HANDBOOK ON AGRICULTURE, BIOTECHNOLOGY AND DEVELOPMENT

Handbook on Agriculture, Biotechnology and Development

Edited by

Stuart J. Smyth

Research Scientist, Department of Bioresource Policy, Business and Economics, University of Saskatchewan, Canada

Peter W.B. Phillips

Professor of Public Policy, University of Saskatchewan, Canada

David Castle

University of Edinburgh, UK

Edward Elgar
Cheltenham, UK • Northampton, MA, USA

Published by
Edward Elgar Publishing Limited
The Lypiatts
15 Lansdown Road
Cheltenham
Glos GL50 2JA
UK

Edward Elgar Publishing, Inc.
William Pratt House
9 Dewey Court
Northampton
Massachusetts 01060
USA

A catalogue record for this book
is available from the British Library

Library of Congress Control Number: 2013943217

This book is available electronically in the ElgarOnline.com
Economics Subject Collection, E-ISBN 978 0 85793 835 0

ISBN 978 0 85793 834 3 (cased)

Typeset by Servis Filmsetting Ltd, Stockport, Cheshire
Printed and bound in Great Britain by T.J. International Ltd, Padstow

Contents

Contributors

Ademola A. Adenle, United Nations University, Japan

Philipp Aerni, University of Zurich, Switzerland

Corinne Alexander, Purdue University, USA

Julian M. Alston, University of California – Davis, USA

Volker Beckmann, Ernst-Moritz-Arndt University Greifswald, Germany

Julia Bognar, University of Toronto, Canada

Carlos G. Borroto, Center for Genetic Engineering and Biotechnology, Cuba

Derek Brewin, University of Manitoba, Canada

Graham Brookes, PG Economics, UK

Janet Carpenter, J.E. Carpenter Consulting LLC, USA

Yves Carrière, University of Arizona, USA

David Castle, University of Edinburgh, UK

Mao Chen, Ministry of Agriculture, China

Pedro Conceição, United Nations Development Programme, USA

Buwani Dayananda, University of Saskatchewan, Canada

Matty Demont, International Rice Research Institute, Philippines

Koen Dillen, Institute of Prospective Technological Studies, Spain

Derek Eaton, Graduate Institute of International and Development Studies, Switzerland

Edna Einsiedel, University of Calgary, Canada

José Falck-Zepeda, International Food Policy Research Institute, USA

Jorge Fernandez-Cornejo, United States Department of Agriculture – Economic Research Service, USA

George B. Frisvold, University of Arizona, USA

Carol V. Gonsalves, Hawaii, USA

Dennis Gonsalves, United States Department of Agriculture – Pacific Basin Agricultural Research Center, USA

Marnus Gouse, University of Pretoria, South Africa

Gregory Graff, Colorado State University, USA

Richard Gray, University of Saskatchewan, Canada

Aarushi Gupta, University of New Hampshire, USA

William O. Hennessey, University of New Hampshire, USA

Jill E. Hobbs, University of Saskatchewan, Canada

Wallace E. Huffman, Iowa State University, USA

Lee Ann Jackson, World Trade Organization, Switzerland

Calestous Juma, Harvard University, USA

Nicholas Kalaitzandonakes, University of Missouri, USA

Scott Kaplan, University of California – Berkeley, USA

Valerie J. Karplus, Massachusetts Institute of Technology, USA

William A. Kerr, University of Saskatchewan, Canada

George G. Khachatourians, University of Saskatchewan, Canada

Enoch M. Kikulwe, University of Goettingen, Germany

Eunice Kim, University of California – Berkeley, USA

Deepthi Elizabeth Kolady, Cornell University, USA

Stanley P. Kowalski, University of New Hampshire, USA

John Kruse, World Agricultural and Environmental Services, USA

Les Levidow, Open University, UK

Sebastian Levine, United Nations Development Programme, USA

Karinne Ludlow, Monash University, Australia

Xingliang Ma, International Food Policy Research Institute, USA

Alexandre Magnier, University of Missouri, USA

Stavroula Malla, University of Lethbridge, Canada

Ira Matuschke, Organisation for Economic Co-operation and Development, France

Jill J. McCluskey, Washington State University, USA

Alan McHughen, University of California – Riverside, USA

Jennifer Medlock, University of Calgary, Canada

Douglas Miller, University of Missouri, USA

Latha Nagarajan, International Fertilizer Development Center and Rutgers University, USA

Anwar Naseem, McGill University, Canada

Chidi Oguamanam, University of Ottawa, Canada

Marc Ouattarra, Institut de l'Environnement et de Recherches Agricoles, Burkina Faso

Micheal Owen, Iowa State University, USA

Robert Paarlberg, Wellesley College, USA

Peter W.B. Phillips, University of Saskatchewan, Canada

Matin Qaim, University of Goettingen, Germany

Terri Raney, Food and Agriculture Organization of the United Nations, Italy

Jeanne M. Reeves, Cotton Incorporated, USA

Sybil D. Rhodes, Universidad del CEMA, Argentina

Syed Masood H. Rizvi, Dow AgroSciences Canada Inc., Canada

Camille D. Ryan, University of Saskatchewan, Canada

David Schimmelpfennig, United States Department of Agriculture – Economic Research Service, USA

Graham J. Scoles, University of Saskatchewan, Canada

Grace Skogstad, University of Toronto, Canada

Stuart J. Smyth, University of Saskatchewan, Canada

Claudio Soregaroli, Università Cattolica del Sacro Cuore, Italy

David J. Spielman, International Food Policy Research Institute, USA

Alexander J. Stein, International Food Policy Research Institute, USA

Jennifer Thomson, University of Cape Town, South Africa

Jeffrey Vitale, Oklahoma State University, USA

Gaspard Vognan, Institut de l'Environnement et de Recherches Agricoles, Burkina Faso

Gina Waterfield, University of California – Berkeley, USA

Seth Wechsler, United States Department of Agriculture – Economic Research Service, USA

Justus Wesseler, Technical University of Munich, Germany

Alphanso Williams, University of Saskatchewan, Canada

William W. Wilson, North Dakota State University, USA

L. LaReesa Wolfenbarger, University of Nebraska Omaha, USA

Gongyin Ye, Ministry of Agriculture, China

Jose Yorobe Jr, University of the Philippines, Philippines

Douglas Zhihua Zeng, World Bank, USA

David Zilberman, University of California – Berkeley, USA

1 Introduction to agriculture, biotechnology and development

Stuart J. Smyth, Peter W.B. Phillips and David Castle

1 INTRODUCTION

Evidence is playing a greater and more conspicuous role in the shaping and management of public and private organizations and the rules that govern them. Decision-makers rely on evidence and must account for its use, but the processes for creating, normalizing and disseminating knowledge, as well as the standards for the evidence itself, are widely debated, discussed and documented. Apart from the theories and typologies of evidence generation and evaluation, one crucial variable underpinning and defining effective uptake and use of new knowledge is often missing: timing. In the context of emerging technologies where the knowledge base is incomplete, the timing of knowledge generation and use can determine the fate of technologies. This is especially true in the dialogue about when, where and how society should adapt, adopt and use new science-based inventions in the global agri-food system. The emergence of biotechnology in the global system has triggered a broad range of individuals and groups asserting that the technology is, or is not, appropriate to the context and needs of global agriculture and development. Careful consideration of the quality and provenance of evidence will always matter, but so too does the vital question of when is the best time to disseminate knowledge, especially as it relates to innovative concepts and products?

While every case of agri-food biotechnology innovation is different, some general rules about when and how to disseminate information apply. For example, knowledge acquired in the early years of a new technology's development, testing and uptake should, for a variety of reasons, be carefully documented and disseminated. Ultimately, this evidence will contribute to a complete story of the benefits and risks of the technology. In some instances, this may be at the first point of commercialization (what economists might call the *ex-ante* evaluation); at other times it may follow a lengthy period of sustained adoption (what economists call *ex-post* evaluation). In the context of agriculture, biotechnology and development, in 2013 both forms of evaluation exist. Readily available is extensive, detailed evidence of the impacts and effects of a few first generation technological adoptions on the actors, the debates, the economy and society. Less readily available are experiences supporting estimates of the impact of second and third generations of technologies and products, many of which might fundamentally change the structure and function of the global agri-food system.

Knowledge about iterative and transformative technologies, such as the recent technological changes in agriculture, accumulates incrementally over many years. Articles in the early to mid-1980s described the process of how one might genetically transform plants. These scientific publications have since provided more information and greater details about how to insert or activate new traits into plants. By the mid to late 1990s new articles began to appear that attempted to estimate the effects of this technology on consumers, producers and industry, both in developed and developing nations. This was matched by a flurry of work on consumer responses, intellectual property, regulatory frameworks, international trade impacts, biosafety assessments, adoption benefits and many other topics.

The application of biotechnology to agriculture has precipitated, if not the largest change in the history of agriculture, certainly the largest change since the move to mechanized agriculture. Responses to this innovation span a wide spectrum of applications and impacts. Much knowledge about the state of agri-food technology and its socio-economic impact on global agriculture, biotechnology and development has accumulated over the last quarter century of production history. Specific studies on the effect of agricultural biotechnology now provide a rich history and offer grounded thoughts on the future for the technology in this sector. The experiences and impacts of agricultural biotechnology (agbiotech) provide a unique perspective on how an innovation is able to provide global benefits in the face of some of the most hostile responses to commercialization experienced in recent memory. As this knowledge base has grown, drawing from an increasingly diverse set of sources, the timing is opportune for a handbook that gathers together the evidence on the main issues in agbiotech.

2 OBJECTIVES OF THE *HANDBOOK*

This book provides the reader with a diverse, but concentrated, perspective of the global application of biotechnology to plant agriculture. Readers will be able to gain rich insights into specific aspects of agbiotech (that is, impacts of genetically modified traits in an array of products, ranging from such large-area crops as corn and soybeans to niche products like papaya), and will also be provided with an assessment of the overarching structure that governs the trade and regulation of agbiotech processes (that is the role of national and international governance systems, such as the United States and European Union regulatory process and the World Trade Organization). These perspectives provide an evidence base for the reader to compare and contrast the results within the different applications. The book offers the reader detailed evidence of both the products and the processes that are part of, and important to, agbiotech. Readers will gain new insights into why agbiotech has been more successful in some geographic locations than others and why some products have been more successful than others and what this says about both the technology and its application to agriculture and the broader question of technology as a part of the development agenda.

3 THE ORGANIZING HEURISTICS FOR THE VOLUME

To realize the objectives of the book, the editors have applied three interrelated, three-pronged approaches to assessing the relationships between agriculture, biotechnology and development. In the first of the approaches, which might be labelled the 'epistemic' approach, the editors drew on previous work about the codification and organization of knowledge in an effort to lay the foundation for contributors to sort between: (a) what is accepted or 'Kuhnian' normal science; (b) what is the focus of current research, where theory, methods and evidence are being advanced and tested; and (c) speculative pursuits, where standard research practices have yet to emerge. For the second approach, the editors adopted Elinor Ostrom's institutional analysis and decision (IAD) framework as the superstructure for the *Handbook*. This framework supports rigorous analysis of: (a) the exogenous variables (that is, the actors); (b) the action arenas (that is, issues and decision points); and (c) the outcomes. The IAD is a useful organizing framework for categorizing the chapters of the *Handbook* into somewhat discrete categories. As a third approach, the chapter authors were asked to structure their contributions by self-consciously considering how they use: (a) established analytical or normative theories and models; (b) existing methods; and (c) standard or experimental metrics. The models–methods–metrics approach was encouraged as a way for authors to reflect on relationships that could be articulated in the IAD framework, and to speculate on knowledge gaps and research opportunities, thereby integrating the epistemic approach into their chapter. Combining these three approaches helps to give individual chapters, parts and the entire *Handbook* an integrated structure that bridges different sources of knowledge, institutional settings and dynamics and the tools and techniques used to understand the many facets of agbiotech. In the following three sections the approaches are described more fully.

3.1 Knowledge Framing, Priority Setting and Triage

Because this *Handbook* is a compendium of knowledge drawn from many complex and overlapping systems, it is important to specify the grounding criteria used to separate the grains of interest from the large volume of chaff. Four criteria were used. The first is straightforwardly epistemic. For any given question about agbiotech, a matrix can be constructed in which theory and evidence relevant to the question comprise the axis, and each axis is divided into 'known' theory or evidence and 'unknown' theory or evidence. Authors were asked to consider their topic in light of the 'known evidence and known theory', the inverted pair of 'unknown–known' theory and evidence, and cases where both theory and evidence are unknown but may be under discussion. This approach helps to categorize some questions as 'normal research' questions, in the sense defined by Kuhn (1970); other questions require future research, as uncertainties dominate.

Second, any investigation of knowledge must start with some concept of how 'deep' to go with a topic. Authors in this *Handbook* have been asked to address specifically both the 'depth' and the 'breadth' of the knowledge they surveyed. Some chapters, such as those examining the impact of specific technologies, are inherently fine-grained analyses. Others, such as the general operation of the regulatory or trade system, are more coarse-grained. Similar levels of granularity have been imposed within each of the three parts of the book, but between parts, divergence has been allowed. One might look at parts I, II and III as moving from coarse to medium to fine granularity, as the scale, scope and impact of the technology is increasingly delimited from the exogenous environment it engages with, to the adaptations in the action arenas and finally into specific applications in specific markets.

A third consideration is the 'stopping rule' used to circumscribe the range of topics considered in this *Handbook*. In the first instance, the basic topic – biotechnology applications in plant agriculture and its impact on development – narrows the range of considerations. At the chapter level, authors have been directed to focus on their core topic and only secondarily to address overlapping considerations considered in other chapters. Thus circumscribed, this *Handbook* does not explore in depth many worthy but increasingly peripheral issues where 'green' agbiotech intersects with 'red' (medical), 'blue' (marine), and 'white' (industrial) biotechnology. These points of intersection could themselves comprise a lengthy and interesting *Handbook* in coming years.

Fourth, chapter authors were asked to differentiate between the iterative and transformative effects of the technology. Iterative change usually involves marginal improvements, resulting in technologies and products that substitute for, or complement, existing technologies. Iterative change is relatively modest and short-lived, but is easily assimilated by existing regulatory systems, commercial actors and ultimately by citizens and consumers. In contrast, transformative change involves new technologies that generate a wide array of new production, consumption and political and social opportunities, consequently challenging accepted concepts, technologies, products and organizational structures. The impact on governance systems can be acute and enduring, and tends to precipitate discourse and conflict. While the rate, scale and scope of the change will vary depending on whether the technology involves small, iterative adjustments or poses large, transformative modifications, the challenge remains the same. Institutions need to respond and to adapt to the new circumstances (Phillips, 2007).

3.2 The Institutional Analysis and Development Framework[1]

This *Handbook*, and many of its chapters, uses Ostrom's (2005) institutional analysis and development (IAD) framework to structure the analysis of the theoretical and the institutional underpinnings of the complex system that has delivered a range of new technologies and products in the global agri-food system. The IAD framework helps to separate the nested layers of organizational environment, rules, actors and outcomes in governance systems. The framework focuses on more than just the organizations; it also directs how one might interpret interactions between and among actors and the institutional rules and norms that govern their exchanges, thereby offering insights into the behavioural underpinnings of many developments. Importantly, the IAD framework accepts that interactions can be simple or complex. While some critics are concerned that this adds to the already complex and dynamic picture of institutional analysis, applied consistently it can help resolve much of the complexity surrounding any given problem without resorting to the reductionist simplification of many other approaches.

The IAD framework is a systems approach to policy processes, specifically focusing on inputs, decision-makers, outputs, outcomes, evaluative criteria and feedback effects (Figure 1.1). Ostrom argues that any complex system can be viewed as being composed of subsystems (she calls them 'holons') that interact with an overarching system. Each subsystem 'can be "dissected" into its constituent branches on which the holons represent the nodes of the tree, and the lines connecting them the channels of communication, control or transportation' (Ostrom, 2005, p. 12). While this notion of nearly decomposable complex subsystems has been around since Simon (1955), the IAD framework advances the approach by more clearly articulating the constituent parts and offering a framing for understanding how the components integrate into the meta-system. The

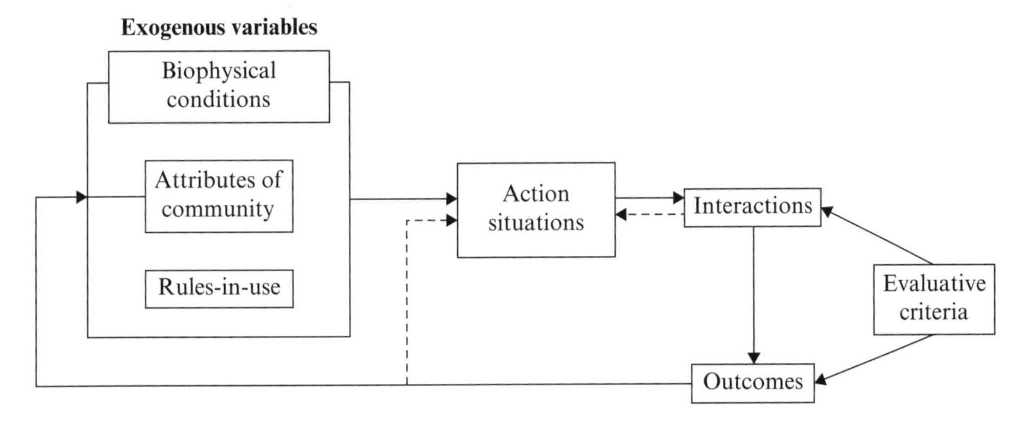

Source: Ostrom (2005, p. 15).

Figure 1.1 Components of the IAD framework

approach highlights that many of the interactions within and across the subsystems occur simultaneously and at multiple levels. The IAD framework therefore provides analysts with the luxury of either analysing the system as a composite or of focusimg on selected subsystems independently or jointly. This very flexibility highlights the importance of the exogenous variables, the action arenas and the rules and linkages between them.

One key analytical feature of the IAD framework is that it helps to frame and interrogate the possibilities of polycentric governance – 'a system of governance in which authorities from overlapping jurisdictions (or centers of authority) interact to determine the conditions under which these authorities, as well as the citizens subject to these jurisdictional units, are authorized to act as well as the constraints put upon their activities for public purposes' (McGinnis, 2011, p. 171). Most theories posit that knowledge-based economies, such as agbiotech, rely on polycentric systems of governance (for example Porter, 1990; Nelson, 1993; Lundvall, 1992; Leydesdorff and Etzkowitz, 1998). The IAD supports the analyses of these systems by targeting attention on the governing forces which exert influence on each subsystem and in turn on the overall system. In the context of research-based innovation systems, multi-sectoral and multi-functional polycentric governance dominate.

At its most basic level, the IAD framework consists of three elements: (1) exogenous variables; (2) an action arena; and (3) the interactions that generate outputs and outcomes. Ostrom defines exogenous variables to include biophysical or material conditions (for example the physical and biological constraints and challenges in different growing regions), attributes of community (for example the industrial structure and political systems governing agriculture) and rules (for example the overarching legal and institutional norms and practices that delimit choices). The action arena is composed of action situations and participants – variously defining problems, issues, policy areas and networks or communities of individuals and organizations. Stone (1989) asserts that action areas are not so much found as constructed through causal stories. Interactions between action arenas and exogenous variables determine the outputs and outcomes, which are evaluated using criteria adapted from external systems or developed explicitly for the circumstances. Outcomes are continuously fed back onto the action arenas (and at times change the exogenous factors), which results in ongoing transformations in the system (either at the level of the holons or subsystems or at the comprehensive system level). In this way each of these variables are connected to each other, which leads to extensive learning-by-doing and change.

In applying Ostrom's IAD framework to the examination of biotechnology in the crops-based agri-food sector, authors of each chapter were asked to examine an exogenous variable (Part I), an action arena (Part II) or a set of interactions and outcomes (Part III). Each author was asked to apply appropriate knowledge filters to consider the models, methods and metrics people have used, or ought to use, to understand the role of biotechnology. In effect, they were tasked with examining a specific set of institutions, issue areas or applications, identifying the models and analytical methods that are commonly used to define and examine the action area (what Ostrom might call the evaluative criteria) and then to critically survey and assess the outcomes observed. This approach is meant to identify gaps in the theory, methods or evidence needed to fully understand the implications of this ongoing technological transformation.

3.3 Models, Models and Metrics

Authors of the chapters were also asked to examine critically all aspects of the known social and natural science associated with their topic. In this context, clearly delimited and articulated models, methods and metrics are important tools for understanding the dynamics between contexts, organizations and actors reflected in the IAD framework. Because of the plurality of types and uses of models, methods and metrics, authors were encouraged to be explicit about the tools that they were using. For example various types of models can be used to represent the causal relationships in a system, and given the diversity of the actors, issues and applications, there are inevitably many facets of a system that can be represented. While this diversity adds significant explanatory power, it often is at the root of disagreements and disputes. While resolving these disagreements and disputes is not an appropriate task for a *Handbook*, this *Handbook* offers a compendium of a wide range of models that appeal to and bridge theoretical concepts from sociology, psychology, political science, policy and economics.

Models of a system rely on underlying methods to construct and validate them. Many disciplines and authors approach the validation of models inductively, with micro-level detail filling in the gaps of the theory and conforming to the implied structure, without explicitly testing the models in a scientific manner. In short, much of the work takes the theory as given and then applies it to estimate the impacts implied by the theory. Other work is more deductive in nature, using theory to derive refutable, testable hypotheses. Both methods have their place, but it is vitally important to know which is being undertaken. Standard performance indicators and conventional modelling can deliver useful information about how effectively closed-system technical problems have been solved, but they may be less effective in explaining how science and technology perform in an unbounded social context.

With respect to metrics, authors were asked to summarize and assess metrics that comprise the evidence base used in decision-making about agbiotech innovation. In some cases metrics refer to qualitative or expert judgement; in other cases, empirical study is required. It is important to document data collection methods, whether involving primary or secondary data. The provenance of data can often have a major impact on general acceptance – data from statistical agencies, firms, community groups and farm-level or consumer surveys will have differing import. It is similarly important to disclose analytical methods; it is vital to know whether the focus is on economic, social, institutional or political factors, or whether the analysis is based on comparative static, partial equilibrium or computable general equilibrium approaches. Finally, every discipline has norms or standards to differentiate between data that is included or excluded; moreover, statistical practices and norms to validate statistics vary. It is important to keep in mind that methodological pitfalls exist in underlying fields such as econometrics (Leamer, 1983; McCluskey, 1983).

3.4 The Editorial Process

How this bespoke volume arose deserves comment. The editors, having considered how the IAD framework could be adapted to the context of agbiotech innovation, developed the basic structure of the book's sections and range of topics to be covered in each section. A list of potential authors was identified from the editors' networks and in some cases by researching the topic. Each author was asked to deliver a chapter of approximately 6000 words, including figures, tables and references; the goal was to produce the drafts within three months of commissioning. Many authors met the deadlines but exceeded the somewhat arbitrary word limit. In an attempt to assist readers to compare and to some extent contrast the diversity of biotechnology applications within agriculture, the chapters follow a standard set of basic headings. The value of this regular structure is twofold: first the authors had a clearer understanding of what they were to address within the body of their chapter; and second, the readers of the book have the ability to make assessments from one chapter to another. Authors were given a template that included a context-specific introduction, a survey of the theoretical models used, a review of the quantitative and qualitative methodological approaches used and a summary of the metrics, including the aggregate ranges of applications and impacts, the scale and scope of effects and the regional and distributional effects. Furthermore, the authors were asked to critically assess future theoretical developments, convergences, divergences and development of methods, and the shortcomings in data that need to be considered and remedied. Each chapter was reviewed by all of the editors, who provided detailed content and stylistic comments, suggestions and questions, which frequently initiated a second round of writing by the authors and review by the editors.

4 FINDINGS: THEMES, LESSONS AND GAPS

To frame the discussion in the many chapters of the *Handbook*, we invited Graham Brookes to provide an overview of the state of the industry in the period 1995–2010 for the first substantive chapter of the *Handbook* (Chapter 2). Although the first commercial genetically modified (GM) crop was planted in 1994 (tomatoes), 1996 was the first year in which a significant area of crops containing GM traits were planted (1.66 million hectares). Since then there has been a dramatic increase in plantings; by 2010/11, the global planted area reached over 139 million hectares (and ISAAA estimates it rose to 170 m ha in 2012), equivalent to 71 per cent of the total utilized agricultural area of the European Union (James, 2013). GM traits have largely been adopted in four main crops – canola, corn, cotton and soybeans – although small areas of GM crops in sugar beet (adopted in the USA and Canada since 2008), papaya (in the USA since 1999 and China since 2008) and squash (in the USA since 2004) have also been planted. In 2012, 28 countries planted GM crops, more than half of which were developing nations. Six countries accounted for 94 per cent of total production: US, Brazil, Argentina, Canada, India and China (in declining order of area). Two traits dominate: herbicide-tolerant crops account for 65 per cent of the total GM area, while insect-resistant crops account for 35 per cent of global plantings. GM seeds account for 70 per cent of the global soybean acreage, 52 per cent for cotton, 26 per cent for corn and 20 per cent for canola. In those countries adopting GM varieties, GM seed market share has risen above 80 per cent. GM crops have also been pro-trade, in that adoption and production is concentrated in leading export nations. Brookes estimates that biotechnology producers account for between 72 per cent of cotton and 95 per cent of soybean global trade. These statistics suggest that there are *highly differentiated* actors, issues and impacts of the development and commercialization of biotechnology in plant agriculture.

4.1 Part I: Exogenous Variables: The Environment, Actors and Rules

The 18 chapters in this part examine a wide array of exogenous factors, external constraints and rules that govern the interactions of the key actors in the biotechnology, agriculture and development communities. The overriding message of these chapters is that there is a divergence in the various subsystems between those who are developers, adaptors, adopters and users and all the rest. Those with a stake in the effective, efficient and economically relevant use of biotechnology and agriculture are generally prepared to optimize their value. Those who are, or believe they are, disenfranchised from the technology, either because they do not own any of it, cannot productively use it or find it does not meet their social-economic interests, are actively engaged in opposing the technology. As with most controversial and contentious issues, the vast majority of actors in the agricultural and development communities are either silent bystanders or they are simply unaware or insufficiently invested to embroil themselves in disputes about the technology.

As befits a *Handbook* examining the current state of affairs in agbiotech, this part of the book offers a range of observations and analyses of how the lead developers and adopters and the pro-science institutions have accommodated and managed the introduction of new GM crops. The leading adopters, including Canada and the US (Bognar and Skogstad), Argentina and Brazil (Rhodes) and China (Karplus), and the partial or selective adopters, including India (Qaim), Oceania (Ludlow and Yorobe) and Africa (Thomson), have followed a common path. For the most part they have directed their public research capacities (Gray and Dayananda) towards advancing the science and applications and have at a minimum tolerated and in most cases partnered with private industry (Hobbs) to accelerate and focus the technology and its uses. These actors and jurisdictions have also proactively used existing regulatory authorities and structures and drawn on the longstanding policies and procedures in the global intellectual property system (Oguamanam), the multilateral risk analysis framework (Jackson) and the international trade system (Kerr) to effectively advance the technology in the agri-food setting (Bognar and Skogstad).

Juxtaposed against this juggernaut of investors, developers, promoters and users, there is a smaller but still significant group of sceptics, doubters and campaigners. While one might be tempted to point a finger at individual actors such as the European Union (Levidow), developing world non-adopters (Paarlberg and Falck-Zepeda), non-governmental organizations (NGOs) (Aerni) and consumer advocacy groups (Smyth and Castle), none of these actors alone appears to have enough power and scope to have much effect on the evolution of the sector. Together, however, they have effectively stalled, or at least slowed, the introduction of the technology into regions and markets that represent more than half of the global agri-food sector and about one third of the world's consumers.

Gray and Dayananda (Chapter 3) assert that the advent of biotechnology profoundly impacted agricultural research. Advances in molecular science vastly increased the scope of what is understood about crop, weed and pest genetics. Explosive growth of knowledge changed the possible, both in terms of the speed of genetic development and the scope of outcomes. But the potential was only realized when new intellectual property rights to genes, microbes, plants and varieties attracted the attention of the agri-chemical and pharmaceutical industry. Beginning in the mid-1980s, firms shifted focus and resources to exploit these new opportunities. Hobbs (Chapter 4) investigates the resulting periods of rapid investment, industry mergers and acquisitions and the ultimate globalization of research, development and commercialization. The end result is that we now have a highly concentrated, vertically and horizontally integrated, global industry. Somewhat paradoxically for such an innovative technology, the nature of the science and the regulatory challenges in both pro-biotech and other markets has worked to amplify the industrial concentration, to the point that there is little true entrepreneurial activity in the research and development (R&D) stages of the sector. Meanwhile most of the world's public research capacity grappled with how to respond. Gray and Dayananda (Chapter 3) note that while some public research organizations (PROs) withdrew in the face of the competition from multinational enterprise (MNE) research programmes (for example in the UK), others sought to partner with them to accelerate the development and adoption of the technology. They report that recently PROs are testing new models of engagement in order to sustain research on common pool problems such as genomics, elite germplasm and agronomy.

As these investments were proved up, they presented a challenge to regulators and the supply chain. The first and undoubtedly most important accommodations came in the United States. Bognar and Skogstad (Chapter 5) assert that American dominance in plant biotechnology can be attributed to explicit government policies to promote GM products. The promotional policies included strong protection for intellectual property rights and a generally permissive regulatory framework that does not discriminate against biotechnology. This approach, for the most part emulated in Canada and to a lesser extent in Mexico (which has a permissive but cautious approach), was guided by pursuit of competitiveness and economic gain.

This model of permissive and promotional regulation has been adapted and adopted across the globe, first in Latin America (Argentina and Brazil) and then successively in South Africa, China and most recently in India. While Alston, Kalaitzandonakes and

Kruse (Chapter 45) suggest that all adopters do not win equally, and some may barely hold their own, this bloc of adopting nations has held firm and is gaining adherents. James (2013) estimates that in 2012, at least 28 countries, representing the lion's share of the leading exporters for the four main commodities using GM varieties, have adapted, adopted, approved and commercialized at least one GM crop variety.

South America has been a strong adopter of GM soybeans. Rhodes (Chapter 6) asserts that Argentina, and more recently Brazil, have been generally pro-technology, pro-business and pro-international trade, with differences mostly based on the role of small, poor or landless farmers (especially in Brazil) and the role the multinational companies have played in the markets. One interesting recent development is that Brazil, which lagged in adoption due to erratic policy in the 1990s, has proactively cultivated relationships with the MNEs, while Argentina has lagged in locking in any significant MNE investments, at least partly due to its conflict-driven domestic politics and international policies.

A few African countries have been long-time incubators of biotechnology. Thomson (Chapter 7) examines the political economy of adoption of cotton in South Africa and Burkina Faso, maize/corn in South Africa and the ongoing research into cowpea, bananas, cassava and nutritionally enhanced crops in a variety of countries. Thomson concludes that while adoption has been modest, GM crops may soon dominate what is becoming a somewhat more prosperous region (*The Economist*, 2013).

China's role in agricultural biotechnology is indeterminate. While China developed and commercialized the world's first GM crop (herbicide-tolerant tobacco), it delayed making any irreversible decisions until the new millennium. Karplus (Chapter 8) asserts that China has been catching up and, in recent years, has emerged as a leader in crop biotechnology innovation. Karplus notes that China is a hotbed of socio-economic research and innovation, as scholars and practitioners are focused on understanding the transformations underway. This body of work may ultimately define what is examined and how research is undertaken in the many other industrializing and developing nations which have so far avoided making any explicit choices about this new technology.

Qaim (Chapter 9) examines the other enigma – India. While an early and engaged investor in biotechnology, India has engaged in a slow and conflict-ridden process of adoption. While the economic evidence suggests that Indian farmers benefit from new technologies by dramatically raising their cotton yields while containing their costs, there are persistent debates and invective directed at these developments. Farmer hardship and reports of related suicides have caught the attention of both Indians and international NGOs and the development community, generating a vociferous argument about the appropriateness of GM technology in this context.

Physically isolated from other nations and relatively free from external influence, a number of Oceania countries have responded to GM crops with unique, comprehensive GM regulatory schemes. Ludlow and Yorobe (Chapter 10) assert that distinctive regulatory schemes arise from many factors. Japan's absolute trade dependence and cultural sensitivities to change in key products has framed the measured and managed introduction of GM commodities through trade. Physical isolation of Australia and New Zealand has led to the development of highly distinctive flora and fauna and a history of harmful introduction of new species, and thus an overriding concern about the environmental impact of release of any new organisms. That isolation has also provided a modest competitive marketing advantage as their products have been free from a range of pests common elsewhere and, in recent times, free of GM traits. No one wants to discard these benefits without full consideration of the trade-offs. Particularly in the case of New Zealand, the development of a strong organic farming sector provided a counterforce to GM crops. Finally, the Philippines are engaged in developing and testing new GM crops, but have had to wait for appropriate applications for the agro-industrial system.

The antithesis to the dominant model of adoption is the European Union, where a mix of agronomic realities, socio-economic considerations and political pressures has led to a multi-track policy, with some countries able and willing to adapt and use GM technologies (for example Spain), some opposed to them (for example Austria) and some for the most part waiting to see whether a GM or GM-free future offers the greatest potential benefits (for example France, Germany and the UK). Levidow (Chapter 11) notes that while the EU, especially the Commission and a few key member states (such as the UK), were early promoters of the technology, the technology became negatively associated with the hazards and apparent unsustainability of factory farming in the 1990s. Since then regulatory manipulations and prevarications have forestalled all but exploratory plantings – precaution rules in the face of uncertainties about the technology and its impacts. Since then, public and policy support has slipped, investment has declined and focus shifted to other priorities. The development pathway is now broken, with conflicts related to regulation, market access and consumer acceptance blocking widespread introduction of GM crops for the foreseeable future.

Paarlberg (Chapter 12) blames the EU and its policy and regulatory stance, both domestically and internationally, for stifling the more widespread adaption and adoption of GM crops in Africa, where he believes it could have major positive impacts. He asserts that the pattern of persistent regulatory blockage in the EU and uncertainty about market access for African exporters has discouraged both national scientists and international technology providers from making significant biotechnology investments to help farmers in Africa, which he laments. Given extensive channels of influence (for example bilateral assistance, multilateral technical assistance, European-based advocacy, interregional trade flows and cultural linkages), he fears that Africa will lose as a result of their inability to use this technology.

Falck-Zepeda (Chapter 13) offers a complementary story about non-adopters in Latin America, with similar pathways of influence having similar effects in a range of small to large markets that are unambiguously importers of technology. In those instances, a mix of concerns about protecting bio-diverse regions, challenges related to smallholders and the landless and concerns transmitted through environmental NGOs, donor agencies and Campesinos (rural farm workers) all have stalled development of scientific, regulatory and commercial capacity to adapt and adopt the technology. By contrast, Borroto (Chapter 14) offers quite a different picture through his investigation of Cuba and its efforts to nurture an indigenous, stand-alone, research-based industry. Perhaps most instructive, Cuba has demonstrated that lower-income developing countries can acquire and use the technology in ways that fit with their agro-ecological circumstances. The challenge is to get effective leadership and governance systems in place.

While problems and disputes often start at the local or national level in key countries, they quickly escalate to international forums. Jackson (Chapter 15) lays out the baseline multilateral risk analysis framework that virtually all countries adhere to and which underpin dispute management in most forums. The difficulty, as Kerr (Chapter 16) points out, is that the corresponding dispute settlement processes in the multilateral trade system, embodied primarily in the World Trade Organization, various regional trade agreements (such as the NAFTA) and a new set of environmental agreements under the aegis of the Convention on Biological Diversity (especially the Cartagena Protocol on Biosafety, which attempts to manage the risks inherent in the first trans-boundary movements of living modified organisms) have gone some way to adapting the basic trade principles and processes to address the scientifically-based risks, but are unable to effectively address the socio-cultural concerns that motivate many critics. Despite more than two decades of negotiations to harmonize the rules and procedures related to trade in GM products, little has been accomplished. Kerr likens this effort to putting a square peg in a round hole, arguing that none of the international trade organizations are competent to deal with consumer and citizen protectionism – each of these organizations was designed to control industrial rent-seeking. Given that the trade rules cannot seem to bend to fit the new technology, many look to the other major international forums – a favourite next stop is the array of institutions engaged in managing ownership of intellectual property (IP). The WTO, through the Agreement on Trade-Related Aspects of Intellectual Property Rights (TRIPs), has made international IP disputes actionable, which has raised the interest in how the International Union for the Protection of New Varieties of Plants (UPOV), the International Treaty on Plant Genetic Resources and the World Intellectual Property Organization (WIPO) can contribute to resolving disputes. Oguamanam (Chapter 17) offers an insight into the argumentation and strategies that developers, farmers, communities, activists and nation states have adopted to use these organizations.

The flies in the ointment, in many people's eyes, are the NGOs. On the one hand, there are consumer advocacy groups that report on, and attempt to foster, concerns about GMOs in foods. The main focus of consumer advocacy has been on implementing a set of mandatory disclosures through product labels, in an effort to advance 'consumer sovereignty'. The challenge has been that the evidence does not support the policy. When consumers are asked, only a small proportion are vehemently against GM foods; a larger portion of people express concerns that, as reported in Smyth and Castle (Chapter 18), can be traded off against other considerations, such as cost. Consumer concerns about GM foods have been on the wane, generally speaking. Whereas a decade ago heightened concern might have warranted changes in labelling regulations, Smyth and Castle conclude that given the shift in consumer attitudes, changes in labelling standards might not have the same warrant as they did years ago. One recent aspect of the labelling issue is further addressed in Chapter 30.

Aerni (Chapter 19) examines the 'uncompromising views' of environmental pressure groups towards the use of genetic modification in agriculture and its consequences for freedom of speech, dissenting views in academia, the empowerment of small-scale farmers and global sustainable development. He argues that environmentalists are not able to agree how to address sustainability problems effectively. One faction believes that technological and economic change is necessary to make the world more sustainable, which has led in many instances to entrepreneurial ventures and experimentation. An opposing faction of the environmental movement believes that new technologies are not able to be part of the solution, but part of the problem of sustainable development.

Drawing on lessons from nuclear technology, these groups focus on potentially irreversible effects of agbiotech on biological diversity. This perspective underlies the powerful, global network of NGOs and campaigners who target the technology wherever it is proposed for introduction, having drawn an immutable line in the sand. For the foreseeable future, Aerni sees little hope of convergence between those two perspectives. At present, they have embedded themselves in national and international policy processes and communities in ways that lead to a boisterous debate but little concrete action.

4.2 Part II: Action Arenas

Given the colourful cast of actors introduced in the first section – albeit necessarily a partial accounting – it is inevitable that there are a large number of action arenas. Each chapter in Part II examines a major arena, defined by different policy processes upstream (for example research management, technology transfer, intellectual property) or downstream (farmer adoption, commercial uptake and use, market effects and environmental effects). As one might expect from our preceding discussion of the actors, each action arena is contested; the adopters and non-adopters have differing views about what can and should be done with the relevant policy instruments. For the most part the developers, adopters and promoters dominated these arenas in earlier years, but their dominance now is contested. For the sake of simplicity, we have parsed the arenas into three main domains: upstream research management; the mixing of technology and markets; and downstream impacts and controversies.

Anticipating future research, or outcomes of current research, remains difficult, particularly as time horizons extend beyond any immediate set of projects. A number of authors in the *Handbook* were tasked with considering the agbiotech innovation pipeline, the transfer of technology from the pipeline to the users and the innovation processes themselves. Phillips (Chapter 20) offers a survey of the methods used for defining and interrogating the research pipeline, offering some insights into what these analyses say about the likely and prospective flows of inventions into the regulatory and commercial systems. In the first instance, this is important for investors seeking the best research prospects, and then for regulators who will have to respond to new GM crops. The message is somewhat troubling, as most capital and effort is directed to a handful of large-area crops and a small number of traits. New crops and new traits seem to be much lower priorities, a fact which will slow the value generation of agricultural biotechnology. Phillips cautions that these predictions are probably good for five years, after which there is little basis for prediction.

Instead of making predictions about the agbiotech pipeline, others consider the conditions under which clusters and innovation systems deliver new technologies, products and processes. Spielman, Zeng and Ma (Chapter 21) note that industry clusters have historically played a central role in economic growth and development, driven significantly by technology spillovers within geographically concentrated innovation systems. They posit that biotechnology clusters are emerging and concentrating around developing-country agriculture, which could accelerate the supply of new tools, products and value for small-scale, resource-poor farmers in the developing world. Their study of clusters in India, Kenya and China highlights the critical role for public investment, the need for innovation-friendly policies to support research and commercialization and the critical role that context plays in success. The conclusion is that clusters are a powerful, but risky, development strategy.

Hennessey, Gupta and Kowalski (Chapter 22) investigate a number of case studies of technology transfer to developing countries, concluding that international tech-transfer is highly specialized and iterative, necessitating trained local human and institutional intermediaries. They note that the international movement of innovation is hampered by whatever proves to be the weakest link in the tech-transfer system (which can include laws and treaties or the owners and financers involved in tech-transfer investment). In the case of agbiotech transfer to developing countries, the weakest links appear to be incomplete IPRs, inadequate tech-transfer competence, poor information access and weak management capacities. Their prescription is that developing countries will need to build some internal capacity and become full participants in the global system; right now they are mostly passive recipients.

Moving along the commercialization pathway, the next major step to consider is the set of structures and process that mix and match new technology with existing capacities and markets. This happens in the supply chain, on farms and in markets themselves. Schimmelpfennig (Chapter 23) examines the structure and operation of the commercial supply chain in North America, ground zero for the new technology. He considers a range of exogenous variables such as the environment in Canada, Mexico and the US, the principal actors and the rules that have influence over them. He concluded that while the technology literature on agbiotech is extensive, several information gaps were uncovered that resulted from network failures in either industry structure, regulation or both. The clearest examples of these gaps were found to come from outside of North America, but these examples were chosen for illustrative purposes, not because information gaps did not exist in North America. The network gaps show large market impacts due to slow-moving or unavailable new agricultural biotechnologies, with related impacts on both producer and consumer well-being.

Farmers are a vital component in the commercialization pathway. Alexander (Chapter 24) notes that farmer choices are based on a mix of factors, including pecuniary rewards (for example yield gains, cost reductions due to saved inputs, higher profit, increased overall productivity and risk management) and non-pecuniary effects (for example simplicity or labour saving). But farmers are not all alike; many have different tolerances for risk and different production contexts that drive different valuation and adoption. Adopters win, but some more than others, based on their individual attributes; non-adopters lose unambiguously. Interestingly, none of the studies shows definitively that the benefits are biased towards larger or more intensive operations. In fact there are winners across the spectrum. While Alexander reports mostly on site-specific, comparative-static studies, Alston, Kalaitzandonakes and Kruse (Chapter 45) offer results from a new, partial equilibrium study that suggests that the effects may not be as straightforward as these single studies imply. What value individuals appear to gain from the farm-level perspective may be clawed away by the interrelationships between technologies, products and markets. Their chapter highlights the complexity of the technology treadmill.

A critical question often raised about the impact of biotechnology is whether GM crops can coexist with other crops. The concern is that if crops cannot be appropriately differentiated, then even small-scale adoptions could cannibalize otherwise valuable market segments. Beckmann, Soregaroli and Wesseler (Chapter 25) examine the economic problem of coexistence, framing it in the context of three factors: consumer and farmer preferences for different production methods; the agro-ecological dynamics, which depend on the biology of the crops concerned and the agro-ecological environment in which they are released; and the broader institutional framework. They assert that while these key factors vary both nationally and internationally, coexistence can and has been achieved, wherever consumers are willing to pay. The cultivation of GM and non-GM crops at world level shows coexistence is possible. They conclude that small-scale studies using gene flow models show that commonly-used threshold levels of 0.9 per cent can be achieved without additional regulatory interventions. In general, whether or not coexistence will be possible depends on the threshold level and where the obligations are imposed. The resulting systems vary based on differing consumer preferences, while the distribution of cost and benefits differ according to how the regulations are structured.

One common concern is about corporate concentration in the biotechnology industry. Naseem and Nagarajan (Chapter 26) discuss whether the inputs industry has a greater degree of concentration due to agbiotech and, if so, what implications this has for market structure. The authors establish that consolidation has taken place in the agbiotech industry, but that cannot be solely attributed to the advent of agbiotech – it is commonplace to all maturing industries. Consolidation has primarily occurred within three sectors: seed development firms; technology firms; and chemical firms. In discussing the implications of this, the authors identify that the literature is inconclusive regarding impacts to innovation.

One of the biggest myths around new technology is that all of the benefits go to the inventor, or at least to the agent that commercializes the innovation. The reality is that even in the most monopolistic context, benefits are widely disseminated in the marketplace. Magnier, Kalaitzandonakes and Miller (Chapter 27) develop an innovative approach to estimating market power in industries with a large number of differentiated products. Their method yielded a worst-case scenario which offers an upper limit on the benefits that flow to inventors. They developed and fitted an econometric model and found that the upper bound on the corn seed mark-up is roughly 15 per cent, which they use to conclude that the amount of market power that was exercised in the US seed corn industry between 1997 and 2008 was rather modest. They also found that the life-cycle effect is concave over time, such that the initial price starts at a lower level, increases until the innovation's fourth year on the market, and then declines until the innovation is removed from the market or replaced by succeeding innovations. This provides empirical support for the notion that product life-cycle dynamics are important when examining patterns of price competition in industries with a large number of differentiated products.

Downstream, a range of controversies are bubbling along, variously commanding headline attention, resources and effort. Some are designed to optimize the operations of the system, while others are more about changing the underlying rules.

The evolution, structure and impact of what Eaton and Graff (Chapter 28) call 'the dynamic IP system' examines the past 30 years of litigation, legislation and treaty making, assesses the direct incentive effects of IPRs, the potential impact on competition and market structure and the theoretical and practical effects on public research, technology transfer, freedom to operate and international transfer. While some effects of the impact of IPRs on large-area GM crops are known, there is only limited evidence related to small area crops and, perhaps more importantly, on the nature and direction that R&D programmes take in response to the evolving private rights for intellectual property.

Wolfenbarger, Carrière and Owen (Chapter 29) examine the environmental impacts of GM crops. Given that agriculture (defined as croplands and pasture) occupies about 38 per cent of the earth's terrestrial surface, the single largest global land use, the environmental effects of agriculture are widespread, with impacts both in production zones and surrounding areas. Which crops are produced, where they are produced and how they are produced determine the potential impacts on biodiversity and the ecological services upon which humans depend. The baseline for considering the impact of biotechnology is the well documented negative circumstances attributable to unsustainable practices of conventional and traditional agriculture. They conclude that herbicide-tolerant (HT) and insect-resistant (IR) crops have not exacerbated the negative impacts of agriculture on biodiversity and ecosystem services; rather, studies indicate that the use of GM crops is reducing the adverse impacts of agriculture by promoting conservation tillage and lower use of long-acting herbicides and highly toxic insecticides. To some extent, those gains are under constant threat. Pest resistance decreases the effectiveness of IR traits as insect species evolve resistance, and selective pressures due to widespread glyphosate use have led to at least 24 weed species evolving resistance. The agricultural practices that are adopted to minimize and manage the evolution of resistance will determine the future effectiveness of these GM crops and their related environmental impacts.

Consumers have responded to the uncertainty and controversy about GM crops in a variety of ways. The economic evidence suggests that consumers do not appear content to accept price signals alone to determine their purchases of GM foods. Many also want proactive, mandatory labelling of any GM content, in order to allow them to make choices based on other, intrinsic criteria. Huffman and McCluskey (Chapter 30) examined the role of labelling of GM foods. While all markets have some consumers expressing interest in differential labelling, only a subset of countries, mostly non-adopters (for example EU) and partial adopters (many Latin American and African countries), have implemented mandatory labelling through regulation. The other markets rely on voluntary labelling (enforced by tort and fraud laws) or standards-based labels (for example Canada with its labelling standard). While the labelling issue has quieted recently, the evidence presented by Huffman and McCluskey suggests that there is no reason to assume it will go away soon.

In the face of endemic food shortages in parts of Africa and Asia and increasing price volatility caused by mismatches in supply and demand and speculation on the market, the global policy community has global food security clearly in its sights. Agriculture was elevated to near the top of the agenda of world leaders in 2008–12 with strong, affirmative positions taken by leaders meeting at the G8, G20, UN, OECD and other multilateral forums. Juma, Conceição and Levine (Chapter 31) present a strong argument that the application of new technologies has played a critical role in past efforts to reduce food insecurity and will need to do so again. They point out that food insecurity is a range of mismatches between supply and demand in the market which leads to undernourishment, chronic malnourishment due to incomplete diets and instability in local food systems due to agronomic challenges. Biotechnology can help to remedy all three problems. They caution, however, that solving world hunger requires more than just producing more food. Strong institutional frameworks will be needed to strengthen public oversight and transparency and empower small-scale producers and consumers alike. Inclusive processes of technology development, adaptation and adoption will ensure that benefits are maximized across society but biased in favour of the poorest and most food-insecure.

Given that many of the world's estimated 800 million food-insecure people are also subsistence farmers, one enduring concern about new technology is that it may affect small landholders, and indigenous farmers' ability to sustain their livelihood. This concern is usually debated in the context of traditional knowledge, farmers' privileges to save and reuse seed and the various national and international laws and treaties that assign and mediate property rights related to genetic materials in seeds. Kolady (Chapter 32) reviews the evolution of this topic, examines the international institutional structure of treaties and concludes that while a system of sorts is now in place, it has differential value depending on the farmer and governance capacity domestically and is a challenge to enforce due to the diversity of actors and venues.

The action arenas considered in this part of the *Handbook* are in many cases driven by and are often linked together by two recent innovative processes. In the past generation the rise of the media, especially electronic sources and the Internet, and the emergence of planned and purposeful efforts by governments to formally engage and to proactively facilitate exchanges with citizens and NGOs have ensured sustained and expansive dialogues about the appropriate choices for agricultural biotechnology. Medlock and Einsiedel (Chapter 33) assert that agricultural biotechnology became an iconic policy issue worldwide in the 1990s, crystallizing growing public concerns about the overall governance of science and technology developments. Concerns included impacts of the commercialization of science, including issues of ownership and control, the adequacy of risk assessment and regulatory processes, the role of technology in advancing globalization, social justice and the voice of citizens in shaping future trajectories of new technologies. In response, governments worldwide embarked on a wave of experimentation with new methods of public engagement, many based on new models of dialogue and deliberation, as a way to rebuild public trust and manage controversy. Medlock and Einsiedel examine the theoretical underpinnings of democratic engagement, review the diversity of methods, focusing on the wide range of citizens' consensus conferences held on the matter of biotechnology, and offer a review of the frameworks for evaluating public participation. They conclude that these processes, while converging on a set of common practices, are not perfect. They often generate significant friction between participants and sponsors, they tend to deliver precautionary, conditional advice and it is unclear what influence, if any, they have on policy outcomes. As a means of examining the application of democratic engagement, Zilberman, Kaplan, Kim and Waterfield (Chapter 34) provide an examination of both sides of the debate in California's recent Proposition 37 on the labelling of GM food products – which was narrowly defeated in the 2012 vote.

The media has become a battleground for ideas and influence. While traditional print and broadcast media remain influential, the emergence of the Internet and all of its attendant activities has fundamentally changed debate and discourse. The 24/7, global system of communications has empowered new actors and enabled new strategies for getting messages and ideas disseminated. Ryan (Chapter 35) addresses these phenomena through the lens of marketing, examining the rise of new actors and the sharpening of messages (what she calls mythmaking) and discussing the response from industry and government. Along the way, all participants – industry, government, NGOs, activists and scientists – have been challenged to adapt. In many ways, it is probably too early to fully comprehend the impact of these changes on our ability to structure action arenas and make evidence-informed decisions.

4.3 Part III: Outcomes

Technology changes our production and consumption possibilities, but only if it is used. The chapters in part III explore the impacts and outcomes of the application of biotechnology in a number of different ways. An array of chapters use comparative statics to assess the actual or potential impact of GM traits on a number of crops, including those which have achieved large areas of cultivation, those that are planned but not yet in the market and those which failed. A second group of chapters looks at the meta-effects of biotechnology in agriculture and development, using comparative statics and partial equilibrium analyses to assess the absolute and relative impacts of the cumulative adoption of an array of first generation traits.

The analyses of the impact of GM technologies on the farm economy vary significantly by crop and region; a wide range of approaches are used, including farm-level surveys, producer and consumer surveys and comparative static econometric models. All of the studies converge on a set of common findings: developers have some market power and when able to get a product to the market, generally make a decent return; farmers who adopt gain a mix of pecuniary (higher yield, lower cost or risk management) or non-pecuniary (convenience) benefits; non-adopting farmers almost always lose, as yield gains by adopters drive down market prices; farmer gains tend to be somewhat transitory, as one-time yield gains are eroded by lower long-term prices; consumers in aggregate unambiguously gain due to lower prices; consumers who have strong preferences for goods differentiated by production methods may gain or lose; and new technologies tend to be pro-trade, as adoption and production of GM crops is concentrated in exporting countries.

Nevertheless, there are some notable features in each crop. Fernandez-Cornejo and Wechsler (Chapter 36) examine soybeans, arguably the most important and most global crop, with more than half of its production in developing countries. A key differentiating feature here is that most of the gains come in the form of non-pecuniary benefits. Alston, Kalaitzandonakes and Kruse (Chapter 45) extend the analysis and use a multi-sectoral, partial equilibrium model to assess the impact of this critical technology both on adopters and others. Their analysis offers compelling evidence of the importance of considering the spillovers between areas adopting technologies and both non-adopters and competitive sectors.

Carpenter, Gouse and Yorobe assess the impact of maize/corn (Chapter 37). *Bacillus thuringiensis* (Bt) corn performs more as an insurance product, so that gains in years of high pest pressures often offset the premiums paid in low pressure years. Stacked corn varieties are on the upswing, heralding a new business model. Vitale, Vognan and Ouattarra (Chapter 38) note that Bt cotton is now a critical crop crossing between the developed and developing world and is a core part of the bioscience strategy in China, India and parts of Africa. Gains to producers come either from lower costs (if farmers were previously intensively managing pests using chemicals) or higher yields (if farmers were not managing pests), but seldom both. Brewin and Malla (Chapter 39) note that canola is somewhat different, in that unlike the other three large-area crops, it was neither developed in, nor is its core market in, the US. This mostly Canadian innovation has a higher degree of competition (in that three MNEs hotly contest for market share for HR varieties), developers have both successfully developed and commercialized second and third generation GM traits and there is a long and successful experience in managing industrial and food varieties in the same landscape. Gonsalves and Gonsalves (Chapter 40) tell the fascinating story of the development and promotion of a technology, virus resistance in papaya, that for many farmers is the difference between operating or not. They discuss the long and difficult route to commercialization, with optimal global uptake and use still stalled by slow or incomplete regulatory assessments in key producing areas. Dillen and Demont (Chapter 41) note that while sugar beets present a good vector for GM traits because of the low risk of gene flow and the potential to generate significant commercial returns, adoption remains limited due to an inability to get approved in the EU. In this sense, their study shows the opportunity cost of ineffective regulatory structures. Demont, Chen, Ye and Stein (Chapter 42) similarly examine rice, which has been promoted for decades as an important crop that could be enhanced through genetic modification but that has only just reached the market in recent years. This is particularly important as GM rice is projected to have significant potential to address poverty reduction and health in food-insecure regions of the world.

A number of authors were invited to undertake more comprehensive studies of the impact of new technologies on the broader economy. The difference between comparative static and meta effects is significant and important for understanding the role of biotechnology in agriculture and development. Frisvold and Reeves (Chapter 43) undertook an analysis of the comparative statics studies in each of the product areas to assess the aggregate effects on adopters, non-adopters, investors and consumers. Their conclusions support the emerging consensus from the product-specific chapters. While critics of agricultural biotechnology have expressed serious concerns that all of the benefits, especially in developing countries, accrue to technology development firms, this clearly is not, and never has been, the case. In some instances, the adoption benefits in developing countries actually lie more heavily with producers than any other stakeholder, but this is frequently due to the lack of rigorous IPRs or inability to enforce IPRs in that specific country. Even when enforceable and reliable IPRs exist in a developing country, the majority of the benefits from adopting GM crops lie with producers. Raney, Adenle and Matuschke (Chapter 44) go on to discuss the policy implications flowing from these findings. Finally, Alston, Kalaitzandonakes and Kruse (Chapter 45) report the results of their effort to develop a partial equilibrium model to estimate the gross and net effects of the introduction of GM traits, both on adopters and targeted crops and on non-adopters and consumers. The conclusion is that the gains from GM crops to date exceed US$40 billion per year, but that number varies depending on the overall market context. Tight markets drive out gains while loose markets amplify gains. Moreover, consumers ultimately gain virtually all of the net benefits, farmers who are early adopters gain, but those gains diminish over time, and crop sectors (for example wheat) and individual farmers that do not use the technology unambiguously are worse off than if the technology had not been used. In short, this technology in some ways has amplified the longstanding technology treadmill that farmers are tied to.

A second category of studies in Part III looks at the 'wannabes' and failures. Wilson

(Chapter 46) discusses wheat, which, as one of the most important food grains, was an early target for genetic modification. When a candidate GM event, Monsanto's Roundup Ready herbicide tolerance, was presented for assessment in Canada and the US, it was roundly criticized by all involved and the technology was suspended. While this was viewed as a victory for anti-GM activists and farm populism over the giant Monsanto, in the longer term it has simply perpetuated the decline in wheat's relative competitiveness with crops that are adapting and adopting GM technologies. Recently this has led to a renewal of interest in investigating how GM technologies in wheat might reverse this trend. Rizvi and Scoles (Chapter 47) tell a somewhat similar story for barley, oats and rye, three other small grains popular in crop rotations around the world. Partly in response to the pushback on GM wheat, and partly due to the smaller scale of these crops, efforts to date on these crops have focused on using biotechnology tools (for example genomics, molecular marker assisted selection) rather than transgenics. Breeders generally prefer mutagenic techniques, which in most markets are not regulated to the same extent as GM technologies. There have been efforts to transform each of these crops using transgene technologies, but none of the results have been presented for regulatory review or commercial evaluation. Kikulwe, Falck-Zepeda and Wesseler (Chapter 48) look at the prospects for bananas in Africa, reporting on significant potential that they expect will be realized shortly. Williams and Kerr (Chapter 49) look at one of the highly touted industrial applications, GM feedstock for biofuels, concluding that in spite of the best of intentions and an array of proscriptive regulations and programming, the promise is unlikely to be realized in quite the way proponents and developers expect. Finally, Khachatourians (Chapter 50) looks forward to the long-touted future of a bio-industrial future, where plants will be used for a range of non-industrial purposes, including: producing industrial proteins, fibres and molecules; phytoremediating contaminated soils and water; and producing critical pharmaceutical proteins and enzymes in plant hosts at a lower cost, greater scale and higher purity than conventional methods. While each opportunity sounds exciting, a subtle subtext to Khachatourians' chapter is that we need to hope for the best but plan for the most likely – benefits from this type of venture are some ways off yet.

Meanwhile there are a few cases of products that were developed, approved by regulators and commercialized, but then withdrawn. Studying failure can sometimes tell us more than studying success. Ryan and McHughen (Chapter 51) examine three headline-making failures: tomatoes, the first large-scale commercial GM crop; potatoes; and flax. Their message is that despite numerous tests for safety of products passing through the rigorous regulatory approval processes, good science is no guarantee for commercial success of GM food or crop technologies. The introduction of innovative technologies for tomato, potato and flax precipitated well-timed anti-technology programmes that essentially rendered these good products and technologies null and void. While it is difficult to calculate the cost of value lost and costs of withdrawal of these products from the market, it is evident that these costs are significant.

5 CONCLUSIONS

Handbooks are not usually intended to lead to grand conclusions and insights because they are not driven by a single thesis in support of which evidence and argument are amassed. Rather, Handbooks consolidate themes and topics to aid further investigation of a larger problem. In this sense, the three interrelated, three-pronged approaches described above create an organizational and interpretive framework for the chapters in this *Handbook*. Reflecting back on the elements of that framework – the epistemic criteria, the IAD framework and the models, methods and metrics – one can say, with some certainty, that biotechnology has engaged an array of purposeful actors, stimulated an array of institutional innovation, precipitated and focused dialogue and research on a number of concerns that, while centred on agriculture, are relevant to a significant number of other areas, and ultimately is generating value. Along the way it creates both winners and losers, which generates policy pressure.

Nevertheless, the impressive contributions in this *Handbook* raise a number of conceptual and methodological issues. In the first instance, while the standard performance indicators and conventional models and methods used to assess agbiotech can deliver useful information about how effectively closed-system technical problems have been solved, they often offer limited insight into how science and technology performs in a broader social context. New methods to provide this composite appraisal are needed to tackle challenging questions about the opportunities, constraints and governance of agricultural biotechnology. A number of emerging methods – such as complexity theory and purposeful foresighting – offer interesting prospects for assisting in understanding the dynamics of change in the field.

There is no certainty regarding which models of coordination and governance are most effective in advancing agricultural biotechnology innovation. Several approaches that assess and encourage social acceptance and use can be found; similarly, a range of public, private and collective models of innovation governance exist. Mostly, these differ along disciplinary lines, according to different interpretations of the actors, their motivations, relationships and normal practices. Various models used in other research contexts have been adapted, tested and incorporated into new models related to the social acceptance, adoption, uptake and governance of agricultural biotechnology. Examples can be found in mathematical models that measure evasiveness or the threats of weediness, simulation models on pest resistance or governance models on risk.

The current narrow set of metrics for the assessment of innovation and the resulting social and economic benefits from agricultural biotechnology research is incomplete and often misleading. These measures simply define the tip of the innovation iceberg; most of the critical processes, outcomes and impacts remain uncharacterized or ignored. The methods and models that have been developed and tested must ultimately be grounded in measures that better capture social causes and effects of innovation. The advent of boundary-crossing science and technology that destabilizes the regulatory environment, together with awareness of the far-reaching social and environmental effects of innovation, increases the demand for meaningful measures of innovation (Phillips et al., 2012).

NOTE

1. This section draws on the work of Munim (2011); this is a partial précis of his analysis of the IAD applied to biotechnology.

REFERENCES

Economist, The (2013), 'The world's fastest-growing continent: aspiring Africa', available at: http://www.economist.com/news/leaders/21572773-pride-africas-achievements-should-be-coupled-determination-make-even-faster, accessed 17 April 2013.
James, C. (2013), 'Global status of commercialized biotech/GM crops: 2012', *International Service for the Acquisition of Agri-Biotech Applications (ISAAA) Brief*, 44–2012, ISAAA, Ithaca, NY: Cornell, available at http://www.isaaa.org/resources/publications/briefs/44/executivesummary/default.asp, accessed 8 April 2013.
Kuhn, T. (1970), *The Structure of Scientific Revolutions*, 2nd edn, Chicago, III: University of Chicago Press.
Leamer, E. (1983), 'Let's take the con out of econometrics', *American Economic Review*, **73**(1), 31–43.
Leydesdorff, L. and H. Etzkowitz (1998), 'The Triple Helix as a model for innovation studies', *Science and Public Policy*, **25**(3), 195–203.
Lundvall, B-Å. (ed.) (1992), *National Systems of Innovation: Towards a Theory of Innovation and Interactive Learning*, London: Pinter.
McCluskey, D. (1983), 'The rhetoric of economics', *Journal of Economic Literature*, **31**(2), 482–504.
McGinnis, M.D. (2011), 'An introduction to IAD and the language of the Ostrom workshop: a simple guide to a complex framework', *Policy Studies Journal*, **39**(1), 169–83.
Munim, A-U. (2011), 'An exploration of an institutional and behavioural analytic framework for technology transfer and commercialization partnerships', unpublished MPP thesis, University of Saskatchewan, available at: http://ecommons.usask.ca/handle/10388/ETD-2011–08–64, accessed 17 April 2013.
Nelson, R. (1993), *National Innovation Systems: A Comparative Analysis*, Oxford: Oxford University Press.
Ostrom, E. (2005), *Understanding Institutional Diversity*, Princeton, NJ: Princeton University Press.
Phillips, P. (2007), *Governing Transformative Technological Innovation: Who's in Charge?*, Cheltenham, UK and Northampton, MA, USA: Edward Elgar Publishing.
Phillips, P., G. Webb, J. Karwandy and C. Ryan (2012), *Innovation in Agri-food Research Systems: Theory and Case Studies*, Wallingford: CABI.
Porter, M. (1990), *The Competitive Advantage of Nations*, New York: Free Press.
Simon, H.A. (1955), 'A behavioural model of rational choice', *The Quarterly Journal of Economics*, **69**(1), 99–118.
Stone, D. (1989), 'Causal stories and the formation of policy agendas', *Political Science Quarterly*, 104(2), 281–300.

PART I

EXOGENOUS VARIABLES: THE ENVIRONMENT, ACTORS AND RULES

2 Global adoption of GM crops, 1995–2010

Graham Brookes

INTRODUCTION

Although the first commercial genetically modified (GM) crops were planted in 1994 (tomatoes), 1996 was the first year in which a significant area of crops containing GM traits were planted (1.66 million hectares). Since then there has been a dramatic increase in plantings and by 2010/11, the global planted area reached over 139 million hectares. This is equal to 71 per cent of the total utilized agricultural area of the European Union or two and a quarter times the EU 27 area devoted to cereals.

GM traits have largely been adopted in four main crops – canola, corn, cotton and soybeans – although small areas of GM crops in sugar beet (adopted in the USA and Canada since 2008), papaya (in the USA since 1999 and China since 2008) and squash (in the USA since 2004) have also been planted.

In terms of the share of the four main crops in which GM traits have been commercialized, GM traits accounted for 42 per cent of the global plantings to these four crops in 2010.

CONTEXT OF THE TECHNOLOGY AND DEFINITIONS

All crops grown in the world are the product of thousands of years of breeding by man to improve the quality and yield of the end product. Crop biotechnology is a modern extension of plant breeding techniques that allows plant breeders to select genes with desirable or beneficial traits for expression in a new variety. It represents a new step in the evolution of plant breeding because it allows for the transfer of genes with desirable traits between unrelated species (that is, it allows for the transfer of genes between species that are unlikely to have been possible using traditional plant breeding techniques). It is also a more precise and selective process than traditional cross-breeding for producing desired agronomic crop traits.

The main traits so far commercialized have essentially been derived from bacteria and convey:

- Herbicide tolerance (HT) to specific herbicides (notably to glyphosate and glufosinate) in all four crops of corn, cotton, canola (spring oilseed rape) and soybeans (and sugar beet in North America). The technology allows for 'over the top' spraying of HT crops with broad-spectrum herbicides that target both grass and broad-leaved weeds.
- Resistance to specific insect pests (often called insect-resistant or IR crops): genes coded for the *Bacillus thuringiensis* (Bt) toxin have been introduced into corn and

cotton to offer resistance in the plants to major pests such as corn borers, corn rootworm, cotton bollworm and cotton budworm.

In terms of specific 'jargon' relating to crop biotechnology, it is important to explain and clarify two words: 'trait' and 'event'. A trait, as described above, is a desirable or target attribute such as pest resistance or more specifically resistance to corn rootworm or corn-boring pests. An event is the specific genetic change that delivers the trait. For example, in relation to the trait delivering resistance to corn-boring pests, there are several events delivering this trait, with specific names including Mon 810 (developed by Monsanto), DAS 1507 (developed by Dow AgroSciences) and Bt 11 (developed by Syngenta).

PLANTINGS BY CROP AND TRAIT

Almost all of the global GM crop area is found in the crops of soybeans, corn, cotton and canola (Figure 2.1).[1] In 2010, GM soybeans accounted for the largest share (51 per cent), followed by corn (30 per cent), cotton (14 per cent) and canola (5 per cent).

In terms of the share of total global plantings for these four crops, GM traits accounted for the majority of soybean plantings (70 per cent) in 2010. For the other three main crops, the GM shares in 2010 were 26 per cent for corn, 52 per cent for cotton and 20 per cent for canola (Figure 2.2).

The trend in plantings to biotech crops (by crop) since 1996 is shown in Figure 2.3. This shows a general upward trend in adoption levels over the 15-year period, with the share of total plantings of these four crops in aggregate accounted for GM traits rising steadily (for example, from 15 per cent in 1999, to 25 per cent by 2005 and 40 per cent by 2009). Soybean has consistently been the crop with the largest area planted to GM traits,

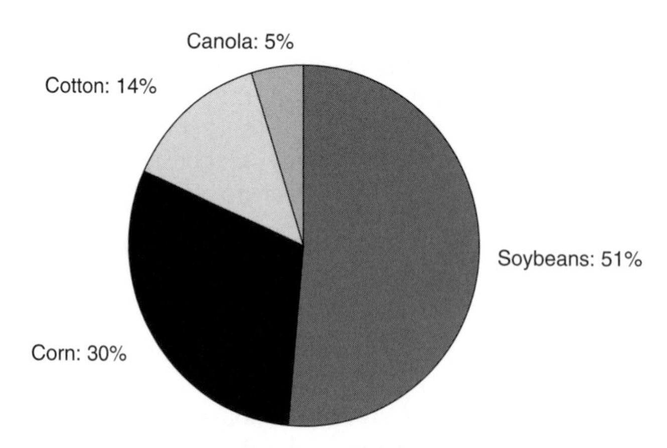

Sources: Various, including ISAAA, Canola Council of Canada, CropLife Canada, USDA, CSIRO, ArgenBio, National Ministries of Agriculture (Mexico, Philippines, Spain), Grains South Africa.

Figure 2.1 GM crop plantings 2010 by crop (base area of the four crops: 139.3 million hectares (ha))

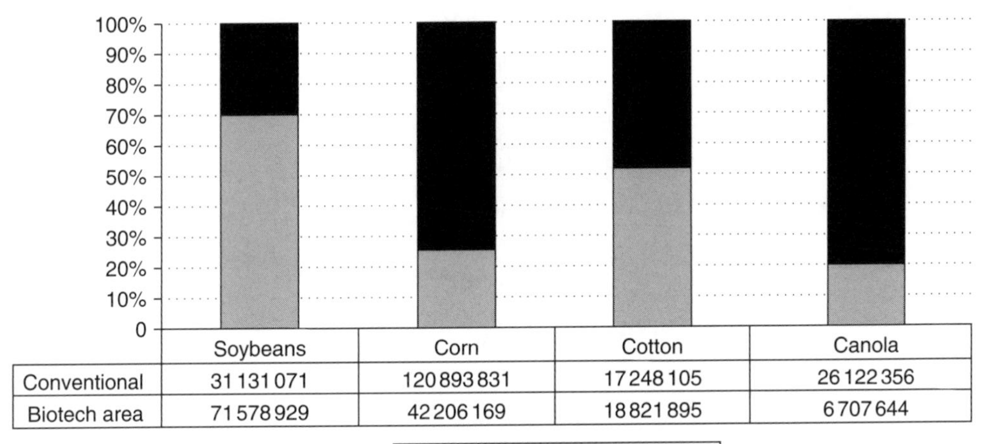

Sources: Various, including ISAAA, Canola Council of Canada, CropLife Canada, USDA, CSIRO, ArgenBio, National Ministries of Agriculture (Mexico, Philippines, Spain), Grains South Africa.

Figure 2.2 2010's share of GM crops in global plantings of key crops (ha)

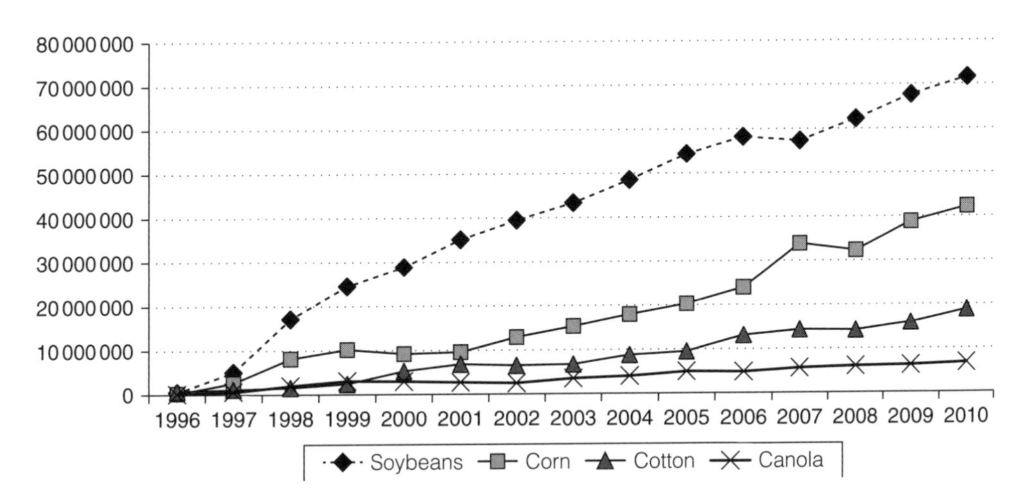

Sources: Various, including ISAAA, Canola Council of Canada, CropLife Canada, USDA, CSIRO, ArgenBio, National Ministries of Agriculture (Mexico, Philippines, Spain), Grains South Africa.

Figure 2.3 GM crop plantings by crop, 1996–2010 (ha)

although the crop showing the largest increases in GM plantings in the last five years has been corn.

Figure 2.4 summarizes the breakdown of the main GM traits planted globally in 2010. GM herbicide-tolerant soybeans dominate, accounting for 42 per cent of the total, followed by GM insect-resistant (largely Bt) corn, herbicide-tolerant corn and insect-resistant cotton with respective shares of 24 per cent, 16 per cent and 10 per cent.[2] In

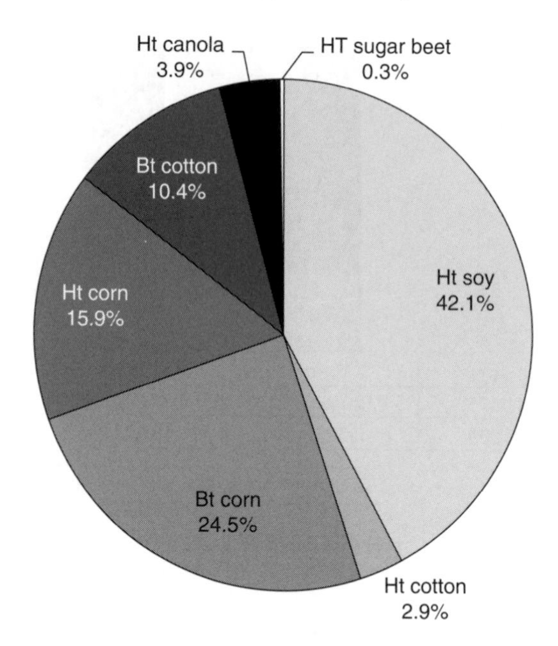

Sources: Various, including ISAAA, Canola Council of Canada, CropLife Canada, USDA, CSIRO, ArgenBio, National Ministries of Agriculture (Mexico, Philippines, Spain), Grains South Africa.

Figure 2.4 Global GM crop plantings by main trait and crop: 2010

total, herbicide-tolerant crops account for 65 per cent and insect-resistant crops account for 35 per cent of global plantings.

GM HT soybean was one of the first technologies widely adopted in countries like the US, Canada and Argentina and, as a result, over 50 per cent of world soybean production used this technology by 2002. From the late 1990s adoption spread to other countries including South Africa, Paraguay, Uruguay and Romania, before Brazil legalized GM HT soybean use in 2003. In 2010, over 70 per cent of the total world production of soybeans used GM HT technology. In almost all adopting countries, the share of the total crop using this technology has reached over 80 per cent and in some cases, notably Argentina and Uruguay, the adoption rate is over 99 per cent.

In corn, GM IR technology has been adopted in several countries from the mid to late 1990s (for example, the US, Canada, Spain, South Africa, Argentina, Uruguay), with the Philippines and Brazil adopting the technology in more recent years (2003 Philippines, 2008 Brazil). GM HT corn was first commercialized in 1997 in the US. Its adoption in other countries has been slower than the adoption of GM HT soybeans, with Canada adopting GM HT corn in the late 1990s, Argentina, South Africa and the Philippines adopting in the early to mid-2000s and Brazil and Colombia adopting in 2010. The largest increase in adoption of GM HT corn technology, outside the US, has coincided with the increased availability of stacked-traited corn, containing both HT and IR technology, in the last five years.

The adoption of GM technology in cotton has followed a similar pattern to corn. IR technology was the first to be adopted in the mid to late 1990s in the US, China,

Table 2.1 GM crops commercially grown in 2010–11: by event

Crop	Number of events grown commercially
Soybeans	12
Corn	33
Cotton	21
Canola	3
Others	2

Australia, South Africa, Mexico and Argentina. Its adoption then spread to India and Colombia in the early 2000s, and more recently to Brazil and Burkina Faso. GM HT cotton was first adopted in the US in 1997, before being adopted in Australia, Argentina and South Africa in the period 1999–2002, and more recently (since the mid-2000s) in Mexico, Colombia and Brazil.

GM HT canola adoption has been more limited, being adopted in Canada in 1996 and the US in 1999. Since then adoption levels in both countries have reached over 90 per cent of the total crop. The only other country to adopt GM HT canola has been Australia, in 2008.

PLANTINGS BY EVENT

Since GM crops were first permitted for commercial planting, approximately 130 events have been granted approval for commercial planting somewhere in the world. In 2010–11, 71 of these events could be found in commercial crops.[3] A breakdown of these is shown in Table 2.1. The crops with the largest number of GM events are corn and cotton, reflecting the availability of both herbicide tolerance and insect resistance traits, often in stacked formats.

PLANTINGS BY COUNTRY

The US had the largest share of global GM crop plantings in 2010 (45 per cent), followed by Brazil (19 per cent). The other main countries planting biotech crops in 2010 were Argentina, India, Canada and China (Figure 2.5). Most of these countries have been using GM traits for many years, with the US, China, Argentina and Canada being some of the earliest adopters of the technology (mid to late 1990s). The widespread planting of GM crops in Brazil and India has been more recent, with insect-resistant cotton first used in India in 2002, and GM crops (herbicide-tolerant soybeans) not legalized in Brazil until 2003. GM cotton and corn are also more recent adoptions in Brazil, with the first GM cotton grown commercially in 2006 and the first GM corn in 2008.

In terms of the biotech share of production in the main adopting countries, Table 2.2 shows that, in 2010, the technology accounted for significant shares of total production of the four main crops in several countries. More specifically:

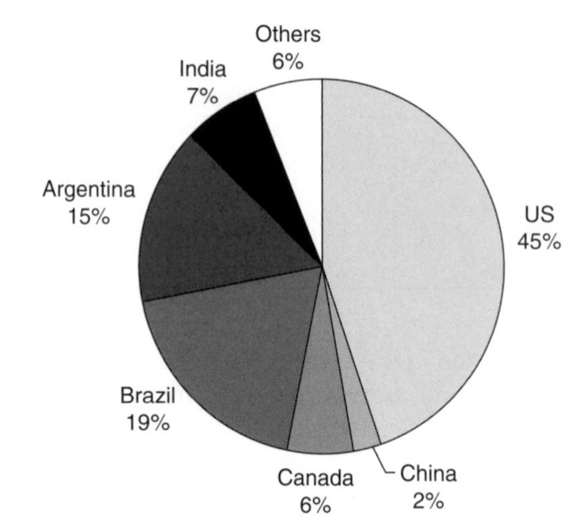

Sources: Various, including ISAAA, Canola Council of Canada, CropLife Canada, USDA, CSIRO, ArgenBio, National Ministries of Agriculture (Mexico, Philippines, Spain), Grains South Africa.

Figure 2.5 Global GM crop plantings 2010 by country

Table 2.2 GM share of crop plantings in 2010 by country (% of total plantings)

	Soybeans	Corn	Cotton	Canola
USA	93	86	93	88
Canada	70	94	N/A	93
Argentina	99	86	98	N/A
South Africa	85	69	100	N/A
Australia	N/A	N/A	99	8
China	N/A	N/A	67	N/A
Philippines	N/A	22	N/A	N/A
Paraguay	96	N/A	N/A	N/A
Brazil	76	55	27	N/A
Uruguay	94	95	N/A	N/A
India	N/A	N/A	85	N/A
Colombia	N/A	8	87	N/A
Mexico	24	N/A	47	N/A
Bolivia	79	N/A	N/A	N/A

Note: N/A = not applicable.

- *The US*: was one of the first countries to adopt the technology in 1996 for traits in soybeans, corn and cotton, and from 1999 in canola, hence the very high adoption levels that have been reached in 2010. Almost all of the US sugar beet crop (96 per cent) used GM herbicide-tolerant technology in 2010, a level of almost total dominance of the crop achieved in just three years (it was available commercially to US farmers only in 2008).

- *Canada and Argentina*, like the US, were early adopters, with the technology now dominating production in soybeans, corn and canola in Canada, and corn, cotton and soybeans in Argentina.
- *South Africa* was the first, and remains the primary African country[4] to embrace the technology, with commercial cultivation beginning in 2000. The technology is widely used in the important crops of corn and soybeans, and in 2010 accounted for all of the small cotton crop (about 13000 ha in 2010).
- *Australia*: was an early adopter of GM technology in cotton (1996), with GM traits now accounting for almost all cotton production. Extension of the technology to other crops, however, did not occur until 2008 when herbicide-tolerant canola was allowed in some Australian states.
- In *Asia*, three countries used GM crops in 2010. China was the first Asian country to use the technology commercially, starting in 1997 with GM insect-resistant cotton technology. This technology rapidly expanded to about two-thirds of the total crop within five years and has remained at this level ever since. GM virus-resistant papaya has also been used in China since 2008. In India, insect-resistant cotton was first adopted in 2002, and its use increased rapidly in subsequent years, so that by 2010 this technology dominates total cotton production (85 per cent of the total). Lastly in the Philippines, insect-resistant corn was first used commercially in 2003, with herbicide-tolerant corn adopted from 2006.
- In *South America*, there are interesting country examples where the adoption of GM technology in one country resulted in a spread of the technology, initially illegally, across borders into countries which were first reluctant to legalize the use of the technology. Thus GM herbicide-tolerant soybeans were first grown illegally in the southernmost states of Brazil in 1997, a year after legal adoption in Argentina. It was not until 2003 that the Brazilian government legalized the commercial growing of GM HT soybeans, by which time it was estimated that more than 10 per cent of the country's soybean crop had been using the technology illegally (in 2002). Since then, GM technology use has extended to cotton in 2006 and corn in 2008. A similar process of widespread illegal adoption of GM HT soybeans occurred in Paraguay and Bolivia before the respective governments authorized the planting of soybean crops using this GM trait.

Notable for their absence from the list of prominent countries adopting GM technology are member states of the European Union. Here adoption of the technology has been limited due to a history of negative attitudes towards the technology,[5] coupled with limited and restrictive authorizations. More specifically:

- In the *European Union*, only two traits have been given approval for commercial planting. An insect-resistant (to corn-boring pests) trait in corn was approved in 1998 (before an effective moratorium on adoption was operational between 1999 and 2003) and in 2010 it was planted on 88500 ha, equal to only 1.1 per cent of the total EU corn area. The majority of this corn was planted in Spain (86 per cent). A high starch content potato was also approved in 2010, when a small area of 245 ha was planted to this trait in the Czech Republic, Sweden and Germany.
- In *Romania* GM HT soybeans were approved in 1999 and became widely adopted.

In 2006, 131 000 ha of soybeans used this technology (69 per cent of the total crop). When Romania joined the EU in 2007, it was no longer permitted to allow the planting of GM HT soybeans. Largely as a direct result of this ban and the associated loss of significant farm income gains that had arisen from use of this technology (equal to an average of US$104/ha between 1999 and 2006), the total area of soybeans has fallen by 77 per cent to only 41 000 ha in 2010.

In addition, in the *Ukraine* the planting of GM crops is banned and no GM traits have been approved for use. Despite this, about 25 per cent of the soybean crop in 2011 was estimated to be planted to illegally imported varieties containing the GM HT trait (tolerance to glyphosate) and a small area of corn is also estimated (1–2 per cent) to be planted to illegally imported varieties containing GM resistance to corn-boring pests.

GM TECHNOLOGY ADOPTION AND SIZE OF FARM

In relation to the nature and size of biotech crop adopters, there is fairly clear evidence that the size of the farm has not been a factor affecting use of the technology. It has been adopted by both large and small farmers, with size of operation generally not a barrier to adoption. In 2010, 15.4 million farmers were using the technology globally (James, 2011), 90 per cent of which were resource-poor farmers in developing countries. For example the average size of cotton farms using GM IR technology in China and India is 0.4 ha and 1.6 ha respectively.

GLOBAL TRADE AND GM CROPS

Global trade in the four main biotech-trait-using crops and their derivatives is heavily influenced by the presence of GM-derived materials. Looking at the extent to which the leading biotech-producing countries are traders (mostly exporters) of these crops and key derivatives, Table 2.3 and Table 2.4 show the following:

- *Soybeans*: in 2010/11, 35 per cent of global production was exported and 98 per cent of this trade came from countries which grow biotech soybeans. As there has been some development of a market for certified conventional soybeans and derivatives (mostly in the EU, Japan and South Korea), this has necessitated some segregation of exports into biotech versus conventional supplies or sourcing from countries that do not use biotech soybeans. Based on estimates of the size of the certified conventional soy markets in the EU and SE Asia (the main markets),[6] about 3.3 per cent of global trade in soybeans is probably required to be certified as conventional, and if it is assumed that this volume of soybeans traded is segregated from biotech soybeans, then the biotech share of global trade is 95 per cent. A similar pattern occurs in soymeal, where 85 per cent of globally traded meal probably contains biotech material.
- *Corn*: 11 per cent of global production was internationally traded in 2010/11.[7] Within the leading exporting nations, the biotech growers of the US, Argentina,

Table 2.3 Share of global crop trade accounted for biotech production 2010/11 (million tonnes)

	Soybeans	Corn	Cotton	Canola
Global production	264.2	828.0	25.1	60.2
Global trade (exports)	93.45	91.3	7.75	10.34
Share of global trade from biotech producers	91.66 (98.1%)	76.5 (83.7%)	5.56 (71.7%)	8.46 (81.8%)
Estimated size of market requiring certified conventional crops (in countries that have import requirements)	3.1	4.5	Negligible	Negligible
Estimated share of global trade that may contain biotech (i.e. not required to be segregated)	88.56	72.0	5.56	8.46
Share of global trade that may be biotech	94.8%	78.9%	71.7%	81.8%

Notes: Estimated size of market requiring certified conventional crops in countries with import requirements excludes countries with markets for certified conventional crops for which all requirements are satisfied by domestic production (for example, corn in the EU). Estimated size of certified conventional crops market for soybeans (based primarily on demand for derivatives used mostly in the food industry): EU 2.1 million tonnes bean equivalents, Japan and South Korea 1 million tonnes.

Sources: Derived from and updated – USDA & Oil World statistics, Brookes (2008).

Table 2.4 Share of global crop derivative (meal) trade accounted for biotech production 2010/11 (million tonnes)

	Soymeal	Cottonseed meal	Canola/rape meal
Global production	177.8	19.7	33.6
Global trade (exports)	60.0	0.5	4.8
Share of global trade from biotech producers	53.6 (89.3%)	0.236 (47.2%)	2.8 (58.3%)
Estimated size of market requiring certified conventional crops (in countries that have import requirements)	2.5	Negligible	Negligible
Estimated share of global trade that may contain biotech (i.e. not required to be segregated)	51.1	0.236	2.8
Share of global trade that may be biotech	85.2%	47.2%	58.3%

Notes: Estimated size of certified conventional market for soymeal: EU 2.25 million tonnes, Japan and South Korea 0.25 million tonnes (derived largely from certified conventional beans referred to in above table).

Sources: Derived from and updated – USDA & Oil World statistics, Brookes (2008).

Brazil, South Africa and Canada are important players, accounting for 83 per cent of global trade. As there has been some limited development of a biotech versus certified conventional corn market (mostly in the EU, and to a lesser extent in Japan and South Korea), which has necessitated some segregation of exports into biotech versus certified conventional supplies, the likely share of global trade accounted for by biotech corn exports is about 79 per cent.

- *Cotton*: in 2010/11, 31 per cent of global production was traded internationally. Of the leading exporting nations, the biotech cotton-growing countries of the US, Australia, India, Brazil and Burkina Faso are prominent exporters, accounting for 72 per cent of global trade. Given that the market for certified conventional cotton is very small, virtually all of this share of global cotton trade from biotech cotton-growing countries is probably not subject to any form of segregation and almost any shipment of cotton may contain biotech-derived material.[8] In terms of cottonseed meal, the biotech share of global trade is 47 per cent.
- *Canola*: 17 per cent of global canola production in 2010/11 was exported, with Canada being the main global trading country. The share of global canola exports accounted for by the three biotech canola-producing countries (Canada, the US and Australia) was 82 per cent in 2010/11. As there has been only a very small development of a market for certified conventional canola globally (the EU, the main market where certified conventional products are required, has been largely self-sufficient in canola and does not currently grow biotech canola), non-segregated biotech exports from Canada and the US probably account for 82 per cent of global trade. For canola/rapemeal, the biotech share of global trade is about 58 per cent.

CONCLUSION

GM traits have largely been adopted in four main crops – canola, corn, cotton and soybeans – and derive from two main categories of trait: insect resistance in corn and cotton, and herbicide tolerance in all four crops. In the countries where these technologies have been available to farmers, adoption has tended to be rapid and has often accounted for the majority of production within several years. The rate of adoption has followed a fairly consistent upward trend, although the increasing availability of stacked-traited seed, containing both IR and HT technology in corn and cotton, has made an important contribution to adoption rates in the last 5–7 years.

NOTES

1. In 2010 there were also additional GM crop plantings of papaya (410 ha), squash (2000 ha) and sugar beet (445 000 ha) in the USA. There were also 4500 ha of papaya in China and 20 000 ha of sugar beet in Canada.
2. The reader should note that the total plantings by trait produces a higher global planted area (162.4 million ha) than the global area by crop (129.4 million ha) because of the planting of some crops containing the stacked traits of herbicide tolerance and insect resistance.

3. The balance represents a combination of events withdrawn from the market for commercial reasons (for example, being superseded by improved events) and others that, whilst given regulatory approval for planting, were not subsequently brought to market by the companies that developed them.
4. The only other African country where commercial GM crops were planted in 2010 was Burkina Faso. First used commercially in 2008, insect-resistant cotton now accounts for two-thirds (260 000 ha) of the total crop (2010).
5. In the media, amongst some citizens and politicans, with anti-GMO activist groups very prominent and vocal in the EU.
6. Brookes (2008) and updated from industry sources and own research.
7. Corn is an important subsistence crop in many parts of the world and hence the majority of production is consumed within the country of production.
8. We consider this to be a reasonable assumption; we are not aware of any significant development of a certified conventional versus biotech cotton market and hence there is little evidence of any active segregation of exports from the US and Australia into these two possible streams of product. This includes the exports from other biotech-growing countries such as China and Argentina.

REFERENCES

Brookes, G. (2008), 'Economic impact of low level presence of not yet approved GMOs on the EU food sector', Brussels: GBC Ltd, for CIAA.

James, C. (2011), 'Global status of commercialized biotech/GM crops: 2010', *International Service for the Acquisition of Agri-Biotech Applications* (*ISAAA*) *Brief*, 42, Ithaca, NY, Cornell.

3 Structure of public research

Richard Gray and Buwani Dayananda

1 INTRODUCTION

Biotechnology has profoundly impacted agricultural research. Molecular science has vastly increased the scope of what is understood and what can be understood about crop and pest genetics. This explosive growth of knowledge changes the possible, both in terms of the speed of genetic development and the scope of outcomes. Importantly, biotechnology has also enabled better protection of genetically related intellectual property, changing the incentives for private crop breeding. Beginning in the mid-1980s, these new incentives attracted significant new private investment followed by a period of industry mergers and acquisitions as several firms acquired the pools of intellectual property required to become leaders in agricultural biotechnology (Fulton and Giannakas, 2001; Graff et al., 2003). In the mid-1990s there was a widespread acquisition of seed firms as the firms heavily invested in biotechnology realized they needed access to locally-adapted germplasm and seed marketing channels (Howard, 2009).

Private firms tend to dominate breeding in those few genetically modified (GM) crops (for example maize, soybeans, cotton and canola) in those jurisdictions where GM cultivars can be commercialized – mainly USA, Brazil, Argentina, India and Canada (James, 2011). Private investment is also largely a developed country phenomenon: the private sector is responsible for over one half of total research expenditures in developed countries while private investment makes up less than 10 per cent of total agricultural investment in developing countries (Pardey et al., 2006). Notably, because of the large fixed costs associated with the commercialization of GM crops, small-area crops have yet to receive significant biotechnology investment from either the public or private sectors in either the developed or developing world.

This chapter focuses on how agricultural biotechnology has altered the role and structure of public crop research. Crop research is important both in terms of resource costs and research outcomes. The examination of how the structure of public research has changed for crops and countries where biotechnology has been widely adopted provides insights into how and where biotechnology might be adopted in the future. From a normative perspective, a review of successful structures can provide models for implementation elsewhere.

The remainder of this chapter is organized in four sections consistent with Ostrom's (2011) Institutional Analysis and Development (IAD) framework. Section 2 deals with models. Methods (section 3) describes how selected jurisdictions have organized the structure of public and producer research related to biotechnology. Metrics (section 4) examines evidence of the performance of various public research structures and outcomes, being mindful of the youthfulness of the biotechnology sector and the long lags involved in R&D. The critical assessment (section 5) summarizes the chapter, assessing how the structure of public research is changing and the challenges it faces in supporting biotechnology-driven innovation.

2 MODELS

2.1 Public Research as Part of a National Innovation System

The understanding of the forces that drive innovation have evolved significantly over the past 50 years. Noisi (1999) argues that we are now in the fourth generation of innovation frameworks. First generation models were linear in nature, where research led to knowledge that led to new products. In the second generation, the importance of feedback between each stage of the model was recognized, leading to chain link models of innovation. In the third stage it was recognized that innovation in a firm required a more integrated approach, where research programmes incorporated feedback from all points in the supply chain, including product end users.

The fourth generation, modern conception recognizes innovation as a complex system with interactions and feedback between firms and many institutions. Success of private R&D was dependent on the public research, infrastructure and so on. Nelson (1993) was among the first to articulate a National Innovation System approach, where an array of institutions are needed for successful development. Etzkowitz and Leydesdorff (2000) built on this with the Triple Helix model that focuses on the vitally important linkages and feedback between the government, universities and private enterprise in R&D commercialization. This fourth generation concept of innovation recognizes not only the relationship between public and private enterprise but also the array of relationships within the private sector including industry organization, strategic licensing, research consortiums, IP sharing and knowledge acquisition strategies and institutions.

The development of these frameworks both created ways of conceptualizing how innovation is organized and, in many cases, changed how the innovation system actually operates. In turn, developments and functional changes within the innovation system have changed our understanding of innovation. This co-evolution of the innovation framework and the development of the innovation system, which was refined during the recent era of biotechnology, have profoundly affected the structure of public research and the role it plays.

2.2 Economic Theories Related to the Structure of Public Research

A number of economic theories based in neoclassical and new institutional economics have been used to give insight into the structure of public research in the context of biotechnology. Neoclassical economic theory describes the role of spillovers, property rights, toll goods and industrial organization in the failure of markets to provide adequate incentives for innovation. New Institutional Economics (NIE) is a field of economics that recognizes the presence of transaction costs in an economy. Information is costly to create, acquire and process. Because information is costly, agents often operate and interact in an environment with incomplete information and incomplete contracts, which increase the scope for opportunistic behaviour. In the NIE framework organizations and institutions matter because they determine how information is acquired and used. This NIE framework has spawned a number of theories relevant to the structure of public research and biotechnology. Three of these theories relate to transactions costs associated with the proliferation of IP and one is related to the theory of voice and effective governance. Although we recognize and appreciate that theories from other academic disciplines are relevant for determining the structure of public research, space and knowledge limitations restrict us from exploring the other theories in this chapter.

2.2.1 Inadequate private research investment

A central tenet of neoclassical economics is that when property rights are defined, such that individuals must pay for the goods they consume, prices will reflect social marginal cost and benefits, and markets will operate to maximize benefits to society. The powerful economic result, often referred to as 'the invisible hand', hinges on whether or not a good is excludable. The right to exclude others from using a good is central to the notion of private property. Without the ability to exclude others, goods are essentially in the public domain because private individuals have no incentive to pay for or provide a non-excludable good. If a good is non-excludable, the ability to use a good 'spills over' to others who are not obliged to pay for the good, and market demand will not reflect the good's full social value. The market demand will reflect only those benefits that can be excluded if payment is not made. Economists use the term 'spillovers' to refer to the benefits that are received but not paid for by the recipients in a market transaction.

The use of property rights as a mechanism to eliminate spillovers was elucidated by Ronald Coase in the Theory of Social Cost (1960). Coase argued that in the absence of transactions costs, spillovers can be eliminated and private incentive corrected through the assignment of property rights. In their most basic form, intellectual property rights (IPRs) grant exclusive rights for individuals to use and protect knowledge from their innovation. In industry, copyright and patent systems have conferred exclusive rights to inventors for more than a century in many countries and patents have been a very important form of protection for biotechnological knowledge and GM crops.

While governments have gone some distance in creating IPRs for agricultural research, they are often limited in nature and difficult to enforce. As a result, many forms of agricultural intellectual property have incomplete protection and spillovers persist. Notably, IPRs are often strongest where technology is difficult or costly to copy, such as a hybrid crop or a pesticide, and weakest when there are widespread low-cost means of reproduction (such as for open pollinated seed). As discussed later, governments have often been reluctant to create strong IPRs for agricultural knowledge, especially for plant varieties. In some cases, this reluctance is driven by the inability to enforce the IPRs and in other cases it is based on a conviction that some forms of knowledge created in the public domain should remain within the public sphere.

2.2.2 Market power and monopolistic pricing of IP

In addition to excludability, which is dependent on property rights, neoclassical economic theory recognizes that goods can also differ in the extent that they are rivalrous (often referred to as subtractability). Most economic goods are 'rival', such that if they are used by one individual they cannot be used or consumed by another. For example, a sandwich is only eaten once, or a litre of gasoline is burned only once. However, some goods, including new crop varieties, new genetic markers, or agronomic knowledge, are non-rival and are not diminished by use. Once created, these non-rival goods can be used any number of times and shared without incurring a significant marginal cost. When protected by IPRs, knowledge becomes a toll good (Lesser, 1998; Fulton, 1997), which is non-rival yet excludable.

When toll goods, including protected IPRs, are a key input into a production process, they are likely to result in significant market concentration (Fulton, 1997). Because the toll-good input is non-rival, it only has to be purchased or created once. This fixed cost is incurred only once for each such good – for example, a new variety of canola – and the same genetic material can be used again and again without additional costs or without reducing its availability to others. This means that the average cost of producing the final output (that is, seed using this genetic material) decreases with the quantity produced because the cost associated with purchasing, or creating, the non-rival input (the new variety) is spread over more units of output. The declining average cost implies that large firms will always have a cost advantage over smaller firms. The lowest industry average cost can be achieved if the good is supplied by a single monopoly.

Toll-good industries for which fixed costs represent a large share of total costs, such as railways, software companies, or electrical distribution networks, are often referred to as natural monopolies. The theory suggests that conditions of perfect competition cannot exist in a toll-good industry. As Lesser (1998) and Fulton (1997) point out, given the nature of the cost curve, if a firm were to price its product equal to marginal cost, which is a condition of perfect competition, the firm would not be able to cover the fixed cost of the research and would not be viable.

2.2.3 Knowledge sharing and duplication of research effort

The biotechnology research revolution created whole new platforms for crop research and genetics advancement, spawning vast quantities of IP with patent claims that can increase transactions costs. To illustrate the vast number of patents, on 12 July 2012 a simple search of the US patent database revealed 6004 patents that are related to 'stress and tolerance and corn'. Searching this large database and identifying which patents are potentially useful, determining what patents are enforceable, and what IP can be safely used without violating other patents, is a time-consuming and costly undertaking. The resulting outcome is often called the 'patent' thicket (Wright and Pardey, 2006), which economic theory suggests adds to the cost of protection and use of IP, which in turn can discourage innovation and limit the use of available knowledge. Public research can be structured to protect IP and add to the growing number of patents or be published in the public domain.

When a single product of biotechnology is made up of several independently-owned pieces of IP, a phenomenon known as the anti-commons can limit the ability of researchers to access existing knowledge if the requisite IP is jointly held by several independent owners (Graff et al., 2003; Heller, 1998; Wright and Pardey, 2006). If many firms own complementary IP, the *ex-post* bargaining behaviour of the individual owners may make it difficult and sometimes impossible to reach an incentive-compatible sharing agreement among all of the requisite owners (Parisi et al., 2005; Smyth and Gray, 2011). The classic case of this is GoldenRice™, which was estimated to contain 44 pieces of IP in the United States that were owned by at least 12 different organizations (Kryder et al., 2000).

For all of the above reasons, firms often have not licensed their IP and have opted to develop their own research platforms, which duplicate effort and drive up the industry cost curve. Other solutions include the creation of research consortiums with *ex ante* IP sharing arrangements.

2.2.4 Coordination and voice within the innovation system

Within the scope of new institutional economics, Picciotto (1997) adds an important dimension to the classification of goods, arguing that some goods require voice for effective governance. These goods fall between those provided by the public sector through hierarchical government institutions and those that are excludable and rival and can effectively be provided by the private sector. In many cases, these goods are effectively provided by a participatory sector, with or without a supporting policy framework. Elements of this distinction were made by Coase (1974) when he pointed out that lighthouses, a classic non-rival, non-excludable public good, were not provided by the government but had been effectively provided by Trinity House, an association of ship owners in England for more than 300 years. Coase pointed out that ship owners had a better sense of how many lighthouses to build and where they were needed than the government. By enabling Trinity House to tax ship owners at port, the government effectively gave voice and control to those with a direct interest in the operation of lighthouses. In dispelling the myth of the tragedy of the commons, Ostrom et al. (1999) point out that the use of the commons is often actively controlled by the communities that use them, who make up rules and practices that preserve the valuable pasture resources. Picciotto (1997) argues that non-profit organizations coming from civil society can play an important role in providing those goods that require industry or local voice and are usually inefficiently provided or not provided at all by the private sector.

In agriculture, levy-based funding is common, and monies collected with industry sales are reinvested in market development, research and other industry activities. The funds are commonly established with a vote of farmers whose representatives are on the board of directors that determine spending allocation, giving voice to those who will fund the research and use the knowledge generated.

Alston et al. (1995) model the economic incentives for levy-funded research and show that under some conditions the incidence of a levy that leads to a per-unit reduction in costs is shared between producers and downstream consumers in a proportion equal to the benefits of research. Later research by Alston and Fulton (2012) shows that this only holds if producers are homogeneous. In a more general case, producers will invest too little in research when often super-majority rules are used in the governance process. Other research shows that voluntary or refundable levies are subject to free-riding. The lack of excludability and the threat of additional free-riding reduces the levy to suboptimal levels, resulting in the underfunding of levy-funded research.

In sum, economic frameworks, theories and models are somewhat prescriptive in terms of how effectively markets, non-profits and the public sector operate. This theory suggests that non-excludability, non-rivalry and the need for voice will affect the efficacy of national innovation systems and more specifically the role of the public sector within these systems. These theories and outcomes have been an important source in shaping the role of the public, private and producer (industry) sectors in agbiotech research.

3 METHODS

The purpose of this section is to describe the structure of public research with respect to agricultural biotechnology, which varies considerably by country and often by crop.

A major catalyst for change in the structure of public research has been the entry of the private sector in agricultural research and crop breeding, which in part has been driven by biotechnology. Given the potential scope of this task, we have confined our focus to an overview of global change in public expenditure levels, a summary of some of the structural changes in some of the larger agricultural countries, an overview of the impacts on private investment and finally a general description of how these investments have changed public roles and the development of public–private partnerships.

The development of biotechnology has significantly changed the structure of public agricultural research. These impacts, however, differ considerably by crop and by jurisdiction. In general, where strong intellectual property rights exist for large crops, the private sector has seed revenues exceeding US$100 million per year, and have developed large and sophisticated breeding and pre-breeding research programmes, leaving only basic science for the public sector. Where IPRs are somewhat weaker, such as the EU crop sector protected by plant breeders' rights (PBR), the private sector is considerably smaller and by necessity focuses primarily on crop breeding and relies to a far larger extent on the public sector for pre-breeding research and basic science. Where property rights are very weak or non-existent, private sector research is not viable and the innovation system depends on the public sector for basic science, pre-breeding and applied breeding activities.

3.1 Global Public Agricultural R&D Spending Levels

Boettiger et al. (2004) argue that R&D in agriculture is unique among other industries because of the global reach of R&D and the historical success of this largely public enterprise. In 2009 global public agricultural and food R&D expenditure was around $33.5 billion (2005 PPP$) (Pardey et al., 2012), having grown on average (inflation-adjusted) at 3.3 per cent per year since 1960 (Pardey et al., 2012). As illustrated in Figure 3.1, four

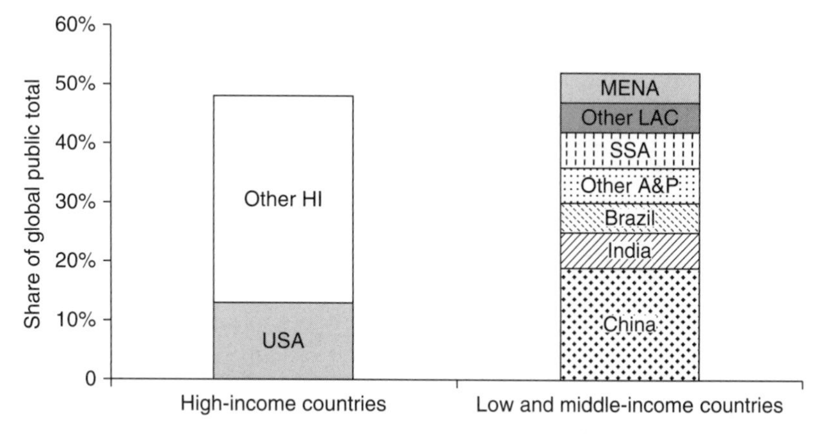

Notes: MENA = Middle East & North Africa (excluding Qatar & United Arab Emirates); SSA = sub-Saharan Africa; LAC = Latin America & Caribbean; A&P = Asia & Pacific (excluding Japan & Singapore).

Source: Pardey et al. (2012, p. 40).

Figure 3.1 Global public agricultural and food R&D spending in 2009

countries – China, USA, India and Brazil – play a major role in global public agricultural R&D. Interestingly the Asia-Pacific countries' share has increased from 20 per cent in 1960 to 31 per cent in 2009, mainly due to increased investments of China and India.

High-income countries[1] as a group decreased their share of global public agricultural R&D spending from 58 per cent in 1960 to 48 per cent in 2009 (Pardey et al., 2012). However, private agricultural R&D spending increased considerably from $6.4 billion in 1981 to $12.8 billion in 2000 (Alston et al., 2010).

3.1.1 The US

The US plays the major role in agricultural R&D and therefore it is worthwhile to examine the changes in US crop research funding over time. In 2006 total global private sector expenditure for food and agriculture R&D was $18.6 billion, and the US accounted for one third of that amount (Fuglie et al., 2011). Schimmelpfennig and Heisey (2009) have carried out a comprehensive analysis on changes in funding sources and shifts in emphasis of US public agricultural research during 1980–2005 and found that funding levels from the various sources that support public agricultural research have changed since 1980 and shifts in funding from these various sources have resulted in constant or slowly increasing overall expenditures on public agricultural research. Figure 3.2 illustrates that private-sector investment in agricultural R&D in the US surpassed public-sector investment by the early 1980s.

Schimmelpfennig and Heisey (2009) further suggest that there is an increased attention in the US public agricultural research portfolio towards applied research than basic research.

3.1.2 China

As explained by Hu (2012), agricultural research has been an engine of growth in China. China's total factor productivity (TFP) has increased over 1995–2004 in a number of crops including rice, wheat, maize, soybean and cotton, at least partly due to increased

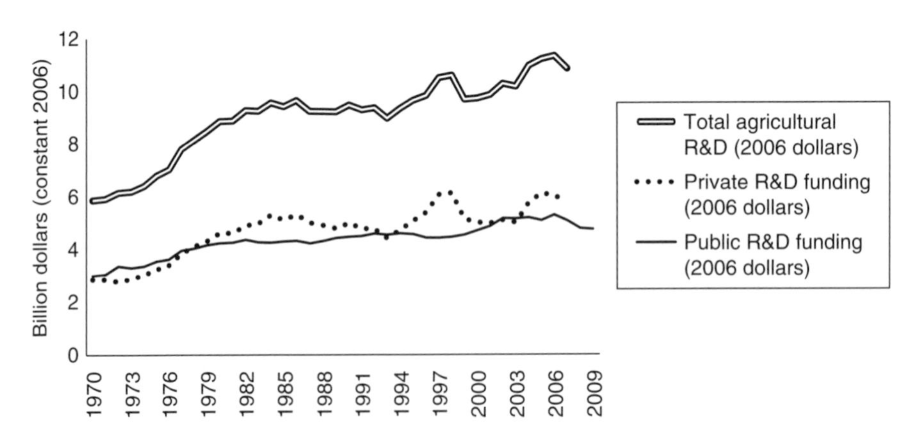

Source: Schimmelpfennig and Heisey (2009).

Figure 3.2 Agricultural research funding in the public and private sectors in the US, 1970–2009

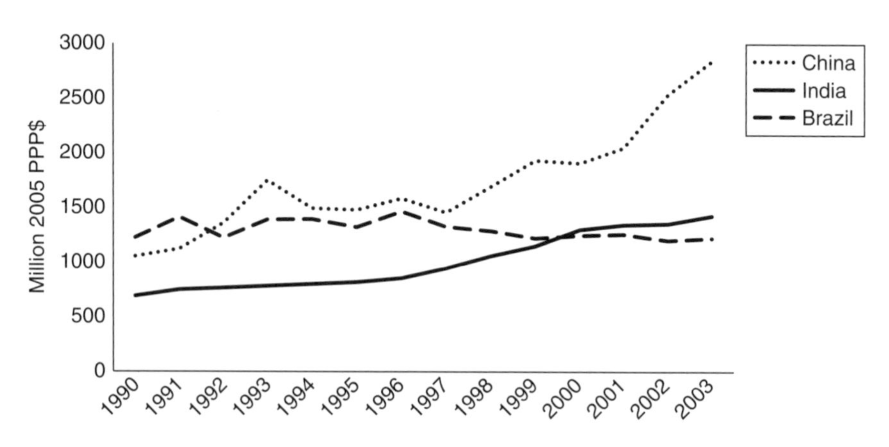

Source: ASTI (2012).

Figure 3.3 Total agricultural R&D spending: public sector (million 2005 PPP$)

public sector investment in agricultural R&D. From 2000–2009 China's fiscal investment in agricultural R&D has increased at an annual growth rate of nearly 15.9 per cent (Hu, 2012). As shown in Figure 3.3, in 2008 public investment in agricultural R&D in China was US$3.6 billion (2005 PPP$) (ASTI, 2012). Furthermore, China has the world's largest agricultural extension system with one million staff members who bring innovation information even to the furthest corners of China (Hu, 2012).

Reforms in China's agricultural research system in the 1980s and 1990s were aimed at encouraging and stimulating the private sector to take a greater role in applied research (Hu et al., 2011). In 2006, private investments in agricultural R&D accounted for nearly 17 per cent of the total agricultural R&D in China (Hu et al., 2011) and was three times higher than the share in other developing countries (6 per cent).

In 2008, China initiated 'The breeding and cultivation of new GM varieties' project with a total investment of US$3.5 billion. The funding for this project comes from central and local governments as well as investments by companies (USDA, 2011a). The project aims to integrate GM technology with traditional methods to improve crops such as rice, wheat, maize, cotton and soybean to obtain excellent quality, high yield and tolerance to abiotic and biotic stresses (Huang et al., 2011).

3.1.3 India

Public sector investment in agricultural R&D in India has increased gradually over the last four decades. Pal and Byerlee (2006) show that funding increased in real terms from $284 million (in 1999 PPP$) in 1961 to $2.9 billion (in 2000 PPP$). According to Pal and Byerlee (2006), this trend shows a continuing, strong political commitment to invest in agricultural R&D despite continuous challenges related to changes in governments and shifts in public-investment priorities. Bt cotton is the only biotech crop currently approved for commercial cultivation in India, and 90 per cent of the total cotton area uses GM varieties (USDA, 2011b). Most of the approved Bt cotton hybrids in India are from two Monsanto events; other approved events include one sourced from China and the three locally developed ones (USDA, 2011b). In addition to cotton, private and public sector research institutions in India are working on the development of various biotech crops mainly for traits like pest resistance, nutritional enhancement, drought tolerance and yield enhancement. The crops currently being developed by public sector institutions include banana, cabbage, cassava, cauliflower, chickpea, cotton, eggplant, rapeseed/mustard, papaya, pigeon pea, potato, rice, tomato, watermelon and wheat (USDA, 2011b).

3.1.4 Brazil

Brazil's large and complex agricultural research system, composed of public institutes, universities, private companies and non-governmental organizations, is considered as one of the most comprehensive and efficient R&D systems in the tropics (Lopes, 2012). The Brazilian Agricultural Research Corporation (Embrapa), the largest actor in the Brazilian agricultural R&D system, is mostly funded by the federal government. In 2006, public sector investment in agricultural R&D totalled $1.224 billion (2005 PPP$), accounting for 41 per cent of Latin America's total agricultural R&D (Stads and Beintema, 2009). However, in the case of agbiotech the private sector plays a leading role in Brazil, and 78 per cent of the biotech companies get public funding for their R&D in biotechnology (BRBIOTEC Brasil and Apex-Brasil, 2011). The main biotech crops grown in Brazil are cotton, corn and soybean.

3.1.5 Africa

Karembu et al. (2009) have analysed the status of biotech crops in Africa and shown that public-sector R&D is low in agbiotech in Africa while different innovative, public–private partnerships have been adopted recently. Through these partnerships, various GM varieties of cotton, maize, cassava, sweet potato, banana, soybean, groundnut and cowpeas have been tested under confined field trials in different countries in Africa during the last few years. However, cotton, maize and soybean are the only commercially producing GM crops in Africa, with South Africa, Burkina Faso and Egypt being the only African countries growing biotech crops commercially as of 2011 (James, 2011). Monsanto has the largest share of the agbiotech market in Africa, being the proprietor of most of the commercially grown GM events.

3.1.6 The European Union

In the European Union, crop breeding is generally undertaken by the private sector, while upstream crop research is almost exclusively a public endeavour. The private sector breeding programmes are modest in size and scope and funded through royalties on the sale of seed. While there is a good deal of expenditure on public genomics-related research, transgenic research remains very limited as GM events are not approved for general use and have yet to be commercialized in any significant way.

The extent to which public expenditures on genomics and agricultural biotechnology are linked to the needs of the private breeding industry differs by country. In France, which has the largest public research programme, it appears to be well integrated with the downstream private breeding sector. The UK government sold PBI (Plant Breeding Institute), a highly successful public wheat breeding institute at Cambridge, to Unilever in 1987, ending the public role in wheat breeding. Subsequently, many UK public scientists shifted to basic science, far removed from applied pre-breeding wheat research, leaving a small private wheat breeding sector without significant public research support. Over the past decade, public researchers have been carrying out more applied wheat research, a move brought about by public–private research consortiums and changes in public research funding formulas that encourage public researchers to collaborate with the private sector.

3.1.7 Canada

Canada has a modest sized agbiotech industry supported by both private and public sector research activities. The private sector is made up of several large multinational firms (Bayer, BASF, Monsanto, Pioneer, Dow and Cargill) that have invested in crop breeding and trait development, while many smaller firms continue to explore novel processes, traits and functional food attributes. Agbiotech research continues at many public institutions, and federal and provincial governments invest in a wide spectrum agricultural research.

Biotechnology's impact on the public sector is most evident in the canola sector, where very successful transgenic traits and hybridization have transformed the structure of the industry. The early development of the canola sector had its roots in public research. In the late 1950s public scientists began working with the Rapeseed Association of Canada to create an oilseed crop for Canada. By 1974 they had successfully bred varieties of rapeseed with low levels of erucic acid and glycosynolates, which were eventually trademarked as a new crop called canola. In the early 1980s public scientists were developing new breeding techniques and exploring hybridization. By the late 1980s there was a large influx of private companies conducting research, exploring transgenic technics and undertaking canola breeding. The first transgenic herbicide-tolerant canola was introduced in 1995. Using patent protection Monsanto was able to earn licensing revenue for its Roundup Ready varieties, and AgrEvo (now Bayer) was able to earn rents from the sale of herbicide required for their HT system. These rents were reinvested in private breeding. As private breeding increased, Agriculture Canada withdrew from varietal development and moved upstream to support the private sector through trait and germplasm development. In the late 1990s the first canola hybrids were commercialized. By 2004 the industry was dominated by the sale of superior hybrid, HT varieties. In 2012, with many hybrid seed varieties being sold at \$18 per kg, plus other licensing costs, the 2012 Canadian canola seed industry sales approached \$1 billion.[2] Given historical investment patterns, this would provide about \$100 million for reinvestment in breeding and research activities (CSTA, 2007). The private sector also has a significant but far smaller presence in soybean and hybrid corn research, where patents and commercial transgenic crop production also created a return to biotechnological research. In other crops, notably wheat, barley, flax and pulse crops, where transgenic varieties are not licensed for commercial production in Canada, private crop research has remained limited.

The public sector continues to actively fund agricultural biotechnology research in Canada. Public researchers undertake basic scientific research contributing to both public knowledge and the pool IP protected knowledge. Genomics research is coordinated through Genome Canada, which provides approximately \$100 million per year project funding for matching from industry and other sources. Of this total, approximately 20 per cent is used for agriculture-related genomic research. This level of funding enables public researchers to develop expertise in some crops, which is then leveraged through international collaboration, to improve the understanding of crop genomics and function.

3.2 Emerging Realms for the Private and Public Sectors in Research and Crop Breeding

The role of the private sector is largely determined by the ability of firms to earn a return on investment, which depends on the strength of IPRs, the size of the market and the cost of doing research. Not surprisingly private firms have invested the most where IPRs are strong and the market is large. For instance, prior to biotechnology, hybrid corn had a very substantial private research presence, as seed companies could sell farmers seed each year and capture a return on their investment.

The strength of IPRs varies a great deal by crop and jurisdiction. In many developing countries, the lack of IPR enforcement makes it nearly impossible to capture a private return from self-pollenated crop breeding. PBRs offer some protection. In the weakest form of PBR established in the International Union for the Protection of New Varieties of Plants (UPOV) (1978), producers retain the farmers' privilege or right to save seed for reseeding. Perrin and Fulginiti (2009) estimate that the farmers' exemption limits breeders to capturing about 11 per cent of the value of their innovations. In the UK a 50 per cent royalty levied on farm-saved seed increases breeders' returns somewhat. In Australia, PBRs include the right to charge 'end point royalties', effectively conferring complete hybrid-like protection for plant breeders.

The development of biotechnology had a large impact on breeder IPRs. First of all, in jurisdictions where they are permitted to be commercialized, the introduction of GM crops strengthened IPRs for seeds as patented traits allowed firms to earn royalties or 'technical use fees' each time a variety is grown. Second, a large number of the tools, processes, and even genetic traits developed using biotechnology can be patented and protected by their owners. This upstream protection allows private firms to earn a private return from biotechnology research. A good deal of biotechnology IP is now owned by the private sector and is used as part of cross-licensing agreements with rival firms (Smyth and Gray, 2011).

The changing realm of private research and breeding activities has directly impacted the realm of public crop research and breeding. In general, governments and universities have withdrawn from breeding in those crops with a large private presence in breeding (for example US corn, Canadian canola and UK wheat). In addition, a number of public–private partnerships have developed in Australia (for example InterGrain Pty Ltd, Australian Grain Technologies Pty Ltd, and HRZ Wheats Pty Ltd). Some countries have created quasi-government institutions for breeding and commercialization of varieties. Finally, the public sector continues to operate breeding programmes in some crop sectors – often where there are weak IPRs, or for small crops that have not attracted private breeding investments.

In contrast to breeding, the public sector continues to fund upstream research for nearly all commercially important crops. In most crops there is little private investment in upstream research and this lack of private research creates a demand for public research. Firms in the industry often lack enough revenue to invest individually in research beyond breeding. These firms and industries argue they require the public research expenditure to continue to be competitive in downstream breeding activities. There is some evidence that when the public support for biotech research decreased in Sweden, the firms relocated to where there was public research, to utilize in their research programmes. In the very large crops with well protected IPR, the private sector does invest heavily in many forms of upstream research, including biotechnology. Despite significant private investment, the public sector continues to undertake research in some topics as a means of supporting and complementing the private research. The public sector is the least active in research where the governments lack the fiscal capacity and/or the crop is too small to warrant research investment.

Now that the public sector is the source of upstream knowledge for the private breeding sector, a whole new set of partnerships, institutions and knowledge-sharing structures has emerged. When breeders and researchers were both in the private sector, a two-way flow of knowledge existed, where researchers and breeders worked together to solve strategic issues. With the development of the private breeding sector and the withdrawal of the public sector from these activities, a whole new set of public–private relationships is required. In general, basic science is funded through systems of competitive grants with limited input from the downstream private sector. As the research becomes more applied and fits into the category of pre-breeding research, the public and private industry recognize the need to create institutions to develop a better two-way knowledge flow.

Finally, in countries with large public breeding systems, there is an increasing awareness that the private sector owns a great deal of potentially useful IP which can be utilized with the appropriate property rights, partnerships arrangements and investment incentives. For example, China, Brazil and India all have partnership arrangements with multinational companies while maintaining very active public research and breeding programmes.

3.3 Knowledge Sharing and the Structure of Public Research

Public sector research has been influenced by several issues related to IP protection and knowledge sharing. The growth in the pool of patented IP, most of it in the private domain, has increased the cost and limited access to available knowledge. The existence of thousands of patents has made the task of determining what knowledge is protected and what can be protected an onerous and somewhat expensive task. Obtaining access to protected IP is also problematic for many public institutions. Firms are often reluctant to license their IP until a specific end use is apparent. The result is that public discoveries can sometimes sit on the shelf and not be commercialized. As a result of the cost and the uncertainties surrounding the use of external IP, many public institutions have developed and maintain their own research platforms, even when this involves a duplication of existing IP.

The IP landscape has also profoundly affected how most public institutions manage their own IP. Prior to biotechnology, most public institutions freely shared their agricultural assets. This changed abruptly when the US recognized the ability to patent biotech IP, and encouraged public institutions to protect and commercialize IP. This created a domino effect where public institutions around the world began to protect their IP in part as a defensive strategy and in part to capture any windfalls from discovery. In the early days of biotechnology, public institutions typically protected nearly all of their IP and would exclusively license their IP for commercial use if the opportunity arose. Over time governments have come to realize that the desire and ability to protect IP negatively impacts the sharing of knowledge between public institutions and between the public and private sector.

As the benefits of sharing knowledge became more apparent to policy makers, the public sector has taken measures to share more IP. This involved more scrutiny at the institutional level, including the development of Technology Transfer Offices to target only the most valuable IP for protection and commercialization. Public institutions have also collaborated with other public institutions – bilaterally through standardized material transfer agreements and multilaterally through the establishment of Public Intellectual Property Resource for Agriculture (PIPRA). Internationally, governments have collaborated to fund the research institutions of the Consultative Group on International Agricultural Research (CGIAR) system. In 2006 the most important gene bank collections of the world's key food and forage crops came under the International Treaty on Plant Genetic Resources for Food and Agriculture. The gene bank ensures that plant breeders, farmers and researchers will be able to access these plant genetic resources under standard conditions and share in the benefits arising from their use. More recently in 2011 the Agriculture Ministers from the G20 group of nations adopted the International Research Initiative for Wheat Improvement (IRIWI) with the goal to coordinate worldwide research efforts in wheat genetics, genomics and agronomy. At the national and sector level, public funding agencies are increasingly making the sharing of IP a requirement to receive funding. One result is that a good deal of genomics IP is published and lies in the public domain.

3.4 Industry Organizations and Public Research

The structure of public research has been impacted through cooperation with industry organizations that have given voice to downstream knowledge users. Industry organizations vary a great deal. In some cases the public sector research institutions deal with research consortiums made of the downstream breeding firms that commercialize technology. In other cases industry councils and advisory bodies provide input and advice into public research priorities. The Rapeseed Association of Canada played a major role in coordinating the research resources required to develop the canola industry. Levy-funded producer-directed research organizations established through legislation have had a large impact in some countries. In Australia, the Grain Research Development Corporation (GRDC) established in 1988, funded by a 1 per cent levy matched 0.5 per cent by government, has had a major impact on public grains research. With over $100 million per year in revenue, the GRDC is a major of funder of crop research and has provided leadership in transforming the role of public research. The GRDC worked with state governments to create three new corporations with public, producer and private shareholders that are now responsible for wheat breeding, while the GRDC and public resources have moved upstream to fund pre-breeding and other research activity. In Canada, the Saskatchewan Pulse Growers are funded through a 1 per cent levy on pulse sales and have worked with governments to undertake the research and breeding required to develop a vibrant export-based pulse industry.

4 METRICS

In the prevailing framework of innovation systems, the structure of public research is an important determinant of how effective the agricultural innovation systems will be in fostering productivity growth. Economic theory suggests that the inherent excludable, and non-rival nature of knowledge creates a number of challenges for the development of effective innovation systems. As outlined in section 2, there are four major challenges: (1) the presence of spillovers reduces incentives and results in *inadequate private research investment*; (2) in the presence of strong IP, market concentration arising from the economies of size can lead to *monopolistic pricing of IP inhibiting adoption*; (3) strategic IP ownership, patent thickets and anti-commons issues can result in *inhibited knowledge sharing and duplication of research effort*; and (4) spillovers and non-rivalry of knowledge creates the potential for synergies at the industry, national or international level that are unexploited without *sufficient coordination and voice within the innovation system*. Economists and other social scientists have made some effort to measure how effective the existing structures public research has been in addressing these issues.

4.1 Addressing Inadequate Private Investment

Economic theory suggests that if IPRs are incomplete, spillovers will reduce incentives for private research. Achieving the socially optimal level of investment involves creating stronger IPRs, undertaking public research, or subsidizing private research. The rate of return to research is a key metric for the level of investment. High marginal internal rates of return or high marginal benefit–cost ratios indicate a situation where larger investment would create net benefits for society. Unfortunately, there is a large, compelling, and growing body of empirical evidence that shows persistently high rates of returns to crop research. This implies that the measures taken to increase crop research have been ineffective in achieving the socially optimal level of research, even in an era of biotechnology.

Alston et al. (2000) compiled a comprehensive data set of studies that assess the rates of return to agricultural research. This meta analysis assembled 292 studies with available evidence on the returns to investment in agricultural R&D published since 1953. The average rate of return across 1128 observations used in the regression was 65 per cent per year with a mode of 28 per cent and median of 42 per cent. The overall mean rate of return across 1852 observations was 81 per cent per year with a mode of 40 per cent and median of 44.3 per cent. The findings of Alston et al. (2000) prove that the rate of return to investment in agricultural R&D was quite large.

In 2011 CGIAR estimated rates of return on their investment in all crop improvement research and reported a 39 per cent rate of return in Latin America and a more than 100 per cent rate of return in Asia, the Middle East and North Africa (CGIAR, 2011). Furthermore, CGIAR (2011) claim that for every $1 invested in CGIAR research, $9 worth of additional food is produced in the developing world.

Alston et al. (2010) modelled state-specific US agricultural productivity for the period of 1949–2002 as a function of public agricultural research and extension (R&E) investment over 1890–2002. Perhaps the most remarkable result of this study is the finding of very long research lags showing that the maximum impact of a dollar spent today occurs 24 years into the future. The finding of the study shows that marginal investments in agricultural R&E by 48 US states generated an average benefit of US$21 per research dollar invested. When the spillover benefits into other states are considered, the national average benefit–cost ratio turned out to be US$32 per dollar invested (at 3 per cent real discount rate). These large benefit–cost ratios confirm that there have been and continue to be high rates of return to agricultural research.

Alston et al. (2009) found an economically significant slowdown in agricultural productivity growth in most of the world excluding China and Latin America since 1990. The authors speculate that an earlier slowdown in agricultural R&D spending growth might be the contributing factor of the recent slowdown in productivity growth. Pardey et al. (2006) emphasize the importance of increasing investment on productivity-oriented R&D by the public sector to handle growing hunger and food security problems. Heisey et al. (2011) also argue that public spending on agricultural R&D will need to rise to maintain historical rates of productivity growth in US agriculture. Pardey and Alston (2012) argue that not only is there a slowdown in productivity growth rates in the agricultural sector but also there is increasing competition from non-farm sectors for water and land and new competition from biofuel demand, which spells slower growth in the supply of food. Pardey et al. (2006) argue that diverting the interests of developed country research away from productivity-enhancing technologies in agriculture will have long-term consequences, especially for the world's poor countries. It is also worth noting that most of the biotechnology research investment has occurred within the past 20 years and, given the lags discovered by Alston et al. (2010), it is too early to clearly understand the full impact of these technologies on productivity growth.

4.2 Market Power and Monopolistic Pricing of IP

There is clear evidence that the introduction of IPRs acted as an incentive for private agricultural R&D. However, as Fulton and Gray (2007) point out, once protected by IPR, knowledge acts as a toll good, and toll-good industries are often referred to as natural monopolies. Fuglie et al. (2011) compared five categories of agricultural input prices in the US – fertilizer, crop seed, farm machinery, animal feed and agricultural chemicals – and found that the largest change during 1994–2010 was in crop seed prices. Crop seed prices more than doubled relative to the prices received by farmers during that period. The fact that seeds are priced well above the marginal cost of seed production is consistent with the toll-good nature of the industry (Moschini and Lapan, 1997). The economic significance of seed cost has attracted the attention of economists. A study by Stiegert et al. (2010) found that the pricing of traits is correlated with measures of market concentration. Wilson and Dahl (2010) and Fernandez-Cornejo and Caswell (2006) argue that, on balance, the economies of size realized by concentration more than offset the higher pricing incentives. As recently as 2007 the Anti-Trust Division of the US Department of Justice held an inquiry into Monsanto's pricing behaviour (Wilson and Dahl, 2010). While unresolved, this issue continues to be a concern for policy makers.

4.3 Knowledge Sharing and Duplication of Research Effort

The ability to patent the knowledge of biotechnology has also affected how private and public firms operate with respect to their IP (Galushko et al., 2012). A proliferation of

IP has led to patent thickets, strategic licensing behaviour, and some anti-commons, which collectively reduced knowledge sharing, increased transaction costs, and reduced the freedom to operate for both private and public researchers (Graff and Zilberman, 2001).

There are many signs that the situation is improving for the private sector. Smyth and Gray (2011) show that after a significant period of very little cross-licensing activity, a flurry of cross-licensing activity took place between 2003 and 2006, which greatly enhanced knowledge sharing among the larger agricultural research firms, facilitating the stacking of key traits.

In the public sector, standardized material transfer agreements have facilitated greater exchange of germplasm. Some granting agencies have made knowledge sharing a requirement for funding. In the Australian wheat sector the GRDC requires that any knowledge created from GRDC grants be made available to all potential downstream users. In the UK the Biotechnology and Biological Science Research Council (BBSRC) requires publication of research results. As outlined in section 3.3, the public sector has played and continues to play a very important role in the international sharing of germplasm.

4.4 Coordination and Voice within the Innovation System

As innovations systems get larger and more complex, there is a need for greater coordination and voice in the system. Innovation systems now often involve public, private and industry organizations, which creates a need to coordinate research activities so that upstream research serves the needs of downstream users, duplication of effort is minimized, and important industry-wide needs are met.

Previous research has shown that institutions that share knowledge and give voice to downstream users can have a positive impact on the innovation system. While there is growing recognition of the need, these institutions have often been slow to develop. The drive to make the research more effective in creating innovation often leads to closer ties with downstream users, collaborative funding, research consortiums, or more general industry councils. Producer-run, levy-based research organizations have been very effective.

5 CRITICAL ASSESSMENT

Policy makers increasingly recognize innovation as important for national competitiveness and productivity improvement. For agriculture, it is also recognized as vital for the sector's ability to feed a growing population under increasingly difficult environmental conditions. The widespread application of biotechnology to agricultural research over the past 25 years has created a new dynamic that is fundamentally changing what research is undertaken, how it is funded, who undertakes the research, how IP is managed, and how products are commercialized and adopted.

As a result of this recognition, social scientists continue to actively study agricultural innovation processes as the industries and countries evolve and employ new strategies to enhance the effectiveness of the innovation system.

The fourth generation of innovation frameworks, which conceives innovation as a complex system, is especially appropriate for describing the current agricultural innovation system. Interactions between private and public actors are increasingly complex as each supply chain and sometimes each product involves a number of critical discrete activities and relationships. Unfortunately, this framework of innovation often lacks any general prescriptive power without very precise and nuanced understanding of the particular innovation processes at play. For more complete understanding, general descriptions need to be replaced by more in-depth studies of particular systems. Undoubtedly, with the long research lags and slow processes of adoption, many dynamics are at play within innovation systems that are not well understood or articulated in existing models. More work is needed to understand how these systems evolve over time.

Economic theory and models provide valuable insight into innovation systems. The non-rival and often non-excludable nature of knowledge and related transaction costs create market failures that can be addressed through policy. A good deal is understood about research spillovers and the potential for property rights and government expenditure to deal with inadequate private incentives. But the pervasive high rates of return to agricultural research suggest that research continues to be underfunded. Governments could address this underfunding with either larger expenditure or stronger IPRs, but are reluctant to do either, leaving parts of the innovation system starved of resources. The strong competition of public resources has severely constrained the resources available for agricultural research, which could get worse given the fiscal situation unless higher food prices drive governments to action.

Far less is understood about managing the toll good nature of research. At this point the agbiotech industry is less than two decades old. As this youthful industry evolves, the adverse effects of concentration and market power are likely to become more apparent, requiring more policy consideration.

Knowledge sharing is important for efficient research. The explosion of biotechnology has resulted in a patent thicket that is increasingly costly to utilize. More research is needed to understand the optimal patent policy and how public research should operate within the existing IP framework.

Agricultural innovation will continue to involve many private and public institutions. The persistence of spillovers and the non-rival nature of knowledge creates demand for goods that will not be produced by a competitive private sector – they must be produced either by the public sector or by industry organizations as industry goods. Given there are some novel, levy-funded institutional structures that have resulted in successful innovation, more research is needed to understand these organizations, how they operate, how they interact with the public and private sectors and how they impact the innovation system.

To be effective, public sector research must link into private research. The public sector must learn how to provide research and applied research that will be used by a downstream private research industry and ultimately producers further downstream. The public research sector must learn how to complement the activities of the private sector and where possible use resources provided by the crop research industry and producers to benefit the sector.

NOTES

1. Classified according to average per capita incomes in 2009 (Alston et al., 2012).
2. This estimate is based on the authors' calculation given observed seed prices of $8.50/lb, 5 pound per acre seeding rates, $15/ac TUA on 40 per cent of area, and a seeded area of 21.3 million acres.

REFERENCES

Alston, J.M. and M. Fulton (2012), 'Sources of institutional failure and underinvestment in levy-funded agricultural research', paper presented at the Australian Agricultural & Resource Economics Society (AARES) meeting in Fremantle, Western Australia, 7–10 February.

Alston, J.M., J.M. Beddow and P.G. Pardey (2009), 'Agricultural research, productivity, and food prices in the long run', *Science*, **325**(4)

Alston, J.M., R.S. Gray and K. Bolek (2012), 'Farmer-funded R&D: institutional innovations for enhancing agricultural research investments', CAIRN Working Paper, available at: http://words.usask.ca/cgpc/files/2012/05/New-CAIRN-Working-paper.pdf.

Alston, J.M., G.W. Norton and P.G. Pardey (1995), *Science Under Scarcity: Principles and Practice for Agricultural Research Evaluation and Priority Setting*, Ithaca, NY: Cornell University Press.

Alston, J.M., M.A. Andersen, J.S. James and P.G. Pardey (2010), *Persistence Pays: US Agricultural Productivity Growth and the Benefits from Public R&D Spending*, New York: Springer.

Alston, J.M., C. Chan-Kang, M.C. Marra, P.G. Pardey and T.J. Wyatt (2000), 'A meta analysis of rates of return to agricultural R&D: ex pede Herculem?', IFPRI Research Report, No. 113 Washington, DC.

ASTI (Agricultural Science and Technology Indicators) (2012), 'Internationally comparable data on agricultural R&D investments and capacity for developing countries', available at: http://www.asti.cgiar.org/data/?exportgeo1=IND.

Boettiger, S., Gregory D. Graff, Philip G. Pardey, Eric Van Dusen and Brian D. Wright (2004), 'Intellectual property rights for plant biotechnology: international aspects', in Paul Christou and Harry Klee (eds), *Handbook of Plant Biotechnology*, Chichester: John Wiley.

BRBIOTEC Brasil and Apex-Brasil (2011), 'Brazil Biotech Map 2011', available at: www.cebrap.org.br/v1/upload/pdf/Brazil_Biotec_Map_2011.pdf.

CGIAR (2011), 'The CGIAR at 40 and beyond: impacts that matter for the poor and the planet', available at: http://www.cgiar.org/our-research/.

Coase, R.H. (1960), 'The problem of social cost', *Journal of Law and Economics*, **3,** 1–44.

Coase, R.H. (1974), 'The lighthouse in economics', *Journal of Law and Economics*, **17**(2), 357–76.

Etzkowitz, H. and L. Leydesdorff (2000), 'The dynamics of innovation: from National Systems and "Mode 2" to a triple helix of university–industry–government relations', *Elsevier Research Policy Journal*, **29**(2000), 109–23.

Fernandez-Cornejo, J. and M.F. Caswell (2006), 'The first decade of genetically engineered crops in the United States', *Economic Information Bulletin*, No. 11. April, US Department of Agriculture–Economic Research Service, Washington, DC, available at: www.ers.usda.gov/publications/eib11/eib11.pdf, accessed 24 March 2011.

Fulton, M.E. (1997), 'The economics of intellectual property rights: discussion', *American Journal of Agricultural Economics*, **79**(5), 1592–94.

Fulton, M. and K. Giannakas (2001), 'Agricultural biotechnology and industry structure', *AgBioForum*, **4**(2), 137–51.

Fulton, M.E. and R. Gray (2007), 'Toll goods and agricultural policy', *CAIRN Policy Brief*, No. 9.

Fuglie, Keith O., Paul W. Heisey, John L. King, Carl E. Pray, Kelly Day-Rubenstein, David Schimmelpfennig, Sun Ling Wang and Rupa Karmarkar-Deshmukh (2011), 'Research investments and market structure in the food processing, agricultural input, and biofuel industries worldwide', Economic Research Report no.ERR-130, US Dept. of Agriculture, December.

Galushko, V., R. Gray and E. Oikonomou (2012), 'Operating in an intellectual property world: knowledge sharing among plant breeders in Canada', *Canadian Journal of Agricultural Economics/Revue Canadienne d'Agroeconomie*, **60**, 295–316.

Graff, Gregory and David Zilberman (2001), 'An intellectual property clearinghouse for agricultural biotechnology', *Nature Biotechnology*, **19**, 1179–80.

Graff, G., G. Rausser and A. Small (2003), 'Agricultural biotechnologys' complementary intellectual assets', *Review of Economics and Statistics*, **85**, 349–63.

Heisey, Paul, Sun Lin Wang and Keith Fuglie (2011), 'Public agricultural research spending and future US agricultural productivity growth: scenarios for 2010–2050', Economic Research Report no. EB-17, US Department of Agriculture, July.

Heller, M.A. (1998), 'Tragedy of the anticommons: property in the transition from Marx to markets', *Harvard Law Review*, **111**(3), 621–88.

Howard, Philip H. (2009), 'Visualizing consolidation in the global seed industry: 1996–2008', *Sustainability*, **1**(4), 1266–87.

Huang Dafang, Zhang Tian, Yue Tongqing and Zhang Hongxiang (2011), 'Beyond technology: initiatives to popularize genetic modification', in Mariechel J. Navarro and Randy A. Hautea (eds), *Communication Challenges and Convergence in Crop Biotechnology*, Ithaca, NY: International Service for the Acquisition of Agri-biotech Applications (ISAAA) and Los Baños, Philippines: SEAMEO Southeast Asian Regional Center for Graduate Study and Research in Agriculture (SEARCA).

Hu, Ruifa (2012), 'China's agricultural innovation system: issues and reform', Proceedings of the OECD Conference on *Improving Agricultural Knowledge and Innovation Systems*, 15–17 June, Paris: OECD Publications. pp. 63–74.

Hu, Ruifa, Qin Liang, Carl Pray, Jikun Huang and Yanhong Jin (2011), 'Privatization, public R&D policy, and Private R&D investment in China's agriculture', *Journal of Agricultural and Resource Economics*, **36**(2), 416–32.

James, C. (2011), 'Executive summary: global status of commercialized biotech GM crops: 2011', *International Service for the Acquisition of Agri-biotech Applications* (ISAAA) Brief 43–2011, Ithaca, NY: Cornell, available at: http://www.isaaa.org/resources/publications/briefs/43/executivesummary/default.asp

Karembu, M., F. Nguthi and H. Ismail (2009), 'Biotech crops in Africa: the final frontier', Nairobi, Kenya: ISAAA AfriCenter.

Kryder, R.D., S.P. Kowalski and A.F. Krattiger (2000), 'The intellectual and technical property components of Pro-Vitamin A Rice (GoldenRice™): a preliminary freedom-to-operate review', *ISAAA Briefs*, no. 20.

Lesser, W. (1998), 'Intellectual property rights and concentration in agricultural biotechnology', *AgBioForum*, **1**(2), 56–61.

Lopes, M.A. (2012), 'The Brazilian agricultural research for development (ARD) system', Proceedings of the OECD Conference on *Improving Agricultural Knowledge and Innovation Systems*, 15–17 June, Paris: OECD Publications, pp. 323–43.

Moschini, G. and H. Lapan (1997), 'Intellectual property rights and the welfare effects of agricultural R&D', *American Journal of Agricultural Economics*, **79**, 1229–42.

Nelson, R.R. (ed.) (1993), *National Systems of Innovation: A Comparative Analysis*, Oxford: Oxford University Press.

Noisi, J. (1999), 'Fourth-Generation R&D: from linear models to flexible innovation', *Journal of Business Research*, **45**, 111–17.

Ostrom, E. (2011), 'Background on the institutional analysis and development framework', *The Policy Studies Journal*, **39**(1).

Ostrom, E., J. Burger, C.B. Field, R.B. Norgaard and D. Policansky (1999), 'Revisiting the commons: local lessons, global challenges', *Science*, **9–284**(5412), 278–82.

Pal, S. and D. Byerlee (2006), 'India: the funding and organization of agricultural research in India: evolution and emerging policy issues' Chapter 7 in P.G. Pardey, J.M. Alston and R.R. Piggott (eds), *Agricultural R&D in the Developing World: Too Little, Too Late*?, Washington, DC: International Food Policy Research Institute.

Pardey, P.G. and Julian M. Alston (2012), 'Global and US trends in agricultural R&D in a global food security setting', proceedings of the OECD Conference on *Improving Agricultural Knowledge and Innovation Systems*, 15–17 June, Paris: OECD Publications, pp. 19–40.

Pardey, P.G., J.M. Alston and C.C. Kang (2012), 'Agricultural production, productivity and R&D over the past half century: an emerging new world order', invited plenary paper presented at the International Association of Agricultural Economists (IAAE) Triennial Conference, Foz do Iguaçu, Brazil, 18–24 August.

Pardey, P.G., N.M. Beintema, S. Dehmer and S. Wood (2006), 'Agricultural research: a growing global divide?', Food Policy Report, Washington, DC: International Food Policy Research Institute, available at: http://www.ifpri.org/sites/default/files/pubs/pubs/fpr/pr17.pdf.

Parisi, F., N. Schulz and B. Depoorter (2005), 'Duality in property: commons and anticommons', *International Review of Law and Economics*, **25**(4), 578–91.

Perrin, R.K. and L.E. Fulginiti (2009), 'Pricing and welfare impacts of new crop traits: the role of IPRs and Coase's conjecture revisited', AgBioForum, **11**(2), article 7.

Picciotto, R. (1997), 'Putting institutional economics to work: from participation to governance', in C. Clague (ed.), *Institutions and Economic Development: Growth and Governance in Less Developed and Post-Socialist Countries*, Baltimore, MD: Johns Hopkins University Press, pp. 343–67.

Schimmelpfennig, David and Paul Heisey (2009), 'US public agricultural research: changes in funding sources and shifts in emphasis, 1980–2005', Economic Research Report no. EIB-45, US Dept of Agriculture, March.

Smyth, S.J. and R. Gray (2011), 'Intellectual property sharing agreements in gene technology: implications for research & commercialization', *International Journal of Intellectual Property Management*, **4**(3), 2011.
Stads, G.J. and N.M. Beintema (2009), 'Public agricultural research in Latin America and the Caribbean: investment and capacity trends', ASTI Regional Report, International Food Policy Research Institute and Inter-American Development Bank.
Stiegert, K.W., Shi Guanming and Jean Paul Chavas (2010), 'Innovation, integration, and the biotechnology revolution in US seed markets', *Choices*, **25**(2), 1.
USDA (2011a), 'China: agricultural biotechnology annual 2011', GAIN report no. CH11050.
USDA (2011b), 'India: agricultural biotechnology annual 2011', GAIN report no. IN1187.
Wilson, W.W. and Bruce Dahl (2010), 'Dynamic changes in Market Structure and competition in the corn and soybean seed sector', Agribusiness & Applied Economics Report no. 657, Dept of Agribusiness and Applied Economic Agricultural Experiment Station, North Dakota State University.
Wright, B.D. and P.G. Pardey (2006), 'Changing intellectual property regimes: implications for developing country agriculture', *International Journal of Technology and Globalisation*, **2**(1/2), 93–114.

4 The private sector: MNEs and SMEs

Jill E. Hobbs

1 INTRODUCTION

Private sector contributions to the development and growth of agricultural biotechnology (agbiotech) are comprised of those arising from a relatively small number of large Multinational Enterprises (MNEs) together with the output of numerous innovative Small and Medium-Sized Enterprises (SMEs). This chapter explores the major drivers of industry structure and firm behaviour in the agbiotech sector. Key themes include mergers, acquisitions and increasing industry concentration among MNEs, entrepreneurship and the management of knowledge resources, and an ever-evolving pattern of alliances and networks. The chapter explores various models and methods that have been used to examine these issues.

Multinational Enterprises, also commonly called Multinational Corporations (MNCs), are large firms that operate across national boundaries in more than one country, often with a diverse portfolio of products. In the context of agbiotech, the dominant MNEs acted strategically in the 1990s to position themselves as life-science companies operating across pharmaceutical, agriculture-related, and nutrition or consumer portfolios, with significant investments in the application of biotechnology to seeds, crop protection, animal health and human health. In contrast, Small and Medium-Sized Enterprises (SMEs) are firms whose number of employees and turnover fall below certain limits. Although there is no commonly accepted international definition of an SME, typically it includes firms with a limited number of employees – less than 250 in the EU and UK or below 500 in Canada and the US – with an annual turnover below a certain level – for example, less than €50 million in annual turnover according to the definition established by the EU Commission (OECD, 2005). SMEs have been an important source of research and development (R&D) and innovative activity in the agbiotech sector.

Section 2 describes the evolution of the agbiotech sector, the key exogenous influences on industry development, and the methods and metrics by which analysts have measured industry structure. Section 3 explores the role of alliances and networks of private sector firms, and the models and lenses through which these relationships have been examined, including transaction cost economics and contributions from the strategic management literature. Interwoven through this discussion is consideration of several drivers of the relationship between MNEs and SMEs, including exogenous influences (institutional, regulatory and technological), inter-firm structures (alliances, licensing agreements and networks) and internal strategic management drivers (entrepreneurial and knowledge resources). The chapter concludes with a prospective look at developments that can be expected in the coming decade.

2 EVOLUTION OF THE PRIVATE SECTOR

2.1 An Industry in Transition

Private sector involvement in agbiotech was influenced by two key exogenous developments: the emergence of a new transformative technology, together with changes to the institutional environment governing intellectual property rights (IPRs). The discovery of deoxyribonucleic acid (DNA), and subsequent development of gene and DNA sequencing methods, paved the way for the development of transgenic varieties of crops with desirable traits. Initially these were primarily input traits such as herbicide tolerance and insect resistance, with the promise of 'second generation' output-trait crops with enhanced nutritional and end-use qualities, and 'third generation' crops with industrial or pharmaceutical properties.

In addition, changes to plant breeders' rights and intellectual property protection facilitated the application of biotechnology techniques to the agricultural sector, as well as proving a major driver of merger, acquisition and alliance activity in the private sector. Patenting of higher-life forms was enshrined in the 1930 US Plant Patent Act, which allowed patent protection in the US for new varieties of plants produced through traditional vegetative methods (Smyth and Gray, 2011). The UPOV Convention (Union Internationale pour la Protection des Obtensions Végétales, also known as the International Union for the Protection of New Varieties of Plants) was the first international treaty concerning intellectual property in agriculture. UPOV was drawn up in Paris in 1961 to harmonize existing national systems of plant breeders' rights and came into force upon its ratification by the UK, the Netherlands and West Germany in 1968 (Dutfield, 2008). By 2012 membership of UPOV had grown to 70 countries. With the advent of biotechnology, greater clarity was needed over what could and could not be patented and, through a series of regulatory reforms and patent case law during the 1980s and 1990s, the patentability of cells, genes and plants was given formal legal status in a number of countries (for a more detailed discussion see Smyth and Gray, 2011).

Product Life-Cycle theory (or the theory of Dominant Design) infers a parallel relationship between innovation and the life-cycle of an industry; it offers a lens through which the early stages of evolution of the agbiotech sector can be viewed (Kalaitzandonakes and Hayenga, 2000; Marks et al., 1999; Hobbs and Kerr, 2003). During the early phases of innovation (the 'fluid phase') immediately following a technological breakthrough, rapid entry into the industry occurs as firms innovate. The crop biotechnology sector in the 1980s had many of these features, with a variety of potential technologies competing for dominance, including transgenic plants and genetically engineered micro-organisms that could be applied to a seed in the form of an inoculant (Kalaitzandonakes and Hayenga, 2000). Both new and existing firms compete for the dominant product design, with industry structure in a state of flux. As the product matures and a dominant design emerges, entry barriers rise and the industry becomes more concentrated (the 'transitional phase'). Process innovation replaces product innovation as the industry transitions and consolidates around a dominant design. In many ways, this reflects the industry consolidation in agbiotech that occurred through the 1990s once transgenic plants had emerged as the dominant design in the form of first generation input-trait crops with built-in pest resistance and herbicide tolerance.

The Product Life-Cycle lens, although a useful depiction of the industry's evolution in its early stages, appears less applicable to the later stages of development in the agbiotech sector, in part perhaps because biotechnology is not simply one technology or one product but rather a collection of increasingly diverse technologies, innovations and applications rooted in the genetic revolution. According to the Product Life-Cycle view of the world, an industry eventually enters a 'specific phase', with product and process innovation tapering off as an industry consolidates around a few dominant firms and a dominant design, leaving the remaining firms to compete on cost, volume and capacity efficiencies rather than through product innovation. While the agbiotech industry did consolidate around a few dominant firms from the early 2000s onwards, innovations and new product developments have continued; as such the characterization of a 'specific phase' does not appear to mesh closely with the current reality of the sector. Other models provide a more complete picture of the subsequent drivers and consequences of industry structure and are examined in section 3 of this chapter.

A relatively small number of agbiotech multinationals now dominate the private sector, both in terms of product market share as well as the share of intellectual property currently owned or controlled by these firms. The life science MNEs emerged from the 1990s when multinational pharmaceutical and agricultural chemical companies repositioned themselves at the interface between biology, agriculture and human health, investing in biotechnology and seed enterprises. Considerable merger and acquisition activity occurred through the 1990s, amid a period of rapid technological change and a shifting landscape of intellectual property protection. For example, Monsanto purchased biotechnology companies Calgene (1996) and Agracetus (1995), as well as seed businesses (for example DeKalb in 1996 and Asgrow in 1997); Dow purchased several biotechnology companies in the mid-1990s, including Mycogen and Ribozyme; DuPont purchased Pioneer Hi-Bred International; and Norvatis Agribusiness and Zeneca Agrochemicals merged to form Syngenta in 2000 (Hobbs and Kerr, 2003; Gaisford et al., 2001).

Since the early 2000s industry structure has been relatively stable, although mergers and acquisitions continue to occur, albeit on a much smaller scale. Pray et al. (2005) report a peak of 22 mergers and acquisitions in 2000, with a notable decline thereafter, and indeed some divestiture of crop science units and agricultural businesses in the early 2000s. For example, Monsanto merged with Pharmacia in March 2000 but was spun off entirely in 2002, while Aventis sold its Crop Science Unit to Bayer in 2001 (for a more detailed overview of merger and acquisition activity and evolving industry structure through the 1990s and 2000s see Pray et al., 2005). Six large MNEs, three each headquartered in Europe and the US, eventually emerged as the dominant private sector firms in agbiotech: BASF, Bayer, Dow, DuPont, Monsanto and Syngenta. The first four hold diverse portfolios across a broad range of sectors, with agriculture or crop protection divisions representing only a portion of their businesses, while agriculture represents a core business focus for Monsanto and Syngenta. In January 2012 BASF moved the headquarters of BASF Plant Science from Germany to North Carolina in the US, declaring its intention to concentrate its plant biotechnology activities on the main markets in North and South America, given challenges created by the continued lack of acceptance of the technology in the European market (BASF, 2012). Table 4.1 provides a snapshot overview of the 'big six' MNEs that dominate agbiotech.

Table 4.1 A snapshot view of the 'big six'

Company	Key facts
BASF	• Headquartered in Europe (Germany). Founded in 1865 • Major divisions (% sales, 2010): Performance Products (19%); Chemicals (18%); Oil & Gas (17%); Plastics (15%); Functional Solutions (15%); **Agricultural Solutions (Crop Protection) (6%)** • Employees, 2010 (BASF total): 110 000 • Net sales, 2010: €63.9 billion • Net sales, Crop Protection Division, 2010: €4.033 billion • R&D expenditures, 2010 (BASF total): €1.5 billion
Bayer	• Headquartered in Europe (Germany). Founded in 1863 • Major divisions (% sales, 2010): Bayer Healthcare (48%); Bayer Material Science (29%); **Bayer Crop Science (Crop Protection & Environmental Science, Bioscience) (19.5%)**; Bayer Business Services and Bayer Technology Services (4%) • Employees, 2010 (Bayer Group): 111 000 • Employees, Bayer Crop Science, 2010: 20 700 • Net sales, 2010 (Bayer Group): €35.09 billion • Net sales, Bayer Crop Science, 2010: €6.8 billion • R&D Expenditures, 2010 (Bayer Crop Science): €722 million
Dow	• Headquartered in the USA. Founded 1897; Dow AgroSciences is a division • Major divisions (% sales, 2010): Plastics (22%); Performance Products (20%); Performance Systems (12%); Hydrocarbons (10%); Electronic & Speciality Materials (9%); Coatings and Infrastructure (10%); **Health and Agricultural Sciences (9%)**; Chemicals and Energy (7%) • Employees, 2010 (Dow total): 49 400 • Net sales, 2010 (Dow total): US$53.67 billion • Net sales, Health and Agricultural Sciences (2010): US $4.87 billion • R&D Expenditures, 2010 (Dow total): US$1.66 billion
DuPont	• Headquartered in the USA. Founded 1802 • Major Divisions (% net sales, 2011): **Agriculture (24%)**; Performance Chemicals (20%); Performance Materials (18%); Performance Coatings (11%); Safety & Protection (10%); Electronics & Communications (8%); Nutrition & Health (6%); Industrial Biosciences (2%) • Employees, 2010 (DuPont Total): 67 000 • Net Sales, 2011: US$37.96 billion • Net sales, Agriculture, 2011: US $9.17 billion • R&D Expenditures, 2011 (Dupont total): $1.96 billion
Monsanto	• Headquartered in the USA. Founded in 1901. Prior to 1997 operated as combined chemical, pharmaceutical and agricultural business. Divested as separate agricultural business in 2002 • Two primary areas of business focus: seeds and traits; crop protection • Employees, 2011: 20 600 • Net sales, 2011: US$ 11.82 billion • R&D Expenditures, 2011: $1.39 billion

Table 4.1 (continued)

Company	Key facts
Syngenta	• Headquartered in Europe (Switzerland). Formed in 2000 through the merger of Norvatis and Astra-Zeneca. Historical origins date back to 1758 with the formation of Geigy • Core focus: agribusiness, including crop protection & seeds • Employees, 2010: 26 179 • Net sales, 2010: $11.64 billion of which Crop Protection sales of $8.88 billion and Seeds sales of $2.80 billion • R&D Expenditures, 2010: $1.03 billion

Sources: BASF Factbook 2011 (BASF, 2011); Bayer 2010 Annual Report (Bayer, 2011); The Dow Chemical Company 2010 Annual Report (Dow, 2011); DuPont website (www.dupont.com); Monsanto 2011 Annual Report (Monsanto, 2011a); Monsanto 2011 Form 10-K (Monsanto, 2011b); Syngenta Annual Review 2010 (Syngenta, 2011).

2.2 Measuring Industry Concentration

The dominance of a relatively small number of MNEs in agbiotech has been well documented in the literature. High levels of industry concentration raise concerns over market power, competitiveness, and the long-run incentives for innovation. Analysts have used a variety of measures and metrics to track changing industry structure and concentration levels. Industry concentration measures generally relate market power to market share. In output (final goods) markets this is typically measured as the share of industry sales weighted by the number of firms in the industry. In the context of agbiotech, however, the concept of innovation markets is a popular means of evaluating industry structure, wherein the share of innovation activity accounted for by the largest firms is the relevant measure. Concentration in innovation markets is measured by the ownership and control of intellectual property (patents), field trial data and crop variety registration (deregulation) activity.

Field trial data provide an indicator of biotechnology research activity by individual firms and have been used to track changing industry structure over time (for example Phillips and Khachatourians, 2001; Pray et al., 2005; Brennan et al., 2005; Oehmke and Wolf, 2003). From the first transgenic field trial in 1987, US data show that the number of firms engaged in field trials climbed rapidly to a peak of 55 firms in 1996, with that number levelling off in the low to mid-40s for the following decade (Pray et al., 2005). This field trial activity was heavily concentrated among a few firms; using a Four-Firm Concentration Ratio (CR-4), Brennan et al. (2005) estimate that 80 per cent of US field trial activity in 2001 can be attributed to the four largest firms.

The Herfindahl–Hirschman Index (HHI), widely regarded as a more comprehensive measure of concentration because it gives more weight to larger firms, relates the market shares of the firms to the size of the industry. Defined as the sum of the squares of the market shares of the N firms in an industry, it ranges from 1/N to 1, with a smaller index indicating a high degree of competiveness and 1 indicating a pure monopoly (Hirschman, 1964; Rhoades, 1995). Applying the HHI method to field trial data confirms very high levels of concentration from the mid-1990s onwards, particularly after accounting for the mergers that occurred in the sector during the 1990s (Brennan et al., 2005).

Patent ownership and control represents another basis on which to evaluate levels of concentration in the private sector. Although canola was developed by public institutions in Canada, an examination of patent distribution for Canadian canola and oilseed technology in 2006 revealed 38 identified patents, of which 32 were held in the private sector, with only seven of these held by SMEs and the remaining 25 held by large US or European MNEs (Smyth and Gray, 2011). A similar pattern of dominance by the private sector is found in agbiotech patents granted by the European Patent Office and the United States Patent and Trademark office between 2002 and 2009, with approximately 62 per cent of patents held purely in the private sector, 40 per cent of those by MNEs (Frisio et al., 2010).

Applying the two concentration measures (HHI and CR-4) to this patent data charts a consistent story of the dominance of the six large MNEs, particularly in cultivar patents (Frisio et al., 2010). Indeed, by 1998 six firms accounted for over 80 per cent of genetically modified crop field trials for regulatory release in the US and by 2002 the same six firms controlled over 40 per cent of private sector agbiotech patents issued in the US. Interestingly, subsidiaries of these firms accounted for the majority (70 per cent) of the patented technology controlled by the 'big six' (King and Schimmelpfennig, 2005). This suggests that SMEs probably played an important role in the development of these technologies, only subsequently becoming affiliated with larger firms through acquisitions, sub-contracts and cross-licensing agreements.

In a nuanced approach to measuring concentration in the agbiotech sector, Oehmke and Wolf (2003) develop a measure of concentration in innovation markets that takes explicit account of inter-firm alliances in the form of cross-licensing agreements and subcontracts to utilize technologies. Inter-firm alliances to test a single gene construct are not counted separately as independent trials; the measure of concentration (which the authors term GR4) focuses on ownership of gene constructs rather than on the number of firms performing R&D. Applying this approach to data on transgenic field trials and deregulations[1] in the US, the authors find notably higher levels of concentration compared with using a traditional CR-4 measure. For all herbicide-tolerant crops the gene-construct-based measure of concentration (GR4) ranged from 85 per cent to 99 per cent between 1991 and 2000; that is, the four largest firms accounted for virtually all field trials and deregulation requests during that period. For transgenic insect resistance, GR-4 concentration levels are in the 95 per cent to 99 per cent range, compared with 48 per cent to 92 per cent using a firm-based CR-4 measure that does not account for inter-firm technology-sharing agreements (Oehmke and Wolf, 2003).

Concentration in innovation markets has implications for long-run competitiveness and innovation because future R&D activity is often contingent on access to existing intellectual property, including patented gene constructs. It therefore matters who owns the IPRs, as well as the extent to which the development of new gene constructs is concentrated among a few firms, either through ownership or through technology licensing agreements, contracts and joint ventures.

Three take-home messages are apparent from this overview of the evolution of the agbiotech private sector. First, the sector evolved fairly rapidly to a situation in which a small number of very large multinational enterprises dominate the sector. Second, measures of industry concentration need to be cognizant of both concentration in output markets as well as in innovation markets, and the extent to which networks, alliances and technology licensing agreements further concentrate the control of IPRs. Finally, although multinationals dominate the ownership and control of intellectual property, the innovative activity generating the IPRs was often created elsewhere, in many cases through small and medium-sized firms. The relationship between MNEs and SMEs, including the role of alliances and networks, is of specific interest in this chapter and the subject of the next section.

3 ALLIANCES, NETWORKS AND INNOVATION

While large multinationals dominate the private sector in terms of market share in both output and innovation markets, small and medium-sized firms have played a key role in the development of the sector, from the generation of ideas, to the development of nascent technologies, testing of concepts, and the exploration, and sometimes commercialization, of new applications and products. Access to these technologies is managed through a variety of inter-firm relationships, both horizontal and vertical, from strategic alliances, to joint ventures, licensing agreements, and other forms of contract. In some cases inter-firm relationships have metamorphosed into outright ownership through vertical and horizontal integration. Several theoretical lenses shed light on these inter-firm relationships, including transaction cost economics and contributions from the strategic management literature using resource-based views of the firm.

3.1 Supply Chains, Uncertainty and Hold-ups

Transaction cost economics (TCE) recognizes that transactions do not occur in a frictionless environment; there are costs to inter-firm transactions arising from searching for information, negotiating the terms of a transaction and monitoring or enforcing a transaction (Williamson, 1979). Transaction costs influence the governance structure that emerges: from contracts, to strategic alliances, licensing agreements, or vertical integration (ownership). If the costs of enforcing a contract become too great, or a licensing agreement becomes too complex to negotiate, firms will seek an alternative means of securing access to the technology, such as through mergers or acquisition, thus internalizing the transaction. The TCE lens posits that transaction costs are driven by the characteristics of a transaction, including uncertainty, asset specificity, frequency and complexity. Asset specificity arises when a party has made an investment that is specific to a relationship, with little or no value in an alternative use or to an alternative user. Asset-specific investments are vulnerable to opportunistic behaviour wherein another party attempts to capture rents. Vulnerability to opportunistic behaviour deters investment in specific assets. This is also known as the 'hold-up' problem (see Klein et al., 1978 for a discussion of asset specificity). In the presence of uncertainty, and when asset-specific investments are in place (or contemplated), firms will seek to safeguard their investments through more formalized exchange relationships such as long-term contracts, joint ventures or vertical integration. Increasing complexity compounds this situation. In environments characterized by considerable uncertainty, complexity and highly idiosyncratic (specialized) investments, the transaction costs of executing enforceable contracts are likely to be high, providing an incentive for firms to vertically integrate as a solution to the hold-up problem.

Uncertainty, complexity and asset specificity characterize many of the relationships along agbiotech supply chains (Gaisford et al., 2001; Hobbs and Kerr, 2003). With advances in genetic research, gene stacking is increasingly commonplace such that plants can contain numerous patentable traits. Often one patentable trait is developed on the basis of other patentable traits or processes. Platform technologies, such as the sequencing or mapping of genomes and transgenic methods, have formed the basis of numerous subsequent innovations (Pray and Naseem, 2005). Technological complexity is compounded by the institutional complexity surrounding the ownership of intellectual property in these traits and processes. Overlapping patents and multiple pieces of intellectual property incorporated into a single crop variety have created an incredibly complex environment of IPRs within the agbiotech sector (for example, several new transgenic crop varieties are reported to incorporate as many as 30 separate pieces of IP; agreements on over 70 pieces of intellectual property were needed prior to the commencement of the research that led to Golden Rice (Smyth and Gray, 2011)). Perhaps not surprisingly, numerous patent disputes have arisen over the past 20 years between firms, as well as between private firms and public sector research institutions (Hayenga, 1998; Pray and Naseem, 2005). In one example, patent interference proceedings in the US between four parties (three private sector and one public sector) that claimed the IPRs to the *Agrobacterium* platform technology went on for 20 years (Pray and Naseem, 2005).

Along with technological and institutional complexity, uncertainty characterizes the R&D landscape in the high-technology sector. Uncertainty over the ownership of intellectual property notwithstanding, the process of scientific discovery, by its very nature, is fraught with uncertainty. Time lags between an initial research idea and eventual commercialization of a product are typically lengthy due both to the process of scientific discovery, as well as the time required to gain regulatory approval, with a small fraction of initial research ideas eventually making it through to commercialization. Writing full contingent contracts to coordinate biotechnology and seed assets years before a product might reach commercialization is extremely difficult and often not economically justified, thereby creating an incentive for vertical integration to internalize these transaction costs (Gaisford et al., 2001).

Finally, asset specificity arises if an investment in a biotechnology innovation is specific to the germplasm (seed) through which it will be delivered, or is dependent upon a platform technology or other patentable trait. Firms may be deterred from pursuing an otherwise promising avenue of research due to uncertainty over access to the necessary intellectual property, or the risk of opportunistic rent extraction by the holder of the IP. Vertical integration is one solution to this problem, and the series of mergers and acquisitions that occurred through the 1990s were in part a response to this situation. More recently, the hold-up problem has been addressed through cross-licensing agreements between the large MNEs, as well as between MNEs and innovative SMEs. Cross-licensing agreements take the form of *ex post* agreements that allow for shared access to established technology platforms, as well as *ex ante* agreements that facilitate the development of future traits (Smyth and Gray, 2011).

In summary, transaction cost economics provides insights into the nature of inter-firm relationships and the incentives for firms to collaborate through long-term alliances. Under conditions of extreme uncertainty, asset specificity and complexity, TCE predicts that firms will vertically integrate to address the hold-up problem through internalizing the transaction. In practice, anti-trust concerns may limit the extent to which further consolidation is possible within the agbiotech sector. Cross-licensing agreements and alliances often present a more flexible alternative to full vertical integration, enabling the private sector to be more nimble in its response to new market and technological opportunities. The strategic management literature, including the Resource-Based View (RBV) of the firm, and literature on entrepreneurship and networks offer additional insights into the motivations for MNEs and SMEs to form networks and alliance partnerships.

3.2 Knowledge Assets, Entrepreneurship and Networks

The relationship between MNEs and SMEs can also be examined from the perspective of managing knowledge assets and the contribution of knowledge spillovers to innovation processes. Literature on the Resource-Based View of the firm, intrapreneurship and the market versus entrepreneurial orientation of SMEs informs this perspective. The *Resource-Based View* of the firm characterizes the firm as a collection of resources and capabilities. Subsets of this approach focus on the knowledge resources of the firm, including human capital, knowledge management, and intellectual property rights. *Intrapreneurship* refers to entrepreneurial and innovation processes within an organization, such as an MNE. The intrapreneurship literature examines the extent to which internal organizational norms and access to knowledge resources encourage or stifle innovation *within* a firm. The recognition that innovation processes often occur as a result of *inter-firm* collaborations led to an interest in so-called 'interpreneurial' networks or 'inter-firm intrapreneurship' (Antoncic, 2001; Bonnafous-Boucher and Lamy, 2010; Peña, 2009).

Finally, the *market orientation* literature shows that SMEs can be characterized alternatively by a market-pull approach (market orientation) or a science-push approach (entrepreneurial orientation), and examines technological capabilities as a determinant of SME success (Renko et al., 2009). All three strands of literature, discussed in more detail below, draw upon the management of knowledge resources within and between firms and the role of knowledge spillovers in fostering innovation, providing a lens through which to view the relationship between MNEs and SMEs in the agbiotech sector.

In the resource-based view of the world, firms engage in strategic alliances, joint ventures or vertical integration to expand access to core capabilities including knowledge resources. Transferring knowledge between independent firms across a market interface may be ineffective, particularly in the case of tacit knowledge. A firm's access to competencies (capabilities) allows it to create new products and technologies in response to market signals and forms the source of competitive advantage for the firm (Teece and Pisano, 1994). The capabilities of a firm set limits on its boundaries and provide an incentive for mergers, acquisitions and strategic alliances to expand that knowledge resource base. In the context of agbiotech, securing access to knowledge resources has been a strong motivation for the establishment of strategic relationships between MNEs and SMEs.

The sources of innovation rarely lie exclusively within company boundaries, nor are new ideas always traded in markets. They often arise from relationships within networks of high-tech firms, downstream users and scientists in public sector research institutions. Business networks among knowledge-intensive companies became commonplace in many high-tech sectors in the 1990s, from semiconductors, to nanotechnology and medical biotechnology, as these companies sought to gain an innovation advantage (Niosi and Queenton, 2010; Peña, 2009). Evidence from the human health biotechnology sector in the US and Canada suggests that proximity to 'star scientists' (as measured by patent and publication success) has a positive effect on the performance of biotechnology firms (Niosi and Queenton, 2010; Zucker et al., 1998). Geographical clusters of firms, institutions and individuals have arisen to drive knowledge-based innovation in the life sciences. Phillips and Ryan (2007) examine the role of innovation clusters in the biotechnology sector, finding a significant private sector presence as well as a number of identifiable stars from public sector research institutions associated with each major cluster. The ability and willingness to foster knowledge spillovers and manage access to IP, both within the cluster and across international boundaries, are identified as key components of successful innovation strategies (Phillips and Ryan, 2007).

While geographic clusters of firms, institutions and individuals can be a source of innovation, collaborations also occur beyond local boundaries and it is the nature of the relationship between these collaborators that is of interest. In the biotechnology sector, as in other high-tech sectors, inter-firm cooperative agreements, or 'interpreneurial' networks, are formed for the express purpose of developing groundbreaking technologies. Since firms often lack all the necessary knowledge resources to develop their technologies completely in-house, the formation of interpreneurial strategic alliances with technology-intensive small firms can be an effective means of pooling complementary knowledge assets to enhance innovation capacity (Peña, 2009). Of course, networks can become too large, with managerial diseconomies of scale arising from coordination and control problems when the number of strategic alliance partnerships becomes too great. Studies of the relationship between interpreneurial alliances and the rate of new product development in the biotechnology sector confirm an initial positive relationship between alliance partnerships and new product development, but the relationship eventually reverses as the number of alliances increases beyond a threshold level (Deeds and Hill, 1996; Peña, 2009).

Given the technological and institutional uncertainty (especially with respect to IPRs) that characterizes the biotechnology sector, firms may prefer a strategy of gradual investment through intrepreneurial R&D alliances rather than investing heavily in outright acquisition of technologies and firms at an early stage. A 'real options' strategy of developing alliances with SMEs to develop promising technologies provides multinationals with the opportunity to gain a first-mover advantage in a new technology without committing resources outright to the purchase and development of the technology in-house. In an examination of interpreneurial networks in the US and European life science and medical biotechnology sectors in the 1990s, Peña (2009) suggests that major biotechnology firms following a 'real options' approach to interpreneurial R&D alliances tended to exhibit stronger financial performance compared to MNEs who took an intrapreneurial approach to innovation and made larger in-house R&D investments (Peña, 2009). However, presumably the long lags between R&D and successful commercialization could also partly explain poorer financial performance by firms with large in-house R&D investments.

Participation in interpreneurial networks and alliances is a recurring theme in the long-run survival and success of biotechnology SMEs, facilitating strategic use of their knowledge assets and enabling small firms to capitalize on knowledge spillovers within the network. The business models and internal managerial orientation of the SME influence the success with which small firms engage in networked research and commercialization activities. The concept of 'market orientation' in the marketing literature relates to the values, beliefs or actions that a firm exhibits: putting the 'customer first' or taking a demand-pull approach to business. Behavioural manifestations of a market orientation encompass the gathering and dissemination of market intelligence and a firm's responsiveness to this information. Numerous scholars have explored the relationship between market orientation and various measures of business performance, including growth and market share (Renko et al., 2009). While there is empirical evidence of a causal link between market orientation and innovation (for example Han et al., 1998; Lukas and Ferrell, 2000; Kirca et al., 2005), the relationship is more nuanced in a high-technology environment where it is the ability to respond to latent rather than expressed market (customer) needs that characterizes a successful transformative innovation (Renko et al., 2009).

In contrast, an entrepreneurial orientation emphasizes aggressive innovation, risk-taking and a tendency to take a proactive approach to anticipating future market needs and changing the competitive landscape. SMEs with an entrepreneurial orientation therefore tend to generate, disseminate and respond to technological knowledge rather than market intelligence. A science-push rather than demand-pull approach characterizes the approach to innovation. The underlying resource constraints, capabilities and competencies of the firm are tightly interwoven with the tendency towards a market or an entrepreneurial orientation and may also inform the choice of alliance partners to complement (or counteract) the firm's natural tendency.

Renko et al. (2009) construct an empirical analysis of the relationships between market orientation, entrepreneurial orientation, technological capabilities and two outcome measures: product innovativeness and capital invested in the company. Using a sample of 85 US and Scandinavian medical biotechnology SMEs, the authors find evidence of a causal link between technological capability and innovativeness. While market orientation appeared to drive higher levels of capital investment in the firm, there was no evidence of a positive relationship between either market orientation or entrepreneurial orientation and product innovativeness. Whether these results are specific to the medical biotechnology sector and would hold across other industry segments is in need of further investigation. In an analysis of the business models used by agbiotech SMEs, Blank (2008) finds a strong focus on product development rather than market development, along with a portfolio approach to product development and active engagement in collaborative partnerships as a means to manage risk.

The entrepreneurship literature offers a useful lens through which to view the drivers of successful business models in small and medium-sized firms; an examination of the market versus entrepreneurial orientation among agbiotech SMEs is a fruitful avenue for future research. The extent to which a market orientation or an entrepreneurial orientation (or some combination of the two) characterizes firms leading the charge in the development of second and third generation agricultural biotechnologies is worthy of investigation.

In summary, the strategic management literature provides further insights into the nature of inter-firm relationships and the incentives for firms to collaborate through long-term alliances. Many of these insights are grounded in resource-based views of the firm with recognition, whether explicit or implicit, of the strategic use of knowledge resources and access to knowledge spillovers as drivers of collaborative relationships to enhance innovation and firm performance. Evidence from the literature on entrepreneurship, as well as clusters, suggests that networks encompassing both multinationals and biotechnology SMEs, together with public sector research institutions, are a core component of the life science sector. Unpacking the drivers of survival and long-run commercial success for SMEs is more complex, particularly given the diversity of firms that operate in this space. A variety of business models and risk management strategies exist, with the market versus entrepreneurial orientation remaining a useful distinction worthy of further investigation. The chapter concludes with some thoughts on external drivers that will continue to define the structure of the agbiotech private sector.

4 THE FUTURE

Reflecting on the issues discussed in this chapter, the factors defining the composition of the private sector can be conceptualized at three interrelated levels: exogenous factors (institutional, regulatory and technological), inter-firm factors (alliances, licensing agreements and networks), and internal strategic management factors (entrepreneurial and knowledge resources). The dominance of a few large multinationals in both output and innovation markets looks set to continue, with SMEs operating within the commercial sphere of influence of these large firms through strategic alliance and licensing arrangements. The primary institutional, regulatory and technological constraints that will continue to define the nature of inter-firm relationships and strategic management decisions include the complexity of intellectual property rights, the constraints of anti-trust regulations, and the technological uncertainty embedded in the innovation process.

Securing access to the necessary IP is a prerequisite for innovations that build on patented platform technologies, or when plants contain multiple stacked traits, and as such is a major driver of decisions by MNEs to pursue specific lines of research, form alliances, negotiate cross-licensing agreements, or acquire firms holding the IPRs. Challenges in obtaining access to IPRs have been characterized as a 'tragedy of the anti-commons', wherein firms have the ability to exclude others from using a resource but cannot use the resource themselves, so that a resource is not used to its full economic value. In this context, investment in innovation is suboptimal, or firms are unable to commercialize a viable technology, and the anticipation of costly bargaining processes result in hold-up at the commercialization stage (Smyth and Gray, 2011). Assuming that bargaining solutions to the hold-up problem are too costly, there is an incentive for vertical (horizontal) integration of the parties holding the IPRs. Thus the IPR landscape creates ongoing pressure for further consolidation within the private sector.

Mitigating this pressure, however, is the potential for moves toward further consolidation to run afoul of various anti-trust and competitive regulatory frameworks in the host countries and core markets of the life-science MNEs. Further consolidation raises concerns over the effects of market concentration on market efficiency, including consumer welfare, barriers to entry and the incentives for innovation. The industry consolidation that occurred through the 1990s and early 2000s triggered the interest of the US Federal Trade Commission and led to a number of rulings requiring divesture of strategic assets by the large agbiotech companies (Smyth and Gray, 2011). Rather than significant merger and acquisition activity, therefore, cross-licencing agreements and other more nuanced forms of inter-firm alliance are likely to characterize the sector over the coming decade.

Finally, the technological uncertainty embedded in innovation and discovery processes suggests that there will always be a role for small and medium-sized firms as nimble risk-taking innovators, the vast majority of whom will fail while a few achieve radical breakthroughs. Providing sufficient breathing space for small innovators to explore new avenues of discovery through cross-licencing agreements and other manifestations of interpreneurial networks should be an attractive strategy to MNEs.

Models, methods and lenses from the economics and strategic management literature offer rich insights into the changing nature of strategic relationships within the private sector. The drivers of inter-firm alliances and implications for innovation, competition and growth in the agbiotech sector remain fruitful avenues for further enquiry.

NOTE

1. After conducting field trials and having selected a genetically improved plant, to obtain legal approval to use the new gene in breeding programmes a firm applies to the United States Department of Agriculture (USDA) for deregulated status for the plant. APHIS (the Animal and Plant Health Inspection Service of USDA) maintains a publicly available database of deregulated crops (Pray et al., 2005).

REFERENCES

Antoncic, B. (2001), 'Organisational processes in intrapreneurship: a conceptual integration', *Journal of Enterprising Culture*, **9** (2) 221–35.

BASF (2011), 'BASF factbook 2011: we create chemistry', Ludwigshafen, Germany: BASF SE, Investor Relations, available at: www.basf.com. accessed 29 January 2012.

BASF (2012), 'BASF to concentrate plant biotechnology activities on main markets in North and South America', BASF News Release, 16 January 2012, Ludwigshafen, Germany BASF SE: available at: www.basf.com/group/pressrelease/P-12–109, accessed 1 October 2012.

Bayer (2011), 'Bayer 2010 Annual Report', Leverkusen, Germany: Bayer AG, available at: www.bayer.com, accessed 29 January 2012.

Blank, S.C. (2008), 'Insiders' views on business models used by small agricultural biotechnology firms: economic implications for the emerging global industry', *Agbioforum*, **11** (2), 71–81.

Bonnafous-Boucher, M. and E. Lamy (2010), 'Hybridisation of the profiles of entrepreneurs and intrapreneurs: the case of French biotechnology SMEs', *International Journal of Entrepreneurship and Small Business*, **10** (3), 410–27.

Brennan, M., C. Pray, A. Naseem and J.F. Oehmke (2005), 'An innovation market approach to analyzing impacts of mergers and acquisitions in the plant biotechnology industry', *Agbioforum*, **8** (2&3), 89–99.

Deeds, D.L. and C.W.L. Hill (1996), 'Strategic alliances and the rate of new product development: an empirical study of entrepreneurial biotechnology firms', *Journal of Business Venturing*, **11** (1), 41–55.

Dow (2011), '*The right formula for growth: the Dow Chemical Company 2010 annual report*', Midland, MO: The Dow Chemical Company, available at: www.dow.com/financial, accessed 29 January 2012.

Dutfield, G. (2008), 'Turning plant varieties into intellectual property: the UPOV convention', in G. Tansey and T. Rajotte (eds), *The Future Control of Food: A Guide to International Negotiations and Rules on Intellectual Property, Biodiversity and Food Security*, London: Earthscan, pp. 27–47.

Frisio, D.G., G. Ferrazzi, V. Ventura and M. Vigani (2010), 'Public vs. private agbiotech research in the United States and European Union, 2002–2009', *Agbioforum*, **13** (4), 333–42.

Gaisford, J.D., J.E. Hobbs, W.A. Kerr, N. Perdikis and M. Plunkett (2001), *The Economics of Biotechnology*, Cheltenham, UK and Northampton, MA, USA: Edward Elgar Publishing.

Han, K.J., N. Kim and R.K. Srivastava (1998), 'Market orientation and organizational performance: is innovation a missing link?' *Journal of Marketing*, **62** (4), 30–45.

Hayenga, M. (1998), 'Structural change in the biotech seed and chemical industry complex', *Agbioforum*, **1** (2), 43–55.

Hirschman, A.O. (1964), 'The paternity of an index', *The American Economic Review*, **54** (5), 761.

Hobbs, J.E. and W.A. Kerr (2003), 'Biotechnology and the changing structure of agri-food supply chains', *International Journal of Biotechnology*, **5** (3/4), 297–310.

Kalaitzandonakes, N. and M. Hayenga (2000), 'Structural changes in the biotechnology and seed industrial complex: theory and evidence', in W.H. Lesser (ed.), *Transitions in Agbiotech: Economics of Strategy and Policy*, Storrs, CT: Food Marketing Policy Center, University of Connecticut, pp. 217–27.

King, J.L. and D. Schimmelpfennig (2005), 'Mergers, acquisitions, and stocks of agricultural biotechnology intellectual property', *Agbioforum*, **8** (2&3), 83–8.

Kirca, A.H., S. Jayachandran and W.O. Bearden (2005), 'Market orientation: a meta-analytic review and assessment of its antecedents and impact on performance', *Journal of Marketing*, **69**(2), 24–41.

Klein, B., R.G. Crawford and A.A. Alchian (1978), 'Vertical integration, appropriable rents, and the competitive contracting process', *Journal of Law and Economics*, **21**(2), 297–326.

Lukas, B.A. and O.C. Ferrell (2000), 'The effect of market orientation on product innovation', *Academy of Marketing Science Journal*, **28**(2), 239–47.

Marks, L.A., B. Freeze and N. Kalaitzandonakes (1999), 'The agbiotech industry – a US–Canadian perspective', *Canadian Journal of Agricultural Economics*, **47**(4), 419–31.

Monsanto (2011a), 'Leading in innovation: 2011 annual report – Monsanto', St Louis, MO: Monsanto, available at: www.monsanto.com, accessed 29 January 2012.

Monsanto (2011b), 'Monsanto 2011 Form 10-K', St Louis, MO: Monsanto, available at: www.monsanto.com, accessed 29 January 2012.

Niosi, J. and J. Queenton (2010), 'Knowledge capital in biotechnology industry: impacts on Canadian firm performance', *International Journal of Knowledge-Based Development*, **1**(1/2), 136–51.

OECD (2005), *OECD SME and Entrepreneurship Outlook*, Paris: Organisation for Economic Co-operation and Development.

Oehmke, J.F. and C.A. Wolf (2003), 'Measuring concentration in the biotechnology R&D industry: adjusting for interfirm transfer of genetic materials', *Agbioform*, **6**(3), 134–40.

Peña, I. (2009), 'Intrepreneurial strategic behaviour and business performance', *International Journal of Entrepreneurship and Innovation Management*, **10**(1), 3–19.

Phillips, P.W.B. and G.G. Khachatourians (eds) (2001), *The Biotechnology Revolution in Global Agriculture: Innovation, Invention and Investment in the Canola Sector*, Wallingford: CABI Publishing.

Phillips, P.W.B. and C. Ryan (2007), 'The role of clusters in driving innovation', in A. Krattiger, R.T. Mahoney and L. Nelson (eds), *Intellectual Property Management in Health and Agricultural Innovation: A Handbook of Best Practices*, Oxford: MIHR and Davis, CA: PIPRA, pp. 281–94.

Pray, C. and A. Naseem (2005), 'Intellectual property rights on research tools: incentives or barriers to innovation? Case studies of rice genomics and plant transformation technologies', *Agbioforum*, **8**(2&3), 108–17.

Pray, C., J.F. Oehmke and A. Naseem (2005), 'Innovation and dynamic efficiency in plant biotechnology: an introduction to the researchable issues', *Agbioforum*, **8**(2&3), 52–63.

Renko, M., A. Carsrud and M. Brännback (2009), 'The effect of market orientation, entrepreneurial orientation, and technological capability on innovativeness: a study of young biotechnology ventures in the United States and Scandinavia', *Journal of Small Business Management*, **47**(3), 331–69.

Rhoades, S.A. (1995), 'Market share inequality, the HHI and other measures of the firm-composition of a market', *Review of Industrial Organization*, **10**(6), 657–74.

Smyth, S.J. and R. Gray (2011), 'Intellectual property sharing agreements in gene technology: implications for research and commercialisation', *International Journal of Intellectual Property Management*, **4**(3), 179–90.

Syngenta (2011), '*Syngenta annual review 2010*', Basel: Syngenta International AG available at: www.syngenta.com, accessed 29 January 2012.

Teece, D.J. and G. Pisano (1994), 'The dynamic capability of firms: an introduction', *Industrial and Corporate Change*, **3**(3), 537–56.
Williamson, O.E. (1979), 'Transaction-cost economics: the governance of contractual relations', *Journal of Law and Economics*, **22**(2), 233–61.
Zucker, L., M. Darby and M. Brewer (1998), 'Intellectual human capital and the birth of US biotechnology enterprises', *American Journal of Economics*, **88**(1), 290–306.

5 Biotechnology in North America: the United States, Canada and Mexico

Julia Bognar and Grace Skogstad

Biotechnology in North America, as it is globally, is dominated by the United States. American biotechnology firms vastly outnumber those elsewhere in the world, they spend more on research and development, and they are responsible for two-fifths of biotechnology patents worldwide (OECD, 2011). In the specific field of plant biotechnology, the US also dominates. In 2011, American farmers dominated global and North American production of GM crops, with 45 per cent and 87 per cent of production respectively (ISAAA, 2011). Canadian farmers have also embraced GM crops, particularly GM canola, positioning Canada in the top five GM nations in the world (ibid.). Mexico, the third member of the North American Free Trade Agreement (NAFTA), presents a different biotechnology story. Although it was one of the first countries to permit field trials of GM crops, it does not allow any commercial planting of GM varieties of native white corn, its most important cereal for human consumption. Moreover, pilot projects of GM varieties of yellow corn were only approved under limited conditions in early 2011. Mexico currently ranks seventeenth in the world in GM production (ibid.).[1]

American dominance in plant biotechnology can be attributed to explicit policies of its governments to promote GM products. These promotional policies include strong protection for intellectual property rights as well as a permissive regulatory framework that does not discriminate against biotechnology. The American regulatory framework authorizes GM products on the basis of principles of scientific risk assessments, substantial equivalence of GM and traditional products, and industry self-regulation. This permissive approach to genetically modified organisms (GMOs) contrasts sharply with the restrictive and precautionary approach taken in the European Union (EU) (see Chapter 11 of this *Handbook*).

Canadian governments have also promoted GM crops and foods by adopting policy instruments and principles similar to their American counterparts. These similarities include intellectual property rights protection for companies that develop GMOs, government funding for GMO research and development, and a regulatory framework based on principles of substantial equivalence and science-based risk assessment. Although industry self-regulation is not as predominant in the Canadian regulatory framework as in the American, greater government oversight of the regulatory process in Canada does not result in appreciable differences in policy outcomes when it comes to the approval and marketing of GM crops and foods.

As major food exporters, including of GM crops and foods, the United States and Canada have used their membership in international organizations such as the Organisation for Economic Co-operation and Development (OECD) and World Trade Organization (WTO) to establish rules and guidelines that do not discriminate against GM products and act as templates for cross-national harmonization. The two countries helped to define the OECD's 1986 guidelines on how to manage the safety risks of GMOs from the laboratory to confined field trials to unconfined commercial planting, as well as its 1992 guidelines on GM foods. In addition, the US was a major proponent of the WTO's Trade-Related Intellectual Property Rights (TRIPS): an agreement sought by its biotechnology companies and one that protects their intellectual property rights. Canada and the US have both been ardent supporters of the WTO Agreement on Sanitary and Phytosanitary Measures (SPS Agreement), an agreement that precludes countries permanently barring imports of GM products on grounds other than scientific evidence of their human health and environmental risks. The SPS Agreement proved an effective lever for Canada and the US pressuring the European Union to end its de facto moratorium on approvals of new GMO varieties between 1998 and 2004.

Mexico's GMO institutional framework, policy principles and policy outcomes differ from the American and Canadian in that they are a mixture of the promotional and permissive elements of American and Canadian GMO regulation as well as the precautionary and protectionist features of EU GMO regulation. There are large public investments in biotechnology R&D in Mexico, and Mexico's legislation to protect the property rights of GMO developers is in compliance with the WTO TRIPS. Mexico permits field/experimental trials of most GM plants and seeds, as well as commercial-scale planting of GM cotton and GM soy. The regulatory approach is also permissive in posing no trade barriers to imports of GM products. However, Mexico's GM approach is precautionary when it comes to genetically modified *maize* seeds and plants. No GM maize has yet been approved for commercial planting. Besides allowing special regimes to be established to protect maize, other precautionary and restrictive measures in Mexican GMO law enable states within Mexico to declare GMO-free zones, and require non-processed imports containing 5 per cent genetically modified material to be labelled. These precautionary measures are consistent with Mexico having signed and ratified the UN Cartagena Protocol on Biosafety, an agreement which regulates trans-boundary movement of live GMOs and one that Canada and the United States have opposed and not signed or ratified.

What explains the substance of American, Canadian and Mexican plant biotechnology policies? Domestic factors are most important in explaining US GMO policies while a combination of both domestic and international factors better explains Canadian and Mexico GMO policies. In terms of American biotechnology policies, five explanatory factors are uppermost: first, the cohesion and formidable organizational and political resources of the pro-biotechnology coalition; second, the privileged access of this coalition to policy makers; third, the congruence of the goals of the pro-biotechnology coalition with America's pro-business and pro-science political culture; fourth, the willingness of American farmers to embrace the technology; and fifth, the American public's comparatively few concerns about GMOs and its high trust in government authorities.

A combination of similar domestic political, institutional and cultural factors, including a belief in innovative technologies as a way to enhance the competitiveness of the agricultural sector, also explains Canadian GMO policies. International factors – in the form of trade dependence on US markets as well as obligations to respect the terms of international treaties – also affect Canadian biotechnology policies, and help explain their considerable convergence with American policies. In terms of Mexico, international factors loom large in explaining its permissive approach, especially with respect to GM

Table 5.1 US, Canadian and Mexican GMO regulatory principles and policies

	United States	Canada	Mexico
GMO specific legislation	No	No	Yes
Government role in authorizing GM crop trials	Permits or notification	Notification	Approval compulsory
Government role in GM commercial planting	Permits or notification	Approval compulsory	Approval compulsory
GM-free zones	No	No	Yes
Principle of substantial equivalence for food	Yes	Yes	Yes
Govt notification or approval of GM foods	Voluntary	Mandatory notification	Mandatory public opinion notification
GM food labelling	None	Voluntary: provision to label GM-free if less than 5% GM content	Compulsory for non-processed GM foods; imports above 5% threshold labelled
Recognition and application of precautionary principle	No	No	Formally recognized; application is discretionary

imports which come overwhelmingly from its most important trading partner, the US. By contrast, domestic political and cultural factors are most important in explaining Mexico's precautionary approach with respect to GM maize. At the same time, this precautionary approach has been facilitated by the international treaty on biosafety.

The chapter discusses in turn the GMO regulatory frameworks of the United States, Canada and Mexico, describing and explaining their fundamental features, as well as their similarities and differences. The latter are summarized in Table 5.1.

1 THE UNITED STATES

The United States is the global leader in the development and production of GM crops. It has allowed more herbicide-tolerant and insect-resistant crops to be tested, planted and introduced into the market than any other country. Over 90 per cent of soybeans and in excess of 70 per cent of cotton and corn grown in the US are genetically modified.[2]

The unparalleled growth of the American biotechnology industry, and the longstanding status of the United States as the globe's pre-eminent producer of GM crops, can be traced to a permissive regulatory framework that does not discriminate against GM products and expedites their approval in a timely fashion. Among the key principles that guide regulatory procedures for the approval of GM crops and foods are the following. First, the *process* of genetic engineering does not pose any inherent risks and hence GM *products* can be regulated by the same procedures that govern similar products derived from traditional production methods. Second, insofar as GM foods are substantially equivalent to traditional foods, there is no need to regulate them more stringently. Third, industry self-regulation is a reliable instrument for ensuring the safety of GM foods. And fourth, scientific risk assessments affirming the environmental and health safety of GM products are the sole criterion for authorizing their sale. Ethical and socio-economic considerations are not considered in the formal regulatory regime.

Adhering to the OECD 1986 guidelines, the 1986 Coordinated Framework for the Regulation of Biotechnology (United States, 1986) defined procedures for the approval of GM crops and foods, and articulated the principle that genetic engineering posed no particular risks for human health or the environment that warranted the specific regulation of this production method. Because genetic engineering was deemed to be not inherently risky, it was only necessary to regulate the *products* of biotechnology and this could be done under existing legislation and by existing organizations of the executive branch.[3]

The regulatory agencies are the Animal and Plant Health Inspection Service (APHIS) of the US Department of Agriculture (USDA), responsible for regulating potential plant pests and noxious weeds from being introduced into the environment; the Environmental Protection Agency (EPA), responsible for regulating the safe use of pesticides and establishing allowable levels of pesticides in food; and the Food and Drug Administration (FDA), responsible for ensuring the safety of food and feed. Their collective task is to ensure that GM crops and foods derived from or containing GM material meet standards of environmental and human health safety.

The regulatory process and role of government agencies varies depending upon the characteristics of the GM product. A GM corn that has had the *Bacillus thuringiensis* (Bt) gene inserted to produce an insecticidal protein/pesticide requires the pre-marketing approval of the EPA. Before the Bt corn can be marketed commercially, the developer of the Bt corn is required to demonstrate to the EPA that it presents no environmental risks, and, if the crop is to enter the food or feed chain, no health risks. It meets this safety requirement by providing data from field tests that have been registered with the EPA and conducted according to EPA guidelines. The company may also, but need not, consult the FDA with respect to any possible changes in nutritional composition or levels of toxics that could affect food or feed safety. Once the EPA is satisfied that the pesticide-containing Bt corn will not have adverse environmental or human health effects, it can be registered for commercial sale under the conditions established by the EPA for its safe use.

Developers of other GM crops – those not encoded with a pesticide – must also provide government-specified data affirming that there will be no significant risks from widespread planting of the GM crop before it can be approved for commercial planting. Such approval can come in one of two ways: following application to APHIS/USDA for a permit to conduct field testing or, since 1993, by simply notifying APHIS that such testing will take place. While a permit imposes restrictions on planting (to prevent the escape of plant material that may pose a pest risk), a notification does not. The notification procedure applies to all genetically modified plants not considered a noxious weed, a category into which 99 per cent of genetically modified plants fall. Under the notification procedure, government officials review risk assessment data provided by the biotechnology company, including information that affirms the eligibility of the GM plant for the notification procedure (for example, it is a plant species that APHIS has determined may be safely introduced). In 1993 six transgenic crops were approved for commercialization under the notification procedure; by 1997, the majority of genetically modified crop varieties grown and used in the US were approved. The notification procedure has reduced the approval process from 10–18 months (National Academy of Science, 2002) to 30 days (Pew Initiative on Food and Biotechnology, 2001).

The FDA has interpreted its regulatory responsibility – to ensure the safety of plants that will be consumed as food or feed – in a way that also expedites approval of GM foods. It regulates substances added to food as either food additives, which require pre-market review and approval as safe by the FDA, or as substances that are 'generally recognized as safe' (GRAS) and which do not require prior FDA approval before marketing. In a 1992 policy statement, the FDA indicated its intention to consider foods derived from genetically modified plants in the same way it had traditionally treated foods containing additives developed from more traditional forms of plant breeding. It also indicated that most foods derived from genetically modified plants were presumptively GRAS. Only GM foods that were not substantially equivalent – in structure, function or composition – to substances currently found in food, would need FDA pre-market approval. At the same time, the FDA also made it voluntary for biotechnology developers to consult with the FDA regarding the safety of GM foods prior to marketing them. The rationale for voluntary notification is that US food safety law makes those who develop new products legally responsible for their safety. Since 1992, then, it has been at the discretion of food developers to provide FDA scientists with 'summary information' on safety and nutritional aspects of transgenic plants used in food. They are reported to have done so and consulted the FDA on all GM foods and feeds currently marketed in the US, despite such notification not being required.

The FDA has not invoked other authority it possesses to regulate GM products. For example, it can require food producers to provide specific information to consumers, including labelling if the GM product is no longer substantially equivalent to the corresponding conventional product. However, given the principle of substantial equivalence of GM and traditional foods, the FDA has not introduced either mandatory or voluntary labelling to signal to consumers that a product has been produced using GM plants or foods.

To reiterate, the US science-based risk assessment biotechnology regulatory process has not handicapped expeditious approval of GM crops for cultivation and sale. Once having satisfied APHIS/USDA regulators that GM crops are safe, there is limited regulatory oversight of their subsequent cultivation. The exception is Bt corn crops, where EPA monitoring rules require companies to have stewardship programmes that include, first, managing refugia (natural habitats) or setting aside areas in an effort to sustain Bt resistant bugs; and, second, collecting and submitting data that assess increased risks of insect resistance or adverse health or environmental effects.

What explains the American permissive and promotional approach to biotechnology? The account begins with recognition that this regulatory approach is the one favoured by American biotechnology companies. In so far as biotechnology is a high-cost and high-risk industry, investors in the technology have sought protection for their investments. From their perspective, such protection is best promoted by intellectual property rights protection and a regulatory approach that assures consumers of the safety of GMOs but also facilitates their commercialization by expeditious approval for new products. Such a regulatory framework, the biotechnology sector argued to policy makers in advance of the 1986 Coordinated Framework for the Regulation of Biotechnology, would enable the biotechnology sector to be a major driver of economic growth. It would also allow the United States to gain an internationally competitive edge in the agricultural sector. These arguments resonated with farmers, accustomed as they were to adopting the latest technological innovations as a way to enhance their productivity. They were also persuasive with the deregulatory-oriented Republican administrations of Presidents Regan and Bush in the mid-1980s and early 1990s (United States, 1985, 1991) that established the US biotechnology regulatory framework.

The success of the American pro-biotechnology lobby in persuading American governments to adopt their preferred regulatory approach can be attributed to several factors (Bernauer and Meins, 2003; Bernauer, 2003; Moore, 2000; Vogel, 2003; Newell, 2003; Skogstad and Moore, 2004). First is the cohesion and formidable organizational and political resources of the pro-biotechnology coalition. The private sector members of the pro-biotechnology coalition include biotechnology developers, represented by the Biotechnology Industry Organization; farmers in the American Farm Bureau; and food processors in their umbrella organization, the Grocery Products Manufacturers of America. The public sector members of the pro-biotechnology coalition include officials in federal and state governments as well as academics in public universities. Second is the privileged access this coalition has enjoyed to policy makers in the American political institutional framework of decision-making. The latter is comprised of elected politicians who are particularly sensitive to rural (farmer) interests as well as to organized interest groups who finance their election campaigns. As noted above, during the pivotal period when the US biotechnology regulatory framework was established, the fact that it was dominated by deregulatory Republicans significantly enhanced the influence of businesses, including plant biotechnology firms, over regulatory policies like those pertaining to GMOs. When it comes to the implementation of the biotechnology regulatory framework, the very limited statutory authority and scarce resources of regulators also gives plant biotechnology firms considerable scope to influence regulatory policy-making (Moore, 2000).

A third factor explaining the capacity of the pro-biotechnology coalition to prevail with its preferred regulatory approach is the compatibility of its goals with America's political culture. This political culture is pro-business (Vogel, 2003) and is deferential to the scientific authority that is relied on to assure Americans that the benefits of GMOs outweigh their risks (Brossard and Nisbet, 2006). A fourth factor, American farmers' early embrace of GM crops as a way to enhance their productivity and profits, has also been crucial. Alongside the American public's comparatively few concerns about GMOs and its high trust in government authorities (Bernhauer and Meins, 2003), this supportive political culture of biotechnology isolated the consumer and environmental groups that argued the need for a more precautionary American approach to managing the risks of GMOs. It was virtually impossible for these critics to get political traction.

2 CANADA

Also based on OECD guidelines, the Canadian biotechnology regulatory approach is very similar to the American approach. Its analogous principles include, first, the belief that it is the products of genetic engineering and not the process that should be regulated; second, the principle that genetically modified products can be regulated with the same legal instruments and institutions used for non-GM products; third, the principle of the substantial equivalence of GM foods to traditional products; and fourth, science-based risk assessments are the only relevant criteria for GM products' approval. While industry self-regulation is also present in the Canadian biotechnology regulatory system, it appears to a lesser extent than in the US. As in the US, subscription to these principles has resulted in a regulatory framework that promotes GM products.

The Canadian biotechnology regulatory approach was established in the 1993 Canadian Regulatory Framework for Biotechnology. It stipulated that genetically modified plants should be regulated in the same way as all *plants with novel traits*, that is, plants containing traits not present in existing plants of the same species and that result from either traditional breeding or genetic engineering. Canada's use of the concept 'novelty' distinguishes its regulatory system from all other countries, and results in different products being reviewed in Canada than elsewhere. For ease of comparability across the three NAFTA countries, however, the discussion here will substitute the terms 'GM plants' and 'GM foods' for those in Canadian legislation: 'plants with novel traits' and 'novel foods'. Canadian legislation requires a GM (novel) product to be evaluated on the basis of its own merits and characteristics, not in terms of the process by which it had been developed. The 1993 Framework also stipulated that existing legislation would serve as the basis for developing regulations to govern biotechnology.[4] A second document, *Guidelines for the Assessment of Novel Foods*, issued in 1994 and also modelled on OECD guidelines, established responsible agencies and procedures to execute the 1993 Framework.

As in the United States, the Canadian decision to approve or not to approve a genetically modified plant is taken solely on the basis of scientific evidence supplied by the biotechnology developer regarding the product's environmental and human health safety. Social, economic and ethical issues are immaterial to GM product regulation. Biotechnology companies seeking authorization for a GMO field trial notify the Canadian Food Inspection Agency (CFIA), which then assesses the plans and carries out inspections to ensure the field trial complies with regulatory conditions. While the CFIA is responsible for ensuring that GM plants have no adverse environmental impacts, Health Canada is entrusted with ensuring the GM plant's safety as a food. Like the CFIA, it relies on risk assessment data submitted by the developer to make this decision. Similar to their American counterparts, regulators in the CFIA and Health Canada decrease or waive the requirements for risk assessment information for products judged to be 'familiar' or 'substantially equivalent' to conventional plant or food species with a history of safe use or previously approved plants with novel traits. However, in an important contrast to the US, it is mandatory for GM food manufacturers (or importers) to submit data confirming a GM food product's safety before it can be sold in Canada.

Following the publication in 1998 of a Biotechnology Strategy document, whose declared objective was to make Canada capture a significant share of the presumed benefits of biotechnology, the approval process for novel foods was clarified. Amendments to the Canadian Food and Drug Law in 1999 confirmed the principle of substantial equivalence as the primary means for deciding whether a crop will be regulated as a conventional crop or as a novel product (Isaac, 2002). Another amendment stipulated that

Health Canada would not be reviewing *all* foods new to the Canadian market, but rather only those that fell under their definition of 'novel'. This policy introduced the concept of 'prior safe use as a food', a concept which allows Health Canada to exclude foods that have been accepted as safe for food use by other countries from being subject to novel food notifications (Canada, 1999; Isaac, 2002). So far there is no evidence that this provision has been used to approve new GM foodstuffs in Canada without undertaking the appropriate domestic assessment.

An explanation of Canada's promotional biotechnology regulatory framework, and its striking similarity to that in the United States, takes us to domestic factors similar to those that explain American plant biotechnology policies. These factors include, first, a shared belief by biotechnology companies, Canada's largest farmers' organizations, and Canadian federal and provincial governments in plant biotechnology as an instrument of enhanced agricultural productivity and competitiveness; second, an institutional framework of decision-making that has given privileged access to this well organized and resourceful pro-biotechnology coalition and posed formidable obstacles to GMO critics; and third, a Canadian political culture marked by faith in science and technology generally and deferential to scientific experts and government regulators to make decisions on product safety. As in the United States, these factors together have made it impossible for critics of plant biotechnology – who have included many diverse groups of smaller and organic farmers as well as environmentalists and consumer groups – to derail the Canadian government's promotion of GM crops.

Goals of economic competitiveness have loomed large in Canadian biotechnology policy. A major concern of Canada's biotechnology sector was that it should not be put at a competitive disadvantage compared to its foreign, especially American, counterparts. A framework that licensed GM crops and foods solely on the basis of their scientifically-determined risks was such a competitive framework; one that paid heed to the broader social, economic and ethical issues raised by GMOs was not. The decision of the Canadian government to use existing legislation and supplementary guidelines to regulate GM products – rather than introduce new legislation – proved pivotal. It eliminated the possibility for these broader issues to arise in parliamentary and public debates, and meant that decisions around how to regulate GMOs were taken in closed policy networks constituted by government regulators, scientific experts and representatives of biotechnology companies and farmers (Moore, 2000).

International factors, in the form of economic competitiveness concerns and conformity with provisions of trade agreements, also account for other features of Canadian plant biotechnology policy. One example is GMO labelling policies. The Canadian government has rejected opinion polls showing strong consumer demand for mandatory labelling of GM foods, and rebuffed the argument of environmental and other anti-GMO groups that labelling is needed to give consumers the right to know. It has done so out of the belief that mandatory labelling in Canada would put at risk both trade and investment with the US where there is no mandatory labelling of GM foods. Mandatory labelling of GM products could be challenged under NAFTA as a restrictive trade barrier. Beyond concern about maintaining access to the American market, there has been the recognition that were Canada to introduce labelling to identify the process by which a product has been made (that is, from genetically modifying a seed), such labelling would seriously compromise the country's ability to oppose initiatives by other countries to do the same.

In so far as the Canadian government regards food labelling that is based on process and production methods to be a non-tariff barrier to trade, it would compromise the county's trade interests by adopting such measures itself. The Canadian government's compromise to advance its trade concerns without ignoring citizens' desire for GM food labelling was to ask an industry-dominated body to develop a voluntary GM food labelling scheme. This strategy has not proved efficacious for consumers because Canadian manufacturers have not taken it up. Nonetheless, a voluntary GM food labelling scheme is consistent with policy initiatives in the US, Canada's most important trade partner (Skogstad, 2008, pp. 230–34).

A second example where international competitiveness concerns, owing to external market dependence, have shaped Canadian GMO policies is the debate over licensing a GM wheat developed by Monsanto. A series of market reports and analyses by the Canadian Wheat Board indicated that Canadian wheat farmers could lose many of their most important markets – which extend well beyond the US – if GM wheat varieties were licensed in Canada. To ward off this possibility, Canadian wheat farmers mounted a broad coalition that included longstanding GMO opponents, including health, environmental and anti-globalization groups, to oppose the approval of GM wheat for cultivation. They succeeded when Monsanto announced in May 2004 that it was deferring its efforts to introduce its GM wheat variety in Canada. Monsanto also withdrew a similar application in the US, where it had been likewise opposed by American farmers, millers and bakers (Skogstad, 2008, pp. 234–38).

Although plant biotechnology is not without controversy in Canada, those who oppose it or seek to broaden the debate to include discussion of the social, economic and ethical issues of GMOs have been frustrated by their inability to shift the terms of debate or alter the regulatory framework. The 1998 Biotechnology Strategy created the Canadian Biotechnology Advisory Committee (CBAC), an expert advisory panel with a mandate to provide independent advice to the government of Canada on issues related to the health, environmental, ethical, social, economic and regulatory aspects of biotechnology. A sub-committee of CBAC created in early 2001 to provide advice on GM food included representation from civil society groups, including those opposing GMOs. However, most of the latter boycotted CBAC's consultation process. They did so out of the belief that CBAC's process was too narrow to address adequately the balance of GMO risks and benefits, and, further, that CBAC itself was not impartial. Rather, it was a mechanism to further the Canadian government's goal of fostering the biotechnology industry in Canada (Abergel, 2012).

A closed policy network envelops Canadian biotechnology regulation and limits the influence of those who are sceptical of the net balance of the technology's risks and benefits. Even when this policy network opens up – as it has when parliamentary committees have debated GM product labelling and GM wheat – elected politicians' support falls overwhelmingly on the side of the agri-food sector that is the major user of the technology (Skogstad, 2008, pp. 232–8). GMO sceptics cannot count on the Canadian public for support. Canadians trust regulatory authorities and defer to the scientific expertise upon which biotechnology regulators rely (Miller et al., 1997, p. 110; Gaskell and Jackson, 2005). As in the US, surveys of Canadians reveal that a majority believes that biotechnology regulations should be based upon expert advice and scientific evidence, rather than moral and ethical criteria or public opinion (Gaskell et al., 2005).

3 MEXICO

In contrast to the promotional and permissive approach to biotechnology in the US and Canada, Mexico's approach is more ambivalent, combining promotion and precaution. On the one hand, the Mexican government has been permissive in approving field trials of GM products and their subsequent cultivation. Mexico was one of the first developing countries to allow field trials of GM plants, and it has subsequently allowed large-scale planting of Bt cotton. In addition, imports of GM products, including GM maize/corn, are allowed. On the other hand, Mexico has taken a precautionary approach when it comes to the approval of genetically modified *maize* seeds and plants. It allowed pilot and experimental tests of GM maize only in 2011 and has not yet approved any GM maize crops for widespread cultivation. As elaborated further below, Mexico's precautionary approach can be explained by the cultural and economic significance of maize.

The permissive side of the Mexican biotechnology approach is reflected in its approval since 1988 of field tests of GM crops, by international companies like Monsanto but also by academic institutions and national research centres. The first law governing transgenic crops was passed in 1995. Among other things, NOM-056-FITO-1995 required a government certificate to be obtained prior to releasing GM material into the environment for experimental programmes or pest control processes. Under this law, 373 release permits were issued between 1988 and 2006, including permits for GM maize, tomatoes, cotton and soybeans (González, 2010, p. 77).

The precautionary side of Mexican GM policy manifested itself when opposition to the release of transgenic maize grew from 1997 onward as evidence surfaced of gene flow from GM maize to native varieties. In 1998, on the recommendation of the environment and agriculture ministries, the Mexican government stopped the approval of field trials with commercial varieties of GM maize. The moratorium on the release of GM maize did not, however, quell controversy. A 2001 study published in *Nature* reported evidence suggesting that GM traits had spread to and persisted in native varieties of maize in the Mexican state of Oaxaca. The authors alleged that the gene flow was due to the Mexican government's failure to implement its own policy prohibiting field trials of GM maize. Further investigations by local organizations discovered evidence of introgression: the spread and persistence of GM traits in local maize varieties in 33 different locations. While the Mexican Ministry of Agriculture and biotechnology industry representatives argued that gene flow between GM varieties and native maize was part of a natural and beneficial process of hybridization, GM critics argued that it was a form of 'genetic pollution'. The Mexican National Ecology Institute and the Biodiversity Commission, whose tests corroborated the findings of the *Nature* study, attributed the contamination to soaring GM maize imports from the US. Mexico imports 89 per cent of the corn it needs for animal feed, and 99.8 per cent of that yellow corn comes from the US (USDA, 2008).

Evidence of contamination of local varieties of maize with GM sequences led affected communities to request the Secretariat of the Commission for Environmental Cooperation (CEC) to prepare a report on the potential direct and indirect environmental impacts on biodiversity caused by the release of GM maize in the state of Oaxaca. This report, prepared pursuant to Article 13 of the North American Agreement on Environmental Cooperation and presented in 2004, concluded that there was introgression. However, it found no negative impacts on biodiversity and health. Among its recommendations were the implementation of a monitoring strategy to guarantee the preservation of the genetic diversity of Mexico's maize and the labelling of GM maize imports from the US. Believing that the genetic contamination had occurred when Mexican farmers had planted GM corn imports intended for the food system (and not for cultivation), the CEC also recommended Mexican farmers be educated on handling GM maize (Commission for Environmental Cooperation, 2004).

Against this backdrop of mobilized concern about GM crops contaminating native corn varieties, the Mexican Law on Biosafety and GMOs was passed in 2005. Preceded by the lifting of the moratorium on GM field trials in 2004, the Biosafety Law established the current regulatory framework for GMOs. First, companies must acquire permits on a case-by-case basis for experimental trials of GMOs and their commercial sale. Second, the law recognizes the importance of maize biodiversity and stipulates that it must be protected. Third, there is provision for the establishment of special regimes of protection for maize and other crops originating in Mexico, including a *de facto* moratorium on the commercial planting of GM maize. Fourth, states within Mexico can declare GMO-free zones within their borders. Fifth, the law calls for compulsory labelling of agricultural non-processed GM foods. Labelling information should acknowledge that the seeds are genetically engineered and should include information on the characteristics of the acquired genetic combination, and the implications with regard to special conditions and growing requirements. Lastly, there is a mandatory procedure for the public to express their concerns, within a 20-day period, when a request for authorization is submitted. Public opinions must be based upon scientific evidence.

Consistent with Mexico's membership in and ratification of the Cartagena Protocol on Biosafety, the 2005 Biosafety Law permits application of the precautionary principle. Its reference to the precautionary principle in Article 9 (IV)[5] is, however, accompanied by Article 63. It states that the Secretariat in charge must also take into account administrative procedures provided in this Law, and international trade agreements and guidelines developed by international organizations to which Mexico is party. In so far as the relevant international treaties are not only the Cartagena Protocol on Biosafety but also the WTO Agreement on the SPS Agreement, the Secretariat has discretion to decide whether to apply the precautionary principle according to Cartagena or the principle of substantial equivalence in the SPS Agreement. Critics argue that regulatory agencies have used their discretion *not* to apply the precautionary principle in the event of uncertainty regarding a GMO's risks, but, rather, to promote biotechnology.

As in the US and Canada, GMO approval in Mexico is based on scientific evidence and case-by-case assessments. The Secretariat of Agriculture (SAGARPA) issues permits for testing and growing GMOs, using risk assessment data received from the biotechnology developer to determine potential risks of the GMO to animal, plant and aquatic health. The Secretariat of the Environment and Natural Resources, SEMARNAT, analyses the potential risks of GMOs to the environment and biological diversity, while the Secretariat of Health (SALUD) ensures the food safety of biotechnology-derived agricultural products destined for human consumption. GM foods and crops are evaluated using the principle of substantial equivalence: a crop or food found to be substantially equivalent to its non-GM counterpart is not subject to further testing or more stringent regulation (Fitting, 2011).

As a complement to the 2005 Biosafety Law, SEMARNAT updated the Implementation rules in 2008 to allow developers and research institutions to experiment with GM maize in approved regions of Mexico where native corn varieties typically do not grow. While the Biosafety Law does call for a special regimen for GM maize to prevent the release of transgenic seeds in centres of origin, and a demarcation of GM maize-free zones, there have been questions about its effectiveness. For example, US imports of GM maize for food, feed or processing continue unabated with little monitoring and few oversight mechanisms. Moreover, the Biosafety Law is silent on the specific mechanisms a state can utilize to apply for 'centre of origin status' in order to ban the planting of GM seeds. This lack of clarity helps explain why only two states, Tlaxcala and Michoacan, have to date banned GM maize under the condition that they are centres of origin. Together the two states produce one third of the white corn that is native to Mexico.

An explanation for Mexico's ambivalent approach to plant biotechnology takes us to both domestic and international factors. Looking first at domestic factors, Mexico's precautionary approach to GM maize owes much to the influence of a well-organized anti-GMO lobby. It has consisted of environmental groups such as Greenpeace, as well as farmers and indigenous rights groups mobilized by concerns about introgression to landrace (traditional) maize. Under the umbrella of the Network for the Defence of Maize, these anti-GMO activists raised public awareness of GMO issues through the media and successfully pressed regulators to limit GM maize approvals. They did so by tapping into cultural values that link maize and the production of it by local farmers to the preservation of the Mexican identity and way of life. Mexico is a 'centre of origin' for several varieties of maize, the plant is a staple of the Mexican diet, and it is highly important to Mexico's agriculture economy. More than half of cultivated acreage in Mexico is devoted to corn production. The anti-GMO campaign framed genetic engineering, the multinational corporations who develop and promote it, and the Mexican governments whose lax policies allowed contamination of native varieties of maize by GM varieties, as a threat to Mexican peasant lifestyles and livelihoods as well as to the country's biodiversity and autonomy (Malkin, 2005; Otero, 2008; Fitting, 2008; Poitras, 2008; Goodman, 2009).

The anti-GMO campaign's success in galvanizing peasant and indigenous rights communities around this particular framing of GM corn (Fitting, 2008) and in persuading regulators not to approve GM maize commercialization invites comparison with Canada and Mexico where anti-GMO campaigns have failed. At least three differentiating factors, besides the symbolic and cultural importance of GM maize, appear to be important. One is Mexicans' low level of trust in government and industry when it comes to issues of biotechnology (Poitras, 2008). A second factor is division within the Mexican scientific community regarding the safety of GM products as compared to substantial scientific cohesion in Canada and the US.[6] While Mexican government scientists espoused the safety of GM products, scientists working in academic and private research centres were less cohesive in their conclusions, with many expressing concerns over the effects of biotechnology on Mexico's biodiversity (Antal, 2008). A third is the lesser insulation of Mexican regulators from organized groups representing not only industry but also environmental and peasant organizations (ibid.).

Adherence to the terms of international treaties also explains Mexico's precautionary approach. Mexico has signed and ratified the Cartagena Protocol on Biosafety, an agreement which regulates trans-boundary movements of GMOs and enables countries to insist on advance notice of shipments containing GM material. As noted above, the Biosafety Protocol was strongly opposed by the governments of the United States and Canada. A 2004 trilateral agreement negotiated with its NAFTA partners enabled Mexico to require, consistent with the Cartagena Protocol, that bulk commodity shipments coming from the US be labelled to indicate they 'may contain' GMO content when they contain 5 per cent or more GMO material.

Turning to Mexico's permissive approach – permission for imports of GM maize and commercial plantings of GM crops other than maize (such as soy and cotton) – international factors are an important explanation. Mexico's membership in NAFTA required Mexico to remove price support mechanisms for local farmers, a loss which made it imperative for Mexican farmers to become more competitive with their trading partners. To promote their agricultural sector's competitiveness, Mexican governments have emphasized agricultural reforms and adoption of capital-intensive technologies, such as genetic modification (Poitras, 2008). NAFTA also imposed obligations on Mexico when it came to allowing imports – including of GM corn imports – from its Canadian and American trading partners.

4 CONCLUSION

Within the NAFTA bloc, there is both congruence and incongruence in the three member countries' plant biotechnology policies. Canada and the United States, guided by similar competitiveness goals and responsive to the interests of biotechnology developers and most farmers, have adopted permissive plant biotechnology policies. By contrast, responding to a different set of organized interests and concerns, Mexico, the third NAFTA member, has charted a different course. It does not discriminate against imports of GM products but is more precautionary when it comes to approving the cultivation of GM varieties of the country's most important crop, maize.

Mexico's distinctive approach to GM crops is a potent reminder of the gap between countries that have been leaders in developing innovative technologies as a strategy to maximize the efficiency and profitability of their agricultural sectors (Canada and the USA), and those that are laggards in adopting this strategy (Mexico). But there is more to the different regulatory policies of the three countries than simply whether a country is an innovator in technology adoption. Rather, the NAFTA countries' differences are also a potent reminder of the limitations of international agreements that attempt to treat agricultural commodities – including those that have been genetically modified – as similar to other (industrial) commodities. For a host of reasons, agricultural products in some countries are exceptional commodities in the sense of being deemed intrinsic to preserving a culture. For quite legitimate cultural and economic reasons, corn has acquired this status in Mexico. There does not appear to be any commodity that has acquired a similar cultural status in Canada and the USA. Rather, as the example of GM wheat demonstrates, in these countries an argument about the need to treat a commodity as exceptional boils down to its economic significance.

Extrapolating beyond NAFTA to global/WTO regulations, the challenge continues to be to find the appropriate balance between goals of trade liberalization and the equally legitimate goal of preserving indigenous agricultural practices that reflect the consensual values of a society. We have not yet found that equilibrium in international agreements.

NOTES

1. The assistance of Sebastian Baglioni, PhD student at the University of Toronto, with Spanish documents on the Mexico case is gratefully acknowledged.
2. Data obtained from the USDA Economic Research Service, 'Adoption of genetically engineered crops in the US', available at: http://www.ers.usda.gov/Data/BiotechCrops/, accessed 29 December 2011.
3. The existing Federal Acts that were potentially relevant to the regulation of biotechnology included the Federal Plant Pest Act, the Plant Quarantine Act, the Federal Food, Drug and Cosmetic Act, the Federal Insecticide, Fungicide and Rodenticide Act, and the Toxic Substance Act.
4. This legislation is the Feeds Act, the Fertilizers Act, the Seeds Act, Health of Animals Act and the Plant Protection Act.
5. The Article states that 'where there are threats of serious and or irreversible damage, lack of full scientific uncertainty shall not be used as a reason for postponing cost effective measures to prevent environmental degradation.' This Article is consistent with Article 10(6) and 11(8) of the Cartagena Protocol on Biosafety.
6. Scientists in Canada have been critical of Canada's biotechnology regulatory approach but show more unity in terms of their judgements regarding the safety of GMOs (providing certain conditions of their release are met).

REFERENCES

Abergel, Elisabeth (2012), 'The Canadian Biotechnology Advisory Committee: legitimacy, participation, and attempts to improve GE regulation in Canada', in Rod MacRae and Elisabeth Abergel (eds), *Health and Sustainability in the Canadian Food System: Advocacy and Opportunity for Civil Society*, Vancouver: UBC Press, pp. 97–126.

Antal, Edit (2008), 'Interaccion entre politica, cienca y sociedad en biotechnologia: la regulacion de los organismos geneticamente modificados en Canada y Mexico', *Norteamerica*, **3**(1), 11–63.

Bernauer, Thomas (2003), *Genes, Trade and Regulation: The Seeds of Conflict in Food Biotechnology*, Princeton, NJ: Princeton University Press.

Bernauer, Thomas and Erika Meins (2003), 'Technological revolution meets policy and the market: explaining cross-national differences in agricultural biotechnology regulation', *European Journal of Political Research*, **42**(5), 643–83.

Brossard, Dominique and Matthew Nisbet (2006), 'Deference to scientific authority among a low information public: understanding US opinion on agricultural biotechnology', *International Journal of Public Opinion Research*, **19**(1), 24–52.

Canada (1999), 'Ministry of Health: Regulations amending the food and drug regulations (948 – novel foods)', Ottawa, available at: http://www.gazette.gc.ca/archives/p2/1999/1999-10-27/html/sor-dors392-eng.html.

Commission for Environmental Cooperation (2004), 'Maize & biodiversity: the effects of transgenic maize in Mexico', available at: http://www.cec.org/Storage/56/4837_Maize-and-Biodiversity_en.pdf.

Fitting, Elizabeth (2008), 'Importing corn, exporting labor: the neoliberal corn regime, GMOs, and the erosion of Mexican biodiversity', in Gerardo Otero (ed.), *Food for the Few: Neoliberal Globalism and Biotechnology in Latin America*, Austin, TX: University of Texas Press.

Fitting, Elizabeth (2011), *The Struggle for Maize: Campesinos, Workers and Transgenic Corn in the Mexican Countryside*, Durham, NC: Duke University Press.

Gaskell, George and Jonathan Jackson (2005), 'A comparative analysis of public opinion: Canada, the USA and the European Union', in E. Einsiedel (ed.), *First Impressions: Understanding Public Views on Emerging Technologies*, Ottawa, pp. 63–75, available at: http://publications.gc.ca/collections/Collection/Ju199-4-2005E.pdf.

Gaskell, George, Edna Einsiedel, William Hallman, Susanna Hornig Priest, Jonathan Jackson and Johannus Olsthoorn (2005), 'Social values and the governance of science', *Science*, **310**, 1908–09.

González, Alicia Gutiérrez (2010), *The Protection of Maize Under the Mexican Biosafety Law: Environment and Trade*, Göttingen: Universitääsverlag Göttingen.

Goodman, Jessica Lauren (2009), 'Defining symbolic spaces in response to a globalizing world: incorporating GMOs and segregating markets in Mexican agriculture. MA thesis, University of British Columbia.
Isaac, Grant (2002), *Agricultural Biotechnology and Transatlantic Trade: Regulatory Barriers to GM Crops*, New York: CABI Publishing.
ISAAA (International Service for the Acquisition of Agri-Biotech Applications) (2011), 'Executive summary: global status of commercialized biotech GM crops', ISAA Brief no. 43–2011.
Malkin, Elizabeth (2005), 'Science vs. culture in Mexico's corn staple', New York Times, 27 May.
Miller, Jon D., Rafael Pardo and Fujio Niwa (1997), *Public Perceptions of Science and Technology: A Comparative Study of the European Union, the United States, Japan and Canada*, Chicago, IL: The Chicago Academy of Social Science.
Moore, Elizabeth (2000), 'Science, internationalization and policy networks: regulating genetically engineered food crops in Canada and the United States, 1973–1998', PhD thesis, University of Toronto.
National Academy of Science (2002), *Environmental Effects of Transgenic Plants: The Scope and Adequacy of Regulation*, Washington, DC: National Academy Press.
Newell, Peter (2003), 'Globalization and the governance of biotechnology', *Global Environmental Politics*, **3**(2), 56–71.
OECD (2011), *Key Biotechnology Indicators*, Paris: Directorate for Science and Technology, available at: http://www. oecd.org/sti/biotechnology/indicators.
Otero, Gerardo (2008), 'Neoliberal globalism and the biotechnology revolution: economic and historical context', in Gerardo Otero (ed.), *Food for the Few: Neoliberal Globalism and Biotechnology in Latin America*, Austin, TX: University of Texas Press.
Pew Initiative on Food and Biotechnology (2001), 'Guide to US regulation of genetically modified food and agricultural biotechnology products', available at: http://www.pewtrusts.org/uploadedFiles/wwwpewtrustsorg/Reports/Food_and_Biotechnology/hhs_biotech_0901.pdf.
Poitras, Manuel (2008), 'Social movements and techno-democracy: reclaiming the genetic commons', in Gerardo Otero (ed.), *Food for the Few: Neoliberal Globalism and Biotechnology in Latin America*, Austin, TX: University of Texas Press.
Skogstad, Grace (2008), *Internationalization and Canadian Agriculture: Policy and Governing Paradigms*, Toronto: University of Toronto Press.
Skogstad, Grace and Elizabeth Moore (2004), 'Regulating genetic engineering in the United States and the European Union: policy development and policy resilience', *Policy and Society*, **23**(4), 32–56.
United States (1985), 'Biotechnology: the United States Department of Agriculture's biotechnology research effort', Washington, DC: GAO.
United States. Executive Office of the President, Office of Science and Technology Policy (1986), 'Coordinated framework for regulation of biotechnology', Washington, DC: GAO.
United States Congress, House Committee on Agriculture, Subcommittee on Department Operations, Research, and Foreign Agriculture (1991), 'Review of current and proposed Agricultural Biotechnology Regulatory Authority and the Omnibus Biotechnology Act', Washington, DC: GAO.
United States Department of Agriculture (2008), 'NAFTA's effect on corn trade between the United States and Mexico', USDA Grain Inspection, Packers and Stockyards Administration, Washington, DC, available at: http://archive.gispa.usda.gov/rdd/NAFTA_corn_trade.pdf.
Vogel, David (2003), 'The politics of risk regulation in Europe and the United States', *Yearbook of European Environmental Law*, **3**.

6 South American adopters: Argentina and Brazil

Sybil D. Rhodes

INTRODUCTION

Several South American countries have been early adopters in the cultivation of crops using genetic modification (GM) technology. Among the South American adopters, Argentina and Brazil stand out both in terms of their policies favorable to the use of agbiotech and the extent of production. In 2008, for example, according to data from the International Service for the Acquisition of Agri-biotech Applications (ISAAA), Argentina and Brazil were numbered second and third respectively in the ranking of countries according to land under cultivation with GM crops, while Paraguay ranked seventh and Uruguay ninth. Relevant crops include cotton, corn and some others, but the explanation for the early adoption of agbiotech in this region has much to do with the expansion of world demand for soy in the past several decades. GM technology, in combination with the herbicide glyphosate and no-till seeding, have been key developments that have transformed an important swathe of South America into one of the world's largest production sites for soybeans and soy products traded on the international market. Argentina, Brazil and Paraguay provided 55 per cent of the world's soy exports in 2008 (Turzi, 2010).

Most aspects of the soy market are dominated by multinational companies, from upstream research to downstream marketing and processing. GM technology has favored the relative position of multinational technology and seed companies within the global commodity chain. Distributing, processing and trading companies have also benefited. Large economies of scale have led to concentration and vertical integration within the crop biotech sector, exemplified by companies like Monsanto, Bayer, Pioneer and Syngenta. In the context of South America, the transnationalization of the production process of an area encompassing important parts of Argentina, Brazil and Paraguay has been labeled a 'soybean republic' by business groups, academic analysts and critical activists alike.

The international soy trade is without a doubt the single most important contextual feature of agbiotech governance in the South American adapting markets. Growing conditions in Argentina and Brazil, as well as Paraguay, Uruguay and parts of Bolivia, are conducive to soybeans and some of the crops for which technologies have been readily available. The ideas and interests of actors towards GM are linked to their positioning within the international soy trade. Governments have found themselves in the position of playing regulatory catch-up to decisions and demands of the soy export sector. The reality that these countries are contiguous geographically has facilitated the transfer of the technology across a wide area on the continent, sometimes in defiance of the legal rules.

In spite of these commonalities, the variations in regulatory permissiveness in the South American region are greater than the differences between North America and the European Union. The range of policy and regulatory stances runs from outright prohibitions encoded in national constitutions, as in the cases of Bolivia and Ecuador, to extremely tolerant policies that actively promoted the early adoption of technologies, as in Argentina. The first regulatory approval for GM soy in Argentina was granted in 1996, less than two years after the same variety was approved in the United States. Since that time, there has been continuous growth in the number of approvals granted, mostly for corn and cotton.

The permissiveness of formal policy has also varied quite a bit in individual countries over time. Brazil, for example, moved from mildly averse to a much more liberal level of tolerance during the past decade. Brazil approved GM soy in 1998 but then there followed a period of seven years during which it did not issue any other formal approvals. In recent years there have been many approvals, especially for corn and cotton.

In the context of the region as a whole, and in Argentina and Brazil in particular, the governance of GM technologies has been mostly favorable or at least not completely antithetical to the interests of the multinational companies. Both leading GM-adopting nations have developed regulatory structures that are centered around national biosafety and seed regulations. There are, however, some important differences in how the sector is governed. This chapter discusses some of the differences between Argentina and Brazil, arguing that they are largely attributable to their different roles in the agricultural export sector and to the engagement of social interests, such as small farmers and the landless, environmentalists, and consumer advocates, in the national political systems of the two countries.

MODELS

In the past two decades South American countries have experienced increased economic growth and improvement in various economic development indicators. The process has not been smooth, as several countries have suffered from currency crises, such as the one associated with the collapse of convertibility and ensuing debt default in Argentina in 2001. When compared to the 'lost decade' of the 1980s, however, recent South American economic development is notable. It appears even more impressive since 2008, as the region has remained largely insulated from the ongoing problems in the US and European economies. Modernizing, or liberalizing, economic reforms enacted largely in the 1990s, generally prudent fiscal and monetary policies, and, since the mid-2000s, higher prices for its mineral and agricultural exports, such as oil, copper, corn and soy, are three of the most important factors contributing to South America's recovery from the 1980s and its current resilience. The developments in agbiotech have been influenced by and have contributed to this trajectory.

There are two identifiable analytical and normative models that specialists and interested actors apply to the development of agbiotech in Argentina and Brazil. Neither is especially original or unique to the region. Both are influenced by the soybean-dominated context of the GM revolution in the region and both incorporate a nationalist perspective, which is not necessarily shared by the multinational agrochemical companies who are the top players in the soy commodity chain.

The dominant model, which could be called the 'national modernization' model, is generally pro-technology, pro-business and pro-international trade. Adherents believe it is better to produce technologies locally rather than borrow them from others, but they much prefer to import foreign technology than not to use it at all. They reject the old image of Latin American agricultural elites as oppressive and unproductive oligarchs who live off rents from the land in favor of the idea of a productive and dynamic agribusiness sector. Proponents of the national modernization model define successful development primarily as increased exports of crops and, especially, processed products such as soymeal and soybean oil. They are less directly concerned with distributive issues, although there may be both alliances and conflicts among large and small producers who share the national modernization outlook. Advocates of the national modernization credit the growth of GM soy for leading a modernizing revolution of national agriculture, arguing that it has many positive effects for national development. For example, Brazilian studies have shown that the level of education of all kinds of producers is increasing as a consequence of the changing demands of the sector (Gasques et al., 2004). Higher education courses linked to agricultural business are also increasing in number in response to demands from the sector for skilled labor. The modernization of soy production has spilled over into traditional crops such as coffee and sugar cane. Agribusiness, particularly the soy business, is positive for the country; it contributes to 'The Brazil that Works', to quote the title of a book subtitled 'The Saga of Brazilian Soy' (Zancopé and Nasser, 2005). Proponents of the national modernization model take pride in the expansion of agribusiness to neighboring countries and even to African countries, such as South Africa, Mozambique, Zambia, Swaziland and Zimbabwe (Bertello, 2010).

The second model, referred to here as the 'national protection' model, has several variants, but is united in its criticism of the national modernization approach. The most important two perspectives are those of small, poor and landless farmers and environmental preservationists, who are often aligned. Consumer advocates are also important players in some instances. The national protection model is most concerned with protecting vulnerable groups from abuses associated with agricultural biotechnology. Its proponents may sometimes have very different assumptions about the nature of the threat as well as the groups that need protecting. Small, poor and landless farmers at times may well have the goal of more inclusive access to new technologies, while environmentalists and consumers embrace the precautionary principle out of fear of damage to the environment or human health. The national protection model is associated with more diverse definitions of successful development, which tend to include the concepts of equality and equity, food sovereignty, food security and sustainable ecology.

The two models have an uneasy coexistence in the early-adopting countries. At various points the tension between them has manifested itself as confrontation in the political arena, provoking short-term instability in the governance arena and some costly effects for economic development. However, evaluation of the effects of any technology on development are ultimately subject to judgement through a political process. Even in cases of apparent stability in governance, there is a latent possibility of conflict.

METRICS: APPLICATIONS OF THE MODELS IN ARGENTINA AND BRAZIL

Argentina

The government of President Carlos Menem (1989–99) created the National Advisory Committee on Agricultural Biosafety (CONABIA), an advisory agency within the Secretary of Agriculture, Livestock, Fisheries and Food (SAGPyA) in 1991. Public and private sector entities are represented within CONABIA, which has strong scientific and technical requirements for membership.

The agency granted the first regulatory approval for GM soy in 1996, less than two years after the same approval in the United States. Since that time, there has been continuous growth in the number of approvals granted, mostly for corn and cotton. The permissive policies toward GM technology were consistent with Menem's broadly liberal economic package, which included opening the country to foreign investment.

Unlike Brazil, Argentina has no comprehensive biosafety law. Some members of Congress began to prepare a bill proposing such a law in 2001, but the economic and political crisis of that year subsequently meant it did not advance.

The national modernization model of agbiotech has been dominant in Argentine policy. No specific challenges of GM technology from the national protectionist perspective have succeeded in entering broader national discourse nor have they penetrated the seed governance structure. Virtually no actors have attempted to initiate a discussion of the effects of consuming GM crops on human health. Consequently there has been no real divergence between regulatory approvals for cultivation and consumption, as has occurred in some other countries. Small, poor and landless farmers and environmentalists, such as the rural development NGO *Grupo Reflexión Rural*, have criticized various aspects of the expansion of the soybean cultivation, but rather than framing their concerns as opposition to GM technology they tend to criticize the broader phenomena of globalized agribusiness and soy monoculture.

The most salient issue in Argentine governance of agbiotech has involved intellectual property protection. Because of Argentine law, Monsanto has been unable to patent its Roundup Ready seeds or to prevent farmers from saving the seeds for the next harvest. Successive governments' protection of Argentine farmers from royalty payments amounts to a subsidy that increases their comparative advantage on the world market. Monsanto has tried a variety of strategies to reverse this policy. The company announced it might stop all its activities in Argentina in 2004. In 2005 it convinced several European courts to detain Argentine soy shipments that it alleged contained unlicensed Roundup Ready soy meal in violation of those countries' patent laws. In 2010 the European Court of Justice ruled that the protections applied to whole soybeans but not processed products (Case C-428/08, European Court of Justice, 2010).

Argentine policy has been less favorable than Brazilian policy to multinationals' concerns about intellectual property. At least part of the explanation for this is that the Argentine National Agricultural Technology Institute (INTA) plays only a limited role in research, which means it, and the state, has little direct state interest in intellectual property protection (Turzi, 2010). The Argentine seed industry is concentrated in the private sector. While the association of national private seed companies largely supported

Monsanto's position in the royalty dispute, their support was not enough to change the government's position favoring the growers.

From Argentina, the GM seeds were illegally smuggled into Brazil and Paraguay. While this ostensibly was a violation of Monsanto's intellectual property, it also had the effect of introducing the seeds in Brazil, in keeping with Monsanto's commercial interest.

The dominance of the national modernization model does not mean that there is no politicization of soybean farming in Argentina. There have been fierce political battles, especially following the Menem years and the economic and political crisis of 2001. In 2003, Néstor Kirchner (like Menem, of the Justicialist or Peronist party) became president without an electoral majority. Kirchner increased the level of state intervention in the economy and promoted a left-leaning, populist discourse. His populist politics were continued and deepened by his wife, Cristina Fernández de Kirchner, who was elected president in 2007 and re-elected in 2011.

In spite of the new political rhetoric, which included criticism of Menem's neoliberalism, the national modernization model continued to dominate the governance of agbiotech under both Kirchner presidencies. The tension with the national protection model regarding GM technology has not entered national political discourse in a direct way. In the 2000s, however, the protectionist model began to challenge the dominant model indirectly. The opposition came in two forms: environmental concern about deforestation and damage to small farmers caused by agrochemical spills. Both of these concerns were associated with opposition to the agribusiness or export-oriented agricultural pattern.

Under the Kirchners, the Argentine national government has had sustained conflict with agricultural producers. The conflict stems from the nature of Argentine political cleavages. The ruling Peronist party generally taxes agricultural exporters in order to provide subsidies to urban and industrial sectors, where the party gains support for its populist policies. Unlike Brazil, Argentina charges export taxes for soybeans. Implemented in 2002 via an emergency law, the tax structure is designed to promote the export of processed soy products rather than the bean, reinforcing Argentina's position as the world's leading exporter of processed soy products such as soybean meal.

In 2008, the government of Cristina Kirchner proposed to revise the tax system by introducing a sliding scale of export taxes based on world soy prices. Producer groups organized and protested against this measure as an example of the government's uncontrolled fiscal voracity. Ultimately there was a dramatic showdown vote in the Senate where the Vice President cast a deciding vote against the sliding tax, which meant that the export tax remains fixed at 35 per cent.

The bitter political clash of 2008 revealed that national politicians looked favorably upon some criticisms of producers. News stories began to appear in the pro-government media, criticizing producers for engaging in deforestation and the violent expulsion of poor farmers from their traditional lands in order to grow more soybeans as well as for allowing dangerous agrochemical spills. In 2006 in Córdoba, the provincial court had prohibited spraying near urban areas and a pending Supreme Court case that would ban glyphosate spraying in the country has producers worried. The motion (no. 262/09) was filed by a group of environmental lawyers (the *Asociación Argentina de Abogados Ambientalistas*, or AAdeAA) in April 2009.

The movement against spraying is composed of local groups of affected people and supported by medical and scientific professionals and environmental lawyers. Perhaps the most prominent individual is scientist Andrés Carrasco, who published a study in 2010 claiming to show that glyphosate is harmful to frog embryos. Attempts to publicize his findings have been met with local hostility, even a reported incident of physical violence. Representatives of producer interests criticize the study as being politically motivated.

In spite of the partial advance of concerns about spraying and deforestation, Argentina is still characterized overall by what Peter Newell (2009) calls 'bio-hegemony', whereby export-agriculture's reliance on GM technology is mostly unquestioned. Politicians and state officials often have an economic stake in the sector, which helps perpetuate the national modernization model and keep national protectionist concerns off the agenda. News coverage of motion 262/09 in the most recent few years has been minimal.

Although the national modernization model remains dominant for now, thorny relations between the agribusiness sector and the urban populist political coalition and the lack of an institutionalized space for negotiation between proponents of the two dominant models raises uncertainty about the future of the Argentine governance system.

Brazil

Brazil has a state-led research agency dedicated to agbiotech research. Brazilian scientists working for the Brazilian Agricultural and Livestock Research Company (Embrapa) developed the country's first GM plant in 1986 and have continued to develop GM crop varieties, particularly for tropical agriculture. The existence of Embrapa ensures that the Brazilian government has more of a direct incentive to protect intellectual property than Argentina.

But the different levels of state involvement in GM research do not explain the differences in levels of overall caution reflected in the policies of the two countries in the late 1990s and early to mid-2000s. During that period, while Argentina was a permissive pioneer, Brazil was the most important large agricultural exporting country to resist approving Monsanto's Roundup Ready soy.

President Fernando Henrique Cardoso (1995–2002) signed a relatively strict Biosecurity Law in 1995. This Law specifically required environmental impact studies be carried out before GM crops could be grown at a commercial scale. Cardoso also issued a regulatory decree creating the National Biosafety Technical Commission (CTNBio) as part of the Ministry of Science and Technology, which was given the final regulatory authority over decisions regarding GM crops. Representatives from various ministries as well as civil society were included on the CTNBio.

In 1997, the CTNBio approved Monsanto's request to conduct field tests of its GM soy. That same year, Monsanto signed an agreement with Embrapa regarding research on GM soy. In response, the Consumer Defense Council (Idec), which had a representative in the CTNBio, opposed the decision to approve Monsanto's request to conduct field tests, arguing that environmental impact studies performed abroad were not sufficient, and separate laboratory tests needed to be carried out in Brazil before open-field trials. When the CTNBio decision was sustained, Idec's representative resigned from the CTNBio, arguing that the body had succumbed to pressure from large biotechnology companies to approve GM products before defining the rules for environmental evaluation.

Together with other consumer protection and environmental NGOs (including Greenpeace) and the Landless Rural Workers' Movement (MST), Idec went on over the next few years to lead the Campaign for a Transgenic-Free Brazil, engaging in battles against the cultivation and commercialization of GM soy and other GM crops in Brazilian courts, the national and state legislatures, and public opinion. In response, a federal court granted an injunction, officially prohibiting the commercialization of the GM soy in Brazil. The decision was upheld on appeal in 1999.

In the meantime, agbiotech became a heated point of contention within the Cardoso government. The Ministers of Agriculture and Science and Technology openly supported the rapid acceptance of genetic engineering, but others disagreed. In 1998, Minister of Health José Serra declared that he favored a moratorium on GM technology and, in 1999, environmental minister José Sarney Filho signed a document in support of the MST's campaign against GM crops.

Individual Brazilian states began to issue their own regulations, some specifically prohibiting the commercialization of GM crops, and others, including the important state of São Paulo, requiring mandatory labeling of all foods with GM content. The southernmost state of Rio Grande do Sul, which at the time was governed by the opposition Workers' Party (PT), began a campaign to encourage farmers not to use GM technology. This would turn out to be a futile quest, as farmers began to import contraband GM seeds from Argentina. The prohibition campaign ended when the PT lost control of the governor's office in 2003. The GM debate was not a decisive factor in the election (the public perceived its overall performance as radical and divisive), but the PT governor's decision to incorporate the issue into party politics had a polarizing effect on the GM debate in Rio Grande do Sul (Amaro da Silveira and Almeida, 2005). After 2003, the Rio Grande do Sul state government would become a strong supporter of GM technology. By the time the public returned the PT to the governor's office in 2011, polemics over agbiotech had largely disappeared from political discourse in the state.

At least two other states, Piauí (under PT governor Wellington Dias) and Santa Catarina (under PMDB governor Luiz Henrique) also attempted to develop as GM-free zones during the years of the Cardoso presidency. A few years later, neighboring Paraná also attempted but failed to gain market niches for conventional soy by restricting the use of GM technology. In Paraná the attempt was led by PMDB governor Roberto Requião, who prohibited the transport of GM crops within the state and refused to allow their exportation through the Porto de Paranaguá. The dispute between Requião, the national Ministry of Agriculture, and the pro-GM forces in other states became very heated – Requião at one point asserted that the declarations of one agricultural ministry official 'better equipped to make him a manager for Monsanto than minister of agriculture' (Jornal da Ciência, 2003). Agriculture Minister Roberto Rodrigues threatened to sue Requião for libel. Requião responded with demands that the Ministry of Agriculture reveal the names of 580 producers from Paraná presumed to be growing GM soy in defiance of the state law prohibiting it, but the ministry refused to reveal the information. Exporters and agricultural representatives, including the Federation of Agriculture of Paraná, went so far as to ask the federal government to intervene to overturn the ban and to facilitate the export of GM soy through the port. The conservative PFL party and the governor of Mato Grosso do Sul Blairo Maggi (also the largest soy producer in the country) filed legal actions against the law in Paraná and eventually Requião's ban was overturned by an order from the Supreme Court.

The members of the Campaign for a Transgenic-Free Brazil had reason to believe they had won a key victory with the triumph of PT candidate Luiz Inácio 'Lula' da Silva in the 2002 presidential elections, but they were ultimately disappointed. After Lula took office, the GM soy issue began to split his coalition, with pro-GM proponents gaining ascendancy. Lula's government ended up supporting a legal challenge by Monsanto against Idec and Greenpeace. Lula also pushed through two separate decree laws specifically permitting farmers with contraband seeds to grow and sell GM soy. The most important legislative battle of Lula's administration concerned the bill that would become the new biosecurity law of 2005. The original bill the government submitted to the legislature in December 2003 was relatively strict regarding GM technology, as it would have allowed the health and environmental ministries to require studies before GM crops were grown or commercialized in the country. The version that was ultimately passed maintained the authority of the CTNBio to rely on evidence produced in other jurisdictions, such as the United States.

Thus, eventually Brazil developed permissive regulations similar to those in Argentina, but the process was much slower because of greater political opposition. The governance structure in Brazil is somewhat more open to national protection advocates than in Argentina (although the government agenda is still clearly dominated by the national modernization model). In Brazil the governance structure was created through a comprehensive and generally open political process that led to the biosafety law passed by the Congress, while in Argentina the governance structure for GM crops was created through a relatively closed process within the executive branch.

Explaining the Differences between Argentina and Brazil: Interests and Institutions

The main factors that explain the influence of the national protectionist viewpoint in Brazil compared with Argentina are the relative strength of various social interests in Brazilian politics, including small, poor and landless farmers, producers in the agribusiness sector, environmentalists and consumers, as well as the nature of the countries' political institutions and overall political contexts.

Small, poor and landless farmers

In order to understand the role of small, poor and landless farmers it is necessary to consider the historical context of debates about agbiotech in Brazil. Inside and outside the country, the development of GM technologies is often likened to, and even considered an extension of, the green revolution, but in the Brazilian context there are equally compelling parallels with the struggle for land property rights. Specifically, all three of these debates have pitted forces claiming to represent modern, scientific, large-scale, export-oriented agriculture against other actors claiming to represent justice for small-scale farmers with stronger ties to the land and national markets.

In Brazil, the national protection perspective is more important in rural and agricultural debates generally. Scholars of Brazilian history and politics argue that the deepest roots of rural conflict in the country are to be found in some combination of the unequal distribution of land and the lack of clear property rights in rural areas. The fear of the demand of land reform has been a major cause of rural oppression in Brazil. Land reform still is the central *raison d'être* of the MST, perhaps the best known and largest social movement in Latin America, and MST is one of the national coordinators of the 'Brazil Free of Transgenes' campaign. Argentina, in contrast, historically has had less land concentration, and demands for land reform have not held the same political salience.

The Brazilian rural union confederation (Contag) views land invasions with sympathy, is concerned about the power of the agribusiness sector, and was not sympathetic to introduction of GM crops. Contag was a party in lawsuits against the decree laws granting amnesties for farmers growing crops with contraband GM soy seeds.

In Brazil and in Argentina, small farmers and even MST members appear to have been divided by the debates over GM technology, however. Certainly they were not part of the initial market for GM grains, but some have hopes that they eventually would benefit from genetic improvement of smaller-scale crops. Some former landless workers who had received land even demonstrated in favor of the right to grow GM crops.

Agribusiness: multinationals, the state sector and national producers

The biotech revolution in recent decades has largely been driven by the private sector, specifically the multinational companies that dominate the world agribusiness sector, including Monsanto, Aventis, Dow AgroSciences, DuPont and Syngenta. The multinationals' interest in national regulatory decisions has not always been united, however. As one example, some Brazilian newspapers reported that some of Monsanto's competitors were accused of financing the NGO campaigns against GM technology (Escobar, 2004).

Even so, as mentioned earlier, the state sector is more important for agbiotech development in Brazil than in Argentina. The single most important entity in Brazilian agricultural science is the state-owned agricultural technology company Embrapa. By the mid to late 1990s the lack of clear regulation had become an obstacle to further development and Embrapa scientists had grown very frustrated over the bureaucratic delays of their research projects. The scientists argued that most GM research in Brazil is focused on small producers who cultivate crops such as bean, mamão, sweet potato, tomato, maracuja and banana. Embrapa scientists' frustration led them to argue in favor of increasing the autonomy of the CTNBio and reducing the power of the environmental ministry.

There has been confusion over whether the broader scientific community is united in a clear position on the GM debate in Brazil. As the national biosafety law was being debated, the National Biosecurity Association (ANBio), a pro-biotechnology organization created in 1999, stated that it had the support of twelve other entities for an open letter it sent to the Senate suggesting changes in the version of the Bill voted on by the chamber of deputies. The *Folha de São Paulo* revealed that at least two of the named scientific societies (the Sociedade Brasileira de Genética and Sociedade Brasileira de Microbiologia) had not discussed the document internally, although the ANBio said delegations including the respective presidents of these organizations had expressed their support for the text of the letter.

Most crucially, in Brazil large national grain producers and other agribusiness interests are better organized for collective action within the broader political system than their Argentine counterparts, largely because of the ongoing struggles over land reform. The leading organization exerted significant influence as it could mobilize congressional power through the rural caucus (*bancada ruralista*) of the legislature as well as through the agricultural ministry on the executive branch – this was especially helpful to the pro-GM movement.

Consumer protection advocates

A consumer reaction against the introduction of GM crops occurred in Brazil, but scarcely at all in Argentina. Brazil's most prominent consumer protection NGO, the Idec, was key in the legal battle against GM soy. This is unsurprising if we take into account the history of Idec and the consumer protection movement in the country. While consumer activism dates from the 1970s, the transition to democracy in the mid-1980s emboldened activists' demands for greater consumer protection at the national level. Since then, Brazil has had perhaps the strongest consumer protection movement in Latin America. Because of the efforts of this movement, the 1988 constitution contained a clause calling upon the national congress to pass a consumer protection code. The congress passed the law in August of 1990, and the strongest consumer protection legislation in Latin America went into effect in early 1991. This legislation provides the legal justification for the decrees requiring the labeling of foods with GM content. In Argentina, consumer advocacy is relatively weaker, less interested in the issue of GM labeling, and lacks the legislative access available to their Brazilian counterparts.

As a science advisor for Greenpeace's international GM campaign put it in an interview with the author, labeling can be 'a tool' for anti-GM campaigns in specific countries but is not an integral part of the international strategy (Stabinsky, 2006). After 2005, the Brazilian anti-GM campaign shifted its overall strategy to put more emphasis on preventing deforestation caused by the expansion of agribusiness. This change in strategy was partly in response to emerging evidence that consumption of GM crops does not pose any measurable dangers to human health.

Environmental advocates

Just as it has an important consumer movement, Brazil also has a relatively strong domestic environmental movement and a constitutional clause and other legislation that provide the movement with legislative leverage. The majority of the social, political and legal resources mobilized in the country are leveraged by concerns about the Amazon region in international environmental debates. Environmental advocates in Argentina, in contrast, have fewer institutional resources and the environmental movement is not buttressed by internationally prominent environmental symbols.

Along with Idec and the MST, the Brazilian branch of the international environmental NGO Greenpeace was the other leader of the Campaign for a Transgenic-Free Brazil. Some environmental activists actually entered and gained power in the environmental ministry of President Lula. Most notably, before she became Lula's environmental minister, Senator Marina Silva had openly favored a five-year moratorium on GM technology. Another high-ranking environmental official, Marijane Lisboa, had coordinated the anti-GMO campaign of Greenpeace for several years, and American David Hathaway, the official translator for Lula's entourage in the United States and India, had been a leader in the anti-GM campaign.

Political institutions: federalism

Brazilian federalism allows much more autonomy than the Argentinian system to subnational units. The constitutional division of powers is an important reason Brazil was relatively slow to develop a national system for the regulation of GM technology. However, in the absence of clear national rules, attempts to establish GM-free zones at the state level were undermined by the use of contraband seeds as much as by the laws governing interstate commerce. Grain purchasers were aware that if farmers in Rio Grande do Sul could smuggle contraband GM seeds from Argentina then it surely was possible for Paranense or Catarenense farmers to acquire and plant seeds brought across state lines. The 'stealth' nature of the technology (Herring, 2007) combined with the ultimate national authority at the federal level, as enforced by the courts, to prevent the institutional 'ratcheting up' of regulations governing GM technology that Bernauer (2003) describes in Europe.

CRITICAL ASSESSMENT

The debate about GM technology in Argentina and Brazil is embedded in the debates about soybeans. Whether GM technology benefits development depends in part on which analytical and normative model is applied. In recent years, governance in both countries has converged around the national modernization model. Both have succeeded at attracting foreign investment and have become known worldwide as agricultural powerhouses, especially because of their soybean exports. The Argentine case in particular demonstrates the economic benefits of early adoption: one study estimated the overall cumulative value to the country of GM techniques in soy, cotton and corn during the period 1996–2010 at US$72 645.52 million (Trigo, 2011). A Brazilian study, in contrast, found that the delays in Brazil implied an unrealized potential benefit of nearly US$28.4 billion from 1996/97 to 2008/09 (Celerehs, 2010). Many analysts in both countries point to continued economic and environmental benefits if they continue to take rapid advantage of developments in agbiotech.

The Argentine and Brazilian experiences have not been without costs, which include, in particular, deforestation, agrochemical spills, and the dislocation of small farmers. It is likely that the non-adoption of GM techniques may have had even worse effects (requiring, for example, the cultivation of additional lands and the use of more harmful chemicals); however, this counterfactual does not prevent these costs from translating into ongoing political tensions. In Argentina this is evident in the recent political attention to the effects of agrochemical spills on human health. The issue is mostly used instrumentally in the conflict over export taxes that continues between the populist political coalition and the agribusiness sector.

Although Brazilian policy appeared more erratic in the 1990s and early 2000s, in more recent years Brazil has come to be regarded as more predictable for GM soy producer interests than Argentina. In spite of the historical salience of rural and agricultural issues in Brazilian politics, relations between governments and producers are characterized by less confrontation than in Argentina. Nonetheless, concern about the relationship between GM crops and deforestation continues to have the potential to trigger important political mobilization in Brazil.

Other South American countries that are adopting or considering the adoption of agbiotech face different challenges because of different combinations of possible benefits and costs. These challenges are filtered through existing political contexts. Where those contexts are more polarized domestically, the debates about agbiotech tend to be confrontational as well. The tensest situations are likely to occur where both national-protection and national-modernization coalitions are at least somewhat evenly matched politically, in which there is a history of confrontation over other issues dividing these two groups, and in which political institutions allow for more decentralized involvement of diverse social and economic interests. Participation by international actors may further aggravate conflictive situations.

Countries that already adopted or may soon adopt agbiotech also face the challenge of convincing customers that their crops are not of lesser quality than non-GM crops. Regional cooperation is one possible response to this difficulty. Some growers' organizations promote the idea of cooperation within the hemisphere in order to prevent customers from paying lower prices for GM crops.

Regional cooperation could allow other countries to benefit from the experience of the early adopters regarding an array of biosafety governance issues. Indeed, many analysts representing both of the main approaches discussed here at least partly agree that the importance of agbiotech to development (and the difficulties in controlling the spread of the technology across national borders) in South America merits more work to facilitate regional cooperation. However, thus far there has been minimal to no coordination of decisions regarding agbiotech in the Common Market of the South (Mercosur) or other South American regional arrangements, at least in part because Argentina, conscious of the benefits it has achieved from fast adoption and fearing any possible regulatory obstacles, has resisted attempts to include biosafety in environmental agreements, while other countries, subject to the political dynamics discussed in this chapter, are unwilling to accept what they view as overly permissive standards. In the meantime, national policy differences continue and conflicts will emerge from time to time.

REFERENCES

Amaro da Silveira, Cristiane and Jalcione Almeida (2005), 'Biossegurança e democracia: entre um espaço dialógico e novos fundamentalismos', *Sociedade e Estado*, **20**(1)(April).

Bernauer, Thomas (2003), *Genes, Trade, and Regulation: The Seeds of Conflict in Food Biotechnology*, Princeton, NJ: Princeton University Press.

Bertello, Fernando (2010), 'Sudáfrica, los primeros campeones. Francisco Lafuente, Pío Silveyra y Juan Carlos Soldano Deheza integran una legión argentina que puso un pie allí', *La Nación*, (20 February). Buenos Aires.

Celeres (2010), 'The economic benefits from crop biotechnology in Brazil: 1996–2009', study prepared for Associação Brasileira de Sementes e Mudas (ABRASEM), available at: http://www.cib.org.br/estudos/Rel_BiotechBenefits_2009_Economico_Eng.pdf, accessed 27 August, 2012.

Escobar, Herton (2004), 'Ideologias e interesses alimentam a polêmica dos transgénicos', *Estado de São Paulo*, (7 March).

European Court of Justice (2010), Case C-428/08 *Monsanto Technology LLC v Cefetra BV and Others* [2010] I-06765.

Gasques, José García, Gervásio Castro de Rezende, Carlos Monteiro Villa Verde, Mario Sergio Salerno, Júnia Cristina P.R. da Conceição and João Carlos de Souza Carvalho (2004), 'Desempenho e crescimento do agronegócio no Brasil', Discussion paper no. 1009, Brasília: IPEA. available at: http://www.ipea.gov.br/, accessed 15 November 2008.

Herring, Ronald J. (2007), 'Stealth seeds: bioproperty, biosafety, biopolitics', *Journal of Development Studies*, **43**(1), 130–57.
Jornal da Ciência (2003), 'Transgênicos: requião faz crítica em carta a Lula', *Jornal da Ciência*, 13 November.
Newell, Peter (2009), 'Bio-hegemony: the political economy of agricultural biotechnology in Argentina', *Journal of Latin American Studies*, **41**(February), 27–57.
Stabinsky, Doreen (2006), Science advisor for Greenpeace GM Campaign, telephone Interview, 15 November.
Trigo, E.J. (2011), '15 years of genetically modified crops in Argentine agriculture', ArgenBio report, available at: http://gmopundit.blogspot.com/2011/12/15-years-of-genetically-modified-crops.html#!/2011/12/15-years-of-genetically-modified-crops.html, accessed 26 August 2012.
Turzi, Mariano, (2010), 'The political economy of soybean production in Brazil, Argentina, and Paraguay', PhD dissertation, Johns Hopkins University.
Zancopé, Gilberto J. and José Monir Nasser (2005), *O Brasil que Deu Certo*, Curitiba: Triade.

7 Africa

Jennifer Thomson

1 INTRODUCTION

Most African countries have been slow to adopt genetically modified (GM) crops, largely for the reasons outlined in Chapter 12 by Robert Paarlberg. In this chapter the main crops (cotton and maize) and the main traits (insect and herbicide resistance) will be discussed. So far South Africa, the first adopting country in Africa, has commercialized cotton (insect resistance), maize (insect resistance and herbicide tolerance), and soybeans (herbicide resistance). Egypt, the second, grows insect-resistant maize, and Burkina Faso, the most recent, has introduced insect-resistant cotton (James, 2011). Upcoming traits (virus resistance, drought tolerance and nutritional enhancement) and new crops (cowpeas, bananas and cassava) will also be addressed. Finally a few public perceptions which are also hindering the acceptance of these crops will be mentioned.

When considering GM crops it is worth noting what Bill Gates wrote in the 2012 Annual Letter from the Bill and Melinda Gates Foundation: 'We can help poor farmers sustainably increase their productivity so they can feed themselves and their families. But that will only happen if we prioritize agricultural innovation.'

2 COTTON: INSECT RESISTANCE

The single largest GM crop by area in Africa is insect-resistant cotton. Cotton is a shrub native to tropical and subtropical regions around the world, including Africa, India, Australia and the Americas. It was independently domesticated in the Old and New Worlds and is an important traded commodity in many African countries. Unfortunately the larvae of certain lepidopterans, such as the cotton bollworm, *Helicoverpa armigera*, bore into cotton bolls causing extensive damage which can often reach staggering proportions, with farmers losing most of their crops. Most insecticides are rather ineffective against these pests which burrow into plant tissues where they are protected against the spray. The biotechnological approach, which has been extremely effective in many cotton-growing countries, is now being used in major cotton-growing regions, including in parts of Africa. *Bacillus thuringiensis*, a naturally occurring soil bacterium that produces proteins called Bt proteins, are toxic to certain insects and have been expressed into a variety of commercial crops (Schnepf et al., 1998). They cause little or no harm to most non-target organisms, including humans and wildlife, due to the specificity of their mode of action. The toxin binds to specific cells in the lining of the insect's gut where it causes lysis and rapid death. It has been used in sprays in conventional and organic agriculture for decades with little or no ill effects on human health or the environment. It is not very effective in spray form as the toxin protein is sensitive to UV light and hence decays on exposure to sunlight. In organic farming, where Bt toxin is relied upon, it must therefore be used in large quantities. Scientists have therefore taken the gene(s) encoding one or more of the more than 30 Bt toxins and introduced them into crops such as cotton to produce plants that can protect themselves from larval attack.

2.1 Bt Cotton in South Africa

Bt cotton, first introduced into the US in 1996, has been grown commercially in South Africa since 1999. An analysis of the benefits of adoption by both small- and large-scale farmers was published by Gouse et al. in 2004. They found the yield increase of large-scale farmers who irrigated was 18.5 per cent, 13.8 per cent for non-irrigated farms and 45.8 per cent for small-scale farmers. Besides the yield benefits, the adoption of Bt cotton also caused a decrease in the volume of insecticides sprayed, with associated cost benefits. Because small-scale farmers do most of their spraying by hand, this reduction usually meant more time for weeding and other farm management activities.

According to Gouse et al. (2004, p. 192) 'a high percentage of large-scale farmers have indicated that peace of mind about bollworms is a very important benefit of Bt cotton'. This peace of mind gave farmers managerial freedom to devote time to other crops or general farming activities. These farmers also noticed increased populations of beneficial insects, such as ladybirds and lacewings, in Bt cotton fields, indicating a possible environmental benefit due to reduced insecticide applications.

The area in South Africa where most of the Bt cotton is grown by small-scale farmers is the Makhathini Flats of KwaZulu-Natal. This region is rich in indigenous plants and weeds that act as natural host plants for all the bollworm species. They therefore act as alternative refuges for the moths and have helped to prevent the build-up of insects resistant to Bt (Green et al., 2003). This is in contrast to the case with Bt maize (see below).

In recent years plantings of cotton, whether Bt or not, have decreased partly due to the global drop in cotton prices. Dry-land cotton, produced predominantly by small-scale farmers with yields of 12 to 15 tons per hectare (ha), is not competitive. The cotton that was planted in 2010 was expected to be almost entirely Bt (James, 2011).

2.2 Bt Cotton in Burkina Faso

In 2009 Burkina Faso became the third country in Africa, after South Africa and Egypt, to commercialize a genetically modified (GM) crop, Bt cotton, but they did it in grand style. By planting just over 125 000 ha of insect-resistant cotton they introduced the largest biotechnology crop onto the continent (Vitale et al., 2010 and Chapter 38 in this book). But this did not happen overnight; the commercial release was made possible through a joint collaboration between their national cotton companies and Monsanto that began in May 2000. The genetic construct introduced is Bollgard II® (BGII), which encodes Cry1Ac and Cry2Ab from *B. thuringiensis*, and Monsanto helped to transfer it to the two regional cotton varieties, STAM 59 and STAM 103. But even before then the various cotton stakeholders in the country had been coordinating their efforts to satisfy the technical, legal and business requirements before any commercial release could be effected.

Burkina Faso's national agricultural research centre, Institut de l' Environnement et de Recherches Agricoles (INERA), played an important role as they have been testing the efficacy of Bt cotton since 2003 by conducting environmental assessments as part of the inputs required to satisfy biosafey protocols and to monitor the socio-economic impacts of BGII (Vitale et al., 2010). They carried out three years of confined field trials and then in 2006 the National Biosafety Committee approved an additional trial outside the INERA research farm environment. This was also the first trial using STAM 59 and STAM 103 varieties. The following year INERA continued these trials on 20 testing sites under the control of three major cotton companies. These 2007 trials provided encouraging results, with average yield increases of 20 per cent. On the basis of these extensive studies the National Biosafety Committee authorized the commercial planting of BGII in June 2008. Production during the first year was limited to 15000 ha as the supply of BGII seed was limited and a seed multiplication year was needed to provide the anticipated demand for the first broad deployment in 2009.

To determine the impact of this first year, INERA conducted a survey of 160 rural households in 10 villages during the summer and fall (Vitale et al., 2010). There are three main cotton-growing zones in the country and each is controlled or administered by a separate company: SOFITEX in the west, the traditional producing zone, SOCOMA in the centre and Faso Cotton in the east, the latter two representing more recently introduced cotton production. Villages were selected randomly to represent typical conditions in each of the zones. The area planted averaged 3.2 ha and the average household contained 14.1 persons, with 8.6 actively engaged in the family's farming operations. It was found that the Faso Cotton zone had the highest yield advantage (36.6 per cent), followed by SOFITEX at 16.5 per cent and SOCOMA with 14.3 per cent. The yield differences could have been due to a combination of factors including environmental characteristics, pest pressure, and secondary pest spray differences (Vitale et al., 2010).

Conventionally treated cotton requires a regimen of six sprays: four targeting Lepidoptera (the primary pest), and the last two against secondary pests such as aphids and jassids. Bollgard II can only protect against the primary pest. However, a majority of producers surveyed (78 per cent) did not carry out the recommended late season sprays, leaving fields unprotected from secondary pests (Vitale et al., 2010). The effect on the yields of those farmers who did undertake late season sprays was not uniform across the zones and further studies will be needed to determine the role of these interventions.

However, yield increase is not the only factor that could improve cotton farmers' lives in Burkina Faso. The overall economic benefits are what really count and in the period under review BGII increased incomes by an average of US$61.88 per ha (Vitale et al., 2010). More significantly, Bt cotton enabled producers surveyed to move from a negative return of −$22.89 per ha generated by conventional cotton to a positive return of US$39.00 per ha. Vitale et al. (2010, p. 329) concluded that 'in relative terms, BGII's economic impact in 2009 corresponded to more than a doubling of the income that would have been earned by conventional cotton, a 270% increase in cotton income'. This was partly due to the fact that production costs had no significant effect on cotton income as costs were nearly the same between BGII and non-Bt cotton.

It is of interest to note that in 2010, out of a total of 400000 ha planted to cotton, 65 per cent were planted to Bt by 80000 farmers. Thus the increase from 2009 (115000 ha) to 2010 was a dramatic jump of 126 per cent, making it the highest proportional increase in GM hectares of any country in the world (James, 2011).

As pointed out by Vitale et al. (2010), continued monitoring will be required to determine whether Bt cotton will be technically and economically viable in the long term. They concluded that 'experience from other parts of the world suggest that benefits can change significantly from one year to another due to differences in weather, pest density, and economic conditions' (ibid., p. 334). An extremely important aspect of such monitoring will be the assessment of farmer compliance with the use of refugia to prevent the build-up of pest resistance. If the BGII experience in Burkina Faso is positive, it will be interesting to see whether neighbouring cotton-producing countries such as Mali and Benin will follow suit.

3 GM MAIZE

Maize, although technically a grain, is used in cooking as a vegetable or starch. In many African countries it is the staple food, where people eat it up to three times a day. White maize is consumed by humans while yellow maize is fed to livestock and poultry.

Most of the maize produced in South Africa is consumed locally, with commercial farmers producing about 96 per cent of the crop and small-scale farmers 4 per cent. Yields over the past 20 years have steadily increased. The average for the five years from 1990/91 to 1994/95 was 1.9 tons/ha while for the most recent five years (2005/6 to 2009/10) it rose to 3.8 tons/ha. Both these periods included a season of drought. GM maize was introduced in 1997 but only became commercially adopted on a major scale in 2000. Since then GM plantings have increased dramatically. In the 2009/10 production season, GM maize contributed 78 per cent of the total commercial area planted to maize (79 per cent of white corn and 77 per cent of yellow corn) (South African Agricultural Business Chamber: www.agbiz.co.za). Although the increase in overall yields is due to a number of factors, one of these factors is likely to be the increase in GM maize.

3.1 Bt Maize

Maize can be severely damaged by the larvae of the maize stalk borer, *Busseola fusca*, and, as with cotton, there are genes coding for varieties of the Bt toxin that can be introduced to this crop to protect it.

In South Africa in 2010, GM maize comprised about 75 per cent of the total maize area and of this about half was planted to Bt maize. It has been estimated that between 2001 and 2010 the yield benefit of this crop to farmers has been in the order of US$376 million (James, 2011). A 2009 study of 80 farmers planting this crop reported significant advantages associated with these plantings, including convenient management (88 per cent) and increased productivity (42.5 per cent); 42.5 per cent also indicated that they perceived Bt technology to be environmentally friendly (Kruger et al., 2009).

The first report of resistance of the maize stalk borer to Bt maize came in 2007. In order to limit such resistance farmers are required to plant refugia (Monsanto, 2007). Refuges are defined as habitats in which the target pest is not under selection pressure because of the toxin and it therefore provides a sustainable habitat for pest development. The principle underlying the high dose and refuge strategy is that any resistant insects emerging from the Bt crop are more likely to mate with one of the much larger number of susceptible pest insects emerging from refugia than with each other, thereby decreasing the selection of Bt resistance alleles (Bourguet, 2004).

Kruger et al. (2009) found that initial levels of refuge compliance were low – even though farmers were obliged to plant a refuge area for each Bt maize field, only 77.7 per cent did so during 1998 – but increased to 100 per cent during 2008. Although the evolution of resistance of *B. fusca* can probably be ascribed to several factors, including rainfall and humidity, the low initial levels of compliance to refuge requirements were most likely to have played an important role (Kruger et al., 2011a). Interestingly farmers remain positive about the technology in spite of resistance development (Kruger et al., 2011b).

A different story emerged when small-scale farmers were surveyed. Of the 78 farmers interviewed, just over half (59 per cent) had more than 10 years' experience in cultivating maize and were well aware of the key production constraints. Their knowledge of GM maize production practices were, however, very poor and knowledge of the risks associated with this technology was completely lacking. None of the farmers interviewed understood the refuge strategy properly. In addition most were illiterate and were therefore unable to read and understand the information in the user guides (Assefa and van den Berg, 2010). This is clearly an issue that needs to be addressed if small-scale farmers are to cultivate Bt maize.

South Africa is the only country in the world where small-scale subsistence farmers have been producing a genetically modified crop for a relatively long period of time. In a study of adoption among farmers in the KwaZulu-Natal province from 2003 to 2010 it was found that although the initial uptake had been for limited amounts of Bt maize, by the end of the period few still planted conventional maize and the rest planted herbicide-resistant varieties or those containing both the stacked Bt and herbicide-resistant genes (M. Gouse, personal communication).

The only other African country to plant Bt maize is Egypt, which was also the first of the Arab countries to adopt GM crops when it planted this crop in 2008. In 2010 farmers planted 2000 ha of Bt yellow maize which represented a 100 per cent increase over 2009. Increases in yield plus insecticide savings net of the additional cost of seed have led to an estimated benefit of US$281 per ha (James, 2011).

3.2 Herbicide-tolerant Maize

Weeds compete with crops for moisture, nutrients and light. Uncontrolled weed growth can thus result in significant losses in yield. Farmers have therefore been spraying herbicides on their crops for decades. As with insecticidal sprays, this is often done using aeroplanes, with the result that a great deal of the spray drifts away from the target sites.

The best known example of transgenic herbicide-tolerance (HT) is Monsanto's RoundupReady®. The active ingredient in the herbicide Roundup is glyphosate, which acts on an enzyme found in many plants, including maize and its weeds. Using Roundup on conventional maize fields is a tricky operation as the herbicide must not make contact with the crop. RoundupReady® maize produces a naturally occurring form of the target enzyme, 5-*enol*pyruvylshikimate-3-phosphate synthase (EPSPS) that is resistant to glyphosate and hence to the herbicide. The gene encoding the glyphosate-resistant form of EPSPS was derived from *Agrobacterium tumefaciens*, coincidentally the bacterium that is used to genetically manipulate plants.

In the Republic of South Africa, the only country in Africa where HT maize is grown, the crop accounted for 13.4 per cent of the GM maize planted in 2010. An additional 41 per cent of the maize acreage used stacked Bt and herbicide-tolerant genes (James, 2011).

One of the positive environmental impacts of HT maize is the use of no-till cultivation or conservation tillage. With conventional maize, farmers would till the soil to allow weeds to grow, spray with herbicide and then wait a sufficient time for it to degrade before planting. Now they can allow weeds and maize to grow together before spraying. This means less soil erosion and better retention of moisture in the soil. In addition, Roundup is more readily degradable by bacteria than many other herbicides (Balthazor and Hallas, 1986).

A somewhat different HT maize is the case of *Striga* resistance. *Striga*, or witchweed, is a major parasitic weed that infests about 20 million hectares of arable land in sub-Saharan Africa. The seeds lie dormant in the soil until a maize plant germinates and chemical compounds then trigger the germination of the weeds. As the *Striga* plants grow, their roots become intertwined with the roots of the maize plants, making hand weeding extremely difficult. BASF has developed, not by genetic engineering but by a variety of breeding processes, maize that is resistant to the herbicide Imazapyr. If seeds are planted, coated with the herbicide, the developing *Striga* plants are killed. The African Agricultural Technology Foundation (AATF), based in Nairobi, has facilitated the delivery of StrigAway® seeds to farmers in Kenya and Tanzania, where commercial seed production began in 2006 and 2010 respectively. Variety testing is ongoing in Uganda and Nigeria. One of the reasons for the early success of this product is that it is not a GM variety and so none of the attendant regulatory hurdles had to be met. However, a lesson was learnt from the early deployment of StrigAway® seeds that could be useful in the future use of GM seeds. It was found that if farmers who intercrop their maize with crops such as legumes do not wash their hands in between planting the two, the legumes could be affected. The AATF therefore distributed information leaflets to farmers to explain this requirement.

3.3 Virus-resistant maize

Despite maize being a crucial staple food crop in Africa, the average maize yield per hectare on the continent is the lowest in the world, leading to food shortages and famine. A major contributing factor to these low yields is maize streak virus (MSV). MSV is the most significant pathogen of maize in Africa, resulting in crop yield losses of up to 100 per cent. Transmitted by tiny leafhopper insects, *Cicadulina mbila*, it is indigenous to Africa and neighbouring Indian Ocean islands. A group at the University of Cape Town has partnered with PANNAR Seed Ltd to develop maize resistant to MSV. The group is using two strategies for introducing resistance. First, the gene coding for the replication associated protein (Rep) is one source for resistance. MSV has a very simple genome, with only three genes. One of them is *rep*, which is the first gene to be expressed when the virus infects a plant. The Rep proteins form a multimer and bind to the virus DNA origin of replication. This is a very specific sequence that, as the DNA is single stranded, forms a hairpin structure. The Rep complex initiates DNA replication at this point, using host proteins to complete the process. Thus, if the host plant produces many copies of mutated Rep proteins these would compete with the virus Reps and prevent them from initiating replication. The second strategy is to use siRNA (small interfering RNA) molecules. These small pieces of RNA can cause genes to be silenced, or inactivated, hence interfering with the expression of specific genes. If the siRNA targets essential parts of the MSV transcriptome they can act as antiviral agents. Both these strategies are undergoing trials and time will tell which, if either, is successful.

3.4 Drought-tolerant Maize

In recent years severe droughts have been affecting many parts of Africa, particularly countries such as Somalia, Ethiopia and Kenya in East Africa. This is threatening the livelihood of millions of people, resulting in a huge refugee problem. With the onset of climate change and more climate variability, this situation can only get worse. Many groups worldwide are working to develop drought-tolerant crops. One of these is a partnership called Water Efficient Maize for Africa (WEMA), funded by the Bill and Melinda Gates and Howard G. Buffet Foundations.

The gene being used in this research is *csp*B, donated royalty-free by Monsanto, which encodes a cold shock protein from the bacterium *Bacillus subtilis* (Castiglioni et al., 2008). This has been shown by Monsanto research to confer tolerance to dehydration in transgenic maize in the USA. The gene has been introduced into African maize varieties by The International Maize and Wheat Improvement Center (CIMMYT) and these are being subjected to confined field trials by the national agricultural research services in Kenya, Mozambique, South Africa, Tanzania and Uganda. The project is being managed by the AATF.

The first South African WEMA trials took place in 2010 in Lutzville, a tiny village up the west coast about four hours drive from Cape Town. It is predominantly a wine-growing area where little, if any, maize is grown. This, together with the fact that it is usually reliably dry in the summer months, was why it was chosen as the location for field trials. At a community meeting in the town in December 2010 local farmers reported that the technology was not relevant to their needs – they largely grew vegetables and cared nothing about maize and even less about GM maize. As the meeting progressed it became increasingly clear that the farmers were extremely angry with the Agricultural Research Council (ARC) for not caring about their problems, and for foisting on them something they neither needed nor wanted. Quite naturally, this translated into anti-GM sentiment. The overriding message from this event was that researchers and commercializers, particularly in developing countries and subregions thereof, should 'do their homework on their target audience'. Otherwise the whole exercise can be derailed by factors having little or nothing to do with the technology itself. At a more recent community meeting (November 2011) the ARC included an expert on vegetable growing and the atmosphere was far more positive (personal communication with BioSafety SA).

In addition to the WEMA project a research group based both at the University of Cape Town and Kenyatta University in Nairobi are using genes isolated from a 'resurrection plant', *Xerophyta viscosa*. These plants are unique in that they are able to tolerate almost complete desiccation. They can lose 95 per cent of their water content and remain in a dormant stage, looking completely dead, for months on end. Upon addition of water, the plants can literally 'resurrect' in a matter of days. *X. viscosa* are often found growing in cracks in rocks in mountains where they grow in very little soil and can dry out rapidly. Day temperatures there are often as high as 40°C while temperatures at night can drop to below freezing. These plants turn out to have genes coding for some very interesting proteins which enable them to survive in these environments. Some of these genes are being introduced into plants to code for the synthesis of osmoprotectants, cell membrane signalling proteins and antioxidants, driven by a stress inducible promoter. This work is still experimental but offers significant potential to address specific agronomic challenges in Africa.

4 COWPEA: INSECT RESISTANCE

Cowpeas are one of the most important food legume crops in the semi-arid tropics. They are rich in protein and are consumed by approximately 200 million people in Africa. In addition, they are drought tolerant and hence well adapted to the drier regions of the tropics. As a legume, the plant fixes atmospheric nitrogen through its root nodules, but unlike many other legumes its green leaves and pods can also be eaten before crop maturity, which helps to bridge the hunger gap between harvests. Cowpea grows well in poor soils with more than 85 per cent sand and with less than 0.2–2 per cent organic matter and low levels of phosphorus (Singh, 2003). In addition, the plant is shade tolerant and can therefore be used in intercropping, a farming method popular in many parts of Africa. Cowpeas are intercropped with maize, millet, sorghum, sugarcane and cotton.

Unfortunately, however, cowpeas are often infested by the borer, *Maruca vitrata*, which decreases the yield from a potential 2–2.5 tons/ha to 0.05–0.5 tons/ha. Fortunately, *Maruca*, being an insect, is sensitive to the Bt toxin. The Commonwealth Scientific and Industrial Research Organization (CSIRO) in Australia has transformed cowpea with the Bt *cry1Ab* gene for *Maruca* resistance and has formed a partnership with the AATF to conduct field trials in West Africa, where cowpeas are a major crop.

The first Bt cowpea field trials were held in Puerto Rico in 2009, mainly because the regulatory regime in that country is such that permission was given in a relatively short time. The problem, however, came with the insects, or the lack thereof. That year there just wasn't the normal infestation of *M. vitrata*. As a result the team decided to use artificially reared larvae for the trials in Nigeria and Burkina Faso. The Puerto Rico trials did, however, enable the researchers to determine which of the lines performed the best agronomically. The African trials have had excellent cooperation from the Nigerian and Burkina Faso authorities and, as a result, the AATF has established an office in Nigeria.

The high level of insect pressure introduced into the field trials has so far shown very encouraging results. There is a clear and striking difference between the non-transgenic and transgenic plants. The level of floral, pod and leaf damage is pronounced in the former and no damage has been observed on the latter. Only dead, first instar larvae (those that were between their first and second moult) were observed inside the flowers of transgenic events, showing that these plants are resistant to *Maruca* infestation (personal communication with H. Mignouna, AATF).

5 BANANAS: DISEASE RESISTANCE

To most Westerners bananas are soft and sweet, the so-called 'dessert banana'. In Africa most banana cultivars grown are of a firmer, starchier consistency and are called plantains or 'cooking bananas'. They are a major food and income source for small-holder farmers in East Africa. In countries such as Uganda, Burundi and Rwanda per capita consumption has been estimated at 45 kg per year, the highest in the world.

Cultivated bananas are parthenocarpic, which literally means 'virgin fruit'; the fruit are produced without fertilization and are thus seedless and sterile. Propagation involves farmers removing and planting a sucker, a vertical shoot that develops from the base of the plant. If the sucker is removed before it has elongated, the process is even easier as these suckers can be left out of the ground for up to two weeks.

The main risk is banana bacterial wilt, which is caused by the bacterium *Xanthomonas campestris*. It threatens the livelihood of millions of farmers in the Great Lakes region who rely on this crop as a staple food and for income generation. The disease was first identified in Uganda in 2001 and subsequently reported in the Democratic Republic of Congo, Rwanda, Tanzania and Kenya. It is very destructive, infecting all banana varieties. Prospects of developing varieties with resistance to bacterial wilt through conventional breeding are limited, as no source of germplasm exhibiting resistance against the disease has been identified (personal communication with H. Mignouna, AATF).

The Academia Sinica in Taiwan has isolated two genes from sweet peppers which, when transformed into bananas, protect them from this disease. The first, *pflp*, codes for the ferredoxin-like amphipathic protein and the second, *hrap*, for a hypersensitive response-assisting protein (Chen et al., 2000; Huang et al., 2004). This project is also being managed by the AATF who have negotiated a royalty-free licence of the two genes. Confined field trials are being carried out in Uganda using lines that showed enhanced resistance in screen house trials (personal communication with A. Kiggundu, NARO).

Another problem is Black Sigatoka, a fungal leaf spot disease, which impedes photosynthesis by blackening parts of the leaves, eventually killing the entire leaf. Starved for energy, fruit production falls by 50 per cent or more, and the bananas that do grow ripen prematurely. The fungus is particularly resistant to treatment by antifungal sprays, which, in any case, are usually too expensive for small-holder farmers in East Africa. As *pflp* and *hrap* are defence genes and can provide resistance against a broad spectrum of pathogens, transgenic bananas will also be assessed for resistance to this fungal disease (personal communication with A. Kiggundu, NARO).

A very interesting study has been carried out on the social benefits, costs and consumer preferences of fungal-resistant bananas in Uganda (Kikulwe, 2010). The author found that urban consumers were less in favour of their introduction than rural ones, who are closer to the producers. He pointed out that the inclusion of benefits for producers as an attribute in consumer preferences towards GM food often has not been considered in studies on consumer willingness to pay. He also cautions that although the GM banana technology is likely to improve the overall welfare in Uganda, it is necessary to think carefully about those who may lose from the introduction of this technology. As these will mainly be urban consumers, by thinking about this beforehand the negative impact of this opposing segment could be reduced. Finally his results show that with each year of delay in the introduction of these GM bananas, Uganda loses about US$179–365 million. As the maximum incremental social tolerable irreversible costs are about US$176, considerably lower than the cost of delay, he urged the National Agricultural Research Organization of Uganda to work harder to push the GM banana through the biosafety protocols as promptly and efficiently as possible.

6 CASSAVA: VIRUS RESISTANCE

Cassava, or manioc, is extensively cultivated as an annual crop in parts of Africa for its edible starchy tuberous root, a major source of carbohydrates, with the crop accounting for up to 30 per cent of the daily caloric intake of Ghanaians. The plants do well on poor soils and with low rainfalls and because it is a perennial it can be harvested as required. Its wide harvesting window allows it to act as a famine reserve. The importance of cassava to many Africans is epitomised in the Ewe (a language spoken in Ghana, Togo and Benin) name for the plant, *agbeli*, meaning 'there is life'.

Cassava mosaic virus is related to maize streak virus and causes the leaves to wither, limiting the growth of the root. The virus caused a major African famine in the 1920s. Sometime in the late 1980s a mutation occurred in Uganda that made the virus even more harmful, causing the complete loss of leaves. This mutated virus has been spreading at a rate of 50 miles per year and is currently encroaching into Nigeria via the Democratic Republic of Congo and neighbouring countries. Nigeria, as the world's largest producer of cassava, is starting to become seriously concerned.

Cassava mosaic disease (CMD) is being tackled by Claude Fauquet and his team at the Donald Danforth Plant Research Institute in St Louis, Missouri. They are using RNA interference (RNAi) technology via the replication associated *AC1* gene from the East African strain of cauliflower mosaic virus (CMV) isolated in Uganda. This technology can be described as a natural defence mechanism of plants and other organisms, and consists of 'teaching' the plant to recognize virus genetic sequences in advance, so that the plant is ready to act when the real virus attacks. Indeed, it can be thought of as a type of plant vaccination.

The transgenic lines have been tested extensively under glasshouse conditions in the USA. The National Agricultural Research Organisation of Uganda is now undertaking confined field trials in Namulonge, a known hotspot for CMD (personal communication with A. Kiggundu, NARO).

7 NUTRITIONALLY ENHANCED CROPS

More than half a billion people around the world rely on sorghum as a dietary staple. Its tolerance to drought and heat make it an important food crop in Africa, where it is indigenous to Ethiopia and Sudan. However, it lacks certain essential nutrients. In order to give it added nutritional value, the Bill and Melinda Gates Foundation funded the African Biofortified Sorghum project, run by an international consortium under the leadership of Africa Harvest, an African-based international non-profit organization. African biofortified sorghum contains the gene for a high-lysine storage protein from barley and has increased levels of Vitamin A, iron and zinc.

Vitamin A deficiency is common in developing countries. One of its earliest manifestations is night blindness, which is often found in malnourished pregnant women and children. Deficiencies in vitamin A have been estimated to be responsible for 0.6 million deaths in children annually (Black et al., 2008). In an effort to help overcome this problem in Asian countries where rice is the staple food, Ingo Potrykus and Peter Beyer genetically engineered a variety of rice that produced ß-carotene, a precursor of vitamin A. The first version only produced low levels of the micronutrient, but a subsequent version contains sufficient amounts to provide the entire dietary requirement of the nutrient to people who eat about 75g of this 'golden rice' per day (Paine et al., 2005). It is called 'golden rice' because ß-carotene has a yellow colour.

In 2006, scientists from the Council for Scientific and Industrial Research (CSIR), the South African partner in the consortium, applied to the Registrar of the Directorate for Genetic Resources Management in the National Department of Agriculture, the body that administers the GMO Act, to undertake greenhouse trials for biofortified sorghum. This application was denied by the Executive Committee on the grounds of potential risks pertaining to environmental impact (as a result of gene flow) as the experiment was being conducted on an indigenous species. They also took note that the possibility of the CSIR obtaining a trial or general release authorization some time in the future, were they to apply, would be extremely low. They finally expressed concerns regarding the containment levels of the facilities that would be involved in the proposed activities and indicated that such activities should be conducted in at least a Level-3 containment facility.

In its appeal, the CSIR pointed out that the South African Biotechnology Strategy of 2002 had stressed the importance of value-addition to indigenous crops. However, the decision of the Executive Committee could be interpreted to mean that no research on indigenous crops should be allowed. The CSIR also noted that this committee was prejudging future applications for field trials and general release and was turning down a glasshouse trial in case of a possible future application. Finally, the appeal noted that the CSIR did, indeed, have a Level-3 containment facility that had been approved by that very Directorate for Genetic Resources.

Two appeals were turned down but, finally, in 2009 permission was granted. However, the damage had already been done by this slow and complicated process. The Gates Foundation moved the R&D for this project to Kenya in 2008 where approval for GM sorghum greenhouse trials was obtained within three months and trials began within five months.

8 PUBLIC PERCEPTION PROBLEMS

Public perceptions of the potential dangers of GM crops have in part hampered their adoption in Africa. Food safety is one issue that many Africans raise against eating food derived from GM crops. Despite assurances from organizations such as the British Medical Association (2004), the European Union Research Directorate (2001) and the German Academies of Science and Humanities (Helt, 2004), there is still the belief in many countries that food safety has not been sufficiently tested. Indeed, one argument used by the anti-GM crop lobby is that if people eat food derived from such crops they will become sterile. The origin of this is some unsubstantiated research on mice. Unsubstantiated or not, the mere thought of decreased fertility is anathema to an African man.

Another of the criticisms often levelled against the planting of GM crops is that farmers will be forced to buy seed every season. What these critics omit to acknowledge is that many farmers have been buying seed ever since hybrids were developed in the 1930s. In hybrid seed production, elite inbred varieties with well documented and consistent characteristics, such as yield, are crossed and the resulting hybrid seed is collected. An important parameter in the selection of the elite male and female lines is their combining ability. This is the term used to describe the level of heterosis, or hybrid vigour, that the parents will generate in the resultant seed. Higher combining ability between parents results in increased performance in the resulting hybrid seed.

Hybrid seed has been one of the main contributing factors to the dramatic rise in agricultural output in the developed world during the last half of the twentieth century. This is despite the fact that such seed is more expensive than conventional seeds due to the investment in their production and the built-in intellectual property protection that allows seed merchants to extract higher prices. In Africa many farmers still choose to plant open-pollinated varieties. These are cheaper but do not have the trait advantages of hybrids. As all GM crops in Africa are so far in hybrids, farmers still have the same choice – plant hybrids, whether GM or not, or plant open pollinated varieties. Farmers are smart economists who do their sums – if the hybrids pay, they will buy them; if they do not, they won't.

9 CONCLUSIONS

Where to from here? Among the existing GM crops it will be interesting to watch the development of Bt cotton in Burkina Faso, and to see whether this crop will be adopted in neighbouring countries, especially landlocked Mali. Kenya will also be a country to watch. Will any of their field trials translate into commercial crops? And what will happen in some of the neighbouring East African countries who are also involved in field trials?

Of the upcoming GM crops, the development of Bt cowpea in Nigeria from field trials to commercial release is a possibility. And of particular interest will be bananas and cassava in Uganda, the former resistant to both bacterial wilt and Black Sigatoka, and the latter to Cassava Mosaic Disease.

Finally the development of drought-tolerant maize, under the auspices of the WEMA project: will the field trials currently underway in South Africa, Kenya, Mozambique and Uganda be successful? This is probably the single most important trait for the future of food security in Africa.

REFERENCES

Assefa, Y. and J. van den Berg (2010), 'Genetically modified maize: adoption practices of small-scale farmers in South Africa and implications for resource poor farmers on the continent', *Aspects of Applied Biology*, **96**, 215–23.

Balthazor, T.M. and L. Hallas (1986), 'Glyphosate-degrading microorganisms in industrial waste treatment biosystems', *Applied and Environmental Microbiology*, **51**, 432–34.

Black, R.E., L.H. Allen, Z.A. Bhutta, L.E. Caulfield, M. de Onis, M. Ezzati, C. Mathers and J. Rivera (2008), 'Maternal and child undernutrition: global and regional exposures and health consequences', *The Lancet*, **371**, 243–60.

Bourguet, D. (2004), 'Resistance to *Bacillus thuringiensis* toxins in the European corn borer: what chance for Bt maize?', *Physiological Entomology*, **29**, 251–56.

British Medical Association (2004), 'Genetically modified foods and health: a second interim statement', London: BMA, March.

Castiglioni, P., D. Warner, R.J. Bensen, D.C. Anstrom, J. Harrison, M. Stoeker, M. Abad, G. Kumar, S. Salvador, R. D'Ordine, S. Navarro, S. Back, M. Fernandes, J. Targolli, S. Dasgupta, C. Bonin, M.H. Luethy and J.E. Heard (2008), 'Bacterial RNA chaperones confer abiotic stress tolerance in plants and improved grain yield in maize under water-limited conditions', *Plant Physiology*, **147**, 446–55.

Chen, C.H., H.J. Lin, M.J. Ger, D. Chow and T.Y. Feng (2000), 'cDNA cloning and characterization of a plant protein that may be associated with the harpinPSS-mediated hypersensitive response', *Plant Molecular Biology*, **43**, 429–38.

European Union (EU) Research Directorate (2001), 'GMOs: are there any risks?', Brussels: EU Commission, press briefing, 9 October.

Gouse, M., C. Pary and D. Schimmelpfennig (2004), 'The distribution of benefits from Bt cotton adoption in South Africa', *AgBioForum*, 7(4), 187–94.

Green, W.M., M.C. de Billot, T. Joffe, L. van Staden, A. Bennett-Nel, C.L.N. du Toit and L. van der Westhuizen (2003), 'Indigenous plants and weeds on the Makhathini Flats as refuge hosts to maintain bollworm population susceptibility to transgenic cotton (Bollgard™)', *African Entomology*, **11**(1), 21–9.

Helt, H.W. (2004), 'Are there hazards for the consumer when eating food from genetically modified plants?', Union of the German Academies of Science and Humanities, Commission on Green Biotechnology. Göttingen: University of Göttingen.

Huang, S.N., C.H. Chen, H.J. Lin, M.J. Ger, Z.I. Chen and T.Y. Feng (2004), 'Plant ferredoxin-like protein AP1 enhances Erwinia-induced hypersensitive response of tobacco', *Physiological and Molecular Plant Pathology*, **64**, 103–10.

James, C. (2011), 'Global status of commercialized biotech/GM crops: 2010', *International Service for the Acquisition of Agri-Biotech Applications (ISAAA) Brief*, **42**, Ithaca, NY: ISAAA.

Kikulwe, E. (2010), 'On the introduction of genetically modified bananas into Uganda: social benefits, costs and consumer preferences', PhD thesis, Wageningen University.

Kruger, M., J.B.J. van Rensburg and J. Van den Berg (2009), 'Perspective on the development of stem borer resistance to Bt maize and refuge compliance at the Vaalharts irrigation scheme in South Africa', *Crop Protection*, **28**, 684–9.

Kruger, M., J.B.J. van Rensburg and J. Van den Berg (2011a), 'Resistance to Bt maize in *Busseola fusca* (Lepidoptera: Noctuidae) from Vaalharts, South Africa', *Environmental Entomology*, **40**(2), 477–83.

Kruger, M., J.B.J. van Rensburg and J. Van den Berg (2011b), 'Transgenic Bt maize: farmers' perceptions, refuge compliance and reports of stem borer resistance in South Africa', *Journal of Applied Entomology*, **X**, 1–12.

Monsanto (2007), 'User guide for the production of Yield-Gard, Roundup Ready and YieldGard with Roundup Ready maize', available at: www.monsanto.com.

Paine, J.A., C.A. Shipton, S. Chaggar, R.M. Howells, M.J. Kennedy, G. Vernon, S.Y. Wright and E. Hinchliffe et al. (2005), 'Improving the nutritional value of Golden Rice through increased pro-vitamin A content', *Nature Biotechnology*, **23**, 482–7.

Schnepf, E., N. Crickmore, J. van Rie, D. Lereclus, J. Baum, J. Feitelson, D.R. Zeigler and D.R. Dean (1998), '*Bacillus thuringiensis* and its pesticidal crystal proteins', *Microbiology and Molecular Biology Reviews*, **62**, 775–806.

Singh, B. (2003), 'Improving the production and utilization of cowpea as food and fodder', *Field Crops Research*, **84**, 150–69.

Vitale, J.D., G. Vognan, M. Ouattarra and O. Traore (2010), 'The commercial application of GMO crops in Africa: Burkina Faso's decade of experience with Bt cotton', *AgBioForum*, **13**(4), 320–38.

8 China

Valerie J. Karplus

1 INTRODUCTION: CHINA AS A CENTRE OF INNOVATION IN CROP BIOTECHNOLOGY

Over the past three decades, China has emerged as a leader in crop biotechnology innovation. China's strong, public-led research enterprise is by many measures the largest outside North America (Chen, 2006; Karplus and Deng, 2008). Guided by central government emphasis on balanced economic development, state-sponsored research is in large part focused on challenges of national food security and improving the livelihoods of the country's smallholder farmers. Outlays of resources and support for education and training in crop biotechnology have increased in lock-step with public funding since the early years of China's reform and opening in the late 1970s. The result has been the successful generation of thousands of transgenic varieties spanning crops as diverse as rice, maize, potato, cotton, sweet potato and eggplant.

While the number of transgenic varieties under development in laboratories and field trials has grown over the past two decades, biosafety approvals and commercial planting have proceeded more slowly. Advances in the laboratory and field trial stages have in many cases not yet achieved success on a commercial scale. In the wake of global anti-transgenic sentiments in the early 2000s, regulatory approvals in China slowed to a standstill, before resuming again in late 2009. Several major crops have received approval since then, including varieties of transgenic rice and maize (Waltz, 2010). Biosafety approval does not, however, guarantee successful commercial application. In many respects, diffusion of transgenic crops in China has been slow when compared to Argentina, Australia, Canada, the United States, and other nations with large investments in transgenic crops.

While transgenic varieties have been the primary target of public opposition globally, translating into slower biosafety approvals in China, the application of non-transgenic biotechnology techniques to develop advanced crop varieties in China has continued apace. Advances in hybrid breeding, tissue culture, and other biotechnology techniques have produced a broad array of crop improvements. Crops developed with non-transgenic techniques do not face the same level of biosafety concern and scrutiny, nor are they subject to labelling requirements. A large number of advanced biotechnology crops are not transgenic, which suggests that studies focused exclusively on transgenic varieties may overlook some of China's most significant crop biotechnology achievements, including the application of molecular techniques to hybrid rice breeding (Wang et al., 2005). It is important to recognize that China's crop biotechnology enterprise encompasses far more than transgenics, which helps to explain why broader efforts to promote biotechnology development have been undeterred by opposition abroad and slow regulatory approvals for transgenic crops early on. Recent advances utilizing advanced biotechnology have built on a long history of experience in hybrid rice breeding that predates China's reform and opening period that began in the late 1970s (Lin and Yuan, 1980).

Crop biotechnology outcomes in China will exert strong influence on the future of both transgenic and non-transgenic applications of the technology worldwide, both through the channels of global scientific research exchanges and in end-use markets for transgenic crops and products. In recent decades, scholars have produced a large number of quantitative and qualitative studies of the evolution of the industry in China, attempting to explain outcomes and to extract lessons for policy, both domestically and abroad. This chapter considers how these efforts have collectively produced insights into the drivers of industry development in China and identifies the gaps and puzzles that remain to be explained. By taking stock of the evolution of scholarship applied to understand the drivers of outcomes in China's crop biotechnology sector, this chapter reviews current models and metrics applied based on various disciplinary approaches, with an eye to identifying promising existing as well as untapped directions for future research.

2 MODELS DEVELOPED TO EXPLAIN PLANT BIOTECHNOLOGY OUTCOMES IN CHINA

Previous scholarship has focused on characterizing factors that shape the contours of China's plant biotechnology research enterprise, attempting to explain the translation of outcomes from the laboratory to field trials, market commercialization and international trade. This section first describes in detail the outcomes observed across China's complex plant biotechnology innovation system, offering a review of the major achievements and remaining challenges to widespread realization of benefits and stewardship of the risks. To the extent possible, this description focuses on facts and leaves a discussion of causal mechanisms for later sections. Then, models used to explain plant biotechnology outcomes in China are reviewed, focusing on explanations related to biophysical or material conditions, community attributes, legal structures, and international trade and governance.

2.1 Outcomes in China's Plant Biotechnology System

The accomplishments of China's complex system for plant biotechnology innovation have been quantified using a variety of metrics. These outcomes are perhaps most visible in the scale of research underway, the variety and number of novel varieties generated using plant biotechnology, the number of these varieties approved for commercial planting, and the area planted to these varieties as a fraction of total acreage by crop. Here we focus mainly on outcomes for transgenic crops, as it can be difficult to find a standard definition of a crop produced using advanced biotechnology techniques. Yet the role of broader investment in agricultural biotechnology tools and training and the strength of the traditional agricultural research and extension system are also important outcomes in their own right, but due to space are not considered in more depth here (see also Karplus and Deng, 2008).

2.1.1 Research and development

The fruits of China's investment in crop biotechnology are perhaps most visible and measurable at the laboratory and contained field trial stages. One of the most striking features is the extent of public sector involvement – China has the largest publicly-funded crop biotechnology investment outside of North America, with research funding doubling approximately every five years (Chen, 2006). Over 80 per cent of applications for plant variety protection in China originate from public research institutes (Koo et al., 2006). The Chinese Academy of Sciences (CAS), Chinese Academy of Agricultural Sciences (CAAS) along with their provincial and prefectural branches, China Agricultural University, Huazhong Agricultural University, and Peking University rank among the leading institutions involved in crop biotechnology in China. Many universities and institutes are home to State Key Laboratories that focus on various aspects of agricultural biotechnology. Over the past two decades, laboratories have at times partnered with multinational companies seeking to enter the Chinese market. For example, the Monsanto Company originally partnered with the Chinese National Cotton Research Institute of the CAAS at Anyang, Henan, in the mid-1990s (Huang et al., 2002b). A more recent example is the Beijing Weiming Kaituo Agriculture Biotech Co., which was formed by a partnership of Peking University, Institute of Genetics and Developmental Biology of CAS, Institute of Biotechnology of CAAS, and Beijing Academy of Agriculture and Forestry Sciences, and now collaborates with the Pioneer Overseas Corporation (a DuPont Company) (BWK, 2012).

The range of transgenic and non-transgenic crops under development in China is vast. Rice alone has been engineered with resistance to insects, blight, fungal rust, salt and drought and with herbicide tolerance. Research has also involved nutritional improvements, including the well-known golden rice fortified with precursors to Vitamin A to address nutritional deficiencies (Dawe and Unnevehr, 2007). These new varieties address a number of concerns, such as vulnerability of agriculture to accelerated climate change (Pray et al., 2011).

Research funding has also increased substantially since the earliest research initiatives were started in the 1980s. For example, in the case of insect-resistant rice, public research expenditures increased from 8 million yuan in 1986 to 195 million yuan in 2003 (Chen et al., 2011). In late 2008 a new programme focused on transgenic plants started with 26 billion yuan, with emphasis on wheat, soybean and maize as well as rice (Xia et al., 2011). Wheat research has focused on drought tolerance, pest resistance, and aspects of grain quality (Xia et al., 2011).

2.1.2 Biosafety approvals

China has established a biosafety regulatory system to certify newly developed varieties prior to their introduction in fields. Approvals are required both for small-scale experimental planting as well as large-scale field trials; requirements are more stringent for the latter. Pre-production trials are sometimes required for crops planted over large areas or deemed essential to national food security. China's biosafety system was initially modelled in the 1990s after the regulatory framework in the United States, which embraced the principle of 'substantial equivalence' in the final product as the criterion for determining whether heightened regulatory scrutiny should be applied. In the early 2000s, new regulations requiring labelling and tracing of transgenic crops were introduced in China, incorporating features of regulations in several European countries that embodied a precautionary stance towards the technology. Today China's regulatory procedures reflect a blend of the two systems and have been applied to evaluate hundreds, if not thousands, of new crop varieties. It should be noted that this system is separate from the varietal registration procedure, which applies to both transgenic and non-transgenic varieties.

The earliest crops to receive biosafety approval in China included varieties of insect-resistant (Bt) cotton, colour-altered petunia, and a virus-resistant sweet pepper (Huang et al., 2002b). Virus-resistant tobacco was also commercialized before biosafety regulations were formally established in China, but later withdrawn from the market amid concerns about acceptance in export markets (Karplus and Deng, 2008). Of these crops, only Bt cotton is planted commercially in China on a large scale. Biosafety concerns were somewhat allayed by the fact that cotton is a non-food crop. Varieties developed by domestic research institutes and the multinational company Monsanto have both received approval and are under cultivation.

After 2001, for a long period no additional transgenic crops were approved for commercial planting in China, although research and field trials (including extended pre-production trials for transgenic varieties of rice and maize) continued apace. At the end of 2009, the Ministry of Agriculture issued safety certificates for varieties of phytase-enhanced maize (for use in animal feed) and transgenic insect-resistant rice developed by scientists at Huazhong Agricultural University, conditional on the successful completion of additional pre-production trials (Waltz, 2010; Lu, 2010). China was not the first country to approve insect-resistant rice for commercial planting – indeed, regulators in the United States approved a variety of transgenic insect-resistant rice developed by Bayer CropScience in 1999, but the company never commercialized it amid fears that it would not be accepted in export markets (Waltz, 2010).

Transgenic poplar trees and a virus-resistant papaya variety have received the green light for commercialization in China (James, 2011). Wheat, by contrast, has not yet received commercial approval and there has been only slow progress on research for a range of traits. As of 2009 only 62 varieties of wheat had entered into the biosafety regulatory process (Xia et al., 2011).

2.1.3 Biotechnology crops grown commercially

The greatest success of any transgenic crop in China in terms of planted area and benefits to date is Bt cotton, which was planted across over 65 per cent of the country's cotton-growing areas in 2008, while adoption reached 100 per cent in some of the eastern provinces (Huang et al., 2010). The most successful varieties were developed by the Chinese Academy of Agricultural Sciences and the provincial Academy of Agricultural Sciences in Henan Province. Early on, varieties developed by the Monsanto Company were very successful when they were first adopted, but gradually lost market share to domestically developed varieties.

The benefits of growing Bt cotton in China are estimated to be large. Benefits, which mainly take the form of increased yields, decreased production costs, and health and environmental benefits due to reduced pesticide use, have been sustained since the crops' initial introduction, owing in part to continued development of new varieties. These benefits have accrued primarily to the smallholder farms that historically have accounted for a large share of total cotton acreage (Huang et al., 2002a and 2010).

It is important to note that biosafety approval does not mean a crop will be successfully commercialized. In fact a few of the first transgenic crops approved for commercial planting were not commercialized successfully, including a variety of virus-resistant sweet pepper. The large-scale commercial potential of transgenic rice has also not been realized, and it is not clear how extensively new transgenic rice varieties will be planted, either in China or internationally.

2.2 Explaining Outcomes in China's Biotechnology Sector

Accounting for the crop biotechnology outcomes observed in China as well as other countries is a complex task that has engaged researchers across a wide array of disciplines. This chapter focuses on three categories of drivers: (1) biophysical and material conditions; (2) attributes of the community; and (3) rules and overarching legal structure, including biosafety regulations, intellectual property rights, and the strength and mandates of governing institutions. Paralleling the structure of section 2.1, each of these drivers is investigated to identify how they help explain outcomes at various stages of crop biotechnology development. The objective is to illustrate how different studies applying a diverse range of methodologies have linked observed outcomes to underlying drivers.

2.2.1 Biophysical and material conditions

Geography and the nature of localized agricultural challenges such as soil nutrient profiles, water availability, pest populations and other features of China's agricultural landscape have played an important role in the development of crop biotechnology in China. Resource constraints in China have shaped the country's agricultural development throughout history. Food security and overcoming geographical limitations to agricultural productivity and expansion has long been a concern of government policy architects over several thousands of years.

In the case of crops that have been successfully commercialized in China, environmental and rural challenges have been documented as a key driver. In particular, the rapid uptake of Bt cotton has been driven by the declining efficacy of conventional pesticides and the health and environmental effects of intensification of chemical fertilizers, pesticides and other inputs (Pray et al., 2006; Huang et al., 2002a). Development of transgenic insect-resistant rice was also a response to increasing stem borer infestations after 1993 and, in particular, a severe outbreak in 1996 (Chen et al., 2011). Rice yield losses due to stem borer infestations have been documented since the Song Dynasty (960–1279 AD); however, these infestations were effectively controlled and caused negligible yield losses until the 1990s when pest resistance to synthetic pesticides began to increase. The commercialization of transgenic rice in 2009 was also suggested to be driven in large part by needs specific to China's domestic geography as well as cropping conditions and occurred in spite of concerns about rejection in export markets (Waltz, 2010).

The practices employed in China are also shaped by the biophysical and material conditions. Farmers in China often have limited education or training in sophisticated pest management practices developed in research institutes or universities, and mechanisms for encouraging these practices are still weak in many parts of the country. Chen et al. (2011) have suggested that these weaknesses explain why the dissemination of advanced seed varieties was more effective in raising yields, relative to encouraging a shift to more modern practices that would require behavioural change among farmers.

2.2.2 Attributes of the community

Some research has identified how attributes of the stakeholder community in China have enabled crop biotechnology development (Paarlberg and Pray, 2007; Keeley, 2006; van Zwanenberg et al., 2010). Stakeholders involved in crop biotechnology decisions in China are quite diverse and have exerted significant influence upon different parts of the innovation system. Researchers at institutes and universities, government officials focused on science and technology, environmental and public health and agricultural policy, and consumers, farmers and rural development advocacy communities have played very different roles in influencing the direction and pace of crop biotechnology development. Agbiotech development in China has benefited from strong state support and stewardship, led mostly by the Ministry of Science and Technology and the Ministry of Agriculture. This feature of China's agbiotech sector has allowed research, development and demonstration to proceed in spite of periodic opposition from the environmental community and consumer sentiment in and outside China (Keeley, 2006; van Zwanenberg et al., 2010). The strong and persistent emphasis on promoting crop biotechnology research and development, even as commercial approvals slowed through the first decade of the 2000s, reflects the unique roles of these communities in China.

Strong government support for publicly-funded research and development benefits from both deep historical roots and broad contemporary support at the intersection of the country's science and technology and rural development agendas. China had a well-established and respected network of government-funded agricultural and fundamental scientific research institutes prior to the beginning of reforms in the late 1970s. Many of these research institutes benefited from close ties to the government, both to the Ministry of Science and Technology and to the Ministry of Agriculture. With the first applications of agricultural biotechnology globally, China's leaders saw an opportunity to catch up in this advanced technology area that could simultaneously offer benefits to the large consumer population and smallholder farmers. With support from both ministries, policies to strengthen research on crop biotechnology and set up enabling biosafety regulations were among the first in the world to gain ground in the early to mid-1990s.

It is also important to understand how advanced biotechnology fitted into China's overall reform-era (post-1978) national science and technology strategy. Significant resources were directed towards upgrading China's laboratories and research capabilities, particularly in areas where China was deemed to be capable of quickly operating at the global frontier. A large number of students were supported to study abroad in the United States, Europe and Japan. Only in recent years are a large number of these former students returning to China (see Figure 8.1). Research funding has also increased, particularly in the eight research areas identified in the early science and technology development plans.

The introduction in the mid-1990s of the first transgenic crops developed in China coincided with increasing opposition to transgenic methods elsewhere in the world, particularly in parts of Europe. In China environmental and public health officials, influenced to some extent by developments abroad, sought to exert increased influence over the evaluation of transgenic crops nearing final biosafety approval. These communities took a more precautionary stance, aided by several unauthorized releases of transgenic seed that were detected and widely publicized by international advocacy organizations. At the same time, anti-transgenic sentiments were growing within China as environmental

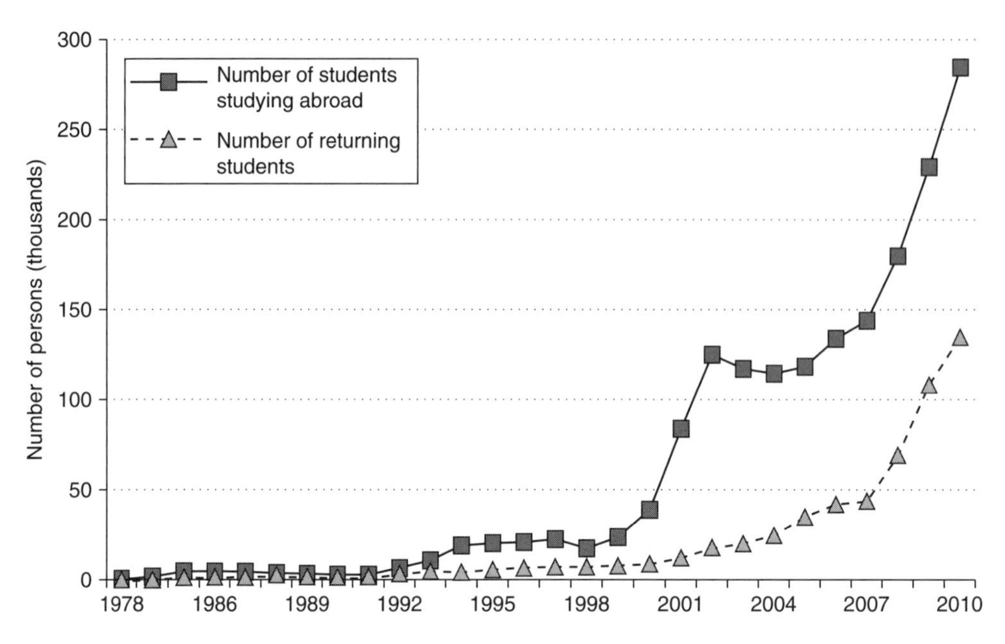

Source: *China Statistical Yearbook* (2011).

Figure 8.1 Chinese students studying abroad and oversees returnees since reforms began in 1978

and health concerns inched higher on the national agenda, encouraged by growing evidence of the adverse environmental consequences of China's rapid economic expansion.

The role of farmers in the development of China's crop biotechnology system has been limited mostly to the evaluation and adoption of new varieties, with their interests represented by agricultural and social policy-makers within the government. Many studies have evaluated the benefits of crops to farmers. Huang et al. (2002a), for example, evaluated the early impact of the first transgenic cotton varieties in north-eastern China, reporting significant economic and health benefits and satisfaction among farmers. Positive reports of the technology's acceptance and widespread adoption have encouraged continued support for the technology among policy communities. Early studies of the impact of transgenic rice have also demonstrated benefits to rural farmers. The studies by Huang et al. (2005) employed econometric techniques based on samples of rural households, and drew comparisons between adoption and non-adopting households. Other studies using qualitative interviews and extensive observation of users of Bt cotton over a single growing season yielded different conclusions about farm-level benefits (Berge and Ricroch, 2010). Still other studies pointed out that an increase in pesticide use was needed to control secondary pests, which emerged once broad-spectrum pesticides were no longer being used (Ho et al., 2009). China's economic expansion has coincided with rapid urbanization, and today over half of the country's population lives in urban areas. The diets of urban populations increasingly rely on dairy and meat, decreasing the dietary share of basic staples such as rice, potato and millet. Several studies have considered how consumer opinion and values are changing with rising wealth and urbanization, and these have focused on urban centres such as Beijing, Shanghai, Nanjing and Tsingdao (Li et al., 2002; Bai et al., 2008; Ho and Vermeer, 2004). Consumers are also beginning to favour foods made from organic or locally grown ingredients, mirroring trends in many developed countries, and to become more interested and engaged in food safety debates. In Shanghai news reports documented a case of a woman who filed a lawsuit in Shanghai alleging that chocolate milk power she had purchased for her son contained transgenic additives (*Shanghai Daily*, 2005). This case is perhaps an exception – for the most part public surveys have indicated broad support for the adoption of transgenic crops and are willing to consume foods derived from them (Tan et al., 2011). As China's population grows richer, it is unclear how tastes will shift. Researchers have suggested that the lack of strong consumer opposition in China to date helps to explain why crop biotechnology has continually moved forward (Cheng, 2011).

2.2.3 Rules and overarching legal structure

Features of the institutions that govern China's crop biotechnology have also allowed the technology to move forward rapidly over the last two decades, particularly at the stage of research and development. Scholars argue that several institutions have played a key role: the biosafety and regulatory approval process; intellectual property rights; industrial policy and trade concerns; and conflicts among the interests of stakeholder groups. Characteristics of these institutions have been used, primarily relying on methodologies from political science, economics, sociology and law, to explain many of the unique outcomes in China's crop biotechnology sector.

Since being implemented in the mid-1990s, China's biosafety regulatory system has evolved from a relatively permissive system to one that embodies a more precautionary approach. A key point of change was 2002, when a second round of regulations was issued. This shift has been documented by Paarlberg (2001), Karplus and Deng (2008), and Pray et al. (2006). Figure 8.2 shows the progression of China's biosafety regulations, and names some of the key triggering events and broader influences. The biosafety approval system has in practice continuously allowed the field testing of crops, and it is only at the final approval stage that several extensively-tested varieties been stalled, sometimes for many years (perhaps the best-known case is transgenic rice).

Scholars have linked the strength and patterns of oversight of intellectual property rights for novel crop biotechnology innovations to decisions by the public and private sectors to invest in crop biotechnology in particular countries. Paarlberg (2001) documented weaknesses in China's intellectual property protection that led to widespread copying of transgenic insect-resistant Bt cotton seed when proprietary varieties were first introduced by the multinational company Monsanto. This disregard for intellectual rights occurred despite a government guarantee of legal protection, and deterred Monsanto from broadening its involvement in China. At that time, farmers had to pay a premium 'technology fee' for the use of the Monsanto seed, but fees for the domestically developed varieties were much lower or non-existent. Since that time, recognition of the importance of intellectual property rights protections in China has grown, which is judged an important factor in the commercialization decisions of both domestic and multinational market entrants. While the problem of intellectual property rights is still not fully resolved, it has not uniformly impeded technology diffusion, as it benefits domestic developers of the technology and shelters them from international competition. Meanwhile, some domestic developers who are advancing new traits rather than simply

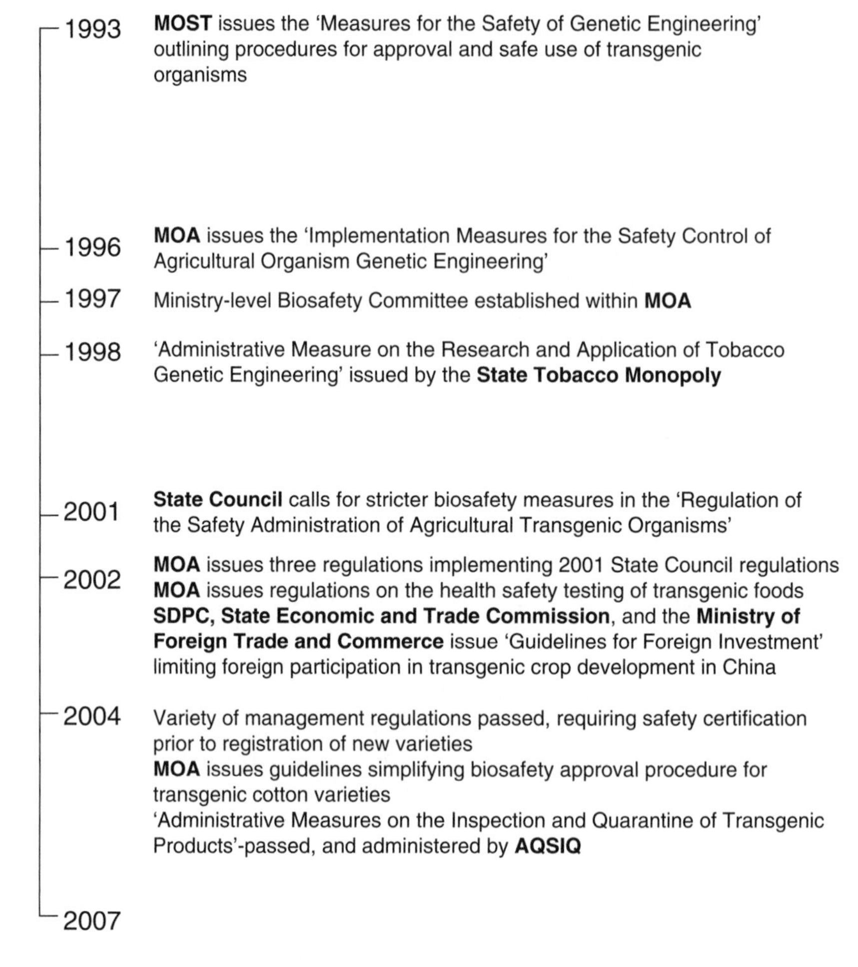

Source: Reprinted from Karplus and Deng (2008).

Figure 8.2 Major developments in the biosafety regulation of transgenic crops in China

replicating other technologies are increasingly appealing to intellectual property protection to protect their own profit margins (Karplus and Deng, 2008).

Finally, scholars have noted that conflicting institutional incentives have meant that China lacks mechanisms to support commercialization of crop biotechnology, even once they have received final biosafety approval. Some of this weakness has been attributed to the fragmentation in the seed market and a lack of effective oversight, which allowed pirated seed to spread widely in the early days of Bt cotton introduction (Ho et al., 2009). While the proliferation of pirated seed may be related to the weakness of intellectual property rights, it is not the full story. Given the large population involved in smallholder farming and small rural businesses and the perception that agbiotech development should lead to improved rural conditions, it is perhaps not surprising that intellectual property rights were not enforced at the expense of access to the technology.

3 METHODS

The discussion above has described the key developments in China's crop biotechnology sector over the past two decades and summarized the insights of studies that have attempted to explain apparent paradoxes or unexpected phenomena. This section takes stock of the methods that have generated these insights and considers the relationships between the conclusions and the methods used to generate them. The objective is to create broader awareness of the disciplinary and inter-disciplinary foundations of previous studies and to consider what remaining questions these or other new methods could answer.

Economics is well represented among the studies used to probe the impacts of both investment in research and development as well as impacts of transgenic crops in farmers' fields. Dr Huang Jikun and Dr Hu Ruifa, Prof. Carl Pray and collaborators have generated a rich body of literature, applying these methods in the Chinese context (Huang et al., 2002a, 2002b, 2003, 2005, 2010; Pray et al. 2006; Hu et al. 2011). Tan et al. (2011) have also used quantitative techniques to measure impacts.

A political economy perspective has also been applied to study China's experience with crop biotechnology, particularly transgenic crops. Fok and Xu (2011), Keeley (2003), Paarlberg (2001), and Zhang et al. (2012) have offered institutional explanations for the outcomes observed across all stages of China's crop biotechnology development.

Finally, reflecting the strong involvement of scientists, particularly Chinese scientists, in policy-making in China for crop biotechnology, a number of scientists have offered descriptive and prescriptive studies of the successes and limitations facing development in China. Chen et al. (2011) provide an assessment of the impacts of transgenic insect-resistant rice based on scientific assessment of its potential. Karplus and Deng (2008) also review developments in China's agricultural biotechnology sector, basing their analysis on a survey of studies as well as detailed interviews with scientists and policy-makers involved with the creation of China's crop biotechnology enterprise. Chen (2006) reviews developments in China's agricultural biotechnology sector broadly, based in part on his early and deep involvement in the promotion of related activities in China. Xia et al. (2011) provide further reflections on the challenges to commercialization of transgenic wheat, based on direct experience.

4 METRICS

Metrics used to assess the success of crop biotechnology in China span the full range of methodologies and stages in the development process. Here we use the stages from section 2 to summarize metrics that originate in a variety of disciplines but describe a common subset of outcomes in the plant biotechnology sector. From the survey of metrics it becomes apparent that metrics applied stem largely from the quantitative methods in economics, public policy, as well as political economy. Some of these studies span long time horizons, as in the case of studies of the adoption and impact of insect-resistant transgenic cotton and to some extent transgenic insect-resistant rice in China. Fewer studies employ qualitative methods, regardless of discipline.

4.1 Early-stage Research and Field Trials

Several metrics commonly used to assess the extent of activity in China's laboratories and early-stage field trials (which face comparable biosafety scrutiny, but are not subject to the more stringent regulations governing large-scale or pre-production trials) include government spending, number of personnel employed, number of biotechnology varieties under development, and the number of crops in field trials. Chen (2006) estimates that China conducted over 700 field trials for transgenic crops by 2006, although exact numbers are not publicly reported, and the number has continued to grow.

Scientists involved in crop biotechnology development have themselves contributed to rich narratives on the early and persistent challenges to crop biotechnology in China. Chen et al. (2006) describe how the research system for plant biology came of age in China, focusing on metrics such as overseas returning students, expenditures and observed shifts in government priorities, as broader reforms affected the structure of Chinese industries and research institutes. While not easy to quantify, researchers have remarked that ongoing reforms left ever more research institutes – including agricultural research institutes – without guaranteed funds, galvanizing a shift toward more profit-oriented activities. Chen et al. (2006) have noted that funding for essential forms of basic research is still a major shortcoming of China's crop biotechnology research system.

4.2 Biosafety System Performance

Indicators of the performance of China's biosafety system range from the number of biosafety approvals to the frequency and severity of detected biosafety breaches. Both are imperfect measures of system performance (a lack of breaches detected could mean either strong performance or limited potential for detection). Pray et al. (2006) have found it useful to draw international comparisons, for instance to the case of India, when interpreting metrics related to China's biosafety system performance.

Biosafety approvals in China have long been a measure of the effectiveness of China's regulations governing the safe introduction of the products of advanced biotechnology into the agricultural system. Indeed, measures such as the number of transgenic varieties approved for commercial planting in China in each year, as well as the number of total applications for field releases, are signals that the system is functioning despite barriers of cost or administrative hurdles, and is successfully attracting applications from scientists and breeders. China's biosafety regulatory system has reviewed a large number of applications for field testing since it was formally set up in the late 1990s. As of 2005, the government had received 1044 applications for field trials or commercial release and 777 of these applications had been approved (Pray et al., 2006; Chen, 2006). These applications predominantly covered 60 crops, as well as several animals and a large number of micro-organisms (Pray et al., 2006).

4.3 Commercialization of Transgenic Crops: Magnitude and Distribution of Benefits

Perhaps the greatest body of work has been focused on the agronomic and economic impact of transgenic crops in the field to date, despite limited examples of large-scale adoption. The Center for Chinese Agricultural Policy and its international collaborators

Table 8.1 Estimated adoption and benefits from Bt cotton in China

Quantity	Value
Commercialized since	1997
Adoption rate (%)	66
Change in pesticide cost (%)	−65
Change in yield (%)	24
Change in seed cost (US$/ha)	32
Change in net income (US$/ha)	470
Farmers' benefit share (%)	94
Companies' benefit share (%)	6

Source: Pray et al. (2002), reported in Qaim (2005).

have documented the impacts for both Bt cotton and Bt rice varieties through a series of studies that rely on farm-level observation of crop impacts. Key metrics include the adoption rate (in terms of the fraction of total planted area), change in crop yield, change in farm income, and the distribution of benefits between farmers and companies (developers) (see Table 8.1). Another oft-cited metric is the number of small farmers growing transgenic crops in China, estimated to total 7 million in 2011 (James, 2011; Anthony and Ferroni, 2011). More recent estimates provided in Huang et al. (2010) suggest that earlier studies may in fact have underestimated benefits to smallholder farmers. In particular, the authors found that the economic benefits (a combination of the input cost savings and productivity gains) were greater when estimated using a larger sample over a longer time horizon. The results also suggest that pest resistance has not emerged in spite of the fact that refugia were not required for most small farms in the north and central parts of China, due to the small scale and diversity of neighbouring crops (Huang et al., 2010).

5 CRITICAL ASSESSMENT

In China a large and growing body of work has quantified the impacts and extracted lessons from the country's experience with agricultural biotechnology. This chapter has provided an overview of how China's crop biotechnology development has unfolded. While the work cited includes diverse claims of strengths and weaknesses at different points in the system, several insights are widely shared among studies to date. The first recognizes China's significant progress in developing a large, publicly-funded research enterprise that has so far delivered crops aimed at meeting the needs of smallholder farmers. The second is that while China's publicly-funded research in crop biotechnology is strong, the commercial success of many crops is still limited. This apparent paradox has been attributed to weak intellectual property rights early on, and more recently to gaps in incentives created by state involvement in the biotechnology sector that discourage research institutes from taking risks on new varieties, many of which may be highly capital intensive to develop and deliver. A third insight relates to the fact that crop biotechnology in the most comprehensive sense is continuing to play an important role in China's future, even if transgenic varieties are adopted relatively slowly. Long-term continued diffusion is to some extent the result of large state investment in the tools and techniques of biotechnology, the influence of scientists in the public policy process, and basic public trust in the promise of science to solve national challenges. However, future trends for each of these factors are difficult to predict.

The scholarly community has so far converged on a range of empirical methods in economics, political science and the natural sciences to monitor and elicit lessons from the unfolding story of crop biotechnology in China. The number of studies of the impact of China's transgenic crops in the field will also increase with experience, for instance, to assess the performance of insect-resistant transgenic rice and fortified maize varieties since commercialization. More detailed quantitative analysis of the biosafety system in China as well as the potential and limitations of policy intervention to stimulate domestic private sector activity in the seed industry will be of interest in the future. Finally, continued study of the empirical landscape may eventually lend itself to the construction and validation of improved models, both qualitative and quantitative, extending and grounding efforts to estimate benefits and costs using large-scale national or regional economic and natural systems models that reflect a rich understanding of the complex inter-relationships in China's crop biotechnology innovation system.

REFERENCES

Anthony, V.M. and M. Ferroni (2011), 'Agricultural biotechnology and smallholder farmers in developing countries', *Current Opinion in Biotechnology*, **23**, 1–8.

Bai, J., T.I. Wahl and J.J. McCluskey (2008), 'Consumer choice of retail food store formats in Qingdao, China', *Journal of International Food & Agribusiness Marketing*, **20**(2), 89–109.

Berge, J.B. and A.E. Ricroch (2010), 'Emergence of minor pests becoming major pests in GE cotton in China: what are the reasons? What are the alternative practices to this change of status?', *GM Crops*, **1**(4), 214–19.

BWK 'Beijing Weiming Kaituo' (2012), website at: http://www.bwkcrop.com/EN/Intro/.

Chen, H., V. Karplus, H. Ma and X.W. Deng (2006), 'Plant biology research comes of age in China', *The Plant Cell*, **18**(11), 2855–64.

Chen, M., A. Shelton and G. Ye (2011), 'Insect-resistant genetically modified rice in China: from research to commercialization', *Annual Reviews of Entomology*, **56**, 81–101.

Chen, Z.L. (2006), 'Chinese agricultural biotechnology in the field', in A. Eaglesham and Ralph W.F. Hardy (eds), *Agricultural Biotechnology: Economic Growth Through New Products, Partnerships and Workforce Development*, Ithaca, NY: National Agricultural Biotechnology Council.

Cheng, Y. (2011), 'Public has doubts over modified food', *People's Daily Online*, 23 February, available at: http://english.peopledaily.com.cn/90001/90776/90882/7296799.html.

China Statistical Yearbook (2011), Beijing: China Statistics Press.

Dawe, D. and L. Unnevehr (2007), 'Crop case study: GMO golden rice in Asia with enhanced Vitamin A benefits for consumers', *AgBioForum*, **10**(3), 154–60.

Fok, M. and N. Xu (2011), 'Variety market development: A Bt cotton cropping factor and constraint in China', *AgBioForum*, **14**(2), 47–60.

Ho, P. and E.B. Vermeer (2004), 'Food safety concerns and biotechnology: consumers' attitudes to genetically modified products in urban china', *AgBioForum*, **7**(4) 158–75.

Ho, P., J.H. Zhao and D. Xue (2009), 'Access and control of agro-biotechnology: Bt cotton, ecological change and risk in China', *The Journal of Peasant Studies*, **36**(2), 345–64.

Hu, R., Q. Liang, C. Pray, J. Huang and Y. Jin (2011), 'Privatization, public R&D policy, and private R&D investment in China's agriculture', *Journal of Agricultural and Resource Economics*, **36**(2), 416–32.

Huang, J., R. Hu, S. Rozelle and C. Pray (2005), 'Insect-resistant GM rice in farmers' fields: Assessing productivity and health effects in China', *Science*, **308**(5722), 688–90.

Huang, J., J.W. Mi, H. Lin, Z. Wang, R. Chen, R. Hu, S. Rozelle and C. Pray (2010), 'A decade of Bt cotton in Chinese fields: assessing the direct effects and indirect externalities of Bt cotton adoption in China', *Science China: Life Sciences*, **53**(8), 981–91.

Huang, J.K., R.F. Hu, C. Fan, C.E. Pray and S.S. Rozelle (2002a), 'Bt cotton benefits, costs, and impacts in China', *AgBioForum*, **5**(4), 153–66.
Huang, J.K., S. Rozelle, C. Pray and Q. Wang (2002b), 'Plant biotechnology in China', *Science*, **295**, 674–7.
Huang, J.K., R.F. Hu, C. Pray, F. Qiao and S. Rozelle (2003), 'Biotechnology as an alternative to chemical pesticides: a case study of Bt cotton in China', *Agricultural Economics*, **29**, 55–67.
James, C. (2011), 'Executive summary, Global status of commercialized biotech/GM crops: 2011', *International Service for the Acquisition of Agri-Biotech Applications (ISAAA) Brief* 43, Ithaca, NY: Cornell.
Karplus, V.J. and X.W. Deng (2008), *Agricultural Biotechnology in China: Origins and Prospects*, New York: Springer.
Keeley, J. (2003), 'Regulating biotechnology in China: the politics of biosafety', IDS Working Paper no. 208, Brighton: Institute of Development Studies.
Keeley, J. (2006), 'Balancing technological innovation and environmental regulation: an analysis of Chinese agricultural biotechnology governance', *Environmental Politics*, **15**(2), 293–309.
Koo, B., P.G. Pardey, K. Qian and Y. Zhang (2006), 'An option perspective on generating and maintaining plant variety rights in China', *Agricultural Economics*, **35**, 35–48.
Li, Q., K.R. Curtis, J.J. McCluskey and T.I. Wahl (2002), 'Consumer attitudes toward genetically modified foods in Beijing, China', *AgBioForum*, **5**(4), 145–52.
Lin, S.-C. and L.-P. Yuan (1980), 'Hybrid rice breeding in China', in G. Argosino, V.S. Durvasula and W.H. Smith (eds), *Innovative Approaches to Hybrid Rice Breeding Selected Papers from the 1979 International Rice Research Conference*, Laguna, Philippines: International Rice Research Institute, pp. 35–51.
Lu, C. (2010), 'The first approved transgenic rice in China', *GM Crops*, **1**(3), 113–15.
Paarlberg, R.L. (2001), *The Politics of Precaution: Genetically Modified Crops in Developing Countries*, Baltimore, MD: Johns Hopkins University Press.
Paarlberg, R. and C.E. Pray (2007), 'Political actors on the landscape', *AgBioForum*, **10**(3), 144–53.
Pray, C.E., L. Nagarajan, J. Huang, R. Hu and B. Ramaswami (2011), 'Impact of Bt cotton, the potential future benefits from biotechnology in China and India', *Food and Global Welfare*, available at: http://www.isid.ac.in/~bharat/Research/prayetal.pdf.
Pray, C.E., B. Ramaswami, J. Huang, R. Hu, P. Bengali and H. Zhang (2006), 'Costs and enforcement of biosafety regulations in India and China', *International Journal of Technology and Globalisation*, **2**(1/2), 137–57.
Qaim, M. (2005), 'Agricultural biotechnology adoption in developing countries', *American Journal of Agricultural Economics*, **87**(5), 1317–24.
Shanghai Daily (2005), '3rd GM food suit filed in Shanghai', 13 December, available at: www.china.org.cn/english/China/151770.htm.
Tan, T., J. Zhan and C. Chen (2011), 'The impact of commercialization of GM rice in China', *American-Eurasian Journal of Agricultural & Environmental Sciences*, **10**(3), 296–9.
Van Zwanenberg, P., A. Ely, A. Smith, C. Chen, S. Ding, M-E. Fazio and L. Goldberg (2010), 'Regulatory harmonization and agricultural biotechnology in Argentina and China: critical assessment of state-centred and decentred approaches', *Regulation & Governance*, **5**(2), 166–86.
Waltz, E. (2010), 'China's GM rice first', *Nature Biotechnology*, **28**(1), 8.
Xia, L., Y. Ma, Y. He and H.D. Jones (2011), 'GM wheat development in China: current status and challenges to commercialization', *Journal of Experimental Botany*, **63**(5), 1785–90.
Zhang, F., P. Cooke and F. Wu (2012), 'State-sponsored research and development: a case study of China's biotechnology', *Regional Studies*, **45**(5), 575–95.

9 Agricultural biotechnology in India: impacts and controversies

Matin Qaim

1 INTRODUCTION

India's agricultural research system had embarked upon modern biotechnology early on, with sizeable investments made by strong public and private sector research organizations. India was also one of the first countries in Asia to commercialize a genetically modified (GM) crop, namely insect-resistant cotton with inbuilt toxins from *Bacillus thuringiensis* (Bt). Bt cotton was first commercialized in India in 2002, and was rapidly adopted by cotton farmers in the following years. By 2012, over 7 million farmers had adopted Bt cotton on 27 million acres (James, 2012). Today, India has the largest area under Bt cotton worldwide. Since the technology's introduction, national cotton production has increased tremendously, mostly as a result of higher yields (Rao and Dev, 2010; Cotton Corporation of India, 2012). This sounds like a big success story for Bt cotton. Indeed, several peer-reviewed studies that analysed impacts confirmed sizeable benefits for adopting farmers (Morse et al., 2005; Bennett et al., 2006; Qaim et al., 2006; Subramanian and Qaim, 2010).

Nevertheless, public attitudes remain sceptical about the merits of Bt cotton technology. Anti-biotech interest groups reports have highlighted incidents such as failures of this technology in farmers' fields, exploitation of smallholder farmers through seed companies, and disruption of traditional cultivation practices (Sahai and Rahman, 2003; Shiva et al., 2011). There have also been repeated claims of a causal link between Bt cotton and farmer suicides observed in cotton belts of India (Coalition for a GM-Free India, 2012). Such claims contradict the evidence, but they are nonetheless perpetuated by the mass media in India and elsewhere. Negative opinions have also shaped public attitudes and policy-making processes. This goes far beyond the specific case of Bt cotton and also affects other potential GM crop applications. In 2009, the Indian Environmental Minister suspended the commercialization of Bt eggplant for an indefinite period of time. This moratorium was imposed even though Bt eggplant was declared safe after a careful review and could have produced large benefits for vegetable producers and consumers (Krishna and Qaim, 2008).

The controversy about GM crops in India continues, partly due to the success of anti-biotech groups in terms of lobbying political authorities. In August 2012, a high-level parliamentary panel launched a report concluding that GM crops would harm rather than benefit Indian agriculture (Committee on Agriculture, 2012). This report called for an immediate halt of all GM field trials. It builds heavily on NGO statements and claims about the alleged negative social consequences of Bt cotton for smallholder farmers (Bagla, 2012). A technical expert committee was appointed to further review the evidence with inputs from scientists before further political decisions are made.

2 BT COTTON ADOPTION AND IMPACTS

Bt cotton was commercialized in India in 2002. The varieties approved provide resistance to the cotton bollworm complex – a major insect pest. In the first year, around 90000 acres were planted with Bt hybrids (Qaim et al., 2006). In subsequent years the area under Bt increased substantially, reaching 27 million acres in 2012 – equivalent to 93 per cent of the total Indian cotton area (James, 2012). Also, the number of Bt cotton varieties increased considerably over time. In 2002, only three Bt hybrids, which were developed by the Indian seed company Mahyco and contained Monsanto's Bollgard I technology, were approved by the national regulatory authorities. In 2004 and 2005, three other Indian seed companies, which had sublicensed the Bollgard I technology, received approval for the commercialization of several additional Bt hybrids.

In 2006, the number of approved Bt hybrids increased further. In addition, new Bt events were deregulated by the national authorities, including Monsanto's Bollgard II technology and a range of technologies developed by public research institutes (Table 9.1). These were then backcrossed into hybrids from several local seed companies. In 2008, India's Central Institute for Cotton Research (CICR) released the first open-pollinated Bt variety based on an event developed by the Agricultural University in Dharwad. By 2012, over 1000 different Bt cotton varieties marketed by around 40 companies were available in India (James, 2012). This increase in the number of Bt events and varieties is positive for farmers, as it should offer greater effectiveness against a broader spectrum of insect pest species and better adaptation to different agroecological conditions.

In India, cotton is primarily grown by smallholder farmers with farm sizes of less than 15 acres and cotton holdings of 3–4 acres on average. Several studies have shown that Bt cotton adoption is associated with significant benefits in terms of reduced pesticide use and higher yields and profits (Qaim and Zilberman, 2003; Morse et al., 2005; Bennett et al., 2006; Qaim et al., 2006; Sadashivappa and Qaim, 2009; Rao and Dev, 2010). This is consistent with studies on the impacts of Bt cotton in other countries (Pray et al., 2002; Qaim, 2009; Carpenter, 2010; see also Chapter 38 in this volume).

Many of the early studies on Bt cotton impacts have important shortcomings, which may be one reason why the controversies continue. First, most of the evidence is based on data from the first few growing seasons after the commercial release of Bt varieties.

Table 9.1 Commercially approved Bt cotton events in India (2002–12)

No.	Event	Developer	Year of approval
1	MON-531 (Bollgard I)	Mahyco/Monsanto	2002
2	MON-15985 (Bollgard II)	Mahyco/Monsanto	2006
3	Event-1	JK Agri-Genetics	2006
4	GFM Event	Nath Seeds	2006
5	BNLA-601	CICR and Agricultural University Dharwad	2008
6	MLS-9124	Metahelix Life Sciences	2009

Source: Adapted from James (2012).

This is unsatisfying because it does not allow analysis of longer-term developments. For instance, resistance build-up in pest populations or growing importance of secondary pests may potentially lower Bt benefits over time. Second, most impact studies do not properly control for selection bias, which may occur when more successful farmers adopt the new technology earlier or more widely (Crost et al., 2007). As these successful farmers may have higher crop yields and profits anyway, this can lead to inflated benefit estimates. Third, most studies focus on agronomic impacts of Bt, such as yield and pesticide use effects, while economic effects, such as profit changes, are not analysed at all, or only based on simplistic comparisons. Fourth, and related to the previous point, many existing studies concentrate on impacts at the plot level, without considering possible broader welfare effects for farm households and rural development (Glover, 2010; Stone, 2011).

3 ADDITIONAL EVIDENCE FROM PANEL DATA

Most studies on Bt cotton impacts build on cross-section surveys of farmers with and without the technology. With cross-section data, impact dynamics cannot be analysed, and selection bias is difficult to control. This is different with panel data, where the same farms are surveyed in several rounds over a number of years.

A panel survey of Indian cotton farm households was carried out in four rounds between 2002 and 2008 using a multistage sampling procedure (Kathage and Qaim, 2012; Krishna and Qaim, 2012; Kouser and Qaim, 2011). At first, four states were purposively selected, namely Maharashtra, Karnataka, Andhra Pradesh and Tamil Nadu. These four states cover a wide variety of different cotton-growing situations and they produce 60 per cent of all cotton in central and southern India. In the four states, 10 cotton-growing districts and 58 villages were randomly selected, using a combination of census data and agricultural production statistics. Within each village, farm households were randomly selected from complete lists of cotton producers. The first-round survey interviews took place in early 2003, shortly after the cotton harvest for the 2002 season was completed. The same survey was repeated at two-year intervals in early 2005 (referring to the 2004 cotton season), early 2007 (referring to the 2006 season), and early 2009 (referring to the 2008 season). To our knowledge, this is the only longer-term panel survey of Bt cotton farm households in a developing country.

In total, 533 households were interviewed during the seven-year period. Most of these households were visited in several rounds. The total sample consists of 1431 household observations, 1085 Bt-adopting and 346 non-adopting households. During face-to-face interviews in all four rounds, household heads were asked to provide a wide array of agronomic and economic information, including input–output details on their cotton plots. Farmers who grew Bt and non-Bt cotton simultaneously provided details for both alternatives, so that the number of plot observations is somewhat larger than the number of farmers surveyed. The total number of cotton plot observations is 1655 over the four rounds.

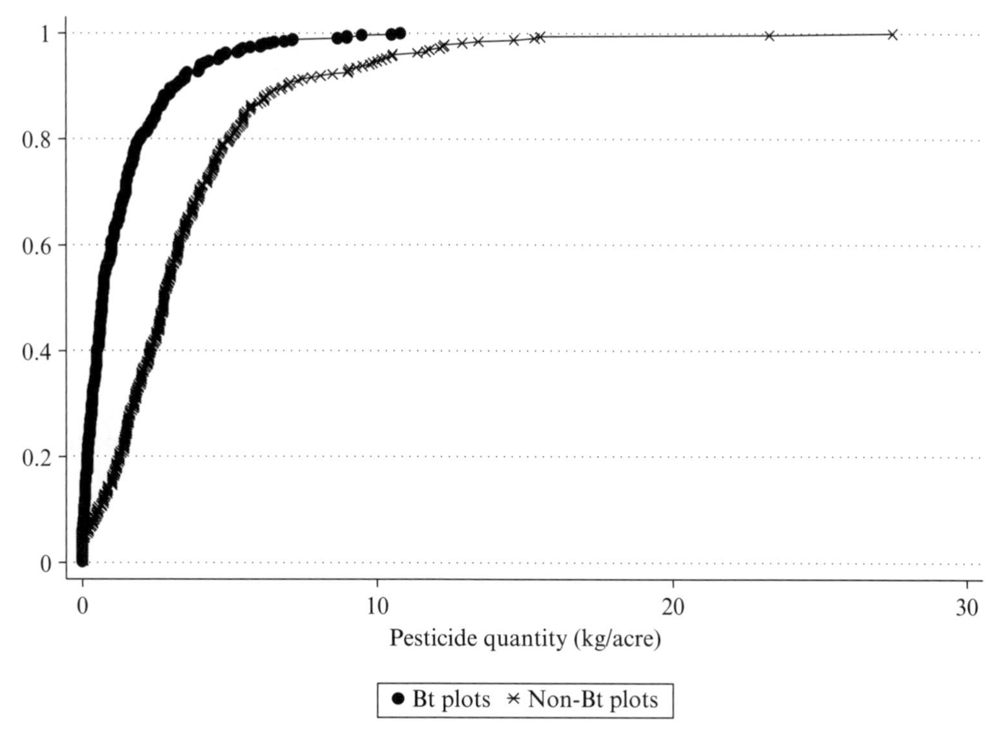

Source: Kouser and Qaim (2011).

Figure 9.1 Cumulative distribution of pesticide quantity used on Bt and non-Bt cotton plots

3.1 Impact of Bt on Pesticide Use

Building on these panel data, Figure 9.1 shows that farmers use significantly lower pesticide quantities on Bt plots than on non-Bt plots. This suggests that the technology is effective in reducing bollworm populations, so that the need for chemical pest control diminishes on Bt cotton plots.

Simply comparing pesticide use on Bt and non-Bt plots may be misleading, because there may be systematic differences in terms of local conditions (for example, irrigation or agroecological factors). Likewise, Bt-adopting and non-adopting households may differ systematically in terms of education, experience and other socio-economic variables. If this is the case, a simple comparison could overestimate the technology's net impact. This can be accounted for in regression analysis. Standard regression models can control for systematic differences to the extent that these differences are observed. However, unobserved heterogeneity between adopters and non-adopters can still lead to selection bias. In this connection, panel data are of particular advantage. Bt adoption is a time-variant variable; in the panel many households were surveyed before and after Bt adoption. Hence, a fixed effects estimator can be used. With fixed effects models, differencing within households eliminates time-invariant unobserved factors, so that they can no longer bias the impact estimates.

Table 9.2 *Net effects of Bt cotton technology on pesticide use per acre*

	(1) Pesticide quantity (kg a.i.)	(2) Pesticide cost (Rs)
$Bt^{2002-04}$ (dummy)	−0.954***	−879***
$Bt^{2006-08}$ (dummy)	−1.299***	−1284***

Notes:
Estimates are based on fixed effects panel regressions. Values shown can be interpreted as marginal effects. Control variables included in estimation are not shown here for brevity.
*** Statistically significant at the 0.01 level.

Source: Krishna and Qaim (2012).

Krishna and Qaim (2012) used fixed effects models to explain determinants of pesticide use. In one specification, pesticide quantity per acre measured in terms of active ingredients (a.i.) was used as a dependent variable, while in another specification pesticide cost measured in Indian Rupees (Rs) was considered. To identify net effects of Bt technology, Bt was included as an explanatory variable next to a number of other covariates. Two dummy variables represent Bt adoption: $Bt^{2002-04}$, which takes a value of 1 if Bt was adopted on a particular plot in the 2002–04 period; and $Bt^{2006-08}$, which takes a value of 1 if Bt was used in 2006–08.

The estimation results are summarized in Table 9.2. Column (1) confirms that Bt has reduced pesticide use significantly and that this effect has actually grown over time. While the net Bt impact was a reduction of 0.95 kg of pesticide a.i. per acre in 2002–04, the reduction was around 1.3 kg/acre in 2006–08. Relative to what has been sprayed without Bt in 2002–04, the net reduction was 37 per cent and 52 per cent in 2002–04 and 2006–08, respectively. The results in column (2) confirm the same patterns also when measuring pesticide use in terms of monetary costs. We conclude that Bt resistance development is not yet an issue of practical relevance in India.

When disaggregating pesticides by target pests, Krishna and Qaim (2012) found a slight increase in sprays against sucking pests, which are not controlled by the Bt toxins. This is not surprising because Bt reduced the use of broad-spectrum pesticides that harm both bollworms and sucking pests. So far, the pesticide-reducing effect of Bt is stronger than the increasing effect through secondary pests. It should also be stressed that broad-spectrum pesticides and those used specifically against bollworms are often much more toxic for the environment and human health than specific pesticides against sucking pests. Indeed, the largest pesticide reductions through Bt occur in the most toxic products categorized in hazard categories I and II (Kouser and Qaim, 2011).

Another interesting finding arises on non-Bt plots, where pesticide use was also reduced significantly over time (Krishna and Qaim, 2012). This can be interpreted as a spillover effect of Bt technology; widespread adoption and thus area-wide suppression of bollworm populations has allowed significant chemical pesticide reductions not only on Bt but also on non-Bt plots. Similar spillover effects of Bt technology have also been reported in China and the USA (Wu et al., 2008; Hutchison et al., 2010).

Smallholder cotton farmers in India apply pesticides manually with little or no protective clothing. In this situation, acute pesticide poisonings are commonplace. Using the same panel data and fixed effects regression models, Kouser and Qaim (2011) showed that Bt adoption has notably reduced the incidence of acute pesticide poisoning among cotton growers, especially in the 2006–08 period. Bt cotton helps to avoid several million cases of pesticide poisoning in India every year, which also entails sizeable health cost savings.

3.2 Impact of Bt on Yield and Profit

Comparing yield levels on the cotton plots in the panel data, Bt strongly outperformed non-Bt cotton throughout the different survey rounds. The higher yields are not due to higher yield potentials of Bt hybrids, but to more effective pest control and thus lower crop losses. Higher yields are also the main reason for much higher profits on Bt cotton plots. But again, this simple comparison between Bt and non-Bt plots cannot establish net effects of technology adoption, because there may be systematic differences between plots and farmers that need to be controlled for. Fixed effects panel models were used to control for observed and unobserved heterogeneity (Kathage and Qaim, 2012).

In these models, Bt was included in a slightly different way. A Bt dummy was included as treatment variable, which is 1 for a Bt plot in any particular year and zero otherwise. In addition, a change-over-time dummy was used, which takes a value of 1 if Bt was used in 2006 or 2008. While the Bt dummy indicates whether or not the technology has a positive net effect on cotton yield and profit, the change-over-time dummy reveals whether there are impact dynamics; if the Bt coefficient is positive and significant, and the change-over-time effect is statistically insignificant, then the technology causes benefits that do not change over time. A negative change-over-time effect would indicate shrinking benefits, whereas a positive change effect would reveal increasing benefits.

The estimation results are shown in Table 9.3. The positive and significant effect of Bt in column (1) indicates that Bt has a positive net impact on cotton yield. Controlling for all other factors, Bt has increased cotton yield by 126 kg per acre, which is equivalent to a 24 per cent gain over mean yields on non-Bt cotton plots. The change-over-time effect is

Table 9.3 Net effects of Bt cotton technology on yield and profit per acre

	(1) Yield (kg)	(2) Profit (Rs)
Bt (dummy)	125.90***	1877**
Bt change over time (dummy)	3.59	−260
2004 (dummy) [a]	125.39***	2066***
2006 (dummy) [a]	297.03***	5007***
2008 (dummy) [a]	208.61***	2332**

Notes:
Estimates are based on fixed effects panel regressions. Values shown can be interpreted as marginal effects. Additional control variables included in estimation are not shown here for brevity.
, * Statistically significant at the 0.05 and 0.01 level, respectively.
[a] The reference year is 2002.

Source: Kathage and Qaim (2012).

insignificant, indicating that the Bt yield effect was stable over time and did not increase or decrease in the later as compared to the earlier period. Also shown are year dummies that are all positive and significant, implying that overall yield levels in cotton have increased since 2002 on top of the Bt effect. This may be due to other innovations in the cotton sector. Since Bt adoption increased significantly over time, there is a close correlation between the Bt and the year dummies. When not including these year dummies in the model, the Bt change effect increases and turns significant, pointing at possible increases in Bt yield gains over time (Kathage and Qaim, 2012). The Bt effects shown in Table 9.3 with year dummies can therefore be interpreted as conservative estimates.

The estimates in column (2) of Table 9.3 indicate that Bt has increased profit by 1877 Rs per acre (US$ 38), equivalent to a 50 per cent profit gain over non-Bt cotton. The Bt impact per acre has not changed significantly over time, when year dummies are included. Similar to the yield effects, the profit gains increase significantly when the year dummies are not included in estimation. In any case, total cotton profits per farm rose over time, because farmers increased their Bt cotton acreage. Applying the per-acre effect of 1877 Rs to the average number of Bt acres on adopting farms, Bt added 5307 Rs (US$ 107) to annual farm-level cotton profits during 2002–04 and 10 524 Rs (US$ 213) during 2006–08. Nationwide, for the 27 million acres under Bt in 2012, this implies an annual net gain of over 50 billion Rs (US$1 billion) in cotton profits (see Chapter 45 in this volume for gains on soybeans in the US and Chapter 38 in this volume for cotton gains in Burkina Faso).

3.3 Impact of Bt on Household Living Standards

Cotton is often the major crop for cotton-producing households in India, so that profit gains through Bt technology are also likely to increase household living standards. A common way of measuring living standards in the development literature is to look at household consumption expenditures. Kathage and Qaim (2012) used a fixed effects specification of a consumption expenditure model. Their results suggest that Bt had no significant effect in the early adoption period, but it increased household living standards significantly in the 2006–08 period. This is plausible. While Bt-adopting households also had higher cotton profits during 2002–04, they did not immediately change their consumption behaviour but waited until they realized that the profit gains were sustainable. In the 2006–08 period, Bt-adopting households increased their annual consumption expenditures by 15 841 Rs (US$ 321) on average. Compared to non-adopters, this implies a net increase of 18 per cent, which confirms that Bt cotton has significantly raised living standards of smallholder farm households (Kathage and Qaim, 2012). Bt-adopting households spend significantly more on food than non-adopting households. Qaim and Kouser (2013) confirmed that Bt cotton technology has improved food security and dietary quality in India.

Using the first three rounds of these panel data and a village social accounting matrix, Subramanian and Qaim (2010) showed that Bt both increased the living standard of adopting farm households and had positive multiplier effects for landless households in the rural economy. These positive multiplier effects occur mainly through the rural labour market. Cotton is a labour-intensive crop. Higher yields through Bt technology entail employment generation on cotton farms and in downstream sectors such as transportation and trade. Subramanian and Qaim (2010) estimated that each dollar of direct profit gains for farmers through Bt adoption is associated with over 80 cents of indirect gains in the local economy. They also showed that 60 per cent of the total gains are captured by households with per capita incomes of less than $2 a day (a common World Bank threshold for moderate poverty). Bt technology has contributed to poverty reduction in the cotton-growing belts of India.

4 PERSISTENT CRITICISM

The panel data we collected in India are unique, and the results regarding the impacts of Bt are very robust. But they do not please the anti-biotech lobby – there has been significant effort to discredit these findings. A standard argument we hear is that our results could not be believed because the research was funded by biotech companies. This allegation is easy to spread, but it is wrong. Our research was entirely funded by public money and was not influenced by any private sector interests. Another argument is that Bt may work well in irrigated environments or for farmers with good access to input markets and extension services, but not for poorer farmers in marginal environments. This is a valid argument for some of the earlier impact studies that used cross-section data, but it does not apply to the more recent analyses with the panel data. The panel fixed-effects models not only account for observed differences between Bt and non-Bt plots and adopters, but they also control for any time-invariant unobserved factors. Fixed effects models explicitly consider differences within farms and households so that potentially confounding factors cancel out. Hence, the treatment effects reported are net impacts of Bt technology.

Critics also mentioned that the sample size of 533 cotton-producing households is too small for meaningful analyses in a large country like India. We agree that a larger sample size is always desirable, but also point out that these data with four rounds is probably the largest panel data base that has ever been compiled to analyse GM crop impacts in any developing country. The sampling framework was carefully developed, involving random selection in relatively diverse agroecological environments. Like any random sample, it is not able to represent every single situation, but it is representative on average. This is confirmed by comparing key variables with other studies and statistics.

Also related to data limitations is the argument that our observations stop in 2008, so that longer-term developments could not be assessed. It is true that simple extrapolation should be done with caution. On the other hand, we have shown that the Bt effects were stable or even increasing over a period of seven years, so it is unlikely that the benefits would abruptly stop after that period of consideration. Of course, potential problems like development of Bt resistance and secondary pest outbreaks need to be monitored and managed in the long run, but there are no indications that they would have become major issues in India up until now.

Looking at the statistics from the Cotton Corporation of India (2012), cotton yield levels seem to have stagnated since 2008 (Figure 9.2). These statistics have been used by some to argue that Bt cotton failed after a certain period of success (Coalition for a GM-Free India, 2012). Interestingly, data from the Agricultural Statistical Yearbook are different, suggesting that yields have further increased since 2008 (Ministry of Agriculture, 2011). Obviously, there are different statistics and it is unclear which ones are more

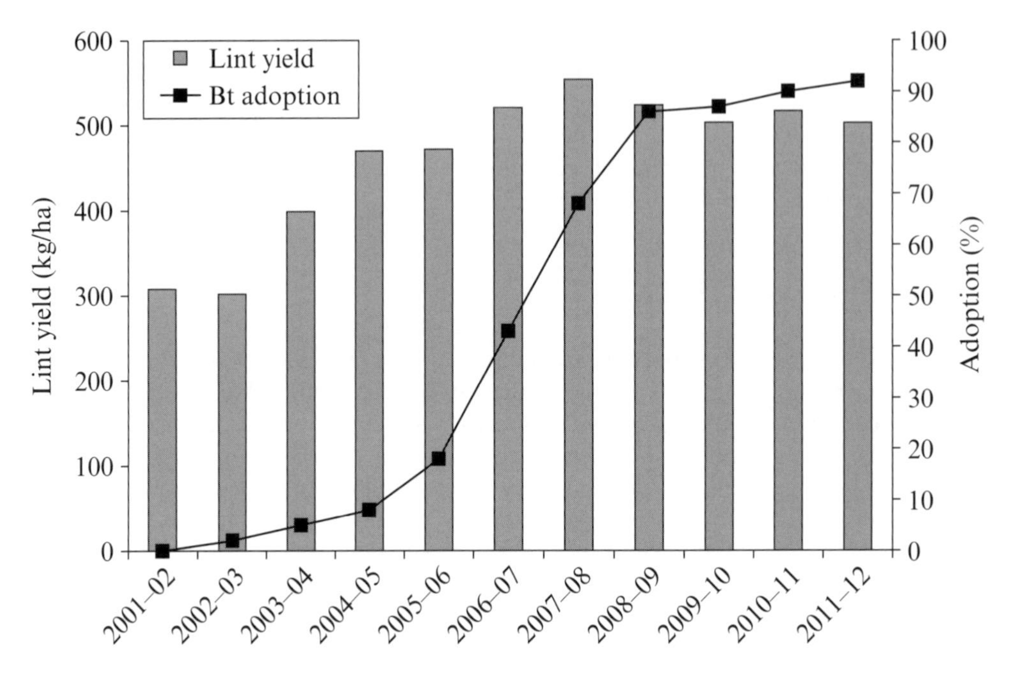

Source: Own presentation with data reported in Cotton Corporation of India (2012) and James (2012).

Figure 9.2 Developments of yields and Bt adoption in the Indian cotton sector

reliable. But even if yield levels have stagnated since 2008, this does not prove a failure of Bt technology. In 2008, almost 90 per cent of the cotton area in India was already under Bt, so that the yield gains over non-Bt cotton had already been realized (Figure 9.2). In this situation, Bt can hardly be expected to increase yields further, unless new technologies are being introduced. In connection with Table 9.3 above, it was shown that in the period between 2002 and 2008 there were also yield gains that were unrelated to Bt technology, probably due to other innovations in the cotton sector. Stagnating yields after 2008 could well suggest that such other innovations have not been sufficiently promoted in recent years. Sustainable yield growth certainly requires broad-based technological progress and cannot rely on Bt technology alone.

Finally, the farmer-suicide argument is still used widely and seems to be very powerful in the public debate. Indeed, thousands of suicides are observed every year among Indian farmers, many of which occur in the cotton-growing belts of the country. These suicides are tragic, but they seem to be caused by a variety of factors. Erratic rainfalls and related crop failures and indebtedness seem to be important components (Sheridan, 2009; Patel et al., 2012). But suicides among farmers in India have been reported long before Bt cotton was introduced, and the rates have not increased since 2002 (Figure 9.3). Hence, there is no direct link between suicides and Bt technology. What is also not widely known is that suicide rates in India are among the highest in the world. Suicides among farmers account for less than 10 per cent of all suicides in India (Gruere and Sengupta, 2011; Patel et al., 2012), suggesting that suicide rates among farmers are actually lower than in other population groups. The overall problem surely needs attention, but this is an issue

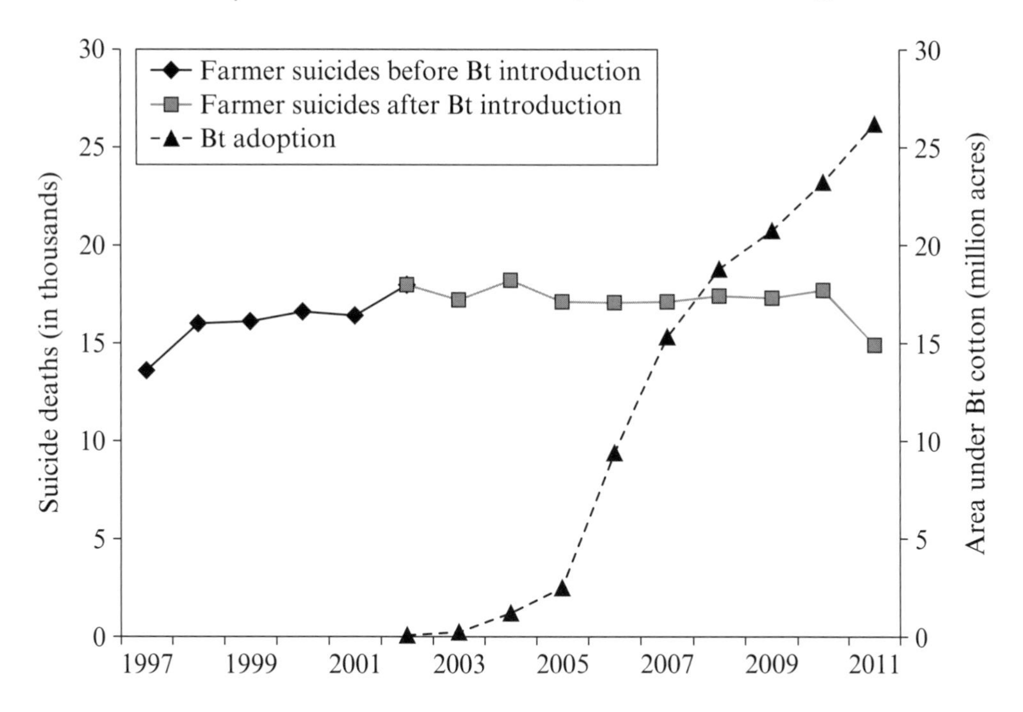

Source: Own presentation based on data reported in Gruere and Sengupta (2011), James (2012) and Hesselbarth (2013).

Figure 9.3 Bt cotton adoption and farmer suicides in India

that predates and extends far beyond GM crop policies. The argument in connection with Bt cotton is pure anti-biotech propaganda.

These facts have not yet changed the minds of anti-biotech groups, which are not open to dialogue and are not therefore willing to engage with reasoning and new scientific evidence of the kind presented in this chapter. Openness to evidence appears to be conditional on whether it purports to show negative effects of GM crops. The research reported here has been published in renowned academic journals after rigid peer-review, and has thus passed a widely recognized, if not definitive standard. As for any research, if there is justifiable criticism after publication, the usual procedure is to write a comment or research paper, which will also be peer-reviewed and published if deemed relevant. None of the direct criticism of our papers that we are aware of was published as a comment or peer-reviewed article in an academic journal; in our estimation, it would not have passed a peer-review process. Instead, the criticism and allegations are posted in Internet forums and newsletters without any quality control. Unfortunately, formation of public opinion does not differentiate between peer-reviewed research findings and unsubstantiated claims, and the Internet is a very powerful medium in reaching people worldwide.

If scientific evidence is not successful in terms of changing public opinion, one could at least trust farmers to make the right decisions for themselves (Herring and Rao, 2012). If farmers were not benefiting from Bt cotton, how could one explain that adoption rates have increased year after year, rapidly reaching a level of over 90 per cent (James, 2012)? Anti-biotech groups know that their line of argumentation is undermined with more and more farmers deciding in favour of GM crops. What these groups do as a countermeasure is to take advantage of individual farmers who have suffered crop failures due to drought or other problems and depict these failures as outcomes of Bt cotton adoption. The misfortune of individuals often makes good stories – so good that one easily forgets about the millions of other farmers who are extremely satisfied with Bt cotton and with how the technology has improved their lives.

5 CONCLUSION

The scientific literature shows that Bt cotton adoption has significantly reduced chemical pesticide use in India, while increasing farmers' yields and profits. Studies also show that farm household incomes and living standards have increased as a result of Bt cotton adoption. A substantial share of these benefits accrues to poor farm families. Moreover, there is no correlation between farmer suicides and Bt cotton adoption. Also, there is no indication that Bt causes any differential harm to the environment or human health. On the contrary, the reduction in chemical pesticide use has brought about sizeable positive environmental and health effects.

We acknowledge that cotton farmers in many parts of India suffer from erratic rainfalls and other conditions that can contribute to crop failures and social hardship. We also acknowledge that Bt technology is not a silver bullet for all these problems. But Bt is a solution to some long-standing serious pest problems, and farmers are clearly better off with this technology than without it. The scientific evidence for this conclusion is so strong that it can hardly be ignored. The benefits for poor farmers and consumers will further increase when GM food crops are developed and commercialized. This is currently being debated in India, and because India is one of the few countries where GM crops are widely used by smallholder farmers, many other countries are carefully observing developments and political decisions there. Abiding by the demands of activists and banning further GM crop trials in India could have serious negative implications for development-oriented agricultural biotechnology research worldwide.

REFERENCES

Bagla, P. (2012), 'Negative report on GM crops shakes government's food agenda', *Science*, **337**, 789.

Bennett, R., U. Kambhampati, S. Morse and Y. Ismael (2006), 'Farm-level economic performance of genetically modified cotton in Maharashtra, India', *Review of Agricultural Economics*, **28**, 59–71.

Carpenter, J.E. (2010), 'Peer-reviewed surveys indicate positive impact of commercialized GM crops', *Nature Biotechnology*, **28**, 319–21.

Coalition for a GM-Free India (2012), '10 Years of Bt cotton: false hypes and false promises: cotton farmers crisis continues with crop failures and suicides', Coalition for a GM-Free India, available at: http://indiagminfo.org/.

Committee on Agriculture (2012), 'Cultivation of genetically modified food crops: prospects and effects', New Delhi: Ministry of Agriculture.

Cotton Corporation of India (2012), '*Cotton statistics*', New Delhi: Cotton Corporation of India.

Crost, B., B. Shankar, R. Bennett and S. Morse (2007), 'Bias from farmer self-selection in genetically modified crop productivity estimates: evidence from Indian data', *Journal of Agricultural Economics*, **58** 24–36.

Glover, D. (2010), 'Is Bt cotton a pro-poor technology? A review and critique of the empirical record', *Journal of Agrarian Change*, **10**, 482–509.
Gruere, G. and D. Sengupta (2011), 'Bt cotton and farmer suicides in India: an evidence-based assessment', *Journal of Development Studies*, **47**, 316–37.
Herring, R.J. and N.C. Rao (2012), 'On the "failure of Bt cotton": analyzing a decade of experience', *Economic and Political Weekly*, **XLVII**(18), 45–54.
Hesselbarth, K. (2013), 'The deadly myth of Bt cotton', *AgriFuture*, Spring, 8–10.
Hutchison, W.D., E.C. Burkness, P.D. Mitchell, R.D. Moon, T.W. Leslie, S.J. Fleischer, M. Abrahamson, K.L. Hamilton, K.L. Steffey, M.E. Gray, R.L. Hellmich, L.V. Kaster, T.E. Hunt, R.J. Wright, K. Pecinovsky, T.L. Rabaey, B.R. Flood and E.S. Raun, (2010), 'Areawide suppression of European corn borer with Bt maize reaps savings to non-Bt maize growers', *Science*, **330**, 222–5.
James, C. (2012),. 'Global status of commercialized biotech/GM crops: 2012', *International Service for the Acquisition of Agri-Biotech Applications (ISAAA) Brief*, 44, Ithaca, NY: ISAAA.
Kathage, J. M. Qaim (2012), 'Economic impacts and impact dynamics of Bt (*Bacillus thuringiensis*) cotton in India', *Proceedings of the National Academy of Sciences USA*, **109**, 11652–6.
Kouser, S. and M. Qaim, (2011), 'Impact of Bt cotton on pesticide poisoning in smallholder agriculture: a panel data analysis', *Ecological Economics*, **70**, 2105–13.
Krishna, V.V. and M. Qaim (2008), 'Potential impacts of Bt eggplant on economic surplus and farmers' health in India', *Agricultural Economics*, **38**, 167–80.
Krishna, V.V. and M. Qaim (2012), 'Bt cotton and sustainability of pesticide reductions in India', *Agricultural Systems*, **107**, 47–55.
Ministry of Agriculture (2011), 'Agricultural statistics at a glance, 2011', New Delhi: Directorate of Economics and Statistics, Ministry of Agriculture.
Morse, S., R.M. Bennett and Y. Ismael (2005), 'Genetically modified insect resistance in cotton: some farm level economic impacts in India', *Crop Protection*, **24**, 433–40.
Patel, V., C. Ramasundarahettige, L. Vijayakumar, J.S. Thakur, V. Gajalakshmi, G. Gururaj, W. Suraweera and P. Jha (2012), 'Suicide mortality in India: a nationally representative survey', *The Lancet*, **379**, 2343–51.
Pray, C.E., J. Huang, R. Hu and S. Rozelle (2002), 'Five years of Bt cotton in China – the benefits continue', *Plant Journal*, **31**, 423–30.
Qaim, M. (2009), 'The economics of genetically modified crops', *Annual Review of Resource Economics*, **1**, 665–93.
Qaim, M. and S.Kouser (2013), 'Genetically modified crops and food security', *PLOS ONE*, **8**(6), e64879.
Qaim, M., A. Subramanian, G. Naik and D. Zilberman (2006), 'Adoption of Bt cotton and impact variability: insights from India', *Review of Agricultural Economics*, **28**, 48–58.
Qaim, M. and D. Zilberman (2003), 'Yield effects of genetically modified crops in developing countries', *Science*, **299**, 900–902.
Rao, N.C. and M. Dev (2010), *Biotechnology in Indian Agriculture: Potential, Performance and Concerns*, Hyderabad: Center for Economic and Social Studies.
Sadashivappa, P. and M. Qaim (2009), 'Bt cotton in India: development of benefits and the role of government seed price interventions', *AgBioForum*, **12**, 172–83.
Sahai, S. and S. Rahman (2003), 'Performance of Bt cotton in India: data from the first commercial crop', New Delhi: Gene Campaign.
Sheridan, C. (2009), 'Doubts surround link between Bt cotton failure and farmer suicides', *Nature Biotechnology*, **27**, 9–10.
Shiva, V., D. Barker and C. Lockhart (2011), *The GMO Emperor has No Clothes*, New Delhi: Navdanya International.
Stone, G.D. (2011). 'Field versus farm in Warangal: Bt cotton, higher yields, and larger questions', *World Development*, **39**, 387–98.
Subramanian, A. and M. Qaim (2010), 'The impact of Bt cotton on poor households in rural India', *Journal of Development Studies*, **46**, 295–311.
Wu, K.M., Y.-H. Lu, H.-Q. Feng, Y.-Y. Jiang and J.-Z. Zhao (2008), 'Suppression of cotton bollworm in multiple crops in China in areas with Bt toxin-containing cotton', *Science*, **321**, 1676–78.

10 Oceania: Australia, New Zealand, Japan and the Philippines

Karinne Ludlow and Jose Yorobe Jr

Physically isolated from other nations and relatively free from external influence, the Oceania countries have responded to GM crops with comprehensive and individualistic GM regulatory schemes. These distinctive regulatory schemes arise from many factors. The physical isolation of their land masses led, particularly in Australia and New Zealand, to the development of highly distinctive flora and fauna and a history of concern about environmental releases of new organisms. That isolation also provided a competitive marketing advantage for their products as being free from particular pests, or in more recent times GMOs, which was not to be discarded easily. Particularly in the case of New Zealand, the development of a strong organic farming sector provided a push against GM crops. Finally, there is an expectation in these nations of public participation in GM policy development and that governments address socio-economic concerns. Internationally, all are members of the WTO, and Japan, the Philippines and New Zealand are members of the Cartagena Protocol on Biosafety; Australia is not. All except the Philippines also adhere to either UPOV 1991 or 1999, and the Philippines has implemented its own plant variety protection legislation.

AUSTRALIA

Australia and the other Oceania nations are island countries in the Pacific Ocean. Australia, the largest in landmass of the four nations, lies in several climatic and geographic zones, with agricultural production possible in all except the desert middle zones. The crop industries make a significant contribution to the Australian economy, and Australia is a net exporter of food and feed. Per annum, the grains and oilseeds industry (subject to seasonal variation) produces about 35–45 million tonnes of grain from an area of more than 20 million hectares, with an annual gross value of production (GVP) of around A$9–13 billion (US$9.39–13.56 billion) (Australian Government, DAFF). Australia is the world's third largest canola exporter, and cotton is also a major contributor to its economy, with a GVP exceeding A$2 billion (Australian Government, DAFF). Australia's place in the southern hemisphere also means some plant breeders can use the opposite production season to accelerate seed production and make up seasonal shortfalls in their own production (New Zealand Grain and Seed Trade Association).

Australia is a constitutional monarchy with a parliamentary democracy. It is a federation of eight states and territories with three tiers of government: federal, state and over 600 local governments. The federal model of government used in Australia means that the state parliaments have general powers on all matters except where the Australian Constitution has expressly withdrawn a specific power from the states or vested the power exclusively on the federal parliament. Responsibility for agriculture is shared between the different levels of government, with the federal parliament responsible for setting general policy on international trade and the state parliaments responsible for most other matters. This means that to achieve a nationally consistent approach to regulation on a state matter, such as the regulation of gene technology, the state, territory and federal governments must reach agreement. Federal legislation is then enacted, which is mirrored in counterpart legislation in each state and territory. Australian politics is dominated by two main political parties, although the number of parliamentarians from green and independent backgrounds has risen in recent elections. Voting is compulsory in Australia, with members of the lower house, called the House of Representatives, elected through preferential voting and proportional voting for the upper house, which is called the Senate.

English is the predominant language in Australia. The Australian indigenous people, Aborigines and Torres Straits Islanders, make up about 3 per cent of the Australian population and speak a number of different indigenous languages. There is no provision for Australian indigenous people to be specifically represented in parliament or for their views to be specifically considered under the GM regulatory scheme.

History of GM Adoption

Australia approved the commercial release of its first GM plant, GM carnations with novel flower colours, in 1995. GM cotton has been grown in Australia since 1996 and nearly all Australian cotton is now GM, except for the small acreage of non-GM cotton needed to provide refuges and buffer zones. Other GMOs have been approved for experimental or commercial release, including GM herbicide-tolerant canola in 2003. As discussed below, however, moratorium legislation in the canola-growing states blocked commercial plantings until 2008, when most moratoriums were lifted. Approvals for commercial release of GM corn and soy varieties have not been sought.

GM Regulatory Scheme

Australia had a voluntary self-regulatory system for GM research from the mid-1970s until the national GM regulatory scheme was introduced on 21 June, 2001. The centrepiece of that scheme is the national Gene Technology Act 2000 (GT Act) (adopted in each state and territory, other than the Northern Territory, by mirroring state legislation) which establishes the Gene Technology Regulator (GTR). The GTR heads the national Office of the GTR (OGTR), a federal regulatory agency.

All dealings with GMOs in Australia are prohibited *unless* authorized under the GT Act. The term 'dealings' includes most uses of GMOs, including research, release, import and commercial use. The Act, amongst other things, also creates a publicly available database of all GMOs and GM products approved in Australia. GMO dealings are authorized only if they fall within one of four defined categories and relevant regulatory conditions are satisfied. Approval conditions become more onerous as perceived risk increases.

The most rigorous process for authorization is where a licence involving intentional release of GMOs into the environment is sought. Such cases require, for example, consultation with the public, the states and other federal government authorities such as the Environment Minister. The GTR must also be satisfied that any risks posed by the proposed dealing can be managed in a way that protects human health and safety and the environment. Whilst it was originally envisaged that social and ethical issues would also be considered during this process, and public consultation during the scheme's creation showed the Australian public considered ethical issues equally important to safety and environmental concerns, the final regulatory scheme does not address these matters. Importantly, possible trade and marketing ramifications of GMO releases on other forms of agriculture, such as non-GM agriculture, are not considered. With respect to the assessed risks, they are assessed in the context of the risks posed by the non-modified parental organisms. Accordingly if the risks posed by a GMO are no greater than those posed by the conventional version of the organism, approval is likely.

Since the establishment of the Australian national scheme, there have been calls from some states for socio-economic considerations to be added to the GTR's risk assessment process. The scheme was also independently reviewed in 2006. That review concluded that the existing scope of the Act should be maintained. The limited scope of the GTR's assessment process and the proposed commercial release of GM canola in 2003 were important in the states' decision to introduce their own legislation dealing with socio-economic concerns. This action resulted in moratoriums on certain GMO releases in every state and territory except Queensland and the Northern Territory where canola cannot be grown. The purpose of these moratoriums was to preserve the identity of non-GMOs for marketing purposes and provide time to consider the socio-economic ramifications of GMO releases. That legislation remained in force in 2012 but the moratoriums had expired in all states except South Australia and Tasmania, where the ban is expected to continue until at least 2014. In Tasmania, all dealings with GMOs, including research, are expressly prohibited unless licensed by the GTR *and* authorized under state legislation. The South Australian ban applies, in contrast, only to cultivation of GM food crops, with 'food crop' being defined to include crops intended for animal consumption. A new statutory liability provision designed to compensate those harmed by the cultivation of a GM crop has also been created in South Australia and Western Australia. The differing legislative response and uptake of GM crops between adopting states partly reflects the differing needs of the states – Western Australia, for example, has a much greater weed problem than other states, which leads to greater concern about the development of herbicide tolerance – and because states differ in the crops grown and their export markets.

Food and Feed

Australia and New Zealand share a common food regulator, Food Standards Australia New Zealand (FSANZ). Since May 1999 FSANZ must assess GM foods prior to initial commercial release onto the Australian or New Zealand market. There are also specific regulations concerning labelling of such food.

'Standard 1.5.2 – Food Produced Using Gene Technology' of the Australia New Zealand Food Standards Code applies to all food produced using gene technology. Such food is defined as food derived or developed from an organism modified by gene technology. Gene technology is defined as 'recombinant DNA techniques that alter the heritable genetic material of living cells or organisms'. This does not include food derived from animals which have been fed feed produced using gene technology, unless the animals are themselves products of that technology. Neither does it include food from organisms that have been given GM agents, such as GM veterinary products, via non-genetic routes.

The Standard has two Divisions: Division 1, which concerns the sale and use of foods covered by the Standard; and Division 2, which provides for the labelling requirements. It is unlawful to sell or use as an ingredient or component any foods produced using gene technology unless specifically approved. Approval requires a risk-based case-by-case safety assessment by FSANZ and approval by the high-level intergovernmental committee which oversees food regulation, the Legislative and Governance Forum on Food Regulation. The food must also comply with any listed special conditions on the sale of the food, such as special labelling. For example, to get approval for sale, the GM food must have been determined to be at least as safe as its traditional counterpart. However, if it contains a factor known to cause an allergic reaction in some part of the population, it may need to be labelled appropriately. The GM trait involved may also raise significant ethical, cultural or religious concerns regarding the origin of the genetic material used in the GM. In that case additional labelling requirements may be specified.

The general labelling requirements for GM food are set out in Division 2. Food that is, or contains as an ingredient, including a processing aid, a food produced using gene technology which contains novel DNA and/or novel protein or has altered characteristics must be labelled. Novel DNA and/or novel protein is DNA or protein which, as a result of the use of gene technology, is different in chemical sequence or structure from DNA or protein present in counterpart non-GM food. Altered characteristics are present if:

- the GM has resulted in one or more significant changes in composition or nutritional parameters of the food, outside the normal range of values for existing non-GM counterpart food;
- there are significant differences in levels of anti-nutritional factors or natural toxicants in the food compared to its existing non-GM counterpart;
- the food contains a new factor known to cause an allergic response in particular sections of the population; or
- the intended use of the GM food is different from the existing non-GM counterpart food.

From 7 December 2001 all GM food package labels must include the statement 'genetically modified' in conjunction with the name of that food, ingredient or processing aid. For GM foods not sold in packaging, such as fruit and vegetables, the required information must be displayed on or in connection with the display of the food. For non-GM foods, no statement as to genetic status is required. If producers choose to label such products, for example as being GM-free, they must be able to verify the truth of that statement or risk penalties under consumer protection laws. This may require the method used to verify that status to be disclosed on the label.

Exemptions from the labelling requirements apply to:

- highly refined food without altered characteristics where the refining process removes the novel DNA and/or novel protein;
- processing aids and food additives except where novel DNA and/or novel protein from them is present in the final food;
- flavours present in concentrations less than or equal to 0.1 per cent in the final food; and
- food intended for immediate consumption prepared and sold from premises and vehicles, such as take-away food, food in restaurants or airline food.

Food, ingredients and processing aids can contain up to 1 per cent of *unintended* presence of approved GM product without requiring labelling. However, there is a zero tolerance for unapproved GM food, even where a GM food is exempt from labelling requirements.

Feed is regulated separately but GM and non-GM feed is not distinguished. Provided the feed is safe and will not cause harm to the animal (and the GM regulatory scheme regarding import is satisfied), it can gain entry.

NEW ZEALAND

New Zealand is the southernmost of the Oceania nations. This position means that crops such as canola, soy or cotton are not produced to any significant extent. New Zealand is a net exporter of food and feed, and agriculture is a major part of the New Zealand economy, with NZ$25.9 billion (US$21.06 billion) of agricultural products' exports, particularly dairy products and meat, for the year ending June 2011 (Statistics New Zealand, 2011). Pasture seed and vegetable seed are also important export items, particularly to the United States, Australia, Europe and Japan (New Zealand Grain and Seed Trade Association). Grain and oilseeds form a very small part of New Zealand's agriculture. Like Australia, its place in the southern hemisphere means some plant breeders can use the opposite production season to accelerate seed production and make up seasonal shortfalls in their own production (New Zealand Grain and Seed Trade Association).

New Zealand is a constitutional monarchy with a parliamentary democracy. In addition to the national parliament, it has 11 regional councils and 67 territorial authorities for local government purposes. New Zealand has a proportional representation form of government, called Mixed Member Proportional (MMP), whereby each voter has two votes – one for the electoral seats and the second for a party. It has been suggested that the MMP form of government means political parties are less able to dominate the legislative process, and political focus on discrete issues is increased. This has allowed a sustained focus on GM issues and ultimately led to the New Zealand Royal Commission on GM, discussed below (Hope, 2002).

English is the predominant language in New Zealand although the Maori language is also an official language and the second most widely spoken (Statistics New Zealand). Maori make up 14.6 per cent of the New Zealand population (Statistics New Zealand) and have seven seats guaranteed for them in the parliament.

History of GM Adoption

There has been only one approved commercially released GMO in New Zealand, a GM equine influenza vaccine. Because there is no commercial GM crop production in New Zealand, there has been no real experience in the functionality of the New Zealand regulatory scheme for GM crop releases. The low number of approved GM events is attributed to 'the costly, lengthy and unproven nature of the regulatory approval process' (USDA, New Zealand, 2011). A further explanation for the low number of approvals is thought to be due to some GM trials being conducted offshore, in Australia or other places rather than attempting to comply with the relatively untested New Zealand regulatory approval process. Furthermore, the important GM crops such as corn, soybeans and cotton are not commercially grown in New Zealand and the relatively few pests in the country translates into less uptake of first generation GM crops. There have, however, been some field tests, imports and development of GMOs in containment. As of July 2011, 19 contained field trials had been approved.

GM Regulatory Scheme

In 1978 an indefinite moratorium on GM experiments outside strict containment was imposed on public institutions. In 1986 the Minister for Science and Technology established a working party to report on how best to regulate GM field testing and release – it recommended a statutory committee (Hope, 2002). The Minister for the Environment was given responsibility for establishing this committee because he was already acting to reform legislation relating to hazardous substances and new organisms, GMOs being considered a class of new organism. An interim committee was established in 1988 and the moratorium lifted. The committee only had authority over public sector researchers but private sector researchers were asked to comply voluntarily (Hope, 2002). The first GMO field trial, for GM potatoes, occurred in 1988–89 (Hope, 2002). The committee was replaced in 1998 by the legislative scheme discussed below.

Growing public opposition to GMOs, particularly GM crops and food, and dissatisfaction with the response to this opposition by government agencies, led to the establishment of a Royal Commission on Genetic Modification (Hope, 2001) which reported in 2001 (New Zealand, Royal Commission, 2001). The Commission has been described as an international-high water mark for public participation in GM policy development (Hope, 2001), with more than 10000 written submissions being made. Public concern focused on environmental safety and economic concerns about possible commercial impact of any 'contamination' of non-GM crops, both in a physical sense and in consumer perception (Hope, 2001). While the Commission completed its work, there was a two-year voluntary moratorium both on GMO releases into the environment (except vaccines and some human health products) and on field testing (Hanley and Elborough, 2004). The moratorium ended in October 2003.

Since 1 July 1998 the development, import, field testing or intentional release of GMOs has not been permitted in New Zealand without approval by the Environmental Protection Agency (EPA). Approvals are given on a case-by-case basis under the Hazardous Substances and New Organisms Act 1996, although the risk assessment details are generally set out in the Hazardous Substances and New Organisms (Methodology)

Order 1998. Some low-risk contained activities can be carried out under more streamlined processes. Two types of releases are permitted – conditional releases where controls are imposed and unconditional releases where no controls are imposed if there are no potential risks needing management. Applications with '(a) significant cultural, economic, environmental, ethical, health, international, or spiritual effects; or (b) significant effects in an area in which the Authority lacks sufficient knowledge or experience' may be called in and decided by the Minister for the Environment.

The New Zealand GM regulatory scheme is relatively unique because adverse effects are weighed against beneficial effects on a case-by-case basis. 'Effects' includes potential, probable and future effects. This contrasts with the other Oceania schemes where only risks are considered rather than the potential benefits. During the 2006 review of the Australian scheme most submissions thought expansion to include consideration of benefits would be impracticable or undesirable, first, because 'the existence or scale of many benefits did not become apparent for some years after the GMO was released' and secondly because 'it would be very difficult to construct a calculus for measuring risk and benefit in the same time frame and dimension' (IRGTA, 2006, chap. 3, p. 2). GMO opponents also argued that consideration of benefits may result in presumed benefits outweighing risks, while GMO proponents expressed concerns that it would be seen to compromise the GTR's scientific approach to risk assessment (IRGTA, 2006). The review concluded that the Australian GTR's risk assessment process should not be expanded to a risk–benefit assessment (IRGTA, 2006).

The New Zealand GM regulatory scheme also provides for social and cultural as well as scientific and economic values to be considered (as does the Philippines' scheme, below) and makes input by Maoris and the consideration of their spiritual beliefs when making decisions mandatory. Whakapapa (genealogy) is a central notion in Maori culture and defines the 'basic structure of relationships between generations and between species'. Science and industry's emphasis of the importance of genes in determining an organism's characteristics to illustrate the importance of GM technology may have unwittingly generated opposition from Maoris on the basis of interference with whakapapa (Hope, 2002). There have also been difficulties in consulting with Maoris because no one organization can speak on behalf of all.

The EPA begins with consideration of the scientific evidence and then examines evidence relating to other values and matters. To illustrate these other values, in an application regarding the development of GM cows (NZ Environmental Risks Management Authority Decision, 2001), ethical risks such as risks to animal welfare, risks arising from using companion animals for GM experimentation and risks arising from contravening the Bible were discussed. Economic risks such as potential damage to New Zealand's clean green image and risks to tourism were also considered. Benefits were also widely defined, including increased scientific knowledge and skills and indirect benefits 'arising from capacity building due to retention of intellectual capital in [NZ] and the enhancement of [NZ's] international reputation in science' (Wheen, 2004, p. 162).

Both the Australian and New Zealand schemes refer to precaution – but New Zealand legislation explicitly states that a precautionary approach shall be taken into account where there is relevant uncertainty, while the Australian legislation provides that precautions to prevent environmental degradation can be taken provided they are cost-effective and serious or irreversible environmental damage is threatened.

Food and Feed

New Zealand shares a food regulator with Australia. Feed does not have to be labelled as GM or non–GM although GM feed must comply with the regulatory scheme described above.

JAPAN

Only 15 per cent of Japan's land is suitable for cultivation and agriculture is subsidized and protected. Small amounts of canola, soy, corn and rice are grown there for domestic markets but Japan is both a net importer and the world's largest per capita gross importer of GM food and feed (USDA, 2010).

Japan, another constitutional monarchy with a parliamentary democracy, in addition to its national Diet, has 47 prefectures and many administrative sub-prefecture regions, adding to its administrative complexity. The Diet has authority regarding GMO approval. Article 7 of the Basic Law on Food Safety No. 48 provides that '[l]ocal governments shall be responsible for formulating and implementing policies to ensure food safety that corresponds to the natural, economic, and social conditions of the area under their jurisdiction, on the code of the basic principles and on the basis of an appropriate sharing of roles with the state.' Therefore at the prefecture and even at the municipal level there are local rules relating to agricultural biotechnology. Most, if not all, of these rules are political responses to popular concerns, and are not science based (USDA, 2010). These can set larger minimum planting distances between GM and conventional crops or require additional approval for planting GM crops from the prefecture. Such local government rules are seen as further discouraging Japanese farmers from growing GMOs (USDA, 2011).

Japan's population is nearly entirely culturally and linguistically Japanese. Japanese consumers are given responsibility, by law, to express their opinions on food safety policies (Basic Law on Food Safety).[1]

History of GM Adoption

Japan was an early adopter of agricultural biotechnology products. The first food from a GM plant was approved for sale in Japan in 1996; GM roses have been cultivated there since 2009. The import of GM papaya was approved in 2011, with the first shipment arriving in December 2011 (see Chapter 40 in this volume for more on this). However, although over 95 events in 7 crops have been approved for environmental release (USDA, 2011; the following is largely drawn from the USDA report), as of September 2011 no GM food crop has been cultivated in Japan. GM crop research is undertaken in Japanese public institutions, particularly for rice and soybeans but, due in part to regulatory costs and delays, that research is unlikely to be commercialized in Japan. Public institutions do not have the regulatory expertise or money for approval and the size of the seed market does not justify Japan-specific biotech product development.

Whilst large quantities of GM and non-segregated corn are imported by Japan, this has mostly been used for feed. Of the food corn, most is used in ways not requiring labelling of the end product, mostly as processed ingredients such as starch and sweeteners. Similarly for soybeans, soybean oil does not require labelling and therefore GM soybeans are imported for oil. But soy products for direct consumption, such as tofu and natto, must be labelled; as a result, food use of GM soybeans is small and limited to products not subject to mandatory labelling, such as soy sauce. Nevertheless, more recent data shows declining demand for non-GMO corn for food because of 'high premiums for segregated "Non-GMO" corn and a lack of end-user opposition to biotech ingredients' (USDA, 2011). Even where mandatory labelling requirements apply, non-segregated food-use corn is being used quite widely (USDA, 2011).

GM Regulatory Scheme

The Japanese GM regulatory system is commonly seen as complex and costly, but also predictable and functional. Consumer concerns are frequently raised as the reason for this complexity (USDA, 2010). Nevertheless, the expense of compliance with the regulatory scheme is judged by many as the reason that Japan GM research has not progressed to the field trial stage (USDA, 2010).

Commercialization of GM crop plants requires separate environmental, food and feed approvals. Four ministries are involved in the regulatory framework: the Ministry of Agriculture, Forestry and Fisheries (MAFF) (responsible for feed and environmental safety); the Ministry of Health, Labour and Welfare (MHLW) (responsible for food safety of GM products); the Ministry of Environment (MOE) (responsible for biodiversity protection, together with MAFF); and the Ministry of Education, Culture, Sports, Science and Technology (MEXT) (responsible for approval of experimental developments). The Consumer Affairs Agency (CAA) established in 2010, protects and enhances consumer rights.

Any introduction or development of new GMOs into Japan first requires approval from the MAFF and is regulated under the Law Concerning the Conservation and Sustainable Use of Biological Diversity through Regulations on the Use of Living Modified Organisms (Law no. 97 2003). Under this Law, LMO use is divided into two groups: Type 1 and Type 2 (Yamanouchi, 2005). Type 2 concerns use where measures are taken to prevent escape into the environment and is not generally relevant to field testing, or the cultivation or import, of GM crops (these are usually contained experiments). Type 1 concerns GMO use without measures preventing environmental dispersal (Yamanouchi, 2005).

Impact on biodiversity is assessed by the Committee for Evaluation of Biological Diversity Effects (CEBDE), within the MAFF. The CEBDE considers competitive superiority, potential production of toxic substances and cross-pollination (USDA, 2010). As part of the biodiversity impact assessment seed companies must conduct field testing in Japan in isolated plots (USDA, 2011), an aspect of Japanese regulation criticized by some as being unnecessary (USDA, 2011). In any case, lack of cultivatable land makes it difficult to access such plots (USDA, 2011). A 2004 MAFF coexistence guideline also requires detailed information to be made public through web pages and meetings with local residents and the establishment of buffer zones (USDA, 2011).

MAFF also monitors the quality and safety of imported feed ingredients at the ports. A sampling programme of import shipments and processed food products at the retail level is used to ensure compliance with all the domestic rules about introduction and use of GM crops and foods. Imported grains, oilseeds and foods are tested at ports by the Ministry of Health, Labour and Welfare (MHLW). Testing is done at ISO 17025 US and Japanese accredited laboratories by the government agency, Food and Agricultural Materials Inspection Center (FAMIC).

Food and Feed

All GM foods and feed are assessed for safety prior to being granted certification for distribution to Japanese domestic markets. The safety assessment of GM food (and feed) has been mandatory since April 2001, following a revision of the standards for foods and food additives under the provisions of the Food Sanitation Law (FSL). Pursuant to the FSL, the Ministry of Health, Labour and Welfare (MHLW) is responsible for food safety of all GM food products. The food safety risk assessments are performed by the Food Safety Commission (FSC). The standards used are set out in the 'Standards for the Safety Assessment of Genetically Modified Foods (Seed Plants)' and are based on the Codex standards. Japan has a zero tolerance for unapproved GM food.

All GM feed must be approved for feed safety by the Ministry of Agriculture, Forestry and Fisheries (MAFF). Feed safety is regulated by the Law Concerning the Safety and Quality Improvement of Feed and safety assessments consider significant changes in feed use compared with existing traditional crops and the potential for the production of toxic substances. MAFF may set an exemption of a 1 per cent tolerance for unintentional commingling of GM products in feed if that product is approved in other countries, but not yet in Japan. However, the exporting country must be recognized by the MAFF minister as having a safety assessment programme equivalent to or stricter than Japan's (USDA, 2011). Issues such as environmental, ethical, moral and socio-economic aspects of the research and development, production and marketing of GM foods or feeds are irrelevant to the food and feed safety assessments.

Two laws (with essentially identical requirements) regulate labelling of GM foods: the Food Sanitation Law (FSL) and Japan Agricultural Standards (JAS). Food labelling falls under the authority of the Consumer Affairs Agency (CAA). There are three possible GM food label claims: Non-GMO, GMO and Non-Segregated. Non-GMO labelling is voluntary, but to use the label the commodity must be able to be shown to have been identity preservation (IP) handled from production through to processing. Under the JAS law, Japan has an informal tolerance of 5 per cent for GM ingredients in products labelled Non-GMO. However, this only applies to events approved in Japan.

Under FSL, if the GM content of any of the top three ingredients by weight in a food, or if the weight of the GM ingredient is more than 5 per cent of total weight of the food, they must be labelled with either the phrase 'Biotech Ingredients Used' or 'Biotech Ingredient Not Segregated' (USDA, 2011, section IV). Non-segregated products are those in which the identity has not been preserved and therefore are assumed to be primarily derived from GM varieties (USDA, 2010, section II).

THE PHILIPPINES

The Philippine economy is predominantly agricultural, devoting 32 per cent of the total land area to agriculture and employing 33 per cent of the population. Rice and corn are the major staple crops cultivated for food and feed, but production remains insufficient for domestic needs.

The Philippines has a democratic system of government with three equal branches: executive, legislative and judiciary. Its political system is pluralist and based on a US-style presidential system characterized by a strong adversarial culture in politics where powerful interest groups, mostly family clans, fight for political influence and scarce public resources (Aerni, 2001). The executive branch has national responsibility for agricultural biotechnology under the Department of Science and Technology (DOST). Local government units, such as provinces and municipalities, have a legal mandate to pursue policies on GMOs independent of the national thrusts.

The Philippines has a rich cultural diversity distinguished by race, religion and ethnicity and is influenced by its colonized past, having been both a US and Spanish colony (http://www.philippine country.com/Philippine-culture/Philippine-society-html). The indigenous people, estimated at about 12 per cent of the population, are very much involved in GM crop production in the country. Their participation in GMO decision making is specifically embodied in sections 5 and 7 of Executive Order No. 514, called the National Biosafety Framework of the Philippines, requiring 'public participation and conduct of social, economic, ethical, cultural, and other assessments prior to decisions to commercialize products of modern biotechnology'.

History of GM Adoption

The first GMO approved for commercial release in the Philippines was *Bacillus thuringiensis* (Bt) corn in 2002 after seven years in the regulatory approval process. Initial planting covered more than 10 000 hectares strategically located in major corn-growing areas in the country. The area devoted to GM corn has increased substantially, reaching more than a half million hectares in 2010. More events with insect resistance and herbicide-tolerance have been approved for commercial propagation since 2002: herbicide tolerant corn in 2005 and a stacked herbicide tolerance and insect-resistant corn in 2005 and 2010. The adoption of stacked-trait corn started in 2006, rising to more than 400 000 hectares in the first five years (the Philippines, Department of Agriculture, 2010). Farmers' preference for this seed type has been overwhelming, considering the burgeoning cost of farm labour.

By 2010, 93 transformation events had been approved for direct use for food, feed and processing in the Philippines. These events cover seven agricultural crops, but mainly corn, cotton, potato and soybean. Sixteen corn transformation events have been approved for field trials and seven events of GM corn have been approved for commercial release. Corn seeds with insect tolerance and stacked traits (insect resistance and herbicide tolerance) are the more popular types among corn farmers in the country. There are also local and international research institutes developing GM crops. Those crops currently in the regulatory process include Golden Rice, Bt eggplant, papaya with delayed ripening and ringspot-virus resistance, Bt cotton and virus-resistant varieties of sweet potato and abaca (James, 2010).

GM Regulatory Scheme

The Philippines was the first in the ASEAN region to establish a GM regulatory system and became the model for other countries in the region. The system, formalized by Executive Order (EO) No. 430 in 1990, involves the National Committee on Biosafety of the Philippines (NCBP) which was given the mandate of identifying and evaluating potential hazards in the introduction of GMOs into the country and formulating and reviewing national biosafety policies and guidelines (National Committee, 2006). The Committee is responsible for the research and development stages of the GM regulatory process, leaving the regulation of commercial production and use to the Bureau of Plant Industry of the Department of Agriculture. During the introduction of GM Bt corn in 2003, small foreign-funded groups and local non-governmental organizations conducted misinformation campaigns to block field testing, resulting in some local government units passing resolutions that blocked the field testing of GM crops (Halos, 2000). The campaigns included public protests and hunger strikes and, sometimes, destruction of field trials (Cabanilla, 2007).

To further secure safety in handling GM crops, the functions of the NCBP were further expanded under Executive Order No. 514 issued in 2006, making regulatory evaluation more consistent with the Cartagena Protocol on Biosafety and including socio-economic, ethical, cultural and other considerations in biosafety decisions. The NCBP, an inter-departmental committee with the Department of Science and Technology as Chair and with representatives from the Departments of Agriculture, Environment and Natural Resources and Health, scientists, and community representatives, must approve all GM applications. Other regulatory functions are delegated to its member departments.

Submission and review of activities relating to GMOs or hazardous biological work are categorized into three types (National Academy on Science and Technology, 2008; National Committee, 2006; Department of Agriculture, Philippines (2007)). For contained facility research, proposals are considered by the relevant institution's Institutional Biosafety Committee (IBC), a group in the research institution tasked to evaluate and monitor biosafety aspects of biological research, which then passes appropriate proposals onto the NCBP for review. The proposal is scientifically evaluated by the Scientific and Technical Review Panel (STRP), created by NCBP for that purpose. For field testing, a public information campaign is required at the test sites and in print media. The number and locations of multi-location field trials are approved and monitored by NCBP. Commercial releases are monitored and approved by the Department of Agriculture following the issuance of Administrative Order No. 8 of 2002 on the 'Rules and Regulations for the Importation and Release into the Environment of Plants and Plant Products Derived from the Use of Modern Biotechnology'. This next level of risk assessment includes short and long-term effects on the environment, ecology and human health. With AO 8, field testing and commercialization of GM products become the responsibility of the Department of Agriculture after certification of completion of contained facility testing is secured from NCBP. Part of regulatory policy is the development of transparency and public participation through a sustained public information campaign on regulatory procedures and status of GM applications. This is at least in part a response to membership in the Cartagena Protocol, which requires all exporting and importing member states to declare the presence of GMO in products for transit, which may then be subject to biosafety regulation (NAST).

Food and Feed

The policy on the direct use (importation and release into the environment) of GM products for food and feed is governed by the Department of Agriculture's Administrative Order No. 8 series of 2002.

The order stipulates that no regulated GM product shall be allowed for food, feed or processing unless approved by the Bureau of Plant Industry, Department of Agriculture. It must also pass the risk evaluation of the STRP and have been assessed through the science-based risk assessment analysis performed by the Bureau of Agricultural and Fisheries Product Standards (BAFPS) for food and the Bureau of Animal Industry (BAI) for feed (The Philippines, Department of Agriculture, Process Flow for food and feed Assessments).

Although not binding to the Philippines, the biosafety policy on food follows the guidelines of the Codex Alimentarius. It has adopted Annex 3 of the Codex Plant Guideline, that is, 'Food Safety Assessment in Situations of Low-Level Presence of Recombinant-DNA Plant Material in Food', when conducting food safety assessment in situations of low-level presence of recombinant-DNA plant materials in food and feed. This is embodied in the DA's Administrative Order No. 1 series of 2009 (USDA, 2009). As a net importer of food and feed grains, this policy minimizes trade disruptions as a result of low-level presence or minimal GMO traces in trade shipments.

The existing regulatory policies on GMOs do not cover labelling. However, there are discussions and studies on potential guidelines and benefits of labelling on commercial food and feed with GMO content (De Leon et al., 2004).

CONCLUSIONS

The Oceania nations provide interesting contrasts in both their GM adoption and in their regulation of GM crops. For example, although both Australia and New Zealand are exporters of food and feed, Australian agriculture is adopting agricultural GM technology at an increasing rate. New Zealand, on the other hand, is not. The Philippines and Japan are both unable to produce enough food and feed to satisfy their own domestic needs and so are net importers of these things, including of GM food and feed. However, the Philippines is strongly adopting GM technology in its agriculture while Japan has been slow to do so.

All four Oceania nations have created regulatory regimes specifically for GM. All have government department(s) responsible for approval of GM crops prior to their release into the environment. They all also require GM foods to undergo safety assessment before sale. Although none require GM feed to be labelled, all, except the Philippines, require GM food to be labelled as such.

NOTE

1. Article 9 'Consumers shall play an active role in ensuring food safety by endeavoring [sic] to improve their own knowledge and understanding of food safety and by making efforts to express their opinions about policies to ensure food safety.' See also Article 13 regarding promoting the exchange of information and opinions.

REFERENCES

Aerni, P. (2001), 'Public attitudes towards agricultural biotechnology in developing countries: a comparison between Mexico and the Philippines', Science, Technology and Innovation Discussion Paper No. 10, Cambridge, MA: Center for International Development.

Australian Government, DAFF (Dept of Agriculture, Fisheries, and Forestry), accessed 23 November 2011 at: http://daff.gov.au/agriculture-food/crops/about_australian_crop_industries.

Basic Law on Food Safety, Law no. 48, Japan, available at: http://www.fsc.go.jp/sonata/fsb_law 1807.pdf.

Cabanilla, L.S. (2007), 'Socio-economic and political concerns for GM foods and biotechnology adoption in the Philippines', *AgBioForum*, **10**(3), 178–83.

De Leon, A., A. Manalo and F. Guilatco (2004), 'The cost implications of GM food labeling in the Philippines', *Crop Biotech Brief*, **4**(2).

Department of Agriculture, Philippines (2007), Administrative Order no. 8, February 2007, accessed 10 December 2011, at: www.da.gov.ph/images/PDFFiles/LawsIssuances /A0/2007A0/a0_08.pdf.

Halos, S.A. (2000), 'Defining the agricultural biotechnology policy of the Philippines', Policy Notes no. 2000–06, Philippines: PIDS.

Hanley, Z. and K. Elborough (2004), 'Making genetic modification safe for New Zealand – and vice versa', available at: http://greenbio.checkbiotech.org/news/making_genetic_modification_safe_new_zealand_and_vice_versa.

Hope, J. (2001), 'New Zealand Royal Commission on genetic modification', *Environmental and Planning Law Journal*, **18**(5), 441.

Hope, J. (2002), 'A history of biotechnology regulation in New Zealand', *New Zealand Journal of Environmental Law*, **6**, 1.

Independent Review of the Gene Technology Act 2000 (IRGTA) (2006), *Statutory Review of the Gene Technology Act 2000 and the Gene Technology Agreement*, Canberra: Commonwealth, Department of Health and Ageing.

James, C. (2010), 'Global status of commercialized biotech/GM crops: 2010', *International Service for the Acquisition of Agri-Biotech Applications (ISAAA) Brief* 42, Ithaca, NY: ISAAA.

National Academy on Science and Technology (2008), 'Biosafety regulations in the Philippines: a review of the first fifteen years, preparing for the next fifteen', January, Taguig City.

National Committee on Biosafety of the Philippines (2006), 'Executive order no. 514: Establishing the National Biosafety Framework, prescribing guidelines for its implementation, strengthening the NCBP, and for other purposes', monograph, Taguig City: NCBP.

New Zealand Environmental Risks Management Authority Decision (2001), 'Approval code GMF000026 on application code GMF98009', 23 May, available at: http://www.epa.govt.nz/search-databases/pages/applications-details.aspx?appID=GMF98009#.

New Zealand, *Report of the Royal Commission on Genetic Modification: Report and Recommendations 2001*, available at: http://www.mfe.govt.nz/publications/organisms/royal-commission-gm/.

New Zealand Grain and Seed Trade Association, accessed 18 July 2013 at: www.nzgsta.co.nz/about-us/.

Statistics New Zealand (2011), 'Global New Zealand – international trade, investment, and travel profile: year ended June 2011', Wellington: Statistics New Zealand.

Statistics New Zealand, '2006 census', accessed 23 November 2011 at: http://www.stats.govt.nz/Census/2006CensusHomePage.aspx.

The Philippines, Department of Agriculture, Bureau of Plant Industry (2010), 'Areas devoted to GM corn in the Philippines', accessed at: 15 January 2012 at: http://biotech.da.gov.ph.

The Philippines, Department of Agriculture, Bureau of Plant Industry, 'Process Flow for Food and Feed Assessments' accessed at: 17 January 2012 at: http://biotech.da.gov.ph.

USDA Foreign Agricultural Service (2009), 'Agricultural biotechnology annual update on Philippine biotechnology situation', GAIN (Global Agricultural Information Network) Report no. RP9030, Washington, DC.

USDA Foreign Agricultural Service (2010), 'Japan: biotechnology – GE plants and animals. Biotechnology

Annual Report 2010', GAIN (Global Agricultural Information Network) Report no. JA0025, 19 August, Washington, DC.

USDA Foreign Agricultural Service (2011), 'Japan: agricultural biotechnology. Biotechnology annual report 2011', prepared by Suguru Sato, GAIN (Global Agricultural Information Network) Report no. JA0039, 19 September, Washington, DC.

USDA Foreign Agricultural Service (re. New Zealand) (2011), 'New Zealand Agricultural biotechnology annual report prepared by David Lee-Jones, GAIN (Global Agricultural Information Network) Report no. NZ1111, 15 July, Washington, DC.

Wheen, N. (2004), 'Genetic modification, risk assessment, and Maori belief under New Zealand's Hazardous Substances and New Organisms Act', *Asia Pacific Journal of Environmental Law*, **8**, 141.

Yamanouchi, K. (2005), 'Regulatory considerations in the development and application of biotechnology in Japan', *Revue Scientifique et Technique*, **24**(1), 109–15.

11 European Union policy conflicts over agbiotech: ecological modernization perspectives and critiques

Les Levidow

INTRODUCTION

In the early 1990s agricultural biotechnology was being promoted as a symbol of European progress by all EU institutions, especially the European Commission. According to proponents, agbiotech provides a clean technology for enhancing eco-efficient agro-production, while also addressing the agrochemical problems of industrial agriculture. By the late 1990s, however, 'GM food' became negatively associated with factory farming, its hazards and unsustainable agriculture. GM products have generally faced commercial and/or regulatory blockages to market access in Europe. Commercial use has been largely limited to animal feed from soya imports and from Bt insecticidal maize cultivation in Spain (at least until 2013).

This article discusses four key questions:

- How did the EU promote agbiotech as an eco-efficient innovation?
- What societal and policy conflicts arose?
- Despite support from EU institutions, why did agbiotech encounter such great obstacles?
- How does this case signal future difficulties of claims for eco-efficiency solutions?

To answer those questions, this chapter draws on perspectives from ecological modernization (EM), whose relevance to EU policy is explained in the next section. Subsequent sections analyse the agbiotech case through the following sequence: early EU promotion of agbiotech; the 1990s agbiotech controversy; and policy changes in response. The conclusion summarizes the relevance of EM for explaining the EU-wide conflict. The text says 'biotech/biotechnological' wherever policy frameworks refer to an overall technology; otherwise it refers to 'agbiotech' for this specific sector.

1 ECOLOGICAL MODERNIZATION: PERSPECTIVES ON ECO-EFFICIENCY

In the academic literature, ecological modernization (EM) generally refers to a perspective that eco-efficient innovation can (or even will) reconcile environmental and economic sustainability. This perspective has had an ambiguous status – as a normative standpoint, as policy frameworks, and as a means to explain or analyse such frameworks (Weale, 1992, 1993). In the latter sense, EM has sought to illuminate policy changes which integrate economic with environmental criteria. According to sociologist Joseph Huber, ecological problems arise from the industrial techno-system colonizing the socio-sphere and eco-sphere. In his view, the remedy lies in an eco-social restructuring of the techno-system: 'the dirty and ugly industrial caterpillar will transform into an ecological butterfly' (Huber, 1985, p. 20; translated in Mol, 1995, p. 37; also see Mol, 1996, pp. 312–13).

Such improvements have been seen as overcoming the conflict between economy and ecology – theorized as interdependent features of innovation. EM emphasizes the potential for institutionalizing ecology in decisions on production and consumption. This could be done through government measures stimulating self-regulation of industry, thus transferring responsibilities from the state to the market (Mol, 1996, p. 306). This focus implicitly accepts a 'market failure' diagnosis from neoclassical economics – in contrast to a 'state failure' diagnosis which could justify state measures directing innovation along more environmentally better pathways.

Capitalist modernization is seen as potentially benign, hindered mainly by markets failing to provide appropriate incentives, thus warranting state assistance or incentives to correct the failure. 'Eco-efficiency' is reduced to an input–output efficiency of resource usage (Buttel, 2000). EM is limited by 'the preoccupation with efficiency and pollution control over broader concerns about aggregate resource consumption and its environmental impacts' as well as by an uncritical stance towards the transformative potentials of modern capitalism (ibid., p. 64). Amidst significant variations in EM approaches, they all presuppose technological change decoupling economic growth from environmental burdens (Davidson, 2012).

Eco-efficiency discourses generally promote techno-fixes, as if no other diagnosis were thinkable. According to Hajer (1995, p. 38), ecological modernization 'uses the language of business and conceptualises environmental pollution as a matter of inefficiency, while operating within the boundaries of cost-effectiveness and administrative efficiency . . . [EM] is . . . basically a modernist and technocratic approach to the environment that suggests that there is a techno-institutionalist fix for the present problems.' In that sense, EM is predominantly 'techno-corporatist', dependent upon agreements with industry for technological improvements (ibid., pp. 38, 281).

Despite that limitation, EM perspectives can help to illuminate tensions among different environmentalist frameworks. Often citizens protest against claims for environmental improvements through techno-fixes, especially when linked with imperatives of economic competitiveness. According to Hajer and Versteeg (2005, p. 179) 'the late 1990s showed how citizens not so much opposed eco-modernist governmental policies but conceived of the environmental problem in different, more culturally loaded terms . . . Furthermore, governments could be seen to strengthen the ties between eco-modernist thinking and neo-liberal economic discourse.' Such critical perspectives can illuminate how eco-modernist discourses are linked with specific policy agendas and cultural meanings which may drive societal conflict. Hajer's perspective can help identify socio-political tensions around an 'eco-efficient' innovation.

Since at least the early 1990s EU policy has emphasized eco-efficiency measures reconciling economic and environmental sustainability. As an ongoing tension, the EU makes symbolic declarations on 'sustainable development' while adopting EM strategies which are more compatible with the European integration project. According to Susan Baker, '[EM] is in keeping with its key tenet, namely the construction of a neoliberal free-market economy in support of industrial competitiveness'. Sustainable development is largely understood as market-based eco-efficiency measures decoupling economic growth from environmental harm. Nature is still framed as a 'standing reserve' of exploitable resources (Baker, 2007, pp. 302–3).

European agbiotech controversy has already been analysed by elaborating EM perspectives in various ways. An early account linked EM with efforts at more stringent regulation of agbiotech (Gottweis, 1998, pp. 232–5). By contrast, other accounts have associated EM with pro-biotech agendas. In the UK, New Labour supported GM crops as a means to maintain economic competitive advantage, while enhancing sustainability through more intensive, eco-efficient agri-production (Barry and Paterson, 2004). UK agbiotech supporters made claims for environmental advantages over conventional agriculture. Opposition groups rejected that framework along with industrial agriculture altogether, while appealing to natural characteristics of alternatives, especially organic farming (Toke, 2002, 2004). Yet such alternatives have been theorized as a different form of ecological modernization (Marsden, 2004). While drawing upon all those insights, this article will associate EM mainly with biotechnological claims for eco-efficiency.

2 AGBIOTECH PROMOTED AS SUSTAINABILITY VIA ECO-EFFICIENCY

Within an overall policy emphasizing eco-efficient innovations, the European Commission identified biotechnology as necessary for societal progress. Its 1993 White Paper on *Growth, Competitiveness, Employment* advocated a 'clean technology' base to facilitate a positive relation between the environment and economic growth. The policy also counselled European adaptation to inexorable competitive pressures: 'The pressure of the market-place is spreading and growing, obliging businesses to exploit every opportunity available to increase productivity and efficiency' (CEC, 1993a, pp. 92–3). This imperative was linked with innovations such as biotechnology: 'The European Union must harness these new technologies at the core of the knowledge-based economy' (ibid., p. 7).

Likewise the 5th Environmental Action Programme (EAP) elaborated claims for technology which could provide efficiency gains towards environmental objectives:

> Many of the new clean and low-waste technologies not only reduce pollution substantially, but also economise on the consumption of raw materials to such extent that cost savings can more than offset initial higher investment costs and thereby reduce unit production costs. A case in point is represented in the development and use of new techniques in the field of genetic engineering and biotechnology; these offer considerable potential for useful applications in agriculture, food processing, chemicals, pharmaceuticals, environmental clean-up and the development of new material and energy sources. (CEC, 1993b, p. 28)

Thus sustainable development was framed as resource efficiency in the image of agbiotech: greater economic efficiency would help to minimize environmental pollution. Indeed, such claims pre-dated European cultivation of GM crops: the Economic and Social Committee, officially representing the Commission's link with civil society, asserted that they are 'guaranteeing yields, helping to cut the use of plant health products in combating pests and diseases, and creating quality products' (EESC, 1998). Such statements accepted claims from agbiotech companies that GM crops simultaneously achieve economic and environmental objectives, especially by reducing input costs and waste (for example Monsanto, 1997, p. 16).

The biotech industry anticipated and emphasized greater market pressures on European agriculture, as both an imperative and an opportunity for a biotechnological solution. Under reforms of the Common Agricultural Policy, EC agricultural subsidies would be reduced and would lose their former link with production. In the view of many company managers, market liberalization and subsidy reduction would continue, thus offering greater opportunity to sell inputs to farmers, as well as to finance future development of GM crops (Chataway et al., 2004, p. 1053).

EU policies also extended proprietary claims on genetic resources but that provoked opposition. At issue was the concept of 'biopiracy', whether this meant unauthorized use of GM seeds, or rather 'Patents on Life', that is, patent rights on mere discoveries of common resources. The 'biopiracy' issue raised doubts among those on the political Left and trade-union groups which were otherwise inclined to support technological innovation as societal progress. After a decade-long conflict, an EC Directive extended patent rights to 'biotechnological inventions' (EC, 1998). This broadened the scope of discoveries or techniques which could be privatized. As public controversy continued afterwards, several EU member states failed or refused to incorporate the Directive into national law. Nevertheless the Directive strengthened incentives for the use of GM techniques.

EU R&D policies also strengthened incentives for agbiotech, especially by blurring the boundary between the public and private sectors. In many EU member states, public-sector agricultural research institutes were allocated fewer state funds than before and were expected to substitute income from the private sector or from royalties on patents. The EU's R&D funding priorities complemented that shift towards marketizing hitherto 'public-sector' research (Levidow et al., 2002). By 1990 EC funds for biotech research became conditional upon industry partners committing resources to a project. Research was given a clear economic role, with 'more careful attention to the long-term needs of industry', according to managers of the DG Research Biotechnology Division (Magnien and de Nettancourt, 1993, p. 51). In their view, 'The most vital resource for the competitiveness of the biotechnology industry is the capacity to uncover the mechanisms of biological processes and figure out the blueprint of living matter' (ibid., p. 53). Nature was conceptualized as an information machine whose deficiencies had to be corrected, as an essential means towards European industrial regeneration and competitive advantage.

Along those lines, biotech was given prominence in the Commission's Framework Programmes for research, which comprised half the EU's total budget. The agenda emphasized 'technologies needed to design and develop processes and produce "clean", high-quality products'. GM techniques were promoted in the name of 'pre-competitive' research whose results could later be developed into marketable products. Framework Programme 5 (1998–2002) included a large programme on 'Life Sciences and Biotechnology' which echoed industry's discourse on synergies between pharmaceutical and agbiotech research, also complementing the industrial mergers of the mid-1990s (Tait et al., 2002).

In appropriating the language of sustainable agriculture, EU policy attributed agri-environmental problems to genetic deficiencies, while attributing eco-efficiency benefits to inherent properties of GM crops. Eco-efficiency was also linked with the productivity imperatives of international competitiveness. These imperatives informed the Commission's policy of 'risk-based regulation', as shown next.

3 REGULATING AGBIOTECH: AGRI-INDUSTRIAL EFFICIENCY IN QUESTION

After much disagreement among EU institutions on the appropriate regulatory framework for agbiotech, the result was a compromise linking product safety with the wider project to overcome trade barriers among member states (Gottweis, 2005). According to the 1990 Directive on the Deliberate Release of GMOs into the Environment, 'completion of the internal market would be based on a high level of protection for the environment and human health'. Member states had a duty to ensure that GMOs do not cause 'adverse effects' (EEC, 1990). The Directive left open the definition of what would count as harm.

The initial political response was largely hostile to the implicit precautionary basis of the Directive. Soon after enactment, the Directive came under attack by a new consortium of chemical multinationals which were buying up seed companies. They blamed European 'over-regulation' for obstructing European economic competitiveness. This blame was incorporated into Commission policy in several ways, including 'risk-based regulation' (CEC, 1993a), especially 'the need for balanced and proportionate regulatory requirements commensurate with the identified risks' (CEC, 1994). This language implied that any regulatory burdens must be justified by prior evidence of risk.

Conveniently for agbiotech companies, EU regulatory procedures identified no risks by adopting a narrow definition of environmental harm. For a long time, critics had warned that the familiar 'pesticide treadmill' could be supplemented by a genetic treadmill; in particular, insecticidal- or herbicide-resistant crops could generate resistant pests. But these scenarios were dismissed as merely agronomic problems which also sometimes resulted from conventional pesticide techniques. Some member states sought more evidence or control measures regarding such potential effects of GM crops, but their requests were marginalized in the EU-wide procedure (Levidow et al., 1996, 2000). National differences over regulatory standards were levelled down through Commission decisions to approve GM crops for commercial cultivation in 1996–97. Thus safety claims accepted the normal hazards of intensive monoculture.

Claims for beneficent eco-efficient products were turned into an ominous prospect, however, especially through analogies with the 'mad cow' scandal. This originated in the UK, where expert advice downplayed risks and uncertainties in order to protect the beef industry from public panic. Instead the consequent scandal undermined the credibility of official safety claims for food products, while also aggravating suspicion towards intensive agricultural methods. When the Commission approved an insecticidal (Bt) maize in January 1997, despite opposition from member states, criticism came from a wide range of organizations, including the European Parliament. The mass media echoed attacks on the Commission for 'recidivism', suggesting they were repeating the crimes over approval of British beef. Extending such analogies, activists cited the unpredictable effects of agbiotech as grounds for a moratorium on GM products.

Agbiotech opponents warned against various risks: GM crops would impose 'uncontrollable risks', would spread 'genetic pollution', would extend 'unsustainable' intensive agricultural methods which had already generated agri-food hazards, and would extend corporate power over the agri-food chain (Levidow, 2000). Activists eventually generated broad opposition among civil society: according to opinion surveys, the public distrusted the role of the biotechnology industry in developing new products (Gaskell and Bauer, 2001, p. 71).

Thus critics challenged the biotechnological promise that more efficient agri-inputs would remedy environmental problems. They also counterposed different agriculture models, including organic farming and alternative cultivation methods such as Integrated Pest Management. Protest was driven mainly by activists from environmentalist and farmer groups. In some countries, such as France and Italy, mass organizations of small-scale farmers played political roles in generating mass opposition to GM crops. In response to public controversy, some governments devised a more cautious regulatory approach, went beyond 'risk' to 'sustainability' issues and promoted alternative development pathways. Four national examples briefly illustrate those pressures and responses.

In the late 1990s the French agbiotech debate expanded from 'risk' to sustainability issues, framing agriculture as common good linking producers with consumers. Some industrial-type farmers had sought access to GM crops as a means to enhance their economic competitiveness. Representing smaller-scale producers, the Confédération Paysanne attacked such products as a multiple threat to their economic independence, to high-quality French products, to consumer choice and even to democracy. They counterposed their own *paysan savoir-faire* as a basis for a different societal future (Heller, 2002). They advocated de-intensification measures, based on 'remunerative agricultural prices and sustainable family farming, with multiple benefits for society' (CPE, 2001). Supporting agbiotech in principle, the French government initially led EU-wide approval of a GM herbicide-tolerant oilseed rape, but soon reversed its stance and blocked approval. This regulatory blockage responded to expert concerns about GM crops creating a genetic treadmill, while also accommodating public anxieties and peasant opposition.

Italian anti-agbiotech opponents sought to protect the agro-food chain as an environment for craft methods and local speciality products, known as *prodotti tipici*. The Italian parliament had already allocated subsidies to promote such products and foresaw these being displaced by GM crops. According to a parliamentary report, the government must 'prevent Italian agriculture from becoming dependent on multinational companies due to the introduction of genetically manipulated seeds'. Parliamentarians suggested that local administrations applying EU legislation on sustainable agriculture should link these criteria with a requirement to use only non-GM materials. Parliament was adopting arguments from Coltivatori Diretti, a million-strong union of mainly small-scale farmers who opposed GM crops (Terragni and Recchia, 1999). Implicitly responding to these pressures, Italian authorities obstructed regulatory approval of GM crops.

The Austrian government was already promoting organic agriculture in the early 1990s; agbiotech opponents turned GM crops into a symbolic threat to that alternative economic strategy. Austrian regulators unfavourably compared potential environmental effects of GM crops to methods which use no agrochemicals, among other grounds, to oppose commercial approval (Torgersen and Seifert, 2000). Civil servants drew links between sustainable development and the Precautionary Principle which had emerged in the Convention on Biological Diversity in 1992. In their risk–benefit analysis of GM crops, risks were always uncertain while benefits were restricted to sustainability, understood mainly as organic farming (Torgersen and Bogner, 2005).

In the UK, agbiotech critics drew an analogy between GM crops, industrialized agriculture and the market pressures which led to the BSE crisis. These suspicions were echoed widely in civil society and even in government agencies. The Consumers' Association attacked the agro-food industry for its 'unshakeable belief in whizz-bang techniques to conjure up the impossible – food that is safe and nutritious but also cheap enough to beat the global competition' (McKechnie, 1999). As official advisors to the government, nature conservationists warned that broad-spectrum herbicides could increase harm to wildlife habitats in or near agricultural fields. According to a report of the UK Environment Agency, agbiotech products became controversial because they are designed for an 'increasingly intensive monoculture'; the report argued that GM crops should therefore be evaluated in a wider debate about sustainable agriculture, 'not just relative to today's substantially less-than-sustainable norm' (Everard and Ray, 1999).

In various ways across European countries, agbiotech was turned into a symbol of agri-industrial threats, misguided efficiency and neoliberal globalization. Playing on consumer anxieties about food safety, radical environmentalists built an anti-agbiotech coalition counterposing various alternatives, including food sovereignty, local control of food production and 'natural' food. Those issue-framings relate to rival political economies: for example, radical greens advocated decentralizing the economy, in opposition to a globalized agri-industrial production (Hines, 2000; Woodin and Lucas, 2004).

Also contentious were rules for labelling products as 'GM'. Protesters demanded comprehensive GM labelling, so that consumers would not be unwittingly 'force-fed' GM food and could make their own judgements on product safety. In the late 1990s the food industry and likewise the EU eventually accommodated demands for more stringent rules. Soon European food retail chains excluded GM grain from their own-brand products rather than apply a 'GM' label. Consequently, GM grain found a market only for animal feed, whose products required no GM label (Levidow and Bijman, 2002). By the late 1990s agbiotech products were being blocked through a commercial boycott more than by any regulatory obstacles. Lacking a market for GM grain, European farmers were deterred from cultivating GM crops, except in some maize fields in Spain, which otherwise would have a shortage of animal feed.

4 REGULATORY CHANGES VERSUS AGRI-EFFICIENCY

In response to public controversy, environmental risk criteria became broader and thereby less amenable to the agri-efficiency for which GM crops were designed. Some governments restricted or even banned GM products which already had EU approval. In 1999 the UK announced a voluntary moratorium on commercial cultivation of GM crops, citing the scientific uncertainties about how herbicide sprays on herbicide-tolerant crops could affect farmland biodiversity. This move aimed to allow time for commercial-scale testing of herbicide usage and its effects, thus delaying an awkward political decision.

Moreover, in 1999 the EU Council had a significant minority blocking consideration of any more GM products. Several members demanded that the EU regulatory procedure must first incorporate more stringent precautionary criteria. This procedural blockage, which became known as the de facto moratorium, represented both a policy impasse over regulatory criteria and difficulties in addressing public concerns.

The moratorium stimulated changes towards broader regulatory criteria for agbiotech. In consultation with member states, the Commission revised the 1990 Directive. The revision encompassed a broader range of risks, including the wider effects of herbicide usage and delayed or indirect effects. It also required that scientific uncertainty be made explicit about any 'identified risk', potentially as a basis to require commercial-stage monitoring for specific risks. The Directive now required public consultation (EC, 2001). Together these changes potentially required greater evidence of safety before and even during commercialization, along with greater public accountability for regulatory decisions. In parallel with these legislative changes, the Commission extended earlier arguments about biotechnological imperatives. Given the 'revolution taking place in the knowledge base of life sciences and biotechnology', this could provide a major contribution to the EC 'becoming a leading knowledge-based economy'. At the same time, the Commission emphasized the scope to shape innovation for greater societal benefit:

> Europe is faced with a major policy choice: either accept a passive and re-active role, and bear the implications of the development of these technologies elsewhere, or develop pro-active policies to exploit them in a responsible manner, consistent with European values and standards. The longer Europe hesitates, the less realistic this second option would be. (CEC, 2002, p. 7)

Its strategy document celebrated the prospects for GM crops to generate 'more sustainable agricultural practices', which were equated with eco-efficiency measures such as pesticide reduction (CEC, 2002, pp. 6, 15). Technological progress carries socio-ethical implications which 'cannot be adequately addressed within the narrow context of regulatory product approvals'; consequently, EU procedures need transparency, accountability and participatory approaches. Nevertheless, regulatory oversight 'is the expression of societal choices': rules should ensure that market mechanisms function effectively, so that safe products become available to accommodate consumer preferences (ibid., pp. 14, 21). Consequently, regulatory oversight bore the burden of societal conflict over agbiotech, in the absence of any procedure for evaluating agricultural development pathways.

Rather than await the changes in EU law, the Commission sought to restart the EU-wide regulatory procedure, which had been stalled since 1999. In its view, precaution could not justify permanent blockages or bans on the GM products under consideration. When the regulatory procedure finally resumed in 2003, member states applied even more stringent criteria than in the late 1990s; these stances led to more disagreements over new GM products, again mainly involving herbicide-tolerant and insect-resistant traits. More member states than before challenged the available evidence of safety and raised extra uncertainties; some criticisms came from governments which anyway opposed GM crops in general (for example, Italy, Greece and Austria). Environmental NGOs continued to attack safety claims while counterposing alternative agricultures. Whenever the Commission sought EU approval for a specific GM product, there was little support from member states, so the Commission lacked legitimacy when approving the product (Levidow et al., 2005).

Also in response to EU Council demands and with strong support from the Parliament, new legislation further broadened the criteria for 'GM' labelling. Formerly this requirement depended upon the presence of detectable DNA or protein in GM food. Under a new law, 'GM' labelling was now required for any food or feed product containing GM material, regardless of detectability, and its presence must be traceable throughout the agro-food chain (EC, 2003a, 2003b). This effectively required labelling for a broader range of products than before. With this extra weapon, NGOs made further efforts to deter food companies from using GM grain, whose use by now was limited to animal feed.

The inadvertent spread of GM material became another contentious issue. European agbiotech opponents had warned against the prospect that GM crops would irreversibly 'contaminate' the environment and non-GM crops, which could impose a financial loss for non-GM farmers. Yet segregation measures to limit the spread could impose economic burdens on GM farmers and so constrain crop cultivation (Levidow and Boschert, 2011).

To regain control over the 'contamination' issue, the Commission developed a 'coexistence' policy for ensuring farmers' free choice to cultivate GM, conventional or organic crops. This policy sharply distinguished between environmental issues, which were appropriate for risk regulation under the Directive, and merely economic damage from the spread of 'safe' GM material to non-GM crops (CEC, 2003). But this key distinction was challenged, blurred and undermined. Again under pressure from the Parliament, the Commission agreed to amend the Deliberate Release Directive so that 'member states may take appropriate measures to avoid the unintended presence of GMOs in other products' (EC, 2003a, p. 20).

Such segregation measures were being developed by many national or regional authorities. For some, 'coexistence' policy increasingly meant segregation measures which would marginalize or preclude GM crops. A Europe-wide charter of regional authorities discursively linked 'GMO-free zones' with food sovereignty, 'quality' labels on food products and regional biodiversity. The charter identified GM crops as a threat to 'sustainable and organic farming and regional marketing priorities for their rural development' (FFA, 2005).

Ostensibly about 'GM contamination', this conflict expressed rival development pathways – GM crops as eco-efficiency measures for agri-industrial development versus alternatives through agrarian-based rural development (Levidow and Boschert, 2008; Marsden, 2004, 2008; Marsden et al. 2002). Agbiotech opposition was joined by the Assembly of European Regions, which proposed that coexistence would be based on in-depth feasibility studies examining the environmental, socio-economic and cultural impact of GMOs. Areas could be designated as 'GMO free' in order to protect any added value of certified quality products (AER/FoEE 2005). This proposal generalized from stringent rules already being devised by some regional authorities, especially as a means to protect product 'quality' as recognized by consumers.

5 CONCLUSION: CONTENDING DEVELOPMENT PATHWAYS

Returning to the original questions about the EU policy conflicts, agbiotech was a contentious development pathway. Some EM perspectives can help to explain how agbiotech was promoted as a benign eco-efficient tool essential for both environmental improvement and economic competitiveness, as a basis to gain government support, especially from the European Commission. Agbiotech was promoted as both an instrument and beneficiary of neoliberal policies – by extending patent rights, by making the public sector more dependent upon private finance, and by elaborating 'risk-based regulation', so that narrowly-defined risks could be the only basis for questioning EU-wide commercial approval of a GM product. Safety standards were linked with the imperative to create a single market, in turn necessary for global competitiveness.

By the mid-1990s EU-wide disputes arose over the appropriate regulatory criteria. The dominant policy accepted industry claims for eco-efficient, safe GM products, taking for granted the normal hazards of intensive monoculture such as pest resistance. Yet some policy-makers sought to test claims for eco-efficiency through better evidence, before and during commercial use of GM crops, especially regarding the prospect of a genetic treadmill. As in other regulatory areas, Commission policy featured tensions and shifts between these two different environmental bases for an EU internal market (cf. Weale and Williams, 1993).

Meanwhile NGO campaigns were opposing GM crops as the supposed solution for agro-environmental sustainability problems. Through mass protest since the late 1990s, agbiotech was popularly stigmatized as a threat to the environment, sustainable development and democratic accountability. Critics diagnosed agri-industrial efficiency as a threat to alternative pathways, understood along lines of an agrarian-based rural development. Such agendas undermined the EM perspective that 'the dirty and ugly industrial caterpillar will transform into an ecological butterfly' (Huber, 1985), at least in the agbiotech case.

Societal conflict arose partly from an overall EU policy seeking innovation which combines greater economic efficiency with less pollution, while representing this policy as 'sustainable development' (cf. Hanf, 1996; Baker, 2007). For agbiotech in particular, Commission policy relegated societal choices to regulatory oversight (CEC, 2002), which thereby bore the burden of societal conflict and so remained contentious. Such conflict was translated into disputes over the appropriate regulatory criteria – for example, for risk assessment, labelling and segregation. All those criteria became more stringent and thus more amenable to civil society efforts at blocking market access to GM products. By 1999 food retail chains were boycotting GM inputs, which found a market only in animal feed.

Those societal conflicts and their outcomes cannot be readily explained by EM perspectives which attribute sustainability problems to market failure, in turn due to inadequate state incentives for eco-efficient innovation. The EU case is better understood through critical perspectives on EM: namely, state failure (or perhaps refusal) to evaluate various options for a more environmentally-sustainable agriculture. EU institutions embraced techno-fixes amenable to privatizing knowledge for agri-industrial systems, so governments were seen 'to strengthen the ties between eco-modernist thinking and neoliberal economic discourse' (Hajer and Versteeg, 2005).

Moreover, Europeans were told that they had no choice but to accept agbiotech as a development pathway towards a better future. In response, opponents turned agbiotech into a test of democratic accountability for societal choices (Levidow and Carr, 2010). Eventually opposition networks counterposed alternative development pathways, thus opening up a broader debate about societal futures.

ACKNOWLEDGEMENTS AND METHODS

For its information sources and research methods, this chapter draws mainly upon three research projects on agbiotech regulation:

- 'From precautionary to risk-based regulation: the case of GMOs', funded by the ESRC, 1995–96.
- 'Safety regulation of transgenic crops: completing the internal market?', funded by the European Commission, DG XII/E5, Ethical, Legal and Socio-Economic Aspects (ELSA), Biotechnology horizontal programme, during 1997–99.
- 'Precautionary expertise for GM crops (PEG)', funded by the European Commission, DG-Research, Quality of Life programme, during 2002–04.

REFERENCES

AER/FoEE (2005), 'Safeguarding sustainable European agriculture: coexistence, GMO free zones and the promotion of quality food produce in Europe', Friends of the Earth Europe and Assembly of European Regions conference, 17 May, Brussels.

Baker, S. (2007), 'Sustainable development as symbolic commitment: declaratory politics and the seductive appeal of ecological modernisation in the European Union', *Environmental Politics*, **16**(2), 297–317.

Barry, J. and M. Paterson (2004), 'Globalisation, ecological modernisation and New Labour', *Political Studies*, **52**, 767–84.

Buttel, F. (2000), 'Ecological modernization as social theory', *Geoforum*, **31**, 57–65.

CEC (1993a), 'Growth, competitiveness, employment: the challenges and ways forward into the 21st century', *Bulletin of the European Communities*, supplement 6/93, esp. pp. 100–103, Brussels: Commission of the European Communities.

CEC (1993b), 'Towards sustainable development', Fifth Environmental Action Programme; also in *Official Journal of the European Communities*, C 138, 17 May, 5–98.

CEC (1994), 'Biotechnology and the white paper on growth, competitiveness and employment: preparing the next stage', *Ebis Newsletter*, **4**(2), 2–5, Brussels: European Biotechnology Info Service.

CEC (2002), 'Life Sciences and Biotechnology: A Strategy for Europe', Luxembourg: Office for Official Publications of the European Communities available at: http://europa.eu.int/comm/biotechnology.

CEC (2003), 'Commission Recommendation of 23 July on guidelines for the development of national strategies and best practices to ensure co-existence of GM crops with conventional and organic farming', Brussels: Commission of the European Communities.

Chataway, J., J. Tait and D. Wield (2004), 'Understanding company R&D strategies in agro-biotechnology: trajectories and blind spots', *Research Policy*, **33**(6–7), 1041–57.

CPE (2001), 'To change the CAP', Brussels: Coordination Paysanne Européenne, available at: www.cpefarmers.org.

Davidson, S. (2012), 'The insuperable imperative: a critique of the ecologically modernizing state', *Capitalism Nature Socialism*, **23**(2), 31–50.

EC (1998), 'Directive 98/44/EC of the European Parliament and of the Council on Protection of Biotechnological Inventions', *Official Journal of the European Union*, 30 July, L 213, p. 13.

EC (2001) 'European Parliament and Council Directive 2001/18/EC of 12 March on the deliberate release into the environment of genetically modified organisms and repealing Council Directive 90/220/EEC', *Official Journal of the European Union*, L 106, pp. 1–38.

EC (2003a), 'Regulation 1829/2003 of 22 September 2003 on genetically modified food and feed', *Official Journal of the European Union*, L268, 18 October, pp. 1–23.

EC (2003b), 'Regulation 1830/2003 of 22 September 2003 concerning the traceability and labelling of GMOs and traceability of food and feed produced from GMOs and amending Directive 2001/18', *Official Journal of the European Union*, L268, 18 October, pp. 24–28.

EEC (1990), 'Council Directive 90/220/EEC on the deliberate release to the environment of genetically modified organisms', *Official Journal of the European Communities*, L 117, pp. 15–27.

EESC (1998), 'Opinion of the European Economic and Social Committee on genetically modified organisms in agriculture – impact on the CAP', *Official Journal of the European Communities*, C 284, pp. 39–50.

Everard, M. and D. Ray (1999), *Genetic Modification and Sustainability*. 2020 Vision Series No.1. Bristol: Environment Agency/Cheltenham: The Natural Step.
FFA (2005), 'Berlin manifesto for GMO-free regions and biodiversity in Europe', Foundation for Future Farming, Assembly of European Regions, January, available at: www.are-regions-europe.org.
Gaskell, G. and M. Bauer (eds) (2001), *Biotechnology 1996–2000*, London: Science Museum.
Gottweis, H. (1998), *Governing Molecules: The Discursive Politics of Genetic Engineering in Europe and the United States*, Cambridge, MA: MIT Press.
Gottweis, H. (2005), 'Transnationalizing recombinant-DNA regulation: between Asilomar, EMBO, the OECD, and the European Community', *Science as Culture*, **14**(4), 299–308.
Hajer, M. (1995), *The Politics of Environmental Discourse*, Oxford: Oxford University Press.
Hajer, M. and W. Versteeg (2005), 'A decade of discourse analysis of environmental politics: achievements, challenges, perspectives', *Journal of Environmental Policy & Planning*, **7**(3), 175–84.
Hanf, K. (1996), 'Implementing environmental policies', in A. Blowers and P. Glasbergen (eds), *Environmental Policy in an International Context – Prospects for Environmental Change*, London: Arnold, pp. 197–221.
Heller, C. (2002), 'From scientific risk to paysan *savoir-faire*: peasant expertise in the French and global debate over GM crops', *Science as Culture*, **11**(1), 5–37.
Hines, C. (2000), *Localization: A Global Manifesto*, London: Earthscan.
Huber, J. (1985), *Die Regenbogengesellschaft. Ökologie und Sozialpolitik*, Frankfurt am Main: Fisher Verlag.
Levidow, L. (2000), 'Pollution metaphors in the UK biotechnology controversy', *Science as Culture*, **9**(3), 325–51.
Levidow, L. and J. Bijman (2002), 'Farm inputs under pressure from the European food industry', *Food Policy*, **27**(1), 31–45.
Levidow, L. and K. Boschert (2008), 'Coexistence or contradiction? GM crops versus alternative agricultures in Europe', *Geoforum*, **39**(1), 174–90.
Levidow, L. and K. Boschert (2011), 'Segregating GM crops: why a contentious "risk" issue in Europe?', *Science as Culture*, **20**(2), 255–79.
Levidow, L. and S. Carr (2010), *GM Food on Trial: Testing European Democracy*, London: Routledge.
Levidow, L., S. Carr and D. Wield (2000), 'Genetically modified crops in the European Union: regulatory conflicts as precautionary opportunities', *Journal of Risk Research*, **23**(3), 189–208.
Levidow, L., S. Carr and D. Wield (2005), 'EU regulation of agri-biotechnology: precautionary links between science, expertise and policy', *Science & Public Policy*, **32**(4), 261–76.
Levidow, L., S. Carr, R. von Schomberg and D. Wield (1996), 'Regulating agricultural biotechnology in Europe: harmonization difficulties, opportunities, dilemmas', *Science & Public Policy*, **23**(3), 135–57.
Levidow, L., V. Søgaard and S. Carr (2002), 'Agricultural PSREs in Western Europe: research priorities in conflict', *Science and Public Policy*, **29**(4), 287–95.
Magnien, E. and D. de Nettancourt (1993), 'What drives European biotechnological research?', in E.J. Blakelely and K.W. Willoughby (eds), *Biotechnology Review no.1: The Management and Economic Potential of Biotechnology*, Brussels: Commission of the European Communities, pp. 47–8.
Marsden, T.K. (2004), 'The quest for ecological modernisation: respacing rural development and agro-food studies', *Sociologia Ruralis*, **44**(2), 129–47.
Marsden, T.K. (2008), 'Agri-food contestations in rural space: GM in its regulatory context', *Geoforum*, **39**(1).
Marsden, T., J. Banks and J. Bristow (2002), 'The social management of rural nature: understanding agrarian-based rural development', *Environment and Planning A*, **34**, 809–25.
McKechnie, S. (1999), 'Food fright', *The Guardian*, 10 February (director of the Consumers Association).
Mol, A.P.J. (1995), *The Refinement of Production: Ecological Modernisation Theory and the Chemical Industry*, Utrecht: Jan van Arkel/International Books.
Mol, A.P.J. (1996), 'Ecological modernisation and institutional reflexivity: environmental reform in the late modern age', *Environmental Politics*, **5**(2), 302–23.
Monsanto (1997), *Report on Sustainable Development*, St Louis, MO: Monsanto Company.
Tait, J., J. Chataway and D. Wield (2002), 'The life science industry sector: evolution of agro-biotechnology in Europe', *Science and Public Policy*, **29**(4), 253–8.
Terragni, F. and E. Recchia (1999), 'Italy: precaution for environmental diversity?', report for 'Safety Regulation of Transgenic Crops: Completing the Internal Market', DGXII RTD project coordinated by the Open University, available at: http://technology.open.ac.uk/cts/srtc/index.html.
Toke, D. (2002), 'Ecological modernisation and GM food', *Environmental Politics*, **11**(3), 145–63.
Toke, D. (2004), *The Politics of GM Food*, London: Routledge.
Torgersen, H. and A. Bogner (2005), 'Austria's agri-biotechnology regulation: political consensus despite divergent concepts of precaution', *Science & Public Policy*, **32**(4): 277–84.
Torgersen, H. and F. Seifert (2000), 'Austria: precautionary blockage of agricultural biotechnology', *Journal of Risk Research*, **3**(3), 209–17.

Weale, A. (1992), *The New Politics of Pollution*, Manchester: Manchester University Press.
Weale, A. (1993), 'Ecological modernisation and the integration of European environmental policy', in J.D. Liefferink, P.D. Lowe and A.P.J. Mol (eds), *European Integration and Environmental Policy*, London: Belhaven, pp. 202–16.
Weale, A. and A. Williams (1993), 'Between economy and ecology? The single market and the integration of environmental policy', in D.Judge (ed), *A Green Dimension for the European Community*, London: Frank Cass, pp. 45–64.
Woodin, M. and C. Lucas (2004), *Green Alternatives to Globalisation – A Manifesto*, London: Pluto.

12 African non-adopters

Robert Paarlberg

The future of agricultural biotechnology in Africa will depend most on the choices African governments make regarding regulation of this technology. High regulatory costs and the uncertainty of a final regulatory approval by national biosafety committees are currently the limiting factors blocking uptake of locally adapted biotechnology applications for African farmers.

In nearly all countries in sub-Saharan Africa today it is still not legal for farmers to plant any GMO crops, and in most countries on the continent it is still not legal even to do research on GMOs. Only two countries in sub-Saharan Africa have approved any GMO crops for commercial planting: the Republic of South Africa and Burkina Faso. GMO crops in wide use by small farmers elsewhere, such as Bt cotton and Bt maize, are waiting to be used by African farmers, but national regulatory authorities have not yet given farmers permission to plant these seeds. This pattern of persistent regulatory blockage and uncertainty has discouraged both national scientists and international technology providers from making significant biotechnology investments to help farmers in Africa.

In Africa, the regulatory approval process for GMO crops is typically held up at one of two sequential choke points: non-approval to conduct research on GMOs (for example in labs, greenhouses, or in physically confined field trials, or CFTs), or a subsequent non-approval to plant these crops in unconfined settings such as farmers' fields (known as an 'environmental release'). Technology developers cannot gain a full commercial release, enabling ordinary farmers to plant the seeds, until national regulatory committees have given affirmative approval first for CFTs and then for an environmental release. At both of these key approval points, the applicant must present to a national biosafety committee (NBC) a carefully prepared 'dossier' of information. At the CFT stage this dossier must include details regarding the molecular biology of the GMO in question, the experiments that are being proposed, and the confinement steps that will be taken to ensure the biological safety ('biosafety') of those experiments, such as the physical security and isolation of the research site, plans for disposing of plant materials, and procedures for monitoring and accidents. At the subsequent environmental release stage, a second dossier must be submitted which provides evidence gathered from the field trials demonstrating that uncontained planting will present no significant new biological risks, plus evidence that consumption by humans and animals will present no significant new health risks. This will be in addition to all the normal requirements for demonstrated efficacy in farmers' fields in national performance trials, sanitary and phytosanitary (SPS) clearance, and formal seed certification.

Simply getting this regulatory approval started is usually time consuming in Africa. A first source of delay can be the slow pace of parliamentary action in passing the separate national laws ('biosafety laws') that Africans believe GMO crops must require. In Kenya, it took five years to move from a draft law to a final act of parliament. A second source of delay then becomes drafting and publishing separate regulations and guidelines for handling GMOs under the new law, a process that took three more years in Kenya. A third source of delay is often the formation of a functional national biosafety committee (NBC) needed to review applications and make case-by-case decisions to approve (or not) the import and planting of GMO seeds. Slowing all of these processes further has been an atmosphere in Africa of strong social and political suspicion toward GMOs, expressing itself either as a worry about the impacts of GMOs on the environment (biosafety), or on human health (food safety), or on the rural poor (socioeconomic safety). These worries are persistently enflamed by activist groups – many of which are non-African – that reject GMOs categorically, seeing them as a dangerous manifestation of corporate globalization.

The ability of a GMO crop to run this regulatory gauntlet in Africa depends, country by country, on the presence or absence of at least three critical factors. First, the presence of a functioning national biosafety committee, legally constituted under a separate act of government. Second, a strong agronomic performance for the crop (compared to available alternative varieties), demonstrated through in-country field trials. Third, organized in-country political support, sufficient to overcome the organized opposition to all GMO crops that has already formed in Africa.

So far, few GMO crops have been able to get through this gauntlet. Consider the case of insect-resistant Bt maize in Kenya. Under an Insect Resistant Maize for Africa (IRMA) project funded by the Syngenta Foundation initiated in 1999, scientists at Kenya's national agricultural research institute (KARI) were given permission to conduct two years of confined field trials for a variety of Bt maize designed to protect against stalk borers. The technology presented no evidence of new risks to human or animal health, and no new risks to the environment, yet it was never granted an environmental release. In this case, all three of the above factors were missing: prior to 2009 Kenya did not have in place a national biosafety law; the Bt gene incorporated in the crop was an older technology that controlled stalk borer pests in only part of the country; and (partly as a consequence) there were few champions for the technology inside Kenya, beyond the scientific community.

Bt cotton has also failed to make it through the regulatory system in Kenya, for slightly different reasons. There is strong political support in Kenya for Bt cotton and relatively weak political opposition (as cotton is an industrial rather than a food crop), and confined field trials of Bt cotton have been underway at the Kenya Agricultural Research Institute (KARI) since 2004. Yet as of 2011 there has not yet been an environmental release of this GMO technology. Prior to 2011, the explanation given by Bt cotton supporters had been the absence of openly published GMO biosafety regulations based on a separate national biosafety law, but by 2011 the law had been passed and the regulations published ('gazetted'). The remaining explanation was that the germplasm used by KARI was Monsanto germplasm, which is not ideally suited to the Kenyan environment. So, in this cotton case, success factors 1 and 3 were present, but factor 2 was not strong enough. Kenya is one of the African governments farthest along in working with GMOs, yet it is still not legal for farmers in Kenya to plant any GMO crops.

CHOOSING A REGULATORY SYSTEM FOR GMOS

When they started working with GMOs, African governments could have chosen either one of two quite different regulatory approaches: the approach used by the European Union, or the approach used by the United States. The decision was made by nearly all to adopt Europe's approach, and this remains the principal reason so few regulatory approvals have been given in Africa. The European approach has produced almost no approvals at all in Europe since 1998, so it should be little surprise it is producing so few approvals in Africa.

There are four key differences between the European and American approaches:

- The regulatory approach used in Europe requires new and separate laws that are specific to genetically engineered (GMO) foods and crops. In contrast, the United States regulates GMOs for food safety and for environmental safety using the laws that were already in place to govern non-GMO foods and crops.
- The European approach also requires the creation of new institutions (for example, national biosafety committees) and a separate screening and approval process for GMOs. In the United States the institutions that screen and approve GMOs (the Food and Drug Administration, the Animal and Plant Health Inspection Service, and the Environmental Protection Agency) are the same institutions that screen and approve non-GMO foods and crops.
- The European approach also differs because regulators can decline to approve a new technology on grounds of 'uncertainty' alone, without any evidence of risk. Even a hypothetical risk not yet validated can be sufficient reason for blockage. This is known as the precautionary approach. In the United States, if standard tests for known risks such as toxicity, allergenicity and digestibility have been successfully passed, there is usually no further regulatory barrier to commercial release.
- Finally, in Europe products in the marketplace with GMO content must carry identifying labels and operators in the food chain must 'trace' the GMOs they handle by maintaining a documentary record of purchases and sales. In the United States the Food and Drug Administration (FDA) does not require labels on any approved GMO foods and there are no traceability requirements imposed on GMOs.

Which of these two approaches is better? An optimal regulatory system is one that allows new technologies to be used while preventing new risks to human health or the environment. Using this standard it is clear that the US approach has so far done a better job. Most of the GMOs actually put on the market over the past dozen years have been approved under the risk-based American approach rather than the highly precautionary, uncertainty-based European regulatory approach, yet the safety record for the technology has remained essentially unblemished. If the European approach had been followed everywhere, the safety record would not be improved, but many fewer productive technologies would have been available to farmers.

It is essential to know that there has not yet been any documented evidence that approved GMOs have posed new risks either to human health or the environment. This finding of 'no new risks' has long been the official view of scientific authorities even in

Europe. European science academies took a number of years to study the impacts of GM crops on human health and the environment following the first commercializations in 1995, and by 2001–04 a consensus had emerged, even in Europe, that no new risks from these seeds had been documented.

In 2001, the Research Directorate General of the EU released a summary of 81 separate scientific studies conducted over a 15-year period (all financed by the EU rather than private industry) aimed at determining whether GM products were unsafe, insufficiently tested, or under-regulated (Kessler and Economidis, 2001). The EU Research Directorate concluded from this study that 'research on GM plants and derived products so far developed and marketed, following usual risk assessment procedures, has not shown any new risks on human health or the environment' (EU Research Directorate, 2001). National Academies of Science in Europe began drawing this same conclusion within the year. In December 2002, the French Academy of Sciences stated that 'all the criticisms against GMOs can be set aside based for the most part on strictly scientific criteria' (French Academy of Sciences, 2002, p. xxxviii). At the same time the French Academy of Medicine announced it had found no evidence of health problems in the countries where GMOs had been widely eaten for several years (French Academy of Medicine, 2002). In the UK in May 2003, the Royal Society presented to a government-sponsored review two submissions that found no credible evidence GM foods were more harmful than non-GM foods, and the Vice-President and Biological Secretary of the Royal Society, Professor Patrick Bateson, expressed irritation at the undocumented assertions of risk that continued to come from anti-GMO advocates:

> We conducted a major review of the evidence about GM plants and human health last year, and we have not seen any evidence since then that changes our original conclusions. If credible evidence does exist that GM foods are more harmful to people than non-GM foods, we should like to know why it has not been made public. (Royal Society, 2003)

In March 2004, the British Medical Association (BMA), which had earlier withheld judgment, endorsed these Royal Society conclusions (BMA, 2004). In September 2004, the Union of the German Academies of Science and Humanities produced a report which concluded, 'according to present scientific knowledge it is most unlikely that the consumption of the well characterized transgenic DNA from approved GMO food harbours any recognizable health risk' (Helt, 2004, p. 4). This report added that food from insect-resistant GM maize was probably healthier than from non-GM maize due to lower average levels of the fungal toxins that insect damage can cause.

A consensus has also emerged at the global scientific level of no new risks linked to any of the GM crops and foods to have reached the market so far. In March 2000, the Organisation for Economic Co-operation and Development (OECD) in Paris organized a conference with 400 expert participants from a variety of backgrounds. These experts announced their agreement that 'no peer-reviewed scientific article has yet appeared which reports adverse effects on human health as a consequence of eating GM food' (OECD, 2000, p. 2). In August 2002, the Director-General of the World Health Organization (WHO) endorsed consumption of GM foods, saying, 'WHO is not aware of scientifically documented cases in which the consumption of these foods has negative human health effects. These foods may therefore be eaten.' (Mantell, 2002).

Some accept that GM foods are probably safe to eat yet still question their safety for other living things in the biological environment (their 'biosafety'). All farming disturbs and changes nature, so it is difficult to agree on exactly what level of disturbance can be considered acceptable. For example, planting a GM variety of beet or rapeseed can help farmers control weeds in the field (compared to conventional beet or rapeseed), and as a result there may be fewer insects in the farm field (insects use weeds for food and shelter) and fewer weed seeds for some farmland birds to eat. Some might see this as a damaging disturbance of nature. Yet by most conventional definitions of biosafety, the GM crops currently on the market have not disturbed nature (beyond farm fields) any more than conventional crops. A 2003 study conducted by scientists from New Zealand and the Netherlands published in *The Plant Journal* examined data collected worldwide up to that time and the authors concluded from this data that the GM crops approved to that point had been no more likely to worsen weed problems than conventional crops, no more invasive or persistent, and no more likely to lead to gene transfer. There was no evidence that GM crops had transferred to other organisms (including weeds) new advantages such as resistance to pests or diseases or tolerance to environmental stress (Connor et al., 2003). Later in 2003 the International Council for Science (ICSU) examined the findings of roughly 50 different scientific studies that had been published in 2002–03 and concluded 'there is no evidence of any deleterious environmental effects having occurred from the trait/species combinations currently available' (International Council for Science, 2003, p. 3) In May 2004, the United Nations Food and Agriculture Organization (FAO) issued a 106-page report summarizing evidence that 'to date, no verifiable untoward toxic or nutritionally deleterious effects resulting from the consumption of foods derived from genetically modified foods have been discovered anywhere in the world'. On the matter of environmental safety, this FAO report found the environmental effects of the GM crops approved so far, including effects such as gene transfer to other crops and wild relatives, weediness, and unintended adverse effects on non-target species (such as butterflies), had been similar to those that were already existing from conventional agricultural crops (FAO, 2004). Finally in 2007, a study carried out for the journal *Advanced Biochemical Engineering/Biotechnology* surveyed ten years of research published in peer-reviewed scientific journals, scientific books, reports from regions with extensive GM cultivation, and reports from international governmental organizations and found that, '[t]he data available so far provide no scientific evidence that the cultivation of the presently commercialized GM crops has caused environmental harm.'(Sanvido et al., 2007).

In 2010, the Directorate-General for Research and Innovation of the European Union produced yet another reassuring report on GMO safety. This report said:

> the main conclusion to be drawn from the efforts of more than 130 research projects, covering a period of more than 25 years of research, and involving more than 500 independent research groups, is that biotechnology, and in particular GMOs, are not *per se* more risky than e.g. conventional plant breeding technologies. (EU, 2010)

Skeptics who remain fearful sometimes respond that 'absence of evidence is not the same thing as evidence of absence'. Yet if you look for something for 15 years and fail to find it, that must surely be accepted as *evidence* of absence. It may not be *proof* that risks are absent, but proving something is absent (proving a negative) is known to be logically impossible.

The explanation for Europe's highly precautionary regulatory approach toward

GMOs goes beyond risks. It is a policy posture that reflects not a presence of new risks for Europeans, but instead an absence, for most Europeans, of new benefits. The first generation of GM crops provided benefits to farmers, but almost no discernible benefit at all to food consumers.

The first generation of GM crops that came to the market in 1995–96 provided benefits mostly to farmers growing cotton, maize and soybean, in the form of lower costs for the control of insects and weeds. Yet Europe does not have many cotton, maize and soybean farmers, so the new technology had few champions. For the 99 per cent of Europeans who were not maize, cotton or soybean farmers, the new technology offered no apparent direct benefit at all. For consumers in Europe, the new GM products did not taste any better, look any better, smell any better, prepare any better, or deliver any improved nutrition and while prices in these crops might have been marginally lower than before the technology, the changes were minor. Because the vast majority of Europeans saw little or no direct benefit from the technology, they felt they had nothing to lose by keeping it out of farm fields and out of their food supply. They welcomed a highly precautionary regulatory approach as one way to ensure that outcome.

To demonstrate that it was a benefit calculation rather than a risk calculation that mattered most to Europeans in this case, look at the quite different way Europe regulates GMOs in medicine, versus GMOs in agriculture. In the case of medical drugs, Europe does not hesitate to permit the commercial sale of medicines developed with genetic engineering. By 2006, the European Medicines Agency had actually approved 87 recombinant drugs, derived from genetically engineered bacteria or from the ovary cells of genetically engineered Chinese hamsters. Significantly, these drugs were not free from new risks; it had been learned from clinical trials that many of these drugs actually increased risks of heart disease, malignancy, and gastric illness, but European regulators approved them just the same, because of the benefits the drugs could deliver to so many Europeans. While fewer than 1 per cent of Europeans stood to benefit directly from GM agricultural crops, 100 per cent were vulnerable to the diseases these GMO drugs could help treat – so the regulatory treatment of the GMO drugs was far less precautionary. There were both known risks from clinical trials and plenty of uncertainties surrounding long-term exposures, yet these risks and uncertainties were not allowed to block the commercial release of a technology that could bring significant benefits to Europeans.

AFRICA IS NOT EUROPE

At first glance it might seem puzzling for Africa to follow this European approach to the regulation of agricultural biotechnology. In Africa, the proportion of the population that might benefit directly from agricultural GMOs is much higher than in Europe, because 60 per cent or more of all Africans are still farmers who depend directly on agriculture for income and subsistence. Some GM crop traits now widely commercialized outside Africa, such as Bt crops (for example, maize and cotton) that resist insect damage with fewer chemical sprays, could have wide benefits if planted in Africa today. Other GMO traits soon to come out of the research pipeline, including abiotic stress tolerance traits such as drought resistance, could provide even wider benefits in the future.

Drought-tolerant maize is one of the more interesting new GM crop technologies now emerging from the research pipeline. Maize is a staple food for more than 300 million people in sub-Saharan Africa, many of whom are themselves growers of maize. These Africans remain poor and food insecure because the productivity of their labor in farming is so low. Population growth has been pushing maize production into marginal areas with little and unreliable rainfall, and only 4 per cent of cropland in sub-Saharan Africa is irrigated. These factors, combined with human-induced climate change, are expected to increase drought risks to maize growers in Africa in the years ahead. The development of maize varieties better able to tolerate drought is one important response to this growing challenge.

Not all drought-tolerant maize varieties in Africa must be GMOs. CIMMYT's Drought Tolerant Maize for Africa (DTMA) initiative, funded in 2007 by the Bill and Melinda Gates Foundation and the Howard G. Buffet Foundation, is designed to accelerate the breeding of non-GM drought-tolerant varieties of maize, both hybrids and open pollinated varieties (OPVs) in 13 countries in sub-Saharan Africa. This initiative will use conventional and marker-assisted selection breeding but no transgenic techniques. A second initiative does use GM techniques. This is the Water Efficient Maize for Africa (WEMA) project, funded in 2008 by the Bill and Melinda Gates Foundation and operated in Africa by the African Agricultural Technology Foundation (AATF). CIMMYT is a partner in this project, as is the Monsanto Company. This initiative will use transgenic techniques in addition to conventional and marker-assisted selection.

Regulatory requirements in Africa for GMOs emerge as a critical consideration here. WEMA's genetically engineered (GE) varieties of drought-tolerant (DT) maize will deliver benefits to African farmers only if African regulators first allow the technology to be tested in confined field trials in Africa, and then approve the technology for environmental release and commercial release to farmers. The regulatory gauntlet for this technology will be long and difficult. As of 2011, only three countries in Africa (Republic of South Africa, Kenya and Uganda) had yet given permission for confined field trials of these new GMO varieties. Even with no arbitrary regulatory slowdowns or blockages, it will still require years for GM varieties of drought-tolerant tropical white maize to make their way through the regulatory system in Africa. The first year of WEMA field trials of GM white maize hybrids were completed in 2009 at two sites in the Republic of South Africa: these were subtropical varieties intended to be used by smallholders in RSA and in parts of Mozambique, but field trials in Mozambique itself are not yet approved. Later in 2009, national biosafety regulators in both Kenya and Uganda followed RSA and approved confined field trials for WEMA maize, but regulators in Tanzania did not. Moving beyond the confined field trial stage will be a special challenge for Uganda, since the parliament there has not yet passed a national biosafety bill, a measure that GMO critics insist is needed before a full environmental release of any GMO is permitted. In Kenya, moving beyond the confined field trial stage for WEMA maize will be difficult as well. Field trials of GMO cotton have been underway in Kenya since 2004, yet as of 2011 approval for commercial planting had yet to be given. So even if this new technology performs well, it will not be in the hands of farmers until perhaps 2015 at the earliest, and in a number of countries participating in the project the institutions that are needed to approve planting by farmers are not yet even in place.

WHY DOES AFRICA FOLLOW EUROPE?

Why have so many governments in Africa chosen to follow this highly precautionary European approach toward regulating GM foods and crops, despite the absence of documented evidence of new risks so far, and despite the technology blockages and extended delays that result from the European approach? Five separate channels of external influence on Africa have led to this choice (Paarlberg, 2008).

Bilateral foreign assistance is the first channel of external influence on Africa. Governments in Africa are still significantly dependent on foreign assistance, on average four times as aid-dependent relative to GDP as the rest of the developing world. For this reason, much that takes place in Africa today remains 'donor driven'. Since Africa's official development assistance (ODA) from Europe is three times as large as ODA from the United States, the voice of European donors in Africa tends to be more dominant than any American voice. Governments in Europe have used their ODA to encourage African governments to draft and implement European-style regulatory systems for agricultural GMOs.

A second channel of external influence has been multilateral technical assistance through the UNEP/GEF Global Project for Development of National Biosafety Frameworks (NBFs). Of 23 African governments that had completed a NBF under this UNEP program by October 2006, all but the Republic of South Africa and Zimbabwe had no regulations in place for agricultural GMOs, so UNEP was in effect writing on a blank slate. In the end, 21 of these 23 countries embraced the strongest possible approach (the 'Level One' approach), requiring regulations through binding legal instruments approved by the legislative branch of government (parliament), parallel to the European approach. Europe had greater influence than the United States over this UNEP/GEF program because European governments contribute roughly three times as much to the GEF trust fund as does the United States.

A third channel of external influence has been advocacy campaigns against GMOs from international non-governmental organizations (INGOs), the most active of which are headquartered in Europe. Greenpeace International and Friends of the Earth International, both based in Amsterdam, have campaigned heavily in Africa against agricultural GMOs. Zambian officials were told by Greenpeace that if GMOs were let into their country, organic produce sales to Europe would collapse. A United States organization named Genetic Food Alert warned Zambia in 2002 of the 'unknown and unassessed implications' of eating GM foods, and a British group named Farming and Livestock Concern warned them that GM corn could form a retrovirus similar to HIV. These assertions, which were not backed by any evidence, frightened the Zambians into banning GMOs completely.

A group of mostly European NGOs continued this campaign against GMOs at the 2002 World Summit on Sustainable Development in Johannesburg. Led by Friends of the Earth International, they coached their African partners into signing an open letter warning that GMOs might cause allergies, chronic toxic effects, and cancers. At this same meeting in 2002, two Dutch organizations, HIVOS and NOVIB, joined with partner groups from Belgium, Germany and the UK to finance a 'small farmers' march' on Johannesburg (led by a non-farmer) that ended with a pronouncement that Africans 'say NO to genetically modified foods'.

A fourth channel of external influence has been commercial agricultural trade. Africa's farm exports to Europe are six times as large as exports to the United States, so it is European consumer tastes and European regulatory systems that Africans most often must adjust to. In 2000, private European buyers stopped importing beef from Namibia because it had been fed on GM maize from the Republic of South Africa, and then in 2002, Zambia rejected GM maize as food aid in part because an export company (Agriflora Ltd) and the export-oriented national farmers union (ZNFU) were anxious that exports of organic baby corn to Europe not be compromised. The risks of export rejections from African countries that plant GMOs are actually quite small, as evidenced by the continued growth of food sales to Europe from the Republic of South Africa, yet anxieties surrounding export loss play a political role.

The final channel of external influence is cultural. Most policy-making elites in Africa have much closer cultural ties to Europe than to the United States, so they are naturally inclined to view European practices as the best practices. For example, the Kenyan author of a 2004 article (published by a European-financed NGO, PELUM) which was titled 'Twelve reasons for Africa to reject GM crops', later explained to a newspaper reporter, 'Europe has more knowledge, education. So why are they refusing [GM foods]? That is the question everybody is asking.' Policy-making elites in Africa have often been educated in Europe, they send their children to European schools, and they travel to Europe frequently both on official and unofficial business. It is not surprising that they would be inclined to adopt European-style regulations for GMOs, despite the fact that Africa's needs and circumstances are so different from those of Europe.

External influence of this kind is not unique to Africa, of course. In Latin America, within the sphere of influence of the United States, government policies toward GM crops have usually been closer to the American approach than to the European approach. As of 2010, seven out of the top eleven countries around the world with significant plantings of GMOs were Western Hemisphere countries. It is also telling that the only Asian country to have fully approved GMO maize, the Philippines, was a former American colony.

Political leaders in Africa pay a price in this case by simply 'doing what Europeans do'. Europe has placed stifling regulations on GM foods and crops because Europe itself has little need for this new technology. European farmers are already highly productive without it and European consumers are already well fed. Indeed, like consumers in the United States, Europeans are increasingly over-fed. In Africa, where farmers are not yet productive and where so many consumers are not yet well fed, the potential gains that GM crops can provide are more costly to do without.

Rather than deferring to outsiders, either Europeans or Americans, Africans might usefully look for ways to make independent judgments of their own regarding how to regulate GM crops. Other countries in the developing world that still have large farming sectors and operate relatively free from external influence – such as the People's Republic of China – have so far seen a strong value in this technology, and have been investing significant public budget resources of their own to develop this technology, for their own distinct and independent benefit.

REFERENCES

British Medical Association (BMA) (2004), 'Genetically modified foods and health: a second interim statement', March, London: British Medical Association.
Connor, Anthony J., Travis R. Glare and Jan-Peter Nap (2003), 'The release of genetically modified crops into the environment. Part II. Overview of ecological risk assessment', *The Plant Journal*, **33**, 19–46.
European Union (EU) Research Directorate (2001), 'GMOs: are there any risks?', press briefing, 9 October, Brussels: EU Commission.
European Union (EU) Research Directorate (2010), *A decade of EU-Funded GMO research (2001–2010)*, EUR 24473 EN, Brussels: European Commission.
Food and Agriculture Organization (FAO) of the United Nations (2004), *The State of Food and Agriculture 2003–04: Agricultural Biotechnology: Meeting the Needs of the Poor?*, Rome: FAO.
French Academy of Medicine (2002), 'OGM et santé', recommendations (Alain Rérat), communiqué adopted on 10 December.
French Academy of Sciences (2002), 'Genetically modified plants', Institut de France, Academie des Sciences, Report on Science and Technology, 13 (December).
Helt, Hans Walter (2004), 'Are there hazards for the consumer when eating food from genetically modified plants?', Union of the German Academies of Science and Humanities, Commission on Green Biotechnology, Göttingen: University of Göttingen.
International Council for Science (2003), 'New genetics, food and agriculture: scientific discoveries – societal dilemmas', available at: www.icsu.org.
Kessler, Charles and Ioannis Economidis (eds) (2001), *EU-Sponsored Research on Safety of Genetically Modified Organisms: A Review of Results*, Luxembourg: Office for Official Publications of the European Communities.
Mantell, Katie (2002), 'WHO urges Africa to accept GM food aid', Science and Development Network, 30 August, available at: www.scidev.net/News.
Organisation for Economic Co-operation and Development (OECD) (2000), 'GM food safety: facts, uncertainties, and assessment: rapporteurs' summary', The OECD Edinburgh Conference on the Scientific and Health Aspects of Genetically Modified Foods, 28 February – 1 March.
Paarlberg, Robert (2008), *Starved for Science: How Biotechnology is Being Kept Out of Africa*, Cambridge, MA: Harvard University Press.
Royal Society (2003), 'Where is the evidence that GM foods are inherently unsafe? asks Royal Society', Press Release, Royal Society, 8 May.
Sanvido, Olivier, Jorg Romeis and Franz Bigler (2007), 'Ecological impacts of genetically modified crops: ten years of field research and commercial cultivation', *Advanced Biochemical Engineering/Biotechnology*, **107** (31 March), 235–78, published online.

13 Non-adopters of GM crops in Latin America

José Falck-Zepeda

INTRODUCTION

Producers in a number of countries in Latin America have embraced genetically modified (GM) crops for commercial cultivation and consumption. Countries that have made this choice include Argentina, Brazil, Bolivia,[1] Colombia, Costa Rica, Honduras, Paraguay and Uruguay. The area cultivated with GM crops in these Latin American and Caribbean (LAC) countries represents approximately 63 million hectares, which is roughly 40 per cent of global adoption (James, 2011). Yet this adoption is largely confined to three crops and two traits. Crops include cotton, corn and soybeans, which embody use of insect protection and herbicide tolerance, or combinations of both traits.

To date all of the products have been developed and commercialized by multinational companies. It is not surprising that the adopted GM technologies come from multinationals as they have been the only readily available technologies to farmers in the region. In fact, multinational technologies may be the only ones available commercially for a long time in spite of the many investments in GM technologies by the public sector in LAC countries and elsewhere, as there are very few public sector technologies in advanced stages of the regulatory pipeline (Stein and Rodríguez-Cerezo, 2009). This includes those GM technologies in the regulatory pipeline – generated *by* developing countries *for* developing countries – which are in a relatively advanced state of compliance with biosafety regulations but are not advancing to the deployment to farmers stage (Atanassov et al., 2004). This situation will continue unless countries proactively address and overcome the regulatory hurdles that are slowing down or stopping technology deployment (Nature Biotechnology, 2012).

Most LAC countries have invested significant financial and human resources in a broad range of biotechnology innovations. Investments range from those made to develop traditional biotechnology applications such as plant breeding and tissue culture, to advanced biotechnology techniques such as DNA transformation and functional genomics (Falck-Zepeda et al., 2009). These investments have been the result of existing public policies and private efforts in response to the many challenges typical of the multiple agro-ecological zones and environments. Challenges include the existence of many mega-biodiverse[2] and culturally diverse countries. In addition to the challenges posed by biological and cultural diversity, the Latin American science, technology and innovation system has faced challenges imposed by dual agricultural production systems co-existing and in most cases competing with each other for state investments. Different stakeholders devise positions and policies addressing these issues in different policy arenas, yielding substantially different outcomes.

Alongside long-term investments in biotechnology R&D and the impressive increases in cultivated area with GM crops in the continent, there are several Latin American countries who have not adopted these technologies to date. Some countries have even passed informal or formal bans or moratoriums on the cultivation, use or importation of GM crops. The countries who have not adopted include Peru, Ecuador, Bolivia and Venezuela in South America. In Central America, Guatemala, El Salvador, Nicaragua and Belize and all countries in the Caribbean – with the exception of Cuba – have not adopted these technologies.

The existing policy milieu has shaped Latin American countries' political and decision-making landscape. Two policy questions relevant to examining the adoption of GM crops in LAC countries are: what has driven investments in biotechnology innovations and why have countries in Latin America and the Caribbean taken different policy paths regarding GM biotechnology adoption, use and importation? This chapter draws from the existing literature to explore these questions and provides a general framework discussing the temporal policy development pathway that may help explain existing GM biotechnology R&D investment policies in broad terms, while explaining the apparent paradox of some countries not adopting the technology.

This chapter's discussion will be kept as general as possible following the Institutional Analysis and Development (IAD) framework described by Ostrom (2005) and used elsewhere in this book. The discussion identifies the models, metrics, actors and policy arenas for discussion for broad policy decisions, but abstains from discussing specific (sovereign) country decisions. A country-specific analysis will require in-depth study at the national level, which falls outside the scope of this chapter.

MODELS

We have identified three knowledge strands in the literature that will be useful in the application of the IAD framework to help explain biotechnology innovation in the region and the subsequent pathways taken by countries in Latin America and the Caribbean (LAC). The three strands include: (1) productivity and endogenous growth theory and models; (2) technological dualism and the structure of the agricultural sector; and (3) political economy of agricultural policy and innovation.

Productivity and Endogenous Growth

Productivity and endogenous growth theories and subsequent models focus on the role of technology in promoting growth and externalities as a source of endogenous growth (Solow, 1957; Lucas, 1988). The original conceptual theories were later refined to consider frictions in technology R&D and transfer. The induced innovation model (Hayami and Ruttan, 1985; Binswanger and Ruttan, 1978) showed that technical innovation is most often directly tied to the relative incentives faced by farmers.

The technology treadmill model (Cochrane, 1958; Gardner, 2002) indicated that productivity gains depend critically on the price elasticity of demand for the product under analysis. In small, open economies, typical of most developing countries, price elasticity tends to be almost or even perfectly inelastic, thus most gains from innovation tend to flow to consumers. In the long run, producers tend not to gain from technological change, and only pursue newer technologies to capture short-term gains in productivity. This perpetual search for profitable innovations that are then bid away by the market is the basis for the technological treadmill.

Schultz (1964) proposed the 'poor but efficient hypothesis' which suggested that most developing country farmers lack technological and financial resources to induce gains in their production possibilities frontier that would allow welfare gains and thereby improve their situation. Nevertheless, most developing countries' producers operate at the most efficient production frontier given existing resources. The implication of this hypothesis is that the only way to improve farmer welfare in developing countries would be to make substantial investments in technology and complementary inputs to help producers move their production possibilities frontier.

The role that agricultural technology may have in promoting agricultural and economic growth triggered subsequent discussion about the process of technology adoption and the potential impacts from adoption. The seminal paper by Griliches (1957) introduced the concept of technology adoption dynamics while other papers introduced plausible explanations why, for example, larger farms or households may be first adopters of new technologies (Feder et al., 1985; de Janvry, 1981; Feder, 1985).[3] The later discussion in the literature is important in the context of the technology dualism co-existing in many LAC economies, the focus of our next sub-section.

These models and hypotheses provide the intellectual framework that supported the significant investments by multilateral development agencies and developed and developing countries that gave rise to the Green Revolution in the 1950s to 1970s, and for the continued investments in agricultural innovation and R&D in LAC countries and other developing regions.

Technology and Sectorial Dualism and the Structure of the Agricultural Sector

Models in the development economics literature have introduced the concept of dualism. Lewis (1954) considered that growth in developing economies could be described and might be limited by the existence of dual sectors in the economy. One can find modern and capitalistic activities coexisting with traditional activities. Often rural farmers tend to be subsistence farmers, who are frequently poor and excluded from the formal market driven economic activity system.

A quite related model to the sector model is the technology dualism model proposed by Eckaus (1955) and Higgins (1956). These models focus on technology dissimilarities to explain the differences between the two tracks in the economy. Technological dissimilarities may be a consequence of divergent production possibilities functions and frontiers based on biased technical change and uneven technological endowments. Alternative explanations focus on differences in input or factor endowments between the dual streams.

Dualism has also been expanded to explain dissimilarities between developed and developing countries' pathways and the deteriorating terms of trade observed by Singer (1970), Myrdal (1968) and Prebisch (1950), amongst others. These models have in the past guided economic development efforts and public policy decisions in Latin America and the Caribbean, leading to import substitution policies, heavy concentration on industrial development and a significant decrease in agricultural investments. As a result, agriculture has declined in importance for investors as an engine of growth and as a causal agent determining social welfare in developing countries. This downward trend has not turned around, although there are some signs that investors are re-examining the role that agriculture can play as a source of economic growth (see World Bank, 2008).

From the standpoint of biotechnology innovation and policy development in Latin America, these models are important as the region has one of the most unequal economic distributions in the world (Lopez and Perry, 2008; Puryear and Malloy Jewers, 2009) and wide divergences between the dual sectors of the economy. The modern sector may have access to modern technology, financing and credit, supplementary inputs and other productivity-enhancing inputs, as well as robust connections to markets and marketing channels. Institutional constraints including the availability of critical inputs, which may include biotech seeds, often determine the level and distribution of benefits to farmers vis-à-vis other actors in the economic system (Smale et al., 2009).

In turn, the traditional sector may not have such factor endowments and market access, and tends to be economically poor and under-developed. Furthermore, in LAC countries, the largest share of economic poverty in the population falls disproportionately on the rural poor, and in some countries, on indigenous and non-white ethnic communities. These segments of the population often have other concerns such as protection of norms, customs and ancestral lands which work to shape their positions vis-à-vis the modern sector and technological innovations. In many situations, modern biotechnologies and technologies in general may not have been relevant to their crops or traits of interest. As a result, traditional producers face binding productivity constraints.

The persistent co-existence between both the modern and traditional sectors and the relentless perseverance of economic inequality in Latin America and the Caribbean is perhaps one of the drivers for policy and decision-making positions, pressure groups, and other formal and informal organizations that have opposed GM biotechnologies. The latter groups may in fact oppose most 'modern' technologies because they seem to reinforce duality and poverty. At times they assert that the new technologies may be irrelevant or not economically justifiable under the set of decision-making parameters available to traditional producers. These groups include indigenous, land-rights-based and landless movements on the continent. As poverty tends to be associated with ethnicity and race, duality can also be a driver for the indigenous and ethnically-based organizations and movements.

Political Economy and Institutional Analysis Considerations

The Institutional Analysis and Decision (IAD) Framework proposed by Ostrom (2005) and the political economy approaches[4] are quite similar in their focus and objectives. In fact, it may be more productive to recognize the similarities between both approaches in their implementation. Duncan and Williams (2010) outline some of the central characteristics which can be considered common between both approaches, including that both consider the centrality of politics while focusing on institutions to describe the incentives frameworks that guide behavior. In addition, both approaches focus on understanding the intricacies of the systems or country situations and thus base analysis and potential strategies for improvement on a strong contextual base. Furthermore, both approaches describe the factors that shape the political and institutional processes and the arenas for interaction. Whether a study emphasizes more politics or institutions may determine whether it is labeled one or the other. We identify in the next sections relevant metrics, actors and policy arenas that may become part of a more in-depth systematic analysis.

METRICS: APPLICATIONS TO NON-ADOPTERS AND SOME ADOPTERS IN LATIN AMERICA

A study commissioned by the Inter-American Development Bank (IADB) examined 208 public and private sector R&D organizations in 18 Latin American countries (Falck-Zepeda et al., 2009). This study showed that all of these countries were implementing a wide range of agricultural biotechnology techniques.[5] Yet most of the applications (71 per cent) were traditional techniques such as tissue culture or marker-assisted selection. As Table 13.1 shows, there are significant differences in terms of countries and sectors, although most of the effort has been made by the public sector.

The IADB study showed that approximately 89 per cent of annual investments in

Table 13.1 Number of agricultural biotechnology techniques used by sector and country

Country	Private sector	Public sector	TOTAL
Adopters			
Argentina	57	182	239
Brasil	NA	288	288
Chile	10	170	180
Colombia	8	150	158
Costa Rica	15	53	68
Honduras	4	10	14
Mexico	22	64	86
Paraguay	NA	15	15
Uruguay	17	100	117
Non-adopters			
Bolivia	NA	61	61
Ecuador	14	27	41
El Salvador	3	11	13
Guatemala	12	12	24
Nicaragua	NA	7	7
Panama	NA	19	19
Peru	12	186	198
Dominican Republic	8	16	24
Venezuela	NA	53	53
Total for LAC	182	1424	1606

Notes:
NA = Not available.
The case of Mexico is very particular because the study focused more on the type of techniques across organizations performing R&D. In this case, at the aggregate level Mexico has functional capacity across all types of techniques.
Techniques include cell and tissue culture, molecular markers, diagnostic kits, recombinant DNA, genetic transformations, functional and structural genomics. Counts include agricultural applications for crops, animals and micro-organisms but do not include pharmaceuticals.
The Caribbean region is only represented by Dominican Republic in this table. This table does not include Cuba, which has significant investments in biotechnology.

Source: Falck-Zepeda et al. (2009) based on country surveys part of a study commissioned by IADB, 2006–2007.

Table 13.2 Total investments in agricultural biotechnology in Latin America by region and/or country (thousands US$)

Region/Country	Private	Public	Total
Mexico	–	24 775	24 775
Central America and Dominican Republic	–	6 309	6 309
Southern Cone	4 500	8 322	12 822
Brazil	13 761	55 046	68 807
Andean Region	5 716	14 545	20 261
Total	23 978	108 996	132 974

Note: Columns for public and private sector do not add to the 'total' column as in some countries the authors did not know the distribution between sectors, and/or the confidentiality promises included in the surveys limit how much data we can show about individual countries.

Source: Falck-Zepeda (2009) based on surveys done in study commissioned by the IADB, 2006–2007.

2007 in the region were made by the public sector. Total annual investments in agricultural biotechnology in 2007 were estimated to be approximately US$132 million (see Table 13.2). As the authors of the study are careful to point out, this estimate implies a significant underestimation of the actual level of investments due in particular to the under-reporting in Brazil and Mexico caused partly by the sensitive nature of biosafety and biotechnology policy discussions at the time. Since the study, the situation in Brazil has changed substantially with the biosafety system re-structuring and with new national priorities related to energy independence and on the potential for agriculture to supply different types of fuels and biomass.

Based on the results of the IADB study and other data collected, Trigo et al. (2010) proposed a set of quantitative and qualitative determinants that would allow categorization of different policy situations in each of the 18 countries included in the IADB study. The policy situation covers those policies relevant to biotechnology innovation in the region. Policies include public sector investments, intellectual property management, biosafety regulations, food and feed safety, consumer protection and public sector support to technology transfer and seed systems.

Table 13.3 provides a summary of the policy situations for adopters and non-adopters in the region. As shown in the table, non-adopters tend to have policies that work to prevent or are at most neutral towards biotechnology innovation. In some cases the lack of adoption may be more related to low investments and relatively lower innovative capacity compared to those countries that are adopters. Obviously one needs to examine in much more detail the specific policy situations for each country to help explain differences in terms of science, technology and innovation – such as potential market size, explicit public sector policies and the trends over time for many of these and other determinants – to ensure a full appreciation of the realities of GM biotechnology adoption in LAC countries.

Table 13.4 introduces a classification of countries by innovation category and potential market size. As described in Trigo et al. (2010), even though the authors of the study used several quantitative and qualitative indicators to situate a country in a specific cell in the table, there is always a high degree of subjectivity in categorizing each specific

Table 13.3 Summary of policy situations in selected non-adopter and adopter countries in Latin America

	Public sector investments in biotechnology applications	Intellectual property management	Biosafety regulations	Food/feed safety and consumer protection	Support for public sector participation and tech transfer including seed systems
Non-adopters					
Bolivia	0	0	–	–	0
Ecuador	0	0	–	–	0
El Salvador	0	–	0	–	–
Guatemala	0	–	0	0	–
Nicaragua	0	–	0	0	–
Peru	0	–	–	0	0
Venezuela	+	–	–	0	0
Adopters					
Argentina	+	0	0	+	+
Brazil	+	–	0	0	+
Costa Rica	+	–	0	0	+
Honduras	0	–	0	0	–
Mexico	+	0	0	0	+
Uruguay	+	0	0	0	+

Notes:
+ signifies promotional policies, 0 denotes neutral policies, – reflects preventive policies.
Brazil was categorized as having a preventive biosafety policy in Trigo et al. (2010), but is reclassified here as neutral, based on recent developments in the country.

Source: Selected countries from Trigo et al. (2010).

Table 13.4 Policy situation classification and market size distribution for selected countries in Latin America

	Small markets	Medium markets	Large markets
Non-selective importers of technology	*El Salvador*, *Guatemala*, Honduras, *Nicaragua*, *Panama*	*Bolivia*, *Ecuador*	
Selective importers of technology	Costa Rica, Uruguay	Paraguay, *Peru*	*Venezuela*
Tool users	–	Colombia, Chile	Argentina, Mexico
Innovators			Brazil

Note: Non-adopters are listed in italic text.

Source: Trigo et al. (2010).

Table 13.5 *Country bans and/or moratoriums on the use of GMOs in Latin America*

Country	Relevant formal and informal policies
Peru	2011 Law No. 29811 'Law which establishes a moratorium to the introduction and production of living modified organisms to the national territory for a 10 year period.' This Law annuls Supreme Decree 003-2011-AG which described and authorized a biosafety regulatory system. Law No. 29811 seems to be under review.
Ecuador	Article 401 of the 2008 Constitution: 'Ecuador is declared free of transgenic seeds and crops. Exceptionally, and only in those cases of national interest properly documented by President's Office and approved by the National Assembly, the introduction of genetically modified seeds and crops will be allowed. The State will regulate under strict regulatory norms the use and development of modern biotechnology and its products, in addition to experimentation, use and commercialization. It is forbidden the application of risky or experimental biotechnologies.'
Bolivia (Plurinational Republic of)	• The 2008 Bolivian Constitution (ratified 2009) – Article 255 No. 8 'Food security and sovereignty for all the population; the importation, production and commercialization of genetically modified organisms and other toxic elements which damage health and the environment is forbidden.' – Article 409 'The production, importation and commercialization of transgenics will be regulated by Law.' • Administrative Resolution No. 016/2005 *approving* cultivation of RR soybeans • Law 071 of the Plurinational State December 2010 'Law of Rights of Mother Earth'
Venezuela (Bolivarian Republic of)	In 2004 President Chavez launched a moratorium on the use and cultivation of GM crops in its territory.

Note: Author compilation and free translation from original documents.

country. There is also some degree of uncertainty as the political or policy arena is dynamic. For example, Brazil in the Trigo et al. (2010) paper was originally classified as a large market tool user. Here it is classified as an innovator as the biosafety policy has been clarified significantly since the source paper was published. In another example, Costa Rica and Uruguay could arguably be classified as tool users, in spite of small potential markets.

Table 13.4 shows that most non-adopters tend to be non-selective importers of technology. However, there are some non-adopting countries that have evolved into selective importers of technology, including Peru, Venezuela, Bolivia and Ecuador. Furthermore Honduras is a bit of an outlier as a non-selective importer with a small market but also an adopter. To help explain some of these differences, Table 13.5 provides a compilation of relevant formal or informal moratorium or ban policies in Latin America. These laws and policies can at times completely limit the potential deployment of a GM biotechnology

Table 13.6 Actors and organizations and their positions with regard to the adoption and use of GM biotechnologies

Tend to favor GM technology	Tend not to favor introduction	Tend to be neutral
• Private sector • Large and medium producers • Rural producers • Public R&D and technology transfer sector • Commercial producer associations • Government/State • Donor and multilateral investment agencies	• Smallholder and landless producers • Indigenous communities • Urban and/or higher income consumers • Pressure groups: Environmental, indigenous movement and land rights • 'Campesino'/Peasant/ Landless • Government/State • Donor and multilateral investment agencies	• Regulatory agencies • Government/State • Donor and multilateral investment agencies

Note: Author's compilation.

in these countries. To determine why these countries chose this specific policy option one would need to examine in more detail the internal policy arenas and their actors.

INTERESTS AND INSTITUTIONS

Groups

Table 13.6 includes a list of potential groups of actors and their key positions with regard to GM biotechnologies. As Kikulwe et al. (2010, 2011) remind us, there may be significant heterogeneity amongst different groups in society such as producers and consumers of a specific commodity. These differences may shape perceptions and influence policy. In some cases, the proximity of a particular group to the decision-making center(s) may help explain how much they influence policy outcomes. In Kikulwe's papers, urban consumers in Uganda tended to have a negative view of GM crops in general, whereas rural producers, who were the largest beneficiaries from the potential adoption of a GM banana technology, were mostly positively disposed to the potential release of such technology. This is a good reminder that there may be significant variation in and among individual groups as to their views on technology.

Groups favoring the deployment of a GM technology tend to be those who participate in the 'modern' sector of the economy. Those that do not favor or even oppose the introduction of a GM biotechnology often belong to the 'traditional' sector of the economy. The latter group probably includes those who have been marginalized by government policies; these groups frequently have a high share of the poor and the ultra-poor that may be hampered by critical institutional failures including poor access to productive inputs and information.

As shown in Table 13.6, different government or state actors in the same country may be found in each the categories. This reflects the fact that different ministries within a specific country may have different positions with regard to the adoption and use of GM biotechnologies, in part due to differences in their objectives and time horizons for decision making. For example, Ministries of Agriculture or Science and Technology, which are usually tasked with providing appropriate technologies to producers and to improve productivity, tend to favor GM biotechnologies. In turn, Ministries of Environment or Natural Resources may not favor the use of GM biotechnologies, in part due to the need to preserve biodiversity and or in response to other environmental considerations.

The seemingly conflicting positions between ministries and their objectives deserve a much more detailed examination and further study. Certainly the different drivers pushing for increases in agricultural production and productivity will continue putting pressure on the agricultural and policy systems. Yet governments increasingly recognize the need to address environmental concerns to ensure long-term sustainability of the agricultural system. In turn, protecting the environment, particularly those ecosystems too fragile for agricultural production and mega-diverse regions, may best be achieved by improving agricultural productivity to ensure that no more land will be needed to satisfy biomass needs. In some instances, this might involve withdrawing from those areas which are too fragile for agricultural production.

Supranational Policy Arenas

National policy outcomes are often shaped by discussions in supranational policy arenas. In the case of LAC countries, several policy arenas may be relevant in terms of shaping national policies towards GM biotechnologies. At the international/multilateral level, international treaties such as the Convention of Biological Diversity and the Cartagena Protocol on Biosafety have framed and in some cases have been the shaper of national policies towards biosafety and biotechnology adoption. The precautionary principle has been a major driver for the implementation of biosafety regulatory systems worldwide.[6] Biosafety regulatory systems, particularly in those countries that are signatories to the Cartagena Protocol on Biosafety, have for the most part incorporated the precautionary principle, albeit using different interpretations ranging from weak to strong. The strongest interpretations, of the precautionary principle can lead to bans or moratoriums as it requires an assurance of no risk, which is demonstrating impossibility.

The World Trade Organization (WTO), designed to facilitate trade amongst nations by articulating the rules for conducting international trade, presents real boundaries for national policies. Trade issues may arise when countries have different procedures or requirements to test and approve the use or importation of GM crops or in the context of trade related to materials protected by intellectual property. The WTO has already had an effect on countries and their policies related to GM biotechnologies. Given that the WTO mission is to facilitate trade, the point of intersection with the Cartagena Protocol on Biosafety will be explicitly related to trade issues. So far, the only trade dispute related to GM biotechnologies examined in the WTO has been the EU moratorium, which does not provide too much guidance in terms of the relationship between the existing biosafety framework at the Cartagena Protocol on Biosafety, national regulations and trade.

Several regional trade and cooperation blocs may also become relevant in supporting

GM biotechnology decision-making, including the Central American Integration System (SICA), Andean Integration System and the Andean Pact, and the Southern Cone/MERCOSUR. To date these blocs have not developed a strong position related to GM biotechnologies. Another relevant bloc is the Bolivarian Alliance for the Peoples of Our America, better known by its Spanish acronym ALBA, for Alianza Bolivariana para los Pueblos de Nuestra América, which is an international cooperation organization and treaty whose member countries, many of which can be categorized as social democratic governments, include Antigua and Barbuda, Bolivia, Cuba, Dominica, Ecuador, Nicaragua, Saint Vincent and the Grenadines, and Venezuela. Suriname and Saint Lucia have been admitted as guest countries.

ALBA is based on socialist ideals with the objective of pursuing social welfare, redistribution of wealth and mutual economic aid. The organization seeks an all-inclusive social, political and economic integration process amongst its members, including the introduction of the Sucre, a new regional currency for trading purposes amongst the bloc partners. The organization is designed to be a model for developing country integration as an alternative to the liberal model based on trade, markets and economic liberalization embodied in the WTO and the Free Trade Alliance for the Americas (Al Attar and Miller, 2010). The ALBA bloc has been closely associated with the resurgence of a strong environmental movement, indigenous rights groups, and pressure groups and the 21st Century Revolution, initially supported by Venezuela and Cuba. Although ALBA does not seem to have developed a specific policy towards GM biotechnologies, it has stated anti-corporate, anti-capitalist, anti-competitive and anti-neoliberal sentiments in the past, which may put it in opposition to the currently available technologies in LAC countries. Most of its country members have passed moratoriums or bans on GM biotechnologies, with the exception of Cuba.

CRITICAL ASSESSMENT

Most of the LAC countries that have not adopted GM biotechnologies tend to have insufficient scientific, technology transfer and institutional capacity for biotechnology innovation. In many cases, relatively poor biotechnology innovation capacity coincides with significant political pressures and other policy developments which have worked to shape official policies reflecting a negative approach towards GM technologies. To establish how effective these political pressures and policy developments have been in mobilizing government action (especially moratoriums or bans) may need further in-depth analysis of the leading groups mobilizing against adoption, the policy arena itself and the overall political and institutional processes in non-adopting countries. As Takeshima and Gruère (2011) suggest, anti-GMO lobbying efforts may be more successful in those countries where conditions may not be favorable for the introduction of a GMO in the first place. In fact, the results from that paper indicate that where conditions are favorable for the introduction of GMOs, anti-GMO lobbying has not been as successful.

All adopters and non-adopters in Latin America have made some investments in biotechnology applications but most have been for traditional techniques used by the public sector directed at a wide variety of crops, animals and micro-organisms. This policy stance is in line with the productivity and endogenous growth models which recommend investments in science, technology and innovation as a prerequisite to promote economic growth – but the theory is clear that these investments are necessary but not sufficient for optimal growth. In spite of these R&D investments in GM biotechnology, to date there have been very few public or private products resulting from these investments that have been deployed to farmers, at least partly due to binding biosafety regulatory hurdles and other regulatory requirements imposed by LAC countries.

The implementation of stricter regulatory hurdles, in some cases beyond what can be considered necessary to guarantee a level of safety based on current evidence, implies seemingly contradictory governance processes. LAC governments may incur such contradictory processes if they are promoting improvements in agricultural productivity through biotechnology while at the same time putting obstacles in the way of specific biotechnology techniques. This policy outcome may be due to the lack of coordination between the agricultural, environmental and natural resources ministries in the region. Tensions between those ministries are often connected to their roles and international obligations. For example, Ministries of Environment tend to represent their governments at the Cartagena Protocol on Biosafety and at the Convention of Biological Diversity, which leads them to focus primarily on biodiversity preservation. In turn, Ministries of Agriculture, Science and Technology or Trade usually focus more on improving productivity or maximizing trade only.

In turn, LAC adopters tend to have better investments in biotechnology scientific capacity or to have facilitated the commercialization and regulatory approval processes so that they may take advantage of technologies developed elsewhere. An example of this is Honduras, which has little or no domestic biotechnology innovative – in fact, Honduras has little overall R&D capacity at all – yet it has an explicit policy to improve agricultural productivity. This has led to the development of a science and evidence-based regulatory approval process especially focused on assessing those technologies approved elsewhere. The resulting outcome has been the approval of at least six GM events so far in the country.

Latin America and the Caribbean is a region of contrasts. On the one hand it holds one of the most diverse income inequality conditions in the world. At the same time it has some countries which have invested significantly in science, technology and innovation as a notable public policy. LAC faces major challenges including production of biomass and the protection of mega-biodiverse ecologies and cultural diversity. These challenges will become even more difficult with increases in climate change variability and changes in biomass production for other than food and feed purposes. There is no doubt that LAC countries will need to increase their food production and productivity to ensure food security and to improve economic conditions for their populations.

Such improvements will require binding productivity constraints to be addressed through the best available options and through multi-faceted and synergistic approaches to science, technology and innovation. This will undoubtedly require focused effort on specific crops, traits and binding production and productivity constraints, while at the same time ensuring environmental protection and sustainability over time. This future will necessitate the prudent use of biotechnologies, including those that produce genetically-modified transformations and genomics, amongst others. LAC cannot afford to disregard any option that can prove to be valuable over time. This is the prudent course of action, especially if efforts are focused on the crops, traits and productivity constraints critical to the region.

NOTES

1. Bolivia is a very special case of adoption and use. Herbicide-tolerant soybeans were introduced illegally into the country – without any formal regulatory review or approval – and were massively adopted especially in the Santa Cruz Province. The Bolivian government developed a policy to ban the cultivation of all GM crops. This policy was to be later included in the new 2008 Bolivian constitution. After heavy resistance from producers in Santa Cruz who have adopted the herbicide tolerance technology in significant numbers, the Bolivian government authorized in 2005 the legal commercial cultivation through an administrative resolution (No. 016/2005), valid for that event only.
2. This of course includes mega-biodiversity for pests and diseases, typical of tropical ecologies (Roca, personal communication 2012).
3. See Barrett et al. (2010) for a more detailed description of this literature.
4. Approaches such as those used and/or proposed by SIDA (Powers of Change), DFID (Drivers of Change), World Bank (Poverty and Social Impact Analysis). See Collinson (2003), DFID (2009) and Duncan and Williams (2010) for a more extensive description of these approaches to political economy analysis.
5. This study considered only agricultural biotechnologies. The study did not include medical or industrial applications, although capacity in developing the latter may be used to develop agricultural applications. Thus there may be some underestimation of existing innovative capacity in the region. Another important consideration is the fact that Cuba was not included in the study. Cuba has a quite innovative biotechnology sector which may be considered in many ways cutting edge (*Nature*, 2009).
6. There are several definitions of the precautionary principle or approach. One general definition is that if there is a potential risk that can cause damage resulting from a policy or action, and even though there is no scientific consensus about the risk and its likelihood, then the burden of proof to determine that it is not harmful falls on those doing the action or the policy.

REFERENCES

Al Attar, M. and R. Miller (2010), 'Towards an emancipatory international law: the Bolivarian reconstruction', *Third World Quarterly*, **31**(3), 347 – 63.

Atanassov, A., A. Bahieldin, J. Brink, M. Burachik, J.I. Cohen, V. Dhawan, R.V. Ebora, J. Falck-Zepeda, L. Herrera-Estrella, J. Komen, F.C. Low, E. Omaliko, B. Odhiambo, H. Quemada, Y. Peng, M. J. Sampaio, I. Sithole-Niang, A. Sittenfeld, M. Smale, Sutrisno, R. Valyasevi, Y. Zafar and P. Zambrano (2004), 'To reach the poor: results from the ISNAR-IFPRI next harvest study on genetically modified crops, public research, and policy implications', EPTD Discussion Paper 116, Washington, DC: International Food Policy Research Institute.

Barrett, C.B., M.R. Carter and P. Timmer (2010), 'A century-long perspective on agricultural development', *American Journal of Agricultural Economics*, **92**(2), 447–68.

Binswanger, Hans and V.W. Ruttan (1978), *Induced Innovation: Technology, Institutions and Development*, Baltimore, MD: Johns Hopkins University Press.

Cochrane, Willard W. (1958), *Farm Prices: Myth and Reality*, Minneapolis, MO: University of Minnesota Press.

Collinson, Sarah (2003), *Power, Livelihoods and Conflict: Case Studies in Political Economy Analysis for Humanitarian Action*, London: Overseas Development Institute, available at: http://www.odi.org.uk/resources/download/241.pdf.

De Janvry, Alain (1981), *The Agrarian Question and Reformism in Latin America*, Baltimore, MD: Johns Hopkins University Press.

DFID (2009), 'Political economy analysis: how to note', A DFID Practice Paper, Department for International Development, London, available at: http://www.odi.org.uk/events/2009/07/23/1929-dfid-note-political-economy-analysis.pdf.

Duncan, Alex and G. Williams (2010), 'Making development assistance more effective by using political economy analysis: what has been done and what have we learned?', The Policy Practice, Brighton, available at: http://www.thepolicypractice.com/papers/16.pdf.

Eckaus, R.S. (1955), 'The factor proportions problem in underdeveloped areas', *American Economic Review*, **45**(4), 539–65.

Falck-Zepeda, Jose B., C. Falconi, M.J. Sampaio-Amstalden, J.L. Solleiro Rebolledo, E. Trigo and J. Verástegui (2009), 'La biotecnología agropecuaria en América Latina: Una visión cuantitativa', IFPRI Discussion Paper 860SP, Washington, DC: International Food Policy Research Institute (IFPRI), available at: http://www.ifpri.org/sites/default/files/publications/ifpridp00860sp.pdf.

Feder, G. (1985), 'The relation between farm size and farm productivity: the role of family labor, supervision and credit constraints', *Journal of Development Economics*, **18**(2–3), 85–101.
Feder, G., R.E. Just and D. Zilberman (1985), 'Adoption of agricultural innovations in developing countries: a survey', *Economic Development and Cultural Change*, **33**(2), 255–98.
Gardner, Bruce L. (2002), *American Agriculture in the Twentieth Century: How it Flourished and What it Cost*, Cambridge, MA: Harvard University Press.
Griliches, Z. (1957), 'Hybrid corn: an exploration in the economics of technological change', *Econometrica*, **25**(4), 501–22.
Hayami, Yujiro and V. Ruttan (1985), *Agricultural Development: An International Perspective*, Baltimore, MD: Johns Hopkins University Press.
Higgins, B. (1956), 'The dualistic theory of underdeveloped areas', *Economic Development and Cultural Change*, **4**(2), 99–115.
James, Clive (2011), 'Global status of commercialized biotech/GM crops: 2011', *International Service for the Acquisition of Agri-Biotech Applications (ISAAA) Brief*, **43**, Ithaca, NY: ISAAA.
Kikulwe, E.M., J. Wesseler and J. Falck-Zepeda (2011), 'Attitudes, perceptions, and trust: insights from a consumer survey regarding genetically modified banana in Uganda', *Appetite*, **57**(2), 401–13.
Kikulwe, E.M., E. Birol, J. Wesseler and J. Falck-Zepeda (2010), 'A latent class approach to investigating demand for genetically modified banana in Uganda', *Agricultural Economics*, **42**(5), 547–60.
Lewis, Willliam A. (1954), *Economic Development with Unlimited Supplies of Labor*, Manchester: Manchester School of Economic and Social Studies.
Lopez, J.H. and G. Perry (2008), 'Inequality in Latin America: determinants and consequences', Policy Research Working Paper 4504, Washington, DC: The World Bank.
Lucas, R.E. (1988), 'On the mechanics of economic development', *Journal of Monetary Economics*, **22**, 2–42.
Myrdal, Karl G. (1968), *Asian Drama: An Inquiry into the Poverty of Nations*, New York: Pantheon.
Nature (2009), 'Cuba's biotech boom: Editorial', *Nature*, **457**(7226), 130.
Nature Biotechnology (2012), 'Agnostic about agriculture: Editorial', *Nature Biotechnology*, **30**(3), 197.
Ostrom, Ellinor (2005), *Understanding Institutional Diversity*, Princeton, NJ: Princeton University Press.
Prebisch, Raul (1950), *The Economic Development of Latin America and its Principal Problems*, New York: United Nations.
Puryear, Jeffrey and M. Malloy Jewers (2009), 'How poor and unequal is Latin America and the Caribbean', Inter-American Dialogue Policy Brief 1, Washington DC, USA: Inter-American Dialogue.
Roca, Maria M. (2012), personal communication.
Schultz, Theodore W. (1964), *Transforming Traditional Agriculture*, New Haven, CT: Yale University Press.
Singer, H.W. (1970), 'Dualism revisited: a new approach to the problems of the dual society in development countries', *Journal of Development Studies*, **7**(1), 60–75.
Smale, Melinda, P. Zambrano, G. Gruere, J.B. Falck-Zepeda, I. Matuschke, D. Horna, L. Nagarajan, I. Yerramareddy and H. Jones (2009), 'Measuring the economic impacts of transgenic crops in developing agriculture during the first decade: approaches, findings, and future directions', Food Policy Review 10. Washington, DC: International Food Policy Research Institute (IFPRI), available at: http://www.ifpri.org/sites/default/files/publications/pv10.pdf.
Solow, R.M. (1957), 'Technical change and the aggregate production function', *Review of Economics and Statistics*, **39**(3), 312–20.
Stein, A.J. and E. Rodríguez-Cerezo (2009), 'The global pipeline of new GM crops: implications of asynchronous approval for international trade', JRC Technical Report EUR 23486 EN, Luxembourg: Office for Official Publications of the European Communities, available at: http://ipts.jrc.ec.europa.eu/publications/pub.cfm?id=2420.
Takeshima, H. and G. Gruère (2011), 'Pressure group competition and GMO regulations in Sub-Saharan Africa: Insights from the Becker model', *Journal of Agricultural & Food Industrial Organization*, **9**(1), available at: http://dx.doi.org/10.2202/1542-0485.1325.
Trigo, E., J. Falck-Zepeda and C. Falconi (2010), 'Biotecnología agropecuaria para el desarrollo en América Latina: oportunidades y retos', Working Paper LAC/01/10, Rome, Italy: Programa de Cooperación – FAO/Banco Interamericano de Desarrollo – Servicio para América Latina y el Caribe-División del Centro de Inversiones.
World Bank (2008), *World Bank Development Report 2008: Agriculture for Development,*: Washington, DC: The World Bank.

14 The Cuban context for agriculture and innovation

Carlos G. Borroto

1 INTRODUCTION

1.1 Cuban Geography, Climate and Population Characteristics

Cuba is an archipelago located in the Caribbean Sea, with a total area of 110 860 km^2, making it the seventeenth largest island in the world by land area. The main island of Cuba constitutes most of the nation's land area. Terrain is mostly flat to rolling plains, with rugged hills and mountains in the south-east. The lowest point is at sea level and the highest point is Pico Turquino at 1974 m, which is part of the Sierra Maestra mountain range, located in the south-east of the island.

According to the modified Köppen classification, the predominant climate of Cuba is mild-hot tropical with a rainy summer season. It has a high maritime influence. The country has an annual average rainfall of 1200 mm, with around 30 per cent of the precipitation in the winter period (November to April) and the remaining 70 per cent in the summer (May to October); in general rains are more abundant in the occident of the country than in the east. The average temperature is 23°C in January and 27°C in July. Cuba lies in the path of hurricanes, which are most common in September and October. During recent years, extreme weather conditions, as a consequence of global climate change, have been more common in Cuba. These changes involve longer drought periods, higher frequency of hurricanes and flooding and elevation of sea levels with associated saline intrusions.

Cuba has approximately 11.2 million inhabitants, with 101 inhabitants per km^2. During the last 50 years there has been a big urban population increase with a declining rural population. This trend is expected to continue according to official forecasts. As with many other countries, the Cuban population is rapidly aging.

1.2 Education and Scientific Environment

After the triumph of the Cuban revolution in 1959, special efforts were undertaken in education, culture and science in an effort to convert human resources into a key asset of the country. This has been the cornerstone of Cuba's scientific development ever since. Most of the present research centres in Cuba were born of research groups, or started out as institutes, in the decades immediately following the Cuban revolution in 1959. Some of these centres, like the National Centre for Scientific Research (CENIC) founded in 1965, played an essential role in training young science students at home and in the eventual establishment of many other research institutions (Clark, 2010).

By 2000, Cuba was perceived as a proficient country in terms of scientific capacity, despite having experienced more than four decades of a trade embargo and restrictions on scientific exchanges imposed by successive US administrations (Pastrana and Clegg, 2008).

Human health and agriculture have been among the main priorities for research and development (R&D) in Cuba. Biotechnology development is well recognized and some assert that Cuba has the developing world's most established biotechnology industry (*Nature* Editorial, 2009). Cuba is a member of the World Trade Organization (WTO), and hence a signatory of the TRIPs Agreement which provides for strict rules of intellectual property, according to the international practices. Due to the beneficial impacts of several scientific efforts over the past decades, Cuban citizens have, in general, an excellent perception about the national science capacity and the role of national scientists in socio-economic development.

1.3 Social Context and Agriculture Changes

Before 1959, Cuban agriculture was characterized by mono-crop exploitation of sugar cane, with a high concentration of land ownership and high presence of foreign companies, mainly from the US.

Following the 1959 revolution, two profound agrarian reforms changed the situation dramatically. By the mid-1980s, 70 per cent of the agricultural land was state run. During this period the living condition of the rural population increased significantly in terms of economic development, health, education and general welfare. Nevertheless, the highly technological model of agriculture implemented was highly dependent on foreign inputs and created several cases of serious environmental damage.

In 1991, with the disappearance of the Soviet bloc, Cuba's main economic alliance, main source of agricultural inputs and the main export market all vanished, with an enormous impact on the Cuban economy and particularly on Cuban agriculture, which experienced an abrupt decline.

Notwithstanding that shock, this was a good period to apply more sustainable forms of agriculture. Instead of chemical products, the use of biopesticides and biofertilizers and other environmental friendly practices were promoted; Cuban agriculture was converted to one of the more organically-oriented and sustainable in the world (Oppenheim, 2001; Febles et al., 2011).

Also remarkable was the massive development of urban agriculture, also with an organic approach (Companioni et al., 2001). This was the result of a unique combination of highly developed scientific capacity and the ecological approach of various programmes and groups, with the lack of resources to continue an input-oriented agriculture. This set of biophysical and social contexts and the resulting 'rules in use' largely defined how biotechnology has been applied to food production in Cuba.

2 AGRICULTURAL BIOTECHNOLOGY IN CUBA

Very often, the term 'modern biotechnology' refers only to transgenic technologies. For the developing world, other applications of biotechnology are extremely important too. The range includes: the development of new bioproducts with better formulation and known action mechanisms; mass plant propagation based on tissue culture techniques; marker assisted selection to improve the efficacy of plant and animal breeding programmes; use of molecular technologies for the diagnosis of plant and animal diseases;

growth enhancement of animals with better feed conversion rates; and new biotech-based vaccines and drugs for animal health. In this chapter, both uses of biotechnologies in agriculture will be included, as it is strongly believed that they should be implemented in practice.

Plant biotechnology tools used to improve food security in Cuba can be classified into two main fields according to the kind of technical support used: various cell and tissue culture-based technology, not based on the use of molecular biology; and gene identification and manipulation technology based on molecular biology.

These can then be grouped into six categories: micropropagation; bioproducts used as biopesticides, biofertilizers and bioregulators; germplasm conservation, characterization and exchange; genetic improvement assisted by molecular markers (MAS) and genomics-proteomics; advanced disease diagnosis based on immunology and molecular biology; and genetic engineering for plant genetic improvement based on transgenic crops.

2.1 Micropropagation

In vitro propagation has many advantages, such as higher rates of multiplying clean (pest and disease-free) planting material and the small amount of space required to multiply large numbers of plants. Micropropagation allows: the production of large numbers of plants from small pieces of the stock plant in relatively short periods of time, as needed when introducing a new cultivar; plant improvement through somaclonal variation and early selection in tissue culture; production of disease-free plants including the cleaning up of disease-infected material; and *in vitro* culture which is essential for genetic transformation technology and to propagate any transgenic plant obtained.

In Cuba, protocols are available and have been implemented to micropropagate all the main vegetative propagated crops, including: banana and plantain; pineapple; sugar cane; cassava; sweet potato; yam; dasheen, taro and cocoyam; potato; papaya; forest trees such as eucalyptus, sesbania and acacia; opuntia; agave; cocoa; coconut; and ornamentals like athurium, syngonium, spathiphylum, orchids, ficus, rose, alstromeria, carnation and chrysanthemum.

By focusing on tissue culture, the skills necessary to maintain and to manage a biotechnology laboratory have been developed (Pérez et al., 2000). In the pipeline there are important aspects related to technology improvement, making it more efficient, economically feasible and safe. Among these new developments are more automated technologies, like temporary immersion bioreactor systems and cloning via somatic embryogenesis, that will enhance the use of plant micropropagation (Escalona et al., 1999; 2003; Gómez et al., 2001; Lorenzo et al., 2001; Pérez et al., 2001).

2.2 Bio-products used as Biopesticides, Biofertilizers and Bioregulators

The continued need for pest management in agriculture is self-evident, with increasing pressure on agriculture to produce more from less land. Pests (which include invertebrates, pathogens and weeds) are estimated to cause between 27 per cent and 42 per cent production losses for major crops around the world, but this would rise to a staggering 48–83 per cent without crop protection (Oerke and Dehne, 2004). Food production increases in the past 40 years have been based on massive increases in the use of synthetic pesticides (15–20 times), which is not sustainable. There is ample evidence of the adverse effects on the environment and human health associated with the use of synthetic pesticides. This has led to a strong consumer push for more environmentally benign pest control methods, coupled with regulatory actions to reduce risk, such as the withdrawal of many synthetic pesticides (Glare et al., 2012).

Biopesticides (using a broad definition) represent about 3.5 per cent of the overall global pesticides market (US$1.6 billion in 2009) (Lehr, 2010). The relatively low proportion of biopesticides used in comparison with agrochemicals is generally due to more technical approaches needed for their use, higher prices, traditional use of agrochemicals supported by strong advertising and technical services of the agrochemicals-producing companies. However, in recent years as the public are increasingly concerned about ecology and the consequent spread of organic agriculture, there is a trend to increase the use of the bioproducts in agriculture. The biopesticides market has been increasing more than 15 per cent annually in recent years.

Most of the products available in the market are based on the bacteria *Bacillus thuringiensis* as a biopesticide or *Rhizobium* as a biofertilizer. However, every year new products arrive on the market, and more are in the pipeline at several companies. Products based on different bacteria, fungi, viruses and nematodes are used as biopesticides, biofertilizer or bioregulators.

Although organic agriculture is often shown as an alternative to or against biotechnology, in fact it is based in great measure on the use of bioproducts, which is the use of living organisms or parts of living organisms to provide new products and services, coinciding with the definition and scope of biotechnology.

Taking into account the importance of these products, there is an urgent need to conduct fundamental as well as problem-oriented basic and applied research, especially on the occurrence and distribution of strains and their isolation, maintenance and preservation, physiology, genetics, biochemistry and economic utilization, including better formulations. It is clear that modern biotechnology can contribute to the optimal use of bioproducts in agriculture.

The first attempt to manage pests by means of biological control in Cuba was in 1927 when Bruner and Arango successfully introduced the Australian beetle (*Rodolia cardinalis* Mulsant) from Florida to control *Icerya purchasi* (cottony cushion scale). At the end of the 1960s and beginning of the 1970s, the first development of technologies and facilities for the simple production of *Bacillus thuringiensis* occurred. These results constituted the basis for the creation of a network of bio-laboratories or Centers for the Reproduction of Entomophagous and Entomopathogens (CREE) (Fernandez-Larrea, 2007).

The fall of the Soviet bloc in 1991 led to a collapse in the supplies that had fed the high input agriculture conducted. During this period the network of CREE was increased and reinforced. According to Fernandez-Larrea, 195 centres were finally in operation, producing in the order of 2000 t of biopesticides per year (Fernandez-Larrea, 2007).

In order to contribute to the biological control of phytonematodes in the soil, the Center for Genetic Engineering and Biotechnology of Cuba (CIGB) developed the bionematocide Hebernem, based on the soil bacteria *Tsukamurella paurometabola*. This bacterial strain was isolated from the soils of Cuba as a good candidate for the biological control of phytonematodes; its action mechanism involves the combining effect of desulfurase and chitinase on nematodes and nematodes eggs (Mena, 2004; Hernández et al., 2007; Fernández, 2007).

There is a continuous increase in the availability of biopesticide products in Cuba for agricultural use. Some of the most important ones available today are: Bt (*Bacillus thuringiensis*) formulations to control lepidopterae species but also coleoptera and other genera; *Trichoderma sp*. mainly to control soil fungi; *Verticillium lecanii* to control whiteflies, aphids, thrips and other insects; *Beauveria bassiana* to control thrips and other insects; *Paecelomyces fumosoroseus* controlling different insects, mainly beetles, fire ants and nematodes; *Tsukamurella paurometabola* to control nematodes; and *Metarhizium anisopliae* to control locust, termites, various coleoptera insects, leafhoppers and aphids.

There has also been a profound change to a more ecological approach to fertilization technology in Cuba. The excessive use of chemical fertilizers to increase crop productivity led to a high level of water contamination, eutrification of water reservoirs, high emissions of nitrogen oxide, high use of non-renewable energy sources and other deleterious effects.

Advances in soil microbiology and in applied agricultural biotechnology, including bioengineering, are allowing the exploitation of selected soil micro-organisms that have a high potential for fixing atmospheric nitrogen, solubilizing fixed phosphorus in the soil, and synthesizing biochemical substances that, when interacting with plants, stimulate metabolism. Worldwide over the last half century, the potential of different micro-organisms to have beneficial effects on crop productivity has been demonstrated in diverse cultivation systems (Martinez-Viera et al., 2010; Terry et al., 2002).

The importance of mycorrhizal symbiosis for plants is also widely recognized by the international scientific community as increasing the plants' capacity for water and nutrient absorption, producing a degree of tolerance for root diseases and pests, and at the same time improving some of the soil's physical properties.

In the early 1990s, a full-scale research programme was started in Cuba with the objective of establishing a solid scientific understanding of how mycorrhizal symbiosis can be managed as a positive element of agro-ecosystems. With such knowledge, it has been possible to develop mycorrhizal products to be applied at low doses, evaluating them at the production scale. A mycorrhizal inoculant in solid form, known as EcoMic, has been developed to colonize roots. Its excellent adhesive properties improve the retention of propagules on the seed surface and make it appropriate for seed-dressing technology (Rivera and Fernàndez, 2006).

Biofertilizers represent a vital component of sustainable systems in Cuba, constituting an economically attractive and ecological acceptable practice. The main biofertilizers used are: *Rhizobium* and *Bradyrhizobium*, the most widely used nitrogen-fixing bacteria around the world; *Acetobacter diazotrophicus*, a significant N-fixing bacterium under the right conditions, which inhabits the intercellular spaces and vascular system of sugar cane and other monocot plants; *Azotobacter* species, which has the ability to fix atmospheric nitrogen by converting it to ammonia in the free-living state without entering symbioses with plants; *Azospirillum*, which can fix the nitrogen compound associated with the roots of gramineous plants and more recently in non-gramineous ones; P-solubizer bacteria, which promotes the better use of soil phosphorus, reducing the contaminant effect of excess insoluble phosphorus in soils; and Mycorrhizal inoculants, which increase plants' capacity for water and nutrient absorption, producing a degree of tolerance for root diseases and pests, and at the same time improving some of the soil's physical properties.

In the field of biostimulators and bionutrients, two main products are been used very successfully in the country: Brassinosteroids and Fitomas E. Brassinosteroids (BRs); these products have relatively recently been recognized as a new class of phytohormones which play an important role in growth regulation (Clouse and Sasse, 1998). There have been several reports describing the relationship between BRs and plant stress responses such as activation of cold resistance and induction of ethylene biosynthesis, suggesting that BRs may play a role in stress-responding systems (Nakashita et al., 2003).

In 1987 the Study Center for Natural Products at the Havana University started to synthesize Brassinosteroids analogues and several scientific institutions have been involved in evaluating their action in plant growth and development, particularly related to their anti-stress effects. This has led to the initial introduction of a Brassinosteroid-analogue-based product called Biobras 6 in the field (Izquierdo, 2011; Rodríguez et al., 2003; Nuñez and Mazorra, 2001).

FitoMas E is a new by-product of the Cuban sugar cane industry with a marked anti-stress effect. This bionutrient contains neither growth hormones, plant exogenous stimulators, nutrient fixing, nor solubilizing micro-organisms. It contains only substances of the natural plant metabolism that significantly favour soil–plant exchange. FitoMas E, a mixture of mineral salts and high energy biochemical components (amino acids, nitrogenous bases, active saccharides and polysaccharides) (Montano, 2008), is now commercially used on several crops with outstanding results in Cuba.

Other future approaches would be the combination of bioproducts with transgenic plants. At present, certified organic agriculture does not accept GMOs. In Cuba, many promoters of the agro-ecological growing system are radically opposed to the use of transgenic plants and, as in other countries, there is a passionate debate on the matter.

The author of this chapter has a profound ecological and sustainable approach in all his scientific work and strongly believes that the use of transgenic plants resistant to biotic or abiotic stress, provided they are demonstrated to be safe for humans and the environment through a complete risk assessment, in combination with other agro-ecological practices could have a transcendental impact on food security worldwide, especially in the third world. A lot of work should be done in risk assessment and management, on the combination of different approaches, on the evaluation of the social impact and on the information related to transgenic crops, before it can become accepted. It is also crucial that this sustainable approach not be contaminated with the policy of some transnational companies, whose final goal with transgenic plants is to promote greater use of their agrochemicals.

2.3 Advanced Disease Diagnosis Based on Immunology and Molecular Biology

All programmes to control plant or animal diseases start with a reliable and accurate diagnosis of the disease-causing organism. The earlier this is done, the greater the possibilities to establish successful control. Even more important is the role of the diagnosis in the implementation of effective quarantine and other prevention measures for devastating exotic diseases.

In Cuba, there is a network of Territorial Plant Protection Stations with diagnostic facilities and trained personnel to run the continuous pest and disease monitoring programmes, as well as central diagnostic facilities at the National Institute of Plant Health (INISAV) and at the Center for Plant and Animal Health (CENSA) (Fiallo-olivé et al., 2009a, 2009b; González, 2007; Stefanova and García, 2007). In these centres there are up-to-date protocols with modern molecular tools, which can be used as reference labs. For almost all the most important diseases of plants and animals of the region there are suitable diagnosis methods based on immune enzymatic or molecular procedures (Fiallo-olivé et al., 2009a, 2009b).

2.4 Genetic Modification and Transgenic Plants

The advent of genetic modification over the past 20 years has enabled plant breeders to develop new varieties of crops with improved traits and at a faster rate than was possible using traditional methods, with huge potential for further beneficial developments. The genetic modification of organisms by recombinant DNA techniques can range from enhancing or suppressing the performance of existing genes to transferring genetic information from one organism to another. Genetic modification simplifies the identification of the genes responsible for particular desirable traits and allows scientists to transfer single traits between species precisely.

While there is little controversy about many aspects of biotechnology and its application, genetically modified organisms (GMOs) have become the target of a very intensive and, at times, emotionally charged debate. The FAO recognizes that genetic modification has the potential to help increase production and productivity in agriculture, forestry and fisheries. It could lead to higher yields on marginal lands in countries that today cannot grow enough food to feed their people. The FAO is also aware of the concern about the potential risks posed by certain aspects of biotechnology. These risks fall into two basic categories: the effects on human and animal health, and the environmental consequences. FAO strongly supports a science-based evaluation system that would objectively determine the benefits and risks of each individual GMO. This calls for a cautious case-by-case approach to address legitimate concerns for the biosafety of each product or process prior to its release (FAO, 2000).

In Cuba there has been scientific work with transgenic plants for more than 20 years and there are a range of facilities, technologies and high-level scientific groups in this field (Cockcroft et al., 2003). Cuba is a signatory of The Convention on Biological Diversity (CBD) and of the Cartagena Protocol and has a whole package of laws and instructions to regulate the research and production of genetically modified organisms. There is no moratorium in place, but there are strict rules on research and release of GMOs. The National Center for Biosafety of the Ministry of Science, Technology and Environment is the regulatory agency empowered to provide all the authorizations and licences for the different phases of research, development and releasing of genetically modified organisms to the environment. The National Institute of Food Hygiene of the Ministry of Health regulates and authorizes any kind of food for human consumption. Finally the Ministry of Agriculture regulates and authorizes the use of any new variety for agricultural production in the country.

There are several ongoing projects focused on transgenic plants in different phases of development in crops such as maize, soybean, potato, sweet potato, banana, tomato, rice and sugar cane (Fuentes et al., 2004, 2008; Moran et al., 1998; Doreste et al., 2002). Maize and soybean crop research is more advanced, and one Bt maize has obtained the first licence for large-scale use in the field (Aldridge, 2009). The main traits involved are resistance to biotic stress (pests and diseases) and abiotic stress (drought and salinity).

A significant effort has also been made in the use of plants as bioreactors to produce molecules related to human health (plant made pharmaceuticals, or PMPs). In 2000 the first 'plantibody' in the world to be used in the immunopurification of a human vaccine was registered in Cuba (Knablein et al., 2007; Pujol et al., 2005, 2007; Valdés et al., 2003a, 2003b, 2007). The review of PMP technology published in *Science* in 2008 recognized that the plantibody of the CIGB is the only one already on the market (Kaiser, 2008).

One key point that characterizes the work with transgenic plants in Cuba is the use of this technology in the framework of sustainable agriculture with an ecological approach. In fact, the use of GM plants in Cuba is integrated in a package with other bioproducts and with the use of as few chemicals as possible. This characteristic differentiates the use of this technology in Cuba from the strategy and market interests of several transnational companies of the sector.

2.5 Functional Genomics

Upon completion of the *Arabidopsis* and rice genome sequences and following the collection of vast amounts of expressed sequence tags (ESTs) from crops, medicinal plants and trees, plant science has entered into a qualitatively new development phase. Model systems have played a crucial role in understanding biological processes at genetic, molecular and systems levels.

The ultimate goal is to gain an integrated knowledge on the temporal and spatial regulation of transcription, translation, protein interactions, proteolysis, metabolism, transport and so on, and to learn how these pathways control events through the plant life-cycle. Plant science thus is moving from an analytical approach to integrative multidisciplinary approaches. New powerful technologies and knowledge are becoming equally accessible to plant breeders, geneticists, physiologists, anatomists, ecologists, pharmacologists and botanists. There are many intriguing developments in progress that exploit the results of genome projects leading to the gradual conversion from classical breeding to molecular breeding (Koncz, 2003).

In fact, as a result of the rapid accumulation of DNA sequence information, including the complete genome sequences for several plant species, plant research is focusing increasingly on analysis of gene function. The sequence information in itself cannot provide significant knowledge of the biology of the organisms, but the application of the high-throughput technologies will dramatically increase knowledge about complex biological networks.

Knowing when and where a gene product (RNA and/or protein) is expressed can provide important clues to its biological function. The high throughput approach might be used to study gene expression, enabling study of the level of regulation. The main tools of plant functional genomics used in Cuba include: subtractive cDNA libraries; differential expression; representational difference analysis (RDA); amplified fragment length polymorphism (cDNA–AFLP); suppression subtractive hybridization (SSH); cDNA microarrays; super serial analysis of gene expression (Super SAGE); and identification of the function of genes, including functional complementation, RNA interference (RNAi)/ RNA silencing/virus induce gene silencing (VIGS) and gain of function.

Marker assisted selection (MAS) offers clear advantages in genetic terms over traditional selection in many circumstances. The increasingly widespread application of PCR-based diagnostics and marker assisted breeding are enabling new and more intensive research at a lower cost. The application of DNA marker technology is likely to facilitate the attainment of new objectives in crop research and genetic improvement that have proven to be difficult to achieve using classical techniques.

In Cuba, DNA polymorphism of sugar cane, revealed by molecular marker techniques (RAPD, RFLP and AFLP), has demonstrated that it is a useful complement to traditional methods for the identification and diversity assessment of the germplasm that constitutes the genetic foundation of the sugarcane-breeding programme (Cornide et al., 1999). More recently, at the Center of Genetic Engineering and Biotechnology, an intensive study with wide international collaboration has identified potentially useful genes and targets to be used as molecular markers in order to generate transgenic plants or to produce useful bioproducts able to trigger systemic acquired immunity (SAR).

In 2006, to identify tobacco specie (*Nicotiana megalosiphon*) genes involved in broad-spectrum resistance to tobacco blue mould (*Peronospora hyoscyami* f. sp. *tabacina*), suppression subtractive hybridization was used to generate cDNA from transcripts that are differentially expressed during an incompatible interaction. Virus-induced gene silencing (VIGS) of the lipid transfer protein gene or glutamate decarboxylase gene in *Nicotiana megalosiphon* did not affect blue mould resistance, while silencing of an EIL2 transcription factor gene and a glutathione synthetase gene was found to compromise the resistance of *Nicotiana megalosiphon* to blue mould (Borras et al., 2006). Also in that year cDNA-AFLP rice genes differentially expressed in a tolerant rice variety upon water-deficit stress were identified, allowing the reporting of new genes involved in the response to water-deficit stress in a tolerant rice variety (Rodríguez et al., 2006).

In 2008, a novel sugar cane ethylene responsive factor (ERF), named SodERF3, was reported to enhance salt and drought tolerance when over-expressed in tobacco plants (Trujillo et al., 2008). Meanwhile in 2009, a glutathione S-transferase (GST) gene was amplified from cDNA of *Nicotiana tabacum* roots infected with *Phytophthora parasitica* var. nicotianae. The gene was cloned in sense and anti-sense orientation in RNAi vector for induced gene silencing. A statistically significant increase in resistance of *N. tabacum* to infection following gene silencing was found for GST-silenced plants compared with control plants. This was the first evidence of the role of GST as a negative regulator of defence response.

This gene seems to be crucial for an efficient defence against *P. parasitica* var. nicotianae during compatible interaction and this might now be exploited in strategies to develop durable resistance in cultivated tobacco and other plants through either marker assisted breeding or other biotechnological approaches (Hernández et al., 2009).

In 2010, to identify *Nicotiana tabacum* genes involved in resistance and susceptibility to *Rhizoctonia solani*, suppression subtractive hybridization was used to generate a cDNA library from transcripts that are differentially expressed during a compatible and incompatible interaction. This allowed the isolation of a protein kinase cDNA that was down-regulated during a compatible interaction and up-regulated during an incompatible interaction. Several genes previously associated with resistance in tobacco, were up-regulated in tobacco plants over-expressing the protein kinase cDNA (Chacón et al., 2010). Potentially, the protein kinase gene could be used to engineer resistance to *R. solani* in tobacco and other pathogen–plant interactions.

Also in 2010, a novel cDNA encoding a defensin named NmDef02 was reported. The purified recombinant protein expressed in *Pichia pastoris* was found to display antimicrobial activity *in vitro* against important plant pathogens. Constitutive expression of the NmDef02 gene in transgenic tobacco and potato plants enhanced resistance against various plant pathogens, including the oomycete *Phytophthora infestans*, a causal agent of the economically important late blight disease of potatoes grown under greenhouse and field conditions (Portieles et al., 2010). Ongoing experiments have demonstrated very useful potential applications of this gene in other very important crops such as soybean.

Recently, protection against a fungal disease (*Rhizoctonia solani*) and two dangerous oomycetes (*Phytophthora parasitica* var. nicotianae and *Peronospora hyoscyami* f. sp. *tabacina*) by a gene encoding a polygalacturonase-inhibiting protein (PGIP) was reported. These results show that expression of PGIP is a powerful way of engineering broad-spectrum disease resistance (Borras-Hidalgo et al., 2012).

A significant threat to Bt-based insect control is the potential of key pests to develop resistance to Cry toxins. One key goal of the Environmental Biotechnology group of CIGB is to understand the mechanism conducting insect resistance and to develop products and technologies to avoid it. One of the main approaches of the lab is to find the serine-protease gene related to insect resistance (Rodríguez et al., 2008, 2010). These results give important clues for insect resistance management.

3 MAIN OUTCOMES

There are several outcomes from biotechnology applied to agriculture in Cuba. They can be summarized as follows.

Cuba has successfully developed and implemented an alternative agricultural system with deep ecological roots. This approach made it possible to continue producing food during the serious economic constraint that followed the collapse of the Soviet bloc.

Cuba has developed a full network of production facilities to supply the whole package of bioproducts that this organic-oriented agriculture demands. These facilities range from simple, but efficient, Centers for the Reproduction of Entomophagous and Entomopathogens up to sophisticated industrial plants fulfilling higher international standards.

Commercial tissue culture labs (biofactories) built in all Cuban territories have a total potential capacity of more than 50 million vitroplants per year, produced by means of micropropagation.

There is a network of Territorial Plant Protection Stations with diagnostic facilities and trained personnel to run the continuous pest and disease monitoring programmes as well as central diagnostic facilities at the National Institute of Plant Health (INISAV) and at the Center for Plant and Animal Health (CENSA). In these centres there are up-to-date protocols with modern molecular tools, which can be used as reference labs. All this full diagnostic coverage of early monitoring, together with the strict quarantine systems that are in place, have made it possible for Cuba to be free from many serious plant pests and diseases. The combination of this monitoring system with the use of biocontrols has made integrated pest management (IPM) possible.

There is an integrated system of biosafety controls in place with a full package of laws and regulations to regulate the application of current and future applications from the labs to the field. Without this regulatory system, it would be impossible to develop the full array of agbiotech in the country.

There is also a network of labs in place to perform all the toxicology and ecotoxicology studies needed to register new products. The protocols of the trials are internationally approved.

For nearly 30 years a National Program of Agricultural Biotechnology has existed, which promotes, makes research calls, peer review proposals and approves execution of projects in this field. Throughout this time, the main focus has been on projects related to an ecological and sustainable agriculture.

At the Center of Genetic Engineering and Biotechnology (CIGB), other research centres and universities, well-trained scientists, facilities and modern labs are able to carry out research and applications, using frontier technologies in genomics, bioinformatics, genetic modification and related fields.

One might argue that the main outcome of this biotechnology revolution in Cuba has been the change in attitude among scientists, farmers, officers and decision-makers: the prevailing view now is that sustainable agriculture is both possible and needed.

The main current challenge is how to effectively and safely combine the new technologies of agbiotech with the agro-ecology approach. An inclusive strategy to improve agriculture will need to mobilize different actors and opinions, rather than exclude them. In fact, to meet the challenge of feeding the Cuban population and to attain real food sovereignty in a sustainable manner will require all types of knowledge and technology to be utilized, and the involvement of all parties.

REFERENCES

Aldridge, S. (2009), 'Cuba's first GM corn', *Nature Biotechnology*, **27**, 110.

Borras, O., B.P.H.J. Thomma, C. Collazo, O. Chacón, C.J. Borroto, C. Ayra et al. (2006), 'EIL2 transcription factor and Glutathione Synthetase are required for defense of tobacco against tobacco blue mold', *Molecular Plant–Microbe Interactions*, **19**, 399–406.

Borras-Hidalgo, O., C. Caprari, I. Hernandez-Estevez, G.D. Lorenzo and F. Cervone (2012), 'A gene for plant protection: expression of a bean polygalacturonase inhibitor in tobacco confers a strong resistance against Rhizoctonia solani and two oomycetes', *Frontiers in Plant Science*, **3**(268), 1–6.

Chacón, O., M. González, Y. López, R. Portieles, M. Pujol, E. González et al. (2010), 'Over-expression of a protein kinase gene enhances the defense of tobacco against Rhizoctonia solani', *Gene*, **452**, 54–62.

Clark, I. (2010), 'Cuba', chapter 6 in *UNESCO Science Report 2010*, Paris: UNESCO.

Clouse, S. and J. Sasse (1998), 'Brassinosteroids: essential regulators of plant growth and development', *Annual Review of Plant Physiology and Plant Moecular Biology*, **49**, 427–51.

Cockcroft, C.E., L. Herrera-Estrella and C.G. Borroto (2003), 'Agricultural biotechnology in Latin America and the Caribbean', in P.Christou and H. Klee (eds), *Handbook of Plant Biotechnology*, John Wiley, pp. 1–47.

Companioni, N., Y. Ojeda, E. Páez and C. Murphy (2001), 'La agricultura urbana en Cuba', in F. Funes et al. (eds), *Transformando el Campo Cubano: Avances de la Agricultura Sostenible*, La Habana: ACTAF, pp. 93–109.

Cornide, M.T., O. Coto, E. Canales, A. Sigarroa, J.E. Sánchez, M. Ramos et al. (1999), 'Desarrollo de las aplicaciones de los marcadores moleculares en el mejoramiento de la caña de azúcar en Cuba', *Biotecnologia Aplicada*, **16**, 186–9.

Doreste, V., E.L. Ramos, G.A. Enríquez, R. Rodríguez, R. Peral and M. Pujol (2002), 'Transgenic potato

plants expressing the potato virus X (PVX) coat protein gene developed resistance to the viral infection', *Phytoparasitica*, **30**, 177–85.
Escalona, M., G., Samson, C. G. Borroto and Y. Desjardins (2003), 'Physiology of effects of temporary immersion bioreactors on micropropagated pineapple plantlets', *In Vitro Cellular & Developmental Biology – Plant*, **39**, 651–6.
Escalona, M., J.C. Lorenzo, B. González, M. Daquinta, J.L. González, Y. Desjardins et al. (1999), 'Pineapple (Ananas comosus L. Merr) micropropagation in temporary immersion systems'. *Plant Cell Reports*, **18**, 743–8.
FAO (2000), 'FAO statement on biotechnology', March, pp. 1–3.
Febles, J.M., A. Tolón, X. Lastra and X. Acosta (2011), 'Cuban agricultural policy in the last 25 years: from conventional to organic agriculture', *Land Use Policy*, **28**, 723–35.
Fernández, E. (2007), 'Manejo de fitonematodos en la agricultura Cubana', *Fitosanidad*, **11**, 57–6.
Fernandez-Larrea, O. (2007), 'Pasado, presente y futuro del control biológico en Cuba', *Fitosanidad*, **11**, 61–6.
Fiallo-Olivé, E., Y. Martinez-Zubiaur and R.F. Rivera-Bustamante (2009a), 'Tomato yellow leaf distortion virus, a new bipartite begomovirus infecting tomato in Cuba', Tobacco yellow crinkle virus, a new bipartite begomovirus infecting tobacco and pepper in Cuba. *Plant Pathology*, **58**(4), 785.
Fiallo-Olivé, R.F. Rivera-Bustamante and Y. Martínez-Zubiaur (2009b), 'Tobacco yellow crinkle virus, a new bipartite begomovirus infecting tobacco and pepper in Cuba', *Plant Pathology*, **58**(4), 785.
Fuentes, A., P.L. Ramos, C. Ayra, M. Rodriguez, N. Ramirez and M. Pujol (2004), 'Development of a highly efficient system for assessing recombinant gene expression in plant cell suspensions via Agrobacterium tumefaciens transformation', *Biotechnology and Applied Biochemistry*, **39**, 355–61.
Fuentes, A.D., P.L. Ramos, Sánchez, Y. D. Callard, A. Ferreira, K. Tiel et al. (2008), 'A transformation procedure for recalcitrant tomato by addressing transgenic plant-recovery limiting factors', *Biotechnology Journal*, **3**, 1088–93.
Glare, T., J. Caradus, W. Gelernter, T. Jackson, N. Keyhani, J. Köhl, et al. (2012), 'Have biopesticides come of age?', *Trends in Biotechnology*, **30**, 250–58.
Gómez, R., L. del Sol, M. Reyes, M. Freire, L. Posada, I. Herrera et al. (2001), 'Embriogénesis somática en bananos y plátanos partiendo de flores masculinas inmaduras', *Biotecnología Vegetal*, **1**, 29–35.
González, G. (2007), 'Desarrollo del diagnóstico de virus, viroides y fitoplasmas en Cuba y su aplicación práctica', *Fitosanidad*, **11**, 11–15.
Hernández, A., F. Weekers, J. Mena, E. Pimentel, J. Zamora, C. Borroto et al. (2007), 'Culture and spray-drying of Tsukamurella paurometabola C-924: stability of formulated powders', *Biotechnology Letters*, **29**, 1723–28.
Hernández, I., O. Chacón, R. Rodriguez, R. Portieles, Y. López, M. Pujol et al. (2009), 'Black shank resistant tobacco by silencing of glutathione S-transferase', *Biochemical and Biophysical Research Communications*, **387**, 300–304.
Izquierdo, H. (2011), 'Actividad biologica de los brasinoesteroides y sus anaílogos en las plantas', *Temas de Ciencia y Tecnología*, **15**, 45–50.
Kaiser, J. (2008), 'Is the drought over for pharming?', *Science*, **320**, 473–5.
Knablein, J., M. Pujol and C.G. Borroto (2007), 'Plantibodies for human therapeutic use', *Bioworld Europe*, **1**, 14–17.
Koncz, C. (2003), 'Plant biotechnology: from genome projects to molecular breeding: editorial overview', *Current Opinion in Biotechnology*, **14**, 133–5.
Lehr, P. (2010), 'Biopesticides: the global market', available at: http://www.bccresearch.com/market-research/chemicals/biopesticides-market-chm029c.html.
Lorenzo, J.C., E. Ojeda, A. Espinosa and C. Borroto (2001), 'Field performance of temporary immersion bioreactor-derived sugarcane plants', *In Vitro Cellular and Developmental Biology – Plant*, **37**, 803–806.
Martinez-Viera, R., B. Dibut and Y. Rios (2010), 'Efecto de la integración de aplicaciones agrícolas de biofertilizantes y fertilizantes minerales sobre las relaciones suelo–planta', *Cultivos Tropicales*, **31**, 27–31.
Mena, J. (2004), 'Determinación de cepas bacterianas con actividad nematicida', PhD thesis, Centro de Ingeniería Genética y Biotecnología de Camagüey, Universidad Central de las Villas, Santa Clara.
Montano, R. (2008), 'FitoMas-E, bionutriente derivado de la industria azucarera', Instituto Cubano de Investigaciones de los Derivados de la Caña de Azúcar (ICIDCA) La Habana.
Moran, R., R. Garcia, A. Lopez, Z. Zaldua, J. Mena, M. Garcia et al. (1998), 'Transgenic sweet potato plants carrying the delta-endotoxin gene from Bacillus thuringiensis var. tenebrionis', *Plant Science*, **139**, 175–84.
Nakashita, H., M. Yasuda, T. Nitta, T. Asami, Fujioka S. and Y. Arai (2003), 'Brassinosteroid functions in a broad range of disease resistance in tobacco and rice', *Plant Science*, **33**(5), 887–98.
Nature editorial (2009), 'Cuba's biotech boom', *Nature*, **457**, 1079.
Nuñez, M. and L.M. Mazorra (2001), 'Los brasinoesteroides y la respuesta de las plantas al estres', *Cultivos Tropicales*, **22**, 19–26.
Oerke, E.C. and H.W. Dehne (2004), 'Safeguarding production: losses in major crops and the role of crop protection', *Crop Protection*, **23**, 275–85.

Oppenheim, Sara (2001), 'Alternative agriculture in Cuba', *American Entomologist*, Winter, 216–27.
Pastrana, S.J. and M.T. Clegg (2008), 'US–Cuban scientific relations', *Science*, **322**, 345.
Pérez, J.N., M. Suárez and P. Orellana (2000), 'Posibilidades y potencial de la propagación masiva de plantas en Cuba', *Biotecnología Vegetal*, **1**, 3–12.
Pérez, N., M. de Feria, E. Jiménez, A. Capote, M. Chávez and E. Quiala, (2001), 'Empleo de sistemas de inmersión temporal para la producción a gran escala de tubérculos *in vitro* de *Solanum tuberosum* L. var. Atlantic y estudio de su comportamiento en el campo', *Biotecnología Vegetal*, **1**(1), 17–21.
Portieles, R., C. Ayra, E. Gonzalez, A. Gallo, R. Rodriguez, O. Chacón, et al. (2010), 'NmDef02, a novel antimicrobial gene isolated from Nicotiana megalosiphon confers high-level pathogen resistance under greenhouse and field conditions', *Plant Biotechnology Journal*, **8**, 678–90.
Pujol, M., J.V. Gavilondo, M. Ayala, M. Rodriguez, E.M. Gonzalez and L. Perez, (2007), 'Fighting cancer with plant-expressed pharmaceuticals', *Trends in Biotechnology*, **25**, 455–9.
Pujol, M., N.I. Ramirez, M. Ayala, J.V. Gavilondo, R. Valdes, M. Rodriguez, et al. (2005), 'An integral approach towards a practical application for a plant-made monoclonal antibody in vaccine purification', *Vaccine*, **23**, 1833–7.
Rivera, R. and Fernández, F. (2006), 'Inoculation and management of mycorrhizal fungi within tropical agroecosystems', in N. Uphof et al. (eds), *Biological Approaches to Sustainable Soil Systems*, Boca Raton, FL: Taylor & Francis, pp. 479–89.
Rodríguez, C.R., Y.I. Villalobos, E.A. Becerra and F.C. Manchado (2003), 'Synthesis and biological activity of three new 5a-hydroxy spirostanic brassinosteroid analogues', *Journal of the Brazilian Chemical Society*, **14**, 466–9.
Rodríguez, L., D. Trujillo, O. Borrás, D.J. Wright and C. Ayra (2008), 'Molecular characterization of Spodoptera frugiperda – Bacillus thuringiensis Cry1Ca toxin interaction', *Toxicon*, **51**, 681–92.
Rodríguez, L., D. Trujillo, O. Borrás, D.J. Wright and C. Ayra (2010), 'RNAi-mediated knockdown of a Spodoptera frugiperda trypsin-like serine-protease gene reduces susceptibility to a Bacillus thuringiensis Cry1Ca1 protoxin', *Environmental Microbiology*, **12**, 2894–903.
Rodríguez, M., E. Canales, C.J. Borroto, E. Carmona, M. Pujol and O. Borra (2006), 'Identification of genes induced upon water-deficit stress in a drought-tolerant rice cultivar', *Journal of Plant Physiology*, **163**, 577–84.
Stefanova, M. and A. García (2007), 'El servicio de diagnóstico de las bacterias fitopatógenas en Cuba: desarrollo y alcances', *Fitosanidad*, **11**, 5–10.
Terry, E., Z. Teran, R. Martinez-Viera and M.A. Pino (2002), 'Biofertilizantes, una alternativa promisoria para la producción hortícola en organopóonicos', *Cultivos Tropicales*, **23**, 43–6.
Trujillo, L.E., M. Sotolongo, C. Menendez, M.E. Ochogavía, Y. Coll I. Hernandez et al. (2008), 'SodERF3, a novel sugarcane ethylene responsive factor (ERF), enhances salt and drought tolerance when overexpressed in tobacco plants', *Plant Cell Physiology*, **49**, 512–25.
Valdés, R., R. Biunayki, T. Alvarez, J. García, M. Jose, A. Figueroa et al. (2003a), 'Hepatitis B surface antigen immunopurification using a plant-derived specific antibody produced in large scale', *Biochemical and Biophysical Research Communications*, **310**, 742–7.
Valdés, R., L. Gómez, S. Padilla, J. Brito, B. Reyes, T. Alvarez et al. (2003b), 'Large-scale purification of an antibody directed against hepatitis B surface antigen from transgenic tobacco plants', *Biochemical and Biophysical Research Communications*, **308**, 94–100.
Valdés, R., D. Geada, W. Ferro, M. del C. Abrahantes, J. Cremata, G. Cabrera et al. (2007), 'Effects of tobacco extract and temperature on the stability of the monoclonal antibody CB. Hep-1 expressed in transgenic tobacco plants', *Bioprocessing Journal*, Spring, 16–24.

15 Risk assessment frameworks in the multilateral setting

Lee Ann Jackson

1 INTRODUCTION

International trade agreements uniformly recognize that countries have the right to take actions to protect domestic health or environmental safety even when these actions will have direct or indirect effects on trade. Indeed, to the extent that imports can act as a vector for particular risks, interventions to manage risk are likely to restrict trade. The challenge for multilateral trade rules is to find the appropriate balance between constraining a country's ability to use measures for protectionist ends while also ensuring that they maintain the right to protect the health of their citizens and their environment. Since the early years of the GATT (1947) multilateral trade rules have included some constraints on the ability of countries to use health as a justification for implementing trade-distorting policies. Nevertheless over time as successive trade rounds reduced tariffs and more attention was focused on the potential impact of non-tariff measures, it became clear that the existing disciplines embedded in the GATT did not adequately limit health-related non-tariff measures.

During the Uruguay Round (UR) negotiators were struggling to develop disciplines on non-tariff, health-related measures that would not interfere with the basic right of states to protect the health of their citizens and their environment. With respect to certain types of health risks, negotiators agreed on a solution that recognized the important role played by science in distinguishing between authentic measures implemented to manage risk and measures imposed with protectionist intent. This decision was likely to have been influenced by the evolution of formalized risk assessments methods during the 1970s and 1980s and a growing confidence that risk assessment could play an important role in many areas of regulatory policy setting (Button, 2004). At the conclusion of the UR the WTO was created and WTO members adopted a new set of rules. These rules included the Agreement on Sanitary and Phytosanitary Measures (SPS Agreement) which established, for the first time, specific disciplines with respect to regulatory interventions impacting trade for food safety and animal and plant health.

The SPS Agreement specifies the relevant commitments for countries in relation to measures that they may take to manage possible risks related to food safety and animal and plant health. In the SPS Agreement the right of countries to intervene in markets in order to control potential risks associated with food safety and plant and animal health is balanced by an obligation to ensure that these measures are based on science. In particular, according to the SPS Agreement WTO member countries should ensure that their SPS measures are based on an assessment of the risks associated with particular imports.

The goal of this chapter is to describe the general framework for risk assessment in the multilateral context, highlighting, where appropriate, aspects that are specifically relevant to biotechnology. This chapter describes how the multilateral trading system, in particular the SPS Agreement, the conclusions of WTO disputes related to the SPS Agreement and relevant international standards-setting bodies, creates the context for risk assessments at the national level. The chapter does not explore the role of risk assessment within risk analysis frameworks more generally. In particular the chapter does not examine the ways in which risk assessment informs risk management decisions made by regulatory authorities – that process is well articulated in the Red, Orange and Blue Books produced by the US National Research Council (NRC, 1983, 1994 and 1996) and codified in a variety of International Standards Organizations standards (for example 9000, 14001 and 32000).

The chapter begins by highlighting key aspects of the SPS Agreement and WTO disputes relating to risk assessment. In parallel, the chapter briefly summarizes the guidance that exists from the relevant international standards-setting bodies (ISSBs). The chapter next examines WTO and the ISSBs' guidance on the choice of quantitative or qualitative risk assessment methods. Finally, the chapter examines what direction is provided by multilateral rules and recommendations regarding risk assessments when scientific information is insufficient.

2 RISK ASSESSMENT IN THE WTO CONTEXT

The SPS Agreement applies to a precise set of measures that are defined based on the type of risk the measure is intended to address. This definition states that any measure taken to address a particular category of risks can be considered to be an SPS measure. Specifically the SPS Agreement, Annex A, states that a sanitary or phytosanitary measure is any measure applied:

a. to protect animal or plant life or health within the territory of the member from risks arising from the entry, establishment or spread of pests, diseases, disease-carrying organisms or disease-causing organisms;
b. to protect human or animal life or health within the territory of the member from risks arising from additives, contaminants, toxins or disease-causing organisms in foods, beverages or feedstuffs;
c. to protect human life or health within the territory of the member from risks arising from diseases carried by animals, plants or products thereof, or from the entry, establishment or spread of pests; or
d. to prevent or limit other damage within the territory of the member from the entry, establishment or spread of pests.

Certain other types of technical measures, for example labelling or packaging, fall under the Agreement on Technical Barriers to Trade (TBT Agreement), a parallel but separate sub-agreement. While a full discussion of the TBT Agreement is outside the scope of this chapter, it is useful to note that the TBT Agreement does not require that WTO members base their TBT measures on risk assessment but does require the concepts of legitimate objectives, proportionality and non-discrimination.

The SPS Agreement, Annex A, goes on to say that sanitary or phytosanitary measures:

include all relevant laws, decrees, regulations, requirements and procedures including, *inter alia*, end product criteria; processes and production methods; testing, inspection, certification and approval procedures; quarantine treatments including relevant requirements associated with the transport of animals or plants, or with the materials necessary for their survival during transport; provisions on relevant statistical methods, sampling procedures and methods of risk assessment; and packaging and labelling requirements directly related to food safety.

The reference to scientific justification of SPS measures arises in several parts of the SPS Agreement. The core rights and obligations captured in the SPS Agreement are laid out in Article 2, which says:

> Members have *the right* to take sanitary phytosanitary measures necessary for the protection of human, animal or plant life or health, . . . [And] . . . Members shall ensure that any sanitary or phytosanitary measure is applied only to the extent necessary to protect human, animal or plant life or health, *is based on scientific principles* and is not maintained without sufficient scientific evidence, except as provided for in paragraph 7 of Article 5 (emphasis added).

The first reference to scientific principles in the SPS Agreement is quite general and the Agreement later identifies risk assessment as the appropriate way of establishing that a country's choice of SPS measures fits the specific circumstances.

The SPS Agreement also recognizes that there may be situations when scientific information is not adequate to conduct a full risk assessment, and regulatory authorities would still want to implement an SPS measure (paragraph 7, article 5). This paragraph provides WTO members with the right to implement trade restrictive SPS measures provisionally when the relevant body of science is insufficient to conduct a risk assessment. Members implementing these types of provisional measures are obliged to seek additional information with a view to conducting a more objective risk assessment and to review the measures within a reasonable period of time. Section 4 of this chapter discusses these types of situations highlighting the conditions under which WTO members have the right to implement provisional measures.

Additional guidance on risk assessments can also evolve through the process of dispute settlement in the WTO. In disputes WTO members can submit complaints on particular measures being implemented by their trading partners for consideration by a panel of external experts, drawn from a pool of independent legal and technical trade experts nominated by member states. Over the course of the dispute process these panellists assess the facts presented by the parties to the dispute and make findings with respect to whether members have violated provisions of the WTO Agreements. Parties to the dispute have the right to appeal, in which case an Appellate Body evaluates the findings from the panel, and makes rulings on legal interpretation. Thus each dispute that considers the SPS Agreement can potentially offer insight into the legal meaning of the Agreement, through interpretation by the panel and the Appellate Body. In several instances WTO disputes have provided additional interpretative guidance on the SPS Agreement specifically related to the role of risk assessment, and these are discussed below.[1]

In all but one SPS disputes the panels have sought the advice of scientific experts.[2] The selection of the scientific experts is done through consultation with relevant international bodies and with the parties to the dispute. The number of experts selected depends upon the number and types of issues on which advice is needed. These groups of experts are called upon to provide their scientific judgement on the issues that parties to the dispute have raised. Hence the types of questions that are explored with the scientific experts are limited to those issues which have been raised by the parties to the dispute. For example, in the *EC–Biotech* case the panel convened a panel of scientific experts in order to obtain information, among other things, on the risk assessment provided by the parties to the dispute. The group of scientists included a crop geneticist, ecologist, biologist, environmental scientists, molecular biologist, and a food engineer. These questions focused on the availability of scientific evidence on a variety of possible risks associated with biotechnology crops including the emergence of resistance, impacts on non-target organisms, antibiotic resistance, and toxicity of biotech crops to humans. The areas of expertise represented by scientific experts and the questions raised depend upon the specific SPS measures that are at issue in each SPS case.

The SPS Agreement sets out a general framework for considering what would normally be included in a risk assessment. The SPS Agreement distinguishes between two types of risk assessment: the first is specific to risks associated with consumption of food, beverages or feedstuffs, while the second is specific to risks associated with the spread of pests or diseases. Risk assessment in relation to food-borne risks is defined as: 'the evaluation of the potential for adverse effects on human or animal health arising from the presence of additives, contaminants, toxins or disease-causing organisms in food, beverages or feedstuffs' (Annex A, para. 4).

In contrast, the SPS Agreement defines risk assessments covering disease or pest risks as: 'the evaluation of the likelihood of entry, establishment or spread of a pest or disease within the territory of an importing Member according to the sanitary or phytosanitary measures which might be applied, and of the associated potential biological economic consequences' (Annex A, para. 4).

The Appellate Body in the *Australia–Salmon* dispute pointed to one fundamental distinction between these two types of assessments. The assessment of food-borne risks requires the evaluation of the *potential* adverse effects on human or animal health, while the assessment of risks related to disease or pests requires an evaluation of the *likelihood* of entry, establishment or spread of a disease and of the associated possible consequences. (*Australia–Salmon*, footnote 69). Furthermore, in this dispute the Appellate Body established a cumulative three-step test for determining what type of risk assessment would satisfy the requirements of the SPS Agreement with respect to assessment of disease or pest risks (*Australia–Salmon*, para. 121). These steps are:

1. identification of the hazards and possible biological and economic consequences of their entry or spreading;
2. evaluation of the likelihood of entry, establishment, or spreading; and
3. evaluation of the impact of SPS measures on this likelihood.

While the SPS Agreement does not give specific technical guidance on what should be included in a risk assessment, the SPS Agreement refers to three specific international organizations as the relevant international standards-setting bodies. For food safety, the standards, guidelines and recommendations established by the Codex Alimentarius Commission (Codex) provide relevant guidance relating to food additives, veterinary drug and pesticide residues, contaminants, methods of analysis and sampling, and codes and guidelines of hygienic practice. For animal health and zoonoses, the relevant standards, guidelines and recommendations are those developed under the auspices of the International Office of Epizootics (OIE). For plant health, the relevant standards, guidelines and recommendations are those developed under the auspices of the Secretariat of the International Plant Protection Convention (IPPC) in cooperation with regional organizations operating within the framework of the International Plant Protection Convention.[3]

The designation of these three ISSBs is crucial to the application of the SPS Agreement. When a member's SPS measure conforms to international standards, guidelines or recommendations from these three organizations they are presumed to be consistent with the SPS Agreement (Article 3, SPS Agreement). Thus the Agreement simultaneously encourages WTO members to harmonize their measures and directs them towards adoption or implementation of international standards guidelines and recommendations. At the same time conclusions of disputes have indicated that SPS Agreement maintains a certain flexibility of choice with respect to SPS regulation. Specifically, in the *Australia–Apples* case the Appellate body noted that:

> while Article 5.1 directs a Member conducting a pest risk assessment to take into account internationally developed risk assessment techniques, this does not mean that a risk assessment must be based on or conform to such techniques. Nor does it imply that compliance with such techniques alone suffices to demonstrate compliance with a Member's obligations under the *SPS Agreement* (para. 246).

The following paragraphs turn briefly to the specific guidance provided by the international standards-setting bodies. Table 15.1 provides a summary of the documentation from the three ISSBs.

At a general level these bodies recommend similar generic risk assessment frameworks. Risk assessments, as recommended by the three international standards-setting bodies should include, in general, an identification of hazards, a characterization of exposure to hazards, an evaluation of likely adverse effects, and an estimation of the risks. While

Table 15.1 Risk assessment guidance from the ISSBs

Codex (2002)	Principles for the risk analysis of foods derived from modern biotechnology
Codex (2003)	Guideline for the conduct of food safety assessment of foods derived from recombinant-DNA plants
Codex (2007)	Working principles for risk analysis for food safety for application by governments
Codex (2008)	Guideline for the conduct of food safety assessment of foods derived from recombinant-DNA animals
Codex (2011)	Working principles for the application of risk analysis in the framework of the Codex Alimentarius
IPPC (2004)	ISPM 11: Pest risk analysis for quarantine pests including analysis of environmental risks and living modified organisms
IPPC (2007)	ISPM No. 21: Pest risk analysis for regulated non-quarantine pests
IPPC (2011)	ISPM No. 2: Framework for pest risk analysis
OIE (2011a)	Aquatic animal health code
OIE (2011b)	Terrestrial animal health code
OIE (2011c)	Guidelines for assessing the risk of non-native animals becoming invasive

Table 15.2 Comparison of terminology in the leading ISSBs

Generic risk assessment process	CAC	OIE	IPPC
Identification of hazards	Hazard identification	–	Pest categorization
Characterization of exposure to hazards	Exposure characterization	Release assessment Exposure assessment	Assessment of probability of introduction and spread
Evaluation of likely adverse effects associated with hazards	Hazard characterization (including dose/ response if available)	Consequence assessment	Assessment of potential economic consequences
Estimation of risks	Risk characterization	Risk assessment	Conclusion of risk assessment

Source: FAO (2007).

the general framework is similar across the three bodies, the specific terminology differs slightly (see Table 15.2).

2.1 Codex Alimentarius Commission (Codex)

The Codex has developed 'Working principles for risk analysis' which represent guidance to the Codex in establishing standards in the area of food safety. The objective of these working principles is to provide guidance to the Codex Alimentarius Commission and the joint FAO/WHO expert bodies and consultations, to ensure that Codex standards are based on the principles of risk analysis (Codex, 2011). Also the Codex has 'Working principles for risk analysis for food safety for application by governments' (Codex, 2007). (FAO and WHO also have published documents in this area. See for example, FAO/ WHO, 2006.)

The Codex has further developed 'Principles for the risk analysis of foods derived from modern biotechnology' (Codex, 2003). This document focuses on the safety and nutritional aspects of foods derived from modern biotechnology. Two guidelines on the conduct of safety assessments, one for foods from DNA-modified plants and the other for foods from DNA-modified micro-organisms, consider intended and unintended effects of the genetic modification and an assessment of possible allergenicity. In terms of risk assessment, the principles note that the assessment should identify whether the hazard is present, and if so gather information on its nature and severity. The assessment should compare the relevant food derived from modern biotechnology with the conventional counterpart taking into account intended and unintended effects, identifying new or altered hazards, and identifying changes relevant to human health in key nutrients. Information for this risk assessment can be collected from a variety of sources including the developer of the product, scientific literature, regulatory agencies and international bodies.

The Codex 'Annex on food safety assessment in situations of low-level presence of recombinant-DNA plant material in food' (2008) addresses assessment in the situation where a GM food has been approved in an exporting country but has not yet been assessed by a country that may be importing the food occasionally at low levels. Codex has also developed a 'Guideline for the conduct of food safety assessment of foods derived from recombinant-DNA plants'.

Sampling and detection methods may also be relevant in the context of assessment risks since sampling methods contribute to overall uncertainty of analytical results. Codex currently is discussing the development of 'Guidelines on criteria for methods for the detection and identification of foods derived from biotechnology'. The current draft of these guidelines includes protocols for the validation of both quantitative and qualitative PCR and protein-based testing methods.

Codex risk assessments are conducted by specialized FAO/WHO expert bodies. For example the Joint FAO/WHO Expert Committee on Food Additives (JECFA) is responsible for risk assessments for food additives, veterinary drug residues and contaminants in food. The work of these specialized bodies typically results in the establishment of an international standard for limits on particular residues or hazards in foods.

2.2 International Plant Protection Convention (IPPC)

The IPPC has developed three standards that focus on risk analysis: ISPM 2 on 'Guidelines for pest risk analysis', adopted in 1996 and revised in 2007 and renamed as 'Framework for pest risk analysis'; ISPM 11 on 'Pest risk analysis for quarantine pests', adopted in 2001 and revised in 2003 and 2004; and ISPM 21 on 'Pest risk analysis for regulated non-quarantine pests'.

In terms of the pest risk analysis described in ISPM 2, three interrelated steps are relevant: pest categorization; assessment of the probability of introduction and spread; and the assessment of the potential economic consequences. The categorization of a pest as a quarantine pest allows for the opportunity to eliminate organisms from consideration before a more in-depth analysis is undertaken. Categorization includes the identity of the pest, the presence or absence in relevant areas, regulatory status, potential for establishment and economic consequences. With regard to regulatory status, if a pest is present at low levels, some type of official control may be expected, including in the case of living modified organisms (LMOs) including measures on parent organisms, donor organisms, and gene vectors.

ISPM No. 11 covers pest risk analysis for quarantine pests, including analysis of environmental risks associated with LMOs. The framework for pest risk analysis under ISPM 11 includes pest categorization, assessment of the probability of introduction and spread, assessment of potential economic consequences and degree of uncertainty. ISPM 11 also includes a section on pest risk management that suggests various options for different types of products and pathways of introduction of pests. ISPM 11 also includes an annexe that describes how to determine the potential for a living modified organism to become a pest. The standard has been referenced in WTO disputes dealing with plant health issues, including *Japan–Apples*, *EC–Biotech* and *Australia–Apples*. Of particular relevance to this chapter is the fact that the standard was important in the *EC–Biotech* case for the evaluation of whether the SPS Agreement covers regulation of GM products. This was important in this case because some of the potential risks identified by

EU regulation as associated with cultivating GM products were not directly attributable to the GM plants, but rather to the potential ecological impact of introducing modified genes into the environment. For example, the text of ISPM 11 states that:

> In cases of phytosanitary risks related to gene flow, the LMO is acting more as a potential vector or pathway for introduction of a genetic construct of phytosanitary concern rather than as a pest in and of itself. Therefore, the term 'pest' should be understood to include the potential of an LMO to act as a vector or pathway for introduction of a gene presenting a potential phytosanitary risk. (Annex 3, p. 27)

2.3 International Office of Epizootics (OIE)

In contrast to the IPPC, the OIE does not include a risk assessment standard that is specifically targeted at GM animals. The risk analysis guidelines that would be relevant are contained in the terrestrial and aquatic animal health codes. The principal aim of import risk assessment according to the OIE code is:

> to provide importing countries with an objective and defensible method of assessing the disease risks associated with the importation of animals, animal products, animal genetic material, feedstuffs, biological products and pathological material . . . This is necessary so that the exporting country is provided with clear reasons for the imposition of import conditions or refusal to import.

The OIE recently adopted 'Guidelines for assessing the risk of non-native animals becoming invasive'. The definition of an invasive, non-native animal is 'an animal that has been introduced and subsequently become established and spread outside its native distribution area, and caused harm to the environment, animal or human health, or the economy' (OIE, 2011c, p. 1). These guidelines may also provide guidance in terms of the relevant methods for determining the risks that could be associated with an LMO that could also be classified as an invasive, non-native species.

A 2005 resolution by the OIE stated that the organization would develop guidelines for international trade in GM-derived animals and products, in particular on the exclusion of unapproved animals and products from the livestock population and segregation from the feed and food supply. The OIE also constituted an ad hoc Group on Biotechnology (which later became the ad hoc Group on Vaccines Related to New and Emerging Technologies) and an ad hoc Group on Diagnostic Tests Related to New and Emerging Technologies to support the work of OIE specialist commissions and related working groups. The ad hoc Group on Vaccines Related to New and Emerging Technologies has been revising certain sections of the Terrestrial Manual in light of developments in biotechnology. These revisions deal primarily with the scientific aspects of biotechnology and the development of vaccines.

3 QUANTITATIVE VERSUS QUALITATIVE RISK ASSESSMENT METHODS

In general risk assessments can include both quantitative and qualitative analysis. In quantitative risk assessment numerical expressions of risk are provided. In qualitative risk assessment the evaluated risk is described in words and the estimate of risk can be ranked or separated into descriptive categories.

Quantitative risk assessments are valuable because they explicitly highlight the relative importance of uncertainty in various decision-making situations. The results of these analyses typically provide information across a distribution of scenarios. Sensitivity analysis using quantitative models provides an in-depth examination of how specific factors influence the outcomes of the risk assessment, and thus provide risk managers with more information regarding which interventions could be most effective for managing risk. The challenge of quantitative assessments is that data necessary for the full specification of the model are rarely available.

Qualitative assessments, on the other hand, provide the opportunity to conduct analysis in the absence of detailed quantitative studies. The resulting risk characterization will usually be descriptive or categorical in nature, rather than tied to a specific quantitative measurement. Similar principles apply to qualitative risk assessments as to quantitative assessments – that is, they should be transparent, well-documented and based on appropriate data. The main differences between qualitative and quantitative risk assessments are the ways that data are synthesized and the communication of results. For example, qualitative risk assessments will usually describe the evaluated risk in words and provide an estimate of risk ranked in descriptive categories, whereas quantitative risk assessments rely on numerical expressions of risk.

While the SPS Agreement does not say anything explicit about the relative merit of quantitative versus qualitative risk assessments, WTO dispute case law has considered the issue. Importantly, the *EC–Hormones* case established that there is no requirement implied in the SPS Agreement for a quantitative evaluation of risks. Rather the report states that while risk assessments can be either quantitative or qualitative, the assessment needs to be specific enough that it addresses the relevant risks at issue (see Pauwelyn, 1999 for a more detailed examination of these topics).

All of the three ISSBs mentioned above recognize that both types of assessments can be valid. Indeed, risk assessments focusing on plant health and environmental risks have often been qualitative. The results of these non-quantitative assessments provide less clear guidance for decision makers with respect to how particular regulatory controls lead to particular levels of health protection (FAO, 2007). Often expert opinion may be necessary to gather information on topics where there is a high level of uncertainty. Techniques are available for eliciting expert opinions in order to ensure that these opinions are as evidence-based as possible. According to IPPC guidelines, assessment of the probability of introduction and spread of LMOs includes assessment of both entry and establishment. In the case of an LMO the analysis will include both the intended and unintended pathways of introduction and intended use. The conclusion on the probability of introduction and spread can be quantitative or qualitative. OIE guidelines recognize that qualitative assessments may include ranking of costs or impacts as high, medium or low.

Codex recommendations and guidelines also recognize that qualitative risk assessment can play an important role. The Codex 'Working principles for risk analysis for food safety for application by governments' (2007) state that 'risk assessment should be based on scientific data most relevant to the national context. It should use available quantitative information to the greatest extent possible. Risk assessment may also take into account qualitative information.' Furthermore these principles state that:

> constraints, uncertainties and assumptions having an impact on the risk assessment should be explicitly considered at each step in the risk assessment and documented in a transparent manner. Expression of uncertainty or variability in risk estimates may be qualitative or quantitative, but should be quantified to the extent that is scientifically achievable.

In relation to biotechnology risk assessment, while the Codex risk assessment guidelines for GM foods do not explicitly refer to qualitative risk assessment, they do note that they should be read in conjunction with the Codex 'Working principles for risk analysis' which, as noted above, refer to qualitative methods.

The estimation of potential non-commercial and environmental consequences suffers from data limitations. ISPM 11 explicitly mentions that when analysing LMOs, potential adverse effects on non-target organisms as well as economic consequences related to pest properties should be considered. The OIE 'Guidelines for assessing the risk of non-native animals becoming invasive' indicate that both direct and indirect consequences should be considered, although the social and biological costs are often difficult to assess. Examples of direct consequences include harm to ecosystems, harm to native species, economic damage and impacts on human health and well-being. Indirect consequences include eradication costs, compensation costs, potential trade losses, and impacts on socio-cultural values.

Impact scoring is one example of a method that has been used in risk assessments to combine quantitative and qualitative assessments in order to reach a conclusion about the risk consequences of particular imports. This method was considered in the WTO *Australia–Apples* dispute where the panel looked at biological and economic consequences as examined in Australia's Import Risk Assessment (IRA). While some expert judgement is necessary to bridge gaps in existing information, the use of expert opinion introduces a certain amount of subjectivity into the final conclusion. So, for example, in the *Australia–Apples* dispute experts consulted by the panel were sceptical about the impact scores assigned by the IRA to some criteria since they were not drawn from scientific measurements. The panel was of the view that choices made in Australia's risk assessment led to an overestimation of the entry, establishment and spread of the pests at issue.

Another issue that arose in the context of the *Australia–Apples* dispute was the importance of documentation in risk assessments that use expert judgement. The ISSBs have also developed guidance in this area. For example, ISPM 11 highlights the importance of documenting where expert judgement plays a role in the risk assessment:

> It is important to document the areas of uncertainty and the degree of uncertainty in the assessment, and to indicate where expert judgement has been used. This is necessary for transparency and may also be useful for identifying and prioritizing research needs. It should be noted that the assessment of the probability and consequences of environmental hazards of pests of uncultivated and unmanaged plants often involves greater uncertainty than for pests of cultivated or managed plants. This is due to the lack of information, additional complexity associated with ecosystems, and variability associated with pests, hosts or habitats.

It may be particularly relevant to know where expert judgement has been used in the analysis and how sensitive the results of the analysis are to parameters based on expert judgement.

Quantitative methodologies are not always preferable, in particular when the use of quantitative methods masks data limitations. In *Australia–Apples* the panel noted:

> that a quantitative methodology should only be used 'when reliable specific numeric data are available' to support the choice of probability ranges and probability shapes. In the absence of sufficient data, and particularly if numbers are chosen in an arbitrary manner, a quantitative method would only give a misleading impression of objectivity and precision (para. 7.441).

4 RISK ASSESSMENTS AND SUFFICIENT SCIENTIFIC EVIDENCE

Risk assessments provide a structure for organizing scientific information, including information on relevant uncertainties. The challenge for risk assessments of new and evolving agricultural technologies is twofold. First, the pace of advancement in agricultural, industrial and food production practices is rapid and arguably increasing. Second, given the complexity of the systems and the fact that some impacts, such as impacts on ecosystems, may evolve over time, the long-term consequences of introduction to the environment are difficult to predict. Thus technological advancement creates continuous pressure for risk assessments so that new products can be put on the market at the same time that the assessment of risks is becoming increasingly complicated.

The SPS Agreement seems to distinguish between two situations in which regulatory authorities may wish to intervene to manage SPS risks. The first situation is one where there is adequate scientific information to conduct a risk assessment that satisfies the requirements of Article 5.1. The second situation occurs when insufficient evidence exists for a risk assessment to be conducted. In this situation, as mentioned in section 2, the SPS Agreement includes guidance on the rights of members to implement SPS measures. Specifically Article 5.7 states that:

> In cases where relevant scientific evidence is insufficient, a Member may provisionally adopt sanitary or phytosanitary measures on the basis of available pertinent information, including that from the relevant international organizations as well as from sanitary or phytosanitary measures applied by other Members. In such circumstances, Members shall seek to obtain the additional information necessary for a more objective assessment of risk and review the sanitary or phytosanitary measures accordingly within a reasonable period of time.

Several WTO SPS disputes have considered the relationship between the obligation to conduct a risk assessment and the existence of scientific evidence. The Appellate Body has explained that the relevant scientific evidence will be considered 'insufficient' for purposes of Article 5.7 'if the body of available scientific evidence does not allow, in quantitative or qualitative terms, the performance of an adequate assessment of risks as required under Article 5.1 and as defined in Annex A to the SPS Agreement' (*Japan–Apples*, para. 179).

WTO dispute panels have clearly recognized the difference between scientific uncertainty (which is an integral part of risk assessments) and insufficient scientific evidence. Specifically, in *Japan–Apples* the Appellate Body stated that 'The application of Article 5.7 is triggered not by the existence of scientific uncertainty, but rather by the insufficiency of scientific evidence. The text of Article 5.7 is clear: it refers to "cases where relevant scientific evidence is insufficient", not to "scientific uncertainty". The two concepts are not interchangeable' (*Japan–Apples*, para. 184). Peel (2004) suggests that this distinction indicates that provisional measures are an option primarily for members where limited scientific research exists on a particular risk, rather than where there is a large body of existing scientific evidence.

In the context of the WTO *EC–Biotech* dispute, the risk assessments were not evaluated in detail. Rather in evaluating whether there was a moratorium in place at the EU Member State level, the panel considered whether there had already been a risk assessment conducted in the context of the EU and whether the member states had followed the conclusions of this assessment. With regard to the EC member state safeguard measures, the panel found that the member states of the European Communities had not based their safeguard measures on existing risk assessments which satisfied the definition of the SPS Agreement. Hence these measures could be presumed to be maintained without sufficient scientific evidence. The panel further argued that since a risk assessment had been conducted by other scientific bodies, the relevant scientific information could not have been insufficient to conduct a risk assessment. Thus the panel found that EU member states' safeguard measures were not consistent with Articles 5.7 and 5.1.

Another aspect to this issue is how the evaluation of whether there is sufficient scientific information available to conduct a risk assessment could differ depending upon the level of protection sought by the implementing member. In *US–Continued Suspension* the Appellate Body noted that:

> where the chosen level of protection is higher than would be achieved by a measure based on an international standard, this may have some bearing on the scope or method of the risk assessment. In such a situation, the fact that the WTO Member has chosen to set a higher level of protection may require it to perform certain research as part of its risk assessment that is different from the parameters considered and the research carried out in the risk assessment underlying the international standard (para. 685).

Some analysts have argued that the existence of a specific standard with an associated risk assessment for a particular product would not necessarily imply that there would be sufficient scientific evidence to justify a higher level of standard for the same product (see Jackson and Jansen, 2010 for a more detailed discussion).

The discussion about how to determine whether there is sufficient science to complete a risk assessment is particularly relevant in the context of the SPS Agreement because the determination of insufficient scientific evidence provides countries with the option of temporarily bypassing the requirement to conduct a risk assessment. The three ISSBs are also operating in the context of rapidly evolving science and have established their own mechanisms for advancing their work while acknowledging the limits on scientific evidence. For example, as noted by Button (2004), Codex has adopted a practice that if scientific data is insufficient or incomplete, then the CAC should not elaborate a standard but rather would develop a text, such as a code of practice, which 'would be supported by available scientific evidence'. In this way standards-setting in the ISSBs can evolve in parallel with the evolution of scientific evidence.

5 CONCLUSIONS

This chapter provides an overview of the multilateral context for risk assessment, with a view to establishing a common understanding of existing rules and guidance in the area of risk assessment relevant for genetically modified crops. In response to advancing agricultural science and evolving trade patterns, many countries actively manage perceived risks by intervening with trade. While countries may be averse to accepting the cultivation of these crops within their national borders, the disciplines embedded in the WTO SPS Agreement oblige countries to justify scientifically their regulatory interventions that have trade impacts. As discussed in this chapter, risk assessment plays a crucial role in this justification. The SPS Agreement and related disputes provide general guidance on the ways that risk assessment is designed and targeted. This guidance develops over time with the rulings of WTO disputes on SPS issues. The ISSBs provide detailed technical guidance depending on their area of competence and this guidance is also evolving in parallel to the development of new technologies.

The ongoing development and trade of genetically modified crops creates new challenges for the trading system due to the diversity of views about these products. The work of the institutions discussed in this chapter is invaluable for national policy makers and regulatory authorities seeking to ensure that risks associated with these products are adequately managed. Still, the work of these institutions represents just one piece of a complex system of governance in this area. More work is needed to develop a deeper understanding of how these rules and guidance influence actors and organizations, and ultimately adoption outcomes of biotechnology products on the ground.

NOTES

1. As of July 2012 the WTO had received 38 complaints to the Dispute Settlement System involving SPS measures; the organization had issued panel or appellate body reports in eight cases (http://www.wto.org/english/tratop_e/dispu_e/dispu_agreements_index_e.htm?id=A19#selected_agreement). This number counts situations where multiple WTO members complain about the same measure as one dispute. In these situations, while each complaint receives a unique number in the dispute settlement catalogue, often a single report is issued covering the issues raised by multiple complaining parties.
2. The panel in the *US–Poultry* dispute did not seek the advice of scientific experts.
3. While the SPS Agreement identifies three specific ISSBs, in the context of the specific focus of this book on biotechnology products, it is useful to mention that the Convention on Biological Diversity also provides guidance on risk assessment in the area of Living Modified Organisms. The Cartegena Protocol, an agreement under the authority of the CBD, describes risk assessment as the 'assessment of the adverse effects of LMOs on the conservation and sustainable use of biological diversity, also taking into account risks to human health.' The crucial point is that the CBD is not explicitly identified as one of the three international standards-setting organizations in the SPS Agreement. Thus if a member based their risk assessment on the guidance of CBD they would not necessarily be considered to be in conformity with the provisions of the SPS Agreement.

REFERENCES

Button, C. (2004), The Power to Protect: Trade, Health and Uncertainty in the WTO, Oxford: Hart Publishing.
Codex (2002), 'Principles for the risk analysis of foods derived from modern biotechnology', CAC/GL 44–2003, Rome: FAO/WHO.

Codex (2003), 'Guideline for the conduct of food safety assessment of foods derived from recombinant-DNA plants', CAC/GL 45–200, Rome: FAO/WHO.
Codex (2007), 'Working principles for risk analysis for food safety for application by governments', CAC/GL 62–2007, Rome: FAO/WHO.
Codex (2008), 'Guideline for the conduct of food safety assessment of foods derived from recombinant-DNA animals', CAC/GL, Rome: FAO/WHO.
Codex (2011), 'Working principles for the application of risk analysis in the framework of the Codex Alimentarius', *Codex Alimentarius Commission Procedural Manual*, 20th edn, Rome: Food and Agriculture Organization.
FAO/WHO (2006), 'Food safety risk analysis: a guide for national food safety authorities', FAO Food and Nutrition Paper 97, Rome: Food and Agriculture Organization.
FAO (2007), 'FAO biosecurity toolkit', Rome: Food and Agriculture Organization.
IPPC (2004), 'Pest risk analysis for quarantine pests including analysis of environmental risks and living modified organisms', ISPM No. 11, Rome: IPPC.
IPPC (2007), 'Framework for pest risk analysis', ISPM No. 2, Rome: IPPC.
Jackson, L. and M. Jansen (2010), 'Risk assessment in the international food policy arena: can the multilateral institutions encourage unbiased outcomes?', *Food Policy*, **35**(6), 538–47.
National Research Council (1983), 'Risk assessment in the federal government: managing the process', Washington, DC: National Academy Press.
National Research Council (1994), 'Science and judgment in risk assessment', Washington, DC: National Academy Press.
National Research Council (1996), 'Understanding risk: informing decisions in a democratic society', Washington, DC: National Academy Press.
OIE (2011a), 'Aquatic animal health code', Paris: OIE.
OIE (2011b), 'Terrestrial animal health code', Paris: OIE.
OIE (2011c), 'Guidelines for assessing the risk of non-native animals becoming invasive', Paris: OIE.
Pauwelyn, J. (1999), 'The WTO Agreement on sanitary and phytosanitary (SPS) measures as applied in the first three SPS disputes', *Journal of International Economic Law*, 641–64.
Peel, J. (2004), 'Risk regulation under the WTO SPS Agreement: science as an international normative yardstick?', Jean Monnet Working Paper June 2004, New York: New York University Law School.
WTO (1995), 'Agreement on sanitary and phytosanitary measures', Geneva: WTO.
WTO (1998), 'Australia – measures affecting importation of salmon', WT/DS18/AB.
WTO (1996), 'European Communities – measures concerning meat and meat products', WT/DS26/R.
WTO (2003), 'Japan – measures affecting the importation of apples', WT/DS245/R.
WTO (2006), 'European Communities – Measures Affecting the Approval and Marketing of Biotech Products', WT/DS291/R.
WTO (2008), 'United States – Continued Suspension of Obligations in the EC – Hormones Dispute', WT/DS/320/R.
WTO (2010), 'Australia – measures affecting the importation of apples from New Zealand', WT/DS367/AB.

16 The trade system and biotechnology

William A. Kerr

INTRODUCTION

Transformative technologies represent significant regulatory challenges both for domestic policy makers and for those charged with putting arrangements in place for their international governance. Transformative technologies move existing ways of doing things and institutional arrangements from states of near equilibrium into states of considerable disequilibrium before settling down to a new altered state of near equilibrium. That is the nature of the transformative process and what sets some new technologies apart from the more common iterative or marginal gains in technological progress. The problem for policy makers is that the end point of transformation is seldom clear and disequilibrium creates both considerable angst in society that they must attempt to allay, as well as losers that will demand protection or that the transformation be halted. Policy initiatives may have unintended consequences or kill the goose that lays the golden egg of technological benefits. Often, it is not clear in the beginning that the technology is transformative and policy makers may attempt to govern it within existing institutional arrangements that become increasingly inappropriate as the state of disequilibrium deepens. For example, in the early 1900s who could have foreseen the changes that harnessing the internal combustion engine for transportation would bring – networks of freeways, extensive fuel distribution systems, huge industrial manufacturing complexes, urban sprawl, parking lots, air pollution, carhops, driver training, junkyards, radial tyres, motels, just-in-time-delivery, 18-wheelers, drive-in churches, motor homes, demolition derbies; the list is almost endless. None of this was foreseen by whoever was making the regulation whereby a flagman had to precede a horseless carriage to warn those with potentially skittish horses of its arrival.

The automobile transformed every country in the world, but far from uniformly. In part, these differences arise from geographic realities and levels of economic development, but also, in part, due to regulatory divergence. While there has been considerable international cooperation in devising automobile regulations, somehow the world ended up with half of it driving on the left and half of it driving on the right. The result of the latter is that international trade in automobiles is considerably inhibited, with cars that trade between countries that drive on different sides of the road having to be engineered with the capability to place the controls on either side of the vehicle – a non-trivial increase in cost. This absence of international driver-side harmonization has also led to acrimonious trade relations. For example, US car makers complained for decades that non-tariff trade barriers inhibited the sales of American cars in Japan. The reality was, of course, that almost no American cars were engineered to have the controls on the right-hand side of the car to accommodate Japanese drivers (Kerr and Perdikis, 2003). Of course, lack of international regulatory harmonization for transformative technologies is not confined to automobiles but has been widespread – and with accompanying disincentives to trade.

The absence of harmonization in domestically distributed electricity current means that small electric appliances do not trade; or must be designed to accept a number of alternative currents: with the exception of laptop computers, electric shavers and hair dryers, most are not. Television standards differ, railway gauges differ, cellphone configurations differ. Of course, there has been a considerable degree of international harmonization which acts to enhance trade: agreement of English as the international language for air traffic control, the size of shipping containers, the internet, to name a few. The point is that one should not expect that the governance of transformative technologies will evolve in ways that produce international harmonization and, as a result, the benefits of international trade are often forgone.

The reasons why international regulatory harmonization does not necessarily arise for transformative technologies are multifaceted and, to some extent, particular to each individual technology. It is important to realize that a failure to harmonize is likely to mean that the potential societal benefits that the technology can yield will be reduced. Biotechnology, while it has been identified as a transformative technology (Phillips et al., 2006), is probably at the regulatory stage of the automobile in the first decade of the twentieth century, with flagmen posted to warn of its approach. Policy makers are attempting to grapple with how to regulate the technology where its future evolution is opaque, disequilibrium is manifest and increasing, angst is mounting in some quarters and vested interests fear losses. International governance of biotechnology is very much a work in progress.

INTERNATIONAL TRADE INSTITUTIONS

As with any major change to technology, everything from terminology to the measurement of efficacy will have to be determined. Attempts to fill these regulatory gaps will initially be made by existing national and international institutions. This is no different in the case of biotechnology, where a range of international institutions vie to take the lead. The Organisation for Economic Co-operation and Development (OECD) took the lead in establishing agreed terminology and measurement methods so that confusion in international communication was reduced. Existing scientific organizations such as the International Plant Protection Convention (IPPC), The World Organization for Animal Health (OIE) and the Codex Alimentarius (Codex) are involved in developing standards for genetically modified (GM) organisms and products. The IPPC has addressed the international regulation of GM crops through several International Standards for Phytosanitary Measures (ISPMs) while the OIE has developed standards for diagnostic reagents, sera, and vaccines for animals in the International Animal Health Code. There has been significant effort in Codex to develop, for example, a standard for labelling food products derived from biotechnology (Phillips and Kerr, 2002).

As with other transformative technologies, the use of biotechnology became a contentious issue worldwide, but it proved to be more controversial in some countries than others. This led to it being regulated in different ways in individual countries and these divergent regulations resulted in limitations on market access for the products of biotechnology in some countries. Thus, the multilateral trade institutions, particularly the

World Trade Organization (WTO), became the focus of international regulatory conflict (Kerr, 1999).

The multilateral governance system for international trade arose in the wake of the Great Depression of the 1930s, when there were no multilateral rules, and the Second World War. Based on their experience from the Great Depression and the slide into the second global war in a century, the victors – particularly the US and the UK – were convinced that there was a need for a set of multilateral institutions to reduce political and economic conflicts between nation states (Kerr, 2010a). In the late 1940s a number of multilateral institutions were negotiated to deal with sources of country-to-country conflict: the United Nations to deal with political disputes; the International Monetary Fund to deal with strategic currency devaluations; and the World Bank to deal with inequalities in levels of economic development. A fourth multilateral organization, the International Trade Organization (ITO) was proposed (US Department of State, 1945) and negotiated (Kerr, 2010b), but never put in place due to protectionist interests in the US Congress. One of the ITO's sub-agreements, the General Agreement on Tariffs and Trade (GATT) was, however, ratified by the US Congress and came into being in 1947.

The GATT–1947 proved to be a good vehicle for removing the very high Depression-era tariffs but, unlike the ITO, was too narrow in scope to deal with other trade-restricting measures that came to the fore as tariffs were removed (Gaisford and Kerr, 2001). It was also a consensus-based organization, a characteristic that extended to its disputes mechanism. Further, trade in agricultural products was largely excluded from the GATT disciplines due to waivers granted to the US in the organization's early days. These waivers allowed countries to retain: (1) high tariffs on agricultural products; (2) the continued use of import quotas and variable levies to inhibit agricultural imports; (3) the extensive use of export subsidies; and (4) the paying of large domestic subsidies to farmers without the fear of countervailing measures being applied (Kerr, 2000). The absence of disciplines on trade in agricultural products led to considerable trade distortion in the sector and, eventually, a beggar-thy-neighbour subsidy war between the EU and the US in the 1980s and early 1990s. The subsidy war in agriculture threatened to spread to other sectors and began to spill over into other areas of international relations.

It was agreed that a return to the GATT disciplines for agricultural products should be included in the negotiating agenda for the Uruguay Round of GATT negotiations that began in 1986. The Uruguay Round negotiations were long and acrimonious, with it being particularly difficult to arrive at a set of disciplines on agriculture. One of the concerns of the negotiators was that once disciplines on the use of trade barriers and subsidies were imposed on agriculture, countries would simply switch to providing support through the use of trade barriers justified on nefarious sanitary and phytosanitary grounds. In an attempt to eliminate such ploys, a new GATT sub-agreement was negotiated – the Agreement on the Application of Sanitary and Phytosanitary Measures (SPS). All member states, including the European Union, agreed that science should be the only justification for imposing SPS-based trade barriers. The Uruguay Round was concluded in 1994, prior to any major commercialization of GM crops. In addition to the SPS agreement, the end of the Uruguay Round saw the most significant revamping of the multilateral trade system since its inception in 1947. The Agreement on Technical Barriers to Trade (TBT) was strengthened, new agreements on Trade in Services and Trade Related Aspects of Intellectual Property (TRIPS) were negotiated and a new organization, the World Trade Organization (WTO), put in place to oversee these agreements and the revised GATT, the GATT–1994. The new WTO was endowed with a binding disputes settlement mechanism.

As most attempts to restrict trade in GM organisms and products have been justified on an SPS basis, it has been at the centre of the subsequent international controversies regarding these products, although aspects of the TBT (on labelling of GM foods in particular) and the broader GATT–1994 are also applicable. In addition, given that much of agricultural biotechnology was developed in the private sector, the international protection of intellectual property has also been a major issue, which comes under the TRIPS agreement.

THE AGREEMENT ON THE APPLICATION OF SANITARY AND PHYTOSANITARY MEASURES

The economic model upon which the GATT disciplines are premised does not expect any group in society to ask for protection through the imposition of trade barriers besides producers of goods, such as firms or farmers (Kerr, 2010c). This premise continues in the WTO and the SPS. Hence, there is no provision to allow governments to impose trade barriers in response to other groups in society that are seeking to exclude goods from their country's market. As a result, governments are forced to seek alternative justification for trade barriers when faced with protectionist pressure from, for example, consumers or environmentalists (Kerr, 2010c; Perdikis and Kerr, 1999).

The SPS deals with trade barriers put in place for reasons of human health, animal health, plant health and protection of the natural environment. SPS measures to restrict trade must be consistent with the general WTO principle of non-discrimination, meaning that foreign suppliers should be treated no differently than domestic suppliers and that there should be no discrimination among foreign suppliers (Isaac, 2007). Discrimination on either basis is allowed, however, if diseases and pests are specific to some countries but not to others or if foreign regulations, personnel or facilities are not able to effectively manage the safety of exported food and agricultural products (Kerr, 2004).

It is the decision-making criteria enshrined in the SPS, however, that set it apart from other agreements and sub-agreements of the WTO. In the rest of the WTO the decision to impose trade barriers is largely political, with the constraints on political decisions largely economic. For example, certain forms of economic protection are banned (for example import quotas), other forms of protection are constrained (for example tariff rates are bound) and the barriers imposed must be the least trade distorting (for example least cost for foreign suppliers) available to achieve the ends (Kerr, 2010a). If a country cannot live up to its commitments due to domestic political constraints then countries that do not receive the expected benefit are allowed to engage in economic retaliation (for example imposing countervailing duties on goods found to be unfairly subsidized) (Baylis, 2007). In the SPS, while the decision to initiate the imposition of trade barriers remains political, the constraint is scientific legitimacy. Economic factors again enter the decision process only when a political decision to impose trade barriers is made when there is no legitimate scientific justification – when the costs of retaliation must be considered by decision makers.

The legitimacy of science as enshrined in the SPS has two components: (1) a trade barrier can only be imposed if human, animal, plant health or the natural environment would be threatened by imports (for example imported grain could carry fungal diseases that could threaten the health of local production in the importing country); and (2) a risk assessment that determines imports that could threaten human, animal and plant health and the environment actually pose an unacceptable level of risk. The second component is included because there could be circumstances where trade in products with a potential health problem can be imported because the risk is at an acceptable level for the society (for example some individuals are allergic to peanuts but with proper management the risk is within acceptable limits). Thus, for the imposition of trade barriers there must be a scientifically identified health or environmental problem associated with the imported product and some positive assessment that shows that importing those products would lead to an unacceptable level of risk to health or the environment.

The framers of the SPS presumed that there would be a scientific consensus that policy makers could draw upon to determine whether a scientific basis for the imposition of trade barriers existed and that policy makers would accept the statistical basis of risk (for example that guaranteeing zero risk is a statistical impossibility). It was agreed that countries could set their own levels of risk. Those that negotiated the SPS recognized that in the case of new scientific discoveries there could be a lag in garnering sufficient scientific information to form a consensus. In those cases, countries are allowed to temporarily impose trade barriers on a precautionary basis. If trade barriers are imposed on the basis of precaution, then the country imposing the barrier is obligated to proactively seek the information to fill in the gap in scientific knowledge (Kerr, 2003a).

In a fashion that is considered consistent with the scientific method, the scientific community has institutionalized its means of dealing with new discoveries – this is now called the Risk Analysis Framework (RAF) (Isaac, 2007). The RAF was developed to deal with the regulation of advanced technology products (which were characterized by a major information gap between the producers of the innovation and the intended consumers) where the goal was to credibly inject science into public policy development (Isaac and Kerr, 2007; National Academy of Science, 1983). The apparent universality of the RAF made it a suitable framework for building standardized regulatory rules in various domestic jurisdictions as well as in multilateral agreements and treaties. The RAF is also supported by international organizations such as the Organisation for Economic Co-operation and Development (OECD), the World Trade Organisation (WTO) and several United Nations agencies, including the World Health Organization (WHO) and the Food and Agriculture Organization (FAO). It is used by the Codex, the OIE and the IPPC international standards organizations, all of which predated the SPS. These science-based standards-setting organizations were recognized explicitly in the SPS and, thus, the RAF was implicitly accepted in the Agreement. More fundamentally, the RAF has been the foundation of all regulatory approaches to the research, development and commercialization of GMOs in all countries that now have specific regulatory capacity (Dimitrov et al., 2004; Krimsky, 2000; Isaac, 2002).

International standards are a goal in the SPS and standards agreed by the Codex, OIE and IPPC are given safe haven status if they are adopted by member states of the WTO in justifying the imposition of trade barriers – in other words they are not open to challenge by other WTO members (Isaac et al., 2002). Countries are allowed to establish their own standards that are stricter than those of the Codex, OIE or IPPC but they must provide a scientific justification for those stricter standards. In essence, the framers of the SPS attempted to ring fence the Agreement from producers seeking economic protection from their politicians by removing the ability of politicians to interfere in the decision-making process for imposing trade barriers justified on sanitary and phytosanitary grounds.

There have been some pressures to reopen the SPS agreement in subsequent negotiations in recent years. The Uruguay Round produced an Agreement on Agriculture that (re-)applied GATT disciplines to the sector, but only to a limited degree. The member states could not agree on full transition to GATT disciplines, thus threatening the successful conclusion of the entire round. Thus, to save the round they accepted a first step, but mandated renewed negotiations on the Uruguay Round agenda after a five-year break. Thus, the negotiations on agriculture were to start again in 1999 but with an agenda that had largely been determined at the beginning of the Uruguay Round – long before biotechnology had become a trade issue. These separate mandated negotiations were subsequently rolled in 2001 into the wider Doha Round, albeit with the agenda unchanged (Hobbs and Kerr, 2000). In short, the SPS was not re-opened for negotiation in the Doha Round (Perdikis et al. 2001). Thus, the multilateral rules of trade that govern biotechnology were all negotiated before the widespread commercialization of biotechnology and there is no provision to address them in the Doha Round negotiating agenda at the WTO, yet the international trade environment has changed considerably since the trade negotiation agenda for agriculture, including the SPS, was agreed in 1986 (Kerr, 2010c).

Biotechnology became a significant international trade issue at the end of the 1990s – primarily in the European Union. While a few GM crops were approved for cultivation in the EU in the 1990s, angst regarding the appropriateness of embracing the clearly transformative technology rose quickly among some segments of civil society with strongly held preferences (Gaisford et al., 2001). Their vociferous opposition to biotechnology could not be ignored by politicians in the EU, and in 1999 a temporary ban on the licensing of new GM crops was put in place until a revised regulatory regime could be devised (Viju et al., 2011). This was accompanied by a moratorium on imports of GM organisms and products.

Some groups in civil society in other countries also expressed doubts regarding the wisdom of embracing the transformative technology and in some cases they succeeded in stopping or slowing down the acceptance of biotechnology by policy makers. In a number of countries, however, their influence was smaller and registrations of new biotechnology-based products continued in an orderly fashion under the broad scientific umbrella of the RAF. In particular, adoption of GM crops was widespread in major agricultural-exporting countries such as the US, Canada and Argentina. Thus, the stage was set for a major trade confrontation.

The political problem in the EU largely arose from consumers, environmentalists, those with ethical concerns and those worried about the influence of multilateral corporations – the major developers and owners of intellectual property in agricultural biotechnology (agbiotech) – in the food system (Kerr, 2001; Gaisford et al., 2001). It was not, for the most part, farmers or other commercial actors in the EU asking for a ban on the technology. The key exception was organic producers who redefined the *organic* standard to exclude GM crops, at least in part for marketing reasons. As suggested above, there is no recognition in the WTO that groups other than producers would ask their policy makers for protection from imports (Kerr, 2010c). Under the existing WTO arrangements there is no direct mechanism by which the EU could impose trade barriers on the basis of resistance from other groups in civil society. Some of those who objected to the technology were doing so for reasons that fit within the SPS – concerns over the effect of consuming GM products on human health and concerns over risks to the natural environment of releasing GM organisms (Perdikis, 2000). These concerns were dismissed by biotechnology exporters because their licensed GM crops had all been assessed using the RAF framework and, thus, had passed the scientific hurdle.

While having agreed to the *scientific legitimacy* basis of the SPS, the EU began to chafe under its restrictions. The first major test to the SPS came not from crop biotechnology, but rather the use of growth hormones in beef production. Growth hormones became a focus of consumer concerns in the EU after a few children in Italy exhibited abnormal physical development (Roberts, 1998; Kerr and Hobbs, 2002). Their use in the EU was severely restricted and imports of beef produced using them banned. This was in spite of scientific assessments undertaken by the EU that indicated neither health concerns nor unacceptable levels of risk. Basically, some EU consumers refused to defer to the scientific consensus and politicians acquiesced (Roberts, 1998). Growth hormones are widely used in beef production around the world and a case was brought to the WTO by the US and Canada regarding the EU import ban. The WTO Panel sided with the complainants, finding that the EU had not presented sufficient evidence of a threat to human health and had not carried out a risk assessment. The EU claimed that the ban was justified on the basis of precaution because some sub-populations had not been tested for risks to their health – the panel sided with the scientific consensus. If the sub-populations argument had been accepted, trade barriers could always be justified because there are always more sub-populations that could be tested (Smyth et al., 2011). Further, the panel indicated how it would deal with the question of the level of risk. Under the SPS each country is allowed to determine its acceptable level of risk. For beef hormones the EU effectively set the level of risk at zero. The panel did not accept this because the risk tolerated by the EU for other similar products was considerably more lax (Roberts, 1998). Thus, the panel attempted to close off this potential loophole by making its decision on the basis of comparable risks. The panel asked the EU to remove its ban. The EU chose to ignore the WTO Panel's ruling – something very rare – and accept retaliation, as is its right under the WTO. The value of trade affected was small and, hence, the cost of accepting retaliation deemed acceptable. The US and Canada subsequently retaliated (Kerr and Hobbs, 2005). The outcome of the beef hormone case was a clear disappointment for the EU and was considered a test for the rising storm over GM products. Given its failure to win the scientific legitimacy battle it began to alter its stance regarding what had been agreed, suggesting that science should only *inform* decisions in SPS cases and that other political and economic concerns should also inform the decisions (Isaac, 2007). In effect, it walked away from the attempt to ring fence SPS issues from political influence. This set up a major conflict between GM exporters, particularly the US, and the EU (Isaac and Kerr, 2003a).

The 1999 EU ban on imports, while imposed on a temporary basis, continued over a considerable period as the EU attempted to put in place a new domestic regulatory regime for licensing GM products. The issue became one of the most divisive ever faced by the European Commission, and one that they had not been able to fully resolve by 2011 (Viju et al., 2011). The EU's ban on imports was challenged at the WTO by the US, Canada and others in 2003 (Isaac and Kerr, 2003b). Given the transformative nature of biotechnology, the EU attempted to argue that the ban was justified on the basis of insufficient scientific information – the precaution clause in the SPS. This set off a broader debate regarding the nature of the precautionary principle, with some in civil society seeing it as a means to effectively stop the transformative technology being pursued (Van den Belt, 2003). While the precautionary principle has been enshrined in a number of international agreements and national legislations, operationalizing the principle for decision making has proved very difficult and contentious (Kerr, 2003a).

In 2006 the WTO Panel ruled against the EU's temporary import ban and the EU promised to bring its import regime into compliance with WTO commitments, but requested time to do so (Viju et al., 2011). The rudiments of the new EU regime were put in place in 2003 but it has continued as a work in progress as there remains vociferous opposition to biotechnology in some EU member states. The first new GM crop approved in the EU under the new rules was approved only in 2010, based on a 2005 application. The new regime, however, remains consistent with the EU's stance on the SPS that science should only have an advisory role – it leaves the final approval of a GM organism or product to political institutions (Viju et al., 2011). Thus, the EU would appear open to another SPS challenge at the WTO.

In the meantime, trade in the products of biotechnology remains constrained. Given that segregation of GM and non GM crops in major producing and exporting countries is likely to be prohibitively expensive even if it is technically feasible in some cases (Smyth et al. 2004), the world remains divided between those that license and trade GM crops and those that do not. In some cases, countries have failed to adopt the technology due to fears over losing access to non-GM markets, particularly the EU (see Chapter 12 by Paarlberg in this volume for a discussion of the effect in Africa). Further, the development and release of some GM crops, such as wheat for example, have been postponed due to worries over access to non-GM markets. These trade difficulties regarding biotechnology are likely to be inhibiting investment in biotechnology and, thus, reduce ability to realize the potential benefits associated with the technology.

Given the transformative nature of the technology, it has been suggested by some that the WTO is not the appropriate institution to make the trade rules for biotechnology. It was argued that an alternative institution was required to regulate trade in GM products and organisms. As an alternative, the Cartagena Protocol on Biosafety (CPB) was negotiated between 1996 and 2000 by 138 countries under the auspices of the 1992 Convention on BioDiversity (CBD) of the United Nations Environmental Program (UNEP). It has been ratified by the requisite number of countries, including the EU, but not by the US or a number of countries that have been major adopters of GM crops. Thus there are, in effect, competing international rules for governing the trade in the products of biotechnology (Phillips and Kerr, 2000). The CPB explicitly incorporates the precautionary principle, accepts that other considerations than science may be taken into account when putting trade barriers in place and allows countries to unilaterally impose trade barriers without appeal by exporters (Hobbs et al., 2005). It accepts the principle that exporters are liable for any negative consequences that may arise as a result of the import of GM organisms and products, including economic losses. The parties to the CPB have not, as yet, been able to agree on the rules for liability (Holtby et al., 2007). The CPB does not have a binding dispute settlement mechanism. The CPB's rules are, to a considerable degree, in conflict with those of the WTO and it is unclear whether the CPB or the WTO takes precedent in international law (Phillips and Kerr, 2000). This confusion has led some countries to try to incorporate aspects of both regimes in their domestic laws (for example the Philippines as discussed by Ludlow and Yorobe in Chapter 10 in this volume). Given that the WTO has a binding disputes settlement mechanism it seems likely that any trade disputes will be brought to the WTO and decided according to WTO rules. Clearly, the multilateral governance system is struggling to find the appropriate means to regulate international trade in this transformative technology.

THE INTERNATIONAL PROTECTION OF INTELLECTUAL PROPERTY

One of the major changes to the multilateral international institutional arrangements governing international trade was an attempt to harness the sanctioning power of the GATT for the enforcement of foreign intellectual property rights. In the last quarter of the twentieth century the proportion of the value of goods comprised of intellectual property was increasing rapidly (Yampoin and Kerr, 1998). While a number of international agreements existed to foster the international protection of intellectual property and an international coordinating institution, the World Intellectual Property Organization (WIPO) had been put in place, there was no binding dispute settlement mechanism, no particular incentive for countries to join, and no mechanism to sanction those that failed to live up to their commitments (Kerr, 2003b). The result was that many developing countries chose not to join the existing arrangements to protect intellectual property (Gaisford and Richardson, 2000), and those that did made little effort to enforce them (Kerr 2003b). Given the rising technical capacity in some developing countries, particularly in Asia, piracy was leading to significant forgone revenues for those investing in developing new intellectual property, largely located in developed countries.

The restructuring of the GATT into the WTO during the Uruguay Round was undertaken in part to rectify these deficiencies. In particular, the new TRIPS agreement was negotiated. The WTO was created to administer the TRIPS, GATT–1994 and the new General Agreement on Trade in Services (GATS). The WTO was given a binding disputes settlement mechanism. Central to the organizational structure was cross-agreement retaliation whereby trade sanctions under the GATT–1994 could be applied against countries that failed to live up to their TRIPS commitments to enforce the intellectual property rights held by foreign firms. To gain the benefits of the GATT, all countries had to join the WTO and, hence, accept TRIPS disciplines. Developing countries signed up to the TRIPS unenthusiastically and only after being promised better market access to developed country markets for their agricultural products and textiles (Boyd et al., 2003).

Developed country governments, faced with budgetary difficulties, were not willing to shoulder the major portion of the investments required to develop the apparent potential of biotechnology – in contrast to their major efforts using conventional plant breeding in, for example, the Green Revolution of the 1960s and 1970s. To provide the incentive for the private sector to undertake research and development activities in agbiotech, intellectual property rights had to be extended to living organisms (Gaisford et al., 2001; Kerr and Yampoin, 2000). Over time the required changes were made in domestic intellectual property legislation and were incorporated in the TRIPS.

The TRIPS requires that all countries put domestic intellectual property legislation in place that gives equal treatment to foreign owners of intellectual property. Countries must have transparent and judicially-based methods for firms to complain about violations of their intellectual property and countries must commit resources to enforcement – although this provision is rather vague. During the TRIPS negotiations, developing countries expressed wide-ranging concerns regarding patent protection for agbiotech for reasons of food security, anti-competitive practices of agbiotech firms and threats to the environment (Ringo, 1994). They also generally hold reservations relating to the patenting of life forms and agricultural crops in particular. Due to these concerns, the TRIPS contains an exception whereby countries can choose not to enforce intellectual property rights for 'plants and animals (other than micro-organisms) and essential biological processes for the production of plants and animals, other than non-biological and microbiological processes' (TRIPS Article 27.3). Plant varieties, however, must be protected either by patents or a system of plant breeders' rights, and products developed using biotechnology are to be protected.

Developing countries have tended to focus on the costs of protecting intellectual property, particularly the high costs of pharmaceuticals for impoverished citizens and the costs of seed for poor farmers, rather than the benefits of new technology (Boyd et al., 2003). As developing countries produce little intellectual property, most of the benefits of enforcing intellectual property accrue to developed-country firms (Kerr and Isaac, 2005). Thus, developing countries have little incentive to enforce intellectual property rights. Further, as some of the genetic material that provides the potential gains from biotechnology comes from plants that are part of the natural environment or traditional varieties used in developing countries, there has been considerable debate over the appropriateness of assigning intellectual property to this genetic material – the so-called bioprospecting or biopiracy debate over traditional knowledge (Isaac and Kerr, 2004; also see Chapter 32 in this volume). This has raised further questions in developing countries regarding their enforcement of TRIPS commitments.

The result is that enforcement in developing countries has been lax (Cardwell and Kerr, 2008; Loppacher and Kerr, 2004). Theoretical investigations of the likely efficacy of the threat of sanctions incorporated in the WTO suggest that in most circumstances they will not provide sufficient incentive for developing countries to enforce intellectual property rights in biotechnology (Yampoin and Kerr, 1998; Gaisford and Richardson, 2000; Gaisford et al. 2002). In spite of the continued increase in incidents of intellectual property piracy the mechanism of cross-agreement retaliatory sanctioning remains relatively untested. The predictable results of lax enforcement have been two-fold. First, investment in crops appropriate for developing countries is largely absent (Gaisford et al., 2007; Loppacher and Kerr, 2005). Second, firms that invest in biotechnology are seeking alternative mechanisms to protect their intellectual property (Cardwell and Kerr, 2008). In short, the WTO and TRIPS have not provided an institutional mechanism that facilitates maximizing the potential benefits from biotechnology.

CONCLUSIONS

Multilateral international trade institutions have thus far struggled with providing effective and efficient governance for the transformative technology of biotechnology. For both the rules of trade and the protection of intellectual property there has been an attempt to govern biotechnology using existing norms and institutions. In the case of a transformative technology this may be an inappropriate attempt to put a square peg into a round hole. The political pressures faced by policy makers in both developed countries and developing countries are different from those the existing multilateral institutions were designed to deal with. In particular, the rise of consumers, environmentalists and other groups in civil society seeking protection from imports runs counter to a major assumption of the economic model upon which the multilateral rules of trade are premised. In developing countries there has never been any real buy-in to the commitments in the TRIPS and, in particular, the appropriateness of trade sanctions. Furthermore, biotechnology has also proved to be controversial among civil society groups in developing countries, particularly because of the role of multinational companies in providing seed and the question of property rights in traditional knowledge. The one attempt to create a new multilateral institution to deal with trade in the products of biotechnology, the CPB, sets out rules that are largely mirrored opposites to those of the WTO. Hence, the CPB is as unacceptable to those wishing to export GM organisms and products as the WTO is to those who wish to exclude the products of biotechnology from their markets. Thus far, the member states of the WTO have been unwilling to deal with the issue of the new protectionists. As opening the SPS to re-negotation, for example, will require consensus among WTO member states, reform within existing institutions seems remote. As a result, trade in the products of biotechnology is inhibited and investment in the technology is reduced and skewed away from developing country crops. The potential benefits of biotechnology are commensurately reduced. The way forward is not clear. It is, however, early days and just as with other transformative technologies, governance is likely to progress beyond placing a flagman ahead of the horseless carriage.

REFERENCES

Baylis, K. (2007), 'Unfair subsidies and countervailing duties', in W.A. Kerr and J.D. Gaisford (eds), *Handbook on International Trade Policy*, Cheltenham, UK and Northampton, MA, USA: Edward Elgar Publishing, pp. 347–59.

Boyd, S.L., W.A. Kerr and N. Perdikis (2003), 'Agricultural biotechnology innovations versus intellectual property rights – are developing countries at the mercy of multinationals?', *The Journal of World Intellectual Property*, **6**(2), 211–32.

Cardwell, R. and W.A. Kerr (2008), 'Protecting biotechnology IPRs in developing countries: simple analytics of a levy solution', *Journal of Agricultural Economics*, **59**(2), 217–36.

Dimitrov, S., T. Etty, M. Oomens and H. Somsen (2004), 'Regulating biotechnology', *Journal of International Biotechnology Law*, **1**(4), 174–6.

Gaisford, J.D. and W.A. Kerr (2001), *Economic Analysis for International Trade Negotiations*, Cheltenham, UK and Northampton, MA, USA: Edward Elgar Publishing.

Gaisford, J.D. and R.S. Richardson (2000), 'The TRIPS disagreement: should GATT traditions have been abandoned?', *Journal of International Law and Trade Policy*, **1**(2), 137–51.

Gaisford, J.D., J.E. Hobbs and W.A. Kerr (2007), 'Will the TRIPS Agreement foster appropriate biotechnologies for developing countries?', *Journal of Agricultural Economics*, **58**(2), 199–217.

Gaisford, J.D., R. Tarvydas, J.E. Hobbs and W.A. Kerr (2002), 'Biotechnology piracy: rethinking the international protection of intellectual property', *Canadian Journal of Agricultural Economics*, **50**(1), 1–14.
Gaisford, J.D., J.E. Hobbs. W.A. Kerr, N. Perdikis and M.D. Plunkett (2001), *The Economics of Biotechnology*, Cheltenham, UK and Northampton, MA, USA: Edward Elgar Publishing.
Hobbs, A.L. and W.A. Kerr (2000), 'First salvoes: a guide to the opening proposals at the WTO negotiations on agricultural trade', *Agribusiness Paesaggio and Ambiente*, **4**(2), 97–108.
Hobbs, A.L., J.E. Hobbs and W.A. Kerr (2005), 'The biosafety protocol: multilateral agreement on protecting the environment or protectionist club?', *Journal of World Trade*, **39**(2), 281–300.
Holtby, K.L., W.A. Kerr and J.E. Hobbs (2007), *International Environmental Liability and Barriers to Trade*, Cheltenham, UK and Northampton, MA, USA: Edward Elgar Publishing.
Isaac, G.E. (2002), *Agricultural Biotechnology and Transatlantic Trade: Regulatory Barriers to GM Crops*, Wallingford: CABI Publishing Inc.
Isaac, G.E. (2007), 'Sanitary and phytosanitary issues', in W.A. Kerr and J.D. Gaisford (eds), *Handbook on International Trade Policy*, Cheltenham, UK and Northampton, MA, USA: Edward Elgar Publishing, pp. 383–93.
Isaac, G.E and W.A. Kerr (2003a), 'Genetically modified organisms and trade rules: identifying important challenges for the WTO', *The World Economy*, **26**(1), 29–42.
Isaac, G.E and W.A. Kerr (2003b), 'GMOs at the WTO – a harvest of trouble', *Journal of World Trade*, **37**(6), 1083–95.
Isaac, G.E. and W.A. Kerr (2004), 'Bioprospecting or biopiracy? – Intellectual property and traditional knowledge in biotechnology innovation', *The Journal of World Intellectual Property*, **7**(1), 35–52.
Isaac, G.E. and W.A. Kerr (2007), 'The biosafety protocol and the WTO: concert or conflict', in R. Falkner (ed.), *The International Politics of Genetically Modified Food*, Houndmills: Palgrave Macmillan, pp. 195–212.
Isaac, G.E., M. Phillipson, and W.A. Kerr (2002), 'International regulation of trade in the products of biotechnology', Estey Centre Research Papers No. 2, Saskatoon: Estey Centre for Law and Economics in International Trade.
Kerr, W.A. (1999), 'International trade in transgenic food products: a new focus for agricultural trade disputes', *The World Economy*, **22**(2), 245–59.
Kerr, W.A. (2000), 'The next step will be harder: issues for the new round of agricultural negotiations at the World Trade Organization', *Journal of World Trade*, **34**(1), 123–40.
Kerr, W.A. (2001), 'The World Trade Organization and the environment', in H.J. Michelman, J. Rude, J. Stabler and G. Storey (eds), *Globalization and Agricultural Trade Policy*, Boulder, Co: Lynne Rienner, pp. 53–65.
Kerr, W.A. (2003a), 'Science-based rules of trade – a mantra for some, an anathema for others', *Journal of International Law and Trade Policy*, **4**(2), 86–97.
Kerr, W.A. (2003b), 'The efficacy of the TRIPS: incentives, capacity and threats', *Journal of International Law and Trade Policy*, **4**(1), 1–14.
Kerr, W.A. (2004), 'Sanitary barriers and international trade: governance issues for the NAFTA beef market', in R.M.A. Loyns, K. Meilke, R.D. Knutson and A. Yunez-Naude (eds), *Keeping the Borders Open*, Proceedings of the Eighth Agricultural and Food Policy Systems Information Workshop, Guelph: University of Guelph, pp. 26–49.
Kerr, W.A. (2010a), *Conflict, Chaos and Confusion: The Crisis in the International Trading System*, Cheltenham, UK and Northampton, MA, USA: Edward Elgar Publishing.
Kerr, W.A. (2010b), 'GATT-1947: a living legacy fostering the liberalization of international trade', *Journal of International Law and Trade Policy*, **11**(1), 1–11.
Kerr, W.A. (2010c), 'What is new in protectionism?: Consumers, cranks and captives', *Canadian Journal of Agricultural Economics*, **58**(1), 5–22.
Kerr, W.A. and J.E. Hobbs (2002), 'The North American–European Union dispute over beef produced using growth hormones: a major test for the new international trade regime', *The World Economy*, **25**(2), 283–96.
Kerr, W.A. and J.E. Hobbs (2005), 'Consumers, cows and carousels: why the dispute over beef hormones is far more important than its commercial value', in N. Perdikis and R. Read (eds), *The WTO and the Regulation of International Trade*, Cheltenham, UK and Northampton, MA, USA: Edward Elgar Publishing, pp. 191–214.
Kerr, W.A. and G.E. Isaac (2005), 'The international treatment of biological material as intellectual property', *Journal of International Biotechnology Law*, **2**(3), 105–11.
Kerr, W.A. and N. Perdikis (2003), *The Economics of International Business: A Guide to the Global Commercial Environment*, Estey Centre Program in International Trade Education, Saskatoon: Estey Centre for Law and Economics in International Trade.
Kerr, W.A. and R. Yampoin (2000), 'Adoption of biotechnology in Thailand and the threat of intellectual property piracy', *Canadian Journal of Agricultural Economics*, **48**(4), 597–66.
Krimsky, S. (2000), 'Risk assessment and regulation of bioengineered food products', *International Journal of Biotechnology*, **2**(1,2 & 3), 231–8.

Loppacher, L.J. and W.A. Kerr (2004), 'China's biotechnology industry and the international protection of intellectual property rights', *Journal of International Biotechnology Law*, **1**(5), 177–86.
Loppacher, L.J. and W.A. Kerr (2005), 'Developing countries' protection of intellectual property under TRIPS and innovation investment', *The Journal of World Intellectual Property Rights*, **1**(1–2), 1–17.
National Academy of Sciences (NAS) (1983), *Risk Assessment in the Federal Government: Managing the Process*, Committee on the Institutional Means for Assessment of Risks to Public Health, Commission of Life Sciences, Washington, DC: The National Academies Press.
Perdikis, N. (2000), 'A conflict of legitimate concerns or pandering to vested interests: conflicting attitudes toward trade in genetically modified goods – the EU and the US', *Journal of International Law and Trade Policy*, **1**(1), 51–65.
Perdikis, N. and W.A. Kerr (1999), 'Can consumer-based demands for protection be incorporated in the WTO? – The case of genetically modified foods', *Canadian Journal of Agricultural Economics*, **47**(4), 457–65.
Perdikis, N., W.A. Kerr and J.E. Hobbs (2001), 'Reforming the WTO to defuse potential trade conflicts in genetically modified goods', *World Economy*, **24**(3), 379–98.
Phillips, P.W.B. and W.A. Kerr (2000), 'Alternative paradigms: the WTO versus the biosafety protocol for trade in genetically modified organisms', *Journal of World Trade*, **34**(4), 63–75.
Phillips, P.W.B. and W.A. Kerr (2002), 'Frustrating competition through regulatory uncertainty: international trade in the products of biotechnology', *World Competition Law and Economics Review*, **25**(1), 81–99.
Phillips, P.W.B., S.J. Smyth and W.A. Kerr (eds.) (2006), *Governing Risk in the 21st Century*, New York: Nova Science Publishers.
Ringo, F.S. (1994), 'The Trade-Related Aspects of Intellectual Property Rights Agreement in the GATT and legal implications for sub-Saharan Africa: prospective policy issues for the World Trade Organisation', *Journal of World Trade*, **28**, 121–39.
Roberts, D. (1998), 'Preliminary assessment of the effects of the WTO Agreement on sanitary and phytosanitary trade regulations', *Journal of International Economic Law*, **1**, 377–405.
Smyth, S.J., W.A. Kerr and P.W.B. Phillips (2011), 'Recent trends in the scientific basis of sanitary and phytosanitary trade rules and their potential impact on investment', *Journal of World Investment and Trade*, **12**(1), 5–26.
Smyth, S.J., P.W.B. Phillips, W.A. Kerr and G.G. Khatchatourians (2004), *Regulating the Liabilities of Agricultural Biotechnology*, Wallingford: CABI Publishing.
US Department of State (1945), *Proposals for the Expansion of World Trade and Employment*, Department of State Publication No. 2411, Washington: US Department of State.
Van den Belt, H. (2003), 'Debating the precautionary principle: "Guilty until proven innocent" or "Innocent until proven guilty"?', *Plant Physiology*, **132**(July), 1122–26, available at: http://www.plantphysiol.org/cgi/reprints/132/3/1122.pdf.
Viju, C., M.T. Yeung and W.A. Kerr (2011), 'Post-moratorium EU regulation of genetically modified products: trade concerns', CATPRN Commissioned Paper, No. 2011–02, Canadian Agricultural Trade and Competitiveness Research Network, available at: http://www.uoguelph.ca/catprn/PDF-CP/CP-2011–02-Viju.pdf.
Yampoin, R. and W.A. Kerr (1998), 'Can trade measures induce compliance with TRIPS?', *Journal of the Asia Pacific Economy*, **3**(2), 165–82.

17 Developing countries and the legal institutions at the intersection of agbiotech and development

Chidi Oguamanam

1 INTRODUCTION

The establishment, evolution and modus operandi of global regulatory institutions reflects the interests and tendencies of the global balance of power (Braithwaite and Drahos, 2000). While attempts are made to accommodate the complex aggregation of interests amongst actors in the global policy-making processes, the overall outcome is hardly satisfactory (Krasner, 1983).The tendency of stronger states to wield influence over critical global regulatory institutions accounts for the perennial suspicion that characterizes the engagement with, and expectations of developing countries and their local and indigenous communities from those institutions. As such, the permanent challenge of global regulatory institutions is to bridge, rather than escalate, the development gap between developed and less developed countries through targeted policy interventions.

As a consequence, most global regulatory institutions and processes operate at the intersection of North–South tension with often conflicting visions of development (Develtere, 2012). Some of the institutions that operate within the social policy spectra, such as the World Health Organization (WHO), United Nations Educational, Scientific and Cultural Organization (UNESCO), United Nations Development Program (UNDP), the United Nations Environmental Program (UNEP) (via the Convention on Biological Diversity (CBD)), and the UN human rights system, are comparatively more confident about their role and relationship with developing countries but, somewhat unexpectedly, often they have a tense relationship with developed countries.[1]

On the other hand, most other global regulatory and governance institutions, including the Bretton Woods trio[2] and various other regional economic bodies, perceive their mandate within the narrow confines of economic and financial agendas (Gathii, 2001). With limited, if any, role in the socio-political arena, they are often distanced from the mandate of the UN. Having no obligations or primary commitment to the social development orientation of the UN and its specialized agencies, they are perceived as essentially serving the economic agenda of developed countries in ways that aggravate the global development divide. As a result of their historical origins, most of these institutions are organized within the free market economic ideal. It is not surprising that their activities account for 'the ambiguous place of social policy in international economic and political institution in the post Second World War era' (Oguamanam, 2006, p. 420).

Simplistic binary impressions as echoed in the preceding paragraphs may be inaccurate. However, they assist to situate the tensions that characterize the role of global regulatory institutions and their agenda. Perhaps, more importantly, that tension provides the context for critical appraisal of the role of those institutions. Contemporary development discourse is littered with suspicions regarding the outcomes of various institutional and regulatory interventions. For example, the debate is still open as to whether the Agreement on Trade Related Aspects of Intellectual Property Rights (TRIPS)'s one-size-fits-all standard of intellectual property protection serves the interest of developed countries to the detriment of their counterparts in the developing world (Reichman, 2000). The jury is also still out on the impact of Free Trade Agreements (FTAs) on weaker members of the partnerships (McQueen, 2002).

Similarly, the claim of proponents that agricultural biotechnology (agbiotech) is a magic wand for global eradication of hunger remains a matter of credible scepticism (Serageldin and Persley, 2003; Shiva, 2000). Related to all of the above, the overall impact of economic globalization and its collective regulatory frameworks on the development aspirations of weaker states and indigenous and local communities is a sensitive subject that evokes hardened passions on both sides (Santos, 2006; Stiglitz, 2006, 2002).

Building upon the critical ambience of inconclusive or suspect tenor of the role and impact of global regulatory interventions and institutions, this chapter explores a few such institutions that have direct ramifications for agriculture, biotechnology and development. These institutions are constitutive but are hardly the exhaustive set of legal regulatory structures that govern agriculture and biotechnology. The aim is to map these institutions, and to highlight their cross-cutting operational experiences in the context of the tensions that surround the evolution and global uptake of agbiotech. The chapter reflects on the modest successes arising from these institutional dynamics and contemplates how they can be recalibrated to harness agbiotech for equitable development, especially in indigenous and local communities in developing countries and elsewhere. The legal institutions at the intersection of plant agriculture, biotechnology and development are as diverse as they are open-ended. In the present context, however, in the interest of discretion and brevity and in keeping with the framework of this *Handbook*, our analysis is limited to a few such institutions. They are the World Trade Organization (WTO) – through select component agreements, including the TRIPS, the agreement on agriculture (AoA) and the agreement on sanitary and phytosanitary measures (SPS); the Food and Agriculture Organization (FAO) – through the international treaty on plant genetic resources for food and agriculture (ITPGRFA), and the international union for the protection of new plan varieties (UPOV), a crucial multilateral regime on plant agriculture. This framework of convenience does not discount peripheral regime association and diverse issue linkages within the cross-cutting nature of agriculture, biotechnology and development.[3]

2 THE WORLD TRADE ORGANIZATION (WTO)

The WTO resulted from earlier GATT and later Uruguay Round multilateral trade negotiations which culminated in 1994.[4] It administers a global package of twenty international agreements and seven Understandings that provide legal ground rules for international commerce and a framework for opening markets for barrier-free trade. According to the WTO, '[t]he system's overriding purpose is to help trade flow as freely as possible . . . because that is important for economic development and well being' (WTO, 2013).

Arguably, all of the WTO agreements deal with specific subjects at the intersection of economics and development, which are framed as 'economic development' in the WTO's overarching vision. As indicated above, in relation to agriculture and biotechnology, three important WTO agreements are directly implicated more so than others; they are the AoA, SPS and, most importantly, the TRIPS agreement. As a continuing exercise, the WTO system is presently embroiled in over a decade-long process of negotiating the so-called 'Doha Development Round', which aims generally to address peculiar challenges that developing countries face in the implementation of the WTO agreements.[5]

The WTO and its vision of trade liberalization is a post-Cold War economic restructuring effort. In a way, as an institutional and ideological symbol, the WTO is a global attempt at the universalization of free market neoliberal economic model building upon the post-World War II Bretton Woods structures. More than anything else, its attraction to developing countries lies mainly in the realm of trade in agriculture (through the promise of opening up of markets).

Following the carrot and stick of 'bargain-linkage diplomacy' through which all previous rounds of multilateral trade negotiations were accomplished, developing countries were inducted into the free trade system (Picciotto, 2011). For the promise of opening up their markets to third-world produce (mostly agricultural exports), the United States and its developed-country partners convinced developing countries to make critical concessions in areas that have had dire consequences for their economic and development aspirations (Braithwaite and Drahos, 2000). One of those areas, as we will explore later, is intellectual property. For example, developed countries promised to scale down domestic subsidies on their agricultural products in order to enable developing countries' produce to compete in Western markets. Developing countries were also assured that the United States would pull back from its high-handed unilateral trade sanction regime and play by the multilateral trade enforcement process established under the WTO process.

A combination of several factors and circumstances are blamed for the largely unfulfilled expectations of developing countries in regard to trade in agricultural produce pursuant to the WTO system via the AoA. Some of these factors include the dynamics in the evolution of agricultural technology. Such unfulfilled expectations resonate in the development aspirations of many developing countries, especially in the realization of the benefit of further technological transformations, knowledge interactions and market opportunities in agricultural developments.

2.1 The Agreement on Agriculture

The major objectives of the agriculture agreement include the establishment of balance in global trade through the promotion of market access, management of domestic support for agricultural production and reduction of export subsidies in order to ensure a level playing field and overall improved market orientation in international trade. The AoA is one of the agreements at the heart of the bargain-linkage diplomacy referred to above. In a way, long skewed in favour of industrialized countries, agricultural produce export is the irresistible bait through which developing countries have been lured into making critical concessions, especially in the areas of intellectual property.

The implementation of the tarrification and subsidy regimes under the AoA remains a contentious issue (Gonzalez, 2004; Tokar, 2004; FAO, 2012). From a developing countries' perspective, the impression is largely that developed countries have neither delivered on the commitment to reduce domestic agricultural subsidies, nor are they genuinely committed to the opening up of their markets. Rather, the devil in the details of the subsidy regime[6] and other components of the AoA operate to advance the interest of developed countries. For example, the subsidy regime prohibits future subsidies where none were used within the base period (1986–90). This policy discriminates against developing countries – that traditionally have no active practice of creating subsidies – while it tolerates existing subsidy regimes of developed countries (Chaonwa, 2005, p. 26).

Even if developed countries are genuinely committed to addressing the distorting effects of subsidies, their ability to continue several domestic support measures for their local farmers is not impaired. Under the so-called green box policy, the agreement saves measures such as government services in support of agricultural research, disease control, disaster relief and infrastructure provision from being construed as having significant impact on trade.[7] These interventions are outside the total aggregate measurement of support (AMS) for domestic agricultural production which parties committed to reduce in the interest of improved global agricultural market orientation. For most developing countries, there are hardly any dedicated government services in the nature of those encouraged under the AoA's green box policy. To ensure a level playing field, perhaps the WTO system ought to support the development of a more proactive green box system in developing countries.

Another source of apprehension over the sincerity of developed countries to the WTO's bargain-linkage deal is their determination to make new trade deals with developing countries which commit the latter to stronger terms than the WTO framework (El Said, 2005). After the WTO came into effect, developed countries neither let off on continued expansion of bilateral FTAs nor on unilateral implementation of targeted sanction regimes. This development has had a chilling effect on the ability of developed countries to leverage opportunities in the WTO to advance trade in agriculture.

The foregoing sketch does not adequately reflect the operations of the tariffication and subsidy regime under the AoA. However, it highlights the mistrust and negative sentiments that characterize the perception of the AoA regime by many developing countries. A truly equitable tariff and subsidy regime for improved market orientation for agricultural products could benefit developing countries.[8] This is especially important in the context of rapid advancement of agbiotech and any corresponding global demand for non-GMO food products, which potentially creates a niche market for developing countries that have yet to embrace genetic modification.

2.2 The SPS Agreement

While the AoA promotes free trade in agricultural products, the SPS is designed to ensure that all trade allows for the protection of human, animal/plant life and health, and overall food safety.[9] The SPS represents built-in balance in the WTO process by recognizing that unqualified free trade may be prone to creating undesirable side effects that ought to be fixed. Since the WTO, and especially the SPS, came into effect there has been hardly any controversy over the need to ensure that agricultural products meet acceptable standard of human, animal and plant health. What is quite controversial, however, is the appropriate standard for assessing the safety of agricultural export, especially in regards to agbiotech.

An appraisal of the SPS experience shows some paradox that transcends the often-conflicted North–South perspectives on the instruments and highlights the transatlantic muscle-flexing between the United States and Europe (Oguamanam, 2007). On a paradoxical note, the SPS is perceived as a veritable mechanism for erecting sophisticated barriers on developing-country exports (Henderson et al., 1999; Jensen, 2002). Yet developed countries, especially those actively promoting GM foods, such as the US and Canada, capitalize on the SPS process to promote trade in GM foods, using the WTO dispute settlement system to challenge regulatory and trade impediments in countries or regions like the EU and elsewhere where consumers have expressed reservations about the technology (see Chapter 18 in this volume for a discussion of consumer preferences).[10]

The SPS allows countries to implement their own standards of safety for agricultural exports. The problem arises, however, when such standards are inconsistent with an importing country's safety standards. Because of this practical disconnect, there is continual mutual suspicion by transacting parties in regard to each other's tendency to arbitrarily or ostensibly deploy safety standards to erect barriers on trade and perpetuate protectionism contrary to the WTO principles.

Parties have sought to resolve this by seeking a common standard for measuring safety. Article 2 of the SPS Agreement recommends that safety measures be 'based on scientific principles', and does not support any approach not premised on 'sufficient scientific evidence'. It, however, makes a narrow exception outside the scientific paradigm. In Article 5(7), it provides that where scientific evidence is not available, safety measures may be tailored on the basis of available information. This can be done by drawing on practices of relevant international organizations or existing measures in member states. All of these are subject to steps being taken within a reasonable time to objectively assess the suspected risk and to consequently review the ad hoc measures in order to make definitive safety decisions (see Chapters 15 and 16 in this volume for more discussion).

Agbiotech is a critical site for contesting the ramifications of these provisions. Interestingly, while Europe's arguably restrained approach to GMOs seeks to interpret the SPS agreement in the framework of pre-existing international understanding of the precautionary principle (as discussed in Chapters 15 and 16 in this volume), the US, Canada and their allies demand adherence to a strict scientific approach (Menrad, 2007). Recent relevant decisions on the subject from the WTO Dispute Settlement Body have endorsed the Codex Alimentarius Commission's (CAC) model of food safety. The latter is a result of a joint effort of the WHO and FAO to set international food safety standards. The Codex model, which is favoured by the United States' agribusiness and biotechnology industrial complex, adopts a narrow approach to food safety that does not accommodate substantial or more elaborate considerations under the precautionary model (Oguamanam, 2007).

The WTO's Dispute Settlement Body's reliance on the CAC can, to some extent, be justified under Article 5(7) of the SPS. However, it is less accurate to suggest, as the WTO does, that SPS seriously accommodates the tenets of the precautionary principle.[11] Under the SPS, the requirement to resolve any apprehension in regard to the risks of agrobiotech products within a reasonable time is viewed by many as antithetical not only to the generally inchoate nature of agbiotech claims and evidence, but also to the dictates of precautionary principles. The victory of the United States and its allies over Europe on the latter's moratorium on products of agricultural biotechnology underscores the triumph of the WTO free market over the foundational precautionary principles established earlier in the environmental regime.

It is instructive that Canada, which is a party to the Convention on Biological Diversity (CBD), did not sign one of the Convention's important initiatives – the 2000 Cartagena Protocol on Biosafety. The latter 'seeks to protect biological diversity from the potential risks posed by living modified organisms [LMOs] resulting from modern biotechnology' (Cartagena Biosafety Protocol, 2000; Art. 1). The United States is not even a party to the CBD. The Protocol is premised on the precautionary principle which owes its origin to the Rio Declaration on Environment and Development, which, pursuant to principle 12, asserts that 'where there are threats of serious and irreversible damage, lack of full scientific certainty shall not be used as a reason for postponing cost-effective measures to prevent environmental degradation'. Along those lines, under the Cartagena Protocol, 'where there are threats of serious irreversible damage', from LMOs or products of agricultural biotechnology full *scientific certainty* is *not* required as a condition precedent to take or impose a remedial action (Oguamanam, 2007, pp. 420–21). (Chapter 15 in this volume discusses this in more detail.)

The present interpretation of the SPS inevitability will create cross-cutting interactions in the regulation of agricultural biotechnology. As noted, the CAC standard is shaped by two prominent intergovernmental bodies, the WHO and the FAO, while the UN environmental system through the Rio Declaration and the CBD provide the platform for the precautionary principle and its relevance in the regulation of biotechnology. Resolution of this cross-cutting regime tension will be critical for equitable regulation of agbiotech.

It has been observed that the consequences of the transatlantic disagreement over the regulation of agbiotech will be felt more elsewhere than in the two leading trading nations/regions. Europe continues to import GM crops from the US, Canada and elsewhere, albeit mainly for restricted uses. Unlike Africa or Asia, the European food system can handle the diversity of supplies which is presently segmented to meet differentiated demands. In contrast, Africa, South America and Asia are projected as having broken food systems which agbiotech is touted to fix. Accordingly, '"[s]ince the United States has no real hope of boosting sales of [genetically engineered] foods to unwilling Europeans . . . the decision [which forced Europe to lift its self-imposed moratorium on GM food] is clearly an effort to chill other nations from pursuing any regulations on [genetically engineered] foods." African and Asian countries have been identified as "by far the most conspicuous targets"' (Oguamanam, 2007, p. 242).

2.3 The TRIPS Agreement

TRIPS is one of the WTO component agreements with great relevance to biotechnology, agriculture and, arguably, development. Its subject matter is intellectual property, which involves law and mechanisms for the control of knowledge, information and innovation. Inherently, intellectual property rights are an acceptable barrier to free trade because they create temporary monopolies for rights holders by vesting the right to exploit an invention in the inventor or their assignee. In effect, IPRs temporarily ward off competition. Before TRIPS, there was no globalized intellectual property standard. Rather, what existed was a range of national patent and IP laws, integrated through a melange of voluntary multilateral agreements with fairly loose commitments on specific areas of creativity and innovation, administered by the World Intellectual Property Organization (WIPO).

Under the WTO framework, developing countries, led by the United States, legally re-engineered intellectual property by articulating the framework for minimum universal binding standards for intellectual property protection, enforceable through the WTO disputes settlement system. In this new transformation, intellectual property is envisioned as an instrument for the promotion of free trade. Parties must commit to TRIPS in order to benefit from the rights and privileges of free trade under the WTO. Consequently, unlike under the pre-TRIPS order, any country's ability to shape their domestic intellectual property standards to suit their national socio-economic exigencies has been curtailed. For example, most developing countries did not encourage strict patent protection in strategic areas, such as agriculture and medicine. Under TRIPS, that discretion is radically constrained. Indeed, strategically, TRIPS targets the optimization of protection on agbiotech innovations.

The key TRIPS element is Article 27, also known as the biotechnology clause. This article is the most direct attempt to co-opt contemporary agriculture innovations into the intellectual property system. Among other things, it provides that: 'patents shall be available for any inventions, whether products or processes, in all fields of technology, provided that they are new, involve an inventive step and are capable of industrial application'. Further, it extends patent protection to 'micro-organisms, and essentially biological processes for the production of plants or animals as well as for plant varieties' (TRIPS Art. 27(3) (b)).

Exploration of the ramifications of Article 27 is the subject of an active strand of legal scholarship.[12] In the present context, one can highlight the implications of the provision in relation to traditional agricultural practices of indigenous and local communities. First, it extends intellectual property protection to all fields of technology (which arguably includes indigenous knowledge). Secondly, however, it filters against much of collective knowledge by erecting the hurdles of newness, inventive step and industrial application: separately and collectively these are technical terms in patent jurisprudence which have been used to exclude informal processes of knowledge generation outside the conventional scientific platform (Mgbeoji, 2001; Oguamanam, 2004). Thirdly, these provisions reflect no regard to cultural sensitivities against the extension of private rights claims over life forms generally, even though this is antithetical to the communal practice of sharing and exchange of seeds amongst traditional farmers in indigenous and local communities of developing countries and elsewhere (Dutfield, 2008).

Article 27 provisions are understood in the context of the historic struggle to control or suppress (both by law and technological intervention) the inherent propagating character of seeds and genetic materials. Control of propagation of genetic resources is projected as the most effective way for proprietary seed breeders and new biotechnology interests to profit from their enterprises. Dutfield (2008) argues that in Europe and America this struggle pitted small-scale traditional subsistence farmers against their corporately organized seed breeder counterparts. Now, with the advent of agbiotech, there is a convergence of seed breeders and mega transnational chemical and agro-allied concerns (GRAIN, 2012, pp. 24–7). The latter's modus operandi translates into pressures for total control, if not outright elimination of traditional farmers and other stakeholders steeped in the tradition of seed exchange.

3 THE UPOV

In 1961, prior to the TRIPS agreement, developed countries (especially those with a head start in seed breeding), formed an elite club under the aegis of the UPOV. According to Dutfield (2008, p. 31), the UPOV is an 'international regime designed specifically to protect plant varieties whose seeds could otherwise be easily saved, replanted and sold' by farmers. It is the most authoritative framework for the development of intellectual property (that is plant breeders' rights, PBR) in plant agriculture. Save for *sui generis* initiatives such as those under UPOV and specific domestic regimes, intellectual property protection, such as patents, did not extend to life forms or plant genetic resources (PGRs) until such obligations were incorporated into the TRIPS agreement.

Through its evolution, UPOV has created a strategic schism between breeders and farmers even as it privileges the former. While all versions of UPOV (1961, 1972, 1978 and 1991) balance the allocation of rights for breeders of new varieties with both a breeders' exemption and a farmers' privilege to save and reuse seed, UPOV 1991 now allows breeders to use both PBR and patents on the same cultivars, which in most cases would eliminate the breeders' exemption and farmers' privilege. This has radically curtailed existing accommodations granted to farmers to access and use farm-saved seeds. Because of its success, developed countries behind the UPOV model have continued to recruit their developing-country partners directly into the membership of UPOV or indirectly through TRIPS plus FTAs,[13] exacting in these countries a tighter regime of intellectual property that undermines the interest of predominantly small-scale traditional farmers. Unexpectedly, developed countries favour an interpretation of TRIPS provision for plant breeders' rights (PBR) under Article 27 that adopts the UPOV plant-breeding system as a default standard.

Ironically, however, continued improvement in the science and technology of plant breeding has resulted in a more stringent interpretation of the standards for bred products, that is distinctiveness, stability and uniformity. These technical formalities which are recognized as the standard for the granting of PBR under UPOV draw plant breeding closer to plant agricultural innovations under biotechnology. Consequently, they bring PBR closer to patent standards, a situation that challenges the continued relevance of PBR.[14]

4 THE FAO TREATY

The processes that culminated in the 2001 International Treaty on Plant Genetic Resources for Food and Agriculture (ITPGRFA) dates back to the 1980s. The treaty is the result of initiatives to mitigate the inequitable effects of the use of intellectual property to appropriate the benefits of technological innovations on plant genetic resources from global *ex situ* seed banks. One of the treaty's major goals is the promotion of the concept of 'farmers' rights' which is a counterpoise to the intellectual property system that only privileges formal scientific and biotechnological agriculture at the expense of traditional and informal farming practices.

Under the treaty, farmers' rights are associated with 'the enormous contribution that the local and indigenous communities and farmers of all regions of the world . . . make for the conservation and development of plant genetic resources which constitute the basis of food and agriculture production throughout the world' (ITPGRFA Treaty 2001, Art. 9.1). Those PGRs in *ex situ* seed banks were pooled from the centres of global genetic diversity in indigenous and local communities of the global south and managed by the Consultative Group on International Agricultural Research (CGIAR) system under the principle of common heritage of mankind. However, the idea of free access to the benefits of agro-technological research endeavours deriving from the commonly-held PGRs has historically not found favour with developed countries. Their inclination, as demonstrated by UPOV, the TRIPS Agreement and various national laws, is to retain exclusive proprietary control of those innovations via intellectual property.

Building upon previous initiatives over equitable access to commonly held PGRs, the treaty creates a multilateral system (MLS) of access and benefit-sharing (ABS) in regard to 64 crops and forages through a framework of a standard material transfer agreement. This framework offers options for various stakeholders, including corporate entities, who seek to appropriate genetic resources from the common pool to give back by way of financial contribution to a complementary benefit-sharing fund (BSF). Proceeds from this fund are channelled exclusively to projects and research initiatives that support sustainable traditional and local farming needs in developing countries. This initiative makes the FAO treaty the first fully operational global instrument on ABS for PGRs inspired by the CBD. As well, the treaty is the first significant attempt to operationalize the concept of farmers' rights.

In the context of the treaty (Art 9.2(a)), farmers' rights are associated with traditional knowledge practices. As already indicated, it is an integral part of ABS under the CBD framework. For instance Article 1.1 specifies the objectives of the treaty as 'the conservation and sustainable use of plant genetic resources for food and agriculture and the fair and equitable sharing of the benefits arising out of their use, in harmony with the Convention on Biological Diversity . . .'.The ABS narrative and the intellectual property system are both part of an elaborate cross-cutting relationship between traditional knowledge, including agricultural practices, and various forms of innovation including agbiotech, which is captured by the United Nations Declaration on the Rights of Indigenous Peoples.

Despite the significance of the FAO treaty, like other initiatives on farmers' rights, the treaty is hamstrung in many regards. First, contributions to the BSF are mostly voluntary. Second, farmers' rights are conceptually vague and weak in terms of jurisprudence (Oguamanam, 2007). Third, commitment of parties to farmers' rights is moderated by their national interests and existing legal obligations. The national interest of most developed countries is for proprietary control of PGR innovations as opposed to the free and equitable access favoured by their developing counterparts. As indicated above, most of the existing national and multilateral legal obligations of developed countries run contrary to the idea of farmers' rights. Fourth and perhaps most important, the ITPGRFA text reflects a decisive victory for developed countries in a very crucial respect: in regard to the scope of application of intellectual property over PGRs under the MLS. That question is at the heart of the treaty and the idea of farmers' rights. Article 12.3(d) forecloses intellectual property claims on PGRs 'for food and agriculture, or their genetic parts or components' only when they are in 'the form received from the Multilateral System.' This creates an important loophole whereby artificial manipulation or forms of genetic modification via agro-biotechnological insights, including isolation and purification of genes, transforms PGRs from 'the form' in which they were 'received' or deposited in the common pool, thereby enabling private rights to be claimed and exploited.

On balance, the ITPGRFA has moved the conversation on farmers' rights and ABS to an operational level. But the core weakness of the instrument works against any significant change in the status quo. As a process and instrument, the ITPGRFA demonstrates the resilience and enduring influence of developed countries in the marginalization of the interests of developing countries at the intersection of agriculture, biotechnology and development.

5 CONCLUSION

The intersection of agriculture, biotechnology and development involves diverse legal and regulatory institutions and processes in a cross-cutting horizontal relationship typical of global governance dynamics. A select list of these institutions is examined in this chapter. Given the economic and development ramifications of agriculture and biotechnology, the evolutionary and institutional dynamics of the regimes and organizations reflect the tensions between developing and developed countries. Within the WTO framework, specifically in regard to the AoA, the SPS and TRIPS, developed countries have continued to pursue their interests in various ways. While there are considerable attempts to accommodate the interests of developing countries, on balance and in practice the overlapping and cross-cutting regimes at the intersection of agriculture, biotechnology and development do not, for the most part, meet the expectations of the developing world.

Agbiotech continues to advance; increasingly it is projected as a panacea for eradication of hunger in the developing world. Current legal systems and regulations, especially those examined in this chapter, generally support the potential transformation of the global food supply through the application of agbiotech. Within these technological and institutional transformations, there is only limited and peripheral accommodation of alternative forms of agricultural knowledge and production practices in indigenous and local communities of the developing world. The objective of equitable market access for agricultural exports from the developing world under the WTO free trade regime is undermined by the legal frameworks for the governance of agricultural innovations which do not adequately account for agricultural practices of indigenous and local communities.

So far, agbiotech is dominated by technologically powerful nations. For the most part, it is corporate or private sector-driven and operates within an industrial production model as opposed to the cultural and communal experiential model of agriculture found in most indigenous and local communities of developing countries. Consequently, one of the challenges of the current, even if moribund, Doha Development Round of the WTO process is how best to reposition the legal regulatory intervention around agriculture and biotechnology to serve the development aspirations of indigenous and local communities and developing countries. The weaknesses in the legal regulatory institutions identified in this chapter are worthy of attention in any such initiative. Whether or not agbiotech becomes the main source of global food supply, its ramification for, and the nature of its engagement with alternative agricultural practices that are now the base of food supply for the majority of hungry people globally, especially in indigenous and local communities of the developing world, needs to be constructively addressed.

NOTES

1. For instance, the United States' relationship with the WHO and UNCEF is often tense and characterized by threats of actual and real withdrawal of financial support and suspension of membership.
2. The institutional foundations of the World Bank, the International Monetary Fund and the GATT were laid in Bretton Woods, NH, in 1944 primarily to finance the reconstruction of war-ravaged Europe and to address the economic causes of war. The WB and IMF have since metamorphosed into global technical agencies and principal lenders and financiers of developing projects in the third world – a status by virtue of which they influence the economic policy in developing countries. While the IMF's focus is on monetary management, the WB focuses on development finance and the two are now complemented by the WTO as the institutional framework for global harmonization of economic regulation and trade.
3. The CBD and its role in regard to the protection of indigenous knowledge and by extension traditional knowledge-based agricultural practices are respectively symbolic of such regimes and issues.
4. The Final Act Embodying the Results of the Uruguay Round of Multilateral Trade Negotiations was signed by world trade ministers in Marrakesh on 15 April 1994 and the WTO was established as a singular governance institution encompassing the GATT (as modified by the Uruguay Round), all constitutive agreements and arrangements that resulted from the Uruguay Round.
5. Started since late 2001 at the instance of the Doha Ministerial Declaration, the focus of the Doha Development Round is on agriculture, services and intellectual property rights. The Development Round has been a site of intrigue between developed and developing countries and it is characterized by serial stalling on key issues, such as the revision of Article 27 of the TRIPS agreement among others.
6. Article 9 of the AoA contains the details of the subsidy regime.
7. See Annex 2 of the AoA for the detailed list of domestics support activities that are exempt from reduction commitments.
8. Clearly, the AoA text contains significant concessions aimed at supporting agricultural export potential of developing countries, but the problem lies in the loopholes within the same text through which those concessions are undermined by unwilling developed countries.
9. Unlike the SPS Agreement, the Agreement on Technical Barriers to Trade (TBT) focuses on non-discrimination measures.
10. Several decisions of the WTO Dispute Resolution Body including the EU–USA Beef Hormone, Australian Salmon and Japanese Agriculture decisions, are inclined toward a permissive interpretation of SPS standard against objections based on other more prudent approach to food safety.
11. The WTO claims that 'Article 5.7 of the SPS allows for temporary "precautionary measures"'. See 'Understanding the WTO: the agreements – standards and safety', available at: <http://www.wto.org/english/thewto_e/whatis_e/tif_e/agrm4_e.htm>.
12. Those include the contentious subject of patentability, the issue of disclosure of origin or source of genetic materials used in subject matters of patent applications, not to mention diverse issues at the intersection of intellectual property in biotechnology, biodiversity and traditional knowledge.
13. In 1968, UPOV had only four exclusively European members (Denmark, Germany, the Netherlands and UK). At the time of writing, the membership stands at 70 countries drawn from all geographic regions. Most FTAs commit developing-country parties to the UPOV as standard of PGR protection.
14. Dutfield (2008, p. 32) notes that 'the increasing strength of the PVP right of recent years is beginning to approximate to that of patent'.

REFERENCES

Secondary Resources

Braithwaite, J. and P. Drahos (2000), *Global Business Regulation*, Cambridge: Cambridge University Press.
Chaonwa, A. (2005), 'The WTO Agreement on agriculture: impact on farmers', Harare, Zimbabwe: *Trade and Development Studies Centre* Issue no. 32, 2005.

Consultative Group on International Agricultural Research (n.d.), available at: <http://www.cgiar.org/>.
Develtere, P. (2012), *How Do We Help? The Free Market in Development Aid*, Leuven: Leuven University Press.
Dutfield, G. (2008), 'Turning plant varieties into intellectual property: the UPOV convention', in Geoff Tansey and Tasmin Rajotte (eds), *The Future Control of Food*, London: Earthscan, pp. 27–47.
El Said, M. (2005), 'The road from TRIPS-minus to TRIPS to TRIPS-plus: implications of IPRs for the Arab world', *Journal of World Intellectual Property*, **8**, 53–65.
FAO (2012), 'The implications of the Uruguay Round Agreement on agriculture for developing countries', available at: FAO <http://www.fao.org/docrep/004/w7814e/W7814E08.htm>.
Gathii, J.T. (2001), 'Re-characterizing the social in the constitutionalizaton the WTO: a preliminary analysis', *Widener Law Symposium Journal*, **7**, 137.
Gonzalez, C. (2004), 'Trade liberalization, food security, and the environment: the neoliberal threat to sustainable rural development', *Transnational Law and Contemporary Problems*, **14**, 419.
GRAIN, *The Great Food Robbery*, Barcelona: GRAIN and Pambazuka Press.
Henderson, S. et al. (1999), 'The impact of sanitary and phytosaniatry measures on developing country export of agricultural and food products', paper presented at the Conference on Agriculture and the New Trade Agenda in the WTO 2000 Negotiations, 1–2 October, 1999, Geneva.
Jensen, M.F. (2002), 'Reviewing the SPS Agreement: a developing country perspective', CDR Working Paper no 02.3.
Krasner, S.D. (1983), *International Regimes*, Ithaca, NJ: Cornell University Press.
McQueen, M. (2002), 'EU's Free Trade Agreement with developing countries: a case of wishful thinking?', 25:9 *The World Economy*, **25**(9), 1269.
Menrad, K. (2007), 'The regulatory regime and its impact on innovation activities in agro-food biotechnology in the EU and US', in Jay P. Kesan (ed.), *Agricultural Biotechnology and Intellectual Property: Seeds of Change*, New York: CABI, pp. 231–43.
Mgbeoji, I. (2001), 'Patents and TK of the uses of plants: is a communal patent regime part of the solution to the scourge of biopiracy?', *Indiana Journal of Global Legal Studies*, **9**, 163.
Oguamanam, C. (2004), 'Localization of intellectual property in the globalization epoch: the integration of indigenous knowledge', *Indiana Journal of Global Legal Studies*, **11**, 153.
Oguamanam, C. (2006), 'Regime tension in the intellectual property rights arena: farmers' rights and post-TRIPS counter regime trends', *Dalhousie Law Journal*, **29**, 413.
Oguamanam, C. (2007), 'Agro-biodiversity and food security: biotechnology and traditional agricultural practices in international intellectual property regime complex', *Michigan State Law Review*, 215.
Picciotto, S. (2011), *Regulating Global Corporate Capitalism*, Cambridge: Cambridge University Press.
Reichman, J.H. (2000), 'The TRIPs Agreement comes of age: conflict or cooperation with the developing countries?', *Case Western Reserve Journal of International Law*, **32**, 441.
Santos, B. de Sousa (2006), 'Globalizations', *Theory, Culture & Society*, **23**, 389.
Serageldin, I and G.J. Persley (eds.) (2003), *Biotechnology and Sustainable Development: Voices of the South and North*, New York: CABI.
Shiva, V. (2000), *Stolen Harvest: The Hijacking of the Global Food Supply*, Cambridge: South End Press.
Stiglitz, J.E. (2002), *Globalization and Its Discontents*, New York: W.W. Norton & Co.
Stiglitz, J.E. (2006), *Making Globalization Work*, New York: W.W. Norton & Co.
Tokar, B. (ed.) (2004), *Gene Traders: World Trade, Biotechnology and Globalization of Hunger*, Vermont: Toward Freedom.
WTO (2013), "Understanding the WTO: Who we are", available at: <http://www.wto.org/english/thewto_e/whatis_e/who_we_are_e.htm>.

International Legal Materials

Agreement on Agriculture, Apr. 15, 1994, Marrakesh Agreement Establishing the World Trade Organization, Annex 1A, Legal Instruments – Results of the Uruguay Round, 1867 U.N.T.S. 410 (1994) (Hereinafter 'AoA').
Agreement on Trade-Related Aspects of Intellectual Property Rights, 15 April 1994, 1869 U.N.T.S. 299: 33 I.L.M. 1197 (Hereinafter, 'TRIPS Agreement').
Cartagena Biosafety Protocol to the Convention on Biological Diversity, 29 January 2000, 39 I.L.M. 1027 (entered into force 11 September 2003), Article 1.
Declaration on the Rights of Indigenous Peoples, GA Res. A/61/295, 107th Plen. Mtg., (2007).
International Treaty on Plant Genetic Resources for Food and Agriculture, Opened for Signature on Nov. 3, 2001, FAO Res. 3/2003, online: FAO <ftp://ftp.fao.org/ag/cgrfa/it/itpgre.pdf>.
International Convention for the Protection of New Varieties of Plants 2 December 1961, 33UST. 2703, 815 U.N.T.S. 89 (as revised at Geneva on 10 November 1972, 13 October 1978) the 1991 revision (Hereinafter 'UPOV 1991').

Rio Declaration on Environment and Development, June 1992, 32 I.L.M. (1992) 874.
The WTO Agreement on the Application of Sanitary and Phytosanitary Measures, 25 December 1993, 33 I.L.M. XXX (entered into force 1 January 1995) (Hereinafter 'SPS Agreement').
The General Agreement on Tariffs and Trade, 15 April 1994, 1867 U.N.T.S. 187; 33 I.L.M. 1153.

18 Consumer attitudes and preferences for GM products

Stuart J. Smyth and David Castle

1 INTRODUCTION

In Canada, it has been estimated that genetically modified (GM) foods and food ingredients are detectable in 11 per cent of foods consumed and might be present (but often not detectable) in up to 75 per cent of the processed foods sold in stores. Examples range from GM papaya and GM sweet corn that are directly consumed, to sucrose and fructose from GM corn that are used as ingredients in products like chewing gum. Although regulators around the world have ruled that there is no scientific evidence to support claims that these foods involve any new or magnified risks, many civil society groups and a large portion of consumers are simply not convinced. In absence of any definitive long-term studies showing these foods are safe, and in response to heightened apprehension about food safety issues, civil society groups and consumers seek mandatory labelling for GM foods. The reasons offered in defence of mandatory labels include: consumers' right to know what is in their food; giving consumers the ability, at point of sale, to choose or avoid GM foods; and enhancing long-term monitoring and surveillance of GM foods.

Food concerns in industrial countries have evolved from food security concerns in the middle of the past century to provenance concerns at the present. Corresponding to this is the lengthening of supply chains, which in many markets has removed the direct relationship between the producer of food products and consumers. The demand from consumers to know what is in their food is not a stand-alone issue, but part of a greater societal movement pertaining to our proximity to food. Witness the recent concern about horsemeat contamination in Europe, mandatory nutrition and country of manufacture labelling in many countries, the inclusion of calorie counts for meals in restaurants and the rise of urban gardening as a means of shortening food chains. All of these examples indicated that consumers are increasingly concerned about potential risks related to their food consumption habits, and perhaps more importantly these examples indicate that consumers want recourse to accountability in food systems that traceability and other documentation are intended to support.

While advances have been made regarding the labelling of many of the above examples, labelling for GM content has been more problematic. The European Union requires labelling on products that contain greater than 0.9 per cent GM ingredient content. Even something as seemingly straightforward as putting a label on a product becomes very problematic as considerable challenges become apparent when one stops to consider questions about how to label and the extent to which labelling should be applied throughout a food system: does it apply from papaya to processed corn to enzymes and yeasts? Enzymes used in the production of cheese are often from genetically modified bacteria, as are yeasts that are used in most baked products (GMO Compass, 2009, 2010). This uncertainty has resulted in varieties of cheese being labelled as GM in some EU countries and not labelled as GM in others. The demand to know precisely what is in our food ultimately needs some boundaries, as often labels are simply not big enough to contain all of the ingredients.

At root, labelling is about providing accessible and meaningful information to consumers. Information is like money, in the sense that people will generally respond that they would prefer to have more of it rather than less. An interesting aspect of money, however, is that happiness does not always follow affluence. After a certain point, as reports from late-stage industrialized countries indicate, more money is actually associated with less happiness, and many people will trade money for other things they value, like health or time. Following this analogy, a question can be raised about access to information. Does a similar pattern exist for information, such that the desire for more information, and the ability to access that information, reaches a saturation point where new information does not improve welfare and perhaps even undermines will-being? A further question can be raised about whether the impulse to want more information needs to be considered in the sobering light of how useful the information is in the context of daily life, particularly when the information might be traded off for other considerations deemed more valuable?

When Canadians (and most citizens in OECD countries) are asked if they would prefer to have information about whether their food is GM, or contains GM ingredients, nearly all will say they want the information. This result is consistent with research into the public's information needs regarding new technologies, for example in the energy, health care and biotechnology sectors. The issue of labelling, whether it is mandatory or voluntary, spans the full spectrum of opinions. Environmental groups and critics of biotechnology claim that more than 95 per cent of consumers responding to surveys indicate that they want GM content to be labelled, but other surveys show that only 2 per cent of unprompted consumers ask for GM labelling. The real demand for labelling lies somewhere between the two; determining whether it is greater or less than 50 per cent should help to determine what type of labelling is optimal.

With respect to GM foods, the consistency with which survey respondents state they want information about GM foods suggests these foods should be labelled. In the United States, for example, consumers state they would prefer a text description accompanied by a symbol to indicate the presence of GM food or food ingredients (Harrison and McLennon, 2003). Labels are thought to be the right way to convey information in so far as they have the potential to give point-of-sale control to consumers who want to make decisions about the products they consume. One approach is that if autonomous decision making by consumers is the issue, a label indicating whether the product is GM (or GM-free) should be able to provide enough information for a consumer to opt out of consuming GM foods. When given the opportunity to vote directly on this issue, however, voters in California rejected the mandatory labelling of GM foods in the fall of 2012 under Proposition 37 (see Chapter 34 in this volume for further discussion).

Suppose two things can be taken for granted. First, people generally want more information about their food so they can decide if they should consume GM foods. Second, it is possible to develop a label that conveys enough information to allow consistent, autonomous decision making by consumers to opt out of buying GM foods. These conditions could be interpreted as counting in favour of a mandatory GM food labelling policy, in the sense that if no barrier exists to giving information, then why not give it? The question, however, is whether the ability to label GM foods provides information to consumers that they always desire. That is, are blanket statements about preferring more information, in principle, reflective of consumers' contextualized uses of information, or is it true only in light of the broad presumption that more information is better than less? Returning to the analogy between information and money, perhaps it really is the case that information about GM foods is desirable in the way that more money is desirable, but only in the narrow sense that it can be traded for other valuable things. If this is true, then there is less presumptive support for mandatory labels than it would first appear.

2 IF LABELS, THEN WHAT INFORMATION?

Because labelling is one among several potential sources of costs, producers and packagers have an interest in whether the effect of labelling GM contents in foods will meet consumers' desire to have this information, while at the same time generating adequate revenues to recoup the extra costs associated with any label change. Marketing products to consumers involves a complicated mix of label messages, consumer knowledge and product context. All these factors can influence the extent to which consumers will be willing to pay for GM products or to avoid them. To discern the relative impact of different product attributes, including information on the label, conjoint analysis can be used to assign values to a series of 'discrete choices'. In this approach, research participants make pair-wise trade-offs between a fixed set of product attributes. The results of these trade-offs can be analysed and assigned fixed values or ordinal rankings in order to identify the relative importance of various product attributes. The art to this method is that partial values can be discerned with the researchers revealing which of the variables is of principal research interest.

To understand how the labelling of GM food might influence consumers' willingness to pay, and how label information would interact with other product attributes, three experimental hypotheses are frequently used in these types of studies. First, conjoint analysis can be used to decide between different offerings of information on a label. For example, consumers might be more willing to pay for information about benefits to them as consumers, but might not be as interested in paying more to learn about the impact of the product on environmental sustainability. Second, conjoint analysis can be used to understand how the same information, such as country of origin, presence of transgenes, or consumer benefits, will appeal to consumers across different product categories, such as fresh produce versus processed foods. Third, conjoint analysis can be used to evaluate hypotheses about the relationship between the extent of consumers' knowledge about GM and their response to the limited capacity of a label to communicate any more than merely indicative information.

The major challenge in determining how much consumers would be willing to pay to have increased labelling information is that wherever possible, people want information at little or no cost to themselves. Frequently critics of biotechnology make the statement that 'labelling for GM is costless – all that is required is to put a label on the products'. Obviously, nothing is free, but the challenge is to determine what the costs are and who should bear them. Consumer studies can assist in determining the aggregate value individuals place on this information, and should allow us to determine the optimal amount of information that should be provided.

Market surveys by McGarry Wolf et al. (2004) showed that many US consumers do not understand the term 'GM-free' when used in product label information. Using simulated test markets for salty snack food and fresh packed vegetables, eight characteristics were offered to inform the consumer about the product. The characteristic 'free of genetically modified ingredients' was the lowest rated. This experiment also revealed that consumers, who before the experiment indicated that GM-free ingredients were 'extremely' or 'very' desirable when making a purchase decision, did not express more interest in the food products labelled GM-free than in products labelled GM when faced with actual purchase choices.

Huffman et al. (2004) attempted to determine whether mandatory or voluntary labelling produces a more efficient economic outcome. An experimental bid auction in the USA, involving products that had basic labels, voluntary labels and mandatory labels for cooking oil, chips and potatoes, found that participants discounted the GM-labelled oil more than the other products. One interesting result from the auction, in which participants were given a fixed amount of money to spend on products with an array of attributes, was that in the mandatory labelling experiment the plain labelled products were perceived to be the non-GM products, yet in the voluntary labelling experiment the plain labelled product was perceived to be the GM product. It was concluded from this study that for the US market, voluntary labelling would be a more efficient policy than mandatory labelling.

Some literature suggests that consumer willingness to pay for products labelled as GM can vary widely (Rousu et al., 2007). Consumer preference for products labelled with a 1 per cent tolerance level versus a 5 per cent tolerance level were studied and this survey found that consumption of products labelled as GM would drop 7–13 per cent regardless of whether the tolerance level was 1 per cent or 5 per cent. The authors reported that there is no statistical support for US consumers having a preference for a 1 per cent tolerance level over a 5 per cent tolerance level. They concluded that if the USA wanted to adopt a tolerance level for the labelling of GM food products, 5 per cent would be the socially optimal level.

The willingness of consumers to pay in the US and UK was assessed by offering consumers the choice between two identically priced boxes of breakfast cereal, one box labelled GM and the other labelled non-GM (Moon and Balasubramanian, 2001). When asked which cereal they would choose if priced the same, 71 per cent of UK respondents chose non-GM, 2 per cent chose the GM cereal, 23 per cent were indifferent and 4 per cent did not know. In the USA, 44 per cent chose the non-GM product, 6 per cent preferred the GM product, 28 per cent did not know and 22 per cent had no preference. The preference to consume non-GM cereal dropped considerably, however, when a premium was placed on the non-GM product: the preference for non-GM cereal dropped to 56 per cent among the UK respondents and to 37 per cent among the US respondents. The number of consumers who were indifferent (that is not willing to pay a premium) was constant at 22 per cent in both countries when a premium for non-GM was introduced.

Other studies on GM food labelling explore the contribution that label information could make to consumers' ability to make decisions based on 'reliable' – which is to say verifiable as opposed to speculative or specious – information. The presumption of these studies is that consumers generally lack reliable information about genetic modification, and do not have sufficient background knowledge about food production, regulatory approval, nutrition sciences and so on to make unaided choices. With reliable information and the background knowledge, they might be in a better position to make informed judgements about the risks and benefits of GM food (Bredahl et al., 1998; Castle et al., 2005; Lambraki, 2002; Poortinga and Pidgeon, 2004). Labelling is regarded as one way to increase consumer knowledge and provide consumers with reliable information, thereby reducing their uncertainty about GM products (Runge and Jackson, 2000). Some authors suggest that providing reliable information on food labels is an important way to avoid consumer assessments of risk and benefits that may be driven by heuristic-based reasoning that can lead to prejudicially negative attitudes toward GM foods (Bredahl et al., 1998; Grunert et al., 2004).

Research on labelling aimed at determining consumers' needs for information suggests that labels should be relatively simple (Finlay et al., 1999), should include admissions of risk or uncertainty about the consequences of GM technology (Bettman et al., 1986), and should provide information about the reasons for the presence of genetic modification (Kutznesof and Ritson, 1996). Other research suggests that consumer understanding and evaluation of information on these labels may be driven by overall knowledge about the subject (Castle et al., 2005; Huffman et al., 2004). Bredahl et al., (1998) suggest that consumers with greater knowledge and expertise about genetic modification may be less motivated to process label information as thoroughly as would consumers who are less knowledgeable. Several researchers have also found that the type of organism being manipulated (animal, vegetable or micro-organism) affects attitudes towards GM products (Frewer et al., 1997; Hossain and Onyango, 2004). Consumers also tend to be less accepting of organisms that are directly manipulated (such as fruits) than of processed foods that contain GM ingredients (such as chips).

The debate about labelling has raised two challenges. First, it is not clear if there are economic incentives for firms to provide GM labelling information voluntarily. If there is no economic incentive, the market will not spontaneously provide this information. Instead, firms will provide what is most profitable and least risky, which could mean that only GM or only GM-free products would be available, depending on the country, or that precautionary labelling claims (such as 'may contain') would be used. Both of these alternative outcomes would not necessarily improve consumer information. If GM labelling is perceived to be of political value, governments have the option of requiring mandatory labelling; in this case the cost of labelling for GM content would be shared between the industry and consumers, depending on the elasticities of supply and demand in each food market. Given that each food category faces different price elasticities, the incidence of the costs would vary widely across the food basket. There is growing evidence that consumers might not derive enough value from the added information to justify this cost – at least in some markets where consumers might end up bearing most or all of the incremental costs. Even where economic incentives might exist, it is not clear how the various supply chains would provide greater information about GM food products to consumers. A US-based survey (McGarry Wolf et al., 2004) shows that terminology such as 'GM', 'non-GM' or 'GM-free' is not overly effective in providing product information to consumers.

The second challenge is how to manage and communicate information to consumers. The way in which a product is differentiated has a great bearing on how it is accepted by consumers. The three systems currently used to provide information to consumers are segregation, identity preservation and traceability. Segregation systems are imposed by regulatory agencies to handle real and identifiable risks, identity preservation systems are initiated by private firms in the supply chain in the search of product differentiation, and profit and traceability systems are usually operated by downstream companies (often retailers) to satisfy consumer demands about the safety or provenance of a product, and correspondingly handling the risks of product recalls and harms that might lead to tort cases. Each system communicates a different message to the consumer. Segregation and identity preserved systems are initiated at the start of the supply chain and are used to manage the flow and/or quality of the food product, whereas traceability is initiated by downstream firms in the supply chain to provide the ability to detect origins of food safety problems and to rectify problems in the cheapest and most efficacious way (Smyth and Phillips, 2002).

The calls for, and information about, labelling can be put in context when consumer shopping habits are examined. The Produce Marketing Association (PMA) closely examines shopping trends in North America, and a survey by the PMA of shopping habits in 2001 at the height of consumer concern regarding GM food products found that the primary factors for consumers when making produce purchases are: expected taste (87 per cent); appearance (83 per cent); cleanliness (74 per cent); degree of ripeness (70 per cent); and nutritional values (57 per cent). This survey highlighted that when consumers are in a retail store faced with a purchase decision, concern over whether the product is GM, non-GM or GM-free might be relatively unimportant for the vast majority of North American consumers. Labelling systems alone therefore appear to be a poor way to communicate with these consumers and would be unlikely to substantially enhance post-market monitoring and surveillance.

Post-market monitoring and surveillance about the long-term consumption effects has been advocated as an important element of food safety concerns regarding GM foods. On one hand, it undermines the regulatory approval process that was used to approve the food product, while on the other, there are always segments of the population that are at increased risk of many consumables (that is, the elderly, infants, immune-suppressed). From a science-based perspective, both the OECD and the UK's Medical Research Council found no justification for undertaking and conducting post-market monitoring and surveillance (OECD, 2000; Medical Research Council, 2000). From a market-based perspective, the problem of implementing such a system is the lack of comparable data and benchmarks. To have a functional post-market monitoring and surveillance system, data would need to be collected on the sale and consumption of all food products, thus providing the information necessary to determine if the consumption of GM food products triggered any different health effects than the consumption of organic food products. 'The implementation of such a system . . . [would be] complex and costly.' (CBAC, 2002, p. 31).

3 GLOBAL PREFERENCES TO CONSUMER ATTITUDES OF GM FOOD PRODUCTS

One of the early assessments of the literature on consumer attitudes in developing countries was undertaken by Curtis et al. (2004), in the early part of the last decade.

The authors identified two major impediments to the adoption and acceptance of GM crops in developing countries. The first involved concerns about GM crop impacts on domestic crop diversity. Many developing countries have inadequate regulatory systems to be able to properly assess the safety of GM crops, and concerns existed, and still do in many countries, about the impact of GM crops on domestic crop diversity but also about the system-level effects of GM crops on biodiversity. The second concern identified was that countries that have strong trade connections, especially with Europe, but also Japan, would be in jeopardy of losing these export markets should GM crops be commercialized domestically. Having to ensure that all crop exports would be free of GM materials can be cost prohibitive – costs arise in having to deal with positive detections of GM presence at points of import entry and then in having to find an alternative market at a potentially reduced price. These trade implications have had a substantial impact on acceptance of GM crops in developing countries (see Chapter 12 in this volume).

While many consumer acceptance studies focus on GM food products in general, one study of interest focused on the specific product of GM bananas and the consumer interests in Uganda (Kikulwe, 2010). Based on the results of a survey of 421 banana-consuming households, overall acceptance to purchase and consume GM bananas was found to be 75 per cent. When respondents were asked about purchasing GM bananas that were the same price as other non-GM bananas, but more nutritious, approval soared to 92 per cent. Respondents were also asked purchase questions about discounts and premiums for GM bananas. If GM bananas were offered at a price discount, acceptance was 78 per cent, while if sold at a premium to other bananas, 39 per cent indicated that they would purchase GM bananas, which is a surprisingly high percentage given that the survey respondents live in a developing nation.

Consumer acceptance studies focused on Africa are far too infrequent, with another study examining the purchasing of GM corn in Kenya (Kimenju and De Groote, 2008). Based upon the results of a survey of 604 consumers in Nairobi, Kenya, in 2003, the researchers found that 38 per cent of the respondents were aware of GM crop technology. In spite of the relatively low level of awareness, consumers were quite willing to embrace GM corn meal products, with 68 per cent indicating that they would buy GM products if priced the same as competing products. While it is not possible to extrapolate the results of these two studies over the entire African content, they do demonstrate that at least in these two examples, consumers are willing to purchase GM products, if and when they are available.

The commercialization of GM eggplant was widely anticipated in India in 2010. In February, 2010, the Indian Minister for Environment and Forests made a surprise announcement that it would ban the commercial release of GM eggplant, in spite of the approval for commercial release from the Indian regulatory system of GM crops (Gardner, 2010). Krishna and Qaim (2008) examined the consumer responses. India is a very large producer of vegetables and one of the leading health concerns is the level of pesticide residues on vegetables that are purchased for consumption. A survey of 645 households was conducted regarding consumers' willingness to pay for GM vegetables that would be free from pesticide residues. The results showed that 68 per cent of respondents would be willing to buy GM vegetables free of pesticides if they were offered at the same price as existing vegetables, with a further 11 per cent indicating that they were indifferent. The remaining 21 per cent were not willing to purchase GM vegetables.

Assessing consumer acceptance of GM products in a country as large and diverse as China is obviously an important and challenging issue, and there have been some studies undertaken that begin to reveal some of the attitudes and preferences of Chinese consumers. The first of these studies dates from 2002, which revealed that while Chinese consumers had virtually no knowledge about biotechnology, they were supportive of the technology (Li et al., 2002). This study surveyed 599 consumers in Beijing and found that 54 per cent had no knowledge of biotechnology, with a further 44 per cent having little knowledge. When asked about their opinion regarding biotechnology, 62 per cent responded favourably, 7 per cent were neutral, with a further 22 per cent undecided. Only 9 per cent responded with a negative opinion. When asked if they would be willing to purchase GM rice if it was priced the same as non-GM rice, 80 per cent indicated that they would be willing to make this purchase. When asked if they would pay a premium to purchase GM rice, 44 per cent of consumers indicated that they would be willing to pay more. Only 14% of respondents were not willing to purchase GM rice, even with a discounted price.

A 2004 study shows how quickly consumers were becoming informed about biotechnology and GM crops, as it found that the percentage of consumers who indicated that they had never heard of biotechnology or GM foods had dropped to between 20 and 35 per cent (Zhang et al., 2004). In a survey of 1005 people, Chinese consumers expressed stronger support (60 per cent) for non-food applications of biotechnology, such as GM cotton. Overall support for GM plants and food products was found to be 52 per cent in this study. Subsequently, in 2006, a study revealed that while awareness about GM foods had risen still further to 67 per cent, there was not a great depth of knowledge about biotechnology, with half or more of respondents incorrectly answering questions about biotechnology (Huang et al., 2006). This premise is supported by a subsequent study (Ho et al., 2006) that found 71 per cent of respondents had heard of GM products, while only 18 per cent could correctly answer questions about biotechnology.

Huang et al. (2006) is unique in that the researchers followed up one year later with the respondents to the previous study, and resurveyed 666 of the previous 1005. When asked about acceptance of a variety of GM food products, support for these products increased by as much as 10 per cent in some instances, with overall support registering at 64 per cent. Meanwhile, there was a noticeable decline in the number of individuals willing to pay a premium to purchase a GM food product than previously reported, with only 3–5 per cent willing to do so. There was also a slight increase in the number of consumers that would not be willing to purchase GM food products under any condition, even with a price discount: 20 per cent of respondents indicated this to be their preference.

Much of the subsequent literature on consumer acceptance of GM plants and food products was undertaken by environmental groups and none of it peer reviewed, but acceptance of GM food would appear still to be strong in China. In late 2009, the Chinese government approved the production of GM rice (Shuping and Miles, 2009). The developers of GM rice expect it could reduce pesticide use by as much as 80 per cent and increase yields by 8 per cent. Large-scale production was expected to begin to occur within 2–3 years of introduction. This would indicate that consumer acceptance for GM rice must be relatively strong – even an undemocratic government is not likely to proceed in the face of strong opposition to something as sensitive as food supply.

If acceptance of GM crops and food products was based on producer adoption, then much of South America would be considered to be strongly in favour of biotechnology. Brazil and Argentina are globally ranked second and third in terms of GM crop production, with 37 and 24 million hectares, respectively (James, 2013). Numerous other nations in Central and South America have adopted GM crops, including: Bolivia, Chile, Columbia, Costa Rica, Honduras, Mexico, Paraguay and Uruguay. Consumer acceptance has not been examined in many of these nations, in part because some of the adoption is for non-food crops like GM cotton and, in others, the production is largely export driven, such as GM soybeans.

One of the challenges between consumer acceptance of GM foods and food security is highlighted in a study of consumer acceptance in Columbia (Pachico and McGarry Wolf, 2004). A survey on consumer acceptance undertaken just after the turn of the millennium shows low levels of consumer awareness, with only 13 per cent of consumers having any familiarity with GM foods and 77 per cent having no awareness. Notably, 40 per cent of consumers routinely do not have enough food to feed their families; as a result, low prices are the most important factor when making a food purchase decision. Seventy-five per cent of the survey respondents reported concerns about purchasing GM foods, but two-thirds of these respondents were willing to buy GM foods due to the more pressing concern of food security.

A survey of 256 consumers in Buenos Aires, Argentina, in 2004 showed that negative perceptions of GM products were higher than positive impressions (Mucci et al., 2004). In this study, 61 per cent indicated that they believed GM crops and foods provided a greater potential for harmful effects than beneficial effects.

A final study from the same period (Hoban, 2004), using data from 2000, shows the acceptance level for a variety of Latin American nations. Table 18.1 illustrates the variation in consumer acceptance of biotechnology. The challenge of accurately assessing consumer acceptance is highlighted in this data: results for Argentina are relatively close to what was reported above, while those for Columbia differ greatly.

It is worth considering, finally, the international context in which any country or set of countries would undertake GM food labelling. The Codex Alimentarius Commission

Table 18.1 Benefits of biotechnology outweigh the risks

Country	Agree (per cent)	Disagree (per cent)
Argentina	44	31
Brazil	55	32
Chile	47	37
Columbia	66	26
Cuba	79	4
Dominican Republic	69	25
Mexico	62	24
Panama	59	34
Peru	58	26
Uruguay	46	23
Paraguay	64	17

Source: Hoban (2004).

(Codex) develops international food standards that identify a processed food product and its essential composition and quality factors, identify additives and potential contaminants, set hygiene requirements, provide labelling requirements and establish the scientific procedures used to sample and analyse the product. Jackson and Jansen (2010) provide a detailed discussion of the science-based risk assessment process for food safety and its relationship to WTO dispute cases. It commonly takes in excess of six years to develop a Codex standard. For a Codex standard to be adopted, each member country is encouraged to incorporate it into any relevant domestic rules and legislation, but they may unilaterally impose more stringent food safety regulations for consumer protection, provided the different standards are scientifically justifiable.

Codex plays an important role in agri-food trade because its standards, guidelines and recommendations are acknowledged in the Sanitary and Phytosanitary (SPS) and Technical Barriers to Trade (TBT) Agreements of the World Trade Organization (WTO). There are currently no Codex standards in place for products of biotechnology; however, there has been significant effort on behalf of Codex members to develop a standard for the labelling of food products derived from biotechnology. The Codex Committee on Food Labelling was tasked in 1993 to initiate work on the development of a standard on the labelling of GM-derived foods and for nearly 20 years the Committee efforts were gridlocked. However, in 2011, the US relented on its opposition to the labelling of GM food products, and in 2012 Codex adopted the principles for a risk analysis of foods derived from biotechnology, which establishes that if a risk is identified, labelling is an appropriate management strategy. Codex stresses that any risk analysis of biotechnology-derived foods has to be science based and that these principles do not address 'environmental, ethical, moral and socio-economic aspects' (Codex, 2012, p. 1). It is important to note that this is a Codex principle on risk analysis of foods derived from biotechnology and not the standard on the labelling of GM foods that the Committee was tasked with 20 years ago.

4 SUMMARY

Issues of consumer acceptance of GM foods would appear to have been a greater concern around the turn of the millennium than they are today. If the availability of literature on this topic can be taken as evidence of an increase in consumer acceptance, the lack of recent articles on the topic suggests that no new significant issues have arisen that are driving researchers and research funders to delve into the topic once again. If there are simmering issues about current labelling standards for GM products already being sold, they are by and large unresolved. How important these issues are to individuals is difficult to ascertain. In a study context people might react to questions about information provision differently than they do in other contexts where finding information can be burdensome. For example, people living in countries that either do not have GM crops or food products, or that live in countries that are not supportive of the technology, from time to time will travel to North America. The fact that they do so and probably eat foods produced in both Canada and the US, 75 per cent of which contain GM ingredients, indicates that consumers are probably more supportive of biotechnology than they are willing to report in consumer acceptance surveys.

The low level of North American consumer willingness to pay for labelling information regarding GM, non-GM or GM-free products suggests that consumers might not perceive a benefit from labels. Consumers, when asked in polls, express a high level of desire to have these products labelled so that they can be differentiated, but when the purchasing decisions are taking place within grocery stores the perceived value of this kind of labelling rapidly diminishes. This might be because the majority of consumers want to go into a grocery store and purchase their food products as quickly as possible and as cheaply as possible. Fresh produce and meats are purchased primarily based on appearance, and processed foods based on brand recognition.

Although mandatory GM labelling does not appear to be economically justifiable in all countries, some alternative is needed to provide consumers with the information that they are demanding. It is apparent that trying to develop systems for tracing product ingredients back to their source and facilitating removal of dangerous elements from the market, based on current definitions of GM foods, would not be feasible because there are too many contrasting market signals for this to work. Ideally, the best route would be for representatives of the biotechnology industry, the food processing industry and government departments and agencies to come together with parts of civil society to develop a strategy for post-market monitoring and surveillance. The strategy should provide labelling information regarding GM content to consumers that is valid and meaningful, keeping in mind the way people shop for and consume foods. To function and meet the needs of all consumers, it is likely that any resulting system will need to operate in a way that enables a wide range of types of information to flow between the supply chain and the consumers. Ultimately, the success or failure of any resulting system should be judged on whether it facilitates an increase in the amount of product information flowing down the supply chain, and at the same time enables commercialization and optimal production and use of safe food. Overly complex and inflexible labelling systems that impede commercial activity will not enhance the delivery of healthy, nutritious and safe food to consumers.

REFERENCES

Bettman, J.R., J.W. Payne and R. Staelin (1986), 'Cognitive considerations in designing effective labels for presenting risk information', *Journal of Public Policy and Marketing*, **5**, 1–28.

Bredahl, L., K.G. Grunert and L.J. Frewer (1998), 'Consumer attitudes and decision making with regard to genetically engineered food products: a review of the literature and a presentation of models for future research', *Journal of Consumer Policy*, **21**(3), 251–78.

Canadian Biotechnology Advisory Committee (CBAC) (2002), *The Regulation of Genetically Modified Foods*, Ottawa: Canadian Biotechnology Advisory Committee.

Castle, D., K. Finlay and S. Clark (2005), 'Proactive consumer consultation: the effect of information provision', *Journal of Public Affairs*, **5**(3–4), 200–217.

Codex Alimentarius Commisssion (2012), 'Principles for the risk analysis of foods derived from modern biotechnology', available at: www.codexalimentarius.org/input/download/standards/. . ./CXG_044e.pdf.

Curtis, K.R., J.J. McCluskey and T.I. Wahl (2004), 'Consumer acceptance of genetically modified food products in the developing world', *AgBioForum*, **7**(1&2), 70–75.

Finlay, K., S. Morris, J. Londerville and T. Watts (1999), 'The impact of information and trust on consumer perceptions of biotechnology', *Canadian Journal of Marketing Research*, **18**, 15–30.

Frewer, L.J., C. Howard, D. Hedderly and R. Shepherd (1997), 'Consumer attitudes towards different food-processing technologies used in cheese production: the influence of consumer benefit', *Food Quality and Preference*, **8**, 271–80.

Gardner, H. (2010), 'India bans GM crop after green protests', *The National*, 10 February, available at:

http://www.thenational.ae/news/worldwide/asia-pacific/india-bans-gm-crop-after-green-protests, accessed 20 January 2012.

GMO Compass (2010), 'Chymosin', available at: http://www.gmo-compass.org/eng/database/enzymes/83.chymosin.html.

GMO Compass, (2009), 'Yeast', available at: http://www.gmo-compass.org/eng/database/ingredients/124.yeast.html.

Grunert, K.G., T. Bech-Larsen, K. Lahteenmaki, O. Ueland and A. Astrom (2004), 'Attitudes towards the use of GMOs in food production and their impact on buying intention: the role of positive sensory experience', *Agribusiness*, **20**(1), 95–108.

Harrison, R.W. and E. McLennon (2003), 'Analysis of US consumer preferences for labelling of biotech foods', available at: https://www.ifama.org/events/conferences/2003/cmsdocs/Harrison.PDF.

Ho, P., E.B. Vermeer and J.H. Zhao (2006), 'Biotechnology and food safety in China: consumers' acceptance or resistance?', *Development and Change*, **37**(1), 227–54.

Hoban, T.J. (2004), 'Public attitudes towards agricultural biotechnology', ESA working paper no. 04–09, available at: http://www.croplifeasia.org/ref_library/biotechnology/public_att_biotech_hoban.pdf, accessed 17 January, 2012.

Hossain, F. and B. Onyango (2004), 'Product attributes and consumer acceptance of nutritionally enhanced genetically modified foods', *International Journal of Consumer Studies*, **28**(3), 255–67.

Huang, J., H. Qiu, J. Bai and C. Pray (2006), 'Awareness, acceptance of and willingness to buy genetically modified foods in urban China', *Appetite*, **46**, 144–51.

Huffman, W.E., M. Rousu, J.F. Shogren, and A. Tegene (2004), 'Who do consumers trust for information: the case of genetically modified foods?', *American Journal of Agricultural Economics*, **86**, 1222–9.

Jackson, L.A. and M. Jansen (2010), 'Risk assessment in the international food safety policy arena: can the multilateral institutions encourage unbiased outcomes?', *Food Policy*, **35**, 538–47.

James, C. (2013), 'Global status of commercialized biotech/GM crops: 2011', *International Service for the Acquisition of Agri-Biotech Applications (ISAAA) Briefs*, 44, Ithaca, NY: ISAAA.

Kikulwe, E.M. (2010), *On the Introduction of Genetically Modified Bananas in Uganda: Social Benefits, Costs and Consumer Preferences*, Wageningen: Wageningen University Press.

Kimenju, S.C. and H. De Groote (2008), 'Consumer willingness to pay for genetically modified food in Kenya', *Agricultural Economics*, **38**, 35–46.

Krishna, V.V. and M. Qaim (2008), 'Consumer attitudes toward GM food and pesticide residues in India', *Review of Agricultural Economics*, **30**(2), 233–51.

Kutznesof, S. and C. Ritson (1996), 'Consumer acceptability of genetically modified foods with special reference to farmed salmon', *British Food Journal*, **98**(4/5), 39–47.

Lambraki, I. (2002), 'An exploratory qualitative and quantitative study on consumers' attitudes towards genetically modified foods', *Dissertation Abstracts International – MAI*, **40**(06).

Li, Q., K.R. Curtis, J.J. McCluskey and T.I. Wahl (2002), 'Consumer attitudes toward genetically modified foods in Beijing, China', *AgBioForum*, **5**(4), 145–52.

McGarry Wolf, M., A. Stephens and N. Pederazzi (2004), 'Using simulated test marketing to examine purchase interest in food products that are positioned as GMO free', in R.E. Evenson and V. Santaniello (eds), *Consumer Acceptance of Genetically Modified Foods*, Wallingford: CABI Publishing, pp. 53–9.

Medical Research Council (2000), *Report of a Medical Research Council Expert Group on Genetically Modified (GM) Foods*, London: Medical Research Council.

Moon, W. and S. Balasubramanian (2001), 'Public perceptions and willingness-to-pay a premium for non-GM foods in the US and UK', *AgBioForum*, **4**(3&4), 221–31.

Mucci, A., G. Hough and C. Ziliani (2004), 'Factors that influence purchase intent and perceptions of genetically modified foods among Argentine consumers', *Food Quality and Preference*, **15**, 559–67.

Organisation for Economic Co-operation and Development (OECD) (2000), *Report of the Task Force for the Safety of Novel Foods and Feeds*, Paris: OECD Printer.

Pachico, D. and M. McGarry Wolf (2004), 'Attitudes toward genetically modified food in Colombia', in R.E. Evenson and V. Santaniello (eds), *Consumer Acceptance of Genetically Modified Foods*, Wallingford: CABI Publishing, pp. 155–62.

Poortinga, W. and N.F. Pidgeon (2004), 'Trust, the asymmetry principle, and the role of prior beliefs', *Risk Analysis*, **24**(6), 1475–86.

Rousu, M., W.E. Huffman, J.F. Shogren and A. Tegene (2007), 'Effects and value of verifiable information in a controversial market: evidence from lab auctions of genetically modified food', *Economic Inquiry*, **45**, 409–32.

Runge, C.F. and L.A. Jackson (2000), 'Labelling, trade and genetically modified organisms: a proposed solution', *Journal of World Trade*, **34**(1), 111–22.

Shuping, N. and T. Miles (2009), 'China gives safety approval to GMO rice', *Reuters*, 27 November, available at: http://www.reuters.com/article/2009/11/27/idUSPEK37812, accessed 20 January, 2012.

Smyth, S.J. and P.W.B. Phillips (2002), 'Product differentiation alternatives: identity preservation, segregation and traceability', *AgBioForum*, **5**(2), 30–42.
Zhang, C., J. Bai, J. Huang, W. Hallman, C. Pray and H. Aquino (2004), 'Consumer acceptance of genetically modified foods: a comparison between the US and China', paper presented at the American Agricultural Economics Association Annual Meeting, 1–4 August, Denver, Colorado, available at: http://ageconsearch.umn.edu/bitstream/20026/1/sp04zh10.pdf, accessed 20 January, 2012.

19 The motivation and impact of organized public resistance against agricultural biotechnology

Philipp Aerni

Fifteen years of experience with the commercial cultivation of genetically modified (GM) crops and countless national and international risk assessments of genetically modified organisms (GMOs) suggest that the risks related to this new technology are not any different from those already known in conventional agriculture. Despite these reassuring findings, public distrust toward GMOs has not decreased. Europe has even further tightened its de facto ban on genetic engineering in agriculture and most African countries continue to be reluctant to approve any GM crops for commercial cultivation, even if they may prove to be particularly beneficial for small-scale farmers. In order to understand this puzzling situation, we have to look at the global controversy on GMOs in the larger historical context. Professional pressure groups against GMOs have their roots in the environmental movement of the 1970s. At that time they criticized the negative environmental consequences of the Green Revolution. By assuming that the current Gene Revolution would largely represent a repetition of the mistakes of the Green Revolution, they were able to shape the risk narrative of genetic engineering in agriculture to a great extent. As an alternative to GMOs, they advocate the concept of 'food sovereignty' which envisions a type of agricultural system that helps countries to ensure food security without having to rely on agricultural trade and the use of new technologies in agriculture. In this chapter we argue that this kind of bipolar world view of good and evil agriculture is unlikely to be helpful in addressing the multiple sustainability challenges of the twenty-first century because it tends to burn rather than build bridges between the actors that would be most suitable to join forces in the fight against hunger and climate change.

ENVIRONMENT AND TECHNOLOGY IN THE TWENTIETH CENTURY

The societal discourse on the global environmental crisis gained momentum in the 1970s with the publication of the report of the Club of Rome called 'Limits to growth' (Meadows et al., 1972). Its basic, Malthusian message that exponential population growth and industry's hunger for natural resources would ruin the planet chimed well with the growing public concerns about the oil crisis at that time. There was, however, not just despair but also hope, thanks to many breakthroughs in science and technology as well as a growing willingness of governments to promote energy conservation, invest more in agricultural research, and alternative energy systems (especially solar energy) (EIA, 1998). Governments also started to create ministries of environment and pass legislation on environment protection. At the same time, the education system responded by introducing environmental awareness programmes in high schools and by establishing the disciplines of environmental studies or sciences as new research and teaching subjects at universities. These trends inspired many young people to study for solutions to cope with the environmental crisis and eventually build a more sustainable global economy (Otis and Graham, 2000). Yet some of the leaders in the environmental movement regarded business and new technologies to be part of the problem (Ehrlich, 1970) while others, though condemning unsustainable business practices, saw the need to transform the economy not by fighting it but by renewing it (Carson, 1962; Lovins, 1976). While followers of the former were primarily fond of nature conservation and saw the human being as a major threat to nature, the latter tended to see the human being as being part of nature and therefore also as part of the solution; as techno-optimists, they believed in sustainable technological change through academic entrepreneurship. As a result, many university spin-off firms were created in the 1970s that aimed at conquering markets with new technologies. These new technologies included information and communication technologies (ICT), solar technology and biotechnology. Yet, ultimately only ICT companies managed to win the tacit approval of environmental NGOs (despite the problem of electronic waste) and to transform the way people communicate and do business (for example Apple, Microsoft and Sun Microsystems). The biotechnology revolution in the 1970s also produced many start-up companies, mainly spin-offs from universities (Cetus Corp., Genentech, Calgene), but were later acquired by the larger established corporations in the pharmaceutical and agribusiness industry (Henderson et al., 1999; Hughes, 2011; Martineau, 2001). In both red and green biotechnology,[1] the key driver of concentration was increasing costs of complying with uncertain, complex and time-intensive regulatory framework (especially in large markets such as Europe and Japan) rather than intellectual property rights (IPRs). It allowed incumbents to continue to dominate their respective markets. In the case of agricultural biotechnology the situation was even more difficult for start-up companies due to the negative public image in Europe that induced many investors to refrain from investing in the emerging companies (Aerni, 2011b).

The solar technology business enjoyed great government support in the 1970s but then had to face falling oil prices and decreasing public and private investments in the 1980s. This led to a general setback in business and technology development in the solar industry until recently, when it enjoyed a revival through subsidies for clean technology (Bradford, 2006).

At any rate, the commitment to mitigate the impact of industrial activity on the environment through technological change was great at the end of the 1970s, no matter whether environmentalists decided to become active in politics, government, science or business. The diverging trend between technology-friendly and technology-sceptic environmentalists increased, however, particularly in the field of agricultural biotechnology. This divide of environmentalists regarding the role of agricultural biotechnology increasingly reflected two meta-visions of sustainable agriculture and how to achieve it. As a consequence, the public debate on sustainable agriculture became increasingly polarized (Aerni, 2009).

Economic and technological change in agriculture was discussed controversially even before the advent of the Green Revolution. Resistance to the use of Mendelian genetics in plant breeding before World War I caused many alternative forms of seed production and farming systems (Kingsbury, 2009). The breeding of landraces continued to be preferred by farmers in Europe until the late 1930s as the advantage of using hybrid

F1 seeds was not yet obvious. Once farmers and the food industry were convinced of the value of modern hybrids, adoption rates of purchased seeds increased rapidly. At the same time, scepticism towards industrial agriculture increased especially because of hybrid corn that tended to encourage monoculture and increased the dependence of farmers on seed companies. Even though the Green Revolution after World War II was largely a public sector initiative that did not rely on hybrids, it further added to the controversy. It was a US-driven initiative to reduce hunger and malnutrition in non-aligned developing countries and must therefore also be understood in the context of the Cold War. Farmers in developing countries were encouraged to adopt high-yielding varieties, fertilizer and irrigation. It gave a big boost to agricultural productivity and thus averted the predicted worldwide food crisis in India in the 1970s (Kingsbury, 2009) but it also had many unintended side-effects on human health and the environment (Pingali and Roger, 1995). The bred high-yielding varieties grew well in favourable agricultural areas but did little to promote sustainable agriculture or address the concerns of small-scale farmers in more marginal areas (Byerlee and Morris, 1993). Even though genetic engineering now allows breeders to address these concerns much more effectively, because the particular traits can be directly inserted into the locally preferred and well-adapted variety, it continues to be associated with corporate power and societal risks. As a result, policy makers focused on regulating corporate agribusiness and, in response to the end of the Cold War, decreased support for the public sector agricultural R&D. The cut in public funding especially affected public research in plant biotechnology. Policy makers in affluent countries preferred to support the type of agriculture that urban consumers considered to be appropriate to safeguard the environment, protect the livelihood of farmers and consumer health. Organic agriculture was closely linked to sustainable agriculture as conceived by the non-farming affluent, largely urban, majority. In return, opponents of agricultural modernization increasingly framed technological change in agriculture in general and agricultural biotechnology in particular as a threat to sustainable agriculture. This largely helps explain why national agricultural research institutes in Europe and the United States that were originally designed to facilitate product innovation in agriculture are now largely considered to be controlling institutes ensuring that farmers are able to comply with public and private environmental and food safety standards (Aerni and Bernauer, 2006; Kingsbury 2009).

Especially in Europe, this dualistic and reductionist view of agricultural biotechnology is shaping the environmental discourse today in schools, politics and academia (Aerni and Oser, 2011) and thus has become a persistent narrative, which can also be found in the United States (Herring, 2009). In turn, advances in plant biotechnology have been revolutionary over the past two decades, adoption rates of GM crops continuing to increase rapidly outside Europe. Fifteen years of experience with GM crops and countless risk assessment studies (EC, 2010) were, however, unable to offset the fears of opponents.

Whereas the dispute over GMOs is largely portrayed as a transatlantic dispute in recent political science literature, with the United States as the major producer and proponent of GM crops and the EU as the major opponent (Falkner, 2007; Schreurs et al., 2009), public resentment toward genetic engineering in agriculture seems to have become widespread all over the developed world. Ironically, anti-biotechnology activists in the affluent regions of the world also regard themselves to be advocates of the rights of small-holder farmers in developing countries. Their baseline assumption is that these farmers are normally self-sufficient and sustainable but are now threatened by trade and technology. As a consequence they lobby on the national and international levels to ban GM crops (Aerni, 2011a). Over the past decade they indeed succeeded in convincing most retailers, research foundations, financial institutions and governments to promote organic agriculture as an alternative to genetic engineering, implying that the former would be sustainable while the latter would be unsustainable. Yet can we assume that the world is becoming more sustainable thanks to such campaigns? The ongoing global food crisis and growing impact of agriculture on global climate change suggest that this is not the case.

The dominant view among the main actors involved in the global environmental debate is nevertheless that technological and economic change is the problem rather than part of the solution to the food and environmental crisis. In this sense, the debate has become narrower and less imaginative compared to the 1970s when many still believed that technology, innovation and entrepreneurship must be part of a sustainable solution.

The trend among environmental NGOs toward more narrow but more media-relevant advocacy agendas based on the simple dualist mindset of 'people versus profits' reflects the increasing influence and professionalization of environmental pressure groups in organized public protest. Even though there is still a great diversity of environmental NGOs, and many of them still assume a critical watchdog function in society, the need to join the agenda of well-known international pressure groups that are able to mobilize effective on- and off-line public protest campaigns, has increased. As a consequence, protest topics increasingly reflect the priorities of large pressure groups such as Greenpeace and Friends of the Earth that focus on popular symbolic campaigns linking the risks of genetic engineering to the risks of global capitalism. Specific local environmental problems that cannot be linked to already well-established global symbolic campaigns tend to be ignored even though they may matter most to local people (Aerni and Bernauer, 2006; Nelkin, 1996; Bob, 2005; Meera, 2005).

HOW NGOS EMERGED AS REPRESENTATIVES OF CIVIL SOCIETY

The term NGO usually includes all types of organizations that pursue social, economic and political objectives but were not established by governments and have no profit motive (Willets, 1982). The number of NGOs with international reach (INGOs), has increased substantially since the 1970s. At the beginning of the twenty-first century there are about 50 000 organizations recognized as INGOs and they have held around 5000 annual world congresses. A disproportionate number have their main offices in the EU and Switzerland and were established in the 1970s (Keane, 2003). They are often portrayed as a politically well-organized force representing the interests of global civil society. Civil society is again conceived as consisting of non-profit organizations that represent different public concerns that are connected neither to business nor government interests. This view tends to ignore that these organizations are increasingly supported by government and business institutions that would also like to be seen as being supportive of a more sustainable world. It is estimated that around US$7 billion are disbursed as government funds annually to INGOs (Keane, 2003). As for business linkages, many representatives of INGOs have joined the boards of large multinational corporations (especially in the energy and retail industry) as they seek to emphasize their commitment to corporate social responsibility. Moreover, INGOs pursue a wide range of agendas ranging from highly commendable objectives (to promote sustainable development) to criminal (for example Mafia) or terrorist activities (for example terrorism justified as religious, ideological or environmental self-defence) (Mittelman, 2000). One additional challenge is that the more global NGOs become, the more they may have an accountability problem with regard to those they claim to represent (Kaldor, 2002).

As for professional environmental advocacy activities, the term INGOs is defined more narrowly here as professional international pressure groups or intervener groups (Ottway and Peltu, 1985) that seek political outcomes that are favourable to their view of environmental protection and sustainable agriculture. Such groups include Greenpeace, Friends of the Earth, Sierra Club, ETC Group and GRAIN. Even though most of these organizations have their roots in the social movement of the 1970s, the movement back then was too heterogeneous to fit any particular type of political organization. The movement in Europe and the United States was primarily united by the resistance against state paternalism and the military-industrial complex that dominated political priorities during the Cold War (Radkau, 1995).

At the same time, the emergence of many underground resistance movements against authoritarian communist or feudalist regimes in the developing world produced another type of civil society that primarily fought against US or Soviet support for corrupt domestic regimes and focused their advocacy work on issues related to social and economic empowerment, political participation, individual rights, better access to health and education and agrarian reform (Schuurman, 1992; Castells, 2004).

Many of the targeted political regimes have been toppled in the 1980s. In the course of widespread democratic transition, underground movements became legal and were invited to assume responsibility as representatives of civil society in the young and fragile democracies. Many national and international donors decided to support these new actors as watchdogs and political opposition groups that represent the interest of the poor and the environment. The movements themselves were often divided in factions; the hard-liners distrusted the new democratic system and still believed in communist revolution while the pragmatic wing was willing to compromise and participate in the newly established political system in pursuit of peaceful reforms rather than revolution (Aerni, 1999). International donors primarily focused their support on the less radical group. In order to be able to get support as partners for development, these public interest groups needed to register as legal not-for-profit entities. Eventually the term non-governmental organization (NGOs) was found to be appropriate (Dine, 2005). Western funding for this new type of organization had two great consequences. First, the number of registered NGOs increased massively in developing countries and, second, the priorities in their advocacy work shifted from topics of local concern to topics of international concern. In that way the agenda of local NGOs in developing countries was increasingly set by their partners in developed countries. The case of the Philippines in the 1990s illustrates this well (Alegre, 1996).

NGOS IN THE PHILIPPINES AND THEIR ATTITUDE TOWARDS IRRI

In the 1960s, two crucial international events influenced civil society in the Philippines: first, the global student movement generated by the growing discontent with state paternalism, and, second, the Second Vatican Council, which resulted in a more progressive Catholic Church that began to show more concern for social and rural development. The rise of Liberation theology in Latin America was one of the consequences of this (Schuurmann, 1992).

The national repression in the Philippines under the Marcos dictatorship combined with the new trends on the international level led to a student movement at national universities. The movement was linked to the underground Communist Party (CPP) and eventually formed the National Democratic Front (NDF) (Rocamora, 1994). What united NDF member organizations was the general discontent with the US support for the Marcos regime. Ferdinand Marcos was an important anti-communist ally and a firm supporter of the US-funded Green Revolution. He gave the International Rice Research Institute (IRRI) the status of an international organization and to some extent granted the institute immunity under national jurisdiction. The main purpose was to promote agricultural modernization and make the Philippines self-sufficient in rice production (Aerni, 1999). Due to his intolerance of democratic change, Marcos found himself facing ever-increasing international isolation as well as domestic opposition not just from NDF but also from the entrepreneurial middle class who were fighting incompetent state intervention and corruption. All these developments made it increasingly difficult for the conservative Reagan administration to continue to support the regime. Marcos was finally dethroned by the People's Power Revolution (EDSA Revolution) in 1986. It led to a shift from a dictatorship to a weak and pluralist democratic state that continued to be largely based on a traditionally entrenched patron–client system (Rocamora, 1994).

Corazon Aquino became the new president of the young democracy in 1986. The new constitution was based on the American-style presidential system and came into force in February 1987. It contained well-founded declarations on labour rights, women's rights and the rights of indigenous communities. It was inspired by the Brundtland report on sustainable development. The international donor community welcomed this transition and funding was not just given to the Aquino government but also to NGOs that were involved in grassroots activities. Increasing financial support for registered NGOs had the effect of increasing their numbers from 16000 in 1986 to 26000 in 1992 (Alegre, 1996).

In 1992, Ramos, a former military general, was narrowly elected President of the Philippines. He did not adopt a policy of 'business as usual', but drafted an ambitious plan for economic development. The so-called 'Vision Philippines 2000' was designed to encourage Filipinos to unite efforts to make the Philippines a Newly Industrializing Country (NIC) by 2000. The promotion of agricultural biotechnology was part of his plan and so was greater involvement of NGOs as representatives of civil society. NGOs, however, were opposed to agricultural modernization because they linked the Green Revolution with IRRI and its collaboration with the Marcos regime. They also received increasing financial support from anti-GMO organizations located in Europe (Aerni, 1999).

Finally, by promoting the influence of the mass media (rather than suppressing it as Marcos did), president Ramos aimed at the gradual elimination of the influence of patron–client ties which form the basis of the oligarchic structure in Philippine national politics (Rocamora, 1994).

While Ramos' primary goal was to stimulate economic growth, his programme was also committed to sustainable development. In this respect, his priorities included political stabilization, decentralization and rural development. His commitment to sustainable development led him to sign Agenda 21 for Sustainable Development at the UNCED conference in Rio in 1992 and to seek wide cooperation with national NGOs and farmer organizations in development strategies. One result of all this is that the Philippines was one of the first Asian states to create National Biosafety Guidelines. In this context, a National Biosafety Committee was created that also included NGOs, with a goal to ensure the safe use of the technology.

This ensured the continuing political importance of NGOs and their ties to various sectoral movements (including the peasantry, church, trade unions and women) (Alegre, 1996). National NGOs also developed strong links to well-endowed and media-savvy INGOs. These INGOs included well-known European and American pressure groups. They were focused primarily on the organization of symbolic protest that mainly resonated with their affluent constituency back home. As such they did not just shape the agenda of local partner NGOs but also influenced the allocation of funding of official development assistance. Greenpeace, the largest and best-known of these advocacy groups, started to exert its influence in the Philippines in the mid-1990s when it collaborated with local NGOs to intercept transgenic rice seed that was sent from ETH Zurich to IRRI in Los Banos. Greenpeace alleged that a particular paragraph of the Philippine Biosafety Guidelines had been disregarded in this case. This paragraph stipulates the obligation to hold public consultations on such biosafety issues. Greenpeace and local NGO activists eventually succeeded in convincing the congressional committee on ecology to call for hearings in the House of Representatives on the health and ecological risks of genetically engineered rice. Thus, this operation created a new political arena regarding genetic engineering in agriculture in the Philippines. But one could argue that it was an elite discussion that hardly affected the public at large. At any rate, Greenpeace was able to shape the civil society agenda in the Philippines by mobilizing many of the local NGOs to oppose IRRI as a stooge of Western interests. A local NGO called MASIPAG has developed an alternative rice breeding programme which was claimed to be more popular and successful with local rice farmers than Western technology from IRRI. Even though MASIPAG refuses to collaborate with IRRI because it is viewed as a foreign institution, MASIPAG would probably not exist without the support of Western donors either. A stakeholder survey showed that most of the opponents of GM crops were funded by European NGOs and development agencies (Aerni, 1999). Stakeholders confirmed the same trends in Mexico and South Africa (Aerni and Bernauer 2006). The almost neo-colonial approach applied by professional Western pressure groups such as Greenpeace was confirmed in a conversation with stakeholder representatives in the Philippines and Mexico. For example, a Filipino academic and MASIPAG representative frankly said that focusing his advocacy work on resistance to GM food ensures that he gets invited to Europe to talk about the victims of Western agribusiness firms in developing countries. The local representative of Greenpeace in Mexico complained about the top-down approach in the priority-setting process of Greenpeace International in the Netherlands (Wahl, 1998). The Mexican Greenpeace representative tried once to convince the headquarters that

Mexico has more important local issues that affect people and the environment much more than GMOs. She was rebuffed and told to look for another job if she could not put up with the Greenpeace agenda. In South Africa it was interesting to observe how secretive European development agencies are about their support for anti-GMO NGOs. When GTZ/GiZ (the German Technical Cooperation) was asked how much and for what kind of projects they fund Biowatch South Africa, an anti-GMO NGO, the answer was that they are not obliged to report on such issues to ordinary people but would only be accountable to their own government (Aerni and Bernauer, 2006).

As for the potential of biotechnology to address challenges related to climate change mitigation and adaptation, anti-biotechnology advocacy groups managed to ensure that the issue of agricultural biotechnology is not raised in global discussions on how to make agriculture in developing countries more productive and sustainable in the face of global climate change. The potential of the technology to address growing biotic and abiotic stress factors in crop cultivation is, however, undeniable. When Lin Erda, member of the CGIAR commission on climate change and agriculture, was asked at an IFPRI conference on Climate Change and Food Security in Beijing on 5 November, 2011,[2] whether sustainable intensification in agriculture would be possible without the use of biotechnology, he first did not want to respond and then finally said, 'I do not know anything about biotechnology'. On an earlier occasion, Robert Watson gave a talk on 30 March 2009. 'The interlinkage of global climate change, economy and poverty' at ETH Zurich. He concluded his presentation by pointing out the future challenge of mankind to grow more food on less land and with fewer greenhouse gas emissions. Yet he could not see any role for modern biotechnology to contribute to this daunting challenge. In the subsequent discussion he denounced modern biotechnology as a potential threat to the environment and regarded advocates of agricultural biotechnology as being merely naive or, worse, stooges of industry. His vehement rejection of biotechnology as an important platform technology may be related to the fact that Watson is not just a scientist (he was Chief Scientist of the World Bank) but also an influential stakeholder in the global debates on climate change, agricultural biotechnology, technology transfer and biodiversity. He was former Chairman of the Global Environment Facility (GEF) in the early 1990s, then became Chairman of the Intergovernmental Panel on Climate Change (IPCC) and Co-chair of the Millennium Ecosystem Assessment. Finally he was also appointed as Director of the International Assessment of Agricultural Science & Technology for Development (IAASTD).[3] Most of the reports he edited start from the same baseline assumption, namely that the success of increasing food production in the twentieth century came at tremendous costs to the environment. Considering that the areas that suffer most from environmental degradation are found in poor countries with high populations but almost no industry (Beckerman, 2002), this baseline assumption might be wrong despite being replicated in countless international environmental reports.

THE EMERGENCE OF PROFESSIONAL PRESSURE GROUPS IN EUROPE AND THE UNITED STATES

Pressure groups such as Greenpeace are a product of the environment movement that emerged in the 1970s in Europe and North America. Their most prominent protests were directed against the use of nuclear technology, commercial whaling and toxic waste. Later on they extended their risk discourse from nuclear technology to agricultural biotechnology. Agricultural biotechnology could hardly be associated with the military-industrial complex that promoted nuclear technology during the Cold War but served as a perfect proxy for grievances about economic globalization and its presumed negative impact on the environment and the poor. Moreover, even though the currently approved GM crops proved to be as safe as conventionally bred crops (EC, 2010), cognitive psychology showed that the public perceives the risks of agricultural biotechnology to be similar to the risks of nuclear technology. They are perceived to be uncontrollable, unobservable, involuntary and potentially catastrophic (Slovic, 1992). This made it easy to scare people with images of 'Frankenfood' and get the attention of the media (Nelkin, 1996).

The German sociologist Niklas Luhmann (1991) called these pressure groups 'Protestbewegungen' (protest movements or 'intervenor groups'). They play a facilitator role in the formation of public resistance to high technologies by means of public stunts, protest campaigns and position papers that attract the interest of the mass media (Peltu, 1985). Media attention helps these groups to gain influence in formation of public opinion as trustworthy actors that fight for the public interest. Public trust provides them with political legitimacy and access to essential material resources. As presumed advocates of nature and the poor, they take up arms against the powerful institutions in business, politics and science which are portrayed as objects of distrust that pursue private interests at the expense of the public at large. The representatives of the targeted institutions may have political power, well elaborated PR strategies and financial resources at their disposal, but if they lack public trust, their expensive campaigns become ineffective or sometimes even counterproductive. In the case of large agribusiness companies such as Monsanto, aggressive PR campaigns to assure regulators and the public of the safety of GM crops in the 1990s led to more public distrust of the company, additionally fuelled by protest campaigns organized by well-known pressure groups.

As a result, regulation of GMOs in Europe became gradually more restrictive even though an increasing amount of risk research suggested that potential risks have been exaggerated. The fact that national and international biosafety regulation became almost exclusively focused on preventing the potential risks of agricultural biotechnology had two major consequences. First, costly and lengthy approval processes tended to push small innovative biotech firms either out of business or into the arms of the big companies, which led to more concentration in industry. Second, the highly preventive regulation of agricultural biotechnology suggested to the public that there must indeed be high risks involved in the use of genetic engineering in agriculture. As a consequence, biotechnology was no longer considered to be an environmentally-sound technology and essentially banned from the public discussion on sustainable agriculture (Aerni, 2011b).

Many of the large agribusiness firms learned their lessons from ill-conceived PR strategies and focused more on collaborating with moderate national and international NGOs that work with small-scale farmers on the ground, especially those interested in helping to introduce agricultural innovation that may help the farmers to cope better with risk and increase yields (in most cases it does not involve genetic engineering but other tools of modern biotechnology) (Aerni, 2006). Even though many examples of successful public–private partnerships have emerged since then (Ferroni and Castle, 2011), they are poorly covered by the mass media and not addressed in the international debate on the global challenge of climate change and the food crisis.

RESENTMENT AGAINST AGRICULTURAL MODERNIZATION AND THE FOOD SOVEREIGNTY MOVEMENT

The reason many pressure groups tend to resent public–private partnerships has a lot to do with their highly selective view of agricultural development and innovation. Agriculture in the nineteenth century is described either as a form of class struggle or a centre–periphery system in which European colonial powers exploited the people and the natural resources of their colonies (Friedmann and McMicheal, 1989). However, the nineteenth century could also be described as the century of human empowerment in the countryside. This was made possible through the establishment of agricultural colleges and agricultural research institutes that collaborated closely with farmers and local entrepreneurs in the regions in order to solve agricultural problems, develop innovative products and technologies and help rural regions organize economically and socially (Kingsbury, 2009). It was technological change that resulted in social mobility and enhanced the self-confidence of the country-dwellers. It also helped create an entrepreneurial middle class that wanted to have a say with regard to the allocation of government taxes. Their active participation in politics contributed to the political stabilization of the young democracies by making use of the system of checks and balances.

However, even though successfully replicated today in New Zealand and many emerging economies, these success stories of human empowerment are largely ignored by environmental and agricultural ministries that follow a concept of sustainable agriculture shaped by civil society that advocate the concept of 'Food Sovereignty' and pressure groups that regard technological change, entrepreneurship and innovation to be the problem rather than part of the solution of sustainable agriculture (Aerni, 2009).

The term 'Food Sovereignty' is said to have been coined by an international NGO called La Via Campesina. It was founded in 1993 in Mons, Belgium, and currently counts 148 organizations from 69 countries as its members. Its creation must also be understood as a response to the growing pressure for agricultural trade liberalization after the completion of the Uruguay Round and the establishment of the World Trade Organization (WTO) in 1995. This was a great worry for highly subsidized and well-protected farmers in affluent Europe. In their efforts to create a global resistance movement against agricultural trade liberalization, they found partners in the civil society of developing countries. Their partners in the South were also worried about liberalization but not because of a possible loss of their entitlements. They were instead concerned that open agricultural markets would primarily benefit large domestic farmers but make business more difficult for small-scale local food producers. This, they feared, would lead to a corporate takeover of the domestic food economy and consequently cause an increase in dependence on foreign firms as well as a loss of importance of local food production (Aerni, 2011a).

Via Campesina may sound like a Latin American social movement but since it is largely funded by European stakeholders, the concept of food sovereignty was therefore defined in a defensive rather than a progressive way. Food sovereignty advocates in civil society may acknowledge the large productivity gains in agriculture and the resulting low food prices that were caused by the Green Revolution. Yet they think it was the wrong way to promote sustainable agriculture and denounce it for having led to monoculture practices, the loss of biodiversity and the abandonment of local varieties. In addition, they correctly note that the widespread use of fertilizer and pesticides has caused environmental and public health problems in some areas. Yet they cannot blame the private sector for that, because the Green Revolution was a public sector initiative (Kingsbury, 2009). Instead the movement frames the environmental problems of modern agriculture in the context of the work of Rachel Carson, who wrote the iconic book *Silent Spring* about the negative consequences of the use of chemicals in agriculture (Carson, 1962). If Food Sovereignty advocates read her book carefully, however, they would notice that she was not opposed to business and technology. She praised public and private sector researchers who jointly developed insect sterilization techniques, as well as the first microbial insecticides based on the effect of Bacillus thuriginensis (Bt). She was strongly in favour of bacterial warfare in agriculture because, in contrast to chemicals, insect pathogens are harmless to non-target insects. She was also a scientist who wanted to reach out to all parties to find a joint solution. She wanted to make agriculture more sustainable by facilitating technological change rather than preventing it.

The Food Sovereignty movement comprises countless NGOs that are concerned about global food production, food safety, as well as food culture. Some are focused on practising new forms of alternative agriculture and quality cuisine (slow food); others are entirely concerned with advocacy work and therefore mostly work as pressure groups. The movement is therefore very heterogeneous. It is largely united by what it opposes rather than what it stands for and what kind of changes it envisions for the future. In view of their defensive posture one might wonder whether the countless self-appointed Food Sovereignty advocates ranging from Prince Charles to Vandana Shiva to José Bové are just anxious to defend their privileged lifestyles, which they consider sustainable, against the forces of change, which they consider unsustainable. In spite of the rhetoric of Food Sovereignty advocates, the current world food system has grown over centuries and is not the product of a deliberately enforced global ideology that aims at enriching the powerful and exploiting the poor. Numerous columnists in the big national dailies all over the world have, however, largely adopted the narrative of well-known pressure groups such as Greenpeace that can be summarized in the slogan 'profits versus people'. This slogan has also been embraced by many high school teachers that would like to tell their students who are the good and the bad guys in the global debate on sustainable agriculture. This dualist and reductionist view is very convenient because it reduces complexity and offers the appearance of certainty. People that use the slogan in public do not really have the intention to say something reasonable about today's global agricultural problems, but to express concern and say something that sounds meaningful within their own peer group (Crano and Prislin, 2006; Crano, 2001; Converse, 1964). Successful pressure groups have realized a long time ago that it is not necessary for the people to bother about facts and history; it is enough to learn who stands for the corporate ('evil') system and who stands for the alternative ('good') system. An Internet search will provide you with everything else you need to know. It helps communities of like-minded people to create an echo chamber in which they can feel reassured about their views even though they lack any concrete experience with any of the systems (Sunstein, 2007; Sennett, 1976).

Anti-biotech activists like Vandana Shiva have honed this PR strategy of just repeating the same catchy slogans to perfection, especially true of her false claim that farmers adopting Bt cotton in India are more likely to get into debt and eventually commit suicide than those who use conventional cotton. By continually repeating the message she has created a persistent narrative that became a public truth that no one needed to verify any longer, and it was then also taught in school as an example of the socio-economic risks of genetically modified crops (Herring, 2009). Yet she must be well aware that large-scale surveys have shown that *fewer* and not more farmers commit suicide after they have adopted Bt cotton (Gruère et al., 2008). These empirical studies showed that Bt cotton in India was widely adopted by small-scale farmers not because they were fooled by seed companies but because they had better yields, needed less chemical input and thus generated more revenues and suffered less from health and environmental problems (Kouser and Qaim, 2011). The success of Bt cotton explains why more than 90 per cent of the farmers who have adopted GM crops worldwide are small-scale farmers (Sadashivappa and Qaim, 2009). This also applies to Burkina Faso, the only country in francophone Africa that decided to ignore France's advice and approve GM Bt cotton for commercial cultivation (Vitale et al., 2011). As in India, Bt cotton was a boon in Burkina Faso, especially for small-scale farmers. But these are obviously not the small-scale farmers the food sovereignty advocates want to hear about. They might argue that Bt cotton is not about food anyway. It is true that so far the only transgenic crops that have been approved were not meant for direct human consumption, with the exception of virus-resistant papaya in Hawaii, which has already been consumed in the United States for more than a decade. So why do we still have to wait for GM food crops that have a real value for poor food consumers? The case of vitamin A-enriched 'Golden Rice', developed by the Golden Rice consortium, has already spent 12 years since the first proof-of-concept trying to comply with national and international regulatory requirements (Potrykus, 2010). The costs so far amount to about US$25 million. Only the strong will of the researchers and the generous support of public and private institutions have sustained the project.

Preventive regulation in Europe is also discouraging plant scientists who would like to make use of their research to create viable commercial products. Plant biotechnologists at universities have become rather disinterested in going beyond proof-of-concept to develop useful products for the poor in cooperation with public and private institutions. This stands in strong contrast to the original purpose of the Cartagena Protocol on Biosafety, which has its foundation in Article 19 of the UN Convention on Biological Diversity (CBD). In Article 19, the purpose of the protocol was described as enabling the safe transfer of biotechnology. But the way it has ultimately been interpreted serves the opposite purpose: it prevents technology transfer (Juma, 2011b). Some are asking how it was possible to entirely reframe the purpose of the Biosafety Protocol. Some scholars argue that this was largely due to the lobbying, advocacy and pressure activities of NGOs. They successfully influenced the formation, interpretation and implementation of the protocol (Arts and Mack, 2007).

While the US government may have been in a position to do some capacity building globally on the risks and benefits of GM crops in view of its considerable experience with commercial cultivation over the past decade, instead it is starting to question its own relatively permissive regulation of GM crops. EPA largely responded to the activism of the growing Food Sovereignty movement in the United States which got a great boost due to the popularity of the books published by food writers such as Michael Pollan. This led in 2011 to a protest letter addressed to the US Environment Protection Agency (EPA) by the leading researchers in the field (Tribe, 2011). The researchers that wrote the letter were concerned that the anti-science attitude of the Bush administration was continuing in the Obama administration, but this time not against the recognition of man-made climate change but agricultural biotechnology. The Food Sovereignty advocates, however, have celebrated this as a victory against the corporate food regime in the United States.

Many Third World activists such as Vandana Shiva and her NGO Navdanya seem to have a more significant influence on European regulation of GM crops than any sort of empirical studies conducted by European scientists. Among others, she is credited for having changed the mind of the Germans about GM crops that eventually led to a ban of the only approved transgenic variety (MON 810) by the German government (Schäffer, 2009). Shiva knew that catering to Western anxieties and stereotypes about the 'corporate' regime and its victims in developing countries can be more rewarding in terms of media attention than fighting for the real local concerns as expressed by the hard-working poor Indian population, whose main wish is to earn a decent living and have better access to basic resources. Their bottom-up social movements usually demand land rights, protection from abuse, access to knowledge, and finance and technology: banning GMOs is *not* one of their priorities unless they get paid by Europeans to advocate for such a ban.

Shiva is not an exception but represents a new type of political entrepreneurship in developing countries. The model seems to involve first creating an NGO to advocate for genuine local concerns (for example land rights) but once in the limelight of the mass media, they become focused on their celebrity status in the West (or simply need funding from Western NGOs) and eventually abandon the local struggle in favour of fervent speeches before Western audiences, embracing a narrative that addresses Western concerns and lifestyles (Bob, 2005; Maxwell, 2003).

Looking at the history of the meaning of 'Food Sovereignty' a similar shift in the narrative can be observed – reflecting the need to cater to the political preferences of affluent donors. The original principles of Food Sovereignty as defined by La Via Campesina at the World Food Summit in Rome in 1996 still referred to the particular grievances of marginal farmers in developing countries (for example agrarian reform, social peace and political participation). But increasingly the meaning of food sovereignty was associated with multi-functional agriculture, sustainable food systems and community food security (CFS) in highly subsidized western countries (Patel, 2011). All kinds of intellectual acrobatics and conspiracy theories are then applied to explain why such a costly approach would also be worth adopting in developing countries. The authors themselves have never carried out empirical research on the problems farmers face in developing countries, but are mostly quoting the Food Sovereignty literature to underpin the validity of their arguments.

The belief that poor small-scale farmers in developing countries would share the lifestyle view of farmers in affluent countries ignores the fact that many of these poor countries must first address the productivity leap in agriculture. The big increase in productivity, thanks to technological change, enabled the United States and Europe in the nineteenth century to feed their growing population and facilitate the emergence of an empowered middle class that would create an inclusive and prosperous economy, and a vibrant democracy. Most economic development advisors believe this still needs to happen in Africa.

THE CHANGING NATURE OF PROTEST

NGOs have their roots in empowering and heterogeneous bottom-up movements in developed and developing countries that were largely concerned with concrete local issues of grievance. In the age of mass communication, there is no longer much space for local grievances that are too complex for ordinary people outside the context to understand. Therefore protest has largely shifted toward symbolic activism consisting of emotional tags and catchy slogans that are able to rally people together quickly with many different grievances (Furedi, 2002). Even though symbolic protest does not address the real problems, they help reduce complexity and provide people with meaning, identity and orientation (Aerni and Bernauer, 2006). Modern information and communication technologies further help to quickly mobilize protest using emotional imagery and fun entertainment to attract the masses even though they would have a hard time explaining what they are protesting against (Foer, 2000). Since GMOs have become a negative emotional tag that can be attached to almost any grievance about economic and technological change, it helps to mobilize people quickly for protest campaigns. The term serves as a proxy for all the things that people do not like about the modern system of food and agriculture.

The success of a protest is increasingly linked to indicators such as numbers of people that participated, entertainment value and media resonance. The content has become predictable because the same slogans are repeated again and again and the scapegoats for particular grievances are the usual suspects (the corporate regime). The value of such protest campaigns is primarily of a psychological nature. The dominating narrative provides meaning, identity and orientation through the reduction of complexity. There are, however, still many NGOs that undertake the role of watchdogs of society. Instead of focusing on providing psychological benefits they provide social benefits by identifying wrongdoing in government and industry through meticulous and time-intensive investigations. Yet it is unlikely that they are able to stop the trend towards the professionalization of protest with its focus on symbolic protest that provides a reduction of complexity and therefore psychological benefits. Symbolic protest is cheaper, more popular and also attracts the attention of the mass media. It therefore has several economic and political advantages and increases donations and ensures ongoing public legitimacy (Aerni and Bernauer, 2006).

There are a number of efforts to counteract this trend. In his presidential address to the American Economic Association in January 2007, George Akerlof highlighted the importance of norms in motivation. People have an ideal for how they should or should not behave. This ideal mostly reflects the norms and values of the community they grew up in. People tend to be happy when they live up to these norms and are correspondingly unhappy when they fail to achieve these norms. So those who are able to shape these norms and values are likely to gain public trust and symbolic power in politics (Bourdieu, 2001). In the agricultural debate, INGOs (including professional pressure groups) were able to shape the public view of sustainable agriculture and consequently induced many other influential stakeholders in business and politics to endorse their opposition to GMOs in agriculture as an expression of social values. In an increasingly affluent and urbanized society, farming comes to be seen as a freely-chosen lifestyle rather than a serious business to produce food (Aerni, 2009). Highly dependent on subsidies from urban taxpayers, farmers and retailers embrace the rustic stereotypes that urban people prefer. Instead of investing their resources in improving productivity in agriculture and enhancing the quality of food, they tend to spend more on political lobbying and marketing. They increasingly see the viability of their business as being dependent on taxpayers willing to pay their subsidies and on consumers willing to buy 'value-added' domestic products that mostly refer to certified labels that are meant to assure the consumers that the food they consume is safe from an ethically, environmentally and health-related perspective. Even though there is no evidence that GM crops cannot meet these requirements, GMO-free agriculture has become a condition for sustainable agriculture. As a consequence lobbying and marketing campaigns by farmers tend to focus on preserving the status quo and reject technological and economic change as a threat to their existence (Aerni, 2009). The consequence is self-censorship in affluent democracies. Researchers involved in plant biotechnology research at universities or employed by industry are not judged by their ability to develop sustainable solutions for crop protection but by their presumed motives that are felt to be opposed to the social values that shaped the public view of sustainable agriculture. Instead the public increasingly considers green politicians, environmental NGOs and hobby organic farmers to be experts and representatives of sustainable agriculture. Since they are trusted to have good motives, they shape the public debate on food as well as sustainable agriculture, without having to worry that someone could challenge their views or question their motives. As a consequence, advocates of the sustainable use of biotechnology are finding few supporters in business and politics. Most actors fear falling into disgrace with popular NGOs and professional pressure groups that are perceived to act in the public interest. The alliance against genetic engineering in agriculture has therefore steadily grown beyond the realm of NGOs. In Europe it now includes political bodies from all levels of governance and even businesses in the food chain that are close to urban consumers (including food writers). With the exception of the UK Government report on the future of farming,[4] only agribusiness companies that invested in biotechnology and research institutes concerned with the genetic improvement of crops still dare to defend the value of GMOs in public.

Ultimately, the power of these anti-biotech alliances and broad antagonism to GMOs induced many European governments to hand over more competence to regulate GMOs to environmental ministries to give the impression that they care about sustainable agriculture (Paarlberg, 2008). Since national environmental ministries and environmental advocacy groups are also the ones involved in the design of Multilateral Environmental Agreements (MEA), it is not surprising that most of the recent MEAs do not link biotechnology to sustainable development, even though biotechnology was largely portrayed as an environmentally sound technology in Agenda 21 and the Rio 1992 Declaration (Aerni, 2011b). Yet all the self-censorship does not eliminate the fact that environmental policies that aim at preventing rather than facilitating technological change proved to be rather ineffective in addressing the global sustainability challenges (Nordhaus and Shellenberger, 2007).

Even firm advocates of the use of biotechnology would caution that the technology is far from being a golden bullet. Instead they argue that it can only succeed if embedded in an integrated approach to farming (Garnett 2011; Juma 2011a). Yet the global debate has become so polarized that opponents have embraced an ethics of purity linking the use of genetic engineering to the contamination of nature (as if all other agricultural techniques would be a product of nature rather than human culture). This ethics of purity among the more radical opponents has also led to alliances with xenophobic and nationalist right-wing movements which also insist on racial purity and the protection of national heritage (Nanda, 2003). On the other hand, companies that have greatly invested in the use of modern agricultural biotechnology have largely withdrawn from the public debate, while other private sector players that are largely concerned about their public image are actively boasting about their 'GMO-free' products and try to link up with anti-biotech NGOs.

All these trends highlight the ambiguous role of NGOs in the global debate on the risks and benefits of agricultural biotechnology. They used to have an important watchdog function, when agricultural biotechnology was still in an early stage, uncertainty about its potential risks was still great and arrogance and aggressive marketing and lobbying by big agribusiness firms was fuelling significant distrust. Yet once industry started to respond to their demands and became more transparent and open to collaboration with civil society, there was no effort made to recognize the positive change and acknowledge the potential benefits of tailoring the technology to particular needs among small-scale farmers in developing countries. Industry became a scapegoat and even government institutions and academic research institutes that wanted to or had to collaborate with industry became objects of distrust. As a consequence, constructive dialogue with industry has become almost impossible. Big companies like Monsanto largely withdrew from the public debate and have even abandoned some of the efforts to transfer its technology to developing countries via public–private partnerships because it was interpreted in the public debate as a mere attempt to increase acceptance of biotechnology in developing countries. In other words, NGOs may have been a force of good in triggering an early awareness for the potential risks of abusing a powerful new platform technology but they have not played a very constructive role when it comes to outlining a clear agenda for how to employ the technology for the benefit of mankind as a whole. In the course of time, their influence has grown enormously and now INGOs increasingly shape the agenda of environmental, development and agricultural research at universities and have a major influence on related public policy. The resulting self-censorship in academia and government could eventually become a problem for academic freedom and efforts to find integrated, effective and sustainable solutions for the global sustainability challenges in future.

CONCLUSIONS

Environmentalism in the 1970s was part of the heterogeneous social movement in Europe and the United States that primarily protested against the paternalism of the state and its support for an opaque and unaccountable military–industrial complex. Environmental problems, the Vietnam War, the nuclear arms race as well as the indifference toward public objections to the construction of nuclear plants were all considered to be part of a larger problem of lack of government accountability. Environmentalists were, however, not able to agree how to address sustainability problems effectively. One faction believed that technological and economic change is necessary to make the world more sustainable. Many young people who shared this view eventually became entrepreneurs and set up university spin-off firms in the field of information technology, solar technology and biotechnology to develop new products and services that more efficiently use renewable resources. Those concerned with the environmental problems in agriculture were enthusiastic about the potential of agricultural biotechnology (including the environmentalist Rachel Carson who endorsed microbial pesticides and the sterilization of insects) because they were convinced that this technology will help to make modern agriculture less energy intensive and less dependent on polluting pesticides and fungicides.

A significant faction of the environmental movement believed, however, that new technologies are not part of the solution but part of the problem of sustainable development. Activists in this group were also concerned about the potentially irreversible effects of agricultural biotechnology on biological diversity, especially when releasing genetically modified organisms into the environment. Most of these conservationists eventually joined one of the numerous environmental non-government organizations that were founded at that time to fight environmental pollution. Their protest campaigns greatly contributed to more public awareness of the environmental problems of industrialized countries in general and industrial agriculture in particular. The governments in Europe and North America responded to the public pressure created by these NGOs by passing more strict environmental regulation and by investing more in alternative clean technologies that could substitute for old polluting technology. Despite the techno-critical attitude in this early stage, there was still a large consensus about the need to search for effective ways to cope with the environmental challenges. This search was based on a system of trial and error. Yet, over time, many environmental NGOs morphed into media-savvy international pressure groups that largely opposed the trial and error approach (which might be risky by itself) and focused on symbolic protest against the capitalist system that they viewed as the root of environmental problems. They started to insist that environmentalism is primarily about banning, controlling and regulating activities in industry. Biotechnology, which used to be considered a clean technology that could facilitate sustainable change in agriculture, was suddenly considered to be inappropriate and linked to the dangers of nuclear technology, even though no evidence has emerged so far that genetic engineering in agriculture is inherently more risky than conventional agriculture. Resistance against agricultural biotechnology by environmental NGOs led many governments to regulate the technology separate from other crop breeding techniques and made the approval process of GM crops very time intensive and costly. As a result, concentration in industry increased substantially because small companies no longer had the means to comply with the regulatory barriers that were raised in the name of the precautionary principle. The precautionary principle, which was never meant to prevent technological change but to make it more safe, became a policy instrument that allowed governments to impose de facto bans on agricultural biotechnology without the need to justify it by means of scientific evidence. The precautionary strategy in the face of political uncertainty was one of 'better safe than sorry'. Yet despite all the strict regulation of agricultural biotechnology, trust in safe use of GMOs in affluent countries decreased. People thought that there must indeed be something wrong with this technology if it has to be regulated to such an extent.

The situation in developing countries was quite different. Preserving the status quo was not an option – people wanted to participate in business and industry and tended to endorse technological change as a source of new economic opportunities and a way to better cope with existing social and environmental risks. NGOs in developing countries were also influenced by the ideas of social empowerment as they manifested themselves in the social movement of the 1970s in Europe and the United States. Yet their priorities were different. They were less focused on environmental issues but on freedom from political repression. Local NGOs in these countries eventually became part of underground resistance movements against authoritarian national regimes that were supported either by the West or the Soviet Union. Once many of the regimes were toppled at the end of the Cold War, international donor agencies encouraged and supported NGOs to participate in the young democracies as representatives of civil society. As a consequence, the number of NGOs in developing countries increased exponentially and with it their dependence on foreign grants and policy agendas. Environmental NGOs and particularly the internationally-minded environmental pressure groups in affluent countries were eager to support these NGOs in developing countries and, at the same time, lobby with their national donor agencies to set priorities in accordance with what they considered to be sustainable development.

These international environmental pressure groups were consequently not just shaping the advocacy agendas of NGOs in developing countries but also influencing donor priorities and research priorities in the field of development and environmental studies back home. They became increasingly opposed to new technologies and international trade in agriculture and started to advocate for food sovereignty, which was largely defined as support for local, GMO-free agriculture. As a result, their views about sustainable agriculture became defensive rather than progressive. Even though the concept of food sovereignty is not new and previous efforts to design national agricultural policies in accordance with the principles of food sovereignty largely failed, it nevertheless gained in popularity in civil society, business and politics. The increasing ability of international environmental pressure groups to set the policy agenda of sustainable development through documentary movies, social media activities, well-coordinated protest campaigns at gatherings of international decision makers and regular symbolic protest events influenced both the media and the environment and development curriculum in the national education system. Teachers started to frame technological change as a sustainability problem, which then influenced the sustainability agenda of business and government institutions that primarily adopt policies that are in line with public views. These developments have essentially led to self-censorship in academia, business and politics when it comes to the role of agricultural biotechnology in addressing the global problems of climate change, hunger and malnutrition. This is especially true for highly exposed academics in the field of environmental sciences that feel the pressure of remaining popular within their own like-minded constituency, as well as the important donors in the field of environmental research. This pressure to be politically correct often goes at the expense of the quality of science and ultimately the public interest.

What does the foregoing mean for the future of agricultural biotechnology? Currently, the popularity of international environmental pressure groups and their anti-biotechnology activities continues to be high in developed and developing countries alike. At the same time, the pressure to act in the face of increasing global sustainability problems increases every year; there is a general acceptance in many quarters that the sustainable intensification in agriculture needed to feed a growing world population cannot be achieved without the use of biotechnology. The most promising development would be a change of mind among a moderate wing within the powerful environmental pressure groups that then might join forces with public plant science researchers. This coalition would need to question the baseline assumption of the current view of global sustainability and reframe the dialogue about biotechnology – ultimately, biotechnology needs to be seen as a solution to the growing sustainability problems in our global economy and society. In the face of the growing pressure to go beyond the ineffective and well-tried defensive, conservation policies, the public might become more open to new approaches. Once there is a change in public opinion there will also be a change of mind in business and politics.

NOTE

1. Red biotechnology is applied to medical processes. Green biotechnology is biotechnology applied to agricultural processes.
2. See http://ccafs.cgiar.org/events/06/nov/2011/international-conference-climate-change-and-food-security-icccfs.
3. IAASTD (2009), www.globalarchitecture.org.
4. http://www.bis.gov.uk/assets/foresight/docs/food-and-farming/11–546-future-of-food-and-farming-report.pdf.

REFERENCES

Aerni, P. (1999), 'Public acceptance of transgenic rice and its potential impact on rice markets in Southeast Asian countries', PhD dissertation, Swiss Federal Institute of Technology, ETH Zurich, Switzerland.
Aerni, P. (2006), 'Mobilizing science and technology for development: the case of the Cassava Biotechnology Network (CBN)', *AgBioForum*, **9**(1), 1–14.
Aerni, P. (2009), 'What is sustainable agriculture? Empirical evidence of diverging views in Switzerland and New Zealand', *Ecological Economics*, **68**(6), 1872–82.
Aerni, P. (2011a), 'Food sovereignty and its discontents', *ATDF Journal*, **8**(1/2), 23–40.
Aerni, P. (2011b), 'Lock-in situations in the global debates on climate change, biotechnology and international trade: evidence from a global stakeholder survey', NCCR Trade Working Paper no. 2011/21, World Trade Institute, University of Bern.
Aerni, P. and T. Bernauer (2006), 'Stakeholder attitudes towards GMOs in the Philippines, Mexico and South Africa: the issue of public trust', *World Development*, **34**(3), 557–75.
Aerni, P. and F. Oser (eds) (2011), *Forschung Verändert Schule: Erkenntnisse aus den Empirischen Wissenschaften für Didaktik, Erziehung und Politik*, Zurich: Seismo Verlag.
Aerni, P. and P. Rieder (2001), 'Public policy responses to biotechnology', Chapter BG6.58.9.2 in *Our Fragile World: Challenges and Opportunities for Sustainable Development*, Encyclopaedia of Life Support Systems (ELOSS), Paris: UNESCO.
Alegre, A.G. (1996), *Trends and Traditions, Challenges and Choices. A Strategic Study of Philippine NGOs*, Manila: Ateneo Center for Social Policy and Public Affairs.
Arts, B. and S. Mack (2007), 'NGO strategies and influence in the biosafety arena, 1992–2005', in R. Falkner (ed.), *The International Politics of Genetically Modified Food*, New York: Palgrave Macmillan.
Beckerman, W. (2002), *A Poverty of Reason: Sustainable Development and Economic Growth*, Oakland, CA: The Independent Institute.
Bob, C. (2005), *The Marketing of Rebellion: Insurgents, Media, and International Activism*, Cambridge: Cambridge University Press.
Bourdieu, P. (2001), *Language and Symbolic Power*, Cambridge, MA: Harvard University Press.
Bradford, T. (2006), *Solar Revolution*, Cambridge, MA: The MIT Press.
Byerlee, D. and M. Morris (1993), 'Research for marginal environments: are we underinvested?', *Food Policy*, **18**(5), 381–94.
Carson, R. (1962), *Silent Spring*, New York: Houghton Mifflin.
Castells, M. (2004), *The Power of Identity*, 2nd edn, Malden, MA: Blackwell Publishing.

Converse, P.E. (1964), 'The nature of belief systems in mass publics', in D. Apter (ed.), *Ideology and Discontent*, New York: Free Press.

Crano, W.D. (2001), 'Social influence, social identity, and ingroup leniency', in C.K.W. de Dreu and N.K. de-Vries (eds), *Group Consensus and Minority Influence: Implications for Innovation*, Oxford: Blackwell, pp. 122–43.

Crano, W.D. and R. Prislin (2006), 'Attitudes and persuasion', *Annual Review of Psychology*, **57**, 345–74.

Dine, J. (2005), *Companies, International Trade and Human Rights*, New York: Cambridge University Press.

Ehrlich, P. (1970), *Population Bomb*, New York: Ballantine Books.

Energy Information Agency (EIA) (1998), '25th anniversary of the 1973 oil embargo', available at: http://www.eia.gov/emeu/25opec/anniversary.html.

European Commission (EC) (2010), *A Decade of EU-funded GMO Research*, EUR 24473 EN, Brussels: European Commission.

Falkner, R. (2007), *The International Politics of Genetically Modified Foods*, New York: Palgrave Macmillan.

Ferroni, M. and P. Castle (2011), 'Public–private partnerships and sustainable agricultural development', *Sustainability*, **3**(7), 1064–73.

Foer, F. (2000), 'Meet the new new Left: bold, fun, and stupid. Protest too much', *The New Republic*, May.

Friedmann, H. and P. McMichael (1989), 'Agriculture and the state system: the rise and decline of national agricultures, 1870 to the present', *Sociologia Ruralis*, **29**(2), 93–117.

Furedi, F. (2002), *Culture of Fear: Risk Taking and the Morality of Low Expectation*, London: Continuum International Publishing Group.

Garnett, T. (2011), 'Where are the best opportunities for reducing greenhouse gas emissions in the food system (including the food chain)?', *Food Policy*, **36**, 23–32.

Gruère, G.P., P. Mehta-Bhatt and D. Sengupta (2008), 'Bt cotton and farmer suicides in India: reviewing the evidence', IFPRI Discussion Paper 00808, Washington DC, available at: http://www.ifpri.org/sites/default/files/publications/ifpridp00808.pdf.

Henderson, R., L. Orsenigo and G. Pisano (1999), 'The pharmaceutical industry and the revolution in molecular biology: interactions among scientific, institutional, and organizational change', in D.C. Mowery and R.R. Nelson (eds), *Sources of Industrial Leadership*, New York: Cambridge University Press.

Herring, R. (2009), 'Persistent narratives: why is the "Failure of Bt cotton in India" story still with us?', *AgBioForum*, **12**(1), 14–22.

Hughes, S.S. (2011), *Genentech: The Beginnings of Biotech*, Chicago, IL: University of Chicago Press.

International Assessment of Agricultural Knowledge, Science and Technology for Development (IAASTD) (2009), *Agriculture at a Crossroads*, The Global Report, Washington, DC: Island Press.

Juma, C. (2011a), *The New Harvest: Agricultural Innovation in Africa*, Cambridge, MA: Harvard University Press.

Juma, C. (2011b), 'Science meets farming in Africa', *Science*, 334, 1323.

Kaldor, M. (2002), 'Civil society and accountability', background paper for the *UN Human Development Report*, New York: UNDP.

Keane, J. (2003), *Global Civil Society?*, Cambridge: Cambridge University Press.

Kingsbury, N. (2009), *The History and Science of Plant Breeding*, Chicago, IL: University of Chicago Press.

Kouser, S. and M. Qaim (2011), 'Impact of Bt cotton on pesticide poisoning in smallholder agriculture: a panel data analysis', *Ecological Economics*, **70**(11), 2105–13.

Lovins, A.B. (1976), 'Energy strategy: the road not taken?', *Foreign Affairs*, **55**(1), 65–96.

Luhmann, N. (1991), *Soziologie des Risikos*, Berlin: Walter de Gruyter.

Martineau, B. (2001), *First Fruit: The Creation of the Flavr Savr Tomato and the Birth of Biotech Foods*, New York: McGraw-Hill.

Maxwell, K. (2003), *Naked Tropics: Essays on Empire and Other Rogues*, New York: Routledge, pp. 219–42.

Meadows, D.H., D.L. Meadows, J. Randers and W.W. Behrens (1972), *The Limits to Growth*, New York: Universe Books.

Meera, N. (2005), *The Wrongs of the Religious Right: Reflections on Science, Secularism and Hindutva*, Gurgaon, Haryana: Three Essays Collective.

Mittelman, J.H. (2000), The Globalization Syndrome: Transformation and Resistance, Princeton, NJ: Princeton University Press.

Nanda, M. (2003), *Prophets Facing Backward: Postmodern Critiques of Science and Hindu Nationalism in India*, New Brunswick, NJ: Rutgers University Press.

Nelkin, D. (1996), *Selling Science*, revised edn, New York: W.H. Freeman.

Nordhaus, T. and M. Shellenberger (2007), 'A manifesto for a new environmentalism', *The New Republic*, **24**, September 30–33.

Otis, L. and J.M. Graham (2000), *Environmental Politics and Policy 1960s to 1990s*, University Park, PA: Pennsylvania State University Press.

Ottway, H.J. and M. Peltu (1985), *Regulating Industrial Risks*, Waltham, MA: Butterworth-Heinemann.

Paarlberg, R. (2008), *Starved for Science: How Biotechnology is Being Kept Out of Africa*, Cambridge, MA: Harvard University Press.
Patel, R. (2011), 'What does food sovereignty look like?', in H. Wittman, A. Desmarais and N. Wiebe (eds), *Food Sovereignty: Reconnecting Food, Nature and Community*, Oakland, CA: Food First Books.
Peltu, M. (1985), 'The role of communications media', in H. Ottway and M. Peltu (eds), *Regulating Industrial Risks*, Cambridge: Cambridge University Press, pp. 128–47.
Pingali, P. and P.A. Roger (1995), *The Impact of Pesticides on Farmer Health and the Rice Environment*, Norwell, MA: Kluwer Academic Publishers.
Potrykus, I. (2010), 'Regulation must be revolutionized', *Nature*, **466**, 561.
Radkau, J. (1995), 'Learning from Chernobyl for the fight against genetics?', in M. Bauer (ed.), *Resistance to New Technology Nuclear Power, Information Technology, Biotechnology*, Cambridge: Cambridge University Press.
Rocamora, J. (1994), *Breaking Through. The Struggle within the Communist Party of the Philippines*, Manila: Anvil Publishing.
Sadashivappa, P. and M. Qaim (2009), 'Bt Cotton in India: development of benefits and the role of government seed price interventions', *AgBioForum*, **12**(2), 172–83.
Schäffer, A. (2009), 'Gentechnik und CSU: ein Wunder der politischen Logopädie', *Frankfurter Allgemeine (FAZ)*, 3. May, available at: http://www.faz.net/artikel/C30923/gentechnik-und-die-csu-ein-wunder-der-politischen-logopaedie-30123883.html.
Schreurs, M.A., S.D. VanDeveer and H. Selin (eds) (2009), *Transatlantic Energy and Environmental Politics: Comparative and International Perspectives*, Aldershot: Ashgate Press.
Schuurman, F.J. (1992), *Social Movements and NGOs in Latin America*, Nijmegen Studies in Development and Cultural Change, Saarbrücken: Verlag Breitenbach.
Sennett, R. (1976), *The Fall of Public Man*, New York: W.W. Norton.
Slovic, P. (1992), 'Perception of risk: reflection on the psychometric paradigm', in D. Golding and S. Krimski (eds), *Social Theories of Risk*, New York: Praeger.
Sunstein, C. (2007), *Republic 2.0*, Princeton, NJ: Princeton University Press.
Tribe, D. (2011), 'Fedoroff Letter to EPA raises serious concerns over EPA blundering' *Biology Fortified*, Inc., 28 September, available at: http://www.biofortified.org/2011/09/fedoroff-letter-to-epa-raises-serious-concerns-over-epa-blundering/.
Vitale, J., M. Ouattarra and G. Vognan (2011), 'Enhancing sustainability of cotton production systems in West Africa: a summary of empirical evidence from Burkina Faso', *Sustainability*, **3**(8), 1136–69.
Wahl, P. (1998), 'NGO-multis, McGreenpeace und Netzwerk-Guerilla', *Peripherie*, **7**, 55–68.
Willets, P. (ed.) (1982), *Pressure Groups in the Global System*, London: Frances Pinter.

PART II

ACTION ARENAS

20 The research pipeline

Peter W.B. Phillips

1 INTRODUCTION

For the first time in a long while, agriculture has been elevated to the top of the agenda of world leaders, with strong, affirmative positions taken by leaders meeting at the G8, G20, UN, OECD and other multilateral forums in 2008–12. Agricultural development offers prospects to address a number of pressing global issues: new technology should enhance food security for a projected world population of 9 billion in 2050; new applications could accelerate economic and social development for an estimated half of the world's population currently employed in agri-food production; new technologies and new farming methods offer opportunities to improve the global environment, reduce soil degradation, improve local water quality and accelerate carbon sequestration to offset climate change; and new crops could trigger industrial development, with prospects ranging from biofuels, to bioproducts (for example replacements for plastics or as artificial fibres) and proteins and enzymes for health and industrial development.

To realize these goals, farmers, the agri-food sector, policy makers, regulators and the financial community need a better sense of what technologies and opportunities might emerge, where they may be first introduced and when this might happen in order to have the requisite policy and infrastructure in place to facilitate their uptake and use. This leads to strong demand for projections or forecasts of what the future might look like. The research pipeline is the focal point for much of this effort. While the pipeline is not formally defined, one should generally consider it to include the upstream basic and applied scientific research, the application of that science to develop new cultivars and complementary technologies (for example chemicals, fertilizers, inoculants and related machinery) and the capacity of the supply chain to accommodate new technologies and products.

While we have been forecasting harvests since the beginning of recorded history, our methods and results remain suspect. This chapter examines the array of quantitative and qualitative methods developed, adapted and adopted to the task of anticipating the future (section 2) and then assesses a range of recent forecasts and projections (section 3). The overarching lesson from this survey is that while methods are being enhanced, outputs remain contingent and partial at best. The high degree of uncertainty in the forecasts poses particular problems for investors and policy makers as they contemplate allocating resources to research, development, regulatory or commercial systems: there are no clear signals to direct attention and command the resources that will undoubtedly be needed to realize the promise of biotechnology in the global agri-food system.

2 METHODS

Kings, individual farmers, merchants and citizens have always wanted to know more about the future, particularly the prospects for crop production. In the prehistoric and ancient worlds seers, sibyls or shaman were presumed to have access to the spirits. Cicero (Div. 2.130) asserted that divination involves 'the power to see, understand, and explain premonitory signs given to men by the gods'. These special people entered trances, threw bones and consulted the spirits through a wide range of practices, ultimately offering advice to supplicants. Some of these practices were entirely mystical while some had a scientific underpinning. In many societies high-priests acquired knowledge of the movement of the sun and stars and erected monuments to track the spring and fall equinox, couching those annual events into rituals that would then trigger seeding and the harvest.

In some ways, our modern crop forecasting systems are natural extensions of this history. Farmers, commodity traders and policy makers continue to seek guidance on how the current or next crop will do, blending a mix of science – for example weather forecasting, climatology and *in situ* analysis of explicit factors of production – with a mix of mysticism (for example the Farmers' Almanac) and sibylesque prognostication based on what are often highly judgemental black-box forecasting models or systems. Most of this prognostication was focused on the forthcoming season, but in some cases more ambitious advisors asserted they had found cycles, such as the biblical seven-year cycle, which then led to new practice, such as fallowing fields.

The distinction between the prehistorical or ancient forecasters and modern-day efforts is that the earlier practitioners were focused on stochastic change in a static system (that is one without any technological change) while modern forecasters are beginning to formally include dynamic effects of technological change. Endogenous growth theorist Paul Romer (1994) likened the earlier effort to weather forecasting, where the best we can do is rearrange our limited resources to minimize the negative effects, while the modern effort is more like climatology, where we can fundamentally change the operating parameters and outcomes. While short-term forecasting has particular value for rulers and traders, longer-term forecasts offer more fundamental opportunities to realize prosperity. Depending on one's perspective, the long-term could be 3–5 years ahead or could span more than a generation (25–30 years). The short to medium-term outlook can be of particular strategic and tactical importance for farmers and firms, as getting it right can offer enhanced profit. Governments and policy makers are also particularly concerned about the 3–5-year window, as that is the period when they must take action – but at best any responses will simply ameliorate the effects of the underlying forecast. In contrast, the long-term, intergenerational view is vitally important for public policy, as it offers the greatest opportunity for governments and their partners to fundamentally solve problems. Recent work by Alston et al. (2010), as discussed in Chapter 3 of this volume, has revealed that the maximum impact of every dollar invested in agriculture occurs 24 years into the future, which means there is a strong link between current-day decisions and long-term prospects.

While the earlier practices provided a foundation for modern-day forecasting, we have added a range of theoretical and methodological enhancements. We now use both objective, inductive, observation-based approaches and a range of artificial or theoretically structured systems to extend our capacity. Bernstein (1996) posits that the most important development in forecasting was the transformation of fate into probabilistic risk. This transition occurred over a half millennium. The introduction of the Indian number system (supplemented with zero, negative numbers, algebra and statistical inference) transformed our ability to represent the here-and-now and to anticipate potential futures in a formal, structured way. When the scientific revolution opened our eyes to the linkages between inputs and outputs, we were able to characterize them through greater mathematical precision.

Thus, we now have a range of methods of forecasting, some based on intuition, dreams and revelations (often called expert judgement) and others based on formally structured, analytical precepts. Many forecasting efforts involve a mixture of these methods, developing probabilistically-based extrapolations or forecasts, which are both based on and often modified by expert or practitioner intuition. While econometrically-derived forecasts have a high degree of quantitative structure, they all involve extensive operator judgement and intuition – first in the identification of the key dependent variables, second in the articulation of the causal pathways, third through the specifications chosen to represent the relationships and finally in the calibrating of the models to deliver stable results. All of these involve extensive qualitative judgement, which is sometimes complemented and justified with Delphi-like surveys of competitive forecasts done by other organizations. Explicit methods range from qualitative surveys, focus groups or Delphi experiments to elicit expert and practitioner opinions to a set of successively more sophisticated quantitative approaches, including random-walk models, time-series analysis and causal, multi-sectoral, behaviourally-based econometric forecasting models.

Popper (2008) undertook a study of the commonly adopted methods of technology forecasting and foresighting used in the context of more than 886 studies. He identified 33 key methods used to delimit and structure the analysis of the future, 15 which he characterized as qualitative, six as semi-qualitative, three as uniquely quantitative and the rest as blended quantitative and qualitative (Table 20.1). He then assessed them based on their ability to gather or process information based on four factors: evidence, expertise, interaction or creativity, and plotted them in what he called a 'foresight diamond'. He asserted that all processes ultimately involved all four attributes: they simply varied in the balance. For example, he suggested that generic expert panels would probably be based 10 per cent on evidence, 70 per cent on expertise, 10 per cent on interaction and 10 per cent on creative aspects. In contrast, citizen's' panels or juries would put a greater weight on interaction. Generally, evidence and creativity are traded off while interaction and expertise offset each other. The evidence-based spectrum includes mostly quantitative and blended methods, such as economic modelling, extrapolation, bibliometrics, literature reviews, benchmarking, patent analysis and structural analysis. The expert domain involves such methods as expert panels, roadmapping, Delphi, relevance trees/logic diagrams, morphological analysis, quantitative scenarios and interviews. Interactive methods include largely qualitative methods, such as brainstorming, workshops, conferences, panels, voting and polling. The creative domain includes such methods as wild cards, science fictioning, simulation gaming, essays, scenarios, acting/role playing and genius forecasting. At the core of the diamond are SWOT analyses, surveys and multiple methods.

This data can be assessed in a number of ways. In the first instance, the data in Table 20.1 shows that these methods are quite unevenly used. The average number of times a method was used was 133, with a standard deviation of 127.5, which means that

Table 20.1 Forecasting and foresighting methods commonly in use in technology assessment

Method	Style*	Evidence (+) Creativity (−)	Interaction (+) Expertise (−)	Number of uses in 886 studies
Literature review	Qual	7	−1	477
Expert panels	Qual	0	−8	440
Scenarios	Qual	5	−6	372
Trend extrapolation/megatrends	Quant	6	−2	223
Futures workshop	Qual	−2	5	216
Brainstorming	Qual	−3	5	198
Other methods	Blend	0	0	157
Interviews	Qual	3	−5	154
Delphi	S.Qual	−1	−2	137
Key technologies	S.Qual	1	−7	133
Questionnaires/surveys	Qual	−1	0	133
Environmental scanning	Qual	6	3	124
Essays	Qual	−5	−1	109
SWOT analyses	Qual	−3	0	101
Technology road mapping	S.Qual	−1	−6	72
Modelling and simulation	Quant	8	0	67
Backcasting	Qual	−3	−5	47
Stakeholder mapping	S.Qual	2	4	46
Cross-impact/sectorial analysis	S.Qual	3	0	36
Bibliometrics	Blend	5	−3	22
Morphological analysis	Qual	0	−4	21
Citizen panels	Qual	−1	8	19
Relevance trees	Qual	−2	−7	17
Multi-criteria analysis	S.Qual	1	0	11
Gaming	Qual	−6	−1	6
Pearson's r correlation coefficient between indicator and number of times used (not significant at 95% confidence, d.f. = 23)		0.3169	−0.1936	–

Note: * Qual = qualitative; S.Qual = semi-qualitative; Quant = quantitative; Blend = mixed methods.

Source: Author's calculations using data from Popper (2008).

we are only statistically confident (at the 90 per cent level) that literature reviews, expert panels and scenarios are used more than others and only that gaming is statistically used less. Similarly, we can calculate the Pearson's r correlation coefficient to determine whether there is a statistically significant bias in use by factor. There appears to be a weakly significant, moderately positive bias towards using evidence-based methods. The marginal negative correlation on the engagement scale suggests there is a minor preference for lay input, but it is not robustly significant. Figure 20.1 shows the results in Table 20.1 graphically, suggesting there is a slight preference to pair creative processes with experts and evidence-based efforts with more interactive and open processes. While

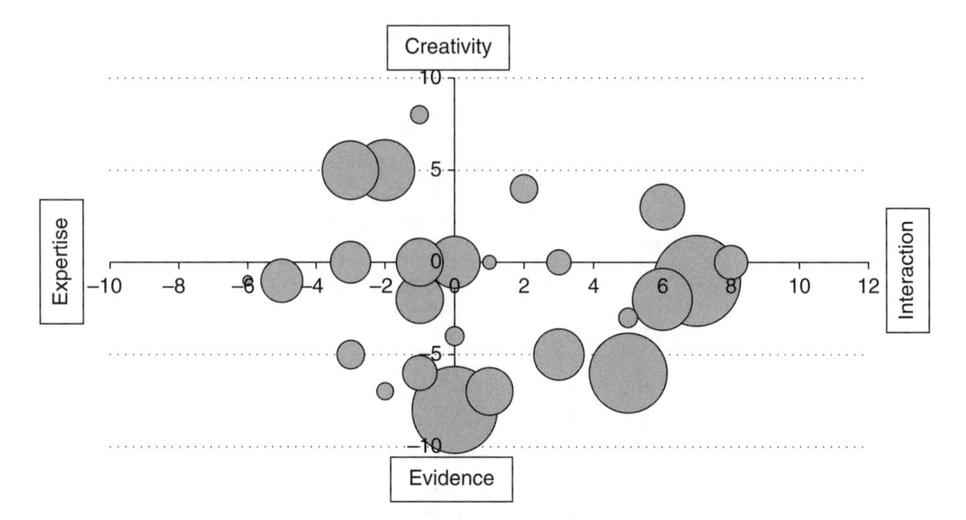

Source: Author's calculations based on data from Popper (2008).

Figure 20.1 Relative popularity of different forecasting and foresighting methods

the relationship is statistically insignificant, one can intuitively see the value in those pairings, as evidence-based elements might ground interactive processes, tempering the flights of fancy of lay persons, while creative processes might challenge the group-think of experts.

Another way to look at the methods is in terms of their purpose. All prognostication efforts have a client, formal or implied. In some cases the efforts are focused on a positivist statement of what is (or could be without any individual or policy responses). At other times forecasts are constructed to frame discussions about how to change the future. These normative efforts are generally highly purposeful, sometimes to the point of ignoring the evidence in favour of more extreme conclusions. An alternative way to characterize the literature would be as either extrapolation or foresighting. Extrapolation tends to be highly grounded and often bounded by the evidence from the present or recent past. Most technology modelling and trend analysis fundamentally depends on what is known or knowable, often relying on a range of artefacts in the literature (for example bibliometric citations and records on research grants), intellectual property system (for example patent filings or trademark registrations), regulatory system (for example applications for confined field trials) or markets (for example sales, revenues or acres seeded). The limitation of this approach is that the contemporary evidence can be incomplete and developments frequently are non-linear. Meanwhile, more qualitative foresighting often engages a range of actors in efforts to set aside the current evidence and hypothesize and speculate about what might occur well beyond the horizon for extrapolation. Given these two dichotomies, one can imagine the literature could be parsed into four quadrants: positivist extrapolations (for example Graff et al., 2009); positivist foresighting (for example Dunwell, 2010), normative extrapolation (for example Stein and Rodríguez-Cerezo, 2010) and normative foresighting (for example WEF, 2012) (see Table 20.2).

Table 20.2 Mapping the forecasting and foresighting literature

		Method	
		Extrapolation	Foresight
Purpose	Positive	Graff, Zilberman and Bennett (2009)	Dunwell (2010)
	Normative	Stein and Rodriquez-Cerezo (2009)	WEF (2102)

3 INSIGHTS

For the purposes of this analysis, the insights from the literature are divided into three phases, consistent with Alston et al. (1995), where there is a set of inputs, processes and outputs from research (for example funded by governments, business and philanthropic foundations and undertaken in universities, public labs and some private companies), there are a range of gestational activities (for example proof of concept, reduction to practice, scale up, regulatory compliance and business planning) and the supply chain and consumers in multiple markets adapt, adopt and use the technology. This framing extends both upstream and downstream to encompass what many assert is the research pipeline, or what Graff et al. (2010) call the 'research sieve'. They converged on this term as more descriptive of the triage process that takes place within the gestational stage. In their report they identified 560 biotechnology traits at the 'proof of concept' stage which led to 383 early-stage field trials, but only 47 advanced trials, 14 regulatory applications, 5 market introductions and only two sustained products.

3.1 Research Phase

The research phase is usually judged to be the most speculative and uncertain. While most of the transformative changes in technology that cause step adjustments in productivity and quality in the agri-food system come from the research phase, there is ample evidence that most discoveries and inventions from research are never commercialized and, more daunting for forecasting purposes, many fundamental technological transformations involve little or no basic scientific research or discovery – they are often simply re-combinations of existing technology that generate new opportunities.

Nevertheless, when one is looking beyond the next 3–5 years, it is absolutely essential to have a sense of where the technology is going, where the greatest needs or opportunities are, where the most important science and technology is being developed and who might have an interest in shepherding any inventions or innovations to the market.

Analysis focused on the research phase tends to look for artefacts of the research system and engage experts in processes to elicit their judgements about where and when new developments might emerge. This analysis is complicated as there are two somewhat discrete key sectors: the public domain, populated largely by universities and public labs, and private firms, dominated by a handful of multinational corporations, and a large number of small and often informal contributors, including entrepreneurial firms and individual farmers. James (2003) summed up the biotechnology research effort in 2001, concluding that in that year there was about $4.4 billion invested, 70 per cent by private firms and 30 per cent in the public domain. Pardey et al. (2006) reported that global agri-food R&D (pre-, on- and post-farm effort, including biotechnology-related R&D), totalled US$36.5 billion in 2000. About 37 per cent of that was conducted by private firms, virtually all of it in developed countries. While this data gives us insight into part of the sector, it offers little information on what the informal actors are doing.

Disproportionately, the public sector invests in upstream research and discovery work. The key artefacts of that system include databases of competitive research grants, research outputs such as journal articles, books or conference papers and patents or invention disclosures. There has been extensive effort in recent years to mine the key national and international grant databases, including the US NIH and USDA, the European Union Framework Programs, Genome Canada and the Canadian scholarly granting councils, the Australian Research Council, the commodity-based research check-off organizations (such as the Grains Research Development Corp (GRDC) in Australia and the Western Grains Research Foundation (WGRF) and Saskatchewan Pulse Growers (SPG) in Canada), the philanthropic foundations (for example Bill and Melinda Gates, Rockefeller and Ford Foundations) and the efforts in the Consultative Group of International Agricultural Research centres. As for research outputs, the key artefacts are conference papers, journal articles, book chapters or books, which initially were tracked and could be analysed by the Institute for Scientific Investigation (ISI) but now can be tracked through the Google Scholar search engine. Ultimately, technologies and applications that are candidates for commercialization are frequently subject to invention disclosure in most public and private universities and then are often codified and protected through patents. So one could look in the national patent databases, in individual institutional records and the Association for University Technology Managers (AUTM) reports to identify where inventions are emerging. While there is no hard and fast relationship, in the life sciences there tends to be a rule of thumb that research artefacts solidify research intentions, which lead to discoveries, inventions and innovations with a variable lag (from months to decades). Graff et al. (2009) estimated that the lag between proof of concept and publication was about 0.6 years in 1987–97, dropping to 0.2 years in 1998–2004.

A number of investigators have worked with this and other data to identify trends in research that may influence the agri-food sector. Phillips and Khachatourians (2001) made one of the earlier attempts, assembling the 1945–99 data for rapeseed and canola to track the evolution of the sector. This data was then used by Malla et al. (2004) to undertake a lifecycle gains-to-research analysis of the creation of canola and by Phillips (2003) to assess the economic impact of the introduction of herbicide-tolerant canola (see Chapter 39 in this volume). The challenge is that the full lifecycle data does not always give an accurate picture of what might emerge. As Graff et al. (2009) showed, the research sieve is ruthless, culling all but a vanishingly small percentage of inventions. The implication is that one would need a large flow of effort upstream to realize even modest uptake and use. More on the prospective side, Pardey et al. (2006) used similar data to develop a longitudinal database of research into agriculture. Both those and other related analyses concluded that while private sector investments in research and development accelerated in the 1990s, public sector research remained relatively flat, which many forecast is having a significant dampening effect on productivity in primary agriculture.

The private sector is a key actor as one moves from the upstream discovery research to more applied research, including the development of specific technologies and breeding of specific traits into crops. One way of getting insight into firm plans is to ask them. For example, Statistics Canada has undertaken a biannual Biotechnology Use and Development Survey since 1997. The challenge is that these surveys, while seeking to add insight into emerging technologies, are infrequent and often lag developments. Another way this activity can be tracked and analysed is through the 'research pipeline' reports issued by multinational firms to shareholders and customers to signal their research intents and efforts. Phillips (2011), for example, undertook a review of the financial statements and publicly revealed research plans, concluding that the top five private investors (Monsanto, Bayer, DuPont Pioneer, Syngenta and Dow Agrosciences) currently invest about 6 per cent of gross revenues on research and development (see Table 20.3). Monsanto is at the high end, investing about 11.5

Table 20.3 GM crops in research pipelines for five key multinationals, 2011

	Bayer	Dow	DuPont	Monsanto	Syngenta	Total
Sales ($ millions)	35088	53674	32733	10502	11641	143638
R&D ($ millions)	3053	1660	1652	1205	1032	8602
R&D (% revenues)	8.7%	3.1%	5.0%	11.5%	8.9%	6.0%
R&D (% ag)	24%	Na	Na	~100%	Na	Na
Crop targets						
Corn	1	Yes	14	11	Yes	~14
Soybean	up to 7	Yes	11	12	1	~12
Canola	up to 7	Na	2	4	1	~7
Cotton	3	Yes	Na	6	Na	~6
Alfalfa	Na	Na	Na	2	Na	~2
Wheat	Na	Na	Na	2	Na	~2
Sugar cane	Na	Na	Na	1	Na	~1
Rice	2	Na	3	Na	Na	~3
Other	Cucumber Melons Onions Tomato	7 traits	–	Broccoli Cucumber Lettuce Pepper Sweet corn Tomato	Sunflower and other vegetables	–

Sources:
Adapted from Phillips (2011).

- Financial data from http://moneycentral.msn.com/investor/invsub/results/statemnt.aspx?symbol={company symbol}.
- Syngenta: http://www.syngenta.com/GLOBAL/CORPORATE/EN/PRODUCTS-AND-INNOVATION/RESEARCH-AND DEVELOPMENT/BIOTECHNOLOGY/Pages/biotechnology-pipeline.aspx.
- Monsanto: http://www.monsanto.com/products/Documents/pipeline-flash/pdfs/TechPipelineALL.pdf.
- Bayer: http://www.bayer.com/en/Bayer-CropScience.aspx.
- Pioneer Hi-Bred: http://www.pioneer.com/home/site/about/research/pipeline/specification-sheets/.
- Dow: http://www.dowagro.com/innovation/pipeline/index.htm.

per cent; at the other end, Dow invests about 3.1 per cent of gross revenues (the lower rate may be due to Dow's more diversified portfolio of activities). A review of the publicly-stated research pipelines of the five key multinationals shows that the bulk of their effort remains focused on corn, soybeans and canola (also cotton), with what appear to be only tentative efforts related to other large-area field crops or vegetables. In Canada, for example, in 2010 all but 15 of the 979 field trials were conducted on canola, corn and soybeans; one single alfalfa trial (by Monsanto) was the exception that made the rule.

A complementary approach is to engage experts and laypersons in the articulation of research priorities, identification of non-linear research efforts and to add probabilities and weights to the developments and their timing.

A range of efforts have taken this approach. One of the earlier and more popular examples was the Delphi process used by Daar et al. (2002) to identify the top ten biotechnologies for improving health in developing countries. Experts were provided with a series of questionnaires about what technologies then existed and might emerge over the coming years that might influence health, with each successive questionnaire based upon the results of the previous questionnaire until a consensus was reached. Their top ten included a number of agri-food-related technologies, including modified molecular technologies, recombinant technologies, technologies for environmental improvement (for example sanitation, clean water and bioremediation) and GM crops with increased nutrients to counter specific deficiencies. Since then *Science*, the *World Economic Forum* and *The Economist* newspaper have all issued annual or quarterly technology surveys that attempt to identify important new and emerging technologies.

More recently, a range of mixed-method assessments have been released. Miller et al. (2010) cited the results of an American Seed Research Summit that identified nine research priorities or policy goals to accelerate seed science and crop improvement. The list included greater effort on conservation of germplasm, more effort to understand basic genetic mechanisms, development of efficient, high throughput analysis systems, improved data analysis tools, managing complex traits, deciphering the genetic basis of plant-environment responses, increasing plant efficiency and quality, improving seed health, quality and performance and developing cost-efficient risk analysis systems for products of new technology. Dunwell (2010) uses a combination of literature analysis, patent searches and technology mapping to highlight a range of plant technologies, especially for soybeans, that might emerge in coming years. Lusser et al. (2012) undertook an extensive analysis of new plant-breeding techniques, conducting a keyword search of the ISI bibliographic database and searches of three key public databases for patents (WIPO, EPO and USPTO) and combined these with a written survey of plant-breeding companies and an international workshop. They identified seven new plant-breeding techniques and applications – ZFN, ODM, cisgenesis intrangenesis, RdDM, grafting on GM rootstock, agro-infiltration and reverse breeding – that could be used to bring new varieties to the market in the next 3–20 years.

While there is an emerging consensus on what science is being done where, and the relative importance of these developments to crop productivity and differentiation, there is no obvious consensus on when these technologies will emerge. In response, there has been increased effort to understand the opportunities and barriers embedded in the gestational and market systems.

Table 20.4 Research sieve in agricultural biotechnology R&D pipeline

Innovation introduced in period eventually advancing . . .	% advancing		Average lag (years)	
	1987–97	1998–2004	1987–97	1998–2004
to publication	37.2	48.3	0.6	0.1
to initial field trials	75.1	55.5	0.5	0.1
to late-stage trials	6.3	2.8	2.9	2.9
to regulatory	4.6	0.6	4.0	3.0
to market	1.7	0.3	4.5	4.0

Source: Graff et al. (2009).

3.2 Gestational Phase

In many ways, the gestational phase is more transparent than the research stage, as it generally encompasses reduction to practice, protecting inventions via intellectual property mechanisms, compliance with regulation and scale-up. Each of these activities creates artefacts, for example field trials, patent filings and regulatory actions. Many scholars and practitioners use these as a base for extrapolations and models, which are often then adjusted based on expert opinion. As one might expect, extrapolations and models are most often used for the short to medium term, as beyond five years the artefacts in the system become highly variable.

Three recent studies use a mix of these datasets and methods and offer converging outlooks.

Graff et al. (2009) identified a set of 558 potential quality-enhancing transgenic product innovations though a mixed-method approach. The primary backward-looking survey involved a literature search and examination of the regulatory records for field trial and regulatory filings. The investigators then undertook a secondary, forward-looking survey of expectations, involving meetings with and presentations by company representatives, review of company annual reports, websites and other publications and personal communications with analysts and research managers. They then used a range of evidence and sources to identify how far and how fast the innovations have progressed in the gestational process. Once the dataset was complete, they calculated the probabilities of advancement and the average lag between steps (Table 20.4) for the technologies in the database and applied those parameters to those product innovations that had not yet been abandoned. They concluded that while only five of the innovations had reached the market by 2006, a further 49 innovations could make it through the gauntlet by 2015.

Arundel and Sawaya (2009) undertook a survey for the OECD of the prospects for biotechnologies in agriculture and related natural resources to 2015. In their study they combined a literature survey with a mix of field trial data, annual reports and research pipeline reports to estimate the approximate date and type of new GM traits that could be introduced by crop. Extrapolating from 2000–06 field trial data, they concluded that up to 59 new traits (mostly agronomic, herbicide tolerant, pest resistant and product quality) could be introduced in the 2008–15 period, involving 27 species of agricultural crops and trees. They then compared this to the firm-based plans revealed in their public

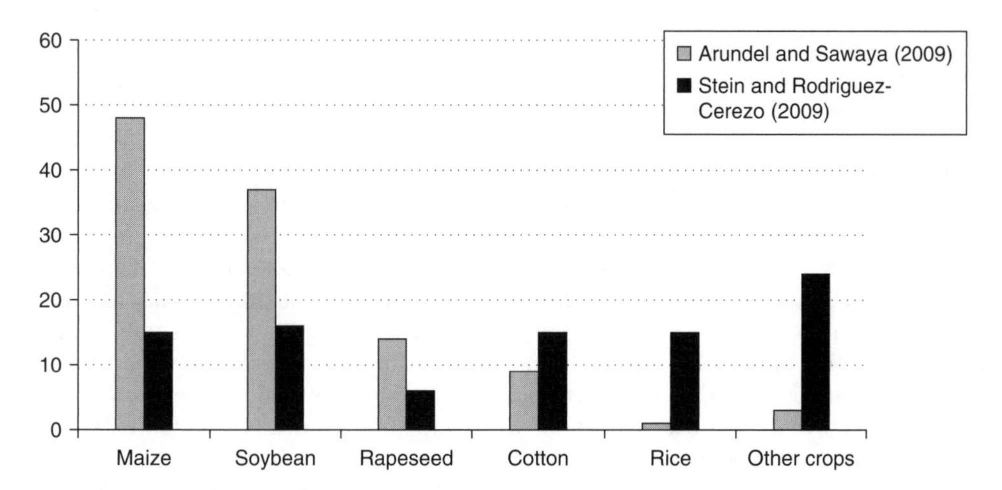

Figure 20.2 Forecasts of new GM events by 2015

reports. That body of evidence suggested that up to 112 new events in seven key agri-food crops could be forthcoming in 2008–18. To put that into context, they reported on an undated FAO study that identified more than 1675 non-GM applications of biotechnology to agriculture in developing countries alone (for example bio-pesticides, bio-fertilizers, diagnostics, fermentation, plant breeding and micro-propagation), noting that more than 143 were already commercialized.

Stein and Rodríguez-Cerezo (2010) offered a complementary approach. They derived an upper-limit estimate of the number of new events in commercial GM crops by 2015 by aggregating the data from the commercial and regulatory pipelines and those in advanced development. Specifically, they undertook a detailed analysis of every trait and crop in the commercial or regulatory pipeline (for example in field trials or under intensive review), looking at the specific country, crop, trait and developer for each event. To identify likely candidates for commercialization in advanced development, the Institute for Prospective Technological Studies convened a workshop of experts and practitioners in November 2008 to build the basis for the forecast. They then constructed a dataset of events in advanced development, using their judgement to sort and present the most likely candidates. They concluded that an additional 91 new GM crop events might be realized by 2015, on top of the 33 events already commercialized in 2008 (see Figure 20.2).

3.3 Market Phase

Given the uncertainty of where the science is going and what might emerge from the gestational phase, forecasting market prospects is particularly problematic. As discussed earlier, the short to medium term can be handled through extrapolation, while the longer term requires often heroic efforts or assumptions.

The simplest approach is to assume a form of random walk, with new technologies and products entering various markets at different times, each then following in some sense a relatively standard S-curve of adoption, reaching peak adoption at some finite point that can be calculated based on the early stages of adoption. In some cases, the S-curve is very steep, and almost total adoption may occur quickly (as seen with HT canola in Canada). In other cases, such as Bt corn, adoption may be slower, with peak levels of uptake far short of 100 per cent of the seeded area.

One difficulty is that data on adoption and use is not readily accessible. While the USDA and Agriculture Canada undertake annual or semi-annual surveys of producers to determine levels of adoption of GM seed in key crops, this level of effort is not standard. As a result, most analysts are forced to fall back on the data produced by Clive James at ISAAA. He uses publicly available data and his extensive contacts with the biotech seed companies, the industrial supply chain and regulators to construct a comprehensive estimate of uptake and use of the technology around the world, which is published as an annual report and is publicly available at the institution's website. The latest release for 2011 reported that GM crops were planted on about 160 million hectares by 16.7 million farmers in 29 countries (James, 2012). The top ten countries each grew more than 1 million hectares of biotech crops. James (2012) estimated that about half of the GM crops were grown in developing countries and that about 60 per cent or around 4 billion people live in the 29 countries planting biotech crops. GM soybean was the principal biotech crop in 2011, planted on 75.4 million hectares, followed by GM maize on 51.0 million hectares, GM cotton on 24.7 million hectares and GM canola on 8.2 million hectares. The key traits were herbicide tolerance (59 per cent of area) and pest resistance (15 per cent), with stacked trait (especially HT/PR) varieties now grown on 26 per cent of the GM area.

James (2012) also offered a forecast of the adoption of biotech crops for the 2012 to 2015 period. He asserted that adoption will depend on timely implementation of regulatory systems, strong political will, enabling financial and material support and a continued supply of improved GM crop varieties to meet the varied needs of the global market. He forecast that up to 10 more countries may approve and adopt GM crops for the first time, bringing the total number of biotech crop-producing countries to 40 by 2015. These new producers could include three more countries in Asia, up to seven countries in sub-Saharan Africa and possibly some additional countries in Latin or Central America and Western or Eastern Europe. Even with those countries that have currently approved production using GM seeds, there remains considerable potential for increasing the adoption rate of the four current large-area GM crops (maize, soybean, cotton and canola), given that there are an estimated 150 million hectares planted to these crops not using the technology. He asserts that in contrast to first generation GM crops that offered higher yields by protecting crops from losses caused by pests, weeds and disease, a number of new cultivars should help to improve product quality. He specifically noted a number of new GM crops that could have particular effect on adoption rates, including: drought-tolerant maize planned for release in North America in 2013 and in Africa by 2017; Golden Rice to be released in the Philippines in 2013 or 2014; and Bt rice in Asia.

While James' (2012) numbers are heavily cited and extensively used by others to undertake economic impact analyses and forecasts, they do not conform to the norms of statistical sampling. Moreover, the data is compiled in a non-transparent way. Users should exercise caution in accepting these numbers as fact as there is the potential that bias (in this case wishful thinking) may underlie some of the results.

In much the same way, proprietary reports from business consultants are another important but relatively inaccessible and non-transparent source of information. Quite a few companies appear to make significant earnings out of producing annual or less frequent reports on biotechnology and its market prospects. While it is difficult to gauge the methods and accuracy of these works, we can at least assume that they must generate some value, both to their clients (who can pay between $500 and $25 000 for these reports) and to the consultants themselves, who would not produce the range and frequency of reports if they did not realize a profit. Current providers, for example BizAcumen, Inc., Global Industry Analysts, Inc., Burrill & Company and RNCOS E-Services Pvt Limited, each produce a range of reports that seem to be sustained for extended periods, suggesting they generate some value.

For longer-term outlooks, the extrapolation method is less effective. A number of scholars and practitioners have used a range of quantitative and qualitative approaches to identify key drivers, scenarios, trends and conflicts between supply and demand that might emerge.

In the first instance, a number of studies have looked at the impact of new technological development on productivity growth. Huffman and Jin (2012), for instance, developed an econometric model for productivity, calibrated using annual public agricultural research efforts in 48 states and total factor productivity for the country, using data from 1970–2004, and then used the model to develop annual forecasts for productivity in the 2005–12 period, offering an upper and lower bound to the forecast based on the 90 per cent confidence interval. They concluded that their forecasts show a significant slowing in growth in total factor productivity for a large number of states and that the forecast for productivity in 2012 was below the actual value in 2004 in some states. Also, the 90 per cent confidence interval for the forecasts includes the possibility of negative growth in productivity from 2004–12 in quite a few states.

Large-scale, long-term modelling efforts somewhat surprisingly tend to take current trends in productivity, as modified by contemporary technological change analysis by crop, as assumptions that underlie the rest of their modelling effort. In this sense, technological change is more exogenous than endogenous in the models. Two major efforts dominate in this area. The annual Agricultural Outlook, prepared jointly by the Organisation for Economic Co-operation and Development (OECD) and the Food and Agriculture Organization (FAO) of the United Nations, is perhaps the most authoritative and often is the base for individual national forecasts. Their projections are derived from a modelling system that incorporates the OECD Aglink and FAO Cosimo models and is assessed and modified based on consultations with national experts in OECD countries. The latest forecast for 2011–20 assumes that wheat and rice yields will rise about 1 per cent per year while coarse grains (including corn) will advance about 1.1 per cent per year (OECD, 2011). USDA also produces a 10-year forecast based on a composite of model results and judgement-based analyses. As with the OECD–FAO effort, productivity is more of an assumption than a derived variable. In the latest forecast (USDA, 2012) covering the 2012–21 period, the model incorporates the assumption that yields per acre will rise an average 1.9 per cent per year for maize, 0.9 per cent for wheat and 1.1 per cent for soybeans. Interestingly, while the adoption of new technology, such as biotechnology and GM crops, will be a key determinant of the ultimate productive capacity of US and global agriculture, neither report makes any mention of their explicit assumptions about what crops will be modified and which ones will be adopted. The conclusion from both efforts is that while productivity is assumed to continue to rise, yield growth will begin to slow, causing markets to tighten, leading to higher prices and more likelihood that supply may not be adequate to meet demand in some years. The resulting world will be more volatile.

At least partly in this response to the over-reliance on models with exogenous assumptions about technological change, Haggui (2004) developed a world wheat market model with GM adoption endogenously generated based on supply and demand and consumer acceptance of GM traits. The model was used to estimate global food security, allowing dynamic economic responses to production shocks, such as climate change. Haggui estimated that under ideal conditions the adaptation and adoption of GM wheat varieties could reduce food insecurity, but that in the face of both consumer concerns and regulatory barriers producers in a number of market areas would hesitate to adopt the technology, which would exacerbate the long-term uncertainties and volatility forecast by the OECD–FAO and USDA.

For all these reasons, the World Economic Forum in 2012 issued a new vision for agriculture (WEF, 2012), and Bill Gates (2012) devoted a significant part of his annual letter from the Bill and Melinda Gates Foundation to the importance of adapting, adopting and using biotechnology to realize maximum productivity growth in agriculture.

4 LESSONS

This chapter has dwelt more on the models and methods of investigating the research pipeline than on the resulting metrics. This was by design. To paraphrase and misappropriate an aphorism from Benjamin Franklin, forecasts, 'like fish, begin to smell after three days'. The actual point or range estimates of science, technology, products and markets are seldom accurate or even that important. The general trend and directions may remain sound, but either by design or by happenstance, the specific forecasts and foresights are seldom realized in quite the way they were divined.

McCloskey (1992) offered a good way to think of the research pipeline. She asserted that true forecasts that go beyond the simple extrapolations of what is already in the system require prescience. But 'prescience is an oxymoron' because it implies we can know pre-science. She asserts that this fundamentally cannot be done.

Thus, we are left with a number of well delimited and extensively used models and methods that use the artefacts of the research and gestational systems to extrapolate from the here-and-now into the near future (with a technical maximum 3–5 years ahead). Beyond that, our collective crystal balls grow misty. While econometric models and expert analyses offer some light, they are highly problematic and variable and should be used cautiously.

REFERENCES

Alston, J., G. Norton and P. Pardey (1995), *Science under Scarcity: Principles and Practice of Agricultural Research Evaluation and Priority Setting*, Ithaca, NY: Cornell University Press.

Alston, J.M., M.A. Andersen, J.S. James and P.G. Pardey (2010), *Persistence Pays: US Agricultural Productivity Growth and the Benefits from Public R&D Spending*, New York: Springer.

Arundel, A. and D. Sawaya (2009), 'Biotechnologies in agriculture and related natural resources to 2015', *OECD Journal: General Papers*, 2009/3, 7–111.

Bernstein, P. (1996), *Against the Gods: the Remarkable Story of Risk*, New York: Wiley.
Daar, A., H. Thorsteinsdóttir, D. Martin, A. Smith, S. Nast and P. Singer (2002), 'Top ten biotechnologies for improving health in developing countries', *Nature Genetics*, **32**(2), 229–32.
Dunwell, J. (2010), 'Foresight project on global food and farming futures – Crop biotechnology: prospects and opportunities', *Journal of Agricultural Sciences*, 1–11.
Gates, B. (2012), '2012 annual letter from Bill Gates', Seattle, WA: Bill and Melinda Gates Foundation, available at: http://www.gatesfoundation.org/annual-letter/2012/Pages/home-en.aspx, accessed 20 April 2012.
Graff, G., D. Zilberman and A. Bennett (2009), 'Correspondence: the contraction of agbiotech product quality innovation', *Nature Biotechnology*, **27**(8), 702–704.
Graff, G., D. Zilberman and A. Bennett (2010), 'The commercialization of biotechnology traits', *Plant Science*, **179**, 635–44.
Haggui, F. (2004), 'Cost of the EU opposition to genetically modified wheat in terms of global food security', unpublished PhD thesis, Department of Agricultural Economics, University of Saskatchewan, available at: http://library.usask.ca/theses/available/etd-09212004–133502/unrestricted/thesis.pdf, accessed 20 April 2012.
Huffman, W. and Y. Jin (2012), 'Reduced funding of public agricultural research stalls future growth of agricultural productivity in the United States', paper presented at the 2012 NC1034 Symposium, The Future of Agricultural Research, Washington, 15–16 March.
James, C. (2003), 'Global review of commercialized transgenic crops: 2002 feature: Bt maize', *International Service for the Acquisition of Agri-Biotech Applications (ISAAA) Brief*, 29, Ithaca, NY: ISAAA.
James, C. (2012), 'Global status of commercialized biotech/GM crops: 2011', *International Service for the Acquisition of Agri-Biotech Applications (ISAA) Brief*, 43, Ithaca, NY: ISAAA, available at: http://www.isaaa.org/resources/publications/briefs/43/default.asp, accessed 20 April 2012.
Lusser, M., C. Parisi, D. Plan and E. Rodríguez-Cerezo (2012), 'Deployment of new biotechnologies in plant breeding', *Nature Biotechnology*, **30**, 231–9.
Malla, S., R. Gray and P. Phillips (2004), 'Gains to research in the presence of intellectual property rights and research subsidies', *Review of Agricultural Economics*, **26**(1), 63–81.
McCloskey, D. (1992), 'The art of forecasting: from ancient to modern times', *Cato Journal*, **12**(1), 23–48.
Miller, J., E. Herman, M. Jahn and K. Bradford (2010), 'Strategic research, education and policy goals for seed science and crop improvement', *Plant Science*, **179**, 645–52.
Organisation for Economic Co-operation and Development (OECD) and Food and Agricultural Organization (FAO) (2011), 'OECD–FAO agricultural outlook 2011–20', OECD and FAO, available at: http://www.agri-outlook.org/pages/0,3417,en_36774715_36775671_1_1_1_1_1,00.html, accessed 20 April 2012.
Pardey, P., J. Alston and N. Beintema (2006), 'Agricultural R&D spending at a critical crossroads', *Farm Policy Journal*, **3**(1), 1–9.
Phillips, P. (2003), 'The economic impact of herbicide tolerant canola in Canada', in N. Kalaitzanondakes (ed.), *The Economic and Environmental Impacts of Agbiotech: A Global Perspective*, New York: Kluwer, pp. 119–40.
Phillips, P. (2011), 'LLP impacts on global trade, agricultural innovation and commodity prices', paper presentated at GM Co-existence Conference, Vancouver, Canada, 27 October.
Phillips, P. and G. Khachatourians (2001), The Biotechnology Revolution in Global Agriculture: Invention, Innovation and Investment in the Canola Sector, Wallingford: CABI Publishing.
Popper, R. (2008), 'How are foresight methods selected?', *Foresight*, **10**(6), 62–89.
Romer, P. (1994), 'Beyond classical and Keynesian macroeconomic policy', *Policy Options*, July/August, available at: http://www.colorado.edu/economics/courses/econ2020/6550/readings/romer1.html, accessed 15 August 2013.
Stein, A. and E. Rodríguez-Cerezo (2010), 'International trade and the global pipeline of new GM crops', *Nature Biotechnology*, **28**, 23–5.
USDA (2012), 'Agricultural projections for 2012–21: US crops', Washington, DC: USDA/ERS, available at: http://www.ers.usda.gov/Publications/OCE121/OCE121d.pdf, accessed 20 April 2012.
World Economic Forum (WEF) (2012), 'Realizing a new vision for agriculture: a roadmap for stakeholders', Geneva: World Economic Forum, available at: www.webforum.org/issues/agriculture-and-food-security.

21 Clusters, innovation systems and biotechnology in developing country agriculture

David J. Spielman, Douglas Zhihua Zeng and Xingliang Ma

COORDINATION FAILURES, CLUSTERS AND BIOTECHNOLOGY

Industry clusters have historically played a central role in economic growth and development – a point that was initially recognized by Adam Smith ([1776] 1993) and expanded on by economists and historians ever since. Clusters are highly interdependent production and distribution networks that are populated by specialized firms and allied business services and are concentrated across a geographical or thematic dimension. Clusters are often driven by strategic alliances between private firms or non-commercial or quasi-commercial actors such as universities, public research organizations, technology transfer programs, knowledge brokers and other similar agents. In effect, a cluster is a concentration of enterprises that generates collective efficiencies or 'agglomeration effects' by leveraging complementarities, economies of scope and scale, technology spillovers and other efficiency gains and positive externalities (Porter, 2000; Schmitz and Nadvi, 1999; Schmitz, 1995).

Theoretical and empirical inquiries into clusters cover a range of topics, including the conceptualization and measurement of net gains from labor pooling, technology spillovers and market access (Krugman, 1991); division of production tasks and reductions in capital barriers to entry through intra-industry credit linkages (Ruan and Zhang, 2009); accumulation and application of appropriate technologies (Caniëls and Romijn, 2003); spillovers of complex, tacit knowledge between firms (Cohen and Levinthal, 1989); spillovers that are external to firms and appropriable by individuals (Audretsch, 1998); and cluster-based firm-level learning modalities (Parrilli and Asheim, 2012).

In many of these inquiries, science, technology and innovation (STI) are central to a cluster's performance. This is the case in, for example, Silicon Valley in California, Research Triangle in North Carolina, Cambridge in Massachusetts, Cambridge in the United Kingdom and Wageningen UR in the Netherlands. These localities host clusters that specialize in the research, development and marketing of scientific tools and technology products. Other, more virtual clusters similarly depend on STI through the exchange of knowledge, information and materials that is enabled by an expanding number of information, communications and logistical technologies.

Less cited, but of no less interest to many scientists, investors and policy makers, are clusters in developing countries where new tools, products and value are being produced for the agricultural sector (see Theus and Zeng, 2012). This chapter examines the nature and design of several geographic and virtual clusters that have emerged in the field of developing-country agriculture and agricultural biotechnology. The chapter describes the underlying innovation systems and processes that have enabled these clusters to emerge.

It highlights the complex interactions between and among heterogeneous economic agents in the agricultural sector and explores the economic institutions, organizational learning processes and technological opportunities that condition their behaviors and relationships. Finally, the chapter emphasizes the relevance of biotechnology clusters to small-scale, resource-poor farmers who constitute a majority of the poor in developing countries across Asia, Africa and Latin America.

One way of thinking about clusters is to understand their role in overcoming coordination failures in the marketplace and at a more systemic level. From a conceptual standpoint, clusters emerge to overcome imperfections within an innovation system and enable both firms and individuals to interact more efficiently. The evidence of their efficiency and effectiveness in this context is an empirical question in many different sectors of the economy.

These cluster attributes can be demonstrated in the context of biotechnology in developing-country agriculture. Biotechnology has the potential to dramatically increase agricultural productivity and reduce poverty in smallholder farming systems (Pingali and Traxler, 2002). The case of Bt cotton in India, China and Burkina Faso already provides early evidence of these effects (see Sadashivappa and Qaim, 2009; Smale et al., 2009; Huang et al., 2006). There are significant impediments to ensuring that the benefits of biotechnology in developing countries reach smallholders (Spielman, 2007). These impediments can be generally classified as market, institutional and system failures, each of which is explained further below.

In the context of developing-country agriculture, market failures exist where end-users of a technology – specifically, the smallholder – have limited access to input and commodity markets, limited purchasing power due to poverty, and limited access to financial services with which to manage the production and market risks inherent in agriculture (see Alston et al., 1995). In such markets, innovators – specifically, firms that develop and market biotechnology products – are generally unable to recoup R&D investment costs fully, thus requiring public investment to bridge the market failure in a manner that equates social and private returns.

A classic example of market failure in developing-country agriculture is evident in the market for improved planting materials for 'orphan' crops of marginal commercial value such as sorghum, millet, groundnut, pigeonpea, cassava, or sweet potato. Where farmers can replant saved seed and capture the gains conferred by science, and where private firms cannot prevent them from doing so through legal or technological means, the profit-maximizing private firm will optimally choose not to invest in the discovery of new techniques to improve these crops, in the development of new and improved crops, or in the market, distribution, or delivery of new planting materials embodying these improvements. This results in a chronic undersupply of improved planting material, thus requiring public-sector intervention in the market, typically through the financing and management of basic research, plant breeding and agronomy programs and agricultural extension services. While there is ample evidence to demonstrate that public investment in agricultural R&D of this nature generates high social returns (see Alston et al., 2000; Evenson and Gollin, 2003; Raitzer and Kelley, 2008; and Renkow and Byerlee, 2010) and significantly favorable impacts on agricultural productivity (Evenson and Rosegrant, 2003; Pratt and Fan, 2010), private sector investment tends to be insignificant throughout much of the developing world (see Beintema and Stads, 2011; Pardey and Beintema, 2001).

In addition to these market failures, institutional failures weigh heavily on developing-country agriculture. Institutions are conventionally defined here as commonly-accepted policies, laws, rules, practices, or norms and as solutions designed to govern exchanges where markets do not exist, do not perform effectively, or incur transaction costs higher than the actual outlays and opportunity costs associated with the exchange (see seminal works by Coase, 1937, 1960; North, 1971; and Williamson, 1975, 1985). Institutions that reduce these transaction costs are viewed as essential to reducing inefficiencies in production and increasing welfare gains.

For example, in the agricultural knowledge market, intellectual property rights (IPRs) are institutions developed to reward innovators for their successful investments in cultivar improvements. IPRs such as plant variety protection laws aim to reduce the transaction costs associated with contested claims over the rights to innovation rents while simultaneously incentivizing innovative behavior. Similarly, quality assurance systems for planting materials (for example, seed certification or truth-in-labeling) are institutions designed to address information asymmetries resulting from the inability of farmers to make *ex ante* assessments of seed quality when such information is known only by the seller. But where these types of legal institutions do not sufficiently facilitate knowledge exchanges, other institutions can emerge. Examples include the vertical integration of an agricultural supply chain – from seed to sales – by a single company to protect the IPRs of technologies used throughout the chain.

A broader set of systemic failures also characterize developing-country agriculture. These failures stem from the inability of agents engaged in knowledge production to learn about each other, identify areas of complementarity and synergy, build and sustain trust through interpersonal or organizational relationships, communicate and exchange ideas effectively, or respond to leadership (see, for example, seminal works on national systems of innovation by Lundvall, 1985, 1992; Freeman, 1987, 1995; Edquist, 1997; Metcalfe, 1995; Nelson, 1993; and Dosi et al., 1988). Systemic failures impede processes of knowledge exchange and create inefficiencies in knowledge production. Consider, for example, the public university scientist who learns about an advanced breeding technique developed by a private company. The technique can dramatically reduce the time needed to identify desirable characteristics in a plant, thus making improved planting materials more readily available to farmers. The company, however, is hesitant to share the technique with a public scientist without the assurance of direct remuneration or intellectual property protection that provides opportunity for remuneration. The university scientist, on the other hand, is unsure how to enter into an agreement with a company without violating university rules and regulations. Nor does she know exactly how to interact with her private sector counterparts: what questions can she ask? What labs can she visit? What materials can she use? This scenario requires legal frameworks to facilitate university–industry partnerships, experienced managers to oversee the collaboration and some goodwill on both sides to see that the science ultimately find its way to farmers.

POLICYMAKING AND SPONTANEOUS VERSUS ORCHESTRATED CLUSTERS

When clusters form spontaneously, their design evolves as the involved firms and individuals accumulate experience and expertise. When clusters are deliberately planned or orchestrated, their designs are shaped by public policies and incentive mechanisms. In practicality, many clusters exist on a continuum between self-emergent and deliberately planned, and most are based on long-term, multifaceted investments. To build an understanding of how clusters develop, we examine the policy dimensions of cluster development and the underlying principles of efficacy and selectivity.

Cluster investments are effective only where minimum conditions of macroeconomic and physical stability, hard and soft business infrastructure and basic institutions for supply-side functions are met (Theus and Zeng, 2012). Government policies play an important role in this regard through economic reforms such as the deregulation of domestic markets, the removal of explicit and implicit trade barriers, the introduction of non-distortionary exchange rates and taxation policies and the development of sound property rights regimes. The public good character of infrastructure, including road, water and telecommunications networks, also makes government involvement an imperative (Zeng, 2010).

Given the government's potentially deep involvement in cluster development, clusters offer policymakers significant leverage over industrial development and economic growth. A visionary policymaker could seize on this opportunity to improve the efficacy of public investment while imposing a degree of selectivity on the allocation of scarce public resources. There are several ways the policymaker might pursue this.

First and foremost, policymakers can invest in cluster development as a realistic approach to removing impediments to competitiveness and innovation. By encouraging a critical mass of firms to invest in a concerted manner, these firms can become an effective vehicle for catalyzing policy reform in support of wider growth and development. Second, policymakers can invest in cluster development as a lower-risk solution to allocating scarce and highly contested public funds. Clusters are, by definition, delineated by geography or sector and are thus less risky than more traditional national investment projects that often incur excessive costs and risks, develop immovable coalitions of rent-seekers, or have a proven track record of failure. Third (and related to the previous point), policymakers can invest in cluster development as a potentially high-impact option for allocating public funds. Because clusters tend to be larger than value chains based on a single commodity or product, and because they tend to generate high levels of spillovers, the intended and unintended outcomes of a cluster can be significant. Finally, investments by policymakers in cluster development have a higher probability of success if the cluster's development and ownership is effectively distributed among diverse agents. Because clusters depend acutely on partnerships between private investors, entrepreneurs, universities, public research agencies and regulators, frequent and durable interactions between these actors are necessary. By engaging these agents in formulating, sequencing and expediting supportive policies and investments, policymakers can potentially realize a greater likelihood of success than many of the traditional, top-down government investment programs in industrial development.

Clusters are not necessarily the perfect investment instrument for policymakers. The long-term success of a given cluster depends acutely on complementary policies, programs and financial mechanisms that can foster new linkages and create opportunities for sustained growth. Without these links and opportunities, crises and challenges that beset clusters can lead to stagnation and decline (Zeng, 2010). Thus, policymakers must maintain a sustained, long-term commitment to cluster development – a commitment potentially challenged by politics.

KNOWLEDGE EXCHANGES AND INNOVATION FINANCING

The long-term success of a given cluster depends on whether firms continue to operate in a relatively competitive market. While innovators traditionally recoup their R&D investments through non-competitive incentive mechanisms (for example, IPRs), the long-term sustainability of a cluster is tied to inter-firm competition and its ability to drive innovation. Necessarily, competition is curtailed where a firm is able to secure exclusive control over resources, assert some form of durable market power, or influence policymaking for its own benefit. Moreover, clusters depend on at least two elements that governments cannot always effectively manage: first, the financing of innovation or mechanisms to reallocate risk across different innovation actors; and exchanges of knowledge, involving iterative learning processes that encourage innovation.

Financing for knowledge-intensive industries is, in many developing countries, characterized by high levels of risks, uncertainty of collateral, high transaction costs and stringent repayment schedules. Sophisticated financial markets are needed to support cluster development. This may encompass a combination of mechanisms, for example, public loan guarantee schemes, development bank financing, business development services, corporate intelligence services, vibrant equity markets, financing through industry taxes or levies and venture capital funds.

Learning and learning cultures are also a key element of success. Industry-specific modes of learning and training, formal and informal, are important means of acquiring and disseminating knowledge and technology in clusters because they influence the clusters' potential for innovation and competitiveness. Modes of learning and training can include anything from formal higher and vocational education programs, apprenticeships, on-the-job training and expert contracting and consulting, to more informal practices such as learning-by-doing and mentoring (Zeng, 2006). The capacity of the firms and individuals in a cluster to learn, solve problems and innovate is essential to success and not always resident in the norms or cultures that influence the cluster.

Sharing, and the institutions designed to promote sharing of knowledge and information, is also critical to a cluster's success. Systems that encourage the transfers and exchanges of tools, materials, products, processes or ideas are a uniquely strategic investment often overlooked by both governments and firms. While conventional means of sharing may exist through scientific publications or disclosures in patent filings, a cluster is likely to require more proactive mechanisms. These may include area or industry development councils, professional networks, industry exhibitions, locally targeted media or other similar mechanisms that encourage a collective ownership of, and commitment to, the cluster's development.

An important aspect of learning and sharing is driven by inter-firm linkages. In the context of the knowledge-intensive industries being examined in this chapter, Hagedoorn (2002) describes inter-firm linkages as different modes of collaboration where two or more firms operate as independent economic agents and organizations but share some of their R&D activities. A typology of inter-firm R&D linkages building on Haagedorn (2002) and others would describe these collaborations as:

1. *R&D partnerships* that are primarily contractual in nature, such as R&D pacts, joint development agreements, contract research, product testing, expert consulting or licensing;

2. *equity-based joint ventures*, where firms invest jointly in a commercial project and assume the associated risks; and
3. *strategic firm behavior*, which primarily involves mergers and acquisitions of one firm by another.

Another way to frame this is to classify inter-firm linkages in terms of horizontal and vertical linkages. Horizontal linkages are collaborations between similar firms, for example, to conduct R&D or develop a particular product. Vertical linkages are collaborations between firms involved in different points of the scientific discovery or technology development. Both horizontal and vertical linkages may exist within the typology described above, depending on the strategic objective of the firms involved.

Of similar importance are the linkages between private sector firms and public research organizations. Conventionally described as public–private partnerships, these linkages cover a range of topics. A functional typology (see Spielman et al., 2010) describes these linkages as follows, each of which addresses the specific roles played by each partner, the distribution of risk across partners and the overall goals of the partnership.

1. *Resourcing partnerships*: Public research organizations receive funding from philanthropic foundations associated with private firms or receive scientific expertise from private firms.
2. *Contracting partnerships*: Public research facilities or expertise are contracted to private firms, or public research organizations contract private firms to conduct research.
3. *Commercializing partnerships*: Public research organizations transfer research findings and materials to private firms for commercialization, marketing and distribution.
4. *Frontier research partnerships*: Public research organizations jointly undertake research activities characterized by some unknown probability of success.
5. *Sector or value chain development partnerships*: Public research organizations collaborate with networks of public, private and civil society partners to develop a commodity subsector or its associated value chain.

AGRICULTURAL BIOTECHNOLOGY CLUSTERS IN DEVELOPING COUNTRIES

Following from the basic principles and logic described above, this section examines the leading biotechnology clusters in developing-country agriculture. The discussion following these examples focuses on three evaluative elements: spontaneity versus orchestration, the role of policy and the long-term influence of these clusters.

Hyderabad, India

Hyderabad's agbiotech cluster has resulted from the high concentration of public and private agriculture-related organizations in the city and its environs. On the public sector side, the city has been a center for both state- and national-level agricultural education and training since the 1970s, thus providing the region with a critical mass of agricultural technicians and scientists. It hosts the Acharya N. G. Ranga Agricultural University (formerly the Andhra Pradesh Agricultural University), which combines educational programs (ranging from the vocational to doctoral levels) with research and extension activities. Allied institutions located adjacent to the university include the National Academy of Agricultural Research Management (NAARM, established in 1976), which is under the umbrella of the Indian Council on Agricultural Research and the National Institute of Agricultural Extension Management (MANAGE, established in 1987), which is under the Ministry of Agriculture.

Hyderabad has also been closely linked to the international agricultural research community through the International Crops Research Institute for the Semi-Arid Tropics (ICRISAT), established in 1972 under the auspices of the Consultative Group on International Agricultural Research (CGIAR). ICRISAT supports an expansive agenda on crop breeding and cultivar improvement, agronomy, crop management and germplasm conservation. ICRISAT has been instrumental not only in improving pearl millet and sorghum cultivars, but also in supplying hybrid parent lines to seed companies in the region and strengthening the private sector's R&D capacity to such an extent that, as of 2005, an estimated 80 per cent of commercial seed sales of pearl millet and sorghum were made by the private sector (Pray and Nagarajan, 2010). Further, ICRISAT has been a thought leader in the area of public–private research partnerships (see Spielman et al., 2010), contributing significantly to the creation of its own science park, agribusiness incubator, technology funds and other mechanisms to encourage enterprise development in science-based agriculture.

Hyderabad is also the center of India's seed industry, a status that is largely credited to policy reforms designed to encourage private investment in cultivar improvement and India's seed market – policies that eventually opened the door for investments in agbiotech. Policy reforms such as the New Policy on Seed Development (1988) and the economy-wide New Industrial Policy (1991) encouraged private-sector participation in higher-value segments of the seed market, first in vegetable hybrids, then with hybrids of sorghum and pearl millet, and more recently with hybrids of maize, cotton and rice (Pal et al., 1998; Pray and Ramaswami, 2001; Ramaswami, 2002). Importantly, these policies also leveled the playing field by limiting the role and government support of state-run seed companies. These policies encouraged rapid growth in the seed market: India's seed industry is estimated to be growing at an average rate of 12–13 per cent per year. In 2008–09, it generated revenues between US$1.3 billion and $1.5 billion, making it the world's fifth largest seed market (Rabobank, 2006).

A significant segment of the seed market is deeply vested in the development and distribution of Bt cotton. Figures from a Biospectrum-ABLE survey (2010) estimate that the value of that market is US$400 million, 72 per cent of which is accounted for by the top four companies (Nuziveedu Seeds, Rasi Seeds, Mahyco and Monsanto). The survey also suggested a growth rate of approximately 26 per cent over the next five years (Biospectrum-ABLE, 2010). The only licensor of Bt technology in India – Mahyco Monsanto Biotech Ltd. (MMB) a joint venture between Mahyco, a domestic seed company, and Monsanto, the world's leading crop-science company – is not headquartered in Hyderabad, but it does maintain facilities in the city alongside many of its licensees (totaling approximately 40).

Today, Hyderabad is home to dozens of small seed companies that multiply, distribute and market publicly-developed seed for a range of crops and, in some cases, operate modest breeding programs. Many of these companies are also MMB licensees and active players in the Bt cotton seed market. Hyderabad also hosts several larger, more intrepid seed companies (including Shiram Bioseed and Advanta, India's own multinational seed company) that run R&D programs and operations in seed multiplication, distribution and marketing, again inclusive of Bt cotton. But Hyderabad is more than just a center of the seed industry. In recent years, it has evolved into a *bona fide* agbiotech cluster with a diversified portfolio of companies working on the frontiers of agricultural science. A number of Hyderabad firms (for example, Metahelix and Avesthagen) operate on knowledge-intensive business models built around the production and licensing of proprietary materials, including transgenic events, for other agriculture and pharmaceutical firms.

Similarly, Hyderabad has emerged as a base for multinational crop-science firms, including most of the Big Six multinational firms (BASF, Bayer CropScience, Dow Agrosciences, Monsanto, Pioneer Hi-Bred International, and Syngenta). These firms have (a) integrated interests in seed, agbiotech and agrichemicals; (b) substantial levels of R&D capacity both in India and abroad; or (c) varying degrees of vertical integration that bring together upstream operations in product development (for example, traits, chemicals) with downstream operations in product marketing (for example, seed, chemicals). These multinationals operate directly in the Indian market through wholly-owned subsidiaries, firms in which they hold an equity stake or licensees of their materials and technologies.

Importantly, several of these companies, including Dupont and its subsidiary Pioneer Hi-Bred International, have built expansive biotech R&D programs in Hyderabad to serve their global business portfolios beyond the Indian market. The presence of these biotech R&D programs in Hyderabad is a highlight of Genome Valley, a strategic initiative of the Government of Andhra Pradesh launched in the early 2000s to attract private sector investment and expertise in biotechnology to a specific Hyderabad suburb. Built on an initial public investment of US$30 million, Genome Valley now hosts over 100 foreign and domestic companies in the fields of pharmaceuticals, agriculture, genomics and allied services.

Support for the Genome Valley's resident companies – and the larger cluster of seed and agbiotech companies in Hyderabad – extends beyond state sanctioning. The cluster owes its operational and commercial existence to the engagement of corporate philanthropies, charitable foundations and public–private partnerships. These include the ICICI Bank, a leading financial services company in India and a central player in Genome Valley through its IKP Knowledge Park, which provides ready-to-use multi-tenanted modular laboratory blocks, common facilities and support services and developed land for customized R&D facilities. A related initiative is ICRISAT's Agri-Science Park, a unique venture that began in 2003 and now hosts six separate projects including the institute's Ag-Biotech Innovation Center, the Agri-Business Incubator and several other projects centered around public–private partnerships and collaborative research. The Agri-Science Park provides start-up firms, including several working in agbiotech, with ready access to scientific infrastructure, materials and expertise.

As with many biotechnology clusters in the developing world, a main draw in Hyderabad is its human resource base. Hyderabad (and India more generally) offers employers with a large pool of highly qualified, relatively low-cost and readily employable workers. This includes not only scientists but also skilled workers, technicians and marketing agents. Access to the growing Indian market and other emerging markets in Asia, the Middle East and Africa is also a distinct advantage for the cluster. And while India's connectivity has long been a challenge to both foreign and domestic investors, the government's aggressive investment programs and policy reforms over the last decade have significantly improved Hyderabad's road, air and communications infrastructure.

Despite extensive government support, significant private investment and a strong sense of partnership within Hyderabad's biotechnology cluster, agbiotech's future in India has received mixed signals in recent years. IP protection in the agricultural sector, whether under the 2001 Protection of Plant Varieties and Farmers' Rights Act or under amendments to the 1970 Patents Act, is relatively untested. Most firms still rely on a pipeline of improved germplasm from the public sector and hybrids to protect their IP protection, albeit with little capacity to combat competition from spurious and copycat seeds in the market. Regulatory uncertainty – brought to light with the ad hoc moratorium on the release of Bt eggplant in 2010 despite the crop's successful passage through the regulatory process in 2009 – is likely to disincentivize significant growth in private investment in other crops and traits. Distortionary interventions by government, such as the imposition of price controls on Bt cotton seed in 2006, create further disincentives for investment. Until the uncertainty surrounding policies and regulations on agbiotech are clarified in India, the contributions of Hyderabad's biotech cluster to agricultural development, economic growth and poverty reduction remain unclear.

Nairobi, Kenya

Nairobi is a biotechnology cluster in the making. The cluster is emerging out of a unique combination of geographic concentration and virtual interaction that is contributing to the development of a relatively extensive agbiotech research platform. Nairobi's emergence as a center of innovative activity is due to a number of factors that have converged over the last two decades.

One key factor dates back to the early 1990s when fertilizer and maize markets were liberalized, resulting in a significant decrease in fertilizer marketing costs and distances traveled by farmers to purchase fertilizer and a corresponding significant increase in fertilizer use (Ariga and Jayne, 2010; Ariga et al., 2006; Freeman and Kaguongo, 2003; Freeman and Omiti, 2003; Omamo and Mose, 2001). This liberalization was relatively novel in the region; most neighboring governments continued to directly manage the procurement, distribution and pricing of inputs to smallholders. The increase in fertilizer use and the related increase in demand for improved seed were probably key drivers behind the government's sustained investment in maize improvement and, more generally, agricultural R&D.

In fact, the Kenyan government spends a relatively high level of public spending on agricultural R&D when compared to most other countries in the region. Public agricultural R&D in Kenya is concentrated around the Kenya Agricultural Research Institute (KARI). In 2008, the country employed over 1000 FTE agricultural researchers and spent KSH 4.5 billion or US$154 million on agricultural R&D (both in 2005 constant prices), reversing a period of significant decline during the 1990s (Flaherty et al., 2010). With its headquarters in Nairobi, KARI is an important center for innovation in Kenya's agricultural sector.

KARI's contribution to the formation of Nairobi's agbiotech cluster is closely tied to several influential international research centers and programs. Two centers, the World Agroforestry Center (ICRAF) and the International Livestock Research Institute (ILRI), are headquartered in Nairobi and provide an umbrella infrastructure for additional programs led by the International Maize and Wheat Improvement Center (CIMMYT), International Crops Research Institute for the Semi-Arid Tropics (ICRISAT) and other centers. These organizations represent an important source of improved germplasm for the public research system and, both directly and indirectly, private seed companies are also key proponents of innovative thinking in the field of agricultural research for development. Importantly, centers such as CIMMYT and ICRISAT are also leaders in the design and implementation of public–private research partnerships, a core modality for innovation in this cluster (see Spielman et al., 2010).

Several initiatives leverage this dense international research presence in Nairobi. One initiative that stands out as a potential hub within the Nairobi cluster is the Biosciences Eastern and Central Africa (BECA) project, a regional center of excellence that is headquartered at ILRI to facilitate the development and application of bioscience research and expertise for agriculture in the region. BECA's mission is to increase access to affordable, world-class research facilities and strengthen human resources in biosciences and related disciplines in Africa (see BECA, 2011).

Another initiative is the African Agriculture Technology Foundation (AATF), a non-profit organization that facilitates public–private partnerships around the use of proprietary agricultural technologies. AATF's work on such crops as maize, cowpea and banana has been instrumental in accessing patented materials from foreign countries and companies to help researchers working in Africa improve the quality and potential impact of their work. Yet another Nairobi player is the Alliance for a Green Revolution in Africa (AGRA) and its Program for Africa's Seed Systems (PASS) (see AGRA, 2009). While AGRA explicitly eschews agbiotech in its project portfolio, PASS and its other programs are potentially critical to increasing Africa's capacity to breed, produce and disseminate quality seed of staple food crops and to building seed markets through which other firms and organizations will deliver agbiotech products.

The presence of these organizations – and related organizations in the public, private and civil society sectors – provides a potentially strong foundation for a research-driven, knowledge-intensive agbiotech cluster. But research is only as good as the policy environment in which it occurs. Thus, the Nairobi cluster is also host to a number of organizations engaged in advocacy and capacity strengthening in support of agbiotech in the country and region. This is particularly relevant given that no genetically modified crops have been commercially released to date in the region and the social and political discourse around agbiotech in Nairobi is often heated.

The city's political corridors are a focal point for interventions by initiatives such as the Program on Biosafety Systems (PBS) which aims to develop capacity to implement effective biosafety regulations to accelerate the availability of agbiotech applications for farmers in the country (see PBS, 2012). It also hosts the Africa office and program of the International Service for the Acquisition of Agri-biotech Applications (ISAAA), which serves the region as a global knowledge-sharing network on agbiotech issues (see ISAAA, 2012).

As a cluster with sizeable investments in research, advocacy and capacity strengthening, Nairobi is necessarily host to several significant projects that target the development of genetically modified crops. These include a now completed project on virus-resistant sweet potato as well as current projects on Bt maize, water-efficient maize and other activities that capitalize on agbiotech research tools for livestock vaccine development and tissue culture micropropagation (Spielman et al., 2010; Hall, 2005). These projects bring together KARI, the above-mentioned international research organizations, the multilateral and philanthropic donor community, several leading multinational firms with significant investments in the life sciences, local seed companies and non-governmental organizations.

Necessarily, these projects and organizations would not be able to invest in the Nairobi agbiotech cluster without support from a diverse community of donors who are interested in accelerating agricultural productivity growth and poverty reduction in Kenya. This is where Nairobi differs from many other agbiotech clusters around the world. In essence, Nairobi's experience has been driven by the international donor community – multilateral donors, bilateral donors and charitable philanthropies – that has invested heavily in building the basic infrastructure needed for cluster take-off. This infrastructure dates back to a long-standing donor commitment to building domestic research capacity and facilities at KARI and its affiliated centers and benefits from the sustained (if sometimes volatile) donor funding to the Nairobi-based international centers and programs of CIMMYT, ICRISAT, ILRI and others. Donors have invested heavily in building a favorable regulatory environment, for example, through Rockefeller Foundation support for the establishment of AATF and through United States Agency for International Development (USAID) funding to PBS. On the research side, the Bill and Melinda Gates Foundation has invested in building capacity in the seed system through AGRA, while the Syngenta Foundation for Sustainable Agriculture has funded the CIMMYT–KARI partnership on Bt maize. Many other donors have provided additional support to various agbiotech and seed industry projects, while multinational companies such as Monsanto, BASF and Bayer CropScience have provided in-kind support to finance their involvement in the research.

Unique to Nairobi is the cluster's significant reliance on virtual linkages. Many of the multinational firms involved in these projects do not have substantive research programs in Kenya: Monsanto and Pioneer Hi-Bred International, for example, site their research work in Southern Africa and at their US research hubs. Nonetheless, these technology leaders engage quite closely with their public sector partners by using a range of information and communications technologies to collaborate.

Also unique to Nairobi is the fact that it is a virtual cluster that serves the region as much as it serves Kenya. Most of Nairobi's agbiotech activities (for example, BECA, AATF and AGRA) are regional in their mandate and mission. Regional bodies such as the Association for Strengthening Agricultural Research in Eastern and Central Africa (ASARECA) or the international agricultural research centers' (IARCs') own networks are designed to ensure that technology spillovers from the Nairobi cluster are strategically translated into solutions to similar problems in countries throughout the region.

Related to this is the potential ability of Nairobi's agbiotech cluster to tackle the small-country problem in which limited market size insufficiently incentivizes private investment in agricultural R&D, despite the existence of socially or economically desirable technology solutions such as agbiotech. Although Kenya's maize seed market – the largest of all its seed markets and the market that is most likely to deliver agbiotech products to Kenya's farmers in the future – chalks up annual revenues of less than US$500 000, it has the potential to serve regional markets with new and better products. This potential may be realized as the harmonization of regional seed trade laws among neighboring countries moves forward under the auspices of the Common Market for Eastern and Southern Africa (COMESA) (see Waithaka et al., 2011).

Beijing, China

The agbiotech cluster in Beijing is firmly grounded in the country's most prominent universities (Peking University, Tsinghua University and the Chinese Agricultural University, among others) and numerous research centers (the Chinese Academy of Science (CAS), Chinese Academy of Agricultural Science (CAAS) and their affiliated institutes). Of the 29 institutes conducting agbiotech research in China, 13 are located in Beijing (Huang et al., 2002).

Beijing's agbiotech cluster emerged quickly in response to economic reforms in the 1980s. In March 1978, the National Congress passed the 1978–85 National Development Plan for Science and Technology, which reestablished and strengthened support for national agricultural research institutes like CAS and CAAS. The 1980s and 1990s witnessed a series of further reforms to China's agricultural policy, including the gradual establishment of an IPRs system. A patent law was enacted in 1983, followed by laws on technology contracts in 1987 and copyright laws in 1990. The first IPRs Court was established in 1994 and numerous revisions to the country's IPRs system were pursued thereafter to meet the requirement of accession to the World Trade Organization. The Chinese government also launched a series of programs to boost science and technology, including the Spark Program in 1985, the Harvest Program in 1987 and the High Tech Research ('863') Program in 1986. Of particular note is the '863' Program, which has a profound impact on China's science and technology development through funding on the order of 33 billion RMB (US$4.9 billion[1]) between 1986 and 2005 and through support for several key agricultural developments such as hybrid rice and Bt cotton (see High Tech Research ('863') Program, 2012).

The agbiotech cluster in Beijing has a geographical base in Zhongguancun, China's first and largest science park. In the early 1980s, as China just started its economic reform, several start-up companies – effectively spin-offs from nearby research centers and universities – established themselves in the Zhongguancun locality. In 1986, the government gave official recognition to these companies, and in 1988 the State Council officially declared Zhongguancun the 'New Technology Industry Development Zone', furnishing it with tax reductions, preferential access to government credit and other supportive policy liberalizations that favored the private firms operating in the cluster (State Council of PRC, 1988). Moreover, the government implicitly encouraged cluster development by introducing policies that encouraged research organizations to generate revenues from commercial activities (Pray, 2001).

The total number of firms sited in Zhongguancun increased from 148 in 1988 to more than 1300 in 1991; by 1999 there were over 4000 firms in Zhongguancun generating total revenues on the order of 45 billion RMB (US$6.8 billion). The same year, the Beijing government enlarged Zhongguancun from about 100 km^2 to about 280 km^2 and improved road, information and communications infrastructure to support this expansion. Further investments were made in Zhongguancun, including the development of a life science park that would host multinational firms such as Monsanto and Syngenta and domestic private firms such as Origin Agritech. The government also designed specific policies to attract overseas Chinese experts and entrepreneurs to work or invest in Beijing and set up a foundation with an annual budget of 200 million RMB (US$24 million) to invest in R&D.

Specific emphasis was placed in Zhongguancun on agbiotech. In 1986, Zhongguancun hosted the Biotechnology Research Center, a relatively small institute under the CAAS system. By 1999, the center was upgraded to become the Biotechnology Research Institute and greatly enlarged to include eight research centers and 104 research staff. In 2003, the Institute of Crop Science was added to the Zhongguancun cluster, extending the research focus on seed and seed-based technologies with a staff of 179 researchers and leadership from scientists trained abroad. Five years later, the Institute of Crop Science added a lab for research focused specifically on genetically modified rice, soybean, wheat and corn with a total investment of about 40.8 million RMB (US$5.9 million).

The scientific and technological outputs from the Zhongguancun cluster have supported the emergence and growth of many private firms in the agricultural sector. For example, Biocentury Transgene, a Shenzhen-based firm, got its start as a spin-off of the Biotechnology Research Institute and was the entity responsible for developing China's version of Bt cotton in 1998. In 2009, the firm's earnings from cotton seed sales (accounting for 95 per cent of its revenue) totaled 120 million RMB (US$18.2 million) (Jia, 2011; Sina Finance News, 2010; CNPVP, 2004).

The Da Bei Nong Group, founded in Beijing in 1994 by a PhD graduate from the Chinese Agricultural University, has grown from a local company specialized in animal feed to a national company doing business in products such as animal feed, seed, fertilizer and agbiotech. Origin Agritech, founded in 1997 in Beijing by two overseas-educated agricultural scientists, has become another top player in agbiotech R&D and the seed industry in China, and was listed on NASDAQ in 2004. Even older state-owned companies based in Beijing have benefited from the cluster, including the China National Seed Group Company (CNSGC), a giant in China's seed industry (Seed China News, 2011b).

While the emphasis of research in the Zhongguancun cluster varies by firm and institute, it is likely that the cluster's research mirrors the overall R&D investment trends in China. The majority of agricultural R&D investment in China is concentrated on crop research, which accounts for about 50 per cent of all agricultural R&D spending among public research institutes in China and 28 per cent of all spending by university research programs (Chen et al., 2012). Within the area of crop research, a specific focus is placed on China's major crops: cotton, rice, wheat and maize (Chen and Zhang, 2011).

The contribution of the Zhongguancun cluster (and Beijing, more broadly) to China's agricultural sector has been so significant that it has encouraged several imitation clusters. Science parks in Hunan and Hainan provinces have been built to support hybrid rice development and distribution, while similar science parks are now found throughout the country, including the Zhangjiang Hi-Tech Park in Shanghai, BioBay in Suzhou and Optics Valley in Wuhan.

But the road to success in the Zhongguancun cluster has not been straight. The strong role played by the government in promoting and regulating the agbiotech cluster has been a topic of concern among some policy analysts and researchers. This includes concerns about the crowding out effects of the public R&D system on private R&D investment (Huang et al., 2002), government-led efforts to consolidate the seed industry and force out smaller firms (Seed China News, 2011b), the persistence of issues relating seed labeling and certification, the marketing of spurious seed to farmers and the enforcement of IPR protections (Seed China News, 2011a).

Related to the issue of IPRs and seed distribution in China is the process by which foreign firms have become players in the Zhongguancun cluster and China's wider seed and agbiotech industry. Prior to 2000, the Chinese government had strategic concerns about introducing foreign companies into its agriculture sector and seed industry (Pray, 2001). In 2000, however, the government changed its position and passed a set of seed laws granting to foreign firms extensive access to the Chinese economy. Since then, foreign investment in agriculture and the life sciences, including investment in the Zhongguancun cluster, have expanded significantly. Surveys conducted by Hu et al. (2011) in 2004 and 2009 suggest that most foreign seed and biotech firms view the current system as appropriate enough to justify their investments in developing plant varieties for the Chinese market. A reflection of this confidence is found in Monsanto's establishment of an agbiotech research center in Zhongguancun, as well as investments in similar centers and marketing or operational headquarters in Beijing by crop-science industry leaders such as Syngenta and Bayer CropScience, chemical industry leaders such as BASF, FMC and firms further up the value chain such as AroFresh. One reflection of this is that one of China's most popular maize hybrids, Xianyu 335, is a product of Pioneer, a subsidiary of the multinational DuPont (Seed China News, 2011a).

COMPARATIVE ANALYSIS

Comparing and contrasting these three agbiotech clusters provides an opportunity to understand the implications of any similarities or differences for future agbiotech research in developing countries. We examine three key attributes here: the cluster's rationale, the nature of its emergence and the modalities of knowledge exchange. While this discussion is largely qualitative, it opens the door for further consideration of how cluster performance and impact might be evaluated in the future.

Across all three cases, the implicit rationale for investing in an agbiotech cluster is consistent with the coordination costs hypothesis described above. Public investments in these clusters were largely driven by recognition of the market, institutional and systemic failures associated with the risks and uncertainties of a new technology for resource-poor, small-scale farmers in a developing-country context. Left to its own devices, a purely competitive market would not have encouraged strategic investments in agbiotech. Smallholders were, at the outset, unfamiliar with the technology, unaware of its potential, or unwilling to pay for the end products. Innovators were unfamiliar with the markets, averse to investing under conditions of uncertainty and incomplete information and unwilling to risk their intellectual property in markets without complete enforcement of IPRs. Yet the potential and real benefits of agbiotech, as documented in the extensive literature on *ex ante* and *ex post* impacts (see Smale et al., 2009 for a review), suggested that the greater social benefits of agbiotech exceeded the private benefits, necessitating some form of government intervention. Thus, governments and bilateral and multilateral donors made the investments necessary to jump-start these three clusters and encourage synergies in science, technology and innovation.

It remains to be seen whether, or the extent to which, these clusters actually achieve the goal of overcoming the coordination failures described above. While all of these clusters succeed in generating new knowledge, information and technology where the market would have otherwise failed, the actual success in application and use of these outcomes is largely unanswered. Certainly, the cluster experience in China has generated significant outcomes in the area of Bt cotton commercialization and marketing alongside other, lesser known agbiotech applications of commercial importance. The experiences in India and Kenya, on the other hand, are still too nascent to determine whether their clusters have ultimately overcome coordination failures in a commercially remunerative manner.

Beyond similarities in rationale and outcome, however, the nature of each cluster's emergence differs significantly. The Hyderabad cluster was effectively a government-led initiative that built on a fairly active public infrastructure of research centers and universities, a set of incentivizing policy reforms and a nascent but emerging private sector in India's high-tech sector. While orchestrated by the Government of Andhra Pradesh, the evidence suggests that private investors played a key role in supporting the cluster's development, particularly at later stages. The Beijing cluster is similar in many ways, although the evidence suggests that public investment and policy reforms played a somewhat larger role in creating the cluster and that the private sector role was initially driven by the need of public institutes to generate revenues through commercial spin-offs. The Nairobi cluster, on the other hand, is similarly driven by public investment, but largely from bilateral and multilateral donors who have a long-term vision of an agbiotech cluster without a more statist plan for generating synergies across spatial and virtual dimensions. The notable limitations on the Kenyan government's leadership role in developing this cluster and the challenges associated with attracting investment from neighboring countries in recognition of the cluster's potential spillover effects, may limit the public sector's role in orchestrating the cluster's growth.

Each cluster – and each country – embodies a different style of knowledge exchange. The Nairobi cluster, for example, depends acutely on facilitated knowledge exchanges through donor-funded organizations, projects and partnerships, including AATF for IPRs management, IARCs and their own-data sharing platforms and public–private research collaborations involving multiple actors to develop and deploy new technologies. The Beijing cluster depends more on formal commercial practices – such as spin-off technology-based companies, foreign direct investment, mergers and acquisitions and IPRs licensing arrangements – that are geared toward the rapid realization of the growing market opportunities in China's vast agricultural sector, most notably around Bt cotton. The Hyderabad experience shares similarities with both: a high level of strategic corporate behavior through licensing, foreign direct investment and IPRs, but also a dependence on public investment from both the government and the international donors that support various organizations, projects and collaborations in the cluster.

Implicit in these three points of comparison is divergent financing experiences. The Beijing cluster has benefited primarily from public spending on agricultural R&D and from public institutes' investments in their own commercial enterprises. The Nairobi cluster, on the other hand, has benefited from donor largesse which, history suggests, is subject to a high level of volatility and prolonged periods of neglect (see Stads, 2011 and Beintema and Rahija, 2011). The Hyderabad cluster again falls somewhere in between, and judging from overall trends in private R&D investment in India's agricultural sector (see Spielman et al., 2011), this particular cluster may benefit from a growing share of private financing relative to public financing. These contrasting financing modalities, combined with the evidence on how the clusters emerged and how knowledge exchanges are managed, indicate varying levels of long-term sustainability and differential impact on smallholders in each cluster's country.

POLICY IMPLICATIONS AND CONCLUSIONS

Industry clusters have historically played a central role in economic growth and development, driven significantly by technology spillovers with geographically or virtually concentrated innovation systems. There is evidence of several biotechnology clusters concentrated around developing-country agriculture, although they are still fairly nascent and the body of evidence on their performance is incomplete. Three clusters, in India, Kenya and China, offer insights into the nature of how clusters emerge, the modalities of knowledge exchanges among actors in each cluster and the sustainability of financial investments made in support of clusters.

The policy implications of these three cluster experiences – notwithstanding the still-emerging evidence of their impact on productivity and welfare in smallholder farming systems – are significant. First, there is an obvious and continuing role for public investment in agricultural R&D. Indeed, China and India have historically demonstrated strong commitment to such investments. However, experiences from Kenya and other small- and medium-sized developing countries suggest that policymakers' and donors' commitment to public spending on agricultural R&D has been inconsistent and, at times, quite volatile. This can have significantly negative implications for any efforts to promote biotechnology research and cluster growth for developing-country agriculture. The development and commercialization of new varieties – whether with or without the use of biotechnology tools – is a process that requires at least a decade of concerted investment and scientific effort, but policymakers' and donors' timelines are simply too short to carry these investments from start to finish.

Second, there is a need for policies that create strong innovation incentives in support of cluster development. The growth of biotechnology clusters depends on a research and business environment that provides strong legal and financial incentives for innovators. Incentives should be designed to encourage innovators in both the public and private sectors to commercialize, license, exchange or otherwise utilize their inventions for economic gains. Incentives should also be designed to ensure that innovators can recoup their investment in research, including provisions for legal recourse for intellectual property theft. Furthermore, incentives should be designed to protect the consumers – in this case, small-scale, resource-poor farmers or food-insecure consumers – from unreasonable welfare losses associated with a monopolistic firm's market power or non-competitive marketing practices. Commercialization programs, intellectual property rights, competition policies and functioning legal enforcement are among the key elements of an effective innovation incentive system.

Finally, consideration should be given to the role that context specificity plays in the success of any cluster. Policymakers are often keen on duplicating a successful cluster experience with similar investments in other localities or industries. Given the scarcity of public resources in developing countries, the risks associated with such investments are often non-trivial, especially when driven by site-specific demonstration effects. Rigorous analysis of the causes of success in a given cluster and the potential for replication in other contexts needs to be at the forefront of decision-making around cluster policies and investments.

In conclusion, this chapter demonstrates that there are several emerging examples of biotechnology clusters that are successfully leveraging public and private resources for the benefit of developing-country agriculture. Their potential to supply new tools, products and value for small-scale, resource-poor farmers is an important rationale for further investment and evaluation of these clusters. Finally, while the analysis set forth in this chapter is largely qualitative in nature, it suggests areas for further research on how cluster performance and impact might be evaluated and how policies to support cluster design and development might be formulated in other developing countries.

NOTE

1. The exchange value was computed by using the average nominal exchange rate from 1986 to 2005. All the following conversions follow the same procedure.

REFERENCES

Alliance for a Green Revolution in Africa (AGRA) (2009), 'Alliance for a Green Revolution in Africa', available at: http://www.agra-alliance.org/, accessed 5 April 2012.

Alston, J.M. and W.J. Martin (1995), 'Reversal of fortune: immiserizing technical change in agriculture', *American Journal of Agricultural Economics*, **77**(2), 251.

Alston, J.M., C. Chan-Kang, M.C. Marra, P.G. Pardey and T.J. Wyatt (2000), *A Meta-Analysis of the Rates of Return to Agricultural R&D: Ex Pede Herculem?*, IFPRI Research Report No. 113, Washington, DC: International Food Policy Research Institute.

Ariga, J. and T.S. Jayne (2010), 'Private-sector responses to public investments and policy reforms: the case of fertilizer and maize market development in Kenya', in D.J. Spielman and R. Pandya-Lorch (eds), *Proven Successes in Agricultural Development*, Washington, DC: International Food Policy Research Institute.

Ariga, J., T.S. Jayne and J. Nyoro (2006), 'Factors driving the growth in fertilizer consumption in Kenya, 1990–2005: sustaining the momentum in Kenya and lessons for broader replicability in sub-Saharan Africa', working paper, Nairobi: Tegemeo Institute of Agricultural Policy and Development, Egerton University.

Audretsch, D.B. (1998), 'Agglomeration and the location of innovative activity', *Oxford Review of Economic Policy*, **14**(2), 18–29.

Beintema, N. and M. Rahija (2011), 'Human resource allocations in African agricultural research: revealing more of the story behind the regional trends', paper presented at the conference 'Agricultural R&D: investing in Africa's future: analyzing trends, challenges and opportunities', organized by the Agricultural Science and Technology Indicators (ASTI) Initiative, International Food Policy Research Institute (IFPRI) and Forum for Agricultural Research in Africa (FARA), Accra, Ghana, 5–6 December.

Beintema, N.M. and G.J. Stads (2011), *Africa Agricultural R&D in the New Millennium: Progress for Some, Challenges for Many*, Washington, DC: International Food Policy Research Institute.
Biosciences Eastern and Central Africa (BecA) Project (2011), 'Biosciences Eastern and Central Africa', available at: http://www.ilri.org/BECA, accessed 4 April 2012.
Biospectrum-ABLE (Association of Biotechnology Led Enterprises) (2010), 'The eighth survey of Indian biotech industry', available at: http://ableindia.in/able_biospectrum_surveys.php, accessed 4 April 2012.
Caniëls, M.C.J. and H.A. Romijn (2003), 'Firm-level knowledge accumulation and regional dynamics', *Industrial and Corporate Change*, **12**(6), 1253–78.
Chen, K.Z. and Y. Zhang (2011), *Foresight Project on Global Food and Farming Futures Regional Case Study: R2 Agricultural R&D as an Engine of Productivity Growth: China*, London: Government Office for Science.
Chen, K.Z., K. Flaherty and Y. Zhang (2012), 'China: recent developments in agricultural research', Agricultural Science and Technology Indicators (ASTI) country note, Rome/Washington, DC: International Food Policy Research Institute.
Coase, R. (1937), 'The nature of the firm', *Economica*, **4**, 386–405.
Coase, R. (1960), 'The problem of social cost', *Journal of Law and Economics*, **3**, 1–44.
Cohen, W.M. and D.A. Levinthal (1989), 'Innovation and learning: the two faces of R&D', *Economic Journal*, **99**(397), 569–96.
CNPVP (The Information Network of China Plant Varieties Protection) (2004), 'The information network of China plant varieties protection' [In Chinese], available at: http://www.cnpvp.com/show.asp?ClassCode=0012&n_id=3250, accessed 18 April 2012.
Dosi, G., C. Freeman, R. Nelson, G. Silverberg and L. Soete (eds) (1988), *Technical Change and Economic Theory*, London: Pinter.
Edquist, C. (1997), *Systems of Innovation: Technologies, Institutions and Organizations*, London: Pinter.
Evenson, R.E. and D. Gollin (eds) (2003), *Crop Variety Improvement and Its Effects on Productivity: The Impact of International Agricultural Research*, Wallingford: CABI Publishing.
Evenson, R.E. and M. Rosegrant (2003), 'The economic consequences of crop genetic improvement programmes', in R.E. Evenson and D. Gollin (eds), *Crop Variety improvement and its Effects on Productivity: the Impact of International Agricultural Research*, Wallingford: CABI Publishing.
Flaherty, K., F. Murithi, W. Mulinge and E. Njuguna (2010), *Kenya: Recent Developments in Public Agricultural Research*, Rome/Nairobi: Agricultural Science and Technology Indicators (ASTI) Initiative/Kenya Agricultural Research Institute (KARI).
Freeman, C. (1987), *Technology Policy and Economic Performance: Lessons from Japan*, London: Pinter.
Freeman, C. (1995), 'The national systems of innovation in historical perspective', *Cambridge Journal of Economics*, **19**(1), 5–24.
Freeman, H.A. and W. Kaguongo (2003), 'Fertilizer market liberalization and private retail trade in Kenya', *Food Policy*, **28**(5–6), 505–18.
Freeman, H.A. and J.M. Omiti (2003), 'Fertilizer use in semi-arid areas of Kenya: analysis of smallholder farmers' adoption behavior under liberalized markets', *Nutrient Cycling in Agroecosystems*, **66**(1), 23–31.
Hagedoorn, J. (2002), 'Inter-firm R&D partnerships: an overview of major trends and patterns since 1960', *Research Policy*, **31**(4), 477–92.
Hall, A. (2005), 'Capacity development for agricultural biotechnology in developing countries: an innovation systems view of what it is and how to develop it', *Journal of International Development*, **17**(5), 611–30.
High Tech Research ('863') Program (2012), 'Guojia gao jishu yanjiu fazhan jihua' [In Chinese], available at: http://www.863.gov.cn/, accessed 5 April 2012.
Hu, R., Q. Liang, C. Pray, J. Huang and Y. Jin (2011), 'Privatization, public R&D policy and private R&D investment in China's agriculture', *Journal of Agricultural and Resource Economics*, **36**(2), 416–32.
Huang, J., S. Rozelle, C. Pray and Q. Wang (2002), 'Plant biotechnology in China', *Science*, **262**(5132), 377–8.
Huang, J., H. Lin, R. Hu, S.Rozelle and C. Pray (2006), 'Eight years of Bt cotton in farmer fields in China: is the reduction of insecticide use sustainable?', Beijing Center for Chinese Agricultural Policy, Academy of Science.
ISAAA (International Service for the Acquisition of Agri-biotech Applications) (2012), 'International service for the acquisition of agri-biotech applications', available at: http://www.isaaa.org/, accessed 5 April 2012.
Jia, H. (2011), 'Newsmaker: biocentury transgene', *Nature Biotechnology*, **29**(1), 12.
Krugman, P. (1991) 'Increasing returns and economic geography', *Journal of Political Economy*, **99**(3), 483–99.
Lundvall, B. (1992) *National Systems of Innovation*, London: Pinter.
Lundvall, B. (1985), *Product Innovation and User-Producer Interaction*, Äalborg: Äalborg University Press.
Metcalfe, J.S. (1995), 'The economic foundations of technology policy', in P. Stoneman (ed.), *Handbook of the Economics of Innovation and Technological Change*, Oxford: Blackwell.
Nelson, R.R. (1993), *National Innovation Systems: A Comparative Analysis*, Oxford: Oxford University Press.
North, D.C. (1971), 'Institutional change and economic growth', *The Journal of Economic History*, **31**(1), 118–25.

Omamo, S.W. and L.O. Mose (2001), 'Fertilizer trade under market liberalization: preliminary evidence from Kenya', *Food Policy*, **26**(1), 1–10.
Pal, S., R.P. Singh and M.L. Morris (1998), 'India', in M.L. Morris (ed.), *Maize Seed Industries in Developing Countries*, Boulder, CO: Lynne Rienner.
Pardey, P.G. and N.M. Beintema (2001), 'Slow magic: agricultural R&D a century after mendel', Technical Report no. 36, Washington, DC: Agricultural Science and Technology Indicators/ International Food Policy Research Institute.
Parrilli, D. and B.T. Asheim (eds) (2012), *Interactive Learning for Innovation: A Key Driver within Clusters and Innovation Systems*, London: Palgrave Macmillan.
Pingali, P.L. and G. Traxler (2002), 'Changing locus of agricultural research: will the poor benefit from biotechnology and privatization trends?', *Food Policy*, **27**, 223–8.
Porter, M.E. (2000), 'Location, competition and economic development: local clusters in a global economy', *Economic Development Quarterly*, **14**(1), 15–34.
Pratt, A.N. and S. Fan (2010), 'R&D investment in national and international agricultural research', IFPRI Discussion Paper no. 986, Washington, DC: international Food Policy Research Institute.
Pray, C.E. (2001), 'Public–private sector linkages in research and development: biotechnology and the seed industry in Brazil, China and India', *American Journal of Agricultural Economics*, **83**(3), 742–7.
Pray, C.E. and L. Nagarajan (2010), 'Pearl millet and sorghum improvement in India', in D.J. Spielman and R. Pandya-Lorch (eds), *Proven Successes in Agricultural Development*, Washington, DC: International Food Policy Research Institute.
Pray, C.E. and B. Ramaswami (2001), 'Liberalization's impact on the Indian seed industry: competition, research and impact on farmers', *International Food and Agribusiness Management Review*, **2**(3–4): 407–20.
Program for Biosafety Systems (PBS) (2012), 'Program for Biosafety Systems, International Food Policy Research Institute', available at: http://pbs.ifpri.info/, accessed 5 April 2012.
Rabobank (2006), 'Indian seed industry: market overview and outlook', Industry Note 184-2006, prepared by the Global Department of Food and Agribusiness Research and Advisory, Rabobank International, Utrecht: Rabobank.
Raitzer, D.A. and T.G. Kelley (2008), 'Benefit–cost meta-analysis of investment in the international agricultural research centers of the CGIAR', *Agricultural Systems*, **96**, 108–23.
Ramaswami, B. (2002), 'Understanding the seed industry: contemporary trends and analytical issues', *Indian Journal of Agricultural Economics*, **57**(3), 417–29.
Renkow, M. and D. Byerlee (2010), 'The impacts of CGIAR research: a review of recent evidence', *Food Policy*, **35**(5), 391–402.
Ruan, J. and X. Zhang (2009), 'Finance and cluster-based industrial development in China', *Economic Development and Cultural Change*, **58**(1), 143–64.
Sadashivappa, P. and M. Qaim, (2009), 'Bt cotton in India: development of benefits and the role of government seed price interventions', *AgBioforum*, **12**(2), 172–83.
Schmitz, H. (1995), 'Collective efficiency: growth path for small-scale industry', *Journal of Development Studies*, **31**(4), 529–66.
Schmitz, H. and K. Nadvi (1999), 'Clustering and industrialization: introduction', *World Development*, **27**(9), 1503–14.
Seed China News (2011a), 'Analysis into corn seed market 2011', Vol. 1, January.
Seed China News (2011b), 'CNSGC benefits much from government policy', Vol. 1, January.
Sina Finance News (2010), 'A city without farmers: a base for advanced agriculture' [in Chinese], available at: http://finance.sina.com.cn/roll/20101102/23303505792.shtml accessed 18 April 2012.
Smale, M., P. Zambrano, G. Gruère, J. Falck-Zepeda, I. Matuschke, D. Horna, L. Nagarajan, I. Yerramareddy, J. Jones and J. Falck-Zepeda et al. (2009), 'Measuring the economic impacts of transgenic crops in developing agriculture during the first decade', *Food Policy Review*, **10**, Washington, DC: International Food Policy Research Institute.
Smith, A. '[1776] 1993', *An Inquiry into the Nature and Causes of the Wealth of Nations*, reprint, New York: Oxford University Press.
Spielman, D.J. (2007), 'Pro-poor agricultural biotechnology: can the international research system deliver the goods?', *Food Policy*, **32**(2), 189–204.
Spielman, D.J., F. Hartwich and K. Von Grebmer (2010), 'Public–private partnerships and developing-country agriculture: evidence from the international agricultural research system', *Public Administration and Development*, **30**(4), 261–76.
Spielman, D.J., D. Kolady, A. Cavalieri and N.C. Rao (2011), 'The seed and agricultural biotechnology industries in India: an analysis of industry structure, competition and policy options', IFPRI discussion paper 1103, Washington, DC: IFPRI.
Stads, G. (2011), 'Africa's agricultural R&D funding rollercoaster: an analysis of the elements of funding volatility', paper presented at conference on 'Agricultural R&D: investing in Africa's future: analyzing trends,

challenges and opportunities', organized by the Agricultural Science and Technology Indicators (ASTI) Initiative, International Food Policy Research Institute (IFPRI) and Forum for Agricultural Research in Africa (FARA), Accra, Ghana, 5–6 December 2011.
State Council of the People's Republic of China (PRC) (1988), 'Interim regulations of the Beijing Municipality concerning the experimental area for developing new-technology industries' [in Chinese], in *Laws and Regulations of the People's Republic of China Governing Foreign-Related Matters*, the Bureau of Legislative Affairs of the State Council of the People's Republic of China (eds).
Theus, F. and D. Zeng (2012), 'Agricultural clusters', in *Agricultural Innovation Systems: An Investment Sourcebook*, Washington, DC: World Bank.
Waithaka, M., J. Nzuma, M. Kyotalimye and O. Nyachae (2011), 'Impacts of an improved seed policy environment in Eastern and Central Africa', discussion paper, Entebbe: Association for Strengthening Agricultural Research in Eastern and Central Africa.
Williamson, O.E. (1975), *Markets and Hierarchies*, London: Collier Macmillan.
Williamson, O.E. (1985), *The Economic Institutions of Capitalism*, New York: Free Press.
Zeng, D.Z. (ed.) (2006), *Africa: Knowledge, Technology and Cluster-Based Growth*, World Bank Institute Development Study, Washington, DC: World Bank.
Zeng, D.Z. (2010), *Building Engines for Growth and Competitiveness in China: Experiences with Special Economic Zones and Industrial Clusters*, Directions in Development, No. 56447, Washington, DC: World Bank.

22 Practice driving policy: agbiotech transfer as capacity building

William O. Hennessey, Aarushi Gupta and Stanley P. Kowalski

INTRODUCTION

In the twenty-first century, as arable land and water resource availability stagnates or decreases and populations increase, advancing the development of agricultural technologies (agbiotech) will become increasingly important to meet global food security challenges. This chapter addresses the international technology transfer (tech-transfer) of agbiotech to developing countries within the context of an integrated global innovation system comprised of interlinked technology information networks. Since international tech-transfer is highly specialized and iterative, trained local intermediaries (human and institutional) are indispensable innovation accelerators. The efficiency of agbiotech transactions needs to be increased, thereby lowering the costs expended and the time consumed in the tech-transfer process. For example, intellectual property rights (IPRs) are both assets and tools in the transaction process; they are bought, sold, leased and, in some cases, donated for free in the global technology marketplace. Identifying such assets and utilizing them effectively involves highly sophisticated, knowledge-intensive, interdisciplinary activities.

The international movement of innovations is hampered by whatever proves to be the weakest link in the tech-transfer system or network (which includes the laws and treaties, owners and financers involved in tech-transfer investment). In the case of agbiotech transfer to developing countries, among the weakest links in the system are inadequate IPR, tech-transfer competence and information access and management capacities. Without eliminating such inadequacies, via focused capacity-building at both the human and institutional levels, access to the agbiotech innovations that developing countries increasingly need for food security and economic development will be problematic (at best) and potentially catastrophic (at worst).

Therefore, developing countries must build their own capacities and become full participants in this system rather than remain passive recipients of its benefits, as is largely the case today. This new awareness should enable them not only to foster access to agbiotech innovations necessary to sustain their agricultural/food security, but to enhance their overall economic development and standards of living by creating their own innovations. As they become contributors to agbiotech innovations, such innovations are much more likely to suit local conditions and facilitate follow-on transfers.

The body of the chapter provides case studies of some of the ways agbiotech has been transferred to developing countries in recent years. It highlights the lessons learned, both positive and negative. Finally, it proposes a way forward: human and institutional intermediaries within developing countries gradually mastering tech-transfer skills by doing it themselves instead of having outside experts do it for them over and over again. Many professionals in developing countries have both the aptitude and the motivation to learn effective agbiotech transfer, as well as the local knowledge to make the transfer truly a success.

AGBIOTECH AND DEVELOPING COUNTRIES

Agbiotech, when appropriately applied, stewarded and managed, offers considerable promise for addressing the challenges of developing countries' long-term food and agricultural security, challenges that conventional and traditional agricultural and agronomic practices will no longer be able to address adequately. For example, in Africa planned and projected land expansion and restoration will be inadequate, contributing, at most, 30 per cent of the additional production needed to meet anticipated demand by 2025. Fresh water resources, too limited to make a major contribution even at present, are dwindling. The remainder of the necessary production increases will have to come from adoption of a variety of more productive technologies, including improved crop varieties and agbiotech. Policies that will help widespread adoption of such technologies include improvements in infrastructure, better utilization of educational systems, and focused capacity building (Wiebe et al., 2001).

The positive impact of agbiotech is demonstrable. Over the past two decades, innovative agbiotech adoption has steadily increased in developing countries, having seen more than half of global production in those areas in 2011 (see Figure 22.1), contributing towards poverty alleviation with clear benefits to smallholder farmers. For example, small-scale farmers have benefited significantly from the shift to Bt cotton in South Africa: over 90 per cent of South African cotton is now produced from Bt genetically-engineered varieties. Bt cotton yielded up to a 46 per cent increase per hectare over conventional cotton; farmers paid 42 per cent less for pest control treatments (spraying decreased from an average of 10 applications to 4 per season); women and children have greater time for family and educational pursuits (with a corresponding drop in indirect costs, especially labor inputs); lower production costs coupled with greater yields have increased farmers' revenue margins and improved their overall economic condition. Future innovations (including stacking of Bt, herbicide tolerance, drought and heat tolerance traits) will further benefit farmers and their families (Gouse et al., 2004; Karembu et al., 2009). Similarly in the Philippines and India, the introduction and application of Bt maize, eggplant and cotton have benefited small farmers and reduced poverty (Kolady and Lesser, 2008b; Subramanian and Qaim, 2010; Yorobe and Smale, 2012). Indeed, in general, genetically engineered crops have been shown to benefit smallholder farmers in developing countries. Empirical evidence suggests that the overall impact has been favorable, with multiple studies showing improvements in economic performance for farmers cultivating less than 10 ha in China, Colombia, Mexico, India and South Africa (Carpenter, 2010). Furthermore, developing countries increasingly view agbiotech as crucial for achieving stable and sustainable food security. For example, by supporting strategic objectives to develop/acquire and disseminate improved technologies (including agbiotech), the Nigerian national policy on biotechnology seeks to drive the development of suitable mechanisms and activities to foster the biotechnology enterprise (Chikaire et al., 2012).

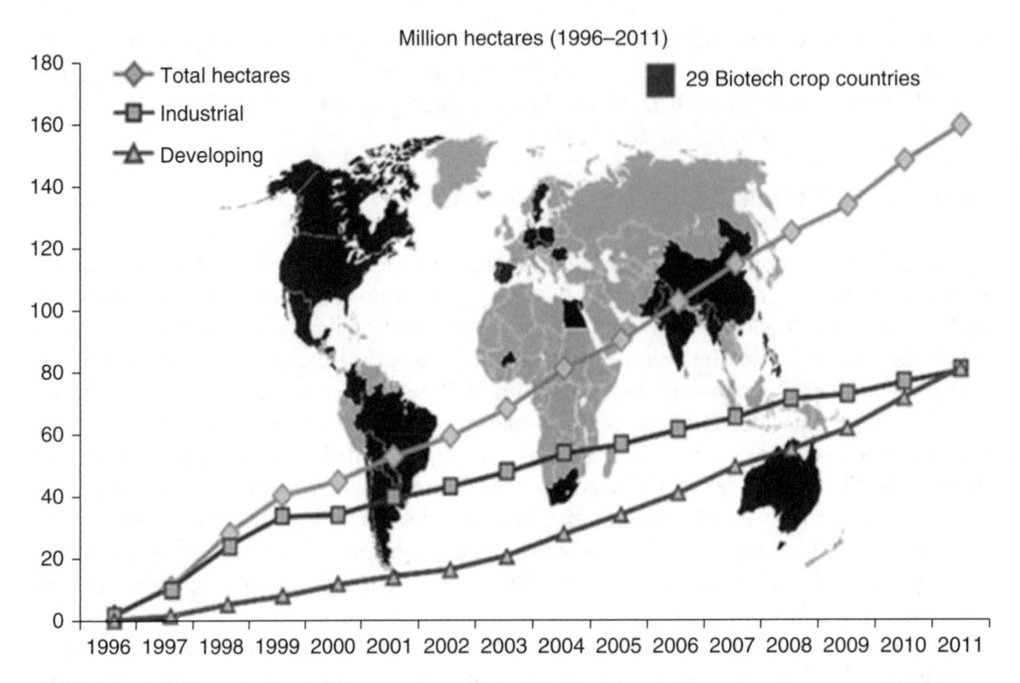

Note: A record 16.7 million farmers, in 29 countries, planted 160 million hectares (395 million acres) in 2011, a sustained increase of 8% or 12 million hectares (30 million acres) over 2010.

Source: Courtesy of Clive James, ISAAA (2011), with permission from ISAAA.

Figure 22.1 Global area of biotech crops

A combination of conventional and innovative technologies in agriculture will be necessary to advance productivity, as countervailing trends including drought, heat, urbanization, salinization, crop pests and diseases negatively impact agricultural efficiency. Access to developed country agbiotech, both directly driven by global agbiotech companies and to a greater extent via tech-transfer, will be key to closing the food security gap in developing countries around the globe (Moussa, 2002). Most advanced agbiotech innovations are, however, owned by developed country entities, often subject to numerous forms of IPR protection; agricultural patent portfolio data suggest that the private sector owns at least 70 per cent of the key IPRs, with an even greater proportion held in the areas of agbiotech, including genetic engineering and crop cultivars (for example, maize varieties). Hence, as most major agbiotech owners are located in developed countries (Rubenstein et al., 2006), international transfer of agbiotech to developing countries is necessary. However, this requires a suite of skills and institutional infrastructure (for example, information management and legal access mechanisms) that are still largely non-existent (United Nations Ministerial Conference of the Least Developed Countries, 2007).

TECH-TRANSFER (IN A NUTSHELL)

Tech-transfer is the sharing of knowledge and facilities among and between industries, universities, governments and other institutions to ensure that scientific and technological developments are accessible to a range of users who can then further develop the technology into new products, processes, materials or services (Kansas State University, 2011). As a technology moves from one entity to another, tech-transfer works effectively if the receiver (transferee) successfully assimilates, applies and utilizes the transferred technology. Transfer might involve tangible and/or intangible assets, including physical property, as with material transfer agreements (MTAs), as well as other IP assets such as patent rights, proprietary processes, know-how and plant variety protection (PVP).

Tech-transfer can occur in a number of directions: from public to private sectors (for example, university to industry for commercialization); from private to public sector (for example, licensing of a patented technology from companies to government and non-governmental agencies); and cross-border from developed to developing countries. Tech-transfer is necessarily transactional, in that the movement of technology is conditioned on agreements and exchanges, which can even be non-monetary, as is the case with some humanitarian licensing provisions. Yet regardless of the direction of the flow of technology, the constant goal remains the maximization of the innovative potential for exploitation of a technology; for example, to transfer and then utilize an advanced heat- or drought-resistant agbiotech for maize in East Africa (Ramanathan, n.d.).

International movement of agbiotech increasingly occurs in a global innovation marketplace. In order to successfully engage, navigate, shop and negotiate in this market, ever greater interdisciplinary expertise by all, including actors in developing countries, is required. Ideally, interdisciplinary expertise should be focused in institutional technology accelerators in developing countries. Such 'innovation accelerators' should combine the functional and operational characteristics of modern tech-transfer offices (TTOs) with a fundamental capacity-building agenda that prepares developing countries to engage rapidly in the global agbiotech innovation market.

PUBLIC–PRIVATE PARTNERSHIPS

Throughout this chapter there is repeated mention of public–private partnerships (P3s). Comprised of partners from the public and private sectors, P3s are joint ventures directed towards a common objective (for example, tech-transfer of drought-, insect- or heat-resistant biotech crops to Africa). P3s are

> any collaborative effort between public and private entities in which each contributes to planning, commits resources, shares risks and benefits and conducts activities to accomplish a shared objective . . . aimed at marrying the efficiency, flexibility and competence of the private sector with the accountability, long-term perspective and social interests of the public sector. (Muraguri, 2010)

Whereas motivations and expectations of the individual partners might vary, the shared P3 purpose drives participation. Partners share decision making, expertise, resources, contributions and risk (Widdus, 2005). P3s can have a significant impact by facilitating and accelerating access to agbiotech in developing countries. However, it is also important to mention that P3s in the context of developing country agriculture, albeit increasing, are still evolving as managerial strategies for advancing agricultural innovation, particularly with respect to private sector collaborations (Spielman et al., 2010).

Participants in a P3 may include entities all along the value chain, including developing-country institutions, multinational corporations, government laboratories and agencies, universities, suppliers, purchasers, international agricultural research centers and philanthropic foundations (Gregory et al., 2008). Familiarity with the many parameters that influence IP management tactics and strategy (for example, assessment of markets, evaluation of candidate technologies, knowledge of regulatory landscapes, complexities of tech-transfer and expertise in the practicalities of delivery) enables P3s to provide dynamic and flexible business models that pool skills, focus funding, formulate strategy to identify challenges and overcome bottlenecks. This reduces transaction and opportunity costs that might otherwise be prohibitive.

For P3s, IPRs are viewed as a tool to facilitate assembly and access to innovation (Wheeler and Berkley, 2001). Hence, for a P3 the aim is not managing IPRs in the standard context as legal, proprietary means to exclude, but rather IPRs as a means for establishing control, assembling partners, mitigating risk, defining objectives, and accelerating access to agbiotech. P3s, therefore, reconcile the possible divergent IP practices and paradigms of partners, such as a developing-country research center and a multinational agbiotech company (for example, the Egypt AGERI, Pioneer Hi-Bred case study presented below). Surprisingly, common space can be found wherein shared objectives and related activities are possible, facilitating strategic application of IPRs. For example, since IP is one driver of pricing, market segmentation can be anticipated as a solution (for example, the Bt eggplant (Brinjal) case study presented below).

Ultimately, the P3 can benefit the industry participant, by reducing risks associated with emerging markets, along with less tangible incentives such as enhancing the corporate public image through demonstration of societal responsibility. For the developing country P3 participants, management of IPR attached to agbiotech can advance food security, public health and economic development (Taubman and Ghafele, 2007).

AGBIOTECH TRANSFER TO DEVELOPING COUNTRIES: CASE STUDIES

The following agbiotech transfer case studies (see also Table 22.1) illustrate the range of actors involved, including but not limited to upstream inventive communities, technology companies seeking new markets and their subsidiaries in developing countries, commercial and government research institutions in developing countries acting to adapt global technologies to regional and local conditions, and ultimately the downstream end-users of transferred technologies (both local seed companies and farmers); these also include the multiplicity of government, international, and nongovernmental organizations acting as broker intermediaries throughout the process (Gregory et al., 2008). From these studies, lessons can also be learned about what works, what problems and blockages hinder effective tech-transfer, what avoidable mistakes were made, and what are the most important issues that need to be addressed going forward.

Table 22.1 Case studies on transfer of agbiotech to developing countries

Project/Case Study	Developing country transferee	Developed country partner/affiliates	Agbiotech transferred	Dates of project	Crops involved	Tech-transfer highlights	Outcomes
Case study 1: Golden Rice	Bangladesh, Philippines, China, India, Indonesia, Vietnam	ISAAA, Greenovation, Syngenta	'Golden Rice': genetically engineered expressing Pro-vitamin A	1990s to present	Rice	P3, Freedom to operate (FTO), assembly of IPR, Humanitarian licensing	'Golden Rice' under development: introgression of phenotype into adapted rice cultivars
Case study 2: Bt insect-resistant eggplant (the Brinjal project)	India, Bangladesh, Philippines	ABSP II, Mahyco (partially owned by Monsanto)	Bt-mediated insect resistance	2000s to present	Eggplant	P3, Licensing with market segmentation between public and private sectors	Commercial development of transgenic eggplant underway
Case study 3: Bt insect-resistant corn	Egypt	ABSP, Michigan State University, Pioneer Hi-bred, USAID	Bt-mediated insect resistance	1990s	Maize	P3	Commercialization being pursued
Case study 4: Bt insect-resistant cotton	India	Mahyco (partially owned by Monsanto)	Bt-mediated insect resistance	2000s to present	Cotton	Multinational Corporate driven	Commercialization
Case study 4: Bt insect-resistant cotton	China	Monsanto, DPL, Singapore EDB	Bt-mediated insect resistance	1990s	Cotton	'Administrative Monopoly' in Hebei Province	Commercialization, but loss of control of technology
Case study 4: Bt insect-resistant cotton	South Africa	Monsanto, DPL	Bt-mediated insect resistance	2000s	Cotton	Multinational Corporate driven	Commercialization

Table 22.1 (continued)

Project/Case Study	Developing country transferee	Developed country partner/affiliates	Agbiotech transferred	Dates of project	Crops involved	Tech-transfer highlights	Outcomes
Case study 5: Bt insect-resistant cabbage	India	Universities of Melbourne, Greenwich and Cornell, Bayer and Dutch Nunhems, AVRDC	Bt-mediated insect resistance	2000s to present	Cabbage	P3, with Bayer providing Bt technology gratis with AVRDC as licensing agent	Commercial development of transgenic brassicas underway
Case study 6: Kenyan agbiotech projects	Kenya	International Atomic Energy Agency (UN affiliate)	Drought resistance	2000s to present	Wheat	R&D collaboration	Varietal development underway
Case study 6: Kenyan agbiotech projects	Kenya	ABSP, Monsanto, ISAAA (provided training)	Virus resistance	1990s to present	Sweet potato	P3: R&D collaboration	Varietal development underway
Case study 6: Kenyan agbiotech projects	Kenya, Mozambique, S. Africa, Tanzania, Uganda	CIMMYT, Monsanto (AATF as intermediary)	Drought resistance	2000s to present	Maize	P3	Varietal development underway
Case study 6: Kenyan agbiotech projects	Kenya	Cornell University, Gates Foundation, DFID	Stem rust resistance	2000s to present	Wheat	R&D collaboration	Varietal development underway

Case study 7: Africa Biofortified Sorghum (ABS)	Kenya (along with several African countries including Nigeria and Burkina Faso)	UC Berkeley and Pioneer Hi-Bred (Africa Harvest as intermediary)	Nutritional enhancement traits/genes	2000s to present	Sorghum	P3	Field testing and introgression of genetically engineered traits into adapted local germplasm
Case study 8: papaya in South-east Asia	India, Indonesia, Malaysia, Philippines, Thailand, Vietnam	ISAAA, University of Hawaii, Monsanto	Virus resistance	1990s to present	Papaya	R&D collaboration	Field testing: varietal release pending approval
Case study 8: papaya in South-East Asia	Indonesia, Malaysia, Philippines, Thailand, Vietnam	ISAAA, Syngenta, University of Nottingham	Delayed ripening	1990s to present	Papaya	R&D collaboration	Field testing: varietal release pending approval
Case study 8: papaya in South-East Asia	Thailand	Cornell University, University of Hawaii	Virus resistance	1990s to present	Papaya	Segmentation of market, with commercial and humanitarian licensing	Field testing: varietal release pending approval
Case study 9: Insect-resistant corn	Indonesia	ISAAA, Syngenta	Insect resistance	1990s	Tropical maize	P3	Project did not succeed

Note: Refer to text for details of case studies and cited pertinent references.

Whereas these case studies, for the most part, illustrate successful short-term transfer and adoption of agbiotech from developed country owners to developing country recipients, they are also likely to represent unsustainable models, which if continued into the current century present the risk for long-term failure. For example, to date, many P3s for delivery of advanced agbiotech to developing countries have been based on a market segmentation model targeting commercially lower-end products in non-competing (that is, humanitarian-based) markets, or, alternatively, a unidirectional model driven by multinational corporations and their local agents (Byerlee and Fischer, 2001; Khush, 2007). In other words, little to no developing-country public sector participation or involvement in the tech-transfer process is evident. Freedom to operate (FTO), brokering, licensing and IP management all appear to be largely undertaken by multinational corporations, their affiliates and subsidiaries, developed-country universities, international broker organizations and developed-country governmental agencies. Where are the developing-country actors, and why are they not more active?

Case Study 1: Golden Rice

Golden Rice, as perhaps the most widely publicized and thoroughly articulated case study, shows how a creative IP management strategy overcame potential obstacles and facilitated subsequent successful delivery of a critically important agbiotech product (Kryder et al., 2000; Kowalski and Kryder, 2002). Upon closer examination, however, it also provides insights into the longer-term limitations of such a strategy.

In developing countries, vitamin A deficiency (VAD), a major problem primarily affecting children under age five, leads to permanent blindness, with up to 250000 children losing their sight every year. VAD also depresses immune systems, with a resulting increase in the incidence and severity of infectious diseases and infant mortality rates. VAD-afflicted children die at nine times the rate of healthy children, with over 1 million children dying of infections every year. As a means to mitigate VAD, relevant genes were genetically engineered into rice, with the goal to enhance beta-carotene in local rice varieties and thereby in local diets: hence, Golden Rice. A great deal of technical complexity was embedded in Golden Rice. Genes (enzymatic activities) were required to catalyze several steps in the carotenogenic biosynthetic pathway. This involved multiple inputs, of both methods and materials (that is, process and product patents).

A systematic analysis (product deconstruction/FTO analysis) of the Golden Rice product was conducted to determine the IP and tangible property (TP) issues potentially embedded. The FTO analysis of Golden Rice tentatively identified 15 TP components and 70 patents (with 31 assignees) of potential relevance. It was subsequently determined, however, that only a few patents pertaining to Golden Rice were applicable in developing countries, along with several TP issues (for example, MTAs).

In effectuating the Golden Rice tech-transfer, assembly of IPR was still deemed expedient, as there were a number of potentially patented inputs in the final genetically engineered beta-carotene rice. This was done efficiently and sequentially, with potential IPR constraints resolved by an IP management, P3 strategy comprising:

1. identification of major IP/TP components (the FTO analysis conducted by ISAAA);

2. inventors of the Golden Rice product assigning their IPR (patent) to Greenovation, which then licensed these IPRs to Zeneca (Syngenta);
3. interpretation, by ISAAA with Zeneca, of the relevance of the FTO review to the proposed humanitarian use in developing countries;
4. licensing in, for humanitarian use, led by and into Zeneca, IPR components it did not already own, including from Bayer AG, Monsanto Company, Novartis AG, Orynova BV and Zeneca Mogen BV; and
5. licensing by Zeneca (Syngenta), via the inventors, of the assembled (or bundled) IPR to public sector institutions in developing countries that could use the rights for the benefit of resource-poor farmers, and others, deficient in vitamin A (Krattiger and Potrykus, 2007).

Licenses, defined for and limited to humanitarian use, were granted back to the inventors of Golden Rice by Syngenta with humanitarian sub-license provisions; Syngenta, however, retained all commercial rights. The inventors could then transfer sub-licenses to developing countries and related institutions (such as the CGIAR, for example, IRRI), for further humanitarian development and access. Licenses to this Golden Rice package were granted to Bangladesh, the Philippines, China, India, Indonesia and Vietnam for introgression of the technology (Golden Rice genotype) into local rice varieties important in VAD areas. Golden Rice is currently still under development for delivery to VAD-afflicted countries, with successful field trials in Louisiana and introgression into local rice varieties in the Philippines and India, where bioavailability trials are underway (Khush, 2007).

It is also important to note that IPRs associated with Golden Rice would not only have related to patents. Germplasm and know-how or show-how might also require consideration. 'The previously described [research proof of concept] lines were generated using the *Japonica* rice line *Taipei 309* because it is relatively easy to transform. *Indica* cultivars, although much less amenable to transformation, are the most predominantly grown and consumed in countries where VAD prevails.' (Hoa et al., 2003). *Taipei 309* was found to be entirely unsuitable for agricultural production, as it was used primarily as a laboratory strain and was not adapted to field conditions of South and South-East Asia. Hence, the Golden Rice phenotype needed to be introduced (via introgression of the complex genotype) into suitably adapted rice cultivars. Whether or not IPR issues related to germplasm are of potential concern needs to be considered whenever moving agricultural technologies from the laboratory to the field; in the case of Golden Rice, this involved the movement of several proprietary genetic constructs from the laboratory rice strain to the locally adapted varieties.

With Golden Rice, as with other agbiotech innovations in general, it will most likely also be necessary to access ancillary IPR not found in patent specifications, such as know-how and show-how possibly protected as trade secrets. As these often make up the bulk of IPRs embedded in advanced innovations, for example, an agbiotech product (Jorda, 2007), access to such trade secrets might also need to be a priority consideration. Furthermore, without understanding this type of IPR, transfer and application of advances in agbiotech to the needs of developing countries could remain attenuated. Comprehensive IP management, which aims to accelerate global access to agbiotech, will therefore require sophisticated capability to identify and evaluate agbiotech innovations and (all of) the possible IPR attached thereto, with a concomitant ability to establish research collaborations, confidentiality agreements, licenses, and/or P3s.

This case study offers a number of lessons. First, in spite of the rhetoric, IPRs did not delay the development and introduction of Golden Rice into developing countries wherein VAD is endemic. Notwithstanding this, the resolution of the potential IP constraints could not be ignored. The inventors of Golden Rice ultimately assigned exclusive rights in the technologies to Syngenta. Syngenta, as the driver, facilitated access to a number of integral technologies, whether owned by Syngenta or other entities, and then granted a humanitarian license which facilitated sub-licensing to public research institutions and low-income farmers in the target developing countries. The P3 between the inventors of Golden Rice and Syngenta was the critical link. All the partners had clearly-defined competence and expertise, and hence, each had its specific role in accelerating the global delivery of Golden Rice. The process reduced the number of patents involved to the necessary minimum, secured humanitarian licenses, established managerial and marketing structures and, ultimately, developed plants optimized both to meet regulatory requirements and to express high levels of desired traits (accumulation of beta-carotene in rice endosperm). Golden Rice shows how public and private sector innovations can be put to work directly to help a developing country prosper via focused and coherent IP management strategy: Golden Rice access was facilitated by dynamic, creative analysis, partnerships and agreements. Furthermore, the FTO review of Golden Rice, in particular before commercial analysis, served to call attention to IP management as a powerful tool for accelerating access to critical innovations in agbiotech (and health).

Golden Rice, as the vanguard for a new class of agbiotech innovations, is a model case study of the need for IP management to facilitate effective, efficient and equitable distribution to developing countries. The Golden Rice story is complex, with early failures in awareness and management of IPRs rescued by a successfully implemented creative licensing approach. Equally important to note is that the VAD-afflicted developing-countries' involvement with the transfer of Golden Rice technologies and the related IPR management was largely downstream, after the earlier crucial steps of identifying IPR constraints, developing strategies to overcome these and then forming the requisite organizational capacity (for example, the P3) to accelerate Golden Rice towards deployment and commercial availability where most needed. Early steps in IP management were carried out by developed-country entities, including Syngenta, ISAAA and the Golden Rice Humanitarian Board.

As IP management was largely undertaken and facilitated by developed countries (that is, non-VAD-afflicted countries), crucial, and perhaps uncomfortable, questions become apparent. Is this a sustainable system for accelerating global access to critical innovations in agbiotech (and health)? Should the developing countries consider becoming themselves the drivers of innovation access and deployment instead of bystanders, observers and possibly (and only sometimes) downstream facilitators? Can it be assumed that in the twenty-first century, developing countries can continue to rely on external, predominantly developed-country actors to intervene with similar humanitarian impulses? Or should organizations such as developing-country National Agricultural Research Systems (NARS) assume a more active role in the delivery and commercialization of crucial innovations in agriculture from the start? If so, and given that IPRs are central to the process of tech-transfer, how critical will be the role of ongoing capacity building to achieve a sustainable system?

In sum, in order to foster sustainable improvement of absorptive capacity for advanced innovations in agriculture, serious and committed capacity building in all aspects of IPR protection, management and transfer expertise should be a priority of developing countries. The Golden Rice story, albeit a success in many ways, is also a wake-up call; it should serve as an alarm to developing countries, that reliance on external humanitarian entities to address critical food security issues is, at best, limiting, and at worst dangerous. Therefore, capacity building to move developing countries from passive bystanders to active participants in addressing their food security must become national priorities. Tech-transfer does not automatically happen; the infrastructure must be built and maintained in order to accelerate access to ever more advanced innovations.

Case Study 2: Bt Insect-resistant Eggplant (the Brinjal Project in India, Bangladesh and the Philippines)

Eggplant, also knowm as Brinjal, is one of the most important vegetables consumed in South and South-East Asia. Bangladesh, the Philippines, India and China are the major producers, accounting for nearly 90 per cent of world production. Throughout South and South-East Asia, however, production of a marketable eggplant is compromised due to herbivorous pests, the most destructive being the eggplant fruit and shoot borer (EFSB) *Leucinodes orbonalis*. This often necessitates exceptionally high rates of pesticide application. Estimates suggest that the crop losses due to pests and diseases in India, Bangladesh and the Philippines range from 54 to 70 per cent and could be 100 per cent if losses from fruit and shoot borer are included. Therefore, there exists a high level of interest in developing fruit and shoot borer resistant (FSBR) eggplant (Gregory et al., 2008; ABSP II, 2011).

In this case, participants included:

1. in India, Hyderabad-based Sathguru Management Consultants Pvt Ltd;
2. the Agricultural Biotechnology Support Project II (ABSP II as a broker), funded by the United States Agency for International Development (USAID);
3. Indian institutional partners, including the Indian Institute of Vegetables Research (IIVR), Tamil Nadu Agricultural University (TNAU), and the University of Agricultural Sciences (UAS, Dharwad);
4. in the Philippines, the Institute of Plant Breeding at the University of the Philippines at Los Banos (IPB-UPLB);
5. in Bangladesh, private sector partner East-West Seeds, a leading seed producer with established market positions in all of the South and South-East Asian target markets and public sector Bangladesh Agricultural Research Institute (BARI); and
6. Mahyco, a private Indian company partially owned by Monsanto, for hybrid Bt eggplant.

Mahyco developed hybrid Bt eggplant varieties resistant to EFSB. The project was to transfer, via conventional plant breeding, this Bt resistance into eggplant varieties for India, Bangladesh and the Philippines. In the case of the Brinjal Project, segmenting markets was feasible–poor farmers would have free access only to the open pollinated varieties while the hybrid varieties would be sold for commercial production. ABSP II initiated and facilitated negotiations between Mahyco-Monsanto and each of the partner countries, which led to licensing agreements that established FTO for all partners. Monsanto, owning the patent rights to the *Cry1Ac* gene and other genetic components such as promoter sequences, licensed these to Mahyco. Mahyco would then sub-license these technologies to the downstream partners. Sathguru, in conjunction with ABSP II, partnered with Mahyco to facilitate the process of tech-transfer to the downstream partner countries, for example, the several public sector institutions. Mahyco sub-licensed the technology, royalty free, to IIVR, TNAU, UAS, Dharwad in India, BARI in Bangladesh, and UPLB in the Philippines. Mahyco also licensed the technology to East-West Seeds on contractual, commercial and royalty-bearing terms.

The major lesson from this case is that market segmentation is a way for the public and private sectors to split an agbiotech asset in such a way that both the commercial interests of the private sector and the development interests of the public sector might be simultaneously accomplished. In other words, both commercial producers and resource-poor farmers will have access to an advanced agbiotech, albeit in different forms, including an open-pollinated variety for the poor farmers as contrasted with an elite hybrid variety for the commercial seed producers (Byerlee and Fischer, 2001). Segmenting markets is feasible when conditioned on varying levels of production capacity, market access and financial factors (Kolady and Lesser, 2008a).

Case Study 3: Bt Insect-resistant Corn (Egypt)

The application of biotechnology to develop transgenic crop plants resistant to pathogenic viruses, bacteria, fungi and insect pests, as well as environmental stresses such as salinity, drought and heat is of crucial importance to Egyptian agriculture and food security. Established in 1990 to promote the transfer and application of agbiotech, the Agricultural Genetic Engineering Research Institute (AGERI) seeks to identify, access and adapt agbiotech to address these challenges (Madkour, 2000). Throughout the 1990s, AGERI collaborated with Pioneer Hi-bred Seeds (Pioneer) to introduce agbiotech appropriate to the needs of Egyptian agriculture. Bt strains were characterized and then patented by AGERI. Pioneer, as part of the technical collaboration, was granted access to these patented Bt genes for further evaluation and research. The AGERI project involved a total of five parties: AGERI, the United States Agency for International Development (USAID), Pioneer Hi-Bred, the Agricultural Biotechnology Support Project (ABSP) and Michigan State University (MSU). Supported by USAID, the project was brokered by ABSP, managed by MSU, with the MSU Office of IP (OIP) and patent counsel and tech-transfer specialists from Pioneer facilitating negotiations and providing critical inputs (Madkour, 2000; Khush, 2007). Negotiations took nearly a year to complete (Maredia et al., 1999).

Regarding the relationship between AGERI and Pioneer, IPR issues were key components to the success of the partnership and the project. ABSP addressed IPR issues in Egypt via training, workshops and the services of a consultant (from Stanford University) who focused on licensing and related IP issues arising from the international transfer of agbiotech. In addition, ABSP assured Pioneer that confidentiality would be observed; IPR protection would need to be delineated and IP management protocols established even before research proceeded (Lewis and Johanson, 2000). Pursuant to the agreement, IP arising from the project would be shared. However Pioneer Hi-Bred and AGERI would solely continue to own their hitherto proprietary genes, germplasm and Bt genes, with the caveat that AGERI would furnish MTAs (the transfer of possession of tangible property but not title, that is, bailment, Bennett et al., 2007) in order to provide Pioneer access to Bt toxin proteins for evaluation (Madkour, 2000).

The goal of the AGERI–Pioneer collaborative project was to develop Bt transgenic maize, resistant to corn borers endemic to Egypt, as well as other Middle East and nearby corn production areas; ABSP provided support, policies and direction to facilitate tech-transfer (Lewis and Johanson, 2000). This case study illustrates several important points, suggesting that it might serve as a prototypical best practices template. First, the P3 between Pioneer Hi-Bred International and AGERI was notable because it was between a developed country private-sector multinational corporation and a developing country public-sector agricultural research institution. Second, AGERI owns IPR to technology arising from the project, filing patent applications on several broad-spectrum Bt toxins in Egypt and the United States, which it subsequently licensed to Pioneer Hi-bred (Madkour, 2000; Khush, 2007). Third, it is particularly notable, and perhaps unique to the case studies presented in this chapter, that efforts were made not only to build human capital in IP management and tech-transfer, but to take the next step: building institutional infrastructure via the establishment of a Technology Transfer Office at AGERI for management of IP issues (Brink, 2002). Hence, the IPR and tech-transfer capacity building initiatives were dynamic, that is, moving beyond the time-honored workshop and white paper format towards sustainable investment in human capital and institutional infrastructure. However, it is also important to note that the fundamental management of IPR and related tech-transfer was still facilitated by developed-country entities (for example USAID, ABSP and MSU OIP) and not the developing-country partner (Egypt).

The P3 embodied notable and innovative features. AGERI came to the table with its own property, that is, a number of locally isolated strains of *Bacillus thuringiensis* with insecticidal activity which Pioneer would be likely to find of interest. This was complemented by Pioneer's contributions to the project, which included marketing and legal expertise of value to AGERI. The exchange of proprietary materials and rights was reciprocal and not only a linear, one-directional tech-transfer; in this respect, the P3 was unlike many others, typically involving a developed-country technology donor and developing-country recipient (Lewis and Johanson, 2000). The AGERI–Pioneer collaboration therefore illustrates not only tech-transfer but, more importantly, co-development of advanced agbiotech applicable to the needs of Egyptian agriculture; central to the project, both parties contributed, with reciprocal movement of proprietary assets (IPR from Pioneer, Bt samples from AGERI) and concomitant application for IPR protection (Lewis and Johanson, 2000; Madkour, 2000; Brenner, 2004). Collaboration between Pioneer and AGERI was also somewhat novel, in that the private sector partner (Pioneer) demonstrated a continued interest in maintaining a stake in the further development of these technologies, which is not necessarily typical, as illustrated in other case studies in this chapter (Brenner, 2004).

Case Study 4: Bt Insect-resistant Cotton (India, China and South Africa)

For the transfer of Bt cotton technology to India, the Indian Council of Agricultural Research (ICAR) and the Department of Biotechnology, Government of India (DBT) initiated a dialogue with Monsanto. Subsequently, Mahyco, partnering with its parent company Monsanto, pioneered the introduction of Bt cotton into India in 2002. Bt cotton usage in India is substantial, with over 3.5 million ha under cultivation (Asia-Pacific Consortium on Agricultural Biotechnology, 2007).

In 1996 in China, Monsanto and Delta & Pineland (DPL) and the Singapore Economic Development Board (EDB) established a joint venture with the Hebei Provincial Seed Company for producing and marketing Monsanto's Bt cotton technology through Ji Dai, a newly formed company. Variety 33B carrying Monsanto's Bt gene had been approved for commercial use in Hebei province. The Monsanto Bt gene was neither patented in China, nor protected by PVP as laws had not yet been passed. Monsanto decided that they would be able to protect their variety through Hebei Provincial Seed Company's administrative monopoly on cotton seed sales in Hebei Province. However, outside of Hebei Province the situation was different. In Anhui, Shandong and Henan provinces, where Monsanto did not have an administrative monopoly, Chinese farmers and seed companies were able to freely plant and reap the benefit of the 33B Monsanto/DPL varieties. Weak IPRs and the absence of administrative monopolies outside Hebei therefore limited Monsanto's ability to control its Bt cotton technology in other provinces of China (Pray et al., 2001).

In contrast, in South Africa Monsanto Corporation licensed and commercialized, via its seed supplier DPL, its Bt cotton technology. This was conducted via the private and public seed sector companies in South Africa directly, as the basic R&D had already been carried out in the USA (Gouse et al., 2004).

The Bt cotton case studies illustrate the key role multinationals can play in tech-transfer. The introduction of Bt cotton in India, China and South Africa was entirely driven from within Monsanto and its partners, sometimes in partnership with public agencies and other times directly through its commercial subsidiaries (Mahyco and DPL). In contrast to Bt papaya and Bt Brinjal, Bt cotton is one of Monsanto's core commercial interests. For obvious reasons, Monsanto was not interested in humanitarian licenses to subsistence farmers. The Chinese case study also illustrates the importance of enforceable IPR in international agbiotech-transfer; the administrative monopoly with Hebei was predicated on compliance, which was a challenge from the start.

Case Study 5: Bt Insect-resistant Cabbage

India is one of the major producers of *Brassica oleracea* (in the cabbage family) worldwide, with approximately 0.5 million ha under production, producing about 6.3 million tons annually. Diamondback moth (*Plutella xylostella*) infestation causes an annual loss estimated at US$16 million (or approximately 25 per cent damage), forcing frequent insecticide application with additional production inputs reaching 38 per cent. Agbiotech, via Bt technology, offers a solution for durable, cost-effective insect resistance; in this case this was realized via transgenic plants with a dual Bt gene construct, expressing two different Bt endotoxins. A P3 for the development of Bt Brassicas was established, called the Collaboration on Insect Management for Brassicas in Asia and Africa (CIMBAA).

CIMBAA, formally headed by AVRDC (The World Vegetable Center, Taiwan), includes the Centre for Environmental Stress and Adaptation Research (CESAR) of the University of Melbourne, Cornell University, and the National Resources Institute (NRI) of the University of Greenwich (UK) as public partners. In addition, a number of research partners are involved that address specific research aspects of the project. Nunhems India Pvt Ltd, a major Indian seed company that is part of the Dutch Nunhems Seeds Corporation (a subsidiary of Bayer CropScience), is the only private partner in the consortium. Financially, the project is supported by Nunhems (which pays for the development of the gene constructs, the transformation, field trials and part of the safety analyses). Bayer CropScience, the patent owner in the CIMBAA consortium, provided its Bt technology gratis. Transformed germplasm ownership will be transferred to a public partner in the consortium (for example, AVRDC) to sub-license the technology to Nunhems. Nunhems is then to transfer (through introgression breeding) the Bt resistance into elite germplasm, considering possible niche markets for commercialization (Vroom, 2009).

The CIMBAA case study once again illustrates the effectiveness of complex P3s for agbiotech-transfer. The driver was AVRDC, an international non-profit research and development institution, with significant cooperation from Nunhems, a subsidiary to Bayer, as well as universities in the UK and USA. As with other case studies, the principal drivers and participants are not the principal beneficiaries, that is, the developing countries, but rather the owners and developers of the technology.

Case Study 6: Kenyan Agbiotech Projects

There are four recent examples of efforts to address Kenya's current crises in agriculture, but none of them are Kenyan-driven.

First, the International Atomic Energy Agency (IAEA) drought-resistant wheat project seeks to develop drought-tolerant wheat varieties for Kenya. Eastern Africa experiences periodic food shortages which are partly due to intermittent but severe drought. In Kenya, semi-arid lands represent 83 per cent of the total land area (56.9 million ha). Although some 300000 ha of this area is arable, it is not fully utilized due to unfavorable weather patterns and a lack of drought-resistant wheat varieties. The project set out to develop improved wheat varieties which would be tolerant of drought, utilizing limited rainfall more efficiently. In addition, the wheat was developed to have moderate resistance to stem rust and yellow rust, improved yield, and good baking quality (International Atomic Energy Agency, 2007).

Second, the Water Efficient Maize for Africa (WEMA) project is focused on South and East Africa. Much of the African continent is drought-prone, making farming risky for millions of small-scale farmers who rely on rainfall to water their crops. Maize is the most widely grown staple crop in Africa – more than 300 million Africans depend on it as their main food source – and it is severely affected by frequent drought. Climate change will only worsen the problem. Drought tolerance has been recognized as one of the most important crop improvement programs, and biotechnology has been identified as a powerful tool to achieve significant drought tolerance. The African Agricultural Technology

Foundation (AATF) is leading a P3 with WEMA to develop drought-tolerant African maize using conventional breeding, marker-assisted breeding and biotechnology. The benefits and safety of the maize varieties are to be assessed by national authorities according to the regulatory requirements in the partner countries: Kenya, Mozambique, South Africa, Tanzania and Uganda. WEMA's other institutional partners include the International Maize and Wheat Improvement Center (CIMMYT) and the Monsanto Corporation (African Agricultural Technology Foundation, 2012).

Third, there is the virus-resistant sweet potato project in Kenya. The sweet potato feathery mottle virus (SPFMV) can lead to 80 per cent yield loss in many parts of Africa where sweet potato is a staple. ISAAA developed and brokered a research partnership which includes the Kenya Agricultural Research Institute (KARI), Monsanto, USAID's ABSP and the Mid-American Consortium. The virus resistance technology, owned by Monsanto, was donated via a royalty-free license for application to sweet potato. Genetically engineered, SPFMV-resistant sweet potato cultivars have been field tested in Kenya (Khush, 2007).

Fourth, the Durable Rust Resistance in Wheat Project (DRRWP) is dedicated to halting the global spread of a potentially devastating strain of wheat stem rust (Ug99), a threat to food security in developing countries which has reached danger level. Ug99 is a particular threat to Kenyan wheat agriculture. In response, the Bill and Melinda Gates Foundation, with the UK Department for International Development (DFID), have supported this project, which seeks to address this crisis (Bill & Melinda Gates Foundation, 2011). Begun in April 2008 by 17 research institutions around the world and led by Cornell University, DRRWP seeks to mitigate that threat through coordinated activities to replace susceptible varieties with durably resistant varieties; advanced agbiotech could be part of the solution to this complex challenge. The strategy will include accelerated multilateral plant breeding and delivery through optimized developing-country seed sectors (Cornell University, 2012).

Similar to the Golden Rice case above, even though these projects are vitally important to Kenya, direct Kenyan intervention has not been notable in any of them; indeed, as illustrated by the majority of case studies presented in this chapter, dynamic, proactive developing-country involvement has been, at best, tangential. At worst, involvement of developing countries in the process has been virtually non-existent. In the case of the WEMA project, negotiations towards formation of the P3, often involving confidentiality, publication rights, public goods and IP, were facilitated solely by American entities, UC Davis (specifically Public Intellectual Property Resource for Agriculture 'PIPRA') and the San Francisco law firm Morrison and Foerster, both based in California. Perhaps this illustrates most clearly that Kenya, as an exemplary African developing country, must build, foster and strengthen its public sector human resource and institutional capacities in IP management and tech-transfer and take the lead in accessing and deploying agricultural technologies critical for its food security, economic growth and general prosperity.

Case Study 7: Africa Biofortified Sorghum (ABS)

As the world's fifth most important cereal crop, grain sorghum, albeit primarily used for feed in the developed countries, is an important food source throughout the semi-arid regions of Africa, indeed the staple for more than 300 million people (UC Berkeley Press Release, 2006). Sorghum has both advantages and disadvantages as a staple crop in Africa: advantages include greater tolerance to environmental stresses (drought, heat and water logging) than maize; disadvantages include inadequate levels of iron, zinc, vitamin A and other minerals, poor nutritional quality due to an amino acid imbalance (inadequate levels of the essential amino acids lysine and tryptophan) and indigestibility of protein and starch in humans due to an abundance of disulfide protein linkages that interfere with enzymatic digestion, the presence of non-waxy starch and the structure of the starch/protein matrix (Case Studies for Global Health, 2009; Kaur et al., 2007). Currently millions in Africa are impacted by serious food shortages, leading to widespread malnutrition and famine (The ABS Project, 2013; Africa Harvest, 2013). The development and deployment of sorghum strains with improved nutritional profiles would therefore be of significant benefit to millions of undernourished Africans who depend on it as their principal staple food.

To address these nutritional deficiencies inherent in sorghum, the Africa Biofortified Sorghum (ABS) Project was launched. A coalition of partners, private/public and international in scope, the ABS Project seeks to develop and deploy advanced and adapted sorghum cultivars with improved nutritional characteristics, towards sustainable mitigation of malnutrition and famine in Africa; this is an increasingly critical issue as climate change and population growth continue to impact food security across wide geographical regions (Blaine, 2011). With funding from the Bill and Melinda Gates Foundation's Grand Challenges in Global Health grant program (as a primary supporter of the ABS Project, approximately $17 million, UC Berkeley Press Release, 2006), Africa Harvest (an African-based, US/Canada incorporated NGO) as the coordinating entity established a consortium, partnering with 11 institutions to conduct research and development on biofortified sorghum (The ABS Project, 2013):

1. DuPont through its business Pioneer Hi-Bred (technology development), USA (which donated $4.8 million in agbiotech to the project, Blaine, 2011);
2. Council for Scientific and Industrial Research (CSIR) (technology development), South Africa;
3. African Agricultural Technology Foundation (AATF) (enabling environment and product distribution), Kenya;
4. The International Crops Research Institute for the Semi-Arid Tropics (ICRISAT) (product development), India;
5. The University of Pretoria (UP) (product development), South Africa;
6. The University of California Berkeley (UC Berkeley) (technology development), USA;
7. The Agricultural Research Council of South Africa (ARC) (product development);
8. The Environmental and Agricultural Research Institute (INERA) (product development), Burkina Faso;
9. The Kenya Agricultural Research Institute (KARI) (product development);
10. The Institute for Agricultural Research (IAR) (product development), Nigeria; and
11. The West African Council for Agricultural Research and Development (CORAF/WECARD) (enabling environment and product distribution), Kenya, Senegal.

Principal agbiotech donors to the ABS Project P3 include Pioneer Hi-Bred, CSIR and UC Berkeley (which has since left the partnership, Case Studies for Global Health, 2009). Pioneer's contribution included sorghum containing 50 per cent more lysine than commonly grown sorghum cultivars and varieties (Speckman, 2008) and in conjunction with CSIR the development of enhanced genetic transformation techniques applicable to sorghum genetic engineering (Sciencescope, 2009). UC Berkeley's contribution included technologies for 'breaking disulfide bonds of plant storage proteins in sorghum grain' (Mimura et al., 2011), via an increase in the levels of TRX proteins in the starchy component of the sorghum grain, thereby to increase digestibility (UC Berkeley Press Release, 2006). Subsequent downstream ABS Project collaborators would then introgress the enhanced traits into locally adapted germplasm, that is,

> [Africa Harvest's] breeder partners in Kenya, Egypt, South Africa and Burkina Faso were provided with sorghum engineered with traits developed at Berkeley (and elsewhere [e.g., possibly Pioneer Hi-Bred alone or via a collaboration with CSIR]), in order to introgress traits (through cross-breeding, or genetic hybridization and repeated back-crossing) into varieties of local interest and then provide the resultant seed to local farmers. (Mimura et al., 2011, pp. 7)

ABS Project progress has shown improved grain sorghum protein quality: 100, 20 and 20 per cent increases in the essential amino acids lysine, tryptophan and threonine respectively, phytate reduction and beta-carotene increase. In addition, there has been optimization and improvement of sorghum genetic transformation technologies, increased bioavailability of zinc and iron, successful field trials in several African countries and backcrossing of ABS traits into African sorghum varieties (introgression into locally adapted germplasm) (AHBFI, 2011, 2012).

Intellectual property management with regard to the ABS Project has involved jointly filed patent applications on innovations arising from the research and development aspects of the project (CSIR and Pioneer Hi-Bred), as well as socially responsible licensing approaches (UC Berkeley).

> UC Berkeley received funding from AHBFI through a sub-award to support its research contribution to the project. . . . In return, UC Berkeley granted to AHBFI a non-exclusive, royalty free license to IP developed with the funding, as well as a license to related research materials. The licenses allow AHBFI ultimately to sell or distribute bio-fortified sorghum (or to allow sublicensees to do so). The improved 'Africa Biofortified Sorghum' varieties will provide more protein, vitamins, and minerals to consumers than existing strains, and the enhanced nutrients will be more 'bio-available' to humans. The license is restricted by a defined field-of-use and, by extension, the charitable objective. UC Berkeley granted the free license prospectively, in advance of IP being created in the collaboration. This approach, and the collaboration structure, preserve incentives so as to induce funding, research, and development by each party as appropriate. (Mimura et al., 2011, pp. 6–7)

Two patent applications filed under the ABS Project relate to improving genetic transformation methods applicable to sorghum (Sciencescope, 2009):

1. a method to improve gene transfer efficiency by activating genetic sequences in the Agrobacterium that are responsible for transferring the T-DNA (Mehlo and Zhao, 2009a), and

2. a process to convert normal cells into a form of progenitor cell line, which is undifferentiated and developmentally flexible (Mehlo and Zhao, 2009b).

'These patents are being filed to protect the technology being developed to enable it to be made available to the poor in line with the policy of the project funders (global access strategy)' (Sciencescope, 2009, pp. 42). Hence, the prevailing IP management strategy for the ABS Project appears to have been focused on enhanced access in the regions where biofortified sorghum is most needed.

Whereas the ABS Project has been repeatedly publicized as an African-led initiative (Speckman, 2008; Case Studies for Global Health, 2009), the make-up of the consortium and the source(s) of funding may belie this assertion. Non-African entities providing significant financial support (either to the ABS Project or the project coordinator Africa Harvest) include the Bill and Melinda Gates, Rockefeller and Ford Foundations, The McLaughlin-Rotman Center for Global Health and USAID (among others, AHBFI, 2012). In addition, two (of the three) technology developers are US-based (UC Berkeley and Pioneer Hi-Bred), with the third (CSIR) working in apparent close collaboration with Pioneer Hi-Bred. The ABS Project's efforts, and indeed early successes, to develop and deliver a crucial innovation in agbiotech to those most in need are noteworthy. However, the complex financial and technological donor structure of the P3 presents questions as to whether this represents a sustainable model for agbiotech-transfer to developing countries.

As with other case studies presented in this chapter, the ABS Project suggests a dearth of leadership by African governmental authorities, that is, the parties who should be most engaged with national food security issues. Muraguri (2010), in a study examining the dynamics of agbiotech P3s in Kenya, has noted that P3s are predominantly science led, supply driven, donor dominated and aid dependent rather than demand or user drawn (by Kenyan institutions), and that a 'wider systems failure in Kenya' is the root cause for this disengagement. National agricultural policy does not support P3s. Kenyan partners have limited capacity and financial resources. With no guiding framework to initiate and promote P3s, they tend to 'evolve spontaneously', in a vacuum devoid of an 'effective conceptual and operative framework', not due to, but 'in spite of the lack of support from national policies'. This situation is not limited to Kenya, but is common throughout the developing world, for example, Latin America (Muraguri, 2010).

Case study 8: Papaya in South-East Asia

Papaya Ringspot Virus (PRSV) causes significant, at times devastating losses to crops in South-East Asia and severely impacts the livelihood of subsistence and smallholder farmers. PRSV resistance technology was owned by Monsanto, in collaboration with the University of Hawaii. ISAAA, mentioned above, brokered and facilitated an agreement with Monsanto to donate the PRSV resistance technology to a network of five countries (Vietnam, the Philippines, Malaysia, Thailand and Indonesia) and also to the Tamil Nadu Agricultural University in India. As part of the program, scientists from the five participating South-East Asian Network countries received training in the fundamentals of IP management. ISAAA likewise brokered a deal for the Network with Syngenta

Seeds and the University of Nottingham for access to their delayed ripening technology (ISAAA, 2011). Due to bureaucratic bottlenecks, progress towards commercialization has been slow. Thailand has field-tested several PRSV-resistant papaya cultivars and Malaysia has field-tested delayed ripening papaya, but varieties are yet to be released for commercial production (Khush, 2007).

In addition, the Papaya Administrative Committee (PAC), a research and marketing group supported by local papaya farmers, facilitated access to PRSV agbiotech, with Cornell University negotiating IPR and FTO on its behalf. Among the key biotechnological inputs necessary for product development were antibiotic resistance genes, the 35S cauliflower mosaic virus promoter and the biolistic gene gun genetic transformation delivery technology. The Cornell Research Foundation (CRF) filed patent applications on PRSV viral coat protein sequences on isolates derived from Jamaica, Venezuela and Brazil. As part of an arrangement for licensing of these (third-party) technologies to the University of Hawaii, CRF also arranged for the technologies to be available to other countries, including developing countries where papaya is an important crop. For example, in a memorandum of understanding with Thailand, humanitarian and commercial uses of the PRSV resistance technology are distinguished: small-scale farmers would have access to royalty-free use of this technology (Davidson, 2008).

This case illustrates that although papaya is a critically important food crop in South-East Asia, as well as in many other countries, with large domestic and export markets and now substantial investment in agbiotech-based improvements (Mendoza et al., 2008), the primary management of IP and tech-transfer related to both PRSV resistance and delayed ripening technologies was handled by a gaggle of developed-country entities, including ISAAA, Cornell University, the University of Hawaii and large, multinational corporate owners of key patents. Developing-country participation was largely downstream and ancillary, for example scientists from the five participating South-East Asian Network countries receiving training in the fundamentals of IP management at ISAAA.

Case Study 9: Insect-resistant Corn in Indonesia

In Indonesia, ABSP attempted to broker a collaboration between Syngenta and the Central Research Institute for Food Crops (CRIFC). The goal was the development of tropical maize resistant to the African corn borer. This project was not successful, largely due to a lack of IP awareness and management capacity in the Indonesian partner. This case clearly illustrates that IP management capability is vital to advance successful P3s. Lack thereof can either stall or derail an otherwise promising project. In other words, efficiency in the system is negatively affected by this constraint, thereby increasing transaction costs beyond what is acceptable to the partners (Khush, 2007). The better the IP management capacity and capability of the developing-country partner, the more efficient will be the transfer of agbiotech innovations. This case clearly illustrates that the weakest link in the agbiotech-transfer chain, in this case IP management in Indonesia, prevented any tech-transfer.

THE GLOBAL AGBIOTECH SYSTEM: THE INNOVATION MARKETPLACE AND ROLE FOR INTERMEDIARIES

A system is a set of interdependent components forming an integrated whole, involving elements and relationships wherein there is movement and interaction (for example, the oceanic ecosystem). Agbiotech increasingly exists in a global innovation system, with its structure defined by:

1. the players, including the producers, owners, broker intermediaries, transferors, transferees and consumers (mostly farmers) of agbiotech;
2. the various laws and treaties (including IP laws such as patent, PVP, trade secrets and germplasm) that regulate the protection, flow of, and access to agbiotech; and
3. the elements (technological components and processes which comprise agbiotech) which includes crop varieties, germplasm resources, plant genetic resources, biodiversity and advanced agbiotech (for example, genetic engineering inputs, tools, such as genomics, gene maps and banks).

The interconnectivity of these components forms a global innovation marketplace. The functioning of every component in this system determines its overall efficiency – the costs of transacting between the components is key. With adequate human resource capacity and capability, a developing country should itself be able to identify and access multiple pieces of agbiotech in this open innovation market.

The concept of a global agbiotech system is consistent with the open-innovation paradigm. Open innovation stresses that organizations should use external as well as internal knowledge to drive innovation and advance technology towards commercialization (Chesbrough, 2003). Under the open-innovation concept, innovation moves in multiple directions as it flows through the global system. In the rapidly emerging global knowledge-based economy, organizations must not rely solely on their own research efforts, but should instead buy or license patented processes or articles as inputs when necessary to accelerate their technological progress. Hence, a rapidly emerging technospace in agbiotech is increasingly driven by economic opportunities for development and new applications, leading to an omni-directional and global tech-transfer ecosystem, yielding spillovers of technologies from the developed countries to be absorbed by the developing countries (Hennessey, 2005).

Golden Rice illustrates this concept, in that it is a complex assemblage of agbiotech inputs (Kowalski and Kryder, 2002), comprising multiple property owners from multiple countries with, potentially, layers of proprietary rights embedded in them, including patents, trade secrets, plant variety protection and TP (for example, MTA). Complex agbiotech products can be assembled and commercialized more rapidly if there exists capability to systematically access the global open innovation system and shop for the best processes and components available. This will increasingly be the case in both agricultural and health innovation.

To operate in this global agbiotech system effectively and efficiently, established intermediaries are necessary to act as accelerators of innovation flows. In an open innovation system, intermediaries function as hubs for connecting technology providers with technology absorbers, via the various mechanisms of tech-transfer, including licensing, collaborative research agreements and MTAs. This is particularly critical in agbiotech, where multiple inputs, potentially owned by government, university and/or corporate entities across the world are involved. Identifying, connecting and then facilitating networked innovation will lower transactions costs, combine optimal inputs and accelerate development of critical innovations (Yang and Shyu, 2010).

In the context of developing countries establishing such intermediaries, investment in human and institutional tech-transfer must be focused, ideally in centralized tech-transfer offices (TTOs) located in key S&T research universities and institutes. While the central mission of these TTO intermediaries will be to accelerate local absorption and adaptation of agbiotech, their operations might also include education, awareness-building, advocacy and networking, as well as the many aspects of IP management (for example, licensing, agreements, deciding what to patent and where and the use of patents and trademarks) (Lim and Park, 2010).

A cost-effective approach could be to gather talent and resources into a centralized TTO that can serve, and also provide training to, multiple institutions and coordinate other national programs building IP awareness, management and capacity, toward the goal of creating a national *culture* of innovation management (that is, a proto-innovation ecosystem). A hub and spoke model might be an optimal initial investment: the hub being a centralized facility that focuses expertise and operations, and then serves as an anchor and driver for further development in the country. Additionally, as Wolson (2004, p. 9) makes clear:

> for IPRs to achieve their objective of incentivizing innovation in developing countries, it is essential that developing countries build capacity to manage IP 'intelligently' – that is, with sufficient knowledge of how the system works, with a comprehensive understanding of the options available, and in a manner appropriate to the surrounding circumstances and environment. . . . Concerted efforts must be made to integrate the full suite of relevant policy interventions and support measures as seamlessly as possible for maximum impact. (See also Wolson, 2007; Payumo, 2011.)

For developing countries to connect to the global agbiotech innovation market and advance tech-transfer, human resource development is key. This could include adequate training of personnel and investment in concomitant institutional infrastructure (Dartey-Baah, 2010). For sustainable success, capacity building must be ongoing rather than one-time, haphazard, or sporadic, as Payumo (2011, pp. 114) articulates:

> Implementation of a sound IP management program by these [developing country public sector R&D] institutions . . . needs continuing capacity building. Researchers and scientists of these institutions should be aware of the potential value of IPRs and of the interest of their laboratories and institutions; be knowledgeable on how to protect their own interests; and be familiar with the rules governing the fair and proper use of IP. Governments and institutions in developing countries have to invest in building the necessary IP and legal capacities.

This, in turn, decreases risk, transaction costs and associated delays in the development and deployment of appropriate agbiotech. This will increasingly be the case with non-commodity subsistence crops: agbiotech, albeit initially deployed in major international commodity crops such as cotton, maize and soybean, will also have considerable potential for improving crops of localized importance, such as cassava, eggplant and pearl millet (Payumo, 2011).

Capacity development is multidimensional. It will necessitate linkages between developed-country producers of knowledge, owners of IP and potential users in developing countries. In particular, emphasis needs to be focused less on producing new knowledge and more on access and productive use of existing knowledge. Building innovation capacity will include 'skills, knowledge and experience (scientific, but also entrepreneurial), institutional structures and linkages or networks connecting science, consumers, entrepreneurs, intermediary organizations and policy bodies.' (Hall, 2005, pp. 616). This is something that should not be pursued haphazardly, but rather in a strategic and thoughtful manner, to focus and maximize resources, talent and expertise in any developing country (Hall, 2005). In addition, there must be full and active involvement on the part of the developing countries themselves, with their own people taking the lead. They cannot, and should not, completely rely on foreign institutional, professional and other help; the latter path is unreliable and likely to become a dangerous practice in the decades to come. This realization also underscores the importance, and necessity, of learning by doing (Davis, 2005); perhaps the irritatingly trite aphorism 'practice makes perfect' is apt here. But it truly does highlight the need for active involvement and investment on the part of the developing countries in order to connect into the global innovation and knowledge-based economy. Actual hands-on experience in tech-transfer and IP management thus becomes the driver of even greater capacity building, which then feeds back, informing strategy and policy, to further energize the process.

CONCLUSION

As the case studies presented herein illustrate, movement of agbiotech from developed to developing countries has been predominantly driven by developed-country entities: governmental agencies and their projects (USAID, USDA, ABSP, ABSP II,), non-governmental organizations, NGOs (ISAAA), brokering agents and advisors (various university TTOs, legal professionals), or multinational agbiotech corporations and/or their local affiliates in developing countries (Monsanto, DPL, Syngenta or DuPont's Pioneer Hi-Bred International). In addition, agbiotech has frequently been transferred via donation or discounted access. Whereas this has certainly led to successful transfer of agbiotech to developing countries, evidenced both by the case studies in this chapter (see Table 22.1) as well as by the aggregate data compiled by ISAAA (see Figure 22.1), the serious concern relates not to *what* was transferred but *how* it was transferred, and whether this represents an adequate and sustainable system for global agbiotech transfer. This concern is all the more urgent in light of the significant challenges that global agriculture is likely to face over the coming decades: heat, drought, population growth and urbanization (particularly in developing countries), pests and disease.

For developing countries to continually depend on, and indeed expect, the largesse and humanitarian impulses of developed-country entities to facilitate and expedite the transfer of crucial innovations in agbiotech might not be an optimal strategy to foster sustainable agricultural systems and food security in the twenty-first century. Hence, to accelerate agbiotech-transfer to developing countries, strategically focused, politically supported, economically committed capacity building, in both human capital and institutional infrastructure, is now necessary; these are the bricks, mortar and reinforced concrete of knowledge-based economic development, fundamental for the emerging economies of Africa, Asia and Latin America – a new development model for a new century. Developing countries need to prioritize investment in these areas, as this is the foundation for accelerating access to critical innovations in health and agriculture.

Unfortunately, governments (in both developed and developing countries), intergovernmental organizations, philanthropic foundations, development banks and NGOs appear to be still largely operating according to a model of development firmly rooted in post-colonial preconceptions of the past century. This structured, paternalistic, project-driven paradigm of development fails to recognize that an innovation revolution is sweeping the globe, that critically important innovations in agriculture, health, energy and communications will need to move along a virtual infrastructure in order to reach those who need them most, that investments are urgently needed to build the systems for accelerating tech-transfer, and that global food security, with all the potential ramifications connected thereto, will be affected by what is done and what is left undone.

Pragmatically speaking, developing countries need personnel trained in tech-transfer, IP management and related business, technical and legal disciplines. These personnel need to be focused in institutional entities, whether called ITECs, TTOs or TISCs:

> A supportive legal environment is necessary but not sufficient for . . . effective technology transfer . . . must be supplemented by the establishment of an Innovation and Technology Entrepreneurship Center (ITEC) to handle . . . spinning-in, adapting for local use, and spinning-out technology. This organization can either be a newly established entity or an existing unit within an established organization (Inclusive Innovation Center or university technology transfer centers), retrofitted to carry out new functions. (Watkins and Mandell, 2010, pp. 20–23)

> A framework to allow technology transfer to the public institutes of developing countries must be stimulated and developed. This has been addressed in some countries by the establishment of . . . TTOs. TTOs are often located in a governmental unit associated with some aspect of agriculture. These offices work with researchers, allowing them to develop new crop varieties, and with government officials to develop appropriate laws and policies for intellectual property protection. They develop means for providing plant variety protection, biotechnology invention protection and intellectual property management. TTOs can play multiple roles in research and development (R&D) institutes, [including] protection of intellectual property . . . revenues through licensing of intellectual property . . . education and awareness, networking . . . creation of new start-up companies . . . institutional policies related to technology transfer [and] service to society. (Maredia et al., 2000, pp. 16–17)

> Technology Innovation Support Centers (TISCs) act as service-oriented providers to: allow local users to benefit effectively from the increased accessibility of intellectual property information offered by internet searches through direct personal assistance; assist local users in creating, protecting, owning and managing their intellectual property rights; strengthen the local technological base by building up or reinforcing local know-how; and to increase technology transfer, e.g. by investigating the possibilities of licensing, joint ventures, etc. In short, TISCs are established so as to act as local drivers of innovation. The training of TISC so as to be able to assist local users and deliver these services is one of the most important elements . . . and while initial training may be focused on searching patent and non-patent technology databases . . . further training in other areas of intellectual property rights is considered particularly useful, as it not only continues to develop staff knowledge and their personal development, but also offers a one-stop-shop as regards other elements of intellectual property rights and of innovation support. (Takagi and Czajkowski, 2012, pp. 32–3)

As the case studies also illustrate, agbiotech exists in a global innovation marketplace, where 'pieces' of a potentially vital technology are scattered, with multiple owners. For example, the numerous inputs necessary to develop Golden Rice needed to be identified, assembled, accessed and ultimately cleared for commercial development. For developing countries, when considering agbiotech innovations and their application, the closed linear approach (as practiced by many developed country research universities and institutes) of invent-disclose-patent-license is insufficient and probably dangerous to adhere to, as many agbiotech innovations are already in the global innovation marketplace. Hence a broader skill-set in tech-transfer, along with an IP toolbox (IP Handbook of Best Practices, 2007) will facilitate transactions in this marketplace, a marketplace that will increasingly be driven by an open innovation paradigm (Chesbrough, 2003).

IPRs stimulate tech-transfer, especially advanced innovations such as agbiotech. This is facilitated via a self-reinforcing cycle of technological absorption, adaptation and diffusion. To occur efficiently, there are critical complementary factors that must be recognized and fostered: quality knowledge institutions (for example, universities and national research centers), financial systems (providing access to research funding), and human capital and networks for research and development interactions (public–private sector and international collaborations) (Park and Lippoldt, 2008). In sum, to accelerate development and growth, there must be competent and motivated individuals and institutions within developing countries to make the tech-transfer happen. As a result, countries will be able to introduce agbiotech from other countries into their markets, and eventually technologically improved agbiotech innovations may even 'spill back' to the developed countries (Hennessey, 2005).

REFERENCES

ABS Project (2013), available at: http://biosorghum.org/abs_project.php, accessed 15 January 2013.

Africa Harvest Biotech Foundation International (AHBFI) (2011), *Africa Harvest Annual Report 2010*, Nairobi, Kenya: AHBFI.

Africa Harvest Biotech Foundation International (AHBFI) (2012), Africa Harvest Strategic Plan 2012-22. Nairobi, Kenya: AHBFI.

Africa Harvest (2013), 'Africa Harvest helping Africa fight hunger and malnutrition with biofortified sorghum', available at: http://africaharvest.org/files/ABS.pdf, accessed 15 January 2013.

African Agricultural Technology Foundation (2012), 'Water efficient maize for Africa', available at: http://www.aatf-africa.org/wema, accessed 7 February 2012.

Agricultural Biotechnology Support Project (ABSP) II (2011), 'Global partnership develops pest resistant Brinjal to enhance rural farm income and livelihood', available at: http://www.absp2.net/index.php?option=com_content&view=article&id=30&Itemid=31, accessed 21 December 2011.

Asia-Pacific Consortium on Agricultural Biotechnology (2007), 'Brainstorming session on models of public-private partnership in agricultural biotechnology – highlights and recommendations', New Delhi, India: Asia-Pacific Consortium on Agricultural Biotechnology and Trust for Advancement of Agricultural Sciences.

Bennett, A.B., W.D. Streitz and R.A. Gacel (2007), 'Specific issues with material transfer agreements' in A. Krattiger, R.T. Mahoney, L. Nelsen et al. (eds), *Intellectual Property Management in Health and Agricultural Innovation: A Handbook of Best Practices*, Oxford, UK: MIHR and Davis, USA: PIPRA, pp. 697–716.

Bill & Melinda Gates Foundation (2011), 'Armed with US$40 million, global research team to fight Ug99; wind-borne wheat pathogen endangers food security worldwide', available at: http://www.gatesfoundation.org/press-releases/Pages/fighting-ug99-pathogen-110227.aspx, accessed 14 March 2012.

Blaine, S. (2011), '"Super" sorghum shows biotechnology helps food security', *Business Day (South Africa)*, available at: http://www.bdlive.co.za/articles/2011/09/08/super-sorghum-shows-biotechnology-helps-food-security, accessed 15 January 2013.

Brenner, C.T. (2004), 'Transgenic technology tales: lessons from the Agricultural Biotechnology Support Project (ABSP) experience', ISAAA Briefs no. 31, Ithaca, NY: ISAAA, available at: https://www.isaaa.org/resources/publications/briefs/31/download/isaaa-brief-31-2004.pdf.

Brink, J. (2002), 'Agricultural Biotechnology Support Project, Egypt Project final report', Lansing, MI: Michigan State University, available at: http://pdf.usaid.gov/pdf_docs/Pdacf932.pdf.

Byerlee, Derek and Ken Fischer (2001), 'Accessing modern science: policy and institutional options for agricultural biotechnology in developing countries', *IP Strategy Today*, no. 1-2001, available at: www.biodevelopments.org/ip/ipst1n.pdf.

Carpenter, Janet E. (2010), 'Peer-reviewed surveys indicate positive impact of commercialized GM crops', *Nature Biotechnology*, **28**(4), 319–21.

Case Studies for Global Health (2009), 'Alliance for case studies for global health, available at: http://www.casestudiesforglobalhealth.org/assets/content/Case%20Studies,%202009%20edition/Sorghum_Malnutrition_GHCS_09.pdf, accessed 15 January 2013.

Chesbrough, Henry (2003), *Open Innovation: The New Imperative for Creating and Profiting from Technology*, Boston, MA: Harvard Business School Press.

Chikaire, J., F.N. Nnadi, N. Ejiogu-Okereke and J.A. Echetama (2012), 'Agricultural biotechnology and biosafety: tools for attaining food security and sustainable industrial growth in Nigeria', *Continental Journal of Agricultural Science*, **6**(1), 6–22.

Cornell University (2012), 'Durable rust resistance in wheat', available at: http://www.wheatrust.cornell.edu/, accessed 14 January 2012.

Dartey-Baah, Kwasi (2010), 'Technology transfer and human resource constraints and challenges: a note to the developing world', Academic Leadership, **8** (1), available at: http://www.academicleadership.org/article/Technology_Transfer_and_Human_Resource_Constraints_and_Challenges_A_note_to_the_developing_world.

Davidson, Sarah Nell (2008), 'Forbidden fruit: transgenic papaya in Thailand', *Plant Physiology*, **147**(2), 487–93.

Davis, Kevin E. (2005), 'Regulation of technology transfer to developing countries: the relevance of institutional capacity', *Law & Policy*, **27**(1), 6–32.

Gouse, Marnus, Carl Pray and David Schimmelpfennig (2004), 'The distribution of benefits from Bt cotton adoption in South Africa', *AgBioForum*, **7**(4), 187–94.

Gregory, Peter, Robert H. Potter, Frank A. Shotkoski et al. (2008), 'Bioengineered crops as tools for international development: opportunities and strategic considerations', *Experimental Agriculture*, **44**(3), 277–99.

Hall, Andy (2005), 'Capacity development for agricultural biotechnology in developing countries: an innovation systems view of what it is and how to develop it', *Journal of International Development*, **17**(5), 611–30.

Hennessey, Bill (2005), 'Changing traffic patterns in technospace', *Michigan State Law Review*, **2005**(1), 201–18.

Hoa, Tran Thi Cuc, Salim Al-Babili, Patrick Schaub, Ingo Potrykus and Peter Beyer (2003), 'Golden Indica and Japonica Rice Lines Amenable to Deregulation', *Plant Physiology*, **133**(1), 161–9.

International Atomic Energy Agency (2007), 'Developing crop resistant wheat for Kenya's drylands', available at: http://tc.iaea.org/tcweb/publications/projectprofiles/AFR-AGR-KenyaWheat.pdf, accessed 21 January 2012.

IP Handbook of Best Practices (2007), 'The IP toolbox summary and overview', available at: http://www.iphandbook.org/handbook/ch04/summary/, accessed 12 September 2012.

ISAAA (2011), 'ISAAA SEAsiacenter: Papaya', available at: http://www.isaaa.org/inbrief/regionalcenters/seasiacenter/default.asp.

James, Clive (2011), 'Global status of commercialized biotech/ GM crops: 2011', *International Service for the Acquisition of Agri-Biotech Applications (ISAA) Brief*, 43, Ithaca, NY: ISAAA.

Jorda, K.F. (2007), 'Trade secrets and trade-secret licensing', in A. Krattiger, R.T. Mahoney, L. Nelsen et al. (eds), *Intellectual Property Management in Health and Agricultural Innovation: A Handbook of Best Practices*, Oxford, UK: MIHR, and Davis, USA: PIPRA, pp. 1043–57.

Kansas State University (2011), 'Technology transfer and development', available at: www.k-state.edu/cecd/technologydev/, accessed 3 November 2011.

Karembu, Margaret, Faith Nguthi and Ismail Abdel-Hamid (eds) (2009), *Biotech Crops in Africa: The Final Frontier*, Nairobi, Kenya: ISAAA AfriCenter.

Kaur, R., J. Wong, J. Singh, J.F. Pedersen, P.G. Lemaux and B.B. Buchanan (2007), 'Toward development of nutritionally enhanced sorghum', meeting abstract, available at: http://www.spcru.ars.usda.gov/research/publications/publications.htm?seq_no_115=215567&pf=1, accessed 15 January 2013.

Khush, Gurdev S. (2007), 'Biotechnology: public–private partnerships and intellectual property rights in the context of developing countries', in Charles R. McManis (ed.), *Biodiversity and the Law*, Sterling, VA: Earthscan, pp. 179–91.

Kolady, Deepthi and William Lesser (2008a), 'Can owners afford humanitarian donations in agbiotech – the case of genetically engineered eggplant in India', *Electronic Journal of Biotechnology*, **11**(2), 7–11.

Kolady, D.E. and W. Lesser (2008b), 'Potential welfare benefits from the public–private partnerships: a case of genetically engineered eggplant in India', *International Journal of Food, Agriculture and Environment*, **6**(3–4), 333–40.

Kowalski, Stanley P. and R. David Kryder (2002), 'Golden rice: a case study in intellectual property management and international capacity building', *RISK: Health, Safety & Environment*, **13**, 47–67.

Krattiger, A. and I. Potrykus (2007), 'Golden Rice: a product-development partnership in agricultural biotechnology and humanitarian licensing', in A. Krattiger, R.T. Mahoney, L. Nelsen et al. (eds), *Executive Guide to Intellectual Property Management in Health and Agricultural Innovation: A Handbook of Best Practices*, Oxford, UK: MIHR and Davis, USA: PIPRA, pp. CS11–CS14.

Kryder R.D., S.P. Kowalski and A.F. Krattiger (2000), 'The intellectual and technical property components of pro-vitamin A rice (golden rice): a preliminary freedom-to-operate review', *ISAAA Briefs* no. 20, Ithaca, NY: ISAAA.

Lewis, J. and A. Johanson (2000) 'The role of biotechnology policies and regulations in technology transfer to developing countries', in William H. Lesser (ed.), *Transitions in Agbiotech: Economies of Strategy and Policy*, Storrs, CT: University of Connecticut Food Marketing Policy Center, pp. 425–35.

Lim, Hyojeong and Yongtae Park (2010), 'Identification of technological knowledge intermediaries', *Scientometrics*, **84**(3), 543–61.

Madkour, M.A. (2000), 'Egypt: biotechnology from laboratory to the marketplace: challenges and opportunities', in G.J. Persley and M.M. Lantin (eds), *Agricultural Biotechnology and the Poor: Proceedings of an International Conference*, Washington, DC: Consultative Group on International Agricultural Research (CGIAR), pp. 97–9.

Maredia, K.M., F.H. Erbisch, C.L. Ives and A.J. Fischer (1999), 'Technology transfer and licensing of agricultural biotechnologies in the international arena', *AgBiotechNet*, **1**(May), pp. 1–7.

Maredia, K.M., F.H. Erbisch and M.J. Sampaio (2000), 'Technology transfer offices for developing countries', *Biotechnology and Development Monitor*, no. 43, 15–18.

Mehlo, Luke and Zuo-Yu Zhao (2009a), 'Method for regulating agrobacterium-mediated transformation', US Patent Application 20090263902, filed 21 January 2009 and published 22 October 2009.

Mehlo, Luke and Zuo-Yu Zhao (2009b), 'Production of progenitor cereal cells', US Patent Application 20090293157, filed 22 January 2009 and published 26 November 2009.

Mendoza, E.M.T., A.C. Laurena and J.R. Botella (2008), 'Recent advances in the development of transgenic papaya technology', in M.R. El-Gewely (ed.), *Biotechnology Annual Review*, Amsterdam: Elsevier, pp. 423–62.

Mimura, C., J. Cheng and B. Penhoet (2011), 'Socially responsible licensing, Euclidean innovation, and the valley of death', *Stanford Journal of Law, Science and Policy*, UC Berkeley Public Law Research Paper, no. 1928837, available at: http://papers.ssrn.com/sol3/papers.cfm?abstract_id=1928837, accessed 15 January 2013.

Moussa, Sami Zaki (2002), 'Technology transfer for agriculture growth in Africa', *Economic Research Papers*, **72**.

Muraguri, L. (2010), 'Unplugged!: an analysis of agricultural biotechnology PPPs in Kenya', *Journal of International Development*, **22**(3), 289–307.

Park, Walter G. and Douglas C. Lippoldt (2008), 'Technology transfer and the economic implications of the strengthening of intellectual property rights in developing countries', *OECD Trade Policy Working Papers*, no. 62.

Payumo, Jane G. (2011), 'An analysis of the implications of strengthened intellectual property rights to agriculture of developing countries and responses of selected public research institutions in Southeast Asia, unpublished PhD dissertation, Washington State University, The Graduate School.

Pray, Carl E., Ann Courtmanche and Ramu Govindasamy (2001), 'The importance of intellectual property rights in the international spread of private sector agricultural biotechnology', report to WIPO, Rutgers, the State University of New Jersey.

Ramanathan, K. (no date), 'An overview of technology transfer and technology transfer models', available at: http://www.business-asia.net/Pdf_Pages/Guidebook%20on%20Technology%20Transfer%20Mechanisms/An%20overview%20of%20TT%20and%20TT%20Models.pdf, accessed 21 December 2011.

Rubenstein, Kelly Day, Paul W. Heisey and John L. King (2006), 'Public sector technology transfer through patents and licensing: the case of US agriculture', *International Journal of Technology Transfer and Commercialisation*, **5**(4), 401–20.

Sciencescope (2009), 'Nutritionally enhanced sorghum for improved African health', November 2009, pp. 40–43, available at: www.csir.co.za/publications, accessed 15 January 2013.

Speckman, A. (2008), 'African scientist's work on sorghum is showcase of science's contribution to development', *The Council for Scientific and Industrial Research (CSIR) South Africa*, available at: http://www.csir.co.za/news/2008/11/Luke.html, accessed 15 January 2013.

Spielman, D.J., F. Hartwich and K. Von Grebmer (2010), 'Public–private partnerships and developing-country

agriculture: evidence from the international agricultural research system', *Public Administration and Development*, **30**(4), 261–76.
Subramanian, Arjunan and Matin Qaim (2010), 'The impact of Bt cotton on poor households in rural India', *Journal of Development Studies*, **46**(2), 295–311.
Takagi, Y. and A. Czajkowski (2012), 'WIPO services for access to patent information: building patent information infrastructure and capacity in LDCs and developing countries', *World Patent Information*, **34**(1), 30–36.
Taubman, A. and R. Ghafele (2007), 'Public sector IP management in the life sciences: reconciling practice and policy – perspectives from WIPO', in A. Krattiger, R.T. Mahoney, L. Nelsen et al. (eds), *Intellectual Property Management in Health and Agricultural Innovation: A Handbook of Best Practices*, Oxford, UK: MIHR and Davis, USA: PIPRA pp. 229–46.
UC Berkeley Press Release (2006), 'African staple crop gets a boost', available at: http://www.berkeley.edu/news/media/releases/2006/04/10_sorghum.shtml, accessed 15 January 2013.
United Nations Ministerial Conference of the Least Developed Countries (2007), *Globalization, Agriculture and the Least Developed Countries*, Istanbul: United Nations Development Programme.
Vroom, Wietse (2009), *Reflexive Biotechnology Development: Studying Plant Breeding Technologies and Genomics for Agriculture in the Developing World*, Amsterdam: Wageningen Academic Publishers.
Watkins, A. and J. Mandell (2010), *Global Forum Action Plan: STI Capacity Building Partnerships for Sustainable Development*, Washington, DC: The World Bank.
Wheeler, Craig and Seth Berkley (2001), 'Initial lessons from public–private partnerships in drug and vaccine development', *Bulletin of the World Health Organization*, **79**, 728–34.
Widdus, R. (2005), 'Public–private partnerships: an overview', *Transactions of the Royal Society of Tropical Medicine and Hygiene*, **99S**, S1–S8.
Wiebe, Keith D., Meredith J. Soule and David E. Schimmelpfennig (2001), 'Agricultural productivity for sustainable food security in sub-Saharan Africa', in Lydia Zepeda (ed.), *Agricultural Investment and Productivity in Developing Countries*, Peru: FAO Economic and Social Development Paper no. 148, pp. 55–74.
Wolson, Rosemary (2004), 'Intellectual property tools, innovation and commercialization of R&D: Options to assist developing countries in positioning themselves to reap the benefits of a stronger intellectual property regime, with special reference to the role of intellectual property management in research organisations', ICTSD / UNCTAD / TIPS Regional Dialogue 'Intellectual Property Rights (IPRs), Innovation and Sustainable Development' in Eastern and Southern Africa, 29 June – 1 July 2004; Cape Town, South Africa, available at: http://www.iprsonline.org/unctadictsd/dialogue/2004-06-29/2004-06-29_Wolson.pdf, accessed 29 October 2011.
Wolson, Rosemary (2007), 'The role of technology transfer offices in building the South African biotechnology sector: an assessment of policies, practices and impact', *Journal of Technology Transfer*, **32**, 343–65.
Yang, Chia-Han and Joseph Z. Shyu (2010), 'A symbiosis dynamic analysis for collaborative R&D in open innovation', *International Journal of Computational Science and Engineering*, **5** (1), 74–84.
Yorobe, J.M. and M. Smale (2012), 'Impacts of Bt maize on smallholder income in the Philippines', *AgBioForum*, **15**(2), 152–62.

23 The North American crop biotech environment, actors and rules

David Schimmelpfennig[1]

INTRODUCTION

By 1996, after ten years of experience with transgenic crops, there were eight countries in the world with applications for commercialization or actual approvals for sale of transgenic crops, counting the European Union as one country. Three of the seven non-EU countries were Canada, Mexico and the US (James and Krattiger, 1996). More revealing about the differences in attitudes towards these seed technologies between North American countries and the rest of the world is that Canada approved for commercial use all 10 applications it had in 1996. Mexico approved five out of five applications, but four were for imported crops only and one of those was for use only as feed. Of the 23 applications in the US by 1996, 20 were approved for sale and two of the remaining three were for crops that already had other biotech varieties approved for sale. The one remaining crop with an application was papaya and this virus-resistant variety developed by Cornell University and the Hawaii Growers' Association has recently been approved. In contrast to the record in North America, the other five countries had a total of 36 applications and only 11 were approved for sale.

Approvals for sale of biotech varieties have led to their adoption: the countries with the highest rates of approvals in 1996 had among the largest areas under cultivation in 2010. The US is first, with 66.8 million hectares of eight crops and Canada is fifth with 8.9 million hectares of four crops (James, 2010). All four crops in Canada also have biotech varieties in the US (canola, corn, soybeans and sugar beets). Canada is distinguished by having almost three times the hectares of biotech crops under cultivation as China, the next biggest adopter. The countries with areas under biotech cultivation between the US and Canada are Brazil, Argentina and India. These top five countries are all large agricultural producers, but the most recent adoption data for the US shows why it has more hectares under biotech crops than Brazil, Argentina and India combined. For 2011, biotech adoption rates on the two largest crops in the US, corn and soybeans, were 88 and 94 per cent respectively. Upland cotton, which only falls behind wheat in hectares planted in the US, shows 90 per cent adoption rates (Fernandez-Cornejo, 2011). Adoption rates on total crop areas planted (NASS, 2011) for these three crops alone adds up to most of the biotech hectares planted in the US. The other five US biotech crops are canola, sugar beets, alfalfa, papaya and squash (James, 2010).

The rest of this chapter considers what made North America so conducive to the use of biotech seed technologies. The answers lie in two basic directions: the structure of the private biotech industry and the regulatory frameworks that were in place. The organization of these two realms will turn out to depend on two substantially independent networks of factors, and while some of the constituent factors appear in both networks (for example characteristics of consumers and commodity supply chains), the structure of the industry and regulatory networks themselves will be portrayed as fundamentally different. The importance of network structure for technology adoption in general is made persuasively by Guinnane et al. (2004). To highlight differences between the North American agbiotech industry and regulatory networks, subsequent sections discuss theories of how parts of each of the networks are structured, methods that have been used to quantitatively and qualitatively test these theories, and metrics that have been developed from successful applications of theory to cases. An application of the network approach highlights gaps that appear in the two networks and an attempt is made to value the missing information in some rudimentary, illustrative way. To begin, the next section lays out the general structure of the frameworks.

NETWORK MODEL FOR INDUSTRY STRUCTURE IN NORTH AMERICAN BIOTECH

Two network frameworks are presented in this section that will be used as points of reference for the later sections that provide more detail on the theoretical background of the networks, methods for testing these theories, and resulting metrics. The first framework on North American biotech industry structure comes from a metric commonly used to quantify the degree of concentration in an industry. The four-firm concentration ratio is the percentage of sales made by the four largest firms in the industry. This ratio declined for the global crop seed industry from 1989–93 (Schimmelpfennig et al., 2004) and in 1994 it was only 21 per cent. But by 2000 it had grown to 33 per cent and in 2009 it stood at 54 per cent (Fuglie et al., 2011). Monsanto changed its business-model from crop chemicals to become a seed/biotechnology company, and made many large acquisitions of seed and related companies, including Holden Foundation Seeds in 1997, a corn seed/germplasm company. Monsanto still has significant sales of glyphosate, but weed tolerance to this herbicide has been increasing. DuPont acquired Pioneer in 1999, then the world's largest seed company. Dow also made a large stake in the crop biotechnology business when it purchased Eli Lilly's share of Dow Elanco in 1997 and formed Dow Agrosciences. Another seed/biotechnology company, Aventis Crop Science, originally AgrEvo, was acquired by Bayer in 2002. Astra-Zeneca's seed business became part of the Advanta Seed Group, but Advanta's seed enterprise was broken up, with parts acquired by Syngenta, Limagrain and others in 2004 and 2005. Details of the consolidation of Syngenta and DuPont into their present forms are presented in Fuglie et al. (2011, Figure 2.2). BASF has been using research collaborations with several seed companies and technology licensing with Monsanto to develop future biotechnologies.

Table 23.1 shows how these companies form an international network with six of the eight largest seed/biotech companies in terms of sales, also having substantial agricultural chemical sales. These Big 6 of US and European-based multinationals also make the largest investments in R&D by several orders of magnitude over the largest companies doing seed/biotech R&D only. In crop seeds, just eight companies were responsible for 76 per cent of 2010 R&D spending – larger than their share of sales. In agricultural chemicals, five companies (each with over US$2 billion sales in 2010) accounted for over 74 per cent of the R&D in this industry (Fuglie et al., 2011). These Big 6 companies form

Table 23.1 Companies with largest crop seed sales in 2009, also largest in agricultural chemicals

Company	Country of incorporation	Crop seed & biotech sales	Agricultural chemical sales	Agricultural R&D (est. 2007)
		Millions US dollars		
Monsanto	USA	7297	3527	770
DuPont/Pioneer	USA	4806	2320	633
Syngenta	Switzerland	2564	8491	830
Limagrain	France	1370	0	171
KWS AG	Germany	996	0	104
Bayer	Germany	699	7535	978
Dow	USA	633	3708	294–380
BASF[1]	Germany	*small*	5065	655

Note: 1. In January 2012 BASF announced its intention to move all of its biotech R&D to the US.

Source: Fuglie et al. (2011), http://www.ers.usda.gov/Publications/ERR130/ERR130.pdf.

a biotechnology research network with the American multinationals having principal R&D facilities located in one or more EU country with the same (in reverse) being true of the European multinationals. These multinational companies also have R&D facilities around the world, establishing global research networks that allow them to develop and adapt new technologies to local conditions, meet national regulatory requirements, and achieve cost economies in their R&D activities while also increasing the rate of their international technology transfer.

The next biggest R&D performer is from the farm machinery sector where US-based John Deere has an estimated US$461 million in agriculturally related R&D and is the world's largest manufacturer of farm machinery with sales of over US$16.5 billion in 2008. In farm machinery, four companies (over US$5 billion sales each) had over 57 per cent of total industry R&D. About 40 per cent of these sales were to markets outside the US and Canada. Deere has research joint ventures in India and Israel.

The Big 6 are supported by substantial North American public research organizations. The US Department of Agriculture Agricultural Research Service (ARS) spent US$456 million on crop science in 2007 and US$171 million on animal science (Fuglie et al., 2011, Table 1.11). More details on the elements of the US agricultural research system including state and federal components are in Schimmelpfennig and Heisey (2009). Agriculture and Agri-Food Canada (AAFC), headquartered in Ottawa, is Canada's federal agricultural entity and includes 19 federal agricultural research organizations (http://en.wikipedia.org/wiki/Canadian_government_scientific_research_organizations). The budget for all of AAFC was almost C$3 billion (gross Canadian dollars) in 2011 but early versions of the 2012 budget had spending cuts of C$418 million, with agricultural research funding taking a C$150 million hit (Heppner, 2011).

While Canada invested US$19.6 billion (constant 2000$) in R&D in 2009, Mexico spent only US$4.4 billion in 2009, placing it at the bottom (with Chile) of the 35 countries surveyed in both R&D as a percentage of GDP (0.4 per cent) and in researchers per thousand workers in the country (0.9 per cent) (OECD, 2011). Positively for Mexico, it hosts the International Maize and Wheat Improvement Center (commonly known by its Spanish acronym CIMMYT for Centro Internacional de Mejoramiento de Maíz y Trigo), one of the largest of the 15 non-profit, research and training institutions in the Consultative Group on International Agricultural Research (CGIAR) which is the world's largest public international agricultural research consortium. CIMMYT focuses on the development of improved varieties of wheat and maize, and on introducing improved agricultural practices to farmers. CIMMYT has some biotechnology capacity but has to be extremely careful to avoid contamination of Mexico's world-leading *in situ* genetic diversity of maize. Mexico's farmers plant 77 per cent of their acreage to landraces. *Ex situ*, CIMMYT maintains a gene bank that in 1995 held 17000 accessions of maize and 123000 accessions of wheat (Smale, 1998).

NETWORK MODEL FOR REGULATION IN NORTH AMERICAN BIOTECH

The previous section discusses how the network of agricultural input sales and R&D in North America is increasingly dominated by seed and chemical technologies from an increasingly narrow group of large companies (the Big 6). These technologies are used by millions of North American agricultural producers (the US has over 3 million farmers alone; Census of Agriculture, USDA, 2007, Table 49), which contrasts with the regulatory network for genetically modified (GM) crops in the US, which has the reverse structure. In the regulatory network the various data requirements and safeguards enforced by several government agencies are intended to ensure that the implementation of regulatory decisions, including approval of field tests and eventual deregulation of approved biotech crops, does not adversely impact human health or the environment. Each of the three regulatory agencies has responsibility for a different aspect of safety based on their expertise in regulatory oversight. Because of the potential for coordination failures between the different organizations, the US government developed a Coordinated Framework for the Regulation of Biotechnology in 1986 to provide an acknowledged structure for the regulatory oversight of organisms derived through genetic modification. This structure is depicted as a network in Figure 23.1 and shows how the information is funneled down to a few outcomes within the coordinated framework. The framework was new but the laws regulating safety, efficacy and environmental impacts of biotechnology inputs were the same as those that were applied earlier to agricultural inputs derived by other methods (OSTP, 1986). The motivation for the framework is at least partially to promote public trust by ensuring adequate legal authority to assess and manage potential risk from new GM technologies across agencies and to increase transparency, clarity and public participation in the process.

Three US government agencies have provided primary guidance for the regulation of GM crops to date. Experimental testing, approval and eventual commercial release of GM crops and other organisms are controlled by the USDA's Animal and Plant Health Inspection Service (APHIS), the Department of Health and Human Services' Food and Drug Administration (FDA), and the Environmental Protection Agency (EPA). APHIS has primary initial responsibility coming from the Plant Protection Act (PPA) that governs plant pests and noxious weeds. The regulations provide for a petition process

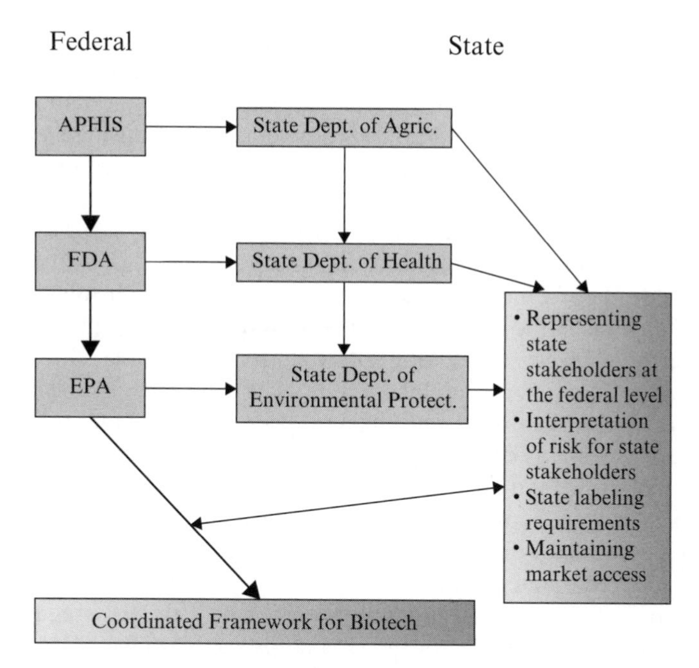

Notes: APHIS: USDA Animal and Plant Health Inspection Service; FDA: Food and Drug Administration, Department of Health and Human Services; EPA: Environmental Protection Agency.

Source: Discussion in Pew Initiative on Food and Biotechnology (2004).

Figure 23.1 US regulation of GE crops

for the determination of non-regulated status, which then allows the GM organism (and its offspring) to be released into US production and consumption without continuing APHIS review.

FDA is responsible for the safety of food and animal feed at The Center for Food Safety and Applied Nutrition (CFSAN) and The Center for Veterinary Medicine (CVM). When drugs and pharmaceutical products are developed from GM crops and animals The Center for Drug Evaluation and Research (CDER) and The Center for Biologics Evaluation and Research (CBER) have asserted responsibility. EPA regulates pesticidal substances produced by GM crops under the Federal Insecticide, Fungicide, and Rodenticide Act (FIFRA) and some other environmentally important chemical substances and GM micro-organisms under the Toxic Substances Control Act (TSCA) (Pew Initiative on Food and Biotechnology, 2004). EPA developed a new regulation for GM plant-incorporated protectants (PIPs) when the regulated GM 'event' could be construed to be a plant, pest or pesticide.

Even though the Coordinated Framework is based on the premise that the potential risks associated with GM organisms fall into the same general categories as those for traditionally bred organisms, USDA rules require notification and permits for field trials of GM plants, neither of which are required for non-GM plants. Likewise, FDA encourages submission of safety data before marketing food that contains a new GM crop variety (Pew Initiative on Food and Biotechnology, 2004). This is within a framework generally considered conducive to the development of GM crops which is reflected in the adoption rates discussed earlier. Other important regulatory issues that are not as well-defined include labeling and consumers' right-to-know, animal welfare, international regulations and trade, and economic-liability and insurance coverage following inadvertent mixing of GM and non-GM crops. Waltz (2011) reads a lot into the APHIS decision that herbicide-tolerant lawn grass falls outside its regulatory authority. Interpreting this as a signal that USDA 'would prefer not to regulate new GM plants anymore' is surprising since USDA is also trying to revise biotech regulations to reflect the changing landscape of innovation.

The right-hand side of Figure 23.1 shows the analogous state involvement compared to the federal functions in the regulation of plant biotechnology in the US. Even though state regulation of GM crops is taking place at the comparable state agencies, their primary motivations are fundamentally different. The state agencies rely on the benefits to their states of rigorous enforcement of nationwide standards and regulations by the federal agencies. As long as basic requirements for food and feed safety, plant health and environmental protection are being handled adequately at the federal level there are actually bigger inter-state coordination problems in regulation and enforcement at the state level that are avoided by having a national coordinated framework. State regulatory agencies have preferred to remain informed about new permits and approvals rather than influencing the basis for them. The primary motivations of the state regulatory agencies are in making sure the benefits of growing GM crops accrue to their GM producers and that their non-GM farmers are not negatively impacted. This usually means helping to preserve access to markets for all their agricultural crops – conventional and organic products as much as the products of biotechnology (Taylor et al., 2004). EPA has closely scrutinized and regulated Bt crops that are genetically modified to be insect resistant, but if the insects themselves become resistant to Bt it could adversely affect organic growers that use naturally occurring Bt primarily in spray form. The technically feasible response to insect resistance is to maintain non-Bt refuges and allow the propagation of populations of non-resistant insects, but the maintenance of refuges has been difficult to enforce in many states.

In fact, state regulators have to work even more diligently than their federal counterparts that have primary regulatory responsibility to respond to the concerns of local citizens and stakeholders. This responsibility for reducing perceived risks from biotechnology can take state agencies beyond the borders presently acknowledged by federal agencies and into labeling, traceability and market segmentation questions within state boundaries and also considerations for maintaining biodiversity and reducing impacts on non-target species. State agencies can cross over into this difficult territory by trying to influence federal regulatory decisions, carrying out local enforcement of federal rules, and even setting and enforcing different state standards than the federal guidelines. Other strategies that have been used include expanding public participation in state decision-making, and considering crop or technology bans at the state level (Taylor et al., 2004).

INDUSTRY NETWORK APPLICATIONS: MODELS

The subsequent sections of this chapter will discuss the operation of the previous networks in detail using insights in the literature on the operation of biotechnology in North

America, starting with theories of how segments of the industry network function. Before addressing the issue of consolidation that appears to be driving the formation of the Big 6 multinationals discussed in association with Table 23.1, a useful model for evaluating company R&D strategies and the major long-term costs involved, that in turn influence industry structure, is the model of technological trajectories. These are the kind of strategic long-term decisions that a little over ten years ago were motivating some of the Big 6 to pursue research synergies by integrating their agricultural and health R&D efforts. This strategy failed partially because of the impact on company share prices of commingling perceived lower pay-off agricultural research with more highly valued health research. A decade later, biofuels and agriculture are considered to have as much or higher potential than the possibly lengthening drug pipeline that is running up against fixed-length patents and pending patent expirations in the pharmaceutical industry. Where we might be now if some of those synergies had been allowed to develop ten years ago is open to speculation, but that ship sailed a while ago. Chataway et al. (2004) use the model of technological trajectories to also consider how public policy can impact strategic R&D decisions and how public reaction was a 'blind spot' to many company managers. Chataway et al. find that 'company distinctiveness', not homogeneity, drove some research agendas, and single-mindedness was certainly a factor driving Monsanto Inc. to pursue biotech input traits that have generally not excited consumers.

Given this seemingly incongruous aspect of agbiotech R&D, how do we get to a Big 6 industry structure? The simple answer is that each of the Big 6 has exploited their own, unique successful view of what would work to enhance their market power. It can be argued that when other companies had trajectories that seemed ready and able to exploit new market niches, the Big 6 or their relatives were poised to merge with or acquire their potential competitors. This should not be particularly surprising given the substantial initial and fixed costs required to operate in the agbiotech input industry. Several models evaluate existing industry structures in agbiotech. Carew's (2005) model of the Canadian agbiotech industry is based on the number and composition of agbiotech field trials and patents registered in Canada and at the US Patent and Trademark Office. The model uses seven technology clusters, including transformation methods, specific trait technologies and germplasm and cultivar development. He notices a production orientation in the patents found during 1996 to 2001 with plant metabolism and production efficiency accounting for roughly one-third of the patents. His interpretation of these results is that biotech companies in Canada are growing and improving the 'agri-food landscape'. Schimmelpfennig et al. (2004) use a research profit function model that also includes field trials and patents for 1989 to 1998 and find that in the US seed industry concentration for corn, cotton and soybeans has reduced biotech research intensity or R&D per unit of seed sales.

These somewhat different results for the US and Canada point to market size as making a difference in how an individual country's science and technology (S&T) policy will influence the extent and desirability of consolidation in that country. Ryan and Smyth (2010) consider factors facilitating innovation in Canadian agbiotech and develop the 'Ag-West Biotech Model'. The factors they discuss have already been mentioned: patents, returns to research, politics, and consumer acceptance. The model offers the novel insight that a non-profit, publicly funded organization like Ag-West can add leadership and direction for companies and regulators to improve outcomes from agbiotech research. Ryan and

Phillips (2004) are also interested in modeling support for a knowledge-intensive industry like agbiotech and find that in Canada agglomeration, or the extent of clustering, promotes innovation when knowledge is diffuse, complicated and actively protected by patents. In Canada, where the proprietary seed markets are smaller than in the US, the emphasis in models of agbiotech research is on research synergies and promotion.

In the US, S&T policy is focused more on the strength and breadth of patents and approving or not approving mergers and acquisitions (M&A). The Department of Justice has been involved in several anti-trust cases with agbiotech firms, and patenting policy has wound through district, federal and in one case even the Supreme Court on the patenting of seed technologies. The case of JEM Ag Supply, Inc. v. Pioneer Hi-Bred International, Inc., 534 US 124, in the Supreme Court in 2001 allowed the enforcement of a utility patent awarded to a hybrid seed variety, opening the way for biotech-seed utility patents; a stronger form of patent protection than either plant patents or plant variety protection certificates. Justice also closely examined Monsanto's separate acquisitions of DeKalb and Delta and Pineland Inc. and required separate anti-trust concessions from the company before allowing both to proceed. Underlying agbiotech in both countries, as mentioned previously and included in the models discussed in this sub-section, is public-sector agricultural research. Sonka and Pueppke (1999) provide justifications for public support of biotechnology research even with the just mentioned strength of relevant intellectual property protection and the growth in size of private agbiotech firms. Rebolledo and Lui (1995) consider many of these same forces operating in Mexican agbiotech and conclude that the technologies are likely to benefit only high-tech producers and to bypass small farmers.

REGULATORY NETWORK APPLICATIONS: MODELS

Additional details to the discussion of Figure 23.1 in the previous section and an evaluation of the Coordinated Framework can be found in Belson (2000). The framework is treated there as one possible model of regulation and discusses a comprehensive interagency review of environmental regulations pertaining to agbiotech carried out in January 2001 by The White House Council on Environmental Quality (CEQ) and the Office of Scientific Policy (OSTP). Representatives from the EPA, FDA, APHIS and other federal agencies participated in this review and the end product was a series of case studies examining how the framework would handle different biotech products, including Bt corn and herbicide-tolerant soybeans. The review stated 'no significant negative environmental impacts have been associated with the use of any previously approved biotechnology product'.

An expanded model of the regulatory process that more closely reflects the state's role on the right-hand side of the regulatory network (Figure 23.1) is discussed in Jackson and Villinski (2002). They concentrate on consumers' apprehensions about GM food, how agbiotech might lead to market failures with distributional equity consequences, and the policy significance of scientific uncertainty. As mentioned earlier, regulatory transparency and rigor can help, but uncertainty can be continually addressed by procedures like those included in the January 2000 Cartagena Protocol on Biosafety, also known as the Convention on Biological Diversity. This agreement supports trade in GM commodities with pledging countries agreeing to post information online about possible GM content in their exports. Trading partners may then choose to accept or reject shipments depending upon their national perception of the risks associated with those GM crops. Crop genetic diversity can increase or decrease in a country due to GM crops as Raney and Pingali (2004) model the question, depending on how they are regulated. When done appropriately, regulation can increase overall availability of plant genetic resources but the resulting impact on the resilience of ag-ecosystems is hard to determine using their approach. A related modeling topic concerns field-level pest resistance. Frisvold and Reeves (2010) consider why herbicide-tolerant GM varieties might be experiencing a bigger problem with resistance than insect-resistant trait varieties. They find that integrated pest and weed management may be more or less compatible with the different technologies, and regulatory and institutional settings can impact resistance outcomes.

INDUSTRY AND REGULATORY NETWORK APPLICATIONS: METHODS

This 'methods' section extends the closely-related previous two sections on models. The distinction here is that the modeling approaches in this section involve qualitative or quantitative approaches whereas the models discussed up to this point have tended to be more conceptual. In the same way that the conceptual models were related to aspects of the networks in Table 23.1 and Figure 23.1, the same will be done here for quantitative and qualitative approaches. Considerable quantitative information was actually used to motivate the networks in the introduction, and this R&D and adoption data will not be reintroduced.

One model that practically requires quantitative assessments and has been applied to many environmental policy questions is benefit–cost analysis (BCA). Formerly known as cost–benefit analysis, the limitations of BCA for policy considerations, treated thoroughly by Schmid (1989) include uncertainty, distributional effects and rationales for public investment. But rather than discrediting the BCA model, the factors they raise look quite familiar in the context of the previous sections. Suitably equipped, BCA may be a natural for evaluation of agbiotech in the networks of Table 23.1 and Figure 23.1, but there are actually very few studies using this approach. It may be that the quantitative requirements have cut back on the number of studies that follow the BCA model explicitly for agbiotech. A specific focus on BCA related to US federal regulations was carried out by Farrow (2011) and focused on the lack of specific guidance from the Office of Management and Budget for analyzing distributional impacts, particularly when benefits are coming from risk reduction of some kind. Zilberman (2006) argues qualitatively that if BCA were applied to biotech regulation both through *ex ante* testing as in Figure 23.1 and *ex post* monitoring, then regulatory requirements would be reduced.

Quantitative assessment of labeling policies (discussed previously with respect to the right-hand side of Figure 23.1), especially their impact on trade in biotech commodities, is considered by Marchant and Song (2005). The North American Free Trade Agreement (NAFTA) countries have voluntary labeling, while in China, the EU and Japan it is mandatory. While trade with Europe has slowed, USDA negotiations and education programs in China, together with a soybean shortage, have kept exports to

China continuing. In June 2011 the European Commission began to allow up to 1 tenth of 1 per cent GM products in animal feed aimed mainly it appears at alleviating feed shortage fears (Meldolesi, 2011) and made possible by newer high-sensitivity testing procedures. Sexton and Zilberman (2011) quantify the impacts of GM crops in reducing tensions between global food supplies, energy production from biofuels, and demands for environmental protection from agriculture. One benefit of qualitative approaches is that they can infer more nuanced insights on the development of agbiotech. Pray et al. (2007) show how political response to biofortified crops (mainly through the actions of states, as represented on the right-hand side of Figure 23.1) depends on the agronomic traits of the improved variety. They consider GM maize, potato, papaya, rice, cassava and sorghum and show how crop qualities that benefit consumers and producers still need the support of political and economic interest groups. When only small farmers with little political clout are the beneficiaries, like potato growers in Mexico, the potential for biofortified GM varieties has seemed more remote.

INDUSTRY AND REGULATORY NETWORK APPLICATIONS: METRICS

This final section applying the networks in Table 23.1 and Figure 23.1 deals with metrics that have resulted from the application of elements of both networks, often in a single paper. Whereas this might sound like the pinnacle of success, the reality is that the range of confounding factors introduced at this level of analysis makes survey approaches most likely to show up in this section and some of the intricacies of the two 'models' sections may be lost. A good example of what can be accomplished at this level of abstraction can be seen in Gusta et al. (2011) who make use of surveys and focus on producers of one crop in one region. Herbicide-tolerant canola in Western Canada is shown to have generated over C$1 billion in added value from 2005 to 2007, when producer experiences with the input industry are evaluated. Lower input costs and better weed control drive this outcome in the context of tillage practices and herbicide use. In the same survey, but on the regulation side, resistance management (discussed previously at the end of the section on regulation network 'models') was reported not to have created many headaches for producers.

A database search is used by Dawson et al. (2009) to consider if underutilized crops in the tropics and sub-tropics have been improved by biotechnology to increase their cultivation. They are mainly interested in whether tissue culture and micropropagation that have benefited major crops have also been applied to underutilized minor crops. The data shows that regulation can interfere with the balancing act taking place between the limited resources available to develop underutilized crops using biotechnology and steps that can be accomplished using traditional methods. They also find that appropriate technologies can be centralized in a few firms and difficult to make use of effectively under license or cooperative agreement. Both of these results touch on the network structures shown in Table 23.1 and Figure 23.1. An example from another source of the types of problems that arise with minor crops comes from the South African Sugar Research Institute, which invested many years of high priority work inserting a Bt construct donated to them by Monsanto Inc. into local sugar cane varieties that proved to be ineffective. Once the problems had been resolved the African Center for Biosafety filed objections to the GM sugar cane (http://www.biosafetyafrica.org.za/index.php/2005021834/Sugarcane-with-a-bacteriozide-protein-Bt-Sugarcane-/-SARSI/menu-id-100023.html). Although the borers that infect maize and sugar cane are different species they were both recently found to be similarly susceptible to the Bt toxin (http://www.isaaa.org/kc/cropbiotechupdate/article/default.asp?ID=8004).

COMPOUNDED INFORMATION GAPS – REGULATION AND INDUSTRY STRUCTURE TOGETHER

The most serious gaps in information evident from both the networks discussed in this chapter are almost surely the result of missing innovations and technologies. Some empirical data shows that for the US agbiotech market leaders in corn, cotton and soybeans, consolidation between 1989 and 1998 led to less research intensity (R&D per unit of seed sales) (Schimmelpfennig et al., 2004) and probably fewer new crop innovations. Global market concentration or the share of global industry sales earned by the largest firms, in most agricultural input industries has risen since the 1990s. This has been most noticeable in the crop seed-biotechnology industry, where four-firm market share more than doubled.

The regulatory network may be partially responsible for this increase in market concentration. To help spread the costs of meeting regulatory approvals, companies have merged or made acquisitions (M&A) while also gaining access to technologies and serving larger markets. The agricultural chemical sector has been heavily affected by stricter government regulations governing the health, safety and environmental impacts of new and existing pesticide formulations. Larger firms may have more staff with greater expertise dealing efficiently with these requirements. A process of reinforcement between regulatory and industry structure networks can lead to more consolidation as patents that protect the intellectual property in newly acquired technologies that might have been licensed by smaller firms are more often exploited in-house by large companies. The result can be patent thickets for complementary technologies that large firms attempt to bring together under one roof, resulting in barriers to entry for other firms. More discussion of agricultural biotechnology patents, strategic patenting, hold-ups and patent thickets can found in Schimmelpfennig (2004).

The number of small and medium-sized enterprises (SMEs) in the crop seed-biotechnology industry has probably been impacted by M&A and regulations. The number of SMEs grew in the 1980s and again in the 2000s, but recently SME exits have outnumbered entrants, and by 2008 only 30 SMEs specializing in crop biotechnology were still active. Of 27 crop biotechnology SMEs that were acquired between 1985 and 2009, 20 were acquired either directly by one of the Big 6 or by a company that itself was eventually acquired by a Big 6 company (Fuglie et al., 2011). The burden of regulatory approvals may have prompted some of these SMEs to seek acquisition as they neared the product approval stage.

One study explicitly linking industry structure and regulation does this in a largely conceptual model. Heisey and Schimmelpfennig (2006) consider the number of approved agbiotech innovations and lobbying efforts of political action committees and find influences from both regulation and industry structure. Their sources of data are private R&D, sales, campaign contributions, and lobbying expenditures, but the empirical results are presented only as a set of stylized facts. Galushko et al. (2010) give intellectual property and private investment in plant breeding research a more in-depth treatment with a survey of canola breeders. They consider the role of public sector plant breeding and private firm freedom-to-operate (FTO) in generating difficulties obtaining access to necessary agbiotechnologies. They show a movement since the 1990s away from M&A as a method to increase FTO, towards more cross-licensing. Another study by Aerni (2004) uses BCA for Mexico to tie together industry characteristics and regulatory environment. Even though adoption has been limited in Mexico, as discussed in the introduction, the 52 stakeholders surveyed see agricultural industry benefits and few problems with human health, but noted an increased need for good biosafety guidelines that are properly implemented in this biodiverse country.

CONCLUSIONS

This chapter has considered exogenous variables in North American agbiotech such as the environment that exists in Canada, US and Mexico, the principal actors, and the rules that have influence over them. To do this in a structured way, the chapter developed network models for industry structure in North America and regulation in US agbiotech from the background data presented in the introduction. The result of this approach to the review of the existing literature on agbiotech is that several information gaps were uncovered that resulted from network failures in either industry structure, regulation or both.

Increasing industry concentration since the 1990s has led to a Big 6 of US and European-based multinationals that make the largest investments in seed/biotech and agricultural chemical R&D. Eight companies in crop seeds had 76 per cent of 2010 industry R&D spending; larger even than their share of sales. Five companies in agricultural chemicals carried out over 74 per cent of the R&D in this industry and each had over US$2 billion sales in 2010 (Fuglie et al., 2011). These Big 6 companies form a biotechnology research network with its primary backbone extending between the US and EU. These multinational companies have specialist R&D facilities around the world, allowing them to increase the rate of their international technology transfer. They do this by developing and adapting new technologies to local conditions, meeting national regulatory requirements from within different countries, while also achieving cost economies in their R&D activities.

The result of multi-layered regulatory networks and the concentration of industry structure dominated by networked multinational corporations is that both the pace and direction of innovation is being decided by these companies. Management and the scientists working for these multinationals to develop new biotech seed technologies and agricultural input chemicals may agree with their company annual reports that they are pushing the limits of the science as far as they can, but it seems that if larger sales were available in presently underserved markets, R&D could be refocused to try to meet those opportunities. It may be revealing to see how widely drought-tolerant corn will be adopted in North America and around the world, and how many market niches make use of the new seed technology.

NOTE

1. The views expressed are not necessarily those of the US Department of Agriculture.

REFERENCES

Aerni, Philipp (2004), 'Stakeholder perceptions towards agricultural biotechnology in Mexico', *International Journal of Agricultural Resources, Governance and Ecology*, **3**(1–2), 95–115.

Belson, Neil A. (2000), 'US regulation of agricultural biotechnology: an overview', *AgBioForum*, **3**(4), 268–80.

Carew, Richard (2005), 'Science policy and agricultural biotechnology in Canada', *Review of Agricultural Economics*, **27**(3), 300–316.

Census of Agriculture (2007), National Agricultural Statistics Service, US Department of Agriculture, Washington, DC, available at: http://www.agcensus.usda.gov/Publications/2007/Full_Report/Volume_1,_Chapter_1_US.

Chataway, Joanna, Joyce Tait and David Wield (2004), 'Understanding company R&D strategies in agro-biotechnology: trajectories and blind spots', *Research Policy*, **33**(6–7), 1041–57.

Dawson, Ian K., Peter E. Hedley, Luigi Guarino and Hannah Jaenicke (2009), 'Does biotechnology have a role in the promotion of underutilised crops?', *Food Policy*, **34**(4), 319–28.

Farrow, Scott (2011), 'Incorporating equity in regulatory and benefit–cost analysis using risk based preferences', *Risk Analysis*, **31**(6), 902–7.

Fernandez-Cornejo, Jorge (2011), 'Adoption of genetically engineered crops in the US' data sets, Economic Research Service, US Department of Agriculture, Washington, DC, July, available at: http://www.ers.usda.gov/Data/BiotechCrops/.

Frisvold, George B. and Jeanne M. Reeves (2010), 'Resistance management and sustainable use of agricultural biotechnology', *AgBioForum*, **13**(4), 343–59.

Fuglie, Keith O., Paul Heisey, John L. King, Carl E. Pray, Kelly Day-Rubenstein, D. Schimmelpfennig, Sun Ling Wang and Rupa Karmarkar-Deshmukh (2011), 'Research investments and market structure in the food processing, agricultural input and biofuel industries world-wide', *Economic Research Report*, no. ERR-130, US Department of Agriculture, December, available at: http://www.ers.usda.gov/Publications/ERR130/ERR130.pdf.

Galushko, Viktoriya, Richard Gray and Stuart Smyth (2010), 'Resolving FTO barriers in GM Canola', *AgBioForum*, **13**(4), 360–69.

Guinnane, Timothy W., William A. Sundstrom and Warren Whatley (eds) (2004), *History Matters: Essays on Economic Growth, Technology, and Demographic Change*, Stanford, CA: Stanford University Press.

Gusta, Michael, Stuart J. Smyth, Kenneth Belcher, Peter W.B. Phillips and David Castle (2011), 'Economic benefits of genetically-modified herbicide-tolerant canola for producers', *AgBioForum*, **14**(1), 1–13.

Heisey, Paul and David Schimmelpfennig (2006), 'Regulation and the structure of biotechnology industries', in R.E. Just, J.M. Alston, and D. Zilberman (eds), *Regulating Agricultural Biotechnology: Economics and Policy*, New York: Springer Publishers, pp. 421–6.

Heppner, Kelvin (2011), 'Cuts to ag Canada budget', *Portage Online*, 4 March, available at: http://www.portageonline.com/index.php?option=com_content&task=view&id=21150&Itemid=469.

Jackson, Lee Ann and Michele T. Villinski (2002), 'Reaping what we sow: emerging issues and policy implications of agricultural biotechnology', *Review of Agricultural Economics*, **24**(1), 3–14.

James, Clive (2010), 'Biotech crop countries and mega-countries, executive summary: global status of commercialized biotech/GM crops: 2010', *International Service for the Acquisition of Agri-Biotech Applications (ISAAA) Brief, 42*, Ithaca, NY: ISAAA, available at: http://www.isaaa.org/resources/publications/briefs/42/executivesummary/default.asp.

James, Clive and Anatole F. Krattiger (1996), 'Global review of the field testing and commercialization of transgenic plants, 1986 to 1995: the first decade of crop biotechnology', *ISAAA Briefs No. 1*, Ithaca, NY: ISAAA.

Marchant, Mary A. and Baohui Song (2005), 'Assessment of biotechnology policies and international trade in key markets for US agriculture', *Journal of Agricultural and Applied Economics*, **37**(2), 379–91.

Meldolesi, Anna (2011), 'EU tolerates GM', *Nature Biotechnology*, **29**(8), 677.

National Agricultural Statistics Service (NASS) (2011), 'Acreage', Washington, DC: June, US Department of Agriculture.

Office of Science and Technology Policy, The White House (OSTP) (1986), 'Announcement of policy and notice for public comment: coordinated framework for regulation of biotechnology', *Federal Register*, 26 June, **51**(23302).

Organisation for Economic Co-operation and Development (OECD) (2011), *Science, Technology and Industry Scoreboard 2011: Innovation and Growth in Knowledge Economies*, Paris: OECD Publishing.
Pew Initiative on Food and Biotechnology (2004), 'Issues in the regulation of genetically engineered plants and animals', April, Washington, DC, available at: www.pewagbiotech.org.
Pray, Carl, Robert Paarlberg and Laurian Unnevehr (2007), 'Patterns of political response to biofortified varieties of crops produced with different breeding techniques and agronomic traits', *AgBioForum*, **10**(3), 135–43.
Raney, Terri and Prabhu Pingali (2004), 'Private research and public goods: implications of biotechnology for biodiversity', Agricultural and Development Economics Division of the Food and Agriculture Organization of the United Nations (FAO–ESA), Working Papers no. 04-07, available at: ftp://ftp.fao.org/docrep/fao/007/ae062e/ae062e00.pdf.
Rebolledo, Solleiro and Jose Lui (1995), 'Biotechnology and sustainable agriculture: the case of Mexico', OECD Development Centre, Working Papers no. 105.
Ryan, Camille D. and Peter W.B. Phillips (2004), 'Knowledge management in advanced technology industries: an examination of international agricultural biotechnology clusters', *Environment and Planning C: Government and Policy*, **22**(2), 217–32.
Ryan, Camille D., and Stuart J. Smyth (2010), 'Facilitating innovation in agricultural biotechnology: an examination of the Ag-West Biotech model, 1989–2004', *AgBioForum*, **13**(2), 183–93.
Schimmelpfennig, David (2004), 'Agricultural patents: are they developing bad habits?', *Choices*, May, pp. 19–23.
Schimmelpfennig, David and Paul Heisey (2009), 'US public agricultural research: changes in funding sources and shifts in emphasis, 1980–2005', Economic Information Bulletin, EIB-45, Economic Research Service, US Department of Agriculture, March.
Schimmelpfennig, David, Carl E. Pray and Margaret F. Brennan (2004), 'The impact of seed industry concentration on innovation: a study of US biotech market leaders', *Agricultural Economics*, **30**(2), 157–67.
Schmid, A. Allan (1989), *Benefit–Cost Analysis: A Political Economy Approach*, Boulder, CO: Westview Press.
Sexton, Steven and David Zilberman (2011), 'How agricultural biotechnology boosts food supply and accommodates biofuels', National Bureau of Economic Research, Inc. NBER Working Papers no. 16699, available at: http://www.nber.org/papers/w16699.pdf.
Smale, Melinda (ed.) (1998), *Farmers, Gene Banks and Crop Breeding: Economic Analyses of Diversity in Wheat, Maize, and Rice*, Norwell, MA: CIMMYT, Kluwer Academic Publishers.
Sonka, Steven and Steven Pueppke (1999), 'Exploring the public's role in agricultural biotechnology research', *AgBioForum*, **2**(1), 33–6.
Taylor, Michael R., Jody S. Tick and Diane M. Sherman (2004), *Tending the Fields: State and Federal Roles in the Oversight of Genetically Modified Crops*, Washington, DC: Pew Initiative on Food and Biotechnology.
Waltz, Emily (2011), 'GM grass eludes outmoded USDA oversight', *Nature Biotechnology*, **29**(9), 772–3.
Zilberman, David (2006), 'The economics of biotechnology regulation', in R.E. Just, J.M. Alston and D. Zilberman (eds), *Regulating Agricultural Biotechnology: Economics and Policy*, New York: Springer Publishers, pp. 243–61.

24 Adoption decisions

Corinne Alexander

1 INTRODUCTION

This chapter reviews the literature on farmer adoption decisions for genetically modified (GM) crops, updating Alexander (2006). Previous literature reviews of farmer adoption of GM crops include Fernandez-Cornejo and McBride (2002), Marra et al. (2002), Qaim and Matuschke (2005) and Qaim (2005). This chapter will discuss the research approaches used to examine farmer decision-making and GM crops and then summarizes the results from the literature.

Productivity gains and economic growth are often driven by the adoption and diffusion of new technologies; hence there is a vast body of literature in economics and sociology on technology adoption (Feder et al., 1985; Solow, 1994). Farmer adoption of GM crops is distinct from other innovations for two reasons. First, regulatory approval for planting GM crops varies greatly by country and by crop, in part due to an intense debate about the appropriate use of agbiotech among nations, agricultural and consumer groups. Many countries either prohibit the planting of GM crops or have a very slow regulatory approval process (Cohen and Paarlberg, 2004; Thomson, 2004). Second, in countries where there is regulatory approval to plant GM crops, the adoption and diffusion of GM crops has been extremely rapid (Qaim, 2005; James, 2011). For example, GM crops were first introduced in the United States in 1996, and by 2011, 88 per cent of the corn acres, 90 per cent of the cotton acreage and 94 per cent of soybean acres were planted to GM varieties or hybrids (US Department of Agriculture National Agricultural Statistics Service, 2011). Further illustrating the extent to which farmer adoption of GM crops is dependent on regulatory approval, James (2011) documents that while GM crops are grown in 29 countries, the majority of the acres are grown in just five countries: the United States; Brazil; Argentina; India; and Canada.

There is a growing body of literature on crops that do not have regulatory approval. Many of these studies focus on understanding the farmer benefit of GM crops with the goal of providing policy-makers with better information for future regulatory decisions. Other studies focus on identifying early adopters to introduce a new GM crop more effectively. These *ex ante* studies include: GM cooking bananas in Uganda (Edmeades and Smale, 2006); IR cotton and IR maize in West Africa (Vitale et al., 2007); IR rice in China (Huang et al., 2008); VR papaya in Thailand (Napasintuwong and Traxler, 2009); GM crops in Nigeria (Nwankwo et al., 2009); GM crops in Ireland (Keelan et al., 2009); and HT sugar beet and IR corn in Germany (Gyau et al., 2009).

2 MODELS: SURVEY OF THE THEORY

The most basic economic adoption model assumes that the farmer will adopt a new technology if it is more profitable than the farmer's current practices. If the GM crop has higher expected yields or lower production costs relative to the current crop, then the farmer will adopt the GM crop. This model is the basis for much of the research on adoption of GM crops in developing countries that assumes farmers are risk neutral and focus primarily on the difference in expected yields between GM crops and the traditional varieties (Zilberman et al., 2007). According to Qaim and Zilberman (2003) the value of GM crops in developing countries depends on a substantial yield effect, rather than their ability to substitute for chemical pesticides.

However, GM and conventional crop yields are variable and farmers face yield risk. Hence, the most common theoretical model assumes farmers are risk averse and care about yield variance. In this model, farmers make their decisions about whether or not to adopt GM crops in order to maximize their expected utility of net returns. If the GM crop results in a higher expected utility of net returns than the current technology used by the farmer, the farmer will choose to adopt the GM crop.

A third theoretical approach is the agricultural household model which assumes the farm household maximizes utility subject to income, production technology and time constraints. This model is primarily used to model farmer decisions in developing countries where household production and consumption decisions are not independent because households consume what they produce. For example, if a farmer chooses to work more hours off-farm and fewer hours producing maize, the household consumption of maize will decline. Fernandez-Cornejo et al. (2005) use the agricultural household model to examine the interaction between off-farm work and the decision to adopt herbicide-tolerant soybeans.

Useche et al. (2009) and Foltz et al. (2012) developed a theoretical adoption model based on the individual traits of the technology. The model is based on the marketing literature for consumer demand that decomposes products into their specific traits, which is particularly well suited for describing GM crops with stacked traits. The model permits researchers to estimate the market value of individual GM traits either *ex ante* or *ex post*.

Researchers have also developed theoretical models to highlight particular aspects of the GM technology and offer testable hypotheses. Fulton and Keyowski (1999) used the Hotelling model of differentiation to model adoption of HT canola when farmers have heterogeneous tillage systems. They find that the benefits associated with HT canola depend on the tillage system, so that adoption similarly depends on the tillage system. Graff et al. (2006) develop an adoption model that estimates the extent farmer adoption of GM crops depend on the price of the technology. Aldana et al. (2011) developed a Bayesian model that highlights the role of information and uncertainty in the dynamics of the adoption decision as GM crops move from single traits to stacked traits.

Researchers also draw on the sociology theoretical adoption model developed by Rogers (2003). Rogers identified five groups of adopters based on when they adopt in the diffusion process – innovators, early adopters, early majority, late majority and laggards – with divisions based on the inflection points in the adoption curve or the number of standard deviations away from the mean adopter. In this context, many of the GM crop adoption studies have focused on identifying and comparing early adopters to late adopters, grouping farmers into two categories rather than five. In addition, some studies have focused on identifying a sixth group of disadopters (Fernandez-Cornejo et al., 2002; Barham et al., 2004).

3 METHODS

The literature on farmer adoption of GM crops can be categorized by four approaches. First, and the most common approach, researchers have used farmer surveys to document the decision of whether or not to adopt GM crops *ex post*, or to elicit farmers' willingness to adopt the GM crop in *ex ante* analyses. Second, researchers have used yield trial data to estimate the value of GM crops to farmers, and most of these studies have focused on developing countries. Third, researchers have used linear and dynamic programming models to predict farmer adoption of GM crops by estimating the benefits of GM crops relative to non-GM crops. Fourth, researchers have used case studies to document the complex decision process *ex ante* and *ex post* for adopting GM crops.

Barham et al. (2004), Morse et al. (2007) and Stone (2011) emphasize the need to have panel data to distinguish between farmers who are early adopters and those who are late adopters because they are distinct and different groups. Many studies that estimate the value of GM crops and make predictions about the diffusion process are based on the data and experiences of early adopters. However, Barham et al. (2004), Morse et al. (2007) and Stone (2011) caution that early adopters are not a random group but instead are biased toward more successful farmers; thus adoption studies must address this sample selection bias. Other researchers have addressed sample selection bias by gathering longitudinal data where each survey respondent has been asked about their GM crop adoption decisions over multiple years (Aldana et al., 2011; Krouser and Qaim, 2011).

4 METRICS

Before summarizing the findings in the literature, it is useful to categorize GM crops by their benefit: herbicide-tolerant crops; insect-resistant crops; and virus-resistant crops. Herbicide-tolerant (HT) crops have been modified to survive the application of specific herbicides that would have previously killed the crop. Insect-resistant (IR) crops or Bt crops have been genetically modified with a gene from *Bacillus thuringiensis*, a soil bacterium, to produce a protein that is toxic when ingested by Lepidopteran insects. This provides better control of the insects because the toxin is produced inside the plant where its effectiveness does not depend on environmental factors. More recently, virus-resistant (VR) crops have been introduced that protect the plant against devastating viral pathogens.

4.1 Factors that Influence Adoption of GM Crops

The farmer adoption decision will be influenced by any farm or producer characteristic that affects the producer's profit from GM crops. The traditional farm and producer characteristics that influence technology adoption include farm size, human capital, location, labor supply, information, risk preferences, cost of the technology and credit constraints (Feder et al., 1985). Since GM crops affect the producer's production system with respect to weed control in the case of HT crops or insect control in the case of IR crops, producer adoption will also be affected by pest pressure and tillage practices. Due to the consumer controversy over GM crops, uncertainty about demand and international market access for GM crops also influences adoption. Lastly, in the case of IR crops in the United States, there are regulatory requirements that may also influence adoption.

4.1.1 Farm size

The decision to adopt GM crops should be invariant to farm size because the technology is divisible and there is no fixed-cost component (Fernandez-Cornejo et al., 2001). However, the majority of analyses have found a significant relationship between farm size and adoption of GM crops. One possible explanation based on Lindner (1983) is that all farms face a fixed cost of learning about the new technology and thus larger farms face a lower information cost on a per acre basis. Qaim and de Janvry (2003) suggest that the transaction costs associated with the IR cotton purchase contracts in Argentina may bias adoption towards larger farms.

Most studies have found that GM crop adoption is positively related to farm size, including: HT soybeans (Fernandez-Cornejo et al., 2002); HT corn (Fernandez-Cornejo et al., 2001; Finger et al., 2009); IR corn (Fernandez-Cornejo et al., 2001); and IR cotton (Qaim and de Janvry, 2003; Sadashivappa and Qaim, 2009). Only one study captured a non-linear relationship between farm size and GM crop adoption: McBride and El-Osta (2002) found that IR corn adoption was significantly positively related to farm size and negatively related to farm size squared. However, two studies have found that GM crop adoption is negatively related to farm size (Hategekimana and Trant, 2002; Fernandez-Cornejo et al., 2005).

4.1.2 Human capital

Human capital relates to a farmer's ability to learn about and successfully implement a new technology. Adoption studies typically measure farmers' human capital through years of formal education, age and experience. Education and age are the two most commonly used explanatory variables in the GM crop adoption literature. Most studies have found that GM crop adoption or stated willingness to adopt is positively related to higher levels of education (Hubbell et al., 2000; Fernandez-Cornejo et al., 2001; McBride and El-Osta, 2002; Fernandez-Cornejo et al., 2002; Qaim and de Janvry, 2003; Barham et al., 2004; Finger et al., 2009; Gyau et al., 2009). Stone (2011) introduces a more complex human capital measure of 'agricultural skilling' which is a continual 'process of learning to perform with given technologies under changing conditions' (p. 393).

Only a couple of studies include the producer's age. Some of these studies model age as having a linear relationship with adoption (Hubbell et al., 2000; Hategekimana and Trant, 2002; McBride and El-Osta, 2002). Other studies model age as having a quadratic relationship with adoption, using both age and age squared as explanatory variables (Payne et al., 2003; Alexander and Mellor, 2005; Fernandez-Cornejo et al., 2005). The four studies that have found age to be significant in explaining adoption include the three that model age as having a quadratic relationship with adoption. In the linear model,

McBride and El-Osta (2002) found that farmer age was positively related to adoption of HT corn but unrelated to adoption of IR corn and HT soybeans.

4.1.3 Pest pressure

GM crops offer producers a substitute for current pest control technologies. Producers who face more severe pest pressures also have a larger potential to benefit from a new technology if it provides better control. The efficacy and cost of the current pest control technology and the severity of the pest have a significant influence on adoption. Farmers who had experienced pest resistance to chemical insecticides were significantly more likely to adopt Bt cotton (Hubbell et al., 2000; Marra et al., 2001). IR cotton and IR corn adoption is positively related to the price of the chemical insecticides (Fernandez-Cornejo et al., 2002; Qaim and de Janvry, 2003). IR corn adoption is positively related to farmer perception of severe pest pressure (Alexander et al., 2003; Payne et al., 2003). Hyde et al. (1999) used a dynamic programming model to estimate the *ex ante* per-acre value of IR corn in Indiana and found that the value of IR corn is equal to the additional seed cost when the probability of an European corn borer (ECB) infestation is 40 per cent; thus farmers who face a lower probability of ECB infestation are unlikely to adopt IR corn. Farmers with severe weed pressure were significantly more likely to adopt HT soybeans (Fernandez-Cornejo et al., 2002). Gonsalves et al. (2007) found that papaya farmers in Hawaii rapidly adopted VR varieties as the only effective treatment for the virulent and destructive papaya ringspot virus and, even more significantly, the availability of VR papaya increased the total papaya acreage by 26 per cent.

4.1.4 Tillage

The decision to adopt HT crops is not independent of the farmer's tillage decision since both tillage and herbicides are used to control weeds. Fernandez-Cornejo et al. (2002) found that producers who use conventional tillage were significantly less likely to adopt HT soybeans in 1997, which was expected since conventional tillage systems are less dependent on herbicides. Ward et al. (2002) concluded that the HT soybean technology performs better with conservation tillage (CT), based on a linear programming model that identified the technically efficient input set based on combinations of tillage practices and whether the cotton seed was HT or conventional. Banerjee and Martin (2009) and Banerjee et al. (2009) examined the interaction between CT and HT cotton adoption using cross-section survey data from the 2003 Agricultural Resource Management Survey (ARMS) conducted by the USDA. Banerjee and Martin (2009) did not find a significant link between CT and HT cotton adoption, while Banerjee et al. (2009) found that adoption of CT is significant in explaining HT cotton adoption and, correspondingly, HT cotton adoption is significant in explaining CT adoption; both papers caution that these findings are based on one year of cross-sectional data. Overcoming this concern, Frisvold et al. (2009) used state-level data from 1997–2002 to simultaneously estimate the factors driving the diffusion of CT and HT cotton. They found that HT cotton and CT are complementary and mutually reinforcing technologies, in that the diffusion of HT cotton and CT technologies was interdependent. Finger et al. (2009) found that all of the farmers who had adopted HT soybeans in Argentina also adopted no-till.

4.1.5 Location

Farm location provides a lot of information such as soil type, pest pressure, climate, availability of information and other market characteristics that affect the profitability of a new technology and ultimately the adoption decision. Previous studies have found that location is a significant predictor of farmer adoption or willingness-to-adopt (Hubbell et al., 2000; McBride and El-Osta, 2002; Alston et al., 2002; Fernandez-Cornejo et al., 2005; and Useche et al., 2009). Keelan et al. (2009) is the only study to explicitly include soil type; they found that Irish farmers with access to high-quality soils were significantly more willing to adopt GM crops.

4.1.6 Risk preferences

A farmer's risk preferences influence the adoption decision when there is uncertainty about the performance and profitability of the new technology. Uncertainty about performance of a new technology may cause the farmer to delay adoption until there is more information and thus less uncertainty about its performance. Some studies have found that farmers who are more risk averse are less likely to adopt HT soybeans (Fernandez-Cornejo et al., 2001), while others have found that more risk-averse farmers are more likely to adopt HT soybeans and IR corn (McBride and El-Osta, 2002) and a few found no relationship (for example, Alexander et al., 2003). Gyau et al. (2009) found that German farmers who said they are more willing to take risks are more likely to state their intention to adopt HT sugar beets or IR corn. Foltz et al. (2012) found that farmers who purchase insurance are willing to pay more for a drought-tolerant trait which reduces production risk.

4.1.7 Information

GM crop performance information can significantly influence adoption, depending on the farmer's trust and confidence in the information source. A farmer's own experience is often very influential in the adoption decision (Marra et al., 2001; Qaim and de Janvry, 2003; Alexander et al., 2003; and Banerjee and Martin, 2009). Finger et al. (2009) found that seed companies were the most important source of information for early adopters of HT soybeans in Argentina, while late adopters relied on information gained from personal contact with other farmers. Nwankwo et al. (2009) conducted an *ex ante* study of farmer willingness to adopt GM crops in Nigeria and found that farmers are much more willing to adopt GM crops if information about these crops is provided by a cooperative. The authors hypothesize that cooperatives are better able to disseminate information about new technologies because they have social capital and farmer trust. Aldana et al. (2011) found that a farmer's own experience with GM crops is extremely important early in the adoption process but diminishes for subsequent decisions. Furthermore, they found that personal experience has a larger impact on the adoption decision for college-educated farmers, which suggests that education enables farmers to learn faster from their own experience.

4.1.8 Demand uncertainty for GM crops

Because of consumer concerns and the potential for trade restrictions for GM crops which may affect price and market access, farmers consider the demand uncertainty when adopting GM crops (Lusk et al., 2003; Noussair et al., 2004; Sobolevsky et al., 2005; and

Lence and Hayes, 2005). Studies have documented that demand uncertainty can have a significant negative impact on GM corn adoption and diffusion (Fernandez-Cornejo et al., 2002; Alexander et al., 2003; and Alexander and Mellor, 2005). Alexander et al. (2003) found that farmers who agreed with the statement that 'Some consumers will not accept biotech food products' were significantly less likely to adopt GM corn, but there was no relationship with adoption of GM soybeans. Gonsalves et al. (2007) documented the impact of Japanese import restrictions on the adoption and diffusion of VR papaya in Hawaii. VR papaya adoption was extremely rapid, with 73 per cent of the farmers surveyed adopting it in the first year. However, because of Japanese import restrictions some farmers have disadopted VR papaya in order to gain access to the Japanese market (Gonsalves et al., 2007; see Chapter 40 in this volume on GM papaya). Napasintuwong and Traxler (2009) conducted an *ex ante* analysis of GM papaya adoption in Thailand, citing that as of May 2009 Thailand had not commercialized GM papaya due to concerns about access to export markets, health and environment.

4.1.9 Labor supply

Labor supply will affect the farmer's adoption decision if the new technology requires a different amount of labor than the current technology. In particular, one major benefit of HT crops is they generally require less labor and tillage (Carpenter and Gianessi, 1999; Flanders et al., 2003). Fernandez-Cornejo et al. (2005) found that while adoption of HT soybeans did not have a significant effect on farm income, it had a significant impact on total household income, which suggests that farmers may adopt HT soybeans because the simplicity and flexibility of the technology allows them to earn a higher income from off-farm labor. Useche et al. (2009) found that farms which use the most family labor also place the highest value on traits that offer potential labor savings.

4.1.10 Credit constraint

GM crops do not require fixed investments and as a result few researchers have studied whether access to credit influences the adoption of GM crops. Qaim and de Janvry (2003) find that farmers' willingness-to-pay for IR cotton in Argentina is negatively related to a credit constraint. Paredes and Martin (2007) find that access to credit is significantly and positively related to HT soybean adoption among smallholder farmers in Argentina, but had no significant relationship to IR corn adoption. However, access to credit is significantly and positively related to adoption of both HT soybeans and IR corn.

4.1.11 Technology fee or seed cost

The cost of a technology directly impacts the adoption decision and the farmers' gains from adopting the technology. Many of the studies that examine the impact of the seed price or technology fee on farmer adoption of GM crops are *ex ante* analyses (Hubbell et al., 2000; Qaim and de Janvry, 2003; and Vitale et al., 2007). For instance, Vitale et al. (2007) estimate that smallholder West African farmers will adopt IR cotton if the technology fee is less than $60 per hectare. In contrast, farmers will be much more sensitive to the technology fee for maize which is consumed domestically. In this case they will only adopt if the fee is less than $30 per hectare.

Paredes and Martin (2007) examined the IR corn and HT soybean adoption decision for smallholder farmers in Argentina. They found that seed cost is the third most important barrier to adoption, after lack of knowledge about GM crops and lack of locally available GM seed. Useche et al. (2009) found that farmer price sensitivity to GM seed cost decreases as farm revenue increases.

Sadashivappa and Qaim (2009) are the only researchers to document the impact of a change in the price of GM technology on farmer adoption. They utilized panel data on farmer adoption of IR cotton seed in India with observations before and after the 2006 government introduction of price caps on IR cotton seed. Overall, they found that the primary impact of the government price intervention was to increase farmer profits and was likely to have been one cause of reduction in the black market for illegal seeds. The impact on farmer adoption was relatively small because the technology was at a late stage in the diffusion process (Sadashivappa and Qaim, 2009).

4.1.12 Non-pecuniary benefits

Non-pecuniary benefits include increased human safety and environmental safety associated with adopting GM crops. Alston et al. (2002) used US farmer survey data to estimate the value associated with increased human safety and environmental safety associated with planting corn resistant to rootworms instead of using chemical insecticides, as well as the pecuniary values of time saved, potential yield risk reduction and equipment cost savings. Farmers willing to pay more for human safety and environmental safety were more likely to adopt IR corn. In contrast, Useche et al. (2009) found that Minnesota and Wisconsin farmers who are concerned about environmental or safety issues are less likely to adopt GM crops.

Krouser and Qaim (2011) documented that adoption of Bt cotton by smallholder farmers in India reduced the incidence of pesticide poisonings. They used panel data from 2002–08 to control for selection bias and estimated that the value of the reduction in pesticide poisonings among smallholder farmers in India from Bt cotton is worth US$14–51 million.

Edmeades and Smale (2006) is one of the few studies that investigated potential demand for a staple food crop, in their case GM cooking bananas in Uganda, by estimating current trait-based demand for banana varieties. In addition to the typical factors that explain farmer willingness to adopt varieties, such as experience and education, area under banana cultivation, disease problems and visits by Extension agents, they found that household demand factors had significant explanatory power. For instance, larger households are willing to adopt more plants to meet their consumption needs, wealthier households demand fewer banana varieties because their income allows them to substitute other foods, households further from market demand more banana varieties and, lastly, perceived cooking quality increases demand.

4.1.13 Country-specific regulations

In the United States, farmers who plant IR crops face insect resistance management (IRM) requirements. Initially farmers were required to plant a minimum of 20 per cent of their acres to a non-Bt crop. Hyde et al. (2000) found that the additional planting costs associated with implementing the refuge to be minimal for Bt corn resistant to ECB. However, in subsequent studies Hubbell et al. (2000), Alexander and Mellor (2005) and Alexander (2007) found these IRM requirements were a barrier to adoption of IR corn. Alexander (2007) also found that farmers who have experience with IRM report less time and effort to meet the requirements, suggesting that additional training could reduce this barrier to adoption.

In Europe, farmers who plant GM crops face coexistence regulations (see Chapter 25 in this volume for a detailed discussion) which require buffer zones and in some countries *ex post* liability for damages caused by contamination. Consmuller et al. (2009) conducted a case study with eight German farmers who grow IR maize and found that most perceive the buffer zone requirement as a negligible cost. In addition, the farmers believe there is minimal *ex post* liability since GM and conventional maize receive the same price in the market.

In Australia, farmers who plant GM canola face a number of requirements designed to segregate GM and non-GM canola (McCauley et al., 2012). In addition to provisions mandated by Monsanto on farmers seeking to trial the crop, the Department of Agriculture and Food in Western Australia required farmers to: (1) notify neighbors; (2) complete a quality-assurance program; (3) accept responsibility for any costs associated with the trial including remedial action; and (4) provide access to regulators to inspect the trial. McCauley et al. (2012) found that these regulations were effective at segregating the crops, with the exception of one case where an organic grower lost certification. The authors concluded from this case that the country needs 'realistic thresholds' to enable coexistence of different production systems.

4.2 Summary of Yield Trail Results

Most of the research comparing GM and non-GM crop yields has focused on the developing world, especially China and India. In general, these studies find that GM crops benefit most, but not all, farmers. Farmer benefits vary greatly depending on the region or the country, and most of this difference is attributable to pest pressure and the efficacy of existing pest control measures. Some of these studies are conducted on crops that do not have regulatory approval and have the goal of providing policy-makers with better information for future regulatory decisions (Huang et al., 2008).

Klotz-Ingram et al. (1999) analyzed US farmer data on cotton yields and pesticide use in 1997 from the Agricultural Resource Management Survey (ARMS). Comparing the mean adopters and non-adopters of HT cotton, they found that mean yields were not significantly different except in one region where the yields for adopters were significantly lower than yields for non-adopters. For IR cotton, they found that in over half the regions analyzed adopters had significantly higher yields than non-adopters. In addition, they found adopters of Bt cotton significantly reduced the use of insecticides that would normally be used on the pests targeted by Bt. One of the authors' primary conclusions was that the benefits of GM cotton in the US depend on the location.

Huang et al. (2002) analyzed whether Bt cotton adoption in China affected yields and pesticide use. Using data from a 1999 farm household survey, they found that on average the yield of Bt cotton was only 5.8 per cent higher than non-Bt cotton. When the authors accounted for other inputs and human capital variables, Bt cotton yields were 15 per cent higher than non-Bt varieties. While there are many production differences between Bt and non-Bt varieties, the biggest differences relate to pesticide use. Bt cotton farmers apply pesticides 6.6 times on average during the season compared to 20 times for non-Bt farmers, which reduces both their input costs and their labor requirements. The most important finding of this paper is that both Bt and non-Bt cotton farmers over-apply pesticides because their marginal effect is estimated to be near zero. The authors hypothesized that the overuse of pesticides may be due to poor information given to the farmers by pest control station personnel.

Qaim and Zilberman (2003) analyzed on-farm trials for Bt cotton in India and found that average cotton yields increased by 80 per cent compared to non-Bt counterparts and by 87 per cent compared to popular check varieties and that insecticide use declined by 70 per cent. They note that these Bt cotton yield increases are substantially higher for India compared to the US and China. They conclude that this relatively large yield gain can be attributed to average pest-related yield losses in conventional cotton in the order of 50 to 60 per cent in India compared to 12 per cent in the United States and 15 per cent in China.

Qaim et al. (2006) analyzed the yield benefit to Indian farmers of adopting Bt cotton using data from farmer interviews. Through a comparison of means of adopters and non-adopters, they show that adopters of Bt cotton make fewer applications of pesticides, have higher costs, have on average 34 per cent higher yields and have higher profits. They then used an econometric model to decompose the yield difference into two effects, a Bt gene effect and a germplasm effect, and find that the Bt gene has a significant positive effect. As with other studies of GM crop adoption, Qaim et al. (2006) find large interregional differences and a high degree of heterogeneity between farmers in terms of the benefits they receive from adopting Bt cotton. Furthermore, while farmers in Maharashtra, Karnataka and Tamil Nadu benefited from adopting Bt cotton, farmers in Andhra Pradesh who adopted Bt cotton had lower average incomes than those who did not adopt. Qaim et al. (2006) conclude that the Bt technology may not benefit all farmers; returns depend on their particular situation in terms of pest pressure and other methods for controlling pests.

Bennett et al. (2006) used field data from farmer surveys and analyzed the yield benefit to farmers adopting Bt cotton in Maharashtra, India. They found large and significant spatial variation in yields for both Bt and non-Bt cotton. Controlling for insecticide sprays, soil type, irrigation and other factors, Bennett et al. (2006) concluded that Bt cotton in India had a positive yield impact in the order of 33–48 per cent but that this benefit varied greatly between seasons and over space.

Gouse et al. (2006) analyzed three years of farmer yield data to measure yield and other benefits to South African smallholders planting IR white corn. They measured yield as kilograms of grain harvested per kilogram of seed planted and found that the IR seed produced significantly higher average yields in the first two years of adoption (32 per cent and 16 per cent, respectively). The third year was dry with low pest pressure, resulting in yields that were not significantly different from conventional seed; given that IR seed is more expensive, farmers planting IR seed had lower profits than farmers who planted conventional seeds. Overall, Gouse et al. (2006) concluded that small-scale subsistence farmers in South Africa benefit from planting IR corn except in years of low pest pressure.

Morse et al. (2007) analyzed two years of farmer yield data for IR cotton and non-GM cotton in India. The authors emphasized the importance of sampling plots, when possible, from farmers who grow both IR and non-GM cotton in order to minimize the sample selection bias between adopters and non-adopters. They found IR cotton yields were higher than non-GM cotton yields for adopters. More importantly, non-GM cotton yields of adopters were also higher than non-GM cotton yields of non-adopters, which highlights the importance of gathering data that permits researchers to control for selection bias.

Huang et al. (2008) estimated the farm-level benefits of IR rice in China (a crop that did not have regulatory approval for widespread use) by measuring the impact on yields and reductions in pesticide use. They used survey data from 2002–04 for 320 households and 584 rice plots, where some households grew only IR rice, some only non-GM rice, and others both non-GM and IR rice. Huang et al. (2008) found that while the IR rice yields were slightly higher, the difference was not statistically significant. However, farmers growing IR rice had substantially higher profits because they used 85 per cent less pesticides and captured labor savings equivalent to 9 days.

Stone (2011) used panel data from cotton farmers in Warangal, India, comparing villages before and after IR cotton adoption to distinguish between 'field-level' and 'farm-level' yield impacts. The major contribution of Stone (2011) is to demonstrate the value of panel data to address selection biases that resulted from: (1) early adopters being more successful farmers than later adopters; and (2) cultivation biases that resulted from IR cotton being planted in fields with the best irrigation because of the high seed cost prior to 2006. Overall, Stone found that IR cotton increased yields 18 per cent, with the largest increases occurring in the villages that had the lowest yields before the introduction of IR cotton. In addition, pesticide spraying dropped by 55 per cent on average with adoption of IR cotton. Stone (2011) also investigated how the introduction of IR cotton affected farmer skill acquisition, tracking farmers' variety choice. Stone observed that farmers frequently changed IR cotton seed, often because a particular seed became a local favorite; he concluded that the rapid pace of change in IR cotton seed varieties has led to 'agricultural deskilling', which is a 'failure of the ongoing process of learning to perform under changing conditions' (p. 394).

5 CRITICAL ASSESSMENT: WHERE IS THE THEORY GOING? WHAT METHODS ARE WE CONVERGING ON? WHICH METHODS ARE BEING DEVELOPED? WHAT DO THE EMPIRICS TELL US? WHAT IS YET TO BE TESTED?

This chapter reviewed the literature on farmer adoption of GM crops. Farmer decisions to adopt GM crops are influenced by a wide range of factors that vary depending on the GM crop and the country. However, not all farmers benefit from planting GM crops and so understanding who will benefit is important, especially in the case of introducing a new GM crop or trait. In particular, there is growing consensus that researchers need to use panel data in *ex post* adoption studies to account for selection biases and cultivation biases (Barham et al., 2004; Morse et al., 2007; Krouser and Qaim, 2011; Stone, 2011).

The regulatory environment surrounding GM crops continues to change. Most of the GM crops with regulatory approval are primarily used for livestock feed, fiber and fuel. As countries approve GM food crops, consumer and household demand will have a larger impact on a farmer's decision to adopt these crops. Gonsalves et al. (2007), who documented VR papaya adoption in Hawaii, highlight the impact of Japan's import restrictions on farmer adoption. Japan lifted this import restriction in December 2011 and it would be interesting to see how the market has responded. Other studies on GM food crops are *ex ante* with the goal of offering policy makers more information about potential benefits for farmers (Edmeades and Smale, 2006; Huang et al., 2008; Gyau et al., 2009).

The dynamic regulatory and market environment as well as the rapid development of new GM traits means that farmers are continually learning new information about GM crops. GM crops offer researchers an opportunity to study the interaction between farmers' human capital, information and their risk preferences as they make decisions about adopting new varieties. Aldana et al. (2011) introduced a theoretical model that explicitly links uncertainty about the GM crop performance, human capital and personal experience or learning. Because the adoption of GM crops generally does not require any fixed investment that would lock farmers into the technology, a growing number or researchers have documented that farmers may disadopt GM crops as they evaluate the benefits they receive from the new technology or as consumer demand changes (Fernandez-Cornejo et al., 2002; Barham et al., 2004; Gonsalves et al., 2007).

Useche et al. (2009) and Foltz et al. (2012) have introduced a new theoretical and empirical approach to literature on GM crop adoption using methods developed in the consumer demand literature. They modeled GM crops as a bundle of traits which enables them to estimate farmer willingness-to-pay for individual traits. This model enables researchers to understand how farmers may value traits *ex ante* and to understand the relative value of individual traits when they are stacked.

In conclusion, the literature on farmer adoption of GM crops will continue to be important as new traits are developed, as GM traits are added to new crops and as regulations change. The existing literature highlights the importance of context and the need to evaluate each GM trait and each GM crop in the context of farmer benefits and consumer preferences.

REFERENCES

Aldana, U., J.D. Foltz, B.L. Barham and P. Useche (2011), 'Sequential adoption of package technologies: the dynamics of stacked trait corn adoption', *American Journal of Agricultural Economics*, **93**(1), 130–43.

Alexander, C. (2006), 'Farmer decisions to adopt genetically-modified crops', *CAB Reviews: Perspectives in Agriculture, Veterinary Science, Nutrition and Natural Resources*, **1**(45), available at: http://www.cababstractsplus.org/cabreviews.

Alexander, C. (2007), 'Insect resistance management plans: the farmers' perspective', *AgBioForum*, **10**(1), 33–43.

Alexander, C. and T.V. Mellor (2005), 'Determinants of corn rootworm resistant corn adoption in Indiana', *AgBioForum*, **8**(4), 197–204.

Alexander, C., J. Fernandez-Cornejo and R.E. Goodhue (2003), 'Effects of the GM controversy on Iowa corn-soybean farmers' acreage allocation decisions', *Journal of Agricultural and Resource Economics*, **28**(3), 580–95.

Alston, J., J. Hyde, M.C. Marra and P.D. Mitchell (2002), 'An ex ante analysis of the benefits from the adoption of corn rootworm resistant transgenic corn technology', *AgBioForum*, **5**(3), 71–84.

Banerjee, S.B. and S.W. Martin (2009), 'A binary logit analysis of factors impacting adoption of genetically modified cotton', *AgBioForum*, **12**(3&4), 218–25.

Banerjee, S.B., S.W. Martin, R.K. Roberts, J.A. Larson, R.J. Hogan Jr, J.L. Johnson, K.W. Paxton and J.M. Reeves (2009), 'Adoption of conservation-tillage practices and herbicide-resistant seed in cotton production', *AgBioForum*, **12**(3&4), 258–68.

Barham, B.L., J.D. Foltz, D. Jackson-Smith and S. Moon (2004), 'The dynamics of agricultural biotechnology adoption: lessons from rBst use in Wisconsin, 1994–2001', *American Journal of Agricultural Economics*, **86**(1), 61–72.
Bennett, R., U., Kambhampati, S., Morse and Y. Ismael (2006), 'Farm-level economic performance of genetically modified cotton in Maharashtra, India', *Review of Agricultural Economics*, **28**(1), 59–71.
Carpenter, J. and L. Gianessi (1999), 'Herbicide tolerant soybeans: why growers are adopting roundup ready varieties', *AgBioForum*, **2**(2), 65–72.
Cohen, J.I. and R. Paarlberg (2004), 'Unlocking crop biotechnology in developing countries – a report from the field', *World Development*, **32**(9), 1563–77.
Consmuller, N., V. Beckmann and C. Schleyer (2009), 'The role of coordination and cooperation in early adoption of GM crops: the case of Bt maize in Brandenburg, Germany', *AgBioForum*, **12**(3&4), 47–59.
Edmeades, S. and M. Smale (2006), 'A trait-based model of the potential demand for a genetically engineered food crop in a developing country', *Agricultural Economics*, **35**, 351–61.
Feder, G., R.E. Just and D. Zilberman (1985), 'Adoption of agricultural innovations in developing countries: a survey', *Economic Development and Cultural Change*, **30**, 59–76.
Fernandez-Cornejo, J. and W.D. McBride (2002), 'Adoption of bioengineered crops', Agricultural Economic Report no. 810, Washington DC: USDA-Economic Research Service, May.
Fernandez-Cornejo, J., C. Alexander and R.E. Goodhue (2002), 'Dynamic diffusion with disadoption: the case of crop biotechnology in the USA', *Agricultural and Resource Economics Review*, **31**(1), 112–26.
Fernandez-Cornejo, J., S. Daberkow and W.D. McBride (2001), 'Decomposing the size effect on the adoption of innovations: agrobiotechnology and precision agriculture', *AgBioForum*, **4**(2), 124–36.
Fernandez-Cornejo, J., C. Hendricks and A. Mishra (2005), 'Technology adoption and off-farm household income: the case of herbicide-tolerant soybeans', *Journal of Agricultural and Applied Economics*, **37**(3), 549–63.
Fernandez-Cornejo, J., C. Klotz-Ingram and S. Jans (2002), 'Farm-level effects of adopting herbicide-tolerant soybeans in the USA', *Journal of Agricultural and Applied Economics*, **34**(1), 149–63.
Finger, R., M. Hartman and M. Feitknecht (2009), 'Adoption patterns of herbicide-tolerant soybeans in Argentina', *AgBioForum*, **12**(3&4), 404–11.
Flanders, A., F.C. White, W.D. Shurley and S.M. Brown (2003), 'Profit and variance analysis of cotton production technologies and rotation crops in Georgia', *Journal of Agricultural and Applied Economics*, **35**(3), 685–94.
Foltz, J.D., P. Useche and B.L. Barham (2012), 'Bundling technology and insurance: packages versus technology traits', *American Journal of Agricultural Economics*, 1–7.
Frisvold, G.B., A. Boor and J.M. Reeves (2009), 'Simultaneous diffusion of herbicide resistant cotton and conservation tillage', *AgBioForum*, **12**(3&4), 249–57.
Fulton, M. and L. Keyowski (1999), 'The producer benefits of herbicide-resistant canola', *AgBioForum*, **2**(2), 85–93.
Gonsalves, C., D.R. Lee and D. Gonsalves (2007), 'The adoption of genetically modified papaya in Hawaii and its implications for developing countries', *Journal of Development Studies*, **43**(1), 177–91.
Gouse, M., C. Pray, D. Schimmelpfennig and K. Johann (2006), 'Three seasons of subsistence insect-resistant maize in South Africa: have smallholders benefited?', *AgBioForum*, **9**(1), 15–22.
Graff, G., D. Roland-Holst and D. Zilberman (2006), 'Agricultural biotechnology and poverty reduction in low-income countries', *World Development*, **34**(8), 1430–45.
Gyau, A., J. Voss, A. Spiller and U. Enneking (2009), 'Farmer acceptance of genetically modified seeds in Germany: results of a cluster analysis', *International Food and Agribusiness Management Review*, **12**(4), 61–80.
Hategekimana, B. and M. Trant (2002), 'Adoption and diffusion of new technology in agriculture: genetically modified corn and soybeans', *Canadian Journal of Agricultural Economics*, **50**(2), 357–71.
Huang, J., R. Hu, S. Rozelle, F. Qiao and C.E. Pray (2002), 'Transgenic varieties and productivity of smallholder cotton farmers in China', *The Australian Journal of Agricultural and Resource Economics*, **46**(3), 367–87.
Huang, J., H. Ruifa, S. Rozelle and C. Pray (2008), 'Genetically modified rice, yields, and pesticides: assessing farm-level productivity effects in China', *Economic Development and Cultural Change*, **56**(2), 241–63.
Hubbell, B.J., M.C. Marra and G.A. Carlson (2000), 'Estimating the demand for a new technology: Bt cotton and insecticide policies', *American Journal of Agricultural Economics*, **82**(1), 118–32.
Hyde, J., M.A. Martin, P.V. Preckel and C.R. Edwards (1999), 'The economics of Bt corn: valuing protection from the European corn borer', *Review of Agricultural Economics*, **21**(2), 442–54.
Hyde, J., M.A. Martin, P.V. Preckel, C.L. Dobbins and C.R. Edwards (2000), 'The economics of within-field Bt corn refuges', *AgBioForum*, **3**(1), 63–8.
James, C. (2011), 'Global status of commercialized biotech/GM crops: 2011', *International Service for the Acquisition of Agri-Biotech Applications (ISAAA) Brief*, 43, Ithaca, NY: ISAAA, available at: http://www.isaaa.org/resources/publications/briefs/43/executivesummary/default.asp.

Keelan, C., F.S. Thorne, P. Flanagan, C. Newman and E. Mulllins (2009), 'Predicted willingness of Irish farmers to adopt GM technology', *AgBioForum*, **12**(3&4), 394–403.
Klotz-Ingram, C., S. Jans, J. Fernandez-Cornejo and W. McBride (1999), 'Farm-level production effects related to the adoption of genetically modified cotton for pest management', *AgBioForum*, **2**(2), 73–84.
Krouser, S. and M. Qaim (2011), 'Impact of Bt cotton on pesticide poisoning in smallholder agriculture: a panel data analysis', *Ecological Economics*, **70**, 2105–13.
Lence, S.H. and D.J. Hayes (2005), 'Genetically modified crops: their market and welfare impacts', *American Journal of Agricultural Economics*, **87**(4), 931–50.
Lindner, R.K. (1983), 'Farm size and time lag to adoption of a scale neutral innovation', mimeograph, Adelaide: University of Adelaide.
Lusk, J.L., J. Roosen and J.A. Fox (2003), 'Demand for beef from cattle administered growth hormones or fed genetically modified corn: a comparison of consumers in France, Germany, and the United States', *American Journal of Agricultural Economics*, **85**(1), 16–29.
Marra, M.C., B. Hubbell and G.A. Carlson (2001), 'Information quality, technology depreciation and Bt cotton adoption in the Southeastern US', *Journal of Agricultural and Resource Economics*, **26**(1), 158–75.
Marra, M.C., P.G. Pardey and J.M. Alston (2002), 'The payoffs to transgenic field crops: an assessment of the evidence', *AgBioForum*, **5**(2), 43–50.
McBride, W.D. and H.S. El-Osta (2002), 'Impacts of the adoption of genetically engineered crops on farm financial performance', *Journal of Agricultural and Applied Economics*, **34**(1), 175–91.
McCauley, R., M. Davies and A. Wyntje (2012), 'The step-wise approach to adoption of genetically modified (GM) canola in Western Australia', *AgBioForum*, **15**(1), 61–9.
Morse, S., R. Bennett and Y. Ismael (2007), 'Inequality and GM crops: a case-study of Bt cotton in India', *AgBioForum*, **10**(1), 44–50.
Napasintuwong, O. and G. Traxler (2009), 'Ex-ante impact assessment of GM papaya adoption in Thailand', *AgBioForum*, **12**(3&4), 209–17.
Noussair, C., S. Robin and B. Ruffieux (2004), 'Do consumers really refuse to buy genetically modified food?', *Economic Journal*, **114**(492), 102–20.
Nwankwo, U.M., K.J. Peters and W. Bokelmann (2009), 'Can cooperative membership and participation affect adoption decisions? Issues for sustainable biotechnology dissemination', *AgBioForum*, **12**(3&4), 437–51.
Paredes, C. and M.A. Martin (2007), 'Adoption of transgenic crops by smallholder farmers in Entre Rios, Argentina', selected paper presented at the AAEA Annual Meeting in Portland Oregon, available at: www.ageconsearch.umn.edu.
Payne, J., J. Fernandez-Cornejo and S. Daberkow (2003), 'Factors affecting the likelihood of corn rootworm Bt seed adoption', *AgBioForum*, **6**(1&2), 79–86.
Qaim, M. (2005), 'Agricultural biotechnology adoption in developing countries', *American Journal of Agricultural Economics*, **87**(5), 1317–24.
Qaim, M. and A. de Janvry (2003), 'Genetically modified crops, corporate pricing strategies and farmers' adoption: the case of Bt cotton in Argentina', *American Journal of Agricultural Economics*, **85**(4), 814–28.
Qaim, M. and I. Matuschke (2005), 'Impacts of genetically modified crops in developing countries: a survey', *Quarterly Journal of International Agriculture*, **44**(3), 207–27.
Qaim, M. and D. Zilberman (2003), 'Yield effects of genetically modified crops in developing countries', *Science*, **299**(5608), 900–903.
Qaim, M., A. Subramanian, G. Naik and D. Zilberman (2006), 'Adoption of Bt cotton and impact variability: insights from India', *Review of Agricultural Economics*, **28**(1), 48–58.
Rogers, E.M. (2003), *Diffusion of Innovations*, 5th edn, New York: Free Press.
Sadashivappa, P. and M. Qaim (2009), 'Bt cotton in India: development of benefits and the role of government seed price interventions', *AgBioForum*, **12**(2), 172–83.
Sobolevsky, A., G. Moschini and H. Lapan (2005), 'Genetically modified crops and product differentiation: trade and welfare effects in the soybean complex', *American Journal of Agricultural Economics*, **87**(3): 621–44.
Solow, R.M. (1994), 'Perspectives on growth theory', *Journal of Economic Perspectives*, **8**(1), 45–54.
Stone, G.D. (2011), 'Field *versus* farm in Warangal: Bt cotton, higher yields, and larger questions', *World Development*, **39**(3), 387–98.
Thomson, J.A. (2004), 'The status of plant biotechnology in Africa', *AgBioForum*, **7**(1&2), 9–12.
US Department of Agriculture (2011), 'National Agricultural Statistics Service: Acreage', June 30, 2011, Washington, DC: USDA, pp. 25–7, available at: http://usda.mannlib.cornell.edu/usda/current/Acre/Acre-06-30-2011.pdf.
Useche, P., B.L. Barham and J.D. Foltz (2009), 'Integrating technology traits and producer heterogeneity: a mixed-multinomial model of genetically modified corn adoption', *American Journal of Agricultural Economics*, **91**(2), 444–61.

Vitale, J., T. Boyer, R. Uaiene and J.H. Sanders (2007), 'The economic impacts of introduction Bt technology in smallholder cotton production systems of West Africa: a case study from Mali', *AgBioForum*, **10**(2), 71–84.
Ward, C.W., A. Flanders, O. Isengildina and F.C. White (2002), 'Efficiency of alternative technologies and cultural practices for cotton in Georgia', *AgBioForum*, **5**(1), 10–13.
Zilberman, D., H. Ameden and M. Qaim (2007), 'The impact of agricultural biotechnology on yields, risks, and biodiversity in low-income countries', *Journal of Development Studies*, **43**(1), 63–78.

25 Coexistence

Volker Beckmann, Claudio Soregaroli and Justus Wesseler

1 INTRODUCTION

Coexistence is defined by the European Coexistence Bureau as:

> the ability of farmers to choose between the cultivation of genetically modified (GM) and non-GM crops, in compliance with the relevant legislation on labelling rules for GM organisms (GMOs), food and feed and/or purity standards . . . Since only GM crops that have passed a rigorous authorisation procedure can be cultivated in the EU, coexistence measures do not concern environmental or health risks. The issues to be addressed in the context of coexistence are in general the potential economic impacts of the admixture of GM and non-GM crops, the identification of workable technical and organisational measures during cultivation, harvest, on-farm storage and transport to minimise admixture, and the cost of these measures. (ECoB, 2012)

In this sense, coexistence of conventional, organic and GM crops can be regarded as (1) an economic issue of (2) preserving the choices of consumer and farmer among different methods of production by (3) controlling agro-ecological dynamics through measures of segregation (4) embedded in a broader regulatory framework of (4a) labeling and (4b) approval for GM crops. The economic problem of coexistence, therefore, consists of at least three framing factors: first, consumers' and farmers' preferences for different production methods; second, the agro-ecological dynamics depending on the biology of the crops concerned and the agro-ecological environment in which they are released; and, third, the broader institutional framework. It should be noted that all these key factors vary internationally, nationally or even regionally. This diversity has implications for achieving practical coexistence. Different preferences on the demand and supply side are fundamental and suspected to affect the institutional environment of GM labeling and approval (for example Gruère et al., 2008; Crespi and Marette, 2003; Kalaitzandonakes and Bijman, 2003; Bernauer and Meins, 2003; for a political economy perspective see Graff et al., 2009). Agro-ecological dynamics are crop and location specific and are extensively addressed in the agronomic literature (for example Heuberger et al., 2010; Jarosz et al., 2005; Ma et al., 2004).

The agro-biological dynamics, in particular cross-pollination and other forms of admixture, by itself do not generate any economic problem. The economic problem emerges from what is considered as GM, conventional or organic, and the related requirements of what needs to be approved or labeled in order to be marketed legally (Bender and Westgren, 2001; Beckmann and Wesseler, 2007). In other words, the issue is related to the legal definition of goods and the segmentation of markets by labels and approval procedures. When a GM crop is approved in the United States (US) or Canada but not in the European Union (EU), adventitious presence may cause an economic loss for the North Amercian non-GM farmer if exporters are no longer able to export non-GM crops to the EU, as the example of adventitious presence of GM flax seeds in exports to Europe illustrates (Falck-Zepeda et al., 2013, Ryan and Smyth, 2012). When GM crops need to be labeled, agro-ecological dynamics may result in an economic loss for a non-GM farmer if he cannot sell his produce as non-GM. Thus, different approval and labeling regimes are fundamental to the regulation of coexistence as an economic problem (Boisson de Chazournes and Mbengue, 2005).

Two distinct approval and labeling regimes emerged, which largely distinguish the US from Europe. In the US, decisions on the approval are made by public agencies such as the Food and Drug Administration (FDA), the US Department of Agriculture (USDA) and the Environmental Protection Agency (EPA) (see Belson, 2000). In Europe, the approval procedures changed significantly between 1990 and 2001. Since the Directive 2001/18/EC on the release of GM organisms (GMOs) into the environment came into force in 2001 (CEC, 2001), risk assessment and risk management at the European level have been divided. Risk assessment based on scientific evidence is prepared by the European Food Safety Agency (EFSA), whereas the approval decision is complex and political, involving the Commission of the European Communities (CEC), committees with representatives of EU Member States and the European Parliament (Wesseler and Kalaitzandonakes, 2011). These different approval regimes have resulted in significant asymmetries with an increasing number of GM crops being approved in the US but not in Europe (Stein and Rodriguez-Cerezo, 2009). For GM crops which are not approved within the EU, currently a zero tolerance policy is applied for its adventitious presence in imported food and a 0.1 per cent threshold is set for feed (CEC, 2011).

Difference also exists regarding labeling regimes (Gruère et al., 2008): whereas the US and Canada currently do not require any GM labeling and allow for voluntary labeling of non-GM food and feed, the EU introduced strict mandatory labeling and traceability of GM food and feed (CEC, 2003a). Whereas under the mandatory labeling and traceability system the GM sector has to bear the costs of labeling, documentation and tracing, voluntary labeling puts the burden on the non-GM sector (Phillips and Isaac, 1998).

In Europe the system of GM labeling has defined goods on the basis of product standards. If a food or feed product contains or consists of a GMO or is produced from or contains ingredients produced from GMOs, it needs to be labeled. If a produce contains more than 0.9 per cent of a single GM event for each single ingredient due to adventitious or technically unavoidable presence it has to be labeled as a GM produce (CEC, 2003a). Since animal products, such as meat or milk, do not contain any GM material even if they are produced with GM feed, these products do not have to be labeled. This standard applies currently for conventional as well as for organic food and feed, although the organic food sector is not allowed to use any GM material within their production processes and sets tolerances lower than 0.9 per cent, while the private sector has introduced GMO-free standards that go beyond the EU regulations (Venus and Wesseler, 2012).

The standards for organic farming in the United States are process based. For organic agriculture this excludes the planting of transgenic crops; however, even if cross-pollination occurs and organic products contain GM material they can still be sold under the US certification scheme (for example Krueger, 2007). The standard set by the US regulators can be seen as a minimum standard. Many private organic certification schemes do have standards that go beyond the federal standard and imply that organic producers bear the additional costs. As Smyth et al. (2010) describe, until now there have been no cases in the United States or Canada where an organic farmer has lost his certification because of adventitious presence of transgenic material.

The following section presents in more detail the coexistence policies in the US and EU. The third section presents selected economic issues of coexistence including labeling and thresholds, property rights and liability rules, cost-effective measures, adoption and regional clustering, implications for international trade and welfare. Section 4 offers some concluding comments.

2 MODELS OF COEXISTENCE REGULATIONS

The institutional differences between the US and Europe are important for understanding and analyzing the regulation of coexistence. Policies governing coexistence can be differentiated into *ex ante* regulations and *ex post* liability rules. This differentiation is useful as the economic implications between the two differ (for example Kolstad et al., 1990; Posner, 2007; Shavell, 1987).

2.1 Regulations in the United States

The United States, as well as Canada, until now has not had explicit regulations addressing the issue of coexistence (Smyth et al., 2010). According to US regulations, if a new crop has been approved for cultivation (that is, deregulated) farmers are free to choose and plant them. Possible damages through the cultivation of transgenic crops are addressed via common law *ex post* liability rules, like nuisance, trespass, negligence or strict liability (Kershen, 2004). However, since labeling of approved GM crops in the US domestic market is not mandatory and non-GM crops are defined in a process-based manner, economic losses due to commingling are difficult to claim and courts have up to now refused compensation claims of farmers for economic damages (Kershen, 2009).

The case, however, is different for partially or non-approved GM crops where economic damages have been successfully claimed by non-GM farmers. One example is StarLink corn which was approved for feed but not for food use and was subsequently detected in the food chain. Aventis CropScience USA was held liable and paid compensation to farmers and other plaintiffs (Kershen, 2009). Another important case is LL601 rice where farmers sued Bayer CropScience for economic damages due to the loss of export markets after non-approved transgenic rice was detected. In both cases, a zero tolerance threshold for non-approved GM crops applied and compensation claims for pure economic losses were admitted by the courts.

For approved GM crops, however, the regulatory standard is essentially 'may contain approved GM material'. Where private standards go beyond this public standard, those following the higher private standards have to bear the additional costs. For example, organic farmers in the US increasingly invest in coexistence measures to meet tightened private standards based on threshold levels for adventitious presence of GM material (for example, Greene and Smith, 2010; Stokstad, 2011). The Non-GM Project, for instance, set a product-based standard of 0.1 per cent GM material for seeds and other propagation material, 0.5 per cent for food, ingredients, supplements, personal care products, and other products that are either ingested or used directly on skin, and 0.9 per cent for animal feed and supplements, packaging, cleaning products, textiles and other products that are not ingested or used directly on skin (Non-GMO Project, 2013). To meet these standards several preventive measures are required or recommended.

Recently, the debate on coexistence reemerged in the US as a consequence of lawsuits filed by the Center of Food Safety and others following GM glyphosate-tolerant sugar beets and alfalfa deregulation by US Department of Agriculture (USDA) Animal and Plant Health Inspection Service (APHIS). The two GM crops were both deregulated in 2005 but cultivation was halted by courts in 2007 for alfalfa (only for new plantings) and in 2010 for sugar beets until new and more complete environmental impact statements (EIS) were provided. Early in 2011, GM alfalfa was completely deregulated (USDA-APHIS, 2011) following a new EIS, and GM sugar beet was partially deregulated as a temporary measure until a full EIS was provided. Finally, GM sugar beet was fully deregulated in July 2012 (USDA-APHIS, 2012).

USDA addressed the coexistence issue by establishing the Advisory Committee on Biotechnology and 21st Century Agriculture (AC21) with the specific objective of suggesting tools and measures to support the practical implementation of coexistence (USDA, 2011a). In particular, AC21 meetings focused on the problems of determining the economic risks for farmers, the eventual compensation mechanisms, the tools and standards to verify eligibility and losses and the question of who pays for the eventual compensation mechanism (USDA, 2011b).

2.2 Regulations in the European Union (EU)

While the risk assessment and approval of GMOs is performed at the EU level, coexistence measures and liability schemes are demanded and implemented at the Member State level. The EU assisted the development of coexistence measures providing general guidelines (CEC, 2003b) and establishing the European Coexistence Bureau (ECoB) having the objective of supporting member states in the development of best practices and technical measures related to coexistence.

Despite the guidelines for coexistence, in 2010 only two GM Crop events were approved for cultivation in the EU (Bt maize MON810 and the potato Amflora) and only eight EU countries planted GM crops (Wesseler and Kalaitzandonakes, 2011). Under EU law approved GM crops may be freely grown in the territory of the EU, provided that farmers respect administrative regulation and agricultural practices. This is often not possible. Some member states, such as Austria, France, Germany, Greece and Hungary for the case of Bt maize, have appealed to article 23 of Directive 2001/18/EC containing the safeguard clause that allows every single state to limit or prohibit the application or the disposal of a GMO (CEC, 2001). This limitation or prohibition can only be used 'where a Member State, as a result of new or additional information made available [. . .] has detailed grounds for considering that a GMO [. . .] contributes a risk to human health or the environment' (CEC, 2001). While this measure is designed to allow for the possibility of banning the planting of GM crops in specific areas, mainly nature reserves, many member states have interpreted it as allowing bans for socio-economic or political reasons.

In 2010, following a number of national bans, the EU Commission reconsidered its coexistence policy with a new Recommendation (CEC, 2010) providing greater flexibility to member states allowing not only the definition of best cultivation practices but also for the possibility of excluding GMOs from large cultivation areas. According to the Commission 'Under certain economic and natural conditions, Member States should consider the possibility to exclude GMO cultivation from large areas of their territory . . . such exclusion should rest on the demonstration by Member States that, for those areas, other measures are not enough to achieve sufficient levels of purity' (CEC, 2010, p. 5). The rationale of this new approach rests on the fact that, in particular areas, avoiding adventitious presence from GM products could be too costly to non-GM operators, also recognizing that local standards might be more stringent with thresholds even below the 0.9 per cent set by the EU legislation. Therefore, with the new approach the Commission is de facto allowing for the introduction of socio-economic evaluations in defining local coexistence regulations, with consequent effects on GM crop adoption.

Member states adopted coexistence regulations under the previous EU general guidelines (CEC, 2003a), where CEC (2009b) reports that, as of February 2009, 15 out of 27 member states adopted specific regulations for coexistence while other countries had legislation at the draft stage. Only a few countries did not develop specific coexistence measures since they are not expecting future GM cultivations in their territory. Table 25.1 summarizes the *ex ante* regulations and *ex post* liability rules by member state using information provided in the CEC report (CEC, 2009b). Draft regulations are distinguished since they are still subject to change, especially during the notification process to the EU Commission.

2.2.1 *Ex ante* regulations

As shown in Table 25.1, *ex ante* regulations in the EU involve prohibitions on cultivation of GM crops in specific areas, cultivation approval procedures, registration systems and information duties requested to GM farmers, as well as technical segregation measures, such as isolation distances, buffer zones, different plant cycles and rotation intervals, and obligatory insurance measures.

Beside the GM ban in certain areas, several member states require farmers to get official approval before they are allowed to plant GM crops. In most cases this is only a formal step following the correct application to authorities by the farmer. In some regions of Austria and Hungary, however, it involves a case-by-case approval of each field. Moreover, Hungarian authorities adopt a case-by-case definition of isolation distances and cultivation conditions and, before the final approval, also require the farmer to get a written approval from the neighboring farmers within the isolation distance. In general, in some countries, the consent from the neighbors is required when the GM farmer cannot respect the minimum isolation distance. A consent from the landowner is also required in Austria, Luxemburg, in two Baltic republics and is foreseen in Italy.

Almost all member states include registration and information responsibilities for GM farmers. These include registration of areas in databases, with individual locations being often publicly available, the duty to inform authorities, neighbors and landowners in advance and requirements for record keeping. Currently, only Spain, France and Sweden do not require prior information to the authorities on the intention to cultivate GM crops.

Technical segregation measures with sexually compatible crops have been adopted in different countries and are quite similar in nature, but their application differs widely.

Table 25.1 Ex ante *regulations and* ex post *liability rules governing coexistence at the farm level among European Union member states*

Policy	EU member states	
	Apply	Intend to apply
***Ex ante* regulations**		
Prohibition and approval procedures		
Prohibition of planting GM crops in specific areas	AT, CY, FI, HU, LV, RO	IT, LT, PL
GM and/or non-GM area on a voluntary base	BE, ES, FR, HU, PT, SV	IT, SI
Case by case approval for each field by local authorities	AT*, HU,	CY, IR
Compulsory training of farmers planting GM crops to be paid by the GM farmer	DK, HU, LT, LV, PT, SK	EE, FI, IR, IT, NL, SI
Consent from landowner needed	AT, LT, LU, LV	IT
Consent from neighbors needed	AT*, BE[a], HU[a], LV, SI[a]	IR[a]
Registration and information duties		
Registration of areas in publicly available database (no information on individual locations and/or restricted access)	CZ, DE, PT, RO, SK, SV	BE, ES, IR
Registration of areas in publicly available database (information on individual locations is released)	AT*, DK, FR, HU, LT, LU, NL,	BE, CY, EE, FI, IT, SI
Informing authorities on the intention to cultivate GM crops prior to cultivation	AT, CZ, HU, LT, LU, LV, PT, SK	BE, ES, IR, SI
Informing authorities on the intention to cultivate GM crops prior to cultivation and at a fixed date of the year	DK, FI, NL, RO	IT
No obligation on informing authorities on the intention to cultivate GM prior to cultivation	ES, FR, SV	
Informing neighboring farmers and/or landowners	AT*, BE, CZ, DE, DK, EE, FI, LT, LV, NL, PT, RO, SK, SV	ES, FR, IR, IT, SI
Record keeping	BE, CZ, DE, DK, ES, FI, HU, LT, LV, PT, RO, SI, SK	EE, IR
Technical segregation measures		
Minimum distance requirements	CZ, DE, DK, HU, LT, LU, LV, NL, PT, RO, SI, SK, SV	EE, ES, FI, FR, IR, IT
Buffer zones	CZ, HU, LT, LV, PT, RO, SI, SK	EE, ES, IT
Plant cycles	HU, PT, RO, SK	ES
Rotation intervals	DK, HU, LT, SK	EE, ES, FI, SI

Table 25.1 (continued)

Policy	EU member states	
	Apply	Intend to apply
Farmer-to-farmer agreements for not applying segregation measures are allowed	BE, DE, DK, HU, LT, LV, PT, SI, SV	EE, ES, FI
Insurance measures		
Compensation fund paid by GM farmers (levy on GM crops) plus support from the central government	BE, DK	
GM farmers need to provide a financial guarantee or a private insurance against damages	AT*, FR, LU	IT
***Ex post* liability**		
Legal liability for damages		
Liability based on civil law (usually fault based)	AT*, BE, CY, CZ, ES, HU, LT, LU, LV, NL, PL, RO, SI, SV	FI, IT
Strict liability for GM farmers	DE, FR	
Penalties		
Fines for non-compliance with *ex ante* regulations	AT, BE, CZ, DK, EE, FR, HU, LT, LU, LV, NL, PT, SK,	FI, IT, SI

Notes:
[a] Only within isolation distance.
AT: Austria, AT*: specific regions of Austria only; BE: Belgium; CY: Cyprus; CZ: Czech Republic; DE: Germany; DK: Denmark; EE: Greece; EL: Estonia; ES: Spain; FI: Finland; FR: France; HU: Hungary; IR: Ireland; IT: Italy; LT: Lithuania; LU: Luxemburg; LV: Latvia; MT: Malta; NL: The Netherlands; PL: Poland; PT: Portugal; SV: Sweden; SI: Slovenia; SK: Slovak Republic; UK: United Kingdom.

Source: Authors' elaboration based on CEC (2009a).

The main instruments are the use of isolation distances, buffer zones, different plant cycles and rotation intervals. The stated objective of these measures is to guarantee that non-GM harvests are below the GM labeling threshold of 0.9 per cent for food and feed.

The definition of the minimum distance requirements differs depending on the country and on the type of crop. Among the different member states, measures are defined for maize, potato, sugar beet, wheat and oilseed rape. Maize is regulated in the highest number of countries. Some countries differentiate additionally between distances to conventional non-GM crops, organic crops and crops for seed production. Isolation distance can be used as the unique requirement, as in Germany and Denmark, or in conjunction with buffer zones that often allow a reduction of the distance requirement. In general, minimum distance requirements are lower for potatoes, followed by sugar beet, maize and oilseed rape and are higher for organic non-GM crops and non-GM seeds. Luxemburg has the largest distances for maize (600 meters), while Hungary imposes a minimum distance of 400 m that can be extended in the case-by-case approval process. Lithuania and Latvia define a distance of 4000 m for GM rapeseed.

A few states consider the use of different plant cycles as a parameter for segregation: this includes using plants with different cycle length and flowering time or simply having a different sowing period. Other countries require attention to the general crop rotation of the farm or to the specific crop interval between GM and non-GM crops. For example, in Slovakia non-GM plants cannot be grown for two years after GM crops of the same species.

Segregation measures may be reduced with written agreements among neighboring farmers on the non-application of such measures. A distinct number of countries allow for this possibility, while a few deny it. With these agreements the neighboring farmer gives up the right to claim compensation for the eventual loss of GM-free status, therefore also reducing the liability risk. In any case, these measures do not overcome the transaction costs of dealing with neighbors.

Not all of the member states impose segregation measures. In Spain, GM farmers need to adopt good agricultural practices and recommendations given by seed companies, mainly consisting in refugia areas and control of volunteers. No minimum distances are defined and the non-GM farmer has the duty to adopt segregation measures if willing to obtain a GM-free harvest. Spain also does not require prior notification to authorities or mandatory information to third parties; it only asks for record keeping. While draft measures for Spain would substantially change the regulatory framework, no coexistence problems have emerged until now and there is no consensus among stakeholders on the need for such regulation.

Even when respecting technical segregation measures, the cultivation of GM crops could cause economic losses to non-GM crops. For this reason some countries require a mandatory financial guarantee from the GM farmer. This could consist of an insurance contract, although at the moment no insurance companies offer such a product since the sufficient information for a proper assessment of risk is lacking. Belgium and Denmark have introduced compensation funds paid for by the GM farmer. These funds are basically a tax on the GM cultivation, and protect the GM farmer from potential financial loss as long as the technical *ex ante* regulations are respected.

2.2.2 *Ex post* liability

In general, most member state legislation refers to civil law and, in particular, to tort law (Koch, 2008; ECTIL, 2007). The GM farmer can be held liable if it can be demonstrated that he was negligent or mismanaged his enterprise. Usually, in this case the GM farmer has to pay compensation for damages and all other expenses. France and Germany, by contrast, adopt a strict liability regime for GM crops wherein the GM farmer is liable for all financial damage independently of his behavior.

It is important to define the occurrence and the level of damage. In general, EU countries define the damage as the need to label a product as GM because of the unwanted admixture with GM material over the 0.9 per cent labeling threshold. The level of the damage is measured as the difference between the price of the GM product and the price of a non-GM product. If there is no price mark-up for conventional non-GM crops, economic damage will be difficult to prove and it is reasonable to expect that no compensation will be required. This actually reduces the problem of coexistence mainly to areas where organic farmers are present. Moreover, it is uncertain how the different legal systems will deal with those cases where the non-GM farmer faces losses originating from delivery contracts voluntarily setting higher standards of purity (less than 0.9 per cent) (ECTIL, 2007).

Liability can also be regulated through insurance or payments for mandatory compensation funds as is the case in Austria, France and Lithuania (all with mandatory insurance) and in Belgium and Denmark (with compensation funds).

Excluding the case of strict liability, compliance with *ex ante* regulations mainly protects farmers against liability claims. *Ex ante* regulation exposes farmers to non-compliance penalties that can be independent of the occurrence of damage. For example, the Czech Republic, Denmark, Hungary and Portugal impose regular controls on GM fields and provide penalties for non-compliance that could include fines or, in some cases, the withdrawal of the authorization to grow GM crops. Other member states do not explicitly state whether the fine for non-compliance is independent of the occurrence of an admixture.

3 SELECTED ECONOMIC ISSUES OF COEXISTENCE

The previous section has illustrated that while the United States and the EU member states do regulate the planting of transgenic crops, the issue of coexistence is explicitly regulated only in the EU. This disparity is primarily grounded in the different institutional framework that requires mandatory GM labeling in the EU and thus induces the economic risk of adventitious presence of GM in non-GM produce, with member states allowed to take appropriate measures to avoid this happening. The economic risk is absent if mandatory labeling is not required and labeling is based on process standards, as in the US domestic market.

Based on the two contrasting cases, a number of economic issues related to coexistence will be elaborated. First, there is an impact arising from (1) differences in the labeling regime and (2) the general assignment of property rights. Then, the (3) effectiveness and costs of different coexistence measures will be assessed including technical and organizational solutions. Successively, the implication of different coexistence regulations on the (4) adoption and (5) clustering of GM crops will be discussed. Finally, we address issue related to (6) international trade and (7) a welfare assessment of coexistence regulations.

3.1 Labeling and Threshold Levels

Labeling should provide consumers with reliable information about valuable attributes of a good which are difficult to detect from visual inspection. Labeling, thus, is an instrument to overcome incomplete and asymmetric information and to enable informed consumer choices.

Process-based labeling of GM and non-GM products circumvents problems related to controlling agro-ecological dynamics. It only matters whether or not authorized GM crops are used within the production process, independent of the adventitious presence of GM material. Thus, even mandatory GM labeling based on process standards would not cause coexistence problems.

For product-based labeling, be it mandatory or voluntary, threshold levels are of crucial importance and agro-ecological dynamics may pose a serious problem. The threshold level, however, is arbitrary and can not be based on scientific reasoning. Whereas in the EU the reference threshold is currently set at 0.9 per cent for food and feed containing approved GM material and at zero for non-approved GM material in the food chain, other countries applying mandatory labeling of GM crops require between zero (for example China) or 5 per cent (for example Japan) (Gruère and Rao, 2007). Private standards usually aim at threshold levels below 0.9 per cent. It can reasonably be assumed that coexistence at the farm level is more costly to achieve, the lower the threshold level (Maltsbarger and Kalaitzandonakes, 2000).

Without labeling, coexistence of approved GM crops would not create any economic problem. Due to the zero tolerance policy, coexistence issues related to differences in the approvals between countries would, however, still remain.

Since labeling and thresholds create segmented markets, their rationale is mainly related to consumer preferences. Different threshold levels may reflect the trade-off between consumer preferences for GM-free produce and the cost of segregation, which is likely to increase with decreasing threshold levels.

3.2 Basic Property and Liability Rules

The allocation of property and liability rights to farmers who plant GM or non-GM crops is a fundamental decision in regulating coexistence (Beckmann and Wesseler, 2007). Property and liability rules determine the rights and duties of farmers in relationship to each other and to the courts (see, for example, Krauss, 1999). The US and the EU case represent two different approaches which are characterized by Beckmann et al. (2011) as:

1. The property right is primarily with the GM farmer. The non-GM farmer has to implement measures to prevent cross-pollination and cannot claim in court compensation for economic damages through cross-pollination.
2. The property right is primarily with the non-GM farmer. The GM farmer has to implement measures to prevent cross-pollination of non GM crops and is liable for economic damages caused by his GM crops on non-GM farmers, which the non-GM farmer can claim using the courts.

In Case 1 the non-GM farmer has to invest in preventive measures by either keeping an appropriate distance from the GM farmer, using buffer strips or other technical measures to reduce cross-pollination, or by talking to the GM farmer and asking him to keep a distance from his non-GM crops, whichever will be a more cost-effective solution. For a given farm size and structure, the potential damage costs for the non-GM farmer will increase with the number of neighboring GM farmers. Under Case 2 the GM farmer has to keep a distance from the non-GM farmer and can do this by either locating his GM field away from the non-GM field, using buffer strips or other technical measures to reduce cross-pollination, or by talking to the non-GM neighbors. Here again, for a given farm size and structure, the costs for the GM farmer can be expected to increase with the number of non-GM neighbors.

Essentially property and liability rules structure responsibilities and the division of costs and benefits. They represent basic entitlements for GM and non-GM farmers. As Coase (1960) has shown, under such circumstances the allocation of property rights does not affect the efficient outcome as long as neighbors are willing to negotiate and information asymmetries are negligible – but the two property rights systems have an effect on the distribution of benefits and costs. Information asymmetries and transaction cost may, however, be significant and at times even prohibitive and thus the distribution of property rights may well affect efficiency. In practice it has proven difficult to determine the allocation of property and liability rights based on efficiency considerations alone. Beckmann et al. (2011), for example, argue that if the costs of using the court are substantial, a property rights system of type 1 is preferable since the number of court cases is reduced. Since the allocation of property and liability always affects the wealth of different actors, the determination will be subject to distributive bargaining and value judgments about entitlements.

Referring to Coase (1960) one may argue that it is more important to create clear property and liability rights and it is less important how they are actually assigned since once rights and duties are settled, farmers can bargain for increasing efficiency.

3.3 Cost-effective Coexistence Measures

The issue of cost-effective measures to achieve coexistence arises independent of the allocation of property and liability rights, but they depend on the labeling system and the private or public standards applied. The question is how cost-effectiveness relates to the objectives.

The literature so far has focused on the European context where thresholds have been set and research concentrates on measures that ensure that farmers stay below the threshold of 0.9 per cent. Other threshold levels exist. Since the agro-ecological dynamics are crop specific and location specific, research results must pay attention to these specificities. The crops that are well researched are maize and rapeseed.

At the farm level, factors affecting cross-pollination include the isolation distance between pollen source and recipient field, the size, shape and orientation of the respective fields, wind directions, pollen viability, male fertility or sterility or synchrony in flowering times (Devos et al., 2005). The coexistence measures most often considered involve spatial isolation distances, pollen barriers and temporal isolation measures.

According to Czarnak-Kłos and Rodríguez-Cerezo (2010) green maize does not require any isolation distance to maintain the 0.9 per cent threshold whereas for grain maize an isolation distance of 15 to 50 m is advised. However, for admixture levels below 0.1 per cent the authors propose an isolation distance of 105 to 500 m for grain maize and 85 to 200 m for green maize. These isolation distances can be reduced by buffer zones or temporal isolation measures, where the latter heavily depend on climatic condition.

Whether or not isolation distances induce any costs depends on the farm-specific adventitious presence risk that is linked to the agricultural structure and the share of the respective crops in the landscape. Munro (2008) discusses in the context of the setting of his model the possibilities of bargaining, and concludes that there are limits to achieving the social optimal level through bargaining, if land cannot be reallocated. This changes if land reallocation is possible. The possibility of land reallocation is a realistic assumption in the EU as the concentration of crops in a given landscape hardly reaches more than 30 per cent (EUROSTAT, NUTS 3 level).[1] The study by Devos et al. (2008) at the smaller regional level indicated for Belgium a concentration of less than 50 per cent in the area with highest maize production density on cultivated land and less than 30 per cent on total area.

Ceddia et al. (2007) in their simulation study on oilseed rape conclude that if no more than 27 per cent of an area planted with oilseed rape is planted with GM seeds the average GM content of the remaining non-GM oilseed rape will be below the 0.9 per cent threshold level. Putting these results in the European context of oilseed rape production, oilseed rape hardly even reaches more than 25 per cent on total land in the EU at NUTS 3 level. Again, in this context coexistence should easily be achieved without minimum distance requirements.

The previous papers cited indicate that achieving coexistence for maize and oilseed rape may not pose a serious problem in the EU and additional policies may not be needed; in essence, the *ex ante* regulations and *ex post* liability rules member states implement or want to implement may be redundant. The conclusion is based on the 0.9 per cent threshold level. But measuring GM content is a difficult task. Numerous studies on cross-pollination indicate that the GM concentration in harvested non-GM crops changes depending on the distance to the source. If field averages are used, GM concentration decreases with field size (Weber et al., 2005). Many non-GM crops such as oilseed rape or maize are collected at a regional level for further handling. If the GM concentration is measured at the elevator, the concentration may further decrease. The reported evidence of GM concentration exceeding the 0.9 per cent threshold level at some measuring points in the field does not imply the average concentration will be above the threshold level (Ceddia et al., 2007; Weber et al., 2005). Studies measuring field level concentrations are scarce and there is no evidence of GM concentration levels above the 0.9 per cent threshold level. Gray et al. (2011) in their study on GM oilseed rape in Australia did conclude that achieving the 0.9 per cent threshold level for oilseed rape does not pose a problem and additional minimum distance requirements are not needed. If targets aim at a lower than 0.9 per cent threshold, distance requirements may be deemed necessary (Czarnak-Kłos and Rodríguez-Cerezo, 2010).

Isolation distances can be mandatory or endogenously determined by the specific agro-ecological dynamics. Gray et al. (2011) discuss this duality, taking into account the allocation of property and liability rights, the farm-specific adventitious presence risk, threshold levels and the opportunity cost for GM or identity-preserved (IP) crops. They conclude that endogenous pollen barriers are superior from an efficiency point of view if transaction costs are minimal. Therefore, flexible policies should be preferable, especially policies which reduce the costs of transactions (that is favoring the possibility of bargaining, land reallocation and availability and use of information by farmers).

3.4 Coexistence Regulation and the Adoption of GM Crops

Theoretically, many of the *ex ante* regulations that are established by EU member states have a negative effect on adoption of GM crops and are not independent of farm size and may cause an irreversibility effect (Beckmann et al., 2010). In particular mandatory minimum distance requirements make it more difficult for smaller farms to adopt GM crops (Soregaroli and Wesseler, 2005) and can cause a substantial domino effect on other farms (Demont et al., 2008), which can be reduced if farmers know what their neighbors are doing and have the opportunity to relocate GM crop cultivation (Groeneveld et al., 2013). Further, communication with neighbors can substantially reduce coexistence regulation compliance costs if the regulations allow for sufficient flexibility (Skevas et al., 2010).

Empirical studies on adoption of Bt maize in the EU have been presented by Gomez-Barbero et al. (2008) for the case of Spain, by Skevas et al. (2009, 2010) for the case of Portugal and by Consmüller et al. (2009, 2010) for the case of Germany. These countries provide quite a range of different *ex ante* regulations and *ex post* liability rules: where Spain has no *ex ante* coexistence regulations in place and relies on civil liability rules, Portugal introduced a range of *ex ante* coexistence measures, which, while mostly mandatory, can be circumvented by private arrangements between farmers, but do not subject farmers to *ex post* liability claims. Farmers in Germany, as in Portugal, face *ex ante* regulations which might by circumvented by private arrangements but are exposed to unlimited and several liability rules. Based on a case study approach of a small number of farms, Skevas et al. (2009, 2010) and Consmüller et al. (2009) analyzed strategies of Bt maize-growing farms coping with coexistence regulations. In both cases, farmers were able to plant Bt maize without significant additional costs due to large farm sizes in Germany and, in Portugal, with the organization of small farms in a cooperative. Although these studies show that coexistence regulations do not necessarily impose additional costs, they cannot be regarded as representative of their respective country. Indeed Consmüller et al. (2010) investigated regional adoption patterns in Germany and found that the amount of maize grown per farm, the occurrence of the European corn borer and the presence of the anti-GMO movement are important determinants for regional differences in the adoption rates. The authors argue that German coexistence regulations did impose fixed costs and therefore discriminate against farms seeking to plant a small amount of maize unless the latter were able to cooperate with their neighbors and organize GM zones as in Portugal. The studies also revealed that the socio-economic environment is an important factor in determining the adoption decision.

The interaction between coexistence regulations, agricultural structures and the socio-economic environment of farms in Europe is also revealed in studies that analyzed potential adoption, like Breustedt et al. (2008) for herbicide-tolerant (HT) oilseed rape in Germany and Areal et al. (2011) for HT maize in Spain, France and Hungary and for HT rapeseed in Germany, Czech Republic and the United Kingdom. These studies show comparatively high potential adoption rates for HTGM crops in the selected EU member states, but found that farmers are concerned about the coexistence regulations, the risk of being held liable for damages on neighboring fields due to adventitious presence and insurance costs. Technical measures, such as buffer zones and isolation distances, are generally not regarded as major obstacles of growing HTGM crops.

Thus, a major critique is that most *ex ante* regulations that are established by EU member states are not proportional to the 0.9 per cent threshold and actually more designed to prevent GM crops from being grown (Beckmann et al., 2006; Demont et al., 2009; Ramessar et al., 2010).

Spain is the only member state of the EU where currently coexistence is not regulated by mandatory requirements. Bt maize in Spain has been growing since 1998. In 2010, Spain had the largest Bt maize acreage in the EU: the total Bt maize area was 76 575 ha, corresponding to 23 per cent of the total maize area in Spain and 80 per cent of the total

Bt maize area in the EU. Despite that, no economic damage for conventional or organic farms has been recorded.

3.5 Spatial Clustering: GM and GMO-free Regions

Theoretical studies have suggested that clustering of GM and non-GM crops may significantly lower coexistence costs and that coexistence regulations may provide farmers with strong incentives to do so (Beckmann et al., 2006; Beckmann and Wesseler, 2007; Beckmann and Schleyer, 2007; Furtan et al., 2007; Beckmann et al., 2010).

Empirical studies on GM or GMO-free zones, however, have been rather limited. Skevas et al. (2009, 2010) present an example of a type of GM zone in Portugal where small-scale farmers reallocate their plots spatially to plant Bt maize. Consmüller et al. (2009) found that cooperation to form GM zones in Brandenburg, Germany, was not deemed necessary due to the large-scale farms. Overall there is little evidence for the establishment of GM zones in Europe currently.

The formation of GMO-free zones, in contrast, has become a widespread phenomenon in Europe and, in particular, in Germany, France, Belgium, Italy and Greece. GMO-free regions may be founded by farmers but also declared by local, regional or national governments. In general, these declarations are not legally binding.

Examples of GM-free zones in the US can be found in cases of seed production. Purity of seed is the starting point for an identity-preserved (IP) product and this becomes even more important in cases of stringent import standards, such as zero tolerance policies in trading partners. Therefore, coexistence practices become more strict and costly in the case of seed production. As Endres (2005) reports, key seed producers in the US contract with neighboring farmers in order to create such areas.

3.6 International Trade

Coexistence between GM and GM-free products also has implications for international trade and downstream to the agro-food supply chain. For example, considering the EU dependency on imports for many feed and food products, EU demand for specific quality attributes, such as GM-free products, and the related EU regulations on labeling can be important drivers of coexistence practices of exporting countries.

Particular attention is directed to the combination of threshold levels and 'asynchronous approval', 'isolated foreign approval' or 'low-level presence' of GM events. Under zero tolerance for unapproved GM events, different authors highlight possible trade disruptions and important consequences on the international and EU market in case of adventitious presence (see Nowicki et al., 2010). The case of LL601 rice is a real example of this impact on trade and firms (Brookes, 2008). Traces of non-EU approved LL601 event were found in long grain rice exported to the EU from the US in August 2006. This produced a series of trade disruptions and commercial difficulties to both EU and US firms.

On the supply side, adoption of GM crops is increasing worldwide and a growing number of new GM crops are expected to be traded in the future (Stein and Rodriguez-Cerezo, 2009). The likelihood of adventitious presence and the capacity to control such events also depend on what happens in exporting areas: countries interested in exporting

IP products could control the degree of asynchrony in their approval process or could create large IP areas in their territory. On the other hand, non-interested countries could create cases of isolated foreign approval and low-level presence of GM events, since the approval process is not activated in the importing country Which route they choose will depend on the attractiveness of GM-free importing markets, a parameter that could vary in the future given the emerging role of new importing countries, such as China (Stein and Rodriguez-Cerezo, 2009).

Firms respond to the risk of adventitious presence using different degrees of vertical coordination and integration. For example, in Brazil some exporting firms own the land where they produce large amounts of non-GM soybeans, and other firms directly control the supply chain from seed production to export delivery to take advantage of the large IP areas in their territory (Pelaez et al., 2010).

Segregated non-GM programs have also been adopted in the US. These programs include interventions to avoid admixture at the different stages of the supply chain, from the seed sector to processing. Zero tolerance results are very difficult to achieve: for example, in 2006/07 in the US a large-scale program involving different actors of the US and international supply chain was activated to segregate the maize event DAS 59122-7 (Herculex Rootworm) from the corn gluten feed exports to the EU due to the fact that in 2006 the event was approved in the US but not in the EU. In spite of their efforts, the coexistence and segregation practices proved not to be sufficient as about half of the samples tested positive for the maize event and, in some circumstances, exports that tested negative at the US export position tested positive at EU ports (Nowicki et al., 2010). Finally, the EU approved DAS 59122-7 in September 2007.

3.7 Welfare Implications

From an economic point of view coexistence at the regional farm level will be preferable if the net benefits of coexistence, namely the aggregated net benefits of cultivation of GM and non-GM crops within the same region, are larger than the aggregated net benefits of cultivating either only GM or non-GM crops. The net benefits depend on how consumers and farmers value GM and non-GM crops and on the costs of separating these crops to maintain their attributes, which relates to the definition of GM and non-GM crops and, thus, the threshold level.

The optimal design of coexistence policies depends on the possible economic damages that may occur and on the additional costs caused by *ex ante* regulations and *ex post* liability. The identification of the minimum costs is not an easy task due to the interdependencies between *ex ante* regulations and *ex post* liability. While the corner solution of only using *ex post* liability needs to be considered, there are a number of arguments against a corner solution. The identification of a single farmer causing the adventitious mixing of GM material in non-GM crops within a region with several farmers is difficult. Further, the distribution of power might be uneven, and moral hazard and adverse behavior could affect the efficiency of using *ex post* liability only (Shleifer, 2010).

Overall, coexistence requirements increase the costs of food and feed production as they require measuring purity as well as additional technical measures to segregate GM and non-GM crops along the supply chain. Neither measurement nor segregation are costless. Whether or not this will be welfare enhancing depends on additional welfare gains among consumers. If consumers express a positive willingness to pay for non-GM food or feed and if this compensates for the additional costs of achieving coexistence, then coexistence can be welfare enhancing. In addition, externality effects on the environment have to be considered. Contrary to the widely held view that GM crop production has a negative impact on the environment, studies show that cultivation of GM crops can generate environmental benefits (for example Smyth et al., 2011; Wesseler et al., 2011; Brookes and Barfoot, 2010). Several studies indicate that GM crop cultivation can also contribute to poverty alleviation (for example Kikulwe et al., 2011; Qaim, 2009). The environmental implications and poverty alleviation issues are relevant as they have an effect on the willingness to pay for non-GM food and feed. Nevertheless, several studies show positive welfare gains for the EU if environmental effects are included and identity preservation (IP) costs excluded (for example Demont et al., 2004; Wesseler et al., 2007). Moschini et al. (2005) include IP costs in their assessment of the welfare benefits of introducing GM crop cultivation within the EU. The authors show the welfare benefits for the EU to become negative due to the IP costs. The differences in the results illustrate one of the fundamental problems of assessing the welfare benefits of coexistence policies. The IP costs are a result of policies governing the production and consumption of GM and non-GM crops. The policies are a response to consumers' willingness to pay for non-GM food products. The consumers' willingness to pay for non-GM food products are based on the fear of negative impacts of GM crops on the environment and human health (Gaskell et al., 1999, 2010). GM crops currently approved for cultivation or processing have passed the environmental and food safety assessment and hence can be considered to be safe. Further, cultivation could even increase environmental benefits. This inconsistency in policy has an effect on the welfare assessment. Much scope for future work remains to fully assess the welfare effects of coexistence.

4 CONCLUSIONS

The cultivation of GM and non-GM crops at world level shows coexistence is possible. Smaller-scale studies using gene flow models show commonly-used threshold levels of 0.9 per cent can be achieved without additional regulatory interventions, such as minimum standards when measured at first point of delivery, such as elevators. At field level, it is also possible to achieve 0.9 per cent thresholds – although within a field, threshold levels may be exceeded at field borders. In general, whether or not coexistence will be possible does depend on the threshold level and where the obligations are imposed. The issue of coexistence is also closely linked with the debate on labeling. The thresholds for labeling of food and feed drive the thresholds for crop production. The major driver in the debate on coexistence is the European Union. The EU labeling policies generate spillover effects on crop production in other countries exporting crops to the EU, increasing the overall costs of crop production.

Despite the possibility of achieving coexistence without additional interventions, EU member states have implemented a set of regulations governing the cultivation of GM crops. When assessing the economic effects at farm level, differentiating between *ex ante* regulations and *ex post* liability is necessary. One of the important results from the literature is that minimum distance requirements discriminate against smaller farms.

Further, the regulations within the EU are very diverse, reflecting the overall view of EU member states on GMOs. Member states with more stringent coexistence regulations are also those member states arguing against approval for cultivation of GM crops. Nevertheless, flexible coexistence regulations allow farmers to respond and to adjust, as seen in Portugal.

Finally, the overall welfare effects of coexistence are difficult to assess. The results will largely depend on the treatment of consumer willingness to pay.

NOTE

1. NUTS stands for Nomenclature of Territorial Units for Statistics and subdivides the economic territory of the European Union into 97 regions at NUTS 1 level, 271 regions at NUTS 2 level and 1303 regions at NUTS 3 level.

REFERENCES

Areal, F.J., L. Riesgo and E. Rodriguez-Cezero (2011), 'Attitudes of European farmers towards GM crop production', *Plan Biotechnology Journal*, **9**, 945–57.

Beckmann, V. and C. Schleyer (2007), 'Neue Formen der Kooperation von Landwirten bei der Befürwortung und Ablehnung der Agro-Gentechnik', in B. Köstner, M. Vogt and B. van Saan-Klein (eds), *Agro-Gentechnik im Ländlichen Raum: Potenziale, Konflikte und Perspektiven. Forum für Interdisziplinäre Forschung*, Berlin: Brandenburgische Akademie der Wissenschaften and Dettelbach: J.H. Röll, pp. 221–43.

Beckmann, V. and J. Wesseler (2007), 'Spatial dimension of externalities and the Coase Theorem: implications for coexistence of transgenic crops', in W. Heijman (ed.), *Regional Externalities*, Berlin: Springer, pp. 215–34.

Beckmann, V., C. Soregaroli and J. Wesseler (2006), 'Co-existence rules and regulations in the European Union', *American Journal of Agricultural Economics*, **88**(5), 1193–9.

Beckmann, V., C. Soregaroli and J. Wesseler (2010), 'Ex-ante regulation and ex-post liability under uncertainty and irreversibility: governing the coexistence of GM crops', *Economics: The Open-Access, Open-Assessment E-Journal*, **4**(2010–9).

Beckmann V., C. Soregaroli and J. Wesseler (2011), 'Coexistence of genetically modified (GM) and non-modified (non-GM) crops: are the two main property rights regimes equivalent with respect to the coexistence value?', in C.A. Carter, G. Moschini and I. Sheldon (eds), *Genetically Modified Food and Global Welfare*, Frontiers of Economics and Globalization, Vol. 10, Bingley: Emerald Group Publishing Limited, pp. 201–24.

Belson, N.A. (2000), 'US regulation of agricultural biotechnology: an overview', *AgBioForum*, **3**(4), 268–80.

Bender, K.L. and R.E. Westgren (2001), 'Social construction of the market(s) for genetically modified and nonmodified crops', *American Behavioral Scientist*, **44**, 1350–70.

Bernauer, T. and E. Meins (2003), 'Technological revolution meets policy and the market: explaining cross-national differences in agricultural biotechnology regulation', *European Journal of Political Research*, **42**(5), 643–83.

Boisson de Chazournes, L. and M-M. Mbengue (2005), 'International legal aspects of the co-existence between GM and non-GM products: approaches under international environment law and international trade law', in A. Messean (ed.), *Proceedings of the Second International Conference on Co-existence between GM and Non-GM based Agricultural Supply Chains*, 14–15 November 2005, Montpellier (France), pp. 15–28.

Breustedt, G., J. Müller-Scheeßel and U. Latacz-Lohmann (2008), 'Forecasting the adoption of GM oilseed rape: evidence from a discrete choice experiment in Germany', *Journal of Agricultural Economics*, **59**(2), 237–56.

Brookes, G. (2008), 'Economic impacts of low level presence of not yet approved GMOs on the EU food sector', briefing document, GBC Ltd, UK, May 2008, available at: http://www.agrodigital.com/images/estudio.pdf.

Brookes, G. and P. Barfoot (2010), 'Global impact of biotech crops: environmental effects, 1996–2008', *AgBioForum*, **13**(1), 76–94.

CEC 'Commission of the European Communities' (2001), 'Directive 2001/18/EC of the European Parliament and of the Council of 12 March 2001 on the deliberate release into the environment of genetically modified organisms and repealing Council Directive 90/220/EEC', *Official Journal of the European Union*, L 106, 17.4.2001, p.1.

CEC, Commission of the European Communities (2003a), 'Regulation (EC) No. 1829/2003 of the European Parliament and of the Council of 22 September 2003 on genetically modified food and feed', *Official Journal of the European Union*, L 268/1-23.

CEC, Commission of the European Communities (2003b), 'Recommendation 2003/556/EC on guidelines for the development of national strategies and best practices to ensure the co-existence of genetically modified crops with conventional and organic farming', *Official Journal of the European Union*, L 189, 29.7.2003, p. 36.

CEC, Commission of the European Communities (2009a), 'Report from the Commission to the Council and the European Parliament: on the coexistence of genetically modified crops with conventional and organic farming', COM(2009) 153, Brussels.

CEC, Commission of the European Communities (2009b), 'Annex to the Report from the Commission to the Council and the European Parliament: on the coexistence of genetically modified crops with conventional and organic farming', Commission Staff Working Document, SEC(2009) 408, Brussels.

CEC, Commission of the European Communities (2010), 'Recommendation 2010/C 200/01 of 13 July 2010 on guidelines for the development of national co-existence measures to avoid the unintended presence of GMOs in conventional and organic crops', *Official Journal of the European Union*, C 200, 22.7.2010.

CEC, Commission of the European Communities (2011), 'Commission Regulation (EU) No. 619/2011 of 24 June 2011 laying down the methods of sampling and analysis for the official control of feed as regards presence of genetically modified material for which an authorisation procedure is pending or the authorisation of which has expired', *Official Journal of the European Union*, L166/9-15. 25.6.2011.

Ceddia, M.G., M. Bartlett and C Perrings (2007), 'Landscape gene flow, coexistence and threshold effect: the case of genetically modified herbicide tolerant oilseed rape (Brassica napus)', *Ecological Modelling*, **205**(1–2), 169–80.

Coase, R. (1960), 'The problem of social costs', *The Journal of Law and Economics*, **3**(1), 1–44.

Consmüller, N., V. Beckmann and M. Petrick (2010), 'An econometric analysis of regional adoption patterns of Bt maize in Germany', *Agricultural Economics*, **41**, 275–84.

Consmüller, N., V. Beckmann and C. Schleyer (2009), 'The role of coordination and cooperation in early adoption of GM crops: the case of Bt maize in Brandenburg, Germany', *AgBioForum*, **12**(1), 47–59.

Crespi, J.M. and S. Marette (2003), '"Does contain" vs. "does not contain": does it matter which GMO label is used?', *European Journal of Law and Economics*, **16**(3), 327–44.

Czarnak-Kłos, M. and E. Rodríguez-Cerezo (2010), 'Best practice documents for coexistence of genetically modified crops with conventional and organic farming: 1. Maize crop production', Seville: European Coexistence Bureau (ECoB), Joint Research Centre (JEC), Institute for Prospective Technological Studies.

Demont, M., J. Wesseler and E. Tollens (2004), 'Biodiversity versus transgenic sugar beets: the one Euro question', *European Review of Agricultural Economics*, **31**(1), 1–18.

Demont, M., W. Daems, K. Dillen, E. Mathijs, C. Sausse and E. Tollens (2008), 'Regulating coexistence in Europe: beware of the domino-effect!', *Ecological Economics*, **64**, 683–9.

Demont, M., K. Dillen, W. Daems, C. Sausse, E. Tollens and E. Mathijs (2009), 'On the proportionality of EU spatial *ex ante* coexistence regulations', *Food Policy*, **34**(6), 508–18.

Devos, Y., D. Reheul and A. De Schrijver (2005), 'The co-existence between transgenic and non-transgenic maize in the European Union: a focus on pollen flow and cross-fertilization', *Environmental Biosafety Research*, **4**, 71–87.

Devos, Y., M. Cougnon, O. Thas and D. Reheul (2008), 'A method to search for optimal field allocations of transgenic maize in the context of Co-existence', *Environmental Biosafety Research*, **7**, 97–104.

ECoB-European Coexistence Bureau (2012), 'Background', available at: http://ecob.jrc.ec.europa.eu/background.html.

ECTIL, European Centre of Tort and Insurance Law (2007), 'Liability and compensation schemes for damage resulting from the presence of genetically modified organisms in non-GM crops', April 2007, available at: http://ec.europa.eu/agriculture/analysis/external/liability_gmo/index_en.htm.

Endres, B.A. (2005), 'Revising seed purity laws to account for the adventitious presence of genetically modified varieties: a first step towards coexistence', *Journal of Food Law and Policy*, **131**.

Falck-Zepeda, J., J. Wesseler and S. Smyth (2013), 'The current status of the debate on socio-economic assessments: positions and policies in Canada, the USA, the EU, and developing countries', *World Review of Science, Technology and Sustainable Development*, **10**(4), 203–27.

Furtan, W.H., A. Güzel and A.S. Weseen (2007), 'Landscape clubs: co-existence of genetically modified and organic crops', *Canadian Journal of Agricultural Economics*, **55**, 185–95.

Gaskell, G., M.W. Bauer, J. Durant and N.C. Allum (1999), 'Worlds apart? The reception of genetically modified food in Europe and the US', *Science*, **285**, 384–7.

Gaskell, G., S. Stares, A. Allansdottir, N. Allum, P. Castro, Y. Esmer, C. Fischler, J. Jackson, N. Kronberger,

J. Hampel, N. Mejlgaard, A. Quintanilha, A. Rammer, G. Revuelta, P. Stoneman, H. Torgersen and W. Wagner (2010), *Europeans and Biotechnology in 2010: Winds of Change?*, Luxembourg: Publications Office of the European Union.

Gomez-Barbero, M., J. Berbel and E. Rodríguez-Cerezo (2008), 'Adoption and performance of the first GM crop introduced in EU agriculture: Bt Maize in Spain', JRC Scientific and Technical Reports. European Commission Joint Research Centre, Institute for Prospective Technological Studies, Seville, Spain.

Graff, G.D., G. Hochman and D. Zilberman (2009), 'The political economy of agricultural biotechnology policies', *AgBioForum*, **12**(1), 34–46.

Gray, E., T. Ancev and R. Drynan (2011), 'Coexistence of GM and non-GM crops with endogenously determined separation', *Ecological Economics*, **70**(12), 2486–93.

Greene, C. and K. Smith (2010), 'Can genetically engineered and organic crops coexist?', *Choices*, **25**(2).

Groeneveld, R., J. Wesseler and P. Berentsen (2013), 'Dominos in the dairy: an analysis of transgenic maize in Dutch dairy farming', *Ecological Economics*, **86**(2), 107–16.

Gruère, G.P. and S.R. Rao (2007), 'A review of international labeling policies of genetically modified food to evaluate India's proposed rule', *AgBioForum*, **10**(1), 51–64.

Gruère, G.P., C.A. Carter and Y.H. Farzin (2008), 'What labelling policy for consumer choice? The case of genetically modified food in Canada and Europe', *Canadian Journal of Economics*, **41**(4), 1472–97.

Heuberger, S, C. Ellers-Kirk, B.E. Tabashnik and Y. Carriere (2010), 'Pollen- and seed-mediated transgene flow in commercial cotton seed production fields', *PLoS ONE*, **5**(11).

Jarosz, N., B. Loubet, B. Durand, X. Foueillassar and L. Huber (2005), 'Variations in maize pollen emission and deposition in relation to microclimate', *Environmental Science and Technology*, **39**, 4377–84.

Kalaitzandonakes, N. and J. Bijman (2003), 'Who is driving biotechnology acceptance?', *Nature Biotechnology*, **21**, 366–9.

Kershen, D. (2004), 'Legal liability issues in agricultural biotechnology', *Crop Science*, **44**, 456–63.

Kershen, D. (2009), 'Three separate and distinct spheres: patents, regulation and liability. The selected works of Drew L. Kershen', available at: http://works.bepress.com/drew_kershen/10.

Kikulwe, E., E. Birol, J. Wesseler and J. Falck-Zepeda (2011), 'A latent class approach to investigating developing country consumers' demand for genetically modified staple food crops: the case of GM banana in Uganda', *Agricultural Economics*, **42**, 547–60.

Koch, B. (ed.) (2008), *Economic Loss Caused by Genetically Modified Organisms*, Vienna: Springer Verlag.

Kolstad, C.D., T.S. Ulen and G.V. Johnson (1990), 'Ex post liability for harm vs. ex ante safety regulation: substitutes or complements?', *American Economic Review*, **80**, 888–901.

Krauss, M.I. (1999), 'Property rules vs. liability rules', in B. Bouckaert and G. De Geest (eds), *Encyclopedia of Law and Economics*, Cheltenham, UK and Northampton, MA, USA: Edward Elgar Publishing.

Krueger, J.K. (2007), *If Your Farm is Organic, Must it be GMO-Free? Organic Farmers, Genetically Modified Organisms, and the Law*, St Paul, MO: Farmer's Legal Action Group, Inc.

Ma, B.L., K.D. Subedi and L.M. Reid (2004), 'Extent of cross-fertilization in maize by pollen from neighboring transgenic hybrids', *Crop Science*, **44**, 1273–82.

Maltsbarger, R. and N. Kalaitzanodakes (2000), 'Direct and hidden costs in identity preserved supply chains', *AgBioForum*, **3**(4), Article 10.

Moschini, G., H. Bulut and L. Cembalo (2005), 'On the segregation of genetically modified, conventional and organic products in European agriculture: a multi-market equilibrium analysis', *Journal of Agricultural Economics*, **56**(3), 347–72.

Munro, A. (2008), 'The spatial impact of genetically modified crops', *Ecological Economics*, **67**, 658–66.

Non-GMO Project (2013), Non-GMO project standard v. 10, available at: http://www.nongmoproject.org/wp-content/uploads/2013/06/Non-GMO-Project-Standard-v10.pdf, accessed 22 July 2013.

Nowicki, P.L, L.H. Aramyan, W. Baltussen, L. Dvortsin, R.A. Jongeneel, I. Perez Dominguez, C.P.A. Wagenberg, N. Kalaitzandonakes, J. Kaufman and D. Miller (2010), 'Study on the implications of asynchronous GMO approvals for EU imports of animal feed products', final report for Contract no. 30-CE-0317175/00-74, Brussels: Directorate-General for Agriculture and Rural Development European Commission, available at: http://ec.europa.eu/agriculture/analysis/external/asynchronous-gmo-approvals/full-text_en.pdf.

Pelaez, V.M., D. Aquino, R. Hofmann and M. Melo (2010), 'Implementation of a traceability and certification system for non-genetically modified soybeans: the experience of Imcopa Co. in Brazil', *International Food and Agribusiness Management Review*, **13**(1).

Phillips, P.W.B. and G. Isaac (1998), 'GMO labeling: threat or opportunity?', *AgBioForum*, **1**(1), 25–30.

Posner, R. (2007), *The Economic Analysis of Law*, 7th edn, New York: Aspen Publishers.

Qaim, M. (2009), 'The economics of genetically modified crops', *Annual Review of Resource Economics*, **1**, 665–94.

Ramessar, K., T. Capell, R.M. Twyman and P. Christou (2010), 'Going to ridiculous lengths: European coexistence regulations for GM crops', *Nature Biotechnology*, **28**(2), 133–6.

Ryan, C.D. and S.J. Smyth (2012), 'Economic implications of low-level presence in a zero-tolerance European import market: the case of Canadian triffid flax', *AgBioForum*, **15**(1), 21–30.
Shavell, S. (1987), *Economic Analysis of Accident Law*, Cambridge, MA: Harvard University Press.
Shleifer, A. (2010), 'Efficient regulation', in D.P. Kessler (ed.), *Regulation vs. Litigation: Perspectives from Economics and Law*, Chicago: University of Chicago Press, pp. 27–43.
Skevas, T., P. Fevereiro and J. Wesseler (2010), 'Coexistence regulations and agricultural production: a case study of five Bt maize producers in Portugal', *Ecological Economics*, **69**, 2402–8.
Skevas, T., J. Wesseler and P. Fevereiro (2009), 'Coping with *ex ante* regulations for planting Bt-maize: the Portuguese experience', *AgBioForum*, **12**(1), 60–69.
Smyth, S.J., A.B. Endres, T. Redick and D. Kershen (2010), *Innovation and Liability in Biotechnology: Transnational and Comparative Perspectives*, Cheltenham, UK and Northampton, MA, USA: Edward Elgar Publishing.
Smyth, S.J., M. Gusta, K. Belcher, P.W.B. Phillips and D. Castle (2011), 'Environmental impacts from herbicide tolerant Canola production in Western Canada', *Agricultural Systems*, **104**(5), 403–10.
Soregaroli, C. and J. Wesseler (2005), 'Minimum distance requirements and liability: implications for coexistence', in J. Wesseler (ed.), *Environmental Costs and Benefits of Transgenic Crops*, Wageningen UR Frontis Series, Vol. 7, Dordrecht: Springer, pp. 165–82.
Stein, A.J. and E. Rodriguez-Cerezo (2009), 'The global pipeline of new GM crops: implications of asynchronous approval for international trade', Joint Research Centre Institute for Prospective Technological Studies report no. 23846 EN, available at: http://ipts.jrc.ec.europa.eu/publications/pub.cfm?id=2420.
Stokstad, E. (2011), 'Can biotech and organic framers get along', *Science*, **322**, 166–9.
USDA (2011a), 'USDA actions to support continued dialogue and constructive coexistence in US agriculture', available at: http://www.usda.gov/documents/USDAContinuedDialogueConstructiveCoexistence.pdf.
USDA (2011b), 'AC21 plenary meeting: draft meeting summary', version 1 9-12-11, 30–31 August, available at: http://www.usda.gov.
USDA-APHIS (2011), 'Glyphosate-tolerant alfalfa events J101 and J163: request for nonregulated status', *Record of Decision*, 27 January, available at:http://www.aphis.usda.gov/brs/aphisdocs/04_11001p_rod.pdf.
USDA-APHIS (2012), 'Glyphosate-tolerant events H7-1 sugar beet: petition for nonregulated status', *Record of Decision*, 19 July, available at: http://www.aphis.usda.gov/brs/aphisdocs/03_32301p_feis_rod.pdf.
Venus, T. and J. Wesseler (2012), 'Bereits doppelt so viel "ohne Gentechnik": Milch als Biomilch in Deutschland: Welche Bedeutung hat GVO-freie Milch für unsere Milchwirtschaft?', *Deutsche Molkerei Zeitung*, **133**(2), 24–6.
Weber, W.E., T. Bringezy, I. Broer, F. Holz and J. Eder (2005), 'Koexistenz von gentechnisch verändertem und konventionellem Mais', *Mais*, 1+2, Special issue, 1–6.
Wesseler, J. and N. Kalaitzandonakes (2011), 'Present and future EU GMO policy', in A. Oskam, G. Meesters and H. Silvis (eds), *EU Policy for Agriculture, Food and Rural Areas*, 2nd edn, Wageningen: Wageningen Academic Publishers, pp. 23–323.
Wesseler, J., S. Scatasta and E. Hadji Fall (2011), 'Environmental benefits and costs of GM crops', in C. Carter, G. Moschini and I. Sheldon (eds), *Genetically Modified Food and Global Welfare*, Frontiers of Economics and Globalization, Vol. 10. Bingley: Emerald Group Publishing, pp. 173–99.
Wesseler, J., S. Scatasta and E. Nillesen (2007), 'The maximum incremental social tolerable irreversible costs (MISTICs) and other benefits and costs of introducing transgenic maize in the EU-15', *Pedobiologia*, **51**(3), 261–9.

26 Biotechnology and the inputs industry

Anwar Naseem and Latha Nagarajan

1 INTRODUCTION

The use of non-labor inputs – such as land and capital, but also intermediate inputs like chemicals, fertilizers and seeds – have been an important driver of productivity gains in agriculture over the last 50 years (Alston et al., 2009). For example, in the US, the use of non-labor inputs has offset the dramatic decline in labor use in agriculture which is estimated to have fallen at a rate of 3.2 per cent per year since 1948 (Fuglie et al., 2007). The reduction in labor use, as well as lower land use in some countries, coupled with the increase of chemicals and machinery has meant that aggregate input use has remained constant. Yet agricultural output has almost trebled since 1961, with an average increase of 2.2 per cent per year (Wik et al., 2008), implying that technological advances have resulted in more productive inputs. Indeed total factor productivity (TFP), which measures the amount of output per unit of input, has increased 2.7 times since the early 1950s (Fuglie et al., 2007). High TFP growth has resulted in lower food prices, saved natural resources (especially land) and has allowed labor to be used in other sectors.

The productivity gains in agriculture are largely a result of advances in biological sciences that have enabled plant breeders to develop crop varieties that are high yielding and of better quality. While productivity gains and increased use of non-labor inputs has dramatically increased global agricultural output, recent trends suggest that output growth has leveled off. For example, from 1960 to 1980, aggregate output grew at 2 per cent per annum but has slowed to 1 per cent annually from 1990 (Alston et al., 2009). The slowdown in output growth is of concern, considering higher demand from population and income growth in the foreseeable future. Agricultural biotechnology (agbiotech) has shown much promise in helping reverse these trends, or at the very least forestalling a further decline in output growth.

The impacts of agbiotech are not limited to simply raising farm productivity, but go beyond the farm gate, affecting different industries and actors across the agricultural value chain. The most significant impact has been and will be, of course, on the seed industry. The application of biotechnology to seeds has allowed plant breeders to develop crops with novel traits which would not have been possible with conventional breeding alone. Since input traits like herbicide tolerance and insect resistance are the most common traits that have been commercialized, their adoption has resulted in changes in the demand for plant protection chemicals, and realignment of the industry as seed and chemical firms seek to take advantage of the complementarities between the two. Traits in development such as those that seek to enhance the efficiency of nitrogen and water use may also result in changes in resource use, which will have ripple effects across different input markets, both sectors that produce intermediate inputs like fertilizer, and also land and labor markets. Besides the secondary impacts through the development of new seeds, biotechnology can also have a direct impact on the chemical industry (through the development of biopesticides) and the fertilizer industry (with biofertilizers), although such impacts are likely to be limited in the short term.

Major innovations such as biotechnology are often regarded as being disruptive in that they transform existing economic institutions and the status quo. Although such changes are to be expected in the course of economic growth and development, their implications for economic efficiency, growth and welfare are highly context specific. The objective of this chapter is to provide a better understanding of the changes that have occurred in the agricultural inputs industry due to the advent of biotechnology. In particular our interest is to examine the changes in structures and relationships of seed and chemical markets and associated institutions as these two have been affected most directly by biotechnology. A focus on them is only natural. For the purposes of this chapter, we ignore the impacts on other input markets such as labor, which are affected by the adoption of (labor-saving) technologies such as herbicide-tolerant crops (what some authors call non-pecuniary benefits).

In the literature on the impact of biotechnology on the inputs industry, two themes and concerns are consistently expressed. First, there is concern about whether input markets, especially seeds, have become more concentrated as a result of biotechnology. In 1995 and before the widespread adoption of genetically modified (GM) crops, for example, the already high four-firm concentration ratios (CR4) for corn and soybean markets in the US were 69 per cent and 47 per cent, but rose to 72 per cent and 55 per cent respectively by 2008 (Fernandez-Cornejo and Just, 2007; Shi et al., 2010). Was this firm consolidation spurred by technological opportunities provided by biotechnology or were other factors at play? Secondly, and related to the issue of market concentration, there are questions about the implications of higher market power? Although there may be static efficiency losses due to concentration, these can be justified if there are dynamic efficiency gains due to greater innovation. A related concern is that since concentration has, in part, been driven by firms seeking to acquire intellectual assets given the strengthening of intellectual property rights (IPRs) for biological materials over the last two decades, it is worth investigating whether IPRs among a few firms stifle innovation as firms limit licensing to protect existing markets. This chapter addresses these issues, first providing the theoretical underpinning of the observed and hypothesized changes and second examining the empirical methods that have been employed and results obtained. More specifically, in the next section we first explore the theories on the determinants of firm consolidation in the inputs industry as well as those related to the impacts of market structure and innovation. In the third section, we review the empirical approaches that have been used to understand the generation of innovations and the impacts of market concentration in agricultural biotechnology. The fourth section presents the results of the studies reviewed in the preceding section along with summary data on the agricultural input industries directly impacted by biotechnology. The concluding section provides a critical assessment of the literature, emphasizing the lessons being learned and directions for future research.

2 MODELS

2.1 Drivers of Consolidation and Concentration

Firm consolidation within industries is a normal evolutionary process of market economies. The industrial organization literature offers different reasons for the factors behind industry consolidation that can be broadly characterized as being motivated either by value maximization or non-value maximization. Value is increased if consolidation results in improvements in efficiencies, greater market power or access to new markets. In research and development (R&D)-intensive industries, consolidation can also overcome the problems of incomplete contracts and ownership, overlapping property rights to innovations, and the distribution of complementarity assets among competing firms. Acquisition is often the least-cost way of overcoming these challenges (for example Grossman and Hart, 1986; Teece, 1986). Although less frequent in the context of consolidation in the agricultural inputs industry, examples of non-value maximizing motives include decisions by managers to pursue their own objectives or when regulatory authorities approve/disapprove certain type of mergers and acquisitions (M&As).

Although there has been a great interest in the study of consolidation and concentration in the agricultural inputs industry since the late 1990s, the industry has been undergoing considerable consolidation for much longer. For example, the number of pesticide firms dropped from 33 to 17 in the period 1972 to 1989, and nearly 120 seed firms have become part of major chemical corporations since the late 1960s (Ollinger and Fernandez-Cornejo, 1995). The interdependence of chemical and seed firms and the role of biotechnology in enabling such a relationship led Just and Hueth (1993) to examine the circumstances under which economies of scope in demand can be a motivator for consolidation. Scope economies in demand occur when two or more products can be combined to attain higher profits for a given cost of production – such as they would if herbicide-tolerant crops are sold together with herbicides. They develop a simple two-stage model of demand-related industry where a multiple output monopolist exploits economics of scope in demand. They find that the industry structure and investment behavior is critically dependent on whether biotechnology products are complements or substitutes regarding a firm's existing chemical product line. Interdependencies of the chemical and seed industry are further explored by Lemarie and Marette (2003), who develop a vertical differentiation model where each differentiated product corresponds to a plant protection solution. Using a multi-period Cournot completion framework, the model is solved for equilibrium quantities of plant protection chemicals – with and without the availability of biotechnology. Simulations based on the theoretical model suggest that the transfer of surplus from chemical firms to biotechnology firms is an important determinant of consolidation activity.

By the late 1990s, the pace of industry consolidation was much more rapid and extensive (in terms of both vertical and horizontal consolidation) which could not be explained by the economies of scope in demand argument of Just and Hueth (1993) alone.

Kalaitzandonakes and Bjornson (1997) emphasize the different strategic motivations and trade-offs for consolidation activities to increase rents in the commercial phase of the innovation continuum. Johnson and Melknonian (1999) formalize these strategic trade-offs by utilizing a game-theoretic framework and find that consolidation can be explained in part due to the benefits of coordinated actions between the players (that is the managers), whether the investment pay-offs are transferable between firms, and the extent to which investment decisions by the firms are substitutes or complements. Game-theoretic modeling to model strategic behavior by biotechnology and seed firms is further studied by Shi (2009) and Wilson and Huso (2008). Wilson and Huso (2008) differentiate between two games, one where two biotechnology firms decide on whether to license or not, and the other where the biotechnology firm has the option to purchase a seed firm. A novelty of their modeling is the consideration of stacking genes as a commercialization strategy.

Strategic motivations aside, the desire for firms to achieve economies of scale and scope remains a compelling reason for consolidation as it implies that larger, more diversified firms have lower average costs (Fulton, 1997; Hayenga, 1999). Fulton and Giannakas (2001) suggest that sources of economics of scale and scope for the seed, chemical and agricultural biotechnology industries are sunk costs in the form of R&D expenditures and regulatory costs. They argue that agricultural R&D is scientific knowledge – a non-rival good – which can be employed in the production process repeatedly without increasing input use, and hence costs, resulting in economics of scale and scope. Similarly costs incurred to meet regulatory approval are sunk costs in that marginal increases in output will not incur additional regulatory costs. As a result, larger firms will have an advantage over smaller firms since they will be able to spread the costs of regulation over more output (Fulton and Giannakas, 2001). Heisey and Schimmelpfennig (2006), while acknowledging the cost of regulation as being a factor in firm decision to consolidate, argue that other costs and revenue factors are likely to be more important. Furthermore they point to regulation being endogenous, in that firms lobby to influence the regulatory process when the marginal benefits of innovation outweigh the marginal costs of lobbying.

A common thread underpinning all the arguments for why one firm acquires another is that the two firms have complementary assets, and acquisition is the least-cost way of combining these assets (Graff et al., 2003; Rausser et al., 1999; Teece, 1986). When complementary assets are held by different agents, market outcomes are sub-optimal as the holders of those assets produce at a level that does not take into account the positive impacts their production has on the owners of complementary assets (Graff et al., 2003). In these situations, coordination of complementary assets may be required and can take several forms, such as the full acquisition and integration of another firm to a more arm's-length relationship such as through a non-exclusive licensing of a patent. Graff et al. (2003) suggest that plant genetic transformation technologies are complementary to trait-specific plant gene technologies and plant germplasm, and that plant gene technologies are complementary to plant germplasm. Having all three mutually complementary technologies would be preferred to just one, so firms will have an incentive to carry out M&As to combine such disparate assets.

The studies reviewed thus far would imply that consolidation due to technological opportunities brought on by biotechnology would be a one-off event and that once a new market structure has evolved no further changes will occur, at least not for the same underlying reasons. Oehmke et al. (2005) take issue with such characterization and suggest that the R&D-based agbiotech industry (and by extension input industries like seeds and chemicals), follow cyclical patterns. They develop a theoretical model of endogenous

R&D investment where R&D is represented as a sequence of stochastic innovation races (see, for example, the models of Grossman and Helpman, 1991; Segerstrom et al., 1990). They test the model empirically and find that it tracks the data on consolidation in the corn, cotton and soybean industries quite well. Their key result is that M&A activity is countercyclical to R&D activity, which contrasts with the analysis of Graff et al. (2003), which suggests that M&A activity increases with the productivity of R&D assets when they are paired with complementary assets. This suggests that R&D activity moves procyclically with M&A activity.

The use of patents to protect intellectual assets in agbiotech has been one of its defining features, but one that has been the subject of considerable controversy. One difficulty has been the overlapping scope of many patents and the uneven control of mutually blocking patents that could result in a low level of innovations and the anti-commons problem (Heller and Eisenberg, 1998). The control of patent rights – either through direct ownership or licensing – can be another motivation for consolidation. Lesser (1998) argues that patents are essentially an entry barrier as the threat of patent litigation discourages small firms from entering the market. Indeed, if one were to examine the aftermath of the consolidation that occurred in the late 1990s, concentration ratios of agbiotech patents held by the market leaders actually rose (Marco and Rausser, 2008). This observation led Anderson and Sheldon (2011) to develop a theoretical model of endogenous sunk costs, licensing and market structure when property rights are well defined by patents. By extending Sutton's 'capabilities' approach (Sutton, 2007, 1998), that firm concentration is a function of endogenous sunk costs in R&D investment rather than deterministically driven by exogenous sunk costs, Anderson and Sheldon are able to show that licensing leads to lower levels of industry concentration than would be implied by Sutton's model.

2.2 Impacts of Consolidation and Concentration

Whatever the reasons behind consolidation in the biotechnology and related inputs industry, the desirability needs to be assessed by studying the welfare impacts. Specifically, one concern is that higher consolidation and concentration could slow innovation as firms use their market power to capture rents. Schumpeter famously hypothesized that higher levels of innovative activity are more likely to occur in industries that are concentrated, reasoning that imperfect competition provided the best environment to internalize the benefits of R&D (Schumpeter, 1950). While some theoretical models have supported the view that innovation is inversely related to the level of competition (Dasgupta and Stiglitz, 1980; Loury, 1979) the modeling of others concludes the opposite to be the case (Lee and Wilde, 1980; Reinganum, 1989). Still others have shown the relationship to be more complex and dynamic, with innovation first increasing then decreasing in competition (Spence, 1984). While there is no consensus on the direction of the relationship, it is generally agreed that technological opportunity, factor markets and appropriability all determine the levels of innovation (Knott and Posen, 2003).

Unlike the theoretical models that have been developed to study the determinants of consolidation, relatively few authors have specifically modeled the theoretical basis to support or refute the Schumpeterian hypothesis for agbiotech. Rather models that are used to explain consolidation are extended to study the implications of innovations. Recall, for example, the analysis of Graff et al. (2003) which implied that innovations should increase in the wake of consolidation as firms take advantage of complementary assets. This is in contrast to the modeling of Oehmke et al. (2005) who suggest that innovative activity moves countercyclically to R&D market concentration, a finding that diverges from the Schumpeterian prediction that the two should move together. While the theory on this issue might be thin, a number of studies have tested the Schumpeterian hypothesis from an empirical standpoint.

3 DATA AND METHODS

Before reviewing the methods employed in studying the agricultural inputs industry, it is helpful to know the data that is commonly available and used in the literature. The development of a GM crop is the culmination of a sequence of events that uses public and private knowledge, much of it proprietary, and research monies to purchase labor and equipment. While the ultimate objective is usually to produce GM crops with desired traits, along the way data is generated which can be useful in analytical work on measuring concentration and innovation. The most common of these metrics is research expenditure, but there exists no standardized process of reporting it and it is especially difficult to obtain disaggregated firm-level data (Fernandez-Cornejo and Just, 2007; Pray et al., 2005). The data on research outputs (both final and intermediate) is relatively richer in the level of detail, and includes measures like field trials of GM crops, patents and plant variety protection certificates. Field trial data in particular has been used extensively to study the market structure of the biotechnology industry as it contains data on firm, crop, gene and phenotype characteristics of the particular trial. Regulatory data on pesticides has also been used to study market structure related to the pesticide industry.

3.1 The Empirics of Industry Consolidation

Evidence for consolidation can come from the number and pace of M&As in the industry over a given period. Howard (2009) presents a graphical representation of the evolving relationships between global seed, chemical and biotechnology firms since 1996, adjusted for firm size and ownership structure. The figures reveal that the pace of the acquisitions was fairly rapid in the late 1990s, slowed in the early 2000s (including a brief pause in 2003), but accelerated again from 2003 to 2008. The changes implied by the visual presentation are confirmed by more quantitative measures of industry concentration, such as the four-firm concentration (CR4) and Herfindahl–Hirschman Index (HHI), reviewed in section 4.

Most of the earlier studies analyzing the impacts of biotechnology on the input industry were limited by data availability given that commercial development was still in its infancy. As such they tended to be qualitative or used the case study approach. Kalaitzandonakes and Bjornson (1997), for example, document nearly 1600 M&As, joint ventures and other forms of collaborative agreements culled from a variety of news sources. This allowed them to determine the number of firms in the industry and their relationships. Their data was particularly useful for Johnson and Melkonian (2003) who used it to carry out an in-depth examination of the strategic reasons for consolidation for the four major players in the agricultural biotechnology industry. In similar fashion, but focused on the strategic behavior of one particular firm, Hennessy and Hayes (2000) study the glyphosate and Roundup Ready soybean seed markets, especially the motivation of Monsanto to pursue tying (or bundling) strategies.

As agbiotech data became more readily available, including patents and field trials, it became possible to test the theories on the firm consolidation spurred by biotechnology. Graff et al. (2003) test their hypothesis that industry consolidation was due to a desire to exploit complementarities, using data on patents and firm characteristic for 76 firms engaged in agbiotech. A positive covariance in the unexplained variation in the patent holds of firms across the industry was used as an indication of statistically significant complementarity. Their results suggest that during the consolidation period firms sought to build a more balanced intellectual asset portfolio than would otherwise be expected.

Although biotechnology patents have been widely implicated in greater concentration of the input sector, studies empirically validating the relationship have been scarce. Towards this end, Marco and Rausser (2008) develop a measure of patent enforceability in order to assess the role of property rights in consolidation. By using survival models they show that patent statistics are useful predictors of the timing of consolidation and that enforceability influences the likelihood consolidation. Patent enforceability increases the likelihood that firms will engage in firm acquisitions while at the same time encouraging them to spin off subsidiaries.

3.2 Market Structure and Innovation in Agriculture Biotechnology

Given the policy and welfare implication of consolidation activity, there is much greater interest in empirically unpacking the innovation–market structure relationship for the case of input industries associated with biotechnology. Despite the vast amount of economics literature examining this relationship, it has remained inconclusive (Gilbert, 2006). Early studies used single-equation models to relate some measure of innovative activity to some index of concentration (Scherer, 1965), but did not find a consistent, robust relationship to support the Schumpeterian hypothesis. Initial studies were also sensitive to the choice of the dependent variable measuring innovation, with support for the Schumpeterian hypothesis weaker when the number of patents rather than the R&D intensity was used (Symeonidis, 1996).

A number of studies have found that while innovation is higher in industries with greater concentration, the relationship is weaker and even non-existent when controlling for industry effects. Scott (1997) examined R&D intensity and four-firm concentration across different lines of business, finding that the effect of concentration disappeared when fixed sector and firm effects were introduced in the regression. Likewise, Levin et al. (1985) find no evidence of a concentration–innovation relationship when using proxies for the conditions underlying the fixed effects, such as differences in appropriabiity and technological opportunity.

One common finding in the literature is that with a rise in concentration, R&D intensity and innovative activity first increase, only to decrease later. Scherer (1967) finds that maximum research intensity occurs at concentration levels of 50 to 60 per cent. Levin et al. (1985) and Scott (1997) corroborate the inverted-U hypothesis, but here too, the results are sensitive to industry effects. The inability to find a robust relationship between innovation and concentration led many to focus on the possibility of a two-way direction of causality. Levin and Reiss (1984) and Levin et al. (1985) use multi-equation models in which concentration and R&D are both endogenously and simultaneously determined by factors such as technological opportunity, appropriability and market size. They find support for the view that both innovation and concentration are simultaneously determined. As a result, Symeonidis (1996) in his survey of the literature concludes that 'recent empirical work suggests that R&D intensity and market structure are jointly determined by technology, the characteristics of demand, the institutional framework, strategic interaction and chances'. The literature therefore is inconclusive as to whether concentrated industries are more innovative. While there is some consensus that the relationship between market structure and innovation might be endogenous, it is important to note that most studies that suggest this have used cross-sectional data that is unable to account for industry-specific factors. This implies that understanding the innovation–concentration relationship requires the study of specific industries over time (Knott and Posen, 2003), where the underlying determinants of innovation such as technological opportunity and appropriability are similar. Agbiotech is such an industry than can be divided into sub-industries (by commodity) that differ primarily on the basis of market structure, allowing one to isolate its impact on concentration.

A relatively straightforward approach to analyzing the impacts of concentration on innovation is to relate some form of concentration measure (CR4, CR8 or HHI) to innovative output, such as patents, field trials or new plant varieties. The study by Brennan et al. (1999) was one of the first to make use of field trial data to establish a connection between concentration and innovation. Although the industry was still nascent at the time of their study, their conclusions raised concerns about a possible slowdown in innovative activity:

> The total number of field trials indicate that innovation concentration by a few firms has not reduced R&D activities for the industry as a whole, but has adversely impacted the R&D activities for firms not in the top four . . . Entry by new firms into the market also appears to have been reduced. Large firm size resulting from merger activity is showing signs of reduced research efficiency. (Brennan et al., 1999, p. 171)

As the industry evolved, and with availability of more data points, Brennan et al. (2005) extend the analysis to construct a mobility index to better capture how market leadership has changed over time. As the same few firms dominate the innovation market from year to year, the mobility indices confirm that concentration in the market has been persistent. They also distinguish between whether firms have the *ability* and *incentive* to limit market R&D investment. They conclude that while there is evidence to support the former, it is unclear whether that resulted in the latter. For example, since the top four firms held almost 50 per cent of the agbiotech patents in the early 2000s, these firms clearly had the ability to decrease their efforts to obtain patents (and by implication, innovation) resulting in lower patent count.

Of course, consolidation by itself is not a determinant of innovative activity, as individual firm characteristics can also have an impact. Using detailed firm characteristics data for 15 important agbiotech firms, Klotz-Ingram et al. (2004) find that small firms in their sample are much more innovative and research intensive than large diversified life-science firms. Interestingly, with only one exception, nearly all the small firms in their sample were acquired by larger firms, raising once again the concern that innovation may slow down if the number of small firms declines due to acquisitions. These findings, based on descriptive and trend analysis of the available data, have been confirmed by more sophisticated econometric modeling. Adopting a simultaneous research profit function, Schimmelpfennig et al. (2004) find that the relationship between firm concentration in the US seed markets of cotton, corn and soybean is inversely related to R&D intensity. That is, higher competition (and hence low concentration) results in greater R&D activity.

Innovation indicators like patents and field trials are useful in determining the level of R&D intensity for a given market structure, but they do not accurately capture underlying productivity effects. That is to say, a concentrated market, while not producing a higher level of innovation, may still be producing innovations that are highly productive. Nolan et al. (2012) explore this relationship for the case of corn in the US. Using a fixed effects model and a dataset based on university trials of corn hybrids, they estimate a production function which enables them to identify the amount of yield that is due to the genetics of the hybrid. By combining the dataset with information on concentration in the corn seed market, they find that there is a positive relationship between industrial concentration and productivity, and by extension innovation. Ceddia (2005) also explores this relationship – in his case for canola in Canada – and finds that as the market has become concentrated, technological lock-in has emerged, such that private firms are unwilling to invest in alternative varieties and diversification. Since he finds evidence that increasing crop diversification results in a stabilizing average canola yield (with no downward pressure on yield), technological lock-in because of concentration would lower diversity and hence yields.

A number of authors have suggested that the public sector has the ability to play a positive, balancing role in concentrated markets (Ceddia, 2005; Oehmke, 2001). In a period when private firms were consolidating and industry concentration was on the rise, Oehmke (2001) found that public sector institutions increased their applied research on crop biotechnology as measured by the number of field trials. Moreover, having a higher number of public sector institutions implies a higher number of commodities to which innovation is applied.

One mechanism whereby concentration could negatively affect innovation is if firms use their accumulated patents to block firm entry into the market by limiting patent licensing (the hold-up problem). Finding evidence for hold-ups is challenging since the alternative outcome is usually unknown. Using a case study approach, Pray and Naseem (2005) asked whether patents on two platform crop biotechnologies – plant transformation technologies and structural genomics – resulted in the patent holder of those technologies holding up further innovation. Through interviews of key stakeholders and additional patent analysis, they found that while there were some instances of patents being held up, the benefits provided by those technologies were significant in generating key commercial GM crops. Furthermore, where accessing firms were unable to license the patent, they eventually found alternative mechanisms for developing their technologies. The empirical question of whether accumulated patent portfolios of consolidated firms increases or decreases innovation is carefully explored by Chan (2011). Using data on patents and plant variety protection certificates, the results of the negative binomial regression reveal that firms with more patents do not exhibit economies of scale in the number of plant varieties developed, nor does a greater diversity in a firm's patent portfolio result in a greater number of varieties. This result would suggest that the complementarities firms seek to exploit through mergers are more illusory than real.

4 METRICS

This section reviews the key indicators on market structure and innovation for the global seed and agro-chemical markets. The goal is to provide an overview of how these markets have evolved as a consequence of biotechnology, with an emphasis on identifying the key actors in this action arena.

4.1 Market Size and Structure

Figure 26.1 shows the trends in the market size for seed and crop protection chemicals. After a period of considerable growth until the mid-1990s, the global crop protection market experienced limited growth from the late 1990s to the mid-2000s. The major reasons for the decline in sales included: stronger regulatory measures towards environmental safety; expiry of patent-protected chemicals (for example Monsanto's Roundup in early 2000); volatility in output markets; and the introduction of GM crops, particularly in North and Central American countries. The market surged partially from the mid-2000s onwards due to favorable cropping conditions from higher commodity prices.

The global seed market has also grown significantly since the early 1970s due primarily to improved yield-enhancing technologies. With the introduction of GM seeds, the combined growth of seed – both conventional and GM – has been relatively higher than agro-chemicals from 2005 onwards.

The effects of the industry consolidation in agricultural inputs can be seen in Table 26.1, which presents the changes in industry concentration and R&D intensity over a 15-year period (1994–2009). As firms began to realize the strong potential for demand

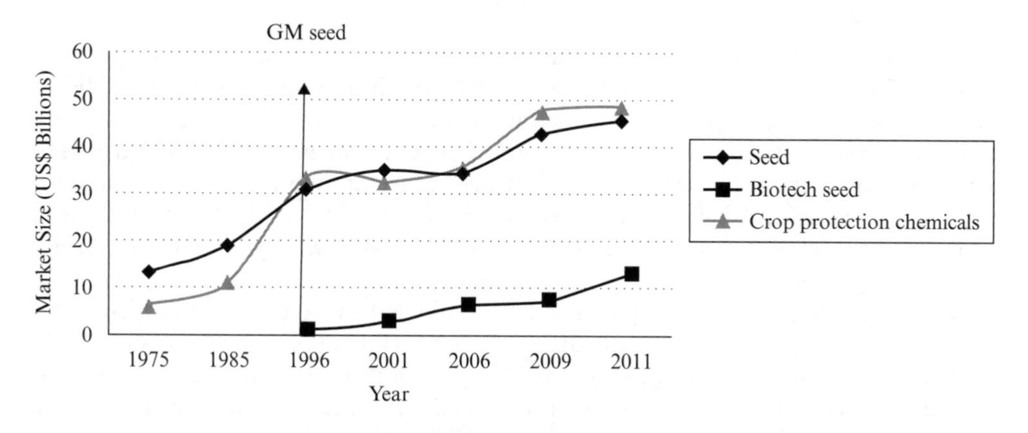

Sources: Fernandez-Cornejo and Just (2007); Fuglie, et al. (2011).

Figure 26.1 Trends in global market size for agricultural inputs (1975–2011)

Table 26.1 Increasing market concentration in global agricultural input industries

Firms	Year	Four-firm concentration ratio Share of global market (%)	Eight-firm concentration ratio Share of global market (%)	Industry R&D intensity (R&D/Sales)
Crop seed and biotechnology	1994	21.1	29.0	11.0
	2000	32.5	43.1	15.0
	2009	53.9	63.4	10.5
Crop protection chemicals	1994	28.5	50.1	7.0
	2000	41.0	62.6	6.8
	2009	53.0	74.8	6.4

Source: Fuglie et al. (2011).

Table 26.2 Top six global ag-input firms market size in 2009

Firms	Seed sales (US$ mill.)	Global market share (%)	Agchemical sales (US$ mill.)	Global market share (%)	Total ag seed, trait, & chemical sales (US$ mill.)	R&D devoted to agbiotech as % of seed-biotech R&D (%)	Ag R&D/ Ag sales (research intensity) (%)
Monsanto	7297	27	2029	10	9326	80	12.5
DuPont	4806	17	2453	5	7259	50	9.6
Syngenta	2564	9	8878	19	11442	15	8.9
Bayer	699	3	7284	17	7983	85	10.5
Dow	633	2	3708	9	4341	85	n.a
BASF			5348	11	5348	100	13.5

Source: Fuglie et al. (2011).

arising from bundling of inputs and the opportunities for 'economies of scope' from biotech innovations through expanded crop portfolios, the industry began to consolidate from the late 1990s onward. The acquisition of seed firms provided market power and control over key technologies. The structural changes in the industry increased overall concentration, with both the four and eight-firm concentration ratios more than doubling over the period for seed and biotechnology firms (Table 26.1). Concentration in crop protection chemicals has also increased, but the market was relatively more concentrated to begin with.

In the late 1990s, ten major global firms dominated the agro-chemical sector, but within the decade the number had dropped to six (Table 26.2). It is useful to note that a high market share in the seed market does not imply high market share for agro-chemicals. For example, both Monsanto and Syngenta have significant presence in both markets, but Monsanto dominates seeds, whereas Syngenta dominates in agro-chemicals.

Table 26.3 Private R&D expenditure (nominal US$ millions)

Year	Crop protection		Seed & biotech	
	US	Global	US	Global
1960	27		4	
1970	126		11	
1979	292		43	
1989	561		272	
1995	751	2390	507	1213
2000	703	2352	1045	2055
2005	612	2678	1095	2254
2010	793	3116	2176	3726

Source: Fuglie et al. (2011).

4.2 Innovation and Research and Development Intensity

Trends in research and development (R&D) intensity within the seed sector can be linked to technological opportunities and the ability to appropriate R&D returns. In particular, three distinct waves of innovation are observed, starting from the 1930s when the commercial application of heterosis in maize allowed seed firms to appropriate the returns of their breeding effort, followed by another wave in the 1970s, largely as a result of the introduction of the Plant Variety Protection Act (PVPA) in the US. Signs of concentrated markets begin to emerge by 1982, with nearly 50 per cent of the PVPA certificates held by 14 firms (Drew, 2010). The third wave began around the mid-1980s when firms began investing in agbiotech R&D, encouraged by the Supreme Court ruling of *Diamond vs. Chakrabarty*. The ability to appropriate returns of research is one reason that the growth of private seed and biotech R&D since 1995 has nearly tripled globally (quadrupled for the US) (see Table 26.3).

By 2010, Fuglie et al. (2011) estimated that eight seed and biotech firms accounted for 76 per cent of all R&D spending. Consequently the ownership of innovations was also concentrated among these firms. By the year 2007, the share of the top ten companies was estimated at about 75 per cent of all patents applications at the USPTO and 43 per cent of all patent applications at the European Patent Office, with two firms, Monsanto and DuPont-Pioneer, holding more than 50 per cent of the patents (Schenkelaars et al., 2011). Economic analysis of the trends in product innovations in the US corn seed industry from 1992–2008 further confirms that increased R&D expenses during this time corresponded with the number of new product offerings and with declining product lifecycle. This trend was also evident in the soybean and cotton seed industries (Kalaitzandonakes et al., 2010).

Between 1997 and 2008, the number of maize hybrids offered annually in the US seed market grew by more than 50 per cent, from 3060 to 4300, and soybean varieties doubled from 650 to 1130 (Kalaitzandonakes et al., 2010). Dooley and Kurtz (2001) also note that the introduction of biotechnology caused a steep increase in the number of hybrids offered in the market. This suggests that the higher levels of concentration in US seed markets for these crops and the introduction of GM varieties have not had any negative

Table 26.4 Number of on-going field trials for GM maize, soybean, cotton and tomato

Firm type	US	EU	India	Australia	Argentina
Crop life firms	10684	1181	25	40	1061
Other private firms	2067	599	6	14	324
Public	3167	449	20	40	86

Source: Schenkelaars et al. (2011).

impacts on the level of innovations in these crops. Kalaitzandonakes et al. (2010) also note that R&D costs of these firms would continue to remain at high level and would be likely to contribute towards further concentration and consolidation in the seed industry on the whole.

Further analysis based on data from the ongoing field trials of GM maize, soybean, cotton and tomato in the US, EU, India, Australia and Argentina by Schenkelaars et al. (2011) indicates that in terms of trial numbers, the ownership and concentration continues to be dominated by the major six crop life companies: BASF, Bayer, Dow, DuPont-Pioneer, Monsanto and Syngenta (Table 26.4). Overall, Monsanto ranks first in numbers of private sector field trials with GM maize, GM soybean, GM cotton and GM tomato in these countries. For the case of GM rice, BASF and Bayer rank first in numbers of private sector field trials, followed by Monsanto.

4.3 Agro-chemicals Industry

The agricultural chemicals sector also underwent major waves of consolidation in the last three decades that resulted in considerable change in R&D investment patterns and ownership of new innovations. Evidence suggests that the concentration ratios of the four dominant firms have steadily increased between 1994 and 2009, from 28 per cent to 53 per cent of global pesticide sales (Table 26.1). By 2009, 12 large firms supplied 86 per cent of the global market for agro-chemicals (Fuglie et al., 2011). There has been a decline in the introduction of new crop protection chemicals, with the mean rate of introduction declining from 12.3 introductions per year to 10.6 across types of chemicals (Table 26.5).

Table 26.5 Trends in new product introductions

Product introductions	1980–89	1990–99	2000–07	In R&D
Herbicides	51	57	32	12
Insecticides	29	37	22	14
Fungicides	36	29	27	17
Others	7	3	4	0
Total	123	126	85	43
Mean rate of introduction/year	12.3	12.6	10.6	8.6

Source: McDougall (2008).

5 CRITICAL ASSESSMENT

Firm consolidation has been a feature of agricultural input industry for many decades, and the advent of agbiotech has led not only to further consolidation within separate input markets but more significantly across the different markets. In particular, agbiotech has primarily led to extensive market integration among the three key areas of the agriculture industry, namely seeds, biotechnologies and agrichemicals, with new and innovative partnerships towards technology development, innovation and in technology transfer mechanisms (UNCTAD, 2006). The global concentration in seed and crop protection industry doubled from 1994 to 2009, signifying the emerging complementarities between these two sectors.

The forces behind the rise in concentration are varied. One suggestion has been that it arises from the inefficiencies of incomplete contracts in the presence of asset specificity. Another possibility is that firms seek strategic advantage in the marketplace in the presence of overlapping property rights. It has also been argued that patents can be used as an entry barrier, limiting competition. Finally the high regulatory costs can also be an entry barrier, encouraging consolidation by incumbents to achieve economies of scale and scope.

Greater concentration is not associated with increased R&D intensities among these two sectors, however. In the case of the seed industry, there was a significant increase in research intensity by the top four firms in the late 1990s and early 2000s as the industry sought to commercialize a number of GM crop varieties (Fuglie et al., 2011). In the case of the ag-chemicals sector, R&D growth has been limited due to environmental policies and regulations, and the rise of generic markets. In recent years, the decline in US firms' agro-chemical research investment due to emphasis on GM crop research has also significantly affected the R&D towards the sector.

REFERENCES

Alston, J.M., J.M. Beddow and P.G. Pardey (2009), 'Agricultural research, productivity, and food prices in the long run', *Science*, **325**, 1209–10.

Anderson, B. and I.M. Sheldon (2011), 'Endogenous R&D investment and market structure: a case study of the agricultural biotechnology industry', paper presented at Annual Meeting of the Agricultural and Applied Economics Association, 24–26 July, Pittsburgh, PA.

Brennan, M.F., C.E. Pray and A. Courtmanche (1999), 'Impact of industry concentration on innovation in the US plant biotech industry', in W. Lesser (ed.), *Transitions in Agro-Biotechnology*, Proceedings of Conference, 24–25 June, Washington, DC. 1999.

Brennan, M., C. Pray, A. Naseem and J.F. Oehmke (2005), 'An innovation market approach to analyzing impacts of mergers and acquisitions in the plant biotechnology industry', *AgBioforum*, **8**, 89–99.

Ceddia, G. (2005), 'The impact of agro-biotechnology on the canola seed industry and canola productivity: technological lock-in?', paper presented at ICABR, 6–10 July, Ravello, Italy.

Chan, H.P. (2011), 'Do firms with larger patent portfolios create more new plant varieties in the US agricultural biotechnology industry?', *Economics of Innovation and New Technology*, **20**, 749–75.

Dasgupta, P. and J. Stiglitz (1980), 'Industrial structure and the nature of innovative activity', *Economic Journal*, **90**, 266–93.

Dooley, F.J. and M.M. Kurtz (2001), 'The effect of a changing market mix in seed corn on inventory costs', paper presented at American Agricultural Economics Association Annual Meetings, 5–8 August, Chicago, IL.

Drew, J. (2010), 'An economic evaluation of the roots and fruits of intellectual property rights for US horticultural plants', Ph.D., The University of Minnesota.

Fernandez-Cornejo, J. and R.E. Just (2007), 'Researchability of modern agricultural input markets and growing concentration', *American Journal of Agricultural Economics*, **89**, 1269–75.
Fuglie, K., J.M. MacDonald and V.E. Ball (2007), 'Productivity growth in US agriculture', USDA–ERS Economic Brief no. 9, Washington, DC.
Fuglie, K., P. Heisey, J. King, K. Day-Rubenstein, D. Schimmelpfennig, S.L. Wang and R. Karmarkar-Deshmukh (2011), 'Research investments and market structure in the food processing, agricultural input, and biofuel industries worldwide', Economic Research Report, Washington, DC: USDA-ERS.
Fulton, M. (1997), 'The economics of intellectual property rights: discussion', *American Journal of Agricultural Economics*, **79**, 1592–4.
Fulton, M. and K. Giannakas (2001), 'Agricultural biotechnology and industry structure', *AgBioforum*, **4**, 137–51.
Gilbert, R. (2006), 'Looking for Mr Schumpeter: where are we in the competition–innovation debate?', in A. Jaffe, J. Lerner and S. Stern (eds), *Innovation Policy and the Economy, Volume 6*, Cambridge, MA: The MIT Press, pp. 159–215.
Graff, G.D., G.C. Rausser and A.A. Small (2003), 'Agricultural biotechnology's complementary intellectual assets', *Review of Economics and Statistics*, **85**, 349–63.
Grossman, S.J. and O.D. Hart (1986), 'The costs and benefits of ownership: a theory of vertical and lateral integration', *The Journal of Political Economy*, **94**, 691–719.
Grossman, G.M. and E. Helpman (1991), 'Quality ladders and product cycles', *The Quarterly Journal of Economics*, **106**, 557–86.
Hayenga, M. (1999), 'Structural change in the biotech seed and chemical industrial complex', *AgBioforum*, **1**, 43–55.
Heisey, P. and D. Schimmelpfennig (2006), 'Regulation and the structure of biotechnology industries', in R.E. Just, J.M. Alston and D. Zilberman (eds), *Regulating Agricultural Biotechnology: Economics and Policy*, New York: Springer, pp. 421–36.
Heller, M.A. and R.S. Eisenberg (1998), 'Can patents deter innovation? The anticommons in biomedical research', *Science*, **280**, 698–701.
Hennessy, D.A. and D.J. Hayes (2000), 'Competition and tying in agrichemical and seed markets', *Review of Agricultural Economics*, **22**, 389–406.
Howard, P.H. (2009), 'Visualizing consolidation in the global seed industry: 1996–2008', *Sustainability*, **1**, 1266–87.
Johnson, S.R. and T.A. Melkonian (1999), 'Policy and technology as factors in industry consolidation', paper presented at: *World Food Security and Sustainability: The Impacts of Biotechnology and Industrial Consolidation*, 6–8 June, University of Nebraska.
Johnson, S.R. and T.A. Melkonian (2003), 'Strategic behavior and consolidation in the agricultural biotechnology industry', *American Journal of Agricultural Economics*, **85**, 216–33.
Just, R.E. and D.L. Hueth (1993), 'Multimarket exploitation: the case of biotechnology and chemicals', *American Journal of Agricultural Economics*, **75**, 936–45.
Kalaitzandonakes, N.G. and B. Bjornson (1997), 'Vertical and horizontal coordination in the agro-biotechnology industry: evidence and implications', *Journal of Agricultural and Applied Economics*, **29**(1), 129–39.
Kalaitzandonakes, N., D. Miller and A. Magnier (2010), 'A worrisome crop?', *Regulation*, **33**, 20–26.
Klotz-Ingram, C., D. Schimmelpfennig, A. Naseem, J. King and C. Pray (2004), 'How firm characteristics influence innovative activity in agricultural biotechnology', in R.E. Evenson and V. Stantaniello (eds), *The Regulation of Agricultural Biotechnology*, Wallingford: CAB International, pp. 171–81.
Knott, A.M. and H.E. Posen (2003), 'Does competition increase innovation? New evidence from old industries', in Working Paper, Philadelphia, PA: The Wharton School, University of Pennsylvania.
Lee, T. and L.L. Wilde (1980), 'Market structure and innovation: a reformulation', *Quarterly Journal of Economics*, **94**, 429–36.
Lemarie, S. and S. Marette (2003), 'Substitution and complementarities in the biotechnology and pesticide markets: a theoretical framework', in N.G. Kalaitzandonakes (ed.), *The Economic and Environmental Impacts of Agbiotech*, New York: Kluwer Academic, pp. 287–306.
Lesser, W. (1998), 'Intellectual property rights and concentrations in agricultural biotechnology', *AgBioforum*, **1**, 56–61.
Levin, R.C. and P.C. Reiss (1984), 'Tests of a Schumpeterian model of R&D and market structure', in Z. Griliches (ed.), *R&D, Patents, and Productivity*, NBER Conference Report, Chicago and London: University of Chicago Press, pp. 175–204.
Levin, R.C., W.M. Cohen and D.C. Mowery (1985), 'R&D appropriability, opportunity, and market structure: new evidence on some Schumpeterian hypotheses', *The American Economic Review*, **75**, 20–24.
Loury, G.C. (1979), 'Market structure and innovation', *Quarterly Journal of Economics*, **93**, 395–410.
Marco, A.C. and G.C. Rausser (2008), 'The role of patent rights in mergers: consolidation in plant biotechnology', *American Journal of Agricultural Economics*, **90**, 133–51.

McDougall, Phillips (2008), 'The global agrochemical and seed markets industry prospects', presentation at CPDA Annual Conference, 21 July, San Francisco.
Nolan, E., P. Santos and G. Shi (2012), 'Market concentration and productivity in the United States corn sector: 2002–2009', paper presented at 2012 Annual Meeting, 12–14 August, Seattle, WA.
Oehmke, J.F. (2001), 'Biotechnology R&D races, industry structure, and public and private sector research orientation', *AgBioforum*, **4**, 105–14.
Oehmke, J.F., C.A. Wolf and K.C. Raper (2005), 'On cyclical industry evolution in agricultural biotechnology R&D', *Journal of Agricultural & Food Industrial Organization*, **3**(1).
Ollinger, M. and J. Fernandez-Cornejo (1995), 'Regulation, innovation, and market structure in the US pesticide industry', Washington, DC: US Department of Agriculture, Economic Research Service.
Pray, C.E. and A. Naseem (2005), 'Intellectual property rights on research tools: incentives or barriers to innovation? Case studies of rice genomics and plant transformation technologies', *AgBioforum*, **8**, 108–17.
Pray, C., J.F. Oehmke and A. Naseem (2005), 'Innovation and dynamic efficiency in plant biotechnology: an introduction to the researchable issues', *AgBioforum*, **8**, 52–63.
Rausser, G.C., S. Scotchmer and L.K. Simon (1999), 'Intellectual property and market structure in agriculture', Working Paper, Berkeley, CA: Department of Agricultural and Resource Economics.
Reinganum, J.F. (1989), 'The timing of innovation: research, development, and diffusion', in R. Schmalensee and R.D. Willig (eds), *Handbook of Industrial Organization: Volume 1*, Amsterdam, Oxford and Tokyo: North-Holland, distributed in the US and Canada by Elsevier Science, pp. 849–908.
Schenkelaars, P., H. de Vriend and N.G. Kalaitzandonakes (2011), 'Drivers of consolidation in the seed industry and its consequence for innovation', report prepared for COGEM, available at: http://www.cogem.net/index.cfm/en/publications/publicatie/drivers-of-consolidation-in-the-seed-industry-and-its-consequences-for-innovation.
Scherer, F.M. (1965), 'Corporate inventive output, profits, and growth', *Journal of Political Economy*, **73**, 290–97.
Scherer, F.M. (1967), 'Market structure and the employment of scientists and engineers', *The American Economic Review*, **57**, 524–31.
Schimmelpfennig, D.E., C.E. Pray and M.F. Brennan (2004), 'The impact of seed industry concentration on innovation: a study of US biotech market leaders', *Agricultural Economics*, **30**, 157–67.
Schumpeter, J.A. (1950), *Capitalism, Socialism and Democracy*, New York: Harper and Row.
Scott, J.T. (1997), 'Schumpeterian competition and environmental R&D', *Managerial and Decision Economics*, **18**, 455–69.
Segerstrom, P.S., T.C.A. Anant and E. Dinopoulos (1990), 'A Schumpeterian model of the product life cycle', *American Economic Review*, **80**, 1077–091.
Shi, G. (2009), 'Bundling and licensing of genes in agricultural biotechnology', *American Journal of Agricultural Economics*, **91**, 264–74.
Shi, G., J-P. Chavas and K. Stiegert (2010), 'An analysis of the pricing of traits in the US corn seed market', *American Journal of Agricultural Economics*, **92**, 1324–38.
Spence, M. (1984), 'Cost reduction, competition, and industry performance', *Econometrica*, **52**, 101–21.
Sutton, J. (1998), *Technology and Market Structure: Theory and Practice*, Cambridge, MA: MIT Press.
Sutton, J. (2007), 'Market structure: theory and evidence', in M. Armstrong and R. Porter (eds), *Handbook of Industrial Organization*, Amsterdam, North-Holland, pp. 2301–68.
Symeonidis, G. (1996), 'Innovation, firm size and market structure: Schumpeterian Hypotheses and Some New Themes', OECD Economics Department working paper no. 161, Paris OECD.
Teece, D.J. (1986), 'Profiting from technological innovation: implications for integration, collaboration, licensing and public policy', *Research Policy*, **15**, 285–305.
UNCTAD (2006), 'Tracking the trend towards market concentration: the case of agricultural input industry', Geneva: UNCTAD.
Wik, M., P. Pingali and S. Brocai (2008), 'Global agricultural performance: past trends and future prospects', Background Paper for the *World Development Report 2008*, Washington, DC: World Bank.
Wilson, W.W. and S.R. Huso (2008), 'Trait stacking, licensing, and seed firm acquisitions in genetically modified grains: a strategic analysis', *Journal of Agricultural and Resource Economics*, **33**(3), 382–401.

27 Market power in the US biotech industry

Alexandre Magnier, Nicholas Kalaitzandonakes and Douglas Miller

1 INTRODUCTION

The advent of agricultural biotechnologies has induced structural changes in parts of the agribusiness industry that have precipitated a heated debate about regulatory oversight of some segments of the agri-food supply chain. Chapters 4, 23 and 26 in this volume address important aspects of these changes.

The US seed industry has been at the center of the structural changes as large multinational firms with significant investments in agricultural biotechnology integrated forward into the seed industry through multiple mergers and acquisitions. This vertical integration changed drastically the ownership and control of the US seed industry as well as its degree of concentration. Over the last fifteen years, the more concentrated and vertically integrated US seed industry has excelled in introducing an ever-increasing number of new differentiated seeds adapted to regional cropping conditions and delivering a large number of new biotech traits. As the number of seeds offered expanded, product life cycles shortened and the value of biotech traits increased, the prices of seeds paid by farmers also escalated. The increasing concentration of the industry and the parallel increases in seed prices have, in some cases, been viewed as evidence of market power that has been exercised in the US seed and biotech industry and has attracted the attention of the US anti-trust authorities. The key question we address in this chapter then is whether market power has been exercised in the US seed/biotech corn industry in recent years, and if so, to what extent.

Measuring market power in industries with a large number of differentiated products is technically difficult. Econometric methods based on structural demand and supply modeling (Appelbaum, 1982; Bresnahan, 1981; Iwata, 1974) were mainly developed for industries with a single homogeneous good (Berry, 1994). Because a large number of cross-price elasticities need to be estimated, structural approaches cannot be easily extended to industries with many differentiated products in the absence of limiting (and questionable) restrictions (Berry and Pakes, 1993). Such structural models may also require additional restrictive assumptions embedded in the parametric specification of the supply and demand functions, making any test of market power jointly dependent on the assumed functional forms and the presence of market power in a given industry (Perloff et al., 2007).

The most commonly used approach for estimating market power in differentiated product industries involves the use of discrete choice models (Berry et al., 1995). Discrete choice models are most suitable for the estimation of demand systems and typically abstract from supply considerations, which must be properly accounted for in the measurement of market power (Di Marina, 2004). They also require assumptions that do not always correspond to the realities of markets. In particular, discrete choice models imply that buyers must be presented with all possible product options from which they choose only one.

Alternative methods for estimating market power in differentiated product industries do exist but they are also subject to empirical challenges of their own. The residual demand approach proposed by Baker and Bresnahan (1988) estimates the product demand of a firm once the conjectures and reactions of all other competing firms have been accounted for. Residual demand, however, can be identified only if relevant cost information exists for each individual product in the market and such information is rarely available, especially when the number of products is large.

Given the inherent empirical challenges, we develop here a second-best approach to measuring market power in industries with a large number of differentiated products. Instead of providing a point estimate of market power, our proposed method calculates an upper bound of the mark-up charged by firms and hence an upper bound to exercised market power. We build on the residual demand approach of Baker and Bresnahan (1988) but we use an imperfect instrumental variable to estimate the elasticity of the residual demand curve and the price mark-up charged by firms. The instrumental variable that we use is an industry-wide cost shifter instead of a product-specific cost shifter, which causes the instrumental variable to fail the exogeneity assumption. The use of the endogenous instrument variable implies that the estimated parameters of the residual demand are biased but we take guidance from economic theory to provide a direction for the bias, thereby establishing an upper bound for the mark-up. We then illustrate how our method can be used in practice by applying it to the US hybrid seed corn industry, which has a very large number of differentiated products. In this context, we provide an estimate of the upper bound of the mark-up charged for hybrid corn seeds from 1997 to 2008 and we find that the market power exercised in the US corn seed market during this period was moderate.

Our study contributes to the literature which seeks to bound parameter estimates when data limitations restrict traditional econometric methods either because of measurement error (Klepper and Leamer, 1984), because instruments are not available (Leamer, 1981), or because instruments fail the exogeneity assumption (Conley et al., 2007; Hahn and Hausman, 2003; Nevo and Rosen, 2008). In this literature, Nevo and Rosen (2008) addressed the issue of identification with imperfect instrumental variables and produced results similar to the estimates we derive. However, we use the framework of the residual demand to isolate the source of the endogeneity of the instrument and provide appropriate inferences regarding the direction of the bias.

Our method builds on the residual demand approach which economizes on data when estimating market power in differentiated product industries and proves to be particularly useful when the degree of product differentiation is uncertain or when the boundaries of the relevant market are unclear. Instead of requiring product-specific instruments which can be intractable when the number of products is large, our method requires only one instrument, which implies that an estimate of the bound on market power can be derived even with limited data.

Differentiated industries where the same industry-wide cost shifter enters the production function of each firm are fairly common. For instance, industries where firms buy at least one production factor from the same input market and there is no discrimination in this factor market fall in this category. The approach we propose can be used in such industries provided that the industry-wide cost shifter employed as the instrument is not correlated with other firm-specific cost shifters. Our proposed approach can also accommodate the estimation of market power for firms producing multiple products under the assumption that, in equilibrium, conjectures about the behavior of other firms and conjectures within the firm are consistent.

Our method yields a worst-case scenario and as such it should be used with caution if the estimate of market power is found to be large. But it should provide a certain degree of reassurance about the competitive structure of a market if found to be relatively small. We illustrate this point empirically by examining the level of market power in the US seed corn industry during a time when the industry underwent meaningful consolidation and came under the scrutiny of the US Department of Justice for anti-competitive behavior.

2 USING THE RESIDUAL DEMAND APPROACH WITH AN IMPERFECT INSTRUMENT

Empirical measures of market power need to account for the varieties of ways in which firms interact in differentiated product industries. Baker and Bresnahan (1988) proposed a framework for estimating the residual market demand while accounting for the supply response of firms providing substitute products. The elasticity of the residual demand curve may then be inverted to obtain an estimate of the mark-up charged by firms.

Briefly, the equilibrium price and quantity of each product is determined by the intersection of a demand relation and a supply relation. Without loss of generality we may first consider the inverse demand of firm 1 for its product:

$$P_1 = P^1(Q_1, \boldsymbol{Q}_{-1}, \boldsymbol{Y}, \boldsymbol{\alpha}_1) \tag{27.1}$$

where P_1 and Q_1 are the price and quantity of product 1, $\boldsymbol{Y}$ is a vector of exogenous variables shifting demand and $\boldsymbol{\alpha}_1$ is a vector of demand parameters, that is the own-price elasticity of demand and the cross-price elasticities. $\boldsymbol{Q}_{-1}$ is the vector of quantities of all products competing with product 1 and does not necessitate any a priori delineation of market boundaries. The vector of inverse demand equations of competing firms is written in a similar way:

$$\boldsymbol{P}_i = \boldsymbol{P}^i(\boldsymbol{Q}_{-i}, Q_i, \boldsymbol{Y}, \boldsymbol{\alpha}_i) \text{ for all } i \neq 1 \tag{27.2}$$

The supply behavior relation is determined by the equality between marginal cost and perceived marginal revenue (PMR). For firm 1, the supply relation is defined as:

$$MC^1(Q_1, \boldsymbol{W}, \boldsymbol{W}_1, \boldsymbol{\beta}_1) = PMR^1(Q_1, \boldsymbol{Q}_{-1}, \boldsymbol{Y}, \boldsymbol{\alpha}_1, \boldsymbol{\theta}_1) \tag{27.3}$$

The supply relation for competing firms is defined identically as:

$$MC^i(Q_i, \boldsymbol{W}, \boldsymbol{W}_i, \boldsymbol{\beta}_i) = PMR^i(\boldsymbol{Q}_{-1}, Q_1, \boldsymbol{Y}, \boldsymbol{\alpha}_i, \boldsymbol{\theta}_i) \text{ for all } i \neq 1 \tag{27.4}$$

where PMR^i is defined as $P^i(.)+Q_i\Sigma_j((\partial P^i/\partial Q_j)(\partial Q_j/\partial Q_i))$ (27.5)

Equations (27.3) and (27.4) indicate that marginal cost for each firm is dependent on a vector of industry-wide factor prices, $\boldsymbol{W}$, that is common to all firms, on a vector of product-specific factor prices, $\boldsymbol{W}_i$, that is not contained in $\boldsymbol{W}$ and on a vector of supply parameters $\boldsymbol{\beta}_i$. Perceived marginal revenue in equations (27.3) and (27.4) depends on conduct variables $\boldsymbol{\theta}_i$ representing the conjectures the firms hold about the other firms' reactions. Each of these conjectures corresponds to the ratios $\partial Q_j/\partial Q_i$ in equation (27.5).

Two steps are then necessary to obtain the residual demand curve for product 1. The first step consists of solving the system of equations defined by the supply and demand relations (27.2) and (27.4) for the quantities of all the products competing with product 1. We may obtain a vector of product quantities $\boldsymbol{Q}_i$ as a function of Q_1 and all the exogenous variables, except for $\boldsymbol{W}_1$, the vector of factor price specific to product 1, as:

$$\boldsymbol{Q}_i = \boldsymbol{E}^i(Q_1, \boldsymbol{Y}, \boldsymbol{W}, \boldsymbol{W}_{-1}, \boldsymbol{\beta}_{-1}, \boldsymbol{\alpha}_{-1}, \boldsymbol{\theta}_{-1}) \tag{27.6}$$

where the vectors of factors and parameters $\boldsymbol{W}_{-1}$, $\boldsymbol{\beta}_{-1}$, $\boldsymbol{\alpha}_{-1}$ and $\boldsymbol{\theta}_{-1}$ represent respectively firm-specific cost factors, supply parameters, demand parameters and conjectures of all the firms in the industry except firm 1. This vector of quantities can therefore be interpreted as best response to a change in.Q_1

The second step consists of substituting the vector of quantities $\boldsymbol{E}^i$ obtained in (27.6) into the demand relation (27.1) and into the supply relation (27.3) which are both specific to firm 1. The first substitution yields the inverse residual demand curve of firm 1:

$$P_1 = P^1(Q_1, \boldsymbol{E}^i(Q_1, \boldsymbol{Y}, \boldsymbol{W}, \boldsymbol{W}_{-1}, \boldsymbol{\beta}_{-1}, \boldsymbol{\alpha}_{-1}, \boldsymbol{\theta}_{-1}), \boldsymbol{Y}, \boldsymbol{\alpha}_1) \tag{27.7}$$

Eliminating the notation redundancies, the residual demand of firm 1 is written as:

$$P_1 = R(Q_1, \boldsymbol{Y}, \boldsymbol{W}, \boldsymbol{W}_{-1}, \boldsymbol{\beta}_{-1}, \boldsymbol{\theta}_{-1}, \boldsymbol{\alpha}_1) \tag{27.8}$$

where α_i represents the union of the demand parameters of all firms. The second substitution yields the supply relation of firm 1:

$$MC^1(Q_1, \boldsymbol{W}, \boldsymbol{W}_1, \boldsymbol{\beta}_1) = PMR^1(Q_1, \boldsymbol{\theta}_1, \boldsymbol{Y}, \boldsymbol{W}, \boldsymbol{W}_{-1}, \boldsymbol{\beta}_{-1}, \boldsymbol{\theta}_{-1}, \boldsymbol{\alpha}_1) \tag{27.9}$$

The residual demand curve (27.8) of product 1 is a function of structural parameters $\boldsymbol{\beta}_{-1}$ and, $\boldsymbol{\alpha}_1$ conjectures $\boldsymbol{\theta}_{-1}$, industry-wide cost factors $\boldsymbol{W}$ and cost factors specific to all firms competing with firm 1, $\boldsymbol{W}_{-1}$. However, the vector of cost factors specific to firm 1, $\boldsymbol{W}_1$, does not appear in the residual demand curve of product 1. The supply relation of product 1 in (27.9) is a function of all the parameters of the model, including $\boldsymbol{W}_1$ on the left-hand side of (27.9). Because $\boldsymbol{W}_1$ is a vector of exogenous variables that appear in the supply relation (27.9) but not in the demand relation (27.8), the residual demand curve of firm 1 can be econometrically estimated if $\boldsymbol{W}_1$ is available.

As long as conjectures are consistent, the mark-up can then be estimated by taking the derivative of the residual demand curve (27.8) in logarithmic form:

$$-\frac{\partial \ln R}{\partial \ln Q_1} = \frac{P_1 - MC_1}{P_1} \tag{27.10}$$

The consistency of conjectures is necessary for (27.10) to hold and it implies that there is an exact correspondence between the conjectural variations of a firm, defined by the ratios $\partial Q_j / \partial Q_i$ in equation (27.5), and the reaction function of competitors (Bresnahan, 1981). This, in turn, suggests that beliefs and actual behavior need to concur in equilibrium.

2.1 Residual Demand Curves of Multiple Product Firms

The framework proposed by Baker and Bresnahan (1988) was proposed for the residual demand of firms producing only one good but their method can be extended to multi-product firms in differentiated product industries. In this case, assumptions about pricing strategies also need to account for the substitution effects of products competing both outside and within the firm. Indeed, when the products of multi-product firms are substitutes in demand, an increase in the price of one product increases the demand for the other products. As a result, the Lerner index of each of the products exceeds the inverse of their own elasticity of the (ordinary) demand (Tirole, 1988). Since the residual demand approach consists of estimating mark-ups using the inverse of the residual demand elasticity, pricing issues of multiple products may invalidate the approach. However, the residual demand approach accounts for those internal effects and remains valid for the multi-products firm. As stressed by Baker and Bresnahan (1988), the key requirement is that conjectures are consistent in equilibrium. In the case of the multi-product firms, this implies that conjectures within the firm and conjectures relative to other firms in the industry are consistent.

To clarify the consistency requirement of the multi-product firm, we may think of two product managers within a firm, each in charge of a differentiated version of the same type of product. The product managers are de facto competitors because the products are substitutes in demand. From the point of view of the aggregate firm, however, the conjecture of the manager of each product ought to be that the best reaction of the other manager is to maximize the joint profit of the firm.[1] In reality, however, there may be some competition between divisions (Ramon and Giralt, 1995) and this should be reflected in the conjectures that managers share about each other (Di Cintio, 2006). But as long as the reaction function of each manager corresponds to the conjecture of the other, the residual demand elasticity of each product will be equal to the mark-up charged for this product as expressed in (27.10).

2.2 Estimation of the Residual Demand with Imperfect Instrumental Variable

The feasibility of the residual demand approach proposed by Baker and Bresnahan depends on the availability of exogenous cost factors. In practice, it may not be possible to obtain product-specific cost factors for the products of interest. Yet the estimates of the parameters of the residual demand curve are likely to be biased if the endogeneity of quantities demanded is not controlled. Furthermore, the estimate of the mark-up is likely to be biased downward since the elasticity of demand is itself often biased upward when quantities in a demand relation are estimated without instruments (Berry, 1994).

In the absence of firm-specific cost factors, industry-wide cost factors may still be readily available to the analyst. This situation is likely to occur for industries where the production technology and the cost of some production factors are similar and available to all firms but subsequent differentiation of products is not properly reflected in the cost structure of firms. An alternative approach would therefore be to instrument quantities with an industry-wide cost shifter even if the exogeneity assumption failed. This solution may be preferable since it is possible under a set of reasonable assumptions to infer the direction of the bias. To demonstrate this point, we first derive an analytical expression for the bias when an industry-wide variable is used as an instrument for quantities.

Consistent with the notation used in the previous section, we initially specify a linear model[2] of the residual demand curve for product 1:

$$P_1 = \beta_0 + \beta_1 Q_1 + \beta_2 W_I + \boldsymbol{\beta}_3 \mathbf{Z} + \mathrm{u}_1 \tag{27.11}$$

where $\mathbf{Z}$ is a vector of demand shifters and u is the error term of the model. In accordance with the residual demand curve (27.8), an industry-wide cost shifter W_1 is included in the demand equation. This industry-wide cost shifter accounts for shifts in the demand of product 1 through the cross-price elasticities when competing firms adjust their quantities in reaction to a change in W_1. We assume that the other industry-wide and firm-specific cost shifters that may have an effect on price of product 1 are exogenous and therefore do not bias the estimates if they are not included in (27.11).

As discussed in the previous section, a factor price specific to product 1, W_1, would be a proper instrument to control the endogeneity of Q_1 in (27.11). If instead Q_1 is instrumented with the industry-wide cost shifter W_I, then by definition W_I should be excluded from equation (27.11). Rewriting (27.11), the model to be estimated would then become

$$P_1 = \beta_0 + \beta_1 Q_1 + \boldsymbol{\beta}_3 \mathbf{Z} + \mathrm{u}_2 \tag{27.12}$$

where

$$\mathrm{u}_2 = \beta_2 W_I + \mathrm{u}_1 \tag{27.13}$$

Here, W_I is now omitted from (27.12), yet W_I shifts the residual demand curve through the interaction of cross-price elasticities and the effect of quantity choice of other firms as indicated in (27.8). Therefore, if used as an instrument for Q_1 in (27.12), W_I would be correlated with Q_1 because it is a cost factor that shifts quantities through the supply relation, but it would fail the exogeneity assumption since it is also correlated with u_2 as indicated by (27.13).

The endogeneity bias that W_I creates when used as an instrumental variable is identical to one that omitted variables create with ordinary least squares estimation when a relevant explanatory variable is not included in the specification of an econometric model. In our case, W_I plays the role of the omitted variable in (27.12) when it is used as an instrumental variable for quantities and omitted from (27.11). Typically, all parameters would be biased when an endogenous variable is omitted from the specification of a model. It turns out, however, that there is a simple relationship between the direction of the bias and the parameters when the two-stage least squares (2SLS) estimator is used to compute the instrumental variable estimate. In this case, the same procedure used to retrieve the direction of the bias created by omitted variables in the linear regression model can indeed be used to derive the bias created by W_I.

In the first regression of the 2SLS procedure, Q_1 is regressed on all the exogeneous variables of the model, including W_I:

$$Q_1 = \alpha_0 + \alpha_1 W_1 + \boldsymbol{\alpha_2 Z} + v_1 \tag{27.14}$$

In the second step, the predicted values for Q_1 obtained in (27.14) are substituted for the quantities in (27.12). As we mentioned, W_I needs to be excluded from (27.11) to avoid perfect multicollinearity between, $\widehat{Q_1}$, $\mathbf{Z}$ and W_I. Instead, we need to use model (12) and the second regression of the 2SLS procedure is therefore

$$P_1 = \beta_0 + \beta_1 \widehat{Q_1} + \boldsymbol{\beta_3 Z} + r \tag{27.15}$$

with $r = u_2 + \beta_1 v_1$ where u_2 corresponds to the error term in (27.12).

Under the exogeneity and relevance assumptions required for instruments to be valid,[3] this second regression would provide an unbiased estimate of the coefficients of the model. However, W_I is omitted from model (27.12) and yet it is correlated with the error term u_2 as (27.13) indicates. As a result, all of the parameter estimators of the model are likely to be biased.

Following Wooldridge (2002), the bias may be obtained by first writing the linear projection of W_I, the missing variable, on all the explanatory variables appearing in (27.15):

$$W_I = \gamma_0 + \gamma_1 \widehat{Q_1} + \boldsymbol{\gamma_2 Z} + v_2 \tag{27.16}$$

Substituting the right hand-side of (27.16) for W_I in (27.11), we then obtain

$$P_1 = (\beta_0 + \beta_2\gamma_0) + (\beta_1 + \beta_2\gamma_1)\widehat{Q_1} + (\boldsymbol{\beta_3} + \beta_2\boldsymbol{\gamma_2})\mathbf{Z} + (u_1 + \beta_2 v_2) \tag{27.17}$$

Equation (27.17) shows that the bias for β_1 is $\beta_2\gamma_1$ and the bias for the coefficient on the demand variables is $\beta_2\boldsymbol{\gamma_2}$ The variance of the estimator could also be affected since the error term of (27.17) is now equal to $u_1 + \beta_2 v_2$.

To obtain the signs of γ_1 and $\boldsymbol{\gamma_2}$, it is easier to work with the predicted values obtained with model (27.14) which are equal to:

$$\widehat{Q_1} = \alpha_0 + \alpha_1 W_I + \boldsymbol{\alpha_2 Z} + \hat{v}_1 \tag{27.18}$$

Rearranging (27.18) we obtain

$$W_I = \left(-\frac{\alpha_0}{\alpha_1} - \frac{\hat{v}_1}{\alpha_1}\right) + \frac{\widehat{Q_1}}{\alpha_1} - \frac{\boldsymbol{\alpha_2 Z}}{\alpha_1} \tag{27.19}$$

Expression (27.16) and (27.19) are therefore equivalent so the bias β_1 for can also be expressed as β_2/α_1 and the bias for the vector of demand parameters $\boldsymbol{\beta}_3$ can be expressed as $-\beta_2\boldsymbol{\alpha}_2/\alpha_1$. Hence, the bias in the estimator of β_1 is $\tilde{\beta}_1 = \beta_1 + \beta_2/\alpha_1$ and the bias in the estimator of $\boldsymbol{\beta}_3$ is $\tilde{\boldsymbol{\beta}}_3 = \boldsymbol{\beta}_3 - \beta_2\boldsymbol{\alpha}_2/\alpha_1$

The sign of the parameters α_1 and $\boldsymbol{\alpha}_2$ can be directly obtained by estimating (27.14) but β_2 cannot be estimated because W_I needs to be omitted from the model (27.11) when it is used as an instrumental variable. As we discuss below, under a set of reasonable assumptions, it is possible to infer the sign of these parameters.

2.3 Sign of the Biases of the Coefficients of the Residual Demand Curve

The patterns of substitutions can be complex in a differentiated oligopoly. However, the direction of the bias can be inferred under the weak assumptions that the effect of an increase in the industry-wide cost shifter W_I is to decrease equilibrium quantities Q_1 in (27.14) and to decrease the quantities of all the other products of the vector $\boldsymbol{E}^i$ in (27.6) while holding Q_1 constant.

The first assumption implies that after accounting for all the substitution patterns and the conjectures of all the firms in the industry, the effect of an increase in an industry-wide cost shifter leads to a decrease in equilibrium quantities of firms as in the standard Cournot model. This is equivalent to $\alpha_1 < 0$ in (27.14). This result may not always hold if more complex patterns of substitutions are taken into consideration. It is possible, for instance, that the increase in the industry-wide cost shifter does not impact the overall cost of all producers equally, which would then make a subset of the products relatively cheaper to produce and lead to an increase in their equilibrium quantities.

The rationale for the second assumption is similar and implies that an increase in the industry-wide cost shifter W_I would reduce the elements of the vector of best response quantities $\boldsymbol{E}^i$ expressed in (27.6). $\boldsymbol{E}^i$ corresponds to the quantities of the competing firms while Q_1 is held constant in their adjustment decisions. These best-response quantities are substituted for the equilibrium quantities in the demand curve to obtain the residual demand curve of product 1 as expressed in (27.8). If products are substitutes, an increase in W_I and a subsequent decrease in the quantities $\boldsymbol{E}^i$ implies that an increase in W_I will shift the residual demand curve to the right and lead to an increase in the price of product 1. As a result, we would expect $\beta_2 < 0$.

Since $\alpha_1 < 0$, this implies that $\beta_2/\alpha_1 < 0$. As result, the sign of the bias in the estimator of β_1 should be negative. Further, since β_1 is negative, the bias estimator $\tilde{\beta}_1$ represent a lower bound for β_1.

Furthermore, recalling (27.10),

$$-\frac{\partial \ln R}{\partial \ln Q_1} = \frac{P_1 - MC_1}{P_1} = |\beta_1| * \frac{Q_1}{P_1} \tag{27.20}$$

Since $|\mathbf{E}(\tilde{\beta}_1)| > |\beta_1|$, the biased estimator of β_1, $\tilde{\beta}_1$, bounds the market power parameter from above, which is the result we have sought to establish.

Using the same reasoning, if any element of the vector of demand parameter $\mathbf{Z}$ in (27.14) has a positive effect on the equilibrium quantity Q_1, then it should also have a positive effect on the price of the residual demand curve in (27.11). As a result, we should expect any element of $\boldsymbol{\beta}_3$ and the corresponding element of $\boldsymbol{\alpha}_2$ to have the same sign. Since $-\beta_2/\alpha_1 > 0$, the bias of any element of $\boldsymbol{\beta}_3$ should be of the same sign as the sign of this element. The effect of the bias is therefore to inflate the values of the demand parameters.

3 MEASURING MARKET POWER IN THE US HYBRID CORN SEED INDUSTRY

In this section we illustrate the empirical relevance of the upper-bound approach of the residual demand we have proposed by examining the degree of market power exercised in the US hybrid corn seed industry over a multi-year period. We selected this industry because firms in this market develop and market a large number of differentiated products and because it is concentrated and has undergone significant consolidation in recent years, to the point that concerns over potential anti-competitive behavior have attracted the attention of the US Department of Justice (Department of Justice, 2010).

Following the development of plant breeders' rights in the 1970s and plant utility patents in the 1980s, and later with the advent of agricultural biotechnologies, the US hybrid corn seed industry went through a series of structural changes which redistributed the control and ownership of most major US firms to large integrated biotech firms. In particular, at the end of the 1990s, multinational biotech firms operating in agricultural input markets integrated forward into the seed industry (Fernandez-Cornejo, 2004). Various efficiency considerations have been advanced to explain these strategic moves (Kalaitzandonakes and Bjornson, 1997; Marks et al., 1999; Rausser et al., 1999), but most boil down to the need to access the best seeds as an optimal delivery mechanism of biotech traits.

Motives aside, the vertical integration of biotechnology firms substantially changed the ownership structure of the US seed industry. A number of medium and large independent seed firms became part of integrated multinational firms with assets in biotechnology and other markets. Parallel changes in the level of concentration of the US seed industry were also realized. The Herfindahl–Hirschman Index (HHI) for the US seed industry in the 1998–2008 period increased, ultimately exceeding 1800 in 2008, a level typically assumed to separate moderate and high levels of concentration in industries (Figure 27.1). Under such circumstances, the chance that market power was exercised in the industry could not be precluded.

Estimating market power in the seed industry, however, is inherently difficult. The first step to any analysis of market power is an adequate demarcation of the market for which firms compete, which should include all the products for which some substitution possibilities exist (Massey, 2000). Yet seed markets typically involve a large number of differentiated products for which it is inherently difficult to clearly delineate market boundaries. In the case of the US hybrid corn seed market, for instance, several thousands of different hybrids are offered in any given year but in reality buyers in any location choose from a more limited set (Magnier et al., 2010). Because hybrids are selected for commercialization based on their yields, disease resistance and other performance characteristics and those vary greatly across geographic regions with different climatic conditions and soils types (Duvick, 1998; Pandey, 1998), the markets of hybrid seeds tend to be limited in geographic scope. Yet the exact geographic boundaries at which the performance of any

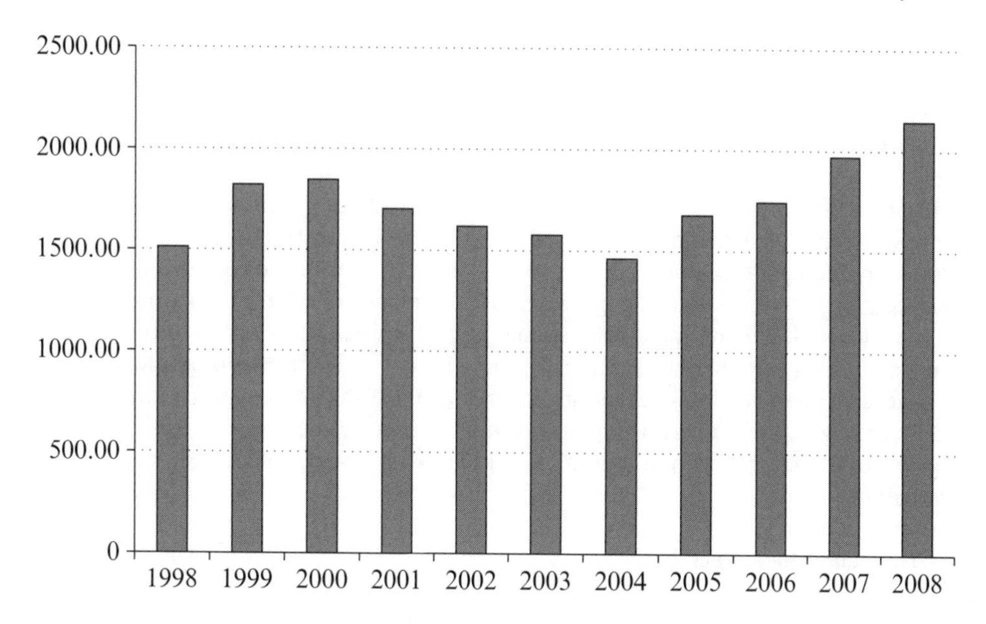

Figure 27.1 Herfindahl–Hirschman Index (HHI) for US seed corn industry, 1998–2008

individual hybrid drops off against alternatives are difficult to delineate for buyers and sellers and even more so for the analyst.

Variable product life cycles also complicate the patterns of market competition among hybrid corn seeds (Magnier et al., 2010) and must be accounted for in the estimation of market power exercised in the industry. Each year, hundreds of new hybrids are brought to the market but as it takes time for buyers to experiment with and adopt new hybrids, new and old vintage products coexist (Dooley and Kurtz, 2001). As a result, there is a time dimension to the substitution patterns of hybrid seeds which is reflected in the pricing strategies of firms. New hybrids may need to be discounted at the beginning of their cycle to encourage adoption and again at the end of their lifecycle when they have to compete with newly introduced hybrids (Morris et al., 2003).

Most significantly, the assessment of market power in the seed industry is inherently difficult because the market is made up of a very large number of differentiated hybrids. Structural modeling approaches cannot be easily extended to industries with such a large number of products. Similarly, discrete choice models require assumptions that do not correspond to the realities of the seed market. In particular, the assumption that each buyer in the market is presented with all available hybrids to choose from is not compatible with the structure of hybrid seed market where buyers typically choose a few hybrids from a limited number that are relevant to their production region. The residual demand approach would require a specific cost variable for each hybrid but this information would be difficult to obtain for such a large number of products, and access to proprietary firm-level cost data is not possible. Hybrids could, in principle, be aggregated to save on the requirement to obtain hybrid-specific instruments but regional differentiation of hybrids with uncertain geographic boundaries prevents meaningful aggregation possibilities.

Not surprisingly, only a few studies have attempted to measure market power in the US hybrid corn seed industry. Fernandez-Cornejo and Spielman (2002) examined the effects of industry concentration on market power and costs (including R&D) in the US cotton and corn seed industries. The authors used a structural econometric model to measure the relative strengths of these effects over the past 30 years. Seed were considered homogeneous and not differentiated products and the preliminary results that the authors presented suggest that increased concentration over the past 30 years resulted in a cost-reducing effect that prevailed over enhanced market power. Shi, et al. (2010) specified a structural model to capture pricing strategies of biotech trait bundling and found evidence that firms engaged in spatial price discrimination and exercised a certain degree of market power in the pricing of some of the biotech traits in hybrid seed corn. To achieve a tractable empirical model and recognizing that hybrid seeds are differentiated by their geographic performance and other characteristics, the authors defined the geographic boundaries of markets for hybrid seeds, a priori, to coincide with selected aggregations of counties.

2.4 Data and Model Specification

Against this background, we use a data set with annual observations of prices and quantities of corn hybrids sold in the US and estimate market power with a price-dependent model. The data set has been collected by a commercial market research company – GFK Kynetec (previously Doane Marketing Research) – through annual surveys of over 5000 US hybrid seed corn buyers between 1997 and 2008. Additional information provided for each hybrid was the states where it was sold and the type of biotech traits bred in each hybrid. Such biotech traits could be singular or stacked together to combine individual trait functionalities. The different types of biotech traits include resistance to insects and tolerance to various types of herbicides such as Roundup, Liberty and Imidazolinone (IMI). However, the data set did not provide for instrumental variables to control for the endogeneity of individual quantities of each hybrid. So in accordance with the methods we proposed in the previous section, we used the same industry-wide cost shifter for all hybrids.

The instrumental variable that we use is the lagged price of corn and serves as a proxy for production cost since the production of seeds occurs during the prior growing season and contract seed growers are paid prices similar to those of commodity corn plus fixed incentive premiums (Krull et al., 1998). Because this production factor affects the costs of all hybrids and represents a significant share of the total costs incurred by seed firms,[4] the lagged price of corn is an industry-wide cost shifter.

The data consists of an unbalanced panel of annual observations for several thousand corn hybrids that were sold in the US over the 1997 to 2008 period. The summary statistics for the corn hybrids used in this study are reported in Table 27.1.

The model that we estimate is specified as

$$\ln(p_{it}) = \beta_0 + \beta_1 \ln(q_{it}) + \beta_2 F_t + \phi(t) + \sum_{r=1}^{K} \gamma_r R_r + \sum_{s=1}^{S} \lambda_s TR_s + u_{it} \tag{27.21}$$

where the time index is $t = 1997, \ldots 2008$, the hybrid index is $i = 1, \ldots$ N with N $= 6170$, p_{it} is the reported market price for hybrid i at time t, q_{it} is the acreage on which the hybrid was planted, $\phi(t)$ is a cubic life-cycle trend that starts with the first year that hybrid i

Table 27.1 Summary statistics for the corn data

Variable	Average	Std. Error	Minimum	Maximum
Seed price (dollars)	97.44	28.04	34.12	278.00
Quantity (acre)	27362.49	80640	12.00	4004987
Mid-Atlantic	0.0559	0.2296	0.00	1.00
South-East	0.0251	0.1564	0.00	1.00
South-West	0.0642	0.2452	0.00	1.00
HERB	0.1927	0.3945	0.00	1.00
CB	0.1452	0.3523	0.00	1.00
RW	0.0098	0.0983	0.00	1.00
CB-HERB	0.1193	0.3241	0.00	1.00
RW-HERB	0.0121	0.1095	0.00	1.00
CB-RW	0.0118	0.1080	0.00	1.00
CB-RW-HERB	0.0386	0.1927	0.00	1.00
CB-HERB2	0.0098	0.0983	0.00	1.00
RW-HERB2	0.0001	0.0118	0.00	1.00
CB-RW-HERB2	0.0021	0.0468	0.00	1.00

appeared on the market, R_r is a regional dummy variable, TR_S is a biotech trait-specific dummy variable and F_t is the expected price of corn just before planting time (that is, the average January–March price of the December corn futures contract on the Chicago Board of Trade).[5]

Quantity is expressed in log form, so β_1 represents the overall price flexibility coefficient for each hybrid and corresponds to the average mark-up across hybrids over the period covered by the data set. We find that the price-dependent specification is relevant in the case of seeds because it implies a direction of causation that runs from quantity to price. Indeed, seeds cannot be stored for long periods of time before they deteriorate in quality, and production typically happens only once a year. As such, quantities cannot be changed during this period, and price is the variable that needs to adjust to clear the market.

Consistent with the upper-bound approach of the residual demand, we estimated the models by OLS and with the two-stage least squares instrumental variable (2SLS-IV) estimator in which the lagged price of corn is used as an instrument for corn seed quantity. The price of corn of the previous year is therefore expected to shift the supply curve of all hybrids since contract seed growers are typically paid prices similar to those of commodity corn the year before seeds are put on the market. We then use the Hausman t-test to compare the OLS and IV estimates of β_1. The observed test statistics are greater than the critical values at conventional Type I error rates, and we reject the null hypothesis that quantity is exogenous. Accordingly, we only report the 2SLS-IV estimates of the corn log-price models in Table 27.2.

The most effective regional specification in the econometric models we estimated was based on multiple state groupings. The three regions with significantly different price levels are the Mid-Atlantic region (New York, New Jersey, Pennsylvania, Delaware, Maryland, Virginia and West Virginia), South-East (North Carolina, South Carolina, Georgia,

Florida, Alabama, Mississippi, Louisiana, Tennessee and Arkansas), and South-west (Kansas, Oklahoma, Texas, Colorado, New Mexico, Arizona and California). Ten distinct biotech and other similar technology traits were aggregated as broad homogeneous trait categories which included herbicide tolerance (Roundup Ready, Liberty Link and IMI), corn borer resistance (CB), and rootworm resistance (RW) plus combinations (stacks) of two or more of these trait categories. Overall, the three corn regions represent about 14.5 per cent of the total observations, and the 10 traits represent about 52 per cent of the observations (that is, 48 per cent of the observations were conventional hybrids).

2.5 Results and Discussion

As indicated in Table 27.2, the fitted model explains roughly 32 per cent of the observed variation in the log-prices of US hybrid seed corn, and the estimated flexibility coefficient

Table 27.2 2SLS-IV estimates of the corn log-price model

Analysis of variance					
Source	DF	Sum of squares	Mean square	F value	Pr > F
Model	18	1001.64	55.64666	548.79	<.0001
Error	21413	2171.25	0.101399		
Corrected Total	21431	1873.335			
Root MSE	0.31843	R-Square	0.31569		
Dependent Mean	4.5373	Adj R-Sq	0.31511		
Coeff Var	7.01809				
Variable	DF	Parameter estimate	Standard error	t-value	Pr > |t|
Intercept	1	5.351704	0.070971	75.41	<.0001
Lnq	1	−0.14623	0.007991	−18.3	<.0001
FuturesP	1	0.000789	0.000028	28.21	<.0001
T	1	0.068863	0.009442	7.29	<.0001
t2	1	−0.01164	0.002153	−5.41	<.0001
t3	1	0.000573	0.000139	4.13	<.0001
midatlantic	1	−0.12124	0.010647	−11.39	<.0001
southwest	1	0.10401	0.009671	10.75	<.0001
southeast	1	0.149652	0.015361	9.74	<.0001
HERB	1	0.200929	0.006298	31.9	<.0001
CB	1	0.234642	0.006689	35.08	<.0001
RW	1	0.293645	0.023009	12.76	<.0001
CB_HERB	1	0.364235	0.00748	48.69	<.0001
RW_HERB	1	0.401493	0.02037	19.71	<.0001
CB_RW	1	0.364098	0.020974	17.36	<.0001
CB_RW_HERB	1	0.531404	0.012679	41.91	<.0001
CB_HERB2	1	0.527478	0.023738	22.22	<.0001
RW_HERB2	1	0.686358	0.18435	3.72	<.0002
CB_RW_HERB2	1	0.779229	0.049882	15.62	<.0001

is −0.14623, which implies that the upper bound on the corn seed mark-up is roughly 15 per cent. Given this result, we can conclude that the amount of market power that was exercised in the US seed corn industry between 1997 and 2008 was rather modest.

The three parameters of the fitted cubic trend component are statistically significant, and the signs on these coefficients indicate that the lifecycle effect is concave over time – the initial price starts at a lower level, increases until the hybrid's fourth year on the market, and then declines until the hybrid is removed from the market. This pattern is consistent with the pricing structure in the US hybrid seed corn market described in Magnier et al. (2010). The result also provides empirical support for the notion that product lifecycle dynamics must be accounted for when examining patterns of price competition in industries with large numbers of differentiated products where one of the differentiating attributes is vintage, as in the case of the US hybrid seed corn industry.

The remaining explanatory variables are semi-flexibilities, and 100 times the estimate is interpreted as the approximate percentage change in the seed corn price given a unit increase in the dummy variable. The regional coefficients imply that seed corn prices in the Mid-Atlantic region are roughly 12.1 per cent lower than the rest of the country, and seed corn prices in the South-west and South-east are roughly 10.4 per cent and 14.9 per cent higher (respectively) than in the rest of the US. These differences may reflect the ability of firms to segment these markets geographically or they may correspond to differences in costs in servicing such markets which are reflected in seed prices.

The values associated with the individual biotech traits are strongly significant and positive: herbicide-tolerant hybrids earn a premium that is roughly 20 per cent higher than the price of conventional seed corn, and corn borer and rootworm-resistant hybrids have premia that are roughly 23 per cent and 29 per cent higher than conventional corn (respectively). We also find that hybrids with stacked biotech traits earn higher premiums, but the values of the combined traits are less than the sum of the individual ones, which provides some evidence of bundled pricing strategies, which appears to dampen the ability to exploit market power.

Our empirical results show that with a rather parsimonious model we can derive an upper bound for the market power that has been exercised in the corn seed industry as well as separate premiums on biotech traits delivered through hybrid corn seeds. This is in a market context of thousands of products differentiated by geography, vintage and other performance characteristics and for which the market boundaries are unclear. Obtaining an upper bound on the seed mark-up and biotech trait premiums charged sheds some light on the amount of market power exercised in the US seed industry during a period of significant structural change and consolidation. We find that despite increasing industry concentration and vertical integration over the period of analysis, the amount of market power exercised in the US seed industry for seeds and biotech traits was modest.

The relative size of the upper bounds estimated for the mark-ups and premiums charged in the US corn seed industry also appear consistent. Firms engaged in hybrid seed breeding activities are expected to charge prices above marginal costs to recoup the fixed costs of breeding. Similarly, firms developing biotech traits are expected to introduce additional mark-ups to recoup biotech R&D expenditures. Without the existence or the prospect of some market power, firms would have no incentive to use more efficient technologies, improve product quality, or introduce new varieties, hybrids and biotech traits. The premium for corn seed is relatively modest while the premiums for the biotech traits, which are more costly to develop, are slightly larger.

4 SUMMARY AND CONCLUDING COMMENTS

Estimating market power is often difficult in industries characterized by a large number of differentiated products. The residual demand method of Baker and Bresnahan (1988) is a powerful approach but requires product-specific instruments to apply and those may not be readily available in practice. In this chapter, we showed that an upper bound for the mark-up and the market power exercised can be derived by using a single industry-wide instrument, which may be easier to obtain.

The methodology we have proposed in this chapter can be applied to any industry with a large number of differentiated products that have fuzzy market boundaries and when cost-specific instruments cannot be obtained for each product. It is also worth noting that while the upper bound of market power that our method provides is a second-best estimate, its accuracy improves as the degree of product differentiation increases. Strategic interactions tend to attenuate as products become less proximate as substitutes (Baker and Bresnahan, 1988) and under such circumstances industry-wide cost shifters, like those we proposed to use as instruments, are less likely to fail the exogeneity assumption.

We have demonstrated the applicability of our proposed method by applying it to the estimation of market power in the US seed corn industry. This industry has undergone fundamental structural changes and consolidation over the last two decades. These have led to complaints about high levels of concentration, unchecked exercise of market power, incidence of anti-competitive behavior and a recent DOJ investigation. Yet, despite these claims there has been little tangible evidence in the literature about the extent of market power in the US seed industry. This gap probably reflects the inherent difficulties in measuring market power exercised within an industry characterized by a very large number of differentiated products with unclear market boundaries. The parsimonious price-dependent model that we specified to obtain an upper bound for the market power exercised in the industry showed that the mark-up on hybrid seeds was, on average, about 14 per cent or less over the 1997–2008 period while premiums earned on biotech traits were higher, ranging from 20 per cent for single traits to 50 per cent for multiple stacked traits.

It has long been recognized that static analysis of market power is not appropriate when applied to innovative industries. Concentration ratios cannot describe the degree and nature of competition that is often present in such industries and static analyses of consolidation measures and price mark-ups fail to account for the large sunk costs that are incurred by innovative firms through R&D investments. Our results would therefore suggest that caution and proper methodology are necessary in the analysis of market power of innovative industries, especially those with a large number of differentiated products like the US seed/biotech industry, since anti-trust regulatory restrictions on the industry's ability to exercise some market power directly influence the industry's capacity to invest in future discoveries and new product development.

NOTES

1. Conjectures within the firm correspond to a subset of the ratios $\partial Q_j/\partial Q_i$ in equation (27.5). The rest of the conjectures correspond to conjectures that a firm holds relative to outside competitors.
2. Model (27.11) could be rewritten in more conventional log-linear form for prices and quantities without changing the analysis but the linear form simplifies the exposition.
3. The exogeneity assumption would imply that $Cov(W_I, u_2) = 0$ where u_2 is the error term in (27.12) and the relevance assumption implies that $\alpha_1 \neq 0$ in (27.14).
4. Our estimates indicate that 30–45 per cent of total costs incurred by seed firms for developing, producing and marketing hybrid corn seeds in the US are attributable to production costs and most of such production is carried out by contract seed production.
5. Because the vast majority of contracts for seed growing between seed firms and independent seed producers are agreed upon months in advance of any growing season, these contracts are typically priced off the CBOT corn futures prices just prior to the planting of the parent lines of the hybrid seeds. Based on interviews we performed with seed firms in the US, the pricing of most production contracts occurs between January and March of every year. In order to evaluate the influence of the choice of corn price on our empirical results we also used alternative averages of CBOT corn futures prices and the USDA annual commodity price. Our empirical results were robust to all alternative specifications.

REFERENCES

Appelbaum, E. (1982), 'The estimation of the degree of oligopoly power', *Journal of Econometrics*, **19**(2–3), 287–99.

Baker, J. B. and T.F. Bresnahan (1988), 'Estimating the residual demand curve facing a single firm', *International Journal of Industrial Organization*, **6**(3), 283–300.

Berry, S. (1994), 'Estimating discrete-choice models of product differentiation', *The RAND Journal of Economics*, **25**(2), 242–62.

Berry, S. and A. Pakes (1993), 'Some applications and limitations of recent advances in empirical industrial organization: merger analysis', *The American Economic Review*, **83**(2), 247–52.

Berry, S., J. Levinsohn and A. Pakes (1995), 'Automobile prices in market equilibrium', *Econometrica: Journal of the Econometric Society*, **63**(4), 841–90.

Bresnahan, T.F. (1981), 'Duopoly models with consistent conjectures', *The American Economic Review*, **71**(5), 934–45.

Conley, T.G., C.B. Hansen and P.E. Rossi (2007), 'Plausibly exogenous', *The Review of Economics and Statistics*, **94**(1), 260–72.

Department of Justice (2010), 'DOJ and USDA hold first-ever workshop on competition issues in agriculture', press release, March, available at: http://www.justice.gov/atr/public/press_releases/2010/256496.htm.

Di Cintio, M. (2006), 'Product differentiation and multiproduct strategies', *University of Lecce Economics Working Paper No. 81/38*.

Di Marina, G. (2004), 'Empirical analysis of markets with differentiated products: the characteristics approach', *Giornale degli Economisti*, **63**(2), 243–88.

Dooley, F. and M. Kurtz (2001), 'The effect of a changing market mix in seed corn on inventory costs', paper presented at the 2001 annual meeting of the American Agricultural Economics Association, Long Beach, CA, July.

Duvick, D. (1998). 'The United States', in M. Morris (ed.), *Maize Seed Industries in Developing Countries* Boulder, CO: Lynne Rienner Publishers and CIMMYT, pp. 193–212.

Fernandez-Cornejo, J. (2004). 'The seed industry in US agriculture: an exploration of data and information on crop seed markets, regulation, industry structure, and research and development', Agriculture Information Bulletin no. AIB786, United States Department of Agriculture.

Fernandez-Cornejo, J. and D. Spielman (2002), 'Concentration, market power, and cost efficiency in the corn seed industry', paper presented at the 2002 annual meeting of the American Agricultural Economics Association, Long Beach, CA, July.

Hahn, J. and J.A. Hausman (2003), 'IV estimation with valid and invalid instruments', Massachusetts Institute of Technology, Department of Economics, working paper no. 03-26.

Iwata, G. (1974), 'Measurement of conjectural variations in oligopoly', *Econometrica: Journal of the Econometric Society*, **42**(5), 947–66.

Kalaitzandonakes, N. and B. Bjornson (1997), 'Vertical and horizontal coordination in the agro-biotechnology industry: evidence and implications', *Journal of Agricultural and Applied Economics*, **29**(1), 129–39.
Klepper, S. and E.E. Leamer (1984), 'Consistent sets of estimates for regressions with errors in all variables', *Econometrica: Journal of the Econometric Society*, **52**(1), 163–83.
Krull, C., J. Prescott and C. Crum (1998), 'Seed marketing and distribution', in M. Morris (ed.), *Maize Seed Industries in Developing Countries* Boulder, CO: Lynne Rienner Publishers and CIMMYT, pp. 125–41.
Leamer, E.E. (1981), 'Is it a demand curve, or is it a supply curve? Partial identification through inequality constraints', *The Review of Economics and Statistics*, **63**(3), 319–27.
Magnier, A., N. Kalaitzandonakes and D.J. Miller (2010), 'Product life cycles and innovation in the US seed corn industry', *International Food and Agribusiness Management Review*, **13**(3), 17–36.
Marks, L.A., B. Freeze and N. Kalaitzandonakes (1999), 'The agbiotech industry: – a US–Canadian perspective', *Canadian Journal of Agricultural Economics/Revue Canadienne d'Agroeconomie*, **47**(4), 419–31.
Massey, P. (2000), 'Market definition and market power in competition analysis: some practical issues', *Economic and Social Review*, **31**(4), 309–28.
Morris, M., K. Dreher, J-M. Ribaut and M. Khairallah (2003), 'Money matters (II): costs of maize inbred line conversion schemes at CIMMYT using conventional and marker-assisted selection', *Molecular Breeding*, **11**(3), 235–47.
Nevo, A. and A.M. Rosen (2008), Identification with imperfect instruments: National Bureau of Economic Research Working Paper no. 14434, Cambridge, MA: NBER.
Pandey, S. (1998), 'Varietal development: conventional plant breeding', in M. Morris (ed.), *Maize Seed Industries in Developing Countries* Boulder, CO: Lynne Rienner Publishers and CIMMYT, pp. 57–76.
Perloff, J.M., L.S. Karp and A. Golan (2007), *Estimating Market Power and Strategies*, New York: Cambridge University Press.
Ramon, F-O. and M. Giralt (1995), 'Competition and cooperation within a multidivisional firm', *The Journal of Industrial Economics*, **43**(1), 77–99.
Rausser, G., S. Scotchmer and L. Simon (1999), 'Intellectual property and market structure in agriculture', Department of Agricultural and Resource Economics, UC Berkeley Working Paper no. 880.
Shi, G., J-P. Chavas and K. Stiegert (2010), 'An analysis of the pricing of traits in the US corn seed market', *American Journal of Agricultural Economics*, **92**(5), 1324–38.
Tirole, J. (1988), *The Theory of Industrial Organization*, Cambridge, MA: MIT Press.
Wooldridge, J.M. (2002), *Econometric Analysis of Cross Section and Panel Data*: Cambridge, MA: MIT Press.

28 The dynamic IP system in crop genetics and biotechnology

Derek Eaton and Gregory Graff

1 INTRODUCTION

Innovation and technological change is governed by a range of institutional frameworks. This chapter concentrates on the intellectual property (IP) system, arguably one of the most important governance mechanisms influencing the evolution of technological trajectories or pathways in agricultural biotechnology. As in other areas of technology, innovation in agricultural biotechnology (agbiotech) has catalyzed changes in the IP system, in this case extending the system into the area of life forms. Novel products of biotechnological research have had to be accommodated by the IP system. This process of institutional change has been driven by a struggle between various actors in the system to influence incremental and major adjustments in the system to their perceived benefit (Graff and Zilberman, 2007; Brousseau et al., 2011). These actors include individuals and organizations involved in research and development (R&D) in agbiotech and plant breeding, in both the public sector and the private sector, as well as seed propagators, farmers, other businesses involved in the agricultural value chain, non-governmental organizations (NGOs), and others. These change agents attempt to influence the components of the IP system, in which a distinction can be made between legislators (who define and create legal instruments), regulators or administrators (IP-granting authorities, who interpret and implement instruments) and the judiciary (which interprets legal instruments as a result of conflicts). Thus, IP both conditions and influences innovation and, conversely, technological change drives the evolution of the IP system by presenting new challenges and disrupting the configuration of varying and opposed interests.

The most important issues that arise now in this area relate to the patentability of technological tools (for example molecular breeding methods), genes and genetic constructs, and the living materials, such as plants, in which these are embodied. The challenges posed to understanding and informing governance frameworks can be placed in a longer-term perspective which recognizes the historical evolution of the IP system and strategies for appropriating returns on investments made in biotechnology R&D and plant breeding.

During the twentieth century, the commercialization of agricultural genetic inputs led to the creation of new IP instruments, including plant patents in the US in 1930 (for asexually reproduced, or vegetatively-propagated, species, other than potatoes and edible tubers) and then the later emergence in Europe of plant varietal protection (PVP), also referred to as plant breeders' rights (PBR), in Europe. (In this chapter, the term 'patent' generally refers to utility patents; plant patents are referred to specifically.) This new form of protection – sometimes described as a combination of copyright and patent protection (for example Janis and Smith, 2006) – was gradually adopted by most industrialized countries. International harmonization led to the International Union for the Protection of New Varieties of Plants (UPOV; Union Internationale pour la Protection des Obtentions Végétales) in 1961. The scope of UPOV protection, both in terms of species coverage and also the rights conferred, was extended in revisions introduced in 1972, 1978 and 1991.

Technological developments in the area of biotechnology induced the introduction of utility patents in the plant-breeding sector. The most notable and earliest shifts in patent policy came in the US through a number of landmark court cases:

- *Diamond v. Chakrabarty* (1980), which involved the first utility patent on a living organism that embodied a biological invention;
- *Ex Parte Hibbard* (1985), which allowed whole plants that embodied biological inventions to be considered as patentable subject matter; and
- *J.E.M. Ag Supply Inc. v. Pioneer HiBred International Inc.* (1995), which confirmed that plant varieties are eligible for utility patents, despite the availability of PVP.

Although other countries followed different paths in their legal evolution, the general trend has been similar to that in the US. In the principal agricultural producing countries of the OECD, patent protection is at least available for components of plants, such as recombinant genes or cell lines, if not for whole plants as an organism. Van Overwalle (2008) provides a description of the European context which is complicated by the existence of IP regimes at both national and supranational levels. European countries, however, have elected not to permit patenting of whole plants as an organism (EC, 1998, 2005). Nevertheless, even where patent protection is not available for whole plants, such plants may effectively be protected by patent protection if they incorporate a patented component, as, for example, ruled by the Supreme Court of Canada in Monsanto vs. Schmeiser in 2004 (see Galushko et al., 2010 for a summary of the Canadian IPR regime). Table 28.1 summarizes the basic differences between the two forms of protection.

Patent and PVP protection is increasingly available in emerging and developing countries. Under the Agreement on Trade Related Aspects of Intellectual Property (TRIPS), which entered into force in 1995, members of the World Trade Organization (WTO) are required to provide patent or some other form of IP protection for plant varieties, as well as patent protection for biotechnological inventions, with extensions having been provided for least-developed countries.

As they involve a reconfiguration of commercial practices and property rights, these changes have been the source of considerable controversy among farmers, companies, governments and other stakeholders, including NGOs (see, for example, Le Buanec, 2005). This is evidenced by the legal cases mentioned above. In terms of governments and legislators, the Dutch parliament has, for example, debated developments in IP protection in 2010 and 2011, partly because of concerns over the implications for the country's substantial private sector presence in plant breeding (Louwaars et al., 2009). The French and German governments have issued legislation intended to prevent a patent on a gene or cell line contained in a plant variety from restricting the exercise of the breeder's exemption under PVP, which allows other breeders to use a PVP-protected variety in further research without requiring permission from the holder of the PVP certificate (Louwaars, 2007). Globally, the trend in agricultural R&D has been characterized by an

Table 28.1 Key characteristics of UPOV plant variety protections as compared to plant and utility patents

Characteristics	UPOV 1978 (Canada, various developing countries)	UPOV 1991 (US, EU and member states)[1]	Plant patents (available in US since 1930)	Utility patents over plant varieties (available in US since 1985)
Protected subject matter	Varieties of species listed by country	Varieties of all genera and species[2]	Varieties of any asexually reproduced plants, except potatoes and edible tubers	Varieties of any sexually reproduced plant[3]
Duration of protection	Minimum 15–20 years (depending on crop)	Minimum 20–25 years (depending on crop)	20 years (from filing date)	20 years (from filing date)
Disclosure	Description of DUS variety otherwise through availability for breeding	Description of DUS variety otherwise through availability for breeding	Description and photographic drawing	Enabling or best mode disclosure plus deposit of novel material
Exclusive rights	Multiplication of variety for commercial purposes	Multiplication of variety for commercial purposes Use of harvested product for planting[1]	Reproduction or sale of patented plant	Multiplication of variety for commercial purposes Use of harvested product Any other commercially-related use (incl. breeding of new variety)

Notes:

1. Under UPOV 1991 the farmers' privilege was removed although a member country is permitted to make exceptions. The US still allows a broad farmers' privilege, while the EU has restricted it considerably (see discussion below).
2. In the US, PVP protection is only available for sexually-reproduced crop species.
3. In the US, as in the EU, and other countries complying with TRIPS Article 27(3)b, utility patents are also available for biotechnological inventions, including genetic transformation events and gene constructs, as well as for tools used in their creation, subject to the regular requirements for patentable subject matter.

ebbing public sector and a rising private sector, aided and abetted by changes to IP. This implies that the costs of developing new plant varieties for many crop species are being internalized within the value chain, rather than being spread across all societal groups through public investment funded through general taxation.

The overarching question from an economic or policy perspective is how IP can best drive innovation as opposed to inhibiting or skewing its advance. A fundamental question concerns the configuration of both rights and responsibilities between farmers, seed companies, public research institutions and governments. This question is still very relevant for many emerging and developing countries (see Chapter 17 in this volume). The introduction of IP in the development of new plants allows a greater role to be played by the private sector and may present other options for public research institutions, such as concentrating on more basic research in plant science, or developing new plant innovations for crop species and market segments in which commercial incentives are lower. After the introduction of IP, particularly PVP (together with the use of trademarks and trade secrets), the further shift to the use of patents also raises new questions about potential effects on the rate and direction of innovation. The prospect that patents in biotechnology could inhibit research through fragmented rights, increased transaction costs and greater uncertainty has been captured by Heller and Eisenberg (1998) in the context of biomedical innovation with their alarm call concerning the 'tragedy of the anticommons'.

This chapter concentrates on research and analysis of the recent increase in the importance of patent and plant variety protections and the dynamics of how this increase is playing out among various actors. Aside from patents and PVPs, other instruments of IP protection are commonly used in this sector, including trademarks, geographical indications and trade secrets. In some jurisdictions, geographical indications have also become available and relevant for the seed sector. Although their importance should not be denied, such forms of intellectual property protection are not addressed in this chapter, partly for reasons of space, but partly because they are arguably not as close to the core of the institutional dynamics being highlighted here. Another limitation is that the chapter concentrates entirely on the field of plant biotechnology, ignoring IPR issues of animal biotechnology, which are concentrated much more on tools and processes, as opposed to protection of final products.

The discussion is organized according to the key issues that are of interest in understanding and refining the IP system. The first issue concerns a basic understanding of the trends in the use of the most important IPRs. The long-term increase in the use of IPRs leads in the third section to an examination of the fundamental incentive effect on innovation that IPs are intended to provide, with a review of both theoretical considerations and empirical findings. A range of indirect, or secondary, effects of IP on the innovation system is then examined, including competition and market structure, the implications for sequential and downstream research, and also publicly funded research actors, such as universities. Toward the end, some space is devoted to international aspects that have been examined. The final section offers a short assessment of overall trends in this field and gaps to be addressed.[1]

2 TRENDS IN THE USE OF IP RIGHTS (IPRS)

It is helpful to ground an understanding of how the IP system has evolved in some basic descriptive data on IP filings in the area of agbiotech. In most OECD countries, information on the application for and/or granting of formal IPRs, particularly PVP certificates, plant patents (where applicable), or utility patents, is generally available, and a number of researchers and investigators have systematically compiled databases that look specifically at developments in crop genetics and biotechnology. For PVP certificates and plant patents, each recorded application or granted title represents a relatively homogeneous data point corresponding to a newly developed crop variety. For utility patents, the situation is considerably more complex, as a wide variety of different types of technologies can be the subject of utility patents and existing systems of patent classification do not fully specify all of the possible technologies that may be utilized in plant breeding or biotechnology. This means that customized searches of classifications and even filtering on technical keywords is required to identify patents in these fields. Such compilations have been done most comprehensively in the US. In the early part of the 2000s, the Economic Research Services of the US Department of Agriculture (USDA-ERS) developed a database of US agbiotech patents (see Heisey et al., 2005 and the USDA-ERS website). Graff et al. (2003a) analysed US Patent and Trademark Office (USPTO) patent data on plant biotechnology patents and divided these into three key categories: patents on genetic transformation tools; patents on genes or genetic characteristics; and patents on elite plant germplasm or plant varieties. The most recent and most complete compilation of IPRs covering plant varieties in the US, including utility patents, PVPs and plant patents, has been compiled by Pardey et al. (2013).

As indicated in Figure 28.1, applications for these three types of IPRs have grown significantly over time. In recent years, applications for plant patents and utility patents have continued to increase, with some fluctuation and slowing in the growth of utility patents between 1999 and 2005. In contrast, applications for PVPs decreased considerably after the late 1990s and only recovered to similar levels by 2008. Pardey et al. (2013) offer possible explanations for these trends, including an increase in the mid-1990s in the average time needed for PVP applications to be processed. It is worth pointing out, though, that despite some variability in the applications, the total of titles in force for all three types of varietal rights in the US has continued to grow.

Pardey et al. (2013) uncover more detailed trends as well. At the crop level, ornamental species account for over half of all varietal rights over the period 1930–2008 in the US, while cereals and oilseeds make up only 13 per cent and 10 per cent respectively. In general, ornamental species, most of which are asexually reproduced, are protected by plant patents, while the use of utility patents (possible only since 1985) has been dominated by corn and soybean.

In Europe, although neither plant patents nor utility patents are available for plant varieties, an analysis of the situation for PVP in Europe is complicated by the fact that protection was initially offered only at the national level. In 1996, a European Union PVP certificate became available with the establishment of the Community Plant Variety Office (CPVO), which now coexists with national level protection. A trend for plant breeders to opt for the broader, but more expensive, CPVO protection has resulted.

Although the UPOV secretariat collates and publishes all PVP titles issued by all member countries, including the CPVO, there is not an available database that easily identifies specific varieties with protection at both the national and European levels, or at the national level but in multiple member countries. Figure 28.2 shows trends in PVP titles granted by the CPVO from 1996–2011 by crop group. The total number of titles granted displays an upward trend. This may be even stronger given that holders of existing national-level rights were initially offered a grace period to register these with the CPVO after introduction of the EU PVP. For most crop groups with the exception

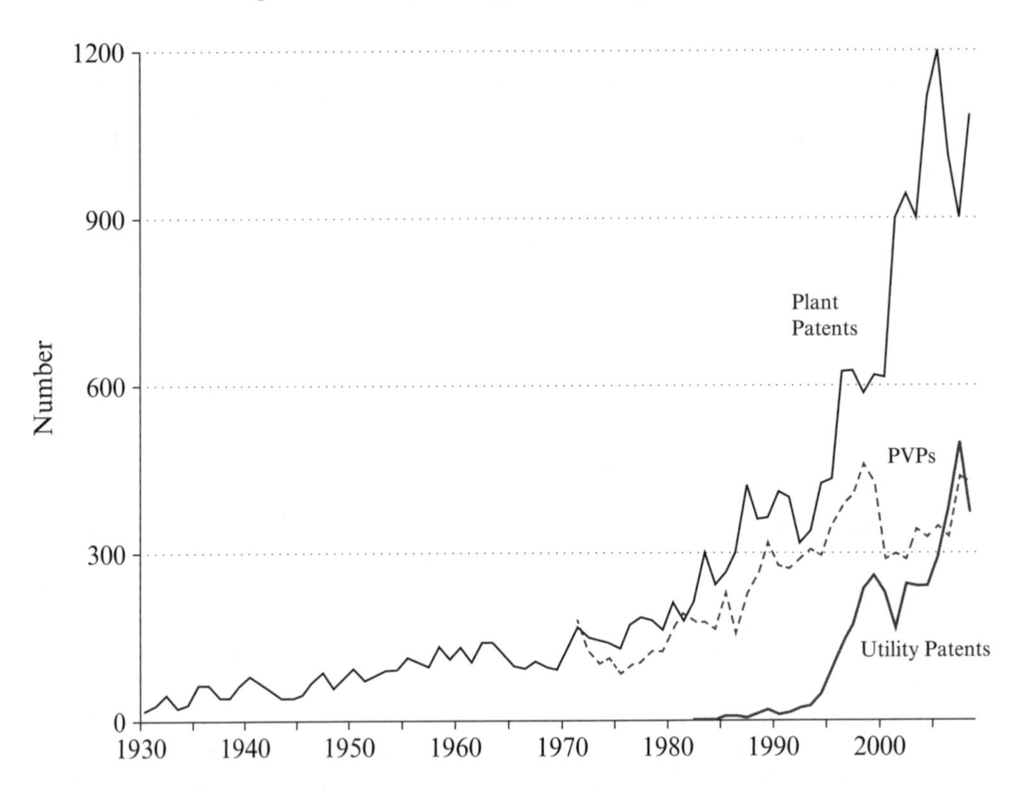

Source and notes: Pardey et al. (2013), Figure 1, who note that 'PVP indicates plant variety protection certificates. All data are reported by year of application. The PVP series represents the number of certificate applications, while the utility and plant patents are the number of granted patents taken here to represent patent applications.' Reprinted by permission of Macmillan Publishers Ltd © 2013.

Figure 28.1 US varietal rights, 1930–2008

of ornamentals, the number of PVPs granted has fluctuated somewhat over the first 10 years, before starting an upward trend during the last 5–6 years. For ornamental crops, which account for more than one-half of the titles granted (similar to proportions observed in the US), the pattern is reversed, with strong growth until 2007 and a stable pattern since then.

The recent trend of increasing numbers of PVP titles at the EU level probably continues an earlier trend in individual member countries. Although a comprehensive assessment is lacking, Louwaars et al. (2009) document a long-term increase in the number of PVP titles granted in the Netherlands (though with data only going back to 1964, not 1941 when PVP was first made available), which currently accounts for one-third of all CPVO applications. They also show how the number of PVPs granted in the country declined considerably in the mid-1990s, coinciding with the introduction of the CPVO right. Total PVPs granted for ornamental crops continued to decline after 1996 while those granted by the CPVO continued to grow. In contrast, the number of PVPs granted in the Netherlands for non-ornamental crops fluctuated at about 100 grants per year over the period 1999–2005, while total CPVO grants for such crops fluctuated between 500

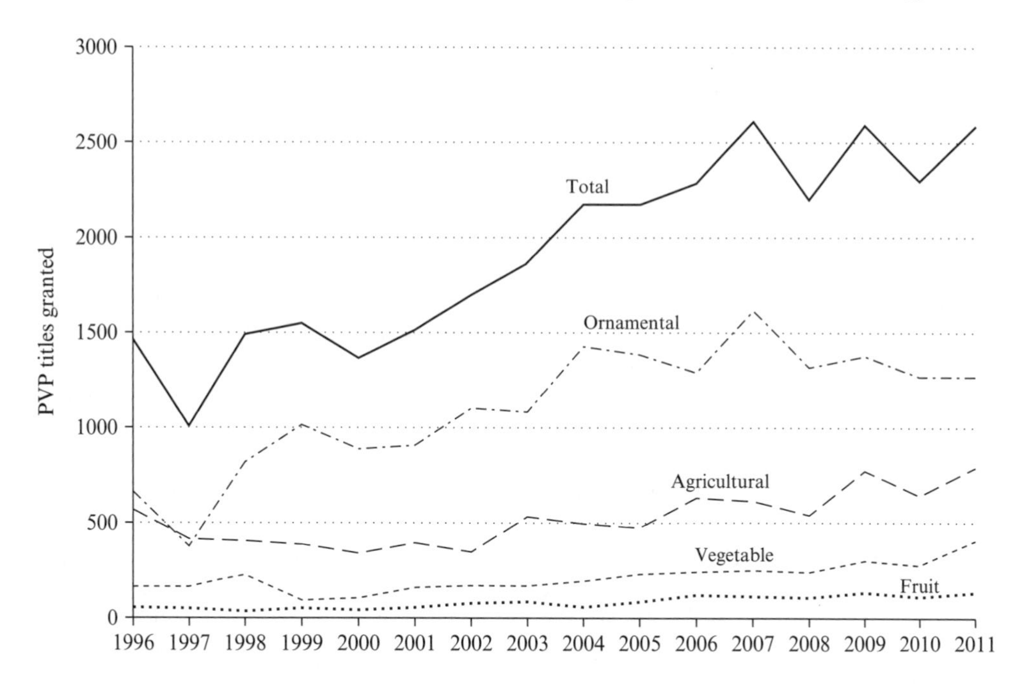

Source: Community Plant Variety Office (www.cpvo.europa.eu).

Figure 28.2 Trends in PVP titles granted by CPVO by crop group, 1996–2011

and 800 per year with no clear trend. Further data collection and analysis is therefore required to assess the trends in Europe over the past 15–20 years.

Describing the basic trends in the use of utility patents for protection of genetic transformation tools and genetic characteristics is more complicated. Graff et al. (2003b) detailed trends in plant biotechnology patents for the USPTO, the EPO and the Japanese Patent Office (JPO) and those registered under the Patent Cooperation Treaty (PCT) from 1980 to 2000. Those results demonstrated an exponentially increasing number of patents issued in the US to 2000, with more moderate numbers and rates of growth of patents granted by the EPO and JPO. They also demonstrated how the public sector was relatively active in patenting in the field of plant biotechnology, compared to other fields of technology.

Buccola and Xia (2004) examined a relatively small subset of USPTO patents issued from 1985 to 2000, restricted to those patents concerning cellular or molecular-level technologies, representing key tools of biotechnology, with the set totaling 691 patents. The data exhibits a marked increase in the number of patents granted from the late 1980s through to the mid-1990s. They also attempted to proxy for the quality, or value, using a count of 'forward' citations, the number of subsequent patents that cite the subject patent, finding that mean citations per patent declined substantially by the late 1990s (even after attempting to account for truncation bias).

The analyses by Graff et al. (2003a), (2003b) and Buccola and Xia (2004) provide important insights about basic trends in biotechnology innovation based on the use of the patent system. Given the long-term nature of the R&D process, it should be considered a priority to carry these analyses forward with updating of data from the first decade of the twenty-first century. (An interesting recent analysis by Agrawala et al. (2012) examines patenting data from 1990–2006 for trends in innovations in crop biotechnology for evidence of R&D to address climate change adaptation challenges in agriculture.) In particular, more effort could be devoted to linking and matching the use of different IPRs in combination or in different jurisdictions. The identification and classification of IPR applicants and their geographic location is also important information for studying trends in which types of organizations (or individuals) apply for patents. This is becoming more relevant, while at the same time more difficult, as patent applications in plant biotechnology are increasingly coming from emerging-economy countries, such as China and India. In addition, there have been only a few studies based on input from breeders and plant scientists (for example Graff et al., 2009; Lei et al., 2009; Tripp et al., 2007), an approach that provides valuable insights; greater collection and use of such data would be useful in understanding the trends, as well as driving forces, in the use of IPRs.

The growing use of IPRs poses the fundamental question as to whether the apparent increase in innovative activity is stimulated in part by the availability of these forms of protection. In addition, questions arise about ownership structure and implications for market power, with possible effects on the rate or direction of innovation in the breeding sector (Graff et al., 2003a; Goeschl, 2005; Graff et al., 2009). In general, it is difficult to distinguish between evolution in the innovation process that might be exogenous from the influence of the IP system and the extent to which changes in the IP system itself are driving shifts in the innovation process. The remaining sections of this chapter make an effort to consider such questions and review contributions that have been made toward answering them. The final section will reflect, however, on how little is still known about the dynamic IP system.

3 DIRECT INCENTIVE EFFECTS

The fundamental question of whether patents and other IPRs enhance the incentive to innovate has not received much systematic attention from researchers, which may seem surprising given how central this presumed effect is to the rationale for an IP system. It is often just taken as common sense that IP protection and the assumed facility to earn an economic return are required for innovators to invest. Still, this view has throughout history been contested in debates concerning national patent systems in particular; for a historical review, see Lerner (2002). There is scant empirical work, even for other technology classes, with the study by Cohen et al. (2000) standing out. As an update of a much-cited paper by some of the same authors (Levin et al., 1987), they conducted a survey of US industry and found that patents were not viewed as being among the most important strategies for earning profits from innovations. More recently, Boldrin and Levine (2008 and 2009) offer a similar, though more contested, analysis of this issue. These critical perspectives emphasize that innovators have a range of means to protect access to their innovations and thus to generate profits (or capture rents). Some commentators have noted that increasing inefficiencies in the operation of the patent or copyright systems may partly undermine the effectiveness of IPRs as a means of stimulating innovation or its diffusion (Jaffe and Lerner, 2004). A recent analysis commissioned by the European Patent Office (2007) indicates that authorities administering the patent system do take such concerns seriously.

An important finding by Levin et al. (1987) was that patent protection was more important for industries, such as pharmaceuticals, with a product that was relatively easy to copy and for which only a limited amount of secrecy is possible (in the case of pharmaceuticals, given the testing and disclosure requirements of health and safety regulations, most the information required to replicate the invention is revealed). This is very much the situation in plant breeding, given the self-reproducing nature of agricultural crops, a fact that was recognized in debates in the 1920s in the US leading to the development of the Plant Patent Act (1930).

3.1 Theoretical Analyses of the Direct Incentive Effects of IPRs

The basic starting point for theoretical models is the non-excludable characteristic of the final product of crop breeding or biotechnology, since new crop varieties are to various extents self-replicating technologies. A basic assumption is that the innovator, the breeder, has no technical means of completely excluding others – whether they are farmers or competitors – from obtaining samples of seed and reproducing it, either for own use (by replanting of seeds), or for sale on the secondary market. The resulting non-appropriability of returns means that incentives to invest in R&D are generally insufficient relative to social needs. Hence legal means in the form of IPRs are required, or other solutions, such as public subsidies or provision of breeding and seeds for farmers (other solutions to this problem used both in agriculture and other industries include public subsidy for research, explicit market monopoly, above and beyond that offered by IPRs, and direct public provision of research).

Still, there are various strategies and technological means by which the developers of new plant varieties can and do enhance their appropriability. Hybridization is probably the most important of these, partly because it is a purely biological mechanism. Griliches (1957 and 1958) conducted the classic study of the nature of this form of excludability within the process of technological change in the case of corn in the US market in the first half of the twentieth century. It is worth noting, however, that the extent of appropriability that can be achieved using hybridization varies considerably depending on the biological characteristics of specific crop species, with general distinctions being drawn between cross-fertilizing (most amenable to hybridization), self-fertilizing, and vegetatively-propagated species (in descending order of the ease with which the crops can be protected by biological means).

Moschini and Lapan (1997) introduced a framework to investigate the distribution of benefits from new agricultural biotechnological innovations, in particular the extent to which the innovators of GM crops are appropriating returns on their investment. Note that this framework can be applied to any kind of agricultural innovation that is purchased by a farmer to enhance productivity. There is nothing about it that is inherently concerning GM crop varieties, but as these were delivered with greater restrictions due to IPRs, it became relevant to examine the resulting impacts in terms of efficiency and equity. Moschini and Lapan adjusted existing analyses of the welfare benefits from agricultural innovations (for example Alston et al., 1998) to explicitly account for the monopoly power arising from IPRs, in particular from patents over GM varieties, and thus allowing for the generation of monopoly rents. Their theoretical model, once calibrated, allows estimates of the distribution of net benefits arising from GM varieties among the innovating seed company, farmers and consumers. This framework was applied subsequently in numerous studies to show varying shares of benefits accruing to different actors (Sobolevsky et al., 2005 for herbicide tolerant soybeans; Falck-Zepeda et al., 2000 for Bt cotton; Pray et al., 2001 for Bt cotton in China). This line of research thus allows a rough estimation of the benefits appropriated by the innovator.

One question is whether some of the net benefits measured using the Moschini and Lapan (1997) framework are due to the earlier evolution of the IP system, especially the protection for GM constructs and varieties. But in most such analyses, no effort is made to disentangle the influence of other factors on the strength of incentives to innovate, such as other appropriation strategies. Such an interpretation would have to recognize that, even in static terms, certain other social costs have not been included in the calculation of net social welfare, in particular the transaction and administration costs of granting and enforcing IPRs, as well as costs arising from regulatory requirements concerning GM crops, which are often covered by innovators using rents appropriated under the IP regime (see Phillips, 2003). In any case, a static welfare analysis does not link the monopoly profits to the incentive to invest in R&D, nor does it examine variations in the extent of appropriability as related to IPRs.

In plant breeding, the potential to earn profits is closely tied to the possibilities for discouraging or restricting farmers from saving seed. This practice, also referred to as planting progeny seed, is still very prevalent in less formal or less commercial seed systems (see Tripp et al., 2006 for a discussion), although it is limited to self-fertilizing and vegetatively-propagated crop species due to the manner of biological reproduction. The development of hybridization for certain species starting in the 1920s and 1930s provided a technological means to encourage farmers to purchase new seed for each cropping season, since progeny seed saved from such varieties, performs relatively poorly and inconsistently. In developing countries, farmers have saved progeny seed from purchased hybrids to cross these with local varieties, which has provided a channel of technology diffusion (Tripp, 2001). Using legal means, patent protection also excludes this practice, as do most countries' PVP systems, with the UPOV 1991 Act establishing the most common and the strongest standards for national PVP laws while allowing a member country to make exceptions in its legislation. Enforcement of IPRs to exclude farmers from saving seed is neither perfect nor costless, and the innovator is generally required to expend resources to identify and pursue legal action against potential infringers. Alternative systems have been established in some European countries under which farmers' associations agree to an additional royalty payment to acquire permission to replant (Gray, 2012). For GM crop varieties, seed companies have typically sold seed to their client farmers only under contracts in which they agree not to save seed, and there have been cases of alleged breach of contract and patent infringement brought before the courts in both the US and Canada concerning GM varieties. In some Latin American countries including Brazil, Monsanto has developed arrangements to collect seed royalties from farmers at the point of sale of the harvested crop, in the case of herbicide-tolerant soybeans (Traxler, 2006).

Though relevant only for non-hybridized crops, this issue has been politically controversial since it entails a shift in legal rights, and is similar in some respects to changes introduced by copyright protection for software and digitalized entertainment media where encryption, 'click-through' user agreements, and other forms of technically or contractually-based copy protection have become important appropriation mechanisms. In the case of crop seed, biotechnology has opened a possibility comparable to software encryption by engineered seed sterility. Such engineered control over seed viability has been referred to as genetic use restriction technologies (GURTs) technology protection systems (TPS), and more figuratively by some civil society organizations opposed to the technology as 'terminator technology'. Planting of progeny seed could be restricted by varietal GURTs, which are distinct from trait-specific GURTs. The latter may be designed to convey agronomic benefits (see Visser et al., 2001). In any case, GURTs have not been commercialized and seed companies have publically announced that they do not intend to develop and commercialize such technologies.

It is important to note, however, that GURTs could serve a separate purpose: to reduce the possibility of unauthorized spread of genetically modified organisms beyond the fields in which they have been intended for planting. This would provide a mechanism to support biosafety regulations and to reduce potential liability for the seed company (Smyth and Phillips, 2002). At an international level, the controversy around GURTs reflects concerns of the potential implications in developing countries, where it can be much more difficult to establish and enforce fully-informed contracts with farmers. Furthermore, the practice of saving seed is much more prevalent and economically important in developing countries, where there has long been a flow of seeds between formal and informal (or less formal) seed systems.

There are few formal analyses of the various incentives and trade-offs that the evolution of the IP system has led to in terms of reducing farmers' rights to save and reuse seed in exchange for the prospect of being able to purchase and use improved seed varieties. An early paper by Heisey and Brennan (1991), predating the commercialization of modern biotechnology, developed an analytical model of demand by farmers for replacement seed. The context for their model was one of developing-country wheat production and efforts by international agricultural research centers (IARCs) such as CIMMYT and national partners to encourage more frequent seed purchasing – effectively a formalization of the seed supply system but still largely involving public sector actors – in order to boost agricultural sector productivity and food security. The core of their work is the trade-off farmers face between the higher cost of more frequently purchased seed and the deteriorating performance of progeny seed; this trade-off is arguably sharper for farmers in many developing countries due to greater difficulties in accessing credit.

This trade-off suggests the application of models for durable consumer goods. Burton et al. (2005) look systematically at alternatives for the innovator to increase its share of profits by enforcing the restrictions on reusing seed in a two-period model. Their model incorporates the basic vertical product differentiation set-up with farmers differing according to how much profit they can generate with the seed. They analyze three alternative options for the monopolist innovator, including two types of contracts and a GURT. This analysis incorporates a probabilistic success of enforcement of contracts by the monopolist and the associated monitoring and transaction costs. Their theoretical results suggest that GURTs would enhance profits for innovators, while farmers (as a whole) would benefit more from contracting. This paper is among the first in the agbiotech field to incorporate elements from the more extensive literature on piracy and copyright protection (for example Shy and Thisse, 1999). Burton et al. (2005) do not incorporate the effects of the differing degrees of appropriability on the incentive by the monopolist to invest.

Alston and Venner (2002) develop a model to examine the incentive that the degree of appropriability accruing to a monopolist, as represented by the royalty rate, has on the incentive to innovate. Less than complete appropriability of returns could be optimal, even for the monopolist, if it generates greater demand. This model is extended by van Tongeren and Eaton (2004) to a two-country setting to assess social welfare effects. Extending this analysis to two countries exploits the potential for price discrimination, which also yields efficiency benefits.

The degree of appropriability can also be interpreted in terms of the potential for infringement and enforcement. Giannakas (2002) develops a theoretical model that assesses the welfare benefits of adoption by farmers of an IP-protected variety in a developing country. The analysis illustrates the idea that permitting a certain amount of infringement on the IPR can have benefits in terms of greater diffusion. This arises naturally in a vertical product differentiation framework with incomplete adoption of a product or innovation. This is a common theme in the economics literature on IP; although monopolist profits may be maximized with very strong IPRs, and innovation may also be faster, social benefits may be greater with less than maximum possible scope or less than perfect enforcement of IPRs. Such a trade-off is similar to the more traditional case of balancing static and dynamic benefits in the length of patent protection (Nordhaus, 1969; Scotchmer, 2004).

3.2 Empirical Analyses of the Direct Incentive Effects of IPRs

The incentive effects of IPRs can be looked at in terms of a number of measures, which in general can be divided into two groups, R&D inputs and R&D outputs. In principle these can be described in terms of quantity and quality, but in practice the former receives the most attention due to relative ease of measurement. R&D inputs are typically assessed in terms of financial investments or the allocation of human R&D resources. R&D outputs are measured in terms of new innovations registered for protection with IP authorities, in particular the granting of patents or plant variety protection registrations. Measures of the quality of R&D outputs would incorporate productivity improvements in downstream production processes, but this possibility has been exploited in only a few studies. Estimates of these benefits have, however, been addressed in studies of returns to R&D in crop breeding and improvement, some of which are mentioned above, where the specific role of IPRs in stimulating the R&D was not the research focus.

Prior to the introduction of modern biotechnology into the plant breeding sector, attention in terms of IPRs focused on the effect of the introduction of PVP, on both inputs and outputs of R&D. PVP offers a discrete event, and historical trends can be examined using statistical and econometric methods to ascertain shifts in innovation that may be attributable to the change in the IP system. Most studies of the effects of PVP introduction have been undertaken in the US but a few have looked internationally.

A recent historical review of the effects of the US Plant Patent Act of 1930 has been undertaken by Moser and Rhode (2011). They concentrate on rose breeding, which accounts for more than half of the plant patents issued in the US between 1930 and 1970, when the Plant Variety Protection Act was enacted (though this did not replace the Plant Patent Act, as explained above). Moser and Rhode demonstrate how the number of rose varieties patented has accounted for only 16 per cent of total rose varieties registered during this period, with the proportion actually declining from about 1950 onwards. They conclude on the basis of this and other circumstantial evidence that the availability of plant patents has had at best a 'secondary effect' on the development of the US rose breeding sector.

Butler and Marion (1985) used data on certificates and a survey of breeders to examine changes in breeders' behavior as a result of the US Plant Variety Protection Act (PVPA) enacted in 1970. The study estimated R&D investments and found a rapid increase in the period leading up to the Act, possibly in anticipation of its passage. The study found evidence of increased investment in a few specific crops, which was also concluded in a follow-up study by Butler (1996), showing an increase in the number of soybean and wheat crop varieties released in the 1970s. These studies are generally perceived as indicating (though not confirming) that the PVPA had a positive effect on R&D incentives for a limited number of crops.

Perrin et al. (1983) surveyed seed companies for data on research expenditures in light of PVP introduction. They also found evidence of a significant increase in research expenditures for a number of crops, including not only hybrid crops. This study also looked at investments as a percentage of seed sales and in terms of dollars of research per dollar of crop value. The general conclusion was that PVP stimulated a marked increase in these investments.

Using econometric techniques, Perrin et al. (1983) also found evidence of a weakly-improving trend in soybean yields in test plots after the introduction of PVP in 1970, an early study assessing the potential effects of PVP on agricultural productivity. On the other hand, Alston and Venner (2002) found no statistically significant evidence that PVP had stimulated improvements in commercial or experimental wheat yields and Venner (1997) found no evidence for increased private sector investments in this crop (other studies are documented by Lesser, 1997). Similarly, Carew and Devadoss (2003) regressed canola yields, for 1995–2001 for 12 growing areas in Manitoba, on a range of variables including the proportion of land planted to varieties protected by PVP, as well as to varieties genetically modified for herbicide tolerance, and to hybrid varieties in general. Whereas those varieties protected by PVP tended to be open-pollinated, the genetically modified varieties were protected mostly by technology use agreements. Carew and Devadoss controlled for other factors, including fertilizer use as well as spatial and temporal diversity of varieties planted in each growing area, but found only weak evidence of an effect of PVP on canola yields. Naseem et al. (2005) conducted a similar study on cotton yields for 10 US states over the period 1950–96. With a longer set of panel data, they were able also to include a trend shift variable for when PVP is expected to start influencing farmers' yields (including a delay for the long-term nature of breeding), and interacted that variable with the proportion of area planted to protected varieties. The results suggest that the combined effect of PVP was to increase yields by about 8–10 per cent. Further work of this sort would help to clarify the seemingly mixed effects of PVP on breeding and productivity. As yield is only one measure of productivity, and not necessarily the most important one to farmers, continued work in this area should incorporate other productivity measures and preferably indicators of economic performance, which also account for the costs of inputs, including seeds. This would allow a more thorough economic assessment.

Taking an international perspective, Diez (2002) assessed the trends in the number of varieties registered in Spain over the period 1974–91, covering the period in which PVP was introduced in the country, first for wheat, barley, rice, potatoes and oats in 1975, with protection extended later to sunflower, alfalfa and maize. Following PVP, the number of varieties registered and protected by private breeders increased, as did the registration of varieties by foreign breeding companies. However, the analysis does not control for other market-related factors such as overall trends in Spanish economic policy which may have played a role.

Jaffe and van Wijk (1995) examined the impact of PVP on R&D in Argentina using a survey of plant-breeding firms. While this study found that investments had increased between 1986 and 1992, survey answers indicated that changing economic policies and market liberalization had provided more of an incentive than the introduction of PVP. Leger (2005) undertook an interview-based survey of effectively all companies and public sector organizations involved in maize breeding in Mexico where PVP was introduced in 1997. There appeared to be little increase in breeding programs that could be attributed to the introduction of PVP, which was primarily being used by public breeding organizations. These results may largely reflect the fact that maize varieties can be protected reasonably well from reproduction through hybridization.

In the case of patents for agbiotech, it is fairly clear that the landmark cases in the US represent the key change in IP policy of relevance that caused patenting applications and grants subsequently to increase. There were similar changes in the interpretation of Canadian patent law which allowed the patenting of genes (described in Galushko et al., 2010), followed by similar possibilities in European countries. This effectively extended the scope of patents to protect uses of the plants and even plant varieties in which such genes had been inserted, as illustrated by the decision in the Monsanto vs. Schmeiser case mentioned above. The sharp rise in the number of patents granted and GM crop varieties subsequently developed and marketed – particularly by the private sector – is generally interpreted as evidence that the changes to the allowable subject matter for patents and associated claims induced such investments. Although this has not yet been analyzed systematically, some anecdotal evidence suggests that additional factors also play a role. In particular, the dominant innovation pathway of genetic transformation of crops has been confined largely to crops and varieties for animal feed and industrial purposes. There has been much lower R&D output in crops destined for direct human consumption as food, at least partly due to the social controversy and the cost of compliance with the attendant regulatory frameworks that have been put in place. The latter include costs of complying with requirements for segregation and traceability, which may well have also influenced the directions taken by research and innovation (see Chapter 25 in this volume).

4 COMPETITION AND MARKET STRUCTURE

The standard economic model views patent protection as a temporary monopoly right given in exchange for the investment of resources in R&D and the disclosure of the resulting invention.[2] What is often overlooked, however, is that monopoly power is limited by the extent to which there are competing products or processes. The economic literature differentiates between drastic and non-drastic (or radical and incremental) innovations as a means to distinguish between those innovations for which few substitutes are available and those innovations whose improvements confer only a relative, as opposed to absolute, advantage for the innovator.

Duopoly models are often employed to examine the effects of changes in IP policy and regime on the incentives facing competing innovators, as well as the likely outcomes in terms of rates of innovation, pricing strategies or market share. In the area of agbiotech, one attempt to look at specific questions of this nature is Eaton and van Tongeren (2006), which proposes a simple static model of vertical product differentiation with interdependence in research costs between two firms. The extent to which the breeders' exemption is limited by patent policy (or a further refinement of PVP policy) affects the extent to which each firm benefits from spillovers from the others' research.

Moschini and Yerokhin (2008) provide a more complete theoretical analysis of the breeders' exemption in a dynamic model employing stochastic game theory, which has become a standard workhorse in the industrial organization literature (Doraszelski and Pakes, 2007). Their model illustrates the intuitive idea of how a research exemption in patent policy, and its parallel, the breeders' exemption in PVP, may reduce the incentive to invest in large, costly breakthrough inventions, since competitors will follow on with smaller improvements that take market share away from the initial inventor. This is the rationale behind the proposition that genetically modified crops would not have provided an interesting enough opportunity for private companies without the more exclusive protection offered by patents. Some actors in the seed sector have indeed argued that the decreasing gains to traditional breeding strategies justify this increase in scope of IP protection, especially when viewed in light of how genetic sequencing has reduced the cost of decoding the make-up of rival crop varieties and thereby the cost of imitating them (Janis and Smith, 2006). As Moschini and Yerokhin (2008) point out, their findings can also be interpreted as an argument for the provision of fundamental, upstream research through public funding (see Malla and Gray, 2003). An additional point to make about this analysis is that it ignores possible differences in the innovation pathway pursued, an area that remains relatively unexplored in the theoretical analysis and which could lead to a qualification of results on the research exemption.

It is worthwhile to pause to consider the value of these models, as it is often not well understood that this area of theoretical work is still young and very much piecemeal. Each specific contribution builds upon the overall body of knowledge and thus is best interpreted in that context. Stylized models necessarily ignore numerous aspects of a question and the results are thus often predicated on assumptions that could usually not be defended in real-world situations. The body of theory emerges as attempts are made to capture different salient aspects into formalized and generalized models. The models developed and described above are inspired by the sector of plant breeding and agbiotech, but their stylized nature means that they are not necessarily specific to the sector. They build on a much broader body of literature which is not reviewed here.

4.1 Empirical Analysis of the Influence of IPRs on Competition and Market Structure

The availability of patent protection for genetic constructs, the tools for creating constructs as well as complementary products such as herbicides (glyphosate), has stimulated increased concentration in the seed and agro-chemical sector. Hayenga (1998) documented the trends in the US at the end of the 1990s which involved a large round of mergers and acquisitions in the sector and precipitated anti-trust cases by the US Department of Justice. The late 1990s also heralded a wave of considerable litigation amongst agbiotech companies, centered on patent infringement, including interpretation of the terms of licensing contracts. The engagement of the Department of Justice indicates that the potential for this technological and IPR-driven process of vertical and horizontal integration to reduce competition in the seed and agro-chemical sector was being taken seriously. International trends in the concentration of PVP rights were analyzed by Srinivasan (2003), providing additional, indirect evidence of these trends.

There have been very few studies that have examined the specific reasons for the restructuring and increased concentration of the US plant breeding and seed industry in the 1990s. Graff et al. (2003b) propose that the restructuring was at least in part due to the potential profits from exploiting complementary intellectual assets, including tools of genetic transformation, genetic constructs and elite germplasm, which were generally assets belonging to different actors holding IP protection. This explanation is based on the reduction in transaction costs and uncertainty in coordinating asset complementarities that acquisition allows (Teece, 1986; Milgrom and Roberts, 1990). Their econometric model is based on the patent production function pioneered by Griliches (1979) and Hausman et al. (1984) in which the number of patents per company is a function of its R&D spending as well as other characteristics. Using the dataset of US patents mentioned above, Graff et al. (2003a) measure complementarities between different types of patents as the positive covariance in the unexplained variation in patent counts, and found that this had increased under the concentrated industry structure observed in 1999 as compared to the less concentrated structure of 1994 before the wave of mergers and acquisitions. The results provide support for the asset-complementarities explanation of industry trends, relative to other proposed explanations that emphasize strategic motivations to erect barriers to entry or engage in strategic patenting (that is accumulating patents as bargaining chips). To date, an update of this study has not been published, nor has a similar exercise been undertaken using EPO data, both of which would be worthwhile at least to inform understanding of evolving trends in the industry and the role of IP.

Marco and Rausser (2008) likewise looked at mergers and acquisitions activity in plant biotechnology in the 1980s and 1990s. They developed an econometric model to explain consolidation in the sector, with US patents being held by firms as one of the explanatory variables. They also included a measure of the enforceability of patents based on the predicted probability that a court would rule a patent valid and infringed. The average enforceability of a firm's patent portfolio was found to be positively correlated with the probability that the firm would be involved in consolidation.

Using tools of industrial organization, a few papers have assessed whether recent trends towards restructuring have led to increased prices for seed. Shi et al. (2010) estimated a mark-up price equation for corn seed in the US market for the years 2000–07 and found evidence that trait bundling was characterized by sub-additive pricing of seed, reflecting either economies of scope in bundling of GM traits or complementarity in demand for those traits. Shi et al. (2011) found similar results for the US cottonseed market with data from 2002 to 2007. Importantly, while they also found evidence of increased seed prices, as theory would predict, they suggested that this is not excessively inhibiting diffusion. Reflecting on this and other evidence, Stiegert et al. (2010) point out that whether the evolving market structure, with increasing concentration and vertical integration, will maintain sufficient competitive pressure in the sector remains a relevant topic for research and policy attention.

5 PUBLIC RESEARCH, TECHNOLOGY TRANSFER AND FREEDOM TO OPERATE

Considerable attention has focused on the effect of IP protection on the reconfiguration of roles between public and private sector actors in agricultural research. One concern is that the increased possibilities for seeking IP protection will alter incentives and shift the relative focus of public research institutions, diverting publicly funded research activities towards more applied topics with greater commercial – although not necessarily social – benefits. In the US, researchers have attempted to assess the impact of the Bayh–Dole Act of 1980 which harmonized IP rules under the different US federal agencies that fund research and encouraged the institutions that receive federal research funds to apply for patent protection. One clearly-stated purpose of this legislation is to stimulate the uptake and commercialization of such publicly-funded research.

For the US, Barham et al. (2002) studied trends in patenting by universities over the 1975–2000 period and noted, starting in 1995, a marked increase both in the number of patents and their citation rates, suggesting increased quality or value of innovations. There is also evidence of path dependence and increased concentration of innovation among fewer universities, although this is mitigated somewhat once the value of patents, in terms of citations, is taken into account.

A wider view of patenting trends was examined by Graff et al. (2003b), using patent data for technologies relevant to the agbiotech sector from the US, Europe, Japan and WIPO. This analysis found that public sector research institutions accounted for 25 per cent of the total stock of patents in the agbiotech field, while the private sector accounted for the remaining 75 per cent. The study also showed that while a handful of major companies enjoyed relatively consolidated patent portfolios, patent holdings in the public sector were, as would be expected, highly fragmented across separate institutions. Under the situation indicated in Graff et al. (2003a), in which a diversified portfolio of different complementary IP appeared to be advantageous or even necessary for access to market, it appeared that any technologies generated by public sector researchers might be dependent upon technologies tied up in the major companies' IP portfolios, making it potentially difficult for public sector institutions to serve the crop genetic needs of their traditional stakeholders using GM technology.

Relatively little attention has been given to the role that IP plays in shifting the research priorities of public sector plant biotechnology researchers and plant breeders. At issue is whether public sector researchers would concentrate more efforts on innovation pathways that are more likely to deliver higher revenue generation opportunities, in effect directly catering to or even competing with the private sector. For example, in an interview survey of the effects of both PVP and patent protection in five developing countries by Tripp et al. (2007), public agricultural research organizations in both Kenya and India reported that PVP protection implied an expectation for these organizations to earn revenues. In their review of the GM canola sector in Canada, Galushko et al. (2010) report findings from a survey of breeders that reveals a lack of strategic coordination of the relative roles of the public and private sectors. In theory, and as argued by Tripp et al. (2007), the development of IP protection possibilities, as part of a larger process of increased commercialization of R&D in the breeding sector, should allow the public sector to relinquish those traditional research pathways that have commercial potential, leaving those efforts to the private sector. The public sector could then concentrate on market and crop segments of less interest to the private sector. In particular, this could include 'orphaned' crops of importance to the low-income smallholder farmers, which is particularly relevant from an international perspective. An alternative role for public sector research is to concentrate on more basic, upstream research with broader potential to generate economic spillovers and stimulate downstream innovation pathways. Or the public sector could simply reduce its agricultural research efforts, freeing up resources to devote to other more pressing public priorities (Malla and Gray, 2003).

A second concern affecting public research organizations is the potential for increased use of IP protection to restrict or shape the research plans of plant scientists and breeders. Lei et al. (2009) surveyed scientists at several US universities regarding what effects the patenting of research tools in the field of plant biology may be having over their research programs. The scientists surveyed reported that IP protection of research tools does have a negative influence on their research. The precise mechanism is, however, typically indirect. Administrators, who perceive potential for revenue from patented research tools, tend to require the use of material transfer agreements (MTAs), a type of inter-institutional contract that stipulates the legal terms under which a proprietary tool can be used in research by the recipient. The negotiation and execution of MTAs impose costly delays on the exchange of research tools, with fully 42 per cent of respondents reporting delays of research projects due to delays in obtaining a requested research tool, and 27 per cent reporting that they ended up using alternative, sometimes less effective, tools than the one they had preferred. Still, most scientists disregard patents altogether, with 91 per cent of scientists reporting that they had never checked whether a research tool is patented.

Galushko et al. (2010) examine the case of GM canola in Canada. Using a survey of breeders in both the public and private sectors, the study seeks to assess the impacts of increased use of IP protection on innovation investments and also the sharing of information, research tools and germplasm among breeders. Interestingly, the most common reason cited for increased protection of research tools and germplasm was as a response to increased patenting by others and thus the desire to ensure freedom to operate. Such a result is not that surprising given the large share of the sample accounted for by public sector breeders, who are likely to have less motivation to seek IP protection for revenue generation purposes. They conclude that ensuring freedom to operate in the canola sector is becoming costly and time-consuming and suggest that the effect of this on the rate of innovation deserves further attention.

The role of the public and private sectors, and the freedom of the former to operate, has additional implications for international agricultural research intended to contribute to agricultural growth in developing countries. Binenbaum et al. (2003) undertook a review for staple crops and concluded that there were few restrictions on developing-country researchers in place at that time. Protection in developing-country jurisdictions had been applied for on only a very limited basis. Furthermore, only a small proportion of developing-country agricultural production is exported to developed country markets where there is greater potential for infringement to incite enforcement and litigation measures. This situation may have changed in the interim as many developing countries have subsequently agreed to IPR provisions in free trade agreements negotiated with either the US or the EU; however, actual patent filing trends may take time to respond to such changes. Regardless, despite such findings, the potential uncertainty regarding freedom to operate may cause reluctance among national agricultural research systems or international agricultural research centers (IARCs) to invest scarce resources in lines of research that could be impacted.

6 INTERNATIONAL TRANSFER

Another premise for the strengthening of IPRs internationally is its potential for stimulating trade in final goods as well as foreign direct investment to transfer higher technologies for economic production. While there are a number of studies that look for correlations between a country's strengthening of IPRs and flows of trade or foreign direct investment, including for the pharmaceutical biotechnology sector (Smith, 2002), there is very little analysis of the agbiotech sector, although Srinivasan (2004) did examine international flow of varieties using PVP data from UPOV.

Eaton (2013) analyzes international trade in seeds over the period 1989–2007 and finds no significant relationship between a country's membership of UPOV and its imports of seeds and planting material, taking into account differences in the enforceability of PVP and IPRs in general. This analysis, which used a gravity model approach common to such analyses (see Smith, 2002), also looked at different crop groups. The study did not focus specifically on GM crop varieties, but does cover the entire seed sector and thus includes both GM crops as well as the use of biotechnology for other purposes in the development of new varieties, such as molecular breeding. There is, however, no systematic study of the relationship between IPRs and FDI in the agricultural sector, which would be of equal, if not greater interest; this lacuna is likely due to data limitations.

With respect to understanding comparisons in IPR use across countries, Maskus (2006) proposes the use of revealed technology advantage, which corresponds to the revealed comparative advantage used in empirical trade analysis but with patents substituting for exports. Revealed technology advantage is thus measured as the share of a country's patents in a given technology field out of all global patents in that field, normalized by that country's share in all patents.[3] Concentrating on the Asia-Pacific region, Maskus (2006) illustrates that countries such as South Korea and China have a lower revealed technology advantage compared to countries such as the US and Japan, as would be expected. This may provide a rationale for maintaining more flexible IPR policies, though well-defined IPRs are still likely to be necessary for countries to encourage technology transfer. Maskus looks exclusively at patents, but his innovative measure could be of use for studying other IPRs.

There is considerable unexploited value in descriptive analysis of IPR data, particularly at an international level, to simply understand trends and inform more systematic analysis. Michiels and Koo (2008) examine patent data from the EPO to assess the development of agrobacterium-mediated transformation technology and its diffusion. Examining data for three years (1996, 2000 and 2004), this study found evidence of a shift in research focus from fundamental to applied research, as well as a rise in applied crop research activity by emerging market countries, in particular China and India.

7 CRITICAL ASSESSMENT

What then is dynamic about the IP system? The availability of formal IPRs for inventions in the realm of plant breeding and biotechnology has changed, generally in discrete intervals, as new forms of IPRs – plant patents and PVP – have been created and as utility patents have evolved through either regulation (as in the EU) or jurisprudence (as in the US). This demonstrates how the IP system shifts to accommodate changes in technology and economic development, while the underlying framework of excludability, as typified by patents, remains relatively stable, with PVP and plant patents representing, in effect, adapted forms of patent protection.

This chapter has attempted to illustrate some of the principal trends in the evolution of IPRs in agbiotech and the important research and policy questions it raises. Much of the research and analysis in this area has concentrated on how policy can improve the system of IPRs in order to maximize the rate of innovation. There are some indications, as suggested in the recent studies by Graff et al. (2009) and Lei et al. (2009) that the risk of an anticommons is a real concern in agbiotech. Agricultural research and technology development has generally been dominated by the efforts of public organizations. Growth in public funding has slowed in recent years and private sector actors have emerged in certain market segments. The effects of the increasing use of various proprietary rights in agricultural research on this complex web of actors, which accelerated in the 1980s and 1990s, are perhaps only now beginning to be seen. Given the challenges of meeting global food security challenges while reducing the environmental impacts of agriculture in the coming decades, these developments should be cause for concern.

Despite some exceptions, a relatively neglected area of research has concerned how the availability and use of IPRs affects the nature and direction that R&D programs take. For instance, the innovation pathway of genetic modification of crops for insect resistance and herbicide tolerance has delivered demonstrable economic, environmental and social benefits, but other innovation pathways might have dominated under different policies. This may benefit from an additional set of methods than has been reviewed here, drawing more on the field of innovations systems analysis, a field which has given attention to historical trajectories and path dependency in other technology areas (see Chapter 21 in this volume). Any research in this area will require improvements in metrics by which the direction of agbiotech research can be measured, as well as the necessary data collection and analysis. In this regard, agbiotech also needs to be seen within the broader context of innovation in agricultural systems, which comprises many more elements than the development of new crop varieties.

Also with respect to methods, there have been relatively few examples of multidisciplinary perspectives as called for by Kesan (2000). Legal research has tended to assess interpretations of law in light of technological developments, including possible adjustments in either legislation or jurisprudence. Economic research, which has been the primary focus of this chapter, has concentrated on understanding the effects of IP policy on outcomes in terms of innovation, market structure and the distribution of benefits and costs. There has been little interaction between these areas of research.

Empirical research on the IP system and biotechnology in agriculture has concentrated on widely-grown grain and oilseed crops. These are indeed the crops that have seen the most cultivation of genetically modified varieties. There appears to be a different array of incentives and dynamics in the high value-added segment of horticultural crops, where biotechnology has become more prominent in the process of breeding, through sequencing and marker-assisted selection techniques. There is less prevalence of genetic modification, particularly as many of these crops are destined for either direct consumption ingestion by humans, but IPRs are still relevant to the breeding process.

Understanding the effects of the dynamic IP system could also benefit from more attention to research on licensing, exchange, enforcement and litigation of these patents. It is arguable that these aspects of the IP system have as much of an effect on the innovation process as legislation and regulations concerning the awarding of patents and their scope. Clearly this research presents both data and methodological challenges that would need to be addressed. Such data may need to be collected through survey methods.

NOTES

1. To a large extent, this emerging area of research was summarized in the monograph edited by Santaniello et al. (2000), which included a number of papers presented at the second conference of the International Consortium on Agricultural Biotechnology Research (ICABR) in 1999. Over more than a decade, this forum (currently titled the International Consortium on Applied Bioeconomy Research) has provided a focal point for presentation and exchange of research results on the social, economic and management issues posed by agricultural biotechnology, including IP. In some respects, this chapter attempts to provide an update on the current state of research in this evolving field.
2. The latter purpose of the patent system, encouraging disclosure of innovations, is often forgotten or ignored in economic analyses of agricultural biotechnology. A more active body of literature among legal scholars has given this social benefit more attention. Landes and Posner (2003), for example, suggest that the benefits from disclosure and its resulting spillovers are greater than the direct incentive effects. It seems plausible that the relative importance of the disclosure function of the patent system, in comparison to the direct incentive effects, is likely to be greater in technological fields where the threat of imitation is not as great. This may be the case with process innovations in agricultural biotechnology, such as the tools of genetic sequencing and engineering, but less so with the product innovations, particularly new crop varieties, due to their self-reproducing nature.
3. Revealed comparative advantage is measured as a share of a country's exports of a particular class of goods out of global exports of those goods, divided by the country's total exports as a share of global exports. Values greater than 1 indicate that the country accounts for a greater share of global exports of that good than the country does for global exports of all goods.

REFERENCES

Agrawala, Shardul, Cecile Bordier, Victoria Schreitter and Valerie Karplus (2012), 'Adaptation and innovation: an analysis of crop biotechnology patent data', environment working paper, Paris: Organisation for Economic Co-operation and Development (OECD).

Alston, J.M, and R.J. Venner (2002), 'The effects of the US Plant Variety Protection Act on wheat genetic improvement', *Research Policy*, **31**(4), 527–42.

Alston, Julian M., George Norton and Philip G. Pardey (1998), *Science Under Scarcity: Principles and Practice for Agricultural Research Evaluation and Priority Setting*, Wallingford: CAB International.

Barham, Bradford, Jeremy Foltz and Kwansoo Kim (2002), 'Trends in university ag-biotech patent production', *Review of Agricultural Economics*, **24**(2) (October), 294–308.

Binenbaum, Eran, Carol Nottenburg, Philip G. Pardey, Brian D. Wright, and Patricia Zambrano (2003), 'South–North trade, intellectual property jurisdictions, and freedom to operate in agricultural research on staple crops', *Economic Development and Cultural Change*, **51**(2) (January), 309–36.

Boldrin, Michele and David K. Levine (2008), *Against Intellectual Monopoly*, Cambridge: Cambridge University Press.

Boldrin, Michele and David K. Levine (2009), 'Does intellectual monopoly help innovation?', *Review of Law & Economics*, **5**(3), 991–1024.

Brousseau, Eric, Pierre Garrouste and Emmanuel Raynaud (2011), 'Institutional changes: alternative theories and consequences for institutional design', *Journal of Economic Behavior & Organization*, **79**(1–2) (June), 3–19.

Buanec, Bernard Le (2005), 'Plant genetic resources and freedom to operate', *Euphytica*, **146**(1–2) (November), 1–8.

Buccola, S. and Y. Xia (2004), 'The rate of progress in agricultural biotechnology', *Review of Agricultural Economics*, **26**(1) (February), 3–18.

Burton, Diana M., H. Alan Love, Gokhan Ozertan, and Curtis R. Taylor (2005), 'Property rights protection of biotechnology innovations', *Journal of Economics & Management Strategy*, **14**(4), 779–812.

Butler, L.J. (1996), 'Plant breeders' rights in the US: update of a 1983 study', in Jeroen van Wijk and Walter Jaffe (eds), *Proceedings of a Seminar on the Impact of Plant Breeders' Rights in Developing Countries*, Santa Fe Bogota, Colombia, 7–8 March, Amsterdam: University of Amsterdam, pp. 17–33.

Butler, L.J, and B.W. Marion (1985), 'Impacts of patent protection in the US seed industry and public plant breeding', North Central Regional Research Bulletin no. 304, Madison, WI: University of Wisconsin.

Carew, Richard and Stephen Devadoss (2003), 'Quantifying the contribution of plant breeders' rights and transgenic varieties to canola yields: evidence from Manitoba', *Canadian Journal of Agricultural Economics*, **51**(3), 371–95.

Cohen, Wesley M., Richard R. Nelson, and John P. Walsh (2000), 'Protecting their intellectual assets: appropriability conditions and why US manufacturing firms patent (or not)', working paper no. 7552, NBER.

Diez, M.C.F. (2002), 'The impact of plant varieties rights on research: the case of Spain', *Food Policy*, **27**, 171–83.

Doraszelski, Ulrich and Ariel Pakes (2007), 'A framework for applied dynamic analysis in IO', in Mark Armstrong and Robert H. Porter (eds) *Handbook of Industrial Organization*, Volume 3: 1887–1966, Amsterdam: Elsevier.

Eaton, Derek (2013), 'Trade and intellectual property rights in the agricultural seed sector', Centre for International Environmental Studies Research Paper no. 20, Graduate Institute, Geneva.

Eaton, Derek J.F. and Frank W. van Tongeren (2006), 'Patents versus plant varietal protection', in R. Evenson and V. Santaniello (eds), *International Dimensions of Agricultural Biotechnology*, Wallingford: CABI.

European Commission (1998), 'Directive 98/44/EC of the European Parliament and of the Council of 6 July 1998 on the legal protection of biotechnological inventions', Official Journal of the European Union no. 213, 30 July.

European Commission (2005), 'Report from the Commission to the Council and the European Parliament Development and implications of patent law in the field of biotechnology and genetic engineering', SEC (2005) 943, 14 July, COM (2005) 312 Final.

European Patent Office (2007), 'Scenarios for the Future', Munich: EPO, available at: http://www.epo.org/news-issues/issues/scenarios.html.

Falck-Zepeda, José B., Greg Traxler and Robert G. Nelson (2000), 'Surplus distribution from the introduction of a biotechnology innovation', *American Journal of Agricultural Economics*, **82**(May), 360–69.

Galushko, Viktoriya, Richard Gray and Stuart Smyth (2010), 'Resolving FTO barriers in GM canola', *AgBioForum*, **13**(4), 1–10.

Giannakas, Konstantinos (2002), 'Infringement of intellectual property rights: causes and consequences', *American Journal of Agricultural Economics*, **84**(2) (May), 482–94.

Goeschl, Timo, (2005), 'Do patent-style intellectual property rights on transgenic crops harm the environment?', in Justus Wesseler (ed.) *Environmental Costs and Benefits of Transgenic Crops in Europe: Implications for Research, Production, and Consumption*, Dordrecht: Kluwer Academic Publishers.

Graff, Gregory D. and David Zilberman (2007), 'The political economy of intellectual property: re-examining European policy on plant biotechnology', in Jay Kesan (ed.), *Agricultural Biotechnology and Intellectual Property*, Wallingford: CAB International.

Graff, Gregory D., Gordon C. Rausser and Arthur A. Small (2003a), 'Agricultural biotechnology's complementary intellectual assets', *Review of Economics and Statistics*, **85**(2), 349–63.

Graff, Gregory D., David Zilberman and Alan B. Bennett (2009), 'The contraction of agbiotech product quality innovation', *Nature Biotechnology*, **27**(8), 702–4.

Graff, Gregory D., Susan E. Cullen, Kent J. Bradford, David Zilberman and Alan B. Bennett (2003b), 'The public–private structure of intellectual property ownership in agricultural biotechnology', *Nature Biotechnology*, **21**(9), 989–95.

Gray, Richard (2012), 'Intellectual property rights and the role of public and levy-funded research: some lessons from international experience', in *Improving Agricultural Knowledge and Innovation Systems*, OECD conference proceedings, Paris: Organisation for Economic Co-operation and Development (OECD).

Griliches, Zvi (1957), 'Hybrid corn: an exploration in the economics of technological change', *Econometrica*, **25**(4) (October), 501–22.

Griliches, Zvi (1958), 'Research costs and social returns: hybrid corn and related innovations', *Journal of Political Economy*, **66**(5) (October), 419–31.

Griliches, Zvi (1979), 'Issues in assessing the contribution of research and development to productivity growth', *Bell Journal of Economics*, **10**(1), 92–116.

Hausman, Jerry, Bronwyn H. Hall and Zvi Griliches (1984), 'Econometric models for count data with an application to the patents–R&D relationship', *Econometrica*, **52**(4) (July), 909–38.

Hayenga, Marvin L. (1998), 'Structural change in the biotech seed and chemical industrial complex', *AgBioForum*, **1**(2), 43–55.

Heisey, Paul W, and John P Brennan, (1991), 'An analytical model of farmers' demand for replacement seed', *American Journal of Agricultural Economics*, **73**(4), 1044–52.

Heisey, Paul W., John L. King and Kelly Day Rubenstein (2005), 'Patterns of public-sector and private-sector patenting in agricultural biotechnology', *AgBioForum*, **8**(2&3), Article 3.

Heller, Michael A. and Rebecca S. Eisenberg (1998), 'Can patents deter innovation? The anticommons in biomedical research', *Science*, **280**(5364), 698–701.

Jaffe, Adam B. and Josh Lerner (2004), *Innovation and Its Discontents: How Our Broken Patent System Is Endangering Innovation and Progress, and What to Do About It*, Princeton, NJ: Princeton University Press.

Jaffe, W. and J. van Wijk (1995), *The Impact of Plant Breeders' Rights in Developing Countries*, Amsterdam: Inter-Am Institute for Co-operation on Agriculture, University of Amsterdam.

Janis, Mark D. and Stephen Smith (2006), 'Obsolescence in intellectual property regimes', Legal Studies Research Paper no. 05-48, Ames, IA: University of Iowa, available at: http://ssrn.com/abstract=897728.

Kesan, Jay P. (2000), 'Intellectual property protection and agricultural biotechnology: a multidisciplinary perspective', *American Behavioral Scientist*, **44**(3) (November), 464–503.

Landes, William M. and Richard A. Posner (2003), *The Economic Structure of Intellectual Property Law*, Cambridge, MA: Belknap Press.

Leger, Andreanne (2005), 'Intellectual property rights in Mexico: do they play a role', *World Development*, **33**(11), 1865–79.

Lei, Zhen, Rakhi Juneja and Brian D. Wright (2009), 'Patents versus patenting: implications of intellectual property protection for biological research', *Nature Biotechnology*, **27**(1), 36–40.

Lerner, Josh (2002), '150 years of patent protection', *American Economic Review*, **92**(2) (May), 221–5.

Lesser, W. (1997), 'Assessing the implications of intellectual property rights on plant and animal agriculture', *American Journal of Agricultural Economics*, **79** (November), 1584–91.

Levin, Richard C., Alvin K. Klevorick, Richard R. Nelson and Sidney G. Winter (1987), 'Appropriating the returns from industrial research and development', *Brookings Papers on Economic Activity* **3**, 783–820.

Louwaars, Niels, P. (2007), 'Seeds of confusion: the impact of policies on seed systems', PhD thesis, University of Wageningen, the Netherlands.

Louwaars, Niels, Hans Dons, Geertrui Van Overwalle, Hans Raven, Anthony Arundel, Derek Eaton and Annemiek Nelis (2009), 'Breeding business: the future of plant breeding in the light of developments in patent rights and plant breeders' Rights', CGN Report no. 2009–14, Wageningen: Centre for Genetic Resources.

Malla, Stavroula and Richard Gray (2003), 'Public research policy for today's agricultural biotech research industry', *Canadian Journal of Agricultural Economics*, **51**(3), 347–69.

Marco, Alan C. and Gordon C. Rausser (2008), 'The role of patent rights in mergers: consolidation in plant biotechnology', *American Journal of Agricultural Economics*, **90**(1) (February), 133–51.
Maskus, Keith E. (2006), 'Intellectual property rights in agriculture and the interests of Asian-Pacific economies', *World Economy*, **29**(6) (July), 715–42.
Michiels, An and Bonwoo Koo (2008), 'Publish or Patent? Knowledge Dissemination in Agricultural Biotechnology', Discussion Paper no. 795, Washington, DC: International Food Policy Research Institute (IFPRI).
Milgrom, Paul and John Roberts (1990), 'The economics of modern manufacturing: technology, strategy, and organization', *American Economic Review*, **80**(3) (June), 511–28.
Moschini, G. and H. Lapan (1997), 'Intellectual property rights and the welfare impacts of agricultural R&D', *American Journal of Agricultural Economics*, **79**, 1229–42.
Moschini, GianCarlo and Oleg Yerokhin (2008), 'Patents, research exemption, and the incentive for sequential innovation', *Journal of Economics & Management Strategy*, **17**(2), 379–412.
Moser, Petra and Paul W. Rhode (2011), 'Did plant patents create the American rose?', National Bureau of Economic Research Working Paper no. 16983. available at: http://www.nber.org/papers/w16983.
Naseem, Anwar, James Oehmke and David Schimmelpfennig (2005), 'Does plant variety intellectual property protection improve farm productivity? Evidence from cotton varieties', *AgBioForum*, **8**(2&3), article 6.
Nordhaus, William D. (1969), *Inventions, Growth and Welfare: A Theoretical Treatment of Technological Change*, Cambridge, MA: MIT Press.
Overwalle, Geertrui van (2008), 'Biotechnology and patents: global standards, European approaches and national accents', in Daniel Wuger and Thomas Cottier (eds), *Genetic Engineering and the World Trade System*, Cambridge: Cambridge University Press, pp. 77–108.
Pardey, Philip G., Bonwoo Koo, Jennifer Drew and Carol Nottenburg (2013), 'The evolving landscape plant varietal rights in the United States, 1930–2008', *Nature Biotechnology*, **31**(1), 25–9.
Perrin, R.K, K.A. Hunnings and L.A. Ihnen (1983), 'Some effects of the US plant Variety Protection Act of 1970', Economics Research Report no. 46, Raleigh, NC: North Carolina State University Department of Economics and Business.
Phillips, Peter C.B. (2003), 'The economic impact of herbicide tolerant canola in Canada', in N. Kalaitzandonakes (ed.), *The Economic and Environmental Impacts of Agbiotech: A Global Perspective*, New York: Kluwer Academic Publishers, pp. 119–40.
Pray, Carl, Danmeng Ma, Jikun Huang and Fangbin Qiao (2001), 'Impact of Bt Cotton in China', *World Development*, **29**(5), 813–25.
Santaniello, V., R.E. Evenson, D. Zilberman and G.A. Carlson (eds) (2000), *Agriculture and Intellectual Property Rights: Economic, Institutional and Implementation Issues in Biotechnology*, Wallingford: CABI.
Scotchmer, Suzanne (2004), *Innovation and Incentives*, Cambridge, MA: MIT Press.
Shi, Guanming, Jean-Paul Chavas and Kyle Stiegert (2010), 'An analysis of the pricing of traits in the US corn seed market', *American Journal of Agricultural Economics*, **92**(5), 1324–38.
Shi, Guanming, Kyle W. Stiegert and Jean P. Chavas (2011), 'An analysis of bundle pricing in horizontal and vertical markets: the case of the US cottonseed market', *Agricultural Economics*, **42**, 77–88.
Shy, Oz and Jacques-Francois Thisse (1999), 'A strategic approach to software protection', *Journal of Economics and Management Strategy*, **8**(2), 163–90.
Smith, Pamela J. (2002), 'Patent rights and trade: analysis of biological products, medicinals and botanicals, and pharmaceuticals', *American Journal of Agricultural Economics*, **84**(2) (May), 495–512.
Smyth, Stuart and Peter W.B. Phillips (2002), 'Product differentiation alternatives: identity preservation, segregation, and traceability', *AgBioForum*, **5**(2), article 1.
Sobolevsky, Andrei, GianCarlo Moschini and Harvey Lapan (2005), 'Genetically modified crops and product differentiation: trade and welfare effects in the soybean complex', *American Journal of Agricultural Economics*, **87**(3) (August), 621–44.
Srinivasan, C.S. (2003), 'Concentration in ownership of plant variety rights: some implications for developing countries', *Food Policy*, **28**(5–6), 519–46.
Srinivasan, C.S. (2004), 'Plant variety protection, innovation, and transferability: some empirical evidence', *Review of Agricultural Economics*, **26**(4), 445–71.
Stiegert, Kyle W., Guanming Shi and Jean-Paul Chavas (2010), 'Innovation, integration, and the biotechnology revolution in US Seed markets', *Choices*, available at: http://www.choicesmagazine.org/magazine/article.php?article=119.
Teece, David J. (1986), 'Profiting from technological innovation: implications for integration, collaboration, licensing and public policy', *Research Policy*, **15**(6), 285–305.
Tongeren, Frank van and Derek Eaton (2004), 'The case for differentiated appropriability in intellectual property rights for plant varieties', in R.E. Evenson and V. Santaniello (eds), *The Regulation of Agricultural Biotechnology*, Wallingford: CABI Publishing, pp. 109–18.

Traxler, Greg (2006), 'The GMO experience in North and South America', *International Journal of Technology and Globalisation*, **2**(1/2), 46–64.
Tripp, Robert (2001), *Seed Provision and Agricultural Development: The Institutions of Rural Change*, London: Overseas Development Institute, in association with James Currey, Oxford and Heinemann, Portsmouth, NH.
Tripp, R., D. Eaton and N. Louwaars (2006), 'Intellectual property rights: designing regimes to support plant breeding in developing countries', Agricultural and Rural Development Department Report no. 35517-GLB, Washington DC: World Bank, available at: http://siteresources.worldbank.org/INTARD/Resources/IPR_ESW.pdf.
Tripp, Robert, Niels Louwaars and Derek Eaton (2007), 'Plant variety protection in developing countries: a report from the field', *Food Policy*, **32**(3) (June), 354–71.
Venner, Raymond J. (1997), 'An economic analysis of the US Plant Variety Protection Act: the case of wheat', PhD dissertation, University of California – Davis.
Visser, B., D. Eaton, N. Louwaars and I. van der Meer (2001), 'Potential impacts of genetic use restriction technologies (GURTs) on agricultural biodiversity and agricultural production systems', FAO Commission on Genetic Resources for Food and Agriculture, Working Group on Plant Genetic Resources for Food and Agriculture.

29 Environmental effects

L. LaReesa Wolfenbarger, Yves Carrière and Micheal Owen

Agriculture (which we have defined as croplands and pasture) occupies approximately 38 per cent of the earth's terrestrial surface with much of the remaining surface unsuited for agrarian practices (Foley et al., 2011). As the single largest land use on the earth, the environmental effects of agriculture are widespread and impact not only the site where production of food and fiber occurs but also offsite areas. What crops are produced, where they are produced and how the crops are produced define the potential environmental impacts. Environmental effects impacted by agricultural systems can be characterized broadly as changes to biodiversity and alterations in the ecological services upon which humans depend. Collectively, the negative environmental impacts attributable to unsustainable practices of agriculture are well documented (Gordon et al., 2010). Cultivation of the soil for planting and for weed management may cause soil erosion onsite and may degrade habitat, air and water resources on and offsite. Use of chemicals to minimize yield losses from weeds and other pest complexes may affect non-target organisms, including desirable ones like pollinators and pest predators that provide benefits to agriculture and humans in general. Concomitant with increasing crop yields, fertilizer use in some regions of the world has dramatically altered global nitrogen and phosphorous cycles and impacted water quality, aquatic ecosystems and marine fisheries (Millennium Ecosystem Assessment, 2005). Similarly, agricultural production accounts for the largest proportion of consumptive freshwater use globally, an estimated 86 per cent of water withdrawals (Foley et al., 2011; Hoekstra and Chapagain, 2007), and in some locations causes water deficiencies for downstream human users and ecosystems (Pfister et al., 2011). Agricultural systems strive to manage pests, including weeds, insects and diseases, to sustain crop yields. However, the management of these organisms has resulted in the evolution of pest populations that no longer respond to these management practices. Lastly, in specific instances the evolution of weeds resistant to herbicides may occur as a consequence of gene flow between introduced crops and sexually compatible wild relatives with weedy characteristics (Ellstrand et al., 1999).

In 2011, genetically modified (GM) crops were established on 160 million hectares globally, which represented approximately 11 per cent of the world's 1.5 billion hectares of cropland (FAOSTAT, 2012; James, 2011). The net environmental impacts of a GM crop depend on its unique characteristics, but also on the production practices that are promoted, replaced or eliminated when the GM crop is adopted. We discuss the environmental impacts of two categories of the most widely adopted GM crops: herbicide resistant tolerant (HT) crops; and insect resistant (IR) crops (James, 2011). Specifically, we present information on HT crops with resistance to the herbicide glyphosate (Roundup™) and IR crops that produce *Bacillus thuringiensis* (Bt) toxins with insecticidal properties that target some insect pests.

Studies on the environmental effects of HT crops and IR crops have focused primarily on how each has changed pesticide use patterns, and the environmental consequences of these changes. Studies have not yet addressed whether the GM crops have altered the amount of agricultural land used globally or water withdrawals. Also, in most cases we lack spatial data to relate or monitor how site-level environmental indicators have changed over the past 16 years with the use of GM crops. In spite of these limitations, we have accumulated numerous studies at local and regional scales that provide insights on how GM crops have impacted organisms and ecological processes in and adjacent to agricultural systems.

ENVIRONMENTAL EFFECTS OF GENETICALLY MODIFIED, GLYPHOSATE TOLERANT CROPS

Herbicide-tolerant crops, including those with modified resistance to glyphosate, facilitate weed management by allowing farmers to spray specific herbicides (for example glyphosate) that would be phytotoxic to the crop without genetic modification. The GM glyphosate-resistant (GR) crops allow the application of glyphosate to kill weeds while the crop is actively growing. Therefore, HT crops in general and GR crops specifically are designed to alter the selection and patterns of herbicides used for weed management. The environmental impacts of herbicide use vary depending on the chemical's toxicity and its persistence and movement when released into the environment. Applying herbicides during the growing season may produce changes in the prevalence of other strategies for weed management, such as pre-season tillage. More effective weed management may reduce the abundance of some pollinators and natural enemies of pests that use weed species for food or habitat (Hawes et al., 2003, 2009).

In parallel with adoption rates of GR soybean, cotton and corn in the United States, glyphosate use on these crops has increased and the use of other herbicides has decreased (National Research Council, 2010). Glyphosate is reported to be more environmentally benign than many of the herbicides relied upon previously in these cropping systems (Cerdeira and Duke, 2006). It binds tightly and rapidly to soil, has low toxicity to mammals, birds and fish and is broken down by soil bacteria in a relatively short period of time compared to other herbicides (Malik et al., 1989). Glyphosate load in field run-off depends on source strength, rainfall and drainage patterns (Coupe et al., 2012). Specific glyphosate formulations containing the surfactant polyoxyethylene amine may negatively affect amphibians and other aquatic organisms if used in violation of the labeling by the US Environmental Protection Agency (Relyea and Jones, 2009).

From 1996 to 2009, a rise in conservation tillage and no-till practices accompanied the adoption of GR soybean and GR cotton in the United States (National Research Council, 2010). The relationship between tillage practices and the adoption of these HT crops is complementary and bi-directional, meaning that conservation tillage practices encouraged the adoption of HT crops and vice versa (Frisvold et al., 2007; Kalaitzandonakes and Suntornpithug, 2003). However, increased conservation tillage and no-tillage practices were less evident for adoption of GR corn (National Research Council, 2010). Conservation tillage and no-tillage practices reduce adverse environmental effects of agriculture. These practices decrease soil erosion and promote carbon sequestration and soil quality onsite, and improve water quality offsite by reducing pollution from agricultural run-off into waterways (reviewed by Uri et al., 1999; Holland,

2004; Blanco-Canqui and Lal, 2009). However, we lack the spatial data needed to assess how GR crops have reduced the environmental impacts of agriculture on and off sites (Marvier et al., 2008).

While GM crops from highly domesticated species are not predicted to persist outside of cultivation, genetic modification of crop species like fruit trees, forage crops and canola that may be able to exist in feral populations may have unforeseen ecological consequences (Gepts, 2002). Schafer et al. (2011) reported volunteer GR canola widespread and abundant along major roadways in North Dakota, and the timing of sampling suggested it persisted outside of cultivation. Environmental impacts from self-sustaining feral populations of GM crops are not reported thus far. Whether environmental impacts occur will depend on how these feral crop populations affect native species and what, if any, weed management strategies might be employed for control.

Hybridization is common among closely related plant species and is a mechanism by which herbicide tolerance might evolve in related weedy species when HT crops co-exist and are able to interbreed with the related weedy species. Crop plant genes (non-transgenic) frequently occur in their wild relatives (Kwit et al., 2011). A few documented cases of transgene movement into feral populations, for example, canola (*Brassica napus*) or landraces exist and include canola (*Brassica napus*), maize (*Zea mays* L.) and creeping bentgrass (*Agrostis stolonifera*) (Warwick et al., 2009; Piñeyro-Nelson et al., 2009a, 2009b; Zapiola et al., 2008). Like the persistence of feral crop populations, environmental impacts will emerge if the presence of transgenes affects the fitness of native species and ecosystems or if weed management strategies required for control of feral populations cause environmental harm. For seven of the world's top crops, hybridization with wild relatives has contributed to the evolution of some weed species (Ellstrand et al., 1999), but no cases are known where aggressive weed species have evolved from the hybridization of GR crops and sexually compatible wild weedy relatives.

EVOLUTION AND MANAGEMENT OF HERBICIDE RESISTANCE

Weeds impact global agricultural profitability and are important pests affecting the well-being of humans. In 2004, weed management cost an estimated $20 billion for farmers in the United States (Basu et al., 2004; Bridges, 1994). Currently, herbicide-resistant (HR) weeds represent an even greater economic threat to United States agriculture (National Research Council, 2010). Herbicide resistance has been a concern since before the introduction of GM crops. However, the incredible global change in the agricultural landscape attributable to GM technology has significantly impacted the evolution of HR weeds. In fact, the first case of herbicide resistance in weeds was reported in the scientific literature in 1970, but the evolution of resistance to herbicides in plants was proposed as early as 1956 (Harper, 1956; Ryan, 1970). Since the original report, over 393 herbicide-resistant weed biotypes in more than 210 different plant species have been reported and the totals increase almost daily (Heap, 2012). Nevertheless, concerns about GM crops, specifically those that are resistant to glyphosate, have dramatically changed how HR weeds are now perceived by farmers, agriculturalists, society and the federal government (Arntzen et al.,

2003; Benbrook, 2009; Gealy et al., 2007; Hails and Kinderlerer, 2003; National Research Council, 2010). The report from the National Research Council (2010), which brought attention to HR weeds in GM GR crops, resulted in the House Oversight Committee on Domestic Policy convening a hearing to debate the implications of evolved weed resistance to glyphosate (http://oversight.house.gov/index.php?option=com_jcalpro&Itemid=1&extmode=view&extid=197).

Historically, HR weeds have not been considered a major economic or environmental problem, except by weed scientists and those growers who have HR weed biotypes in their fields. However, a number of books and review articles have detailed the occurrence of HR weeds at the time of publication and described the evolution of HR weeds as becoming a problem of increasing significance (LeBaron, 1982, 1991). Eight years after the first major review on the topic of HR weeds published in 1982, LeBaron (1991) indicated that the number of reported HR weed species had more than tripled to 107 biotypes and the area of infestations had increased more than ten fold (Heap, 2012). What is concerning is that the major increases in HR weed biotypes involve several mechanisms of action (Heap, 2012). Weeds with evolved resistance to herbicides that inhibit acetolactate synthase (ALS, EC 2.2.1.6) and 5-enolpyruvyl-shikimate-3-phosphate synthase (EPSPS, EC 2.5.1.19) have increased dramatically since 1990 and 2000, respectively, while resistance to herbicides that inhibit photosystem II (like atrazine) has not increased markedly since 1990 (Heap, 2012; Owen, 2008). However, the occurrence of weed biotypes with multiple resistances has also increased dramatically (Heap, 2012). Despite these obvious changes, discussions about the evolution of HR weed biotypes and the resulting risks, economics, and mitigation tactics have not varied greatly in more than three decades (Duke and Powles, 2008; Frisvold et al., 2009).

While GR crops often provide economic and yield benefits to growers, there are concerns about the impact of GR crops through introgression of the GM trait into the wild, the impact on non-target organisms, herbicide use and the general sustainability of agriculture (Andow, 2003; Benbrook, 2009; Ervin et al., 2010; Gura, 2001; Ledford, 2007; National Research Council, 2010). There are also concerns about GR crops becoming pernicious weeds unto themselves (Beckie and Owen, 2007). The controversies and concerns resulting from GR crops and HR weeds raise ethical considerations, grower concerns, questions about how information is transmitted and queries about the effectiveness of extension programs (Llewellyn and Pannell, 2009; Scott et al., 2009). There have been numerous attempts to engage groups representing the industry, government and academics to develop a consensus about these issues (Boerboom and Owen, 2007; Owen and Boerboom, 2004). Despite these efforts, the problems associated with GM crops and HR weeds have gone largely unresolved. This lack of resolution is attributable, in part, to the unwillingness of growers to recognize the implications of their management tactics on weed populations, the marketing perspectives of the herbicide and seed industries, and the potentially erroneous belief that new technologies and tactics will be available in the near future (Owen et al., 2011). As a result, the number of weed species with evolved resistance to glyphosate continues to increase and more importantly, the number of fields with GR weed biotype is rising at an increasing rate (Owen, 2009).

FACTORS INFLUENCING THE RATE OF EVOLUTION OF HERBICIDE RESISTANCE

The recurrent use of any herbicide imparts selection pressures on a weed population and thus creates an ecological advantage for those rare individuals within the population that have one or more mutations conferring the ability to survive the herbicide (Llewellyn et al., 2001). Similarly, the recurrent use of any weed management tactic or crop production strategy, whether cultural or mechanical, will also select for specific weeds or biotypes that are ecologically adapted, thereby providing them with an ecological opportunity to become dominant within the weed community (Owen, 2008). The introduction of crop cultivars with genetic modifications for herbicide tolerance, the subsequent unprecedented adoption of this technology over the last 16 years, and the resulting simplification of grower weed management tactics have enhanced the evolution of GR weed biotypes (Young, 2006). The widespread adoption of GR crops globally and subsequent use of glyphosate has resulted in the evolution of glyphosate resistance in 24 weed species (Heap, 2012).

Weed communities and populations inevitably change due to the agricultural manipulation required to produce crops (Haas and Streibig, 1982). The most important agricultural manipulations, or selective forces that affect changes in weed communities are tillage (disturbance) and herbicide use, which will ultimately cause the composition of the weed communities to change to species that are less affected by these manipulations (Owen, 2008). However, herbicides tend to impart greater and more consistent selective forces on a weed community than tillage, resulting in more temporally immediate changes or shifts in species composition (Heard et al., 2003a, 2003b). Herbicides are useful to sustain crop yield and are important and effective tools to manage weeds (Bridges, 1994). However, despite the economic effectiveness of herbicides, it is telling that no weeds have been 'eliminated' from agroecosystems during more than 60 years of herbicide use (Holt, 1994).

The adoption of GR crops has dramatically affected herbicide use patterns in the United States (Young, 2006; National Research Council, 2010). These herbicide use patterns have, as suggested previously, resulted in changes in weed communities where weeds resistant to glyphosate have become more prevalent. To better manage these changes, more diverse management practices must be used by growers, including alternative herbicides (Owen, 2008; Owen et al., 2010). Importantly, these changes in weed communities have not been only attributable to evolved resistances. For example, *Commelina benghalensis* (tropical spiderwort) and *C. communis* (Asiatic dayflower) have become significant weed problems in the south and midwest United States, respectively, and represent economically important weed shifts not attributable to evolved glyphosate resistance (Culpepper et al., 2004; Ulloa and Owen, 2009). Rather, the change in prominence of these weeds in their respective agroecosystems was indirectly attributable to the use of one herbicide (glyphosate) without the inclusion of other diverse weed management tactics (Owen, 2008). Other changes in weed communities are also documented in response to the adoption of GM crops (Webster and Nichols, 2012).

Evolved resistance to glyphosate in two economically important weed species (for example *Amaranthus tuberculatus* and *A. palmeri*) has resulted in significant changes in crop production systems and impacts on agriculture (Owen et al., 2011; Shaw et al., 2012).

While other weeds have also evolved resistance to glyphosate, these two *Amaranthus* species have become agricultural problems of an unprecedented scale (Owen, 2011). These two species have several important ecological characteristics that make them particularly well-adapted to crop production systems, as well as the demonstrated ability to evolve multiple resistances to as many as five different herbicide sites of action, including glyphosate (Owen, 2011; Tranel et al., 2011). Furthermore, these species are particularly fecund, obligate out-crossers due to dioecious sexual habit, have adapted to conservation tillage systems and can hybridize with other *Amaranthus* species, thus further increasing genetic variability (Gaines et al., 2011; Steckel, 2007; Steckel et al., 2007; Trucco et al., 2009, 2007). Also contributing to the evolved resistance to herbicides is the use of low herbicide concentrations, which growers unfortunately see as an economic opportunity in the short term (Manalil et al., 2011).

CONSEQUENCES OF HERBICIDE RESISTANCE

Modern production practices in agriculture have reportedly caused marked declines in the abundance of some plant and animal species; however, no conclusive evidence exists suggesting that the relatively recent adoption of GM crops has directly affected the biological diversity or species abundance within the fields planted to GM crops (Cerdeira and Duke, 2006; Krebs et al., 1999; Schutte, 2003). As indicated, however, a lack of diversity in weed management for GM crops has resulted in the widespread evolution of HR populations, which have caused crop production practices to change. These changes may include a return to practices associated with environmental concerns, such as more aggressive tillage and use of more and different herbicides (National Research Council, 2010; Ervin et al., 2010; Shaw et al., 2012).

While there are numerous alternative tactics available to growers to manage HR weeds, growers are generally reluctant to adopt these integrated weed management strategies proactively (Green and Owen, 2011; Llewellyn and Pannell, 2009). It has been demonstrated that herbicide mixtures, herbicide rotations, the use of labeled herbicide rates, mechanical tactics and other strategies improve the management of HR weeds (Beckie, 2011; Beckie and Reboud, 2009; Diggle et al., 2003; Owen et al., 2011; Shaw et al., 2012; Wrubel and Gressel, 1994). Mathematical models that demonstrate the impact of a more diverse weed management program on the evolution of HR weeds have been developed (Freckleton and Stephens, 2009; Neve, 2008; Neve et al., 2011a, 2011b). However, growers continue to avoid more diverse programs that manage HR weeds better despite evidence that the alternative programs are economically favorable (Owen et al., 2011; Weirich et al., 2011a, 2011b).

The apparent inability of production agriculture to move beyond the simple and convenient strategies provided by the GR crops threatens the sustainability of GM-based agriculture (Duke, 2011; Gressel, 2011; Owen et al., 2011). It is clear that new models for effective weed management must be adopted to preserve crop production against the threat of evolved resistance to herbicides, especially when there have not been any new herbicide sites of action commercialized within the last 25 years. The latest tactic brought forward to help manage the GR weed populations is new GM crops with tolerance to older herbicides, which has resulted in considerable controversy (National Research

Council, 2010; Mortensen et al., 2012). Importantly, weeds have already evolved resistances to these herbicides, thus compromising the future utility of the GM crops. New and diverse approaches to weed management are needed to combat the evolution of HR weeds and thus ensuring the success of agriculture (Gressel, 2011).

ENVIRONMENTAL EFFECTS OF GM BT CROPS

GM crops producing Bt toxins are tools for managing insect pests and are employed widely in corn and cotton production systems. Through genetic modification, Bt plants produce proteins with insecticidal properties that target some insect pests. Bt crops may produce more than one type of Bt toxin. Like synthetic insecticides, understanding the environmental impacts of a Bt crop requires information on its toxicity to non-target organisms, on exposure of non-target organisms, on the toxin's persistence and movement in the environment, and on how the use of Bt crops may change agricultural practices.

Bt toxins typically target a narrow array of invertebrate taxa and their toxicity is usually confined to insects in the same order as the target pest (Schnepf et al., 1998; Cattaneo et al., 2006) although some exceptions exist (Romeis et al., 2006). Considerable research has quantified the non-target effects of Bt crops and meta-analyses of these studies have produced evidence-based generalizations (Marvier et al., 2007; Wolfenbarger et al., 2008). In cotton and corn, the relative effect of using Bt crops depends on whether or not insecticide treatments are a part of the non-Bt crop comparison or added to the Bt crop. Fields planted with Bt crops had higher abundances of non-target arthropods compared to those with non-Bt crops when insecticides were used only in the non-Bt crop fields, but lower abundances when insecticides were not used on the non-Bt and Bt crop fields (Marvier et al., 2007). With respect to the ecosystem services from biological control of insects, meta-analyses revealed that the adoption of Bt cotton increased abundances of arthropod predators when it replaced insecticide treatments but no consistent effect were observed when insecticide use occurred on Bt cotton (Wolfenbarger et al., 2008). However, Lu et al. (2012) reported substantial increases in predator abundances and concomitant decreases in the cotton aphid (*Aphis gossypii* Glover) between 2001 and 2010 following the adoption of Bt cotton in China, even though insecticide treatments were not eliminated. The increase in biological control occurred not only within cotton fields but was also evident in neighboring corn, peanut and soybean fields (Lu et al., 2012).

As the adoption of Bt cotton rose in China, the total number of insecticide sprays per season declined significantly from a high of more than 16 in 1992 to approximately 11 in recent years, and from approximately 11 to 3 for insecticides targeting the cotton bollworm (*Helicoverpa armigera*), a key pest of cotton in China (Lu et al., 2012). Insecticide use in United States cotton and corn production systems has also declined since the introduction of Bt crops in 1996 (National Research Council, 2010). Reductions in pesticide use, including insecticides, have the potential of improving water quality if fewer agricultural chemicals are transported into waterways or ground water, but we lack data to quantify the extent to which this has occurred.

The adoption of Bt crops affects the population density of insect pests at spatial scales larger than the individual fields where they are deployed. Regional suppression of insect pest populations controlled by Bt cotton and Bt corn has occurred in the United States (Hutchinson et al., 2010; Carrière et al., 2003, 2010) and China (Wu et al., 2008; Wan et al., 2012), and it is probable that additional reductions in insecticide use accompany the regional suppression of insect pest populations.

Because Bt plants produce their insecticidal proteins throughout the plant tissue, a wide variety of organisms may ingest or be affected by Bt toxins. Groups of particular interest are natural enemies (discussed above), pollinators such as honey bees (*Apis melifera*), and soil organisms such as earthworms, nematodes, bacteria and fungi. Pollinators may ingest Bt toxin in the pollen of Bt plants either when pollinating the plant (for example cotton) or when pollen is dispersed by wind from its source (for example corn). No supportable generalizations exist about the effects of Bt crops on pollinators as a group given the lack of scientific research (National Research Council, 2010). Among laboratory studies, no consistent effect of Bt plant pollen or Bt toxin on development time or survival of pollinators has been detected, but effects do vary among pollinator species, particularly for development time (Naranjo, 2009). Laboratory research on Bt toxicity has consistently indicated that Bt pollen and Bt proteins do not decrease survival of adult or larval honey bees, one of agriculture's most important pollinator species (Hendriksma et al., 2011; Duan et al., 2008). Most assessments of the effects of Bt insecticidal proteins on soil organisms, including micro-organisms, have concluded that microbial populations and their measured ecosystem functions are not substantially altered by Bt toxins (Icoz and Stotzky, 2008). Similarly, Bt and non-Bt plant residues have similar decomposition rates, indicating that Bt plants are not altering soil quality or carbon sequestration (Lehman et al., 2008; Lachnicht et al., 2004; Zwahlen et al., 2007; Tarkalson et al., 2008; Kravchenko et al., 2009).

The persistence of IR crops outside of cultivation or the transfer of IR genes to wild relatives could lead to environmental impacts on native species and ecosystems, or to a need for weed management strategies with greater environmental effects depending on whether the feral IR crops were weeds (Letourneau et al., 2003). To date, no known cases of environmental impacts from persistence of commercialized IR crops outside of cultivation or from transgene movement from IR crops to wild relatives have been observed, but transgene movement from GM crops into wild or weedy relatives has been documented for HT crops (Warwick et al., 2009; Piñeyro et al., 2009a, 2009b; Zapiola et al., 2008) as discussed above.

EVOLUTION AND MANAGEMENT OF INSECT RESISTANCE

Bt crops have provided several benefits, including reduced insecticide use, regional insect pest suppression or eradication, and increased or less variable yields (National Research Council, 2010; Carrière et al., 2010; Tabashnik et al., 2010). As the regulatory agency overseeing the use of insecticides in the United States, the Environmental Protection Agency (EPA) concluded that the evolution of pest resistance to Bt toxins is a serious threat to the continued efficacy of Bt crops and Bt sprays used in organic and conventional production systems (Matten et al., 2008; Thompson et al., 2008). Because the EPA considered sustained use of Bt crops in the 'public good', the refuge strategy was mandated to delay the evolution of resistance in major pests targeted by Bt corn and cotton.

THE REFUGE STRATEGY

Mathematical models have been central for development of the refuge strategy. Because many of the observed cases of resistance to Bt crops involve mutations at a single locus (Bravo and Soberón, 2008; Baxter et al., 2011), models typically assume that resistance to a Bt toxin is conferred by mutations at a single locus (Gould, 1998; Carrière et al., 2010). For simplicity, models assume that a single allele per locus confers susceptibility or resistance, even if more than one resistance allele may be present in pest populations (Zhang et al., 2012).

The refuge strategy requires the presence of refuges of non-Bt host plants in or near Bt crop fields. For refuges to be effective, the numerous susceptible insects produced in refuges must mate with the rare resistant pests that survive on Bt crops. When resistance is recessive, most hybrid offspring from such matings are killed by Bt crops, which reduces the heritability of resistance and delays its evolution (Gould, 1998; Carrière et al., 2010). However, resistance to commercialized Bt crops is recessive in many but not all target pests (Tabashnik et al., 2008; Carrière et al., 2010; Brévault et al., 2012).

When refuges are present, the evolution of Bt resistance can be further delayed by reducing the selective differential between individuals with and without resistance alleles (Onstad, 2008; Carrière et al., 2010). The selective differential between resistant and susceptible individuals is affected by crop management practices, such as increasing refuge size that increases relative fitness of susceptible individuals. This selective differential is also affected by pest biology and genetics. Fitness costs associated with resistance to Bt toxins occur in environments that lack Bt toxins if individuals with one or more resistance alleles have lower fitness than individuals without such alleles (Gassmann et al., 2009). Fitness costs of Bt resistance are common in insect populations and select against resistance in refuges, which counterbalances selection favoring an increase in resistance in fields of Bt crops (Gassmann et al., 2009; Carrière et al., 2010). Incomplete resistance occurs when the fitness of resistant individuals is lower on Bt cultivars than on corresponding non-Bt cultivars (Carrière and Tabashnik, 2001). Incomplete resistance is common in insect pest populations and contributes to delaying the evolution of resistance by reducing the selective advantage of resistant individuals (Carrière and Tabashnik, 2001; Tabashnik et al., 2005; Carrière et al., 2010). Manipulation of fitness costs and incomplete resistance has been envisaged to increase the capacity of refuges to delay the evolution of resistance (Crowder and Carrière, 2009; Gassmann et al., 2009; Carrière et al., 2010; Williams et al., 2011). Manipulation of costs and incomplete resistance could be especially useful for improving efficacy of the refuge strategy in delaying resistance in pests with haplodiploid or parthenogenetic modes of reproduction (Carrière, 2003; Crowder and Carrière, 2009).

EMPIRICAL TESTS OF THE REFUGE STRATEGY

Empirical support for the refuge strategy has been provided by short-term laboratory and greenhouse experiments, retrospective analyses of variation in resistance in the field, and large-scale field tests. Laboratory and greenhouse experiments tested the hypothesis that mating between susceptible and resistant individuals delays the evolution of resistance

(Zhao et al., 2005). Such experiments provide only partial support for the effectiveness of the refuge strategy, because they do not consider several factors that affect resistance in the field. Retrospective analyses of variation in resistance evolution in the field have also indicated that refuges are effective (Tabashnik et al., 2008). Such analyses are based on comparisons among species or qualitative comparisons within species. One potential drawback is that factors that vary among species or geographic areas could confound outcomes of these tests. A large-scale field test of the refuge strategy was recently conducted in a single geographic area and insect pest species, the sweet potato whitefly, *Bemisia tabaci* (Carrière et al., 2012). In a first step, spatially-explicit statistical models were developed to describe the association between spatial variation in resistance frequency and the abundance and distribution of refuges and fields where a specific insecticide was used. This provided information on the types of habitats that were effective refuges and their zone of influence, which is the maximum distance at which refuges can provide susceptible individuals to delay the evolution of resistance (Carrière et al., 2004). In a second step, the statistical models were used with independent data to successfully predict spatial variation in resistance, which confirmed that the identified refuges and pesticide-treated fields respectively delayed and accelerated the evolution of resistance (Carrière et al., 2012). Overall, these studies strongly support results from mathematical models, which show that low initial frequency of resistance alleles, recessive inheritance of resistance and abundant refuges of non-Bt host plants are key factors that delay the evolution of resistance.

DEVELOPMENT AND IMPLEMENTATION OF THE REFUGE STRATEGY

Since the introduction of Bt crops in the United States in 1996, refuge strategies mandated by the EPA have evolved, based on a process using scientific knowledge to balance economic and environmental considerations (National Research Council, 2010). As novel knowledge of factors affecting the evolution of Bt resistance in specific insect pests arises, the EPA provides new specifications for the area, configuration, and types of refuges to be used with specific Bt crops. For example, new data and modeling results indicated that cultivated and wild non-cotton host plants could provide sufficient refuges to delay resistance to Bt cotton in the tobacco budworm (*Heliothis virescens*) and cotton bollworm (*Helicoverpa zea*) (US EPA, 2007). This prompted the EPA to abolish the requirement for planting refuges of non-Bt cotton from Texas to the Mid-Atlantic for managing resistance to Bt cotton in these two polyphagous pests (US EPA, 2007).

The first generation of Bt corn and cotton commercialized in 1996 produced the Bt toxins Cry1Ac or Cry1Ab. These cultivars have been progressively replaced by crops that produce two or more Bt toxins, which are known as 'pyramided' plants. One of the factors promoting this replacement is the recognition that the evolution of resistance is more effectively delayed with a pyramid strategy than with one-toxin crops (National Research Council, 2010). The pyramid strategy, under ideal conditions, is based on the principle that insects targeted by two or more Bt toxins are killed as long as they have a susceptibility allele at a resistance locus, a phenomenon called 'redundant killing' (Gould, 1986; Roush, 1998). When the frequency of resistance alleles is low and resistance is recessive, the only genotype that has high survival on a Bt cultivar producing two or more toxins is expected to be extremely rare. Accordingly, the refuge strategy is expected to be more effective in reducing the heritability of resistance when crops produce more than one Bt toxin (Roush, 1998; Gould et al., 2006). However, as mentioned above, resistance to commercialized Bt crops is not always recessive. Furthermore, the concentration of Bt toxins in Bt cultivars typically declines as the growing season progresses, which could invalidate some of the fundamental assumptions of the pyramid strategy in pests less sensitive to Bt toxins, and significantly reduce efficacy of the pyramid strategy (Carrière et al., 2010; Brévault et al., 2012).

It is expected that most Bt crops commercialized in the future by biotechnology companies will produce two or more toxins for the control of individual insect pest species (National Research Council, 2010; http://www.monsanto.com/products/Pages/corn-pipeline.aspx). For example, the complete replacement of one-toxin by two-toxin Bt cotton has occurred in Australia (Baker et al., 2008).

EVIDENCE OF EVOLUTION OF RESISTANCE TO BT CROPS

Reports of field-evolved resistance associated with increased damage to Bt crops currently involve three major pests in the United States (Tabashnik and Carrière, 2012). The cotton bollworm (*Helicoverpa zea*) evolved resistance to the Bt toxins Cry1Ac and Cry2Ab produced by some cotton cultivars (Tabashnik et al., 2008), the fall armyworm (*Spodoptera frugiperda*) evolved resistance to Cry1F corn in Puerto Rico (Storer et al., 2010) and the Western corn rootworm (*Diabrotica virgifera virgifera*) evolved resistance to Cry3Bb corn in Iowa (Gassmann et al., 2011). In other countries, two other pests evolved resistance that resulted in increased damage to Bt crops. The maize stem borer (*Busseola fusca*) evolved resistance to Cry1Ab corn in South Africa (Van Rensburg, 2007), and the pink bollworm (*Pectinophora gossypiella*) evolved resistance to Cry1Ac cotton in India (Dhurua and Gujar, 2011). In many of these cases, evidence indicates that resistance to Bt crops evolved rapidly because refuges were too rare (Tabashnik et al., 2008; Kruger et al., 2012; Tabashnik et al., 2012). Significant increases in the frequency of resistance to toxins produced by Bt cultivars have been detected in other pest populations, although in these cases the frequency of resistant individuals is still too low to translate into significant damage to Bt crops (Downes et al., 2010; Wan et al., 2012; Zhang et al., 2012).

There is usually a lag between the introduction of a novel insecticide and the rise in the number of species that evolve resistance (National Research Council, 2010; Carrière et al., 2010). While no insect pests evolved resistance to Bt crops during the first six years that they were used (Tabashnik et al., 2003), there has been a clear increase in the number of pests that have evolved resistance and the possibility remains that the accumulation of pests resistant to Bt could accelerate (Carrière et al., 2010). The last 16 years have confirmed that Bt crops contribute in reducing the environmental impacts of managing insect pests (National Research Council, 2010; Tabashnik et al., 2010; Hutchison et al., 2010). More than ever, efficient resistance management strategies are needed to sustain the efficacy of Bt crops and to avoid reverting to practices with known negative environmental impacts.

CONCLUSIONS

Studies addressing environmental impacts of HT and IR crops indicate that these commercialized GM crops have not exacerbated the negative impacts of agriculture on biodiversity and ecosystem services. In fact, studies indicate that the use of GM crops has reduced the adverse impacts of agriculture by promoting practices such as no-till planting of GR crops in conjunction with the use of glyphosate, and reduced insecticide use (IR crops). Reduced insecticide use associated with the use of Bt crops has, in turn, enhanced pest management through ecosystem services in some cases. However, the evolution of pest resistance decreases the effectiveness of GR and Bt crops and causes agricultural practices to change in response. At least 24 weed species have evolved resistance to glyphosate since the introduction of GR crops. The number of insect species with evolved resistance to Bt crops is currently lower but increasing. What agricultural practices are implemented to minimize and manage the evolution of resistance will determine the future effectiveness of these GM crops and any change in their environmental impacts.

ACKNOWLEDGEMENTS

This project was supported by Biotechnology Risk Assessment Grant Program competitive grant no 2011-33522-30729 from the USDA National Institute of Food and Agriculture.

REFERENCES

Andow, D.A. (2003), 'UK farm-scale evaluations of transgenic herbicide-tolerant crops', *Nature Biotechnology*, **21**, 1453–4.

Arntzen, C.J., A. Coghlan, B. Johnson, J. Peacock and M. Rodemeyer (2003), 'GM crops: science, politics, and communication', *Nature Reviews Genetics*, **4**, 839–43.

Baker, G.H., C.R. Tann and G.P. Fitt (2008), 'Production of *Helicoverpa* spp. (Lepidoptera, Noctuidae) from different refuge crops to accompany transgenic cotton plantings in eastern Australia', *Australian Journal of Agricultural Research*, **59**, 723–32.

Basu, C., M.D. Halfhill, T.C. Mueller and C.N. Stewart Jr (2004), 'Weed genomics: new tools to understand weed biology', *Trends in Plant Science*, **9**, 391–8.

Baxter, S.W., F.R. Badenes-Pérez, A. Morrison, H. Vogel, N. Crickmore, W. Kain, P. Wang et al. (2011), 'Parallel evolution of *Bacillus thuringiensis* toxin resistance in Lepidoptera', *Genetics*, **189**, 675–9.

Beckie, H.J. (2011), 'Herbicide-resistant weed management: focus on glyphosate', *Pest Management Science*, **67**, 1037–48.

Beckie, H.J. and M.D.K. Owen (2007), 'Herbicide-resistant crops as weeds in North America', *CAB Reviews: Perspectives in Agriculture, Veterinary Science, Nutrition, and Natural Resources*, **44**, 1–22.

Beckie, H.J. and X. Reboud (2009), 'Selecting for weed resistance: herbicide rotation and mixture', *Weed Technology*, **23**, 363–70.

Benbrook, C. (2009), 'Impacts of genetically engineered crops on pesticide use: the first thirteen years', Boulder, CO: The Organic Center.

Blanco-Canqui, H. and R. Lal (2009), 'Crop residue removal impacts on soil productivity and environmental quality', *Critical Reviews in Plant Sciences*, **28**, 139–63.

Boerboom, C. and M. Owen (2007), 'National glyphosate stewardship forum II: a call to action', St Louis, MO, available at: http://www.weeds.iastate.edu/mgmt/2007/NGSFII_final.pdf.

Bravo, A. and M. Soberón (2008), 'How to cope with insect resistance to Bt toxins?', *Trends in Biotechnology*, **26**, 573–9.

Brévault, T., S. Nibouche, J. Achaleke and Y. Carrière (2012), 'Assessing the role of non-cotton refuges in delaying *Helicoverpa armigera* resistance to Bt cotton in West Africa', *Evolutionary Applications*, **5**, 53–65.
Bridges, D.C. (1994), 'Impact of weeds on human endeavors', *Weed Technology*, **8**, 392–5.
Carrière, Y. (2003), 'Haplodiploidy, sex and the evolution of pesticide resistance', *Journal of Economic Entomology*, **96**, 1626–40.
Carrière, Y. and B.E. Tabashnik (2001), 'Reversing insect adaptation to transgenic insecticidal plants', *Proceedings of the Royal Society of London B*, **268**, 1475–80.
Carrière, Y., D.W. Crowder and B.E. Tabashnik (2010), 'Evolutionary ecology of insect adaptation to Bt crops', *Evolutionary Applications*, **3**, 561–73.
Carrière, Y., C. Ellers-Kirk, M.S. Sisterson, L. Antilla, M. Whitlow, T.J. Dennehy and B.E. Tabashnik (2003), 'Long-term regional suppression of pink bollworm by *Bacillus thuringiensis* cotton', *Proceedings of the National Academy of Sciences of USA*, **100**, 1519–23.
Carrière, Y., P. Dutilleul, C. Ellers-Kirk, B. Pedersen, S. Haller, L. Antilla, T.J. Dennehy and B.E. Tabashnik (2004), 'Sources, sinks, and zone of influence of refuges for managing insect resistance to Bt crops', *Ecological Applications*, **14**, 1615–23.
Carrière, Y., C. Ellers-Kirk, K. Harthfield, G. Larocque, B. Degain, P. Dutilleul et al. (2012), 'Large-scale, spatially-explicit test of the refuge strategy for delaying insecticide resistance', *Proceedings of the National Academy of Sciences of USA*, **109**, 775–80.
Cattaneo, M.G., C.M. Yafuso, C. Schmidt, C-Y. Huang, M. Rahman, C. Olson, C. Ellers-Kirk, B.J. Orr, S.E. Marsh, L. Antilla, P. Dutilleul and Y. Carrière (2006), 'Farm-scale evaluation of the impacts of transgenic cotton on biodiversity, pesticide use, and yield', *Proceedings of the National Academy of Sciences of USA*, **103**, 7571–76.
Cerdeira, A.L. and S.O. Duke (2006), 'The current status and environmental impacts of glyphosate-resistant crops: a review', *Journal of Environmental Quality*, **35**, 1633–58.
Coupe, R.H., S.J. Kalkhoff, P.D. Capel and C. Gregoire (2012), 'Fate and transport of glyphosate and aminomethylphosphonic acid in surface waters of agricultural basins', *Pest Management Science*, **68**, 16–30.
Crowder, D.W. and Y. Carrière (2009), 'Comparing the refuge strategy for managing the evolution of insect resistance under different reproductive strategies', *Journal of Theoretical Biology*, **261**, 423–30.
Culpepper, A.S., J.T. Flanders, A.C. York and T.M. Webster (2004), 'Tropical spiderwort (*Commelina benghalensis*) control in glyphosate-resistant cotton', *Weed Technology*, **18**, 432–6.
Dhurua, S. and G.T. Gujar (2011), 'Field-evolved resistance to Bt toxin Cry1Ac in the pink bollworm, *Pectinophora gossypiella* (Saunders) (Lepidoptera: Gelechiidae), from India', *Pest Management Science*, **67**, 898–903.
Diggle, A.J., P.B. Neve and F.P. Smith (2003), 'Herbicides used in combination can reduce the probability of herbicide resistance in finite weed populations', *Weed Research*, **43**, 371–82.
Downes, S., T. Parker and R. Mahon (2010), 'Incipient resistance of *Helicoverpa punctigera* to the Cry2Ab Bt toxin in Bollgard II', *PLoS ONE*, **5**(9), e12567.
Duan, J.J., M. Marvier, J. Huesing, G. Dively and Z.-Y. Huang (2008), 'A meta-analysis of effects of Bt crops on honey bees (Hymenoptera: Apidae)', *PLoS ONE*, **3**(1), e1415.
Duke, S.O. (2011), 'Comparing conventional and biotechnology-based pest management', *Journal of Agricultural and Food Chemistry*, **59**, 5793–8.
Duke, S.O. and S.B. Powles (2008), 'Glyphosate: a once-in-a-century herbicide', *Pest Management Science*, **64**, 319–25.
Ellstrand, N.C., H.C. Prentice and J.F. Hancock (1999), 'Gene flow and introgression from domesticated plants into their wild relatives', *Annual Review of Ecology and Systematics*, **30**, 539–63.
Ervin, D.E., L.L. Glenna and R.A. Jussaume Jr (2010), 'Are biotechnology and sustainable agriculture compatible?', *Renewable Agriculture and Food Systems*, **25**, 143–57.
FAOSTAT (Food and Agriculture Organization of the United Nations), FAO Statistics resources database, Economic and Social Development Department, Food and Agriculture Organization of the United Nations, available at: http://faostat.fao.org/site/377/DesktopDefault.aspx?PageID=377#ancor, accessed June 29, 2012.
Foley, J.A., N. Ramankutty, K.A. Brauman, E.S. Cassidy, J.S. Gerber, M. Johnston et al. (2011), 'Solutions for a cultivated planet', *Nature*, **478**, 337–42.
Freckleton, R.P. and P.A. Stephens (2009), 'Predictive model of weed population dynamics', *Weed Research*, **49**, 225–32.
Frisvold, G., A. Boor and J.M. Reeves (2007), 'Simultaneous diffusion of herbicide tolerant cotton and conservation tillage', in *Proceedings of the 2007 Beltwide Cotton Conference*, New Orleans, LA, 9–12 January 2007, National Cotton Council of America.
Frisvold, G.B., T.M. Hurley and P.D. Mitchell (2009), 'Overview: herbicide resistant crop-diffusion, benefits, pricing, and resistance management', *AgBioForum*, **12**, 244–8.
Gaines, T.A., S.M. Ward, B. Bukun, C. Preston, J.E. Leach and P. Westra (2011), 'Interspecific hybridization

transfers a previously unknown glyphosate resistance mechanism in *Amaranthus* species', *Evolutionary Applications*, **5**, 29–38.
Gassmann, A.J., Y. Carrière and B.E. Tabashnik (2009), 'Fitness costs of insect resistance to *Bacillus thuringiensis*', *Annual Review of Entomology*, **54**, 147–63.
Gassmann, A.J., J.L. Petzold-Maxwell, R.S. Keweshan and M.W. Dunbar (2011), 'Field-evolved resistance to Bt maize by Western corn rootworm', *PLoS ONE*, **6**, e22629.
Gealy, D.R., K.J. Bradford, L. Hall, R. Hellmich, A. Raybold, J. Wolt and D. Zilberman (2007), 'Implications of gene flow in the scale-up and commercial use of biotechnology-derived crops: economic and policy considerations', CAST issue paper no. 37, Ames, IA: CAST.
Gepts, P. (2002), 'A comparison between crop domestication, classical plant breeding, and genetic engineering', *Crop Science*, **42**, 1780–90.
Gordon, L.J., C.M. Finlayson and M. Falkenmark (2010), 'Managing water in agriculture for food production and other ecosystem services', *Agriculture Water Management*, **97**, 512–19.
Gould, F. (1986), 'Simulation models for predicting durability of insect-resistant germ plasm: a deterministic diploid, two-locus model', *Environmental Entomology*, **15**, 1–10.
Gould, F. (1998), 'Sustainability of transgenic insecticidal cultivars: integrating pest genetics and ecology', *Annual Review Entomology*, **43**, 701–26.
Gould, F., M.B. Cohen, J.S. Bentur, G.G. Kennedy and J. Van Duyn (2006), 'Impact of small fitness costs on pest adaptation to crop varieties with multiple toxins: a heuristic model', *Journal of Economic Entomology*, **99**, 2091–9.
Green, J.M., and M.D.K. Owen (2011), 'Herbicide-resistant crops: utilities and limitations for herbicide-resistant weed management', *Journal of Agricultural and Food Chemistry*, **59**, 5819–29.
Gressel, J. (2011), 'Global advances in weed management', *Journal of Agricultural Science*, **149**, 47–53.
Gura, T. (2001), 'The battlefields of Britain', *Nature*, **412**, 760–63.
Haas, H. and J.C. Streibig (1982), 'Changing patterns of weed distribution as a result of herbicide use and other agronomic factors', in H.M. LaBaron and J. Gresse (eds), *Herbicide Resistance in Plants*, New York: John Wiley & Sons, pp. 57–80.
Hails, R. and J. Kinderlerer (2003), 'The GM public debate: context and communication strategies', *Nature Reviews Genetics*, **4**, 819–25.
Harper, J.L. (1956), 'The evolution of weeds in relation to resistance to herbicides', in Proceedings of the 3rd British Weed Control Conference, Farnham British Weed Control Council, pp. 179–88.
Hawes, C., A.J. Haughton, D.A. Bohan and G.R. Squire (2009), 'Functional approaches for assessing plant and invertebrate abundance patterns in arable systems', *Basic and Applied Ecology*, **10**, 34–42.
Hawes, C., A.J. Haughton, J.L. Osborne, D.B. Roy, S.J. Clark et al. (2003), 'Responses of plants and invertebrate trophic groups to contrasting herbicide regimes in the farm scale evaluations of genetically modified herbicide-tolerant crops', *Philosophical Transactions of the Royal Society of London B*, **358**, 1899–913.
Heap, I. (2012), 'The international survey of herbicide resistant weeds', available at: www.weedscience.com, accessed 27 November 2012.
Heard, M.S., C. Hawes, G.T. Champion, S.J. Clark, L.G. Firbank, A.J. Haughton, A.M. Parish, J.N. Perry, P. Rothery, R.J. Scott, M.P. Skellern, G.R. Squire and M.O. Hill (2003a), 'Weeds in fields with contrasting conventional and genetically modified herbicide-tolerant crops. I. Effects on abundance and diversity', *Philosophical Transactions of the Royal Society of London B*, **358**, 1819–32.
Heard, M.S., C. Hawes, G.T. Champion, S.J. Clark, L.G. Firbank, A.J. Haughton, A.M. Parish, J.N. Perry, P. Rothery, R.J. Scott, M.P. Skellern, G.R. Squire and M.O. Hill (2003b), 'Weeds in fields with contrasting conventional and genetically modified herbicide-tolerant crops. II Effects on individual species', *Philosophical Transactions of the Royal Society of London B*, 358, 1833–46.
Hendriksma, H.P., S. Härtel and I. Steffan-Dewenter (2011), 'Testing pollen of single and stacked insect-resistant Bt-maize on *in vitro* reared honey bee larvae', *PLoS ONE*, **6**(12), e28174.
Hoekstra, A.Y. and A.K. Chapagain (2007), 'Water footprints of nationals: water use by people as a function of their consumption pattern', *Water Resource Management*, **21**, 35–48.
Holland, J.M. (2004), 'The environmental consequences of adopting conservation tillage in Europe: reviewing the evidence', *Agriculture, Ecosystems and Environment*, **103**, 1–25.
Holt, J.S. (1994), 'Impact of weed control on weeds: new problems and research needs', *Weed Technology*, **8**, 400–402.
Hutchison, W.D., E.C. Burkness, P.D. Mitchell, R.D. Moon, T.W. Leslie, J. Fleischer et al. (2010), 'Areawide suppression of European corn borer with Bt maize reaps savings to non-Bt maize growers', *Science*, **330**, 222–5.
Icoz, I., and G. Stotzky (2008), 'Fate and effects of insect-resistant Bt crops in soil ecosystems', *Soil Biology and Biochemistry*, **40**, 559–86.
James, C. (2011), 'Global status of commercialized biotech/GM crops: 2011', *International Service for the Acquisition of Agri-Biotech Applications (ISAAA) Brief*, 43, Ithaca, NY: ISAAA.

Kalaitzandonakes, N. and P. Suntornpithug (2003), 'Adoption of cotton biotechnology in the United States: implications for impact assessment', in N.G. Kalaitzandonakes (ed.), *The Economic and Environmental Impacts of Agbiotech: A Global Perspective*, New York: Kluwer Academic/Plenum Publishers, pp. 103–18.
Kravchenko, A.N., X. Hao and G.P. Robertson (2009), 'Seven years of continuously planted Bt corn did not affect mineralizable and total soil C and total N in surface soil', *Plant and Soil*, **318**, 269–74.
Krebs, J.R., J.D. Wilson, R.B. Bradbury and G.M. Siriwardena (1999), 'The second Silent Spring?', *Nature*, **400**, 611–12.
Kruger, M., J.B.J. Van Rensburg and J. Van den Berg (2012), 'Transgenic Bt maize: farmers' perceptions, refuge compliance and reports of stem borer resistance in South Africa', *Journal of Applied Entomology*, **136**, 38–50.
Kwit, C., H.S. Moon, S.I. Warwick and C.N. Stewart (2011), 'Transgene introgression in crop relatives: molecular evidence and mitigation strategies', *Trends in Biotechnology*, **29**, 284–93.
Lachnicht, S.L., P.F. Hendrix, R.L. Potter, D.C. Coleman and D.A. Crossley (2004), 'Winter decomposition of transgenic cotton residue in conventional-till and no-till systems', *Applied Soil Ecology*, **27**, 135–42.
LeBaron, H.M. (1982), 'Introduction', in H.M. LaBaron and J. Gresse (eds), *Herbicide Resistance in Plants*, New York: John Wiley & Sons, pp. 1–8.
LeBaron, H.M. (1991), 'Distribution and seriousness of herbicide-resistant weed infestations worldwide', in J.C. Caseley, G.W. Cussons and R.K. Atkin (eds), *Herbicide Resistance in Weeds and Crops*, Oxford: Butterworth-Heinemann, pp. 27–44.
Ledford, H. (2007), 'Out of bounds', *Nature*, **445**, 132–3.
Lehman, R.M., S.L. Osborne and K.A. Rosentrater (2008), 'No differences in decomposition rates observed between Bacillus thuringiensis and non-Bacillus thuringiensis corn residue incubated in the field', *Agronomy Journal*, **100**, 163–8.
Letourneau, D.K., G.S. Robinson and J.A. Hagen (2003), 'Bt crops: predicting effects of escaped transgenes on the fitness of wild plants and their herbivores', *Environmental Biosafety Research*, **2**, 219–46.
Llewellyn, R.S. and D.J. Pannell (2009), 'Managing the herbicide resource: an evaluation of extension on management of herbicide-resistant weeds', *AgBioForum*, **12**, 358–69.
Llewellyn, R., R.K. Lindner, D.J. Pannell and S. Powles (2001), 'Adoption of integrated weed management to conserve the herbicide resource: review and framework', available at: http://www.agrifood.info/review/2001/Llewellyn.html, accessed 6 December 2010.
Lu, Y., K. Wu, Y. Jiang, Y. Guo and N. Desneux (2012), 'Widespread adoption of Bt cotton and insecticide decrease promotes biocontrol services', *Nature*, **487**, 362–7.
Malik, J., G. Barry and G. Kishore (1989), 'The herbicide glyphosate', *Biofactors*, **2**, 17–25.
Manalil, S., R. Busi, M. Renton and S.B. Powles (2011), 'Rapid evolution of herbicide resistance by low herbicide dosages', *Weed Science*, **59**, 210–17.
Marvier, M., C. McCreedy, J. Regetz and P. Kareiva (2007), 'A meta-analysis of effects of Bt cotton and maize on nontarget invertebrates', *Science*, **316**, 1475–77.
Marvier, M., Y. Carrière, N. Ellstrand, P. Gepts, P. Kareiva, E. Rosi-Marshall, B.E. Tabashnik and L.L. Wolfenbarger (2008), 'Harvesting data from genetically engineered crops', *Science*, **320**, 452–3.
Matten, S.R., G.P. Head and H.D. Quemada (2008), 'How governmental regulation can help or hinder the integration of *Bt* crops into IPM programs', in J. Romeis, A.M. Shelton and G.G. Kennedy (eds), *Integration of Insect-resistant Genetically Modified Crops Within IPM Programs, Progress in Biological Control*, Vol. v. 5, New York: Springer, pp. 27–39.
Millennium Ecosystem Assessment (2005), *Ecosystems and Human Well-being: Current State and Trends, Volume 1*, Washington, DC: Island Press.
Mortensen, D.A., J.F. Egan, B.D. Maxwell, M.R. Ryan and R.G. Smith (2012), 'Navigating a critical juncture for sustainable weed management', *BioScience*, **62**, 75–84.
Naranjo, S. (2009), 'Impacts of Bt crops on non-target invertebrates and insecticide use patterns', *Perspectives in Agriculture, Veterinary Science, Nutrition and Natural Resources*, **4**, 1–11.
National Research Council (2010), *The Impact of Genetically Engineered Crops on Farm Sustainability in the United States*, Washington, DC: National Academy Press.
Neve, P. (2008), 'Simulation modelling to understand the evolution and management of glyphosate resistance in weeds', *Pest Management Science*, **64**, 392–401.
Neve, P., J.K. Norsworthy, K.L. Smith and I.A. Zelaya (2011a), 'Modeling glyphosate resistance management strategies for Palmer amaranth (*Amaranthus palmeri*) in cotton', *Weed Technology*, **25**, 335–43.
Neve, P., J.K. Norsworthy, K.L. Smith and I.A. Zelaya (2011b), 'Modelling evolution and management of glyphosate resistance in *Amaranthus palmeri*', *Weed Research*, **51**, 99–112.
Onstad, D.W. (2008), *Insect Resistance Management: Biology, Economics, and Prediction*, San Diego: Elsevier.
Owen, M.D.K. (2008), 'Weed species shifts in glyphosate-resistant crops', *Pest Management Science*, **64**, 377–87.
Owen, M.D.K. (2009), 'Herbicide-tolerant genetically modified crops: resistance management', in N. Ferry

and A.M.R. Gatehouse (eds), *Environmental Impact of Genetically Modified Crops*, Wallingford: CAB International, pp. 113–62.
Owen, M.D.K. (2011), 'Weed resistance development and management in herbicide-tolerant crops: experiences from the USA', *Journal of Consumer Protection and Food Safety*, **6**, 85–9.
Owen, M. and C. Boerboom (2004), 'National glyphosate stewardship forum', St Louis, MO, available at: http://www.weeds.iastate.edu/weednews/2006/NGSF%20final%20report.pdf.
Owen, M., P. Dixon, D. Shaw, S. Weller, B. Young, R. Wilson and D. Jordan (2010), 'Sustainability of glyphosate-based weed management: the benchmark study', *Information Systems for Biotechnology*, Blacksburg, VA: Virgina Tech.
Owen, M.D.K., B.G. Young, D.R. Shaw, R.G. Wilson, D.L. Jordan, P.M. Dixon and S.C. Weller (2011), 'Benchmark study on glyphosate-resistant cropping systems in the USA. II. Perspective', *Pest Management Science*, **67**, 747–57.
Pfister, S., P. Bayer, A. Koehler and S. Hellweg (2011), 'Environmental impacts of water use in global crop production: hotspots and tradeoffs with land use', *Environmental Science & Technology*, **45**, 5761–68.
Piñeyro-Nelson, A., J. Van Heerwaarden, H.R. Perales, J.A. Serratos-Hernández et al. (2009a), 'Transgenes in Mexican maize: molecular evidence and methodological considerations for GMO detection in landrace populations', *Molecular Ecology*, **18**, 750–61.
Piñeyro-Nelson, A., J. Van Heerwaarden, H.R. Perales, J.A. Serratos-Hernández et al. (2009b), 'Resolution of the Mexican transgene detection controversy: error sources and scientific practice in commercial and ecological contexts', *Molecular Ecology*, **18**, 4145–4150.
Relyea, R.A. and D.K. Jones (2009), 'The toxicity of Roundup Original Max® to 13 species of larval amphibians', *Environmental Toxicology and Chemistry*, **28**, 2004–2008.
Romeis, J., M. Meissle and F. Bigler (2006), 'Transgenic crops expressing *Bacillus thuringiensis* toxins and biological control', *Nature Biotechnology*, **24**, 63–71.
Roush, R.T. (1998), 'Two-toxin strategies for management of insecticidal transgenic crops: can pyramiding succeed where pesticide mixtures have not?', *Philosophical Transactions of the Royal Society B: Biological Sciences*, **353**, 1777–86.
Ryan, G.F. (1970), 'Resistance of common groundsel to simazine and atrazine', *Weed Science*, **18**, 614–16.
Schafer, M.G., A.A. Ross, J.P. Londo, C.A. Burdick, E.H. Lee et al. (2011), 'The establishment of genetically engineered canola populations in the US', *PLoS ONE*, **6**, e25736.
Schnepf, F., N. Crickmore, J. Van Rie, D. Lereclus, J. Baum, J. Feitelson, D.R. Zeigler and D.H. Dean (1998), '*Bacillus thuringiensis* and its pesticidal crystal proteins', *Microbiology and Molecular Biology Reviews*, **62**, 775–806.
Schutte, G. (2003), 'Herbicide resistance: promises and prospects of biodiversity for European agriculture', *Agriculture and Human Values*, **20**, 217–30.
Scott, B.A., M.J. VanGessel and S. White-Hansen (2009), 'Herbicide-resistant weeds in the United States and their impact on extension', *Weed Technology*, **23**, 599–603.
Shaw, D.R., S. Culpepper, M. Owen, A. Price and R. Wilson (2012), 'Herbicide-resistant weeds threaten soil conservation gains: finding a balance for soil and farm sustainability', Ames, IA: Council for Agricultural Science and Technology.
Steckel, L.E. (2007), 'The dioecious *Amaranthus* spp.: here to stay', *Weed Technology*, **21**, 567–70.
Steckel, L.E., C.L. Sprague, E.W. Stoller, L.M. Wax and F.W. Simmons (2007), 'Tillage, cropping system, and soil depth effects on common waterhemp (*Amaranthus rudis*) seed-bank persistence', *Weed Science*, **55**, 235–9.
Storer, N.P., J.M. Babcock, M. Schlenz, T. Meade, G.D. Thompson, J.W. Bing et al. (2010), 'Discovery and characterization of field resistance to Bt maize: *Spodoptera frugiperda* (Lepidoptera: Noctuidae) in Puerto Rico', **103**, 1031–8.
Tabashnik, B.E. and Y. Carrière (2012), 'Resistance to transgenic crops and pest outbreaks', in P. Barbosa, D. Letourneau and A. Agrawal (eds), *Insect Outbreaks Revisited*, Chichester: Blackwell, pp. 342–54.
Tabashnik, B.E., T.J. Dennehy, and Y. Carrière (2005), 'Delayed resistance to transgenic cotton in pink bollworm', *Proceedings of the National Academy of Sciences of USA*, **102**, 15389–93.
Tabashnik, B.E., A.J. Gassmann, D.W. Crowder, and Y. Carrière (2008), 'Insect resistance to Bt crops: evidence versus theory', *Nature Biotechnology*, **26**, 199–202.
Tabashnik, B.E., Y. Carrière, T.J. Dennehy, S. Morin, M. Sisterson, R.T. Roush et al. (2003), 'Insect resistance to transgenic Bt crops: lessons from the laboratory and field', *Journal of Economic Entomology*, **96**, 1031–103.
Tabashnik, B.E., M.S. Sisterson, P.C. Ellsworth, T.J. Dennehy L. Antilla, L. Liesner et al. (2010), 'Suppressing resistance to Bt cotton with sterile insect releases', *Nature Biotechnology*, **28**, 1304–7.
Tabashnik, B.E., S. Morin, G.C. Unnithan, A.J. Yelich, C. Ellers-Kirk V.S. Harpold et al. (2012), 'Sustained susceptibility of Pink Bollworm to Bt cotton in the United States', *GM Crops and Food: Biotechnology in Agriculture and the Food Chain*, **3**, 194–200.

Tarkalson, D.D., S.D. Kachman, J.M.N. Knops, J.E. Thies and C.S. Wortmann (2008), 'Decomposition of Bt and non-Bt corn hybrid residues in the field', *Nutrient Cycling in Agroecosystems*, **80**, 211–22.
Thompson, G.D., S. Matten, I. Denholm, M.E. Whalon and P. Leonard (2008), 'The politics of resistance management: working towards pesticide resistance management globally', in M.E. Whalon, D. Mota-Sanchez and R.M. Hollingworth (eds), *Global Pesticide Resistance in Arthropods*, Wallingford: CAB International, pp. 146–64.
Tranel, P.J., C.W. Riggins, M.S. Bell and A.G. Hagar (2011), 'Herbicide resistances in *Amaranthus tuberculatus*: a call for new options', *Journal of Agricultural and Food Chemistry*, **59**, 5808–12.
Trucco, F., T. Tatum, A.L. Rayburn and P.J. Tranel (2009), 'Out of the swamp: unidirectional hybridization with weedy species may explain the prevalence of *Amaranthus tuberculatus* as a weed', *New Phytologist*, **184**, 819–27.
Trucco, F., D. Zheng, A.J. Woodyard, J.R. Walter, T.C. Tatum, A.L. Rayburn and P.J. Tranel (2007), 'Nonhybrid progeny from crosses of dioecious amaranths: implications for gene-flow research', *Weed Science*, **55**, 119–22.
Ulloa, S.M. and M.D.K. Owen (2009), 'Response of Asiatic dayflower (*Commelina communis*) to glyphosate and alternatives in soybean', *Weed Science*, **57**, 74–80.
Uri, N.D., J.D. Atwood and J. Sanabria (1999), 'The environmental benefits and costs of conservation tillage', *Environmental Geology*, **38**, 111–25.
US EPA, (2007), 'Pesticide news story: EPA approves natural refuge for insect resistance management in bollgard II cotton', US Environmental Protection Agency, available at: http://www.epa.gov/oppfead1/cb/csb_page/updates/2007/bollgard-cotton.htm.
Van Rensburg, J.B.J. (2007), 'First report of field resistance by stem borer, *Busseola fusca* (Fuller) to Bt-transgenic maize', *South African Journal of Plant and Soil*, **24**, 147–51.
Wan, P., Y. Huang, B.E. Tabashnik, M. Huang and K. Wu (2012), 'The halo effect: suppression of pink bollworm on non-Bt cotton by Bt cotton in China', *PloS ONE*, **7**, e42004.
Warwick, S.I., J.H. Beckie and L.M. Hall (2009), 'Gene flow, invasiveness, and ecological impact of genetically modified crops', *Annals of the New York Academy of Sciences*, **1168**, 72–99.
Webster, T.M. and R.L. Nichols (2012), 'Changes in the prevalence of weed species in the major agronomic crops of the Southern United States: 1994/1995 to 2008/2009', *Weed Science*, **60**, 145–57.
Weirich, J.W., D.R. Shaw, K.H. Coble, M.D.K. Owen, P.M. Dixon, S.C. Weller, B.G. Young, R.G. Wilson and D.L. Jordan (2011a), 'Benchmark study on glyphosate-resistant cropping systems in the United States. Part 6: timeliness of economic decision-making in implementing weed resistance management strategies', *Pest Management Science*, **67**, 785–9.
Weirich, J.W., D.R. Shaw, M.D.K. Owen, P.M. Dixon, S.C. Weller, B.G. Young, R.G. Wilson and D.L. Jordan (2011b), 'Benchmark study on glyphosate-resistant cropping systems in the United States. Part 5: effects of glyphosate-based weed management programs on farm-level profitability', *Pest Management Science*, **67**, 781–4.
Williams, J.L., C. Ellers-Kirk, R.G. Orth, A.J. Gassmann, G. Head, B.E. Tabashnik et al. (2011), 'Fitness cost of resistance to Bt cotton linked with increased gossypol content in pink bollworm larvae', *PLoS One*, **6**(6), e21863.
Wolfenbarger, L.L., S. Naranjo, J. Lundgren, R. Bitzer and L. Watrud (2008), 'Bt crop effects on functional guilds of non-target arthropods: a meta-analysis', *PLoS ONE*, **3**(5).
Wrubel, R.P. and J. Gressel (1994), 'Are herbicide mixtures useful for delaying the rapid evolution of resistance? A case study', *Weed Technology*, **8**, 635–48.
Wu, K.-M., Y.-H. Lu, H.-Q. Feng, Y.-Y. Jiang and J.-Z. Zhao (2008), 'Suppression of cotton bollworm in multiple crops in China in areas with Bt toxin-containing cotton', *Science*, **321**, 1676–8.
Young, B.G. (2006), 'Changes in herbicide use patterns and production practices resulting from glyphosate-resistant crops', *Weed Technology*, **20**, 301–307.
Zapiola, M.L., C.K. Campbell, M.D. Butler, C.A. Mallory-Smith (2008), 'Escape and establishment of transgenic glyphosate-resistant creeping bentgrass *Agrostis stolonifera* in Oregon, USA: a 4-year study', *Journal of Applied Ecology*, **45**, 486–94.
Zhang, H., W. Tian, J. Zhao, L. Jin, J. Yang, C. Liu et al. (2012), 'Diverse genetic basis of field-evolved resistance to Bt cotton in cotton bollworm from China', *Proceedings of the National Academy of Sciences of USA*, **109**, 10275–80.
Zhao, J.-Z., J. Cao, H.L. Collins, S.L. Bates, R.T. Roush, E.D. Earle et al. (2005), 'Concurrent use of transgenic plants expressing a single and two *Bacillus thuringiensis* genes speeds insect adaptation to pyramided plants', *Proceedings of the National Academy of Sciences of USA*, **102**, 8426–30.
Zwahlen, C., A. Hilbeck and W. Nentwig (2007), 'Field decomposition of transgenic Bt maize residue and the impact on non-target soil invertebrates', *Plant and Soil*, **300**; 245–57.

30 Labeling of genetically modified foods[1]

Wallace E. Huffman and Jill J. McCluskey

INTRODUCTION

The introduction of genetically modified (GM) crops to world markets has created a new division between the crop trading countries. Despite claims from developers, scientists and official institutions that GM products are safe, consumer attitudes toward GM foods are largely negative in many developed countries, especially in the European Union and Japan (Lusk et al., 2005). Consumer skepticism is usually attributed to the unknown environmental and health consequences of genetically modified crops. Such consequences include, but are not limited to, unanticipated allergic responses, the spread of pest resistance or herbicide tolerance to wild plants, and inadvertent toxicity to wildlife. Further, many are concerned with the ethical dimensions of biotechnology. The United States is an exception to this rule, where consumers are largely indifferent about GM foods (Nelson, 2001; Chern et al., 2002). Studies conducted in lesser developed countries (LDCs) find that consumer attitudes toward GM foods are less negative and in many cases positive (Li et al., 2002; Subrahmanyan and Cheng, 2000).

The agbiotech industry cites numerous benefits of GM crops, including food availability, nutrition enhancement and economic advantages; importantly, not all of these advantages are attributed to the same crop. The current world population is approximately 6 billion and growing at a rapid pace. There is a need to ensure an adequate food supply for this burgeoning population in the future. GM crops have beneficial characteristics that can be exploited to meet the growing food demand. They can be pest resistant, disease resistant, can tolerate drought and salinity, and can also possess industrial, nutritional or medicinal qualities. Environmental groups host websites, hold demonstration and hand out fliers that claim that GM food is 'FrankenFood', that it is generally harmful to the environment and may be harmful to human health (Greenpeace International, 2001). The media, looking for sound bites, sometimes carries stories about the benefits and risks of GM foods. The result is that the agbiotech industry and environmental groups inject information into the discussion of and markets for GM crops, creating markets with greatly conflicted information. To the extent that consumers are aware of this information, it makes good decision-making more difficult. Food labels may carry information about foods. Owing to a tendency for the food industry to exaggerate claims about their products, the US government has established guidelines on the nature of claims that can be made and the placement of information on food labels for food sold in retail stores.

Developing countries have potentially more to gain from GM foods, since they have more urgent needs in terms of food availability and food security. If there is decreasing marginal utility in calorie consumption per capita, the marginal utility of the last calorie consumed is higher in most developing countries. For example, the yield-enhancing qualities associated with many GM crops are important for countries with large and growing populations, such as China. These governments must recognize that if they are going to continue to feed their own people, they must find more efficient agricultural production methods. Further, in developing countries, small-scale farmers are likely to experience greater yield increases because many currently do not use chemical pest control (Qaim and Zilberman, 2003). Even so, the introduction of GM crops into developing countries has also been controversial (Pinstrup-Andersen and Schiøler, 2001).

Initially, most GM products contained 'first-generation' genetic modifications, which address production considerations and have input traits that may result in public-good benefits, such as reduced pesticide usage. As more 'second-generation' GM products, which include quality traits or product-enhancing attributes such as nutritional benefits, enter the market, consumers may be more accepting of biotechnology in food because of the inherent private benefits. With benefits directly accruing to the consumers, those consumer-friendly GM products could be accepted without discounts or could receive a premium. Golden Rice, which by genetic modification produces beta-carotene, a precursor to vitamin A, is one of the emblematic cases of second-generation agbiotech products.

In research aimed at explaining the differences between the United States and Europe, Nelson (2001) concludes that European consumers generally focus on the unknown risks associated with genetically modified products rather than the benefits, whereas US consumers focus neither on the risks nor on the benefits. Nelson argues that European consumers treat the potential harm of GM foods as certain until proven otherwise. This approach is basically the 'Precautionary Principle', which dominates European GM labeling policy. The Precautionary Principle calls for preventive measures to be taken when an activity raises threats even if a direct cause–effect relationship has not been scientifically proven. In its strongest and most distinctive forms, the principle imposes a burden of proof on those who create potential risks, and it requires regulation of activities even if it cannot be shown that those activities are likely to produce significant harms (Sunstein, 2005).

ASYMMETRIC INFORMATION AND INFORMATION QUALITY

GM foods are the result of new developments in crop biotechnology. Uncertainties always exist with new technologies, and agbiotech is no different. Initially some groups raised concerns about the long-term impact of growing GM crops on the environment, including biodiversity. Others raised health and ethical issues. In contrast, the biotechnology industry advertised these crops as the best thing that had happened in food production since hybrid corn.

The markets for new GM crops and foods are ones with contradictory information. The agbiotech industry generally provides information bestowing the benefits of the technology while environmental groups claim a wide range of potential harm to the environment and human health. However, these two groups are interested parties in the debate over genetically modified organisms (GMOs). Since the release of GM crops in the mid-1990s, additional research has been undertaken to identify the benefits and costs, including some of the risks. Information from this research has reduced the uncertainties about GM crops and foods. The presence of GMOs in the US food system since the mid-1990s with no significant health impacts reveals that short-term harm or damages are not significant. The distribution of information about GM crops has been aided by advances in electronic information networks, which transmit information at approximately zero marginal cost.

In contrast to public information which is freely available to everyone, private information, which is held by a particular individual or group, is the source of asymmetric information (Molho, 1997). In two-party interactions in which one party possesses private information, the informed party can be expected to exercise an informational advantage whenever it can expect to gain from it (Molho, 1997). When experienced parties develop intuition about situations potentially leading to opportunistic behavior, asymmetric information can destroy opportunities for welfare-improving exchanges (Akerlof, 1970).

Many decision makers often rely on information provided by individuals or groups who are affected by their decisions. Furthermore, these decision makers may not know all of the alternatives available and have no control over the information provided to them by interested parties. Furthermore, these interested parties may manipulate by distorting or concealing information. For example, farmers rely on information, including advertising, provided by biotechnology companies about the expected performance of new GM crop varieties, which is almost certainly selective. However, these biotechnology firms are constrained by US regulatory policies on GMOs (Vogt and Parish, 1999) and by their interest in repeat sales. Even so, it is widely believed that their communications do not accurately reflect the complete picture. Likewise, communications by anti-GM groups may exaggerate the potential damages to the environment, human health and organic producers, and ignore the benefits, including the dramatic reduction in use of environmentally toxic insecticides.[2]

Environmental groups and organic producers are interest groups. Individuals self-select into these groups because of a common interest or goal focused on the environment, and achieving the group's goal is a public good to its members. Each of these groups has resources – largely members' time, financial contributions and voting power – and their impact is affected by organizational efficiency. By choosing relatively narrow objectives, these groups reduce coordination and decision making costs over organizations that have diverse goals. Free-riding is, however, a major organizational problem (Olson, 1965; Sandler, 1992; Cornes and Sandler, 1996). Free-riding arises when an individual relies on the efforts of others to attain a goal, and, in the extreme, no one exerts any effort toward that goal.

Advances in communication and information technologies have greatly reduced the organizational costs of interest groups and have undoubtedly increased their productivity. They are now able to construct low-cost internet sites for displaying their objectives, news releases, short articles, and other information. For example, during the late 1990s the website of Friends of the Earth was used to display the locations or addresses of field testing of GM crops in England (Huffman and Tegene, 2002). These groups can also use email to rapidly distribute communications among members, for example dealing with demonstrations, or letters opposing particular GMO issues and policies.

Consider the situation where there are two interested parties providing information in an attempt to sway the decisions of producers and consumers of GM foods. The particular capsule of information is a signal, which reflects the self-interest and private information of the sending party. Sending information is inexpensive because it requires little action on the part of the sender, and new information technologies, such as email and websites, have reduced the cost and greatly increased the speed of delivery, although there remains some modest fixed cost of preparing a communication. Thus, the marginal cost of distributing information has become approximately zero.

Misinformation can be distributed as easily as positive information: computer viruses are just one example of misinformation. The diversity of human interactions has also been greatly expanded with these new information technologies. Individuals now communicate with others whom they do not know personally and with whom they have greatly different norms and values. Hence, new issues arise in accessing information quality. Since signaling with communications is so cheap, one possible outcome is that they degrade the expected quality of information to the point that consumers ignore communications from interested parties.

Although the marginal cost of distributing information is approximately zero, it remains costly for decision makers to interpret this information, especially contradictory information. Consumers and agricultural producers, however, differ in the long-term consequences of using incorrect information. Consumers maximize utility subject to a resource constraint. When they use wrong information in their decision making, their welfare decreases, but this does not generally cause them to die. In contrast, producers, who can be described as long-run profit maximizers, will suffer reduced profits when they use misinformation. If profits become negative over the long term, they will go bankrupt and be forced to exit the industry. This is one dimension on which producers face stronger incentives to scrutinize information about GM technology compared with consumers. For consumers and producers, the ability to process information and make good decisions is a valuable skill in the sense that it can be welfare or profit increasing, and this ability seems most likely to be related to their years of schooling and accumulated experience as decision makers (Huffman, 2001). If decision makers recognize their vulnerability to a particular source of private information, they may choose to collect information from several interested parties possessing different points of view in order to reach a more informed decision. There are potential costs and benefits of these actions.

Milgrom and Roberts (1986) examine decision making in an environment where only private information is available to decision makers, such as consumers. They have shown that it is possible for decision makers to make fully informed decisions when they rely on one interested party for information but the necessary conditions seem quite restrictive and are unlikely to be fulfilled. For example, the interested parties' preferences must be known to the decision maker, the information must be costlessly verifiable, the decision maker must know the facts about which interested party has information to be able to detect situations in which information is being withheld, and the decision maker must be able to draw accurate inference when information is withheld (sometimes referenced as being sophisticated). These conditions are demanding, but perhaps they are not impossible attributes for successful producers to possess. They may exceed the capacities of most food consumers. The implication is that if producers or consumers rely only on information provided by biotechnology supply firms or environmental groups, this is unlikely to lead to fully informed decisions. Hence, good reasons exist for society to be skeptical of claims about GMOs made by both agbiotech firms and environmental interest groups.

Rather than obtaining information from only one interested party, Milgrom and Roberts (1986) show that there are advantages for decision makers from having competition among interested parties who provide information. In this situation, the number of restrictions or assumptions needed to attain good decision making are greatly reduced. However, the interested parties must be able to convey their information to the decision maker, and the decision maker must listen to all interested parties who want to convey information. In essence, there must be an opportunity for different interests to come out or to be revealed.

Agbiotech firms, environmental groups and other interested parties differ in their ability and effectiveness in conveying information that they have. This ability might be associated, among other things, with training, communication skills, personalities, organizational philosophies, and information technologies. For the last eight years rapid advances in communication and information technologies have occurred. With the associated dramatic fall in the cost of sending messages and storing information through new networks, accessibility to these technologies seems minimally constrained by capital or credit. Biotechnology companies are primarily private companies interested in long-term profits associated with the sales of their products and the value of their company, and the information that they distribute can be expected to be consistent with this long-term objective and to be constrained by it. Environmental groups, in contrast, are primarily pursuing non-monetary goals which seem less constraining on their actions and on the objectivity of their information.

With competition among interested parties in providing information, decision makers can be unsophisticated, having little or no idea of available options and issues bearing on the decision or preferences of the interested parties. He or she must, however, be able to process the information effectively that he or she receives, and the information must be verifiable. Verifiability means that the truthfulness of information can somehow be checked against a reliable source. Under these conditions, fully informed decisions are possible.

Much information is being distributed about GM food, but it is not verifiable. Although knowledge about biotechnology has increased significantly over the past fifteen years, some effects and impacts remain unknown. A coalition of anti-GM interests, for example, Greenpeace, Friends of the Earth, Action Aid, remains in place, more strongly in Europe than in the United States, but they continue to have the objective of slowing the acceptance of GM technology. These groups continue to raise new questions about short- and long-term effects on the environment from using GMOs and consuming GM food. Some of the activities of the anti-GM groups are not beyond disrupting GM field experiments that could lead to advances in the stock of knowledge about GMOs. In response, UK scientists have sent an open letter to these NGO activists asking them not to destroy their GM wheat field trials (Science Media Center, 2012).

Second-generation GM products, including functional foods, are becoming available. They are primarily credence goods (that is consumers cannot detect GM ingredients even after consuming the product). In addition, the food industry may find itself inclined to make health claims about these foods into a marketing device. Policy makers then face significant challenges arising from new claims because of information asymmetries. Issues considered in developing policies might include factors influencing consumer preferences for these products and validity of health claims (Veeman, 2002).

When information is not verifiable, the reliability of information provided by an interested party depends on the congruity between the objectives of the decision maker and those of the interested party. When objectives diverge, optimal decision-making is difficult, and these complex problems have been the topic of a range of incentive schemes in principal–agent or agency theory literature. These models are, however, well suited only to decision problems with few players.

In these circumstances, communications by interested parties might lead to unduly restrictive public policies being adopted or it might degrade the information content to the extent that sophisticated decision makers ignore it. This will be generally welfare reducing relative to fully informed decision making. For example, social cost-reducing inputs for crop production might not be used by farmers, or socially beneficial GM foods might not be consumed. More generally, long-term delays in adopting GM technologies may occur when it is time consuming to verify or refute claims by the environmental groups about agricultural-biotechnology products. For example, the EU has chosen to apply the 'precautionary principle' of proceeding very slowly in approving GM crops. This will reduce the expected social and private pay-off to research and development (R&D) in this area. This has implications for where the private sector places its future R&D investments.

When a large number of decision makers need the same verified information, research to produce this information provides a public good which may have great social value. After the information is generated, the knowledge is non-rival, in that use by one decision maker does not affect the quantity or quality available to others. Since the opportunity cost of an added user of the information is essentially zero and each user's value of the information is private information, private decision making on how much to produce will lead to under-provision. Unless some organizational device is used to internalize these unrealized externalities, tendencies to free-ride will lead to sub-optimal outcomes (Huffman and Tegene, 2002). The price system is of no aid in extracting information on the social value of verified information or a system to manage it. Hence, consistent with consumer education, independent third-party or verifiable information about agbiotech may have considerable value if available and disseminated to consumers.

Many studies have examined how information affects valuation. For example, Frewer et al. (1999) investigated British consumers' response to a number of messages that all stressed the benefits of GM foods but varied in persuasiveness. None of the messages had a significant impact on consumer attitudes. Scholderer and Frewer (2003) tested communication strategies for changing consumers' attitudes towards GM foods in an experimental study involving consumer samples from Denmark, Germany, Italy and the United Kingdom. Different information strategies were tested against a control group for their ability to change consumer attitudes. They found that no attitude change occurred. Unlike attitudes, however, consumers' product choices were sensitive to provision of information. All information materials significantly decreased the probability of consumers choosing GM products.

The scientific consensus is that first-generation GM foods are equivalent to their conventional counterparts. On average, as we discuss in later sections, consumers want a discount in order to choose the GM foods over the conventional product. This reminds us that it is the public's *perception* of risks, rather than scientifically proven risks, that directly affect markets.

FOOD LABELS

Food products in the United States can include claims on their labels about nutrition, safety, organic status, and whether they contain GM ingredients. The 1990 Nutrition Labeling and Education Act (NLEA) provided new guidelines for nutrition labels on packaged foods sold in US supermarkets; unpackaged foods, for example fresh fruits and vegetables, are not affected (Balasubramanian and Cole, 2002). This law requires packaged foods to display nutrition information prominently in a new label format, namely the Nutrition Facts Panel (NFP). It also standardizes to some degree serving size, health claims (that link a nutrient to a specific disease), and descriptive terms (for example, 'low fat') on food packages. The goal of this legislation was to improve consumer welfare by providing nutrition information that would assist consumers in making healthy food choices.

Costs of Effective Food Labeling

As an indication of the costliness of effective nutrition labeling, it is estimated that the US food industry has spent $2 billion in order to comply with the 1990 Nutrient Labeling Act (Silverglade, 1996). Some attributes, such as enhanced calcium and vitamins A and C, are viewed positively by consumers, in that more is better in the case of positive consumer attributes. Other attributes, such as sodium, fats and pesticide residues, are negative, so less is better. Prior to the Nutrition Labeling Act, the emphasis on negative labeling was more muted. Balasubramanian and Cole (2002) suggest that this tendency can be explained by consumers having an asymmetric value function, weighing a dollar of loss more heavily than a dollar of gain, which illustrates one of the key points of prospect theory (Tversky and Kahneman, 1981).

NLEA policies tend to emphasize negative rather than positive labeling. First, permissible health claims are those that associate specific nutrients with reduced risk of specific diseases. Of the seven health claims approved by the US Food and Drug Administration (FDA) at the onset of the new Nutrient Labeling Act, three linked negative attributes exclusively with deadly diseases, such as dietary fat with cancer, sodium with hypertension, and dietary saturated fat with high cholesterol and heart disease, and only one claim featured a positive attribute, namely calcium and osteoporosis. Later claims have, however, been more balanced. In general, regulations on nutrient-content claims tend to focus more heavily on negative attributes (calories, sugar, sodium, fat, fatty acids and cholesterol) than on positive attributes such as fiber and vitamins.

Clearly, with foods made currently from crop varieties that contain GM input traits, adding a label for GM content would be an example of labeling a negative food attribute. GM content has not been proven scientifically to have human health consequences, except for the transport of some known allergens to new locations. Hence, GM food labels today would not meet the Nutrition Labeling Law requirement of a proven nutrient intake leading to a better health outcome.

GM products used for food do have to pass a food safety test. In 1992, the FDA announced its landmark decision that GM food and food products will be regulated in the same way as those created by conventional means. This policy allows new GM foods to be treated as conventional foods as long as they meet three conditions: their nutritional value has not been lowered; they incorporate new substances that are already a part of the human diet; and they contain no new allergenic substances. In January 2001, the FDA issued a Guidance for Industry reaffirming this policy, which stated that the only GM foods that need to be labeled are foods that have different characteristics from the non-GM version, such as elevated vitamin A levels. In the United States, labeling food for GM content is not otherwise required. Firms must notify the FDA at least four months before putting a new GM food product on the market, and the scientific description of the product is posted on the internet for review during this time (Just et al., 2006). Only minor changes have been made in these guidelines since 2001.

The GM labeling policy in the United States can be classified as being voluntary, but the European Union (EU) imposed mandatory labeling for all foods containing more than 0.9 per cent GM ingredients. In the United States, if a voluntary label is affixed, the FDA has mandated that it cannot use the phrase 'genetically modified'. The FDA defines the term 'genetic modification' as all forms of breeding, genetic engineering and conventional modifications (US FDA, 2001). The FDA prefers the phrase 'genetically engineered' or 'made through biotechnology'. In contrast, the European Commission GM food label policy, adopted in 1997, requires each member country to enact a law requiring labeling of all new products containing substances derived from GM organisms. Japan, Australia and many other countries have also passed laws requiring GM labels for major foods.

Regardless of the approach, effective GM labeling involves costs, especially the costs of testing for the presence of GM content, segregating GM and non-GM products, variable costs of monitoring for truthfulness of labeling and enforcement of the regulations that exist, and risk premiums for being out of contract (Wilson and Dahl, 2005; Roe and Sheldon, 2007).

Effective GM labeling policies require effective segregation of GM from non-GM commodities. If, however, identity preservation through the production, marketing and processing chain were required, this system would be substantially more costly (Wilson and Dahl, 2005). To the extent that there is a market for non-GM products, buyers would be expected to specify in their purchase contracts some limit on GM content and/or precise prescriptions regarding production, marketing and handling processes. One can envision a marketplace of buyers with differentiated demand according to their aversion to GM content. To make this differentiation effective, additional costs and risks are incurred. Additional testing involves costs of conducting the tests, for which there are several technologies of varying accuracy. The risk is that GM products will be commingled with non-GM products, so the detection system must test to see that customers' shipments are within contract limits for GM contamination. This is a serious economic problem, as agents seek to determine the optimal strategy for testing and other risk mitigation strategies. 'Tolerances' are an important issue in segregation and identity preservation. GM tolerance refers to the maximum impurity level for GM content that is tolerated in a product that still carries the non-GM label. There are two levels where tolerances apply: one is defined by regulatory agencies such as the FDA, and the other is commercial tolerance. Individual firms can and seem likely to adopt different tolerance levels, subject to any regulation. Moreover, different countries are likely to have different tolerance levels, and this increases the risks and costs of segregation or identity preservation.

While private sector handlers routinely segregate and blend grains and beans as a primary function of their business, new risks arise when handling GM and non-GM products, due to the added risk of adventitious commingling. If GM products are considered to be the inferior product, growers and handlers of GM products have an incentive to mix GM with non-GM products. For US grains, Wilson and Dahl (2005) suggest that this risk may be about 4 per cent at the grain elevator level. Farmer–processor contracting in horticultural or specialty crops, however, could reduce this margin by specializing in the product being delivered, such as non-GM or a positive GM trait. Another source of risk is testing, since no test is 100 per cent accurate. Testing risk, however, varies with the technology, tolerance and variety of the products handled, and seems likely to fall over time, as the technology of testing advances.

GM Labeling Policy

There are three policy options for GM food labeling: a labeling ban, voluntary labeling, and mandatory labeling. A policy of voluntary labeling works well in a niche market where consumers who prefer GM-free products are willing to pay a premium to obtain them. In the United States, organic food, which is process based and has low probability of GM content (due to adventitious presence), is the low GM-content option.

Caswell (1998) argues that mandatory labeling serves the same purposes as voluntary labeling but at a higher cost because the entire market must be segregated and labeled even though only a portion of consumers care that the product is GMO-free. Mandatory labeling becomes more attractive with increases in the number of people who care whether their food has been genetically altered. For mandatory labeling to be welfare enhancing, policy makers must determine whether consumers place enough value on knowing whether their food is GM-free for the benefits of labeling to exceed its costs.

Supporters of mandatory labeling argue that it is needed for informed consumer choice. Critics of mandatory labeling argue that it carries the potential for information overload, and that consumers face both time and knowledge constraints when interpreting labels. Biotech companies are against mandatory labeling. They argue that labeling will unfairly stigmatize products that contain GM ingredients and hurt sales. The most convincing critique of mandatory labeling is the cost of preserving the identity of all foods that could potentially be genetically modified.

The debate over public labeling policy may turn out to be moot because of the expansion of private standards. Public standards do not adjust to changes in consumer preferences or technology as efficiently as private standards. As consumer tastes and preferences continue to change, the private sector has responded to ensure consumer satisfaction. The consolidation in the food retail sector has given the major food retailers the bargaining power to define the private standards at the industry level. Britain's food giants Tesco, Sainsbury's, Morrisons, Marks and Spencer, Somerfield, Aldi and Co-op have maintained a decade-long ban on GM foods.

How the provision of information about the inclusion of GM ingredients affects consumers' food choices and whether consumers value information are important issues for firms and policy makers. Firms typically have more information about the quality of their products than consumers, creating a situation of asymmetric information. It is prohibitively costly for most consumers to acquire information independently of firms.

EMPIRICAL STUDIES OF CONSUMER RESPONSES TO GM LABELS

There is a large body of literature that investigates consumer demand and the use of GM labels. In this section, we first discuss findings of a set of studies that utilize consumer surveys and stated consumer responses to GM labels. We then discuss studies that utilize economic experiments with an emphasis on the effects of information provision.

Consumers' preferences over types and formats of information, their ability to analyze information and time cost of decision making differ. Hence, tendencies to use various types of information vary across consumers (McCluskey and Swinnen, 2004). Labels are likely to be effective when they address specific informational needs and can be processed and used by their target audience (Verbeke, 2005). In this section, we first discuss findings of a set of studies that utilizes consumer surveys and stated consumer responses to GM claims and labels. The survey-based studies reviewed here include those that measure consumers' willingness to pay (WTP) for foods with various labels, including GM content. The second set of articles reviewed and discussed in this section is based on non-experimental data and focuses on estimating consumer GM label valuations on the basis of revealed preferences through consumers' actual purchase behavior. Here the empirical strategy involves reduced-form hedonic empirical models as well as structural demand analyses and estimation of consumers' WTP for foods with a particular label. The last set of papers reviews empirical findings of consumers' WTP for foods using experimental data sets.

Survey-based Studies

Advantages and disadvantages exist to using survey-based evidence to assess consumers' preferences and willingness to pay for GM products. A major disadvantage is that a respondent to a survey question may not truly reflect what he/she would do in a real-world setting. This is the 'hypothetical bias', which represents any differences that may exist between responses from real and hypothetical valuation questions (Cummings et al., 1995). However, in the right setting survey-based results have advantages and are useful. In particular, one can obtain information on consumers' characteristics, such as age and gender, and ask questions about consumers' attitudes towards health, their beliefs about relevant information or product attributes, among others. Further, researchers can elicit consumer valuations for hypothetical products that do not yet exist. This is especially useful in evaluating consumers' WTP for new products developed using new technologies.

The mostly negative consumer response to food products that contain GM ingredients with 'first-generation' traits is well documented by many researchers. Lusk et al. (2005) provides a comprehensive review and a meta-analysis of 25 studies that in total report 57 valuations for GM foods. They found that 87 per cent of the variation in existing value estimates for genetically modified food can be explained by an econometric model that controls characteristics of the sample, elicitation method and characteristics of the food being valued. Given that results are sensitive to elicitation method, a difficult decision is how to inform public policy. To date consumer surveys with valuation questions have been conducted in many countries, and the findings suggest that cultural factors affect consumer acceptance of biotechnology.

A number of examples show how the methods work and the kind of results they are delivering. Some are studies conducted in the United States, where consumers are more accepting of new technology compared with European countries and Japan. Li et al. (2004) examined US consumers' value for GM corn-fed beef products. Using dichotomous choice contingent valuation (CV) methodology, they found that respondents who received information about the science and possible consumer benefits of GM, were willing to pay more for GM corn-fed beef products. Respondents who were more environmentally-conscious and more educated were more likely to choose GM corn-fed beef when they received GM information. However, a mean discount of 8 per cent was required across information treatments to get consumers to choose the GM product. This is small compared to findings from other studies in Europe and Japan. In another US study, Kaneko and Chern (2005) found that most consumers are not knowledgeable about biotechnology or they are indifferent between GM or non-GM foods. They faced consumers with several options: an indifference option in the CV question for vegetable oil, cornflakes and salmon. Respondents were asked to chose either the non-GM food item or GM alternative at the market price. When the respondents choose the non-GM (GM) item, the price of the non-GM (GM) item was randomly discounted. If the respondent was indifferent between these two foods, then he/she randomly received a discount on the GM alternative or non-GM alternative and was asked whether he/she was still indifferent. The authors found that the WTP to avoid GM salmon was the highest in comparison with GM vegetable oil, GM cornflakes and GM-fed salmon.

Heiman and Zilberman (2011) investigate the effects of positively (negatively) framed statements on perceptions about GM and hypothetical choices of Israeli consumers for genetically modified vegetables. Their survey was conducted among students, and the respondents were randomly divided into negative, positive and control groups. The authors suggest that negative framing of the properties of biotechnology not only affected perception but also increased the weight assigned to health and decreased the weight assigned to taste. The main effects of knowledge and non-negative perceptions of the contribution of GM products to health increased consumers' willingness to purchases GM bell peppers. Environmental and moral consideration were found to have little impact on GM selection. The actual choice between GM versus non-GM peppers was significantly affected by price: larger discounts increased demand for GM peppers. Negative framing tended to reduce the likelihood of choosing GM and positive framing tended to increase it, suggesting that wording and presentation of a GM trait matter.

McCluskey et al. (2003) use a dichotomous choice CV model to examine the preferences of Japanese consumers. They found that the discount needed for Seikyou consumers to purchase noodles made with GM wheat and tofu made with GM soybeans is positively affected, that is a greater discount is required, when consumers report higher levels of risk toward GM food, higher levels of concern about food safety and the environment, and higher self-reported knowledge about biotechnology, education levels and income. The mean discount required for the Japanese consumers to purchase the GM products was greater than 50 per cent, so one can infer from that a transformation of

Japanese consumers' perceptions and attitudes are needed for GM food products to successfully enter the Japanese market.

Grimsrud et al. (2004) conducted a consumer survey of 400 shoppers in the Oslo region of Norway in 2002 to identify the factors that affect the discount required for Norwegian consumers to purchase bread made with GM wheat. The Norwegian consumers they sampled were on average willing to purchase GM bread with approximately a 50 per cent discount compared to non-GM bread, indicating that there is strong skepticism toward GM foods in Norway. The strict Norwegian GM food policies are consistent with the consumer consensus antipathy against the inclusion of GM foods in the Norwegian market.

Antonopoulou et al. (2009) conducted a survey in Greece to investigate the impact of political factors on consumer attitudes towards GM food. Findings include that rejection of globalization is strongly related with increased probability of being against GM foods, supporting a ban on GM foods, mandatory labeling on allowed foods, and refusing to purchase them at lower or discount price.

Given the skepticism of European Union consumers toward GM foods, Grunert et al. (2004) investigate the effects of exposing subjects to positive sensory experiences that can only be obtained by using genetic engineered flavor-enhancing bacteria. The subjects were adults in Denmark, Finland, Norway and Sweden, and the food products were hard yellow cheeses, with one labeled as 'produced using GMOs'. Subjects were first asked to taste the cheeses and to rate them for 'liking' and their willingness to buy over two separate sessions. The cheeses were labeled 'conventional' and 'GM', either without or with a health benefit. The two cheeses were picked from the first session so that the cheese labeled as conventional was rated neutrally among the eight alternatives and the GM cheese was a liked option. They found that price and starter bacteria had the largest impact on subjects' purchase intensions. The non-GM starter has a positive value, whereas the GM starter has a negative value, being larger if GM material is still viable in the final product. Subjects who were in the group that received the cheese labeled 'GM' had a less negative attitude toward GM in food products than the control group. Also, the starter used (GM or conventional) had a smaller impact on subjects' buying intensions than for the control group. Specifically, those subjects who thought that they had tasted a GM cheese attached less weight to the type of starter culture in the rating step, while those that believed that they had tasted a GM cheese with a health benefit (low fat) attached still less weight to it. The study emphasizes that consumers are less negative to GMOs if they provide consumer benefits.

Li et al. (2002) conducted a survey of 600 consumers in Beijing, China. Although the majority of surveyed consumers self-reported that they had no knowledge of biotechnology, their attitude toward GM foods was generally positive. Using a CV methodology, they find that Beijing consumers' WTP for GM soybean oil was positively affected by respondents' positive opinion toward GM foods for both products and by higher levels of self-reported knowledge for soybean oil. A few years later, Lin et al. (2006) provided an assessment of Chinese consumers' acceptance of biotech foods based on a large-scale survey. They found that about three-quarters of urban consumers in their sample had never heard of biotech foods. In spite of this low level of awareness, most of the Chinese subjects responded with favorable or neutral attitudes toward biotech foods.

In Taiwan, Chiang et al. (2012) used the CV method to examine consumers' WTP for

GM foods and their prior purchase intentions. They found that participants' risk perception, education level and age have significant effects on Taiwanese consumers' prior purchase intentions. The results of their analysis show that Taiwanese consumers would prefer to buy non-GM salmon and they would be willing to pay a higher premium if they were old, or if they deem GM soybean-fed salmon to be risky to human health.

On the African continent, Kimenju and Groote (2008) investigate Nairobi consumers' acceptance of GM food through surveys. They found that 38 per cent of the 604 respondents were aware of GM products and the majority of them were people in higher education and income groups. Even so, the authors found that 68 per cent of the respondents would be willing to buy GM maize meal at the same price as their favorite brand. However, the authors reported that consumers were concerned about possible side-effects, particularly on the environment and biodiversity.

A subset of stated-preference studies considers consumer valuation of the benefits of second-generation GM food products. Li et al. (2002) concluded that Chinese consumers in their sample, on average, are willing to pay a 38 per cent premium for Golden Rice. Further, Boccaletti and Moro (2000) reported a decrease in the percentage of consumers who refuse to buy GM food products, from 17 per cent to 12 per cent, when confronted with products that directly benefit them, instead of generic GM food products in Italy. They found that consumers better accept GM products with increased nutritional and organoleptic characteristics, and longer shelf life. Anand et al. (2007) found that consumers' acceptance of GM wheat in India changes according to whether its benefits are portrayed as producer-friendly (cost-reducing) or consumer-friendly (disease-fighting).

As more functional foods have been offered on the market, researchers have examined how the benefits associated with functional foods affect consumer acceptance. West et al. (2002) reported that Canadian respondents were willing to pay higher prices for functional food products such as anti-cancer tomatoes, and heart-healthy potatoes and chicken. Maynard and Franklin (2003) reported a similar tendency with 'cancer-fighting' dairy products. Grunert et al. (2004) investigated the effect of sensory experience of Nordic consumers with GM cheeses and their attitudes and purchase intentions. They reported that the functional benefits of the products, such as fewer calories, less saturated fat, more calcium, and more zinc, generally had small effects. Hartl and Herrmann (2009) found that German consumers are less likely to reject GM foods with consumer-oriented benefits than foods produced using farmer traits.

Revealed-preference-based Studies

In Europe and other developed countries, anti-GMO sentiment has been expressed in numerous surveys. Polling of European consumers consistently indicates a high degree of hostility to the presence of GMOs in the food supply. The aversion to GMOs is based on both private considerations, such as health risks or a preference for natural foods, and a social dimension, such as concerns about environmental effects and ethical aspects. While anti-GMO sentiment in surveys could reflect more of a concern for the social dimension of GMOs, private considerations could still dominate actual purchase decisions.

It is well known that individuals' decisions can differ greatly between contingent value studies or other surveys, which are hypothetical, and auctions where there is a real commitment to purchase (Cummings et al., 1995; List and Shogren, 1998; List and Gallet,

2001). Furthermore, given that most surveys do not inquire about actual purchase decisions at specific prices, it has been shown that small contextual cues or small changes in information provided to survey respondents may change results dramatically. In particular, the problem is most severe for willingness-to-pay for public goods. It can be argued that surveys place respondents in the role of citizens, who make judgments from society's point of view, rather the private consumer, who makes optimizing decisions for his or her household (Noussair et al., 2004).

Laboratory Experiments

Experimental economics, which is the application of experimental methods to study economic questions, offers new insights into consumer attitudes. Experiments may be conducted in the lab, such as a classroom or real lab setting, or in the field, that is, a real-world setting, such as a grocery store or farmers' market. As the description suggests, many economic experiments are designed to mimic market behavior. For lab experiments, this requires that experiments be conducted under institutional rules that support the information discovery process. The field of experimental economics was launched in the 1960s and 1970s by Vernon Smith (1962 and 1976), Kagel and Roth (1995) and others. The goal in experimental economics is to develop and implement a controlled economic experiment on subjects. Good experiments make empirical analysis of economic data much easier.

A number of economic experiments have been performed to assess what value is placed on commodities that contain GMOs, under different information and labeling treatments. In these experiments, participants bid in an auction setting on real consumer goods in a laboratory session under an incentive-compatible, information-revealing mechanism. Auctions provide several advantages. First, they provide a common homogeneous unit, money, to measure strength of preferences by any one individual for a commodity or across commodities and are free of endowment effects. For example, $1 has a common meaning to respondents, whereas in a survey, 'strongly disagree' and 'agree', are likely to be interpreted differently across respondents. Second, money is free of endowment effects. Endowment effects arise when an individual in an experiment is endowed with a commodity, say a hamburger sandwich, and asked to upgrade to another hamburger sandwich containing different attributes, such as one containing lower bacteria levels. Third, in an auction, an individual is committing to an actual purchase, unlike in a poll where there is no commitment. Fourth, under a demand-revealing mechanism, the dominant strategy is for individuals to indicate their true value for a commodity. In principle, this allows WTP to be directly measured, rather than inferred, as in surveys. The existence of a dominant strategy also simplifies calculation of a participant's best strategy because it is independent of own-risk attitudes and beliefs about other participants. Fifth, the bid submitted in an auction is a summary of an individual's valuation of the various attributes of the good, including GM content. In contrast, in a survey an individual may express a strong preference for non-GM products, but this information does not accurately reveal the weight of this attribute relative to other attributes, such as appearance, size, color and taste. In addition, a survey response might be expected to give greater weight to public dimensions of a good, such as negative externalities that result from widespread use of the product, than a private value auction (Noussair et al., 2004).

One of the earliest economic auctions of GM foods was conducted by Noussair et al. (2004). They found that information about the benefits of GM technology significantly reduced the compensation that consumers demanded to consume a GM biscuit (or cookie), and the effect varied by type of information and location. Participants in their experiments were a demographically representative sample of consumers in the Grenoble area of France in July 2000. In the main part of the experiment, participants were asked to place bids on biscuits (or cookies) of the type that are typically available in local stores. The products looked similar and were close substitutes. At the start of the auction, each participant received a sample of each of the four products to taste and was asked to mark down how much she/he liked the product. Each of the four products was simultaneously auctioned over five rounds. In each round, information was incrementally revealed about some or all of the products. No sale prices were drawn until the last (fifth) round. The incrementally revealed information included whether products contained GMOs, the percentage content of GM ingredients, and a four-page handout containing background information about GMOs.

Noussair et al. (2004) found that, in contrast to the generally negative public stand of the French against GMOs in 2000, 65 per cent of participants were willing to purchase a biscuit containing GMOs if the price was low enough. Nearly 25 per cent of participants showed no decrease in their WTP in response to learning that a product contained GMOs. Labeling for 0.1 per cent and 1 per cent thresholds for GMO content generated significantly different bids and were perceived as being significantly different. Thus, participants revealed that their demand for biscuits is decreasing in GMO content. Eight-nine per cent of participants were willing to purchase a product satisfying the 1 per cent threshold, the maximum content that the EU exempts from labeling. In essence, injection of information about GMOs into the experiments had little impact on the bidding behavior of these French consumers.

Lusk et al. (2004) conducted incentive-compatible auctions to test the value of information (including environmental, health, and benefits to lesser developed countries) across locations in England, France and the United States in 2002. Each subject was endowed with a non-GM chocolate chip cookie and then asked to bid the minimum amount they had to be paid to exchange this cookie for a GM cookie with full knowledge that the consumption of the cookie was required at the end of the auction. The subjects participated in ten rounds of bidding with a random number generated to determine the binding round. Winners were chosen by the fifth price auction. Four information treatment effects were implemented: (i) no information; (ii) potential environmental benefits of GM technology; (iii) potential health benefits; and (iv) potential benefits from greater abundance of food.

They found that average willingness to accept (WTA) a payment to exchange their cookie was lowest in the US locations, $0.20 higher in England, and $1 higher in France. This is interpreted as heterogeneity of tastes for GM products across consumers in these developed countries (Lusk et al., 2006). The findings on information effects include that the environmental benefit information resulted in lower impact on WTA among US and English subjects but higher impact on WTA impact for French subjects. Subjects who received the health information had a lower WTA in all areas but France. Subjects only in Texas and England had a lower WTA after receiving information about the potential benefit of greater world food abundance.

Rousu et al. (2007) conducted economic experiments to examine the impact on WTP of GM and plain food labels and diverse information with participants from Des Moines, Iowa, and St Paul, Minnesota in 2001. The researchers used a random *n*th price auction. Participants were asked to bid simultaneously on three commodities: a bottle of vegetable oil, a bag of tortilla chips, and a bag of Russet potatoes. For each product, there were two food labels: (a) a plain label giving the generic type of food, its weight, and date of preparation in the case of tortilla chips; and (b) a label containing the information in (a) plus the statement 'This product is made using genetic modification (GM)'. Statements about GM foods representing perspectives of the agricultural biotechnology industry, environmental groups, and third-party sources were then packaged into five information treatments: agricultural biotechnology perspective only; environmental group perspective only; agricultural biotechnology and environmental perspectives; agricultural biotechnology and third party perspectives; environmental group and third-party perspectives; and all three perspectives. Food labels and information treatments were randomly assigned to particular lab sessions.

They found that, on average, bidders discounted foods with GM labels by 14 per cent relative to the same food item having a plain food label. The average WTP for a GM product was highest when participants received (only) the industry's perspective about GMOs and the lowest when they received (only) the environmental group's perspective. Across these two information treatments and across the three commodities, the average differential was 64 per cent. Third-party information tended to moderate bidding when it was combined with the agricultural biotechnology industry perspective and the environmental group perspective, having the largest and statistically significant moderating force when participants received the environmental group perspective. The study concludes that third-party information has large social welfare value in this market with conflicting information.

Huffman et al. (2007) examine the role of consumers' prior beliefs and new information on their WTP for foods that might be genetically modified. They apply a Bayesian-updating model in which prior beliefs can be updated when the decision maker encounters new information. The study used data from Rousu et al. (2007). They found that participants who had informed prior beliefs discounted GM labeled food products more highly than those who had uninformed prior beliefs. Uninformed participants were especially susceptible to information from interested and third parties. In contrast, informed participants were generally not affected significantly by new information. These results show how both skeptics and proponents of new technologies (interested parties) might try to manage information to achieve private objectives. This is most likely to occur when much is unknown scientifically about the impact of new technologies or when third-party information is limited or unavailable. Opponents of a new technology may try to target people who are relatively uninformed about the technology. Proponents of the technology may try to target people who have informative prior beliefs for maximum effectiveness.

Colson et al. (2011) assessed consumers' acceptance of GM nutritionally enhanced vegetables using a series of auction experiments administered to a random sample of adult consumers. They were investigating consumer rather than farm traits. The experiments were conducted in 2007 in Des Moines, Iowa, and Harrisburg, Pennsylvania. Commodities chosen for these experiments were Russet potatoes, beefsteak tomatoes and broccoli. A total of seven different food labels for each product were used in the experiments: (i) plain label giving generic product name and weight; (ii) information in (i) plus the statement 'GM Free Product'; (iii) information in (i) plus the statement 'Intragenic GM Product'; (iv) information in (i) plus the statement 'Transgenic GM Product'; (v) information in (i) plus statements 'Enhanced levels of Antioxidants and Vitamin C' and 'GM Product'; (vi) information in (i) plus statements 'Enhanced levels of Antioxidants and Vitamin C' and 'Intragenic GM Product'; (vii) information in (i) plus statements 'Enhanced levels of Antioxidants and Vitamin C' and 'Transgenic GM Product'. Transgenic GM products are made from raw products that contain genes that were from another species, for example, soil bacteria. Intragenic GM products are made from raw products that contain genes from only one species but some genes have been moved long distances using genetic engineering. Following Rousu et al. (2007), three perspectives on GMOs were prepared. The information treatments were: (i) No information provided, which served as a baseline; (ii) agricultural biotechnology industry perspective (only); (iii) environmental group perspective (only); (iv) agricultural biotechnology industry and environmental group perspectives; and (v) bio-tech industry, environmental group and third-party perspectives. Food labels and information treatments were randomly assigned to particular lab sessions.

Compared to the baseline of no information, they found that consumers' bids were higher for intragenic GM than for transgenic foods when they received the agricultural biotechnology industry's perspective. When the environmental group perspective was injected, the average bid price by GM technology was reduced below the no information case. When commodities were labeled as enhanced with high levels of anti-oxidants and vitamin C, the average bid prices were higher across all three commodities than when no information on nutrient content was provided. Hence, the increase in nutrient content was shown to have positive economic value. Under the agricultural biotechnology industry perspective and with a label indicating enhanced nutrient content using intragenic GM technology, consumers' bids were significantly higher than for transgenic or a plain label. These results suggest that genetic engineering can be used to improve nutrient content of food, which is a consumer attribute, and that consumers will pay more for that food. Moreover, as these foods come on the market, they may improve US consumers' perspective of the biotech industry.

DISCUSSION AND FUTURE RESEARCH DIRECTIONS

The scientific consensus is that first-generation GM foods are equivalent to their conventional counterparts. However, on average, consumers want a discount in order to choose first-generation GM products over conventional products. Thus, it is the public's *perception* of risks, rather than scientifically proven risks, that directly affect markets. This brings up the issue of scientific versus consumer sovereignty (Roberts, 1999). Although the scientific consensus is that GM foods are completely safe for consumption aside from potential allergens, it may still be the case that a majority of the population in some countries prefer to avoid GM foods. The evidence shows so far that information provision affects valuations and that the information source also matters.

As more 'second-generation' GM products, which include quality traits or product-enhancing attributes, such as nutritional benefits, enter the market, consumers appear to be more accepting of biotechnology in food because of the inherent private benefits. With benefits directly accruing to the consumers, those consumer-friendly GM products could be accepted without discounts or even receive a premium. As technologies evolve, the lines between GM, functional foods, and other 'enhanced foods' are expected to blur. At the same time, there is a growing segment of consumers who value 'natural' or organic foods. The establishment of the USDA's organic standard in 2002 provided a standardized reference point for those interested in these products. Fifteen years ago, organic products were marketed largely in farmers' markets and specialty food stores, but now organic foods are carried by large supermarkets and big box stores (megastores). Products made from GM raw materials are also carried in these same large supermarkets and big box stores, so organic and GM food produces are increasingly coming into direct competition. Although the USDA's organic standard is strictly process based, some producers and consumers believe that organic foods are fresher, safer, more nutritious, environmentally friendly and produced by small, local farmers. However, this characterization is broadly untrue. Hence, research could be undertaken to identify the effects of objective information about the attributes of organic food on consumers' willingness to pay. This seems to provide new research opportunities for assessing consumers' demand for food.

We expect that in the future researchers will focus more on new sources of data, such as retail scanner data and experiments, both in the lab and field. In particular, scanner data can be increasingly individualized, as consumers use grocery store 'loyalty' cards. We also expect researchers to adopt more interdisciplinary or multidisciplinary methods to better understand household food choices. This includes insights from sensory sciences about the factors affecting taste and perceptions and how the experience affects perceptions of technology and the product. Psychology and neuro-economics will probably be integrated into studies of how values, social norms and personality affect choice of food made with new technologies. We expect research to draw increasingly upon findings of behavioral economics either to learn about consumer choice or to overcome cognitive limitations.

Ultimately, the new research agenda may consider how increasing affluence of consumers, increasing health concerns, and increasing expectations for customized products, for example, all the possible combinations offered at Starbuck's and the increasing number of stock-keeping units at grocery stores, affect consumers' food choices. This type of research will be critical to determining how to examine the increased differentiation and variety that exists in today's food markets.

NOTES

1. We thank our former graduate students for the collaboration that they provided in the study of consumer responses to biotechnology and information. We also thank the Iowa State Agricultural Experiment and the Washington State Agricultural Experiment Station for Financial Support.
2. Over the past decade, organic producers in the United States have become increasingly opposed to GM agriculture and would like to see a ban on GM crops (Huffman and Strzok, 2012).

REFERENCES

Anand, A., R.C. Mittelhammer and J.J. McCluskey (2007), 'Consumer response to information and second-generation modified food in India', *Journal of Agricultural & Food Industrial Organization*, **5**, Article 8.

Akerlof, G. (1970), 'The market for lemons: uncertainty and the market mechanism', *Quarterly Journal of Economics*, **84**, 488–500.

Antonopoulou, L., T. Papadas and A. Targoutzidis (2009), 'The impact of socio-demographic factors and political perceptions on consumer attitudes towards genetically modified foods. An econometric investigation', *Agricultural Economics Review*, **10**(2), 89–103.

Balasubramanian, S.K. and C. Cole (2002), 'Consumers' search and use of nutrition information: the challenge and promise of the Nutrition Labeling and Education Act', *Journal of Marketing*, **66**, 113–27.

Boccaletti, S. and D. Moro (2000), 'Consumer willingness to pay for GM food products in Italy', *AgBioForum*, **3**(4), 259–67.

Caswell, J.A. (1998), 'Should use of genetically modified organisms be labeled?', *AgBioForum*, **1**(1), 22–4.

Chern, W.S., K. Rickertsen, N. Tsuboi and T. Fu (2002), 'Consumer acceptance and willingness to pay for genetically modified vegetable oil and salmon: a multiple-country assessment', *AgBioForum*, **5**(3), 105–12.

Chiang, J., C. Lin, T. Fu, and C. Chen (2012), 'Using stated preference and prior purchase intention in the estimation of willingness to pay a premium for genetically modified foods', *Agribusiness*, **28**(1), 103–17.

Colson, G., W.E. Huffman and M. Rousu (2011), 'Will consumers pay more for product enhanced attributes: evidence from food experiments', *Journal of Agricultural and Resource Economics*, **36**, 343–64.

Cornes, R.C. and T. Sandler (1996), *The Theory of Externalities, Public Goods and Club Goods*, 2nd edn, New York: Cambridge University Press.

Cummings, R.G., G.W. Harrison and E.E. Rutstrom (1995), 'Homegrown values and hypothetical surveys: is the dichotomous choice approach incentive-compatible?', *American Economic Review*, **85**(1), 260–66.

Frewer, L.J., C. Howard, D. Hedderley and E. Shepherd (1999), 'Reactions to information about genetic engineering: impact of source characteristics, perceived personal relevance and persuasiveness', *Public Understanding of Science*, **8**, 35–50.

Greenpeace International (2001), 'We want natural food!', available at: http://www.greenpeace.org/~geneng/.

Grimsrud, K.M., J.J. McCluskey, M.L. Loureiro and T.I. Wahl (2004), 'Consumer attitudes to genetically modified food in Norway', *Journal of Agricultural Economics*, **55**(1), 75–90.

Grunert, K.G., T. Bech-Larsen, L. Lähteenmäki, Ø.Ueland and A. Åström (2004), 'Attitudes towards the use of GMOs in food protection and their impact on buying intention: the role of positive sensory experience', *Agribusiness*, **20**, 95–107.

Hartl, J. and R. Herrmann (2009), 'Do they always say no? German consumers and second-generation GM foods', *Agricultural Economics*, **40**, 551–60.

Heiman, A. and D. Zilberman (2011), 'The effects of framing on consumers' choice of GM foods', *AgBioForum*, **14**(3), 171–9.

Huffman, W.E. (2001), 'Human capital: education and agriculture', in Bruce L. Gardner and Gordon C. Rausser (eds), *Handbook of Agricultural Economics*, Vol. IA, Amsterdam: Elsevier Science/North-Holland, pp. 333–81.

Huffman, W.E. and J. Strzok (2012), 'The economics of organic and GMO farming systems: interactions and how they might co-exist', Iowa State University, Department of Economics.

Huffman, W.E. and A. Tegene (2002), 'Public acceptance of and benefits from agricultural biotechnology: a key role for verifiable information', in V. Santaniello, R.E. Evenson and D. Zilberman (eds), *Market Development for Genetically Modified Foods*, Wallingford: CABI International, pp. 179–89.

Huffman, W.E., M. Rousu, J.F. Shogren and A. Tegene (2007), 'The effects of prior beliefs and learning on consumers' acceptance of genetically modified foods', *Journal of Economic Behavior and Organization*, **63**, 193–206.

Just, R.E., J.M. Alston and D. Zilberman (2006), *Regulating Agricultural Biotechnology: Economics and Policy*, New York: Springer Science+Business Media.

Kagel, J. and A. Roth (eds) (1995), *Handbook of Experimental Economics*, Princeton, NJ: Princeton University Press.

Kaneko, N. and W.S. Chern (2005), 'Willingness to pay for genetically modified oil, cornflakes, and salmon: evidence from a US telephone survey', *Journal of Agricultural and Applied Economics*, **37**(3), 701–71.

Kimenju, S.C. and H.D. Groote (2008), 'Consumer willingness to pay for genetically modified food in Kenya', *Agricultural Economics*, **38**, 35–46.

Li, Q., J.J. McCluskey and T.I. Wahl (2004), 'Effects of information on consumers' willingness to pay for GM-corn-fed beef', *Journal of Agricultural and Food Industrial Organization*, **2**(2), Article 9.

Li, Q., K.R. Curtis, J.J. McCluskey and T.I. Wahl (2002), 'Consumer attitudes toward genetically modified foods in China', *AgBioForum*, **5**, 145–52.

Lin,W., A. Somwaru, F. Tuan, J. Huang and J. Bai (2006), 'Consumer attitudes toward biotech foods in China', *Journal of International Food and Agribusiness Marketing*, **18**(1/2), 177–203.
List, John and Craig Gallet (2001), 'What experimental protocol influence disparities between actual and hypothetical stated values?', *Environmental & Resource Economics*, **20**(3), 241–54.
List, John A. and Jason F. Shogren (1998), 'Calibration of the difference between actual and hypothetical valuations in a field experiment', *Journal of Economic Behavior & Organization*, **37**(2), 193–205.
Lusk, J.L., M. Jamal, L. Kurlander, M. Roucan and L. Taulman (2005), 'A meta-analysis of genetically modified food valuation studies', *Journal of Agricultural and Resource Economics*, **30**(1), 28–44.
Lusk, J.L., L.O. House, C. Valli, S.R. Jaeger, M. Moore, B. Morrow and W.B. Traill (2004), 'Effect of information about benefits of biotechnology on consumer acceptance of genetically modified food: evidence from experimental auctions in United States, England, and France', *European Review of Agricultural Economics*, **31**, 179–204.
Lusk, J.L., W.B. Traill, L.O. House, C. Valli, S.R. Jaeger, M. Moore and B. Morrow (2006), 'Comparative advantage in demand: experimental evidence of preferences for genetically modified food in the United States and European Union', *Journal of Agricultural Economics*, **57**, 1–21.
Maynard, L.J. and S.T. Franklin (2003), 'Functional foods as a value-added strategy: the commercial potential of 'cancer-fighting' dairy products', *Review of Agricultural Economics*, **25**, 316–31.
McCluskey, J.J. and J.F.M. Swinnen (2004), 'Political economy of the media and consumer perceptions of biotechnology', *American Journal of Agricultural Economics*, **86**(5), 1230–37.
McCluskey, J.J., K.M. Grimsrud, H. Ouchi and T.I. Wahl (2003), 'Consumer response to genetically modified food products in Japan', *Agricultural and Resource Economics Review*, **32**(2), 222–31.
Milgrom, P. and J. Roberts (1986), 'Relying on the information of interested parties', *Rand Journal of Economics*, **7**, 18–32.
Molho, I. (1997), *The Economics of Information*, Malden, MA: Blackwell Publishing.
Nelson, C.H. (2001), 'Risk perception, behavior, and consumer response to genetically modified organisms', *American Behavioral Scientist*, **44**(8), 1371–88.
Noussair, C.N., B. Ruffieux and S. Robin (2004), 'Do European consumers really refuse to buy GM food?', *Economic Journal*, **114**, 102–20.
Olson, M. (1965), *The Logic of Collective Action*, Cambridge, MA: Harvard University Press.
Pinstrup-Andersen, P. and E. Schiøler (2001), *Seeds of Contention: World Hunger and the Global Controversy over GM Crops*, Baltimore, MD: Johns Hopkins University Press.
Qaim, M. and D. Zilberman (2003), 'Yield effects of genetically modified crops in developing countries', *Science*, **299**, 900.
Roberts, Donna (1999), 'Analyzing technical trade barriers in agricultural markets: challenges and priorities' *Agribusiness*, **15**(3), 335–54.
Roe, B. and I. Sheldon (2007), 'Credence good labeling: the efficiency and distributional implications of several policy approaches', *American Journal of Agricultural Economics*, **89**, 1020–33.
Rousu, M., W.E. Huffman, J.F. Shogren and A. Tegene (2007), 'Effects and value of verifiable information in a controversial market: evidence from lab auctions of genetically modified food', *Economic Inquiry*, **45**, 409–32.
Sandler, T. (1992), *Collective Action*, Ann Arbor, MI: The University of Michigan Press.
Scholderer, J. and L.J. Frewer (2003), 'The biotechnology communication paradox: experimental evidence and the need for a new strategy', *Journal of Consumer Policy*, **26**(2), 125–57.
Science Media Center (2012), 'UK scientists call on anti-GM group to reconsider threats', available at: http://www.sciencemediacentre.co.nz/2012/05/02/uk-scientists-call-for-dialogue-with-anti-gm-group/.
Silverglade, B. (1996), 'The Nutrient Labeling and Education Act: progress to date and challenges for the future', *Journal of Public Policy and Marketing*, **15**, 148–56.
Smith, V.L. (1962), 'An experimental study of competitive market behavior', *Journal of Political Economy*, **70**, 111–37.
Smith, V.L. (1976), 'Experimental economics: induced value theory', *American Economic Review*, **66**, 274–9.
Subrahmanyan, S. and P.S. Cheng (2000), 'Perceptions and attitudes of Singaporeans toward genetically modified food', *The Journal of Consumer Affairs*, **34**(2), 269–90.
Sunstein, C.R. (2005), *Laws of Fear: Beyond the Precautionary Principle*, New York: Cambridge University Press.
Tversky, A. and D. Kahneman (1981), 'The framing of decisions and the psychology of choice', *Science*, **211**, 453–58.
US Food and Drug Administration (FDA) (2001), 'Guidance for industry: Voluntary labeling indicating whether foods have or have not been developed using bioengineering', available at:http://vm.cfsan.fda.gov/~dms/biolabgu.html.
Veeman, M. (2002), 'Policy development for novel foods: issues and challenges for functional food', *Canadian Journal of Agricultural Economics*, **50**(4), 527–39.

Verbeke, W. (2005), 'Agriculture and the food industry in the information age', *European Review of Agricultural Economics*, **32**(3), 347–68.
Vogt, D.U. and M. Parish (1999), 'Food biotechnology in the United States: science, regulation, and issues', *US Department of State*, available at: http://fpc.state.gov/6176.htm, accessed January 2012.
West, G.E., C. Gendron, B. Larue and R. Lambert (2002), 'Consumers' valuation of functional properties of foods: results from a Canada-wide survey', *Canadian Journal of Agricultural Economics*, **50**(4), 541–58.
Wilson, W.W. and B.L. Dahl (2005), 'Costs and risks of testing and segregating genetically modified wheat', *Review of Agricultural Economics*, **27**(2), 212–28.

31 Biotechnology and food security

Calestous Juma, Pedro Conceição and Sebastian Levine[1]

INTRODUCTION

The application of new technologies has played a crucial role in reducing food insecurity, as reflected in 'Green Revolutions' that were adopted across Asia and Latin America in the 1960s and 1970s. But many challenges persist, especially as the gains from these revolutions are decelerating in many parts of the world and large regions, notably in Africa, did not reap many of its benefits. To cope with the rising global population, ecological degradation and demand for improved nutrition, the international community is exploring new avenues to enhance food security by harnessing the power of existing and emerging technologies, especially biotechnology.

This chapter argues that advances in biotechnology hold promise in meeting new and persistent food security challenges, especially in developing countries where the majority of poor and hungry live. The chapter stresses that biotechnology should be viewed as a set of tools that can be applied to address specific challenges and that this can only be done effectively if placed in a wider policy context that advances food security. It also notes that biotechnology should be viewed as complementary to other approaches and not a substitute. The focus should be to enlarge the toolbox for addressing food security challenges, not to reduce it.

The chapter is divided into four sections. The first section provides a brief overview of the global significance of addressing food security and the growing political interest in the matter. Section 2 examines the linkages between food security and economic development. Section 3 analyses the role of biotechnology in addressing food security challenges and the last section reviews policy implications arising from the analysis.

1 BACKGROUND

Food security is back at the top of the global policy agenda. Leaders in African countries that have historically been food insecure are increasingly treating the matter as a priority (Juma, 2011a, Chap. 5). In 2008 a High-Level Task Force on the Global Food Security Crisis was set up comprising the heads of international organizations from the UN, World Bank, IMF and OECD to 'promote a comprehensive and unified response to the challenge of achieving global food security, including by facilitating the creation of a prioritized plan of action and coordinating its implementation' (UNHLTF, n.d.). In 2009, food security was elevated to the Summit of G8 leaders who made commitments to increased, predictable and sustainable funding to strengthen food security and agricultural development (G8, 2009). In 2012, UN Secretary-general Ban Ki-moon issued a 'zero hunger challenge' calling for a complete elimination of hunger and child malnutrition (United Nations News Center, 2012).

This renewed focus on food security has been triggered in recent years by high and volatile food prices and food-related riots in many developing countries (Naylor, 2011). Moreover, recurrent food crises in the horn of Africa in 2011, during which Somalia was gripped by the first famine of the twenty-first century, and later in the Sahel Region of West Africa that affected more than 10 million people, have accentuated public concern and jolted aid agencies into action (UNOCHA, 2012).

These crises also reminded the world (again) that while effective humanitarian responses are critical to ease human suffering, longer-term solutions are needed to effectively address the deep-seated vulnerabilities of millions of food-insecure people. There is growing recognition that enhancing food security requires responding today to the challenges of tomorrow, notably those exacerbated by high population growth especially in the poorest regions, the consequences of unabated environmental degradation and the increasingly negative effects of global warming (FAO et al., 2011).

Innovation and the application of new technologies played a crucial role in alleviating food insecurity in the past, especially during the 'Green Revolutions' that swept across Asia and Latin America in the 1960s and 1970s. New breakthroughs in the field of biotechnology hold promise of alleviating food security by accelerating agricultural productivity, especially in developing countries where most of the poor and hungry live (Juma, 2011b; UNDP, 2012). But the benefits could be much wider and improve both access to and the use of food, as well as make food systems more stable. Reaping the potential benefits of biotechnology for food security will place demands on institutional frameworks to manage risks and ensure inclusive outcomes at global, regional and national levels.

Indeed, the rate of biotechnology adoption remains one of the highest of all agricultural technologies (see Chapter 24 in this volume). This is mainly because farmers recognize the benefits of the technology and weigh them against potential risks before adopting it. This is as true of farmers who have adopted the technology as it is of farmers who have not adopted it, as shown in a recent study in Tanzania (Lewis et al., 2010).

There are no silver bullets to tackling the tremendous challenges of global food security, but under the right conditions, biotechnology can contribute significantly to a 'zero-hunger' future.

2 FOOD SECURITY AS A DEVELOPMENT CHALLENGE

The operational definition of food security has evolved over successive World Food Summits (WFS) from 'availability at all times of adequate world food supplies' (UN, 1975) to a situation where 'all people, at all times, have physical and economic access to sufficient, safe and nutritious food to meet their dietary needs and food preferences for an active and healthy life' (FAO, 1996). Enhancing food security is fundamental to the gradual realization of the right to food enshrined in the 1999 International Covenant on Economic, Social and Cultural Rights (Mechlem, 2004). Later (FAO, 2009), it was further specified that the four pillars of food security are: availability, access, utilization and stability (FAO, 2009). These pillars serve as a useful framework for gauging the current state of global food security (the main focus of this section) and for understanding the potential benefits and risks of biotechnology when it comes to food security (next section).

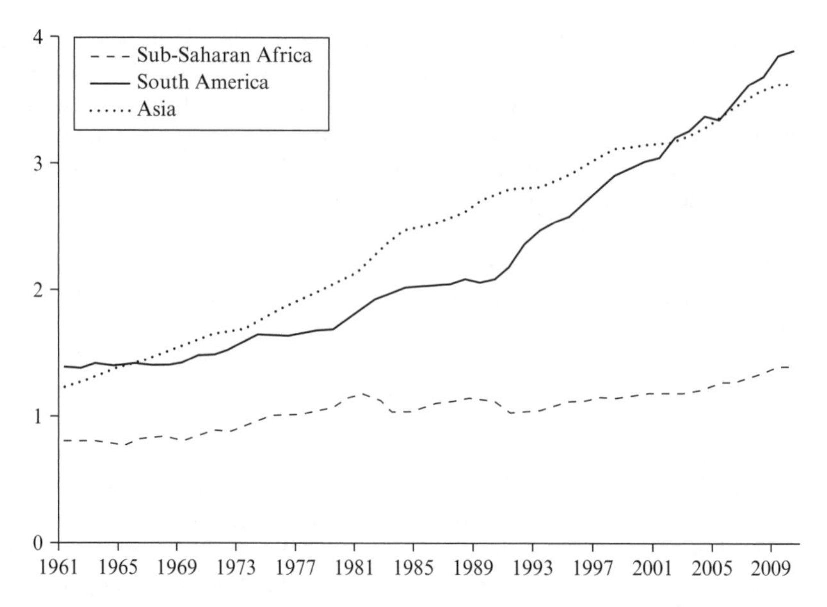

Source: Three-year moving averages computed using data from FAO Statistics Division.

Figure 31.1 Cereal yields (tonnes per hectare)

2.1 Food Availability is Increasing – but not Everywhere

At the global level, food has probably never been more plentiful than it is today. Total food production has increased more than threefold over the past half a century alone. Even when taking population growth into account, a global index on gross production of food is up almost 40 per cent since the early 1960s (FAOSTAT, 2012). Similarly, the total availability of calories from all food stuffs has risen by more than a quarter (29 per cent) from 2190 per capita/day in 1961 to 2830 in 2009. This means that, at least in principle, enough food is available to ensure that the recommended daily food intake of 2100 kilocalories (kcal) can be more than met for all people (Naiken, 2002). Food availability is, however, highly uneven between regions and countries, and even within countries. Figure 31.1, for example, shows one of the causes of this uneven distribution – yields for cereals have risen about fourfold in Asia and South America since 1961, while they have only risen marginally in sub-Saharan Africa, one of the world's most food-insecure regions.

While *total* cereal production has increased more than threefold in sub-Saharan Africa, Asia and South America since the early 1960s, output *per person* fell 13 per cent in sub-Saharan Africa while increasing 44 per cent in Asia and 48 per cent in South America (UNDP, 2012). A similar picture emerges for other food groups such as livestock and fish production. In 2012, out of the 66 countries classified by the FAO as being both low-income (GNI/person is less than US$1905 in 2009) and in food deficit (gross imports exceed gross exports using trade volumes for 2007–09 converted into and aggregated by the calorie content of individual commodities), 37 were from sub-Saharan Africa (FAO, 2012).[2] A further measure of the unavailability of food in sub-Saharan Africa is its reliance on emergency food aid from abroad. Over the past two decades an average of 32 countries in sub-Saharan Africa received emergency food aid every single year.[3]

2.2 Lack of Access to Food Drives Hunger in Asia and Africa

While food availability is a necessary condition for food security, it is not a sufficient condition. 'Food entitlements', in the terminology of Amartya Sen in his pioneering work on poverty and famines, refers to the ability of people to access or acquire food through legal means: *production* for auto-consumption, *exchange* through barter or purchase, or *transfers* such as gifts or humanitarian assistance (Sen, 1981).

In this framework, access to food can be restricted for reasons that go beyond food availability, including: limited purchasing power by individuals and households because of poverty and lack of income-earning opportunities; erosion in informal safety nets, for instance in countries and regions under prolonged environmental and/or health stresses, and the immaturity of formal mechanisms for social protection; and weak physical infrastructure that restricts accessibility of farmers and consumers to markets (for further discussion, see UNDP, 2012). To these entitlement failures, delays, poor targeting and other challenges in implementing humanitarian assistance are sometimes added, representing a *response* failure by policy makers or international aid agencies (Devereux, 2009).

Failures in these entitlements explain why, despite increasing availability of food at the aggregate level, an estimated 850 million people were under-nourished in 2006–08, corresponding to 13 per cent of the total world population (Table 31.1).[4] Two-thirds (568 million) of all under-nourished people in 2006–08 were in Asia (including 225 million in India and 130 million in China) while sub-Saharan Africa had the largest share of under-nourished people (27 per cent of the total population of the region). The share of hungry people in Africa has also declined less (by 13 per cent) than in Asia and Latin America (by 25 per cent and 33 per cent, respectively) since the early 1990s, and the absolute number of hungry people in Africa has actually increased by more than 50 million.

Table 31.1 Undernourishment: number of undernourished persons (in millions), prevalence of undernourishment in total population in brackets (in %)

	1990–92	1995–97	2000–02	2006–08
Asia[a]	607.1 [20]	526.2 [16]	565.7 [16]	567.8 [15]
Latin America and the Caribbean	54.4 [12]	53.4 [11]	50.8 [10]	47.0 [8]
Oceania[b]	0.7 [12]	0.8 [11]	1.0 [13]	1.0 [12]
Sub-Saharan Africa	165.9 [31]	188.2 [31]	197.7 [29]	217.5 [27]
World	848.4 [16]	791.5 [14]	836.2 [14]	850.0 [13]

Notes: [a] = excluding Japan; [b] = excluding Australia and New Zealand.

Source: FAO Statistics Division, http://www.fao.org/fileadmin/templates/ess/documents/food_security_statistics/PrevalenceUndernourishment_en.xls and http://www.fao.org/fileadmin/templates/ess/documents/food_security_statistics/NumberUndernourishment_en.xls, accessed July 2012.

2.3 Uneven Progress in Use of Food: Malnutrition Persists in Africa

In the 2009 WFS definition, nutrition is considered an integral dimension of food security (FAO, 2009). While hunger occurs when a person's diet does not provide adequate calories, malnutrition occurs when the diet lacks either protein or micronutrients (notably vitamin A, zinc, iodine and iron), both of which are necessary for growth and maintenance. The types of food consumed, dietary diversity and the methods used for food processing and preparation all affect the bioavailability of micronutrients. Non-food factors such as disease burden, sanitary conditions and levels of education also interact as determinants of proper food utilization.

An estimated 2 billion people suffer from one or more micronutrient deficiencies (Graham et al., 2006). A diet that lacks proteins, essential fatty acids, and micronutrients such as iron has been shown to lower birth weight, growth and cognitive development and increase child mortality (Pelletier et al., 1993; Habicht, 2008). The damages can be long-term: poor foetal growth or stunting in the first two years from birth leads to irreversible damage, including shorter adult height, lower attained schooling, reduced adult income, and decreased birth weight of offspring (Victoria et al., 2008).

Acute malnutrition, or 'wasting' (low weight for height), occurs as a result of recent rapid weight loss, malnutrition, or a failure to gain weight within a relatively short period of time. Wasting affects 9 per cent of pre-school children or 56 million every year (Table 31.2) and caused an estimated 1.5 million deaths in 2004 (Black et al., 2008). Measures of *chronic* malnutrition, which occurs as a result of deprivation in adequate nutrition over a longer period of time, show that more than one in four (27 per cent) or 171 million of all pre-school children in 2010 were stunted (low height for age) in their growth and almost one in six (16 per cent) or 104 million were underweight (low weight for age).

Over the past two decades child malnourishment has improved markedly in developing Asia and in Latin America; stunting rates are down by more than 40 per cent. Again, for sub-Saharan Africa progress has been much slower. Because of population growth the *actual* number of children that are under-nourished has grown over the period and is forecast to continue to grow over the next decade. In Africa inequities in food availability and access, which produce high rates of malnutrition, are exacerbated by poor access to water and sanitation, and weak health infrastructure in many countries (UNDP, 2012).

While hunger and under-nourishment are significant challenges, mainly in developing countries, obesity and over-nourishment is a growing problem almost everywhere (Popkin et al., 2012). This is caused by a nutrition transition associated with urbanization and changing lifestyles, for example away from manual labour and towards more sedentary types of activities, and dietary habits, including consumption of cheap and energy-dense foods, a process that has been characterized as a clash between human biology and new technologies (Popkin et al., 2012).

People who are overweight and obese are more at risk of developing high blood pressure, high blood cholesterol or other lipid disorders, type 2 diabetes, heart disease, stroke, and certain types of cancer (WHO, 2000). The WHO estimates that in 2008, 1.4 billion adults over the age of 20 were overweight or obese (WHO, 2012). Among children under five, the corresponding number was 43 million children in 2010 (80 per cent of whom were in developing countries), a number that is expected to reach 60 million by 2020 (de Onis et al., 2010).

2.4 Instability in Food Systems is a Growing Concern

Food systems are constantly exposed to change, whether in the form of shocks, trends or cycles. Covariate shocks as a result of natural disasters such as earthquakes, droughts or floods can cause damages that are immediate (for example to food already in storage or the destruction of crops ready for harvesting) or longer term (for example through the destruction of the sector's productive capacity). Man-made disasters also severely affect food security. According to one set of estimates the developing world lost nearly US$121 billion in agricultural output as a result of conflict between 1970 and 1997, enough to lift 330 million people out of hunger in each of the years (FAO, 2002).

The effects of such disruptions go beyond food availability to affect access, such as when infrastructure is damaged and market interactions curtailed, and utilization, with

Table 31.2 Child malnourishment:[a] *number of malnourished children under 5 (in millions), prevalence of malnourishment among all children under 5 in brackets (in %)*

	1990	2000	2010	2020[e]
Wasting (weight-for-height)				
Asia[b]	45.3 [12]	41.3 [11]	38.4 [11]	n/a [n/a]
Latin America and the Caribbean	1.3 [2]	1.1 [2]	0.9 [2]	n/a [n/a]
Oceania[c]	0.1 [6]	0.1 [5]	0.1 [4]	n/a [n/a]
Sub-Saharan Africa[d]	8.2 [9]	10.3 [10]	13.2 [10]	n/a [n/a]
World	57.4 [9]	55.1 [9]	55.5 [9]	n/a [n/a]
Stunted (height-for-age)				
Asia[b]	189.9 [49]	138 [38]	99.5 [28]	68.4 [19]
Latin America and the Caribbean	13.2 [24]	10.2 [18]	7.2 [14]	4.9 [10]
Oceania[c]	n/a [n/a]	0.5 [39]	0.5[38]	n/a [n/a]
Sub-Saharan Africa[d]	38.1 [43]	45.7 [42]	54.8 [41]	59.6 [40]
World	253 [40]	203.8 [33]	171.4 [27]	142 [22]
Underweight (weight-for-age)				
Asia[b]	132 [34]	96.5 [26]	70.5 [20]	49.3 [14]
Latin America and the Caribbean	4.2 [8]	2.8 [5]	1.8 [3]	1.1 [2]
Oceania[c]	n/a [n/a]	0.2 [17]	0.2 [15]	n/a [n/a]
Sub-Saharan Africa[d]	21.5 [24]	24.8 [23]	28.3 [21]	29.5 [20]
World	161.8 [25]	127.4 [21]	103.7 [16]	82.5 [13]

Notes: [a] Each measure is defined as a shortfall in 2 standard deviations or more from the mean of the National Center for Health Statistics/World Health Organization international reference population; [b] excluding Japan; [c] excluding Australia and New Zealand; [d] For sub-Saharan Africa the figures are computed based on sub-regional data; [e] Forecast.

Source: World Health Organization (WHO) (2011).

outbreaks of disease and collapse in public health systems. Moreover, idiosyncratic shocks, such as when a household member loses a job or falls ill, can have particularly severe effects on food security in developing economy settings. Throughout the developing world, but nowhere more than in Southern Africa, the pernicious interaction between malnutrition and HIV/AIDS has reduced the viability of farming as a livelihood, severely increasing the vulnerability of rural communities to a 'new variant famine' (De Waal and Whiteside, 2003, pp. 1234–7). Cyclical or longer-term stresses – such as seasonal harvesting patterns that result in long 'hungry seasons' between harvests or creeping environmental degradation – are slower moving and more predictable (UNDP, 2012). But they devastate communities all the same, especially those that cannot manage their exposure to hazards and protect their livelihoods.

Stresses from population pressure are pervasive and growing: global food production must increase by 70 per cent by 2050 in order to keep up with population growth and changes in demands (Bruinsma, 2009). Unchecked environmental degradation will exacerbate the challenge. The limited data that is available suggests that already in the early 1990s an estimated 23 per cent of agricultural soils were degraded (Oldeman, 1994). Rising temperatures as a result of human-induced climate change, resulting in higher evaporation and increased precipitation, will further destabilize global food systems (Smith et al., 2007).

These factors will cause upward pressures in the prices of the most important agricultural crops (rice, wheat, maize and soybeans) and lead to higher prices for meat and cereals. Projections derived from combining simulations of agricultural production with biophysical crop models show that, as a result of climate change, calorie availability in the developing world in 2050 could be below the level it was in 2000 and that child malnutrition could increase by up to 20 per cent compared to a world without climate change (Nelson et al., 2009).

3 THE ROLE OF BIOTECHNOLOGY IN ENHANCING FOOD SECURITY

In the international arena biotechnology has been defined to mean 'any technological application that uses biological systems, living organisms, or derivatives thereof, to make or modify products or processes for specific use' (UN, 1992). In the agriculture and food sciences biotechnology is considered to straddle a range of research areas, including genomics, marker-assisted selection, tissue culture, cloning, artificial insemination, embryo transfer and other technologies (FAO, 2004).

There are two main types of biotechnological processes. The first is associated with conventional plant or animal breeding methods that have been used by farmers for centuries to speed up and improve desirable biological traits. The second is both more advanced and more controversial as it seeks to genetically modify (GM) the pattern of a plant or animal to express novel traits. The key attraction of genetic modification from a food security perspective is that it makes the process of crop improvement more efficient and timely. In comparison to conventional plant breeding techniques, where it can take years to develop or eliminate traits by selection, GM techniques provide opportunities for manipulation of genetic material with more precision, with greater room for experimentation and with more speed. Both conventional breeding techniques and genetically modified organisms (GMOs) are already transforming agriculture and global food security.

3.1 Biotechnology Can Boost Crop Yields and Food Availability

Innovation and new technologies in agricultural systems, notably new high-yielding crop varieties (HYVs), especially rice and wheat, agro-chemicals and new irrigation techniques, generated the 'Green Revolution' of the 1960s and 1970s, which boosted crop output and helped lift millions of people out of hunger and poverty. The adoption of HYVs occurred rapidly. By 1970, about 20 per cent of the wheat area and 30 per cent of the rice area in developing countries were planted to HYVs, and by 1990, the share had increased to about 70 per cent for both crops, and their yields had virtually doubled (IFPRI, 2002). According to one set of estimates, without this progress crop yields in developing countries would have been 24 per cent lower and prices between 35 per cent and 66 per cent higher in 2000 (Evenson and Gollin, 2003). Moreover, caloric intake would have been 14 per cent lower and the proportion of malnourished children would be nearly 8 per cent higher. The 'Gene Revolution' appears to be outpacing anything that has come before it. An estimated 94-fold increase in area planted from 1.7 million hectares in 1996 to 160 million hectares in 2011 makes GM crops the fastest adopted crop technology in the history of modern agriculture (Brookes and Barfoot, 2012).

To date two major traits of GM crops have gone to market: herbicide-tolerant (HT) crops and insect-resistant (IR) crops (Fukuda-Parr and Orr, 2012). HT crops carry traits that make them resistant to a broad-spectrum herbicide (that is glyphosate or Roundup™, glufosinate or Liberty™ or imidazoline) and are mainly used in maize, canola and soybeans. These modified crops can be sprayed with the herbicide for weed control without being harmed, which makes weed control more cost effective. IR crops, mainly cotton and maize, have been developed to include the bacterium *Bacillus thuringiensis* (Bt) in the genetic make-up of a plant, which makes it poisonous to certain insects. This reduces the quantity of insecticides needed during cultivation, which in turn reduces production costs as well as the risks of crop failure, poisoning and environmental damage. HT and IR traits at times are combined (called 'stacked') in some crops, especially some maize varieties.

Critics of biotechnology are right to question its role in world agriculture – like other agricultural practices, biotechnology is not risk free. Concerns such as the transfer of modified genes to wild relatives and the development of resistance to pests need to be taken seriously and kept under constant review. According to a 2010 European Commission report on GMOs surveying 130 research projects spanning more than 25 years and involving more than 500 independent research groups, 'biotechnology, and in particular GMOs, are not *per se* more risky than e.g. conventional plant breeding technologies' (EC, 2010).

Moreover, studies on public health impacts of GM crops do not give cause for concern either. For instance, in 2004 the US National Research Council found that no adverse health effects could be attributed to GM traits despite American corn consumption rising from 5.9 kilograms per capita annually in 1980 to 15 kilograms annually by 2008, while the portion of the crop that was genetically modified rose from zero to 80 per cent over the same period (Biello, 2011).

3.2 Access to Food Improves when Incomes Rise

Agricultural productivity growth supported by biotechnology can boost access to food and contribute to poverty reduction. There is plenty of evidence that growth underpinned by productivity gains can reduce poverty far more effectively than can growth in the rest of the economy (Thirtle et al., 2003).

There are several reasons why agricultural growth is more effective in reducing poverty than growth in other sectors (Lipton, 2012). First, increased farm production should lead to more and cheaper food (and often more stable access to food, as stocks grow) and to higher income from sales of cash crops and livestock. Both boost purchasing power and, therefore, the ability to buy food. Second, land, the main asset in farming, is typically more evenly distributed in Africa than are the human and physical capital required for non-farm production. Third, small-scale, labour-intensive farming, as in Africa, has productivity advantages that large, capital-intensive mechanized farming does not.[5] The ability of biotechnology to boost the incomes of the poor and their food access will thus depend on its impact in terms of boosting the productivity mainly of smallholder farmers.

What is the evidence related to GM crops so far and what are the potential downsides? Since their first commercial use, the economic benefits of GM crops to small-scale farmers have continued to raise policy interest. Critics of the technology have used episodes of crop failure to argue that the technology does not benefit farmers. Recent evidence from India using unique panel data on the adoption of Bt cotton contradicts these views. It shows that 'Bt cotton has caused a 24% increase in cotton yield per acre through reduced pest damage and a 50% gain in cotton profit among smallholders. These benefits are stable; there are even indications that they have increased over time' (Kathage and Qaim, 2012, pp. 11652–6). The study further 'shows that Bt cotton adoption has raised consumption expenditures, a common measure of household living standard, by 18% during the 2006–2008 period' (ibid.). It concludes that Bt cotton has resulted in large and sustainable benefits, which contribute positively to economic and social development in India. The increase in consumption expenditures is particularly important because it shows the potential increase in the purchasing power of farmers (and thus their exchange entitlements) and their access to food.

Biotechnology research has direct relevance for contributing to current food security in Africa. Take banana cultivation in Uganda as an example. The staple is currently being threatened by a bacterial disease, *Xanthomonas* wilt, causing an annual loss of more than US$500 million. The disease also affects banana production in other Great Lakes countries. There are no resistant banana varieties and no chemical treatment of the disease is available today. Adopting disease-resistant GM banana would benefit consumers and farmers directly and enhance food security. Ugandan scientists are developing a GM banana using a gene from sweet pepper (*Capsicum annuum*) that helps to control the disease (Juma, 2011b). (For a more detailed discussion of GM bananas, see Chapter 48 in this volume.)

Another example of GM technology that could enhance food security in Africa involves the black-eyed pea, a subspecies of the cowpea (*Vigna unguiculata*). The crop is infested by *Maruca vitrata*, a pod-boring insect that results in annual losses of about US$300 million for smallholders in Africa. The continent produces about 5.2 million tonnes of the bean yearly, nearly 70 per cent of global output. Researchers at Ahmadu Bello University in Nigeria have developed a GM cowpea variety using Bt genes that can help control the disease. The only method available today for controlling the disease is spraying expensive pesticides, which cost Nigeria alone over US$500 million annually (Juma, 2011b). Using biotechnology would not only enhance food security but it will also reduce human exposure to toxic pesticides.

There are still many challenges that need to be overcome in order to ensure that the conditions exist for successful use of biotechnology. The institutional shifts related to the economics of GM crops differ in several critical ways from traditional agricultural innovation systems (Fukuda-Parr and Orr, 2012). First, GM crop research today is dominated by the private sector with the exception of the national public research systems of China, Brazil and India. Second, since GM crops are protected by intellectual property (patents) and cannot be reproduced without licence, farmers and mostly local and indigenous seed companies cannot reproduce the seeds freely. Third, they incur higher costs of innovation and certification. These high costs drive out small investors. Fourth, with patenting, they could generate significant profits, particularly when used for crops in high demand globally and therefore attractive to large investors. The unique challenges and opportunities associated with GM crops make institutional shifts essential in order to adapt GM crops to local environments.

The benefits from GM crops appear to be spreading: the US still dominates GM production with 69 million of the world's 160 million hectares of biotech crop area. But the cultivation in developing countries, especially Brazil, Argentina and India, is growing rapidly (their combined GM area in 2011 was 65 million hectares) (Brookes and Barfoot, 2012). Of the 29 countries planting biotech crops in 2011, 19 were developing and 10 were industrial countries. Notably, more than 90 per cent of the farmers using GM varieties were small, resource-poor farmers in developing countries (Brookes and Barfoot, 2012). Still, research and development in biotechnology must continue on the specific challenges faced by the poor while 'regulatory procedures should be strengthened and rationalized to ensure that the environment and public health are protected and that the process is transparent, predictable and science-based' (FAO, 2004).

Technology can empower the poor and vulnerable by building human capabilities and knowledge. But if misapplied, technology can dispossess or marginalize poor people. Narrowly compartmentalized farm science, conducted far from farmers' fields, can produce results that are irrelevant to smallholder farmers, leading to unequal outcomes and wasted opportunities. Success often comes from combining farmers' experiences with research results to build human and social capital and allow technologies to fully inform livelihoods (Lilja and Dixon, 2008). But success is never guaranteed. West Africa's slow uptake of the New Rice for Africa (NERICA), a cross between Asian and African varieties, shows that participatory training sometimes takes too long to replace more traditional extension and seed delivery systems (Kijima et al., 2011).

Gender impacts of biotechnology also need to be considered. Herbicide-tolerant maize (corn), now in use in South Africa and Egypt, reduces the need for weeding – a significant benefit for African farmers, mostly women, who spend nearly 200 hours a year weeding a hectare of land (Juma, 2011b). When a new technology results in a more profitable crop or when a new processing machine increases income, men often move in and take over. Policies thus need a gender perspective to ensure that technologies are developed and applied in ways that shield them from automatic takeover by men (Quisumbing and Meinzen-Dick, 2001).

3.3 Enhancing Nutritional Value through Biotechnology

Bio-fortification of crops, that is, making crops more nutritious through conventional breeding and GM, together with fortification of foods, could increase the nutritional value of food – and its variety. Bio-fortification, though still controversial, is not new. Recent research has produced important breakthroughs (CIAT and IFPRI, 2002). Because bio-fortification tends to focus on unprocessed food staples – foods that poor people eat in large quantities every day – it implicitly targets low-income households that do not consume commercially fortified processed foods.

Bio-fortification also promises low recurrent costs because once self-fortifying seeds have been developed, countries can share them. A bio-fortified crop system can thus be highly sustainable, producing nutrient-strengthened varieties year after year, regardless of policy fads or breaks in the supply of micronutrient supplements. The vitamin A-rich sweet potato developed by HarvestPlus and its partners, now available in Mozambique and Uganda (Syngenta, 2012), hints at the successes possible through bio-fortification. Moreover, after more than a decade, 'Golden Rice', a GM rice that contains enhanced levels of beta carotene, which converts to vitamin A in humans, is advancing towards the completion of its regulatory requirements in the Philippines and Bangladesh and could be released in 2013 (James, 2011).

But there are limits. While bio-fortified crops are still in development, fewer nutrients can be bred into them than can be added through commercial fortification and supplementation, and breeding for micronutrient density might come at the expense of breeding for other traits, such as drought or pest tolerance. In any case, the staple crops targeted by bio-fortification are insufficiently dense in protein and fats to meet the nutritional needs of infants, who suffer the most severe malnutrition. This limitation is partly offset by the health benefits to infants and children of having better-nourished mothers.

Bio-fortified foods are also criticized for concentrating nutrients in a few superior crop varieties bred by a handful of private companies, to the detriment of biodiversity and competition (Shiva et al., 2011). It will take strong research management and socially responsible investments in nutrition by the food industry and private companies for bio-fortification to deliver on its powerful promise. It will also take government regulation, farmer involvement and consumer awareness. The potential of bio-fortification can be realized if plant breeders, nutritionists, researchers, governments, non-governmental organizations and commercial firms collaborate (Bouis and Islam, 2012).

3.4 Making Food Systems More Stable and Resilient

A key criticism of the 'Green Revolution' is that it led to an excessive and inappropriate use of fertilizers and pesticides, which in turn polluted waterways, poisoned agricultural workers, and killed beneficial insects and other wildlife (Conway, 1998). It is essential that agricultural practices become more environmentally sustainable. Currently, food and agriculture are responsible for a large share of human-caused greenhouse gas emissions; agriculture alone contributed 13.5 per cent of global annual emissions of anthropogenic greenhouse gases (GHG) between 1970 and 2004, a slightly higher share than even the transport sector (Pachauri and Reisinger, 2007).

A key challenge to ensure food system stability is therefore to make agriculture a net emissions sink, not a net source of GHG, while at the same time addressing the food needs of a global population that is expected to reach 9 billion by 2050 (Godfray et al., 2010). To meet this challenge focus needs to be not only on raising productivity, but also on preparing the sector – and the people whose livelihoods depend on it – for a warmer world (Naylor, 2011). Biotechnology holds promise in both areas.

The application of biotechnology has a number of unintended ecological benefits. For example, the adoption of GM crops has curbed the release of greenhouse-gas emissions by reducing the use of pesticides, which require energy to manufacture. It has also saved farmers from heavy exposure to these toxic chemicals. In addition, the use of herbicide-tolerant crops enables farmers to cut back on the ploughing and weeding that releases carbon that would otherwise be sequestered in the soil. The combined effects of lower fossil fuel use and increased carbon sequestration as a result of biotechnology applications in agriculture is significant in terms of lowering GHG emissions. For 2010, the impact has been estimated as lowering CO_2 emissions alone by 19 billion kg – the equivalent of the annual emissions of 9 million cars (Brookes and Barfoot, 2012).

Using both conventional breeding methods and GMs, researchers in biotechnology have made great strides in developing varieties that have built-in tolerance mechanisms to water shortage, which makes it possible to deliver high yields even during drought conditions. For instance, under the 'Drought Tolerant Maize for Africa', 34 new drought-tolerant maize varieties have been developed and disseminated to farmers in 13 project countries – Angola, Benin, Ethiopia, Ghana, Kenya, Malawi, Mali, Mozambique, Nigeria, Tanzania, Uganda, Zambia and Zimbabwe – between 2007 and 2011. An estimated 2 million smallholder farmers are using the drought-tolerant maize varieties and have obtained higher yields, improved food security and increased incomes.

4 WIDER POLICY IMPLICATIONS

Biotechnology can only play its role in addressing food security when placed in a wider policy context. This is mainly because biotechnology encompasses a set of tools that need to be integrated into farming systems which are in turn part of overall agricultural development strategies. There are at least three critical policy areas that need to be addressed when dealing with the contribution of biotechnology to food security: fostering food production and nutrition; increasing access to food; and risk management.

4.1 Fostering Food Production and Nutrition

The most urgent task for developing countries is to formulate biotechnology policies and strategies aimed at a diversity of fields associated with biotechnology. So far many developing countries have put considerable effort in developing biosafety policies aimed at regulating importation of biotechnology products as inspired by the Cartagena Protocol on Biosafety to the Convention on Biological Diversity. It has been argued that the implementation of these policies has been slow because of the lack of adequate technical and legislative capacity.

This view, however, misses the critical point that starting with biosafety was a misguided view. What is needed is to design biotechnology policies aimed at advancing the application of life sciences. Biosafety, which will be discussed below, should be part of biotechnology and not the other way round. Biotechnology policies would involve building local capacity in universities and research institutes, promoting international technology cooperation, designing strategies for knowledge management, which include intellectual property protection, and addressing the safety aspects of biotechnology.

The prospects of using biotechnology to improve food production and enhance nutrition are being reinforced by the exponential growth in science and technology and greater international cooperation through enhanced internet connectivity. The phenomenon of abundant technological knowledge is illustrated by the rise of the open access movement and lowering of entry barriers to genomics research (Juma, 2012). For example, in 2001 it cost about US$100 million to sequence the complete human genome. Just over a decade later the price had dropped to US$1000. This drop has put into the public domain enormous amounts of sequence information upon which developing countries can leapfrog into a new age of functional genomics at minimal cost. These advances also usher in a new era of ecologically-sound biotechnology that builds on new understanding of the metabolic pathways of plants (De Luca et al., 2012; Milo and Last, 2012).

Countries will need to set their priorities on how to use biotechnology to promote food production and improve nutrition. This will include improving existing crops and animals as well as increasing the diversification of the food base by adding new sources.

Biotechnology should not be viewed as a substitute for other techniques but as an addition to existing practices. The logic should be to expand the toolbox for agricultural production by promoting technological diversity. For this reason, biotechnology policies should be part of national agricultural policies aimed at promoting the application of science, technology and innovation in development. This approach will require a fundamental review of many of the existing policies that emphasize biosafety as a starting point.

4.2 Increasing Access to Food

Increasing access to food is not limited to producing food, but entails raising income levels. The ability to participate in local, national, regional and international trade is an important aspect of increasing access to food. This is intricately linked to the addition of value to agricultural products. A major limitation on trade is poor infrastructure (transport, energy, irrigation and telecommunications). Without adequate infrastructure it is not possible for farmers to derive much value from their produce. In fact, poor infrastructure tends to limit the scope of production to most peasant systems – farmers tend to grow just enough to use, transport or store (Juma, 2011a, Chapter 5).

In addition to infrastructure, access to food is affected by a wide range of trade-related policies that limit direct or indirect access to markets. One of the most important policies involves restrictions in trade in GM products on safety grounds. Such policies are practised by industrialized and developing countries alike. Not only do such policies restrict access to food by limiting the scope of traded products, but they also inhibit exchange of critical scientific and technological knowledge needed to address food security. Many of these restrictions fall under the broader rubric of biosafety and were promoted in the early stages of the development of the biotechnology industry. There is now sufficient evidence to show that the benefits for biotechnology outweigh the risks. As a result, more favourable policies are needed to foster trade in biotechnology products in general and its application to food security in particular.

4.3 Risk Management

The final major area of concern involves risk management. So far most of the debates surrounding agricultural biotechnology have tended to start off with the presumption of risks associated with the adoption of the technology. A more rational policy, however, should put the risk of adopting new technologies in a broader context. This context would also involve the risks posed by not adopting the technology or not acting in time. Emerging evidence related to long-term impacts of climate management on agriculture now indicate clearly that 'doing nothing' and failing to adopt new technologies in time could have far-reaching implications for global food security.

In this respect, risk management strategies need to balance between a diversity of risks, which include explicitly considering the impacts of 'doing nothing'. Such an approach would require that biosafety, for example, be pursued as part of larger strategies to use biotechnology to advance food safety. It would also require that risk management be science-based while taking into account considerably broader societal consultations. Biosafety policies that focus on narrow biotechnology hazards are therefore more likely to introduce new threats to food security and should be reviewed in light of the need to expand technological options as a strategy for food security.

CONCLUSION

Alleviating world hunger requires more than just producing more food: availability is important for food security but so is access to and utilization of food, and ensuring stability in food systems. Biotechnology has shown promise in terms of accelerating progress across all the dimensions of food security. But as with all new technologies there are relative risks related to the economic, environmental, public health and equity consequences. It is therefore crucial that institutional frameworks are developed to strengthen public oversight and transparency and empower small-scale producers and consumers alike. The aim should be to ensure an inclusive process of technology development, adaptation and adoption where the benefits are maximized across society but biased squarely in favour of the poorest and most food-insecure.

These concerns are not unique to biotechnology and apply to other fields of technological endeavour. The debate over biotechnology provides a unique opportunity for a better understanding of the place of all technologies in society. What is done for biotechnology can serve as a standard that can be applied in other fields. The world is entering a genuine age of technological abundance that opens up new opportunities for more inclusive societies. But such inclusion will require reforms in institutional arrangements. Without such reforms 'food security' may end up serving as a metaphor of depravity in all aspects of socio-economic endeavour. Done right, the world will enter a new age that can translate technological abundance into opportunities for all.

NOTES

1. The chapter presents personal views of Conceição and Levine, and not necessarily those of the United Nations Development Programme.
2. Countries 'graduate' from the list after meeting the two criteria for three consecutive years. A third self-exclusion criterion is also applied when countries that meet the two first criteria specifically make a request to the FAO to that they be excluded from the LIFDC category.
3. Authors' computations using World Food Programme Food Aid Information System. Estimates exclude Somalia and Sudan.
4. FAO is revising the methodology for measuring undernourishment and therefore has not updated its estimates for the number of undernourished people in 2009 and 2010 or estimated the number for 2011.
5. Latin America and the Caribbean are an exception, since large gains in agricultural labour productivity were driven by mechanization, with little impact on rural poverty, in contrast with the experience in Asia (de Janvry and Sadoulet, 2010).

REFERENCES

Biello, David (2011), 'Coming to a cornfield near you: genetically induced drought resistance', *Scientific American*, 13 May.

Black, R.E. et al. (2008), 'Maternal and child undernutrition: global and regional exposures and health consequences', *Lancet*, **371**(9608), 243–60.

Bouis, Howarth and Yassir Islam (2012), 'Biofortification: leveraging agriculture to reduce hidden hunger', in Shenggen Fan and Rajul Pandya-Lorch (eds), *Reshaping Agriculture for Nutrition and Health*, Washington, DC: International Food Policy Research Institute.

Brookes, G. and P. Barfoot (2012), 'Global impact of biotech crops: environmental effects, 1996–2010', *GM Crops and Food: Biotechnology in Agriculture and the Food Chain*, **3**(2), 129–37.

Bruinsma, Jelle (2009), 'The resource outlook to 2050: by how much do land, water, and crop yields need to increase by 2050?', Proceedings of the Food and Agriculture Organization of the United Nations Expert Meeting on How to Feed the World in 2050, 24–26 June, Rome: FAO.

CIAT (International Centre for Tropical Agriculture) and IFPRI (International Food Policy Research Institute) (2002), 'Biofortified crops for improved human nutrition: a challenge program proposal', Washington, DC, available at: http://www.cgiar.org/pdf/biofortification.pdf, accessed 3 October 2012.

Conway, Gordon (1998), *The Doubly Green Revolution: Food for All in the Twenty-First Century*, Ithaca, NY: Cornell University Press.

De Janvry, Alain and Elisabeth Sadoulet (2010), 'The global food crisis and Guatemala: what crisis and for whom', *World Development*, **38**(9), 1328–39.

De Luca, V. et al. (2012), 'Mining the biodiversity of plants: a revolution in the making', *Science*, **336**, 1658–61.

De Onis, M., M. Blössner and E. Borghi (2010), 'Global prevalence and trends of overweight and obesity among preschool children', *American Journal of Clinical Nutrition*, **92**, 1257–64.

De Waal, Alex and Alan Whiteside (2003), 'New variant famine: AIDS and food crisis in Southern Africa', *Lancet*, **362**(9391), 1234–7.

Devereux, Stephen (2009), 'Why does famine persist in Africa?', *Food Security*, **1**(1), 25–35.

European Commission (2010), 'A decade of EU-funded GMO research (2001–2010)', Brussels: European Union, available at: http://ec.europa.eu/research/biosociety/pdf/a_decade_of_eu-funded_gmo_research.pdf, accessed 3 October 2012.

Evenson, R.E. and D. Gollin (2003), 'Assessing the impact of the green revolution, 1960–2000', *Science*, **300**(5620), 758–62.

FAO (1996), 'Rome declaration on world food security and world food summit plan of action', *World Food Summit*, 13–17 November.

FAO (2002), *The State of Food Insecurity in the World (SOFI) 2002*, Rome: FAO.
FAO (2004), 'Agricultural biotechnology: meeting the needs of the poor', in *The State of Food and Agriculture 2003–2004*, part 1 Rome: FAO.
FAO (2009), 'Declaration of the World Summit on Food Security', World Summit on Food Security, 16–18 November.
FAO (Food and Agriculture Organization of the United Nations), IFAD (International Fund for Agricultural Development), IMF (International Monetary Fund), OECD (Organisation for Economic Co-operation and Development), UNCTAD (United Nations Conference on Trade and Development), WFP (World Food Programme), World Bank, WTO (World Trade Organization), IFPRI (International Food Policy Research Institute), and UN HLTF (United Nations High-Level Task Force on Global Food Security) (2011), 'Price volatility in food and agricultural markets: policy responses', Interagency Report to the G20 on Food Price Volatility, available at http://www.oecd.org/tad/agriculturaltrade/48152638.pdf, accessed 3 October 2012.
FAO Country Profiles (2012), 'Low-Income Food-Deficit Countries (LIFDC) – List for 2012', available at http://www.fao.org/countryprofiles/lifdc/en/, accessed June 2012.
FAOSTAT (2012), 'United Nations Food and Agriculture database', available at: http://faostat.fao.org.
Fukuda-Parr, Sakiko and Amy Orr (2012), 'GM crops for food security in Africa: the path not yet taken', Working Paper no. 2012-018, New York: United Nations Development Programme, Regional Bureau for Africa.
G8 Leaders (2009), 'G8 leaders declaration: responsible leadership for a sustainable future', L'Aquila, 10 July, available at: http://www.g8italia2009.it/static/G8_Allegato/Chair_Summary%2c1.pdf, accessed June 2012.
Godfray, H.C. et al. (2010), 'Food security: the challenge of feeding 9 billion people', *Science*, **327**(5967), 812–18.
Graham, R.D.W. et al. (2006), 'Nutritious subsistence food systems', *Advances in Agronomy*, **92**, 1–72.
Habicht, J.P. (2008), 'Malnutrition kills directly, not indirectly', *Lancet*, **24**, 1749–50.
International Food Policy Research Institute (IFPRI) (2002), 'Green Revolution, curse or blessing?', Washington, DC: IFPRI.
James, Clive (2011), 'Global status of commercialized biotech/GM crops: 2011', *International Service for the Acquisition of Agri-Biotech Applications (ISAAA) Brief*, 43. Ithaca, NY: ISAAA.
Juma, C. (2011a), *The New Harvest: Agricultural Innovation in Africa*, chapter 5, New York: Oxford University Press.
Juma, C. (2011b), 'Preventing hunger: biotechnology is key', *Nature*, November, **479**, 471–224.
Juma, C. (2012), 'Technological abundance for global agriculture: the role of biotechnology', Faculty Working Paper Research Series, no. RWP12-008, Cambridge, MA: Harvard Kennedy School.
Kathage, Jonas J. and Matin Qaim (2012), 'Economic impacts and impact dynamics of Bt (Bacillus thuringiensis) cotton in India', *Proceedings of the National Academy of Sciences*, **109**(29), 11652–6.
Kijima, Yoko, Keijiro Otsuka and Dick Sserunkuuma (2011), 'An inquiry into constraints on a Green Revolution in sub-Saharan Africa: the case of NERICA rice in Uganda', *World Development*, **39**(1), 77–86.
Lewis, C. et al. (2010), 'Tanzanian farmers' knowledge and attitudes to GM biotechnology and the potential use of GM crops to provide improved levels of food security: a qualitative study', *BMC Public Health*, **10**, 407–16.
Lilja, Nina and John Dixon (2008), 'Responding to the challenges of impact assessment of participatory research and gender analysis', *Experimental Agriculture*, **44**(1), 3–19.
Lipton, Michael (2012), 'Learning form others: increasing agricultural productivity for human development in sub-Saharan Africa', Working Paper no. 2012-007, New York: United Nations Development Programme, Regional Bureau for Africa.
Mechlem, Kerstin (2004), 'Food security and the right to food in the discourse of the United Nations', *European Law Journal*, **10**(5), 631–48.
Milo, R. and R. Last (2012), 'Achieving diversity in the face of constraints: lessons from metabolism', *Science*, **336**, 1663–67.
Naiken, Loganaden (2002), 'FAO methodology for estimating the prevalence of undernourishment', keynote paper presented at the International Scientific Symposium on Measurement and Assessment of Food Deprivation and Undernutrition, 26–28 June, Rome.
Naylor, Rosamond L. (2011), 'Expanding the boundaries of agricultural development', *Food Security*, **3**(2), 233–51.
Nelson, G.C. et al. (2009), 'Climate change impact on agriculture and costs of adaptation', Washington, DC: International Food Policy Research Institute (updated October 2009).
Oldeman, L. Roel (1994), 'The global extent of land degradation', in D.J. Greenland and István Szabolcs(eds), *Soil Resilience and Sustainable Land Use*, Wallingford: CABI Publishing.
Pachauri, Rajendra K. and Andy Reisinger (eds) (2007), 'Climate change 2007: synthesis report', Geneva: Intergovernmental Panel on Climate Change.

Pelletier, D.L., E.A. Frongillo and J.P. Habicht (1993), 'Epidemiologic evidence for a potentiating effect of malnutrition on child mortality', *American Journal of Public Health*, **83**, 1130–33.
Popkin, B.M., L.S. Adair and S.W. Ng (2012), 'Global nutrition transition and the pandemic of obesity in developing countries,' *Nutrition Reviews*, **70**(1), 3–21.
Quisumbing, Agnes R. and Ruth Meinzen-Dick (2001), 'Empowering women to achieve food security', *2020 Focus 6*, Policy Brief 1, Washington, DC: International Food Policy Research Institute.
Sen, Amartya (1981), *Poverty and Famines: An Essay on Entitlement and Deprivation*, Oxford and New York: Clarendon Press and Oxford University Press.
Shiva, Vandana, Debbie Barker and Caroline Lockhart (2011), 'The GMO emperor has no clothes: a global citizens report on the state of GMOs – false promises, failed technologies', Florence, Italy: Navdanya International.
Smith, P. et al. (2007), 'Agriculture', in B. Metz et al. (eds), *Climate Change 2007: Mitigation*, Cambridge, UK, and New York: Cambridge University Press.
Syngenta Foundation for Sustainable Agriculture (2012), 'Biofortification of staple foods for Africa and Asia – HarvestPlus', available at: www.syngentafoundation.org/index.cfm?pageID=525, accessed 16 January 2012.
Thirtle, C., Lin Lin and Jenifer Piesse (2003), 'The impact of research-led agricultural productivity growth on poverty reduction in Africa, Asia and Latin America', *World Development*, **31**(12), 1959–75.
United Nations (UN) (1975), 'Report of the World Food Conference', 5–16 November 1974, Rome, New York: UN.
United Nations (UN) (1992), 'Convention on biological diversity', New York: UN.
United Nations Development Programme (UNDP) (2012), 'Africa human development report 2012: towards a food secure future', New York: UNDP.
United Nations High Level Task Force (UNHLTF) (n.d.), 'Global food security crisis: background information', available at: http://www.un.org/en/issues/food/taskforce/background.shtml.
United Nations News Center (2012), 'Rio+20: Secretary General challenges nations to achieve zero hunger', 22 June, available at: http://www.un.org/apps/news/story.asp?NewsID=42304, accessed June 2012.
United Nations Office for Coordination of Humanitarian Affairs (UNOCHA) (2012), 'Humanitarian Dashboard – Sahel food and nutrition crisis', February 2012, available at: http://ochanet.unocha.org/p/Documents/Sahel%20dashboard%20Feb%202012.pdf, accessed June 2012.
Victoria, C.G. et al. (2008), 'Maternal and child undernutrition: consequences for adult health and human capital', *Lancet*, **371**(9609), 340–57.
World Health Organization (WHO) (2000), 'Obesity: preventing and managing the global epidemic', Report Series no. 894, Geneva: WHO.
World Health Organization (WHO) (2011), 'Global database on child growth and malnutrition', Geneva: WHO, available at: www.who.int/nutgrowthdb/estimates/en/index.html, accessed 7 July 2012.
World Health Organization (WHO) (2012), 'Obesity and overweight', Fact sheet no. 311, May 2012, available at: http://www.who.int/mediacentre/factsheets/fs311/en/, accessed 10 July 2012.

32 International regimes on plant intellectual property rights and plant genetic resources: implications for stakeholders

Deepthi Elizabeth Kolady

1 INTRODUCTION

Technological advances in agriculture, especially in agricultural biotechnology with its potential to address food and nutritional security in the context of climate change and volatile commodity prices, have brought renewed global focus on agricultural research and development (R&D) investment, and plant intellectual property rights (IPRs). Since biological diversity and access to plant genetic resources are important not only for the development of new and improved varieties, but also for long-term food security, recent debates on IPR protection in agriculture inevitably focus on these topics as well.

We are witnessing a decline in public sector agricultural R&D investment in developed countries (Pardey et al, 2006; Alston et al., 2009). Sufficient evidence exists to suggest that enforcement of plant IPRs, in the form of plant breeders' rights or patents or a combination of both, has attracted private R&D investment into agriculture in developed countries (Perrin et al., 1983; Foster and Perrin, 1991; Fernandez-Cornejo, 2004). But inadequate protection of IPRs in agriculture has been suggested as one of the reasons for the limited participation of the private sector in agricultural R&D in developing countries (Pray, 1992; Kolady et al., 2010).

A number of diverse and cross-cutting arguments are focused on the role of IPRs in developing world agriculture. Some critics of IPRs in agriculture argue that IPRs lead to loss of genetic diversity and others are concerned about increased corporate presence in the seed sector. Others argue that plant breeders are making mostly incremental improvements to the pre-existing plant varieties whose development was facilitated by the farmers in the earlier eras, hence it is important to recognize not only breeders' rights but also farmers' rights (Fowler, 1994; Goodman, 2009). Further, it has been argued that even though IPR holders often use traditional knowledge in the development of protected products or processes, contributions of indigenous and local communities are never acknowledged. This has led to a global debate about the balance between rights and obligations of a plant breeder on the one hand and those of farmers, farming community and indigenous people on the other (Phillips, 2007).

Countries have responded to these concerns by forming various international organizations and negotiating treaties, each with its own specific objectives, scope and guidelines. The International Union for the Protection of New Varieties of Plants (UPOV), the Convention on Biological Diversity (CBD), the Trade Related Aspects of Intellectual Property Rights (TRIPS) Agreement of the World Trade Organization (WTO), and the International Treaty on Plant Genetic Resources for Food and Agriculture (ITPGRFA) are the most relevant treaties in the context of plant IPRs and plant genetic resources.

Even though each of these aims to achieve long-term food security and improvements in social welfare, as will be evident from the sections below, they differ in their objectives and approaches to achieve those objectives, raising the potential for conflicts between them. Because of this, the legal and practical implications of any one of these agreements/treaties in member countries depend not only on their effective implementation but also on each country's membership in other agreements/treaties.

There are concerns about the implications of establishing systems providing multiple rights on the use and exchange of plant genetic resources by various stakeholders such as farmers, plant breeders, agricultural biotechnologists and investors (Ramanna, 2003). Countries differ in their views on what rights should be protected and how they should be protected. This chapter examines the implications of international agreements on plant IPRs and plant genetic resources for farmers and agbiotech industries, with special focus on provisions of farmers' rights and protection of traditional knowledge by UPOV and ITPGRFA, respectively.

This chapter is organized as follows: section 2 briefly lays out the history of IPRs in agriculture; section 3 presents the background of international regimes on plant genetic resources (PGRs); section 4 lists and explains the international treaties/agreements dealing with plant IPRs and PGRs and identifies the main overlapping or conflicting provisions; section 5 identifies the major implications of overlapping or potentially conflicting agreements; and section 6 offers some concluding comments.

2 EVOLUTION OF IPR PROTECTION IN AGRICULTURE

There are two broad approaches for justifying IPRs in general: first, the philosophy that products of the human mind are stamped with the inventor's personality, hence, it is important to endow the inventor with a moral and economic claim to exploit the product to the exclusion of third parties; second, the instrumentalist or economic philosophy posits that an inventor's products are social welfare enhancing, hence it is important to provide adequate incentives for inventors to invest time, financial resources and intellectual capacity to create proprietary products (Lesser, 1997). In the absence of such exclusive rights for inventors, proprietary products would be exploited by the 'free riders'. Legal IPRs address this market failure by legally preventing direct copying through the granting of a temporary property right. Legal IPRs enhance social welfare by placing certain limits on the scope of protection and terms of protection, the requirement to fully describe and disclose the invention along with the application for protection and by recognizing situations in which consumers or other inventors could access and use the protected products for social benefits.

Even though IPR protection has been granted to industrial and manufactured products for centuries, countries until recently excluded plants and other agricultural applications from IPR protection, based on the economic argument that IPR protection in plants could lead to increased prices for essential food commodities, which might decrease social welfare (Kolady and Lesser, 2009). However, as the scope of scientific plant breeding expanded, researchers and policy makers recognized the importance of having some form of intellectual property protection in agriculture so that plant breeders could recoup the cost of researching and developing improvements to the pre-existing biological resources such as landraces or traditional varieties. Policy makers in developed countries argued that because plant varieties are naturally self-replicating, in the absence of such exclusive rights to the breeders, they are susceptible to unauthorized exploitation by third parties, both at home and in foreign countries. Further, they argued that such potential for exploitation and free-riding disincentivizes plant breeders from investing in research and development of new varieties and exporting seeds to other countries, which in turn reduces farmers' access to improved varieties. These discussions and debate led to the formation of the International Union for the Protection of New Varieties of Plants (UPOV) and Article 27.3(b) of the Trade Related Aspects of Intellectual Property Rights (TRIPS) agreement.

3 EVOLUTION OF INTERNATIONAL REGIMES ON PLANT GENETIC RESOURCES

It has been a common practice for farmers to save the seeds of their crops and use them for sowing in their own fields, exchange (usually known as brown-bagging) or to sell to other farmers for sowing. However, as inventors or breeders tried to capture a share of the incremental value generated by their research and development of new improved plant varieties through the creation of IPR mechanisms such as plant breeders' rights or patents, restrictions have been placed on the practice of brown-bagging of improved and protected varieties by farmers. By clearly laying out the exclusive rights of the owners (in this case plant breeders) and exceptions to their rights, the initial IPR regime in agriculture (that is, UPOV) permitted farmers to use, exchange and sell the seeds produced on their farm and called this 'farmers' privilege'. As the debate on IPR protection advanced, there have been concerns about the asymmetric benefits derived by the donors of plant genetic resources (particularly farmers and indigenous people) and the scope and rights of developers of technology. There are also concerns about the conservation and sustainable use of biodiversity and genetic resources.

Traditional knowledge is a broad term that includes subsets such as 'indigenous knowledge', 'folklore', and 'traditional medicinal use'. Traditional knowledge (TK) is embedded in the traditional system that each community has maintained and developed in its local context (WIPO, 2011). It has been argued that even though IPR holders quite often use such knowledge in the development of protected products or processes, contributions of indigenous and local communities are never acknowledged. Castle and Gold (2007) identified two types of traditional knowledge that relate to biodiversity and innovation: first the knowledge about the use of biological resources in health or agriculture based on previous exploitation (for example knowledge about the medicinal uses of neem and turmeric); and second, information about the existence of particular plants or animals that have characteristics that may be of interest to pharmaceutical or chemical companies.

As the demand for more stringent forms of IPRs in agriculture increased, discussions about the rights of farmers, protection of traditional knowledge and mechanisms to share the benefits arising out of their contributions in a fair and equitable way became active, especially at the Food and Agricultural Organization (FAO). These discussions and debate led to the formation of Convention on Biodiveristy (CBD) and International Treaty on Plant Genetic Resources for Food and Agriculture (ITPGRFA). For example, the origins of the International Treaty on Plant Genetic Resources (ITPGR) could be traced to the International Undertaking on Plant Genetic Resources (Undertaking), which argued that 'plant genetic resources are common heritage of mankind and hence should be available without restriction'. FAO Conference Resolution no. 5/89 defined farmers' rights as 'rights arising from the past, present and future contributions of farmers in conserving, improving and making available plant genetic resources, particularly those in the centers of origin/diversity'. It has to be noted that unlike plant breeders' rights or patents, debates on farmers' rights and traditional knowledge are associated more with the conservation and equity objectives.

4 INTERNATIONAL AGREEMENTS AND TREATIES DEALING WITH IPRS AND PLANT GENETIC RESOURCES

4.1 UPOV

Plant breeding industries in Europe recognized the importance of an international system of plant variety protection which attempts to protect the interests of plant breeders while facilitating access to the new plant varieties by farmers in other countries. The initiatives by the International Association for the Protection of Intellectual Property (AIPPI) and the International Association of Plant Breeders (ASSINSEL) were instrumental in establishing an international system of plant variety protection initially aimed mainly at European states. These organizations' initiatives to establish an international plant variety protection (PVP) regime led to diplomatic conferences during 1956–61 which recommended the formation of the International Union for the Protection of New Varieties of Plants (UPOV, its French acronym) in 1961. The UPOV convention came into force on 10 August, 1968, having been ratified by the UK, the Netherlands and Germany (Heitz, 1987; Dutfield, 2011).

The objective of UPOV is to provide and promote an effective system of plant variety protection, with the aim of encouraging the development of new varieties of plants for the benefit of society. Basically, UPOV establishes a framework law that may be adopted into national law. Since its inception, UPOV has adopted Acts in 1961 (amended in 1972), 1978 and 1991. The 1991 Act is the only one presently open for new members. UPOV is an inter-governmental organization under the aegis of the Director General of the World Intellectual Property Organization (WIPO). In 2012 there were 70 members in the UPOV, mostly developed countries.

The 1991 Act adopts core international IPR obligations such as a definition of applicable subject matter (Article 1) and genera and species to be protected (Article 3), identification of the conditions for the granting of plant breeders' rights (Article 5-9) and exceptions and limitations to breeders' rights (14-19).

4.1.1 Definition of applicable subject matter and genera and species to be protected

UPOV protects plant 'varieties', a term open to multiple definitions and understandings. In the 1991 Act (Article 1(vi)) the definition hinges on the existence of specific characteristics which are stable through repeated propagations. The 1991 Act requires that all plant or species are protected within ten years of ratifying or acceding to the Act (Article

3). The Act also permits dual protection, in that a single variety can be protected both by patents and by PVP, and protection of discovered varieties.

4.1.2 Conditions for granting of plant breeders' rights

Plant varieties which are new, distinct, uniform and stable are eligible for PVP protection. New, also referred to as novelty, requires the variety not to have been known or grown publicly for more than one year in the application country or four years elsewhere (six years for trees and vines). Uniformity and stability are technical requirements to document that a variety has been propagated for a sufficient number of generations to be stable. 'A variety shall be deemed to be distinct if it is clearly distinguishable from any other variety whose existence is a matter of common knowledge at the time of the filing of the application'. These three requirements are often abbreviated as DUS. However, these requirements have been criticized as discouraging variability and diversity within cultivated crop varieties, which is essential for long-term food security (Fowler, 1994; Leskien and Flitner, 1997).

4.1.3 Scope of plant breeders' rights

Recipients of PVP certificates may prevent others from using the variety without authorization, including specifically producing, selling, stocking for the purpose of selling, importing and exporting (Article 14.1). These rights extend not only to the variety and its propagating materials but also to 'harvested materials, including entire plants or parts of plants' (Article 14.2) and, at national discretion, to 'products made directly from harvested material of the protected variety' (Article 14.3).

The advent of genetic modification (GM) for crop improvement in the 1990s allowed agricultural biotechnology to insert an altered gene which serves as the distinguishing characteristic, raising the possibility that agbiotech firms could effectively appropriate traditional PVP-protected varieties. To address this issue, the 1991 UPOV Act in Article 14.5 establishes a complex system of 'initial' and 'essentially derived' varieties. In simple terms, if the resultant new variety is deemed to be essentially derived then it cannot be sold without the permission of the owner of the initial variety (Kolady and Lesser, 2009). However, Lesser (2009) argues that even though the 'essential derivation' component of the current version of UPOV is an effort to maintain breeding access and to ensure compensation for using another firm's germplasm, the current approach is not operational because of the proposed rules of delineation. The rules are based on the proportion of genetic materials contributed relative to the initial variety; the challenge is that with genetic modification important changes can involve the addition of proportionally little genetic material. Further, there is the scope to by-pass this approach by adding junk genes with no function during breeding the variety to raise the proportion of new genetic material.

4.1.4 The key limitations and exceptions to plant breeders' rights

A *research exception* (Article 15(1)) allows access to protected varieties for breeding purposes and as such is a general research exception. Due to the incremental nature of much plant breeding wherein the best of the existing varieties are crossed and then selected to produce a new variety expressing the best attributes of the parents, access to existing varieties is considered by many to be important to maintaining ongoing improvements.

The research exception also facilitates access to improved novel varieties by new breeding firms for research purposes.

Farmers' privilege allows farmers to retain the improved and protected crop as a seed source in subsequent years, but only for planting on his or her own land. Unlike the 1971 Act, the revised 1991 Act does not permit exchange or sale of seeds of protected varieties among farmers, an issue that critics have raised as against the traditional practices of farming in developing countries.[1] Further, the 1991 Act (Article 15(2)) makes farmers' privilege a national option and must be exercised 'within reasonable limits and subject to the safeguarding of the legitimate interests of the breeder'. Many interpretations of this provision have been offered. For example, the International Association of Plant Breeders has interpreted the 'reasonable limits' as a state requirement to limit the quantity of seed, acreage and species subject to the farmers' privilege, whereas the 'safeguarding of the legitimate interests of the breeder' is a condition that requires farmers to pay breeders a legitimate share of their returns from the use of the farmers' privilege provision (ISF, 2001). As expected, member states have interpreted and implemented this provision differently. The US allows farmers to save seeds for use on their own holdings while the EU requires 'large' farms to pay a royalty for saved seeds. Because of the research exception and farmers' privilege, plant variety protection offered through plant breeders' rights is considered a weak form of IPR.

4.2 The TRIPS Agreement

Given that the scope of protection offered to plant breeders through UPOV is limited and not strictly enforceable internationally, many developed countries argued for stronger and enforceable IP protection in agriculture: they succeeded to a certain extent with the ratification of the TRIPS agreement.

The TRIPS agreement was adopted as a treaty in 1994 and is administered by the WTO. Since TRIPS is mandated by the WTO, all WTO members have to become TRIPS compliant, but the deadline for compliance varied depending on whether a country is considered developed, developing or least developed. There were 155 member countries in the WTO as of July 2012. It is the first IPR treaty that seeks to establish universal minimum standards of protection across major fields of intellectual property including patents, copyrights, trademarks and trade secrets (Helfer, 2004).

The TRIPS agreement deals with plant variety protection through its Article 27.3(b), which requires member states to provide protection for plant varieties either by patents or by an effective *sui generis* system or by a combination thereof. Article 27.3(b) of TRIPS states that 'Members may exclude from patentability: plants and animals other than microorganism, and essentially biological processes for the production of plants or animals other than non-biological and microbiological processes'.[2] However, members must provide for the protection of plant varieties either by patents or by an effective *sui generis* system or by a combination thereof.

Only the US and three other countries (Australia, South Korea and Japan) allow the routine granting of utility patents for plant varieties. An applicant variety must be new (novel), non-obvious (inventive step in other countries, similar to distinctness) and be examined. Patents must demonstrate utility, essentially a use, if not an economically efficient use. Deposits are typically required and are publicly available but neither breeders' rights nor farmers' privilege apply. Due to those factors, patent protection is considered to provide notably stronger protection for the owner than PVP, even though in practice the standards for a patent are no greater than for PVP (Kolady and Lesser, 2009; Lesser, 2009).

The two important implications of Article 27.3(b) are: TRIPS does not refer to or incorporate any pre-existing IPR agreements such as UPOV Acts and offers flexibility for member states to decide the type of protection they want to provide for plant varieties. Even though UPOV was in force since 1961, its membership grew significantly post-1995 and many attribute this to the TRIPS agreement (Helfer, 2004; Dutfield, 2011).

4.3 Convention on Biological Diversity

The main objective of the Convention on Biological Diversity adopted in 1992 is to conserve biological diversity and to ensure the sustainable use of its components and the fair and equitable sharing of the benefits arising out of the utilization of genetic resources. Currently there are 193 parties to the CBD. Most developing countries view the CBD as a forum to protect the interests of indigenous and local communities. The CBD is linked to IPR legislation through articles 8(j), 15.5 and 16.5.

Article 8(j) directs parties to

> Subject to its national legislation, respect, preserve and maintain knowledge, innovations and practices of indigenous and local communities embodying traditional lifestyles relevant for the conservation and sustainable use of biological diversity and promote their wider application with the approval and involvement of the holders of such knowledge, innovations and practices and encourage the equitable sharing of the benefits arising from the utilization of such knowledge, innovations and practices.

The implications of Article 8(j) are expanded by the requirement that access to this knowledge be made with the 'prior informed consent' of indigenous and local communities (Article 15.5) and by the ratification of the Nagoya Protocol on access to genetic resources and the fair and equitable sharing of benefits arising from their utilization (CBD, 2010).

The CBD is further linked to IPR agreements by Article 16.5 which states that

> The Contracting Parties, recognizing that patents and other intellectual property rights may have *an* influence on the implementation of this Convention, shall cooperate in this regard subject to national legislation and international law in order to ensure that such rights are supportive of and do not run counter to its objectives.

4.4 ITPGRFA

The 27th session of FAO recognized the need to resolve the issues surrounding farmers' rights within the framework of the Undertaking, and negotiations led to the formation of the International Treaty on Plant Genetic Resources for Food and Agriculture (ITPGRFA) in 2001 (Gerstetter et al., 2007). Currently there are 127 parties to the treaty. The treaty, which is in harmony with the CBD, recognizes the contribution farmers and indigenous communities have already made in the conservation and development of plant genetic resources and their potential future contributions.

The treaty recognized farmers' rights by stating that

> the Contracting Parties recognize the enormous contribution that the local and indigenous communities and farmers of all regions of the world, particularly those in the centers of origin and crop diversity, have made and will continue to make for the conservation and development of plant genetic resources which constitute the basis of food and agriculture production throughout the world (Article 9.1).

Article 9.2 states that

> The Contracting Parties agree that the responsibility for realizing farmers' rights, as they relate to plant genetic resources for food and agriculture, rests with national governments. In accordance with their needs and priorities, each Contracting Party should, as appropriate, and subject to its national legislation, take measures to protect and promote farmers' rights, including: (a) protection of traditional knowledge relevant to plant genetic resources for food and agriculture; (b) the right to equitably participate in sharing benefits arising from the utilization of plant genetic resources for food and agriculture; and (c) the right to participate in making decisions, at the national level, on matters related to the conservation and sustainable use of plant genetic resources for food and agriculture.

As evident from Article 9.2 the scope of farmers' rights is much more than the farmers' privilege provision of UPOV. Further Article 9.3 states farmers' rights are to be granted as 'appropriate' and 'subject to national law'. Article 15.2 of UPOV 1991 gives an optional exception to signatories to permit farmers certain uses of harvest obtained from a protected variety (farmers' privilege clause), but does not recognize farmers' rights, traditional knowledge and benefit sharing (Article 9, and Article 13 of ITPGRFA). The 1991 UPOV Act does not permit farmers to exchange or sell seeds of protected varieties, an issue critics have raised as against the traditional practices of farming and farmers' rights recognized in ITPGRFA. A dominant view in developing countries is that, irrespective of the protection status of a crop variety, retaining a variety as a seed source to use either on own farm or to exchange or sell among other farmers is a farmers' right and not farmers' privilege (Sahai, 2002).

Many countries were concerned about the clauses dealing with IPR issues in the ITPGRFA. For example, countries such as Canada, Australia and fifteen countries of the European Union (EU) made their own clarifications about the IPR clauses before ratifying the treaty, while the US and Japan cite these IPR clauses as the reason for not ratifying the treaty. To a great extent, the ITPGRFA tries to avoid tension with IPR provisions of UPOV by recognizing farmers' right not through IPRs but by benefit-sharing mechanisms that compensate farmers for their contributions to plant genetic diversity (Helfer, 2004).

Even though the ITPGRFA is meant to provide facilitated access to plant genetic resources for food and agriculture through its multi-lateral system (MLS), concerns have been raised, especially by inventors, about the interpretations and implications of Article 12.3(d) which states that 'Recipients shall not claim any intellectual property or other rights that limit the facilitated access to the plant genetic resources for food and agriculture, or their genetic parts or components, in the form received from the Multilateral System'.

Article 12.3(d) of ITPGRFA prohibits IPRs on accessions received from the MLS which have not been modified in any form. But then, to earn a PBR under UPOV, the candidate variety should meet the DUS criteria specified in the UPOV Acts. However, the problem arises when a new variety is developed using materials received from the MLS. While the new variety is most likely to meet the DUS criteria of UPOV and be eligible for a PBR, Article 12.3(d) prevents IPRs on such a new variety if such an IPR would 'limit the facilitated access to the PGRFA' (Helfer, 2004). Thus the term 'limit the facilitated access to the PGRFA' is open to interpretation. One could also argue that the research exception provision of UPOV (Article 15) does not restrict facilitated access to the PGRFA as stated in ITPGRFA, and thus there is no conflict between these two treaties.

However, the debate surrounding Article 12.3(d) also focuses on whether the treaty would bar the patenting of isolated genes extracted from germplasm placed in the MLS. Since UPOV does not grant patents, there is no formal conflict between UPOV and ITPGRFA. However, the article is prone to conflict with TRIPS provisions or national IPR laws (as in many industrialized countries) that treat isolated and purified genes as patentable inventions (Helfer, 2004). The UPOV treaty does not stipulate the requirement for benefit sharing (either in the form of cash or technology transfer); thus Article 13 of ITPGRFA is an additional obligation to the parties of ITPGRFA. However, as argued by Helfer (2004), the benefit-sharing provision of the ITPGRFA (Article 13.d(ii)) could be in conflict with the TRIPS provisions (Article 27.1),[3] as this imposes an obligation in connection with the biotechnology patents that is not imposed with other types of patents.

5 IMPLICATIONS OF INTERNATIONAL REGIMES ON IPRS AND PGRS: NATIONAL LEVEL

Even though, as discussed earlier, the rights and exceptions of these treaties avoid a formal legal conflict and provide significant leeway for countries to decide the extent and scope of protection of various rights, there are legitimate concerns about how member countries to these treaties could effectively implement and enforce these provisions. It is clear from the above analysis that farmers' rights and protection of traditional knowledge provisions of CBD and ITPGRFA aimed at conservation and sustainable use of biodiversity and plant genetic resources are in conflict with plant breeders' rights provisions of IPR treaties, making their joint implementation an onerous task for member states (Table 32.1).

Many developing countries are reluctant to join UPOV because of their concern that UPOV protects the interests of the plant breeding industry (mostly located in developed countries) while ignoring the interests and rights of the indigenous farmers, who significantly contributed (and continue to contribute) to the crop improvement for thousands of years (Sahai, 2002). UPOV non-members have more flexibility in enacting national laws that enable them to meet the competing social objectives embodied in the concepts of breeders' rights and farmers' rights. The increased membership of developing countries in ITPGRFA and their limited presence in UPOV suggests that many developing countries plan to make use of the *sui generis* provision of TRIPS in the protection of plants and animals by forming national legislation balancing both farmers' rights and breeders' rights. India, a member of ITPGRFA but not UPOV, provides an example of enacting a *sui generis* system of plant variety protection which tries to balance farmers' and breeders' rights.

Table 32.1 International regimes on plant IPRs and PGRs: an overview

Category	Regimes on plant IPRs		Regimes on PGRs	
Regime	*UPOV*	*TRIPS*	*CBD*	*ITPGRFA*
Year	1961	1994	1992	2001
Main Objectives	Protection of plant breeders' rights	Establish universal minimum standards for IP protection across major fields	Conservation and sustainable use of biological resources	Conservation, sustainable use of PGR, and benefit sharing in the context of access to PGR
Rights protected	Plant breeders' rights	Plant breeders' rights	Acknowledges TK	Farmers' rights, Traditional knowledge
Articles' potential for conflicts	15.2	27.1	8j, 15.5, 16.5	9.1, 9.2, 9.3, 12.3(d), 13
Conflicting provisions	Farmers' privilege	Patent rights enjoyable without discretion against any particular field of technology	Benefit sharing and prior informed consent	Farmers' rights, protection of TK, equitable sharing of benefits
Members (#)	70	155	193	127

Source: Author.

5.1 India's PVP System

The Protection of Plant Varieties and Farmers' Rights Act (2001) of India protects not only new varieties but also extant varieties,[4] which includes already notified varieties, farmers' varieties and varieties of common knowledge.

Chapter 6 of the Act on Farmers' Rights states that

> i) a farmer who has bred or developed a new variety shall be entitled for registration and other protection in like manner as a breeder of a variety under this Act; (ii) the farmers' variety shall be entitled for registration if the application contains declaration as specified in clause (h) of sub-section (1) of section 18; (iii) a farmer who is engaged in the conservation of genetic resources of land races and wild relatives of economic plants and their improvement through selection and preservation shall be entitled in the prescribed manner for recognition and reward from the Gene Fund. Provided that material so selected and preserved has been used as donors of genes in varieties registrable under this Act; and (iv) a farmer shall be deemed to be entitled to save, use, sow, resow, exchange, share or sell his farm produce including seed of a variety protected under this Act in the same manner as he was entitled before the coming into force of this Act, provided that the farmer shall not be entitled to sell branded seed of a variety protected under this Act. (PPV&FR Authority, 2009)

Thus the Indian Act protects breeders and farmers' rights and recognizes a farmer's breeding activities by protecting farmer varieties. Even though countries such as Thailand, Indonesia and Malaysia also enacted laws protecting farmers' rights, the

Indian Act stands apart in allowing farmers to save, sell or exchange seeds of protected varieties. The Indian Act facilitates benefit sharing in two ways: first, individuals or organizations can claim compensation based on their contribution towards the development of protected variety; and, second, individuals or organizations can claim compensation on behalf of a community or communities based on their contribution towards the development of a protected variety. The Act adopts many provisions of the 1978 UPOV by granting plant breeders' rights, while going much beyond the farmers' privilege provisions of the UPOV Acts in granting farmers' rights. Thus the Indian Act tries to address the plant variety protection objective of UPOV along with the conservation, sustainable use and benefit-sharing objectives of ITPGRFA in a unique way. The Indian Act also provides protection for initial or dependent varieties (from 1991 UPOV), and limits the number of genera and species to be protected according to 1978 UPOV, while leaving undefined 'farmer variety'.

Interactions with seed industry stakeholders in India suggest that, overall, they are supportive of the PVP Act and they are applying for protection of their varieties under the Act (Personal communication, 2010). From the industry standpoint, allowing farmers to sell or exchange the seeds of protected varieties may not cost them much, because those farmers are most likely to be small and marginal farmers who have limited resources to purchase new seeds every year. Industry stakeholders also note an overall trend among Indian farmers to purchase seeds annually, even for self-pollinated crops such as rice and wheat, a fact supported by the increasing seed replacement rate for these crops nationally. However, most of the industry stakeholders expect that the Act would prevent theft of their popular varieties by other local seed companies, which they consider a greater threat to their business than brown bagging of improved varieties by small and marginal farmers. However, the effectiveness of the Authority in enforcing the Act, especially in adjudicating infringement cases and enforcing penalties for infringement, has yet to be tested. Further, while the Indian Act is unique in linking breeders' rights and farmers' rights, there are concerns about its effective implementation, its effect on technology transfer, access to PGRs and research and development investment.

Even though developing a *sui generis* system of IPR protection gives flexibility for nations to protect farmers' and breeders' rights, many countries face significant challenges in terms of human and financial resources for enforcing the system. Significant differences in the IPR systems across countries could also create complexities and uncertainties for breeders and investors in development and transfer of agricultural technologies.

5.2 Implications of Farmers' Privilege, and Protection of Farmers' Rights and Traditional Knowledge

Because of farmers' privilege, and the research exception, many developed countries consider UPOV as a weak forum of IPR protection in plants, and PVP as a weak form of IPR protection compared to patents. Even though the 1991 UPOV Act reduced the scope of farmers' privilege, and made it a national option, concerns have been raised about the farmers' privilege provision. Phillips (2007) identified five implications of farmers' privilege and incomplete or weak IPR mechanisms for the agbiotech industry: (i) industry might retrench and the public sector might re-emerge as the main developer of new varieties; (ii) industry might retrench and only develop varieties where effective biological controls are possible (e.g. hybrids); (iii) industry might abandon certain crop sectors, traits or markets especially for small acreage crops; (iv) industry might exploit provisions in legally enforceable international treaties such as TRIPS to collect royalties (for example. Monsanto's plan to use end use license and royalty scheme for Roundup Ready soybeans in Argentina, Brazil and Uruguay); and (v) industry might resort to new marketing strategies such as private contracts, increasing prices and increasing varietal turnover rate.

With the advancement of biotechnology applications in agriculture, as pointed out by Phillips (2007), the chances of the public sector leading the development and commercialization of novel varieties, especially genetically modified varieties, are very limited. However, the increased private sector investment into hybrid crops such as corn and pearl millet in India suggest that when there is lack of or incomplete IPR regimes, the private sector will focus on crop sectors where effective biological control such as hybridization is possible. Further, as may be happening in the case of hybrid rice, the private sector might consider hybrids as a platform for introducing genetically engineered traits in crops that are currently self-pollinated (Kolady et al., 2010). Whether such a strategy by private firms is in the best interest of farmers is a debatable topic. When the private sector in developing countries adopts such a strategy, it will be the responsibility of the public sector to develop similar technologies to meet the needs of the poor and small farmers. But the public sector in many developing countries lacks the human and financial capital to do so. Similarly, as argued by Phillips (2007), when the private sector resorts to marketing strategies such as increased prices and reducing the life cycle of novel varieties, these may not be in the best interests of farmers.

Compared to farmers' privilege, discussion and debate on farmers' rights and traditional knowledge in the context of IPRs is relatively recent. However, as evident from the revocation of US patents on turmeric and basmati rice and European patents on neem, proponents of TK have been successful in pointing out how traditional knowledge and genetic resources have been used by patent applicants (mostly in developed countries) without acknowledging the contribution of local and indigenous communities (in developing countries) in the development of such knowledge and resources. Further, these cases opened a debate on the definition of 'prior art' (a term used for things which are 'not-novel' in patent applications) and 'bio-piracy' (a situation where indigenous or TK, originating with indigenous people, is exploited for commercial gain without permission from and with no compensation to the indigenous, local communities). Under United States patent law, prior existing knowledge that could be used to challenge a patent generally is publication in any journal, but not unwritten knowledge known and available in oral or folk traditions outside the US. However, the recent US Patent Reform Act of 2011 made the following changes on 'prior art':

(a) the invention was known or used by others in this country, or patented or described in a printed publication in this or a foreign country, before the invention thereof by the applicant for patent, or
(b) the invention was patented or described in a printed publication in this or a foreign country or in public use or on sale in this country, more than one year prior to the date of the application for patent in the United States (US (35 USC Sec. 102(a))).

Following the success of India in revoking the patents on turmeric, neem and basmati there has been a demand to modify IPR texts to include traditional knowledge, written or not, as prior art. A Third World Network expert group on implementing TRIPS recommended in 1998 that developing countries apply a broad concept of prior art in their national legislation to ensure that patents are granted only to new inventions, and to require novelty of the inventive process as a condition for granting a patent, to avoid biopiracy and patenting of knowledge or materials developed by and diffused within local and indigenous communities (Raghavan, 2000).

The legal battle surrounding the case of publicly developed genetically-engineered open-pollinated varieties of eggplant (Bt eggplant) in India highlights the problems of implementing and enforcing various provisions of CBD and ITPGRFA and their implications on access to plant genetic resources and on research and development of improved varieties, especially GE varieties. Soon after the Indian government put a moratorium on commercialization of Bt eggplant in India, Environmental Support Group (ESG), a non-governmental organization, filed a complaint with Karnataka State Biodiversity Board stating that the developers of Bt eggplant – which include Mahyco (Mahyco developed the Bt hybrid eggplant and donated the technology to public institutions to develop open pollinated varieties of Bt eggplant; Monsanto has a 26 per cent stake in Mahyco), University of Agricultural Sciences Dharwad and Tamil Nadu Agricultural University – violated the 2002 Indian Biodiversity Act (India is a party to CBD) by accessing and using the local varieties of eggplant for developing Bt eggplant varieties, without prior approval from state or national biodiversity authorities.[5] While Mahyco denied violating any biodiversity protection laws, the state agricultural universities claimed that the law does not apply to them as they are part of the public sector. In its meeting on 20 June, 2011, the National Biodiversity Authority (NBA) decided to bring legal proceedings against Mahyco and Monsanto and all others concerned to resolve the issue (The Hindu, 2011). However, the Karnataka State Biodiversity Board, the regulatory authority charged with implementing the National Biodiversity Act, decided to drop legal actions against Mahyco and US-based Monsanto (DNA, 2012). At this point it is not clear what effects this legal battle will have on accessing plant genetic resources for use in public and private sector breeding programs in the future. This example from India highlights the complexities and challenges of enacting and enforcing effective national legislations in the context of membership in various international treaties. Further, the Indian example highlights the lack of awareness regarding the scope of legislation among various stakeholders.

To address the issues surrounding farmers' rights and TK, two proposals are under discussion at the World Intellectual Property Organization (WIPO): first, a general scheme of IP-like protection of TK and farmers' varieties (that is, Indian model); and, second, a specific proposal to require patent applicants to disclose the origin of their biological resources in their patent application. An intergovernmental committee (IGC) was established in 2000 by WIPO with the objective of reaching an agreement on a text of an international legal instrument (or instruments) which will ensure the effective protection of traditional knowledge, traditional cultural expressions (TCEs), folklore and genetic resources. The reports of the IGC suggest that discussions are still continuing and no consensus has been reached on the definition, objectives and mechanism for protection of TK.

However, the specific proposal to require disclosure of the origin of biological resources in patent applications is gaining traction in various forums. For example, the group led by India and Brazil wants to amend the TRIPS agreement to required patent applicants to disclose the country of origin of genetic resources and traditional knowledge used in the inventions, to present evidence that they have arranged prior informed consent and to offer evidence of fair and equitable sharing. The EU proposed requiring all patent applicants to disclose the source or origin of genetic material, with legal consequences of not meeting this requirement lying outside the scope of patent law. The US has argued for national legislation to include contracts rather than a mandatory disclosure mechanism to meet the objectives of facilitated access to genetic resources and fair and equitable sharing of the benefits (WTO, 2011). Lesser (2009) argues that it will require a unique combination of measures to balance breeders' needs and those of farmers and consumers. He suggests the difficulty in designing a functional system of IP protection for plant varieties arises from: (i) multi-levels of breeding contributions, from incremental annual enhancements to the occasional multi-step advancements; (ii) multiple phenotypic attributes of finished varieties; (iii) a short commercial life of new varieties; (iv) the ability of seeds of most crops to reproduce true-to-form, meaning that farmers can save the crop as future seed source; and (v) the sequential nature of much of plant breeding.

6 CONCLUSIONS AND POLICY IMPLICATIONS

There is sufficient evidence to suggest that granting proprietary rights through plant IPRs has had a positive effect on increasing agricultural R&D investment, especially from the private sector. Nevertheless, there is a consensus among various stakeholders that conservation and sustainable use of plant genetic resources and access to plant genetic resources are necessary for the development of new plant varieties, and that the contribution of indigenous and farming communities in conservation and development of new plant varieties are not properly recognized. However, there is no consensus regarding how to achieve these competing demands for protection of plant breeders' rights, farmers' rights, and traditional knowledge. Countries have responded to these concerns by forming various international organizations and negotiating treaties, each with its own specific objectives, scope and guidelines. The International Union for the Protection of New Varieties of Plants (UPOV), the Convention on Biological Diversity (CBD), the Trade Related Aspects of Intellectual Property Rights (TRIPS) agreement, and the International Treaty on Plant Genetic Resources for Food and Agriculture (ITPGRFA) are the most relevant treaties in the context of plant IPRs and plant genetic resources. It is evident from the above analysis that in particular farmers' rights and protection of traditional knowledge provisions of the CBD and ITPGRFA aimed at conservation and sustainable use of biodiversity and plant genetic resources are in potential conflict with plant breeders' rights provisions of IPR treaties, making their joint implementation an onerous task for member states.

It is clear from this analysis that, overall, the views of developed and developing countries on the scope and potential of existing international regimes differ based on their technological capacity and the traditional knowledge and plant genetic resources that have arisen or reside in their jurisdictions. It is evident from the Indian case that, even at the national level, the proposal to link farmers' rights and traditional knowledge to traditional legal IPR mechanisms can be a lengthy and difficult process. Hence, the efforts to link conservation and sustainable use of PGRs, and equity objectives with traditional IPRs at the international level are undoubtedly going to be even more complex, costly and harder to implement.

The above analysis highlights the challenges of enacting and enforcing a comprehensive policy framework both to stimulate private investment in agricultural R&D, which is expected to increase global social welfare, and to ensure that the rights of farmers and farming communities are recognized. Even though there has been some progress in this direction, the current international system on IPRs and PGRs is far from optimal. While harmonization among various treaties could lead to fewer conflicts, the international negotiations have shown that perfect harmonization among these treaties is going to be difficult to achieve.

It is clear that in regions or countries where IPR regimes are weak or absent, the industry relies on other exclusion strategies, such as marketing contracts, biological controls (for example hybrids) and pricing strategies. Given that legal IPRs are just one among the many tools available for plant breeders or inventors to extract a return on their investments, focusing only on legal IPRs may not lead to a welfare-enhancing IPR or policy regime. Those engaged in debates on farmers' rights and traditional knowledge also need to recognize that those issues are discussed more in the context of conservation and equity objectives, hence narrowly focusing on legal regimes for protection of farmers' rights and traditional knowledge may not achieve its objectives. Policy makers need to take into account the unique combinations of dimensions that apply to plant varieties, cultural differences in farming and value systems, socio-economic conditions of farmers and different levels of R&D capacity that prevail in the world, in their policy dialogue regarding designing a system that is functional, welfare enhancing, easy and cost-effective to implement and enforce.

As the range of technological opportunities expands, it is also important that we examine the economic feasibility of other supply- and demand-related policy interventions to conserve genetic diversity, such as providing funds or subsidies for farmers or local communities to grow traditional varieties or landraces in centers of origin or diversity. More research needs to be conducted to examine the economic feasibility of various alternative supply- and demand-related policy instruments for the conservation of crop genetic diversity.

NOTES

1. For more details on history of farmers' privilege provision and its practical effects, please see Phillips (2007).
2. It should be pointed out here that currently, hybrids, inbred parents and genes, and processes such as hybridization systems and transformation systems are patentable in many countries.
3. Article 27.1 of TRIPS requires members to make 'patents . . . available and patent rights enjoyable without discrimination as to . . . field of technology . . .'.
4. A large number of popular, high-performing varieties of crops notified under the Seeds Act of 1966 continue to be marketed by different agencies. To provide legal protection to these varieties retrospectively, the PPV&FR Act has provisions for their registration within three years of crop notification in the gazette for variety registration for the remaining period. The Extant Variety Recommendation Committee advises the registrar on the suitability of these varieties for registration. Three types of extant varieties are permitted

for certification: (1) varieties notified under the Seeds Act of 1966 that have not completed a 15-year (for annual crops) or 18-year (for biennials and perennials) period; (2) varieties of common knowledge; and (3) farmers' varieties.
5. For more details on the public–private partnership involved in the development of Bt eggplant please see Kolady and Lesser (2011).

REFERENCES

Alston, J.M., J.M. Beddow and P.G. Pardey (2009), 'Agricultural research, productivity, and food prices in the long run', *Science*, **325**(4), 1209–10.

Castle, D. and E.R. Gold (2007), 'Traditional knowledge and benefit sharing: from compensation to transaction', in P.W.B Phillips and C.B. Onwuekwe (eds), *Accessing and Sharing the Benefits of the Genomics Revolution*, Dordrecht, Springer: pp. 65–79.

Convention on Biological Diversity (CBD) (2010), 'The Nagoya protocol on access and benefit sharing', available at: http://www.cbd.int/abs/, accessed December 2011.

DNA (Daily News and Analysis) (2012), 'Biopiracy charges dropped against two US companies', available at: http://www.dnaindia.com/bangalore/report_bio-piracy-charges-dropped-against-two-us-companies_1648647, accessed February 2012.

Dutfield, G. (2011), 'Food, biological diversity, and intellectual property: the role of the International Union for the Protection of New Varieties of Plants (UPOV)', Global Economic Issue Publications, Intellectual Property Issue Paper no. 9, The Quaker United Nations Office, available online at: http://www.quno.org/geneva/pdf/economic/Issues/UPOV%20study%20by%20QUNO_English.pdf, accessed October 2011.

Fernandez-Cornejo, J. (2004), 'The seed industry in US agriculture: an exploration of data and information on crop seed markets, regulation, industry structure, and research and development', Agriculture Information Bulletin no. AIB 786, Washington, DC: US Department of Agriculture Economic Research Service.

Foster, W.E. and R. Perrin (1991), 'Economic incentives and plant breeding research', faculty working paper, Raleigh, NC: North Carolina State University.

Fowler, C. (1994), *Unnatural Selection: Technology, Politics and Plant Evolution*, New York: Gordon and Breach.

Gerstetter, C., B. Gorlach, K., Neumann and D. Schaffrin (2007), 'The International Treaty on Plant Genetic Resources for Food and Agriculture within the current legal regime complex on plant genetic resources', *The Journal of World Intellectual Property*, **10**(3/4), 259–83.

Goodman, Z. (2009), 'Seeds of hunger: intellectual property rights on seeds and the human rights response', Backgrounder no. 2 in the THREAD series, available at: www.3dthree.org/pdf_3D/3D_THREAD2seeds.pdf, accessed February 2012.

Heitz, A. (1987), 'The history of plant variety protection', in UPOV, *The First Twenty-Five Years of the International Convention for the Protection of New Varieties of Plants*, Publication no. 879, LIPOV, 53–97.

Helfer, L.R. (2004), 'Intellectual property rights in plant varieties: international legal regimes and policy options for national governments', FAO legislative study no 85, available online at: ftp://ftp.fao.org/DOCREP/FAO/007/Y5714e/Y5714e00.pdf, accessed October 2011.

International Seed Federation (ISF) (2001), 'Position paper on farm saved seed', available online at: http://www.worldseed.org/cms/medias/file/PositionPapers/OnIntellectualProperty/Archives/Farm_Saved_Seed_(En)_Archives_20080701.pdf, accessed January 2012.

Kolady, D.E. and W. Lesser (2009), 'But are they meritorious? Genetic productivity gains under plant intellectual property rights', *Agricultural Economics*, **60**(1), 62–79.

Kolady, D.E. and W. Lesser (2011), 'Genetically engineered crops and their effects on varietal diversity: A case of Bt eggplant in India', *Agriculture and Human Values*, **29**(1), 3–15.

Kolady, D., D.J. Spielman and A. Cavalieri (2010), 'Intellectual property rights, private investment in research, and productivity growth in Indian agriculture: a review of evidence and options', IFPRI Discussion Paper no. 1031, Washington, DC: International Food Policy Research Institute.

Leskien, D. and M. Flitner (1997), 'Intellectual property rights and plant genetic resources: options for a *sui generis* system', *Issues in Genetic Resources*, no. 6, Rome: IPGRI.

Lesser, W. (1997), 'The role of intellectual property rights in biotechnology transfer under the convention on biological diversity', ISAA briefs no. 3, Ithaca, NY: ISAAA.

Lesser, W. (2009), 'From penury to prodigal: protection creep for US plant varieties', *Virginia Journal of Law and Technology*, **14**(235), 236–72.

Pardey, P.G., J.M. Alston and R.R. Piggott (2006), *Agricultural R&D in the Developing World: Too Little, too Late?*, Washington, DC: International Food Policy Research Institute.

Perrin, R.K., K. Kunnings and L.A. Ihnen (1983), 'Some effects of the US Plant Variety Protection Act of 1970', Economics Research Report no. 46, Raleigh, NC: North Carolina State University.
Phillips, P.W.B. (2007), 'Farmers' privilege and patented seed', in P.W.B. Phillips and C.B. Onwuekwe (eds), *Accessing and Sharing the Benefits of the Genomics Revolution*, Dordrecht: Springer, pp. 49–64.
PPV and FR Authority (Protection of Plant Varieties and Farmers' Rights Authority) (2009), 'Annual report 2008–09', New Delhi: Ministry of Agriculture, Government of India.
Pray, C.E. (1992), 'Plant breeders' rights legislation, enforcement, and R&D: lessons for developing countries', in G. Peters and B. Stanton (eds), *Sustainable Agricultural Development: The Role of International Cooperation*, Proceedings of the 21st International Conference of Agricultural Economists, Brookfield, VT: Dartmouth.
Raghavan, C. (2000), 'Neem patent revoked by European patent office', available at: http://www.twnside.org.sg/title/revoked.htm, accessed October 2011.
Ramanna, A. (2003), 'India's plant variety and farmers' rights legislation: potential impact on stakeholder access to genetic resources', EPTD discussion paper no. 96, Washington, DC: International Food Policy Research Institute.
Sahai, S. (2002), 'Farmers' rights under attack by decision to join UPOV', available at: http://www.genecampaign.org/Publication/Article/Farmers%20Right/Farmers_right_decission_join_upov.pdf, accessed December 2011.
The Hindu (2011), 'NBA to take action against Bt brinjal biopiracy', available at: http://www.thehindu.com/news/national/article2340768.ece, accessed January 2012.
WIPO (2011), 'Intellectual property and genetic resources, traditional knowledge, and folklore', available at: http://www.wipo.int/freepublications/en/intproperty/450/wipo_pub_l450gtf.pdf, accessed December 2011.
WTO (2011), 'TRIPS: reviews, Article 27.3(B) and related issues: background and the current situation', available at: http://www.wto.org/english/tratop_e/trips_e/art27_3b_background_e.htm, accessed January 2012.

33 Engaging publics on agbiotech: a retrospective look

Jennifer Medlock and Edna Einsiedel

INTRODUCTION

Agricultural biotechnology became a policy issue worldwide in the 1990s (Kearnes et al., 2006) because of growing public concerns about the governance of science and technology developments. These concerns include rules around commercial ownership and control; the adequacy of risk assessment and regulation processes; the shifting role of globalization forces; and incorporating citizen and stakeholder voices in shaping trajectories of new technologies. Confronted by increasing levels of controversy related to genetically modified (GM) crops in particular, governments in several countries embarked on experimentation with new methods of public engagement, many based on a model of dialogue and deliberation, as a way to rebuild public trust and manage the controversy (House of Lords Select Committee, 2000).

Much practical and theoretical attention has been given to developing new engagement techniques, putting them into practice, and testing methods to assess their influence. This chapter provides an opportunity to step back and examine the broad trajectory of these participation experiments in the realm of agbiotech from the late 1980s until the present day. We adapt Ostrom's institutional analysis and development (IAD) framework that guides the *Handbook* as a whole (see Figure 33.1) such that our 'Action Arena' consists of public participation initiatives. We have relabelled the 'Exogenous Variables' in Ostrom's model as 'Contextual Factors' as this mirrors more closely the language used in the participation literature.

We first present three areas of theory providing the rationale and impetus for the action arena of public participation. We then describe participation methods emerging from these theoretical approaches, focusing mostly on the consensus conference process, which has been used in more than 15 countries on issues related to agbiotech (The Loka Institute, n.d.), followed by an overview of the research approaches used to assess these experiments. We end with a synthesis of the empirical research on consensus conferences related to agbiotech, examining the evolution of engagement practices and their policy and social learning outcomes, emphasizing how the consensus conference method has been deployed in different ways across different governance contexts.

MODELS: THEORIZING PUBLIC PARTICIPATION

In this section, we review three relevant theoretical lenses for investigating public participation: deliberative democracy; science and technology studies; and policy sciences and governance.

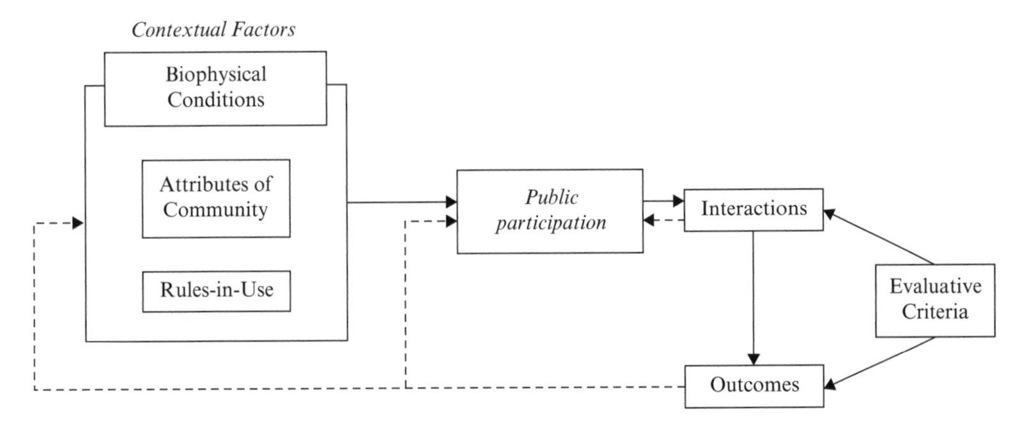

Figure 33.1 Ostrom's (2005) IAD framework adapted for the case of public participation

Deliberative Democracy

In contrast with representative or interest-based approaches to democratic decision-making, deliberative democratic theorists advocate for an ideal of decision-making based upon public reason-giving in the spirit of the common good. Individuals engage in a rational, informed exchange of reasons for or against a particular decision, with the better argument ultimately winning the day (Elster, 1998). Procedurally, this requires creating discussion spaces free from manipulation or strategic behaviour and with equal opportunity for all citizens to participate, and the closer an actual deliberation can come to these ideals, the better (Thompson, 2008). Meeting these conditions of deliberation is critical to ensuring legitimacy, transparency and accountability in decision-making and for the resulting policies to be considered fair and just (see Gastil and Levine, 2005).

In addition, deliberative democracy theorists argue that individuals taking part in deliberation are transformed by the experience, having 'a tendency to broaden perspectives, promote toleration and understanding between groups, and generally encourage public-spirited attitudes' (Chambers, 2003, p. 318). While not replacing traditional electoral and legislative tools of collective decision-making, public deliberation is considered a complementary approach to aid in addressing political deadlock and divisiveness on major policy issues.

Science and Technology Studies

Science and technology studies (STS) provide a second theoretical lens for investigating the participatory turn. Emerging in the 1970s, this field draws attention to the social and political contexts of scientific knowledge production and the consequences for democratic decision-making. Deconstructing expertise has revealed the inherent interests involved in scientific knowledge production and, correspondingly, how expert knowledge is also distributed among communities outside of science (Jasanoff, 2003). Work in this area shows that many contemporary policy debates, ranging from biomedical research to agricultural policy to transnational environmental disputes, revolve around the issue of integrating diverse interests and forms of knowledge in a context where the boundary between science and society is no longer clear-cut. Also, the pervasiveness of science and technology in our everyday experience has led to research on the nature and role of the 'ordinary' citizen in science and technology development (Elam and Bertilsson, 2003).

Other theorists under the umbrella of STS take a macro-level perspective, inspired for example by work on Mode 2 science (Gibbons et al., 1994) or post-normal science (Funtowicz and Ravetz, 1992). They focus broadly on the social embeddedness of knowledge production, the recognition and explicit accounting of uncertainty in decision-making and the recognition and accommodation of social values, all of which could result in more 'socially robust' scientific research outcomes and more sustainable policies. Sheila Jasanoff's (2005) work on civic epistemologies, particularly her case study of biotechnology development across three countries, demonstrates how political communities have particular (historically and culturally-situated) styles of policy reasoning and processes for incorporating and interrogating different forms of knowledge, including scientific and indigenous forms, ultimately leading to different policy outcomes.

Policy Sciences and Governance

A third lens for examining public participation focuses on policy-making and governance, building on the observation that policy-making focused on efficiency, effectiveness and economic considerations is insufficient when confronted with 'wicked problems' – those which were typically interrelated with other problems, which needed consideration of social values, or involved trade-offs among competing values, and where multidisciplinary perspectives were more appropriate (Rittel and Webber, 1973). These wicked problems also generally cross national boundaries and have led to transnational arenas and institutions becoming important for managing cross-boundary policy problems. Along with the influence of social movements that effectively used the media to shift public views and policy agendas (Higgins and Lawrence, 2005), these contexts contributed to musings about the decline of the state. The shift from *government* to *governance* helps to capture the changing model of decision-making from centralized bureaucratic modes to reliance on more distributed modes of expertise (see, for example, Lyall and Tait, 2005).

One arena that illustrates these institutional shifts in policy-making is that of technology assessment. The quintessential model, centred primarily on technical expertise, is the now-defunct US Office of Technology Assessment, considered to be a victim of the internecine battles within the US Congress (Bimber, 2005). Despite its closure, the notion of technology assessment was adopted in Europe through various technology assessment units tied to parliaments. One adaptation in European contexts was the inclusion of citizen engagement, a reflection of an ongoing exploration of different approaches to technology assessment (see Sclove, 2010). An umbrella of policy approaches was offered by constructive technology assessment (CTA), pioneered in particular by policy communities in the Netherlands and Denmark. Dissatisfaction with traditional technology assessment approaches led to an interest in broadening the focus upstream to assessment at the technology design stage rather than solely at the point of deployment (Schot and Rip, 1997). Such a broadened focus was designed to modulate ongoing technology development with implications for technology reshaping and 'strategic niche management' (Schot and Rip, 1997). All of these processes involved dialogue and engagement with a wide variety of stakeholders and members of the public. The theoretical frameworks within which CTA studies were being conducted were based on growing interest in the dynamics of technological development and processes of technological embedding in society. The work of Bijker and his colleagues took account of contingency, of seamless webs between the social and technical, and investigations of patterns of evolution of socio-technical orders (see, for example, Bijker et al., 1997).

The discursive turn in policy analysis provides another avenue for thinking through the theoretical questions around public participation, similarly challenging the dominance of technocratic models of policy analysis and arguing for a broader approach to policy-making: 'the increasing complexity of modern technological society dramatically intensifies the information requirements of modern decision-makers. Policy decisions combine sophisticated technical knowledge with intricate and often subtle social and political realities' (Fischer, 2003, p. 2). The route that had been taken by policy studies in the 1960s and 1970s was strongly empiricist, reliant almost solely on quantitative methods, emphasizing the strict separation of facts and values, and devoid of examination of the social contexts for policy problems. (Fischer, 2003). Recognizing the narrow technical orientation of this approach, some policy sciences scholars have similarly turned to discourse and dialogue as a forum for a fuller understanding of social values underpinning policy preferences and as a normative approach to making better policy.

In this brief overview of theoretical frameworks for representing and understanding the role of public engagement and participation, three common themes emerge: first, a growing dissatisfaction with the explanatory limitations of the linear policy-making process; second, the dissatisfaction with inputs – the limitations of science as a primary base for decision-making and recognition of other sites of expertise; third, the changing contexts of decision-making including the growth of social networks and movements and their implications for more horizontal modes of governance. The inclusion of more voices in the making and shaping of policy, by choice or political pressure, was a response to this changing environment. This led to an interest in the process of public participation – how to design methods of engagement, as well as the outcomes of decision-making – how to make decisions not just technically sound but also politically and socially robust.

METHODS: DEVELOPING NEW TECHNIQUES OF PUBLIC PARTICIPATION

Agbiotech issues have been at the centre of a wide variety of public participation processes, including surveys, focus groups, government-legislated opportunities for public comment, and provision of channels for greater stakeholder participation. In this chapter we focus on the emergence of deliberative models of engagement, as these defined a distinct shift by state governments in modes of public participation over the 1990s from predominantly one-way forms of communication to two-way forms of dialogue (see OECD, 2005). Proponents claim that deliberative approaches are well suited to the shifting governance context for agbiotech issues, characterized by decreasing levels of public trust in decision-makers, increasing levels of controversy, involvement by different sectors of society with competing interests, and increasing public criticism of technocratic policy approaches based primarily, if not solely, on scientific research.

At a practical level, Abelson et al. (2003) argue that deliberative approaches provide time and resources for participants to learn about new technologies and their potential societal implications, and furthermore, to discuss and debate the issues with diverse stakeholders across government, industry and civil society. Deliberative approaches also provide opportunity for participants to talk and reflect with each other before forming recommendations for policy and action, with the expectation that through fair, respectful and rational argumentation, participants can reach consensus on appropriate policy directions.

Consensus Conferences

A variety of techniques fall under the umbrella of deliberative engagement, including citizens' juries, deliberative polling and scenario workshops (see Rowe and Frewer, 2000). By far the most well-known and well-travelled deliberative model is the consensus conference, created by the Danish Board of Technology in the late 1980s (Klüver, 1995). The consensus conference process has been described in detail elsewhere (Grundahl, 1995) so we provide only a short overview. The process is centred on a panel of 12–15 ordinary citizens, generally recruited through random selection or media advertisements. The term 'ordinary' generally refers to people who do not have existing connections to the issue, either through employment or civil society involvement, but is often not explicitly defined (Powell et al., 2011). The panel undergoes an intensive learning process through two preparatory weekends held before a public weekend conference. These weekend meetings provide opportunity for the panelists to get to know each other, to learn the basics of the overarching issue and to develop key questions on the issue to present to an expert panel.

One component that differentiates the consensus conference process from most other deliberative models is the 'scoping' stage of the consultation. In other models, such as the citizens' jury, citizen participants are given more specific direction on the policy issue or question under discussion (such as, should we go ahead with GM food labelling, and if so, how? Or, should GM wheat be approved in Canada?) In the consensus conference, the participants begin with a very general overarching policy area (for example, 'Gene technology in the food chain' in Australia or 'plant biotechnology' in the UK – see The Loka Institute (n.d.) and take on the added task of developing a list of key questions. In consensus conferences that have taken place on GM crops, panel questions have touched on a variety of issues, including: environmental impacts; human health and safety; social justice; public involvement in regulation; and economic implications.

Based on these key questions, organizers invite subject matter experts to participate in the final conference (which is open to the public and the media). On the first day of the conference, the experts give presentations in response to the key questions identified by the panel. Time is available for cross-examination and rebuttal from the citizen panel. Day 2 begins with questions from the audience as well as further queries from the panel. The rest of Day 2 (and sometimes Day 3) is devoted to writing the report. On Day 3 (or 4), the report is presented and the expert panel responds. The management of such a process involves a project management team, along with an advisory committee that

Table 33.1 Citizen consensus conferences on agbiotech

Country	Year	Subject
Denmark	1999	GM foods
	1998	Citizens' food policy
	1992	Transgenic animals
	1987	Gene technology in industry and agriculture
Argentina	2000	GM food
Australia	1999	Gene technology in the food chain
Belgium	2003	GM food
Canada	1999	GM food
France	1998	GM food
Germany	1994	GM crops
Italy	2004	GM crop trials
Japan	2000	GM food
Netherlands	1993	Genetically modified animals
New Zealand	1999	Plant biotechnology
	1999	Biotechnological pest control
	1996	Plant biotechnology
Norway	2000	GM food
	1996	GM food
South Korea	1998	GM food
Switzerland	1999	Gene technology and the food chain
UK	1994	Plant biotechnology
US	2002	GM food
	2003	Future of food (New England)

Sources: Compiled from The Loka Institute (n.d.) and Seifert (2006).

generally consists of representatives from government, industry, civil society, consumers, the scientific community and the project director and manager. Consensus conferences are costly (in comparison to most government-sponsored participation processes) and time-intensive, involving a 6–19-month planning process.

The first consensus conference was held in 1987 on the topic of 'gene technology in agriculture' (Klüver, 1995). Since then, consensus conferences have been held in more than 20 countries on issues such as nanotechnology, fluoride, transgenic animals, ozone depletion and gene therapy. The topic explored most by far through consensus conferences is agbiotech, and especially GM food, (The Loka Institute, n.d.) with more than 20 events held between 1987 and 2004. The vast majority of these were held in the mid-to-late 1990s at the height of controversy around GM food (see Table 33.1). In addition, it has most often been the case that a country's first experimentation with holding a consensus conference was on the topic of GM food (Marris and Joly, 1999).

Expected Outcomes

The potential outcomes of consensus conferences, and other deliberative techniques, are seen as both normative and substantive. They are touted as a good means to: enhance the quality and legitimacy of decision-making; foster public debate on the issue; reduce the 'democratic deficit' by bringing citizens into decision-making processes that affect their lives; build individual self-confidence, deliberation skills and knowledge levels; and develop community capacity for action (Andersen and Jaeger, 1999; Barnes et al., 2003; Delli Carpini et al., 2004).

METRICS: FRAMEWORKS FOR EVALUATING PUBLIC PARTICIPATION

This section describes the chronology of research approaches examining deliberative engagement methods, especially the consensus conference, and corresponds to the 'evaluative criteria' aspect of Ostrom's IAD framework (refer to Figure 33.1).

The earliest research on consensus conferences, undertaken in the early 1990s, was generally descriptive in nature, consisting of single-case studies giving an overview of how the consensus conference methodology was implemented, what the outcomes were and, sometimes, the main lessons learned (Abelson et al., 2003). These studies had a very practical orientation, describing experiences 'from the trenches' and were generally written by people on the organizing team or otherwise closely affiliated with the project (Joss, 1995; Klüver, 1995).

Defining 'Effectiveness' Criteria

In the late 1990s, there was a move to make assessments of deliberative engagements more rigorous by defining and operationalizing 'effectiveness' criteria. These criteria emerged from three main sources (Burgess and Chilvers, 2006): the opinions of researchers and practitioners (for example Rowe and Frewer's (2000) process and acceptance criteria); theoretical and normative principles, particularly those based on deliberative democracy (for example Webler's (1995) fairness and competence criteria); and the perspectives of engagement participants, such as through interviews and questionnaires with lay and expert panelists (Delli Carpini et al., 2004; Mayer et al., 1995). Over time, four main categories of assessment have informed the majority of empirical studies (Abelson et al., 2003): representation; the structure of the process or procedures; the information used in the process; and the outcomes and decisions arising from the process.

Statistical representation is, of course, not possible with the small number of participants in consensus conferences, yet organizers look to achieve different forms of representation, including, for example, geographic, demographic, political or attitudinal types of representation. They also may assess representation through principles of inclusion, for example, in meeting barriers of language, culture, literacy and accessibility or through the perceived legitimacy of the selection process (Ableson et al., 2003; Burgess and Chilvers, 2006).

Evaluation criteria assess the quality of information presented to, and received from, lay panellists within the engagement process. Decisions made about the type, form, volume and accessibility of information frame the issue for discussion in particular ways. In the GM food arena, policy-makers have been accused of narrowly framing the issue as one of scientific risk and obscuring the social and ethical aspects implicit in the technology's development (Wynne, 2005).

Procedural metrics examine the format of the consultation, judging it against principles of fairness and legitimacy. In connection with the policy process, they include criteria like: early involvement (did engagement happen before major technological and political decisions have been made?); relevance of the outputs to current policy discussions; and how the process is linked to decision-makers (who are ultimately receiving the advice and responding back?) This category also includes criteria related to the format of the engagement: do participants have equal opportunity to contribute to the discussion? Was there enough time for discussion? Did participants have the opportunity to challenge the information provided? Was mutual respect maintained throughout the process? (Rowe and Frewer, 2000 and 2004).

Evaluation frameworks sometimes assess consultation outcomes, such as contributions to decision-making and public debate, the levels of learning and satisfaction with the process on the part of participants and the extent of consensus achieved (Delli Carpini et al., 2004). Contribution to decision-making encompasses both the extent to which the lay panel report is taken into consideration by policy-makers (and other stakeholder groups) as well as how policy-makers publicly respond to the input. It also includes perceived decision legitimacy and public trust (Burgess and Chilvers, 2006). Public debate is assessed in a number of ways; the extent of media coverage is a key indicator as is influence on the views of key governance actors (Dryzek and Tucker, 2008). Measuring gains in knowledge and awareness is not limited to the lay panel, but involves other actors involved in the process, including the expert panellists, steering committee and organizing team (Walls et al., 2005). With respect to consensus formation, organizers and researchers have given different levels of importance to coming to a single agreement; in recognizing this, some countries have called the process a citizens' conference rather than consensus conference (Einsiedel and Eastlick, 2000).

A large body of case studies emerged from this perspective over the late 1990s and early 2000s (Einsiedel and Eastlick, 2000; Goven, 2003; Joss, 1994; Marris and Joly, 1999; Mohr, 2003; Nishizawa, 2005; Purdue, 1999; Skorupinski et al., 2007). They are generally designed as single cases and employ qualitative research methodologies, though some have used mixed methods and also incorporated quantitative techniques (Mayer et al., 1995). Common data collection techniques include: interviews and/or written questionnaires with process participants; recordings of conference deliberations; compilation and analysis of planning and outreach documents; tracking and analysis of the extent and tone of media coverage; and tracking and analysis of related policy decisions.

Shifting Attention to Outcomes and Interactions within Broader Governance Contexts

Recent research is critical of this evaluation work, particularly the dominant focus of these assessments on evaluating the *process* aspect of doing engagement, including, for example, operational considerations such as understanding the money and other resources needed to mount events, and understanding the optimal conditions for promoting informed dialogue and debate. The process focus has been accompanied by a paucity of research on the *outcomes* of engagement events. The outcome research that has been done generally centres on the perspectives of participants (lay panel, expert panel and stakeholders) immediately following the event or on the extent and tone of media coverage of the issue and the consensus conference itself. Few studies have addressed longer-term decision-making impacts or the institutional constraints that may impede policy influence (Bickerstaff and Walker, 2005). There are a number of reasons for this. Practically, the timelines for assessments are typically not long enough to account for longer-term impacts (Bickerstaff and Walker, 2005; Rowe and Frewer, 2004). Methodologically, it is difficult to show cause-and-effect of a single initiative when there are myriad inputs into policy decisions (Abelson et al., 2003). The lack of systematic research on outcomes means that claims supporting the links between participatory processes and participatory outcomes remain theoretical and speculative, leaving public engagement initiatives vulnerable to scepticism.

A related criticism of this body of work is the lack of systematic research into the role of contextual factors (see Figure 33.1) in enacting public engagement initiatives. The majority of studies incorporate context in an individual and ad hoc manner, giving, for example, a description of the institutional setting, the goals of the organizers, the nature of public debate, the stage of policy development, and the involved stakeholder groups, followed up with the process adaptations made in response to those conditions (Klüver et al., 2000). More recently, researchers have employed comparative case study approaches to take a systematic look at the interplay between a participatory method and its political, social and institutional context (Dryzek and Tucker, 2008; Hagendijk and Irwin, 2006; Jasanoff, 2005; Nielsen et al., 2007). Concepts such as political culture (Jasanoff, 2005), national participatory styles (Nielsen et al., 2007), and political system types (Dryzek and Tucker, 2008) have been applied to compare consensus conferences on agbiotech issues across a variety of national contexts and have started to identify the most relevant dimensions of context and the mechanisms by which they interact with the deliberative methodology.

A key contribution of this work is making explicit what are often implicit assumptions about the nature of scientific authority, democratic legitimacy and the appropriate role of lay publics in a particular decision context. It is often assumed that the concept of participation has a universal meaning across different contexts (Nielsen et al., 2007), but this research emphasizes how different political communities have unique understandings of 'authority', 'democracy' and 'public knowledge' which create different norms and expectations about the 'legitimate' role of publics in scientific governance and evolve particular institutional structures and practices for including publics over time. Relevant aspects of the institutional configuration for public engagement include: the institutional location for hosting or organizing the event (for example is it an arm's-length technology assessment institution or the department directly responsible for making a policy decision?); the resources available for doing public engagement (financial and human) and organizer experience in putting on these kinds of events; and the way public engagement is linked with the policy decision and other strands of policy input, such as appointed expert advisory committees and other stakeholder engagement processes (Burgess and Chilvers, 2006; Irwin, 2001, 2006).

This strand of research demonstrates the 'interpretive flexibility' of consensus conferences (Horst, 2008) in that there are variations in the perceived rationale for doing a consensus conference, the kinds of information that the lay panel receives and what counts as a successful contribution to the decision-making process. In those contexts with a strong tradition of public deliberation (such as Denmark, the birthplace of consensus conferences), the rationale of a consensus conference follows more directly from its theoretical roots in deliberative democracy as a form of broad social learning and a legitimate and necessary contribution to decision-making. In a context with a tradition of elite-oriented, science-based decision-making, sponsors may perceive the rationale to be one of public education (Joss, 1994), and organizers may shift their goal for the event to that of raising sensitivity and awareness of deliberative models of engagement (Klüver et al., 2000). The EUROPTA project, examining deliberative models of engagement across six countries over the late 1990s, identified a wide variety of contributions that engagement initiatives can make, both direct and indirect. Direct contributions include agenda-setting, filtering of policy alternatives and exploring the nature of policy objectives. More indirect contributions include improving science and public communications, stimulating broader public debate, building awareness of the policy issue and raising sensitivity to new methods of engagement.

SYNTHESIS OF EMPIRICAL RESEARCH: WHERE ARE WE NOW?

In the remainder of this chapter, we explore the body of literature on deliberative engagement specific to agbiotech issues as we have moved from theory to practice. What have we learned from these experiments in deliberative engagement over the past 25 years? A core outcome of the body of work on consensus conferences has been demonstrating that beyond the differences in context described in the previous section, there are also a number of commonalities shared across contexts. Lay people in general take their role in consensus conferences seriously; they are able to learn about and integrate complex information, engage in quality deliberation and work in a mindset geared towards the 'common good' (Delli Carpini et al., 2004; Dryzek and Tucker, 2008). Another commonality across different policy and cultural contexts is the kinds of issues that lay panels raise in conjunction with agbiotech. Questions emerging in lay panel discussions cover the following themes (Mørkrid, 2001; Einsiedel et al., 2001; Goven, 2003; Purdue, 1999):

- What are the purposes of pursuing agbiotech?
- What benefits and risks are implicated?
- Who enjoys the benefits and who experiences the risks?
- How is the technology controlled, monitored and regulated?
- What are the longer-term consequences for human health and the environment?
- Who owns genetically modified plants? Who owns genes?
- What does private ownership mean in terms of access to information?
- How will publics be informed of, and involved in, policy development in an ongoing way?

Furthermore, the conclusions reached in citizen reports on agbiotech issues have generally been precautionary in nature (Dryzek et al., 2009), not proclaiming an absolute 'verdict', but taking a 'yes, but . . .' perspective on governance, focusing on the institutional arrangements surrounding the technological application and investigating the ways in which it can be constructed to maximize social benefits and minimize harm (Marris and Joly, 1999). For example, in Switzerland a panel recommended a moratorium on commercial production and marketing of GM organisms (but not research), suggesting they may have been more supportive had the benefits of the technology been more prevalent (Skorupinski et al., 2007). A French panel also exhibited scepticism of GM technologies, arguing that there had never been consumer demand for GM foods in France and raising the question of whether there was even a need for GM products in a European context (Dryzek et al., 2009). A Canadian lay panel produced a conditionally positive perspective on GM food, contending that it could be a beneficial and safe technology, but only 'if we make it so' (Citizen panel report, p. 7, in Dryzek et al., 2009, p. 275) by considering how it fits within its social, ethical and economic contexts.

Another common finding is that lay panels are aware of the political and institutional context in which their consensus conferences are taking place and are able to respond to it, albeit in different ways in different contexts. Lay panels take their autonomy seriously and are 'extremely wary of not being – or being seen to be – manipulated by stakeholders from any side' (Marris and Joly, 1999, p. 17) and generally have been seen to be not easily manipulated by organizing choices (Drzyek et al., 2009). Thus, despite the highly structured nature of the consensus conference methodology, and suggestions that this leaves open many opportunities for organizers to 'manipulate' the process (through issue framing, selecting kinds of background information and expert selection), research shows that lay panels have generally not been easily swayed by organizational choices (Dryzek et al., 2009).

An extreme example of this is the French consensus conference in 1998. It took place in a highly expert-oriented policy context, where the expectation was that public engagement would serve an education function rather than provide a forum for citizens to contribute recommendations and views to policy-makers. The government announced that the consensus conference would take place after legislation was produced and it framed public views as 'indecisive, insufficiently informed, and incoherent' (Dryzek et al., 2009, p. 269). In this environment, the French lay panel worked hard to maintain their autonomy, interrupting the experts in the preparatory sessions to have them return to points which the panel considered important as well as keeping the scientist-dominated advisory committee out of the room when writing the final report. This behaviour is not the norm in the French context and led to a media article being titled: 'citizens discourteous in the face of specialists' (Marris and Joly, 1999, p. 17).

In a Japanese consensus conference held in 2000, the lay panel responded strategically to the political and cultural context in Japan, particularly the norm for 'conflict-avoidance behavior' (Nishizawa, 2005, p. 483). The panel positioned its recommendations in the context of the expected policy-maker response and a desire to encourage future citizen participation. In the words of a lay panel member, 'if we had insisted on a total ban on GM crops, we were worried that the shell which had started to open after many years would close' (Nishizawa, 2005, p. 483).

The structure of consensus conferences joins citizens and experts in dialogue, which is not the norm in policy development processes. The empirical research shows that, through this interaction, many panels challenge the special status traditionally given to scientific expertise in S&T policy, recognizing that experts cannot fully escape their own biases and that relevant knowledge comes from a variety of sources, including but not exclusive to scientific research (Dryzek and Tucker, 2008). Australian panellists cited the 'recalcitrance' of some experts to avoid directly answering questions posed to them as a major obstacle to the conference process. In a Swiss conference, the panel was sceptical of experts when they made comments outside of their specialist area; in the end, they changed the name from 'expert' to 'informant' (Skorupinski et al., 2007). This challenging does not carry across the board, however, as seen in the UK and New Zealand initiatives. In the 1996 New Zealand conference, characterized as a good-faith attempt (Goven, 2003), a strong deficit-model framing of the issue led to the lay panel perceiving only the scientists involved as 'real' experts and offering the 'facts' to the panel, in comparison to getting 'rhetoric' from non-scientists.

PUTTING DELIBERATIVE MODELS IN PERSPECTIVE

On the whole, there has been extensive academic research and discussion on consensus conferences in general and on public engagement on agbiotech specifically. The high level of attention paid to consensus conferences needs to be placed within the perspective of the broader governance context for biotechnology applications. The total number of consensus conferences is quite small in comparison to other sources of policy input such as public opinion polls and the many expert advisory processes (Hagendijk and Irwin, 2006). Over time, consensus conferences have remained in an 'experimental' frame in most countries (Denmark being a key exception), generally conducted as one-off events and not integrated within a larger public and stakeholder engagement programme.

In addition, there is little evidence of direct influence of consensus conferences on policy decisions related to agbiotech. In some cases, the consensus conference happened after major policy decisions had been made (for example France) while in others, such as Japan, the UK, Canada and Australia, the consensus conference was not explicitly linked to a policy decision (Einsiedel et al., 2001; Nishizawa, 2005; Purdue, 1999; Walls et al., 2005).

Some analysts point to the downstream timing of consensus conferences on agbiotech as part of the problem. Theoretical ideals suggest that consensus conferences are best applied upstream in the technology development process where more choices about a technology's trajectory are open to democratic debate. But in the case of agbiotech, GM crops were already being produced on a commercial scale before the broad experimentation with consensus conferences began in the 1990s (Marris and Joly, 1999). The downstream use of consensus conferences, in many instances after public controversy had erupted (for example France, Switzerland and UK), has led critics to suggest that the broad experimentation with consensus conferences does not represent a groundswell of enthusiasm for deliberative democracy on the part of government, but instead, that the consensus conference is seen as an instrumental tool for managing public controversies that could not be resolved by existing institutional processes (Dryzek et al., 2009; Marris and Joly, 1999). Those processes deployed outside of official processes might be further viewed as efforts to open up more traditional or closed decision-making procedures.

Areas for Further Research

Taken together, the body of research in this area has led to a less romantic and more realistic view of the possibilities and limits of deliberative public engagement. In the remainder of this chapter, we outline key areas of future research related to the institutionalization of deliberative modes of engagement. A number of big questions remain. How are citizen views, which tend to be holistic in nature, and involve social, scientific, political and economic issues in tandem, best incorporated into policy processes dominated by narrower, science-based risk assessments? How are the views of 'unaffiliated citizens', as represented in consensus conferences, best integrated with those of more 'interested' stakeholder groups? How can citizen engagement be managed well in a context of organizational, financial and political constraints?

The UK's most extensive public engagement exercise to date – *GM Nation?* – illustrates the tensions that have precipitated these questions. Taking place over the period of November 2002 to September 2003, *GM Nation?* was a response to public controversy over GM crops, with the aim of providing meaningful input to the UK government 'about the nature and spectrum of the public's views, particularly at the grass roots level, to inform decision-making' (*GM Nation?*, 2003, p. 11, in Irwin, 2006, p. 311). It was a multi-faceted engagement process including: a website that received more than 2.9 million hits; a series of open meetings at the national, regional and local levels, with an estimated 20000 people participating (Rowe and Frewer, 2005); and a series of 'narrow-but-deep' closed discussions (77 'ordinary citizens' in 10 groups) that met on two separate occasions with time between to gather information.

The process and outcomes were challenged on many fronts. The legitimacy of the exercise as a whole was questioned because of the perceived weakness in representation, particularly a concern about 'capture' by 'special interest' groups or 'cranks' (Burgess and Chilvers, 2006). Genuine public opinion was seen to consist of the views of naive, uninterested citizens. The privileging of the 'innocent citizen' (Irwin, 2006) often comes at the expense of other publics' ability to participate in decision-making, most notably 'uninvited' and spontaneous publics, particularly those with fixed, informed views. There are important ramifications from this preference for ordinary citizens: 'Deliberately or not, *invited* public involvement nearly always imposes a frame which already implicitly imposes normative commitments – an implicit politics – as to what is salient and what is not salient, and thus what kinds of knowledge are salient and not salient' (Wynne, 2007, p. 107) while uninvited participation is usually about directly challenging these commitments. Deeper reflection is needed on how to accommodate and connect organized and spontaneous forms of public participation (see Middendorf and Busch, 1997).

A second criticism of *GM Nation?* which exemplifies broader concerns within the public participation field was of how citizen input was integrated with other strands of policy input. The public debate ran in parallel with an economic assessment and review of the available science on GM crops. The economic report was released seven days before the end of the public debate, and the science review three days after. In addition, it was concluded before the publication of results from the UK field trials of GM crops (Irwin, 2006). Beyond inhibiting public discussion of the more technical issues raised by the other strands (Irwin, 2006), Burgess and Chilvers (2006) suggest that 'such compartmentalization upheld fact/value distinctions, undermined any possibility of having an informed debate, and went some way to (re)constructing a "deficit model" of public understanding' (p. 717).

While the trajectory of specific extended models of public participation on agbiotech as exemplified by *GM Nation?* might suggest the demise of deliberative public participation approaches, the application of such approaches to other emerging technologies, from nanotechnology (Powell, 2008; Cobb, 2011) to stem cell research (BMRB, 2008) to synthetic biology (Royal Academy of Engineering, 2009) show 'the deliberative democratic project . . . still very much alive' (see Roberts, 2004, p. 341). The range of practices has been increasing and is projected to continue because of more decentralized approaches to decision-making, networked stakeholder organizations and publics linked by information and communication technologies, and the proliferation of wicked problems (Roberts, 2004). Questions relevant to agricultural biotechnology will continue to be addressed with various forms of public engagement as new (and potentially contentious) applications are planned (see for example Marris et al., 2008), new regulatory demands arise as new GM crops are developed and new driving forces emerge for GM plants (see EPTA, 2009). The changing contexts of governance to address multi-level governance challenges, from climate change to biodiversity, have similarly seen innovation in deliberative approaches to engage the transnational public (see Worthington et al., 2011; Danish Board of Technology, 2012).

The tendency for deliberative exercises to be conducted as one-off experiments (with the exception of ongoing work carried out by Technology Assessment Institutes in Europe), in conjunction with a lack of evidence of their policy impact, makes deliberative engagement vulnerable in an era of global economic recession. For public participation scholars, this calls for putting more attention on the interactions piece of the IAD framework. The field has come a long way in developing new techniques of public participation, and the time is ripe for extending beyond participation exercises as isolated, experimental events, to considering both normatively and practically how they are best positioned within the complex web of governance actors and processes.

REFERENCES

Abelson, J., P-G, Forest, J. Eyles, P. Smith, E. Martin and F-P. Gauvin (2003), 'Deliberations about deliberative methods: Issues in the design and evaluation of public participation processes', *Social Science & Medicine*, **57**, 239–51.

Andersen, I-E. and B. Jæger (1999), 'Danish participatory models: scenario workshops and consensus conferences: towards more democratic decision-making', *Science and Public Policy*, **5**, 331–40.

Barnes, M., J. Newman, A. Knops and H. Sullivan (2003), 'Constituting "the public" in public participation', *Public Administration*, **81**(2), 379–99.

Bickerstaff, K. and G. Walker (2005), 'Shared visions, unholy alliances: power, governance and deliberative processes in local transport planning', *Urban Studies*, **42**(12), 2123–44.

Bijker, W., T. Hughes and T. Pinch (1997), *The Social Construction of Technological Systems: New Directions in the Sociology and History of Technology*, Cambridge, MA: MIT Press.

Bimber, B. (2005), *Politics of Expertise in Congress: The Rise and Fall of the Office of Technology Assessment*, Albany, NY: State University of New York Press.

BMRB (2008), 'Stem cell dialogue', available at: http://www.bbsrc.ac.uk/society/dialogue/activities/stem_cell_final_report.pdf, accessed 22 January 2009.

Burgess, J. and J. Chilvers (2006), 'Upping the *ante*: a conceptual framework for designing and evaluating participatory technology assessments', *Science and Public Policy*, **33**(10), 713–28.

Chambers, S. (2003), 'Deliberative democratic theory', *Annual Review of Political Science*, **6**, 307–26.

Cobb, M. (2011), 'Creating informed public opinion: citizen deliberation about NTs for human enhancement', *Journal of Nanoparticle Research*, **13**, 1533–48.

Danish Board of Technology (2012), 'Results report: worldwide views on biodiversity', Copenhagen.

Delli Carpini, M.X., F.L. Cook and L.R. Jacobs (2004), 'Public deliberation, discursive participation, and citizen engagement: a review of the empirical literature', *Annual Review of Political Science*, **7**, 315–44.

Dryzek, J. and A. Tucker (2008), 'Deliberative innovation to different effect: consensus conferences in Denmark, France, and the United States', *Public Administration Review*, **68**(5), 864–86.
Dryzek, J., R. Goodin, A. Tucker and B. Reber (2009), 'Promethean elites encounter precautionary publics: the case of GM foods', *Science, Technology & Human Values*, **54**(3), 263–88.
Einsiedel, E. and D. Eastlick (2000), 'Consensus conferences as deliberative democracy: a communications perspective', *Science Communication*, **21**, 323–43.
Einsiedel, E., E. Jelsoe and T. Breck (2001), 'Publics at the technology table: the consensus conference in Denmark, Canada and Australia', *Public Understanding of Science*, **10**, 83–98.
Elam, M. and M. Bertilsson (2003), 'Consuming, engaging and confronting science: the emerging dimensions of scientific citizenship', *European Journal of Social Theory*, **6**(2), 233–51.
Elster, J. (1998), *Deliberative Democracy*, Cambridge: Cambridge University Press.
EPTA (2009), 'European Participatory Technology Assessment Network final report: genetically modified plants and foods: challenges and future issues in Europe', Berlin.
Fischer, F. (2003), *Reframing Public Policy: Discursive Politics and Deliberative Practices*, Oxford: Oxford University Press.
Funtowicz, S. and J. Ravetz (1992), 'Three types of risk assessment and the emergence of post-normal science', in S. Krimsky and D. Golding (eds), *Social Theories of Risk*, Westport, CT: Praeger, pp. 251–74.
Gastil, J. and P. Levine (2005), *The Deliberative Democracy Handbook: Strategies for Effective Civic Engagement in the Twenty-First Century*, San Francisco, CA: Jossey-Bass.
Gibbons, M. et. al. (1994), *The New Production of Knowledge: the Dynamics of Science and Research in Contemporary Society*, London: Sage.
Goven, J. (2003), 'Deploying the consensus conference in New Zealand: democracy and de-problematization', *Public Understanding of Science*, **12**, 423–40.
Grundahl, J. (1995), 'The Danish consensus conference model', in S. Joss and J. Durant (eds), *Public Participation in Science: the Role of Consensus Conferences in Europe*, London: Science Museum.
Hagendijk, R. and A. Irwin (2006), 'Public deliberation and governance: engaging with science and technology in contemporary Europe', *Minerva*, **44**, 167–84.
Higgins, V. and G. Lawrence (2005), 'Globalization and agricultural governance', in V. Higgins and G. Lawrence (eds), *Agricultural Governance: Globalization and the New Politics of Regulation*, Abingdon: Routledge.
Horst, M. (2008), 'In search of dialogue: staging science communication in consensus conferences', in D. Cheng et al. (eds), *Communicating Science in Social Contexts*, New York: Springer Science and Business Media.
House of Lords Select Committee (2000), 'Science and Technology – third report', London: UK Parliament.
Irwin, A. (2001), 'Constructing the scientific citizen: science and democracy in the biosciences', *Public Understanding of Science*, **10**(1), 1–18.
Irwin, A. (2006), 'The politics of talk', *Social Studies of Science*, **36**(2), 299–320.
Jasanoff, S. (2003) 'Technologies of humility: citizen participation in governing science', *Minerva*, **41**(3), 223–44.
Jasanoff, S. (2005), Designs on Nature: Science and Democracy in Europe and the US, Princeton, NJ: Princeton University Press.
Joss, S. (1994), 'The UK National consensus conference on plant biotechnology', *Public Understanding of Science*, **4**, 195–204.
Joss, S. (1995), 'Evaluating consensus conferences: necessity or luxury?', in S. Joss and J. Durant (eds), *Public Participation in Science: The Role of Consensus Conferences in Europe*, London: Science Museum, pp. 89–109.
Kearnes, M., R. Grove-White, P. Macnaghtan, J. Wilsdon and B. Wynne (2006), 'From bio to nano: learning lessons from the UK agriculture biotechnology controversy', *Science as Culture*, **15**(4), 291–307.
Klüver, L. (1995), 'Consensus conferences at the Danish Board of Technology', in S. Joss and J. Durant (eds), *Public Participation in Science: the Role of Consensus Conferences in Europe*, London: Science Museum, pp. 41–9.
Klüver, L., M. Nentwich, W. Peissl, H. Torgersen, F. Gloede, L. Hennen, J. van Eijndhoven, R. van Est, S. Joss, S. Bellucci and D. Bütschi (2000), 'EUROPTA: European Participatory Technology Assessment – participatory methods in technology assessment and technology decision-making', Copenhagen: The Danish Board of Technology, available at: www.tekno.dk/europta.
Loka Institute, The (n.d.), 'Tracking Danish-style, citizen-based deliberative consensus conferences on science and technology policy worldwide', *The Loka Institute*, retrieved 1 October 2012 from http://www.loka.org/TrackingConsensus.html.
Lyall, C. and J. Tait (2005), *New Modes of Governance: Developing an Integrated Policy Approach to Science, Technology, Risk, and the Environment*, Aldershot: Ashgate.
Marris, C. and P.-B. Joly (1999), 'Between consensus and citizens: public participation in technology assessment in France', *Science Studies*, **12**(2), 3–32.
Marris, C., P. Benoit-Joly and A. Rip (2008), 'Interactive technology assessment in the real world: dual dynamics in an ITA exercise on genetically modified vines', *Science Technology and Human Values*, **33**(1), 77–100.

Mayer, I., J. de Vries and J. Guerts (1995), 'An evaluation of the effects of participation in a consensus conference', in S. Joss and J. Durant (eds), *Public Participation in Science: the Role of Consensus Conferences in Europe*, London: The Science Museum, pp. 109–124.

Middendorf, G. and L. Busch (1997), 'Inquiry for the public good: democratic participation in agricultural research', *Agriculture and Human Values*, **14**(1), 45–57.

Mørkrid, A. (2001), 'Consensus conferences on genetically modified food in Norway', *Citizens as Partners: Information, Consultation and Public Participation in Policy-making*, Paris: OECD, pp. 223–37.

Mohr, A. (2002), 'Of being seen to do the right thing: provisional findings from the first Australian consensus conference on gene technology in the food chain', *Science and Public Policy*, **29**(1), 2–12.

Nielsen, A., J. Lassen and P. Sandoe (2007), 'Democracy at its best? The consensus conference in a cross-national perspective', *Journal of Agricultural and Environmental Ethics*, **20**, 13–35.

Nishizawa, M. (2005), 'Citizen deliberations on science and technology and their social environments: case study on the Japanese consensus conference on GM crops', *Science and Public Policy*, **32**(6), 479–89.

OECD (2005), *Evaluating Public Participation in Policy-making*, Paris: OECD.

Ostrom, Elinor (2005), *Understanding Institutional Diversity*, Princeton, NJ: Princeton University Press.

Powell, M. (2008), 'Building citizen capacities for participation in nanotechnology decision-making: the democratic virtues of the consensus conference model', *Public Understanding of Science*, **17**(3), 329–48.

Powell, M., M. Colin, D. Kleinman, J. Delborne and A. Anderson (2011), 'Imagining ordinary citizens? Conceptualized and actual participants for deliberations on emerging technologies', *Science as Culture*, **20**(1), 37–70.

Purdue, D. (1999), 'Experiments in the governance of biotechnology: a case study of the UK National Consensus Conference', *New Genetics and Society*, **18**(1), 79–99.

Rittel, H. and M. Webber (1973), 'Dilemmas in a general theory of planning', *Policy Sciences*, **4**, 155–69.

Roberts, N. (2004), 'Public deliberation in an age of direct citizen participation', *American Review of Public Administration*, **34**(4), 315–53.

Rowe, G. and L. Frewer (2000), 'Public participation methods: a framework for evaluation', *Science, Technology and Human Values*, **25**(3), 3–29.

Rowe, G. and L. Frewer (2004), 'Evaluating public participation exercises: a research agenda', *Science, Technology and Human Values*, **29**(4), 512–56.

Rowe, G. and L. Frewer (2005), 'A typology of public engagement mechanisms', *Science, Technology and Human Values*, **30**(2), 251–90.

Royal Academy of Engineering (2009), *Synthetic Biology: A Public Dialogue*, London: Royal Academy of Engineering.

Schot, J. and A. Rip (1997), 'The past and the future of constructive technology assessment', *Technological Forecasting & Social Change*, **54**(2&3), 251–68.

Sclove, R. (2010), *Reinventing Technology Assessment: A 21st Century Model*, Washington, DC: Woodrow Wilson International Center for Scholars.

Seifert, F. (2006), 'Local steps in an international career: a Danish-style consensus conference in Austria', *Public Understanding of Science*, **15**, 73–88.

Skorupinski, B., H. Baranzke, H.W. Ingensiep and M. Meinhardt (2007), 'Consensus conferences – a case study: publiforum in Switzerland with special respect to the role of lay persons and ethics', *Journal of Agricultural and Environmental Ethics*, **20**, 37–52.

Thompson, D. (2008), 'Deliberative democratic theory and empirical political science', *Annual Review of Political Science*, **11**(1), 497–520.

Walls, J., T. Rogers-Hayden, A. Mohr and T. O'Riordan (2005), 'Seeking citizens' views on GM foods: experiences from the United Kingdom, Australia and New Zealand', *Environment*, **47**(7), 22–36.

Webler, T. (1995), 'Right discourse in citizen participation: an evaluative yardstick', in O. Renn, T. Webler and P. Wiedemann (eds), *Fairness and Competence in Citizen Participation: Evaluating Models for Environmental Discourse*, Dordrecht: Kluwer, pp. 35–86.

Worthington, R., M. Rask and M. Lammi (2011), *Citizen Participation in Global Environmental Governance*, London: Earthscan.

Wynne, B. (2005), 'Risk as globalizing "democratic" discourse? Framing subjects and citizen', in M. Leach, I. Scoones and B. Wynne (eds), *Science & Citizens: Globalization and the Challenge of Engagement*, London: Zed Books, pp. 66–82.

Wynne, B. (2007), 'Public participation in science and technology: performing and obscuring a political–conceptual category mistake', *East Asian Science, Technology and Society*, **1**(1), 99–110.

34 Lessons from the California GM labeling proposition on the state of crop biotechnology

David Zilberman, Scott Kaplan, Eunice Kim and Gina Waterfield

1 INTRODUCTION

In the California State election on 6 November 2012, Proposition 37 was introduced to require mandatory labeling of food sold to consumers that was made from plants or animals with genetic material changed in specified ways. This measure failed to pass with 51.5 per cent of Californian voters opposing it and 48.5 per cent in favor after proponents of the legislation led in the polls by a large margin in the early days of the election campaign (see Figure 34.1).

The campaign and the public debate surrounding genetically modified (GM) foods prior to the election was just one battle in a long struggle to shape policies that would impact the fate of crop biotechnologies in the US. The intellectual exchanges, rhetoric and tone of the campaign, as well as the result of Proposition 37 provide insight on biotechnology in the context of agriculture. This chapter analyzes some of these lessons and discuss their implications for the future of agricultural biotechnology.

2 POLITICAL ECONOMY OF AND WILLINGNESS TO PAY FOR TRANSGENIC TECHNOLOGY

Agricultural biotechnology was introduced in the early 1990s and encountered significant resistance, which resulted in the imposition of strict regulation and practical bans in Europe and Africa. Graff et al. (2009) argue that these strict regulations are the result of political economy processes where different interest groups are using their political and/or economic capital to affect policy choices either directly, by influencing policymakers, or indirectly, by affecting public perception to impact political choices, including voting and campaign contributions. The stakeholders involved in the political debate surrounding agricultural biotechnology include supporters such as companies that produce GM-related products, like Monsanto, technology start-ups that develop many of the innovations, scientists and universities conducting research in this area, farming sectors that use GM, and some of the public. Conversely, opponents of GM technology may come from companies without a strong GM portfolio, for example companies that sell pesticides, and environmental groups. Farmers tend to have mixed attitudes. Some farmers support GM because of its potential to alleviate pest pressure, enhance productivity and improve product quality. Other farmers oppose it, especially those that do not benefit from GM directly or because of concern about increases in supply and the resulting decline in prices. Paarlberg (2008) found that because GM technology originated in

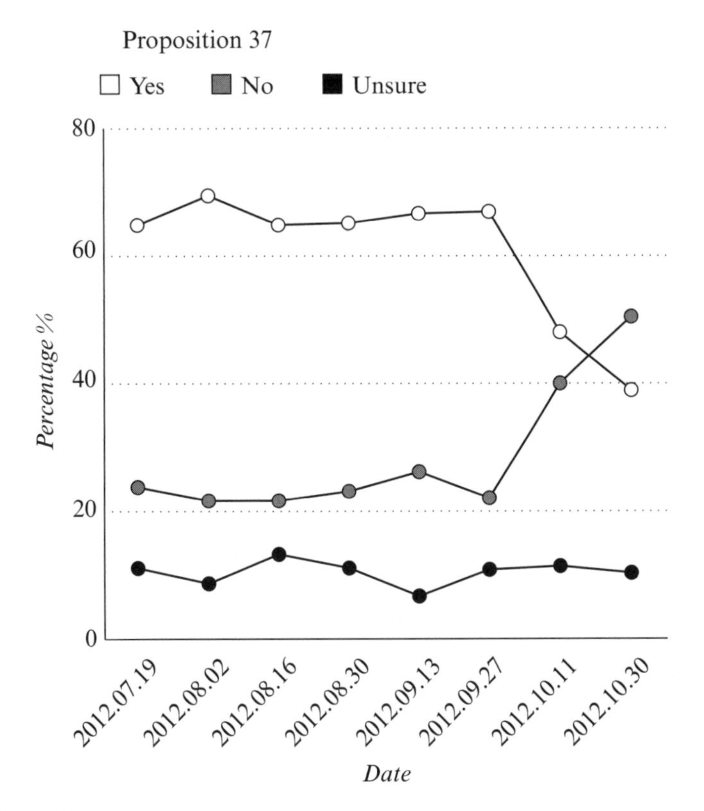

Source: Huffington Post.

Figure 34.1 Change in voter preference for the labeling proposition

the US and benefited American companies, it has more support in the US than in Europe (this is reflected by heavy GM regulation in the EU). Moreover, countries within the US sphere of influence are more likely to adopt GM than countries that lie more in the European sphere of influence.

The political economy literature suggests that the different interest groups will adjust their actions to affect policy choices according to their institutional set-up in various locations. In California, the proposition mechanism can set major policy directions, and different interest groups frequently use it to promote their own benefit. In the past, environmental and other interest groups introduced a proposition to ban pesticides (Proposition 17, 'The Big Green Initiative of 1991'), which failed. The labeling initiative continues along this tradition, and we expect both supporters and opponents of GM to reveal themselves through their monetary and intellectual contributions to this initiative. The ultimate fate of the proposition, however, lies in the decision-making power of the voting public and thus it is important to understand their prior beliefs and the factors that affect their voting behavior.

There is a large body of literature on the public perception of GM. Early studies (Gaskell et al., 1999) emphasized that the big difference in perception in the US versus

Europe is a result of differences in the electorate's trust in the food safety regulatory system and is reflected in differences in media coverage and attitudes towards GM. Heiman et al., (2000) found that gender, education and religious beliefs affect attitudes towards GM. In particular, more religious individuals tended to oppose GM, while individuals with more education were more likely to support it. Overall they identified negative prior attitudes towards GM. The negative prior preferences that consumers have towards GM food were reflected in their willingness to pay a premium for the labeling of GM-free food (Lusk and Coble, 2005). Studies have found that price discounts can sway consumers to choose GM products over traditional food despite negative priors (Huffman et al., 2003).

Consumers are quite heterogeneous, and the amount that they are willing to discount GM food compared to conventional food varies significantly and may be correlated to education, culture, gender, income and the type of GM food considered. Furthermore, the discount for GM products may vary depending on the manner in which consumers' opinions are being solicited (Lusk et al., 2005) as well as the prior information provided to consumers about the impact of biotechnology in general and the products they consider. Colson et al. (2011) suggest that consumers may actually be willing to pay a positive premium for GM products that have traits that are beneficial to health. Hamilton et al. (2003) found that the wide heterogeneity in consumer attitude is also manifested in the case of pesticides – both in terms of willingness to pay for pesticide-free food and to vote for a proposal to ban pesticides. Their findings are consistent with those of Aerni et al. (2011) who compared voting with willingness to pay for GM-free food. They found that individuals who voted to ban the use of chemical pesticides or GMO are not necessarily the ones with highest willingness to pay to avoid GM with 'undesirable' features. This suggests that attitudes towards GM represent attitudes towards food safety and environmental sustainability, and when assessing both the environmental effects and human health risk of GMO, attitudes actually vary among the public.

Kahneman and Tversky (1984) suggested that judgments and decisions are affected by framing of different alternatives. When a technology is presented in a positive way, it will increase the likelihood of a vote in favor of it or increase people's willingness to pay for it. Heiman and Zilberman (2011) used experiments to assess willingness to purchase GM products and found that, as expected, both positive and negative framing affected likelihood of purchase at a given price (or discount compared to non-GM), but negative framing had a stronger impact in deterring consumers. Responses were much more affected by magnitude of the discount for GM products rather than the framing, which suggests that for much of the population, costs may have a stronger impact on decisions regarding GM products than other product features. Heiman and Zilberman's (2011) work implies that negative prior perceptions about GM technology can be amplified by negative information but can be, to some extent, reduced by positive information and price discounts. Thus new information about the impact of GM and the extent that people have to pay for the implementation of labeling, may affect their vote.

3 CONCEPTUAL FRAMEWORK OF VOTING FOR LABELING

Before presenting the specific details of Proposition 37, we develop a conceptual framework that incorporates some of the major features mentioned in the literature, namely heterogeneity of the public in terms of willingness to pay (WTP) to avoid GM products (in this case, through labeling), the cost of implementation of labeling, and the possible comparison of mandatory labeling with voluntary labeling. The California proposition is an almost ideal application of the median voter model, and thus we use it for our analysis (Downs, 1957). The model of course is simple and stylized, but it can provide some insight into the factors that determine the outcome and to what extent the proposition results in efficient resource allocation.

We assume society contains a large number of voters, each of them having a different WTP for labeling of GM products in food. This WTP is the additional amount consumers are ready to pay annually in terms of higher prices of food, inconvenience, and so on. Let WTP be denoted by W, which is distributed from 0 to $\overline{W}$. Let $F(W)$ be the cumulative distribution of the WTP for labeling among a heterogeneous population. Specifically, if the population is ordered according to WTP, the fraction of the population that is willing to pay an amount equal to or greater than a specific W is $F(W)$. Note that $F(\overline{W}) = 1$, at the highest level of WTP the cumulative distribution is equal to 1. If α is a fraction between 0 and 1, then we will define W_α as the maximum amount that α fraction of the population is willing to pay for labeling. If, for example, $W_{0.5} = \$100$, then half the population with a lower WTP for labeling would pay \$100 for labeling. The fraction of the population with WTP above a certain value W is thus, $1 - F(W)$.

First suppose that mandatory labeling will increase the price of food by ΔP_M, and suppose that the public is facing a proposition to introduce mandatory labeling of GM food products. Let us assume that individuals will vote in support of the proposition if their WTP is greater than the increases in the price of food due to labeling. Thus the support for the proposition is $1 - F(\Delta P_M)$ and the proposition will pass if $1 - F(\Delta P_M) \geq .5$ or if $F(\Delta P_M) < .5$, namely, $\Delta P_M < W_{.5}$, the median voter has WTP that is smaller than ΔP_M.

Thus if voters' WTP were fixed, campaigning would serve no purpose. Voters seemed uncertain about both their WTP and the cost of mandatory labeling. Let $F^0(W)$ be the initial distribution of WTP, and ΔP_M^0 the initial estimate of the per capita cost of mandatory labeling. Campaign efforts from both sides aimed to modify the distribution of the WTP and the estimate of ΔP_M. The supporters of the proposition aim to convince the voter to reduce their perceived ΔP_M and increase their WTP. Let $F^1(W)$ be the cumulative distribution of the WTP, and ΔP_M^1 is the estimate per capita cost of the mandatory labeling after the campaign. The proposition has a lower probability of passing if $\Delta P_M^1 > \Delta P_M^0$, namely if the *perceived* cost to the consumer because of the proposition per capita *increased* due to campaign efforts. Another reason for the reduced probability of the proposition passing is that campaign efforts modified the distribution of WTP, in particular, $W_{.5}^0 > W_{.5}^1$, the median voter's WTP for labeling was reduced due to the campaign. A campaign may change the outcome of a vote from support to rejection of the proposition if $W_{.5}^0 - \Delta P_M^0 > 0$ but $W_{.5}^1 - \Delta P_M^1 < 0$.

If the proposition had failed, the consumer could obtain information on GM food through voluntary labeling. In this scenario, the people who do not care much about GM food do not need to pay for the labeling, but the people who want to avoid GM food have the option to do so by paying a higher price ΔP_V. It is plausible that $\Delta P_V > \Delta P_M$ because the cost of mandatory labeling is distributed among more people. If the voters are aware of the option of voluntary labeling, as long as $\Delta P_M < \Delta P_V$, individuals with $W > \Delta P_M$ will vote in favor of the mandatory labeling, but this option is not likely to affect the outcome.[1]

One might ask whether the result of the voting procedure is efficient in an economic sense? We attempt to answer this question partially. First, consider whether the passing of the proposition will improve the welfare of the voters (it is a partial measure of the efficiency effect). Assume that in the initial scenario (benchmark scenario), there is *no mandatory or voluntary labeling*, and the initial voter surplus is zero. The result of the voting is improved efficiency in the narrow sense if it improves the surplus of the voting population. If the proposition passes (Yes), the change in the average voter surplus is:

$$\Delta VS_Y = \int_0^{\overline{W}} (W - \Delta P_M)\, dF(W)^2$$

This surplus change is the difference between the willingness to pay of each individual minus the extra costs multiplied by the measure of the relative size of the population with a given willingness to pay W (by $dF(W)$, which is the differential of the cumulative distribution at W). This change can be decomposed to:

$$\Delta VS_Y = \int_0^{\Delta P_M} (W - \Delta P_M)\, dF(W) + \int_{\Delta P_M}^{\overline{W}} (W - \Delta P_M)\, dF(W) \qquad (34.1)$$

The first element is the surplus of the people that oppose the proposition (their willingness to pay is below ΔP_M, and this term is negative) while the second element is the surplus of the people who support the proposition (their willingness to pay is above ΔP_M). If the sum is positive, the proposition will increase the welfare of the voters. Note that the proposition will pass if:

$$\Delta SY = \int_{\Delta P_M}^{\overline{W}} dF(W) - \int_0^{\Delta P_M} dF(W) > 0 \qquad (34.2)$$

Where ΔSY is the difference between the fraction of those who support the proposition and those who oppose it. Note that the proposition may fail because more people will oppose it, yet it may improve the welfare of the voting public if it passes (namely $\Delta SY < 0$ and $\Delta VS_Y > 0$), if the average gain from the people who support the proposition is sufficiently larger than the average loss of the people who oppose it. Under such a hypothetical situation, failure of the proposition is not improving welfare.

Suppose the proposition fails but increases awareness, leading to a voluntary market. Then the welfare effect on the voting public is:

$$VS_N^{VM} = \int_{\Delta P_V}^{\overline{W}} (W - \Delta P_V)\, dF(W) \qquad (34.3)$$

Compared to the current situation, the people who have high WTP ($W > \Delta P_V$) gain from the voluntary market, but the rest of the population is unaffected. So obviously a voluntary market is an improvement relative to no labeling whatsoever. The passing of the proposition is superior from a welfare perspective to having a voluntary market if:

$$\Delta VS_Y - VS_N^{VM} = \underset{(a)}{\int_0^{\Delta P_M} (W - \Delta P_M) dF(W)} + \underset{(b)}{\int_{\Delta P_M}^{\Delta P_V} (W - \Delta P_V) dF(W)} + \underset{(c)}{\int_{\Delta P_V}^{\overline{W}} (\Delta P_V - \Delta P_M) dF(W)} > 0 \quad (34.4)$$

Equation (34.4) presents the change in the welfare of the voting public from the passing of the proposition *relative to voluntary labeling*. It is the sum of the loss of individuals with low willingness to pay for labeling (group (a) in equation (34.4)), the gain of people with modest WTP for labeling (group (b), with $\Delta P_M \leq W \leq \Delta P_V$), and the gain of the group of people with relatively high WTP for labeling (group (c)). This last group gains less from the passing of the proposition compared to voluntary labeling versus non-labeling because their gain is only the reduction in the per unit cost of labeling as you switch from voluntary to mandatory labeling. Comparing equations (34.1) and (34.4) suggests that the existence of feasible voluntary labeling can affect the intensity of support for mandatory labeling.

The complete impact on social welfare should also consider the impact on groups outside the voting public, but this analysis is beyond the scope of this framework. The debate about the proposition does recognize the existence of these external effects, as we will show below.

4 BACKGROUND ON PROPOSITION 37 TO INTRODUCE MANDATORY LABELING IN CALIFORNIA

California's Proposition 37 required the labeling of certain plant and animal foods containing GM ingredients or processed using GM technology. While the proposition was designed to apply to most types of food, there were several significant exceptions. Generally, raw or processed food made from genetically modified plants or animals requires a label. Meat produced from animals fed GM feed or injected with GM materials, medicines produced using GM technology, foods sold in restaurants for consumption in-house, alcoholic beverages, and raw foods unintentionally produced with GM materials (seeds) were to be exempt from the proposition. These exemptions reflect the reality that most grain-fed meat products are produced from GM feed. Thus the proposition does not label existing traits that are mostly used for animal feed, but erects barriers for possible GM traits in crops for human consumption. The proposition also prohibits labeling GM products as 'natural', and thus is consistent with the exclusion of GM as organic.

Proposition 37 also states that there will be increased costs incurred by the state in monitoring and enforcing this new mandatory labeling system. According to the proposition itself, 'increased annual state costs ranging from a few hundred thousand dollars to over $1 million' will be required in order to 'regulate the labeling of genetically engineered foods' (California Voter Guide, 2012). These regulations range from those on GM research facilities and farm-level monitoring of crop treatment, all the way to processors of food and manufacturers of labels. While these direct costs are significant, indirect costs associated with yield decreases as a result of reduced GM technology used in production may result in additional costs to the government and to the consumer.

Enforcement of the proposition would take place at the county level, with the California Department of Health and Safety bearing the responsibility. The enforcement standards for food labeled under Proposition 37 are actually much stricter than the USDA organic standards. According to Colin Carter (Carter et al., 2012), under the mandatory GM labeling requirement conventional foods would be subject to testing at certain processing levels to check for 'accidental GM' ingredients all the while foods labeled 'organic' are subject only to testing at the farm level, bypassing additional testing thereafter. In addition, the onus of enforcement is largely on the consumer, which may give way for citizens suing 'on behalf of the general public', where companies responsible for the violation would not be given any advanced notice about the allegations nor the opportunity to address the issue before penalties (Cross, 2012). Fines would amount to $1000 per day for companies found to be violating the labeling system (Hiltachk, 2012). The legal implications of this are extremely significant. Much of the involvement of the legal system will not come through lawsuits against labeled GM foods, but rather against conventional food that is not labeled but contains some GM ingredients, even if by accident through processing (Hiltachk, 2012).

Proposition 37 had numerous supporters, headed by the 850000-member Organic Consumer Association, which was the largest support donor with over $1.3 million spent. Other notable donors include Nature's Path (a non-GMO packaged food maker which raised over $650000), Chipotle Mexican Grill, Whole Foods Markets, and the California Democratic Party. Overall, the supporters of Proposition 37 raised $8.7 million. While the supporters were ahead in the polls for much of the debate over the proposition, the ability of the opposition to counter with much larger donations had a significant effect on the final outcome. Monsanto provided the largest donation: its outlays of over $8.1 million nearly matched the entire 'yes' side. Several other food companies donated, including Kraft Foods Global, Hershey Company, Pepsico Inc., General Mills, and the Kellogg Company. Several biotechnology companies, including Dow AgroSciences, Bayer CropScience, BASF Plant Science, each donated $2 million. Both sides were vying for the support of the media; major newspapers such as the *Los Angeles Times*, the *San Francisco Chronicle*, and the *Sacramento Bee* opposed the proposition for various reasons. Michael Pollan and several other leading food activists were very visible supporters of the proposition.

It was clear that given the early lead of those supporting labels in the polls, the opponents had a chance only if they were able to make a large investment. Indeed they did so; the 'no' side spent $45.6 million versus the $8.7 million on the 'yes' side. The final outcome, however, depended on the evolution of the arguments of the two parties, as presented below.

5 EVOLUTION OF THE PUBLIC DEBATE

As the political economic literature suggests, the introduction of a new proposition starts a public debate where two camps gradually introduce arguments that aim to enhance the likelihood that their argument will win. The two camps usually introduce their own arguments to the public arena, respond to their opponent's arguments and assess their impacts via polling. Thus there is an ongoing process of mutual learning that is based on feedback and the ability to come up with new arguments that may sway voters. Our perspective on the evolution of the debate surrounding Proposition 37 is based on observing and participating in it and interacting with key players in the campaign.[3] The debate concerning Proposition 37 had three phases.

Proponents of the proposition supported their stance via three main arguments. First they drew on the basic American principle of freedom of information. They noted that over 40 other countries require labeling for GM foods so their citizens can make informed decisions. This was argued to be especially important as GM food may pose some risk and people have the right to protect themselves. The second line of argumentation aimed to highlight the broadly-held perspective that GMOs do not provide real benefit and simply introduced new risks or enhanced existing ones. Key points included that GM traits are primarily in crops that are not consumed directly by humans, they do not increase yields or replace pesticides and may be dangerous for the environment and humans. The conclusion was that we need more experience to assess their true risks.[4] In addition, the rights to these technologies are held by large agri-businesses and corporations like Monsanto, who capture most of the economic benefits of the technologies and reduce both the freedom to operate and the economic well-being of the small farmer.

Opponents of the proposition focused on addressing each of the proponents' claims. First, they disagreed that the proposition addressed the issue of freedom of information; they tried to recast the debate into whether labeling should be mandatory or voluntary. If there is market demand to avoid GM food, then the supplier will provide it and a certification mechanism will be established to label it. They pointed to other examples of voluntary labeling to accommodate consumer preferences, including fair trade coffee, halal and kosher foods, and even organic foods do not include GM and are certified by the government. The true issue to the 'no' side was what the benchmark would be: would it be non-GM, or GM? Opponents of Proposition 37 suggest that including GM as part of the mainstream food is preferable because of its important contribution to the food system.[5]

The opponents also disputed the assertion that GM provides negligible benefits, citing studies documenting GM's contribution to increasing food availability without jeopardizing food safety. In particular, they argue that GM provides an alternative means to address pest problems by replacing existing chemical pesticides that pose health risks, or to address problems that are not otherwise managed. Ultimately, GM traits increased output. In developing countries the impact of GM on yields is especially pronounced. In particular there are several studies that show that GM cotton yields are more than 50 per cent higher than conventional crops, corn yields may increase by 20 per cent and soybean yields rise by 30 per cent. A key feature is that GM varieties enabled the expansion of acreage by double cropping (Qaim and DeJanvry, 2003; Qaim and Matuschke, 2005; Qaim and Zilberman, 2003). Furthermore, this supply-enhancing effect of GM caused the reduction in the price of major food commodities of 15–30 per cent, and if the

Europeans had removed the de facto ban on GM, some of the recent food price inflations may have been avoided. Thus, the main beneficiaries of GM are poor consumers globally.

Opponents of the proposition cited reports from leading research academies that suggest that GM foods were as safe as conventional foods, both in terms of human health and environmental effects. Furthermore, they argued that one of the main reasons that farmers adopt GM is that it reduces their exposure to toxic pesticides. In response to the allegations that GM primarily benefited big agri-businesses, opponents of Proposition 37 cited results of quantitative studies that reported the benefit of GM foods to be in the billions of dollars, shared among consumers, farmers, suppliers and industry (Alston and Sumner, 2012). They also noted studies that suggested that the excessive regulation of GM actually contributed to the concentration of the industry and prevented the introduction of new traits with even more appealing properties, including varieties that are more drought tolerant and offer enhanced nutritional content, especially in specialty crops. They noted that the industry was supporting new clearinghouse arrangements to assist small farmers to access GM varieties.

The public discourse about Proposition 37 engaged many media outlets, including print, television, social media, and editorial blogging. Each side provided new information to strengthen their argument. For example, the proponents circulated the results of a French study that found tumors in rats fed on GM corn. But the opponents were swift in their response, publicizing the responses of several academies of science including the National Academy of Sciences, American Council on Science and Health, the World Health Organization, and European Food Safety Authority. The opponents benefited from another fortuitous study from Stanford University that found that organic food has no more nutritional value than conventional food.[6]

As Table 34.1 suggests, proponents of Proposition 37 benefited initially from widespread support, with 65 per cent in favor, 25 per cent opposed and 14 per cent undecided; the first phase of the campaign between mid-September and early October actually reduced the support for the proposition to 39 per cent, with many voters reporting they were undecided. The fact that people for the first time heard about some of the benefits of GM and, more importantly, that they were presented with arguments that there is no evidence for its negative effects (and actually no evidence that organic is better) may have caused people to rethink their stance.

The second phase of the public debate began in October 2012. It was clear from the advertisement blitz that the opponents were better financed as their ads were more frequent and were presented through better outlets. In the second phase much of the

Table 34.1 Proposition 37 polling data

Date of poll	In favor	Opposed	Undecided
17–23 September 2012	61%	25%	14%
7–9 October 2012	39%	30%	31%
7–10 October 2012	48.3%	40.2%	11.5%
21–28 October 2012	39.1%	50.5%	10.5%

Source: Poll conducted by the LA Times, available at http://articles.latimes.com/2012/sep/27/business/la-fi-prop37-times-poll-20120927.

arguments became 'personal'. The proponents of Proposition 37 emphasized that this was a case where big businesses were attempting to buy the election. The public was asked, 'who do you trust? Monsanto, Pepsi, or the Consumers Union?'[7] Furthermore, it was insinuated that scientists who opposed Proposition 37 were in the pockets of big business.[8] At the same time, the opponents suggested that the proposition was written by lawyers interested in profiting from litigation arising out of violations of Proposition 37 and that these professionals were simply seeking to erect obstacles to the functioning of markets for personal gain.[9]

From our discussion with the individuals behind 'No on 37', it seems that none of the parties benefited from this particular phase of the debate. Nevertheless, one sub-argument promoted by the opponents gained some traction, namely, that the proposition did not arise out of objective concerns about current products (because 70 per cent of GM foods are exempt from this legislation), rather from a desire to slow the biotechnology sector. The opponents of Proposition 37 stressed that GM varieties may help to address climate change and may be economically valuable for California, which tends to be a leader in provision of technology.

The third phase of the debate emphasized the immediate impact of Proposition 37 on Californians. Alston and Sumner (2012), a study by Northbridge Environmental Consulting and Carter et al. (2012) all presented findings that suggested that Proposition 37 could cause economic hardship. The logic was that its implementation would require the separation of different types of products based on whether or not they contain GM, which would require investment in costly monitoring throughout the supply chain, leading some food retailers to limit their product choices in order to avoid the extra cost of compliance. Thus, it was estimated that the extra food cost per family of four would be $400 per year. Proponents of Proposition 37 had first introduced the cost argument, arguing that voluntary labeling was unfair because it raises prices of non-GM foods and makes it less affordable for the poor. The opponents used these studies to turn the argument around, reframing the discussion into whether people who are indifferent about GM should pay extra for information that they do not value. When Proposition 37 failed at the ballot box, commentators suggested that this final cost argument played a prominent role in its defeat,[10] which supports our conceptual model which states that if the majority of population does not value the labeling more than the extra cost that it imposes, then it will fail.

6 CONCLUSION

Proposition 37 originated from an attempt to stall the advances of GM and biotechnology in California, seeking to drawing on the perception that there is significant public suspicion against the technology and rising awareness and concern about food safety. The initial survey of the public mood presented in this study suggests that there was a potential for the proponents of the proposition to succeed. Furthermore, there is a large body of literature in economics and other fields that suggests that some consumers may be willing to pay significant premiums for non-GM food. However, in the end the proposition failed. Two main arguments seemed to carry the outcome of the vote: first, flaws in the writing of the proposition created suspicion of its intent; and perhaps more importantly, the claim that implementation would raise food prices for Californians by $400 per year per household caught people's attention. To some extent, this was a real experiment on WTP to avoid GM. This experiment showed that among the majority of the populace the WTP was low, suggesting that while some of the perceived objections are widely held, they do not run deeply. Once the public realized the cost of restricting GM, they lost enthusiasm, which suggests that increased education on the benefits of GM and, more importantly, the cost of blocking its use, might bear fruit and help to relax the policies that regulate and restrict GM in other markets.

It seems that if the public faces a serious trade-off and is exposed to sound argumentation as to why a regulatory requirement is excessive, people will vote against restrictions. This bodes well for the future of GM if its proponents can make a strong case for it, given that California has tended to support environmental causes: California is one of the few states that has implemented climate change policies. Another lesson might be simply that money talks, and large contributions to political causes can/may sway the public, possibly even against sound policy. But the elections in 2012 demonstrated that large amounts pending do not always guarantee a win.

NOTES

1. If for some reason some individuals assume that the cost of mandatory labeling is greater than that of voluntary labeling, they may vote against mandatory labeling.
2. This is an average because we weigh the welfare change for each interest group (they may support or oppose the proposition) using the probability weight of that group, that is, the population share of the group.
3. Kathy Fairbanks, 'no' on 37 and Stacy Malkin, 'yes' on 37.
4. http://www.carighttoknow.org/facts
5. Zilberman (2012); http://www.noprop37.com/facts; Alston and Sumner (2012).
6. http://med.stanford.edu/ism/2012/september/organic.html.
7. http://www.carighttoknow.org/.
8. http://blogs.berkeley.edu/2012/06/06/why-labeling-of-gmos-is-actually-bad-for-people-and-the-environment/#comments.
9. http://www.noprop37.com/facts/.
10. http://science.time.com/2012/11/06/prop-37-why-californias-ballot-initiative-on-gm-food-is-about-politics-more-than-science/.

REFERENCES

Aerni, Philipp, Joachim Scholderer and David Ermen (2011), 'How would Swiss consumers decide if they had freedom of choice? Evidence from a field study with organic, conventional and GM corn bread', *Food Policy*, **36**(6), 830.

Alston, Julian M., and Daniel A. Sumner (2012), 'Proposition 37 – California food labeling initiative: economic implications for farmers and the food industry if the proposed initiative were adopted', available at: noonprop37.com.

California Voter Guide (2012), 'California general election', available at: http://voterguide.sos.ca.gov.

Carter, C.A., G.P. Gruère, P. McLaughlin and M. MacLachlan (2012), 'California's Proposition 37: effects of mandatory labeling of GM food', *ARE Update*, **15**(6), 3–8, University of California Giannini Foundation of Agricultural Economics.

Colson, Gregory J., Wallace E. Huffman and Matthew C. Rousu (2011), 'Improving the nutrient content of food through genetic modification: evidence from experimental auctions on consumer acceptance', *Journal of Agricultural and Resource Economics*, **36**(2), 343.

Cross, Rebecca (2012), 'Proposition 37: California's newest challenge for food product companies', *Food Processing*, 20 September.
Downs, Anthony (1957), 'An economic theory of political action in a democracy', *The Journal of Political Economy*, **65**(2), 135–50.
Gaskell, George, Martin W. Bauer, John Durant and Nicholas C. Allum (1999), 'Worlds apart? The reception of genetically modified foods in Europe and the US', *Science*, **285**(5426), 384–7.
Graff, Gregory D., Gal Hochman and David Zilberman (2009), 'The political economy of agricultural biotechnology policies', *AgBioForum*, **12**(4), art.4.
Hamilton, Stephen F., David L. Sunding and David Zilberman (2003), 'Public goods and the value of product quality regulations: the case of food safety', *Journal of Public Economics*, **87**(3), 799–817.
Heiman, Amir and David Zilberman (2011), 'The effects of framing on consumers' choice of GM Foods', *AgBioForum*, **14**(3), article no. 9, available at: http://www.agbioforum.org.
Heiman, Amir, David R. Just and David Zilberman (2000), 'The role of socioeconomic factors and lifestyle variables in attitude and the demand for genetically modified foods', *Journal of Agribusiness*, **18**(3), 249–60.
Hiltachk, Thomas (2012), 'Litigation incentives inherent in Proposition 37', Bell, McAndrews & Hiltachk, LLP, Counsel to 'No on 37' campaign, available at: www. noprop37.com.
Huffman, Wallace E., Jason F. Shogren, Matthew Rousu and Abebayehu Tegene (2003), 'Consumer willingness to pay for genetically modified food labels in a market with diverse information: evidence from experimental auctions,' *Journal of Agricultural and Resource Economics*, **28**(3), 481–502.
Kahneman, Daniel and Amos Tversky (1984), 'Choices, values, and frames', *American Psychologist*, **39**(4), 341.
Lusk, Jayson L. and Keith H. Coble (2005), 'Risk perceptions, risk preferences and acceptance of risky food', *American Journal of Agricultural Economics*, **87**(2), 393–405.
Lusk, Jayson L., Mustafa Jamal, Lauren Kurlander, Maud Roucan and Lesley Taulman (2005), 'A meta-analysis of genetically modified food valuation studies', *Journal of Agricultural and Resource Economics*, **30**(1), 28–44.
Paarlberg, Robert L. (2008), *Starved for Science: How Biotechnology is Being Kept Out of Africa*, Cambridge, MA: Harvard University Press.
Qaim, Matin and Alain De Janvry (2003), 'Genetically modified crops, corporate pricing strategies, and farmers' adoption: the case of Bt cotton in Argentina', *American Journal of Agricultural Economics*, **85**(4), 814–28.
Qaim, Matin and Ira Matuschke (2005), 'Impacts of genetically modified crops in developing countries: a survey', *Quarterly Journal of International Agriculture*, **44**(3), 207–28.
Qaim, Matin, and David Zilberman (2003), 'Yield effects of genetically modified crops in developing countries', *Science*, **299**(5608), 900–902.
Zilberman, David (2012), 'The logic and consequences of labeling GMOs', *ARE Update University of California Giannini Foundation of Agricultural Economics*, **15**(5), 5–8.

35 Biotechnology communications, mythmaking and the media

Camille D. Ryan

1 INTRODUCTION

The art and science of communication involves a 'source' (for example, a company or organization) that is trying to reach a 'receiver' (customer or stakeholder). The goal of communication – from a marketing or public relations perspective – is to be 'understood'; to establish a common frame of reference in order to build value. This involves choosing the right message and appreciating that that message may be interpreted by receivers in any one of a number of ways. Thus, understanding the market and your customers and stakeholders is key in developing and executing a successful marketing and communications plan.

In the context of food and agriculture, traditional business approaches to communication and associated strategies have not always worked within the confines of such a simple formula, particularly in the past two decades in the case of 'big agriculture' and the area of biotechnology. Science has fundamentally changed how farming is done and has significantly increased productivity levels worldwide. Biotechnology, which developed into a bit of a buzzword over 30 years ago with substantive investments in start-up companies (McHughen, 2000), offers up a set of tools to modify organisms for a particular purpose. In the context of agricultural biotechnology, that purpose can include anything from generating higher yields in crops to genetically conferring resistance to certain diseases in plant varieties. The use of genetic engineering techniques (for example cloning, gene-splicing) applied to crop development (in conjunction with traditional plant-breeding techniques) has led to the development of a variety of high-yielding, disease-resistant and more environmentally-friendly crop varieties. As of 2012, there has been a 100-fold increase to 170 million hectares planted to biotechnology crops, up from 1.7 million hectares in 1996. This makes biotech crops the fastest adopted crop technology in the history of modern agriculture (James, 2012).

Despite these touted benefits, biotechnology (bioscience, more broadly speaking) has taken somewhat of a kicking in the media and in the court of public opinion (Marks and Kalaitzandonakes, 2001). Many agree that what is deemed as 'anti-science' or 'anti-GM' (genetic modification) rhetoric has been spawned through the concerted, tactical efforts of special interest groups and/or non-government organizations (NGOs). Others suggest that opposition to GM also gained a foothold in the public psyche due, in large part, to the failings on the part of the agricultural biotechnology industry to consider the 'end user' (consumer) beyond that of its primary target market (producers/farmers) (Pinstrup-Anderson and Cohen, 2000; Gaskell et al., 2000; Moschini, 2001).

In the case of 'big ag', the commercialization of science to develop biotechnology crops has led to the rise of what is viewed by critics as a 'science-industrial complex';

science having evolved into a 'private good', driven by monopolistic interests to promote and develop technology for profit (Maeseele, 2009). Science communication, for the most part, has been largely viewed as 'paternalistic' in terms of a one-way transmission flow from the scientific sphere to the public or, as Scholderer and Frewer (2003) suggest, technology-driven, top-down practices. This traditional communication model interprets public acceptance of science and technology simply by raising awareness through the diffusion of more scientific information, thereby enhancing 'scientific literacy' (Bucchi, 1996; Salleh, 2004).

The perceived paternalistic relationship between science (more broadly) and the public, along with the 'science-industrial complex', sets the stage for the current context. Special interest groups and NGOs sporting a new kind of professional activism now 'combine money and marketing with the growing influence of the Internet to sway public opinion and public acceptance' (Byrne, 2006, p. 144). Anti-corporate organizations effectively interweave well-crafted words and combine them with disarming visual images to create highly influential anti-technology campaigns. These constructed metaphors are blended with well-executed communication strategies, the power of social media and celebrity endorsement making for modern 'myth' creation. And although antagonism for all that is 'genetically modified' runs deep in only a fraction of the population, many consumers 'have become fixed in their opposition' (Chassy, 2007, p. 171). According to McHughen and Wager (2010), 'sometimes the misinformation and fear can become infectious and pathogenic, instigating bad public policy, with substantial negative consequences to everyone' (p. 727).

Unquestionably, the anti-GM movement challenges the assertions of science as a universal and unbiased authority leading to what has been referred to in the literature as 'contentious politics or issues' (Herring and Roberts, 2006) particularly in the context of agricultural biotechnology. According to Boone (2011), this requires an 'issues management' approach to proactively handle these types of complex controversies. At the very least, it is evident that traditional industry-driven communication strategies have not worked to counter the negative impacts of the mythological milieu asserted by the anti-GM movement.

2 COMMUNICATIONS IN AGRICULTURAL BIOTECHNOLOGY: A HISTORY

Traditional communications and marketing in business is about brand-building. It is about constructing value-added around a product or service. It first starts with identifying a gap in the market, then by creating a product or service to address the gap. From that point on, it becomes about beating the competition to the market in order to maximize profits.

The formula laid out here is greatly simplified for illustrative purposes. For the most part, this is how industries (no matter what they ply their trade in) operate to capture and maintain market share and to remain competitive. The biotechnology industry is no different, neither in its historical branding as 'agro-chemical' nor in its more recent organizational format positioned in the market as 'life science' or 'bio-science' oriented. Agricultural biotechnology is simply an industry plying its trade and marketing (and communicating the benefits of) value-added products to a receptive market. Unfortunately, that market has been prone to even more uncertainties due to the activities of the anti-GM movement.

In order to examine the history of communications in biotechnology more thoroughly, it becomes important to understand the roll-out of products (genetically modified crops/organisms), the activities of special interests groups and NGOs (including the role of the media) and the response of industry as a result. Marks and Kalaitzandonakes (2001) argue that media coverage of agricultural biotechnology 'has been both cyclical in tone and event driven' (p. 205). They find that a bias in coverage in the media has emphasized different *frames* at different points in time (on both sides of the Atlantic), depending on unfolding events and scientific and risk management controversies. In this section, the history of the roll-out of GM crops in both North America and the European Union (EU) is highlighted along with key 'events' and activities that define the anti-GM movement in addition to response(s) by industry.

Large acreage sized biotechnology crops of maize, soybean, canola and cotton were commercialized in the US in 1996 (James, 2012). Prior to that, however, Calgene's Flavr Savr tomato was launched in the US in 1994 (see more on this in Chapter 51 in this volume). The industry was ill-prepared for the anti-GM backlash that ensued. Despite being approved for both food and feed safety in the US, the Flavr Savr tomato had a short life on the American market due to the pressure placed on the consumer and retailer by special interest groups. Other mass produced crops introduced a short time later represented huge immediate benefits to the producer, and as discussed in Chapter 2 (this volume) to the consumer over the longer term.

Agricultural biotechnology was 'adopted faster than any other technology in the history of agriculture' (Chassy, 2007, p. 170). So, although the 'maverick' Flavr Savr tomato did not make the market 'cut', production of biotechnology maize, soybean, canola and cotton flourished in the North American market. As of 2012, in the US alone, 93 per cent of soybean, 80 per cent of cotton and 73 per cent of corn acreage were GM (herbicide tolerant). Sixty-seven per cent of corn acreage planted and 77 per cent of cotton was Bt (USDA, 2012).

Despite the adoption of several biotechnology crops, it is fair to assume that most advocates of the technology may have expected more aggressive uptake in the market and were most definitely unprepared for the EU's negative response and the regulatory wrangling that ensued. The backlash of the anti-GM movement has been calculated and relentless, leveraging the media in spreading sensationalized and falsified information. Chassy (2007, p. 170) asserts that 'thousands of consumer, environmental, and charitable NGOs have participated in a well-organized, well-financed, and professionally managed global campaign against GMOs. They have been supported by governments, the organic food industry, the chemical industry, food manufacturing industry, and food retailers among others.'

As previously mentioned, North American markets were largely accepting of new GM varieties (cotton, soybean, maize and canola) that were introduced in the mid-1990s. The EU proved to be a greater problem. From 1996 onward, the anti-GM movement took a firm hold on the EU. In the 1980s, the first groups to get involved were the German Green Party, Greenpeace-Switzerland, the UK Green Alliance, the UK Genetics Forum, the International Coalition for Development Action (following that, the Genetic Resources

Action International, GRAIN), and the Intermediate Technology Development Group in the United Kingdom (Schurman, 2004). By the mid-1990s, new organizations including Friends of the Earth-Europe, Greenpeace International, the British Soil Association, and Confédération Paysannes in France, joined the movement. Greenpeace International devoted fifteen campaigners and a 'highly dynamic campaign coordinator', German journalist and former German Green Party politician Benny Härlin, to its new GM food campaign in 1996 (Schurman, 2004, p. 252).

This coalition of NGOs began their campaign against the technology when the first efforts were made to import genetically modified grains into European ports (Glover, 2008). The Bovine Spongiform Encephalopathy (BSE) crisis of 1996 in the UK only served to fuel the fires of the anti-GM movement. The approval of GM maize and rapeseed varieties in the EU in 1998 set off a series of major events that essentially reframed the debate on agricultural biotechnology from 'a matter of scientific advancement and economic growth to one of public accountability and unforeseen consequences' (Maeseele, 2009, p. 160). In early 1998, Prince Charles in the *Daily Telegraph* publicly criticized genetic engineers, suggesting that they are venturing into 'realms that belong to God and God alone', formally setting science in opposition to faith in the media. Soon after, Arpad Pusztai reported the results of his 'sketchy' research on GM potatoes during an interview on the BBC, and by 1999 UK fast food chains and retailers had banned GM ingredients (Glover, 2008). While many scientists have concluded that Pusztai's research methodology was critically flawed and that no conclusions about the safety of biotech foods could be drawn from his data, it continues to resonate in the public debate. In mid-1999, the general public and policy makers 'reacted in a knee-jerk fashion' (Shelton and Roush, 1999) to the monarch butterfly controversy that ensued after a one-page scientific brief by Losey et al. was published in *Nature* (1999). As a capstone to these events, and to close the decade, a de facto moratorium was proclaimed in the EU on new GM crops.

In an effort to turn public opinion around in the EU, Monsanto developed a €5 million advertising campaign in June of 1998. It was designed to convince the public of the benefits of genetically modified crops (Schurman, 2004). The campaign targeted specific countries, such as the UK and Germany, where opposition to anything genetically modified was deemed highest. According to the late Helen Holder, an avid anti-GMO campaigner who later joined Friends of the Earth Europe, in a speech in Amsterdam in March 1999, the Monsanto strategy was to place a series of full-page advertisements in popular newspapers (*The Guardian* and *The Observer*) normally read by members of 'high socio-economic categories' who were recognized as having influence on key decision-makers (Holder, 1999). The advertisements themselves (comprised more of text than of images) expressed a common theme with some variation according to country: *Food biotechnology is a matter of opinions, Monsanto believes you should hear them all* (the specific UK message).

Despite what could be considered a strategically conceptualized corporate campaign on the part of Monsanto, its critics suggested that it backfired. Thirteen complaints were filed against Monsanto with the Advertising Standards Authority by special interest groups. Four were upheld while nine were dismissed (Gregoriadis, 1999). What followed was a shift in tactics on the part of Monsanto. Robert Shapiro, CEO of Monsanto, addressed a conference organized by Greenpeace in 1999 and was quoted as saying: '[w]e have probably irritated and antagonized more people than we have persuaded . . .

Our confidence in this technology and our enthusiasm for it has, I think, been widely seen – and understandably so – as condescension or indeed arrogance' (Gillis and Swardson, 1999).

The industry soon moved towards more collective action. In early 2000, BASF, Bayer CropScience, Dow, DuPont, Monsanto and Syngenta (along with two trade associations – the Biotechnology Industry Organization and CropLife America) announced the backing of a US$50 million venture called the Council for Biotechnology Information (Lambrecht, 2000; Gillis, 2000; US Newswire, 2000). The mission of the Council for Biotechnology Information (CBI) is to improve understanding and acceptance of biotechnology by collecting balanced, science-based information and communicating it through a variety of channels. The organization currently utilizes videos, Twitter, Facebook, blogs, Internet websites to do this. Although initially CBI was designed to raise awareness and shape opinion in the US and Canada, the organization has since expanded. It now operates in several countries all over the world. Jeffrey Bergau, spokesman for the original campaign, stated: 'The more people are exposed to information from a variety of sources, the more likely they are to embrace the technology. Our goal is to try to link people to information and data that's based on sound science' (Gillis, 2000).

Despite the efforts of the agricultural biotechnology industry (even with the establishment of the CBI), backlash from interest groups has continued. Tactics have evolved over the years. At one point, public demonstrations were the norm. For example, in November 1999 thousands of individuals organized in Seattle in advance of the WTO trade talks to protest what they claimed as the WTO's partiality of corporate interests over those of workers and the environment. With demands for the WTO to 'keep its hands off agriculture', 600 people were arrested, pepper-sprayed, gassed and hit with rubber bullets. Other tactics have included criminal damage to public research institutes. In June 2003, experimental research materials at the John Innes Centre in Norfolk UK were uprooted and destroyed (JIC, 2003). More recently, in July 2011, Greenpeace protesters scaled a fence of the CSIRO experimental station at Canberra and destroyed approximately a half-hectare crop of genetically modified wheat (Tribe, 2011).

Throughout all of this, the agribusiness industry has continued to restructure,[1] and consolidation of the industry has only proved to provide more fodder for anti-GM 'fear mongering'. By blending words and images in creative ways, interest groups have been able to negatively influence public perceptions of biotechnology.

3 'MYTHMAKING': TOOL OF CHOICE FOR THE ANTI-GM MOVEMENT

Myths and mythmaking, and the oral tradition of storytelling, have always been a part of society and social engagement. According to Conko and Miller (2004, p. 19), myths 'illustrate simple moral lessons' and learning from them can be empowering. Levi-Strauss (1966) suggests that myths operate as systems of belief, that they are the mirror to society–a reflection of its social organization. Myths and mythmaking provide context and explanation under conditions of perceived or real uncertainty. They offer pathways for connecting society to a nostalgic past or to a more promising future. Levi-Strauss states that myths are powerful and that mythmaking is, in and of itself, an act of power. Myths possess authority by appealing to the values and beliefs of society through symbolic representations. Words and images, combined, help to position and augment myths in society.

Over the past two decades, the anti-GM movement has systematically employed mythmaking as a way to capture headlines and influence public opinion about agricultural biotechnology products. Armed with a strong anti-corporate bias and artfully blending words and disarming visual images, NGOs and special interest groups employ continuous, well-executed communication strategies and, as a result, they have been very successful in reframing the debate on agricultural biotechnology. As Maeseele (2009) outlines, NGOs have 'eagerly employed the discursive weapon and have communicated many alternative frames for people to interpret this technology' (p. 170).

People think in 'pictures' as a way to visually organize and process information using parts of the brain that pull together the emotional and creative (Bostrom and Clawson, 2000). The media theorist, John Berger, suggests that in human development visualizing or 'seeing' comes before words. Dr Lynell Burmark from the Thornburg Center for Professional Development, an expert on visual literacy, states on her website that: 'unless our words, concepts, ideas are hooked onto an image, they will go in one ear, sail through the brain, and go out the other ear. Images . . . go directly into long-term memory where they are indelibly etched' (Burmark, 2013).

Metaphors and analogies are powerful visual communication aids. There are several examples of these that have been leveraged by the anti-GM movement over the years. Take, for example, the visually provocative term 'Frankenfood'. Paul Lewis (Professor of English at Boston College, Massachusetts) is often credited with coining the term. Lewis wrote a letter in June of 1992 to the *New York Times* in response to the USDA's decision to allow companies to market genetically modified food. In that letter Lewis stated: 'If they want to sell us Frankenfood, perhaps it's time to gather the villagers, light some torches and head to the castle.' The visual component of the word 'Frankenfood' is incredibly powerful and well-recognized in society. It fosters images of Mary Shelley's infamous monster of *Frankenstein* fame. Since 1992, the term has been a pervasive (and quite effective) slogan of the anti-GM movement to vilify GM crops and biotechnology. In addition to this moniker, genetically modified organisms are often referred to as 'dangerous' or 'toxic' 'time bombs', deemed nothing more than sources of 'genetic pollution and contamination'. Mass-produced biotechnology crops are often cited as 'killing fields' or the 'Pandora's box' that, when opened, will avail countless iniquities on Mother Nature and the environment (Maeseele, 2009). This kind of language provides the basis for provocative and emotive story-making.

The anti-GM movement has also spread beyond North American and EU soil and has gained firm (and surprising) footholds in other societies. Myths are readily adopted and, depending upon how they are structured and communicated, can resonate in certain cultures. Take, for example, the case where a judge in the Philippines heard an argument at a divorce trial that a husband's homosexuality was the result of him walking through a GM cornfield. The judge, in question, granted the divorce on that basis and, later, at a science conference posed the technical question to experts as to how homosexuality can be conferred through such a means (McHughen, 2013; McHughen and Wager, 2010). Another illustrative example, again outlined by McHughen (2013) and McHughen and Wager (2010), is the case in Zambia in 2002 where then-President Levi Mwanawasa was convinced by European activists that the GM corn in food aid contributed by the US was 'poison'. It was the height of the famine crisis that year in Zambia. Despite this, Mwanawasa locked up the food, defending his actions by stating that '[s]imply because my people are hungry, that is no justification to give them poison' (BBC News, 2002).

And let's not forget the 'myth re-invented'; a historically significant, well-known and saleable myth – one with a hero and a villain. The 'David and Goliath' story was re-packaged and frequently referred to in the media during, before and after the Percy Schmeiser–Monsanto trial (the trial, in question, was heard from 5 June to 20 June 2000 in the Federal Court in Saskatoon, Saskatchewan). The modern re-invention of a 'sacred text' or story dominated the headlines and created some dramatic imagery around the 'solitary farmer' fighting the 'vile agricultural biotechnology giant'. Schmeiser's personal website still refers to the trial and struggles as 'the classic David and Goliath story'.

Words and images associated with the anti-GM movement, although misleading, are ubiquitous. The media are often the intermediary purveyors of these works and images. Marks and Kalaitzandonakes (2001) suggest that the media have played an agenda-setting role in the framing of the GM debate. At the very least, the authors see that the media raised awareness of agricultural biotechnology in both the EU and the US. The role of the media in shaping public opinion or ideology, however, is often debated. Entman (1989) suggests that the media cannot really tell people *how* to think but it can certainly influence what people think *about*. Entman (1989) further asserts, however, that the media are not the sole source of influence for the framing of issues in the eyes of the consumer or the public: '[People] form and maintain the orientations they use to process information. Their partisan and ideological loyalties arise from socialization in a political culture transmitted, reinforced, and constantly altered by [close personal networks] – most of whom use the media' (p. 366).

Yale Law and Psychology Professor, Dan Kahan (2012), illustrates the influence of personal networks on deeply divided opinions about climate change: 'People acquire their scientific knowledge by consulting others who share their values and whom they therefore trust and understand' (p. 255). Kahan warns, however, that we live in a 'richly-stocked' science-communication environment:

> 'The trouble starts when this communication environment fills up with toxic partisan meanings– ones that effectively announce that *'if you are one of us, believe this; otherwise, we'll know you are one of them*'. In that situation, ordinary individuals' lives will go better if their perceptions of societal risk conform with those of their group'. (p. 255)

It would seem that in the context of the anti-GM movement, leveraging the media and the influential effects of personal networks come together to create avenues for perpetuating myths and mythmaking. This would suggest that this stylized version of the 'Matthew Effect'[2] in the pro-GM/anti-GM communications and mythmaking context just further serves to dichotomize the debate.

With the advent of the Internet and growing use of mobile technology, it is important to note that traditional media (newspaper, TV) no longer operate in a vacuum. For the 'communicator', there is enormous value now in integrating multiple media platforms (what Ilhan (2011) calls 'transmedia storytelling'). Information generated through the selection of traditional media formats is further amplified through social media and other new communication tools. These tools permit the involvement of 'citizen journalists' (Gant, 2007) that employ Facebook, Twitter, LinkedIn, YouTube and Flickr as part of their daily social networking activities. Blogs and online diarizing have also become quite popular over the past several years. Even the casual user can now quickly post and transmit messages, photographs, and/or videos on any subject at all (Gant, 2007).

The anti-GM movement effectively takes advantage of all of the social media tools as well as traditional media formats for 'transmedia storytelling'; and sometimes in very controversial ways. Take, for example, the case of the questionable (and biased) Séralini et al. study published in September 2012 in *Food and Chemical Toxicology*. Prior to this publication, Eric-Gilles Séralini already had a reputation as a GM critic, having published several disputed articles in the past. In this latest study, his results suggest that rats fed on Monsanto's genetically modified corn or exposed to glyphosate, its top-selling weed-killer marketed as Roundup®, suffered tumours as well as multiple organ damage. The study – replete with errors of spelling and grammar, a weak statistical methodology and a lack of appropriate controls – was quite clearly not properly peer reviewed. Moreover, Séralini refused to release his data. But what was particularly compelling about this case was *how* the release of the study was managed. Most certainly, it was an 'exercise in media manipulation' (Ryan and Wager, 2012); in short, a public relations campaign. Sustainable Food Trust (SFT) (UK) was at the forefront in supporting and communicating Séralini's study results. Prior to the press conference in London on the 19 September 2012 where the study was publicly launched, SFT instituted an aggressive, pre-emptive online social media campaign (Facebook, Twitter), circulating tweets with 'fear-mongering' messages such as 'New study warns of dangers of GMOs!' with shocking images of tumour-infested rats. In fact, SFT had a webpage set up with pre-constructed tweet messages for 'tweeps' ('citizen journalists') to copy and paste and circulate on Twitter and other social networking sites. Additionally, there was the intriguing timing of other (related) items: 'It's no coincidence that [Séralini] launched an anti-GM book and a movie [both titled *Tous Cobayes? (All of Us Guinea-Pigs Now?)*] that same week. It appears as if the goal of the study was to "prove" something rather than to objectively "investigate" something' (Ryan and Wager, 2012).

Finally, there was Séralini's unprecedented move to impose a Non-Disclosure Agreement (NDA) on the media prior to the publication of the study. Embargoes are the norm. But an NDA is something no academic scientist would ever think to instigate or endorse. The whole Séralini debacle was a very convoluted example of 'bad science' getting 'good legs' in the media. The study has since been discredited by several food safety organizations, including the European Food Safety Authority (EFSA), Health Canada, CTNBio (Brazil), ViB (Belgium), BfR (Germany), FSANZ (New Zealand) and a number of independent research scientists the world over. But the anti-GM 'myth' lives on, complete with increasingly outrageous headlines and a series of alarming photos that continue to circulate on the Internet and through social media.

And last, but certainly not least, the power of 'celebrity' endorsement is pushing political agendas. Use of celebrities has been a common part of corporate communication strategies for some time where firms select endorsers that are seen as dynamic with likeable qualities and ones that will 'foster . . . a match or connection between the celebrity endorser and the . . . brand' (Costanzo and Goodnight, 2005, p.49). Mass media have brought celebrities into our homes, humanizing them and 'creating a new intimacy between performers and audiences' (Larkin, 2009). Celebrities have been used increasingly by political activists as an integral part of the lobbying effort; for example, actress Pamela Anderson is the official spokesperson for People for the Ethical Treatment of Animals (PETA). According to Hirschman and Thompson (1997) 'motion pictures, television shows and fashion and entertainment magazines present images, icons and stories that give the [public] a [powerful] frame of reference' (p. 43). Society's closer, more 'intimate' connection to celebrity (Larkin, 2009) represents 'buy-in' into celebrity 'lifestyles' or 'belief patterns' and a need to tap into that perceived 'good life' (Hirschman and Thompson, 1997). Interpreted meanings or framing around an issue, or image or myth can readily emerge from this. Celebrities have power. And they often exude that power by endorsing 'junk science' and 'pseudoscience'. When the weight of celebrity is propelling the myth forward, it gains even more momentum.

There have been a number of examples like this. After *The Lancet* published the flawed Wakefield study in 1998 (later retracted in 2012), a movement against vaccinations ensued. Actress Jenny McCarthy spent several years using platforms such as Oprah and Larry King Live TV shows to spread the message of a causal link between MMR vaccines and autism. Due to what can only be assumed the power of celebrity influence, there was a steep decline in vaccination rates, and the incidence of disease skyrocketed. Actress and author Suzanne Somers has found a new level of celebrity status as a critic of GMOs, particularly as it relates to the fuzzy pseudoscience concept of 'leaky gut syndrome'. Dr Oz, whose daytime programme attracts almost 4 million viewers, has exhibited a strong anti-GMO bias. In December of 2010, Oz hosted an episode on genetically modified foods wherein he invited 'experts' to discuss the dangers and benefits of GMOs. One of those experts was Jeffrey Smith, a 'popular junk scientist' who has written several books criticizing modern agriculture and genetically modified organisms. Smith made yet another appearance on the Dr Oz Show in October of 2012 on another episode entitled *The Dangers of Genetically Modified Foods*. It is no coincidence that the programme was aired just after the Séralini study was made public. Again, those unpleasant images of tumour-infested rats were repeatedly shown on Oz's programme.

In the last several years, the anti-GM movement has effectively leveraged the power of social media, celebrity endorsement to create and perpetuate myths and to influence public perceptions about agricultural biotechnology. Myths are powerful things. But when falsehoods and fabrications are deceptively shrouded as reality, 'myths can mislead, misinform, and undermine social goals' (Conko and Miller, 2004, p. 19).

4 WHERE ARE THE EXPERTS?

Where are the experts that can counter erroneous claims and dispel these myths? Who *are* the experts?

According to McKenzie (2007), 65 per cent of Canada's estimated 100 000 PhDs were employed in the public sector while 35 per cent are employed in the private sector. It would seem that university and other public sector research institutes carry the balance of power when it comes to knowledge production and scientific and technical know-how in society. Yet, this expert 'voice' is noticeably muted, particularly in the online realm (Ryan and Doerksen, forthcoming). Studies suggest that public sector researchers have been slow to move beyond more traditional forms of public engagement to using online communication tools such as social media (Powell et al., 2011; Lackes et al., 2009; VALGEN, 2010). According to Quinn (2011), scientists view social media tools sceptically. As a means of communication, they are viewed as 'distractions' or 'passing fads' (Faculty Focus, 2009) while traditional publication is seen as the 'gold standard' and the 'hard currency of science' and communication (Iverson et al., 2003). In a world where 62 per cent of the entire adult population look to online social networks or online sources (for example Google) for information, however, this apathy and lack of online engagement represents a significant shortfall (Rainie and Fox, 2012). The fact that universities employ recognition and reward frameworks that are predominantly structured around traditional forms of scholarly works to the detriment of these new forms of online engagement and public discourse only serves to complicate matters more (Ryan and Doerksen, forth coming).

Fortunately, beyond the boundaries of public sector institutions, there is a burgeoning online presence in the form of intermediary, one-off organizations that offer evidence-based information around bioscience. For example, *Biology Fortified Inc.* is an independent, non-profit organization devoted to providing factual information and fostering discussion about issues in biology, with a particular emphasis on plant genetics and genetic engineering in agriculture. *Sense About Science* is a charitable trust that provides information to the public on scientific and medical claims and *Academics Review* tests popular claims against peer-reviewed science. The problem is that, in many cases, these intermediary entities are often underfunded. They also rely heavily on the voluntary contributions by experts (independent public sector scientists) which – presumably due to the lack of institutional incentives – may be sporadic or inconsistent, at best.

The 1996 National Research Council's *Orange Book* (NRC, 1996) presented a new paradigm for risk analysis suggesting that increased transparency would heighten the public's trust in biotechnology (Wolt and Peterson, 2000). Soon after, in 2000, the National Academy of Science published a report (*The Future Role of Pesticides in US Agriculture*) recommending that efforts be made by public institutions to educate and train scientists about the value of public outreach in agricultural sciences. Even now, thirteen years later, there are few institutional incentives or formalized policies in place. And although outreach activities are often part of the academic trifecta of incentivized activities, these undertakings usually revolve around more traditional forms of public outreach such as presenting at public seminars or participating in community-based committees. It is evident that a deeper level of engagement is required:

> Scientists can no longer content themselves with communicating only with fellow scientists; it is in the best interest of everyone that the public can hear directly from scientists themselves . . . As agonizing and as time-consuming as it may be to respond to media requests for interviews, or even to write for popular magazines, it is utterly important for scientists to attend to this duty with the same level of care and attention directed at the conduct of the scientific study itself. Future policy decisions rest on how the public interprets scientific data. (Berenbaum, 2001, p. 512)

This suggests that there is a place for social media policy and new faculty incentives as part of an integrative, proactive communication strategy at the administrative level in universities and other public sector institutes (Ryan and Doerksen, forthcoming). This needs to be done in order to mitigate reputational risks and to tackle the challenge of mythmaking head-on.

5 CONCLUSIONS

The anti-GM movement has actively used mythmaking as a tool for influencing public opinion on agricultural biotechnology. Mythmaking is accomplished through the clever bundling of images and words, the use of old myths in new contexts, by employing a number of social media tools and propelling the story forward through celebrity endorsement.

The problem with all of this is that it is almost impossible to fight this modern 'mythmaking' with traditional marketing or communication strategies. And trust matters in modern technology-dependent societies:

> there is a very real sense in which, because trust is a substitute for knowledge, information does not build it (as is often argued by advocates of the deficit model) but rather makes it redundant. Hence, in one way, trust cannot be effectively created and maintained through information campaigns. Instead, the food sector must aim to improve its trustworthiness by being more socially responsive. (Hansen et al., 2003, p. 119)

It is evident that traditional industry-driven communication strategies have not worked in countering the negative impacts of the mythological milieu asserted by the anti-GM movement. Tackling 'mythmaking' is not part of the private sector portfolio of product and services-delivery activities. Moreover, from a critic's standpoint, the private sector would be considered neither a trusted nor unbiased source.

The credible expert voice must come from an independent source – the public sector. Yet it would seem that that voice has been largely absent from the dialogue. Institutions that house this expert knowledge are not utilizing appropriate 'issues management' strategies in order to address gaps in information and knowledge. Overall, there appears to be a lack of interest on the part of scientific experts, few or no institutional incentives or policies to incite them to participate, and very little funding earmarked to support such activities. Although 'mythmaking' in some form or another will continue to be part of the anti-GM or anti-technology movement, strategically mobilizing resources and knowledge to counter some of the misleading information would go far in mitigating misperceptions.

NOTES

1. Aventis was formed in 1999 when French company Rhône-Poulenc S.A. merged with the German corporation Hoechst Marion Roussel (which was formed from the 1995 merger of Hoechst AG with Roussel Uclaf and Marion Merrell Dow). In 2002, Bayer AG acquired Aventis CropScience (now part of Sanofi) and fused it with its own agrochemicals division (Bayer Pflanzenschutz or 'Crop Protection') to form Bayer CropScience. In 1996, Ciba-Geigy and Sandoz merged to form Novartis. In 1999, Astra AB and Zeneca merged to form AstraZeneca. Novartis and AstraZeneca then merged to form Syngenta in November 2000. Monsanto entered into a merger in 2000 and changed its name to Pharmacia Corporation. In 2000, the new Monsanto Company, based on the previous agricultural division of Pharmacia, was incorporated as a stand-alone subsidiary of the pharmaceutical company. Pharmacia eventually became a subsidiary of Pfizer in 2003 (Dunne, 2007).
2. The Matthew Effect is a sociological term that refers to 'accumulated advantage'. Coined by Robert Merton in the late 1960s, the term is most often associated with the adage 'the rich get richer, the poor get poorer'.

REFERENCES

BBC News, (2002), 'Zambia refuses GM "poison"' 3, September available at: http://news.bbc.co.uk/2/hi/africa/2233839.stm, accessed on 27 November 2012.

Berenbaum, M.R. (2001), 'Interpreting the scientific literature: differences in the scientific and lay communities', *Plant Physiology*, **125**, 509–12.

Boone, K. (2011), 'Building capacity in issues management in the land-grant system', Proceedings of the Issues Management: Building Capacity in the Land-Grant System National Symposium, 13 June.

Bostrom, Robert P. and Vikki Clawson (2000), 'How people think: human information processing', available at: http://www.terry.uga.edu/~bostrom/How%20People%20think.doc, accessed on 4 January 2012.

Bucchi, M. (1996), 'When scientists turn to the public: alternative routes in science communication', *Public Understanding of Science*, **5**(4), 375–94.

Burmark, L. (2013), 'Why visual literacy?', available at: http://www.lynellburmark.org/a_whyvisualliteracy.asp, accessed on 5 January 2013.

Byrne, J. (2006), 'Deconstructing the agricultural biotechnology protest industry', in Jon Entine (ed.), *Let Them Eat Precaution*, Washington, DC: AEI Press, pp. 144–62.

Chassy, B.M. (2007), 'The history and future of GMOs in food and agriculture', *Cereals Feeds World – Perspective*, July–August, **52**(4), 169–72.

Conko, Greg and Henry Miller (2004), *The Frankenfood Myth: How Protest and Politics Threaten the Biotech Revolution*, Westport, CT: Praeger Publishers.

Costanzo, P.J. and J.E. Goodnight (2005), 'Celebrity endorsements', *Journal of Promotion Management*, **11**(4), 49–62.

Dunne, Cheryl L. (2007), 'Syngenta, Bayer CropScience and Monsanto: three of the top seven agriculture/crop protection companies in the world', available at: http://irrec.ifas.ufl.edu/files/student_work/Firm_Comparison_Dunne.pdf, accessed on 15 December 2012.

Entman, R. (1989), 'How the media affect what people think: an information processing approach', *The Journal of Politics*, **51**(2), May, 347–70.

Faculty Focus (2009), 'Twitter in higher education 2009: usage habits and trends of today's college faculty', available at: https://www.facultyfocus.com, accessed 23 May 2012.

Gant, S. (2007), *We're all Journalists Now*, New York: Free Press.

Gaskell, G., M.W. Bauer, N. Allum, J. Durant et al. (2000), 'Biotechnology and the European public', *Nature Biotechnology*, September, **18**, 935–8.

Gillis, Justin (2000), 'Biotech firms launch food ad blitz: up to $50 million annual pledged to woo consumers', *Washington Post*, 4 April.

Gillis, Justin and Anne Swardson (1999), 'Crop busters take on Monsanto', *Washington Post*, 26 October.

Glover, Dominic (2008), 'Made by Monsanto: the corporate sharing of GM crops as a technology for the poor', *STEPS Centre*, available at: http://steps-centre.org/wpsite/wp-content/uploads/GM-Crops-web-final_small.pdf, accessed on 20 December 2012.

Gregoriadis, Linus (1999), 'Monsanto GM food ads found to mislead', *The Guardian*, 11 August, available at: http://www.guardian.co.uk/news/1999/aug/11/food.foodanddrink, accessed on: 13 December 2012.

Hansen, J., L. Holm, L. Frewer, P. Robinson and P. Sandøe (2003), 'Beyond the knowledge deficit: recent research into lay and expert attitudes to food risks', *Appetite*, **41**, 111–21.

Herring, R. and K. Roberts (2006), 'Contentious politics: science, social science and social protest', theme proposal, available online at: http://www.socialsciences.cornell.edu/0609/contentiouspolitcs_iss_theme_proposal.pdf, accessed on 1 December 2012.

Hirschman, Elizabeth C. and Craig J. Thompson (1997), 'Why media matter: toward a richer understanding of consumers' relationships with advertising and mass media', *Journal of Advertising*, **26**(1), 43–60.

Holder, Helen (1999), 'Selling a revolution: the Monsanto PR campaign', reported by *SpinWatch*, available at: http://www.spinwatch.org/video-information-67/41-corporate-spin/18-selling-a-revolution-the-monsanto-pr-campaign, accessed on 12 December 2012.

Ilhan, Behice Ece (2011), 'Transmedia consumption experiences: Consuming and co-creating interrelated stories across media', thesis, University of Illinois at Urbana-Champaign.

Iverson, Margot, Mark S. Frankel and Sanyin Siang (2003), 'Scientific societies and research integrity: what are they doing and how well are they doing it?', *Science and Engineering Ethics*, Vol. 9, 141–58.

James, Clive (2012), 'Global status of commercialized biotech/GM crops: 2011', *International Service for the Acquisition of Agri-Biotech Applications (ISAAA) Brief*, 44, Ithaca, NY: ISAAA.

John Innes Centre (JIC) (2003), 'Criminal damage to GM research trial at the John Innes Centre, Norwich', available at: http://www.jic.ac.uk/corporate/media-and-public/news-archive/030618.htm, accessed on 29 December 2012.

Kahan, D. (2012), 'Why we are poles apart on climate change', *Nature*, **488**(7411), 255.

Lackes, R., M. Siepermann and E. Frank (2009), 'Social networks as an approach to the enhancement of collaboration among scientists', *International Journal of Web-based Communities*, **5**(4), 577–92.

Lambrecht, B. (2000), 'Biotech rivals team up in effort to sell altered food; Monsanto, others launch campaign in US, Canada', *St Louis Post-Dispatch*, 4 April, p. A1.

Larkin, K.G. (2009), 'Star power: models for celebrity political activism', *Virginia Sports and Entertainment Law Journal*, issue 1, fall.

Levi-Strauss, C. (1966), *The Savage Mind*, Chicago, IL: Chicago University Press.

Losey, John E., Linda S. Rayor and Maureen E. Carter (1999), 'Transgenic pollen harms monarch larvae', *Nature*, **399**, 20 May, 214.

Maeseele, P. (2009), 'On media and science in late modern societies: the GM case study', PhD. thesis, University of Ghent, available at: http://lib.ugent.be/fulltxt/RUG01/001/362/258/RUG01-001362258_2010_0001_AC.pdf, accessed on 15 December 2012.

Marks, L. and N. Kalaitzandonakes (2001), 'Mass media communications about agrobiotechnology', *AgBioforum*, **4**(3&4), 199–208.

McHughen, A. (2000), *Pandora's Picnic Basket: The Potential and Hazards of Genetically Modified Foods*, New York: Oxford University Press.

McHughen, Alan (2013), 'Who's afraid of the big bad GMO?', *C2C Journal*, **7**(1), January, available at: http://c2cjournal.ca/, Accessed on 15 January 2013.

McHughen, A. and R. Wager (2010), 'Popular misconceptions: agricultural biotechnology', *New Biotechnology*, vol. **27**, December, 724–8.

McKenzie, Michael (2007), 'Where are the scientists and engineers?', working paper for Science, Innovation and Electronic Information Division, Statistics Canada, available at: http://www.ccwestt.org/Portals/0/publications/Where%20are%20the%20Scientists%20an.pdf, accessed 24 May 2012.

Moschini, GianCarlo (2001), 'Economic benefits and costs of biotechnology innovations in agriculture', working paper no. 01-WP 264, Centre for Agriculture and Rural Development, Iowa State University, presented at the conference *Agricultural Trade Liberalization: Can we Make Progress?*, Canadian Agri-Food Trade Research Network, Quebec City, Canada, 27–28 October 2000.

National Academy of Sciences (NAS) (2000), *The Future Role of Pesticides in US Agriculture*, Washington, DC: National Academy Press.

National Research Council (NRC) (1996), *Understanding Risk: Informing Decisions in a Democratic Society*, Washington, DC: National Research Council, National Academy Press.

Pinstrup-Andersen, Per and Marc J. Cohen (2000), 'Modern biotechnology for food and agriculture: risks and opportunities for the poor', in G.J. Persley and M.M. Lantin (eds), *Agricultural Biotechnology and the Poor*, Washington, DC: The World Bank, pp: 159–69.

Powell, D., C.J. Jacob and B.J. Chapman (2011), 'Using blogs and new media in academic practice: potential roles in research, teaching, learning and extension', *Innovative Higher Education*, 3 December, pp. 1–12.

Quinn, Audrey (2011), 'Are scientists anti-social?', blog posting on the *Science Online NYC* meeting, 16 May 2011, available at: http://audreyquinnaudio.com/2011/05/17/are-scientists-anti-social/, accessed 22 September 22, 2011.

Rainie, L. and S. Fox (2012), 'Just in time information through mobile connections', *Pew Research Center's Internet & American Life Project. Report*, available at: http://pewinternet.org/~/media//Files/Reports/2012/PIP_Just_In_Time_Info.pdf Retrieved May 8.

Ryan, Camille D. and Kari Doerksen (forthcoming), 'Apathy and online activism: an impetus for science and science communication in universities?', *The Journal of Technology, Knowledge and Society*.

Ryan, Camille D. and Robert Wager (2012), 'You can eat your bugs – and your toxins, too', *Western Producer*, op-ed. 23 November, available at: http://www.producer.com/2012/11/you-can-eat-your-bugs-and-toxins-too/, accessed on: 21 December 2012.

Salleh, A. (2004), 'Journalism at risk: factors influencing journalistic coverage of the GM food and crops debate (Australia, 1999–2001) and prospects for critical journalism', unpublished PhD thesis, University of Wollongong (AU), School of Social Sciences, Media and Communication.

Scholderer, J. and L.J. Frewer (2003), 'The biotechnology communication paradox: experimental evidence and the need for a new strategy', *Journal of Consumer Policy*, **26**, 125–57.

Schurman, R. (2004), 'Fighting "Frankenfoods": industry opportunities and the efficacy of the anti-biotech movement in Western Europe', *Social Problems*, **51**(2), 243–68.

Séralini, G-E., R. Mesnage, S. Gress, N. Defarge, M. Malatesta, D. Henniquin, and J. Spiroux de Vendomis (2012), 'Long term toxicity of a Roundup herbicide and a Roundup-tolerant genetically modified maize', *Food and Chemical Toxicology*, **50**(11), 4221–31.

Shelton, A.M. and R. Roush (1999), 'False reports and the ears of men', *Nature Biotechnology*, **17**, 832.

Tribe, D. (2011), 'Greenpeace destroys GM wheat trial in Australia', Biofortified, available at: http://www.biofortified.org/2011/07/greenpeace-destroy-gm-wheat-trial/, accessed 7 January 2013.

United States Department of Agriculture (USDA) (2012), 'Recent trends in GE adoption', available at: http://www.

ers.usda.gov/data-products/adoption-of-genetically-engineered-crops-in-the-us/recent-trends-in-ge-adoption.aspx, accessed on: 4 January, 2013.
US Newswire (2000), 'Public information program on biotechnology begins April 3', 3 April.
VALGEN (2010), 'Genome Canada ABC Integration Workshop Report', Banff, Canada, 28–29 January, available at: www.valgen.ca/wp-content/uploads/2009/09/VALGEN-Banff-Workshop-Final-Report.pdf.
Wolt, J.D. and R.K. Peterson (2000), 'Agricultural biotechnology and societal decision-making: the role of risk analysis', *AgBioForum*, **3**(1), 39–46.

PART III

OUTCOMES

36 Soybeans

Jorge Fernandez-Cornejo and Seth Wechsler[1]

INTRODUCTION

Despite the relatively high price of herbicide-tolerant (HT) soybean seeds compared to conventional varieties, farmers have rapidly adopted HT soybeans since their commercial introduction in 1996 (Figures 36.1 and 36.2). Benefits from adopting HT crops may include higher yields, lower herbicide costs or savings in management time.

HT soybeans were developed to tolerate herbicides that previously would have destroyed the crop along with the targeted weeds. Adopting HT soybeans enables farmers to apply effective post-emergent herbicides, expanding weed management options (Carpenter and Gianessi, 1999; Fernandez-Cornejo and Caswell, 2006). In the United States, HT soybean adoption has expanded more rapidly and widely than any other genetically modified (GM) crop, reaching more than 70 million acres (94 per cent of planted acreage) in 2011.[2] Outside the US, adoption has also been rapid in Argentina, Brazil, Paraguay, Uruguay and Bolivia (James, 2011).

Despite the benefits of HT soybeans, there are some concerns over their future environmental implications. While most researchers agree that HT soybean adoption induces a shift towards more environmentally-benign chemical products, such as glyphosate, they

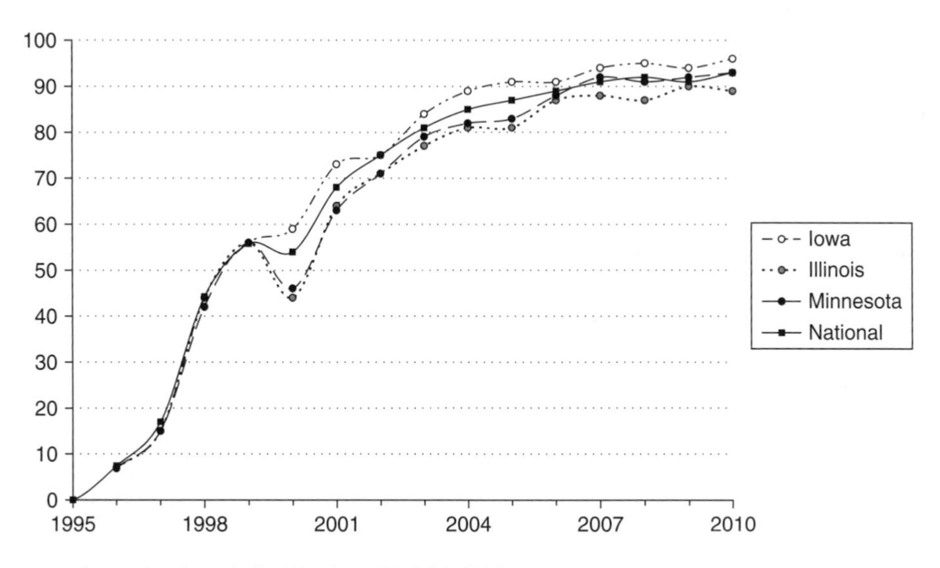

Source: Fernandez-Cornejo (2012) using USDA/NASS data.

Figure 36.1 Adoption of herbicide-tolerant soybeans: percentage of acres, USA and three major states

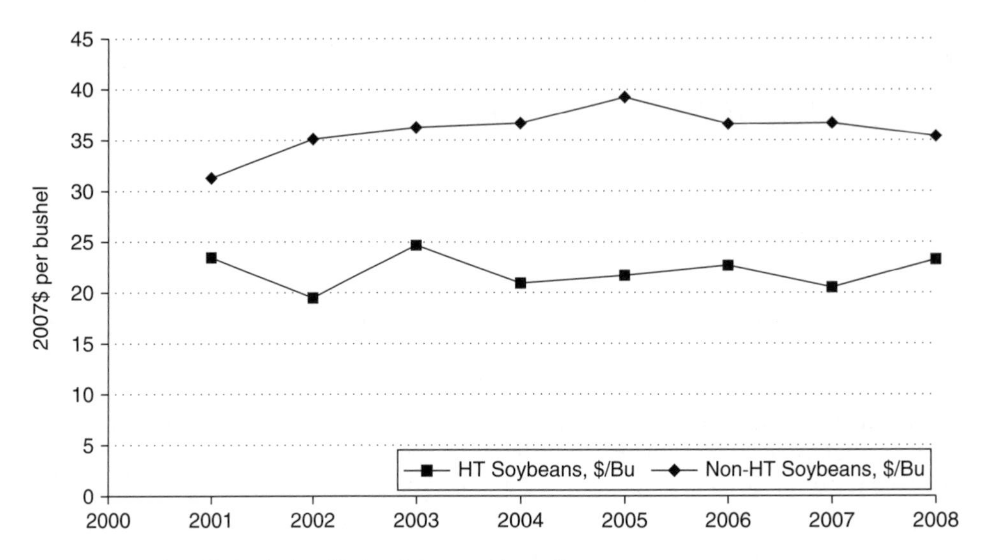

Source: USDA/NASS, Agricultural Prices Summary (Annual).

Figure 36.2 *Prices of HT and conventional soybean seeds*

also believe that using HT soybeans fosters farmers' reliance on these herbicides, possibly reducing their future effectiveness.

The chapter discusses the adoption of HT soybeans. It reviews the models used to analyze the adoption decision, the farm-level impacts of adoption and the welfare implications of adoption. It presents empirical results on the factors influencing adoption as well as on the economic and environmental impacts of adoption, including on yields, herbicide use, conservation tillage, farm household income and welfare.

MODELS AND METHODS

Modeling the Adoption of HT Soybeans

The objectives of this section are: (1) to examine the models used to identify the determinants of HT adoption amongst soybean farmers; and (2) to contrast the relative influence of various factors on the adoption decision, with special emphasis on the role of farm size.

Because adoption involves a binary choice (using a new technology or retaining an old one), a probit model is often used to analyze adoption decisions. Formally, if F denotes the normal distribution, the probability of adopting a seed with HT traits is $P(1_{HT}=1) = F(\delta'_{HT}Z)$, where I_{HT} is an indicator for whether the farmer chooses HT seeds, δ is a vector of parameter estimates, and Z is a vector of explanatory variables. The specification of the adoption equation is: $I_{HT_t} = \delta'_{HT}Z + \varepsilon_{HT}$ where the residuals, ε_{HT} are normally, identically and independently distributed. Elements of Z may include: (i) the relative price of HT seeds; (ii) farm size; (iii) operator experience; (iv) use of operator experience; (v) use of crop insurance (a proxy for risk aversion); (vi) cropping practices; and (vii) operator knowledge about pest infestations.

An alternative to modeling adoption as a binary choice entails analyzing adoption using a Tobit model (Fernandez-Cornejo et al., 2001; Fernandez-Cornejo and McBride, 2002). The Tobit model is preferable when the decision involves the technology choice as well as the extent, or intensity, of adoption (Feder and Umali, 1993).

Models to Examine the Farm-level Effects of Adoption

As Fernandez-Cornejo and McBride (2000, 2002) observe, mean comparisons are often used to analyze results from experiments in which confounding factors are 'controlled' by making them as similar as possible. For instance, field experiments are often used to compare yields (or pesticide use) amongst HT and conventional crops. In these experiments, variables such as soil type, rainfall and sunlight are held constant. One set of soybean plots receives a 'treatment' (for example, HT crops), while the other group does not. As an alternative to controlled experiments, treatment can be randomly administered to respondents.

In 'uncontrolled experiments', such as when comparing means obtained from farm survey data, caution should be exercised when interpreting the results. Conditions other than the 'treatment' vary from respondent to respondent. Consequently, differences in average yields (or pesticide use) cannot necessarily be attributed to the use of HT crops. After all, yields and pesticide use are influenced by factors such as irrigation, weather, soils, nutrient and pest management practices, other cropping practices, operator characteristics and pest pressures (Fernandez-Cornejo et al., 2002; Fernandez-Cornejo and McBride, 2000 and 2002).

Moreover, farmers are not randomly assigned to a treatment group (HT adopters) and a control group (non-adopters); they make the adoption choices themselves. Therefore, adopters and non-adopters may be systematically different from one another. This situation, called self-selection, biases the statistical results unless it is corrected.

The remainder of this section summarizes an econometric model developed by Fernandez-Cornejo et al. (2002) to estimate the farm-level effects of adopting herbicide-tolerant soybeans. The model corrects for self-selection and accounts for simultaneity while analyzing the effect of adoption on herbicide use, crop yields and farm profits. Because adopting HT seeds induces farmers to substitute a particular herbicide (primarily glyphosate) for other herbicides, this process is modeled explicitly. The econometric model accounts for the fact that unobserved variables such as the size of the pest population, pest resistance, farm location and grower perceptions affect both seed choice and pesticide use. Finally, this approach ensures that the econometric model is consistent with farmers' optimization behavior by deriving supply and demand equations from a variable profit function.

The econometric model has two stages. The first stage, which is referred to as the adoption decision model, models the farmer's seed choice (for example, the decision to use HT). This stage is estimated using a probit model. Predicted probabilities of adoption are calculated from the results of the probit model and included as instrumental variables in the second stage of the analysis.

The second stage, which is referred to as the impact model, analyzes the extent to which using HT soybeans affects yields, farm net returns and herbicide use. The impact model includes a profit function, a supply function and derived demand equations for seeds, glyphosate and other herbicides. After specifying a flexible functional form (such as the normalized quadratic for the profit function), the supply function and derived demand equations are calculated and the model is estimated as a simultaneous system (Fernandez-Cornejo et al., 2002).

Modeling the Effect of Global Adoption of GM Crops on Commodity Prices

Over the course of the last 70 years, the rapid adoption of agricultural innovations in US agriculture has contributed to productivity growth and ensured abundance of food (Fernandez-Cornejo, 2004). Many of the gains resulted from a series of biological innovations, beginning with hybrid corn in 1930. More recently, modern biotechnology has facilitated the development of new innovations. As a consequence, the real price (adjusted for inflation) of many agricultural commodities has fallen.

It is difficult to assess the impact of new technologies on commodity prices. As Price et al. (2003) explain,

> once a new technology is introduced and adopted, only the world price that results from the supply shift can be observed. It is not possible to observe the counterfactual price – the price that would have existed, assuming the same supply and demand conditions, if the new technology had not been introduced.

For these reasons, counterfactual prices and quantities are inferred using economic models (NRC, 2010).[3] These models are usually based on a partial equilibrium framework, in which all prices but those of the good under study are kept constant.

As summarized by the NRC report (2010), many researchers (Falk-Zepeda et al., 1999, 2000a, 2000b; Moschini et al., 2000; Price et al., 2003; Qaim and Traxler, 2005) have calculated the effect of GE crops on world commodity prices using the theoretical framework developed by Moschini and Lapan (1997).

Modeling the Welfare Effects of Adoption

Moschini and Lapan (1997) assess the welfare effects of an innovation when a monopolist producer is protected by intellectual property rights in an input (for example, seeds) market (Price et al., 2003). In addition to measuring changes in Marshallian surplus (the sum of producer and consumer welfare), the framework developed by the Moschini and Lapan model calculates the monopoly profits captured by the innovators.

Frisvold et al. (2000) separately determine the benefits for adopters and non-adopters using a mathematical programming model. This model accounts for the impacts of commodity price changes and government price support programs on the stakeholders' welfare.

Falck-Zepeda et al. (2000a, 2000b) specify linear supply and demand curves and assume parallel shifts in the supply curve. These two assumptions impose significant restrictions on the model structure, however. For instance, producer surplus cannot decline with an innovation that causes a parallel supply shift.

Frisvold et al. (2000) and Moschini et al. (2000) used non-linear supply and demand curves and assumed non-parallel supply shifts, which impose fewer restrictions on the model.

Many of the analyses (for example, Falck-Zepeda et al., 2000b; Moschini et al., 2000; Price et al., 2003; Qaim and Traxler, 2005) are based on partial equilibrium models. Some recent studies use a more general computable general equilibrium (CGE) framework (Anderson and Yao, 2003; Anderson and Jackson, 2005; Fernandez-Cornejo et al., 2007).

METRICS

Factors Affecting the Adoption of HT Crops

As discussed in Fernandez-Cornejo and McBride (2002), numerous technology adoption studies have been conducted over the past several decades, beginning with Griliches (1957) and Rogers (1961). Feder et al. (1985) and Feder and Umali (1993) review many of these studies. Rogers (1961 [1995]) hypothesize that innovators or early adopters have attributes different from later adopters or non-adopters. Feder and Umali (1993) distinguish between adoption factors during the early and final phases of adoption; factors such as farm size, tenure, education, information and credit may be significant for the early adopters but not for later adopters.

Several empirical analyses of the factors affecting GM crop adoption have been published. For example, Fernandez-Cornejo and McBride (2000) report (using a probit model) that large operations and educated operators are more likely to adopt herbicide-tolerant (HT) soybeans. They also report that HT soybean adoption is typically associated with the use of conservation or no-till practices. Alexander et al. (2003) examine the role of risk aversion in producer behavior for corn and soybean producers. They find that risk preferences are likely to influence the decision to plant GM corn but not soybeans. More recently, Marra et al. (2004) found that adopters of herbicide-tolerant soybeans tended to be younger and operate more acres than non-adopters (NRC, 2010).

Results of the Tobit analysis for the adoption of genetically modified crops are presented in Fernandez-Cornejo et al. (2001), including the estimated coefficients, standard errors and calculated marginal effects. The elasticities obtained from the Tobit model take into account that a change in an explanatory variable will simultaneously affect the number of adopters and the proportion of acreage under adoption. Among the factors affecting GE adoption, Fernandez-Cornejo et al. (2001) focus on farm size. Since the adoption literature suggests that farm size is often a surrogate for wealth and access to credit, they control for these factors. The econometric results support the hypothesis that farm size does not affect the adoption of herbicide-tolerant soybeans (Fernandez-Cornejo et al., 2002). This result is explained by the fact that HT crop technologies are embodied in variable inputs (for example, seeds), which are completely divisible (unlike technologies embodied in 'lumpy' inputs like tractors or other machinery, which require extensive capital investments).

Table 36.1 Summary of selected studies on the effects of herbicide-tolerant soybeans on yields, pesticide use and returns

Crop/researchers/date of publication	Data source	Effects on		
		Yield	Pesticide use	Net returns
USA				
Delannay et al. (1995)	Experiments	Same	NA	NA
Roberts et al. (1998)	Experiments	Increase	Decrease	Increase
Arnold et al. (1998)	Experiments	Increase	NA	Increase
Reddy and Whiting (2000)	Experiments	Same	NA	Increase
Duffy (2001)	Survey	Small decrease	NA	Same
Fernandez-Cornejo et al. (2002)	Survey	Small increase	Small increase	Same
McBride and El-Osta (2002)	Survey	NA	NA	Same
Bradley et al. (2004)	Experiments	Same	NA	NA
Marra et al. (2004)	Survey	Same	NA	Increase
Developing Countries				
Qaim and Traxler (2005) (Argentina)	Survey	Same	Increase	Increase
Paz et al. (2009) (Bolivia)	Survey	Increase	NA	Increase
Brookes (2005) (Romania)	Interviews	Increase	NA	Increase

Note: NA = not applicable.

Sources: Fernandez-Cornejo and Caswell (2006); Bradley et al. (2004); Marra et al. (2004); Qaim and Traxler (2005); Paz et al. (2009); Brookes (2005).

Some Empirical Results on the Impact of Adoption

As previously discussed, many field-test and enterprise studies have examined the yield, returns and costs of using genetically modified crops (Fernandez-Cornejo and McBride, 2002; Fernandez-Cornejo and Caswell, 2006; Qaim, 2009; NRC, 2010) (see Table 36.1). In addition, many researchers have used econometric models to analyze the farm-level effect of adoption.[4] This section discusses the empirical results of selected studies.

Adoption of HT Soybeans and Yields

As Fernandez-Cornejo and Caswell (2006) observe, currently available GM crops do not increase the yield potential of a hybrid variety.[5] By protecting the plant from certain pests, GM crops can prevent yield losses, particularly when pest infestations are severe. This effect is particularly important for Bt crops.

Most studies suggest that Bt crops produce higher yields than conventional crops, but the case for HT crops is mixed. A 2002 ERS study found that, controlling for other factors, the adoption of HT soybeans led to small but statistically significant increases in soybean yields (Fernandez-Cornejo and McBride, 2002).[6] A survey of Delaware farmers found a three bushel per acre difference between HT soybean adopters and non-adopters (Bernard et al., 2004). More recently, a 2006 soybean USDA survey (with 2204 observations) of 19 major soybean-producing states showed that, on average, HT adopters

Table 36.2 Sample means and definition of main variables: soybean producers, 2006

Variable	Description	All	HT adopters	Non-adopters
Yield	Per acre yields, in bushels	45.51	45.62	40.60
Herbicide use	Pounds Al per acre	1.35	1.36	1.05
Glyphosate use	Pounds per acre	1.21	1.23	0.38
Other herbicide use	Pounds per acre	0.14	0.13	0.66
HT Soybean	Dummy variable = 1 if the operator planted seeds with HT traits	0.97	–	–

Source: NASS/ESR 2006 ARMS Soybean Data.

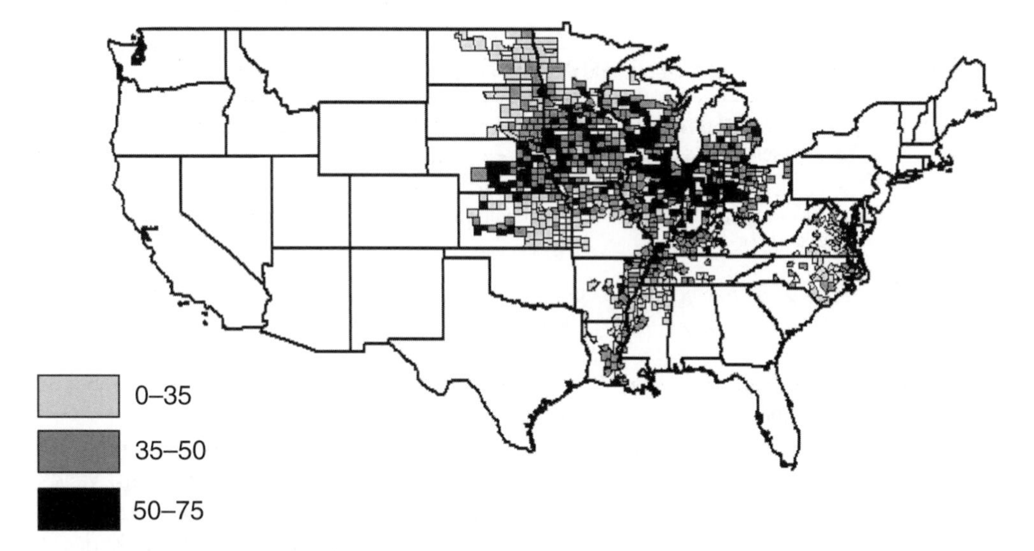

Source: NASS/ERS 2006 ARMS Soybean data.

Figure 36.3 Average soybean yields (in bushels/acre)

had actual soybean yields about five bushels per acre (or 11 per cent) higher than non-adopters (Table 36.2). The geographical distribution of average soybean yields is shown in Figure 36.3.[7]

As Raney and Matuschke (2011) observe, despite the fact that HT soybeans are the main crop in Latin America, there are few farm-level impact studies published for that region. Qaim and Traxler (2005) found that HT soybean adoption had a statistically insignificant impact on yields by analyzing a sample of three provinces in Argentina over a three-year period (yields for HT soybean were 3.01 tons per hectare while yields for conventional soybeans were 3.02 tons per hectare). Similarly, Trigo and Cap (2006) found that, on average, there were no significant differences in soybean yields over a 10-year period.

HT soybeans do appear to offer yield advantages in locations such as Romania and Bolivia where weeds are difficult to control with selective herbicides (Qaim, 2009). Brookes (2005) found HT soybean yields were 31 per cent higher than yields of

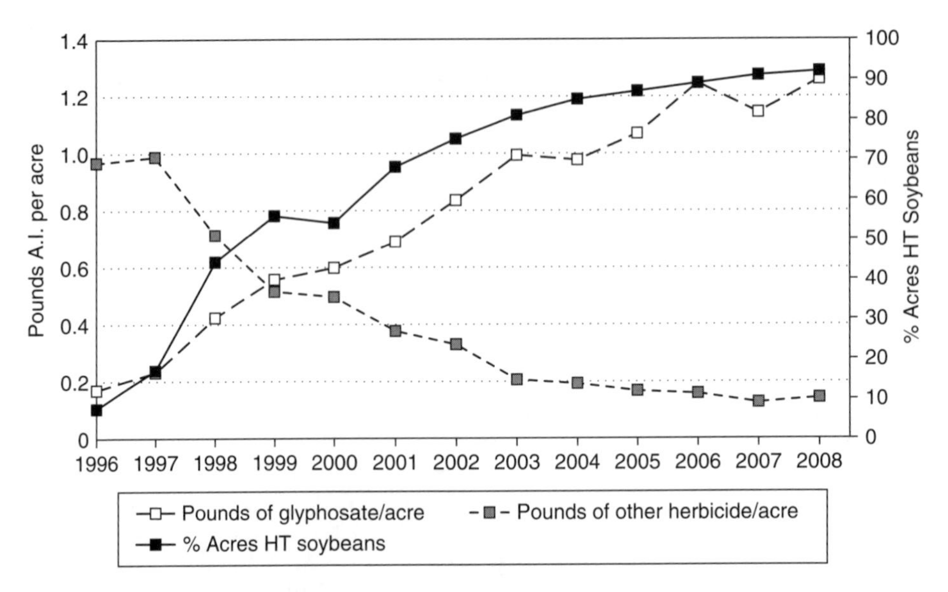

Sources: Fernandez-Cornejo et al. (2009); NASS Agricultural Chemical Usage Summaries; NASS Quick Stats; June Agricultural Survey 2008.

Figure 36.4 Pounds of herbicide A.I. per planted acre and per cent acres of herbicide-tolerant soybeans, 1996–2008

conventional soybeans in Romania. Paz et al. (2009) reported that yields of HT soybeans were 0.44 tons per hectare higher than those of conventional soybeans.

To summarize, in most cases, HT soybeans do not appear to have a significant yield advantage relative to conventional seeds. In fact, in the early years of adoption, yields for conventional soybean varieties were sometimes higher than those of HT varieties. As Qaim and Traxler (2005) and Tassisto (1998) explain, in those years HT traits had not been incorporated into the best available germplasm, leading to what some call yield-drag. Consequently, many farmers used varieties not well adapted to local conditions. These drawbacks have been overcome in recent years.

Adoption of HT Soybeans and Herbicide Use

HT crops enable the substitution of effective broad-spectrum herbicides, such as glyphosate, for other herbicides. Consequently, glyphosate usage has risen rapidly, while the use of other herbicides has declined. Figure 36.4 summarizes trends in HT adoption, glyphosate use and the use of other herbicides on soybeans (Fernandez-Cornejo et al., 2009). Notice that while total herbicide use decreased slightly during the first several years of HT adoption, it has increased in recent years. The results of a 2006 USDA soybean survey bolster these findings. This survey suggests that HT adopters apply herbicides at higher rates than conventional seed users in recent years (Table 36.2). Herbicide expenditures were lower for HT adopters because they pay lower herbicide prices (Table 36.2).

HT adoption appears to have had a more pronounced effect on herbicide use and herbicide costs abroad. Analyzing a survey of Argentinian farmers, Qaim and Traxler

(2005) find that HT soybean adopters applied 107 per cent more herbicide than users of conventional seeds.[8] Herbicide costs were still lower for HT soybean adopters (US$19.10 per hectare) than non-adopters (US$33.64 per hectare) because glyphosate was less expensive than other herbicides.

Similarly, analyzing Bolivian data, Paz et al. (2009) report that herbicide expenditures are nearly US$10 per hectare lower for adopters (US$32/hectare) than for users of conventional soybeans (US$42/ hectare). Analyzing data from Brazil, Pelaez et al. (2004) found an even larger difference in herbicide expenditures. On average, adopters pay US$19.15/hectare while non-adopters pay US$64/hectare.

Environmentally speaking, herbicide-tolerant crops induce farmers to use relatively benign herbicides, such as glyphosate (Fernandez-Cornejo and McBride, 2002; NRC, 2010; Cerdeira and Duke, 2006; Duke and Powles, 2008). Glyphosate has a lower half-life (Wauchope et al., 1993) than the herbicides it commonly replaces. Glyphosate also binds to the soil quickly (which helps prevent leaching) and is rapidly biodegraded by soil bacteria (Malik et al., 1989). Also, glyphosate has very low toxicity to mammals, birds and fish (Malik et al., 1989). Moreover, a recent study has shown that replacing widely used herbicides such as atrazine and alachlor with glyphosate 'can reduce the occurrence of dissolved herbicide concentrations in runoff exceeding drinking water standards' (Shipitalo et al., 2008). As Fernandez-Cornejo and McBride (2002) conclude, glyphosate replaces synthetic herbicides that are at least three times as toxic and that persist in the environment nearly twice as long. Similar results were found in an analysis of Argentinian soybean farmers by Qaim and Traxler (2005). They report that the use of glyphosate replaced herbicides with higher toxicity.

HT soybeans have also recently been shown (Bonny, 2011) to lower the Environmental Impact Quotient index, a well-known measure of the effect pesticides have on human health and the environment (Kovach et al., 1992).

In summary, researchers have shown that HT soybean adoption has led to the use of a more environmentally benign herbicide than the herbicides that it replaced. However, as the NRC (2010) report finds, relying on glyphosate reduces its effectiveness. In fact, recent reports indicate that several weed species have evolved resistance to glyphosate. This may induce US farmers to shift herbicide usage towards conventional herbicides and to return to 'more environmentally damaging practices'. Additionally, the environmental effect of the new stacked HT traits (which make crops resistant to glyphosate as well as dicamba or 2,2-D) have yet to be determined.

Conservation Tillage and HT Soybeans

The environmental impact of conservation tillage (CT), including no-till, ridge-till and mulch-till, is well documented. By leaving substantial amounts of crop residue (at least 30 per cent) covering the soil surface after planting, conservation tillage reduces soil erosion and degradation, increases water retention and reduces water and chemical run-off. In addition, conservation tillage reduces the carbon footprint of agriculture (Fernandez-Cornejo et al., 2010).

Adoption of CT by US soybean growers has risen from approximately 30 per cent in 1996 to 63 per cent in 2006; no-till (the most beneficial of the CT modes) has been adopted even more rapidly.

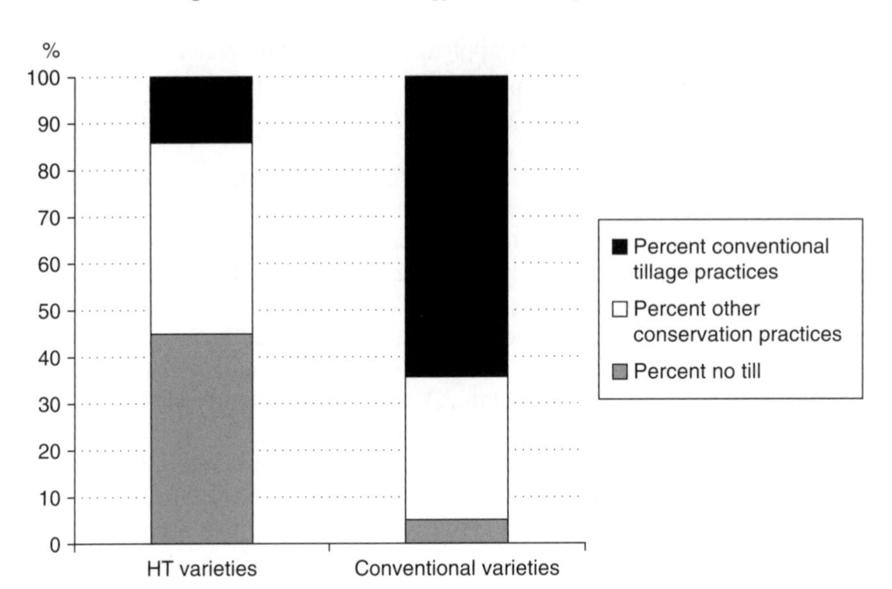

Source: USDA (ARMS data; based on a field-level sample with 2200 observations).

Figure 36.5 Area under conventional tillage, conservation tillage and no-till, soybeans, 2006

Recent survey data illustrates the association between conservation tillage and HT soybean adoption (Figure 36.5). For instance, in 2006, 85 per cent of HT soybeans users adopted CT compared with only 35 per cent of non-adopters. However, it is difficult to isolate the impact of HT adoption on conservation tillage because the direction of causality is not certain. In other words, it is unclear whether HT adoption induces changes in tillage behavior, or vice versa. Therefore, the two decisions must be considered simultaneously. An econometric model developed to address the simultaneous nature of the decisions was presented by Fernandez-Cornejo and McBride (2002). The model was used to determine the nature of the relationship between HT soybean adoption and no-till practices using 1997 national survey data. Farmers using no-till were found to have a higher probability of adopting HT soybeans. The use of HT soybeans did not significantly affect no-till adoption. This result suggests that farmers already using no-till find HT seeds to be an effective weed control mechanism, but that HT soybean adoption does not seem to encourage the adoption of no-till.

Using soybean data from 2002, Mensah (2007) finds a two-way causal relationship between HT soybean adoption and tillage decisions. He finds that farmers who adopt no-till are more likely to use HT soybeans, and that farmers who use HT seeds are more likely to adopt no-till.

More recently, Fernandez-Cornejo et al. (2010) used a state-level panel covering 12 major soybean-producing states over 19 years (from 1988 to 2006) to examine the link between HT adoption and conservation tillage. They quality-adjust prices and quantities using hedonic methods. The results of the analysis suggest that HT adoption induces farmers to adopt conservation tillage practices. In particular, a 1 per cent increase in the adoption of HT soybeans in the US leads to a 0.21 per cent increase in the adoption of conservation tillage.

HT Soybean Adoption and Off-farm Work

Fernandez-Cornejo et al. (2005) analyze how HT soybean adoption affects farm and off-farm income. They develop a model that uses the household rather than the farm business as a unit of analysis and household income rather than farm profits as the metric for economic performance. They estimate the model using data from a nationwide survey of soybean farms collected in 2000. They find that adoption of HT soybeans is positively and significantly related to off-farm household income after controlling for other factors. While farm income was not significantly related to HT adoption, total household income increases significantly with adoption. These findings suggest that the simplicity of weed control programs used by HT soybean farmers provide time savings that enable farmers to earn higher income from off-farm activities.

HT Soybean Adoption and Commodity Prices

The extent to which agricultural biotechnology affects world commodity prices depends on many factors, particularly on where the technology is adopted. HT soybean adoption in the United States induced a soybean price decline between 0.2 per cent (Price et al., 2003) and 1 per cent (Moschini et al., 2000). HT soybean adoption in the United States and Argentina induced a price decline of this commodity of approximately 2 per cent (Qaim and Traxler, 2005). Adoption in the US and South America induced a soybean price decline of 2.2 per cent (Moschini et al., 2000). Global HT soybean adoption induced a price decline of 2.6 per cent (Moschini et al., 2000) (see Table 36.3).

Benefits and their Distribution

Benefits of GM crops accrue to: (1) farmers in adopting countries through potentially higher yields or cost savings; (2) innovators – biotechnology developers and germplasm suppliers – through profits; (3) consumers in the adopting countries through lower commodity prices; and (4) consumers in the rest of the world (ROW), who also gain because of lower world commodity prices. Producers in the ROW realize welfare losses for two primary reasons. First, by producing conventional varieties they do not realize the yield gains or cost savings associated with growing GM crops. Second, they were exposed to the lower world commodity prices caused by the adoption of the technology (Price et al., 2003).

Estimates of the benefits (welfare effects) of GM seeds vary greatly, depending on the year, countries adopting, adoption rates and the models used (Table 36.4). In the case of HT soybeans, estimated annual benefits range from $206 million to $1.2 billion for adoption in the US and Argentina. The share for farmers ranged between 13–29 per cent of the estimated total benefit, innovators captured 34–68 per cent and consumers obtained 5–53 per cent. As Price et al. (2003) observe, the shares are affected by key parameters including the specification of the analytical framework, supply and demand elasticity assumptions, farm-level effects and year-specific variables.

Table 36.3 Effect of global adoption of HT soybeans on commodity prices

Technology and crop	Adopting countries/ regions	Commodity	Price decline	Researchers
HT soybeans	US	Soybeans	0.17 %	Price et al. (2003)
HT soybeans	US	Soybeans	1.0 %	Moschini et al. (2000)
HT soybeans	US and South America	Soybeans	2.2 %	Moschini et al. (2000)
HT soybeans	World	Soybeans	2.6 %	Moschini et al. (2000)
HT soybeans	US and Argentina	Soybeans	1.96 %[b]	Qaim and Traxler (2005)
HT soy & canola + Bt corn[a]	World	Oilseeds	3.08 %	Anderson and Jackson (2005)
Bt corn and HT soybeans	US and Canada	Oilseeds	1.5 %	Fernandez-Cornejo et al. (2007)
Bt corn and HT soybeans	World	Oilseeds	3.87 %	Fernandez-Cornejo et al. (2007)

Notes:
[a] See assumptions in Anderson and Jackson (2005).
[b] Price decline reported is for 2001. Qaim and Traxler (2005) also calculated price declines for other years; e.g., 1.54 per cent for 2000.

Source: NRC (2010).

Table 36.4 Benefits of the adoption of herbicide-tolerant soybeans and their distribution

Study/adopting countries	Year	Total benefits ($ million)	Share of total benefits (Percent)			
			Farmers	Innovators	Consumers	Net ROW[a]
Falck-Zepeda et al. (2000b)/USA	1997-HE[c]	437	29[b]	18	17	28
Moschini et al. (2000)/USA	1999	804	20[b]	45	10	26
Price et al. (2003)/ USA	1997	310	20[b]	68	5	6
Qaim and Traxler (2005)/USA and Argentina	1997	206	16[d]	49	35	NA[e]
Qaim and Traxler (2005)/USA and Argentina	2001	1230	13[d]	34	53	NA[e]

Notes:
NA = not applicable.
[a] ROW = rest of the world (consumers and producers).
[b] US farmers.
[c] HE = high elasticity; assumes a US soybean supply elasticity of 0.92.
[d] Include all soybean producers.
[e] Included in consumers and producers.

Sources: Fernandez-Cornejo and Caswell (2006); Qaim and Traxler (2005); NRC (2010).

CRITICAL ASSESSMENT

As Feder and Umali (1993) hypothesize, factors such as farm size, tenure, education, information and credit may affect the adoption decision of early adopters differently than the adoption decision of late adopters. In other words, the factors affecting adoption are likely to change over time.

The impacts of adoption may also change over time. Unfortunately, as adoption rates rise, accurately quantifying these impacts becomes increasingly difficult. For instance, because HT soybeans have been adopted so extensively in recent years, sample sizes of non-adopters tend to be very small. This makes it difficult to analyze cross-sectional data using traditional econometric techniques. Additionally, many of those few farmers who have not adopted HT soybeans may have different motivations and may use different technologies (for example, organic farming) than traditional non-adopters (that is, laggards). Because these motivations tend to be unobservable, self-selection becomes increasingly problematic with data characterizing the later stages of adoption. That said, the major findings of this chapter are:

1. Adoption of HT soybeans is invariant to farm size.
2. The economic benefits of HT soybeans accrue to farmers in adopting countries, biotechnology developers and germplasm suppliers and consumers. These benefits have increased over time.
3. HT adoption leads to savings in management time, allowing farmers to earn higher total farm household income by allocating time to higher value activities.
4. In most cases, HT soybeans do not appear to have a significant yield advantage relative to conventional seeds.
5. While HT soybean adoption may have increased total herbicide use in many cases, herbicide expenditures are usually lower and HT soybean adoption has led to the use of a more environmentally benign herbicide than the herbicides that it replaced.
6. HT adoption and conservation tillage have a complementary relationship. Consequently, HT soybeans facilitate the benefits of conservation tillage.

One of the most important issues regarding the future of HT soybeans is the evolution of herbicide resistance to weeds and its management. The National Research Council (NRC, 2010) reports that the sustainability of HT crops will be threatened if weeds are managed as they have been. The NRC recommends that government, industry, academia and other stakeholders collaborate to develop cost-effective resistance management practices following Integrated Pest Management (IPM) principles to help to preserve effective weed control in herbicide-tolerant soybeans.

NOTES

1. The views expressed are those of the authors and do not necessarily correspond to the views or policies of the US Department of Agriculture.
2. The most common herbicide-tolerant crops are Roundup Ready crops resistant to glyphosate, a herbicide effective on many species of grasses, broadleaf weeds and sedges (Fernandez-Cornejo et al., 2002). There are also traditionally bred herbicide-tolerant soybeans, resistant to sulfonylurea (STS).

3. Producer and consumer surpluses in the US and international markets and monopoly profits accruing to the biotech developers and seed firms can also be calculated from such models.
4. For detailed results of the econometric model see Fernandez-Cornejo and McBride (2002), and NRC (2010). Unfortunately, there are no reliable results of econometric impact models using more recent data because of the low number of non-adopters in the sample. For example, in the 2006 survey, 97 per cent of the farmers in the sample adopted HT soybean seeds; it was not possible to reliably estimate the impact of HT seed use using the econometric techniques discussed in a previous section.
5. Yields may even decrease if the varieties used to carry the herbicide-tolerant or insect-resistant genes are not the highest yielding cultivars. This yield decrease (yield lag) occurred mostly in early years. HT and Bt genes were introduced into high-yielding cultivars in later years (NRC, 2010).
6. The study used a two-stage econometric model that takes into account simultaneity and self-selectivity.
7. According to the 2006 USDA's ARMS soybean survey, 59.7 per cent of the farmers adopting HT soybeans indicated that they did so in order to increase yields. Other adopters reported that they adopted HT soybeans to decrease herbicide input costs (20.2 per cent), save management time (7.3 per cent), facilitate reduced tillage (6.5 per cent), or for other reasons (6.3 per cent).
8. This large increase was due in part to the fact that HT adoption was associated to the substitution of herbicides for tillage (Qaim, 2009).

REFERENCES

Alexander, C., J. Fernandez-Cornejo and R. Goodhue (2003), 'Effects of the GM controversy on Iowa corn-soybean farmers' acreage allocation decisions', *Journal of Agricultural and Resource Economics*, **28**(3), 580–95.

Anderson, K. and L.A. Jackson (2005), 'Global responses to GM food technology', Rural Industries Research and Development Corporation (RIRDC) publication no. 05/016, February.

Anderson, K. and S. Yao (2003), 'China, GMOs and world trade in agricultural and textile products', *Pacific Economic Review*, **8**, 157–69.

Arnold, J.C., D.R. Shaw and C.R. Medlin (1998), 'Roundup Ready programs versus conventional programs: efficacy, varietal performance, and economics', *Proceedings of the Southern Weed Science Society*, **51**(1998), 272–3.

Bernard, J.C., J.D. Pesek Jr and C. Fan (2004), 'Performance results and characteristics of adopters of genetically engineered soybeans in Delaware', *Agricultural and Resource Economics Review*, **33**(2), 282–92.

Bonny, S. (2011), 'Herbicide-tolerant transgenic soybean over 15 years of cultivation: pesticide use, weed resistance, and some economic issues. The case of the USA', *Sustainability*, **3**(9), 1302–22.

Bradley, K., E. Hagood and P. Davis (2004), 'Trumpetcreeper (*Campsis radicans*) control in double-crop glyphosate-resistant soybean with glyphosate and conventional herbicide systems', *Weed Technology*, **18**(2004), 298–303.

Brookes, G. (2005), 'The farm-level impact of herbicide-tolerant soybeans in Romania', *AgBioForum*, **8**(4), 235–41.

Carpenter, J. and L. Gianessi (1999), 'Herbicide tolerant soybeans: why growers are adopting Roundup Ready varieties', *AgBioForum*, **2**(2)(Spring), 65–72.

Cerdeira, A.L. and S.O. Duke (2006), 'The current status and environmental impacts of glyphosate-resistant crops: a review', *Journal of Environmental Quality*, **35**(5), 1633–58.

Delannaay, X., T.T. Bauman, D.H. Beighley et al. (1995), 'Yield evaluation of a glyphosate tolerant soybean line after treatment with glyphosate', *Crop Science*, **35**, 1461–7.

Duffy, M. (2001), 'Who benefits from biotechnology?', paper presented at the *American Seed Trade Association Meeting*, 5–7 December, Chicago, IL.

Duke, S.O. and S.B. Powles (2008), 'Glyphosate: a once-in-a-century herbicide', *Pest Management Science*, **64**(4), 319–25.

Falck-Zepeda, J.B., Traxler and R.G. Nelson (1999), 'Rent creation and distribution from the first three years of planting Bt cotton', *ISAAA Briefs* no. 14, Ithaca, NY: ISAAA.

Falck-Zepeda, J.B., G. Traxler and R.G. Nelson (2000a), 'Surplus distribution from the introduction of a biotechnology introduction', *American Journal of Agricultural Economics*, **82**, 360–69.

Falck-Zepeda, J.B., G. Traxler and R.G. Nelson (2000b), 'Rent creation and distribution from biotechnology innovations: the case of Bt cotton and herbicide-tolerant soybeans in 1997', *Agribusiness*, **16**, 21–32.

Feder, G. and D.L. Umali (1993), 'The adoption of agricultural innovations: a review', *Technological Forecasting and Social Change*, **43**, 215–39.

Feder, G., R.J. Just and D. Zilberman (1985), 'Adoption of agricultural innovations in developing countries: a survey', *Economic Development and Cultural Change*, **33**, 255–98.

Fernandez-Cornejo, J. (2004), 'The seed industry in US agriculture: an exploration of data and information on crop seed markets, regulation, industry structure, and research and development', *Agricultural Information Bulletin*, AIB-786, Economic Research Service, US Department of Agriculture.

Fernandez-Cornejo, J. (2012), 'Adoption of genetically engineered crops in the US', (2012) data product, US Department of Agriculture, Economic Research Service, available at: http://www.ers.usda.gov/Data/BiotechCrops/.

Fernandez-Cornejo, J. and M. Caswell (2006), 'The first decade of genetically engineered crops in the United States', Economic Information Bulletin no. 11, US Department of Agriculture, Economic Research Service.

Fernandez-Cornejo, J. and W.D. McBride (2000), 'Genetically engineered crops for pest management in US agriculture: farm-level effects', Agricultural Economic Report no. 786, US Department of Agriculture Economic Research Service.

Fernandez-Cornejo, J. and W.D. McBride (2002), 'Adoption of bioengineered crops', Agricultural Economic Report no. 810, US Department of Agriculture, Economic Research Service.

Fernandez-Cornejo, J., S. Daberkow and W. McBride (2001), 'Decomposing the size effect on the adoption of innovations: agrobiotechnology and precision agriculture', *AgBioForum*, **4**(2), 124–236.

Fernandez-Cornejo, J., C. Hendricks and A. Mishra (2005), 'Technology adoption and off-farm household income: the case of herbicide-tolerant soybeans', *Journal of Agricultural and Applied Economics*, **37**(3), 549–63.

Fernandez-Cornejo, J., C. Klotz-Ingram and S. Jans (2002), 'Farm-level effects of adopting herbicide-tolerant soybeans in the USA', *Journal of Agricultural and Applied Economics*, **34**(1), 149–63.

Fernandez-Cornejo, J., R. Lubowski and A. Somwaru (2007), 'Global adoption of agricultural biotechnology: modeling and preliminary results', paper presented at the 10th Annual conference on Global Economic Analysis (GTAP), May, West Lafayette, IN.

Fernandez-Cornejo J., C. Hallahan, R. Nehring, S. Wechsler and A. Grube (2010), 'Conservation tillage, pesticide use, and biotech crops in the USA', selected paper presented at the AAEA Annual Meeting, 26–27 July Denver, CO.

Fernandez-Cornejo, J., R. Nehring, E. Newcomb, A. Grube and A. Vialou (2009), 'Recent trends in pesticide use in US agriculture', selected paper presented at the AAEA Annual Meeting, July, Milwaukee, WI.

Frisvold, G., R. Tronstad and J. Mortensen (2000), 'Effects of Bt cotton adoption: regional differences and commodity program effects', paper presented at the Western Agricultural Economics Association Meeting, 30 June, Vancouver, Canada.

Griliches, Z. (1957), 'Hybrid corn: an exploration in the economics of technological change', *Econometrica*, **25**(1957), 501–22.

James, Clive (2011), 'Global status of commercialized biotech/GM crops: 2010', *International Service for the Acquisition of Agri-Biotech Applications* (*ISAAA*) *Brief*, 42, Ithaca, NY: ISAAA, available at: http://www.isaaa.org/resources/publications/briefs/42/default.asp.

Kovach, J., C. Petzoldt, J. Degni and J. Tette (1992), 'A method to measure the environmental impact of pesticides', *New York's Food and Life Sciences Bulletin* no. 132, Cornell University, Ithaca, NY, updates available at: http://nysipm.cornell.edu/PUBLICATIONS/eiq/).

Malik, J.M., G.F. Barry and G.M. Kishore (1989), 'The herbicide glyphosate', *BioFactors*, **2**(1), 17–25.

Marra, M., N. Piggott and G. Carlson (2004), 'The net benefits, including convenience, of Roundup Ready soybeans: results from a national survey', *IPM Technical Bulletin no.3*, NSF Center, pp. 1–39.

McBride, W.D. and H. El-Osta (2002), 'Impacts of the adoption of genetically engineered crops on farm financial performance', *Journal of Agricultural and Applied Economics*, **34**(1), 175–91.

Mensah, E.C. (2007), *Economics of Technology Adoption: A Simple Approach*, Saarbrücken: VDM Verlag Dr Müller.

Moschini, G. and H. Lapan (1997), 'Intellectual property rights and the welfare effects on agricultural R&D', *American Journal of Agricultural Economics*, **79**, 1229–42.

Moschini, G., H. Lapan and A. Sobolevsky (2000), 'Roundup Ready® soybeans and welfare effects in the soybean complex', *Agribusiness*, **16**(1), 33–55.

National Research Council (NRC) (2010), *The Impact of Genetically Engineered Crops on Farm Sustainability in the United States*, Washington, DC: National Academies Press.

Paz, R., W. Fernandez, P. Zambrano and J. Falck Zepeda (2009), 'GM soybeans in Bolivia', paper presented at the 13th ICABR Conference on The Emerging Bio-Economy, Ravello, Italy.

Pelaez, V., V.L. Albergoni and M.P. Guerra (2004), 'Soja transgenica versus soja covencional: uma analise comparativa de custos e beneficios', *Cadernos de Ciência and Tecnologia*, **21**(2), 279–309.

Price, G.K., W. Lin, J.B. Falck-Zepeda and J. Fernandez-Cornejo (2003), 'Size and distribution of market

benefits from adopting biotech crops', Technical Bulletin no. 1906, Washington, DC: US Department of Agriculture, Economic Research Service.
Qaim, M. (2009), 'The economics of genetically modified crops', *Annual Review of Resource Economics*, **1**, 665–93.
Qaim, M. and G. Traxler (2005), 'Roundup Ready® soybeans in Argentina: farm level and aggregate welfare effects', *Agricultural Economics*, **32**(1), 73–86.
Raney, T. and I. Matuschke (2011), 'Current and potential farm level impacts of genetically modified crops in developing countries', in C.A. Carter and G.C. Moschini (eds), *Genetically Modified Food and Global Welfare*, Bingley, US: Emerald Group Publishing Limited.
Reddy, K.N. and K. Whiting (2000), 'Weed control and economic comparisons of glyphosate-resistant, sulfonylurea-tolerant, and conventional soybean (*Glycine max*) systems', *Weed Technology*, **14**, 204–11.
Roberts, R.K., R. Pendergrass and R.M. Hayes (1998), 'Farm-level economic analysis of Roundup Ready soybeans', paper presented at the Southern Agricultural Economics Association Meeting, 1–4 February, Little Rock, AR.
Rogers, E. (1961 [1995]), *Diffusion of Innovations*, 1st edn, 4th edn, New York: Free Press.
Shipitalo, M.J., R.W. Malone and L.B. Owens (2008), 'Impact of glyphosate-tolerant soybean and glufosinate-tolerant corn production on herbicide losses in surface runoff', *Journal of Environmental Quality*, **37**, 401–408.
Tassisto, R. (1998), 'Control de malezas en sojas RR', in AAPRESID, *Jornada de Intercambio Tecnico de Soja*, Asociación Argentina de Productores en Siembra Directa, Buenos Aires, pp. 77–8.
Trigo, E.J. and E.J. Cap (2006), 'Ten years of genetically modified crops in Argentine agriculture', Argentine Council for Information and Development of Biotechnology–ArgenBio, December.
Wauchope, R.D., T.M. Buttler, A.G. Hornsby, P.W.M. Augustijn-Beckers and J.P. Burt (1993), 'The SCS/ ARS/ CES pesticide properties database for environmental decisionmaking', in G.W. Ware (ed), *Reviews of Environmental Contamination and Toxicology*, vol. 123, New York: Springer Verlag.

37 Maize/corn

Janet Carpenter, Marnus Gouse and Jose Yorobe Jr

INTRODUCTION

Maize (*Zea mays* L. ssp. *mays*), known in North America as corn, is an annually cropped grass that was domesticated in what is now Mexico approximately 6000–7000 years ago. The crop had spread throughout much of the Americas before European explorers in the fifteenth and sixteenth centuries carried seed back for cultivation and introduced maize to other countries. Evolution of maize through selection, and more recently breeding, over thousands of years has resulted in multiple biotypes that can be grown in diverse climates (Carpenter et al., 2002; James, 2003).

Hybrid maize varieties developed from the crossing of inbred lines were introduced for widespread commercialization in the 1930s in the US and quickly dominated maize production, reaching 90 per cent adoption by 1945. By 1999, 94 per cent of maize acreage in developed countries and 54 per cent of maize acreage in developing countries was planted with hybrid varieties (James, 2003). An analysis of the process of adapting and distributing hybrid seed maize and its rate of adoption by US farmers was a key development in the literature on the economics of technological change (Griliches, 1957).

Today maize is among the most important crops grown in the world, accounting for the greatest tonnage produced (1.1 billion metric tonnes in 2009/10), second only to wheat (and followed closely by rice) in terms of area, with nearly 158 million hectares harvested in 2009/2010 (Anonymous, 2011). The US is the leading producer of maize, followed by China, Brazil, the 27 member states in the European Union, Argentina, Mexico, India, France (individually, also included in EU-27), South Africa and Ukraine: all produced over 10 million metric tonnes in 2009/2010 (Table 37.1). The primary use of maize is as animal feed, but substantial amounts are used for ethanol production and human consumption, mainly in developing countries. Major maize exporters are the US, Argentina, Brazil and the Ukraine. Japan is the largest importer, followed by South Korea and Mexico.

In the US, high oil prices have combined with the provisions of the Energy Policy Act of 2005 to increase existing incentives to expand biofuel production. A strong demand for ethanol production resulted in expanded maize acreage after 2005 as corn is the primary feedstock used to produce ethanol in the US. In 2011, 46 per cent of US maize production was used to make alcohol for fuel use, up from just 17 per cent in 2005 (Figure 37.1). The maize area has expanded through adjusting crop rotations in favor of maize, conversion of pasture, reduced fallow, expiring Conservation Reserve Program contracts or shifts from other crops.[1]

China has also experienced major growth in corn production, driven by a surge in industrial uses. Industrial uses are believed to have accounted for nearly all of the growth in corn production in China since 2000. Between 2000 and 2006, industrial uses grew from 20.5 million metric tonness (MMT) to 40 MMT, but receded somewhat

Table 37.1 Maize/corn area, yield and production 2009/10

Country/Region	Area (million ha)	Yield (MT/ha)	Production (million MT)
World	157.8	5.2	813.4
United States	32.2	10.3	332.6
China	31.2	5.1	158.0
Brazil	12.9	4.3	56.1
India	8.3	2.0	16.7
EU-27[a]	8.3	6.9	57.0
Mexico	6.3	3.2	20.4
Nigeria	4.9	1.8	8.8
South Africa	3.3	4.1	13.4
Indonesia	3.1	2.3	6.9
Argentina	2.8	8.5	23.3
Philippines	2.5	2.5	6.2
Romania	2.4	3.2	7.5
Ukraine	2.1	5.0	10.5
Ethiopia	2.0	2.0	3.9
France	1.7	9.1	15.3
Russia	1.4	2.9	4.0
Serbia	1.2	5.3	6.4
Hungary	1.2	6.4	7.5
Canada	1.1	8.4	9.6
Vietnam	1.1	4.1	4.6
Zimbabwe	1.1	0.6	0.7
Thailand	1.0	4.1	4.1
Italy	0.9	9.3	8.7
Egypt	0.8	7.5	6.3
Turkey	0.5	8.0	4.0
Poland	0.3	6.2	1.7
Others	29.7	2.0	60.2

Note: [a] European Union countries also presented individually.

afterwards due to policy changes. The single largest category of use has been starch sugars and alcohol for beverage and industrial use. Fuel ethanol production began in 2004 at three plants and accounted for 3 million metric tonnes of consumption as of 2009 (Gale et al., 2009). Significantly, China has recently become a major importer of corn, believed to be driven by strong feed and industrial demand (Lagos and Junyang, 2011). Figure 37.2 shows a global increase in maize area harvested and total maize production.

Maize is South Africa's most important field crop and annually covers approximately 30 per cent of the total arable land. Maize is a staple food for the majority of the South African population. In recent years white maize made up approximately 60 per cent of the total maize crop and covered 63 per cent of the area under maize. Between 60 and 70 per cent of the South African yellow maize crop is consumed in the broiler production sector and between 2004 and 2007, approximately 80 per cent of the white maize crop went towards direct human consumption as maize flour (mealie meal). South Africa has

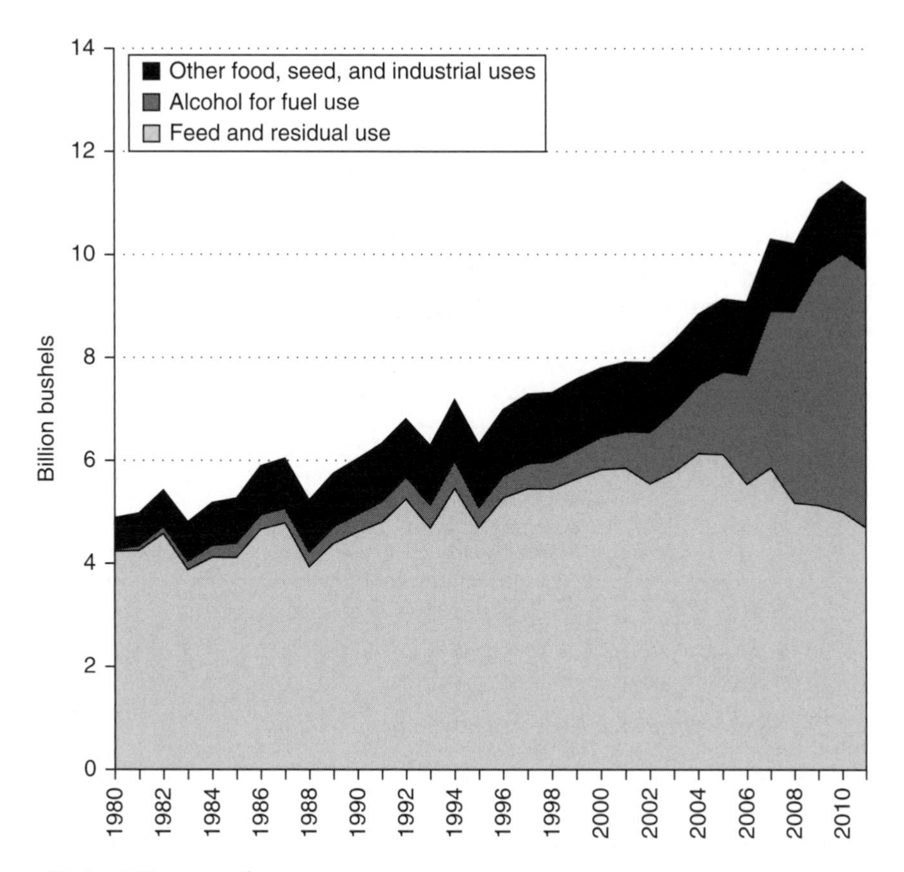

Figure 37.1 US corn utilization

seen a decreasing trend in per capita direct maize consumption as a larger share of the population has moved into the middle class, but this trend was stalled by the 2008 food price spike (BFAP, 2009).

In 2007 the South Africa Cabinet released a biofuels strategy in which maize is excluded as a biofuel feed stock, basing their decision on the fear that the use of maize for ethanol production might threaten food security. This position has been challenged and might be altered in the not-too-distant future, especially in the light of substantial carry-over stocks in recent years, despite exports, and the fact that it is estimated that South African farmers can expand maize production by an additional million hectares. The economic viability of the maize to ethanol process in a fluctuating oil and regional grain market and in the absence of any government support is unclear.

Genetically modified (GM) maize varieties were first commercialized in the US in 1996 and by 2010 were grown in 15 countries around the world, accounting for over a quarter of the global maize area (James, 2010) (see Table 37.2). The first technology to be introduced was insect resistance, which was conferred by transferring a gene from the soil bacterium *Bacillus thuringiensis* (Bt). The first Bt maize events to be commercialized targeted the European corn borer, a pest of primary importance in maize in the US.

Figure 37.2 World maize area harvested and total production

Currently commercialized Bt maize varieties control most stem borers and rootworms as well as some other important pests including corn earworm and fall armyworm (James, 2003). Genetically modified herbicide-tolerant maize varieties were first commercialized in 1998.

As of 2010, seven countries had deployed 'stacked' maize varieties, which embody two or more insect resistance and/or herbicide tolerance traits. Adoption of stacked varieties is already substantial. In the US, 47 per cent of the maize area was planted to stacked varieties in 2010. In Argentina 52 per cent of the maize area was planted to stacked maize hybrids. In Canada, South Africa, the Philippines and Honduras, stacked maize hybrids accounted for 70 per cent, 41 per cent, 76 per cent and 73 per cent of all genetically modified maize plantings, respectively (James, 2010).

Maize has been and will continue to be dominant in the field of GM crops. By 2010, the US Department of Agriculture had granted a total of 83 'determinations of non-regulated status', 26 of which were for maize. Nearly all of these were for insect resistance or herbicide tolerance, with a small number of events with breeding tools such as male sterility and fertility-restored characteristics. Notably, a variety of maize expressing thermostable alpha amylase, which aids in the processing of maize for ethanol, was recently de-regulated as the first industrial technology to be authorized for planting in the US. Drought tolerance is currently under regulatory review and is expected to be commercialized in 2012. Looking into the pipeline, 271 of 672 field trial permits or notifications that were acknowledged or issued in calendar year 2010 by USDA were for corn.

Table 37.2 Year of first genetically modified corn commercialization and adoption rates in 2010

Country	Year of commercialization	Commercialized technology	Adoption rate in 2010
US	1996	IR, HT, IR/HT	86%
Canada	1997	IR, HT, IR/HT	98%
Argentina	1998	IR, HT, IR/HT	86%
South Africa	1998	IR, HT, IR/HT	77%
Spain	1998	IR	24%
Portugal	1999	IR	4%
Honduras	2002	IR, HT, IR/HT	4%
Philippines	2003	IR, HT, IR/HT	42%
Uruguay	2003	IR	83%
Czech Republic	2005	IR	1%
Slovakia	2006	IR	0.50%
Poland	2006	IR	0.50%
Brazil	2007	IR	56%
Romania	2007	IR	<0.5%
Egypt	2008	IR	<0.5%

Source: James (2010).

MODELS AND METHODS

In assessments of the impact of the introduction of genetically modified crops, researchers have employed a range of models. The theory underlying these models is reviewed in Chapter 45 in this volume, and elsewhere (for example Smale et al., 2009), and are therefore only briefly discussed here.

Early studies of the farm-level impacts of the technology employed a standard production function approach, in which yields are modeled as a function of a range of inputs, typically including a dummy variable for the newly introduced technology. Researchers have more recently applied a damage abatement approach, which takes into account the indirect nature of the yield impacts of pesticides, including Bt technology, by reducing the extent of damage in the event of damaging levels of target pest infestation. Another approach to assessing the impact of the introduction of agricultural technology at the farm level is the use of production frontiers to estimate the contribution of various inputs to the efficiency of production. A few studies have also introduced the stochastic production function approach with or without damage abatement specification to determine downside risk effects (or skewness effects) of Bt corn adoption and to evaluate welfare implications using risk premium and certainty equivalent measures. Impacts have also been estimated using a two-stage modeling approach, first modeling probability of adoption then using predicted adoption rates in an impact model to address self-selection bias. More recent models have appeared in the literature addressing second-generation impacts of GM corn, such as the transition from the single to the stacked trait.

At the macro level, partial equilibrium models estimate impacts in a single market while general equilibrium models extend the analysis to other markets that are affected by the introduction of a new technology.

Many of the earliest studies on the economic impact of GM crops based farm-level impact estimates on available information about expected performance of the technology, such as field trials and the expert opinion of extension and industry personnel (Gianessi and Carpenter, 1999, 2000). Farm-level benefits were then extrapolated to the national level based on estimated adoption rates, without consideration of price and production effects.

As adoption grew and the technology spread, farmer surveys were conducted to more definitively assess the impact of the technology. Often, the average outcomes of adopters and non-adopters have been compared, which can over- or understate the impact of the technology if there are any systematic differences between adopters and non-adopters, such a management ability, pest pressure or other factors. Many surveys have attempted to control for this by including within-farm comparisons for partial adopters. Even within-farm comparisons can suffer from bias if growers adopt the technology on fields that are systematically different; this could involve, for instance, adoption of a herbicide-tolerant crop on their weediest fields. In addition, within-farm comparisons may understate the value of the technology by excluding farmers who place the highest value on the technology and fully adopt the technology. The econometric techniques described above have been applied to survey data to address these self-selection biases, endogeneity and censoring issues.

METRICS

The impacts of the introduction of genetically modified maize are relatively well studied. We summarize the literature below, starting with literature addressing global impacts, followed by a discussion of available results by country in alphabetical order. The results of farmer surveys are presented in the Appendix.

Global

Several overarching studies have estimated the impacts of genetically modified maize globally.

Brookes and Barfoot (2008) calculate aggregate economic impacts on farm income using yield, cost and income assumptions from existing literature. Yield increases of between 5 per cent and 32 per cent were assumed for Bt maize and between 0 per cent and 22 per cent for HT maize. Aggregate farm income impacts for Bt maize in 2006 were estimated to be $1131 million and $296 million for HT maize (Brookes and Barfoot, 2008). In a later analysis, these same yield, cost and income assumptions were used in a partial equilibrium model of the impact of biotech corn, canola and soybean on cereal and oilseed markets, finding that corn prices would be 5.8 per cent higher if biotech varieties were not available. The additional production of maize that could be attributed to biotech was estimated at 8.07 million metric tonnes (Brookes et al., 2010).

Using a general equilibrium model, Anderson and Shunli Yao (2003) assume a 5 per cent increase in productivity for GM maize and soybean in the adopting regions (North

America, Southern Cone of South America and SE Asia) with and without China as an adopter. Overall, non-adopting regions see output fall slightly, while adopting regions see an increase. Prices of GM maize and soybean fall much more where the technology is adopted. Adopters increase their net exports, while non-adopters increase their net imports. Since China accounts for a smaller share of global maize and soybean markets than for rice or cotton markets, China's adoption has less effect on changes in world welfare impacts. International prices of maize and soybean are estimated to fall by 0.5 per cent more when China is an adopter (Anderson and Shunli Yao, 2003).

Two studies have compiled results from the existing literature to draw some general conclusions about the impact of currently commercialized technology. Carpenter (2010) compiled the findings of peer-reviewed literature presenting results of farmer surveys that compared yields and other indicators of economic performance of adopters and non-adopters. For Bt maize, 34 results on yields from five countries showed that farmers in developed countries (US, Spain and Canada) achieved a 4 per cent yield increase on average, while farmers in developing countries (South Africa and Philippines) achieved a 16 per cent yield increase on average for yellow maize. For HT maize, a single study from South Africa indicated an 85 per cent yield advantage for adopters. With respect to economic performance, developing-country farmers who adopted the technology compared favorably with non-adopters in 13 out of 19 cases, in four cases the results were neutral and two results showed adopters achieving lower economic performance. In developing countries, farmers achieved improved economic performance in 10 cases and lower economic performance in two cases (Carpenter, 2010). Tables 37.A1 and 37.A2 summarize the results of farmer surveys for insect-resistant and herbicide-tolerant maize, respectively.

Kaphengst et al. (2011) conducted a meta-analysis of the results of both peer-reviewed and non-peer-reviewed publications reporting results comparing the performance of biotech varieties to conventional varieties from both field trials and farmer surveys. With respect to Bt maize, the effect on yields was found to be a 3.9 per cent increase on average across all countries considered. Economic performance was between 10 per cent and 17 per cent greater for adopters of Bt maize compared to non-adopters (Kaphengst et al., 2011).

Argentina

Bt maize was first commercialized in Argentina in 1998, followed by herbicide tolerance in 2004. The economic impact of Bt maize has been estimated, assuming a 10 per cent yield advantage for adopters, resulting from control of the target pest sugarcane borer (*Diatraea saccharalis*). Aggregate benefits of Bt maize were estimated to be $482 million, with 41 per cent of the benefits accruing to technology suppliers, 43 per cent to farmers and 17 per cent to the national government (Trigo and Cap, 2006). No estimates for herbicide-tolerant maize were located.

Brazil

Bt maize was commercialized in Brazil for the first time in 2008/09. A recent survey of 360 farmers from 10 states found that corn accounted for 18 per cent of $3.6 billion aggregate benefits of biotech corn, cotton and soy from 1996/97 to 2008/09, due to reduced production costs and increased yields (Anonymous, 2010). Details on the methodology used for this study were not available.

Canada

Insect-resistant maize was first commercialized in Canada in 1996, followed by herbicide tolerance and stacked varieties. It is estimated that 98 per cent of the Canadian maize area is planted to genetically modified varieties, nearly all herbicide-tolerant varieties (Syngenta, personal communication). Two surveys have examined the impact of Bt maize in Canada, indicating a yield advantage for Bt maize adopters in Ontario of 4.6 per cent and 7.7 per cent in 2000 and 2001 respectively. In Quebec, yields were 12.8 per cent higher and 2.6 per cent lower for adopters compared to non-adopters in 2000 and 2001, respectively (Hategekimana, 2002a, 2002b). Using an adoption model with the 2000 and 2001 data, researchers found that farms with a larger total area and larger area in field crops were less likely to adopt Bt maize, indicating that larger farms were slower to adopt the technology (Hategekimana and Trant, 2002).

European Union

First approved for planting in the EU in 1998, six countries grew Bt maize in 2010: Spain, Portugal, Poland, Slovakia, Romania and Czech Republic. Germany grew Bt maize from 2000 to 2008, but has since discontinued planting due to implementation of a 2005 law covering coexistence and liability (James, 2010).

In a review of the economic impacts of Bt maize across adopting countries in the EU, Brookes (2008) draws from various sources as the basis of estimates, including a farmer survey in Spain, field trials, industry data and estimates of insect damage. Improvements in profitability were estimated between 12 per cent and 21 per cent, but given low adoption have little macroeconomic impact on farm performance (Brookes, 2008).

Philippines

Bt maize was first commercialized in the Philippines in 2003, with herbicide tolerance and stacked Bt/HT in 2005 (James, 2010). The most destructive pest in the major corn-producing regions in the Philippines is the Asian corn borer, with yield losses reported by farmers of between 20 per cent and 80 per cent. Most maize farmers in the Philippines are small, semi-subsistence farmers with average farm size ranging from less than a hectare to about four hectares.

Although a great deal of success was achieved in the production of GM corn in the Philippines in 2003–10, the country remained a net importer of yellow corn in 2012. Three traits of GM yellow corn were commercially produced from 0.54 million hectares in 2010 (James, 2010). Yellow corn is used mainly for feed, while white corn is targeted to food and industrial uses. The country produced 6.38 million metric tonnes in 2010, 66 per cent of which is yellow. The Philippines is the first country in the ASEAN region to commercialize Bt corn in 2003 and to implement a regulatory system for transgenic crops.

The stacked IR/HT corn variety is now the more popular seed type among farmers, planted on more than 300000 hectares.

A survey of maize growers in four provinces (Isabela, Camarines Sur, Bukidnon and South Cotabato) in 2003/04 and 2004/05 found that growers planting Bt maize achieved yields between 20 per cent and 34 per cent higher than non-adopters and had higher gross margins (Gonzales, 2005). Further statistical analysis with the ISAAA 2003/04 corn survey data used a production function to estimate the impact of the technology on yields and a two-stage econometric model to estimate likelihood of adoption and the impact of adoption on net returns. At the margin, a 10 per cent increase in adoption of Bt maize was estimated to result in a 2.25 per cent increase in yields and a 4.1 per cent increase in net farm income. Farmers were estimated to receive 46 per cent of total benefits, with 43 per cent going to technology providers (Yorobe and Quicoy, 2006). Using the same data set and taking into account for endogeneity and selection biases in the Bt corn impact analysis, the inference error has been further minimized through censoring of the pesticide application variable. Different elasticity estimates of the Bt impact result when censoring of the pesticide impact variable is ignored (Mutuc et al., 2012).

Again using the 2003/04 survey data with additional data from 2007/08, researchers used a damage abatement approach to assess the impact of Bt maize on yields and insecticide use. The yield damage abatement effect of Bt maize was found to be statistically significant in both years, but especially in the poor weather conditions of 2007/08, suggesting that insecticide use is less effective in wet weather. Also, the value of insecticide use is reduced when using Bt corn (Mutuc et al., 2011). The same data sets were used to determine the production-risk effects of the Bt corn. Results indicate a strong statistically significant mean-yield increasing effect but no risk-reduction effects. Bt corn farmers were still better off than non-Bt producers, showing a lower probability of loss (Sanglestasawai et al., 2012).

Survey data from 2007/08 was used to explore reasons that adoption of Bt maize might have been stalled despite observed advantages of the technology. A choice experiment was conducted to elicit information about farmers' willingness to pay for attributes of maize seed that were identified as most important, including the Bt trait. Preferences differed between the two survey sites and between the Bt and non-Bt growers at the Isabela site. Farmers valued insect resistance, but not all preferred the Bt trait. The option to pay with credit was valued and a higher WTP was found for information from input suppliers rather than from other farmers (Yorobe et al., 2010).

South Africa

Bt maize was approved for planting for the 1998/99 season and Bt yellow maize was indeed planted in that season. The first planting of Bt white maize was delayed to 2001 as the Bt event first had to be crossed into the better-performing South African white maize varieties. Herbicide-tolerant varieties became available in 2003/04. Table 37.3 indicates the GM maize adoption levels for selected seasons. The popularity of Bt maize and the shift from Bt to HT to the stacked technology is apparent.

South African maize production is dominated by commercial growers who normally produce about 95 per cent of the domestic crop. Smallholders mainly produce on a subsistence level and surplus production depends on the success of the season (highly

Table 37.3 GM maize adoption by South African commercial farmers (%)

	2004/05	2007/08	2009/10
Bt Yellow Maize	23	38	26
Bt White Maize	8	40	60
RR Yellow Maize	1	15	15
RR White Maize	0.3	13	5
Stacked (Bt/HT) Yellow Maize	0	2	20
Stacked (Bt/HT) White Maize	0	3	8

dependent on the amount of rainfall). The adoption figures in Table 37.3 represent adoption by large-scale commercial farmers. GM maize adoption by smallholder farmers is still minimal as smallholders' input investment level and motivations for maize production differ according to their household needs, initial endowments and objectives. Gouse et al. (2008) estimated that in 2008 approximately 10 500 smallholder farmers planted GM maize; that is about 23 per cent of the total number of smallholders who annually buy hybrid seed from the three main seed companies.

Bt maize controls the African maize stem borer (*Busseola fusca*) and the Chilo stem borer (*Chilo partellus*), the most harmful pests to maize and grain sorghum in South Africa. It is estimated that in Southern Africa the African maize stem borer alone causes between 5 per cent and 75 per cent yield reduction and on average 10 per cent across years and regions. As has been found in the US, insecticide use for borer control is limited in South Africa, and nowhere close to the levels of chemical control necessary for bollworm control in cotton. Insecticide use, however, is higher on irrigation schemes where a microclimate and mono-cropping practices are conducive to pest build-up. South African smallholder farmers rarely use insecticides but if borer pressure is high, farmers might apply insecticide granules by hand or use local remedies, like ash or washing powder.

Large-scale commercial farmers, nearly all of whom grew both Bt and conventional maize, were surveyed in 1999/2000 and 2000/01. Yields for both irrigated and dryland farmers were between 7 per cent and 12 per cent higher in Bt fields, and gross margins were higher, due to a combination of savings on plant protection costs and higher yields (Gouse et al., 2005). That commercial maize growers in South Africa have benefited was not surprising, as their operations are similar to those in developed countries such as the US (Gouse et al., 2010).

Gouse et al. (2010) report yield comparisons for smallholder maize producers in the Hlabisa district of KwaZulu Natal who have been planting genetically modified maize since 2001/02. An average yield difference of 12 per cent for the six seasons from 2002/03 to 2007/08 was found, but the yield difference varied between -1 per cent (not statistically significant) in 2003/04 to +32 per cent (statistically significant) in 2004/05. HT maize also rendered a substantial yield increase in the 2006/07 season when a high early-season rainfall and extremely low rainfall during the ear-forming/filling stage resulted in severe weed problems. Chemical weed control by HT adopters was more effective than manual control with hand and hoe, resulting in higher yields for HT adopters. Early indications from South African smallholders are that the labour-saving benefit of chemical weed control might overshadow the insect control benefit of Bt maize (Gouse et al., 2010).

Spain

An early estimate of the impact of Bt maize in Spain comes from Brookes (2002). Based on information provided by a grower cooperative with 500 members, Brookes estimated a yield advantage of 10 per cent for adopters and an increase in gross margin (Brookes, 2002).

In the first statistical analysis of the impacts of Bt maize introduction in Spain, researchers used a bio-economic model to estimate welfare impacts on producers and the seed industry. Assuming an average 9 per cent loss due to corn borers was avoided by adopting the technology, a total welfare gain of 15.5 million euros was estimated for 1998 to 2003, two-thirds of which accrued to farmers, with the rest accruing to the seed industry (Demont and Tollens, 2004).

A farmer survey was conducted in 2002–04 in three regions (Aragon, Catalonia and Castilla-La Mancha) that accounted for approximately 90 per cent of the Bt maize area in Spain in 2006. Bt maize growers achieved higher average yields than conventional growers in all three years, though these gains were significant in only one province. Gross margins were higher in all provinces in all years (Gomez-Barbero et al., 2008).

USA

In 2011, 88 per cent of the US maize area was planted to genetically-modified maize varieties which were either insect resistant, herbicide tolerant or 'stacked' with both insect resistance and herbicide tolerance (Figure 37.3). Farmer surveys have been used to estimate the impacts of Bt and HT corn. The first survey results to be published were from a 1998 survey of Iowa maize growers, a leading maize-growing state. On average, adopters achieved yields that were 8.6 per cent higher than non-adopters and had higher net returns (Duffy and Ernst, 1999). A similar survey conducted in 2000 also found that, on average, yields for adopters were higher by 2 per cent, although net returns were slightly lower (Duffy, 2001).

All other survey results on the impact of Bt corn in the US come from a regular survey conducted by the US Department of Agriculture, the Agricultural Resource Management Survey. Researchers at the USDA Economic Research Service have conducted analysis on the impacts of Bt corn using data for 1998, 2001 and 2005, using a two-stage econometric modeling framework. In the first stage, an adoption decision model is used to determine the factors that influence farmer decisions to use Bt technology. The second stage is an impact model that is used to estimate the impact of adoption on yields, seed demand, insecticide demand and farm profits. An instrumental variables approach is used to correct for self-selection and simultaneity. The results varied from year to year. In 1998, a year of historically low target pest pressure, impacts on yields were not calculated, and the impact on net returns was negative overall (Fernandez-Cornejo and McBride, 2002). In 2001, yields were 9 per cent higher for adopters on average, and the results of econometric modeling showed a difference of 3.9 per cent (Fernandez-Cornejo and Li, 2005). In 2005, adopters' yields were 12.3 per cent higher on average, and the results of the econometric analysis indicated a 1.2 per cent advantage for adopters. Further, economic performance was improved with adoption (Fernandez-Cornejo and Wechsler, 2011).

Researchers have recently confirmed theoretical predictions that adoption of Bt maize

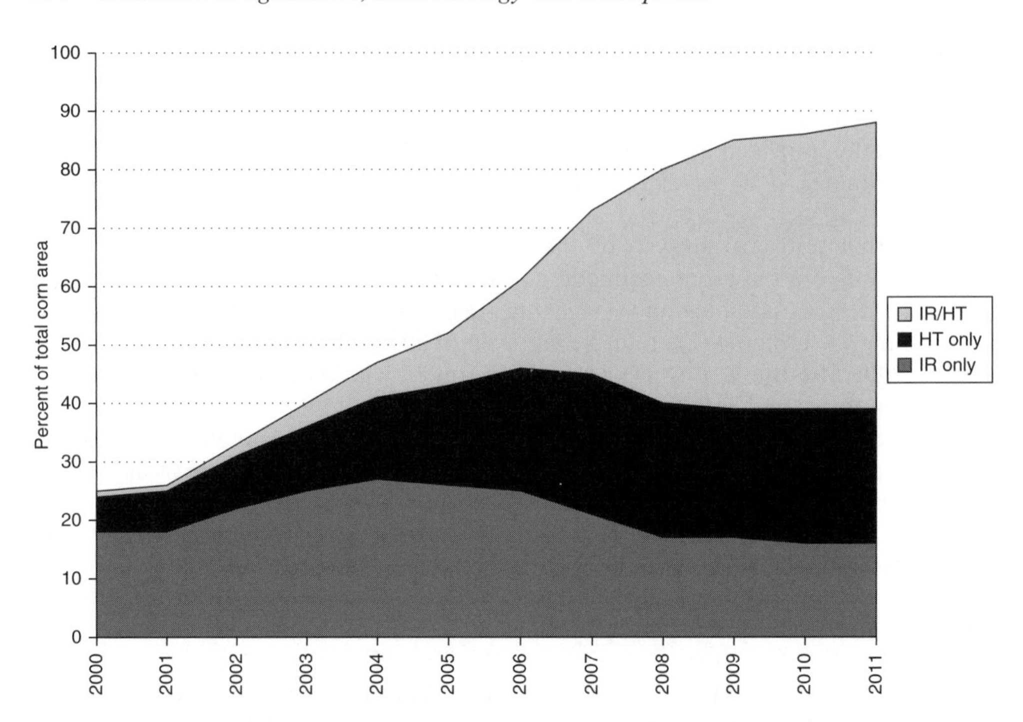

Note: First commercial introduction was in 1996. Data shown from 2000, the first year that USDA published adoption estimates.

Source: http://www.ers.usda.gov/Data/BiotechCrops/alltables.xls.

Figure 37.3 Adoption of genetically modified corn in the US

varieties could suppress populations of the target pest, European corn borer. Using long-term data on larval and moth flight, significantly different per capita population growth rates were found in areas with different levels of adoption. Cumulative benefits of Bt maize adoption were estimated for adopters and non-adopters based on ECB population densities and estimated models of larval stalk tunneling and associated yield loss. Non-adopters were estimated to have gained the majority of benefits, at $2.4 billion of a total $3.2 billion (Hutchison et al., 2010).

CRITICAL ASSESSMENT

The existing literature on the economic impact of genetically modified maize illustrates the wide range of methodological tools that are available in the assessment of new agricultural technology. Outside of a controlled experimental situation, it is likely that researchers will continue to use several tools to gain insight into their observations of farmer and consumer behavior. The methodology employed will depend not only on the availability of data, but on the question that is being asked, which may differ depending on the regulatory environment, for example. The parties to the Cartagena Protocol on

Biosafety are currently negotiating how socio-economic considerations might be incorporated into regulatory frameworks, but the specific parameters that might be considered have not been settled.

The current review exposes some gaps in the assessment of currently commercialized technologies. There are very few results on the impact of herbicide-tolerant maize varieties, which are widely planted in many countries. Further, there are no estimates of the impact of currently commercialized stacked varieties. With high adoption rates for many technologies, new methods will be needed because the opportunities to conduct analysis comparing adopters and non-adopters will no longer be available. New approaches might include the use of farmer surveys to discern willingness to pay where observed variability in market behavior is not available, or time series analysis to estimate the contributions of factors that vary over time but not within a single or a few years of data.

In addition, as we look at the pipeline of technologies (see Chapter 20 in this volume), entirely new models will probably need to be developed to account for quality characteristics and industrial traits. For example, nutritional enhancement targeted to poor populations in developing countries is likely to be assessed by measuring health outcomes. Also, it has been suggested that new cross-price elasticities be developed to account for new market relationships, such as that between corn and oil. The methods that have been developed and used to estimate the impacts of genetically engineered maize to date may inform the analysis of pipeline technologies but will not be entirely adequate to capture the range of possibilities that may emerge in the future.

NOTE

1. http://www.ers.usda.gov/topics/crops/corn/background.aspx.

REFERENCES

Anderson, K. and K. Shunli Yao (2003), 'China, GMOs and world trade in agricultural and textile products', *Pacific Economic Review*, **8**(2), 157–69.

Anonymous (2010), 'The economic benefits of agricultural biotechnology in Brazil: 1996–2009', Altamira, Brazil: Celeres Ambiental.

Anonymous (2011), 'World agricultural production', Washington, DC: US Department of Agriculture Foreign Agricultural Service.

BFAP (2009), 'The South African agricultural baseline, 2009', Bureau for Food and Agricultural Policy, University of Pretoria.

Brookes, G. (2002), 'The farm level impact of using Bt maize in Spain', Canterbury: Brookes West.

Brookes, G. (2008), 'The impact of using GM insect resistant maize in Europe since 1998', *International Journal of Biotechnology*, **10**(2–3), 148–66.

Brookes, G. and P. Barfoot (2008), 'Global impact of biotech crops: socio-economic and environmental effects, 1996–2006', *AgBioForum*, **11**, 21–38.

Brookes, G., T.H. Yu et al. (2010), 'The production and price impact of biotech corn, canola, and soybean crops', *AgBioForum*, **13**(1), 25–52.

Carpenter, J., A. Felsot et al. (2002), 'Comparative environmental impacts of biotechnology-derived and traditional soybean, corn, and cotton crops', Ames, IA: Council for Agricultural Science and Technology (CAST).

Carpenter, J.E. (2010), 'Peer-reviewed surveys indicate positive impact of commercialized GM crops', *Nature Biotechmology*, **28**(4), 319–21.

Demont, M. and E. Tollens (2004), 'First impact of biotechnology in the EU: Bt maize adoption in Spain', *Annals of Applied Biology*, **145**(2), 197–207.
Duffy, M. (2001), 'Study shows no economic advantage for Iowa farmers to plant GMO crops', *Leopold Letter*, **13**(4), 1.
Duffy, M. and M. Ernst (1999), 'Does planting GMO seed boost farmers' profits?', *Leopold Letter*, **11**(3).
Fernandez-Cornejo, J. and J. Li (2005). 'The impacts of adopting genetically engineered crops in the USA: the case of Bt corn', paper presented at *American Agricultural Economics Association Annual Meeting*, 24–27 July, Providence, Rhode Island.
Fernandez-Cornejo, J. and W.D. McBride (2002), 'Adoption of bioengineered crops', Agricultural Economic Report no. 810, US Department of Agriculture, Economic Research Service.
Fernandez-Cornejo, J. and S. Wechsler (2011), 'Revisiting the impact of Bt corn adoption by US farmers', paper presented at *Agricultural and Applied Economics Association and Northeastern Agricultural and Resource Economics Joint Annual Meeting*, 24–26 July, Pittsburgh PA.
Gale, H.F., F. Tuan et al. (2009), 'China is using more corn for industrial products', Washington, DC: US Department of Agriculture Economic Research Service.
Gianessi, L.P. and J.E. Carpenter (1999), 'Agricultural biotechnology: insect control benefits', Washington, DC: National Center for Food and Agricultural Policy.
Gianessi, L.P. and J.E. Carpenter (2000), 'Agricultural biotechnology: benefits of transgenic soybeans', Washington, DC: National Center for Food and Agricultural Policy.
Gomez-Barbero, M., J. Berbel et al. (2008), 'Bt corn in Spain: the performance of the EU's first GM crop', *Nat Biotech*, **26**(4), 384–6.
Gonzales, L.A. (2005), 'Harnessing the benefits of biotechnology: the case of Bt corn in the Philippines', Los Baños: STRIVE Foundation.
Gouse, M., J. Kirsten et al. (2010), 'Insect resistant and herbicide tolerant maize adoption by South African smallholder farmers – making sense of seven years of research', paper presented at *International Consortium for Agricultural Bioeconomy Research*, 15–18 June, Ravello, Italy.
Gouse, M., J.F. Kirsten and W.J. Van der Walt (2008), 'Bt cotton and Bt maize: an evaluation of the direct and indirect impact on the cotton and maize farming sectors in South Africa', commissioned report to the Department of Agriculture: Directorate BioSafety, Pretoria.
Gouse, M., C.E. Pray et al. (2005), 'A GM subsistence crop in Africa: the case of Bt white maize in South Africa,' *International Journal of Biotechnology*, **7**, 84–94.
Gouse, M., J. Piesse et al. (2006a), 'Output and labour effects of GM maize and minimum tillage in a communal area of Kwazulu-Natal', *Journal of Development Perspectives*, **2**(2), 71–86.
Gouse, M., C. Pray et al. (2006b), 'Three seasons of subsistence insect-resistant maize in South Africa: have smallholders benefited?', *AgBioForum*, **9**(1), 15–22.
Gouse, M., J. Piesse et al. (2009), 'Assessing the performance of GM maize amongst smallholders in KwaZulu-Natal, South Africa', *AgBioForum*, **12**(1), 78–89.
Griliches, Z. (1957), 'Hybrid corn: an exploration in the economics of technological change', *Econometrica*, **25**(4), 501–22.
Hategekimana, B. (2002a), 'Genetically modified grain corn and soybeans in Quebec and Ontario in 2000 and 2001', Agriculture and Rural Working Paper Series, working paper no. 54, Ottawa: Statistics Canada.
Hategekimana, B. (2002b), 'Growers of genetically modified grain corn and soybeans in Quebec and Ontario: a profile', Agriculture and Rural Working Paper Series, working paper no. 52, Ottawa: Statistics Canada.
Hategekimana, B. and M. Trant (2002), 'Adoption and diffusion of new technology in agriculture: genetically modified corn and soybeans', *Canadian Journal of Agricultural Economics*, **50**(4), 357–71.
Hutchison, W.D., E.C. Burkness et al. (2010), 'Areawide suppression of European corn borer with Bt maize reaps savings to non-Bt maize growers', *Science*, **330**(6001), 222–5.
James, C. (2003), 'Global review of commercialized transgenic crops: 2002 feature: Bt maize. Ithaca, NY: International Service for the Acquisition of Agri-biotech Applications (ISAAA).
James, C. (2010). 'Global status of commercialized biotech/GM crops: 2010', *International Service for the Acquisition of Agri-biotech Applications (ISAAA) Brief*, 42, Ithaca, NY: ISAAA.
Kaphengst, T., N.E. Benni et al. (2011), 'Assessment of the economic performance of GM crops worldwide', report to the European Commission, available at: http://ec.europa.eu/food/plant/gmo/reports_studies/docs/economic_performance_report_en.pdf.
Lagos, J.A. and J. Junyang (2011), 'People's Republic of China grain and feed annual 2011', Washington, DC: US Department of Agriculture Foreign Agricultural Service.
Marra, Michel E., Nicholas E. Pigott and Olga Sydorovych (2005), 'The impact of herbicide tolerant crops on North Carolina farmers', *North Carolina Economist*, March/April, pp. 1–4, North Carolina Extension Service.

Mutuc, M.E., R.M. Rejesus and J.M. Yorobe Jr (2011), 'Yields, insecticide productivity, and Bt corn: evidence from damage abatement models in the Philippines', *AgBioForum*, **14**(2), 35–46.
Mutuc, M.E., R.M. Rejesus, S. Pan and J.M. Yorobe Jr (2012), 'Impact assessment of Bt corn adoption in the Philippines', *Journal of Agricultural and Applied Economics*, **44**(1), 117–35.
Sanglestasawai, S., R.M. Rejesus and J.M. Yorobe Jr (2012), 'Production risk, farmer welfare, and Bt corn in the Philippines', paper presented at *Agricultural and Applied Economics Association 2012 Annual Meeting*, 12–14 August, Seattle, Washington.
Smale, M., P. Zambrano et al. (2009), 'The economic impact of transgenic crops in developing countries: a note on the methods', *International Journal of Biotechnology*, **10**, 519–51.
Trigo, E.J. and E.J. Cap (2006), 'Ten years of genetically modified crops in Argentine agriculture', Buenos Aires: ArgenBio.
Yorobe, J., E. Birol et al. (2010), 'Farmer preferences for Bt maize, seed information and credit in the Philippines', in J. Bennet and E. Birol (eds), *Choice Experiments in Developing Countries: Implementation, Challenges and Policy Implications*, Cheltenham, UK and Northampton, MA, USA Edward Elgar Publishing, pp. 225–43.
Yorobe, J.M.J. and C.B. Quicoy (2006), 'Economic impact of Bt corn in the Philippines', *The Philippine Agricultural Scientist*, **89**(3), 258–67.

APPENDIX

Table 37A.1 Results of farmer surveys on impact of insect-resistant maize

	Crop	Country	Sub-region	Additional descriptors	Crop year(s)	Survey information	Nature of comparison	Yield (%)	Economic performance (+, 0, −)	Additional notes
Hategekimana (2002a)	Corn	Canada	Ontario		2000	Statistics Canada 2000 November Farm Surveys and 2000 June Farm Surveys	Means	4.6		
Hategekimana (2002a)	Corn	Canada	Quebec		2000	Statistics Canada 2000 November Farm Surveys and 2000 June Farm Surveys	Means	12.8		
Hategekimana (2002a)	Corn	Canada	Ontario		2001	Statistics Canada 2001 June Farm Surveys	Means	7.7		
Hategekimana (2002a)	Corn	Canada	Quebec		2001	Statistics Canada 2001 June Farm Surveys	Means	−2.6		
Gonzales (2005)	Corn	Philippines	Isabela, Camarines Sur, Bukidnon and South Cotabato provinces	Wet season	2003/04	523 farmers	Means	27.0	+	
Gonzales (2005)	Corn	Philippines	Isabela, Camarines Sur, Bukidnon and South Cotabato provinces	Dry season	2003/04	523 farmers	Means	34.0	+	

Gonzales (2005)	Corn	Philippines	Isabela, Camarines Sur, Bukidnon and South Cotabato provinces	Wet season	2004/05	569 farmers	Means	20.0	+
Gonzales (2005)	Corn	Philippines	Isabela, Camarines Sur, Bukidnon and South Cotabato provinces	Dry season	2004/05	569 farmers	Means	24.0	+
Yorobe and Quicoy (2006)	Corn	Philippines	Bukidnon	First cropping	2003/04	407 farmers across all study areas	Means	NS	–
Yorobe and Quicoy (2006)	Corn	Philippines	Bukidnon	Second cropping	2003/04	407 farmers across all study areas	Means	NS	–
Yorobe and Quicoy (2006)	Corn	Philippines	Camarines Sur	First cropping	2003/04	407 farmers across all study areas	Means	37.4	+
Yorobe and Quicoy (2006)	Corn	Philippines	Isabela	Second cropping	2003/04	407 farmers across all study areas	Means	18.3	+
Yorobe and Quicoy (2006)	Corn	Philippines	Isabela, Camarines Sur, Bukidnon and South Cotabato provinces		2003/04	407 farmers across all study areas	Means	34.0	+
Yorobe and Quicoy (2006)	Corn	Philippines	Isabela, Camarines Sur, Bukidnon and South Cotabato provinces		2003/04	407 farmers	Econometric model Cobb–Douglas	22.5	
Yorobe and Quicoy (2006)	Corn	Philippines	South Cotabato	Second cropping	2003/04	407 farmers across all study areas	Means	37.5	+
Gouse et al. (2005)	Corn	South Africa	Mpumalanga	Irrigated	2000/01–2001/02	33 farmers	Means	7.0	+
Gouse et al. (2005)	Corn	South Africa	Mpumalanga	Dryland	2000/01–2001/02	33 farmers	Means	11.0	+
Gouse et al. (2005)	Corn	South Africa	North West	Dryland	2000/01–2001/02	33 farmers	Means	7.0	+

Table 37A.1 (continued)

	Crop	Country	Sub-region	Additional descriptors	Crop year(s)	Survey information	Nature of comparison	Yield (%)	Economic performance (+, 0, −)	Additional notes
Gouse et al. (2005)	Corn	South Africa	Northern Cape	Irrigated	2000/01–2001/02	33 farmers	Means	12.0	+	
Gouse et al. (2009)	Corn	South Africa	KwaZulu-Natal		2006/07	249 farmers	Means	6.0	+	
Gouse et al. (2006b)	Corn (white)	South Africa	Eastern Cape	Mqanduli	2001/02	40 farmers	Means	NS		Yield in kg/kg of seed
Gouse et al. (2006b)	Corn (white)	South Africa	Eastern Cape	Flagstaff	2001/02	60 farmers	Means	34.0		Yield in kg/kg of seed
Gouse et al. (2006b)	Corn (white)	South Africa	KwaZulu-Natal	Hlabisa	2001/02	58 farmers	Means	30.0		Yield in kg/kg of seed
Gouse et al. (2006b)	Corn (white)	South Africa	Limpopo	Venda	2001/02	60 farmers	Means	62.0		Yield in kg/kg of seed
Gouse et al. (2006b)	Corn (white)	South Africa	Mpumalanga	Northern Highveld	2001/02	60 farmers	Means	62.0		Yield in kg/kg of seed
Gouse et al. (2006b)	Corn (white)	South Africa	Mpumalanga	Southern Highveld	2001/02	90 farmers	Means	21.0		Yield in kg/kg of seed
Gouse et al. (2006b)	Corn (white)	South Africa	KwaZulu-Natal		2002/03	104 farmers	Means	16.0		Yield in kg/kg of seed
Gouse et al. (2006a)	Corn (white)	South Africa	KwaZulu-Natal		2003/04	135 farmers	Means	NS		Yield in kg/kg of seed
Gouse et al. (2006a)	Corn (white)	South Africa	KwaZulu-Natal		2003/04	135 farms	Econometric model	NS		Yield in kg/kg of seed
Gouse et al. (2010)	Corn (white)	South Africa	KwaZulu-Natal	Hlabisa	2002/03	67 farmers	Means	31.0		Yield in kg/kg of seed
Gouse et al. (2010)	Corn (white)	South Africa	KwaZulu-Natal	Hlabisa	2004/05	78 farmers	Means	32.0		Yield in kg/kg of seed
Gouse et al. (2010)	Corn (white)	South Africa	KwaZulu-Natal	Hlabisa	2005/06	121 farmers	Means	NS		Yield in kg/kg of seed
Gouse et al. (2010)	Corn (white)	South Africa	KwaZulu-Natal	Hlabisa	2006/07	87 farmers	Means	NS		Yield in kg/kg of seed
Gouse et al. (2010)	Corn (white)	South Africa	KwaZulu-Natal	Hlabisa	2007/08	102 farmers	Means	NS		Yield in kg/kg of seed

Gomez-Barbero et al. (2008)	Corn	Spain	Albacete	2002	69 observations	Means	NS	+
Gomez-Barbero et al. (2008)	Corn	Spain	Lleida	2002	21 observations	Means	NS	+
Gomez-Barbero et al. (2008)	Corn	Spain	Zaragoza	2002	88 observations	Means	12.0	+
Gomez-Barbero et al. (2008)	Corn	Spain	Albacete	2003	77 observations	Means	NS	+
Gomez-Barbero et al. (2008)	Corn	Spain	Lleida	2003	34 observations	Means	NS	+
Gomez-Barbero et al. (2008)	Corn	Spain	Zaragoza	2003	118 observations	Means	10.9	+
Gomez-Barbero et al. (2008)	Corn	Spain	Albacete	2004	88 observations	Means	NS	+
Gomez-Barbero et al. (2008)	Corn	Spain	Lleida	2004	51 observations	Means	NS	+
Gomez-Barbero et al. (2008)	Corn	Spain	Zaragoza	2004	129 observations	Means	11.6	+
Brookes (2002)	Corn	Spain	Huesca	2002	500 farmers	Unclear	10.0	+
Duffy (2001)	Corn	US	Iowa	2000	174 corn fields from USDA Cost and Return Survey	Means	2.0	–
Duffy and Ernst (1999)	Corn	US	Iowa	1998	377 corn fields from USDA Cost and Return Survey	Means	8.6	+
Fernandez-Cornejo and McBride (2002)	Corn	US	Eastern Heartland	1998	USDA ARMS	Econometric model		–
Fernandez-Cornejo and McBride (2002)	Corn	US	Northern Crescent	1998	USDA ARMS	Econometric model		0

Table 37A.1 (continued)

	Crop	Country	Sub-region	Additional descriptors	Crop year(s)	Survey information	Nature of comparison	Yield (%)	Economic performance (+, 0, −)	Additional notes
Fernandez-Cornejo and McBride (2002)	Corn	US	Other regions		1998	USDA ARMS	Econometric model		0	
Fernandez-Cornejo and McBride (2002)	Corn	US	Prairie Gateway		1998	USDA ARMS	Econometric model		0	
Fernandez-Cornejo and McBride (2002)	Corn	US	Western Heartland		1998	USDA ARMS	Econometric model		−	
Fernandez-Cornejo and McBride (2002)	Corn	US			1998	USDA ARMS	Econometric model		−	
Fernandez-Cornejo and Li (2005)	Corn	US	Nationwide		2001	USDA ARMS	Means	9.0		
Fernandez-Cornejo and Li (2005)	Corn	US	Nationwide		2001	USDA ARMS	Econometric model	3.9		
Fernandez-Cornejo and Wechsler (2011)	Corn	US	Nationwide		2005	USDA ARMS	Means	12.3		+
Fernandez-Cornejo and Wechsler (2011)	Corn	US	Nationwide		2005	USDA ARMS	Econometric model	1.2		+

Table 37A.2 Results of farmer surveys on herbicide-tolerant maize

	Country	Sub-region	Crop year(s)	Survey information	Nature of comparison	Yield (%)	Economic performance (+, 0, −)
Gouse et al. (2009)	South Africa	KwaZulu-Natal	2006/07	249 farmers	Means	85.0	+
Gouse et al. (2010)	South Africa	KwaZulu-Natal	2005/06	121 farmers	Means	NS	
Gouse et al. (2010)	South Africa	KwaZulu-Natal	2006/07	87 farmers	Means	184	
Gouse et al. (2010)	South Africa	KwaZulu-Natal	2007/08	102 farmers	Means	NS	
Fernandez-Cornejo and McBride (2002)	US	Eastern Heartland	1998	USDA ARMS	Econometric model		+
Fernandez-Cornejo and McBride (2002)	US	Northern Crescent	1998	USDA ARMS	Econometric model		+
Fernandez-Cornejo and McBride (2002)	US	Prairie Gateway	1998	USDA ARMS	Econometric model		0
Fernandez-Cornejo and McBride (2002)	US	Western Heartland	1998	USDA ARMS	Econometric model		+
Fernandez-Cornejo and McBride (2002)	US		1998	USDA ARMS	Econometric model		
Marra et al. (2005)	US	North Carolina	2001	293 farmers	Within-farm comparison	NS	

38 Cotton

Jeffrey Vitale, Gaspard Vognan and Marc Ouattarra

INTRODUCTION

Insects have traditionally been a major threat to sustaining cotton production over the long term on nearly all continents (Benedict and Altman, 2001). Continuous cotton production accelerates the build-up of insect populations, which can cause substantial production losses on cotton fields (Clay, 2004). Lepidoptera (for example American bollworm) is one of the most pernicious pests found in all of the major cotton-producing areas, including the Americas, Africa and Asia (Benedict and Altman, 2001). Chemical pesticides were developed in the twentieth century to control insect populations but did not provide a sustainable approach to pest control. Over time, pest populations developed resistance to many of the chemical agents, and the toxicity of pesticides created human health and environmental concerns (Goodell et al., 2001).

Research and development of genetically modified (GM) cotton began in the 1980s (Perlak et al., 1990). Transgenic engineering techniques were developed to insert genes that encode and promote the production of Cry proteins (for example Cry1Ac and Cry2Ab), originating from the common soil bacterium *Bacillus thuringiensis* (Bt) (Perlak et al., 1990). The Cry proteins are toxic to certain caterpillar pests common to cotton and are highly effective in killing certain lepidopteran larvae, often called caterpillars (Greenplate et al., 2003). Once ingested, the Cry proteins bind to specific molecular receptors on the lining of the caterpillar's gut, create holes in the gut, and quickly cause death (Hofte and Whiteley, 1989). An attractive feature of GM cotton is that unlike conventional pesticides, many of which kill across a wide spectrum of both targeted and non-targeted (often beneficial) insects, Bt Cry proteins are highly specific to certain caterpillars and do not target other insects (Hofte and Whiteley, 1989; MacIntosh et al. 1990; Sims, 1997).

Although GM cotton has been genetically modified, formulations of microbial Bt fermentation products (containing Cry proteins) have been used for more than 60 years as natural insecticides in spraying programs in agricultural and forestry pest control and recently in organic cotton as 'natural' insecticides (Aronson et al., 1986). While these Bt formulations produced using fermentation methods can be quite effective under certain conditions, the products have never been widely adopted in crops such as cotton because they have short half-lives in the field due to the degradation of Cry proteins by UV light (Zhou, 2011). Many insect larvae may escape control by these Bt products if spray coverage is not optimal, and they are relatively expensive due to production costs incurred by fermentation.

Genetically modified cotton has been one of the more commercially successful GM crops to be introduced. Since its 1996 debut on American cotton farms, the global adoption of GM cotton has spread to 13 countries totaling 27 million ha in 2011 (James, 2012). The leading GM cotton-producing countries, including Argentina, Australia,

China, India and the US, all use the technology. The current use of GM cotton spans a wide range of conditions, from smallholder producers in India and China, to large-scale producers in the US and Australia.

Varieties of GM cotton (Bt) were first available in the US in 1996 and by 2009 use had risen to 65 per cent of total cotton acres (Brookes and Barfoot, 2011). Initial farm-level benefits in the first few seasons (1996–2000) of GM cotton adoption ranged from US$65–115 per hectare, and over the past few seasons (2005–09) have ranged between US$87–128 (Brookes and Barfoot, 2011). Cumulative benefits at the national level reached US$296.79 million in 2009, corresponding to a 6.0 per cent increase in national cotton farm income.

In Australia, insect-resistant cotton (Bt cotton) was approved for production by the Office of the Gene Technology Regulator in 2000. Within three years, adoption rates exceeded 90 per cent and have remained at that level (Cotton Australia, 2011). While production of GM cotton in Australia is relatively modest, with approximately 200 000 hectares on average, the estimated benefits in 2010 were US$18.6 million, with a cumulative benefit at the national level from the initial year of commercial introduction in 2000 through to 2010 of US$31 million (Brookes and Barfoot, 2012).

Cotton that was genetically modified for insect resistance (IR) was approved for production in China in 1997. While there is not a wealth of information available on adoption patterns and the economic impacts, there is enough to provide a general sense of the impacts. However, there is a lag in the data, with the most recent information available from the midpoint of the last decade. In 2005, 66 per cent of China's cotton production was GM cotton, accounting for 3.3 million hectares of production (Gómez-Barbero and Rodríguez-Cerezo, 2006). Most of the benefits are due to the decrease in pesticide applications by smallholder producers. One of the early studies on the adoption of IR cotton in China estimates that in 2001, producers using IR cotton experienced a net revenue gain of US$277 per hectare, compared to a loss of US$225 for non-IR cotton varieties (Huang et al., 2003). A subsequent study using 2005 production data estimated the aggregate economic benefit of GM cotton production in China exceeded US$600 million (Frisvold and Reeves, 2007).

India approved the adoption of GM IR cotton in 2002. These GM IR varieties are grown in nine different Indian states accounting for over 8 million hectares in 2009, with adoption levels ranging from a low of 55 per cent in Gujarat to a high of 100 per cent in Andhra Pradesh and Maharashtra (Arora and Bansal, 2011). Using production data from four different cotton-producing states from four different seasons between 2002–08, Arora and Bansal (2011) estimated that cotton yields increased by 24 per cent with the use of GM IR varieties and per hectare profit increased by 50 per cent. In aggregate, these increases from growing GM cotton were estimated to be in excess of US$200 million in 2005 (Frisvold and Reeves, 2007).

With GM cotton having reached a mature state in many developed countries, accounting for a large proportion of the world's cotton production, over the coming decades most of the potential gains from GM cotton will need to be obtained from less developed regions. Africa accounted for less than 1 per cent of the world's area of GM cotton in 2010 even though it produces 20 per cent of the world's cotton (James, 2010). Africa has been at the center of the biotechnology debate, where opposition by various public interest groups has largely succeeded in delaying the introduction of agricultural biotechnology products (Cohen and Paarlberg, 2002; Paarlberg, 2008). There is reason, however, to be cautiously optimistic about the future use of biotechnology in African agriculture. In 2006, close to 90 per cent of the 10.3 million farmers growing GM crops were small, resource-poor farmers from less developed countries, primarily in Asia (James, 2006). Such progress illustrates that small-scale producers can benefit from GM crops similar to their large-scale counterparts in more developed agricultural contexts.

Burkina Faso established one of the first agricultural biotechnology programs in Africa and has emerged as a leader in the development and use of GM cotton. In 2009, slightly more than 125000 ha of GM cotton were planted by Burkina Faso producers, marking it as the largest ever introduction of GM cotton on the African continent (Vitale et al., 2011). Diffusion of GM cotton has increased rapidly since then, reaching a reported 314000 ha in 2012. In neighboring West African countries, which share similar pest populations, agroecological zones, farming systems and cotton industry structures, the success of GM cotton is expected, with reasonable probability, to be replicated. Hence, the introduction of GM cotton in Burkina Faso could become a pivotal event that may have a substantial impact on the future use of biotechnology in African crop production.

The purpose of this chapter is to present empirical evidence of GM cotton benefits to producers in Burkina Faso following the initial three years of commercial introduction (2009–11) of Bollgard II® (BGII). Results are provided by seasonally conducted household surveys, which document the impacts of BGII vis-à-vis conventional cotton on yields, production costs, cotton income and social factors. The findings presented in this chapter contribute to the biotechnology literature by providing the impact of GM cotton on smallholder producers in Africa, a region where evidence has been limited to only small areas in South Africa. Such empirical evidence will enable greater transparency in the biotechnology debate in Africa by providing empirical evidence of how GM cotton has impacted Burkina Faso smallholder producers throughout the first three years of commercial introduction.

THE COMMERCIALIZATION PROCESS: 2001–08

In Africa, the build-up of pest pressure from long-term cotton production remains one of the most significant and economically important problems facing the West African cotton sector (Baquedano et al., 2010). In West Africa, conventional approaches to control pests through chemical sprays have grown increasingly ineffective as pest populations have developed resistance (Vitale et al., 2011). In a typical year, for example, the Burkina Faso cotton sector may spend over US$30 million for chemically-based pest control products, yet recent studies in Burkina Faso have found significant pest damage on cotton fields that were treated using a standard regimen of six seasonal insecticide sprays, with yield losses ranging between 25 per cent to 85 per cent (Vitale et al., 2011).

Frustrated by conventional pest control methods, key Burkina Faso government and industry stakeholders began working with biotechnology providers in 2001 to introduce GM cotton in locally adapted cotton varieties on a commercial basis (Vitale et al., 2011). Bollgard II had already by that time demonstrated its capacity to have an agronomically-effective and economically-efficient control over *Helicoverpa amerigera*, one of the major cotton pests, as well as some of the other less prominent pests such as *Diparopsis watersi*, *Earias biplaga* and *Syllepte derogate*. Burkina Faso's national agricultural research center, Institut de l'Environnement et de Recherches Agricoles (INERA), began field-testing Bollgard II in 2003. The initial tests were conducted under confined conditions and designed to evaluate efficacy and environmental effects, including pollen-mediated gene flow, and effects on non-target arthropods, including bees. Tests were conducted on two INERA research stations in opposite ends of the country (Farako-Bâ located close to Bobo-Dioulasso in the west and Kouaré located close to Fada N'Gourma in the east). This research evaluated the effectiveness of Bollgard II within field conditions (climate agroecology, insect pressure) to Burkina Faso (Vitale et al., 2008; Hema et al., 2009). In 2006, the National Biosafety Agency approved an additional confined field trial outside of the INERA research farm environment; this Bollgard II (BGII) field trial was located on a seed treatment farm in Boni located about 120 km from Bobo-Dioulasso. The Boni trial represented the first test of Bollgard II technology in the two local germplasm varieties, STAM 59 and STAM 103.

Results of Burkina Faso's first four years of research on Bt cotton were encouraging. On average, BGII cotton increased yields by 35 per cent, reaching as high as 48 per cent in one of the years. The study also found that the number of pesticide treatments required each year could be reduced from six to two when using BGII, eliminating the need for the initial four sprays targeting caterpillar pests. The retention of two late-season sprays to control secondary pests (aphids and jassids) not targeted by the Bt proteins expressed in BGII was recommended by INERA scientists. A reduction in both pesticide use and number of treatments made it possible to save $27.83 per hectare, a 62 per cent cost reduction, based on the trial results.

In July 2007, Bt cotton achieved another important milestone when the National Biosafety Agency gave its approval to conduct several additional large-scale trials much closer to the real operating conditions of the diverse group of cotton growers. That marked the first year of large-scale testing of BGII in the local germplasm varieties. In collaboration with the cotton companies and the cotton growers' union, INERA conducted field trials of these two local varietal versions of BGII on 20 testing sites within the cotton-growing zones under the control of the three major cotton companies SOFITEX, SOCOMA and Faso Coton. All trials were carried out applying appropriate established biosafety protocols. The 2007 test results were also encouraging, with average yield increases of 20 per cent.

In June of 2008 the National Biosafety Agency authorized the commercial planting of BGII in Burkina Faso. This significant accomplishment for Burkina Faso made it the third commercial release of a bioengineered crop in Africa. In the 2008 cotton-growing season, SOFITEX and its contract seed producers planted about 12 000 hectares of the above-mentioned two local varieties containing BGII; this was effectively the seed multiplication year for the anticipated broader commercial deployment in 2009.

Over the past decade, Burkina Faso has emerged as the most progressive country in the West Africa region regarding biotechnology. In 2008, after five years of field testing, monitoring and developing biosafety legislation and protocols, Burkina Faso became only the second African country, following South Africa, to release GM cotton commercially. Adoption of GM cotton has proceeded quickly in Burkina Faso from an initial commercial release of 12 000 ha in 2008 (largely devoted to bulk up commercial BGII seed for the 2009 season) to 125000 ha in 2009 and reaching an estimated 314000 ha in 2012 (approximately 50 per cent of the total cotton acreage).

Conducting the field trials required the development of legal and technical frameworks, including biosafety legislation to formalize regulatory oversight for the research and commercialization of agricultural biotech products. The government Ministries of Agriculture, Environment, and Research and Higher Education became heavily involved. Burkina Faso's national agricultural research center, INERA, was appointed to conduct the primary research needed to test Bt cotton. The Professional Association of Cotton Companies of Burkina (APROCOB) and the national cotton producer cooperative association (the growers' union known as UNPCB), played key roles in the commercialization process, providing technical and managerial assistance as needed.

The Ministry of Environment (MOE) was tasked as the primary legal authority in the commercialization process, and was given the charge of developing a regulatory infrastructure, consistent with the new biosafety laws, to govern testing, development and subsequent environmental release. The National Biosafety Agency within the MOE was established by 2006, and became the competent authority establishing standards for submitted regulatory dossiers and granting approval for field testing and eventual commercialization of Bt cotton. Technology providers contributed by sharing past experiences commercializing Bt cotton in other regions. Monsanto's role included assistance in transferring the Bt genes to two regional commercial cotton varieties, STAM 59 and STAM 103, which are grown as conventional cotton in Burkina Faso.

In addition to the trials, demonstrating the safety and potential value of this technology, the commercial introduction of BGII in Burkina Faso also required the development of a business model linking public and private sectors (through a public–private partnership). The business model required an innovative approach to enable Bt cotton seed to be distributed in Burkina Faso's marketing channels that remained supervised by APROCOB.

IMPACT OF GM COTTON IN BURKINA FASO

Socio-economic analyses were performed for three cotton production seasons, 2009 to 2011, based on household surveys that assessed the impact of BGII on various social, economic and health impact indicators. INERA conducted annual surveys that generated 160, 176 and 544 usable rural household survey responses spread over 10 villages during the summer and autumn of 2009, 2010, and 2011 respectively. The surveys were conducted with representative samples from each of the three main cotton-growing zones, which are controlled or administered by designated cotton companies in Burkina Faso: SOFITEX in the west, SOCOMA in the center, and Faso Coton in the east (Figure 38.1). The survey villages were randomly selected by INERA and represented typical conditions for each of the cotton zones. The annual sampling was intended to include a representative mixture of producers across farm type, with 46.2 per cent large farms (two or more animal draft pairs), 50.6 per cent small farms (one animal draft pair), and 3.1 per cent manual farms. This approached the national typology in which large producers comprise approximately 52 per cent of the farms, small farms 46 per cent, and manual farms the remaining 2 per cent (Vitale et al., 2011).

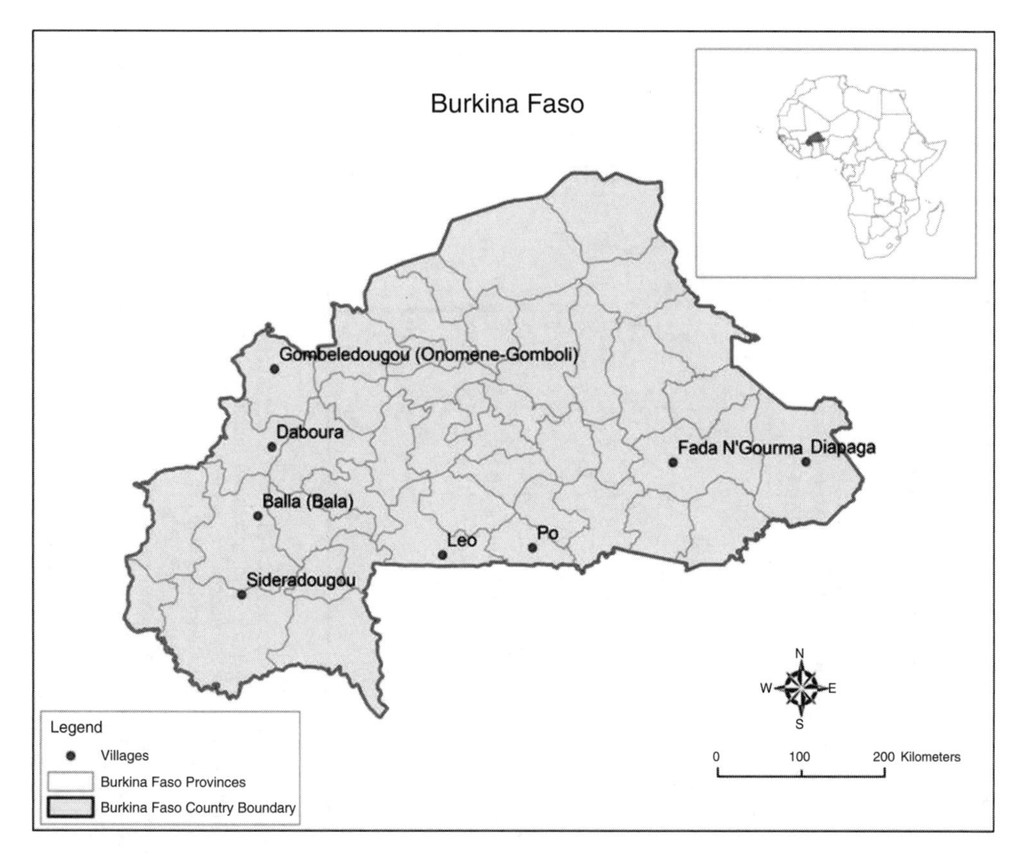

Figure 38.1 Map of study area illustrating the location of surveyed villages

The survey instrument was developed by INERA researchers at the Programme Coton research center in Bobo Dioulasso, Burkina Faso, and administered by local extension workers. Summary statistics for the surveyed households are listed in Table 38.1. The area planted in GM cotton varied between 1.3 ha for the smallest manually-worked farms (hand labor only, no animal assistance) in the Faso Coton zone to 5.1 ha in SOFITEX on large farms. Across all three zones, the average household contained 14.1 persons, with 8.6 of them actively engaged in the family's farming operations which included, but were not limited to, cotton production. The most experienced cotton producers in the survey were from the SOFITEX production zone, with an average farming tenure of 28.0 years (Table 38.1). SOFITEX is the traditional cotton-producing zone with production since the French colonial era, whereas the SOCOMA and Faso Coton zones have been more recently introduced to cotton production beginning in the 1980s (Bassett, 2001). Their longer experience with cotton, the sole cash crop in the region, likely explains why household incomes – including farm and non-farm income – were found to be significantly higher in the SOFITEX zone, with an average household income of US$780 per year (Table 38.1). In the Faso Coton zone, household incomes were found to be US$520 per year, and in the SOCOMA zone, household incomes averaged US$455 per year.

Table 38.1 Summary statistics of the GM cotton producer surveys (2009–11) comparing household demographics, income and farm size across farm types and production zones

Item	SOFITEX[a]				SOCOMA				Faso Coton				All Zones
	Large[b]	Small	Man.	Ave/ Tot	Large	Small	Man.	Ave/ Total	Large	Small	Man.	Ave/ Total	Ave/ Total
GM cotton Households Surveyed[c]													
2009	48	29	3	80	15	25	–[d]	40	11	27	2	40	160
2010	88	56	7	151	14	24	–	38	3	22	4	29	218
2011	66	31	12	109	23	42	8	73	14	61	0	75	257
Total	202	116	22	340	52	91	8	151	28	110	6	144	635
Household size (persons)	16.7[e]	11.0	11	13.9	24.1	10.0	–	18.7	11.5	9.5	11	10.1	14.1
Household farm labor (persons)	10.1	6.4	3.8	8.3	21.6	6.1	–	14.3	5.5	4.4	3.5	4.6	8.6
Area in GM cotton (ha)	5.1	2.9	2.0	4.1	4.1	2.1	1.5	2.8	2.5	1.7	1.3	1.8	3.3
Area in conventional cotton (ha)	4.1	2.6	1.9	3.3	2.8	1.5	1.7	2.0	1.5	0.7	–	0.7	2.6
Distance to GM cotton field (km)	3.7	3.7	8	3.6	2.9	2.8	–	2.8	5.8	5.0	8	5.4	3.8
Experience growing cotton (years)	31.9	25.5	13	28.0	9.1	11.1	–	9.8	8.5	10.8	13.0	10.2	20.4
Household income ($ per year)[f]	924	513	–	780	575	280	–	455	691	471	–	520	655

Notes:

[a] Cotton production zone refers to the areas of operation of the three national cotton companies: SOFITEX, SOCOMA and Faso Coton.

[b] Farm types are defined as follows: Large are farms with two or more animals for assistance in field operations, Small are farms with one animal for assistance in field operations, and Man. are farms where everything is done manually without any assistance of animals.

[c] Sample size reports the total number of GM cotton producers surveyed. Observations with missing data resulted in different sample sizes for each of the variables.

[d] Missing data is represented by '-' and required adjusting the weights used in calculating averages (see note e).

[e] Weighted averages are used when calculating averages across farm type and zone. Weights were determined based on the number of usable observations, i.e. adjusted for missing data. The variables Household size, Experience growing cotton, Distance to GM cotton field, and Household income were collected only in 2009, the first year of the survey.

[f] Household income included farm income from crop and live stock sales, non-farm income and remittances.

Yield Impact

Data collected from grower surveys across all Burkina Faso cotton-growing regions in 2009 (190 surveys), 2010 (170 surveys), and 2011 (540 surveys) were used to construct an analysis of variance (ANOVA) model. The ANOVA modeling activities provide two useful functions. One is to explain cotton yields using the variables collected in the surveys for the purpose of identifying variables possessing significant explanatory power. The other is to conduct a means testing to determine whether/how cotton yields varied within/across an array of explanatory factors. The ANOVA yield model tested several factors including cotton type, zone, year, farm type and pest management practices (number of pesticide sprays applied). The ANOVA yield model was estimated using the PROC GLM statement in the SAS statistical software package.[1]

RESULTS

Genetically modified cotton (BGII) generated significant economic and health benefits for Burkina Faso cotton producers in its initial three years of commercial introduction, 2009–11. Among the surveyed cotton producers and across all three years, BGII cotton yields were on average 22.0 per cent higher than conventional cotton (Figure 38.2). Although production year had a significant effect on cotton yields in the ANOVA model,

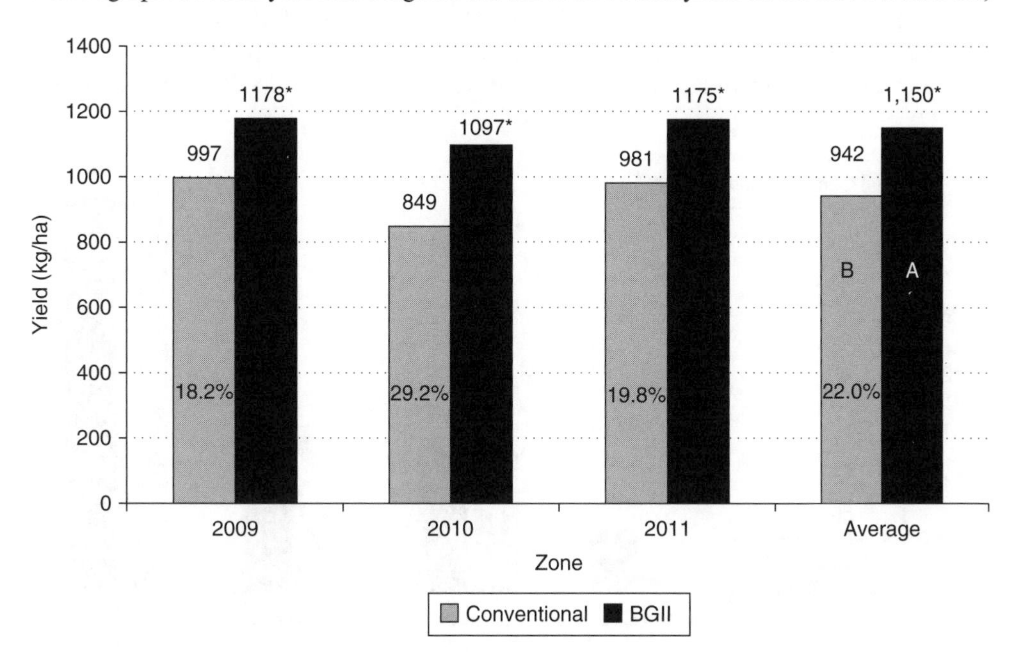

Note: Genetically modified cotton yields were significantly higher than conventional yields in each year and on average across all three years as indicated by the asterisk (*).

Figure 38.2 Comparison of GM versus conventional cotton yields over the first three years of large-scale commercial introduction, 2009–11

the enhanced yield performance of BGII was consistent across the three years. Yield increases obtained from BGII ranged from a low of 18.2 per cent in 2009 to a high of 29.2 per cent in 2010, and within all three years BGII yields were significantly higher (95 per cent confidence level, $p<0.05$) than conventional cotton (Figure 38.2). Across all three years, BGII yields were always significantly higher than conventional cotton, even when comparing between years when conventional yields were above average and BGII yields were below average. Conventional cotton yields were highest in 2009, when yields were above average, reaching 997 kg per ha, yet were still significantly lower than BGII yields in all three years, including when BGII yields were at their lowest in 2010, when yields averaged 1097 kg per ha (Figure 38.2).

Producers were able to retain nearly all of the increased revenue earned by BGII cotton since production costs were nearly identical between BGII and conventional cotton according to survey results (Figure 38.3). Adopting BGII shifted production costs towards higher seed costs, but such increases were largely offset by significantly lower insecticide and labor costs (Figure 38.3). Seed costs averaged US$59.44 per ha for BGII producers across the three years, which was US$46.92 per ha more than the US$12.52 per ha average seed cost paid by conventional cotton producers over the same three-year period (Figure 38.3). With significantly fewer insecticide sprays, BGII producers spent an average of US$11.54 per ha on insecticides across the three years, corresponding to a saving of US$38.25 per ha compared to conventional producers, who spent US$49.79 per ha on insecticides during the same three-year period (Figure 38.3). When insecticide and seed costs are combined with the labor, fertilizer and herbicide costs, BGII producers spent $6.12 per ha more on production inputs than conventional producers, a 1.72 per cent increase in production costs that was not significant (Figure 38.3). Hence, although

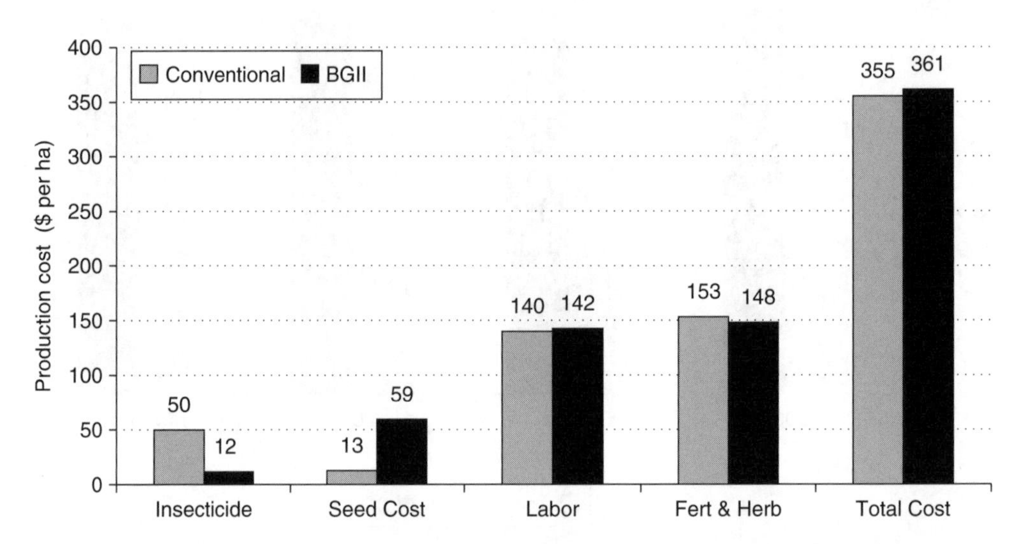

Note: Production costs were not significantly different between GM and conventional cotton.

Figure 38.3 Comparison of production costs for BGII versus conventional cotton averaged across the first three years of large-scale commercial introduction, 2009–11

GM cotton seed costs are often cited as a major constraint to adoption, there is no evidence of unfair seed pricing in Burkina Faso based on the three years of producer surveys.

BGII increased cotton income by an average of \$80.44 per ha relative to conventional cotton over the past three years, corresponding to a 114 per cent increase in cotton income (Figure 38.4). Based on the surveys, an average household producing 3.3 ha of BGII cotton would increase their annual household income, including both farm and non-farm income, by an average of 40.8 per cent from US\$650 per year (US\$1.78 per day) to US\$916 per year (US\$2.51 per day). The impact of BGII on cotton income was stable across time, generating significantly higher income in each of the three years compared to conventional cotton (Figure 38.4). The greatest impact of BGII was in 2011, when cotton income averaged US\$95.35 per ha more than conventional cotton (Figure 38.4). Although the difference was greatest in 2011, this was also the year when both BGII and conventional cotton producers obtained their highest income, benefiting from the substantially higher prices paid by the cotton companies. Between 2010 and 2011, cotton prices increased by 24 per cent from a price of US\$0.50 per lb of cotton lint in 2010 to US\$0.62 per lb of cotton lint in 2011, following a similar increase in cotton prices between 2009 and 2010, when cotton prices increased by 25 per cent from US\$0.40 to US\$0.50 per lb of cotton lint. The substantial price increase explains why conventional cotton producers earned significantly higher cotton income in 2011 than BGII producers in the other two years, 2009 and 2010, even though BGII had significantly higher cotton

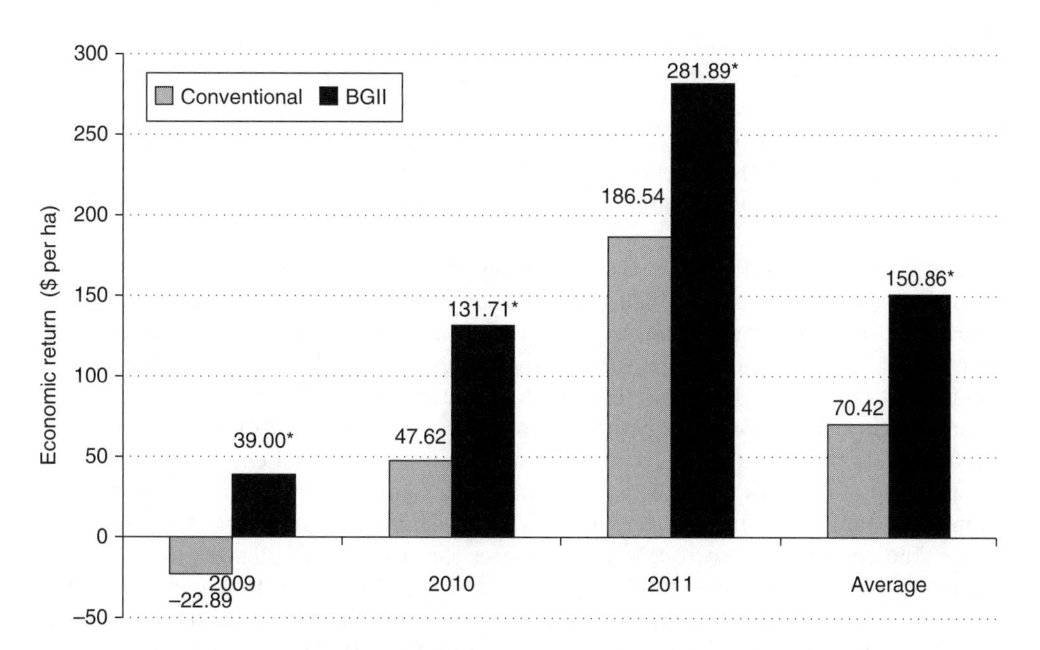

Note: Economic returns from GM cotton were significantly higher than conventional yields in each year and on average across all three years as indicated by the asterisk (*).

Figure 38.4 Comparison of BGII versus conventional cotton economic returns over the first three years of large-scale commercial introduction, 2009–11

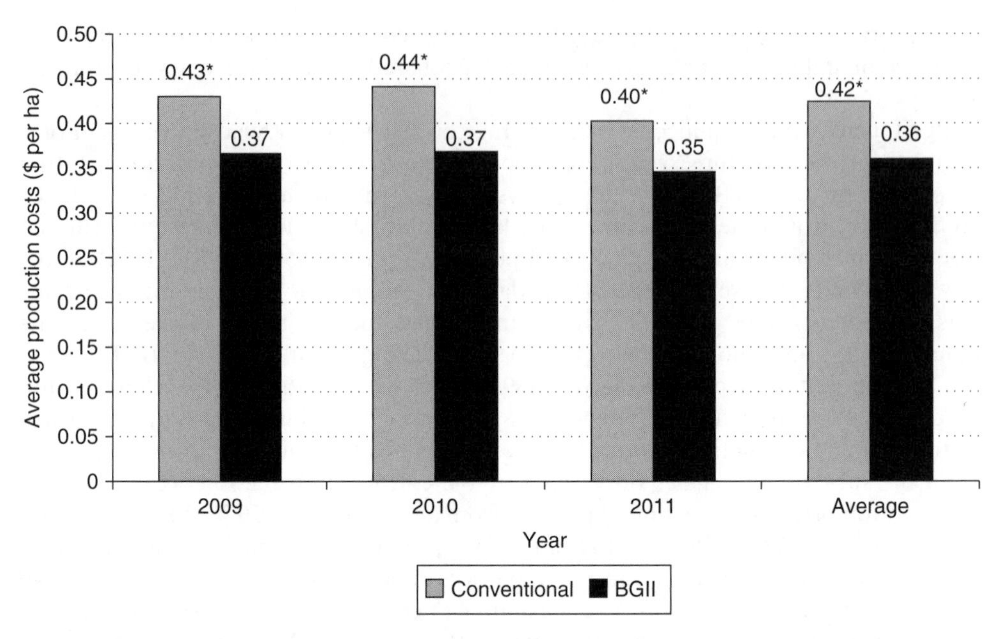

Note: Average production costs were significantly higher for conventional cotton when compared side-by-side with GM cotton in each year and on average across all three years as indicated by the asterisk (*).

Figure 38.5 Comparison of BGII versus conventional cotton average production costs over the first three years of large-scale commercial introduction, 2009–11

yields in 2009 and 2010 than conventional cotton (Figure 38.2). The low price paid to cotton producers in 2009 is particularly evident for conventional cotton producers, whose returns were negative even though yields were the highest in that year (Figures 38.2 and 38.4).

The economic benefits from the producing BGII can be viewed in other ways as well. The average cost of producing a pound of cotton lint was significantly lower for BGII than conventional cotton, averaging 17.9 per cent lower across the three years (Figure 38.5).[2] According to the survey, when assessed over the three years the average production cost of BGII was US$0.36 per lb of cotton lint, which was US$0.06 per lb less than conventional cotton's production cost of US$0.42 per lb of cotton lint (Figure 38.5). Hence, BGII producers earned US$0.06 more than conventional cotton producers for each pound of cotton sold. The year did not have a significant effect on average production costs (Table 38.4). In all three years, BGII had significantly lower average production costs than conventional cotton, with differences that varied only slightly, from a difference of US$0.06 in 2009 and 2011 to a difference of US$0.07 in 2010 (Figure 38.5). The close proximity of the average production costs in 2011 (that is the balancing of lower insecticide costs with increased BGII seed prices) confirms that the primary source of the increase in cotton profit from growing BGII in 2011 was from a combination of the yield increase and the higher cotton price which placed a greater value on output compared to 2009 and 2010.

BGII producers earned significantly higher returns to labor than producers on con-

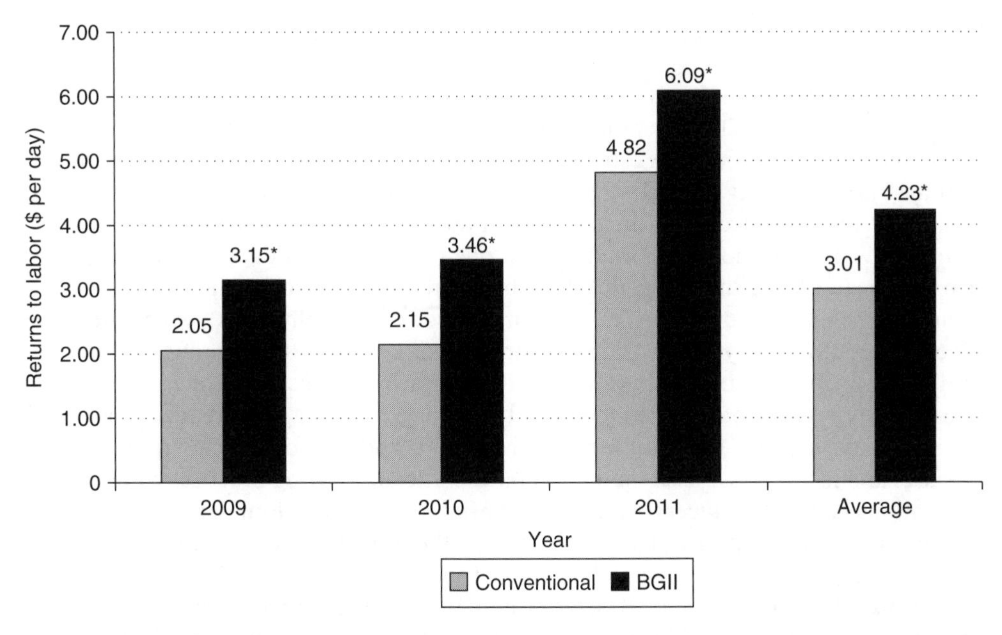

Note: Returns to labor from GM cotton were significantly higher than conventional yields in each year and on average across all three years as indicated by the asterisk (*).

Figure 38.6 Comparison of returns to labor for BGII and conventional cotton across the first three years of large scale commercial introduction, 2009–2011

ventional cotton fields over the past three years (Figure 38.6). In Burkina Faso, where cotton production is labor intensive, households allocate approximately 76 days of labor to each hectare of cotton produced. Households planting BGII were found to have a three-year average return to labor of US$4.23 per day, an increase of 29.0 per cent compared to conventional cotton producers, whose labor returned an average of US$3.01 per day (Figure 38.6). Returns to labor were consistent across the three years, with BGII generating significantly higher returns (95 per cent confidence level, $p<0.05$) to labor than conventional cotton in each year (Figure 38.6). The difference in returns to labor varied modestly across years, from a low of US$1.10 per day in 2009 to US$1.32 in 2010 (Figure 38.6). The largest returns to labor were obtained in 2011 for both BGII and conventional cotton, which were nearly twice as high as they were in the previous two years due to the substantial increase in the cotton price paid to producers and the corresponding increase in economic returns (Figures 38.4 and 38.6). In all three years, returns to labor for BGII were higher than the prevailing institutional wage rate of US$2.33 per day (Figure 38.6). This is a noteworthy finding since in two out of the three years, 2009 and 2010, conventional cotton failed to provide a wage rate commensurate with alternative labor alternatives. In 2009, conventional cotton labor earned a wage of US$2.05 per day, US$0.28 below the institutional wage rate, and in 2010 conventional cotton's return to labor was US$0.18 below institutional wages (Figure 38.6). Conventional cotton's profitability, bolstered by higher cotton prices paid to producers, increased significantly in 2011, with returns to labor US$2.49 higher than institutional wages (Figure 38.6).

Health Impacts and Implications

The introduction of BGII cotton provides safer working conditions by reducing pesticide applications and limiting producers' exposure to potentially toxic agents that are often contained in pesticides. Pesticide poisonings occur frequently on cotton farms in Burkina Faso according to the INERA field surveys conducted over the past three years. Household responses from the three years of producer surveys revealed 50.8 per cent of cotton-growing households reported at least one poisoning incident over the past seven years, 2004–10.[3] More than 81 per cent of the self-reported poisoning incidents occurred during the application of pesticides targeting Lepidoptera (bollworms), which are more effectively controlled by GM cotton and are not sprayed by BGII producers. Planting BGII cotton, for instance, reduced pesticide sprays by 66 per cent (or more in cases when two sprays are not required to control secondary pests), greatly reducing potential pesticide exposure and the resulting negative health and economic costs from medical expenses and lost wages. Based on the standard pesticide spraying practices employed by conventional cotton producers, which targets primary pests with four early season sprays and secondary pests with two late season sprays, the probability of a poisoning incident occurring during a single spray application over the seven years of the household surveys was estimated at 1.04 per cent. Based on the 2011 producer surveys, which found a 75 per cent reduction in pesticide sprays used by BGII, the number of pesticide poisoning incidents, aggregated to the national level, would be reduced by a projected 30 380 cases.

Households reported a range of symptoms and illnesses resulting from the poisoning incidents, including headaches, vomiting, dizziness and flu-like symptoms including coughs and difficulty breathing. The cost of each pesticide incident was assessed based on medical expenses (medicine and doctor bills) and lost wages. This is a conservative approach, since any pain and suffering associated with an illness was not included in the survey. The average cost of an individual poisoning incident, as reported by producers in the INERA surveys, was US$39.22, including US$16.83 in labor costs and US$22.39 in medical expenses. Aggregating the findings from the household surveys to the national level, the use of BGII cotton could generate a positive economic impact of US$1.09 million per year in recouped wages and medical expenses from reduced pesticide poisoning incidents. These findings on health benefits are considered important since pesticide poisonings go largely undocumented, leaving only speculation regarding their frequency and extent in the region.

Concern over health issues associated with pesticide spraying was also found to be a major reason why cotton producers chose to adopt BGII cotton rather than to continue growing conventional cotton. In the most recent survey year, 2011, about one out of every six producers (16.0 per cent) responded that the potential to minimize health risks, by reducing the number of pesticide sprayings, was the single most important reason for adopting GM cotton. A majority (63.5 per cent) of the producers cited a combination of pesticide reduction, including lower cost, along with higher yields, as the most important reasons for adopting BGII. Although health concerns are confounded with higher yields and lower pesticide costs in this response, it is reasonable to attribute the incremental health benefits from reduced pesticide applications as a motivating factor among this group of respondents. Moreover, the surveys indicate that pesticides are a danger not only to producers, but also to the surrounding community. According to surveys,

Table 38.2 Producer characteristics and pesticide application practices among surveyed producers from the 2009 surveys

Item	Sample population		Poisoning cases[a]			
			No poisoning		Yes	
		N	%	N	%	N
Education (Years of schooling)	7.4	191	7.3	59	7.7	132
Attended pesticide safety training workshop						
No training	41%	79	11%	21	30%	58
Training	59	112	20%	38	39%	74
Washed clothes following pesticide application						
No	36%	69	12%	23	24%	46
Yes	64%	122	19%	36	45%	86
Wear protective clothing during application						
No	72%	138	22%	42	50%	96
Yes	28%	53	9%	17	19%	36
Washed body (bathed/showered) following application						
No	19%	37	9%	18	10%	19
Yes	81%	154	21%	41	59%	113
Farm size (ha)	3.89	191	3.62	59	4.01	132

Note: [a] The 'Yes' category refers to households who self-reported the occurrence of at least one pesticide incident during the period 1994–2008.

improper storage and handling led to several poisonings from drinking and eating from emptied pesticide containers, including reported cases of death and suicide. By adopting BGII, Burkina Faso villages could be safer and assume fewer health and environmental risks by eliminating the presence of an estimated 0.96 million pesticide containers.

The INERA surveys collected data on the safety and handling practices of producers in 2009, the first year of the surveys (Table 38.2). The surveys collected information on education level, pesticide training and safety precautions used by individuals responsible for applying pesticides. Wearing protective clothing during the application of pesticides was found to have the greatest effect on reducing pesticide poisonings (Table 38.2). Producers who wore protective clothing had significantly lower pesticide poisoning incidents. Nineteen per cent (of the total sample) of producers who wore protective clothing self-reported a pesticide poisoning incident; in comparison, producers who did not wear protective clothing were more than twice as likely to have incurred a poisoning incident, with a 50 per cent occurrence in pesticide poisonings (Table 38.2). The importance of wearing protective clothing and its beneficial effect on reducing poisonings has been found in other studies and should continue to be emphasized in training programs.

Extension services from the cotton companies provide workshops on safety precautions and best management practices when handling, storing and disposing of pesticides. Attending extension training did not reduce the number of cases of poisoning among the surveyed producers (Table 38.2). Of the producers who self-reported a poisoning incidence, a greater proportion, 39 per cent (of total sample), had attended a safety training workshop compared to 30 per cent (of total sample) producers who did not attend (Table 38.2). This unexpected finding that attending a training workshop implies a *greater* likelihood of poisoning could be due to poor quality and delivery of the safety training programs. Another possible reason might be the high rate of illiteracy among producers in the area, which limits the assimilation and long-term retention and application of the materials taught at the workshop, and perhaps even causes a lack of adequate awareness of the potential health problems associated with pesticides compared to producers who attended training workshops. Perhaps the most reasonable explanation is that while training had no positive effect on reducing pesticide poisonings, the negative effect implied by the surveys is spurious (coincidental).

Bathing and washing clothes following pesticide spray operations are some of the techniques that are taught in extension training programs and are part of the recommended safety practices provided to producers. However, contrary to expectations, neither of those variables had a positive influence on reducing the likelihood of pesticide poisoning incidents (Table 38.2). Among the producers who self-reported a pesticide poisoning incident, 45 per cent had washed their clothes after application, whereas those who did not wash their clothes after application had a lower frequency of pesticide poisoning, 24 per cent (Table 38.2). Likewise, 59 per cent of producers who bathed following pesticide application self-reported poisonings, compared to the much lower incidents self-reported by producers who did not shower following pesticide application, 10 per cent (Table 38.2). As with extension training, the seemingly paradoxical findings that producers who practiced improved hygiene had greater likelihood of poisoning, suggests the negative effect implied by the surveys may be spurious (coincidental) or at least not causally linked.

Cotton growers with the larger farm area are expected to be more likely to be poisoned by pesticides since they generally spend a longer time spraying their farms and are more exposed to pesticide contamination than smaller landholders. The survey results are consistent with expectations: of the 69.1 per cent of the producers who self-reported at least one poisoning incident, farm sizes averaged 4.01 ha, compared to the 3.62 ha average farm size of producers who did not self-report any poisoning incident (Table 38.2).

CONCLUSIONS

Genetically modified cotton may serve as a working example of how African countries can address enhanced sustainability using modern, science-driven technology to increase production levels while reducing input use and improving the health of farm workers. An attractive feature of GM cotton is its potential to increase productivity in the near to short-term, unlike varietal and pest eradication programs that require long-term investment horizons. The Burkina Faso story is emerging as a working model of how biotechnology can be successfully introduced in Africa, providing a working example of how developing countries can overcome challenges in legal frameworks, technocratic bureaucracy and can be supported and sustained by business models that link the private sector to small- and medium-sized producers in developing countries.

If GM cotton continues on its current trajectory in Burkina Faso, its success may create a gateway for the future introduction and development of other biotech crops in Africa. In Burkina Faso, the demand for Bt cotton is driven by the high lepidoptera pest densities and the growing cost of conventional pest-control methods and cotton export markets that strengthen producers' willingness to pay for Bt products. Other cotton-producing countries in the region, such as Mali and Benin, would be likely to benefit as much as Burkina Faso and could be next in line to introduce Bt cotton once legal frameworks are established. While capacity may have been lacking initially, through proper planning, management and partnership with the private sector, Burkina Faso has demonstrated that African countries can successfully introduce GM crops such as BGII cotton and, in so doing, move toward regaining a competitive stance in world markets.

The commercialization of GM cotton in African countries such as Burkina Faso and South Africa can also serve as a gateway to facilitate the other crops. The recent hesitance of certain African countries to accept food aid containing GM maize speaks to the tangible concerns that some African societies currently have over GM crops, particularly when they are intended for human consumption. However, the increased public awareness of the benefits from a GM crop such as BGII could enhance public perception and acceptance of other GM crops. In the long-term, biotechnology is expected to address additional constraints and crops that could benefit large segments of African societies, including consumers.

NOTES

1. The ANOVA model of cotton yield was specified as: $YIELD_i = GENE_{ij} + ZONE_{ik} + TYPE_{il} + SPRAYS_{im} + YEAR_{in} + \varepsilon_i$ where for the ith observation, *j* represents the gene type (GM or conventional), *k* the production zone (SOFITEX, SOCOMA, and Faso Coton), *m* the number of applied pesticide sprays (0–6), *n* the year (2009, 2010 and 2011), and ε the error term. Interaction terms among GENE and the other factors were included but were removed from the model if not significant at the $p<0.05$ level.
2. Average production costs are calculated as total production costs, as listed in Figure 38.3, divided by cotton yield contained in Table 38.1. Average production costs provide a useful measure of profitability since they indicate how much profit is made per unit produced, i.e. profit is given by the difference between selling price and average production cost. Hence, for producers to break even, average production costs must be less than selling price.
3. The producer surveys collected the number of self-reported poisoning incidents that occurred within the household. For each incident reported, respondents listed the type of insecticide used, symptoms experienced, extent of illness, medical expenses, lost wages, and background information on the poisoned individual.

REFERENCES

Aronson, A.I., W. Beckman and P. Dunn (1986), 'Bacillus thuringiensis and related insect pathogens', *Microbiological Review*, **50**,1–24.

Arora, A. and S. Bansal (2011), 'Diffusion of Bt cotton in India: impacts of seed prices and technological development', paper presented at the International Consortium of Applied Bioeconomy Research, 26–29 June, Ravello, Italy.

Baquedano, F.G., J.H. Sanders and J. Vitale (2010), 'Increasing incomes of Malian cotton farmers: is elimination of US subsidies the only solution?', *Agricultural Systems*, **103**, 418–32.

Bassett, T. (2001), *The Peasant Cotton Revolution in West Africa, Côte d'Ivoire, 1880–1995*, Cambridge: Cambridge University Press.

Benedict, J. and D. Altman (2001), 'Commercialization of transgenic cotton expressing insecticidal crystal protein', in J. Jenkins and S. Saha (eds), *Genetic Improvement of Cotton: Emerging Technologies*, Enfield, NH: Science Publishers, pp. 137–201.

Brookes, G. and P. Barfoot (2011), 'GM crops: global socio-economic and environmental impacts 1996–2009', PG Economics, available at: http://www.pgeconomics.co.uk/.

Brookes, G. and P. Barfoot (2012), 'GM crops: global socio-economic and environmental impacts 1996–2010', available at: http://www.pgeconomics.co.uk/publications.php, accessed 8 November 2012.

Clay, J.W. (2004), *World Agriculture and the Environment: A Commodity-by-Commodity Guide to Impacts and Practices*, Washington, DC: Island Press.

Cohen, J. and R. Paarlberg (2002), 'Explaining restricted approval and availability of GM crops in developing countries', *AgBiotechNet*, **4**, 1–6, available at: http://www.agbiotechnet.com/reviews/Abstract.asp?ID=26.

Cotton Australia (2011), 'Facts and figures: natural resource management issues', available at: http://www.cottonaustralia.com.au/facts/factsandfigures.aspx?id=14, accessed 20 December 2011.

Frisvold, G.B. and J. M. Reeves (2007), 'Economy-wide impacts of Bt cotton', Proceedings of the Beltwide Cotton Conferences, January, 2007, available at: http://www.icac.org/cotton_info/tis/biotech/documents/otherdocs/gtap_beltwide.pdf, accessed 21 December 2011.

Goodell, P., L. Godfrey, E. Grafton-Cardwell, N. Toscano and S. Wright (2001), 'Insecticide and miticide resistance management in the San Joaquin Valley cotton for 2001', University of California Davis Division of Agriculture and Nature Resources, publication no. 8033, available at: http://anrcatalog. ucdavis.edu.

Gómez-Barbero, M. and E. Rodríguez-Cerezo (2006), 'Economic impact of dominant GM crops worldwide: a review', Seville: Institute for Prospective Technological Studies, available at: http://ftp.jrc.es/EURdoc/eur22547en. pdf, accessed 8 August 2013.

Greenplate, J., J. Mullins, S. Penn, A. Dahm, B. Reich, J. Osborn, P. Rahn, L. Ruschke and Z.W. Shappley (2003), 'Partial characterization of cotton plants expressing two toxin proteins from *Bacillus thuringiensis*: relative contribution, toxin interaction, and resistance management', *Journal of Applied Entomology*, **127**, 340–47.

Hema, O., H. Some, O. Traore, J. Greenplate and M. Abdennadher (2009), 'Efficacy of transgenic cotton plant containing the Cry1Ac and Cry2Ab genes of *Bacillus thuringiensis* against *Helicoverpa armigera* and *Syllepte derogata* in cotton cultivation in Burkina Faso', *Crop Protection*, **28**(3), 205–14.

Hofte, H. and H.R. Whiteley (1989), 'Insecticidal crystal proteins of *Bacillus thuringiensis*', *Microbiology Review*, **53**(2), 242–55.

Huang, J., R. Hu, C. Fan, C.E. Pray and S. Rozelle (2003), 'Bt cotton benefits, costs, and impacts in China', *AgBioforum*, **5**(4), 153–66.

James, C. (2006), 'Global status of commercialized biotech/GM crops: 2006', *International Service for the Acquisition of Agri-Biotech Applications (ISAAA) Brief*, 35, Ithaca, NY: ISAAA.

James, C. (2010), 'Global status of commercialized biotech/GM crops: 2010', *International Service for the Acquisition of Agri-Biotech Applications (ISAAA) Brief*, 42, Ithaca, NY: ISAAA.

James, C. (2012), 'Global status of commercialized biotech/GM crops: 2012', *International Service for the Acquisition of Agri-Biotech Applications (ISAAA) Brief*, 44, Ithaca, NY: ISAAA.

MacIntosh, S.C., T.B. Stone, S.R. Sims, P.L. Hunst, J.T. Greenplate, P.G. Marrone, F.J. Perlak, D.A. Fischhoff and R.L. Fuchs (1990), 'Specificity and efficacy of purified *Bacillus thuringiensis* proteins against agronomically important insects', *Journal of Invertebrate Pathology*, **56**(2), 258–66.

Paarlberg, R. (2008), *Starved for Science, How Biotechnology is Being Kept out of Africa*, Cambridge, MA: Harvard University Press.

Perlak, F.J., R.W. Deaton, T.A. Armstrong, R.L. Fuchs, S.R. Sims, J.T. Greenplate and D.A. Fischhoff (1990), 'Insect resistant cotton plants', *Bio/Technology*, **8**, 939–43.

Sims, S.R. (1997), 'Host activity spectrum of the Cry2A *Bacillus thuringiensis* subsp. *kurstaki* protein: effects on Lepidoptera, Diptera, and non-target arthropods', *Southwest Entomologist*, **22**, 395–404.

Vitale, J., M. Ouattarra and G. Vognan (2011), 'Enhancing sustainability of cotton production systems in West Africa: a summary of empirical evidence from Burkina Faso', *Sustainability*, **3**(8), 1136–69.

Vitale, J., H. Glick, J. Greenplate, M. Abdennadher and O. Traore (2008), 'Second-generation Bt cotton field trials in Burkina Faso: analyzing the potential benefits to West African farmers', *Crop Science*, **48**, 1958–66.

Zhou, X. (2011), 'Adsorption formulation of the insecticidal protein from *Bacillus thuringiensis*', international conference on Agricultural and Biosystems Engineering, 20–22 February 2011, Hong Kong, *Advances in Biomedical Engineering*, vols 1–2.

39 Canola

Derek Brewin and Stavroula Malla

1 INTRODUCTION

The introduction of biotechnology and intellectual property rights (IPRs) has completely transformed the canola sector in Canada. Currently, private firms dominate in the research investment and control most of the research output both in terms of new varieties and proprietary technology. In the 1970s, most agricultural research was a result of public investment and research output was a public good. The modification of rapeseed into canola and open pollinated canola into herbicide-resistant and hybrid canola through biotechnology was an agronomic revolution. The changes in this crop led to an area increase from less than 1 million hectares (ha) in 1969 to over 7.5 million ha in 2011; from less than 5 per cent of crop land in Canada to over 30 per cent. Canola revenues are forecasted to be over C$5.5 billion in 2012 (AAFC, 2011). While initial public sector research was a vital first step, the role of biotechnology was pivotal in creating herbicide-tolerant (HT) varieties which then facilitated the mass production of hybrid varieties. Together, patentable HT technology and hybrid seed sales fostered a boom in private canola variety development.

Oil pressed from rapeseed has been used for lamps and stoves in Asia and Europe for centuries. This oil was also a valuable lubricant in steam engines in the nineteenth and early twentieth centuries. The first major planting of rapeseed in Canada was for use in these steam engines during World War II (Casséus, 2008). The term 'canola' was first applied in 1978 to a rapeseed developed by western Canadian breeders Baldur Stefansson at the University of Manitoba, and Keith Downey[1] from Agriculture and Agri-food Canada (AAFC) in Saskatoon using traditional breeding techniques with modern testing procedures. The word 'canola' can only be applied to rapeseed varieties that are 'less than 2% erucic acid and . . . less than 30 μmol of glucosinolates per gram of air-dried oil-free meal' (Casséus, 2008).

Canola is now a major edible oil with a seed production of just under 60 million metric tonnes (FAO, 2012). The oil is considered preferable to other major vegetable oils (soy and corn) because of its low level of saturated fat. The five largest seed producers are China (22 per cent of total), Canada (20 per cent), India (11 per cent), Germany (10 per cent), and France (8 per cent). Canadian canola seed exports are 44 per cent of the total rapeseed trade and more than four times higher than their closest competitor, France, at 9 per cent. Canola oil exports from Canadian crushing plants are 39 per cent of the total rapeseed oil trade (FAO, 2012).

Prior to the mid-1980s most of the research investment in canola seed varieties came from public sources. Both Stefansson and Downey were publicly funded. According to Malla et al. (2004) total public investments in rapeseed in the early 1970s were around C$18 million. Improvements in breeders' rights did spark some private sector seed production in the mid-1980s, but the most remarkable shift in canola seed development

Table 39.1 Shares of seeded area, 2008, by HT and breeding systems (%)

Herbicide tolerance	Breeding system		
	Open Pol.	Hybrid*	Total
Clearfield	7.7	3.0	10.7
Liberty Link	0.2	52.1	52.3
Round-Up Ready	6.7	29.6	36.4
None	0.6	0.1	0.6
Total	15.2	84.8	100.0

Note: * Includes synthetic systems.

Sources: Seed Manitoba (2011), MASC (2011), SSG (2011), ASG (2011), Statistics Canada (2011) and authors' calculations.[2]

began with the development of herbicide-resistant varieties through biotechnology. Glysophate resistance was being studied at Monsanto in the mid-1980s (Phillips, 2001a). The first HT canola was introduced in 1995 and some form of herbicide tolerance is now a trait in most of Canada's canola crop. The glysophate-resistant form of HT was a patented trait which was eventually used to compel farmers to pay for Technical Use Agreements (TUAs) to control the HT seed supply and price. This and other forms of HT were also valuable tools in the challenging problem of crossing lines for hybrid canola varieties.

HT is now a key tool in the supply of hybrid canola seed. Hybrid technology does two things. First, it eventually fueled huge increases in yield because of the hybrid vigor that comes from crossing different parent lines: current hybrid varieties normally yield 20 per cent more than open pollinated options. Second, the move to hybrid varieties further supported private incentives to develop new canola varieties, because hybrid seeds are sold after the first crossing and subsequent use of the seed will not lead to the same hybrid vigor yields. Farmers using hybrid varieties require new seed each year, so seed sales are significant. According to our data (see sources for Table 39.1), hybrids now have an 85 per cent share of all canola seed sales in Canada.[3]

Together with the patent-supported TUAs, hybrid seed sales have led to significant returns to the private sector. Revenues from TUAs and seed sales in the canola market are now normally in the neighborhood of C$100/hectare in most of western Canada (SMA, 2011 and MAFRI, 2011) and the area planted has been growing steadily since 1971 to well over 7 million hectares, with the result that gross receipts are in the range of C$700 million in western Canada.

The area planted to canola has grown as varieties with improved yields and herbicide resistance made it a favorite crop anywhere in western Canada that could give it enough moisture. The AAFC forecasted area planted to canola in 2011 was just over 8 million hectares (AAFC, 2011), which was the first time the canola area exceeded spring wheat. Production is forecast to be 15 million metric tonnes (MMT), meeting the industry target of 15 MMT by 2015 two years ahead of schedule (CCC, 2008). The flexibility of herbicide tolerance and the higher yielding hybrids have made canola the most important crop on the Canadian prairies in terms of revenue.

The goal of this chapter is to provide an overview and critical assessment of the impact of GM technologies on the canola industry in Canada. The next section presents the economic models, methods and metrics commonly used to evaluate the impacts of the application of biotechnology in canola, focusing on measures of area, varieties, productivity, returns to research and the distribution of research benefits, particularly as they are affected by intellectual property rights. The last section offers a critical assessment of the impacts of biotechnology on the sector.

2 MODELS, METHODS AND METRICS

2.1 Area, Varieties and Ownership

The first metric of the success of canola is the area of land that farmers were willing to commit to it. As rapeseed turned into canola and open pollinated public varieties gave way to privately controlled hybrid and herbicide-tolerant (HT) varieties, the area of land seeded to canola in Canada exploded over time. Figure 39.1 shows the area of rapeseed/canola from 1943 to 2011 (Statistics Canada, 2011). There are three marked increases in area: around 1970 with the introduction of low erucic acid 'canola' varieties; another jump around 1994 when private breeders entered the market with HT varieties; and lastly starting around 2004 and continuing up to the present, areas started to trend upward due to the introduction of very high yielding hybrids developed using the Liberty Link system.

A second key metric of the major shifts in the canola sector is the ownership and numbers of varieties of canola entering the market place. Table 39.2 lists the 50 leading varieties in terms of area seeded between 1960 and 2008.[4] The dominant varieties from

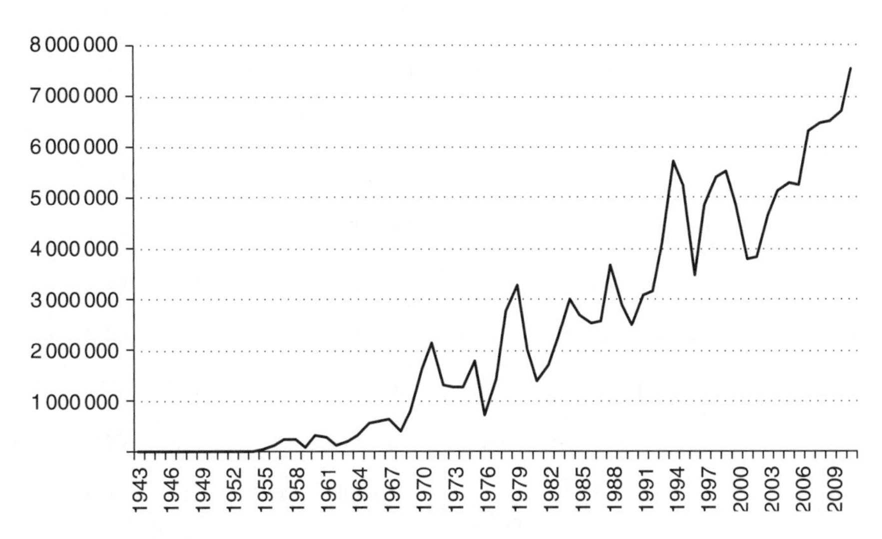

Source: Statistics Canada (2011).

Figure 39.1 Area in hectares of rapeseed and canola, 1943 to 2011

Table 39.2 Top 50 Canadian canola varieties by area, 1965 to 2008

Variety	Released	HT System	Breeder	Arg./Pol.	Pub./ Priv.	Yield Index	OP, HYB	Total Area (Ha)
Echo	1965	Conventional	Agriculture Canada	Polish	Public	109.1	OP	3999714
Target	1967	Conventional	Agriculture Canada (U of MB)	Argentine	Public	102.0	OP	836294
Oro	1968	Conventional	Agriculture Canada	Argentine	Public	99.0	OP	753332
Span	1971	Conventional	Agriculture Canada	Polish	Public	100.1	OP	2255389
Zephyr	1972	Conventional	Agriculture Canada	Argentine	Public	100.0	OP	755219
Torch	1973	Conventional	Agriculture Canada	Polish	Public	100.1	OP	4657145
Midas	1973	Conventional	Agriculture Canada	Argentine	Public	127.0	OP	2067624
Tower	1974	Conventional	Agriculture Canada	Argentine	Public	109.0	OP	3044723
Tobin	1977	Conventional	Agriculture Canada	Polish	Public	90.1	OP	9395614
Candle	1977	Conventional	Agriculture Canada	Polish	Public	87.0	OP	3511164
Regent	1978	Conventional	Agriculture Canada	Argentine	Public	102.0	OP	2933710
Altex	1979	Conventional	Agriculture Canada	Argentine	Public	105.0	OP	1549198
Westar	1984	Conventional	Agriculture Canada	Argentine	Public	122.0	OP	10433193
Legend	1989	Conventional	Svalöf Weibull AB	Argentine	Private	123.0	OP	2724336
Horizon	1990	Conventional	Svalöf Weibull AB	Polish	Private	91.9	OP	1018277
AC Parkland	1990	Conventional	Agriculture Canada	Polish	Public	90.1	OP	765509
Colt	1990	Conventional	Svalöf Weibull AB	Polish	Private	89.2	OP	719485
Delta	1990	Conventional	Svalöf Weibull AB	Argentine	Private	132.4	OP	628745
AC Excel	1991	Conventional	Agriculture Canada	Argentine	Public	119.3	OP	2910293
Vanguard	1991	Conventional	Svalöf Weibull AB	Argentine	Private	118.1	OP	1522612
Profit	1991	Conventional	Agriculture Canada	Argentine	Public	110.7	OP	1344435
Bounty	1991	Conventional	Svalöf Weibull AB	Argentine	Private	132.4	OP	744856
Hyola 401	1992	Conventional	Advanta Canada Inc.	Argentine	Private	121.7	HYB	2869540
Reward	1992	Conventional	University of Manitoba	Polish	Public	92.8	OP	1655547

Table 39.2 (continued)

Variety	Released	HT System	Breeder	Arg./Pol.	Pub./ Priv.	Yield Index	OP, HYB	Total Area (Ha)
Goldrush	1992	Conventional	Svalöf Weibull AB	Polish	Private	92.8	OP	740 873
Garrison	1993	Conventional	Svalöf Weibull AB	Argentine	Private	147.9	OP	977 510
Crusher	1993	Conventional	Svalöf Weibull AB	Argentine	Private	133.6	OP	726 142
Cyclone	1993	Conventional	DLF-Trifollurm	Argentine	Private	137.2	OP	618 065
Legacy	1994	Conventional	Svalöf Weibull AB	Argentine	Private	128.9	OP	1 060 984
Quantum	1996	Conventional	University of Alberta	Argentine	Public	147.9	OP	1 843 006
45A71	1996	Clearfield	Pioneer Hi-Bred Production Limited	Argentine	Private	127.7	OP	1 784 873
Innovator	1996	Liberty Link	Aventis CropScience Canada Co.	Argentine	Private	115.7	OP	900 976
Ebony	1996	Conventional	Monsanto Canada Seeds	Argentine	Private	122.9	OP	705 345
Quest	1997	Roundup Ready	Agricore Cooperative Ltd.	Argentine	Private	126.5	OP	1 469 523
46A65	1997	Conventional	Pioneer Hi-Bred Production Limited	Argentine	Private	133.6	OP	1 179 306
45A51	1998	Roundup Ready	Pioneer Hi-Bred Production Limited	Argentine	Private	130.0	OP	578 132
DKL3235	1999	Roundup Ready	Monsanto Canada	Argentine	Private	120.5	OP	478 773
46A76	2000	Clearfield	Pioneer Hi-Bred Production Limited	Argentine	Private	152.7	OP	1 494 188
DKL34-55	2000	Roundup Ready	Monsanto Canada	Argentine	Private	128.9	OP	1 162 577
InVigor 2663	2000	Liberty Link	Aventis CropScience Canada Co.	Argentine	Private	162.3	HYB	964 443
InVigor 2153	2000	Liberty Link	Aventis CropScience Canada Co.	Argentine	Private	138.4	HYB	603 405
InVigor 2273	2000	Liberty Link	Aventis CropScience Canada Co.	Argentine	Private	143.2	HYB	596 550

Table 39.2 (continued)

Variety	Released	HT System	Breeder	Arg./Pol.	Pub./ Priv.	Yield Index	OP, HYB	Total Area (Ha)
Conquest	2001	Roundup Ready	University of Alberta	Argentine	Public	126.5	OP	499 702
InVigor 2573	2001	Liberty Link	Aventis CropScience Canada Co.	Argentine	Private	158.7	HYB	983 774
InVigor 2733	2001	Liberty Link	Aventis CropScience Canada Co.	Argentine	Private	155.1	HYB	678 581
45H21	2002	Roundup Ready	Pioneer Hi-Bred Production Limited	Argentine	Private	160.4	HYB	1 247 232
InVigor 5020	2003	Liberty Link	Bayer Crop Science	Argentine	Private	169.7	HYB	2 228 406
InVigor 5070	2003	Liberty Link	Bayer Crop Science	Argentine	Private	178.0	HYB	1 509 571
InVigor 5030	2003	Liberty Link	Bayer Crop Science	Argentine	Private	172.4	HYB	1 390 503
DKL71-45RR	2005	Roundup Ready	Monsanto Canada	Argentine	Private	160.4	HYB	639 132

Sources: Seed Manitoba (2011), MASC (2011), SSG (2011), ASG (2011), Prairie Pools Inc. (n.d.) Statistics Canada (2011) and Gray et al. (2006) and authors' calculations.

1965 to 1989 were all bred by public researchers. The varieties with the overall highest total seeded area were Westar and Tobin. Both of these varieties were used to seed over 9 million hectares in aggregate over their lifetime, at a time when the total area of canola seeded in Canada was normally less than 3 million hectares per year. More recently, AAFC introduced AC Excel in 1991, which offered a 19 per cent increase in yield over the 1970s varieties, but it soon faced increased competition from even higher yielding private varieties and had a successful but short lifespan.

The details of private sector seed returns will be discussed in the sections below, but the evolution of breeding rights had a direct effect on variety development and seeded area. These rights were being established in the late 1980s and early 1990s. In the late 1980s a private cooperative from Sweden, Svalöf Weibull AB,[5] began breeding in Canada, attracted to invest in Canada by improved rules in seed registration and thus the returns from seed sales. By 1994, the dominant seeded varieties in terms of area in western Canada were from Svalöf. They had bred several high-yielding varieties (Svalöf's Garrison had a yield index of 1.45) and were starting to be a market leader. Svalöf varieties and AAFC's Excel made up 60 per cent of the seeded area in 1994. Svalöf released Legend in 1989, which went on to be seeded on over 200 000 hectares per year for six years. Between 1989 and 2008 Svalöf released some 75 canola varieties, 10 of which were in the top 50 in terms of total area seeded from 1960 to 2008.

Hyola 401 was the first commercially successful canola hybrid, released in 1992 by Advanta, with a yield index of 1.22. It was seeded on around 6 per cent of the seeded area in 1994. Because hybrid seed is only high yielding in the first generation, every year farmers have to buy the recent crosses to maintain the hybrid yield advantage. Private seed sale returns from hybrid seed was a key factor in attracting private breeders to canola.

In 1996 Monsanto introduced transgenic glysophate-tolerant Ebony and Jewel under the Roundup Ready brand. Competing private firms introduced imazethapyr/imazamox (imaz) and glufosinate-tolerant varieties in 1996 and 1997. Glufosinate-tolerant Innovator, was introduced in 1996 by Aventis and 45A71, an imaz-tolerant variety, was introduced in 1997 by Pioneer (Harker et al., 2000). Herbicide resistance became a key characteristic in the late 1990s. HT canola may have been priced initially at 'nearly' its full benefit, in terms of fewer weed control activities (for example, Fulton and Keyowski, 1999). This may have dampened the HT impact on area, but HT was clearly a production tool that farmers valued and quickly became a standard in the industry. From 1996 to 2007 HT varieties went from 10 per cent to 96 per cent of total area seeded (CCC, 2011).

Technical use agreements pioneered by the early patent holders of the HT traits also facilitated the returns to HT variety breeders. The first HT canola varieties with significant acres were 45A71, Ebony, Independence and Innovator released in 1996, quickly followed by Quest in 1997. The flexibility of these new traits in weed management came to be viewed as a very valuable feature by Canadian producers who were normally willing to pay private breeders for this technology. By 2008, privately-bred, hybrid, HT canola varieties dominated Canadian acreage. No publicly owned variety has seen any significant area in Canada since 2001.

The biotechnology that created HT was also helping some firms, especially Aventis, to develop hybrid varieties, because it facilitated crosses with HT and non-HT parents. The rise of very high yielding hybrids seems to be behind the last big jump in area that started in 2004. By then, herbicide resistance was also a major factor in seed choice. Higher yielding hybrid varieties, especially the Liberty Link varieties, saw a significant increase in market share from 2003 to 2008. By 2006 hybrid varieties with significant yield advantages had come on line.

By 2004 most of the area was going to varieties that were both HT and hybrid crosses, especially the InVigors 5020, 5030 and 5070. All of them had a yield index above 1.69. By 2010, 90 per cent of all seeded areas (around 7 million hectares) was seeded to HT varieties (CCC, 2011). Our own data shows that 85 per cent of these are also hybrid.

Another part of the development of canola seed breeding was the rise of higher quality oils. High erucic acid rapeseed varieties used as industrial lubricants were seeded on several hundred thousand hectares in the mid-1990s. Most of them were developed at the University of Manitoba. Dow Agrosciences released several high omega 9 and low linolenic oils under the brand name Nexera in 2000. Just over 1 million hectares were seeded to Nexera varieties from 2000 to 2008. Cargill introduced competing varieties with similar characteristics to Nexera in 2001. Cargill's Victory varieties were seeded on just over 800000 hectares from 2001 to 2008.

Although still a minor part of the seeded area these new novel traits represent a second generation of biotechnology aimed at meeting consumer needs. Innovation has also moved into the area of medicine. SemBioSys Genetics has developed a technology to use canola to produce the anti-coagulant, hirudin (Phillips, 2001a).

2.2 Partial Productivity Measures: Crop Yields

Productivity is not fully assessed by a single input, but various researchers have used crop yield as a partial productivity measure in the crop sector. The average (actual farm) canola crop yields have increased over time, while acreage-weighted research trial yield indexes (yields in experimental trials) grew unevenly but quite rapidly.

Veeman and Gray (2010) showed that yield trends for the major crops including canola grown in Canada (wheat, barley, canola, corn, soybean, peas) exhibit constant absolute growth but declining proportional growth rate (see Figure 39.2). The yields were based on data from CANSIM and were average actual farm crop yields (yields per seeded acre). Specifically, the yields of canola like the other major crops increased by about 60 per cent from the 1960s to 2007. Veeman and Gray indicated the presence of a linear trend which implies a constant absolute growth and declining proportional growth rate in yield.

The research trial yield indexes for canola (1960–86) are depicted in Figure 39.3. The yield index in experimental trials was based on research trial data and seeded area. According to Veeman and Gray (2010, p. 131)

> Canola yields in experimental trials grew rapidly until 1972, but this growth was reversed from 1975 to 1983 as canola, with low glucosinolate and low erucic acid replaced rapeseed, with the accompanying yield drag of any major crop transformation. Canola trial yields then increased significantly from 1986 to 1994, only to retreat in the late 1990s as herbicide-tolerant varieties were adopted, with major agronomic benefits to growers in terms of weed control but again with the accompanying yield limitations of a major change in the available varieties. Since 1998, canola yields have again grown rapidly, as hybrid varieties have been developed and widely adopted.

Measuring productivity in canola crop in terms of partial productivity such as crop yields is imperative but problematic. As was pointed out by Veeman and Gray (2010) changes in yield have impacts on crop land intensity (such as reduced summer fallow), cropping diversity and cropping technology (like reduced tillage): their impacts on productivity and on each other cannot be easily separated from one another. Furthermore, important quality changes like the canola transformation from rapeseed in the early 1970s, the introduction of HT and hybrid canola varieties in the middle 1990s and the introduction of canola varieties high in oleic acid in the early 2000s are not captured in the yield estimates even though they increase the value of the crop.

2.3 Returns to Research and Distribution of Benefits

Agricultural research typically generates very high returns on investment. Alston et al. (1998) collected 294 studies post-war of returns to R&D investment (involving 1858 estimates). They estimated that the annual rate of return (IRR) averaged 64.2 per cent per year for research only, 46.3 per cent per year for research and extension combined, and 75.6 per cent per year for extension only. They concluded that 'There is no evidence to support the view that the rate of return has declined over time' (p. 27). Furthermore, Brinkman (2004) provided a summary of the Canadian studies on returns to agricultural research (1978–2001) and concluded that 'Research studies in western Canada also show

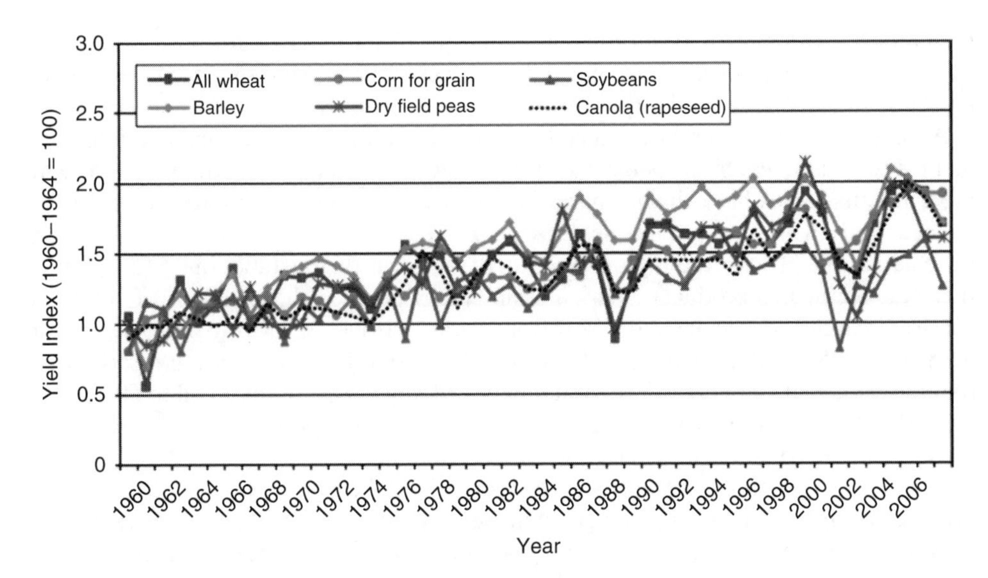

Source: Veeman and Gray (2010) (CANSIM database).

Figure 39.2 Canadian crop yields, 1960–2007 (base 1960:1964 = 100)

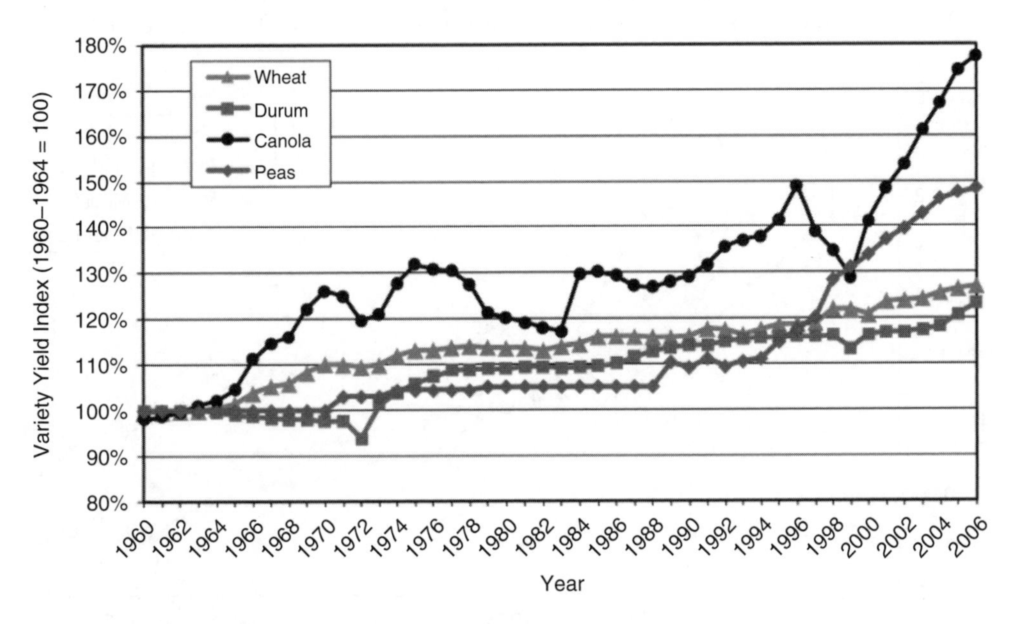

Source: Veeman and Gray (2010) (based on research trial data and seeded area).

Figure 39.3 Research trial yield indexes for selected Canadian crops, 1960–2006 (1960–1964 = 100)

high returns, with benefit–cost ratios ranging from 12.1:1 to 34.1:1 for barley, wheat and rapeseed . . . Overall, it appears that public agricultural research is one of the highest payback uses of public funds.' (p. 132).

A number of studies have examined the return to rapeseed and canola research investment. Previous research has shown very high rates of return to canola research, while more recent studies have shown market rates of returns. Nagy and Furtan (1978) calculated the IRR from improved yield research to be 101 per cent and the benefit to cost ratio equal to 17.64 for the period 1960 to 1974. From their estimation, 47 per cent of this gain accrued to producers. Ulrich et al. (1984) updated the estimates of returns to canola research and calculated the IRR from improved yield research to be 51 per cent for the period 1951 to 1982. Ulrich and Furtan (1985) found by incorporating trade effects that the estimated Canadian IRR from higher yielding varieties was equal to 50 per cent. Ulrich et al. (1984) and Ulrich and Furtan (1985) found that producers could lay claim to 68 per cent and 65 per cent of the gains, respectively.

In a more recent study, Malla et al. (2004) examined the gains to canola research in the presence of intellectual property rights and research subsidies from 1960 to 1999. They show that the rate of return to canola research has diminished since the early 1980s and approached market rates by the mid-1990s (see Figure 39.4: Models 1 and 2). Specifically, they estimated that the average IRR initially exceeded 25 per cent per year, but steadily declined during the period 1970 to 1999, to eventually approach the opportunity cost of capital. Additionally, they found that the IRR for the marginal dollars invested each year showed a much more dramatic decline. During the 1990s the return to these marginal dollars was below a market rate for IRR. Hence, Malla et al. (2004) concluded that

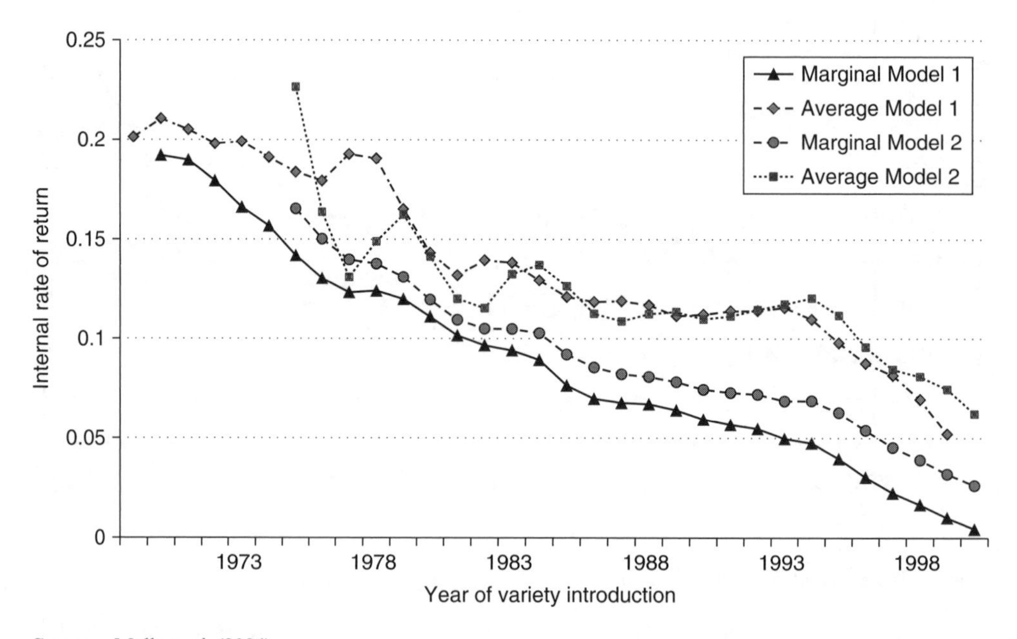

Source: Malla et al. (2004).

Figure 39.4 The marginal and average IRR for canola research, 1968–99

Table 39.3 Economic benefit of GMHT canola (2005–07; all figures in million $CAD)

Year	Acres	Direct	Spillover		Reduced tillage	Cost of volunteer control	Total benefits	
			Low	High			Low	High
2005	12.6M	$141	$63	$103	$153	$14	$343	$383
2006	12.8M	$143	$64	$105	$153	$14	$346	$387
2007	14.8M	$165	$73	$121	$153	$17	$374	$422
Average	13.4M	$150	$67	$110	$153	$15	$354	$397

Source: Gusta et al. (2011).

> the influx of private investment into crop research, which has accompanied the establishment of intellectual property rights (IPRs) and the introduction of biotechnology, suggests high rates of return. The empirical analysis of the canola research sector shows a decline in the total net return to research during a period of increased investment, indicating that net marginal returns to research have been negative . . . the combined effect of IPRs and public incentives has driven the quantity of research beyond the socially optimal. (p. 63)

The Malla et al. analysis suggests that when nearly complete IPRs exist, government incentives for research (for example, research subsidies) could reduce social welfare.

Recently, Gusta et al. (2011) examined the economic benefits of genetically-modified herbicide-tolerant canola (GMHT) for producers. The data for the project was based on an 80-question producer survey from 2007 which comprised six areas of focus: weed control; volunteer canola control; canola production history; specific weed control measures on canola fields and subsequent crops; crop and liability insurance; and general demographics. Their survey focused on three major impacts: 'reduced tillage' (cost of weed control); 'cost of volunteer control' (when GMHT canola becomes an in-crop weed or volunteer); and 'spillover' (multi-year benefits due to fewer weeds or easier weed control on fields next year). Gusta et al. (2011) estimated that 'the new technology generated between $1.063 billion CAD and $1.192 billion net direct and indirect benefits for producers from 2005–07; this is partly attributed to lower input costs and partly attributed to better weed control.' (p. 1) (see Table 39.3).

Previous studies on the producers' benefits from GMHT by the Canola Council of Canada (CCC, 2001) and Phillips (2003) revealed producers' benefits ranging from C$60 to C$70 million respectively in 2000. According to the CCC (2001), improved yield, slightly increased fertilizer usage, increased seed costs, decreased tillage use, improved soil moisture conservation, decreased summer fallow, improved rotation flexibility, lower dockage and decreased herbicide inputs were the main determinants of the producers' benefits. Phillips (2003) estimated the direct and indirect effects on producers with the adoption of the new technology (for example, involving higher seed costs, lower herbicide costs, fewer herbicide applications, lower dockage and earlier seeding) as well as the broad economic impacts on the global economy. The CCC (2001) and Phillips (2003) studies did not take into account the impact of any spillovers or any increased costs from controlling volunteer canola like the study of Gusta et al. (2011). In 2005, the CCC examined the management of volunteer GMHT canola and assessed that there were net benefits to growing GMHT canola instead of conventional varieties.

Furthermore, Smyth et al. (2011a and 2011b) concluded based on a survey (2007) of almost 600 canola farmers in Western Canada that the environmental effects of herbicide-resistant (HR) canola are very important. Specifically, according to Smyth et al. (2011a)

> The commercialization and widespread adoption of herbicide-resistant (HR) canola has changed weed management practices in Western Canada. Before the introduction of HR canola, weeds were controlled by herbicides and tillage as the leading herbicides at that time required tillage to allow for soil incorporation of the herbicide. Much of the tillage associated with HR canola production has been eliminated as 64% of producers are now using zero or minimum tillage as their preferred form of crop and soil management. Additionally, there have been significant changes regarding the use and application of herbicides for weed control in canola . . . The cumulative environmental impact was reduced almost 50% with the use of HR herbicides. (p. 492)

Whereas Smyth et al. (2011b) pointed out that

> a reduction in the total number of chemical applications over the 3-year period was reported, resulting in a decrease of herbicide active ingredient being applied to farmland in Western Canada of nearly 1.3 million kg annually. Fewer tillage passes over the survey period were reported, improving moisture conservation, decreasing soil erosion and contributing to carbon sequestration in annual cropland. An estimated 1 million tonnes of carbon is either sequestered or no longer released under land management facilitated by HT canola production, as compared to 1995. The value of this carbon off-set is estimated to be C$5 million. (p. 403)

Lastly, a different point was raised in the literature regarding producers' heterogeneity and benefits of HT canola. Specifically, Fulton and Keyowski (1999) estimated the producer benefits of herbicide-resistant (HR) canola and argued that

> the pricing and adoption of HR canola in Canada cannot be understood if producers are seen as being homogeneous. We develop a conceptual model of producer heterogeneity that represents the distribution of benefits among producers. In this context, some farmers benefit from the new technology leading to adoption, while others do not. (p. 85) (See their comparisons in Table 39.4.)

Hence, they argued that the differing agronomic, management and technological factors facing each farm are very important determinant of the benefits of HR.

2.4 Intellectual Property (IP), Freedom to Operate and the Tragedy of the Anti-commons

Traditionally, most agri-food research, such as in the early years of canola, was a result of public investment and the products of research (or research output) were put into the public domain as public goods (for example, Malla and Gray, 2003, 2005). Government intervention was justified primarily on the basis that research outcomes were non-excludable (meaning the inventor did not have the ability to exclude others from using, reproducing or selling the new technology or product created from R&D), thus creating limited private R&D incentives. The government intervened to correct the market failure.

The introduction of biotechnology, intellectual property rights (IPR), technical use agreements and hybrid varieties has allowed the creation of excludable non-rival goods in the canola sector. This in turn created incentives for private investment because the inventor could extract most of the economic rents from research investments by retaining ownership over the new technology (Malla and Gray, 2003, 2005). Despite the considerable

Table 39.4 Canola product line: a system comparison from 1999

	Roundup Ready	Smart Open Pol	Liberty Hybrid	Conventional Open Pol
System costs				
Seed cost ($/acre)[a]	$18.70	18.70	24.75 [b]	$13.47
Herbicide cost ($/acre)	$5.00	$26.20	$22.75	$30.00
TUA ($/acre)	$15.00	None	None	None
System cost ($/acre)	$38.70	$44.90	$47.50	$43.47
Gross returns				
Yield (bu/acre)	33.0	31.5	35.7	35.7
Commodity price ($/bu)	$8.00	$8.00	$8.00	$8.00
Expected gross ($/acre)	$264.00	$252.00	$285.60	$285.60
Less system costs ($/acre)	$(38.70)	$(38.25)	$(47.50)	$(43.47)
Gross returns ($/acre)	$225.30	$213.75	$238.10	$242.13

Notes:

[a] Seed cost was calculated assuming a seeding rate of 5.5 lbs/acre.

[b] Recommended seeding rate is 5 lbs/acre for Liberty Hybrids. From Pioneer Grain Company Limited.

Source: Fulton and Keyowski (1999).

growth in private investment, the government continues to make large public investment in research, especially in basic research which may create a positive spillover to private firms as many successful industries are based on breakthrough innovations created in the public sector and, in turn, could result in excess firm entry into the industry (Gray and Malla, 2011). At the same time, the non-rivalrous nature of research outputs (meaning that we can use the technology created from R&D over and over again) has tended to create a concentrated research industry, as firms move to capture economies of scale and scope and exploit the market power offered by property rights to price their technologies above their marginal cost.

A further push towards concentration occurs when firms adopt strategies to preserve their own freedom to operate, either through vertical integration, mergers, acquisitions or joint venture arrangements.[6] Finally, the concentrated nature of the research industry and the exclusive ownership of key pieces of IPRs give research firms some degree of market power, which through higher prices reduces the incentive for product innovation and adoption downstream (Moschini and Lapan, 1997; Malla and Gray, 2003). Furthermore, Galushko et al. (2010) pointed out that due to all these changes in the canola industry, the sharing of knowledge in both the public and private sector has been reduced, increasing significantly the cost of conducting canola breeding research. Hence, the structure of the canola and broad agricultural research industry has been fundamentally altered.

When IPRs are well established, the negotiation of the rights to use innovations can become difficult and expensive, especially when a research product embodies many pieces of IPRs. Specifically, the introduction of single genetic trait into a product of biotechnology can require the use of many separate pieces of IPR. Before the innovator can have 'freedom to operate' the innovator must reach an agreement with each of the other patent owners. If the ownership of the IPRs is dispersed, negotiating FTO agreements to share the proceeds from the innovation is an expensive, time-consuming process that can be subject to hold-up by any of the parties involved (Falcon and Fowler, 2002). The high transaction costs associated with the exchange of IPRs have also adversely affected the structure of the private and public research industries and have created an economic barrier for the commercialization of second generation GM crops.

The resulting problems in freedom to operate could create a tragedy of the anti-commons that results in the under-use of resources and less innovation (for example, Heller and Eisenberg, 1998; Wright, 1998; Buchanan and Yoon, 2000; Falcon and Fowler, 2002; Graff et al., 2003). 'The anti-commons arise whenever individuals have the ability to exclude others from using a resource but cannot use the resource themselves' (Smyth and Gray, 2011). A tragedy of the anti-commons can occur when there are economies of scale in germplasm development but issues with property rights (freedom to operate) lead to less innovatory activity as competing inventors block activity with patent protection (Galushko and Oikonomou, 2007). Too many property rights could lead to less innovation as competing patent rights could actually prevent research activities from being undertaken or products from reaching the market place.

More recently, firms have moved away from mergers and acquisitions towards gene trait cross-licensing agreements to facilitate technology sharing (Galushko et al., 2010; Smyth and Gray, 2011; Stiegert et al., 2010). For example, an important gene trait cross-licensing agreement was made between the largest life science firms in the canola sector, Monsanto and Bayer CropScience, to fight against the tragedy of anti-commons (Smyth and Gray, 2011). Canola varieties with traits owned by these two firms account for more than 85 per cent of the market. Smyth and Gray provided four possible reasons for the recent switch to gene trait cross-license agreements:

> First, anti-trust concerns . . . Second, fostering and maintaining SME [small or medium-sized-enterprises] innovation . . . Third, there is recognition of 'sectoral' benefits to share IP . . . Finally, in perusing cross license agreements, firms will inevitably have some clauses that outline at least the principles of how embodied traits will be commercialized, and how the potential revenue will be shared . . . reassure their shareholders of benefits of the agreement.

These agreements raise additional concerns regarding market concentration.

Overall, the canola industry has witnessed significant changes over time. 'In the 1970s, most research was a result of public investment and the products of research were public goods. Currently, private firms dominate in the canola research investment and control most of the research output both in terms of new varieties and proprietary technology.' (Malla and Gray, 2005, p. 430).

In the early 1980s, varieties were open-pollinated, non-transgenic and there were no effective plant breeder rights (PBR). In the mid-1980s PBRs were established in several countries and there were changes in the seed market and registration rules in Canada. Full PBRs were enacted in Canada in 1990 (Phillips, 2001b). As the technology advanced, the US patent office started to recognize biotech processes which lead to the use of license agreements for some products and processes. This technique was used by the creators of HT canola to compel farmers to continue to pay for the new technology at each planting rather than retaining seed from previous crops. Other technologies facilitated exclusion. Hybrids are a key example as they require the purchase of seeds every year in order to keep the desirable traits.

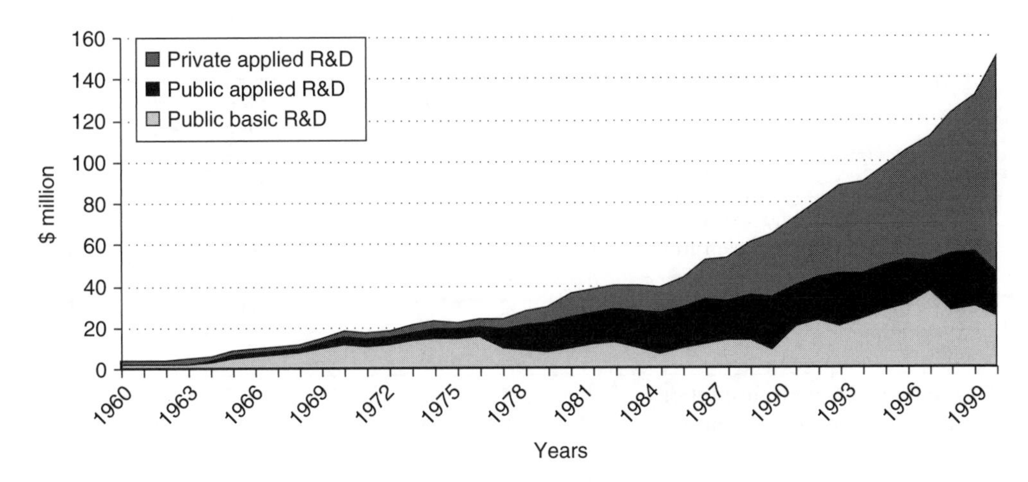

Source: Gray et al. (2001).

Figure 39.5 Research expenditure in the Canola industry (1960–99)

Accordingly, research funding has changed for canola in Canada, which is reflected in the registration of new canola varieties. Over time, research has shifted from a modest public investment to large private investment. In 1970, 83 per cent of the total C$18 million expenditure was public investment and 17 per cent was private investment (Gray et al., 2001; Malla and Gray, 2005). In 1980, research investment was 69 per cent public versus 31 per cent private. By 2000, the private sector share had grown to 70 per cent of the total annual spending on canola research to C$149 million. Today, private research firms dominate the canola industry (see Figure 39.5) with gross receipts over C$700 million.

This is reflected in the array of new varietal registrations. Prior to 1989, the dominant varieties in terms of area were all developed by public institutions. From 1995–98, 88 per cent of the 104 new varieties introduced were private. Today, almost all of the canola varieties are private and over 96 per cent of the total varieties are HT (see Tables 39.5 and 39.6).

The change in the funding of research coincided with a change in the nature of technology and the property rights for the research products. Nowadays, most of the identified canola patents are held by private companies. The public sector has also moved to protect and actively commercialize IP (see Tables 39.7 and 39.8).[7] By the mid-1990s, as a result of these rights, private firms were beginning to capture significant revenue from the sale of their products. In 2001 the private firms earned over C$250 million in revenue, and had effectively crowded out all public sector sales (Gray et al., 2001) (see Figure 39.6). Nowadays, private firms' gross receipts are in the range of C$700 million in western Canada. The canola sector in Canada is fortunate that several forms of herbicide resistance were developed almost simultaneously around 1996. Unlike an unbridled monopoly, these technologies have faced at least some competition, which was not the case in some other markets.

Table 39.5 New varieties developed by institution and by period

	1950–1959	1960–1969	1970–1979	1980–1984	1985–1989	1990–1994	1995–1998
B. napus							
Total varieties by public institutions	1	4	5	4	8	8	10
Total varieties by private institutions	0	0	0	0	12	39	76
Total varieties	1	4	5	4	20	47	86
Number of active institutions	1	2	2	3	11	17	17
B. rapa							
Total varieties by public institutions	1	2	5	1	1	4	2
Total varieties by private institutions	0	0	0	0	3	7	16
Total varieties	1	2	5	1	4	11	18
Number of active institutions	1	2	1	1	3	7	4

Source: Phillips and Khachatourians (2001).

Table 39.6 Adoption rate for HT canola varieties (million acres)

Year	1997	1998	1999	2000	2001	2002	2003	2004	2005	2006	2007
Total canola acres	12.0	13.2	13.6	11.9	9.3	8.9	11.4	11.9	12.6	12.9	15.5
Roundup Ready	0.5	2.8	4.9	4.3	4.0	4.0	5.5	5.7	6.1	5.7	7.0
Liberty Link	1.0	1.6	2.5	1.8	1.5	1.5	2.4	3.7	4.2	5.2	6.2
Clearfield	1.7	2.2	2.5	3.0	1.9	1.4	2.6	2.2	1.8	1.4	1.7
Total HT	3.2	6.6	9.9	9.1	7.4	6.9	10.5	11.6	12.1	12.3	14.9
% HT	26.3%	50%	72.8%	76.5%	79.6%	77.5%	92.1%	97.5%	96%	95.3%	96%

Source: Total canola acres: Canola Council of Canada (2008).

Table 39.7 Ownership of oilseed herbicide tolerance patents in Canada: 2006

Patent Date	Individual	Public		Private					Total
		Can.	US	US MNE*	US small	EU MNE	EU small	Can. small	
Pre 1 October 1989	1	2	0	2	2	2	1	0	10
Post 1 October 1989	0	0	3	16	2	5	0	2	28
Total	1	2	3	18	4	7	1	2	38

Note: * = Multi-National Enterprise.

Source: Smyth and Gray (2011).

Table 39.8 Ownership of canola patents in Canada: 2011

Year	Private	Public
1969–1989	7	3
1990–2006	20	12
2006–2011 (28 July)	14	13

Source: Canadian Intellectual Property Office (CIPO) database; authors' estimates.

Source: Gray et al. (2001)

Figure 39.6 Seed and TUA revenues in the canola industry

3 SUMMARY/ASSESSMENT

The success of canola as a major crop in Canada is linked to the modification of the seed, initially by public researchers and more recently by private researchers mainly through biotechnology. The introduction of biotechnology and the development of IPR protection in plant breeding provided new incentives for massive private investment into canola research, but it also resulted in research fragmentation and patent/IPRs ownership, a series of mergers and acquisitions and more recently a need for gene trait cross-licensing agreements. Evidence of the privatization of the canola industry is seen in the dominance of the private sector in the registration of new canola varieties. This has coincided with a change in the nature of the technology and the property rights for these research products.

While the introduction of biotechnology and the enforcement of IPRs have created incentives to undertake research, it also created a number of complex issues. For instance, additional concerns have been raised with respect to market concentration. Gene trait cross-licensing agreements have led to worries about the close relationships these agreements require between the few dominant firms left in the sector. Furthermore, there have been concerns over the 'freedom to operate' problem that exists in agricultural biotechnology which creates an economic barrier for the commercialization of second generation GM crops and could result in the under-use of resources and less innovation (tragedy of anti-commons). There have also been concerns about the pricing of new canola varieties, the rules of using new technology, and the appropriate government role in today's biotech canola industry. While previous studies have shown very high rates of return to canola research, more recent studies have revealed lower (market) rates of returns to R&D investment.

Overall the industry has experienced significant growth and prosperity. The area seeded to canola varieties, the number of varieties available, and canola crop yields have been on an upward trend for 50 years. The literature indicates that there are considerable benefits from canola research and producer behavior in terms of seeded area and chosen varieties, suggesting they are happy to adopt these innovations. To summarize, the canola industry has benefited enormously from biotechnology but not without some cost. While the problems noted in this chapter regarding concentration might create some challenges, the future of the canola sector in Canada seems bright.

NOTES

1. There is a small amount of controversy over who the 'father of canola' was – Manitoba claims it was Stefansson, who developed low erucic acid varieties in 1968 and Saskatchewan claims it was Downey, who improved the sorting techniques used by several breeders as they chased the low glucosinolates in 1970. Scientific papers clearly show they were working together along with other researchers at the AAFC Research Station in Saskatoon (Downey and Harvey, (1963) and the University of Manitoba (Stefansson et al., 1961).
2. We only give the latest sources in the text. Tables 39.1 and 39.2 are built on data and methods first described in Gray et al. (2006). The data on the percentage of acreage sown of each canola variety were obtained from: Nagy and Furtan (1978) which covered the period 1960 to 1976, various issues of Prairie Pools Inc., *Prairies Grain Variety Survey* (1977–1992), Manitoba Agricultural Services Corporation Variety Survey 1992–2000, Saskatchewan Crop Insurance Corporation (2009), Manitoba Agricultural Services Corporation. Yield Manitoba 2000–2009, www.masc.mb.ca, and Alberta Agriculture, Food and Rural Development, Crop Variety Performance Comparisons. The Prairie variety survey was ended in 1992.

   The relative yield of different canola varieties and the classifications of each canola variety were based on: Alberta Agriculture, Food and Rural Development, Crop Variety Performance Comparisons (n.d.), Saskatchewan Seed Growers (2011), Canadian Food Inspection Agency, *Plant Varieties Journal* (n.d.), Canadian Food Inspection Agency *List of Varieties which are Registered in Canada* (various years), Canola Council of Canada, *Prairie Canola Variety Trials* 2003–2007, Manitoba Agriculture, Food and Rural Initiatives, *Seed Manitoba Variety Guide* (various years), and Wright (2007), 'List of varieties registered for production in Alberta – 2007 and 2008'.

   The area seeded to canola was obtained from Statistics Canada Field Crop Reporting Series – Statistics Canada (Cansim Data: Series 10017, Table 10017). The Minnesota Canola Production Centre (2004) was used to help in the classification of varieties in terms of public or private.
3. We note that some countries have not been as warm to HT adoption as Canada (and the US for a smaller area). Part of this is the power of EU consumers to keep out biotechnology and the fact that most of the world's rapeseed/canola is grown in the EU, Canada or China (FAO, 2012). China is seen as a rapid adopter of biotechnology (ISAAA, 2012).
4. The areas for these varieties were based on Prairie Pools Inc. up to 1997 and after that on producer surveys by western Canadian crop insurance providers (MASC, 2011; SSG, 2011; ASG, 2011). Some varieties could have been missed, especially after 1997, because of poor survey results in some years.
5. Svalöf Weibull AB is a subsidiary of the Swedish Farmers Supply and Crop Marketing Association.

6. The sector has seen mergers. Svalöf Weibull AB created a new company with BASF (Phillips, 2001b).
7. A Statistics Canada survey (2005) of Canadian universities revealed that 75 per cent of University IP license agreements with private firms were exclusive in nature, which is also common practice among US institutions (Smyth and Gray, 2011; Gray and Malla, 2011).

REFERENCES

Agriculture and Agri-Food Canada (AAFC) (2011), 'Canada: grains and oilseeds outlook', 5 July, available at: http://www.agr.gc.ca/pol/mad-dam/index_ e.php?s1=pubs&s2=go-co&s3=php&page=go-co_2011-07-05, accessed 24 August 2011.

Alberta Agriculture, Food and Rural Development (n.d.), 'Crop variety performance comparisons: various soil zones', Edmonton, AB: Alberta Agriculture, Food and Rural Development.

Alberta Seed Guide (ASG) (2011) and various years, 'Alberta Seed Guide 2011', Winnipeg, MB: Alberta Seed Industry Partnership, available at: http://www.seed.ab.ca/index.html, accessed 30 October 2011.

Alston, J.M., M.C. Marra, P.G. Pardey and T.J. Wyatt (1998), 'Research returns redux: a metaanalysis of the returns to agricultural R&D', EPTD Discussion Paper no. 38.

Brinkman, G.L. (2004), 'Strategic policy issues for agricultural research in Canada', *Current Agriculture, Food & Resource Issues*, **5**, 131–47.

Buchanan, J.M. and Y.J. Yoon (2000), 'Symmetric tragedies: commons and anticommons', *The Journal of Law and Economics*, **43**(1), 1–14.

Canadian Food Inspection Agency (n.d.), *Plant Varieties Journal*, various issues, Ottawa, ON: Canadian Food Inspection Agency (CFIA), Plant Breeders' Rights Office.

Canadian Food Inspection Agency (n.d.), 'List of varieties which are registered in Canada', Ottawa, ON: Canadian Food Inspection Agency (CFIA).

Canola Council of Canada (CCC) (2001), 'An agronomic and economic assessment of transgenic canola', Winnipeg, available at: http://www.canola-council.org/gmo_toc.aspx.

Canola Council of Canada (CCC) (2005), 'Herbicide tolerant volunteer canola management in subsequent crops', Winnipeg, available at: http://www.canolacouncil.org/uploads/HT_canola_final.pdf, accessed 20 November 2011.

Canola Council of Canada (CCC) (2008), 'Canola: growing great 2015–in brief', available at: http://www.canolacouncil.org/uploads/1480.CACO.CCC_ 2015shortreportFINAL.pdf, accessed 24 August 2011.

Canola Council of Canada (CCC) (2011), 'Estimated percentage of HT and conventional canola', available at: http://www.canolacouncil.org/ht _conventional_estimates.aspx, accessed 24 August 2011.

Canola Council of Canada (CCC) (n.d.), 'Prairie Canola variety trials', Winnipeg, MB: Canola Council of Canada.

Casséus, L. (2008), 'Canola: a Canadian success story', in Statistics Canada, *Canadian Agriculture at a Glance*, Ottawa: available at: http://www.statcan.gc.ca/pub/96-325-x/2007000/article/10778-eng.htm, accessed 24 August 2011.

Downey, R.K. and B.L. Harvey (1963), 'Methods of breeding for oil quality in rape', *Canadian Journal of Plant Science*, **43**(3), 271–5.

Falcon, W.P. and C. Fowler (2002), 'Carving up the commons: emergence of a new international regime for germplasm development and transfer', *Food Policy*, **27**, 197–222.

Food and Agriculture Organization of the United Nations (FAO) (2012), 'FAOSTAT production and trade data', available at: http://faostat.fao.org/site/291/default.aspx, accessed 23 October 2012.

Fulton, M. and L. Keyowski (1999), 'The producer benefits of herbicide-resistant canola', *AgBioForum*, **2**(2).

Galushko, V. and E. Oikonomou (2007), 'IP protection in Canadian agriculture: a shift to "Tragedy of Anticommons"?', Canadian Agricultural Innovation Research Network *Policy Brief*, no. 4, August.

Galushko, V., R. Gray and S. Smyth (2010), 'Resolving FTO barriers in GM canola', *AgBioForum*, **13**(4), 360–69.

Graff, G.D., S.E. Cullen, K.J. Bradford, D. Zilberman and A.B. Bennett (2003), 'The public–private structure of intellectual property ownership in agricultural biotechnology', *Nature Biotechnology*, **21**(9), 989–95.

Gray, R., and S. Malla (2011), 'Managing public IP with downstream interfirm research spillovers', *Canadian Journal of Agricultural Economics*, **59**, 475–91.

Gray, R. S. Malla and P.W.B. Phillips (2001), 'Industrial development and collective action', in P.W.B. Phillips and G.G. Khachatourians (eds), *The Biotechnology Revolution in Global Agriculture: Innovation, Invention and Investment in the Canola Industry*, Wallingford: CABI Publishing.

Gray, R., S. Malla and K.C. Tran (2006), 'Spillovers and crowding effects in a mixed biotech industry: the case of canola', *AgBioForum*, **9**(1), 31–41.

Gusta, M., S.J. Smyth, K. Belcher, P.W.B. Phillips and D. Castle (2011), 'Economic benefits of genetically-modified herbicide-tolerant canola for producers', *AgBioForum*, **14**(1), 1–13.
Harker, K.N., R.E. Blackshaw, K.J. Kirkland, D.A. Derksen and D. Wall (2000), 'Herbicide-tolerant canola: weed control and yield comparisons in western Canada', *Canadian Journal of Plant Science*, **80**, 647–54.
Heller, M.A. and R.S. Eisenberg (1998), 'Can patents deter innovation? The anticommons in biomedical research', *Science*, **280**(5364), 698–701.
International Service for the Acquisition of Agri-biotech Applications (ISAAA) (2012), 'Global adoption of biotech crops is expected to continue to grow in the future, particularly in developing countries, where there is a promising pipeline of new products' ISAAA press release, Manila, 17 August 2012, available at: http://www.isaaa.org/kc/cropbiotechupdate/pressrelease/2012/default.asp, accessed 23 October 2012.
Malla, S. and R. Gray (2003), 'Public research policy for today's agricultural biotech research industry', *Canadian Journal of Agricultural Economics*, **51**, 347–69.
Malla, S. and R. Gray (2005), 'The crowding effects of basic and applied research: a theoretical and empirical analysis of an agricultural biotech industry', *American Journal of Agricultural Economics*, **87**(2), 423–38.
Malla, S., R. Gray and P.W.B. Phillips (2004), 'Gains to research in the presence of intellectual property rights and research subsidies', *Review of Agricultural Economics*, **26**(1), 63–81.
Manitoba Agriculture, Food and Rural Initiatives (MAFRI) (2011), 'Guidelines for estimating crop production costs – 2011', available at: http://www.gov. mb.ca/agriculture/financial/farm/pdf/copcropproductioncosts-2011westernmb.pdf, accessed 30 October 2011.
Manitoba Agricultural Services Corporation (MASC) (2011), 'Manitoba Management Plus Program: MMPP variety yield data browser', online database, Manitoba: MASC, available at: http://www.mmpp.com/mmpp.nsf/mmpp_browser_variety.html, accessed 30 October 2011.
Minnesota Canola Production Centre (2004), 'Annual report', St Paul, MN: Minnesota Canola Production Centre, available at: http://www.mncanola.org/pdf/2004%20CPC%20Final%20Report.pdf.
Moschini, G. and H. Lapan (1997), 'Intellectual property rights and the welfare effects of agricultural R&D', *American Journal of Agricultural Economics*, **79**, 1229–42.
Nagy, J.G. and W.F. Furtan (1978), 'Economic costs and returns from crop development research: the case of rapeseed breeding in Canada', *Canadian Journal of Agricultural Economics*, **26**, 1–14.
Phillips, P.W.B. (2001a), 'The role of private firms', in P.W.B. Phillips and G.G. Khachatourians (eds), *The Biotechnology Revolution in Global Agriculture: Innovation, Invention and Investment in the Canola Industry*, Wallingford: CABI Publishing.
Phillips, P.W.B. (2001b), 'Regulating discovery', in P.W.B. Phillips and G.G. Khachatourians (eds), *The Biotechnology Revolution in Global Agriculture: Innovation, Invention and Investment in the Canola Industry*, Wallingford: CABI Publishing.
Phillips, P.W.B. (2003), 'The economic impact of herbicide tolerant canola in Canada', in N. Kalaitzandonakes (ed.), *The Economic and Environmental Impacts of Agbiotech: A Global Perspective*, New York: Kluwer Academic Publishers, pp. 119–40.
Phillips, P.W.B. and G.G. Khachatourians (2001), *The Biotechnology Revolution in Global Agriculture: Invention, Innovation and Investment in the Canola Sector*, Wallingford: CABI Publishing.
Prairie Pools Inc. (n.d.), 'Prairie Grain Variety Survey 1977–1992', Regina, SK: Prairie Pools Inc.
Saskatchewan Crop Insurance Corporation (SCIC) (2009), 'Saskatchewan management plus: canola variety data: by crop district, online database, Melville, SK: SCIC, available at: http://www. saskcropinsurance.com/Default.aspx?DN=46abeb04-59a1-46c5-bcdd-3106753216fe.
Saskatchewan Ministry of Agriculture (SMA) (2011), 'Crop planning guide 2011', available at: http://www.agriculture.gov.sk.ca/Default.aspx?DN= 1a1101e3-f7bb-4c4d-ba4c-a4faadff5e80, accessed 30 October 2011.
Saskatchewan Seed Growers (SSG) (2011 and various years), 'SaskSeed', available at: http://www.saskseed.ca/guides.html, accessed 30 October 2011.
Seed Manitoba (2011), 'Seed Manitoba 2012', Carman, MB: published jointly by Manitoba Agriculture, Food & Rural Initiatives, the Manitoba Seed Grower's Association and the Manitoba Cooperator, available at: http://www.seedmb.ca/, accessed 30 October 2011.
Smyth, S.J. and R. Gray (2011), 'Intellectual property sharing agreements in gene technology: implications for research and commercialization', *International Journal of Intellectual Property Management*, **4**(3), 179–90.
Smyth, S., M. Gusta, K. Belcher, P. Phillips and D. Castle (2011a), 'Changes in herbicide use after adoption of HR canola in Western Canada', *Weed Technology*, **25**, 492–500.
Smyth, S.J., M. Gusta, K. Belcher, P.W.B. Phillips and D. Castle (2011b), 'Environmental impacts from herbicide tolerant canola production in Western Canada', *Agricultural Systems*, **104**, 403–10.
Statistics Canada (2011), 'CANSIM Table 10010', Field Crop Reporting Series, available at: http://dc2.chass.utoronto.ca.proxy2.lib.umanitoba.ca/cgi-bin/cansimdim /c2_getArrayDim.pl, accessed 24 August 2011.
Stefansson, B.R., F.W. Hougen and R.K. Downey (1961), 'Note on the isolation of rape plants with seed oil free from erucic acid', *Canadian Journal of Plant Science*, **41**, 218–19.

Stiegert, K.W., G. Shi and J.P. Chavas (2010), 'Innovation, integration, and the biotechnology revolution in US seed markets', *Choices*, **25**(2).
Ulrich, A. and W.H. Furtan (1985), 'An investigation in the rates of returns from the Canadian crop breeding program', unpublished report, Department of Agricultural Economics, University of Saskatchewan, Saskatoon, SK.
Ulrich, A., W.H. Furtan and K. Downey (1984), *Biotechnology and Rapeseed Breeding: Some Economic Considerations*, Ottawa, ON: Science Council of Canada.
Veeman, T.S. and R. Gray (2009), 'Agricultural production and productivity in Canada', *Choices*, **24**(4).
Veeman, T.S. and R. Gray (2010), 'The shifting patterns of agricultural production and productivity in Canada', in J. Alston, B. Babcock and P. Pardey (eds), *The Shifting Patterns of Agricultural Production and Productivity Worldwide*, Ames, Iowa: The Midwest Agribusiness Trade Research and Information Center, Iowa State University.
Wright, B.D. (1998), 'Public germplasm development at a crossroads: biotechnology and intellectual property', *California Agriculture*, **52**(6), 8–13.
Wright, J. (2007), 'List of varieties registered for production in Alberta', Lacombe, AB: Agriculture Financial Services Corporation.

40 The Hawaii papaya story

Carol V. Gonsalves and Dennis Gonsalves

INTRODUCTION

We are currently immersed in the new age of agricultural biotechnology, having consumed trillions of servings of food developed with the aid of molecular technologies unheard of less than half a century ago. This astronomical number was highlighted by Miller (2011), who reported that in the area of North America alone, over three trillion servings of food containing genetically modified (GM) ingredients have been consumed. Billions of dollars and countless hours have been spent in research and development of new and improved agricultural products. Hundreds of millions of acres of genetically modified agronomic food crops such as canola, corn and soybean are being grown and sold in the market place. Horticultural food crops lag behind in acreage, with only papaya, sweet corn and squash available to consumers, while other beneficial products, many of which have already been developed and tested, have yet to be commercialized (Clark et al., 2004). Searching for a clear understanding of how a genetically modified product can make it to market can be a perplexing journey as we navigate through uncharted territory in the cosmos of a biotech universe.

In this chapter we aim to present a clear picture of the process of agricultural biotechnology that was used to transform papayas from those that could not survive an attack by papaya ringspot virus (PRSV) to those that thrive due to engineered resistance to that virus. We discuss efforts to commercialize the new products that were developed, the varieties 'Rainbow' and 'SunUp', so that they could be sold in the United States, Canada and Japan. We also present data from surveys of papaya farmers concerning their satisfaction and adoption of these new genetically modified varieties.

Although considered an orphan crop, one that was not likely to draw the interest of large companies seeking to boost their bottom line, much work has been done throughout the world by many researchers spending lots of money to enhance the value of this crop. Research has been done, for example, in the United States as well as in Bangladesh, Brazil, China, Indonesia, Jamaica, Malaysia, the Philippines, Taiwan, Thailand, Venezuela and Vietnam, in an effort to incorporate virus resistance as well as a variety of other traits into papaya (Gonsalves et al., 2008; Mendoza et al., 2008).

Rather than looking at the worldwide efforts to produce genetically modified papaya, most of which are not in commercial production, we have turned our lens here to our own 'Hawaii papaya story', primarily because it is here that we have first-hand knowledge and experience.

Papaya (*Carica papaya*) is the first horticultural fruit crop on the market today that was produced via agricultural biotechnology. In the 14 years since seeds of 'Rainbow' and 'SunUp' papaya were made available to the papaya farmers and backyard gardeners in Hawaii, reactions from farmers and consumers have been overwhelmingly positive. The news headlines have been almost universally positive: 'Transgenic virus resistant papaya:

from hope to reality for controlling papaya ringspot virus in Hawaii'; 'Farmers say "Yes" to transgenics'; 'Stalked by deadly virus, papaya lives to breed again'; 'Transgenic virus-resistant papaya: the Hawaiian "Rainbow" was rapidly adopted by farmers and is of major importance in Hawaii today'; 'Virus-resistant papaya in Hawaii. A success story'; 'Canada to accept Hawaii's genetically modified papaya'; 'Japan agrees to import GMO papayas from Hawaii Big Island'; and 'Crop savior blazes biotech trail, but few scientists or companies are willing to follow' (Bracken, 2011; Dayton, 2003; Gonsalves, 2000; Gonsalves et al., 2004a 2004b; Voosen, 2011; Westwood, 1999; Yoon, 1999). At a time when papaya should be merely one of many horticultural biotech fruit crops on the market, why is there only one? Possible limitations to development and commercialization will be discussed.

In short, what we have done is to look at the Hawaii papaya story from the very beginning, to try to ascertain what actions were put in place and how they interact in a multicomponent system to arrive at the goals which nested one inside another until the adopted action had finally resulted in a thriving arena of accomplishments. We hope that helpful insights can be gleaned from this exercise and that these insights will greatly help in bringing new and helpful products to the market.

THE MODEL

Papaya Ringspot Virus (PRSV) was first reported in Hawaii in the 1940s on Oahu island, which at that time produced the bulk of Hawaii's commercial papaya (Gonsalves, 1998). In the 1950s, PRSV began to severely affect papaya orchards on Oahu island and thus the papaya industry started to relocate to Puna on Hawaii island (Figure 40.1). Puna was a good region for papaya because of the availability of large tracts of rather inexpensive land to lease or to buy, an abundance of rainfall (over 100 inches per year) and sunshine, the presence of people who wanted to farm papaya, the 'Kapoho' papaya cultivar that was ideally suited for the Puna area and importantly the absence of PRSV. By the 1980s, Puna was producing 95 per cent of the state of Hawaii's papaya, which was Hawaii's second most important fruit crop behind pineapple.

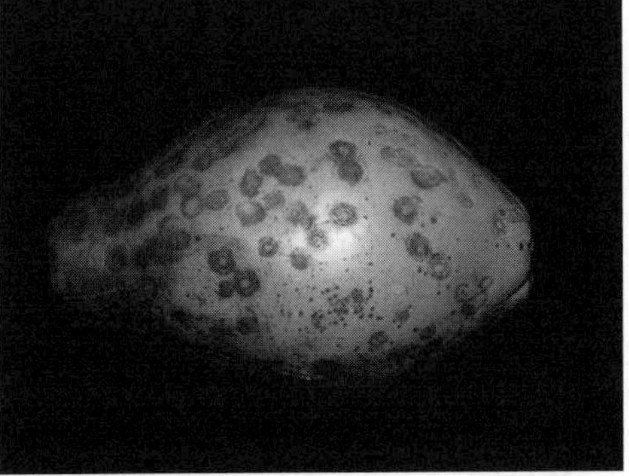

Figure 40.1 Effect of papaya ringspot virus on plant (left) and fruit (right)

However, the threat of PRSV to Hawaii's papaya industry was real since the virus was present in Hilo, which was only about 19 miles away. Hawaii recognized that threat, and the Hawaii Department of Agriculture (HDOA) had an active program to contain PRSV in Hilo and prevent its spread into Puna. Since it was anticipated that PRSV would eventually enter the Puna district, research was started in 1978 to develop strategies to control PRSV on Hawaii island. Cross-protection efforts were investigated; however, the mild mutant, PRSV HA 5-1, was not used widely because its effectiveness was limited to closely-related isolates (Tennant et al., 1994; Yeh and Gonsalves, 1994) and because it induced significant symptoms on fruit of certain papaya cultivars in the winter months (Ferreira and Gonsalves, unpublished observations).

Research was started in 1985 to develop PRSV-resistant transgenic papaya using the concept of pathogen-derived resistance (Gonsalves, 1998). In 1985–86, the concept of pathogen-derived resistance (Sanford and Johnston, 1985) was put forth and the laboratory of Roger Beachy (Powell-Abel et al., 1986) showed that tobacco and tomato expressing the coat protein gene of tobacco mosaic virus were tolerant or resistant to the virus. The concept of pathogen-derived resistance states that a transgenic plant that expressed a transgene of a pathogen would be resistant to that particular pathogen. In applying that concept to papaya, our goal was to develop transgenic papaya that expressed the coat protein gene of PRSV with the expectation that the transgenic papaya would show resistance to PRSV.

Pioneering research had to be done since little information was present on the sequence and cloning of the coat protein gene and tissue culture and transformation of papaya. Very briefly, the coat protein gene of PRSV was cloned and engineered to express a protein in a transformation vector; embryogenic calli of 'Sunset', 'Sunrise', and 'Kapoho' were transformed via the gene gun; and plants containing the coat protein gene of PRSV were regenerated from transformed embryogenic calli. Transformation experiments resulted in the successful regeneration of nine transgenic papaya lines that were screened for resistance to PRSV HA (Fitch et al., 1992). Six transgenic plants were from transformed calli of 'Sunset' and three from 'Kapoho'. 'Sunset' is a commercial red-fleshed Hawaiian 'Solo' papaya that is widely grown in Brazil, but not so much in Hawaii. 'Kapoho' is a yellow-fleshed Hawaiian Solo papaya that was the dominant cultivar being grown in Hawaii. By 1991, greenhouse experiments had shown that RO clones of one transgenic 'Sunset' line, designated 55-1, were resistant to PRSV HA (Fitch et al., 1992). Line 55-1 was a female. Cloned plants were derived from line 55-1 and planted in a field trial in 1992 on Oahu island.

Coincidentally, the long-anticipated PRSV entry was detected in the Puna where 95 per cent of Hawaii's papaya was being grown (Figure 40.2). Since PRSV can be spread by aphids, the virus spread rapidly throughout the Puna district despite efforts to control it. In October 1994, the HDOA declared that PRSV was uncontrollable and stopped the practice of marking trees for roguing. In less than three years, a third of the Puna papaya area was infected. By 1997, Pohoiki and Kahuwai were completely infected. Kalapana was the last place to become heavily infected. Five years after the onset of the virus in Pahoa, the entire Puna area was severely affected (Gonsalves, 1998).

Fortunately, the 1992 field trial confirmed that RO clones of line 55-1 were resistant to PRSV. This same RO field and plants were used to subsequently develop the 'SunUp' and 'Rainbow' cultivars (Manshardt, 1998). 'SunUp' is the transformed line 55-1 that

Figure 40.2 *Healthy papaya in Puna in 1992 (left) compared to severely infected papaya fields in 1994 (right)*

Figure 40.3 *'Rainbow' papaya show excellent resistance (left) while surrounding papaya are severely infected: quality of 'Rainbow' papaya (right) was excellent*

was crossed with non-transgenic 'Sunset' to obtain F_1 seeds, and then selfed for two more generations until the line was homozygous for the coat protein gene and showed resistance. The yellow-fleshed 'Rainbow' is an F1 hybrid from the cross of 'SunUp' with the non-transgenic 'Kapoho'. In 1995, a field trial was established in a severely infected farm in Puna (Figure 40.3). The field trial conclusively confirmed the resistance of 'Rainbow' and 'SunUp' under severe disease pressure and also confirmed the excellent horticultural characteristics of the cultivars (Ferreira et al., 2002).

Starting in 1995, the scientists worked toward regulatory approval of the papaya from Animal Health Inspection Service of USDA (APHIS), Environmental Protection Agency (EPA) and the Food and Drug Administration (FDA). By 1997, line 55-1 and subsequent progenies from this line were deregulated by APHIS and EPA and the FDA consultation was approved. The Papaya Administrative Committee (PAC), a USDA marketing order with the papaya farmers of Hawaii, successfully obtained the licenses to commercialize the new papaya varieties in Hawaii (Westwood, 1999). On 1 May 1998 transgenic seeds

Table 40.1 Fresh papaya production in the state of Hawaii and in the Puna district from 1992–2008

Fresh papaya utilization in Hawaii			
Year	Total (× 1000 lbs)	Puna (× 1000 lbs)	Puna %
1992[a]	55800	52955	95
1993	58200	55785	95
1994	56200	52525	93
1995	41900	29215	93
1996	37800	34195	90
1997	35700	27810	77
1998[b]	35600	26750	75
1999	39400	25455	64
2000	50250	33950	67
2001	52000	40290	77
2002	42700	35880	84
2003	40800	35735	87
2004	34100	29995	87
2005	30700	26910	87
2006	26600	24090	90
2007	31200	28340	90
2008	31500	28500	90

Notes:
[a] PRSV first reported in Puna. Note that figures represent Hawaii County but we assume for nearly all years that much of the papaya was produced in the Puna district of Hawaii County.
[b] Transgenic seed released.

were released to farmers just about seven years after line 55-1 showed resistance to PRSV in greenhouse trials and six years after PRSV was discovered in Puna.

Puna produced 52.9 million lbs of papaya in 1992, the year PRSV was detected in Puna (Table 40.1). By 1998, when 'Rainbow' was released to growers, Puna was totally infected with PRSV and produced only 50 per cent of its 1992 level. Puna produced only 75 per cent of the state's papaya, which was down from 95 per cent in 1992. The release of the transgenic papaya stemmed the production decline in Puna (Table 40.1). Currently, Puna produces 90 per cent of the state's production. The transgenic papaya saved the papaya industry in Hawaii from severe destruction and collapse and it currently represents about 85 per cent of the papaya production in the state (Table 40.2). Furthermore, the resistance of the transgenic papaya has remained durable with no evidence of breakdown. The state's 2008 papaya production is still significantly below that of the 1992 production, however, which is probably due to the difficulty in reclaiming lost markets and because Japan, a major market, would not import the transgenic papaya because it was not deregulated by that country's regulators (Pesante, 2003).

In 1998, Japan had been a major market since the 1980s and Canada was an emerging market. However, the transgenic papaya could not be sold in Japan or Canada until each country assessed and approved it. The scientists again successfully led the efforts to

Table 40.2 Relative percentage of acreage of papaya varieties grown in the state of Hawaii from 2000–09

Variety	2000	2001	2002	2003	2004	2005	2006	2007	2008	2009
'Kapoho'	37	43	42	38	29	30	25	17	16	9
'Rainbow'	32	39	44	46	52	53	58	68	64	77
'Sunrise'	14	11	10	10	11	9	11	8	9	9
Other[a]	7	7	4	6	8	8	6	7	11	5

Note: [a] Transgenic 'SunUp' and other transgenic cultivars derived from 'Rainbow' or 'SunUp' are included in this category, and thus it is reasonable to assume that this percentage refers to transgenic papaya other than 'Rainbow'.

assess the transgenic papaya in these countries. The transgenic papaya was allowed for sale in Canada in 2003 (http://www.hc-sc.gc.ca/fn-an/gmf-agm/appro/papaya_e.html), and in Japan on 1 December 2011 (Sato, 2011). The Canadian market has expanded markedly (from 355 563 fresh kilograms, worth US$1 046 321 in 2004 to 1 522 389 fresh kilograms, worth US$4 264 769 in 2012) (US Trade Online), and the same is expected to occur in the Japanese market. Efforts are now underway to deregulate the transgenic papaya in mainland China. It is noteworthy that the Hawaiian transgenic papayas were the first 'fresh' transgenic US products to be deregulated and commercialized outside of the US.

METRICS

The Hawaiian papaya is often used as a model to illustrate the timely development and commercialization of a public sector developed transgenic crop. Below are the qualitative reasons for this success.

In 1978 PRSV was in Hilo and the potential threat of the virus to the main production area in Puna was real. The scientists took it seriously and started research to control PRSV even though PRSV was not yet in Puna. In fact it took 14 years from the time they started the research before PRSV got into Puna (Gonsalves, 1998).

Carrying out proactive research was crucial because research takes time. Even though the research moved forward at a good pace, it took from 1985–91 to get a PRSV-resistant plant in the greenhouse and another 7 years to get a cultivar, do field trials and to get the papaya deregulated and commercialized. If we had addressed the problem only after PRSV was discovered in Puna in 1992, it is likely that the papaya industry would not exist in Puna today. It would be like the situation in Oahu where commercial plantings were not economical.

In many ways, a characteristic of a good scientist is to know when a selected approach is not worth pursuing. This happened with our cross-protection approach. Then, we took up a completely new approach that could be the ultimate control; that is developing a plant that was resistant to PRSV. Classical breeding for resistance had proven fruitless and very importantly this new approach, if successful, would result in the development of PRSV-resistant papaya that would be the same as the non-transgenic cultivar except for its resistance. This approach was done even though funding was very minimal, because the research team was committed to the cause of developing an effective control method.

A very important factor that contributed to our timely success was the consistent approach to move forward as aggressively as possible, and not to get sidetracked by going for the more appealing 'academic' route. For example, we initially had only nine transgenic plants to test and only one of them showed resistance. We could have gone back and refined the transformation method to get more transgenic lines to test. Since our initial evidence was clear that this one line was resistant to PRSV and resistance was the main objective for our product, we opted to move forward with field trials with this one line.

Additionally, we cloned line 55-1 so we could carry out greenhouse and field tests quickly instead of waiting to get seeds from the transgenic line and subsequent seedlings to the test stage, which would have taken an additional 18 months. Furthermore, the aggressive, practical approach is also reflected in the use of the plant material in the original field trial to develop the 'Rainbow' cultivar, which thus saved significant time. While we did not have much financial support, we pressed forward with the work because of our commitment. In many ways, the papaya industry did not have the financial means to support the work.

Another unusual feature is that the scientists themselves worked to deregulate and help commercialize the product. It is rather unusual that scientists will leave their comfort zone of doing pure research and work to bring a product to the market. Working to deregulate a product is largely the purview of large companies, but in the papaya case the scientists themselves did the work to get the product deregulated. The papaya industry had no financial support nor expertise to get the job done, but somebody had to do this, otherwise the product could not have been released to the farmers. As it turned out, the 'Rainbow' and 'SunUp' papaya varieties offered excellent virus resistance and also a package of horticultural properties that appealed to the farmers, packers and consumers.

In summary, the scientists started the work to help the industry, practiced good science, and never lost sight of their goal to help the farmers.

QUANTITATIVE MEASUREMENT OF TRANSGENIC PAPAYA IMPACT

Farmer Adoption Studies

The crowning glory of all the work thus far was the official 1 May 1998, debut of 'Rainbow' papaya at the Hilo Hawaiian Hotel. Among the hundreds of guests were Peter Dunn, the Assistant Secretary of the United States Department of Agriculture (USDA), scientists, politicians and persons involved in every aspect of the papaya industry, including of course, the farmers.

On this very day, 'Rainbow' and 'SunUp' seeds were made available to the farmers FREE of charge. Interest ran high. However, there was the unanswered question: would the farmers plant it? Our minds were already racing to conceive a method by which we could measure farmer acceptance of these new papaya varieties. Here was a unique and rare opportunity to measure farmer adoption of a horticultural crop produced via the new method of agricultural biotechnology, since at that time 'Rainbow' was the first and only horticultural fruit crop to be deregulated for commercial use in the United States, and indeed in the entire world!

Prior to the 1 May debut, papaya farmers were given the opportunity to register for 'Rainbow' and 'SunUp' seeds through their farmer organization, the Papaya Administrative Committee (PAC). This Committee, established in 1971, was a Marketing Order under the oversight of the USDA. It proactively helps farmers with fruit quality standards, marketing and other farmer issues. In this case this included the work to obtain licenses for the technologies that had been used to develop the transgenic papayas, contracting with the Hawaii Agriculture Research Center (HARC) to produce the seeds and for planning and distributing the seeds to the farmers.

Although the virus-resistant papaya seeds were distributed to the farmers beginning on 1 May, it was not a situation where farmers could come en masse and take as much seed as they wanted. Rather, for the first distributions, seeds sufficient to plant 1000 acres in total were produced, and the distribution of these seeds occurred over a number of months, according to a plan whereby those farmers who were 'currently and historically most affected' by PRSV were the first to receive seed. In the first of four distributions, the top-priority farmers were given two packets of seeds (about 4000 seeds in each packet), which was enough to plant an acre of trees. This plan aimed at preventing the possibility of overproduction, which could occur if all the seeds were distributed at the same time.

Under the PAC system, all growers in the state of Hawaii who produce papayas commercially are required to become members. All these members had to do to be qualified to receive seeds was to register, attend an education session or watch a video which explained what transgenic papayas were all about and how to grow them, and to sign a sublicensing form.

At that time we decided to do a real-time farmer adoption study, concentrating on the adoption of 'Rainbow' papaya in the Puna area, since this was the area where the bulk of the industry was located and where most of the crop loss due to PRSV had occurred. Both 'SunUp' and 'Rainbow' were released to farmers at the same time; however, farmers were keenly interested in 'Rainbow', since it more closely resembled 'Kapoho', the yellow-colored non-transgenic variety that had long been the industry's favorite.

The 1999 survey

There was a sense of urgency and excitement in planning for the first survey (Figure 40.4). This would be an unprecedented adoption study, with the intent to capture adoption data at the earliest possible moment, on the first virus-resistant transgenic fruit to be commercialized in the US. The criteria for 'farmer adoption' of the genetically modified varieties was not determined by whether or not they obtained seeds, but whether they had actually planted their seeds. It was expected that early adopters might already have begun to harvest fruit, since fruit production begins at 9 to 12 months after planting. The interviews would be done during several visits to Hawaii, from June to September 1999, just 13 to 16 months following the 1 May 1998, release of 'Rainbow' and 'SunUp' seeds.

The papaya farmers would play an important role in this survey and, thus, care was taken in selecting the respondents. Since the PAC was organized under the federal Marketing Order of the United States Department of Agriculture (USDA), it was necessary to make a written request, as required by the Freedom of Information Act (FOIA),

Figure 40.4 Carol surveying farmers in adoption study, 1999

in order to obtain names and addresses of the farmers. The first list of 524 farmers received in July 1998 appeared to be highly overstated compared to current Hawaii Agricultural Statistics Service (HASS) production and farm data listing only 241 farms statewide for 1997 (Gonsalves, 2001). Therefore, after consulting with Emerson Llantero, Manager of the PAC, and Donald Martin, of HASS, it was confirmed that this was more of a collective list rather than a current one, since it included farmers who had retired or discontinued papaya farming for reasons not related to PRSV. It also included industry members such as personnel from papaya processing and packing companies. Emerson Llantero suggested that a more current list of active farmers might be found in the PAC list of growers who had registered to receive transgenic seed. He said that nearly all of the active farmers had registered, whether they were planning to plant transgenic seeds or not. This updated list of 256 registered members aligned more closely with the HASS report of 241 farms for 1997. In retrospect, this number was highly consistent with the 262 farms statewide that HASS reported for 1998 (with 208 of them from the island of Hawaii) (Gonsalves, 2001).

Of the 256 registered farmers, 202 resided on the island of Hawaii, with 171 of them farming in the PRSV-devastated Puna area. All of the Puna area farmers were selected for in-person interviews and were contacted by telephone or, if no response or no telephone numbers were found, they were sent letters to the addresses received through FOIA. Table 40.3 shows how the final number of 93 interviewed farmers was obtained (Gonsalves et al., 2004a).

When the virus-resistant papaya seeds were released to the farmers, Carol Gonsalves, who at the time was doing voluntary research in the laboratory of her husband Dennis Gonsalves at Cornell University, expressed a keen interest in carrying out the survey of the papaya farmers to see not only whether or not they were planting the new varieties but also whether the papaya farmers were satisfied with the new varieties.

Table 40.3 Papaya farmers in Puna and surrounding areas who had their farms in Puna selected for personal interviews from the total number of transgenic-registered PAC members

Puna farmer's home address	Total farmers	Number planned	Number interviewed[a]	Final number[b]
Hilo	22	10	6	6
Honokaa	6	1	1	1
Keaau	92	92	56	50
Kurtistown	5	5	0	0
Mountain View	3	3	0	0
Pahala	2	1	1	1
Pahoa	57	57	34	34
Papaikou	1	1	0	0
Pepeekeo	2	1	1	1
Not in Puna area[c]	12	0	0	0
Totals	202	171	99	93

Notes:

[a] Number of farmers interviewed (less those who could not be contacted, did not have time to be interviewed, or declined to be interviewed).

[b] Final number of interviews. Note: six Keaau interviews were disqualified (four due to another family member reporting on the same farm, and two due to conflicting answers).

[c] Two farms in Captain Cook, three in Honaunau and one each in Holualoa, Honomu, Kailua, Kapaau, Oceanview, Paauilo and Waikoloa.

Perhaps it is fitting here to mention that both Carol and Dennis are native Hawaiians whose ancestries also include lineages from Korea, Germany and England (Carol) and China and Portugal (Dennis). Their non-Hawaiian ancestors arrived in Hawaii during the sugar plantation boom days and they were both raised on the Kohala Sugar Plantation, where their fathers worked, about 90 miles from the Puna papaya-growing region on the island of Hawaii. They were used to the rural country life and understood the hard work that it took to farm papayas and they, like most other islanders, enjoyed eating the papaya that grew in their backyards. Their interest in enriching the farmers' lives was not just a passing curiosity.

Although funding for a papaya farmer adoption study was not available, this did not hamper the initiation of this work, for the Gonsalves personally funded all of the expenses for the adoption study. For Carol, it was a great opportunity to fulfill two of her goals: one, to do the adoption study and, two, to use this survey as the basis for her thesis for a Master's degree in Liberal Studies at the State University of New York (SUNY) Empire State College located at Albany, New York. This would be ideal, since most of the courses could be studied via distance learning and a thesis advisor could be selected from virtually any university in the country. Dr David Lee, Professor at the Charles H. Dyson School of Applied Economics and Management at Cornell University, was selected as a thesis advisor for the expertise he could provide, to ensure that the survey would be done properly and that it would be publishable upon completion. Carol would visit the farming area, speak to industry personnel at packing houses, the PAC, the University of Hawaii, the Hawaii Agricultural Research Center (HARC) and the Hawaii Agricultural

Statistics Service (HASS), prior to conducting the interviews. After graduation in 2002, a copy of the thesis was submitted to SUNY Empire State College and to the Cornell University library located at the New York State Agricultural Research Center at Geneva, NY (Gonsalves, 2001). An article based on this work was published online as a Feature Story at APSnet, 'Transgenic virus-resistant papaya: the Hawaiian "Rainbow" was rapidly adopted by farmers and is of major importance in Hawaii today', in the August–September 2004 issue (Gonsalves et al., 2004a).

Of the 93 registered farmers, 84 were men, and nine were women, ranging in age from 22 to 73, with the average sample age of 47. The ethnic heritage of farmers was 91 per cent Filipino and 4 per cent each for Japanese and Caucasian/Other. All farmers were educated either in the United States or the Philippines (29 per cent completed elementary school or partial high school, 43 per cent completed high school and 28 per cent had completed college or other post-graduate school). Most of the farmers interviewed (90 per cent) resided in Keaau or the Pahoa region of Puna; the other 10 per cent farmed there but lived in the surrounding area.

There is no doubt that the papaya farmers are hard-working people who are willing to contend with unpredictability and risk inherent in the act of farming. Many farmers supplemented their income from off-farm jobs (46 per cent) and many of their spouses (47 per cent) worked on the farm, often also working at off-farm jobs. The amount of time spent on off-farm jobs was quite significant, with 29 farmers working 40 hours per week, and 14 of them working 60 or more hours per week, respectively.

All of the 93 farmers interviewed (100 per cent) had registered to receive transgenic papaya seeds and 90 per cent of them (92/93) decided to become 'qualified' to obtain seed, which first became available to farmers on 1 May 1998. Farmer adoption – 'the planting of transgenic papaya seeds' – was high, at 76 per cent. Early adopters (19 per cent) were already harvesting fruit during the time of the interviews. Fully 80 per cent planted less than three months after receiving their seeds, while 20 per cent planted between four to nine months later.

Some time between 1992, when PRSV first entered the Puna growing area, and 1997, all of the farmers' fields had become infected with PRSV. Thus, it is not surprising that the primary reason given for wanting to plant 'Rainbow' and 'SunUp' (91 farmers responding, with multiple responses allowed) was: resistance to PRSV (96 per cent), decrease in production risk (19 per cent), wanting to try the new papaya varieties (16 per cent) and desire for higher profit (13 per cent). The 63 farmers who expressed their reasons for delayed planting (multiple responses allowed) said it was because they did not have land (33 per cent), were not ready (21 per cent), did not have enough seeds (21 per cent) or did not have enough money (17 per cent) (Gonsalves, 2001).

In the Kapoho area of Puna, the area hit hardest by PRSV, analysis of a subset of 34 farmers showed that 94 per cent had obtained seed, 88 per cent were adopters, and 29 per cent were harvesting 'Rainbow' or 'SunUp' at the time of the interviews. Given that it takes 9 to 12 months from sowing seed to fruit production and that the survey was done from 13 to 16 months after seed distribution, annual fruit production data could not be obtained from the interviews; however, it was shown in a five-year field study that 'Rainbow' production in fresh weight pounds is 3.5 times greater than the common non-transgenic papaya (Ferreira et al., 2002).

In addition to the high rate of farmer adoption over a short period of time, a few other interesting data revealed by the survey concern farm size, comparative cost of growing transgenic versus non-transgenic papaya, land cost and an important change in production strategy. Although farm sizes ranged from 1 to 305 acres, only four farms were 50 acres or larger. They accounted for 646 (36 per cent) of the 1796 total transgenic and non-transgenic acres represented in this survey; in contrast, 66 growers (71 per cent of farmers surveyed) had medium (21 to 49.9 acres) or medium-small (6 to 20.9 acres) farms. Twenty growers had small farms (1 to 5.9 acres), for a total of 59.5 acres combined. Three farmers had decided to stop farming. Of the 1796 acres in this survey, 1003 were non-transgenic and 793 were transgenic. Thus, 44 per cent of the trees in the Puna area were the transgenic varieties 'Rainbow' or 'SunUp' at this early stage of farmer adoption (Gonsalves, 2001). In 1999, 1940 acres of papaya were in production in the state of Hawaii according to USDA/NASS (US Trade Online, 2013). Thus, 92.5 per cent of the 1999 farm acres in the state of Hawaii were represented in this study of Puna area farmers. Most farmers felt that it cost about the same to grow transgenic as non-transgenic papayas. One different production strategy was to prepare seedlings for planting in the field, rather than direct-seeding. This way, young seedlings could be nurtured to strength prior to setting them out in the field. Most of the farmers did not own the land that they farmed. Instead, 85 of the 93 respondents (91 per cent) leased land from large landholders at an average of $132 per acre.

The 2011 farmer survey

Given that it has been 14 years since 'Rainbow' and 'SunUp' papayas were first distributed to Hawaii farmers for planting, one might wonder whether PRSV still lingers in papaya fields, whether the non-transgenic varieties still become infected, whether the transgenic varieties still resist infection, and whether farmers are still growing both transgenic and non-transgenic varieties. In order to get answers to these and other questions, 21 papaya farmers were questioned by telephone during the three-week period of 27 October to 16 November 2011, from a standard set of questions. This group included 18 farmers from the Puna area, seven of whom we knew personally due to their activities with the Hawaii Papaya Industry Association (HPIA) and eight who were suggested for interviews by the Calavo and Diamond Head papaya processing companies. In addition, and for the first time ever, three farmers from the island of Oahu were interviewed. Information from the Oahu farmers are discussed, but not included in the tables with the Puna area farmers in order to highlight comparisons between current data collected in Puna and that collected in Puna in 1999 (Gonsalves et al., 2004a). It is interesting to note here that the PAC was inactivated in 2002, by farmer vote. The non-profit HPIA, with voluntary farmer membership which had been in existence since 1965, is currently the major papaya farmer organization today (Information from The Hawaii Papaya Industry Association/ Papaya Info., http://www2.hawaii.edu/~doisteph/Papaya/history.html).

Two of the Puna area farmers had retired within the previous four years; thus, although their data is included in the 18 farmer responses in Table 40.5, they are not included in Table 40.4 (2011) where family participation practices of the 16 currently active farmers are reported. Compared with the earlier 1999 survey, there has been an intensification of families working on and off farms (Table 40.4), often with both spouses working another job besides the farm.

Seventeen of the farmers had established their farms between 1970 and 1994, prior

Table 40.4 Full-time and part time papaya farmers: family participation

On and off farm	Full-time farmers	Part-time farmers
No. of farmers	11	5
No. with off-farm jobs	2	5
Spouse works on farm		
Full-time	4[a]	1
Part-time	3	2
Does not work	3	2
Spouse holds off-farm job		
Yes	2	4
No	8	1
Children work on farm		
Full-time	2	0
Part-time	2/2[b]	1/1[c]
Do not work	5	3

Notes:
[a] One of the full-time farmers did not have a spouse.
[b] The '/2' refers to part-time work by a brother and a mother-in-law.
[c] '/1' refers to a mother-in-law who works part-time.

Table 40.5 Historical adoption data for 18 Puna area papaya farmers

When farm established	No. of farms	'Kapoho Solo' grown	PRSV present	Planted 'Rainbow' or 'SunUp'[a]	PRSV on 'Rainbow' or 'SunUp'
Prior to 1998[b] (1970–94)	17	17	17	16	0
After 1998 (Year 2000)	1	0	0	1	0
Totals	18	17	17	17[c]	0

Notes:
[a] One of the 17 farmers planted only 'Kapoho solo'.
[b] The 'Rainbow' and 'SunUp' seeds became available in 1998.
[c] Adoption rate 17/18 (94%) of farmers growing 'Rainbow' or 'SunUp' papaya.

to the 1998 seed distribution (Table 40.5). All of them were growing the non-transgenic 'Kapoho' cultivar at the time of seed distribution and all of their farms were infected with PRSV at that time. Sixteen of these farmers were 'adopters' of the new 'Rainbow' or 'SunUp' and reported that these varieties have never become infected with PRSV. One of the farmers began his farm in the year 2000, and grows only transgenic varieties. Based on this information, the farmer adoption rate at 94 per cent is very high.

Similar to the 1999 data, there are few large and small farms compared to medium-sized (29 to 49.9 acre) operations (Table 40.6). Currently, 11 farmers grow only transgenic varieties, three grow both transgenics and non-transgenics, and two grow only non-transgenic varieties. The adoption rate of farmers growing transgenic varieties (14/16)

Table 40.6 Puna papaya farm size and varieties grown in November 2011

Acreage grown			No. farmers	
All varieties				
Large	50+ acres		3	
Medium	29–49.9 acres		4	
Med-small	6–29.9 acres		5	
Small	1–5.9 acres		4	
Total farmers			16	
No. of farms growing:			No. of acres	
			GM[a]	Non-GM
'Rainbow' and/or 'SunUp'		11	373	0
'Rainbow'/'SunUp'/Non-GM		3	25	41
Non-GM		2	0	33
Total acres			398	74

Note: [a] GM is genetically modified.

is 88 per cent. The number of transgenic acres (398) was more than five times the non-transgenic acres (74), or 84 per cent of the total acreage grown.

All of the farmers, except for one who 'Never planted "Rainbow"', were clearly aware of the almost certain destruction of their crop if they were to grow non-transgenic papaya. We note here that one other farmer who previously grew 'Rainbow' is currently growing only non-transgenic papaya. A range of quotes speak for themselves: 'Cannot survive without "Rainbow"'; 'More like my security'; 'The only way we could make money'; 'Got tired of cutting down virus-infected trees'; 'Because "Rainbow" is protected from virus'; and 'There's no gamble; risk. Not worth to take the gamble.' It is evident that these seasoned farmers have not forgotten the severity of the damage PRSV did to their crop.

Again, when asked what farmers appreciated most about 'Rainbow', the most common answer is that it is resistant to PRSV. They also cited the juicy, sweet taste and beautiful appearance of the fruit, the productivity and other quality characteristics.

The main problems with growing papaya today included 'PRSV on farms growing non-transgenic "Kapoho" or "Sunrise"', a complaint by those who grow non-transgenic papaya varieties, because when neighboring farmers do not cut down and eliminate infected trees, these become sources of inoculum for aphids to spread to non-infected nearby fields. Other concerns included other pests (for example mealybugs, thrips and mites), other diseases (for example blackspot and white peach scale) and the generally higher cost of growing papayas in the face of uncertain moisture and depressed prices.

Three papaya farmers from the island of Oahu were also interviewed. Like most of the papaya farmers in Puna, these are men with many years of experience, having begun farming 42, 33 and 31 years ago, respectively, and who had to deal with the scourge of PRSV before transgenic-resistant papaya became available. We should remember that the Hawaii papaya industry began on Oahu in the 1940s. Before long, in 1945, PRSV appeared in the production area, eventually leading to the establishment of the major part of the Hawaii papaya industry in Puna, on the island of Hawaii. One of those farmers was a full-time transgenic papaya farmer who has crossed the 'Rainbow' with his 'Kamiya' papaya to produce a resistant line known as 'Laie Gold', also known as 'Kamiya Papaya' ('Kamiya Papaya/the GMO story' at: http://www.kamiyapapaya.com). The other two were full-time farmers with 30–40 per cent of the crop in 'Rainbow' papaya. All three farmers grew up on family farms on Oahu island and were ethnic Japanese. Before transgenic varieties became available, they raised the standard papaya varieties made available by the University of Hawaii (Line 5, Line 8 and Waimanalo Line 77) that were adapted to the climate and elevation in their area, but were not resistant to PRSV. All of the farmers have increased the acreage of papayas grown since the transgenic varieties became available.

Long-term Control of PRSV

The primary rationale for developing transgenic papaya was to control PRSV should it get into Puna. As seen in the above sections and Table 40.1, PRSV caused tremendous damage in Puna starting when it was discovered there in 1992. By 1995, the papaya harvest in Puna had declined to 29 million lbs as compared to about 53 million lbs in 1992. The harvest continued to decrease such that by 1998 the harvest was down to 26 million lbs. Seeds were released in May 1998 and the harvest of papaya started to increase in 2000 when the first full harvest of transgenic papaya was being realized. The harvests increased dramatically to 40 million lbs in 2001 and have fluctuated since between 28 and 35 million lbs. It is safe to say that without transgenic papaya, PRSV would have devastated the papaya industry and eliminated significant commercial production.

Evidence of successful control of PRSV is readily observed in the fields. We have not observed any sign of PRSV overcoming the resistance of the transgenic papaya whereas it has become increasingly difficult and risky to grow non-transgenic papaya because of PRSV. In fact, non-transgenic 'Kapoho' accounted for 95 per cent of the papaya in Puna in 1992 when PRSV was discovered; in 2000, it accounted for about 37 per cent of the papaya but in 2009 it represented only 9 per cent (Table 40.2). The percentage of transgenic papaya was about 39 per cent in 2000 and increased to 83 per cent in 2009.

Regulation and Public Acceptance of Transgenic Papaya

Regulation and consumer acceptance in the US has not been a major issue. While deregulation in the US took some time, it was achieved without any major hold-up. Ever since its release, there have been protests, but transgenic papaya continues to be widely accepted by the consumer.

While special interest groups have been against the transgenic papaya since it was commercialized, it is difficult to evaluate the anti-GM sentiment because formal manuscripts depicting the efforts and responses are not available. It is common knowledge, however, that in Hawaii on three occasions papaya orchards were cut down. While no claim has been made by anti-GM organizations to take credit for this destruction, the targeting of GM plots suggests the attacks are a response to the technology. Undoubtedly, the destruction of the papaya orchards has inflicted personal hardship on the growers. In the latest case, the state and the papaya industry posted a $30 000 reward to apprehend the perpetrators (Anonymous, 2012; Kubota, 2012). Nevertheless, today, probably 95 per cent of papaya sold in the supermarkets in Hawaii are transgenic. The telling figure is also reflected in the fact that transgenic papaya accounts for roughly 83 per cent of the papaya production in Hawaii. Surely, if the public did not want the papaya, the farmers would not grow the crop.

Deregulation and commercialization of transgenic papaya outside the US is another story. The deregulation and commercialization of the papaya in Canada was taken up by the Canada regulators and in 2003 GM papaya was allowed for food consumption (http://www.hc-sc.gc.ca/fn-an/gmf-agm/appro/papaya_e.html). This allowed for a significant expansion of the transgenic papaya market, as noted earlier (US Trade Online). Japan, the major export market for Hawaii papaya since the 1980s, with roughly 20 per cent of Hawaii's production being exported to Japan in 1992, has been a bigger challenge. Even though transgenic papaya became available in 1998, it could not be marketed in Japan because Japan had not deregulated the transgenic papaya. Thus, as PRSV devastated the papaya in Puna, the export of non-transgenic papaya dwindled from a value of $16 million in 1996 to $1 million in 2011 (Sato, 2011).

The natural solution was to gain regulatory approval for the transgenic papaya in Japan. This was a challenge because outside of key commodities such as soybean, corn, cotton and canola, transgenic crops have not been exported from the US to other countries. Moreover, these products are mainly highly processed or used for feed or clothing (as in the case of cotton). Papaya faced two challenges: would the foreign governments allow the deregulation and ultimate commercialization of a whole food like transgenic papaya? Furthermore, given the absence of private investors, the deregulation efforts would be made by public sector scientists and without much funding. In effect, the odds were stacked against deregulation of transgenic papaya in Japan.

Deregulation efforts were started in 1999 and are detailed in another recent review (Gonsalves, 2012). It took 12 years, but the transgenic papaya was deregulated and commercialized in 2011. It should be noted that the length of time to deregulate the papaya was not due to hold-up by the Japanese agencies. It was mainly due to the scientists' lack of manpower to carry out the scientific studies requested by the Japanese; this work was just one of many things that the investigators were doing. By Japanese rules, all transgenic papaya is labeled, which raises questions about whether consumers will accept these products when labeled as GM and how well the transgenic papaya, the first 'fresh' genetically modified product to be labeled and sold in supermarkets, will compete against non-transgenic papaya.

Lastly, the information accumulated for the Japan case provided the background for the scientists to file quickly for deregulation of the transgenic papaya in other countries. A petition to deregulate and market the transgenic papaya in China was submitted in October 2012 and a decision is pending at the time of publication. If successful, China and other markets could offer valuable new opportunities for Hawaiian producers.

In some cases, other countries have asked for help to follow the Hawaii case in development of their own transgenic papaya. The development of the Hawaiian transgenic papaya spurred interest from other countries even before the transgenic papaya was deregulated. In fact, graduate students and scientists from Jamaica, Venezuela, Brazil and Thailand came to Gonsalves' lab to develop transgenic papaya that would be suitable for their respective countries. Technically, each of the countries successfully developed transgenic papaya that showed resistance against strains of PRSV in their country. The details are reviewed elsewhere (Davidson, 2008; Gonsalves et al., 2007, 2008). So far, translating these technical results to deregulation and commercialization of GM papaya in those countries has failed. Possible reasons are discussed in the next section where a critical analysis is made of the transgenic papaya.

One notable exception is a PRSV-resistant transgenic papaya that is commercially grown in Guangdong province in China. While little has been published on the production and impact of this papaya, in a recent review the lead scientist of the group mentioned that 95 per cent of papaya grown in Guangdong province are transgenic, which attests to the apparent great success of their transgenic papaya (Ye and Li, 2010).

A Model System for Translational Biotechnology by the Public Sector

Although the Hawaiian transgenic papaya was deregulated and commercialized 14 years ago, the story is still very popular as judged by the large number of requests that the author has received to give talks and to write reviews on the subject. Why is this? We believe the popularity is because: (1) the story provides a compelling case of impact in that it virtually saved the Hawaiian papaya industry; (2) it involved public sector scientists who did not receive financial support from private companies; (3) it is still evolving from the initial development and commercialization in Hawaii and, more recently, to outside the US; (4) it has triggered papaya efforts in other countries as a direct result of the Hawaiian effort; and (5) the story as documented is the only commercially developed transgenic crop by the public sector in the US.

CRITICAL ASSESSMENT OF THE 'HAWAII PAPAYA STORY'

The success of the Hawaiian transgenic papaya story has been documented above and thus will not be elaborated much further in this section. The facts are there but perhaps it is worth providing a few words on the intangibles that contributed to the success of the story. The Hawaiian story is really a combination of proactive research that combined scientific skill along with luck and the art of aggressive, focused actions. What do we mean by this? First, being proactive was critical – as seen, it took 13 years from the time the transgenic project was started in 1985 to commercialization in 1998. It is safe to say that if work on developing a virus-resistant transgenic papaya started when the virus was discovered in Puna in 1992, the Hawaii papaya industry in Puna would have disappeared by the time a transgenic papaya was commercially available. Another aspect is that we never forgot what we were after – that is, the goal was to develop in a timely manner a virus-resistant plant that would perform under field conditions and with acceptable horticultural characteristics. And, another important aspect was that the work was done in a transparent way in that Hawaii knew what we were trying to do and actually were 'rooting' for us scientists to succeed considering the tremendous damage that was being done to Hawaii's second most important fruit crop. And lastly, the original team members, along with Steve Ferreira who joined the team later, got along well and all contributed in their own way.

Success or failure of events is defined differently by people, depending on the definition or parameters that are chosen. In this section, we define success as a technology being translated to its intended use. It assumes success in developing and testing the potential usefulness of the technology. The translation to its intended use, however, is the definition of success.

As noted above, the Hawaiian transgenic papaya effort resulted in the countries of Jamaica, Thailand, Brazil and Venezuela engaging with our laboratory to produce transgenic papaya for their countries. The clear intention was to translate the technology to a product that would be mobilized in their countries to combat the virus. Technically and academically each of the projects were highly successful. The graduate students and scientists involved in each of the projects performed excellently and they all produced transgenic papaya that were resistant or highly tolerant to PRSV strains in their countries. All of the papaya were taken back to their countries for trials. The technical performance of these papaya has been documented in reviews (Gonsalves et al., 2008). However, none of the papaya have been deregulated and commercialized years after they were brought back to their countries, and thus are not available to the stakeholders. The technological achievement was not translated to its intended use, so we cannot call it a success, by our definition. It would have been a success if these products were made available regardless of whether or not the intended users accepted it.

It is not our intention to go into the potential reasons why this has happened because it would be an academic exercise and we were not directly involved in the decisions made by the governmental agencies. This subject has been discussed in several publications (Davidson, 2008; Gonsalves et al., 2007, 2008). Suffice it to say, our goal of deregulating the transgenic papaya in these countries has not been obtained despite the technical success of developing virus-resistant transgenic papaya.

Papaya is a delicious fruit packed with nutrition for consumers. The Hawaii Papaya Industry Association in its online narrative, 'Papaya information/nutrition – WOW!' (Anonymous, 2006), touts a report by the Center for Science in the Public Interest which places papaya as one of the top five fruit that are high in nutrition, when compared by percentage of the daily value for six nutrients. Papaya is a good source of vitamins A, B, C and E, beta carotene, calcium and potassium. It also provides fiber, has no cholesterol, and is low in fat and calories. It remains our dream that the needy and poor of this world will all have the opportunity to consume papaya in their daily diets.

REFERENCES

Anonymous (2006), 'Papaya information/nutrition – WOW!', Hawaii Papaya Industry Association available at: <http://www.hawaiipapaya.com/info.htm>.

Anonymous (2012), 'Big Island groups offer $30000 reward for information on hacked papaya trees', *Star-Advertiser*, 1 May.

Bracken, S. (2011), 'Japan agrees to import GMO papayas from Hawaii', *Big Island News Center*, 2 September.

Clark, D., H. Klee and A.M. Dandekar (2004), 'Despite benefits, commercialization of transgenic horticultural crops lags', *California Agriculture*, **58**(2), 89–98.

Davidson, S.N. (2008), 'Forbidden fruit: transgenic papaya in Thailand', *Plant Physiology*, **147**(2), 487–93.

Dayton, K. (2003), 'Canada to accept Hawai'i's genetically modified papaya', *Honolulu Advertiser*, 28 January.

Ferreira, S.A., K.Y. Pitz, R. Manshardt, F. Zee, M. Fitch and D. Gonsalves (2002), 'Virus coat protein transgenic papaya provides practical control of papaya ringspot virus in Hawaii', *Plant Disease*, **86**, 101–105.

Fitch, M.M.M., R.M. Manshardt, D. Gonsalves, J.L. Slightom and J.C. Sanford (1992), 'Virus resistant papaya

derived from tissues bombarded with the coat protein gene of papaya ringspot virus', *BioTechnology*, **10**, 1466–72.
Gonsalves, C. (2000), 'Farmers say "yes" to transgenics', *Agriculture Hawaii*, **1**(3), 34.
Gonsalves, C., D. Lee and D. Gonsalves (2004a), 'Transgenic virus resistant papaya: the Hawaiian "Rainbow" was rapidly adopted by farmers and is of major importance in Hawaii today', APSnet Feature, American Phytopathological Society, August–September, available at: http://www.apsnet.org/online/feature/rainbow.
Gonsalves, C., D. R. Lee and D. Gonsalves (2007), 'The adoption of genetically modified papaya in Hawaii and its implications for developing countries', *Journal of Development Studies*, **43**, 177–91.
Gonsalves, C.V. (2001), 'Transgenic virus-resistant papaya: farmer adoption and impact in the Puna area of Hawaii', Master of Arts in Liberal Studies thesis, State University of New York, Empire State College, Albany.
Gonsalves, D. (1998), 'Control of papaya ringspot virus in papaya: a case study', *Annual Review of Phytopathology*, **36**, 415–37.
Gonsalves, D. (2012), 'Hawaii's transgenic papaya story 1978–2012: a personal account', in R. Ming and P. Moore (eds), *Genetics and Genomics of Papaya*, New York: Springer Publishing Company.
Gonsalves, D., S. Ferreira, J. Suzuki and S. Tripathi (2008), 'Papaya', in C. Kole and T.C. Hall (eds), *A Compendium of Transgenic Crop Plants: Tropical and Subtropical Fruits and Nuts*, Vol. 5, Oxford: Blackwell Publishing, pp. 131–62.
Gonsalves, D., C. Gonsalves, S. Ferreira, K. Pitz, M. Fitch, R. Manshardt and J. Slightom (2004b), 'Transgenic virus resistant papaya: from hope to reality for controlling of papaya ringspot virus in Hawaii', APSnet Feature, American Phytopathological Society, August–September, available at: *http://www.apsnet.org/online/feature/ringspot/*.
Kubota, G.T. (2012), 'Decapitation of papaya trees unnerves Hawaii isle farmers', *Star-Advertiser*, 1 May 2012.
Manshardt, R.M. (1998), '"UH Rainbow" papaya', University of Hawaii College of Tropical Agriculture and Human Resources, Germplasm G-1, available at: http://www2.ctahr.haWaii.edu/oc/.
Mendoza, E.M.T., A.C. Laurena, and J.R. Botella (2008), 'Recent advances in the development of transgenic papaya technology', *Biotechnology Annual Review*, **14**, 423–61.
Miller, H. (2011), 'Frankenfish fatuity', *Forbes*, 23 March, available at: <http://www.forbes.com/sites/henrymiller/2011/03/23/frankenfish-fatuity/>.
Pesante, A. (2003), 'Market Outlook Report FRESH PAPAYAS', June, available at: hawaii.gov/hdoa/. . .outlook-reports/papaya%20outlook%20report.pdf.
Powell-Abel, P., R.S. Nelson, B. De, N. Hoffmann, S.G. Rogers, R.T. Fraley and R.N. Beachy (1986), 'Delay of disease development in transgenic plants that express the tobacco mosaic virus coat protein gene', *Science*, **232**(4751), 738–43.
Sanford, J.C. and S.A. Johnston (1985), 'The concept of parasite-derived resistance – deriving resistance genes from the parasite's own genome', *Journal of Theoretical Biology*, 113, 395–405.
Sato, S. (2011), 'Japan approved GM papaya', Gain Report O. JA1048, USDA Foreign Agricultural Service.
Tennant, P.F., C. Gonsalves, K.S. Ling, M. Fitch, R. Manshardt, J.L. Slightom and D. Gonsalves (1994), 'Differential protection against papaya ringspot virus isolates in coat protein gene transgenic papaya and classically cross-protected papaya', *Phytopathology*, **84**(11), 1359–66.
US Trade Online, The Foreign Division of the US Census Bureau, available at: usatrade.census.gov, accessed 27 July 2013.
Voosen, P. (2011), 'Crop savior blazes biotech trail, but few scientists or companies are willing to follow', *New York Times*, 21 September.
Westwood, J. (1999), 'Virus-resistant papaya in Hawaii: a success story', *ISB News Report*, January.
Ye, C. and H. Li (2010), '20 years of transgenic research in China for resistance to papaya ringspot virus', *Transgenic Plant Journal*, **4**, 58–63.
Yeh, S-D. and D. Gonsalves (1994), 'Practices and perspective of control of papaya ringspot virus by cross protection', in (K. F. Harris ed.), *Advances in Disease Vector Research*, Vol. 10, New York: Springer-Verlag, pp. 237–57.
Yoon, C.K. (1999), 'Stalked by deadly virus, papaya lives to breed again', *The New York Times*, 20 July.

41 Sugar beet

Koen Dillen and Matty Demont

INTRODUCTION

Sugar beet (*Beta vulgaris* L.) is a specialized agricultural crop aimed solely at the refined sugar industry. Its root contains a high percentage of sucrose, the basic input for sugar processing. As such, sugar beet directly competes on a global scale with sugar cane, a crop with similar sucrose content. Sugar cane, a C-4 plant, outperforms sugar beet in the warm and dry climates, leading to spatial separation of cultivation. Sugar beet is mainly cultivated in the more temperate and colder climates around the world such as Europe and parts of North America where it is an important agricultural crop. Table 41.1 demonstrates the importance and spatial distribution of sucrose-containing crops worldwide.

The specific properties of sugar beet and its importance in industrialized countries with strong institutions and commercially oriented farmers makes it an interesting crop to be targeted by the biotechnology sector. As early as 1998, approval was granted for both food and feed use and environmental release in the US for genetically modified herbicide tolerant (GMHT) sugar beet (CERA, 2010). In this chapter we will first highlight the particular characteristics of sugar beet which make it suitable for genetic modification and try to explain why the first commercial application of GM sugar beet only took place in 2007, a decade after it was first officially granted permission. Next, we expand on the farmer experiences with cultivation. In the following section different methodologies that were used to assess the socio-economic impact of GM sugar beet are discussed and their respective strengths and weaknesses highlighted. The metrics section continues with

Table 41.1 Global production of sucrose crops in 2009

	Sugar cane	Sugar beet
Area planted worldwide (10^6 ha)	23.8	4.3
Europe (10^3 ha)	–	2911
North America (10^3 ha)	354	476
Rest of Americas (10^3 ha)	11721	15
Africa (10^3 ha)	1537	172
Asia (10^3 ha)	9680	699
Oceania (10^3 ha)	449	–
Yield (ton/ha)	70	53
Production (10^6 ton)	1661	227
Sugar extraction	12%	15%

Sources: F.O.Licht (2005); FAO (2011); USDA (2011).

a presentation of the results of the available socio-economic studies. The final section concludes and raises some topics and consideration for further research.

Sugar, the final product of the sugar beet production chain, consists of 99.7 per cent sucrose. This makes it physically impossible to distinguish GM sugar from conventional sugar as no protein or DNA is present (Klein et al., 1998). As sugar from both sources is chemically identical, human health is not an issue in the case of GM sugar beet. Moreover, from a consumer acceptance point of view, this equivalence made GM sugar beet potentially very interesting. At the time of the development of GMHT sugar beet, the impossibility to detect DNA or protein from GM sugar beet in the final product (sugar) allowed marketing in the EU without the need for labeling. Indeed, Article 8 of Regulation (EC) 258/97 on novel foods and novel foods' ingredients stated that ingredients should be labeled only 'if scientific assessment [. . .] can demonstrate that the characteristics assessed are different in comparison with a conventional food or food ingredient'. However, by the time the EU regulatory framework for GM labeling was revised in 2004 and GM sugar beet could have been deregulated, this regulation had been replaced. The new Regulation (EC) 1830/2003 states that all products produced from GM ingredients should be labeled regardless of the presence of proteins or DNA in the final product. This change in labeling regime affected mainly sugar (and edible oils such as canola) derived from GM crops and further hampered the marketability and hence interest in GM sugar beet in the EU. As a result, technology providers shifted their focus to the remaining sugar beet markets, albeit with a technology initially developed for the EU market.

One of these markets is North America, where labeling regimes are voluntary and consumers are accustomed to GM crops (Gruère, 2007). Nevertheless, an analysis of the US sugar sector's decision of whether or not to allow GM sugar beet demonstrated that public opinion against GM crops was an important consideration for introduction, independent of human health issues (DeVuyst and Wachenheim, 2005). Moreover, the same study showed that the marketability of the by-products was a major constraint to sugar processors accepting GM sugar beet. Pulp and molasses are sold to a variety of outlets, including both food and feed use. These by-products contain DNA and proteins of the GM crops which makes them subject to labeling in a variety of US trade partners. Following the food and feed approval in Australia, New Zeland, Japan and the EU in 2007, the sugar industry has finally opened its door to commercial introduction of GM sugar beet (Dillen et al., 2009b).

Both the end-product and the crop itself possesses specific characteristics that make it interesting for biotechnology. Sugar beet is a biennial plant, flowering and producing seed only every second year. However, sugar beet is harvested after the first growing season, which minimizes the chances of outcrossing between GM and non-GM sugar beet fields and hence facilitating coexistence at farm level. Gene flow with a potential detrimental effect on agricultural productivity can only occur via two pathways (Desplanque et al., 2002). During seed production introgression of the GM trait to conventional seed production may occur. However, this is not a new problem for the seed industry. When breeding different varieties of a conventional variety, breeders also have to assure that the variety is kept as pure as possible. The industry solved this problem through spatial isolation of different seed-producing varieties. According to the OECD, certified sugar beet seed production should be located at least 1000 m from the nearest sugar beet field

(OECD, 2011). In the case of GM sugar beet seed production, the industry is bounded to more stringent regulatory distances and even increased the spatial separation to avoid liability claims. In the US, for example, the legal spatial isolation distance was decided at 4828 m while the industry is requiring a 6437 m isolation distance (McGinnis et al., 2010). Until now this requirement has not created problems in the supply of GM sugar beet seeds, affirming the feasibility of these regulations.

An important concern in sugar beet gene flow is the appearance of 'weed beets'. Occurring since the 1970s, this infestation causes yield losses and mechanical problems during harvest (Longden, 1989). Weed beets are 'hybrids' originated from the introgression of wild beet (*Beta maritima*, *Beta macrocarpa*) genes into fields of cultivated sugar beet seed production. Weed beets are genetic bolters, emerging soon after the sowing of sugar beet, bolting in spring and producing large amounts of seeds before sugar beet is harvested (Sester et al., 2003). As sugar beet may bolter under specific vernalization conditions such as unusually cold springs or due to genetic impurities, cross-pollination with weed beet might occur. If the GM trait transfers to the weed beet, farmers may have severe problems in controlling weed beet population in their fields in the future. Farmers are used to controlling bolters in their fields, as bolters compete with the crop for sucrose allocation and interfere with harvest. Moreover, farmers growing GM sugar beet in the US contractually agree to remove any flowering plants in their field (McGinnis et al., 2010). Hence, compared to some other crops described in this book, the agri-environmental risk from GM sugar beet is rather small. This chapter mainly focuses on genetically modified herbicide tolerant (GMHT) sugar beet for which out-crossing of the GMHT trait to wild relatives is not a major issue as it does not offer a competitive advantage outside of agricultural fields. If in the future GM traits were to be commercialized with competitive advantage outside of the field, the present analysis may have to be updated.

The impossibility to differentiate GM from conventional beet sugar and the manageable environmental and coexistence constraints makes sugar beet an interesting crop for technology providers. Investigating the data on deliberate releases of GM crops in the environment we observe that the biotechnology industry has mainly focused on two specific phenotypes. Since 2004, most field tests were focused on broad spectrum herbicide tolerance. In recent years technology providers have also experimented with a variety resistant to rhizomania, a disease caused by beet necrotic yellow vein virus, which leads to smaller roots and lower sucrose content (European Commission, 2011; Information Systems for Biotechnology, 2011).

Herbicide tolerance is a very interesting trait for sugar beet farmers as economic sugar production is not possible without efficient weed control. Competition from uncontrolled weeds can result in large yield reductions (Märländer, 2005). Therefore, in order to achieve a successful crop, an efficient weed management strategy is required. However, there is no universal solution and the weed management strategies carried out to achieve optimal sugar beet production are expensive, labor intensive and generally complicated. Conventional weed control programs include both pre- and post-emergence herbicide treatments. The most important products in terms of volume are those used pre-emergence. In post-emergence, tank mixes are used in most countries, usually in several split applications at lower dose rates than in pre-emergence. In Europe most farmers use a balanced but very low rate of different active ingredients in the tank mixes in post-emergence. However, these schemes are far from perfect as current herbicides used in Europe are estimated to cause between 5 per cent and 15 per cent yield reduction, mainly as a result of phytotoxicity following application when the sugar beet crop is under stress (Coyette et al., 2002). Typically chemical control is combined with mechanical weeding to fight the bolters and weed beets in the field.

A GM beet tolerant to a broad spectrum herbicide (GMHT sugar beet) gives producers the capability of weed control using only one post-emergence herbicide, applied two or three times during the growing season, of course without neglecting the rules of good agricultural practice. This increases the efficacy of weed control while at the same time reducing complexity and offering higher flexibility, as the timing of applying broad-spectrum herbicide is less stringent. Overall, HT technology reduces the cost of production through lower expenditures for herbicides, labor, machinery and fuel (Qaim, 2009). How this change in production system affects the economic value of the technology is discussed in the sections below. Moreover, Bennett et al. (2006) compare both production systems, GMHT and conventional, in terms of environmental effects through means of a life-cycle assessment. Their findings show that for the case of sugar beet, the GM production system is less of a burden to the environment than the conventional production system.

Against this background, the commercial introduction of GMHT sugar beet took place in 2007 in the US and Canada. The cultivated event is called H7-1, commercialized by Monsanto in cooperation with KWS Saat (CERA, 2010). This specific GMHT crop is tolerant against glyphosate, a commonly used broad spectrum herbicide, often marketed as Roundup™. Note that commercial introduction was preceded by a peak in field trials in the years preceding the decision, as can be seen in Figure 41.1.

Stachler et al. (2010) report on the experiences from US growers in Minnesota and North Dakota. The survey results show that US farmers embraced the GMHT sugar beet (Dobbs, 2011) for its superior weed control, reflected by the swift ramp up in adoption to 49 per cent in 2008, the second year of cultivation, to 88 per cent in 2009 and to 93 per cent in 2010. According to the Animal and Plant Health Inspection Service (APHIS) adoption at the national level was even more rapid, reaching 95 per cent of US acreage

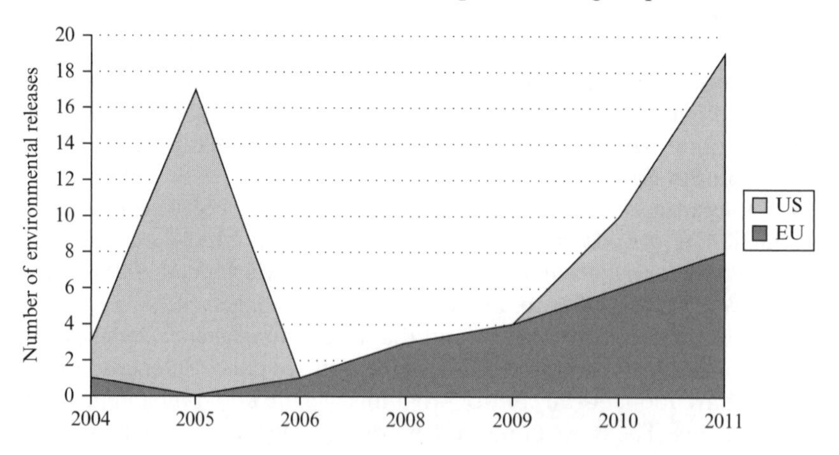

Sources: European Commission (2011); Information systems for biotechnology (2011).

Figure 41.1 The number of field trials with GMHT glyphosate-tolerant sugar beet

in GMHT sugar beet by 2009 (McGinnis et al., 2010). This rate of adoption makes it the most successful GM introduction so far, as predicted by Demont and Dillen (2008). This real life example offers an ideal case study to compare the theoretical agronomic effects with the experience in the field. The regional survey in Minnesota and North Dakota shows that since the introduction of GMHT sugar beet, the average number of herbicide applications decreased from 3.9 to 2.6 during the period 2007–10. Moreover, 71 per cent of the surveyed farmers using GMHT sugar beet reported excellent weed control while among conventional growers only 21 per cent of the farmers made such a statement. The higher efficiency under the broad spectrum herbicide is further confirmed by the significant reduction in the use of the rotary hoe (from 41 per cent to 2.8 per cent during 2007–10) and the area hand-weeded (from 28 per cent to 1 per cent during 2007–10).

Despite the commercial success of GMHT sugar beet in the US, its future is unclear. The initial deregulation of GMHT sugar beet was overturned in 2009 following a lawsuit by some NGOs. As a result, a Californian district court banned the planning in 2010. The court judged that the APHIS failed to take into account that transgenic pollen flow from GMHT sugar beet may hamper the production of conventional and organic sugar beet. However, as a shortage in the sugar market was predicted due to the under-supply of conventional seed (Neuman and Pollack, 2010), APHIS issued a 'partial deregulation' decision as an interim measure pending the completion of the court-ordered environmental impact statement. This statement, due mid-2012, is expected to propose full deregulation; a version open for public consultation is already available on the website (APHIS, 2011). The effect of the court ruling can be clearly observed in Figure 41.1 as field trials in the US restarted in 2009 in order to support the data collection for the new environmental impact statement.

In the EU, the largest sugar beet producer, the future is even more uncertain. As indicated before, H7-1 sugar beet was granted permission for import and feed use in the EU in 2008. Despite the remarkable success in the US and the importance of the crop for EU agriculture, H7-1 is not yet authorized for cultivation in the EU. The application for cultivation started in 2008. This is visible in Figure 41.1 through the increase of field trials which aim at gathering data for the scientific risk assessment conducted by the European Food Safety Authority. At the end of 2011, the assessment had not been finalized as it has to be made compatible with the new EFSA opinion on post-market environmental monitoring (EFSA, 2011).

MODELS

Despite the fact that GMHT sugar beet has been grown in the US for several years, knowledge about the socio-economic impacts of the innovation is limited. Kniss (2010) presents findings from a limited number of fields in Wyoming in the first year of adoption. The limited geographical scope and number of observations makes extrapolations from these results problematic. Dillen et al. (2013), on the other hand, present a simulation model assessing the aggregated effect but have to rely on field trial data as no other data is available. However, several studies have tried to estimate the potential socio-economic impact *ex ante*, that is, before the crop was introduced. In this section we describe the different approaches followed, their particular strengths and weaknesses and the lessons learned.

Table 41.2 Technology fee predictions in the literature on the ex ante *impact of GMHT sugar beet*

Source	Technology fee		Area under research	Methodology
May (2003)	20–30	£/ha	UK	Expert opinion
Märländer (2005)	30–45	€/ha	UK	Expert opinion
Gianessi et al. (2003)	38	€/ha	Denmark, France, Germany, Italy, Netherlands, Spain, UK and Belgium	Expert opinion
Demont and Tollens (2004); Demont et al. (2004)	40	€/ha	EU15	Expert opinion
Dillen et al. (2009a)	88–98	€/ha	EU27	Modeling
Demont and Dillen (2008)	88	€/ha	Czech Republic, Hungary	Modeling
Dillen et al. (2009b)	95	€/ha	EU27	Modeling
Burgener (2001)	123	$/ha	US	Field trials
Rice et al. (2001)	141	$/ha	US	Field trials
Kniss et al. (2004a; 2004b)	245	$/ha	US	Field trials

The results will be summarized in the metrics section. If we start with the US where GMHT sugar beet was adopted first, one *ex ante* impact assessment is available (Kniss et al., 2004a, 2004b). As GMHT sugar beet was not yet commercialized at the time of the study, the authors relied on experimental field trials with near-equivalent cultivars simulating different chemical weed control practices both in conventional and GMHT sugar beet. From the resulting differences in yields and production costs, they calculated potential profits for adopters of the technology. This highly controlled procedure of gathering data leads to very detailed results. The results show that the pay-off from the technology depends on different factors and is not uniformly positive. A first determinant is the price of GMHT sugar beet seed. As the technology is protected by intellectual property rights (IPRs), the technology provider will charge a licensing fee for the use of the technology, which decreases the profit for the adopter. Its direct impact on profits makes the technology fee a crucial parameter in *ex ante* impact assessment of IPR-protected technologies (Table 41.2). Kniss et al. (2004b) predict the technology fee by calculating a break-even price and then assuming the lower bound of the 90 per cent confidence interval to be a realistic estimator for the technology fee.

A second set of determinants consists of the yield and cost effects which are primarily determined by the nature of the 'counterfactual' production practices to which GMHT sugar beet is compared. The study by Kniss et al. (2004b) therefore lists a magnitude of different potential outcomes based on different counterfactuals. However, the set-up of the study does not allow for aggregation of the different scenarios as no information is given on the actual use of conventional treatment programs in the US. Moreover, although a controlled experiment delivers very detailed results on the potential impact of the technology, it does not reflect the situation within farms. The adoption of GMHT sugar beet does not happen in a vacuum. Its introduction on-farm involves a simultaneous decision over a bundle of complementary inputs (for example GMHT seed complemented with a broad-spectrum herbicide, reduced tillage and different crop rotation). Therefore the value of each technology should be calculated as the value of the new production system as a whole taking into account conventional farmer practices (Dillen et al., 2009a).

One possible solution to this problem is to start from aggregated data, an approach followed by several authors (Burgener, 2001; Gianessi et al., 2003; May, 2003). Using average cost and yield data, these studies rely on partial budgeting techniques to estimate the potential value from adopting GMHT sugar beet. This approach has different strengths, including transparency, ease and the low cost of its implementation as it usually relies on existing secondary data. However, following this approach the assessor implicitly assumes the existence of an average farmer representing the whole population. Previous research has showed that the value of a technology is not uniformly distributed among farmers; some realize a profit from the technology and adopt it, while others rationally choose not to adopt. In particular, GM seed technologies will pay off differentially depending on field conditions, pest and weed pressure, crop rotation and environmental conditions. Moreover, the valuation of the technology by any particular farm will depend on managerial expertise and local market conditions that condition the profitability of GM seed technologies relative to alternative technologies (Weaver, 2004). Indeed Demont et al. (2008) demonstrate that not considering this heterogeneity leads to a bias in the *ex ante* impact estimates. Therefore they propose a stochastic simulation model using probability density functions representing the heterogeneity among farmers. Demont and Dillen (2008) and Dillen et al. (2009a) apply this framework to the case of GMHT sugar beet in the EU. Moreover, they build on the framework by explicitly modeling the pricing decision of the technology provider. The value of the technology fee determines to a great extent the adoption by farmers as they weigh it against their valuation for the technology. If farmers consider the value of the innovation to be higher than the technology fee, they will adopt it; otherwise they will reject the technology. Endogenizing this pricing decision adds more realism and consistency to *ex ante* impact assessment and allows one to predict consistent potential adoption ceilings for the GMHT technology, an improvement on previous expert predictions (Coyette et al., 2002).

The former methodologies mainly looked at the isolated impact at the farm level. To gain insight into the welfare creation induced by the adoption, technology effects should be aggregated and market effects incorporated (Frisvold et al., 2003). A first approach, followed by Gianessi et al. (2003) and Park et al. (2011), applies a change in revenue methodology with homothetic extrapolation, approximating the producers' gain in welfare by multiplying the gross margin per hectare with an assumed adoption rate. However, this method implicitly assumes a small, open economy with inelastic supply such that no price effects are generated by the cost-reducing effects of the technology. Demont et al. (2008) assess the impact of GMHT sugar beet through a shift of an inelastic supply curve in two new EU member states. Similarly, the small, open-economy assumption causes prices to be unaffected by the technology. However, large-scale adoption of GMHT sugar beet would affect world prices. The EUWABSIM model, a partial equilibrium model developed by Demont and Tollens (2004), extends this line of reasoning by assuming a large open economy. EUWABSIM assesses the distributional effects of introducing GMHT sugar beet in three regional aggregates, that is, the EU, the Rest of the World (ROW) sugar beet region and the ROW sugar cane region. EU supply is further disaggregated to allow for differences among member states in competitiveness and production practices. Since its conception, the model has evolved to incorporate (i) the change in European sugar policy; (ii) the accession of new member states to the EU; and (iii) the farm-level model by Dillen et al. (2009a) discussed earlier. The model has been used in a variety of studies to address different questions (Dillen et al., 2008, 2009b, 2009c).

Besides market effects, the introduction of a novel technology might also have irreversible and time effects which are not included in the presented studies. As GMHT sugar beet does not show up in the consumed end-product, these would mainly include environmental issues. Many consumers are concerned about potential future irreversible costs of GM crops reducing the overall welfare from introduction. The Bayesian decision analysis of real options, suggested by Morel et al. (2003) and Wesseler (2003) in the context of GM crops, offers a tool to account for these irreversible effects by estimating the maximum incremental social tolerable irreversible costs (MISTICs) for society that would justify the introduction of the technology. Two studies applied this approach using the output from the EUWABSIM model to explain the decision-making process in the EU (Demont et al., 2004; Dillen, 2010).

METRICS

Before turning to a review of the results of different *ex ante* impact assessments, we focus on the technology fees used in the different studies as they largely determine the distribution of welfare. The larger the share of the technology fee in the value creation, the less profit accrues to farmers. The literature overview in Table 41.2 shows the wide range of technology fee predictions that have been used to assess the *ex ante* impact of GMHT sugar beet.

The wide range of estimates can be explained methodologically. The lower end of the spectrum is based on expert opinions while the highest predictions are based on controlled field experiments. The middle ground comes closest to the fee of $131/ha applied in the US (Kniss, 2010) and is a result of modeling attempts to endogenize the pricing decision in the impact assessment. Generally speaking the technology fees are very high compared to other GM food crops (Demont et al., 2007); Bt cotton is the only crop known to reach fees of similar magnitude (for example Qaim and de Janvry, 2003). The mere size of the fee either suggests that the technology is of high value to farmers such that they have a high willingness to pay for it, or that the technology provider is able to extract a large share of the created value, or a combination of both. Dillen et al. (2009a) argue that farmers' heterogeneous valuation of the technology constrains the technology provider's pricing strategy and ability to extract technology rents even under the assumptions of monopolistic market power and price discrimination. The results from different impact assessments in the literature seem to confirm this as farmers are predicted to capture considerable profits from the technology (Table 41.3).

According to this literature, average net profits for farmers from the adoption of GMHT sugar beet are in the range of 50–250 €/ha depending on the geographical scope and the underlying assumptions. Dillen et al. (2013) in their *ex post* analysis estimate an average benefit for the US of $257/ha which falls within this range. Kniss (2010), however, indicates an average benefit of $576/ha, indicating the particularity of the case study. All of the results clearly demonstrate the high economic potential of GMHT sugar beet for farmers. This does not come as a surprise since the adoption by US farmers was

Table 41.3 Net returns generated by the introduction of GMHT sugar beet and yield assumptions

Area under research	Net profit/ha		Type of data	Method	Assumed yield increase	Source
US	141	$/ha	Field trial	Partial budgeting	Case by case	Kniss et al. (2004b)
UK	154	£/ha	Aggregated data	Partial budgeting	2%	May (2003)
Denmark, France, Germany, Italy, Netherlands, Spain, UK and Belgium	111	€/ha	Aggregated data	Partial budgeting	n.a.	Gianessi et al. (2003)
US	149	€/ha	Aggregated data	Partial budgeting	n.a.	Gianessi et al. (2003)
EU27	50–150	€/ha	Secondary data	Review	n.a.	Park et al. (2011)
Czech Republic, Hungary	172–247	€/ha	Aggregated data	Partial budgeting	5%	Demont et al. (2008)
Czech Republic, Hungary	164–174	€/ha	Aggregated data	Partial budgeting	5%	Demont and Dillen (2008)
EU27	116	€/ha	Aggregated data	Partial budgeting	5%	Dillen et al. (2009a)
EU15	217	€/ha	Aggregated data	Partial equilibrium	5%	Demont and Tollens (2004); Demont et al. (2004)

so successful. However, as discussed before, behind these averages are farmers who obtain a heterogeneous pay-off from adopting the technology. Looking at the potential adoption ceilings provides a deeper insight in the farmer segment which is projected to gain from the technology and take part in benefit sharing. The available figures are presented in Table 41.4. The estimated adoption ceilings are high, averaging around 80 per cent for the EU. This indicates that the benefits are not only captured by an exclusive group of farmers with high weed problems but that almost all farmers could potentially gain from the introduction.

Different studies have tried to expand the assessment to the wider economy for which the results are presented in Table 41.5. Depending on the chosen methodology and the geographical coverage, the results vary. The most complete assessment stems from the last update of the EUWABSIM model covering the period from 1996 until 2015 and assessing the benefits forgone from not adopting GMHT sugar beet in the EU when the technology became available (Dillen et al., 2009b). The net present value of these forgone benefits amounts to €15.5 billion worldwide in 2015; 29 per cent of this value would have accrued to the EU producers adopting the technology while 31 per cent of the benefits would have been captured outside the EU. This is a net effect of the gains by consumers from lower sugar prices and the loss of sugar cane growers from the same price decrease.

Table 41.4 Predicted adoption ceilings (EU) and observed adoption rate (US) of GMHT sugar beet

Source	Average adoption	Area under research
Coyette et al. (2002)	84%	Germany, France, UK, Belgium, Netherlands and Spain
Demont et al. (2008)	68–88%	Czech Republic, Hungary
Demont and Dillen (2008)	60–80%	Czech Republic, Hungary
Dillen et al. (2009a)	51–99%	EU27
Demont and Tollens (2004); Demont et al. (2004)	75%	EU15
Dillen et al. (2009b)	47–100%	EU27
McGinnis et al. (2010)	95%	US

Table 41.5 Annual global value generation of GMHT sugar beet and its distribution over the supply chain

Source	Country	Annual global value (10^6 €)	Global benefit sharing			
			Technology provider	Domestic farmers	Domestic consumers	Net ROW
Demont et al. (2008)	Czech Republic	13	34%	66%	0%[a]	0%[a]
	Hungary	11	26%	74%	0%[a]	0%[a]
Gianessi et al. (2003)	Denmark, France, Germany, Italy, Netherlands, Spain, UK and Belgium	390	n.a.	n.a.	n.a.	n.a.
Park et al. (2011)	EU27	73–219	n.a.	n.a.	n.a.	n.a.
Dillen et al. (2009b)	EU27	772	39%	29%	0%	31%

Note: [a] As discussed before, this is a direct effect of the assumption of an open small economy.

The remaining 39 per cent would have been extracted by the technology provider. This simulation shows that even under the situation of strong intellectual property rights and the assumption of a monopolistic technology provider, welfare would have been distributed throughout the supply chain. EU consumers would not have gained from the introduction of GMHT sugar beet as domestic sugar prices are made inelastic by the EU common market organization (CMO). Dillen et al. (2008) in turn demonstrate that the 2006 reform of the CMO for sugar would have significantly altered the demand for the technology in favor of low-cost farmers. On the other hand, the total area planted with sugar beet decreased as a result of the reform, which has eroded the potential market for the technology provider.

Finally, Demont et al. (2004) show that at the aggregated EU level the maximum incremental social tolerable irreversible costs (MISTICs) are high, around €100 million per year. Purely from a rational point of view, this means that if the regulator expects the true irreversible costs of GMHT sugar beet introduction to be lower than this threshold value, the crop should be deregulated. However, until now this has not happened. If we assume that regulators take decisions based on voting incentives and consider the problem from the standpoint of an individual EU household, we can calculate the household-level MISTIC value, which is around €1 per year. Hence it is rational for individual households not to support the introduction of GMHT sugar beet as soon as they expect the irreversible costs to be more than €1 per year. This is easy to understand as EU consumers are hardly capturing any direct benefits from the technology in the context of the actual CMO for sugar. Therefore Dillen (2010) investigates which countries would be most inclined to deregulate GMHT sugar beet if the European Commission decides to delegate the decision on cultivation to the individual member states. His results show that Denmark and Belgium might be most inclined to deregulate while Italy and Portugal will not embrace the technology based on the MISTIC criterion.

CRITICAL ASSESSMENT

During the last 15 years GMHT sugar beet has been the center of attention of different stakeholders in the agricultural complex. Biotechnology providers were interested because of the low risk of gene flow and the potential to generate significant commercial returns on the innovation. Farmers were interested because of problematic weed control in conventional sugar beet and the potential to improve operating margins. The food industry was interested because of the initial absence of labeling and possible consumer acceptance of the equivalent 'GM' sugar, which would come at a lower price. Agricultural economists picked up this interest and applied some of the state-of-the-art methodologies (stochastic modeling, partial equilibrium and real options) to predict the outcome of a commercial introduction on society and assess the distribution of welfare along the supply chain in an *ex ante* setting.

The limited studies available studying the commercial introduction in 2007 in North America, seem to confirm the findings of the *ex ante* assessments which should not come as a surprise given the rapid adoption by US farmers and the size of the technology fee. However, as the area planted with sugar beet in North America only represents 10 per cent of the global sugar beet area, the welfare gains in that region do not compensate in any way the benefits forgone in the EU estimated by the EUWABSIM model. A thorough assessment of the benefits in the supply chain and its spillover effects on the world sugar market is essential. Moreover, the use of a general equilibrium model might be appropriate as the demand for sugar has changed drastically in the last couple of years. Sucrose containing crops are an important input for the biofuel sector. This new outlet for the crop altered the demand and supply of the crops, hence increasing the total value of GMHT sugar beet for society.

As discussed in an earlier section, coexistence is not a major issue in the case of GMHT sugar beet. However, including coexistence measures in socio-economic assessments could fine-tune the results as some EU member states have national coexistence regulations for GM sugar beet in place; for example, Latvia and the Netherlands have established isolation distances (European Commission, 2009). These measures may hamper adoption of the technology and lower welfare creation (Areal et al., 2011; Beckmann et al., 2006; Ceddia et al., 2011; Demont et al., 2009). Finally the existing literature fails to include technological advancements in sugar cane that may reshape the world sugar market. A look at the pipeline for plant biotechnology reveals that the biotechnology sector is developing GM sugar cane with increased yield potential which directly affects the comparative advantage of sugar beet production (CropLife International, 2013).

Furthermore, *ex post* impact assessments would allow for the validation of the different *ex-ante* methodologies that were developed. This would increase the future capacity for evidence-based decision-making by policymakers in *ex ante* settings. Returning to the plant biotechnology pipeline and the results from the field trial analysis, we see that the novel traits from sugar beet will most likely include virus resistance and increases in genetic yield potential. Both of these technologies will have a different socio-economic impact and hence should be studied in due time using the right approaches.

DISCLAIMER

The views expressed are purely those of the authors and may not in any circumstances be regarded as stating an official position of the European Commission.

REFERENCES

APHIS (2011), 'Glyphosate-tolerant H7-1 sugar beets: request for non-regulated status', available at: http://www.aphis.usda.gov/brs/aphisdocs/03_32301p_deis.pdf.

Areal, F.J., L. Riesgo and E. Rodríguez-Cerezo (2011), 'Attitudes of European farmers towards GM crop adoption', *Plant Biotechnology Journal*, **9**, 945–57.

Beckmann, V., C. Soregaroli and J. Wesseler (2006), 'Coexistence rules and regulations in the European Union', *American Journal of Agricultural Economics*, **88**, 1193–9.

Bennett, R.M., R.H. Phipps and A.M. Strange (2006), 'An application of life-cycle assessment for environmental planning and management: the potential environmental and human health impacts of growing genetically-modified herbicide-tolerant sugar beet', *Journal of Environmental Planning and Management*, **49**, 59–74.

Burgener, P. (2001), 'Economics of sugar beet production' in R.G. Wilson (ed.), *Sugar Beet Production Guide*, Lincoln, NE: University of Nebraska, pp. 189–96.

Ceddia, M.G., M. Bartlett, C. De Lucia and C. Perrings (2011), 'On the regulation of spatial externalities: coexistence between GM and conventional crops in the EU and the "newcomer principle"', *Australian Journal of Agricultural and Resource Economics*, **55**, 126–43.

CERA (2010), 'GM crop database', available at: http://cera-gmc.org/index.php?action=gm_crop_database.

Coyette, B., F. Tencalla, I. Brants and Y. Fichet (2002), 'Effect of introducing glyphosate-tolerant sugar beet on pesticide usage in Europe', *Pesticide Outlook*, **13**, 219–23.

CropLife International (2013), 'Plant biotechnology pipeline' available at: http://www.croplifeasia.org/wp-content/uploads/2013/06/PlantBiotechPipeline2013_highresFINAL.pdf.

Demont, M. and K. Dillen (2008), 'Herbicide tolerant sugar beet: the most promising first-generation GM crop?', *International Sugar Journal*, **110**, 613–17.

Demont, M. and E. Tollens (2004), *Ex Ante Welfare Effects of Agricultural Biotechnology in the European Union: The Case of Transgenic Herbicide Tolerant Sugarbeet*, Wallingford: CAB International.

Demont, M., J. Wesseler and E. Tollens (2004), 'Biodiversity versus transgenic sugar beet: the one euro question', *European Review of Agricultural Economics*, **31**, 1–18.

Demont, M., K. Dillen, E. Mathijs and E. Tollens (2007), 'GM crops in Europe: how much value and for whom?', *EuroChoices*, **6**, 46–53.

Demont, M., K. Dillen, W. Daems, C. Sausse, E. Tollens and E. Mathijs (2009), 'On the proportionality of EU spatial ex ante coexistence regulations', *Food Policy*, **34**, 508–18.

Demont, M., M. Cerovska, W. Daems, K. Dillen, J. Fogarasi, E. Mathijs, F. Muska, J. Soukup and E. Tollens

(2008), 'Ex ante impact assessment under imperfect information: biotechnology in New Member States of the EU', *Journal of Agricultural Economics*, **59**(3), 463–86.
Desplanque, B., N. Hautekeete and H. Van Dijk (2002), 'Transgenic weed beets: possible, probable, avoidable?', *Journal of Applied Ecology*, **39**, 561–71.
DeVuyst, C.S. and C.J. Wachenheim (2005), 'American crystal sugar: genetically enhanced sugar beets?', *Review of Agricultural Economics*, **27**, 105–16.
Dillen, K. (2010), 'Opportunities for agricultural biotechnology in the EU: policies and novel constraints', PhD dissertation, Katholieke Universiteit Leuven, Leuven.
Dillen, K., M. Demont and E. Tollens (2008), 'European sugar policy reform and agricultural innovation', *Canadian Journal of Agricultural Economics*, **56**, 533–53.
Dillen, K., M. Demont and E. Tollens (2009a), 'Corporate pricing strategies with heterogeneous adopters: the case of HR sugar beet', *AgBioForum*, **12**, 334–45.
Dillen, K., M. Demont and E. Tollens (2009b), 'Global welfare effects of GM sugar beet under changing EU sugar policies', *AgBioForum*, **12**, 119–29.
Dillen, K., M. Demont and E. Tollens (2009c), 'Potential economic impact of GM sugar beet in the global sugar sector', *International Sugar Journal*, **111**, 638.
Dillen, K., M. Demont, P. Tillie and M. Demont (2013), 'Bred for Europe but grown in the US: the case of GM sugar beet', *New Biotechnology*, **30**(2), 131–5.
Dobbs, M. (2011), 'Excluding coexistence of GMOs? The impact of the EU Commission's 2010 recommendation on coexistence', *Review of European Community & International Environmental Law*, **20**, 180–93.
EFSA (2011), 'Mandate 2008–1034', available at: http://www.efsa.europa.eu/.
European Commission (2009), 'On the coexistence of genetically modified crops with conventional and organic farming', COM (2009) 153 final.
European Commission (2011), 'Deliberate releases and placing on the EU market of genetically modified organisms: GMO register', available at: http://gmoinfo.jrc.ec.europa.eu/gmp_browse.aspx.
FAO (2011), 'FAOSTAT agricultural data', Rome: FAO.
F.O. Licht (2005), *World Sugar Yearbook 2006*, Kent: F.O.Licht.
Frisvold, G.B., J. Sullivan and A. Raneses (2003), 'Genetic improvements in major US crops: the size and distribution of benefits', *Agricultural Economics*, **28**, 109–19.
Gianessi, L.P., S. Sankula and N. Reigner (2003), 'Plant biotechnology: potential impact for improving pest management in European agriculture, sugarbeet case study', Washington, DC: National Center for Food and Agricultural Policy.
Gruère, G. (2007), 'A review of international labeling policies of genetically modified food to evaluate India's proposed rule', *AgBioForum*, **10**, 51–64.
Information Systems for Biotechnology (2011), 'USDA field tests of GM crops', available at: http://www.nbiap.vt.edu/Default.aspx.
Klein, J., J. Altenbuchner and R. Mattes (1998), 'Nucleic acid and protein elimination during the sugar manufacturing process of conventional and transgenic sugar beets', *Journal of Biotechnology*, **60**, 145–53.
Kniss, A.R. (2010), 'Comparison of conventional and glyphosate-resistant sugarbeet in the year of commercial introduction in Wyoming', *Journal of Sugar Beet Research*, **47**, 127–34.
Kniss, A.R., R.G. Wilson, A.R. Martin, P.A. Burgener and D.M. Feuz (2004b), 'Economic evaluation of glyphosate-resistant and conventional sugar beet', *Weed Technology*, **18**, 388–96.
Kniss, A., R. Wilson, P. Burgener, D. Feuz, A. Martin, C. Rice, A. Mesbah and S. Miller (2004a), 'Economic analysis of glyphosate-tolerant sugarbeet', *Zuckerindustrie*, **129**, 171–3.
Longden, P.C. (1989), 'Effects of increasing weed-beet density on sugar-beet yield and quality', *Annals of Applied Biology*, **114**, 527–32.
Märländer, B. (2005), 'Weed control in sugar beet using genetically modified herbicide-tolerant varieties: a review of the economics for cultivation in Europe', *Journal of Agronomy & Crop Science*, **191**, 64–74.
May, M.J. (2003), 'Economic consequences for UK farmers of growing GM herbicide tolerant sugar beet', *Annals of Applied Biology*, 142, 41–8.
McGinnis, E.E., M.H. Meyer and A.G. Smith (2010), 'Sweet and sour: a scientific and legal look at herbicide-tolerant sugar beet', *The Plant Cell Online*, **22**, 1653–7.
Morel, B., S. Farrow, F. Wu and E. Casman (2003), 'Pesticide resistance, the precautionary principle, and the regulation of Bt corn: real option and rational option approaches to decision-making' Washington, DC: Resources for the Future.
Neuman, W. and A. Pollack (2010), 'Duel over sugar beet seeds could create shortage', *New York Times*, available at: http://www.nytimes.com/2010/12/03/business/energy-environment/03sugar.html.
OECD (2011), *OECD Scheme for the Varietal Certification of Sugar Beet and Fodder Beet Seed Moving in International Trade*, Paris: OECD.
Park, J., I. McFarlane, R. Phipps and G. Ceddia (2011), 'The impact of the EU regulatory constraint of transgenic crops on farm income', *New Biotechnology*, **28**, 396–406.

Qaim, M. (2009), 'The economics of genetically modified crops', *Annual Review of Resource Economics*, **1**, 665–93.
Qaim, M. and A. de Janvry (2003), 'Genetically modified crops, corporate pricing strategies, and farmers' adoption: the case of Bt cotton in Argentina', *American Journal of Agricultural Economics*, **85**, 814–28.
Rice, C.A., A. Mesbah and S.D. Miller (2001), 'Economic evaluation of weed management systems in sugarbeets', *Proceedings of the American Society of Sugar Beet Technologists*, **31**, 64.
Sester, M., M. Delanoy, N. Colbach and H. Darmency (2003), 'Crop and density effects on weed beet growth and reproduction', *Weed Research*, **44**, 50–59.
Stachler, J., A. Carlson, J. Luecke, M. Boetel and M. Khan (2010), 'Survey of weed control and production practices on sugarbeet in Minnesota and Eastern North Dakota', available at: http://www.sbreb.org/.
USDA, 2011. 'ERS statistical database', available at: http://www.nass.usda.gov/.
Weaver, R.D. (2004), 'R&D incentives for GM seeds: restricted monopoly, non-market effects, and regulation', in R.E. Evenson and V. Santaniello (eds), *The Regulation of Aricultural Biotechnology*, Wallingford: CABI Publishing, pp. 143–51.
Wesseler, J. (2003), 'Resistance economics of transgenic crops, A real option approach', Washington, DC: Resources for the Future.

42 Rice

Matty Demont, Mao Chen, Gongyin Ye and Alexander J. Stein

INTRODUCTION

Rice belongs to the genus *Oryza* under the family *Poaceae*. This genus comprises more than 22 species distributed across the tropical, sub-tropical and temperate regions of Asia, Africa, central and south America and Australia but only two species are cultivated: *Oryza glaberrima* (Steudel) and *Oryza sativa* (L.) (Lu, 1999), the latter of which comprises the most common rice varieties, indica and japonica. Rice is the most important food crop of the developing world and the staple food of more than half of the world's population, many of whom are also poor and therefore vulnerable to high rice prices. Worldwide, more than 3.5 billion people depend on rice for more than 20 per cent of their daily calorie intake (Seck et al., 2012).

New rice varieties, such as those released during the Green Revolution in Asia, increased farmers' income and reduced the level of under-nutrition (Evenson and Gollin, 2003). Genetically modified (GM) crops are modified through transgenesis or recombinant DNA technology, in which a transgene is incorporated into the host genome or a gene in the host is modified to change its level of expression. Like 'conventional' breeding, genetic modification allows scientists to transfer genes within species but, in particular, it also enables transferring genes that encode for desired traits between species. Key drivers for the application and adoption of rice GM technology come from two main sources: the delivery of higher-yielding, disease-resistant and lower-cost rice production and the provision of nutritionally enhanced rice (Brookes and Barfoot, 2003). It is believed that the commercialization of GM rice will have substantial implications for the alleviation of poverty, hunger and malnutrition, not only for rice-growing and rice-consuming countries, but for all GM crops and their acceptance on a global basis (James, 2005).

Table 42.1 provides an overview of the GM rice traits developed, approved or in the advanced R&D pipeline. As in other GM crops, such as soybeans, maize, canola and cotton, GM rice is dominated by agronomic traits such as herbicide tolerance (HT) and insect resistance (IR) in commercialization and R&D efforts (James, 2005). However, rice traits have already been developed that focus on consumer or industry benefits, so-called second-generation GM traits. These are provitamin A-rich Golden Rice which is expected to convey nutrition benefits to poor populations in rice-eating developing countries who suffer from vitamin A deficiency (VAD), and rice that produces inputs for the pharmaceutical industry; many more traits have been researched to use GM rice for human and livestock health or phytoremediation (Yoshimatsu et al., 2012).

In 2002, proponents of biotechnology were optimistic about the release of GM traits in rice. They forecasted that by 2012, three-quarters of the GM traits would have reached the Asian farmer with a probability of at least 80 per cent (Brookes and Barfoot, 2003,

Table 42.1 GM rice traits approved and in the advanced R&D pipeline

Event, genes or product name	Trait	Developer	First approval or development stage
Rice with agronomic or nutritional traits			
LLRICE62	Herbicide-tolerance	Bayer CropScience, Germany	Fully deregulated in the United States in 2000[a]
LLRICE06	Herbicide-tolerance	Bayer CropScience, Germany	Fully deregulated in the United States in 2000[b]
LLRICE601	Herbicide-tolerance	Bayer CropScience, Germany	Environmental approval in the United States in 2006, approval for food and feed in Columbia in 2008[b]
Tararikhteh B827	Insect-resistance	Iran	Approved by the Agriculture Ministry in Iran in 2005, then revoked, commercialization possible within the next 5 years[a,i]
Huahui 1	Insect-resistance	Huazhong Agricultural University, China	Biosafety certificate for commercial production issued in China in 2009[c, d]
Shanyou 63	Insect-resistance	Huazhong Agricultural University, China	Biosafety certificate for commercial production issued in China in 2009[c, d]
KMD1	Insect-resistance	China	Stage of preproduction testing passed in China in 2009[c]
T1c-9	Insect-resistance	China	Stage of preproduction testing passed in China in 2009[c]
T2A-1	Insect-resistance	China	Stage of preproduction testing passed in China in 2009[c]
Kefeng 6	Insect-resistance	China	Stage of preproduction testing passed in China in 2009[c]
Kefeng 8	Insect-resistance	China	Stage of preproduction testing passed in China in 2009[c]
Golden Rice (GR2)	Beta-carotene content	International Rice Research Institute, Philippines	Preparation of safety information for submission to regulators in the Philippines in 2013[e]

cry1Ac	Insect-resistance	MAHYCO, India	Multi-location research trials[a, f]
cry1Ab, cry1C & bar	Insect-resistance	India	Advanced R&D in India[a]
CP iORF-IV	Virus-resistance	India	Advanced R&D in India[a]
RTBV-ODs2	Virus-resistance	India	Advanced R&D in India[a]
chi11 tlp	Disease-resistance	India	Advanced R&D in India[a]
Glyoxalase I & II	Abiotic stress-tolerance	India	Advanced R&D in India[a]
eOsmotin	Abiotic stress-tolerance	India	Advanced R&D in India[a]
RPD5-11	Yield enhancement	BASF India, India	Field trials authorized for seed production[f]
Bt traits	Insect-resistance	Pakistan	Advanced R&D in Pakistan[a]
Bt traits	Insect-resistance	Indonesia	Advanced R&D in Indonesia[a]
NUE	Nitrogen-use efficiency	Arcadia Biosciences, United States	Advanced R&D in the United States[g]
Rice with pharmaceutical or medical traits			
Lactoferrin	Lactoferrin content	Ventria Bioscience, United States	Clinical trials in the United States[h]

Sources: [a] Stein and Rodriguez-Cerezo (2009), as complemented by Barry (2012); [b] ISAAA (2011); [c] Chen et al. (2011); [d] Lu (2010); [e] IRRI (2012); [f] IGMORIS (2012); [g] Arcadia (2012); [h] Ventria (2012); [i] Ghareyazie (2012).

Table 10, pp. 16). However, in 2012 – despite the wide array of GM rice traits available (Table 42.1) – nowhere in the world has GM rice been commercialized on a large scale.[1] Nevertheless, given the existing approvals and the growing development pipeline, GM rice is nudging slowly towards commercialization. The approval of IR *Bt* (*Bacillus thuringiensis*) rice in China is of major importance to the developing world; this experience can lead the way for other GM food crops, particularly Golden Rice (De Groote, 2011). The expected commercial approval of Golden Rice in the Philippines in 2013/14 will be of significance to China, and also to Vietnam and Bangladesh, which are evaluating the product with a view to deployment (James, 2011).

MODELS

Due to the lack of large-scale commercialization of GM rice, no *ex post* evidence on its impact has been recorded. In the next section, we will discuss the available *ex ante* evidence, but here we will first start with a brief overview of the methods used.

Field trials provide a first impression of the potential impact of GM rice, but they are typically strictly controlled by researchers and do not reflect average farmer practices. In order to account for the latter, researchers have moved experiments from the field to the farm, following two approaches (Smale et al., 2008). First, partial budgeting (PB) techniques attempt to construct the marginal changes in variable costs and benefits per hectare that result from adopting GM crops. In this approach, GM IR rice is hypothesized to increase net yields by reducing insect damage, or to reduce the use and cost of insecticides and related inputs such as labor; GM HT rice is expected to reduce the use of tillage and more toxic herbicides; and drought-tolerant (DT) GM rice is assumed to save water and stabilize yields in times of drought, thereby reducing yield variance. In the second approach, these same hypotheses are tested more rigorously through econometric models with varying specifications.

To date, the most cited and perhaps most reliable data are from the pre-production trials of GM IR rice analyzed by Huang et al. (2008; 2005). The nature of China's pre-production trial system facilitated the analysis of the farm-level impact of GM IR rice before commercialization, while controlling for village and household-specific characteristics. Similar data are lacking for GM HT rice. Regarding GM-DT rice, field trial data are available and on-farm evidence of non-GM DT rice can be used as a proxy for the potential impact of GM DT rice at farm level (Pray et al., 2011).

First-generation GM rice traits typically involve direct productivity and income effects. On the other hand, second-generation GM rice traits, which involve enhanced quality attributes, must be evaluated differently. Quality improvements generally lead to a marginal utility increase and a higher willingness to pay (WTP) among consumers. In developing countries, the situation is different, especially when looking at technologies that are targeted at the poor, such as 'biofortified' GM rice: the process of adding nutritional value to a crop, as was done in the case of the aforementioned Golden Rice, is also called 'biofortification' (Montagnac et al., 2009). Widespread production and consumption of such crops could reduce micronutrient deficiencies, improve health outcomes, and provide economic benefits. However, it is unlikely that they could command higher market prices, because the poor are often not aware of their micronutrient deficiencies, which are also aptly dubbed 'hidden hunger', and anyway may not be able to pay quality premiums. Therefore, social welfare effects of biofortified crops must be evaluated differently (Qaim et al., 2007).

Dawe et al. (2002) quantified likely improvements in vitamin A (VA) intakes in the Philippines due to the consumption of Golden Rice. However, since micronutrient intake is not an end in itself but only a means to ensure healthy body functions, Zimmermann and Qaim (2004) and Stein et al. (2005; 2006; 2008) suggested health benefits be measured directly by comparing the burden of disease of VA with and without Golden Rice – expressed in disability-adjusted life years (DALYs) forgone.

METRICS

Field and Farm-level Impacts

At field and farm level, the anticipated impacts of GM rice (Table 42.2) are in line with classic examples of other first-generation GM crops in developing countries, such as GM IR cotton and maize and GM HT soybeans, reviewed by several authors (Carpenter, 2010; Finger et al., 2011; Qaim, 2009; Raney, 2006; Smale et al., 2009). Despite evidence from field trials (Oard et al., 1996), GM HT rice is generally not expected to increase rice yields in farmers' fields; its major contribution is assumed to be replacement of conventional tank mixes of herbicides by less toxic post-emergence herbicides (Bond et al., 2003; Demont et al., 2009). However, in settings where so far no or very little herbicide was used and where herbicide control was weak, GM HT rice can be expected to boost yields.

In the USA, GM HT rice is further assumed to raise product value as it provides better control of red rice, a weed that cannot be controlled easily with conventional herbicides in rice fields. Removing the red seeds from the commercial rice is necessary but raises costs to the millers, who in turn discount the price they pay to farmers (Annou et al., 2000). Overall cost savings, net of technology licensing fees, are estimated to be in the range of US$17–101/ha.

Stress-resistant or stress-tolerant crops are expected to prevent yield losses whenever and to the extent that the crops are exposed to the stress in question. The impact of GM DT rice, for instance, hinges on: (i) the likelihood and severity of drought stress and (ii) the availability and cost of irrigation in the case of drought. Similarly, the impact of GM IR rice depends on: (i) the likelihood and severity of insect infestation and (ii) the current use of insecticides.

Under total absence of insecticides, yield advantages of 29 per cent (Tu et al., 2000) and 60–65 per cent (Wang et al., 2010) were recorded for GM IR rice relative to its conventional counterpart. The yield advantage disappeared as soon as insecticides were applied often enough and in sufficient quantities to control insect pests. In this case the advantage of GM IR rice consisted in reducing the volume of insecticides that were needed to control insect pests by 50–60 per cent (Wang et al., 2010).

However, recent field trials indicate that the yield advantage of GM IR rice may not be realized when the pressure of target insects in the fields is insufficient. Under normal insect pressure, a yield advantage of 0–68 per cent was recorded, while under low insect pressure, the yield advantage turned into a yield loss of 0–56 per cent (Xia

Table 42.2 Field and farm-level impacts of first-generation GM rice traits

Trait	Region	Yield		Pesticides		Water saving	Cost saving	Adoption potential	Data	Methodology
		Mean	Variance	Volume	Poisoning					
GM HT	USA	+7%							Field trials	ANOVA[a]
GM HT	USA	+0%					$31–101/ha		Survey, assumptions	PB[b]
GM HT	USA	+0%					$17–42/ha[s]		Survey, assumptions	SPB[c]
GM HT	Uruguay	+2.5%					$25/ha	70%	Secondary data	SPB[d]
GM HT	Senegal	+0%		−31%			€22–26/ha	80%	Survey, assumptions	SPB[e]
GM IR	China	+29%		n.a.					Field trials	ANOVA[f]
GM IR	China	+60–65%		n.a.[t]					Field trials	ANOVA[g]
GM IR	China	+0%[u]		−50–60%					Field trials	ANOVA[g]
GM IR	China	+0–68%[v]							Field trials	ANOVA[h]
GM IR	China	−56–0%[w]							Field trials	ANOVA[h]
GM IR	China	+2–4%					$57–84/ha		Field trials	ANOVA[i]
GM IR	China	−0–29%							Field trials	ANOVA[j]
GM IR	China	+6–9%		−80%	−100%				On-farm trials	MR, household effects[k]
GM IR	China	+9–12%		−50%					On-farm trials	MR, village effects[l]

GM IR	China	+0%		−85–90%				On-farm trials	MR, household effects[l]
GM IR	Korea	−0–16%						Field trials	ANOVA[m]
GM DT	China	+0–49%[x]						Field trials	ANOVA[n]
GM DST	China	+18–41%[y]						Field trials	ANOVA[o]
GM DST	India	+25%		0%		−$4.1/ha	50%	Survey, assumptions	PB[p]
GM DST	Bangladesh	+20%		0%	−25%	−$3.0/ha	70%	Survey, assumptions	PB[q]
Non-GM DT	China	+7.8%	−13%		−33%			On-farm trials	MR[r]
Non-GM DT	South India	+28%[z]						On-farm trials	ANOVA[r]
Non-GM DT	South India	+8%[z]						On-farm trials	MR[r]
Non-GM DT	South India	+8.6%[aa]						On-farm trials	ANOVA[r]
Non-GM DT	East India	+29%						Survey	ANOVA[r]

Notes:
GM HT: herbicide-tolerant GM rice; GM IR: insect-resistant GM rice; GM DT: drought-tolerant GM rice; DST GM: drought and salinity-tolerant GM rice; GR: Golden Rice; ANOVA: analysis of variance; PB: partial budgeting; SPB: stochastic partial budgeting; MR: multiple regression; n.a.: not applicable.
[s] Impact for median-cost growers. Acres are converted to hectares (1 ha = 2.471 acres). [t] No insecticides were used on Bt rice and non-Bt rice plots. [u] Bt rice and non-Bt rice plots were protected through insecticides. [v] Normal insect pressure was recorded. [w] Low insect pressure was recorded. [x] Increase of relative yields (ratio of the yield in drought stress to that under normal growth) of GM DT rice compared to the wild type under drought-stress conditions. [y] Based on fresh weights per plant after mild salt stress treatment. [z] Yields of DT rice compared to high-yielding varieties (HYVs). [aa] Yield data compare drought-stressed DT varieties with landraces that had been resown due to the early drought in 2009 which caused poor germination of all varieties.

Sources: [a] Oard et al. (1996); [b] Annou et al. (2000); [c] Bond et al. (2003); [d] Hareau et al. (2006); [e] Demont et al. (2009); [f] Tu et al. (2000); [g] Wang et al. (2010); [h] Xia et al. (2010); [i] Tan et al. (2011); [j] Wang et al. (2012); [k] Huang et al. (2005); [l] Huang et al. (2008); [m] Kim et al. (2008); [n] Xiao et al. (2007); [o] Hu et al. (2006); [p] Ramasamy et al. (2007); [q] Islam and Norton (2007); [r] Pray et al. (2011).

et al., 2010). Yield drags in the same range have been observed in other trials where they were attributed to substantially lower grain filling percentages (Wang et al., 2012; Xia et al., 2010) and phenotypic variations under low infestation levels to the extent that the compensating effect of insect resistance was not reflected in yields (Kim et al., 2008). During 2002–04, GM IR rice was tested in pre-production trials in 17 villages located in 8 different counties in China. Multiple-year household surveys were conducted and the impact of GM IR rice at farm level was estimated through econometric techniques. GM IR rice increased yields by 0–12 per cent while insecticide use declined by 50–90 per cent, depending on whether the regressions controlled for village or household effects.

Similarly, field trial data suggest that GM DT as well as GM drought and salinity-tolerant (GM DST) rice perform substantially better relative to their non-GM counterparts under drought and salt stress conditions (Hu et al., 2006; Xiao et al., 2007). However, these findings have not yet been reproduced in farmers' fields. Non-GM DT cultivars in China, India and Thailand usually yielded more than existing cultivars under normal or high rainfall conditions, but inconclusive results were obtained for severe drought conditions in South India – both the DT and the conventional varieties were either not planted or, if planted, did not yield due to drought stress. In contrast, water use efficiency, as measured by yield per unit of irrigation, was greater than in non-DT varieties in the trials in China and Southern India. In rain-fed conditions of southern and eastern India, crop duration was much shorter with DT varieties than with high-yielding varieties (HYVs) or landraces, also suggesting more effective use of soil moisture (Pray et al., 2011).

The available micro-level impact evidence suggests that agronomic GM rice traits will deliver benefits similar to commercialized first-generation GM crops, namely improved and more stable yields, improved control of biotic or abiotic stresses, reduced pesticide use and related pesticide poisonings of farm workers, and increased water use efficiency (Qaim, 2009).

Consumer, Food Safety and Health Impacts

The potential benefits of second-generation GM traits in rice have been analyzed in depth for Golden Rice in the Philippines (Dawe et al., 2002; Zimmermann and Qaim, 2004) and in India (Stein et al., 2006, 2008). More recently the potential benefits of folate-rich GM rice have also been projected (De Steur et al., 2012a, 2010b). In as far as it reduces pesticide use and related poisonings, GM rice with first-generation agronomic traits can also have a positive health impact. During the above-mentioned pre-production trials in China, 3–11 per cent of non-GM rice farmers reported insecticide poisonings, while there were no reports for GM IR rice farmers, implying a 100 per cent reduction (Huang et al., 2005, 2008). For GM IR cotton Hossain et al. (2004) demonstrated that this can be attributed to the underlying GM technology.

The earlier economic analyses of Golden Rice in the Philippines suggested its development and dissemination could be cost-effective, but the expected impact on vitamin A (VA) intakes or related health outcomes would be insufficient to address VAD. When the first generation of Golden Rice, which was only a proof-of-concept, was replaced by the next generation of Golden Rice with higher provitamin A levels (Paine et al., 2005), new

Table 42.3 Projected health impact of second-generation GM rice traits

Trait	Region	Assumed coverage	Health impact (DALYs saved annually)
Golden Rice	Philippines	40–60%	15311–85137 DALYs[a]
Golden Rice	India	14–50%	204000–1382000 DALYs[b]
Golden Rice	Bangladesh	5–10%	62800–125600 DALYs[c]
Folate-rich rice	China	37–82%	116090–257345 DALYs[d]
Multi-biofortified rice	China	20–60%	1187024–4911814 DALYs[e]

Sources: [a] Zimmermann and Qaim (2004); [b] Stein et al. (2006, supplem.; 2008); [c] own estimation based on various sources (Barba and Cabrera, 2008; FAO, 2010; Stein et al., 2005; WHO, 2009; WHO/FAO, 2004) and personal communications with Fabiana de Moura (HarvestPlus) and Gerard Barry (IRRI); [d] De Steur et al. (2010b); [e] De Steur et al. (2012a).

studies showed that Golden Rice indeed has the potential to address VAD comprehensively and at a very low cost (Stein et al., 2006).

We report the evidence from the literature on the projected health impact of second-generation GM rice traits in Table 42.3. Using various scenarios, Stein et al. (2008) found that if breeding and dissemination of Golden Rice in India is successful, it could reduce the burden of disease of VAD by 60 per cent and, each year anew, prevent the loss of 1.4 million healthy life years (so-called 'disability-adjusted life years' or DALYs). In a 'low impact' scenario, in which breeding, dissemination and consumption of Golden Rice falls short of what could be reasonably expected, Golden Rice would still save over 200000 DALYs – more than 5500 children's lives – each year (Table 42.3).

While at a far earlier stage of research and development, the health benefits of folate-rich rice have been projected using the DALY framework (De Steur et al., 2010b). This analysis was done for the case of China, and in a subsequent study the analysis was expanded to include rice that is bred for – or biofortified with – other micronutrients (De Steur et al., 2012a). Both folate and provitamin A biofortification necessarily needs to be done through genetic engineering and it may be that iron biofortification may only be possible using this approach (Sperotto et al., 2012). Depending on the scenario used, multi-biofortified rice is expected to reduce the Chinese burden of disease of folate deficiency and VAD by 60 per cent and of iron deficiency by 34 per cent (De Steur et al., 2012a).

These benefits cannot materialize unless GM rice is adopted by farmers and accepted by consumers in target countries. For farmers to adopt a new rice variety, what matters most is its agronomic performance and the ability to market the rice (Chong, 2003). Consumers, in turn, are expected to base their purchasing decision primarily on the price of GM rice versus other rice (Corrigan et al., 2009). Surveys indicate that in China as few as 10 per cent of consumers oppose GM rice (Zhang et al., 2010), while almost 80 per cent are willing to pay a premium for more nutritious GM rice (De Steur et al., 2010a), even if they are poor (De Steur et al., 2012b). These results confirm the positive findings of most earlier studies on the acceptance of GM rice in China (see Huang et al., 2006). This general acceptance is also supported by other authors, even if they underline the importance of government endorsement regarding the safety of GM rice or the absence of adverse effects of GM rice for human consumption (Domingo and Gié Bordonaba, 2011; Knight and Gao, 2009; Tan et al., 2011).

Environmental Impacts

Similar to other GM crops, the same set of environmental safety concerns were also cast on GM rice with different agronomic traits, primarily HT and IR. Some of the major concerns include: (i) the capability of the transgenes to escape and potentially outcross to wild populations; (ii) impact on non-target organisms (for example non-target beneficial insects, charismatic species and non-target pest); (iii) the reduction of biodiversity; (iv) the persistence of the transgenes after harvest; and (v) the development of resistance against the engineered genes (Dale et al., 2002).

Outcrossing of transgenes is a major concern for the commercial cultivation of GM rice. Transgene outcrossing from a GM rice variety to its non-transgenic counterparts may cause trade difficulties or even legal disputes between countries and regions (Rong et al., 2012). Transgene outcrossing, especially the ones conveying selective advantageous traits, such as IR and HT, to wild or weedy relatives, may lead to undesired consequences, such as increased weediness or invasiveness of the wild or weedy relatives, and considerable reduction of the genetic diversity in wild populations (Rong et al., 2012). Many studies have been conducted on investigating outcrossing of HT or IR genes from GM rice to non-GM counterparts or weedy rices (Cohen et al., 2008; see reviews by Lu and Snow, 2005). In Asia, transgenes will almost certainly outcross from GM rice varieties to non-GM rice varieties, weedy rice, and two wild species closely-related to *O. sativa*, the perennial *O. rufipogon* and the annual *O. nivara* (Lu and Snow, 2005). Weedy rices, which include various weedy types of *O. sativa* and *O. rufipogon*, are among the most important weeds of cultivated rice in Asia (Baki et al., 2000). Cultivated rice is primarily self-pollinating, but a low level of outcrossing (ca. 0.01 to 1 per cent) occurs among cultivars under field conditions. Rong et al. (2012) found that scale had a significantly negative effect on the frequency of pollen-mediated outcrossing, with decreasing transgene outcrossing at increased scale. The rate of outcrossing declines rapidly with distance (Wang et al., 2006), but weedy rice occurs within rice fields and in some areas *O. rufipogon* is abundant in adjacent irrigation ditches and canals. Little direct evidence is known about the possible consequences of transgene outcrossing to weedy or wild rice. A recent two-year field study reported that GM HT rice Bar68-1 did not have a negative impact on the biodiversity of weeds in paddy fields in comparison with non-GM rice (Jiang et al., 2010).

Impact of GM IR rice on non-target organisms (including non-target herbivores, parasitoids, predators, soil-dwelling detritivores and microbial organisms) and biodiversity has been extensively studied (see reviews by Cohen et al., 2008; Chen et al., 2011). Potential impacts have been assessed under laboratory and field conditions, some over multiple years and sites. Importantly, the field studies have examined not only populations of target herbivores and their natural enemies, but non-target herbivores as well. Generally, few negative effects of GM rice on non-target organisms have been observed, as measured by indicators of fitness, population density and dynamics, and biodiversity indices (Chen et al., 2011). These results are consistent with other crops, as reviewed by Romeis et al. (2006), who compiled data from more than 100 peer-reviewed studies on the impact of GM IR crops on biological control organisms. As expected, some negative effects on natural enemies have been observed when Bt-susceptible, sublethally damaged herbivores are used as prey or hosts (Chen et al., 2011). Current GM IR rice events are primarily targeting lepidopteran insect pests. Planthoppers are the most problematic group of pests in rice ecosystems among non-target herbivores. It has been hypothesized that the reduction in foliage-feeding lepidopterans caused by GM IR rice might lead to reduced populations of generalist natural enemies and consequently to increased populations of non-target herbivores, particularly planthoppers. In a field study conducted at three sites in China over a two-year period with two Bt rice lines, populations of planthoppers and leafhoppers were very similar in GM and non-GM rice fields (Chen et al., 2011). Soil-dwelling detritivores, such as Collembola, play an important role in rice ecosystems, influencing soil structure and nutrient mineralization as well as the activity and composition of the microbial community. Multiple year/site field trials indicated that population levels of the most common soil-dwelling detritivores (for example Scatopsidae, Sminthuridae, Tomoceridae and Ceratopogonidae) were very similar in GM and non-GM rice fields (Cohen et al., 2008).

While transgenic crops offer many unique opportunities for the management of pest populations, one of the major concerns for long-term use of GM crops is the potential for the evolution of resistance. This primarily concerns GM crops with IR traits; however, since the discovery of glyphosate-resistant *Palmer amaranth* weeds in GM HT soybean fields (Sprague, 2012), resistance management for GM HT crops has received more attention. Strategies like 'high dose/refuge' and stacking genes with different mode of action have received strong theoretical support and have been broadly implemented in the USA, Canada and Australia for resistance management. Such strategies will be seriously challenged in GM rice because the insecticidal genes expressed in most of the GM rice events cannot meet 'high dose' standards, and it is difficult to implement a structured refuge for small rice growers. Durability of GM rice may be increased through the development and release of GM rice with stacked genes in which each gene has a different mode of action, especially for GM HT rice. Moreover, regulatory authorities may concurrently release GM rice with different traits; for example, GM HT rice could serve as a refuge for GM IR rice. In addition, ensuring adequate seed supplies of popular non-GM varieties may also help maintain a certain amount of non-GM rice in the field (Chen et al., 2011; Cohen et al., 2008).

CRITICAL ASSESSMENT

The available field and farm-level evidence suggests that the impact of first-generation GM rice traits will be similar to the impacts of other first-generation GM crops recorded in developing countries, such as GM IR cotton and maize and GM HT soybeans. Depending on pest levels and pesticide use, the impact of first-generation GM rice traits will either be reflected through a yield increase or a pest management cost saving. However, before solid conclusions can be drawn, more data will be needed at farm level and impacts should be estimated through econometric techniques in order to isolate the impact of GM rice from confounding factors. The available field trial evidence suggests that GM DT traits have the potential to dramatically increase average yield levels in rice production. However, while their major impact is expected to be found in a reduction of yield variance, little evidence is currently available; farm-level experiments with non-GM DT rice in China suggest that variance could decline by 13 per cent (Pray et al., 2011). Therefore, future research on DT technologies should focus on both first and second order statistics and introduce risk aspects into impact assessment (Konstandini et al., 2009).

On the one hand, unlike the first-generation agronomic traits of GM crops that offered direct benefits 'only' to farmers, biofortified GM rice comes with consumer benefits in the form of better nutrition and health that should facilitate acceptance. On the other hand, even such 'humanitarian' GM crops are embroiled in the larger controversy over modern agricultural biotechnology that is kept alive by various vocal groups of activists that push their agenda in developing countries. Therefore, the introduction of any new GM crop requires careful consideration of the political economy landscape and tailored awareness-building measures. Other factors, such as the different color of Golden Rice, may be less of an issue for consumer acceptance if the recent introduction of orange-fleshed sweet potatoes in Uganda and Mozambique is anything to go by (Hotz et al., 2012a, 2012b). Issues like the targeting, efficacy, food safety and biosafety of the crops, which are sometimes raised, will have to be shown or addressed as a matter of course when the crops reach the corresponding stage of their development, regulation or commercialization; in the case of Golden Rice this has already been done to some extent or such efforts are underway (Nayar, 2011; Tang et al., 2012).

Environmental concerns of first-generation GM rice traits are similar to those raised for other GM crops. Outcrossing of transgenes and the development of resistance are the major concerns for the commercial cultivation of GM rice. Isolation zones between GM and wild rice populations and refuges may offer a potential solution, but they will be particularly challenging for small farms (with 0.5–3 ha under cultivation), which is the dominant scale of rice growing in Asia and sub-Saharan Africa (Seck et al., 2012).

Nevertheless, despite these challenges, the available *ex ante* evidence suggests that GM rice may have a sizeable impact on poverty reduction and health. As widespread commercialization of GM rice may be imminent, the evidence reviewed in this chapter may guide policy makers in their decision-making and in the allocation of financial resources for further R&D.

NOTE

1. GM rice had been commercialized in Iran (James, 2005); however, not on a large scale – after one growing season the approval was cancelled and there was probably little more than seed multiplication going on. Similarly, Huahui 1 and Shanyou 63 have not been fully commercialized in China (Chen et al., 2011), and LL rice in the US has not been (purposefully) commercialized either (Mulvaney et al., 2011). Furthermore, Ventria does not seem to be producing its pharma rice beyond the scale of field trials (Ventria, 2012) (Table 42.1).

REFERENCES

Annou, M.M., E.J. Wailes and G. Cramer (2000), 'Economic analysis of adopting Liberty Link rice', in *Rice Situation and Outlook Yearbook*, no. RCS-2000, Washington, DC: United States Department of Agriculture (USDA), Economic Research Service (ERS).

Arcadia (2012), 'Nitrogen use efficient crops', 14 April, available at: http://www.arcadiabio.com/nitrogen.

Baki, B.B., D.V. Chin and A.M. Mortimer (eds) (2000), *Wild and Weedy Rice in Rice Ecosystems in Asia – a Review*, Los Baños, Philippines: International Rice Research Institute.

Barba, C.V.C. and M.I.Z. Cabrera (2008), 'Recommended dietary allowances harmonization in Southeast Asia', *Asia Pacific Journal of Clinical Nutrition*, **17**, S405–S408.

Barry, G. (2012), 'The future of GM rice and the possible social and economic impact', in M. Lusser, T. Raney, P. Tillie, K. Dillen and E. Rodriguez-Cerezo (eds), *International Workshop on Socio-economic Impacts of Genetically Modified Crops Co-organised by JRC-IPTS and FAO*, workshop proceedings, Luxembourg and Rome: Publications Office of the European Union and the Food and Agriculture Organization of the United Nations (FAO), pp. 129–30.

Bond, C.A., C.A. Carter and Y.H. Farzin (2003), 'Medium grains, high stakes: economics of genetically modified rice in California', *AgBioForum*, **6**(4), 146–54.

Brookes, G. and P. Barfoot (2003), 'GM rice: will this lead the way for global acceptance of GM crop technology?', *International Service for the Acquisition of Agri-Biotech Applications (ISAAA) Briefs*, 28, Ithaca, NY: ISAAA.

Carpenter, J.E. (2010), 'Peer-reviewed surveys indicate positive impact of commercialized GM crops', *Nature Biotechnology*, **28**(4), 319–21.

Chen, M., A. Shelton and G. Ye (2011), 'Insect-resistant genetically modified rice in China: from research to commercialization', *Annual Review of Entomology*, **56**, 81–101.

Chong, M. (2003), 'Acceptance of Golden Rice in the Philippine "rice bowl"', *Nature Biotechnology*, **21**(9), 971–2.

Cohen, M., M. Chen, J.S. Bentur, K.L. Heong and G. Ye (2008), 'Bt rice in Asia: potential benefits, impact, and sustainability', in J. Romeis, A.M. Shelton and G. Kennedy (eds), *Integration of Insect-resistant GM Crops within IPM Programs*, Dordrecht: Springer, pp. 223–248.

Corrigan, J.R., D.P.T. Depositario, R.M. Nayga, X. Wu and T.P. Laude (2009), 'Comparing open-ended choice experiments and experimental auctions: an application to golden rice', *American Journal of Agricultural Economics*, **91**(3), 837–53.

Dale, P.J., B. Clarke and E.M.G. Fontes (2002), 'Potential for the environmental impact of transgenic crops', *Nature Biotechnology*, **20**, 567–74.

Dawe, D., R. Robertson and L. Unnevehr (2002), 'Golden rice: what role could it play in alleviation of vitamin A deficiency?', *Food Policy*, **27**(5–6), 541–60.

De Groote, H. (2011), 'Crop biotechnology in developing countries', in A. Altman and P.M. Hasegawa (eds), *Plant Biotechnology and Agriculture*, Oxford: Academic Press, pp. 563–76.

De Steur, H., X. Gellynck, S. Storozhenko, G. Liqun, W. Lambert, D. Van Der Straeten and J. Viaene (2010a), 'Willingness-to-accept and purchase genetically modified rice with high folate content in Shanxi Province, China', *Appetite*, **54**(1), 118–25.

De Steur, H., X. Gellynck, S. Storozhenko, G. Liqun, W. Lambert, D. Van Der Straeten and J. Viaene (2010b), 'Health impact in China of folate-biofortified rice', *Nature Biotechnology*, **28**(6), 554–6.

De Steur, H., X. Gellynck, D. Blancquaert, W. Lambert, D. Van Der Straeten and M. Qaim (2012a), 'Potential impact and cost-effectiveness of multi-biofortified rice in China', *New Biotechnology*, **29**(3), 432–42.

De Steur, H., X. Gellynck, S. Feng, P. Rutsaert and W. Verbeke (2012b), 'Determinants of willingness-to-pay for GM rice with health benefits in a high-risk region: evidence from experimental auctions for folate biofortified rice in China', *Food Quality and Preference*, **25**(2), 87–94.

Demont, M., J. Rodenburg, M. Diagne and S. Diallo (2009), 'Ex ante impact assessment of herbicide resistant rice in the Sahel', *Crop Protection*, **28**(9), 728–36.

Domingo, J.L. and J. Gié Bordonaba (2011), 'A literature review on the safety assessment of genetically modified plants', *Environment International*, **37**(4), 734–42.

Evenson, R.E. and D. Gollin (2003), 'Assessing the impact of the green revolution, 1960 to 2000', *Science*, **300**(5620), 758–62.

FAO (2010), 'FAOSTAT agriculture data', 19 October, available at: http://faostat.fao.org.

Finger, R., N. El Benni, T. Kaphengst, C. Evans, S. Herbert, B. Lehmann, S. Morse and N. Stupak (2011), 'A meta analysis on farm-level costs and benefits of GM crops', *Sustainability*, **3**(5), 743–62.

Ghareyazie, B. (2012), 'GMOs from Iran in the next 5 years', paper presented at the FAO email conference on 'GMOs in the pipeline', 30 November.

Hareau, G.G., B.F. Mills and G.W. Norton (2006), 'The potential benefits of herbicide-resistant transgenic rice in Uruguay: lessons for small developing countries', *Food Policy*, **31**(2), 162–79.

Hossain, F., C.E. Pray, Y. Lu, J. Huang, C. Fan and R. Hu (2004), 'Genetically modified cotton and farmers' health in China', *International Journal of Occupational and Environmental Health*, **10**(3), 269–303.

Hotz, C., C. Loechl, A. de Brauw, P. Eozenou, D. Gilligan, M. Moursi, B. Munhaua, P. van Jaarsveld, A. Carriquiry and J.V. Meenakshi (2012a), 'A large-scale intervention to introduce orange sweet potato in rural Mozambique increases vitamin A intakes among children and women', *British Journal of Nutrition*, **108**(1), 163–76.

Hotz, C., C. Loechl, A. Lubowa, J.K. Tumwine, G. Ndeezi, A. Nandutu Masawi, R. Baingana, A. Carriquiry, A. de Brauw, J.V. Meenakshi and D.O. Gilligan (2012b), 'Introduction of β-carotene-rich orange sweet potato in rural Uganda resulted in increased vitamin A intakes among children and women and improved vitamin A status among children', *Journal of Nutrition*, **142**(10), 1871–80.

Hu, H., M. Dai, J. Yao, B. Xiao, X. Li, Q. Zhang and L. Xiong (2006), 'Overexpressing a NAM, ATAF, and CUC (NAC) transcription factor enhances drought resistance and salt tolerance in rice', *Proceedings of the National Academy of Sciences*, **103**(35), 12987–92.

Huang, J., R. Hu, S. Rozelle and C.E. Pray (2005), 'Insect-resistant GM rice in farmers' fields: assessing productivity and health effects in China', *Science*, **308**(5722), 688–90.

Huang, J., R. Hu, S. Rozelle and C. Pray (2008), 'Genetically modified rice, yields, and pesticides: assessing farm-level productivity effects in China', *Economic Development & Cultural Change*, **56**(2), 241–63.

Huang, J., H. Qiu, J. Bai and C.E. Pray (2006), 'Awareness, acceptance of and willingness to buy genetically modified foods in Urban China', *Appetite*, **46**(2), 144–51.

IGMORIS (2012), 'Status of GMOs and products: agriculture applications', available at: http://igmoris.nic.in/field_trials2011.asp.

IRRI (2012), 'When will Golden Rice be available to farmers and consumers?', 14 April, available at: http://irri.org/news-events/hot-topics/golden-rice/faq-resources-and-links.

ISAAA (2011), 'Rice (Oryza sativa L.) events', 12 September, available at: http://www.isaaa.org/gmapprovaldatabase/cropevents/default.asp?CropID=16.

Islam, S.M.F. and G.W. Norton (2007), 'Salinity and drought tolerant rice in Bangladesh', in C. Ramasamy, K.N. Selvaraj, G.W. Norton and K. Vijayaraghavan (eds), *Economic and Environmental Benefits and Costs of Transgenic Crops: Ex-ante Assessment*, Coimbatore, India: Tamil Nadu Agricultural University (TNAU) Press, pp. 72–90.

James, C. (2005), 'Global status of commericalized biotech/GM crops: 2005' *International Service for the Acquisition of Agri-Biotech Applications (ISAAA) Briefs*, 34, Ithaca, NY: (ISAAA).

James, C. (2011), 'Global status of commercialized biotech/GM crops: 2011', *International Service for the Acquisition of Agri-Biotech Applications (ISAAA) Briefs*, 43, Ithaca, NY: ISAAA.

Jiang, X., X. Wu and G. Xiao (2010), 'Effects of genetically modified herbicide-tolerant (GMHT) rice on biodiversity of weed in paddy fields', *African Journal of Biotechnology*, **9**(21), 3122–30.

Kim, S., C. Kim, W. Li, T. Kim, Y. Li, M. Zaidi and I. Altosaar (2008), 'Inheritance and field performance of transgenic Korean Bt rice lines resistant to rice yellow stem borer', *Euphytica*, **164**(3), 829–39.

Knight, J.G. and H. Gao (2009), 'Chinese gatekeeper perceptions of genetically modified food', *British Food Journal*, **111**(1), 56–69.

Konstandini, G., B.F. Mills, S.W. Omamo and S. Wood (2009), 'Ex ante analysis of the benefits of transgenic drought tolerance research on cereal crops in low-income countries', *Agricultural Economics*, **40**(4), 477–92.

Lu, B.R. (1999), 'Taxonomy of the genus Oryza (poaceae): historical perspective and current status', *International Rice Research Notes*, **24**, 4–8.

Lu, B.R. and A.A. Snow (2005), 'Gene flow from genetically modified rice and its environmental consequences', *BioScience*, **55**(8), 669–78.

Lu, C. (2010). 'The first approved transgenic rice in China', *GM Crops*, **1**(3), 113–15.

Montagnac, J.A., C.R. Davis and S.A. Tanumihardjo (2009), 'Nutritional value of cassava for use as a staple food and recent advances for improvement', *Comprehensive Reviews in Food Science and Food Safety*, **8**(3), 181–94.

Mulvaney, D.R., T.J. Krupnik and K.B. Koffler (2011), 'Transgenic rice evaluated for risks to marketability', *California Agriculture*, **65**(3), 161–7.

Nayar, A. (2011), 'Grants aim to fight malnutrition: cash boost should help bring fortified rice and cassava to market', 14 April, available at: http://www.nature.com/news/2011/110414/full/news.2011.233.html.

Oard, J.H., S.D. Linscombe, M.P. Braverman, F. Jodari, D.C. Blouin, M. Leech, A. Kohli, P. Vain, J.C. Cooley and P. Christou (1996), 'Development, field evaluation, and agronomic performance of transgenic herbicide resistant rice', *Molecular Breeding*, **2**(4), 359–68.

Paine, J.A., C.A. Shipton, S. Chaggar, R.M. Howells, M.J. Kennedy, G. Vernon, S.Y. Wright, E. Hinchliffe, J.L. Adams, A.L. Silverstone and R. Drake (2005), 'Improving the nutritional value of Golden Rice through increased pro-vitamin A content', *Nature Biotechnology*, **23**(4), 482–7.

Pray, C., L. Nagarajan, L. Li, J. Huang, R. Hu, K.N. Selvaraj, O. Napasintuwong and R.C. Babu (2011), 'Potential impact of biotechnology on adaption of agriculture to climate change: the case of drought tolerant rice breeding in Asia', *Sustainability*, **3**(10), 1723–41.

Qaim, M. (2009), 'The economics of genetically modified crops', *Annual Review of Resource Economics*, **1**(1), 665–94.

Qaim, M., A.J. Stein and J.V. Meenakshi (2007), 'Economics of biofortification', *Agricultural Economics*, **37**(s1), 119–33.

Ramasamy, C., K.N. Selvaraj and G.W. Norton (2007), 'Drought and salinity tolerant rice', in C. Ramasamy, K.N. Selvaraj, G.W. Norton and K. Vijayaraghavan (eds), *Economic and Environmental Benefits and Costs of Transgenic Crops: Ex-ante Assessment*, Coimbatore, India: Tamil Nadu Agricultural University, (TNAU), pp. 7–30.

Raney, T. (2006), 'Economic impact of transgenic crops in developing countries', *Current Opinion in Biotechnology*, **17**(2), 174–8.

Romeis, J., M. Meissle and F. Bigler (2006), 'Transgenic crops expressing Bacillus thuringiensis toxins and biological control', *Nature Biotechnology*, **24**(1), 63–71.

Rong, J., F. Wang, Z. Song, J. Su, R. Chen and B.R. Lu (2012), 'Scale effect on rice pollen-mediated gene flow: implications in assessing transgene flow from genetically engineered plants', *Annals of Applied Biology*, **161**(1), 3–11.

Seck, P.A., A. Diagne, S. Mohanty and M.C.S. Wopereis (2012), 'Crops that feed the world 7: Rice', *Food Security*, **4**(1), 7–24.

Smale, M., P. Zambrano, J.B. Falck-Zepeda, G.P. Gruère and I. Matuschke (2008), 'The economic impact of transgenic crops in developing countries: a note on the methods', *International Journal of Biotechnology*, **10**(6), 519–51.

Smale, M., P. Zambrano, G.P. Gruère, J.B. Falck-Zepeda, I. Matuschke, D. Horna, L. Nagarajan, I. Yerramareddy and H. Jones (2009), *Measuring the Economic Impacts of Transgenic Crops in Developing Agriculture during the First Decade: Approaches, Findings, and Future Directions*, Food Policy Review no. 10, Washington, DC: International Food Policy Research Institute (IFPRI).

Sperotto, R.A., F.K. Ricachenevsky, V.d.A. Waldow and J.P. Fett (2012), 'Iron biofortification in rice: it's a long way to the top', *Plant Science*, **190**, 24–39.

Sprague, C. (2012), 'Glyphosate-resistant Palmer amaranth in Michigan confirmation and management options', available at: http://www.msuweeds.com/assets/ExtensionPubs/Palmer-Glyphosate-Confirmation-Feb12.pdf.

Stein, A.J. and E. Rodriguez Cerezo (2009), 'The global pipeline of new GM crops: implications of asynchronous approval for international trade', JRC Technical Report no. EUR 23486 EN, Luxembourg: Office for Official Publications of the European Communities.

Stein, A.J., H.P.S. and Sachdev M. Qaim (2006), 'Potential impact and cost-effectiveness of Golden Rice', *Nature Biotechnology*, **24**(10), 1200–201.

Stein, A.J., H.P.S. Sachdev and M. Qaim (2008), 'Genetic engineering for the poor: Golden rice and public health in India', *World Development*, **36**(1), 144–58.

Stein, A.J., J.V. Meenakshi, M. Qaim, P. Nestel, H.P.S. Sachdev and Z.A. Bhutta (2005), *Analyzing the Health Benefits of Biofortified Staple Crops by Means of the Disability-Adjusted Life Years Approach: A Handbook Focusing on Iron, Zinc and Vitamin A*, HarvestPlus Technical Monograph 4, Washington, DC and Cali, Colombia: International Food Policy Research Institute (IFPRI) and International Center for Tropical Agriculture (CIAT).

Tan, T., J. Zhan and C. Chen (2011), 'The impact of commercialization of GM rice in China', *American-Eurasian Journal of Agricultural and Environmental Sciences*, **10**(3), 296–9.

Tang, G., Y. Hu, S.A. Yin, Y. Wang, G.E. Dallal, M.A. Grusak and R.M. Russell (2012), 'β-carotene in Golden Rice is as good as β-carotene in oil at providing vitamin A to children', *American Journal of Clinical Nutrition*, **96**(3), 658–64.

Tu, J., G. Zhang, K. Datta, C. Xu, Y. He, Q. Zhang, G.S. Khush and S.K. Datta (2000), 'Field performance of transgenic elite commercial hybrid rice expressing Bacillus thuringiensis δ-endotoxin', *Nature Biotechnology*, **18**(10), 1101–104.

Ventria (2012), 'Phase 2 clinical trial results: VEN100 reduces antibiotic-associated diarrhea by 50% in high-risk patients', 4 April, available at: http://www.ventria.com/content/news/587/phase-2-clinical-trial-results-ven100-reduces-antibiotic-associated-diarrhea-by-50-in-high-risk-patients/.

Wang, F., C. Ye, L. Zhu, L. Nie, K. Cui, S. Peng, Y. Lin and J. Huang (2012), 'Yield differences between Bt transgenic rice lines and their non-Bt counterparts, and its possible mechanism', *Field Crops Research*, **126**(0), 8–15.

Wang, F., Q.H. Yuan, L. Shi, Q. Qian, W.G. Liu, B.G. Kuang, D.L. Zeng, Y.L. Liao, B. Cao and S.R. Jia (2006), 'A large-scale field study of transgene flow from cultivated rice (Oryza sativa) to common wild rice (O. rufipogon) and barnyard grass (Echinochloa crusgalli)', *Plant Biotechnology Journal*, **4**(6), 667–76.

Wang, Y., G. Zhang, J. Du, B. Liu and M. Wang (2010), 'Influence of transgenic hybrid rice expressing a fused gene derived from cry1Ab and cry1Ac on primary insect pests and rice yield', *Crop Protection*, **29**(2), 128–33.
WHO (2009), 'Health statistics and health information systems: Global Burden of Disease (GBD) statistics', Geneva: World Health Organization.
WHO/FAO (2004). *Vitamin and mineral requirements in human nutrition, Second edition*. Geneva, Switzerland: World Health Organization and Food and Agriculture Organization of the United Nations.
Xia, H., L. Chen, F. Wang and B.R. Lu (2010), 'Yield benefit and underlying cost of insect-resistance transgenic rice: implication in breeding and deploying transgenic crops', *Field Crops Research*, **118**(3), 215–20.
Xiao, B., Y. Huang, N. Tang and L. Xiong (2007), 'Over-expression of a *LEA* gene in rice improves drought resistance under the field conditions', *TAG Theoretical and Applied Genetics*, **115**(1), 35–46.
Yoshimatsu, K., N. Kawano, N. Kawahara, H. Akiyama, R. Teshima and M. Nishijima (2012), 'Current status of application and commercialization of genetically modified plants for human and livestock health and phytoremediation', *Yakugaku Zasshi*, **132**(5), 629–74.
Zhang, X., J. Huang, H. Qiu and Z. Huang (2010), 'A consumer segmentation study with regards to genetically modified food in urban China', *Food Policy*, **35**(5), 456–62.
Zimmermann, R. and M. Qaim (2004), 'Potential health benefits of Golden Rice: a Philippine case study', *Food Policy*, **29**(2), 147–68.

43 Aggregate effects: adopters and non-adopters, investors and consumers

George B. Frisvold and Jeanne M. Reeves

1 MEASURING AGGREGATE EFFECTS: THREE NEW DEVELOPMENTS

Economists have been studying the aggregate (market-level) effects of new agricultural technologies for a long time (Alston et al., 1995). Innovations often reduce the cost of production, increasing supply among adopters. This increased production in turn reduces output price. The standard approach of analysis has been to apply methods of welfare economics to calculate estimates of consumer and producer surplus. This is often done in simple models with one commodity and one or two regions. Despite its simplicity, this approach can estimate the distribution of net gains of innovation between producers and consumers and between adopting and non-adopting regions. It can also measure effects of technological spillovers and competition between regions (Edwards and Freebairn, 1984). Analyses have also accounted for interactions with agricultural income support and trade policies (Alston et al., 1988, 1995).

Three factors have changed the way economists assess the aggregate and distribution effects of agricultural innovations. The first pertains to the agricultural innovation process in general and biotechnologies in particular. The private sector has surpassed the public sector in many developed countries in agricultural research and development (R&D) investment. Private technology suppliers usually hold patents or other forms of intellectual property protection, allowing them to exercise some degree of monopoly power – charging higher prices and capturing profits by supplying innovations. Earlier economic studies tended to ignore such monopoly rents in estimating the returns to research and innovation. This may have been a fair assumption in an earlier time when public universities or international research institutions provided many innovations such as improved seed varieties at cost to farmers. Today, however, the private sector brings new biotechnologies to farmers. Technology supply companies and their investors expect a return on their investments. In a seminal paper, Moschini and Lapan (1997) formally modeled the effects of a monopolist bringing an improved agricultural input to market. The new technology increases farmer returns over conventional technology. The monopolist can charge a higher price for the innovation, but that price is limited by the availability of the conventional technology. At too high a price, the innovation is not profitable to farmers and they will not adopt. In this framework, the gains from innovation are divided between producers and consumers (as in earlier approaches), but also between these two groups and the technology supplier. With the rise of large biotechnology supply firms there has been strong policy interest in estimating how much of the gains of innovation are captured as monopoly rents and how much flow to agricultural producers and consumers. The Moschini–Lapan framework has served as the basis for several studies, surveyed here, to formally address these distribution issues.

Second, genetically modified (GM) crops differ from many previous agricultural innovations in that consumers may perceive them as fundamentally different products. In many cases, consumers may judge them inferior to conventionally-grown or organically-grown crops. This complicates analysis because consumers may now have separate demands for GM commodities that interact with their demands for conventional or organic ones. Another complication is the policy response to GM crops. Some countries have imposed or are contemplating product segregation, labeling and identity preservation requirements that increase the cost of supplying GM products. Further, countries may impose production and import bans on GM commodities affecting international trade flows. While economists have long examined interactions of policies and agricultural innovations (Alston et al., 1988, 1995), technology adopters were usually assumed to be selling the same crop as non-adopters. Innovation affected product supply, but not demand. New research on consumer-side impacts represents a significant departure from previous studies (Moschini et al., 2005; Sobolevsky et al., 2005). It illustrates that consumer preferences and labeling and segregation policies have large effects on the size and distribution of benefits of biotechnologies.

The third factor affecting analysis of aggregate effects of biotechnologies is an innovation in economic modeling itself. Many researchers have assessed aggregate market effects of biotechnologies based on data and models of the Global Trade Analysis Project (GTAP) (Hertel, 1997). The project itself is a consortium of scholars and policy analysts. The GTAP model is a global multi-sector, multi-commodity computable general equilibrium (CGE) model calibrated to replicate production, prices, employment and bilateral trade flows between countries in a base year (Hertel, 1997). It also explicitly models various trade policies. While researchers can tailor commodity and regional aggregations to suit their purposes, the GTAP model provides a common data source and framework. It also makes comparison across different CGE analyses more transparent and replicable by other researchers.

GTAP-based studies, surveyed here, have been useful at estimating impacts of adoption of multiple-biotechnologies in multiple regions (which is not possible with simple partial equilibrium models). They have measured the opportunity costs of a country *not* adopting biotechnology and how gains to innovation are affected by technological competition from other countries (spillover effects) and from trade restrictions imposed on exporters of genetically modified products. They can also estimate effects on horizontally or vertically linked markets (for example how adoption of GM cotton affect textile markets, how GM soybeans affect livestock markets, how adoption of GM soybeans or how GM cotton affects other oilseed markets).

This chapter reviews these three main approaches (fixed price, partial equilibrium, and CGE) to estimating the distribution of aggregate gains from agricultural biotechnology. Often, differences in results depend on choice of method used. We have sought to highlight those differences in methods that matter most. For more details about alternative methods and their consequences, Price et al. (2003), Qaim (2009), Scatasta et al. (2006) and Smale et al. (2006) are excellent references.

2 AN INTRODUCTION TO METHODS

Distribution issues have been addressed in three types of studies: (a) single-country, partial-equilibrium (PE) models; (b) multi-country PE models; and (c) multi-country computable general equilibrium (CGE) models. Some additional inferences about the distribution of benefits may also be found by examining farm-level econometric or partial budgeting results. One may compare changes in farm-level net returns (gains to producers) to increased costs of seed (an approximate measure of gains to technology suppliers). Marra (2001), Marra et al. (2002) and the National Research Council (2010) provide good summaries of farm-level effects biotechnologies, as do Chapters 2 and 24 of this volume. Here, the focus is on three types of studies that provide a market-level (as opposed to individual producer) perspective on the distribution of gains from biotechnology.

The three approaches address somewhat different distributional issues. Single-country PE studies often assume that output prices remain fixed, either because the technology-adopting country's production is assumed to be too small to affect world price or because government policies (such as price supports) maintain producer returns at a constant level. This framework focuses on the distribution of gains between producers and technology suppliers. Effects on consumers, government outlays, or other countries are not examined.

In multi-country PE models, output prices are endogenous (that is, they are allowed to fall in response to adoption-driven production increases). These studies consider changes in output price on foreign and domestic producers and consumers. Some studies have examined the role of technological spillovers and the effects of relative rates of adoption and technical progress across regions.

Both single-country and multi-country PE approaches estimate measures of monopoly rents captured by technology suppliers. These are usually measured as the price premium paid for GM seed varieties above the cost of conventional seed. This approach assumes that the seed price premium represents a mark-up above marginal cost for seed production – implicitly assuming (a) the marginal cost of producing GM seed is the same as for conventional varieties and (b) the conventional seed market is competitive so that prices equal marginal costs. This approach ignores the large fixed cost associated with developing and obtaining government approval to commercially release GM seed varieties. Accounting for these up-front costs would reduce the share of returns accruing to technology supplies. In a study of herbicide-tolerant canola, however, Phillips (2003) illustrates the importance of these up-front costs and how the distribution of returns to biotechnology innovations can change over time.

Only a few studies have examined how biotechnology adoption and falling output prices have affected government expenditures for farm income support. This omission is more important in years when commodity prices are low (when effects on government expenditures are likely to be large) than when prices are high (when effects would be small or non-existent).

While the PE models usually examine one market (or possibly a few additional directly related markets), CGE models consider multiple sectors in the economy. Here, all prices are endogenous as are bilateral trade flows of each commodity between multiple countries. All markets are assumed perfectly competitive and CGE studies do not attempt to measure technology supplier monopoly rents. The primary focus here is examination of national or regional welfare (as measured by Hicksian equivalent variation), world prices, production and trade flows. CGE studies focus on the distribution of gains (and losses) between regions, not between producers, consumers and technology suppliers. One can make some inferences about consumer benefits by examining effects on consumer prices (which fall as a result of biotechnology adoption). Studies also often report measures related to producer returns, such as land rents, gross revenues and sector employment. Some studies have also reported effects on production, trade and employment on downstream industries such as rice processors for GM rice and textile and apparel manufacturers for GM cotton (Huang et al., 2004; Frisvold and Reeves, 2007). CGE analyses also often consider how biotechnology adoption interacts with policies to liberalize or restrict trade. Returns to adoption are examined considering whether or not such adoption triggers restrictions on trade in genetically modified products.

3 FIXED OUTPUT PRICE, PARTIAL EQUILIBRIUM MODELS

This section considers studies that explicitly address the distribution of returns to biotechnology between farmers and biotechnology suppliers (Table 43.1). Output price is assumed fixed either because the country is a small, price-taking producer in the world or because national policies guarantee producers a set price. Thus, gains are divided between the biotechnology supplier and farmers. These studies assume technology suppliers have some amount of market power via patents or other intellectual property rights (IPRs) and capture rents through a price premium paid for GM seeds over and above the price of conventional seed (Moschini and Lapan, 1997). The technologies studied are Bt crops, which provide protection against lepidopteron insect pests.

Table 43.1 Distribution of benefits between farmers and innovation supplies from selected studies

Study	Country	Crop/ Technology	Year	Farmer benefit share (%)	Biotechnology supplier share (%)
Traxler et al. (2003)	Mexico	Bt cotton	1997	39[a]	61
			1998	90	10
			Avg.	83	17
Pray et al. (2001)	China	Bt cotton	1999	82–87	13–18
Demont and Tollens (2004)	Spain	Bt maize	1998–2003	60–65	35–40
Vitale et al. (2007)	Mali	Bt cotton	Ex ante	74[b]	26
Vitale et al. (2008)	Burkina Faso	Bt cotton	Ex ante	45–66[c]	19–40
Qaim and de Janvry (2003)	Argentina	Bt cotton	2001	21	79

Notes:
[a] 1997 was a year of light pink bollworm infestation in Mexico.
[b] Simulation assumes a technology fee that maximizes seed company returns. Farmer share decreases as technology fee rises from $45/ha to $100/ha.
[c] Ginning companies receive 15% of benefits under all technology fee scenarios.

In the studies considered, farmers tend to capture the largest share of the benefits of the biotechnology. Two exceptions are Bt cotton in Mexico and Argentina. In 1997, the study area in Mexico faced only light infestations of pink bollworm, the main pest Bt cotton controls there. Farmers paid for pest protection that was not particularly needed that year, with consequently lower benefits. With greater pest pressure in 1998, Bt cotton provided much larger benefits to Mexican farmers. Vitale et al. (2008) also estimated that farmers receive larger benefits under larger pest infestations. Farmers capture a larger share of gains from Bt crops under conditions of greater pest pressure because growers pay a set premium for Bt seeds (so the technology supplier gains are fixed), but receive larger benefits from the technology when pest pressure is greater. For Bt maize in Spain, Demont and Tollens (2004) estimate that producers capture nearly a third of gains each year from 1998 to 2003, but noted that temporal and geographic variation in pest infestations limited both adoption rates and the technology fee that technology suppliers could charge.

The studies by Qaim and de Janvry (2003), Vitale et al. (2007) and Vitale et al. (2008) further illustrate the limits of benefits that even a monopolist technology supplier can extract from farmers. In Argentina, Qaim and de Janvry (2003) estimated that Monsanto charged a price higher than their profit-maximizing price for Bt cottonseed. A lower technology fee would have allowed Monsanto to capture a smaller share of larger benefits, which would have increased their overall profits. In setting a technology fee, the technology supplier faces a trade-off between profits per bag of seed sold (which increase with the fee) and total bags of seed sold (which decrease with the fee). Beyond a certain price threshold, adoption rates plummet, limiting the share of benefits that the technology provider can capture.

One important aspect of the distribution of benefits is the difference in benefits between large-scale and small-scale producers. Evidence from Mexico and Argentina suggests that small farmers adopt and benefit from Bt cotton as much as larger farmers (Traxler et al., 2003, Traxler, 2006). This suggests this particular application of the technology is scale neutral. For China, Pray et al. (2001, p. 819) found that: 'smaller farms and farms which had lower incomes consistently obtained larger increases in net income than larger farmers and those with higher incomes.'

The large relative benefits achieved by farmers, holding output prices fixed, may explain the rapid adoption of GM crops. Among the subset of countries that have approved GM crops, adoption rates are 90 per cent for soybeans, 78 per cent for cotton, 72 per cent for canola, and 60 per cent for maize (Frisvold and Reeves, 2010). The fixed-output price effects may reflect growers' *ex ante* expectations of returns measured in returns per acre. There is a fallacy of composition here, however, as pervasive adoption reduces crop prices and shifts the distribution of gains from producers to consumers. The next section illustrates the importance of output price changes for the distribution of biotechnology benefits.

4 OUTPUT PRICE ENDOGENOUS PARTIAL EQUILIBRIUM STUDIES

In contrast to fixed-price analyses, studies that allow output prices to fall in response to increased production often show large differences in the distribution of net benefits across crops, biotechnologies and regions. Much of the difference is the result of the different modeling and specification choices researchers make in characterizing a given market where innovation occurs. Some readers may be alarmed that (a) results can vary so much from study to study and (b) this variation derives from modeling choices. However, economic theory and previous research often reveals that model results differ in quite predictable ways, given their assumptions. What is important here is not a single, specific model result, but an understanding of *why* a particular model generates a particular result.

4.1 Bt Cotton

Falck-Zepeda et al. (2000b) conducted the first major price-endogenous study of the distribution of gains to Bt cotton. They used a single-commodity, two-region, trade model where the United States is a net exporter of cotton to the rest of the world (ROW). Bt cotton adoption causes a parallel shift in a linear supply curve. Returns to technology suppliers were derived from price premiums paid for Bt seed above the price of conventional seed. They made use of farm survey and experimental plot studies to estimate per-output cost reductions and yield increases for adopters in US states. State-level adoption rates were then used to derive an aggregate downward, vertical shift of the US cotton supply curve. With the model calibrated to 1996 conditions, US producers received 59 per cent of the surplus gains from Bt cotton. Technology suppliers (Monsanto and Delta & Pine Land) captured 26 per cent of the gains. By increasing output, Bt cotton adoption led to a fall in the world price in cotton. US consumers captured 9 per cent of the surplus gains via falling prices. The ROW captured the remaining 6 per cent in net gains, as ROW consumer gains outweighed ROW producer losses.

In a follow-up analysis, Falck-Zepeda et al. (2000a) re-estimated the distribution of Bt cotton adoption gains for 1997, again using the same type of two-region model. They updated yield and cost change estimates to reflect production conditions in 1997. They also accounted for effects of Bt cotton adoption by ROW producers. The spillover effects of ROW adoption were small, however, because ROW Bt cotton acreage was assumed to account for only 8 per cent of world Bt cotton acreage and 0.2 per cent of total cotton acreage. In 1997, total US consumers captured 7 per cent of total surplus gains; the net ROW share (consumer gains minus producer losses) was 6 per cent, similar to 1996. In contrast, the technology-supplier share rose from 26 per cent to 44 per cent, while US producers' share fell from 59 per cent to 42 per cent. Differences in pest pressure accounted for much of the difference between years. In 1995, resistance to pyrethroid pesticides by lepidopteron pests became a severe problem in the US south, especially in Alabama. Bt cotton was an effective substitute for pyrethroids in 1996, conveying large yield advantages (in the order of 32 per cent) and insecticide cost savings in affected areas. In 1996, gains to Alabama producers alone accounted for more than half the gains to US producers (Falck-Zepeda et al., 2000b). In 1997, with lepidoptera under greater control, Bt cotton conferred a more modest yield advantage to Alabama growers (<3 per cent), while technology fees exceeded insecticide cost savings. Estimated Alabama producer surplus actually fell because of higher costs and falling cotton prices in 1997.

Adopters of Bt cotton can potentially gain from insecticide cost savings, yield gains, or both, but suffer as increased output causes output price to fall. Non-adopters, however, only feel the negative effect of falling prices. Falck-Zepeda et al. (2000a, 2000b) did not explicitly measure different impacts between adopters and non-adopters. However, Bt cotton adoption tended to reduce producer surplus in low adoption states, notably California and Texas.

Frisvold and Tronstad (2003) considered Bt cotton impacts from 1996 to 1998. They too used a two-region model with the US as a net cotton exporter to ROW. ROW supply, ROW demand and US demand curves were assumed linear (as in Falck-Zepeda et al., 2000a, 2000b). US supply, however, was modeled in a multi-state mathematical programming model calibrated to annual costs, prices, production and exports in each year. Cotton in the US was modeled using fixed-proportion technologies specific to each region, with different production functions for Bt adopters and non-adopters. Total cotton and Bt cotton acreage constraints were imposed, but calibrated to observed values. Adoption was assumed to alter yield or costs per acre (through changes in insecticide and seed costs per acre). This specification was general enough to represent either divergent or convergent supply curve shifts. Most farm-level and field-plot studies report Bt cotton adoption impacts in terms of change in yield or changes in per-acre costs. Based on these studies, Frisvold and Tronstad (2003) developed low, moderate and high impact estimates of Bt adoption for each cotton-producing state. Changes in per-acre costs generated convergent supply shifts, while yield increases generated more divergent shifts. Convergent shifts steer relative benefits to producers, divergent shifts to consumers (Alston et al., 1995). This approach also allows for separate estimates of impacts on adopters and non-adopters. The analysis also accounted for the fact that, starting in 1998, technology suppliers began to price discriminate regionally, lowering technology fees in areas where Bt cotton provided lower returns. Adoption in the ROW was not considered.

Frisvold and Tronstad (2003) also included the effects of loan deficiency payments (LDPs). Under LDPs, consumers pay the market price for cotton, while US producers receive per-pound output subsidies based on the difference between the world cotton price and a government-set loan rate. LDPs shelter US producers from lower market prices brought about by increased production. World prices were high enough in 1996 and 1997 so that US producers received no LDP payments in 1996 and small payments in 1997. In 1998, however, cotton market prices were lower and LDPs served more as a price floor for US growers.

From 1996 to 1998, under the moderate impact scenario, technology suppliers captured 45–47 per cent of gains, US consumers captured 29–35 per cent, and net ROW 14–21 per cent, with ROW consumer gains outweighing producer losses. US non-adopting producers had *losses* ranging from 5 per cent of *net* surplus gains in 1998 to as high as 45 per cent of net surplus gains in 1996. LDP payments in 1998 significantly protected non-adopters from the effects of falling cotton prices. US adopter gains rose from 44 per cent of net gains in 1996 and 1997 to 59 per cent in 1998. Again, this is the effect of the LDPs.

Government payments to US cotton producers as a share of net surplus rose from 0 per cent in 1996 to 8 per cent and 47 per cent in 1997 and 1998. Of total net US producer gains in 1998, (including adopters and non-adopters), 88 per cent were derived from increased government program payments. Without price support payments in 1996, non-adopter losses exceeded grower gains under the low and moderate impact scenarios. Only under the high impact scenario did US growers gain as a whole. However, this high impact case more closely resembled the high pest pressure faced by producers in 1996.

Frisvold et al. (2006) extended this analysis by updating and re-calibrating their model to 2001 market conditions and splitting out China as a third region, distinct from ROW. They compared the effects of Bt cotton adoption by the US alone, China alone and by both countries. In 2001, the US and China accounted for 95 per cent of global Bt cotton acreage and 40 per cent of world cotton production. Adoption effects in the US were again modeled as yield and per-acre cost shocks in a cotton sector mathematical programming model. Effects in China are modeled as in Falck-Zepeda et al. (2000b) as a parallel shift in a linear supply curve, based on estimated yield and pest control cost changes and adoption rates. US technology supplier returns are estimated as revenues from technology fees. By 2001, regional price discrimination was more extensive, with technology fees ranging from US$12–32/acre. In 1999, Chinese suppliers just covered their costs, while Monsanto/Delta & Pine Land earned less than US$2 million in *gross* revenue (Pray et al., 2001). Monopoly rents in China were suppressed from competition between supply sources and because Chinese farmers saved Bt cottonseed. Frisvold et al. (2006) assumed that technology suppliers captured no rents in China.

With both the US and China adopting, technology suppliers only captured 17 per cent of net world surplus and net ROW captured 8 per cent, while China captured 71 per cent (51 per cent to producers and 20 per cent to consumers). US consumers captured 6 per cent. US producer gains equivalent to 21 per cent of net world surplus were counteracted by government payments (a welfare loss) equal to 24 per cent of world surplus. Absent program payments, US producers would have lost surplus from joint US–Chinese adoption. With joint adoption, technology supplier rents accounted for 83 per cent of US welfare gains. If only the US had adopted, the US would have captured 84 per cent of global surplus. If only China adopted, they would have captured 99 per cent of surplus. This is because technology suppliers capture no rents from China, ROW captures 5 per cent, but the US had welfare losses (resulting from commodity program distortions) of −4 per cent. If only the US had adopted, technology suppliers would have captured 60 per cent of the surplus.

Absolute gains to China were identical – US$595 million both when China adopted alone and when both the US and China adopted. With joint adoption, however, Chinese consumers received larger gains relative to producers. When only the US adopted, Chinese gains fell to just US$3 million. When only the US adopted, US gains were US$198 million compared to US$172 million when both countries adopted. Again, joint adoption shifted relative gains from US producers to US consumers. If only China adopted, the US lost US$26 million of its surplus. Technological competition from another region (that is two countries adopting) appears to have a modest impact on the total welfare of an adopting country, although it shifts relative gains to domestic consumers. Both countries, however, forgo large surplus gains if they fall behind technologically (that is, do not adopt).

Price et al. (2003) adopted the basic model structure of Falck-Zepeda et al. (2000a, 2000b) with some important changes. As in Falck-Zepeda et al. (2000a, 2000b) Bt cotton adoption was modeled as a parallel shift in a linear supply curve, using yield and cost change data from Enhanced Market Data (EMD). Price et al. (2003) also derived supply shift estimates based on the USDA Agricultural Resource Management Survey (ARMS) of US cotton producers. They also used updated USDA data on production and trade flows and different supply and demand elasticities to calibrate the model. They assumed less elastic US cotton supply, more elastic US demand, and less elastic US export demand. They also accounted for adoption in the ROW, but assumed higher rates of ROW adoption than did Falck-Zepeda et al. (2000a).

With supply shifts based on EMD data, US producers captured 39 per cent of new world surplus gains, technology suppliers 25 per cent, US consumers 17 per cent, and net ROW 19 per cent. With supply shocks based on ARMS data, US producers captured 29 per cent of new world surplus gains, technology suppliers 35 per cent, US consumers 14 per cent, and net ROW, 22 per cent. While Falck-Zepeda et al. (2000a) assumed productivity effects of Bt cotton adoption in ROW to be comparable to US effects, Price et al. (2003) assumed ROW effects were only 50 per cent of US effects in their base scenarios. They conducted sensitivity analysis, lowering this rate to 10 per cent and raising it to 100 per cent (as in Falck-Zepeda et al., 2000a). As in Frisvold et al. (2006), increased foreign technological competition had only modest effects on domestic returns of adoption. Falck-Zepeda et al. (2000a) conducted sensitivity analysis allowing ROW adopters to experience supply shifts ranging from 10 per cent of US shifts to shifts equal (in percentage terms) to US shifts. US surplus was changed little as ROW adopter supply shifts increased. In addition, similar to Frisvold et al. (2006), increased technological competition shifted relative gains from US producers to US consumers.

Falck-Zepeda et al. (2008) extend the linear supply–demand/parallel supply shift approach (2000a, 2000b) in several ways in an *ex ante* analysis of potential impacts of Bt cotton adoption in West Africa. They consider adoption in five West African countries (Benin, Burkina Faso, Mali, Senegal and Togo) and the ROW under five scenarios. These countries are net exporters of cotton with low domestic consumption. They examine impacts over 23–24 years, starting in 2012, with adoption reaching 20 per cent in ROW and adoption ceilings in West African countries varying by scenario. They assumed that it takes seven years to reach the maximum adoption level and that the most likely West African technology fee would be US$32/hectare.

In Scenario 1, the ROW adopts Bt cotton, while the five African countries do not, which causes them to suffer producer losses from lower world cotton prices. In Scenario 2, adoption starts first in Burkina Faso in 2012, with the other countries following three years later. West African producers capture 39 per cent of the surplus (14 per cent Burkina Faso, 25 per cent others) and technology suppliers 61 per cent. In Scenario 3, adoption is only permitted for varieties developed in West Africa, following the logic that varieties bred for local conditions may perform better. There is thus a trade-off between the Bt insect-protecting trait and other germplasm traits. Local varieties may also take longer to develop, which Falck-Zepeda et al. (2008) explicitly model. Here, producers receive 40 per cent of the gain, and technology suppliers capture 59 per cent. Thus, gains from developing locally appropriate varieties may outweigh the costs of delay in their development. In Scenario 4, the five countries collectively negotiate a lower technology fee and increase their share to 53 per cent. Scenario 5 assumes adoption, dis-adoption then re-adoption in Benin and Mali. This is meant to show the effects of policy instability on adoption rates. In Scenario 5, producers' share of returns falls to 35 per cent. In all scenarios, West African consumer gains are 1 per cent of the total. Falck-Zepeda et al. (2008) also show how the share of gains that producers receive varies by country and scenario. A major finding is that the countries are worse off without the Bt cotton technology as they face falling world cotton prices, even with a conservative estimate of ROW

Bt cotton adoption. There are also significant costs of dis-adoption that may result from policy instability.

4.2 Herbicide-tolerant Soybeans

Falck-Zepeda et al. (2000a) also used a two-region, linear supply and demand trade model to examine the distribution of gains from herbicide-tolerant (HT) soybean adoption for 1997. Regional yield and per-acre cost changes derived from USDA ARMS data were aggregated based on observed adoption rates and converted to a vertical, parallel supply curve shift. They assumed that ROW HT soybean adoption had the same impacts on yield and costs as in the US. They report results using two different values of the US soybean elasticity of supply. Assuming a supply elasticity of 0.22, US producers capture 76 per cent of net world surplus, technology providers only 10 per cent, net ROW 9 per cent, and US consumers 4 per cent. With more elastic supply (0.92), US producers capture 17 per cent of net world surplus, technology providers only 25 per cent, net ROW 28 per cent, and US consumers 17 per cent. This occurs because the method employed to convert yield increases into vertical shifts in the supply curve are highly sensitive to the supply elasticity parameter (see Alston et al., 1995; Oehmke and Crawford, 2002). The more inelastic the supply, the more a given yield increase translates into a larger cost reduction.

Mensah and Wohlgenant (2010) use a two-region linear supply and demand model to examine HT soybean adoption. The US is a net exporter to ROW and adoption is only assumed for the US. While they do not estimate surplus gains to technology suppliers, they do compare relative returns between a parallel and pivotal supply curve shift. In absolute terms, a pivotal shift cuts US absolute producer gains in half, leaving other surplus measures unaffected. Under a parallel shift, US producers capture 82 per cent, US consumers 13 per cent, and net ROW 5 per cent. Under a pivotal shift, US producers capture 68 per cent of the gains, US consumers 28 per cent, and net ROW 9 per cent.

Moschini et al. (2000) developed a model of HT soybean adoption with three regions: the US, South American adopters (Argentina and Brazil) and the ROW. HT soybean varieties provide better weed control, lowering per-hectare production costs. While HT soybean seed is more expensive, its adoption still lowers per-hectare costs. HT soybeans are assumed to have no effect on yields in the baseline simulation, but this restriction is later relaxed. Yields are endogenous, increasing with the soybean price. Total acreage available for soybean production is increasing with average profits per acre in each region. HT soybean adoption can increase production by increasing farmland planted to soybeans and (later) through yield increases. As with Frisvold and Tronstad (2003) and Frisvold et al. (2006), this approach allows for non-linear and non-parallel shifts in the supply curve. Its specification of technology and land allocation is less restrictive and more realistic. The model is calibrated to match production, price and trade conditions in the 1999–2000 crop year as well as observed HT soybean adoption rates in the US and South America. They assume that HT soybeans have the same impacts on costs and yields in the US and South America, but that technology fees are lower in the latter because of weaker IPRs. Separate demand curves are used for soybeans, soybean meal and soybean oil.

In the base simulation, technology suppliers capture 45 per cent of net global surplus,

US producers 19 per cent, US consumers 10 per cent, South American producers 3 per cent, South American consumers 4 per cent and net ROW 18 per cent. ROW is a net importer of soybean products with consumer gains outweighing producer losses. Moschini et al. (2000) also compared a case of 100 per cent adoption in the US only, with 100 per cent adoption in the US and South America, and 100 per cent in all three regions. As total adoption increases, both global and welfare increases absolutely, with relative gains shifting from US producers to US technology suppliers and consumers. South American welfare declines with increased adoption in ROW, with a 14 per cent decline. Increased technological competition shifts relative gains from producers to consumers.

Moschini et al. (2000) also conducted interesting sensitivity analysis regarding the degree of market power conferred from IPRs in the three regions. From their tables, one can compare the case of 100 per cent adoption in each region where:

a. there is no mark-up in any region (HT seeds supplied competitively in all three regions);
b. where the mark-up for HT seeds is half the US mark-up in South America and one-quarter the US mark-up in the ROW (the base case); and
c. mark-ups in South America and ROW equal those in the US, reflecting the imposition of stronger IPRs in developing countries.

Not surprisingly, increasing mark-ups globally increases technology supplier profits. Moving from case (a) to (b) to (c), technology supplier gains increase while consumer gains in each region decrease. Strengthening IPRs in developing countries from case (b) to (c), so that mark-ups equal US mark-ups, increases US producer gains while reducing South American and ROW producer gains. Producers gain everywhere, with increases continuing until mark-ups fall to zero. Under a scenario of 100 per cent adoption globally, however, there is only a 0.2 per cent gain in total welfare between the competitive and high mark-up pricing. This is due in part to assumptions of inelastic soybean product demand, soybean supply and seed demand. Moreover, the mark-up structure is exogenously set and not necessarily that which maximizes technology supplier profits.

Moschini et al. (2000) also consider the effects of HT soybeans increasing yields as well as reducing per-acre costs of production. As in Frisvold et al. (2006), yield and per-acre cost shocks are specified separately to generate non-linear, non-parallel supply curve shifts, with yield shocks generating a more divergent supply curve shift. Divergent shifts can reduce producer returns because, with inelastic demand, the fall in total revenue can exceed the fall in total costs. This is the case here. With a 5 per cent yield increase from HT soybean adoption, adopter producer returns fall in all scenarios, with the only exception being when the US is the only adopter.

Price et al. (2003) used USDA ARMS data to specify a parallel supply shift from US shocks in a two-region, linear supply–demand system. They found US producers captured 20 per cent of surplus gains, with technology suppliers capturing 68 per cent, US consumers 5 per cent and net ROW 6 per cent. Price et al. (2003) assumed cost and yield changes from HT soybeans adopted in the ROW were 50 per cent of US impacts. Lowering this value to 10 per cent or raising it to 100 per cent had no effect on total US gains. Increasing the efficiency of technology transfer from 10 per cent to 100 per cent did shift absolute gains slightly from US producers to US consumers. It also increased absolute and relative net ROW gains, by reducing ROW producer losses and increasing ROW consumer gains. Price et al. (2003) noted that their results differed from those of Moschini et al. (2000) despite considering the same crop, year and technology. In particular, technology suppliers captured 68 per cent of global gains compared to 45 per cent in Moschini et al. (2000). In a sensitivity analysis, Price et al. (2003) altered their modeling assumptions to be similar to those of Moschini et al. (2000). When they altered costs and elasticity assumptions to match Moschini et al. and similarly assumed no yield advantage from HT soybeans, many of the differences in the size and distribution of benefits between the Price et al. (2003) and Moschini et al. (2000) models were significantly reduced.

Qaim and Traxler (2005) also develop a three-region model (Argentina, US, ROW) for the years 1996–2001 with HT soybeans adopted in the US and Argentina, which are both net exporters, while the ROW is a net importer of soybeans. They assume linear supply and demand functions with adoption creating a parallel supply curve shift. This study highlights differences in IPR protection in the US and Argentina. While HT soybeans benefited from patent protection in the US, they were not patented in Argentina. As a result, Argentine farmers were allowed to save seed, with the result that saved HT soybean seed competed with new market sales. Use of saved and black market (uncertified) HT seed was extensive in Argentina. Thus technology suppliers captured smaller rents for seed in Argentina than they could in the US. Even without patent protection, Argentine farmers were willing to pay a mark-up for certified, Monsanto HT seed. Monsanto also earned price premiums for its Roundup® glyphosate herbicide used with HT seed. In 1996, when adoption only occurred in the US, American producers captured 40 per cent of the surplus, global consumers 47 per cent and technology suppliers 37 per cent. Non-adopting ROW and Argentine producers had losses equal to −25 per cent of net surplus gains. In 1997, however, Argentine growers began adopting HT soybeans rapidly. By 2001, US producers captured 12 per cent of world surplus, Argentine producers 25 per cent, global consumers 53 per cent, and technology suppliers 34 per cent. Non-adopting ROW producers had losses equal to −24 per cent of net global welfare gain. Absolute gains to Argentine farmers grew to more than double US producer gains by 2001. This occurred because adoption rates are higher in Argentina, so US non-adopter losses (from falling output prices) are greater. Absolute losses to non-adopting ROW producers rose from $5 million in 1996 to more than $290 million in 2001, illustrating the large opportunity costs of non-adoption.

Starting with a soybean production specification similar to Moschini et al. (2000), Sobolevsky et al. (2005) extend this analysis in several ways. First, they assumed that consumers treat GM soybeans as weakly inferior to conventional soybean varieties. Although GM soybeans are cheaper to produce, consumers are willing to pay less for them. As long as product segregation costs are not too high (and the discount for GM soybeans is large enough), producers have an incentive to segregate conventional and GM soybeans. Producers must incur costs to segregate the soybean crop. Second, they include effects of US loan deficiency payments (LDPs), which subsidize both producers and consumers of soybeans, but incur government costs. The model has four regions: the US, Brazil, Argentina and ROW, which may all adopt HT soybeans. In equilibrium, the extent of adoption in each region is determined endogenously.

Sobolevsky et al. (2005) report many results by varying key model parameters. With

US LDPs in place and medium segregation costs, all regions adopt. Only Brazil produces conventional soybeans and segregates crops. Technology suppliers capture 54 per cent of global welfare gains. US LDP expenditures (a welfare loss) equal 51 per cent of global gains, however. US producer gains are 27 per cent of this total. Consumers gain in all regions. While producers in the US and ROW gain, those in Brazil and Argentina have welfare losses. Without price supports (but with medium segregation costs), the US becomes the only region not to fully adopt HT soybeans, although adoption does reach 90 per cent of acreage. Removing LDPs reduces welfare to all regions, except ROW, and to all consumer groups. It lowers returns to US producers, but raises returns to producers in all other regions. Without LDPs, US producer returns decline absolutely with global HT soybean adoption. Interestingly, removing LDPs lowers global welfare. Sobolevsky et al. (2005) explain this as a second-best effect. Monopoly pricing of HT soybeans creates over-pricing and under-consumption of HT seed varieties. The LDP subsidies counteract this negative welfare effect.

As segregation costs increase from zero to high levels, global adoption of HT soybeans *increases*. As the costs of growing two types of crops rise, growers specialize in HT soybeans, even though consumers may have a lower willingness to pay. Because it forces producers to specialize in HT crops, increasing segregation costs also increases technology supplier profits. Without LDPs, increasing segregation costs have increasingly negative effects on US producers. With LDPs, however, US producers are unaffected by cost changes. A production and import ban by the ROW benefits ROW producers, but reduces welfare for ROW consumers. It also benefits consumers and hurts producers in all outlying regions by lowering world soybean prices. Technology supplier returns fall 42 per cent under a ban.

4.3 Other Crops

Phillips' (2003) analysis of HT canola in Canada begins with some standard assumptions about how the technology affects producer yields and production costs. HT canola lowers herbicide application costs and can lower dockage costs at elevators because the crop is cleaner. The technology allows earlier planting which can raise yield, but can suffer from yield drag because the new trait is not necessarily in the seed varieties best adapted for local condition. Phillips (2003) introduces some interesting features not included in earlier studies. First, because of pollen drift and outcrossing, HT canola can contaminate organic canola. Organic producers would suffer losses because they could not claim price premiums. Second, while virtually every study assumes that returns to technology suppliers are the gross values of GM seed price premiums, Phillips also considers the discounted, net returns to technology suppliers, accounting for the costs of developing and releasing varieties with the HT trait.

Adopting canola producers gain through lower costs and increased production, but lose through lower market prices (from increased production and reduced organic premiums (although this effect is small). Net gains to producers are overall positive. Phillips reports the distribution of returns from 1997–2000, with the current costs of seed variety development netted out of technology provider gross returns. From 1997–2000, producer share of current returns rises from 6 per cent to 29 per cent, processor and consumer returns rise from 0 per cent to 14 per cent, and technology supplier returns fall from 94 per cent to 57 per cent. Consumer returns are divided among processors, which have some degree of market power, and domestic and foreign purchasers of canola. Phillips notes that processing market power may allow processors to capture much of the downstream consumer benefits and that requirements for identity preservation in export markets could raise the costs of crop segregation, substantially reducing the gains from the technology.

Phillips also traced technology supplier seed development costs back to 1985. In this dynamic context, innovator net gains can be substantially lower than current gross returns, especially in early years of adoption. In each year from 1985 to 1996, development costs outweighed (zero) returns from seed sales. Annual net returns turned positive in 1997, but the net present value of HT canola did not turn positive until 2000.

Price et al. (2003) also estimated the distribution of gains from herbicide-tolerant cotton, using the same model as their Bt cotton analysis, deriving supply shocks from ARMS data. No adoption by ROW was assumed because the US was the only adopter in 1997. US producers captured only 4 per cent of world surplus gains, technology suppliers 6 per cent, US consumers 57 per cent, and net ROW 33 per cent. They report that HT cotton caused more of a price reduction in the world price of cotton than Bt cotton, leading to consumers capturing a large share of the gains.

Dillen et al. (2009) conducted an *ex ante* analysis of potential HT sugar beet adoption by beet producers in the EU and ROW from 1996 to 2014. They focus on the effects of past and current European Common Market Organization (CMO) policy for sugar. Under the current CMO, some countries within the EU face binding production quotas and fixed institutional prices, while others face the world sugar price as their marginal incentive price. Because the CMO fixes prices EU consumers pay, yield increases from HT sugar beet provide no welfare gains from lower prices to EU consumers. Technology suppliers capture 39 per cent of global welfare gains, EU producers 29 per cent and net ROW 31 per cent. ROW consumer gains outweigh ROW producer losses. Among ROW producers, sugar beet producers gain from adopting HT sugar beet, while sugar cane producers have losses due to the absence of any competitive varieties. For ROW producers there is a loss from the falling world price of sugar. Beet, but not cane producers, benefit from higher yields.

Dillen et al. (2009) find that in EU areas facing binding quotas the yield-increasing effect of HT sugar beets reduces technology supplier returns. This occurs because, given a binding quota, higher yields lead to a reduction in acreage planted to sugar beets. With lower acreage come reduced seed sales. Their analysis demonstrates the importance of agricultural and trade policies in determining returns to biotechnologies.

In an *ex ante* analysis of Bt maize adoption in Mali, Vitale et al. (2007) model the domestic demand for maize as price inelastic. Domestic maize consumers are the primary beneficiaries of increased production for adoption. Assuming a zero technology fee, consumers capture 81 per cent of the welfare gain and producers 19 per cent. A technology fee of $10/hectare would maximize seed revenues of a technology supplier. In this case, consumers capture 66 per cent of the gains, technology suppliers 19 per cent, and producers 16 per cent. A $10/hectare technology fee is lower than the fee that suppliers have charged in other regions. Vitale et al. (2007) suggest that innovators may have little incentive to promote Bt maize in Mali.

5 GENERAL EQUILIBRIUM STUDIES

The studies discussed in this section all use variants of the GTAP CGE model (or the GTAP database) to assess international effects of biotechnology adoption. In this framework, all prices are endogenous, and trade, tax and price support policies are explicitly modeled. Technology shocks from biotechnology adoption are modeled as either Hicks neutral or factor-augmenting technical change (Frisvold, 1997). Because all markets are assumed perfectly competitive, monopoly rents accruing to technology suppliers are not measured. The primary focus of these studies is on the effects on national economic welfare and trade flows from biotechnology adoption as well as interactions between technology adoption and trade policies. Separate producer and consumer welfare measures are not reported, but researchers provide indirect measures of how producers are affected, including changes in output prices, land rents, gross revenues and employment.

Anderson and Yao (2003) examine implications of China adopting GM rice, cotton, corn or soybeans in 2005, given prior adoption by other countries, and consider what happens if some countries ban importation of GM crops. In their first three simulations, they compare effects of (a) adoption of a particular GM crop (or crops) in North America, the Southern Cone of South America and South-East Asia with (b) the case where China joins these three regions in adopting the GM crop (or crops). They assume that each adopting region experiences uniform 5 per cent Hicks-neutral productivity growth in the adopting sector. If China does not adopt GM rice, their domestic rice production, price and trade change little. Welfare in China increases slightly from falling world rice prices. However, China gains only 0.5 per cent of the global welfare gain, or US$4 million. Adopting regions (primarily South-East Asia) capture 70 per cent. If China also adopts GM rice, the world price of rice falls further and China's welfare gains increase to US$1.1 billion, 55 per cent of global gains. The share of gains captured by other adopters falls to 28 per cent of the total. All adopting regions still have positive absolute welfare gains when China also adopts. Chinese adoption actually increases South-East Asian welfare further.

In their second simulation, the experiment is repeated for Bt cotton. China is a net importer of cotton, but a net exporter of textile and apparel. If China joins other Bt cotton adopters, cotton prices fall further, China produces more cotton, imports less cotton and exports more textiles and apparel. China's welfare gains rise from US$15 million (<2 per cent of total gains) to US$340 million (23 per cent of the total). Absolute welfare gains in other adopting regions total US$642 million if China does not adopt and US$650 if China does. In contrast, there are welfare losses in other non-adopting cotton-exporting regions, including India, sub-Saharan Africa, Australia and New Zealand. (Anderson and Yao (2003) assumed Australia was a non-adopter of Bt cotton even though it was an early adopter.) These losses increased as China joined adopters. This result is consistent with the many partial equilibrium studies showing high opportunity costs of not adopting agricultural biotechnology. The pattern of results holds for adoption of GM maize and soybeans. Again, China's absolute welfare gains are larger if they become active adopters. Absolute gains to other adopters are affected only modestly by Chinese adoption and all adopting regions gain, even when China adopts aggressively.

In a last simulation, Anderson and Yao (2003) estimated that China would gain US$2.3 billion from adopting all the GM crops (along with the three other regions) but that these gains would fall by <7 per cent if Western Europe banned imports of Chinese agricultural products (ostensibly because they would not want GM products). If North-East Asia joined this import ban, China's gains from GM adoption would fall by two-thirds. A Western European ban has less effect because China's exports to the region are far lower than they are to North-East Asia (primarily Japan and South Korea). The effect of import bans may be overstated, however, because it appears that only Chinese imports are restricted and not those from other GM adopting regions.

Huang et al. (2004) evaluated effects of recent and future adoption of Bt cotton and hypothetical adoption of Bt rice in China. For both crops, they assume non-neutral, factor-augmenting technical change. Chinese trade balances in rice, cotton and textiles all increase, while they decrease in other regions. Welfare gains to China are US$1 billion from Bt cotton adoption, US$4.1 billion from GM rice adoption, totaling US$5.2 billion in annual benefits in 2010 from adoption of both. They estimated that Chinese public R&D on related biotechnology from 1986 to 2000 totaled less than US$0.5 billion (in real 2000 prices) and argue that welfare gains of successful deployment of Bt cotton and GM rice would far outweigh R&D expenditures.

Similar to Anderson and Yao (2003), Huang et al. (2004) consider how a GM crop import ban would affect these gains. They consider an import ban on GM rice (but not Bt cotton) by the EU, Japan, South Korea and South-East Asia. The ban has no effect on overall Chinese welfare. The ban imposes losses on the import-banning regions that are larger than the small losses felt by China. Huang et al. (2004) do not report how the ban would affect land rent in China.

Huang et al. (2004) also examine the implications of China imposing segregation and labeling requirements for soybean imports coming from North and South America. To comply with World Trade Organization regulations, however, China could not impose labeling requirements on imported GM food products without also requiring labeling for its domestically produced GM products, namely rice. Labeling and segregation costs for domestic GM rice are modeled as increased cost of services, while those for imported GM soybeans are modeled through an increase in transport and handling margins. Under the new requirements, Chinese soybean producers would benefit from greater output and higher prices, but the increased rice output from GM rice adoption would be cut in half. Gains from Bt cotton and GM rice adoption would fall by US$1 billion under these conditions.

Elbehri and MacDonald (2004) examine the costs of not adopting Bt cotton in West and Central Africa (WCA) given adoption in other regions. They assume that because of pest resistance problems, productivity in WCA would not just stay constant but would decline without Bt cotton. They assume large productivity gains from adoption in China, India and South Africa (7–10 per cent), moderate gains in WCA (5.3 per cent) and lower gains elsewhere. They assume rather low gains in the US (<2 per cent). Based on adoption rates from 2003, if adoption occurs in other regions but not WCA, then WCA land rents fall, real wages fall and the cotton sector contracts. When WCA adopts too, land rents still fall (but by less), but real wages rise. Land rents rise if technical change is labor-augmenting or labor- and chemical-augmenting. There is also a small downstream increase in textile production. Without adoption, WCA welfare falls by US$88 million, but rises by US$82 million with adoption and assuming neutral technical change, by US$78 million with labor-augmenting technical change, and by US$69 million with labor- and chemical-augmenting technical change. The net gain from adoption is US$150–170 million. They note that WCA cotton production increases more than cotton prices drop, so that gross revenues increase. They consider this a sign that cotton producer welfare and land rents would increase, consistent with producer gains.

When all adopting regions have Hicks-neutral productivity gains, China and India (with the largest productivity gains) capture the largest welfare gains. All cotton-importing, non-adopter regions gain, while Central Asia and Rest of South Asia (non-adopting cotton exporters) face welfare losses. With assumed low productivity growth, US absolute gains are relatively small – smaller even than for South Africa.

Frisvold and Reeves (2007) consider the impacts of adoption and costs of non-adoption of Bt cotton given regional adoption rates and estimated productivity impacts in 2005. Bt cotton adoption is assumed to generate Hicks-neutral technical change in all adopting regions except South Africa, where technical change is assumed to be land augmenting. The size of the productivity shocks varies by region depending on the extent of adoption and estimated effects on adopters. Given that cost/unit output = (cost/acre) / yield, empirical estimates of cost/acre and yield effects of Bt cotton for different countries are used to estimate percentage reductions in average costs. The percentage reduction in average cost is then modeled as a percentage increase in productivity. China experiences the largest productivity gain and the largest gain in domestic welfare. The United States and India follow with lower productivity and welfare gains than China. Mexico has relatively large welfare gains given its level of cotton production, because of large assumed productivity gains and high adoption rates. All regions experience domestic welfare gains, even the non-adopting EU and ROW. These two regions are both net importers of cotton. Textile and apparel net exports expand in China and India, but decline in other net-exporting regions. Textile and apparel imports rise in the EU and ROW. Employment in textile and apparels increases in China and India, but declines elsewhere. Land rents fall in all regions, except Australia, a classic small exporter.

Both the US and China would benefit more if they were the sole adopters of Bt cotton. For China, however, welfare gains are roughly the same whether or not the US adopts. In contrast, if the US did not adopt while China did, the US would forgo about US$200 million in welfare gains. Without adoption, US agricultural land rents fall, and cotton, textile and apparel employment fall. The downstream effects for the apparel and textile industries are quite modest, however.

In another study of global Bt cotton adoption, Anderson et al. (2008) first consider adoption up to 2001. With Hicks-neutral productivity increases in adopting regions assumed to be 5 per cent for the US, Australia and South Africa and 2.5 per cent for China, the US captured 44 per cent of global welfare gains, and China 22 per cent. All seven non-adopting regions also gain, with the exception of the cotton-exporting region of sub-Saharan Africa (SSA, which excludes South Africa in the analysis). The next scenario assumes other regions – except SSA – catch up to the early adopters. US, Australian and South African productivity is held constant (at the post-adoption level), while Chinese productivity grows at 2.5 per cent (to represent their completion of adoption); productivity increases 5 per cent in all other regions, except SSA (0 per cent) and India's share of South Asia (15 per cent). The relative size of the shocks is somewhat arbitrary, but does capture empirical estimates of large yield shocks in India. South Asia captures 48 per cent of the global welfare gain (which is not surprising given the size of its cotton sector and productivity shock). Eastern Europe and Central Asia capture 16 per cent (Central Asia is a large cotton exporter). China, the US and South Africa all post modest gains compared with the first scenario. Australia, a small cotton exporter, loses welfare, as it has no productivity gain in this scenario. SSA, another cotton exporter, loses. If SSA also adopted – thereby obtaining a 15 per cent productivity gain – its welfare would increase by US$187 million. All other regions (except Australia) gain, even with competition from SSA. Again, this illustrates the costs of *not* experiencing productivity gains. Anderson et al. (2008) also conduct sensitivity analysis assuming either labor- or land-augmenting change from Bt cotton adoption instead of neutral change. All shocks are scaled to generate the same unit cost savings. Welfare gains are lower under factor-biased technical change. The reduction is significant for land-saving technical change in many regions.

Bouët and Gruère (2011) use the MIRAGE CGE model to examine Bt cotton adoption impacts. MIRAGE, developed at the Centre d'Etudes Prospectives et d'Informations Internationales in Paris is based on the GTAP database. Their analysis surveys available empirical studies to capture the effects of both Bt cotton and HT cotton, weighing yield and input cost shocks by actual regional adoption rates from 2004/5 and 2008/9 in their first two scenarios. They also explicitly consider how GM cotton affects oilseed markets via changes in cottonseed production. Adoption shocks are factor biased, with the nature of the bias varying by region. Along with yield shocks, pesticide- and labor-saving shocks are assumed. The assumed shocks for these simulations are much larger than in other studies, with absolute welfare effects also being larger. In the 2004/5 scenario, the US captures 40 per cent of global welfare gains, followed by Mexico (13 per cent), China (12 per cent) and India (6 per cent). Non-adopting countries in Asia (which includes Japan and South Korea) and the EU together capture 22 per cent. Compared to the 2004/5 scenario, India's adoption rate in 2008/9 rises from 5 per cent to 70 per cent. In the 2008/9 adoption scenario, India captures 37 per cent of the >US$3.5 billion welfare gain, followed by the US (25 per cent), China (8 per cent), and Mexico (7 per cent). The EU and Rest of Asia capture 16 per cent. In the final scenario, it is assumed that, along with actual 2008/9 adopters, West and Central African (WCA) countries, Senegal, Tanzania and Uganda also adopt at a rate of 50 per cent. Similar to the partial equilibrium results of Falck-Zepeda et al. (2008), the new African adopters avoid welfare losses by adopting Bt cotton. They achieve some modest adoption gains, with Burkina Faso capturing most of the gains. For these countries, Bouët and Gruère find that changing assumptions about adoption rates has more of an impact on their welfare than changes in productivity shocks.

Van Meijl and van Tongeren (2004) examine the effects of the EU Common Agricultural Policy (CAP) and possible GM production and import bans on returns to adoption of Bt corn and HT soybeans. In their Base Scenario (Scenario 0), North American coarse grains and oilseeds sectors experience 5 per cent productivity growth. Productivity growth for coarse grains is modeled as Hicks-neutral technical change. For oilseeds, it is modeled as a combination of chemical- and labor-augmenting technical change. In their Spillover Scenario (Scenario 1), North American adoption has spillover effects encouraging adoption in other regions. Productivity growth in most regions is assumed less than for North America. They model spillovers from the initial adopting region to other regions as a function of trade flows, similarity in farm structure and social acceptance. Scenario 2 adds to Scenario 1 by accounting for the effects of CAP. Scenario 3 assumes that producers in the EU, Australia/New Zealand, Japan (and Newly Industrialized Countries) and the ROW do not accept the biotechnologies. Producers in South America, China and the Rest of Asia adopt. Scenario 4 adds to Scenario 3 the assumption that both production and importation of GM crops are banned in the EU.

In the Base Scenario, with adoption in North America alone, North American welfare increases by US$2.2 billion. EU production of oilseeds and coarse grains declines as North American production increases. EU farm income for each crop falls by about 1 per cent. Falling world prices reduce EU farm income, but benefit EU consumers so that EU welfare increases overall by US$249 million. In Scenario 1, the EU and other countries also adopt biotechnologies. EU oilseeds production increases, but internal market prices fall more. With global biotechnology adoption, farm prices fall more than in Scenario 0. EU farm income falls, but overall EU welfare increases (again because of lower agricultural prices) by nearly US$1.3 billion. Spillovers reduce North American welfare gains by about 3 per cent below Scenario 0 gains. In Scenario 2, which includes the effects of the CAP, EU production of both crops increases. EU farm income from oilseeds falls by a bit more, but coarse grains income (which is protected by CAP) falls by substantially less than under Scenario 1. Compared to Scenario 1, adding effects of the CAP has no effect on North American welfare, although EU welfare gains fall from US$2.2 to <US$0.7 billion because of the distortionary effects of the CAP. If producers in the EU (and some other regions) reject GM crops (Scenario 3), EU production and farm income fall. EU welfare gains from adoption in other countries are only US$152 million. Finally, if the EU bans consumption and importation of GM crops (Scenario 4), EU oilseeds production rises by nearly 20 per cent and coarse grain prices rise 1 per cent. With imports restricted, EU market prices rise. Oilseeds farm income rises by >24 per cent, while coarse grain farm income rises by nearly 3 per cent. Compared to Scenario 0, North American welfare gains are 21 per cent lower. The EU faces welfare losses of >US$1.4 billion. Van Meijl and van Tongeren (2004) contrast their results with an earlier study by Nielsen and Anderson (2001) that estimated that non-adoption in the EU would lead to greater reductions in coarse grain production and farm income from non-adoption. They argue that the CAP's 'price-insulating' feature shelters EU producers from the world price reductions brought on by global biotechnology adoption.

Anderson and Jackson (2003) use the GTAP framework to explain why the EU has adopted a less GM-friendly policy than the United States. They consider five scenarios: (a) adoption of GM coarse grains and oilseeds in all regions; (b) adoption in only Argentina, the EU and North America; (c) adoption in only Argentina and North America; (d) adoption in Argentina and North America, but with an EU moratorium (production and import ban) on GM crops; and (e) the same as (d) except Japan, Korea, Australia and New Zealand join the moratorium. They measure impacts in terms of returns to endowments held by farm households: land, capital and skilled labor.

When all countries adopt, returns to North American and EU farm households fall. When just North America, Argentina and the EU adopt, EU farm income is unchanged; North American farm income still falls, but by less. If only North America and Argentina adopt, EU farm incomes are unchanged, but North American farm incomes fall. North American farm income declines as more regions adopt. EU farmers are little affected by adoption. If the EU imposes a moratorium on GM crops, EU farm income rises, while

North American farm income falls. If the moratorium is extended to more countries, EU farm income rises, while North American losses increase. North American farmers are hurt more by imposing and extending the moratorium than they are from increasing technological competition from other regions.

Anderson and Jackson (2003) use these results to make a political economy argument for differences in farm interest group incentives in the EU and North America. EU farmers are best off under the scenario where the moratorium on GM products is widest, while North American farmers are worst off under this adoption scenario. Thus, while EU farmers stand to gain more if they and other countries reject GM crops, North American producers would be better off with wide acceptance of GM crops (even with technological competition from other regions) than they would be with widespread GM crop bans. Anderson and Jackson (2003), however, do not explain why North American farmers would favor GM crops at all; with global acceptance of GM crops and just North American and Argentina adopting (that is North American technological advantage and widespread consumer acceptance), North American farm income falls! In their model, North American farmers would be *best* off if the GM crop moratorium were extended to *all* countries. Of, course this only applies to farm income. Anderson and Jackson (2003) do not report effects on national welfare. However, all the other GTAP analyses suggest that wider adoption and consumer acceptance would raise aggregate welfare.

6 SOME OBSERVATIONS

6.1 Fixed Price Studies

In many respects, assessment of biotechnology is no different from assessment of other agricultural innovations. Many of the results should be of no surprise given economic theory or the body of literature on returns to agricultural research. While the PE studies generally follow the Moschini and Lapan (1997) approach of including monopoly rents captured by technology suppliers, they conduct the classic 'returns to research' analyses that focus on welfare impacts of supply curve shifts (Alston et al., 1995). An important area where biotechnology is different, however, is in the area of consumer response and potential requirements of product labeling, segregation and identity preservation. In this respect, studies that account for these consumer/policy effects represent a departure from previous 'returns to research' studies.

Studies with fixed output prices have generally estimated that adopting producers capture most of the gains from agricultural biotechnologies, illustrating the limits of monopoly pricing to capture a greater share of the gains of innovation. The technology supplier faces a trade-off because increasing the price of GM seed lowers the quantity sold. Beyond a critical price threshold, adoption plummets. These studies also illustrate that the distribution of gains is sensitive to agronomic conditions in the year of analysis. For example, producers capture a larger share of gains from insect-resistant Bt crops in years with high pest pressure. This also suggests Bt crops have an 'insurance value' that distribution studies have yet to fully explore.

6.2 Price Endogenous Studies

In PE studies with endogenous output prices, the share of net gains (and losses) from biotechnologies varies widely across crop, technology, region and study. However, study results often differ in ways one would expect given economic theory, economic data used and modeling choices made. First, changing demand elasticities affects the distribution of gains between producers and consumers, but has relatively little impact on the aggregate surpluses generated. Consumers capture a greater share of gains (and producers a smaller share) when demand is price inelastic because the increase in supply from biotechnology adoption reduces price more (see Price et al., 2003; Moschini et al., 2000 for example). Second, producers capture more of the welfare gains if the supply curve is more inelastic. Changing the supply elasticity can also increase the estimated total gains of adoption. This has to do with the way that yield increases are converted into vertical supply shifts in linear supply models, assuming parallel supply shifts (Alston et al., 1995; Oehmke and Crawford, 2002).

Third, the effects of spillovers (effect on a home region of adoption by competing producers in other regions) depend on a country's trade position. Edwards and Freebairn (1984) have demonstrated how greater spillovers increase welfare if a region is a net importer of the commodity experiencing a supply shift. If the region is a net exporter, then spillovers can reduce home country welfare, with producer surplus falling and consumer surplus rising. The PE trade models confirm this result. Net-importing regions benefit more as biotechnologies become more widely adopted.

Fourth, effects of spillovers depend on relative rates of adoption. For net-exporting regions, there is a high cost of *not* adopting a biotechnology if other countries are doing so. Even though spillovers tend to reduce gains to producers in the home regions, home producers are often better off with global adoption (for example adoption by themselves and competitors) than with no adoption at all. Producers are not always better off under global adoption, however. They are more likely to experience losses if (a) the new technology has less of a productivity effect at home relative to competing regions, (b) home adoption rates are lower than elsewhere and (c) if the technology generates a divergent, rather than a parallel, supply shift. This leads to a fifth point – the nature of the supply shift affects both absolute and relative producer gains from biotechnology.

6.3 CGE Studies

Because of their complexity and the number of variables changing at once, CGE models provide numerical results where simple PE model intuition may be less helpful. The studies surveyed show that there are high opportunity costs of not adopting biotechnologies if other regions are adopting them. Nations as a whole benefit as adoption becomes more widespread. Consumers clearly gain via falling world prices caused by productivity growth in the adopting sectors. For producers, effects are more ambiguous. Indirect measures of producer income, such as land rents or gross income can often fall with widespread adoption. There appear to be some technological treadmill effects (Cochrane, 1958). Adoption elsewhere puts downward pressure on output price. Home adoption to compete with this lower-cost, lower-price market environment, however, pushes output prices down further. Falling behind technologically, however, (that is not adopting when others do) often is the worst-case scenario. This occurs in part because the function forms used in CGE models often lead to divergent supply shifts in response to productivity growth.

Another key result from CGE studies is that an individual country's gains are often proportional to the size of the productivity shock. This shock in turn is determined by two factors: (a) the size of the yield increase or cost saving of the adopted technology and (b) the national rate of adoption. While earlier studies assume Hicks-neutral technical change, later studies assumed different type of factor-augmenting technical change. The factor bias of technical change appears to matter less than the pure unit cost reduction embodied by the biotechnologies. Yet, in some cases factor bias assumptions alter results significantly. The nature of factor bias in simulations varies by crop, technology and region, and these assumptions are *not* consistent across studies. CGE simulations tend to take results from micro-level analyses and aggregate them up to national level impacts. More econometric research along the seminal work of Binswanger (1974a, 1974b) to estimate the factor-bias induced by biotechnology for national level cost functions would be quite valuable.

Although CGE models estimate impacts of innovation in one sector on related sectors, downstream effects of industries like rice processing or textile markets tend to be small. Given that many of the biotechnologies affect animal feeds, there is scope for looking at the effects of simultaneous adoption of multiple technologies in multiple regions on the livestock sector.

Finally, because of the assumption of perfect competition and general equilibrium closure, CGE studies to date have not attempted to include measures of monopoly rents (*à la* Moschini and Lapan). It might be possible, however, to experiment with partial equilibrium closures in GTAP, where agricultural output prices and land rents are endogenous, while other prices are exogenous. This would allow for imposing exogenous mark-ups in seed price markets.

6.4 Policies Matter

Several studies highlight the important influence that agricultural or trade policies have on the size and distribution of net gains from biotechnology adoption. Price support payments can shelter producers from falling prices that come with technology-induced output increases. Yet this can come at substantial costs to the government and the export-market competitors of the country instituting the policy (Frisvold and Tronstad, 2003; Frisvold et al., 2006; Sobolevsky et al., 2005). They need not be welfare reducing, however, because of second-best effects. Monopoly pricing of GM seeds creates welfare losses that subsidies may counteract.

Import bans on GM commodities tend to reduce welfare of consumers in the regions imposing the ban and GM-exporting countries, while increasing welfare outside the ban area (because the ban exerts downward pressure on prices). Huang et al. (2004) posit that GM bans need not impose significant costs to GM crop-exporting countries. Anderson and Jackson (2003) make political economy arguments that countries impose import bans *because* of their distributional impacts (that is as trade protection for domestic producers).

The studies by Phillips (2003), by Huang et al. (2004) and by Sobolevsky et al. (2005)

illustrate how labeling, crop segregation and identity preservation requirements can greatly reduce the size and alter the relative benefits from biotechnology. The treatment by Sobolevsky et al. (2005) where consumers view conventional and GM crops as different products is an important, novel departure from many earlier approaches that assume biotechnologies affect supply but not demand. A growing body of literature illustrates that given such demand-side effects and policy responses (for example segregation and labeling costs), introducing biotechnology may not improve welfare (Fulton and Giannakas, 2004; Lence and Hayes, 2005; Moschini et al., 2005; Sobolevsky et al., 2005). Interactions with organic food markets are a further complication (for example Phillips, 2003; Moschini et al., 2005). Sobolevsky et al. (2005), however, show that raising crop segregation costs can actually increase specialization in GM crops and increase profits of GM technology providers. This might serve as a cautionary lesson for opponents of GM crops seeking to discourage adoption by increasing segregation costs. The coexistence of differentiated conventional, GM and organic markets will be an important area of continued research as markets for GM crops mature. An important lesson from this review is that 'policy matters' in determining the size and distribution of gains from biotechnology.

REFERENCES

Alston, J.M., G.W. Edwards and J.W. Freebairn (1988), 'Market distortions and benefits from research', *American Journal of Agricultural Economics*, **70**(2), 281–8.

Alston, J.M., G.W. Norton and P.G. Pardey (1995), *Science under Scarcity: Principles and Practice for Agricultural Research Evaluation and Priority Setting*, Ithaca, NY: Cornell University Press.

Anderson, K. and L.A. Jackson (2003), 'Why are US and EU policies toward GMOs so different?', *AgBioForum*, **6**(3), 95–100.

Anderson, K. and S. Yao (2003), 'China, GMOs and world trade in agricultural and textile products', *Pacific Economic Review*, **8**, 157–69.

Anderson, K., E. Valenzuela and L.A. Jackson (2008), 'Recent and prospective adoption of genetically modified cotton: a global computable general equilibrium analysis of economic impacts', *Economic Development and Cultural Change*, **56**, 265–96.

Binswanger, H.P. (1974a), 'A cost function approach to the measurement of elasticities of factor demand and elasticities of substitution', *American Journal of Agricultural Economics*, **56**(2), 377–86.

Binswanger, H.P. (1974b), 'The measurement of technical change biases with many factors of production', *American Economic Review*, **64**(5), 964–76.

Bouët, A. and G. Gruère (2011), 'Refining opportunity cost estimates of not adopting GM cotton: an application in seven Sub-Saharan African countries', *Applied Economic Perspectives and Policy*, **32**, 260–79.

Cochrane, W.W. (1958), *Farm Prices, Myth and Reality*, St Paul. MN: University of Minnesota Press.

Demont, M. and E. Tollens (2004), 'First impact of biotechnology in the EU: Bt maize adoption in Spain', *Annals of Applied Biology*, **145**(3), 197–207.

Dillen, K., M. Demont and E. Tollens (2009), 'Global welfare effects of GM sugar beet under changing EU sugar policies', *AgBioForum*, **12**(1), 119–29.

Edwards, G.W. and J.W. Freebairn (1984), 'The gains from research into tradable commodities', *American Journal of Agricultural Economics*, **66**, 41–9.

Elbehri, A. and S. MacDonald (2004), 'Estimating the impact of transgenic Bt cotton on West and Central Africa: a general equilibrium approach', *World Development*, **32**(12), 2049–64.

Falck-Zepeda, J., D. Horna and M. Smale (2008), 'Betting on cotton: potential payoffs and economic risks of adopting transgenic cotton in West Africa', *African Journal of Agricultural and Resource Economics*, **2**(2), 188–207.

Falck-Zepeda, J., G. Traxler and R. Nelson (2000a), 'Rent creation and distribution from biotechnology innovations: the case of Bt cotton and herbicide-tolerant soybeans in 1997', *Agribusiness*, **16**(1), 21–32.

Falck-Zepeda, J.B., G. Traxler and R.G. Nelson (2000b), 'Surplus distribution from the introduction of a biotechnology innovation', *American Journal of Agricultural Economics*, **82**(2), 360–69.

Frisvold, G.B. (1997), 'Multimarket effects of agricultural research with technological spillovers', in T. Hertel (ed.), *Global Trade Analysis: Modeling and Applications*, Cambridge: Cambridge University Press.
Frisvold, G.B. and J. Reeves (2007), 'Economy-wide impacts of Bt cotton', in D.A. Richter (ed.), *Proceedings, 2007 Beltwide Cotton Conferences*, 9–12 January, New Orleans, LA, Memphis, TN: National Cotton Council of America, pp. 836–42.
Frisvold, G.B. and J.M. Reeves (2010), 'Resistance management and sustainable use of agricultural biotechnology', *AgBioForum*, **13**(4), 343–59.
Frisvold, G.B. and R. Tronstad (2003), 'Economic impacts of Bt cotton adoption in the United States', in N. Kalaitzandonakes (ed.), *The Economic and Environmental Impacts of Agbiotech: a Global Perspective*, Norwell, MA: Kluwer-Plenam.
Frisvold, G.B., J.M. Reeves and R. Tronstad (2006), 'Bt cotton adoption in the United States and China: international trade and welfare effects', *AgBioForum*, **9**(2), 69–78.
Fulton, M. and K. Giannakas (2004), 'Inserting GM products into the food chain: the market and welfare effects of different labeling and regulatory regimes', *American Journal of Agricultural Economics*, **86**, 42–60.
Hertel, T.W. (ed.) (1997), *Global Trade Analysis: Modeling and Applications*, Cambridge: Cambridge University Press.
Huang, J., R. Hu, H. van Meijl and F. van Tongeren (2004), 'Biotechnology boosts to crop productivity in China: trade and welfare implications', *Journal of Development Economics*, **75**, 27–54.
Lence, S. and D. Hayes (2005), 'Genetically modified crops: their market and welfare impacts', *American Journal of Agricultural Economics*, **87**, 931–50.
Marra, M.C. (2001), 'The farm level impacts of transgenic crops: a critical review of the evidence', in P.G. Pardey (ed.), *The Future of Food: Biotechnology Markets in an International Setting*, Baltimore, MD: Johns Hopkins Press and International Food Policy Research Institute, pp. 155–184.
Marra, M., P. Pardey and J. Alston (2002), 'The payoffs to transgenic field crops: an assessment of the evidence', *AgBioForum*, **5**(2), 43–50.
Mensah, E.C. and M.K. Wohlgenant (2010), 'A market impact analysis of soybean technology adoption', *Research in Business and Economics Journal*, **1**, 1–20.
Moschini, G. and H. Lapan (1997), 'Intellectual property rights and the welfare effects of agricultural R&D', *American Journal of Agricultural Economics*, **79**, 1229–42.
Moschini, G., H. Bulut and L. Cembalo (2005), 'On the segregation of genetically modified, conventional and organic products in European agriculture: a multi-market equilibrium analysis', *Journal of Agricultural Economics*, **56**, 347–72.
Moschini, G., H. Lapan and A. Sobolevsky (2000), 'Roundup Ready soybeans and welfare effects in the soybean complex', *Agribusiness*, **16**, 33–55.
National Research Council (2010), 'Impact of genetically-engineered crops on farm sustainability in the United States', Washington, DC: The National Academies Press.
Nielsen, C.P. and K. Anderson (2001), 'Global market effects of alternative European responses to genetically modified organisms', *Weltwirtschafliches Archiv*, **137**(2), 320–46.
Oehmke, J.F. and E.W. Crawford (2002), 'The sensitivity of returns to research calculations to supply elasticity', *American Journal of Agricultural Economics*, **84**, 366–9.
Phillips, P.W.B. (2003), 'The economic impact of herbicide tolerant canola in Canada', in N. Kalaitzandonakes (ed.), *The Economic and Environmental Impacts of Agbiotech*, New York: Kluwer/Plenam, pp. 119–39.
Pray, C., D. Ma, J. Huang and F. Qiao (2001), 'Impact of Bt cotton in China', *World Development*, **29**(5), 813–25.
Price, G.K., W. Lin, J.B. Falck-Zepeda and J. Fernandez-Cornejo (2003), 'Size and distribution of market benefits from adopting biotech crops', technical bulletin no. TB1906, Washington, DC: Economic Research Service.
Qaim, M. (2009), 'The economics of genetically modified crops', *Annual Review of Resource Economics*, **1**, 665–93.
Qaim, M. and A. de Janvry (2003), 'Genetically modified crops, corporate pricing strategies, and farmers' adoption: the case of Bt cotton in Argentina', *American Journal of Agricultural Economics*, **85**, 814–28.
Qaim, M., and G. Traxler (2005), 'Roundup Ready soybeans in Argentina: farm level & aggregate welfare effects', *Agricultural Economics*, **32**(1), 73–86.
Scatasta, S., J. Wesseler and M. Demont (2006), 'A critical assessment of methods for analysis of social welfare impacts of genetically modified crops: a literature survey', report no. 27, Wageningen: Mansholt Graduate School.
Smale, M., P. Zambrano, J. Falck-Zepeda and G. Gruère (2006), 'Parables: applied economics literature about the impact of genetically engineered crop varieties in developing economies', EPT Discussion Paper no. 158, Washington, DC: International Food Policy Research Institute.
Sobolevsky, A., G. Moschini and H. Lapan (2005), 'Genetically modified crops and product differentiation: trade and welfare effects in the soybean complex', *American Journal of Agricultural Economics*, **87**, 621–44.

Traxler, G. (2006), 'The GMO experience in North and South America', *International Journal of Technology Globalization*, **2**, 46–64.
Traxler, G., S. Godoy-Avila, J. Falck-Zepeda and J. Espinoza-Arellano (2003), 'Transgenic cotton in Mexico: a case study of the Comarca Lagunera', in N. Kalaitzandonakes (ed.), *The Economic and Environmental Impacts of Agbiotech*, New York: Kluwer, pp. 183–202.
van Meijl, H. and F. van Tongeren (2004), 'International diffusion of gains from biotechnology and the European Union's common agricultural policy', *Agricultural Economics*, **31**, 307–16.
Vitale, J., T. Boyer, R. Uaiene and J.H. Sanders (2007), 'The economic impacts of introducing Bt technology in smallholder cotton production systems of West Africa: a case study from Mali', *AgBioForum*, **10**(2), 71–84.
Vitale, J., H. Glick, J. Greenplate and O. Traore (2008), 'The economic impact of second generation Bt cotton in West Africa: empirical evidence from Burkina Faso', *International Journal of Biotechnology*, **10**, 167–83.

44 Economic success but political failure? The paradox of GM crops in developing countries

*Terri Raney, Ademola A. Adenle and Ira Matuschke**

1 INTRODUCTION

Genetically modified (GM) crops have been rapidly adopted in many developed and developing countries, including by resource-poor farmers, since they were first commercialized in 1995. An extensive body of peer-reviewed economic literature confirms that farmers benefit from the cultivation of GM crops under a wide range of agro-ecological and socio-economic conditions and that the economy-wide welfare gains are shared among adopting farmers, consumers and technology suppliers. Despite the growing evidence of positive farm-level and economy-wide impacts, the commercialization of new crops and traits has been blocked at the final approval stage (after being cleared on biosafety grounds) in a number of countries, paradoxically in some countries where first-generation GM crops have been the most widely adopted.

While European Union (EU) legislation provides for *ex post* monitoring and reporting on the socio-economic implications of the deliberate release and marketing of genetically modified organisms (GMOs), and the Cartagena Biosafety Protocol provides for socio-economic assessments of the deliberate release into the environment of living modified organisms (LMOs), very few countries explicitly consider *ex ante* socio-economic assessments in their regulatory decision-making processes for GMOs. Nevertheless, many countries do so implicitly. It has been suggested that the explicit incorporation of socio-economic assessments could facilitate the decision-making process for the commercialization of GM crops by separating the science-based, biosafety assessment from the political process of decision-making (Lusser et al., 2012). Based on a survey of peer-reviewed literature on the socio-economic impacts, this chapter considers the potential value of explicitly incorporating socio-economic assessments in the regulatory decision-making processes for GM crops.

A substantial body of evidence on the farm-level economic impacts of GM crops in developing countries has emerged in the past decade. These studies document that the impacts are positive on average but highly variable over time and space and depend crucially on a variety of factors that include but go beyond the agronomic performance of the technology (Raney, 2006). The level of benefits depends, in the first place, on the effectiveness of a given trait in a specific agro-ecological context and its incorporation in locally adapted cultivars; without this first requirement, the technology will not be successful. Similarly, as with any agricultural technology, the ability of farmers to adopt and benefit from GM crops depends on local conditions, adequate infrastructure, functioning markets and extension services. GM crops, perhaps more than other agricultural technologies, also depend on national institutional and regulatory capacity.

2 FARM-LEVEL STUDIES

Farm-level studies aim to identify the benefits that accrue to farmers cultivating GM crops compared to farmers growing conventional crops. Data are often based on *ex post* farm surveys. Partial budget analysis is frequently applied to analyse the changes per unit of land that are associated with the adoption of new technologies. Input costs per hectare are subtracted from the output value per hectare to calculate net benefits.

The farm-level economic impact of transgenic crops depends on the costs and returns of production compared with alternative varieties. Most studies use a partial accounting framework to compare production costs and returns for adopters versus non-adopters of transgenic crops. The key methodological challenges involve ensuring that the technology in question constitutes the only significant difference between the subject and control groups (Lichtenberg and Zilberman, 1986; Kuosmanen et al., 2006; Qaim and Matuschke, 2005).

A more fundamental challenge is that these analyses are too limited to identify the full effect of new technologies. Partial budget analyses are very informative, but they rarely take into account the value of own land or household labour. Also, they often do not account for non-monetary factors that may benefit farmers, such as reduced female or child labour requirements or more flexible management practices. Moreover, partial budget analyses may be biased because of underlying unobservable characteristics of the farm and/or farmers. Studies control for these biases by making plot comparisons between similar plots – one with and one without the innovation – belonging to the same farmer. Other biases in farm-level studies, which are particularly evident in early adoption studies, are related to small sample sizes and selection biases (see Smale et al., 2009). Selection biases are caused by the fact that early adopters of an innovation are generally 'model' farmers who apply modern innovations in a more efficient way. Recent studies (many reviewed in other chapters in this volume) have addressed many of the concerns of earlier studies; for example Kouser and Qaim (2011) use panel data to study the farm-level impacts of Bt cotton in India.

Other approaches applied by researchers to analyse the impact of GM crops in developing countries include production function models that estimate the marginal impact of GM crop yields using regression analysis. Damage abatement models for insect-resistant crops estimate how much potential output is lost to insect pests (Lichtenberg and Zilberman, 1986). The application of insect-resistant technologies decreases crop losses owing to pest infestations and thereby increases potential outputs.

The following review identifies factors influencing the level and distribution of the economic value created by transgenic crops in leading developing countries. The availability of locally adapted transgenic crop cultivars in a country is the most basic requirement for successful adoption by farmers. Institutional factors such as national research capacity and the existence of functioning input markets (FAO, 2004), intellectual property rights (Zilberman and Graff, 2005), environmental and food safety regulatory capacity (Evenson and Santaniello, 2004; Pray et al., 2005), and trade capacity (Tothova and Oehmke, 2005) are crucially important determinants of the level and distribution of gains. The country case studies below illustrate the importance of these factors.

The most extensive *ex post* studies of transgenic crop adoption in developing countries have been conducted for insect-resistant (IR) cotton in Argentina (Qaim and de Janvry,

2003, 2005), China (Pray et al., 2002), India (Qaim et al., 2006), Mexico (Traxler et al., 2003) and South Africa (Bennett et al., 2004b; Morse et al., 2005a). Transgenic herbicide-tolerant (HT) soybeans are being grown in Argentina, Brazil, Paraguay and elsewhere, but Argentina is the only developing country for which peer-reviewed studies have been published (Qaim and Traxler, 2005). Some developing countries also produce HT and/or IR maize, and studies have been published on the impact of IR maize in South Africa (for example Gouse et al., 2005a, 2009).

2.1 China

China, where some 7 million small farmers are growing Bt cotton, represents the most successful case so far in terms of increased yields, increased farm incomes and improved quality of life (James, 2012). As reported by Pray et al. (2002), much of China's success rests on its highly developed public agricultural research system which has independently produced two transgenic constructs that confer insect resistance. These have been incorporated into a large number of locally adapted cotton varieties and compete directly with Monsanto's Bt cotton varieties. As a result, transgenic seed prices are much lower in China than elsewhere and farmers reap substantially higher returns. The role of the public sector in developing and distributing IR cotton varieties has been instrumental in reducing the price premium. Lower costs and marginally higher yields translate into large net profit gains in China. Pray et al. (2011) estimate that almost all GM crop varieties marketed in China have been developed by the public sector. In contrast to the Indian experience most of the cotton area in China is covered with public Bt cotton varieties rather than Bt cotton hybrids.

Chinese farmers experience lower yield gains than in many other countries because pest damage on conventional cotton is controlled by heavy pesticide use. China has been able to significantly reduce its use of chemical pesticides on cotton, with important environmental and farmer health benefits (Huang et al., 2002b).

Pray et al. (2001) analysed the distribution of Bt cotton benefits in China by farm size and found the innovation to be decidedly pro-poor. The smallest farms (less than 0.47 ha) experienced the largest yield gains while mid-size farmers (0.47–1.0 ha) had the largest reductions in total costs due to lower pesticide use. In terms of net income, the gains for the two smaller farm-size categories were more than twice those for the largest farms (over 1.0 ha).

In 2011, more than 70 per cent of China's cotton production used GM varieties, accounting for 3.9 million hectares of production (James, 2012). One of the early studies on the adoption of IR cotton in China estimates that in 2001, producers using IR cotton reported a net revenue gain of US$277 per hectare, compared to a loss of US$225 for non-IR cotton varieties (Huang et al., 2002a). A subsequent study using 2005 production data has estimated that economic benefit of GM cotton production exceeded US$600 million (Frisvold and Reeves, 2007). See Chapter 8 in this volume for further discussion of agbiotech in China.

In recent years, a number of studies reported an increase in secondary pests (mirids) that was directly related to an increase in the area cultivated with Bt cotton (see Pray et al., 2011 for a summary). The Bt gene protects the cotton plant only against certain insects (for example bollworms) and a lower number of pesticide sprays may have led to an increase in these secondary pests. Spraying against secondary pests reduces pesticides savings, and thereby the net income associated with growing Bt cotton (Raney and Matuschke, 2011; Pray et al., 2011). However, as Pray et al. (2011) report, mirid infestations are highly variable with climatic conditions, and efficient pest-management practices can help to address these pests.

2.2 Argentina

The Argentine experience with Bt cotton provides an interesting comparison with the Chinese case in terms of the effect of IPRs. Monsanto has strictly enforced its IPRs on Bt cotton in Argentina and charges significantly higher prices than for conventional cotton seed (Qaim and de Janvry, 2003). As a result, Bt cotton offers relatively small returns and thus has not been widely adopted, in contrast with Bt maize or herbicide-tolerant soybeans which have been enthusiastically embraced by Argentine farmers (Brookes and Barfoot, 2012; James, 2012). The crucial difference is that Monsanto failed to patent its soybean innovation in Argentina and thus has been unable to strictly enforce its IPRs (Qaim and de Janvry, 2003).

Nearly all of the soybeans produced in Argentina are GM and have been modified to be herbicide tolerant (Brookes and Barfoot, 2012; Finger et al., 2009; James, 2012). Brookes and Barfoot (2012) estimated that Argentina HT soybeans generated US$1.16 billion in economic benefits in 2010, while the cumulative impact was US$7.1 billion since 1996.

According to Qaim and Traxler (2005), herbicide-tolerant soybeans have increased total factor productivity by 10 per cent on average in Argentina, with the cost savings being slightly greater for smaller farms (less than 100 ha) than for larger farms, due mainly to lower seed prices among small farmers who are more likely to use uncertified seed than larger farms. They computed aggregate global welfare benefits from herbicide-tolerant soybeans at more than US$1.2 billion, with the largest share going to consumers (53 per cent), followed by seed and biotechnology firms (34 per cent) and farmers (13 per cent). See Chapter 6 in this volume for a detailed discussion of agbiotech in South America.

2.3 South Africa

South Africa provides another important lesson about the role of institutions. The farm-level results for Bt cotton adopters in the Makhathini Flats were positive and the benefits were widely shared by all farm types (Morse et al., 2004). Although the first-year Bt adopters were more likely to be older, larger and male, by the second year smaller farmers of both genders and all age groups were equally likely to adopt. The study found significant pro-poor benefits in that small farmers earned higher per-hectare net returns than larger farmers. This positive experience was not sustained, however (Gouse et al., 2005b). The dissemination of Bt cotton seed to farmers in the Makhathini Flats relied on credit provided by the local cooperative. The cooperative also ran the only cotton gin in the area, so it could rely on collecting its debts from the cotton harvest at the end of the season. When another cotton gin opened in the region, the cooperative was no longer able to collect and soon ceased providing Bt cotton seed on credit (Gouse et al., 2005b).

Subsequent studies have further delved into the impacts on small landholders. Gianessi (2009) reports that to prevent weed damage, maize requires 276 hours of hand weeding per hectare. Gouse (2012) found that smallholder GM maize farmers in South Africa spent on average 267, 177 and 152 hours/ha on manual weeding over three seasons. Vitale et al. (2011) observes that GM cotton production in Burkina Faso between 2009 and 2011 yielded, on average, 22 per cent above those of conventional cotton. The authors observe that higher seed costs were offset by lower pesticide and labour costs, resulting in an overall increase in income by US$80 per hectare. See Chapter 38 in this volume for an in-depth review of GM cotton in Burkina Faso.

2.4 Mexico

The Mexican case provides a clear example of the importance of introducing effective innovations. Traxler et al. (2003) found that the only state in Mexico where Bt cotton adoption exceeded 40 per cent was Comarca Lagunera because the most serious local pests are susceptible to Bt. The major pests in other states are less susceptible to Bt and require continued pesticide use, so Bt varieties are of less value. Traxler and Godoy-Avila (2004) evaluated the distribution of the economic benefits generated by Bt cotton between farmers and the seed suppliers (Monsanto and Delta & Pine Land). Farmers gained 83 per cent of the total economic surplus on average for the two years in the study.

Mexico has had a ban on commercial use of GM corn, which has been costly as corn imports account for 27 per cent of Mexico's trade deficit (Kapstein et al., 2012). These authors posit that if Mexico adopted more innovative corn production technologies, including agbiotech, then yields would rise from the less than 2 tonnes per hectare with landrace corn varieties to over 7 tonnes per hectare. The adoption of higher yielding corn varieties could add as much as US$1 billion to US$6 billion to the Mexican economy.

2.5 India

The Indian experience with Bt cotton has been very controversial. Despite the rapid expansion of Bt cotton and overall positive effects as described in Table 44.1, their economic impact continues to be hotly contested, with some critics charging that Bt adopters are worse off than non-Bt growers (Shiva and Jafri, 2004).

The first economic studies were based on farm-level field trial data, and as such did not reflect actual farm experience with commercial cultivation (Qaim, 2003; Qaim and Zilberman, 2003). These studies estimated potential yield benefits of 80 per cent. A subsequent study shown in detail in Table 44.2 was based on farm-level data from four different states in India (Qaim et al., 2006). This study found large net gains from Bt cotton adoption at the national average level although with significant variation across states, and one state, Andhra Pradesh, experienced negative results. The authors speculated that the lack of locally adapted cultivars was the main reason for poor performance in this state. At the time of the study, the Indian biosafety authorities had approved only four Bt cotton varieties for use throughout the country. By 2010 more than 600 Bt cotton hybrids and varieties had been approved for marketing in the country, and Bt cotton was estimated to cover about 90 per cent of the total cotton area in India (with a substantial acreage of unofficial Bt cotton seeds) (Pray et al., 2011). In 2011, 10.6 million hectares

Table 44.1 Studies showing average farm-level impacts of Bt cotton in India, percentage change versus conventional counterpart

Source	Data	Average agronomic effects		Average economic effects			
		Insecticide sprays	Yields	Seed costs	Insecticide costs	Total costs	Net income
Barwale et al. (2004)	1069 cotton farmers, six states, 2002 season, survey administered by Mahyco	−62	+61	n/a	n/a	n/a	n/a
Qaim et al. (2006)	341 cotton farmers, four states, panel data, 2002–03 season	−62	+34	+221	−69	+17	+69
Morse et al. (2005b)	3496 farmers, Maharashtra state, 2002–03 and 2003–04 season	−59	+51	n/a	−112	n/a	+58
Bennett et al. (2004a)	787 cotton farmers, Maharashtra state, 2002–03 and 2003–04 season	−69	+54	+224	−111	+8	+62
Kambhampati et al. (2006)	2709 cotton farmers, Maharashtra state, 2002–03 and 2003–04 season	−70	+54	n/a	n/a	+9	+62
Morse et al. (2007)	157 cotton farmers, Maharashtra state, 2002–03 and 2003–04 season	n/a	+35–86*	n/a	n/a	+13–32*	+62–144*
Crost et al. (2007)	338 cotton farmers, Maharashtra state, panel data comprising the years 2002 to 2003	n/a	+11–31*	+243	−15	n/a	n/a
Gandhi and Namboodiri (2006)	694 cotton farmers, four states, 2003–04 season	−66	+47	+183	−44	+17	+102
Bennett et al. (2005)	622 cotton farmers, Gujarat state, 2003–04 season	n/a	+29	n/a	n/a	n/a	+132
Sadashivappa and Qaim (2009)	341 cotton farmers, four states, panel data comprising the years 2002–2007	−29	+40	+166	−24	+17	+89

Note: * Refers to lower and upper ranges when accounting for self-selection bias.

Table 44.2 Performance advantage of Bt over non-Bt cotton in main cotton growing states in India, by state, 2002/03, percentage change versus conventional counterpart

	Yield	Revenue	Chemical costs	Total costs	Profits
Maharashtra	32***	29***	−44***	15**	56***
Karnataka	73***	67***	−49***	19**	172***
Tamil Nadu	43***	44***	−73***	5	229***
Andhra Pradesh	−3	−3	−19	13***	40
National average	34***	33***	−41***	17*	69***

Note: *,**,*** Significantly different from zero at 10%, 5% and 1% level respectively.

Source: Qaim et al. (2006).

Table 44.3 Performance advantage of Bt over conventional cotton, Maharashtra, India, percentage change versus conventional counterpart

	Yield	Revenue	Seed costs	Pesticide costs	Total costs	Profit
2002	45	44	232	−48	15	49
2003	63	63	217	−57	2	74
Average	54	53	224	−52	8	62

Note: All results statistically significant at P < 0.001.

Source: Bennett et al. (2004a).

of Bt cotton were largely cultivated by small-scale farmers compared to 50 000 hectares when Bt cotton was first adopted in 2002 (James, 2012).

Researchers from the University of Reading studied Bt cotton adoption in Maharashtra in 2002 and 2003 (Bennett et al., 2004a). Results from Maharashtra are summarized in Table 44.3. Data was collected from two random samples of farmers: 2709 farmers in 2002 and 787 in 2003. The samples covered 16 of 31 districts in the state and 1275 villages. Three Mahyco-Monsanto Bt cotton hybrids were compared with four popular non-Bt varieties, for a total of 7751 plots in 2002 and 1580 plots in 2003. These results, although smaller in magnitude, are consistent with the Qaim and Zilberman (2003) results from the 2001 field trials.

The Reading team also studied the relative merits of two official (MECH 12 and MECH 162) and two unofficial Bt cultivars (F1 and F2 hybrids) with a popular non-Bt counterpart in Gujarat (Morse et al., 2005b). Unofficial varieties have not received government authorization. Their unofficial status has several implications: they are less costly because no royalties are paid; they may be newer varieties because they have not undergone a lengthy approval process; but they may be less reliable than the official varieties. In assessing yields, costs and net returns to the five comparators, the Reading team confirmed that both of the official Bt varieties outperformed the unofficial and non-Bt varieties in terms of yields and net profits, although the difference between

MECH 162 and the F1 unofficial variety was small and statistically insignificant. Chapter 9 in this volume provides greater discussion of the agbiotech experience in India.

3 POTENTIALS AND CHALLENGES OF GM CROPS

According to most recent estimates, 868 million people (12.5 per cent) of the world's population are chronically undernourished, the majority in Asia (563 million) and sub-Saharan Africa (234 million), with the greatest prevalence (26.8 per cent) in the latter (FAO, 2012). In developing countries, global food demand is projected to at least double by the year 2050 when the world population is expected to reach 9.3 billion people, with the majority living in countries where poverty and hunger are widespread (Conforti, 2012; Godfray et al., 2010).

In order to meet global food demand, farmers will have to produce more with fewer resources in a sustainable way. They will be further challenged by increasing abiotic and biotic stresses, like drought, salinity, water-logging, toxic levels, weeds, insects and diseases. Adaptability to climate change will significantly influence the way in which food is produced in the future.

Biotechnology, including GM crops, can play an important role to address these challenges and to boost agricultural productivity by improving crop tolerance to biotic and abiotic stress and enhancing the durability of products during harvesting and shipping (Conway, 1999; Gressel et al., 2004; Sharma et al., 2002; Yang et al., 2009). It is estimated that in the past decade GM crops achieved economic gains of US$78 billion, of which 40 per cent was from reductions in production cost and 60 per cent from substantial yield gains (James, 2012). Lower production costs and higher yields increased farming incomes, also for resource-poor farmers. James (2012) argues that the production of GM cotton in China, India and South Africa and GM maize in the Philippines and South Africa have already improved the livelihood of 15 million resource-poor farmers and contributed significantly to their incomes.

GM technologies could also provide cost-effective solutions to vitamin and mineral deficiencies. For example, vitamin A-enhanced rice (Golden Rice) could assist in alleviating vitamin A deficiencies, which are particularly prevalent among children and women in South-East Asia and Africa (Potrykus, 2010; Stein et al., 2008; Qaim, 2010). Further research on staple food crops like cassava, rice, wheat, sorghum and banana is expected to improve agricultural productivity of farmers in developing economies. However, no genetically modified staple food crop (except for white maize) is currently marketed in developing countries.

Numerous studies have demonstrated positive socio-economic impacts of GM crops in developing countries, but variability in terms of the economic frameworks, methodologies, variables and hypotheses used in these studies means that critical questions remain about the economic impacts on farmers in developing countries. In addition, the review of studies above showed that the socio-economic impacts of GM crops are context specific because of agronomic and institutional variability across locations. Raney (2006) points out that many of the same factors that inhibit the adoption of any technology by poor farmers in developing countries are also relevant for agricultural biotechnologies.

These include lack of access to input and output markets, rural infrastructure, credit, extension services and others. These market factors need to be addressed to unlock the potential of any agricultural innovation, including GM crops.

Institutional weaknesses and political failures have contributed to a lack of a comprehensive framework for the regulation and adoption of GM crops in developing countries (Adenle, 2011; Raney, 2006). Many developing countries do not have the capacity to produce, regulate and apply biotechnology in crop production. Only a few developing countries with considerable scientific infrastructure and clear programs on cutting-edge biotechnology have translated R&D into significant adoption of GM crops and commercialized their products. They also put in place a regulatory framework and monitoring approaches that ensure food and environmental safety.

Many public research institutions in developing countries also lack access to vital information and analytical capacity that could improve their understanding of modern biotechnologies (Adenle, 2011). Some access problems relate to a lack of national capacity in areas such as data collection, management and storage, modern analytical methodologies and tools for accurate decision-making, amongst others. Other problems arise from material transfer agreements which restrict access to biotechnology research materials and further constrain product development, distribution and commercialization (Adenle et al., 2012).

Private sector restrictions on information from the data on the performance of biotechnology products can inhibit the analysis of benefits and challenges. Such restrictions on information sharing stand in sharp contrast to previous epochs of technical change such as the Green Revolution (Conway and Toenniessen, 1999), in which publicly funded research led to publicly available data. Public–private partnerships may be one tool to overcome these constraints. Moreover, regulations in the European Union present a challenge to the development of GMO policies in many developing countries (Stewart, 2009), and also hinder the opportunity of donating new agricultural technologies by multinational firms and other organizations (Adenle, 2011). The most immediate policy need is perhaps to provide transparent and efficient means of data dissemination that can facilitate quantitative assessment of the potential of biotechnology.

4 CONCLUSIONS

The evidence reviewed in this chapter suggests that farmers in developing countries can benefit from transgenic crops, but a fairly high level of national institutional capacity is required to ensure that farmers have access to suitable innovations on competitive terms. The economic impact assessments conducted to date vary considerably in terms of methodological approaches, and the measured impacts reflect these differences as well as the inherent performance variability across agro-ecological zones. Developing countries that lack strong public sector research systems and/or strong commercial seed sectors will be handicapped in the adoption of transgenic varieties.

The enforcement of IPRs clearly influences the farm-level returns to GM crops. Where the public sector provides effective competition with private sector technology suppliers (China), farmers gain a larger share of the economic value produced by GM crops. Where IPRs are strictly enforced (Argentina – cotton), the technology may be priced beyond the reach of farmers, but when there is no IPR enforcement (Argentina – soy beans), the technology supplier cannot earn a return on its investment. Where regulatory capacity is weak (India), there may be significant delays in the development and deployment of suitable innovations. Where credit systems are weak and contracts are unenforceable (South Africa), small farmers may not have access to the new technologies. Institutional factors, including regulatory capacity itself, also influence the level and distribution of socio-economic impacts. The endogenous nature of these relationships suggests that socio-economic assessments of the impacts of GM crops may provide guidance on promising areas for technological innovation, but should not be used as the basis of regulatory decision-making.

NOTE

* The views expressed in this chapter reflect the opinions of the authors and not necessary those of the Food and Agriculture Organization of the United Nations (FAO), the United Nations University-Institute of Advanced Studies (UNU-IAS) and the Organisation for Economic Co-operation and Development (OECD).

REFERENCES

Adenle, A.A. (2011), 'Global capture of crop biotechnology in developing world over a decade', *Journal of Genetic Engineering and Biotechnology*, **9**, 83–95.

Adenle, A.A., S.K. Sowe, G. Parayil and O. Aginam (2012), 'Analysis of open source biotechnology in developing countries: an emerging framework for sustainable agriculture', *Technology in Society*, **34**, 256–69.

Barwale, R.B., V.R. Gadwal, U. Zehr and B. Zehr (2004), 'Prospects for Bt cotton technology in India', *AgBioForum*, **7**(1&2), 23–6.

Bennett, R., Y. Ismael, U. Kambhampati and S. Morse. (2004a), 'Economic impact of genetically-modified cotton in India', *AgBioForum*, **7**(3), 1–5.

Bennett, R., U. Kambhampati, S. Morse and Y. Ismael (2005), 'Perceptions of the impacts of genetically modified cotton varieties: a case study of cotton industries in Gujarat, India', *AgBioForum*, **8**(2&3), 161–71.

Bennett, R., S. Morse, Y. Ismael and B. Shankar (2004b), 'Reductions in insecticide use from adoption of BT cotton in South Africa: impacts on economic performance and toxic load to the environment', *Journal of Agricultural Science*, **142**, 665–74.

Brookes, G. and P. Barfoot (2012), 'GM crops: global socio-economic and environmental impacts 1996–2010', Dorchester: PG Economics Ltd.

Conforti, P. (ed.) (2012), 'Looking ahead in world food and agriculture: perspectives to 2050', Rome: Food and Agriculture Organization.

Conway, G. (1999), *The Doubly Green Revolution: Food for All in the 21st Century*, Londonon: Penguin Books, and Ithaca, NY: Cornell University Press.

Conway, G. and G. Toenniessen (1999), 'Feeding the world in the 21st century', *Nature*, **402**, 55.

Crost, B., B. Shankar, R. Bennett and S. Morse (2007), 'Bias from farmer self-selection in genetically modified crop productivity estimates: evidence from Indian data', *Journal of Agricultural Economics*, **58**(1), 24–36.

Evenson, R. and V. Santaniello (eds) (2004), *The Regulation of Agricultural Biotechnology*, Wallingford: CABI Publishing.

FAO 'Food and Agriculture Organization of the United Nations' (2004), 'The state of food and agriculture 2003–04: agricultural biotechnology: meeting the needs of the poor?', available at: http://www.fao.org/docrep/006/Y5160E/Y5160E00.HTM, accessed 20 May 2011.

FAO (2012), 'The state of food insecurity in the world', Rome: Food and Agriculture Organization.

Finger, R., M. Hartmann and M. Feitknecht (2009), 'Adoption patterns of herbicide-tolerant soybeans in Argentina', *AgBioForum*, **12**(3&4), 404–11.

Frisvold, G.B. and J. M. Reeves (2007), 'Economy-wide impacts of Bt cotton', in Proceedings of the Beltwide

Cotton Conferences, available at: http://www.icac.org/cotton_info/tis/biotech/documents/otherdocs/gtap_beltwide.pdf, accessed 21 December 2011.

Gandhi, V.P. and N.V. Namboodiri (2006), 'The adoption and economics of Bt cotton in India: Preliminary results from a study', working paper, Ahmedabad, Gujarat: Indian Institute of Management.

Gianessi, L. (2009), 'Solving Africa's weed problem: increasing crop production and improving the lives of women', Crop Protection Research Institute, CropLife Foundation, available at: http://croplife foundation.files.wordpress.com/2012/05/Solving-africas-weed-problem-report1.pdf, accessed 21 December 2011.

Godfray, H.C., J.R. Beddington, I.R. Crute, L. Haddad, D. Lawrence, J.F. Muir, J. Pretty, S. Robinson, S.M. Thomas and C. Toulmin (2010), 'Food security: the challenge of feeding 9 billion people', *Science*, **327**, 812–18.

Gouse, M. (2012), 'Farm-level and socio-economic impacts of a genetically modified subsistence crop: the case of smallholder farmers in KwaZulu-Natal, South Africa', PhD dissertation, Department of Agricultural Economics, Extension and Rural Development, University of Pretoria.

Gouse, M., J. Piesse, C. Thirtle and C. Poulton (2009), 'Assessing the performance of GM maize amongst smallholders in KwaZulu-Natal, South Africa', *AgBioForum*, **12**, 78–89.

Gouse, M., C. Pray, J. Kirsten and D. Schimmelpfening (2005a), 'A GM subsistence crop in Africa: the case of Bt white maize in South Africa', *International Journal of Biotechnology*, **7**, 84–94.

Gouse, M., J. Kirsten, B. Shankar and C. Thirtle (2005b), 'Bt cotton in KwaZulu Natal: technological triumph but institutional failure', *AgBiotechNet*, **7**, 1–7.

Gressel, J., A. Hanafi, G. Head, W. Marasas, A.B. Obilana, J. Ochanda, T. Souissi and G. Tzotzos (2004), 'Major heretofore intractable biotic constraints to African food security that may be amenable to novel biotechnological solutions', *Crop Protection*, **23**, 661–89.

Huang, J., R. Hu, C. Fan, C.E. Pray and S. Rozelle (2002a), 'Bt cotton benefits, costs, and impacts in China', *AgBioforum*, **5**(4), 153–66.

Huang, J., S. Rozelle, C. Pray Q. Wang (2002b), 'Plant biotechnology in China', *Science*, **295**, 674–6.

James, C. (2012), 'Global status of commercialized Biotech/GM Crops: 2011', *International Service for the Acquisition of Agri-Biotech Applications ISAAA' Brief*, 43, Ithaca, NY: ISAAA.

Kambhampati, U., S. Morse, R. Bennett and Y. Ismael (2006), 'Farmlevel performance of genetically modified cotton: a frontier analysis of cotton production in Maharashtra', *Outlook on Agriculture*, **35**(4), 291–7.

Kapstein, E., R. Kim and W. Ruster (2012), 'The economic impact of increasing corn yields in Mexico', paper presented at the 16th International Consortium on Applied Bioeconomy Research (ICABR), Ravello, Italy, 24–7 June.

Kouser, S. and M. Qaim (2011), 'Impact of Bt cotton on pesticide poisoning in smallholder agriculture: a panel data analysis', *Ecological Economics*, **70**, 2105–13.

Kuosmanen, T., D. Pemsl and J. Wesseler (2006), 'Specification and estimation of production functions involving damage control inputs: a two-stage, semiparametric approach', *American Journal of Agricultural Economics*, **88**(2), 499–511.

Lichtenberg, E. and D. Zilberman (1986), 'The econometrics of damage control: why specification matters', *American Journal of Agricultural Economics*, **68**(2), 261–73.

Lusser, M., T. Raney, P. Tillie, K. Dillen and E. Cerezo-Rodriguez (2012), *International Workshop on Socio-economic Impacts of Genetically Modified Crops Co-organised by JRC-IPTS and FAO, Workshop Proceedings*, report EUR 25265 EN, EU and FAO.

Morse, S., Bennett R. and Y. Ismael (2005a), 'Bt-cotton boosts the gross margin of small-scale cotton producers in South Africa', *International Journal of Biotechnology*, **7**, 72–83.

Morse, S, R. Bennett and Y. Ismael (2004), 'Why Bt cotton pays for small-scale producers in South Africa', *Nature Biotechnology*, **22**(4), 379–80.

Morse, S., R. Bennett and Y. Ismael (2005b), 'Genetically modified insect resistance in cotton: some economic impacts in India', *Crop Protection*, **24**(5), 433–40.

Morse, S., R. Bennett and Y. Ismael (2007), 'Inequality and GM crops: a case-study of Bt cotton in India', *AgBioForum*, **10**(1), 44–50.

Potrykus, I. (2010), 'Lessons from the "Humanitarian Golden Rice" project: regulation prevents development of public good genetically engineered crop products', *New Biotechnology*, **27**, 466–72.

Pray, C., P. Bengali and B. Ramaswami (2005), 'The cost of biosafety regulations: the Indian experience' *Quarterly Journal of International Agriculture*, **44**, 267–89.

Pray, C., J. Huang, R. Hu and S. Rozelle (2002), 'Five years of Bt cotton in China – the benefits continue', *The Plant Journal*, **31**, 423–30.

Pray, C., D. Ma, J. Huang and F. Qiao (2001), 'Impact of Bt cotton in China', *World Development*, **29**(5), 813–25.

Pray, C.E., L. Nagarajan, J. Huang, R. Hu and B. Ramaswami (2011), 'The impact of Bt cotton and the potential impact of biotechnology on other crops in China and India', in C. Carter, G. Moschini and I. Sheldon

(eds), *Genetically Modified Food and Global Welfare: Frontiers of Economics and Globalization*, Vol. 10, Bingley: Emerald, pp. 83–114.
Qaim, M. (2003), 'Bt cotton in India: field trial results and economic projections, *World Development*, **31**(12), 2115–27.
Qaim, M. (2010), 'Benefits of genetically modified crops for the poor: household income, nutrition, and health', *Nature Biotechnology*, 27, **552**–7.
Qaim, M. and A. de Janvry (2003), 'Genetically modified crops, corporate pricing strategies, and farmers' adoption: the case of Bt cotton in Argentina', *American Journal of Agricultural Economics*, **85**(4), 814–28.
Qaim, M. and A. de Janvry (2005), 'Bt cotton and pesticide use in Argentina: economic and environmental effects', *Environment and Development Economics*, **10**, 179–200.
Qaim, M. and I. Matuschke (2005), 'Impacts of genetically modified crops in developing countries: a survey', *Quarterly Journal of International Agriculture*, **44**(3), 207–27.
Qaim, M. and G. Traxler (2005), 'Roundup Ready soybeans in Argentina: farm level, environmental and welfare effects', *Agricultural Economics*, **32** 2005 73–86.
Qaim, M. and D. Zilberman (2003), 'Yield effects of genetically modified crops in developing countries', *Science*, **299**, 900–902.
Qaim, M. (2010), 'Benefits of genetically modified crops for the poor: household income, nutrition, and health', *New Biotechnology*, **27**, 552–7.
Qaim, M., A. Subramanian, G. Naik and D. Zilberman (2006), 'Adoption of Bt cotton and impact variability: insights from India', *Review of Agricultural Economics*, **28**, 48–58.
Raney, T. (2006), 'Economic impact of transgenic crops in developing countries', *Current Opinion in Biotechnology*, **17**, 1–5.
Raney, T. and I. Matuschke (2011), 'Current and potential farm-level impacts of genetically modified crops in developing countries', in C. Carter, G. Moschini and I. Sheldon (eds), *Genetically Modified Food and Global Welfare: Frontiers of Economics and Globalization*, Vol. 10, Bingley: Emerald, pp. 55–82.
Sadashivappa, P. and M. Qaim (2009), 'Bt Cotton in India: development of benefits and the role of government seed price interventions', *AgBioForum*, **12**(2), 172–83.
Sharma, H.C., J.H. Crouch, K.K. Sharma, N. Seetharama and C.T. Hash (2002), 'Application of biotechnology for crop improvment: prospect and constraints', *Plant Science*, **163**, 381–95.
Shiva, V. and A.H. Jafri (2004), 'Failure of the GMOs in India', *Synthesis/Regeneration*, **33**/Winter, New Delhi, India: Research Foundation for Science, Technology and Ecology.
Smale, M., P. Zambrano, G. Gruère, J. Falck-Zepeda, I. Matuschke, D. Horna, L. Nagarajan, I. Yerramareddy and H. Jones (2009), 'Measuring the economic impact of transgenic crops in developing agriculture during the first decade', *Food Policy Review*, **10**, Washington, DC: The International Food Policy Research Institute.
Stein, A.J., H.P.S. Sachdev and M. Qaim (2008), 'Genetic engineering for the poor: Golden Rice and public health in India', *World Development*, **36**(1), 144–58.
Stewart, R.B. (2009), 'GMO trade regulation and developing countries', working paper, New York University Public Law and Legal Theory.
Tothova, M. and J. Oehmke (2005), 'Whom to join? The small country dilemma in adopting GM crops in a fragmented trade environment', *Quarterly Journal of International Agriculture*, Special Issue, **44**(3), 291–310.
Traxler, G. and S. Godoy-Avila (2004), 'Transgenic cotton in Mexico', *AgBioForum*, **7**, 57–62.
Traxler, G., S. Godoy-Avila, J. Falck-Zepeda and de Jesus J. Espinoza-Arellano (2003), 'Transgenic cotton in Mexico: economic and environmental impacts', in N. Kalaitzandonakes (ed.), *The Economic and Environmental Impacts of Agbiotech: A Global Perspective*, New York: Kluwer-Plenum Academic Publishers.
Vitale, J., G. Vognan, M. Ouattarra and O. Traore (2011), 'The commercial application of GMO crops in Africa: Burkina Faso's decade of experience with Bt Cotton', *AgBioForum*, **13**(4), article 5.
Yang, J., J.W. Kloepper and C.M. Ryu, (2009), 'Rhizosphere bacteria help plants tolerate abiotic stress', *Trends in Plant Science*, **14**, 1–4.
Zilberman, D. and G. Graff (2005), 'IPR innovation and the evolution of biotechnology in developing countries', *Quarterly Journal of International Agriculture*, **44**(3), 247–66.

45 The size and distribution of the benefits from the adoption of biotech soybean varieties

Julian M. Alston, Nicholas Kalaitzandonakes and John Kruse

1 INTRODUCTION

Economists have estimated the annual social benefits from genetically modified (GM) crops to be in the billions of dollars (Brookes et al., 2010; Carpenter, 2010; Falck-Zepeda et al., 2000b; Qaim, 2009; Sobolevsky et al., 2005; Konduru et al., 2008). These measures are generally partial, in that they leave out some elements of costs and benefits, including non-pecuniary benefits to adopting farmers and environmental benefits to others. In addition, they are incomplete in the sense that they represent the 'initial incidence' of benefits from the adoption of the technology, rather than the 'final incidence', which accounts for the shifting of the benefits from farmers to consumers and among different types of farmers as prices are induced to change.

In this chapter we estimate the global benefits from the adoption of biotech soybean varieties, and the distribution of those benefits among technology suppliers, farmers and consumers in a large number of countries, some of which have adopted soybean biotechnologies and some that have not. Our analysis is backward looking, in the sense that we estimate the benefits from the adoption of soybean biotechnologies over the period 1996–2009, based on data of actual adoption. The size of the innovation-induced shift in supply depends on the impact of the innovation on yields and costs.[1] We use information from a comprehensive review of previous studies combined with extensive data from producer surveys to derive estimates of the farm-level consequences of adoption. We combine the estimates of farm-level impacts with information on the rates of adoption to parameterize the implied shifts in market supply functions, in the context of a global multi-market simulation model. The model is used to simulate outcomes for prices and quantities with and without the adoption of biotech soybeans, allowing for the full set of interactions in production and consumption between soybeans and other crops, as well as the impacts of policy-distorted international trade. We use these estimates and other information to infer the net impacts on the welfare of various groups around the world, including producers who adopted the biotech soybean varieties, other producers (non-adopters), consumers and technology suppliers.

2 FARM-LEVEL IMPACTS OF SOYBEAN BIOTECHNOLOGIES

The dominant biotechnology in soybean production during the period of analysis was Roundup Ready™ (RR) soybeans. RR soybeans were genetically modified to tolerate the broad-spectrum herbicide Roundup™ (glyphosate), allowing enhanced weed control at lower cost.[2] The market-wide economic impacts of the RR soybean technology and their distribution are directly related to the determinants of farm-level adoption decisions made by soybean producers. Indeed, to measure the market-wide economic impacts of RR soybeans, the farm-level impacts of the technology must first be quantified. However, this is not straightforward because the counterfactual alternative (what would have happened if farmers had not adopted the technology) may not be easy to infer from observed data.

2.1 The Relevant Counterfactual Scenario

To assess the benefits from RR versus conventional soybeans requires comparing the observed economic outcomes (yields, costs revenues, and so on) against a benchmark that is not observable – the optimal solution that would have been chosen by the adopters in the absence of RR soybean technology. In the case of RR soybean technology the definition of the counterfactual is especially tricky because of the multidimensional nature of the farm-level impacts of the technology.

Some studies have used experimental field trials to assess the farm-level impacts of RR soybeans. Field trials provide useful but partial measures of the comparative performance of RR and conventional soybeans in yield or weed control. Yield trials are designed to maximize varietal yield performance and do not account for the potential advantage of RR soybeans in weed control. Conversely, weed-control trials compare various weed-control programs but overlook yield potential (Carpenter and Gianessi, 2003). Importantly, neither type of trial is designed to minimize unit costs or maximize profits – that is, the sort of behavior that drives producer adoption decisions. Hence, when taken from field trials, measures of yields, the use of herbicide and other inputs, and the implied costs and revenues are typically not representative of actual farm practices and outcomes.

Some studies have compared the economic performance of RR soybean adopters against that of non-adopters or against the population average, which includes both adopters and non-adopters. Such indicators provide useful insights into the relative farm performance of RR soybeans, but they do not account for the influence of (unobserved) differences in land productivity, weed incidence and pressure, differential managerial skills and other factors. Systematic biases can result if the analyst treats the data as though adoption was random when it was, in fact, a result of producer choice influenced by unobserved differences in factors affecting the relative performance of the technological alternatives. It might be possible to control for the influence of management and other unobserved factors by assessing the performance of RR soybeans against that of conventional alternatives on the farms of partial adopters alone (Marra, 2001). This method captures how partial adopters optimally allocate their land resources between RR soybeans and conventional soybeans on the basis of their relative advantages. Nevertheless, such measures can also be subject to biases from other unobserved factors that influence partial adoption, such as risk behavior, learning, and differential quality of land. The key point here is that it is conceptually straightforward but empirically challenging to establish a proper benchmark in an assessment of the farm-level impact of RR soybeans.

2.2 Nature and Sources of Farm-level Impacts

The introduction of RR soybeans augmented input-substitution possibilities in soybean production. Producers can continue to use conventional weed control methods or adopt RR technology, which facilitates substitution of one class of herbicides (for example, broad-spectrum) for another (for example, selective post-emergence) and other inputs, such as management, labor and capital. Over-the-top use of glyphosate on RR soybeans has typically controlled weeds more effectively, thereby extending the application window, increasing convenience, and reducing production risk. Through more effective weed control and reduced production risk, adoption of RR soybeans could, over time, lead to higher yields on average, although not necessarily in every season.[3]

More subtle, but also more intriguing, are the impacts of RR soybeans on agronomic practices and cropping systems. RR soybeans have been associated with increased adoption of conservation tillage and reduced tillage practices (Marra and Piggott, 2006; Trigo and Cap, 2006; Konduru et al., 2008; Fernadez-Cornejo and McBride, 2002). If tillage before planting is eliminated or reduced, weeds must be controlled with herbicides before, at, or after planting, and RR soybeans have been found to be well-suited in such a system.

When adoption of RR soybeans encourages parallel adoption of conservation- and reduced-tillage practices, soil and water conservation can also be affected. Conservation and reduced tillage can improve the availability of organic matter and minerals in the soil, leading to enhanced soil structure and fertility. Soil erosion and water run-off are also reduced, helping to sustain the productive capacity of land and to minimize surface and groundwater contamination. Through improved soil moisture content and retention capacity, efficiencies in water use may also be achieved.

Adoption of RR soybeans has also been associated with shifts towards early planting, increased use of double-cropping, changes in crop rotations and increasing adoption of narrow row planting (Carpenter and Gianessi, 2003; Fernandez-Cornejo et al., 2003; Trigo and Cap, 2006). Such shifts in agronomic practices can result in complex reallocations of inputs – land, labor and capital – and thus significantly broaden the scope of the farm-level impacts of RR technology as well as the difficulty in measuring them.

A distinguishing feature of RR soybean technology is that its use and adoption has also been associated with many non-pecuniary benefits, that is, benefits associated with production aspects that are not priced or traded in the market. These include ease of use and convenience as well as an improved safety profile for those who handle herbicides and for the environment. A number of studies have found that RR soybean technology leads to smaller amounts of active ingredient of herbicide applied (for example, Gianessi and Carpenter, 2000). Also, the toxicity of glyphosate is lower than that of other herbicides and its persistence in the environment is short (Ervin et al., 2000). When comprehensive measures of human and environmental impact are calculated, such as those that take into consideration toxicity and environmental exposure data (for example, the Environmental Impact Quotient), RR soybeans compare well to conventional soybean systems (Brookes and Barfoot, 2005; Nelson and Bullock, 2003). Marra and Piggott (2006) concluded from their national survey that some soybean growers placed significant economic value on the improved operator safety and environmental impact profile of the RR technology.

2.3 Farm-level Impacts and Adoption Decisions

The potential (pecuniary and non-pecuniary) farm-level impacts of RR soybeans frame the incentives for producers who consider adoption. When RR soybeans augment input-substitution possibilities in weed control, producers choose their optimal input mix in view of the relative prices of conventional and RR seeds, relevant herbicides, labor, capital and other inputs. They do similarly when they compare conventional tillage to conservation and reduced tillage systems that are synergistic with RR soybeans. Potential savings in labor, fuel and other variable costs from the adoption of conservation and reduced tillage systems are taken into account. When adoption of RR soybeans is expected to lead to lower input costs per unit of output, producers will be inclined to adopt the technology; likewise, when adoption of RR soybeans is expected to lead to more output per unit of input and higher expected profits. Soybean producers also value the possibility of reduced production risk and increased revenue from potential yield increases. Finally producers who value the convenience as well as the improved human and environmental safety of the technology will, again, be inclined to adopt technology that provides such benefits. These decision elements are reflected in the following simple model of adoption.

Consider the case of an individual price-taking farmer. Farmer i will adopt the RR technology in year t if it is expected to be more profitable than the next best alternative technology (with suitable allowance for a risk premium and for other differences including non-pecuniary aspects). Algebraically, we can represent this behavior as:

$$a_{it} = \begin{Bmatrix} 1 \text{ if } \pi_{it} \geq FC_{it} \\ 0 \text{ if } \pi_{it} < FC_{it} \end{Bmatrix} \text{ where } \pi_{it} = P_{it}\Delta Y_{it} - Y_{it}\Delta P_{it} - \Delta VC_{it} - \Delta S_{it} - D_{it}. \quad (45.1)$$

For farmer i in year t, the variables are defined as follows: a_{it} is an indicator variable that is equal to 1 if farmer i adopts in year t; FC_{it} is the fixed cost per acre associated with the new technology;[4] π_{it} is the total difference in variable profit in dollars per acre between the new technology and the next best alternative technology; P_{it} is the price per bushel of soybeans in year t, and D_{it} is the price discount per bushel for soybeans grown using the new technology, compared with conventional (non-biotech) soybeans; Y_{it} is the average yield, and ΔY_{it} is the difference in yield in bushels per acre between the new technology and the next best alternative soybean technology; ΔVC_{it} is the difference in variable cost of production, in dollars per acre, between the new technology and the next best alternative soybean technology, reflecting both non-pecuniary and pecuniary elements of variable costs other than seed; ΔS_{it} is the difference in seed price per acre between new technology and the next best alternative paid by farmer i in year t.

The benefit per acre on those acres using the new technology, π_{it}, can be expressed alternatively as:

$$\pi_{it} = \left(\frac{\Delta Y_{it}}{Y_{it}} - \frac{D_{it}}{P_{it}} - \frac{\Delta VC_{it}}{P_{it}Y_{it}} - \frac{\Delta S_{it}}{P_{it}Y_{it}}\right) P_{it}Y_{it} = (y_{it} - d_{it} - \theta_v v_{it} - \theta_s s_{it})\, P_{it}Y_{it}, \quad (45.2)$$

where y = the proportional change in yield per acre, d = the proportional price discount (or premium, in which case it is a negative number), v = the proportional change in variable costs per acre other than seed (θ_v is non-seed variable costs as a share of total revenue per acre, or total costs per acre), and s = the proportional change in seed costs – including the associated technology fees or royalties to biotech companies – per acre (θ_s is seed costs as a share of total revenue per acre, or total costs per acre). Then, maintaining the assumption of no market-wide impact on prices, the total annual net benefits from adoption for a particular farmer i, in a particular year t, (NBA_{it}) can be obtained by multiplying the benefits per acre, π_{it} times the number of acres planted to the new variety by that farmer, A_{it}, minus the fixed costs of adoption applicable to that year. Thus:

$$NBA_{it} = (y_{it} - d_{it} - \theta_v v_{it} - \theta_s s_{it})\, P_{it} Y_{it} A_{it} - FC_{it}. \tag{45.3}$$

While the farm-level production and economic impacts of RR soybeans are of interest, we are also interested in how they, in turn, influence aggregate production, consumption, trade and social welfare in different regional and national economies. As a first approximation, farm-level impacts can be extrapolated to regional and national levels by multiplying the micro-level impacts with observed acreage, but these indicators of aggregate impacts do not account for induced price changes and their consequences for soybeans and other crops that compete for farming resources on the supply side or for the consumer dollar on the demand side. Here, we first consider aggregation assuming exogenous prices. Then, we consider the implications of endogenous prices.

2.4 Aggregate Measures of Impacts with Exogenous Prices

If we have data on the elements of equation (45.3) for farmers in different places and corresponding data on the adoption rate and the gross value of production, we can obtain a reasonable measure of aggregate gross annual farmer benefits (FB) from the technical change under the assumption that prices of soybeans are unaffected by the collective impacts of the individual adoption decisions on output:

$$FB_t = \sum_{i=1}^{n} FB_{it} = \sum_{i=1}^{n} (y_{it} - d_{it} - \theta_v v_{it} - \theta_s s_{it})\, P_{it} Y_{it} A_{it}. \tag{45.4}$$

This measure of gross annual benefits – which excludes the element in equation (45.3) associated with the fixed costs of adoption, FC – corresponds to the approximate measure of gross annual research benefits (GARB) introduced by Griliches (1958): $GARB = KPQ$ (see Alston et al., 1995). Implicitly, then, in this representation, adoption of the technology could be modeled as entailing a parallel shift down in the adopter's supply function by a proportion of the initial price and marginal cost, K_{it}, where

$$K_{it} = (y_{it} - d_{it} - \theta_v v_{it} - \theta_s s_{it}). \tag{45.5}$$

The aggregate measure in equation (45.4) entails an equivalent proportional shift in the total supply that is a weighted average of individual supply function shifts.

The measure in equation (45.4) does not include benefits to biotech companies and other technology providers, which might be measured in terms of the element of equation (45.5) associated with a seed premium over the next best alternative technology:

$$BB_t = \sum_{i=1}^{n} \theta_s s_{it} P_{it} Y_{it} A_{it} \tag{45.6}$$

This is a measure of the profits of seed companies and biotech firms, given the pricing strategy that drove the assumed pattern of adoption. It is a gross rather than net benefit to the extent that it has to cover any additional expenses incurred in developing and marketing the new seeds relative to the benchmark alternative.

Combining (45.4) and (45.6) yields a measure of gross annual benefits, including benefits to farmers and technology providers from the adoption of the new technology, the elements of which have been derived under an assumption of no substantial effects on the total quantity of soybeans produced, and thus no effects on the price of soybeans:

$$GB_t = FB_t + BB_t = \sum_{i=1}^{n} (y_{it} - d_{it} - \theta_v v_{it}) P_{it} Y_{it} A_{it}. \tag{45.7}$$

However, if the adoption of the new technology led to an increase in the total quantity of soybeans and this caused a significant reduction in price of soybeans, there would be effects on the welfare of soybean consumers (positive) as well as soybean growers (negative effects on adopters and non-adopters alike, if we assume no segregation costs and no price discounts for biotech varieties). In addition, the innovation might have cross-commodity impacts on the prices of other commodities. Allowing for the endogenous prices would probably not mean a significant change in the overall global impact, but the distribution of benefits could change significantly.[5]

3 MARKET MODELS OF THE BENEFITS FROM RR SOYBEANS

Modeling and measuring the economic impacts of new technologies typically involves a conventional supply and demand framework in which innovations are represented as exogenous supply shifts. This approach, which is discussed in detail by Alston et al. (1995), can be applied to the introduction of RR soybeans with appropriate allowance for the proprietary nature of the technology.

3.1 Standard Single-commodity Market Model

In the standard single-commodity market model, adoption of the new technology causes the commodity supply curve to shift down and out against a stationary demand curve giving rise to an increase in quantity produced and consumed, and a lower price. The benefits are assessed using Marshallian measures of technology-induced changes in consumer surplus for consumer benefits and of technology-induced changes in producer surplus for producer benefits. The total gross annual benefits from the technology depend primarily on the size of the (time-varying) technology-induced supply shift. Other aspects of the analysis typically have limited second-order effects on the measures of total benefits but may have important implications for the distribution of the benefits between producers, consumers and others.

The distribution of the benefits between producers and consumers depends on the relative elasticities of supply and demand, the nature of the technology-induced supply shift and, less importantly, on the functional forms of supply and demand (Alston et al., 1995). The nature of the technology-induced supply shift has been controversial because it matters, especially for findings concerning the distribution of benefits, and is not easy to observe. A common choice is to assume a vertically parallel technology-induced supply shift, with the understanding that this choice has specific implications, compared with alternatives, for measures of the size and distribution of the resulting benefits (for example, see Alston et al., 1995).

The standard model assumes competition in the market for the commodity and the absence of any other market distortions. In the context of the present analysis, the US and global markets for soybeans and related commodities are subject to domestic farm program policies and border policies that, at times, have resulted in distortions in the production, consumption and trade of these commodities. The simulation model used in this chapter incorporates those policies and the simulation results reflect their consequences. However, in the welfare analysis we do not take explicit account of the role of commodity market distortions of these types. Particularly in recent years, because of the relatively high world market prices for agricultural commodities, the effects of such policies have been muted. In any event, even in more normal years, the consequences from ignoring the impacts of these policies on the benefits from technical change are likely to be modest because the quantitative impacts of the policies have been fairly small. For instance, several recent studies (Alston, 2007; Sumner, 2005) estimated that US farm program policies caused increases in production of maize and soybeans in the range of 5–10 per cent (or less) in 2006, when program impacts were more important than they have been in the years since.

The standard model also assumes new technologies are available to the farmers for free. Models that allow for proprietary technology have been developed in recent years (for example, Moschini and Lapan, 1997). The fact that farmers have to pay for proprietary technology affects their incentives to adopt particular innovations and invalidates the conventional surplus measures (for example, as described by Alston et al., 1995). We account for the private nature of the technology in this chapter given the focus of the analysis on the total benefits from the adoption of RR soybeans and the distribution of those benefits between technology developers and others, including farmers and consumers.

The standard model, on which most of the results in the literature are based, is also a single-market, competitive equilibrium model. The introduction of international trade is a straightforward elaboration of the standard model, from which one can obtain measures of welfare impacts for different spatial or market aggregates. More elaborate and complex multi-market models are implied if one wants to disaggregate the market structure horizontally, to represent different geopolitical or spatial markets for a given product, or different products (for example, Freebairn, 1992; Frisvold, 1997; Wohlgenant, 1997; Zhao et al., 2000). Even if we were interested only in a single technology, adopted by a single country, the fact that the technology in question affects the world market prices of multiple commodities means that it is necessary to pay attention to the cross-commodity impacts of price changes on producers and consumers of other commodities. The approach taken here is to use a multi-market model that explicitly represents the supplies and demands for all of the affected commodities as they interact with one another.[6]

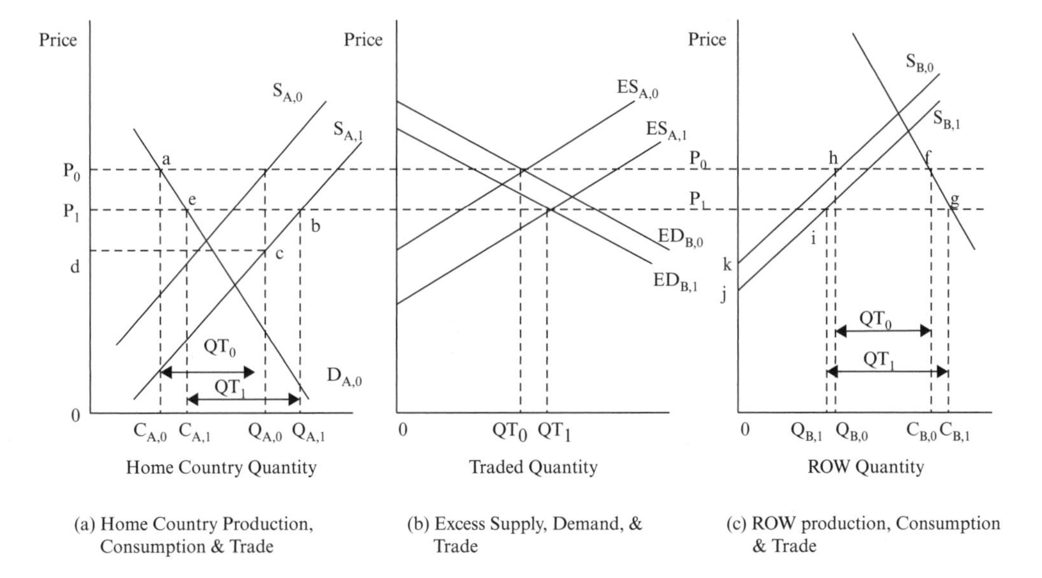

Figure 45.1 A supply and demand model of impacts of innovation by an exporter

3.2 Standard Two-country Model

To illustrate the approach, Figure 45.1 represents a simplified case where RR soybeans are introduced in a large, exporting economy (a single country or region, A). The model allows for technology spillovers in that the rest of the world (region B) is able to adopt RR soybean seeds (with or without payments of technology fees to the developer). It also allows for representation of trade among countries such that market equilibrium is given by equating their excess supply and demand.

In Figure 45.1, panels a, b and c represent respectively (a) the home country where the innovation is initiated (A); (b) the interaction of excess supply and excess demand; and (c) the rest of the world (B). All of the supply and demand curves are assumed to be approximately linear. The international equilibrium price P_0 is obtained from the intersection of the excess supply ($ES_{A,0}$) and excess demand curves ($ED_{B,0}$). The introduction of the innovation in country A results in a parallel shift in domestic supply from $S_{A,0}$ to $S_{A,1}$, and in consequence the excess supply shifts from $ES_{A,0}$ to $ES_{A,1}$. But, because of a technology spillover to the rest of the world, its supply shifts from $S_{B,0}$ to $S_{B,1}$ leading to a reduction in excess demand from $ED_{B,0}$ to $ED_{B,1}$. So, the world price falls from P_0 to P_1. As a result of the innovation and its spillover, producers in country A benefit as long as the overall price reduction (from P_0 to P_1) is smaller than the initial vertical supply shift in country A.

The benefits to producers in country A are given by the area P_1bcd in panel a. Producers in country B are net losers, even after the adoption of the innovation, since the area P_1ij is less than the area P_0hk, but in general they could gain or lose. The benefits to consumers in country A are shown in the figure as the area P_0aeP_1 in panel a and the benefits to consumers in the ROW are shown as the area P_0fgP_1 in panel c. This model can be generalized to include additional countries and commodities, adding realism and detail in the analysis of economic surpluses from innovation and their distribution. Before elaborating on our simulation model, however, we present a brief review of selected estimates of the farm-level impacts and aggregate benefits from RR soybeans that have been presented previously in the literature.[7]

4 PREVIOUS ESTIMATES OF THE ECONOMIC IMPACTS OF RR SOYBEANS

So what do we know about the farm-level impacts and aggregate benefits of RR soybeans? RR soybeans have been grown for over fifteen years, and empirical evidence has continued to accumulate on their impacts at the farm and aggregate levels. A number of studies have measured such impacts at different points in time and locations using different methods and datasets. Their results have begun to paint a picture of the benefits from RR soybeans. A selective review of such studies is presented below.

Fernandez-Cornejo and McBride (2000) used data from the 1997 Agricultural Resource Management Survey (ARMS) of USDA and concluded that adoption of RR soybeans in that year did not have a significant impact on net farm returns. However, they did find that use of RR soybeans was profitable in some US regions (for example, 17 per cent more profitable in the 'Heartland').

Lin et al. (2001) studied the farm-level effects of RR soybeans in the United States, again using the 1997 ARMS survey. They reported that, across all US regions, adopters of RR soybeans spent 1 per cent to 34 per cent less than non-adopters on soybean weed control, and 11 per cent less in the 'Heartland' (where 70 per cent of US soybeans are grown).

Sankula and Blumenthal (2004) and Sankula (2006) studied the farm-level difference in economic performance between RR and conventional soybeans by surveying US experts (state and university weed specialists). They gathered data on weed control and associated expenditures from all US soybean-producing states, thereby representing various production systems. They concluded that RR soybeans allowed 53 per cent and 46 per cent cost savings for weed control in 2003 and 2005, respectively.

Marra and Piggott (2006) used a national producer survey to explore the reasons why US soybean growers chose RR soybeans over conventional varieties and how they benefited from such choices. Their survey evaluated various potential incentives, both pecuniary and non-pecuniary. They estimated substantial net benefits from RR soybeans, along with benefits from a parallel adoption of reduced tillage, up to US$37 per acre for some farmers.

A few studies have also assessed the benefits from RR soybeans at an aggregate level. Some have simply extrapolated from farm-level estimates to estimate the aggregate benefits. Brookes and Barfoot (2005) studied the global economic and environmental impact of genetically modified crops since their commercial introduction in 1996, summarizing farm-level estimates from various countries, and they have presented occasional updates at different points of time (Brookes and Barfoot, 2008, 2011). Their economic impact analyses have focused on farm income effects, while their environmental impact analysis has documented changes in the use of pesticides and the environmental load from crop production, including soybeans. More recently, they have also examined the impact of RR soybeans and other crops on greenhouse gas emissions. They initially reported that, over the period 1996–2004, RR soybeans yielded global economic benefits of approximately US$9.3 billion. More recently, Brookes and Barfoot (2011) reported global economic benefits from RR soybeans over the period 1996–2009 of $18.9 billion ($25.1 billion if second-crop gains in Argentina and Paraguay are included).

Other studies have used market equilibrium models to assess the aggregate economic impacts of RR soybeans at different points in time. Falck-Zepeda et al. (2000a) modeled the welfare effects from the adoption of RR soybeans using a two-region framework (United States and the rest of the world) and estimated a total world benefit of US$1.06 billion from the use of RR soybeans for the year of their calculations. Similarly, Moschini et al. (2002) modeled the global welfare effects of RR soybeans using a three-region model that included a monopolist technology seller as well as consumers and producers. They assumed that the technology resulted in a cost saving of US$20 per acre across all RR soybean adopters, based on their 1997–98 farm-level estimates of such cost savings in Iowa. They found a total efficiency gain worldwide of approximately US$804 million in the 1999–2000 crop year. In another study based on a three-region model, Qaim and Traxler (2005) analyzed the impact of RR soybeans in Argentina, the United States and the rest of the world. They found, from a producer survey, that the RR technology increased total factor productivity by 10 per cent on average. They estimated the global welfare effects of RR soybeans in 2001 at US$1.2 billion.

Konduru et al. (2008) used a multi-region model and estimated the benefits from RR soybeans over the period 1996–2006. Konduru et al. (2008) calculated cost savings not only in weed control but also from the complementary adoption of reduced and minimum tillage practices in the United States, Argentina and Brazil. They estimated the global benefits from RR soybeans to be more than US$31 billion over the period of their analysis.

Trigo and Cap (2006) estimated the total accumulated benefit from RR soybeans in Argentina to be approximately US$19.7 billion during 1996–2005. A large share of the measured gains accrued to producers from double cropping, which, as the authors suggested, was made possible through the RR technology. They analyzed the impact of the increased soybean production in Argentina on consumers worldwide by estimating the impact on soybean price from that additional output. The result was an accumulation of savings in global consumer spending in the range of US$26 billion.

Many of the aggregate economic impact studies have observed that induced shifts in input and output prices affect the size and distribution of the benefits from innovation. For instance, an increase in the price of RR soybean seed transfers a portion of the benefits from the adopters to technology firms and seed suppliers, while a reduction in the output price transfers a portion of the economic gains to consumers. Furthermore, given that input and output prices determine the incentives for producer adoption, shifts in input and output prices cause variations in the rates of adoption and aggregate diffusion and, ultimately, in the market-wide impacts of RR soybeans. Hence, at the aggregate market level, the size and the distribution of the economic impacts are determined jointly with the prices and quantities produced and consumed.

Along these lines, Falck-Zepeda et al. (2000b) estimated the distribution of total world surplus gains among US farmers (76 per cent), the technology supplier (7 per cent), seed suppliers (3 per cent), US consumers (4 per cent) and consumers in the ROW (9 per cent).

The figures obtained for Argentina by Trigo and Cap (2006) were similar. The benefits were distributed among farmers (77 per cent), seed suppliers (4 per cent), herbicide suppliers (5 per cent) and the national government (13 per cent). Moschini et al. (2002) reported the share of the monopolist technology supplier to be as high as 45 per cent of the total surplus. Lin et al. (2001) reported that US farmers received approximately 20 per cent of the total benefits from RR soybeans in 1997.

Most previous studies therefore find that the farm-level and aggregate benefits from the use of RR soybeans have been significant, though their size varies across studies. This variation is, in part, attributable to differences in the methodologies used, the data sets employed, the dimensions of production and other non-pecuniary impacts taken into account, and the time periods and places to which the assessments applied. Year-to-year and place-to-place variations in the benefits from RR soybeans are expected since key underlying factors that determine such impact (for example, severity of weed infestation, input and output prices) vary with time and space, and the adoption process has been cumulative.

We add to the existing literature by evaluating the global economic impacts of RR soybeans over the period 1996–2009. We note here that prior studies have overlooked the fact that innovation in soybean production must have influenced the supply and demand conditions of other oilseeds, such as canola, sunflower and palm oil, as well as other crops, thereby affecting the overall surplus in the market and its distribution. Similarly, previous studies have not accounted for the structure of the soybean complex and the underlying separate demand for oils and meals. Finally, with the exception of Konduru et al., (2008), previous studies have not examined the market effects and benefits from RR soybeans over long periods of time or their year-to-year variation. These shortcomings are addressed in this study by estimating the aggregate worldwide benefits from RR soybeans within the context of a multi-year, multi-crop, multi-sector model. A more granular picture of the benefits associated with the use of RR soybean technology, using a detailed 38 country/region global model, is also presented. In this context it is possible to account for the impacts of market distortions induced by various agricultural policies in different countries (which are explicitly modeled) and to elaborate the distribution of benefits and costs associated with the adoption of RR soybean technology across various countries, adopters and non-adopters, importers and exporters.

5 MODELING THE GLOBAL ECONOMIC IMPACT AND THE DISTRIBUTION OF BENEFITS

To estimate the aggregate economic impacts of RR soybeans over the period 1996–2009, a detailed market simulation model was developed that captures the global interrelationships among oilseeds and competing crops and the rest of the agricultural sector – for instance, the relationship between demand for soy meal or other protein meals and the livestock sector is modeled explicitly. The model allows estimation of price effects, land reallocation patterns, and substitution among oilseeds, as well as between oilseeds and other crops as a result of the adoption of RR soybeans. The model is calibrated such that the baseline simulation replicates the actual values of supply, demand, acreage, stocks and other variables of interest. The scenario describing a world where RR soybeans are not available is simulated by suppressing any changes in yields, costs, profits or other changes caused by the introduction of RR soybeans. The resulting counterfactual prices, quantities, acreages, stocks, imports, exports and other indicators are then compared with the actual historical values (baseline) to account for the effects of RR soybean adoption. The differences between the counterfactual and actual outcomes are used to estimate the benefits from RR soybeans in the form of consumer and producer surpluses.

5.1 Model Structure

The model developed in this study includes a set of supply and demand equations for each commodity of interest. Separate supply and demand functions are specified for the various oilseeds and their derivative oils and meals for each of the 38 countries/regions represented in the model which include the United States, Canada, Mexico, Brazil, Argentina, Paraguay, Uruguay, Bolivia, China, India, Indonesia, Malaysia, Japan, South Korea, EU25, South Africa and Russia. The following equations, which characterize the market clearing conditions for a particular oilseed commodity (and its oil and meal products) in a given country, are indicative of the general nature of the structure of the model:

Beginning Stocks = *Ending stocks*$_{(t-1)}$ *(Oilseeds, Meals and Oils)* (45.8a)

Production = *Harvested Area* * *Yield* *(Oilseeds)* (45.8b)

Production = *Crush* * *Crushing Yield* *(Meals and Oils)* (45.8c)

Total Supply = *Beginning Stocks* + *Production* + *Imports* *(Oilseeds, Meals and Oils)* (45.8d)

Total Demand = *Crush* + *Food Use* + *Other Use* + *Exports* + *Ending Stocks* *(Oilseeds)* (45.8e)

Total Demand = *Food Use* + *Feed Use* + *Industrial Use* + *Exports* + *Ending Stocks* *(Meals and Oils)* (45.8f)

Domestic Use = *Crush* + *Food Use* + *Other Use* + *Ending Stocks* *(Oilseeds)* (45.8g)

Domestic Use = *Food Use* + *Feed Use* + *Industrial Use* + *Ending Stocks* *(Meals and Oils)* (45.8h)

Within the model, for each commodity, the country-specific prices are linked to those of every other trading country using price linkage equations that include import tariffs, taxes and other relevant shifters.

The percentage vertical shift (K) in the supply function of soybeans resulting from the introduction of RR technology is measured as described in (45.4), but in the simulation model only the proportional net change in production costs enters its calculation.[8] These cost savings are net of any technology fees and seed premiums paid for the usage of RR technology.

The model is relatively disaggregated in its treatment of supply and demand for individual countries and regions. The formulas for calculating the change in producer and consumer surpluses follow Alston et al. (1995) and are specified as follows:

$$\Delta PS_{R,S} = P_0 Q_{R,0}(K - Z)(1 + 0.5Z\varepsilon_S)$$

$$\Delta PS_{R,O} = -P_0 Q_{R,0} Z(1 + 0.5Z\varepsilon_O)$$

$$\Delta CS_{R,O} = (P_o - P_1)C_{R,0} + 0.5(C_{R,1} - C_{R,0})(P_0 - P_1)$$

where ΔPS is the change in producer surplus; ΔCS is the change in consumer surplus; R denotes a country or region of interest; S denotes soybeans; O denotes other crops including sunflower, rapeseed and palm oil; P_0 is a counterfactual price and P_1 is an actual price; $C_{R,0}$ is a counterfactual quantity and $C_{R,1}$ denotes an actual quantity; ε_S is the elasticity of supply of soybeans and ε_O denotes elasticities of supply of other crops; Z is the relative price change given by $-(P_1 - P_0)/P_0$.

5.2 Model Calibration and Assumptions

We do not estimate the farm-level impacts of RR soybeans in this study. Instead, we depend on estimates from a large number of previous studies that have measured the farm-level impacts of RR soybeans to infer the changes caused by their introduction.

5.2.1 Influence of RR technology on soybean yields

A number of studies have measured the impacts of RR technology on soybean yields. A few studies have reported yield suppression from the adoption of RR soybeans (Qaim and Traxler, 2005; Duffy and Ernst, 1999). Others have observed that yield reductions in some areas in the early stages of adoption faded over time as the transgene was inserted into more elite varieties and across soybean varieties of different maturities (Marra and Piggott, 2006). Some other studies have reported a positive influence of RR technology on soybean yields (Lin et al., 2001; Gianessi and Carpenter, 2000; Falck-Zepeda et al., 2000a). While drawing firm conclusions from these studies on the impact of RR soybeans on yields is somewhat difficult, it is worth noting that most yield changes, positive or negative, were observed at the very early stages of commercial introduction and adoption of RR soybeans.

Konduru et al. (2008) observed that if significant yield changes were, in fact, caused by the adoption of RR soybeans, a departure from the yield trends observed prior to the introduction of the new technology should be apparent in aggregate data. They tested for structural changes in yield trends in the three countries where RR soybeans have been most broadly adopted (the United States, Brazil and Argentina) and concluded that there was no evidence of a structural shift in yield trends associated with the introduction of the RR technology in any of the countries.

Based on evidence presented in previous studies, it is assumed in the analysis that RR and conventional soybeans have had comparable yields across all years and countries – consistent with similar assumptions made by Moschini et al. (2002), Sankula and

Blumenthal (2004), Sankula (2006), and Marra and Piggott (2006), among others – such that the introduction of RR soybeans had no impact on yields.

5.2.2 Influence of RR technology on production costs

The evidence is strong that use of RR soybeans leads to production cost efficiencies. Cost efficiencies have accompanied several different changes in the standard practices of soybean production associated with the introduction of RR soybeans. Lin et al. (2001) concluded that weed control costs, which include cost of herbicides, herbicide application, scouting and cultivation, were lower for RR soybean adopters than for non-adopters in 1997. They estimated the savings at 11 per cent over conventional varieties in the US Midwest. Along the same lines, Sankula and Blumenthal (2004) and Sankula (2006) estimated that, after accounting for technology fees, the use of RR soybeans saved 53 per cent and 46 per cent of the weed control costs (herbicide costs) relative to those of conventional soybeans in 2004 and 2006. Qaim and Traxler (2005) found similar impacts in Argentina through a producer survey and estimated that adoption of RR soybeans led to 10 per cent savings in variable costs over conventional soybean systems. They also noted that cost savings might be even greater because RR soybeans were found to reduce the time for weed scouting, but such potential cost savings were not included in their estimates.

Konduru et al. (2008) and authors of a number of other studies have noted that adoption of RR soybean led to a parallel increase in the adoption of conservation tillage practices in the United States, Argentina, Brazil and elsewhere. The increased use of conservation tillage has been enabled by the use of more effective post-emergence herbicides, such as glyphosate, that can be applied directly on RR soybeans. Effective control of most weeds through chemical control rather than cultivation has, in turn, led to reduced use of machinery and fuel, and reduced labor requirements. Konduru et al. (2008) estimated the cost savings associated with the increased use of conservation tillage ranging from US$5.50 to almost US$18 per hectare in the United States, Argentina and Brazil.

In addition to the reduction in out-of-pocket costs, the flexibility and simplicity of the production program of glyphosate-tolerant soybeans provides benefits that are not typically included in the accounting of costs. Using non-market valuation techniques, Marra and Piggott (2006) estimated that US farmers gained US$12 per acre (US$30 per hectare) of glyphosate-tolerant soybeans in the form of 'non-pecuniary' benefits of convenience, such as flexibility and time savings in management.

For our simulation analysis we consolidated the estimates of farm-level impacts reported in these and other studies as well as estimates from producer surveys carried out by market research firms in various areas and in different years. If multiple farm-level estimates exist for the same location and year, we use the lowest available. If estimates for a particular location and year do not exist, we interpolate from estimates in previous and subsequent years for the same location or for a comparable location. We omit non-pecuniary effects because too few estimates are available. Similarly, we do not account for cost savings associated with complementary adoption of reduced-tillage practices and double cropping as it is not possible to separate the influence of RR soybean technology from other factors on such adoption. Table 45.1 presents the farm-level cost savings per acre used in the simulation analysis, along with adoption rates for key producing countries.

Table 45.1 Roundup Ready soybeans: farm-level cost savings and adoption rates, 1996/97 to 2009/10

	Units of Measure	1996/97	1997/98	1998/99	1999/00	2000/01	2001/02	2002/03	2003/04	2004/05	2005/06	2006/07	2007/08	2008/09	2009/10
US															
Production cost savings	US $/ Hectare	26.4	22.6	15.6	14.7	30.0	32.8	27.5	39.3	43.4	40.8	24.2	29.9	30.2	24.3
Adoption rate	% of soybean hectarage	7.4%	17.0%	44.2%	55.8%	54.0%	63.0%	75.0%	81.0%	85.0%	87.0%	89.0%	91.0%	92.0%	92.0%
Argentina															
Production cost savings	US $/ Hectare	22.23	23.5	23.5	23.5	23.5	23.5	25.5	25.5	25.5	25.5	24.5	24.5	11.5	15.5
Adoption rate	% of soybean hectarage	0.0%	22.9%	61.2%	90.0%	94.8%	97.3%	99.0%	99.1%	99.3%	99.5%	99.5%	95.2%	99.4%	98.9%
Brazil															
Production cost savings	US $/ Hectare	35.0	35.0	35.0	35.0	35.0	35.0	35.0	35.0	35.0	35.0	29.7	24.4	24.4	25.1
Adoption rate	% of soybean hectarage	0.0%	2.0%	8.0%	15.0%	24.0%	30.0%	32.0%	34.0%	34.0%	49.0%	56.0%	65.9%	64.8%	70.7%
Paraguay															
Production cost savings	US $/ Hectare	0.0	0.0	0.0	23.5	23.5	23.5	26.0	26.0	26.0	26.0	25.0	25.0	25.0	25.0

Adoption rate	% of soybean hectarage	0%	0%	0%	0%	0%	0%	0%	0%	60%	85%	94%	94%	95%	95%
Uruguay															
Production cost savings	US $/ Hectare	0.0	0.0	0.0	23.5	23.5	23.5	26.0	26.0	26.0	26.0	25.5	25.5	25.5	25.5
Adoption rate	% of soybean hectarage	0%	0%	0%	0%	0%	0%	69%	76%	99%	99%	99%	99%	99%	99%
Canada															
Production cost savings	US $/ Hectare	0.0	41.2	35.1	31.6	31.7	29.2	27.4	14.6	17.5	18.9	23.5	24.5	14.3	14.3
Adoption rate	% of soybean hectarage	0.0%	4.0%	12.0%	16.0%	18.0%	27.0%	44.0%	55.0%	55.0%	61.0%	45.0%	45.0%	50.0%	43.0%
Mexico															
Production cost savings	US $/ Hectare	0.0	0.0	0.0	0.0	0.0	0.0	0.0	0.0	82.3	89.4	73.0	66.8	54.1	54.1
Adoption rate	% of soybean hectarage	0.0%	0.0%	0.0%	0.0%	0.0%	0.0%	0.0%	0.0%	0.0%	0.0%	9.0%	11.0%	13.0%	24.0%

Table 45.2 Benefits from RR soybeans: 1996/97 to 2009/10, various countries, in US$2009 million

	Producer surplus	Consumer surplus	Total country surplus
Canada	12.6	166.5	179.2
Mexico	19.8	263.1	282.9
United States	19178.5	3173.2	22351.7
Argentina	6620.3	2005.4	8625.7
Brazil	2565.8	2077.0	4642.8
European Union	−158.1	1381.9	1232.9
China	−968.0	2608.8	1640.8
India	−426.4	593.3	166.8
Japan	−15.0	352.9	337.9
Rest of the World	−1363.5	1885.8	522.3
Crop total	25466.1	14507.8	39974.0
Total technology fees			6638.4
Total crop benefits			46612.3

Note: All reported sums are undiscounted sums of deflated nominal values.

6 MEASURES OF ECONOMIC IMPACT AND THE DISTRIBUTION OF BENEFITS

Table 45.2 reports the estimates of total changes in producer and consumer surplus for the period 1996–2009 for all major producing and consuming countries. The estimated sum of technology fees paid for the RR technology by adopting producers in different countries (that is, the gross revenue of the technology supplier) is also reported in the same table. Together, these figures represent the total estimated benefits from RR technology over the period of analysis (in 2009 dollars). More disaggregated figures on the changes in producer and consumer surplus for selected countries and regions are reported in Table 45.3, in order to clarify the cross-commodity impacts of the innovation.

As the figures in Table 45.2 indicate, RR technology yielded large benefits, exceeding US$46 billion, over the period 1996–2009.[9] Of this total, 55 per cent accrued to producers, 31 per cent to consumers and 14 per cent to the innovator. Hence, unlike most previous studies, we find that consumers benefited significantly from the innovation. Looking at such gains in aggregate, however, masks important nuances in the distribution of gains and losses from the innovation.

For consumers, RR technology reduced the price of soybeans by an average of almost 2 per cent over the period 1996–2009; consequently, the prices of sunflower, canola and palm oil also declined – by an average of 0.7 per cent (Figure 45.2). In contrast, the prices of grains (for example, corn, wheat) increased slightly because of small reductions in their global acreage since the relative profitability of such crops declined against that of soybeans in many producing countries. As a result of all these price changes, global consumer surplus increased by US$14.5 billion, comprising US$13.8 directly from the reduction in soybean prices and the rest from the net effect of the induced changes

Table 45.3 Benefits from the adoption of RR soybeans: 1996/97 to 2009/10, various crops and countries, in 2009 US$ million

	Soybeans		Corn		Canola		Cotton		Sunflower		Palm Oil		Wheat		Country total
	Pro-ducer surplus	Con-sumer surplus	Pro-ducer surplus	Con-sumer surplus	Pro-ducer surplus	Con-sumer surplus	Pro-ducer surplus	Con-sumer surplus	Pro-ducer surplus	Con-sumer surplus	Pro-ducer surplus	Con-sumer surplus	Pro-ducer surplus	Con-sumer surplus	
Canada	57	141	9	−11	−79	43	0	0	−3	2	0	1	28	−9	179
Mexico	−6	277	21	−27	0	13	1	−5	0	1	−1	11	4	−7	283
United States	18829	3398	284	−227	−11	9	44	−16	−37	34	0	12	70	−38	22352
Argentina	6685	1924	17	−6	0	0	1	−1	−101	94	0	0	17	−6	8626
Brazil	2516	2128	42	−39	0	0	11	−10	−2	2	−6	7	5	−12	4643
European Union	−75	1141	58	−60	−153	140	5	−9	−144	165	0	146	151	−140	1224
China	−1088	2597	134	−133	−151	168	63	−80	−43	42	0	139	117	−123	1641
India	−458	450	15	−14	−66	68	37	−35	−36	36	−2	169	83	−82	167
Japan	−16	327	0	−17	0	31	0	−2	0	0	0	20	1	−7	338
Rest of the World	−186	1384	144	−153	−28	23	77	−81	−276	259	−1293	731	229	−277	522
Crop total	26257	13769	695	−686	−487	495	240	−239	−642	635	−1302	1235	705	−701	39974

Note: All reported sums are undiscounted sums of deflated nominal values.

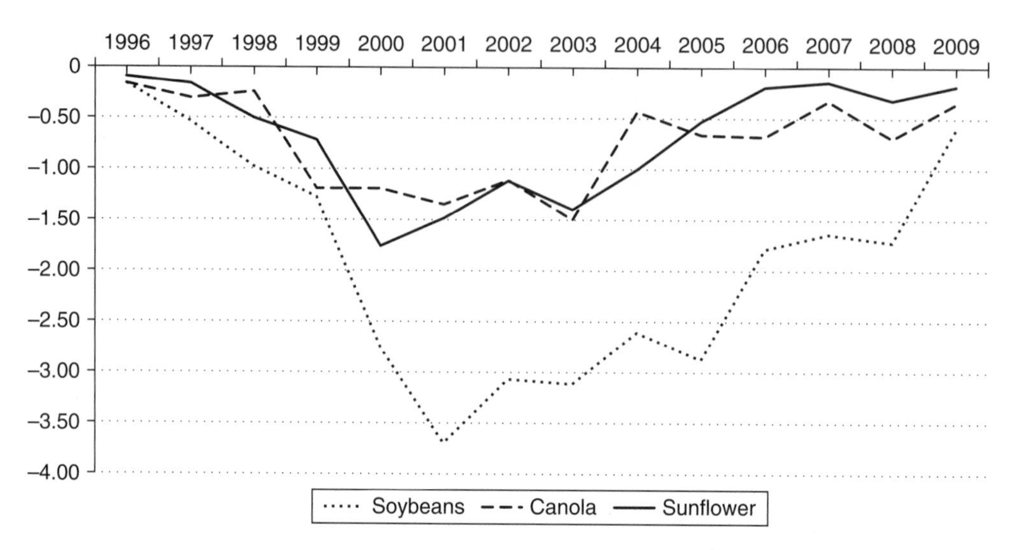

Figure 45.2 Percentage changes in world prices of soybeans, canola and sunflowers caused by the adoption of RR soybeans, 1996–2009

in prices and consumption of all other oilseeds and grains. Large gains in consumer surplus were realized by large consuming countries like the United States, China, Brazil, the EU, India, Japan and Argentina. However, in the case of Argentina, while much of the soybean production is processed domestically, value-added products (for example, soybean meal, soy oil) are exported to many countries, such that much of the consumer gain accrued to consumers in importing countries.

The global change in surplus among soybean producers from the adoption of RR technology is estimated to be US$26.3 billion over the period 1996–2009. The economic impact on soybean producers is driven by the rate of adoption, which determines the total acreage planted to the new technology, and the associated price effects in the global market. Soybean producers in the United States and Argentina have benefited the most from the new technology (gaining US$18.8 billion and US$6.7 billion respectively) because they adopted faster and more extensively. Brazilian producers who began their adoption later have gained US$2.5 billion over the period of analysis, and their share has been increasing in recent years. Together, soybean producers in these three large producing countries captured almost 60 per cent of the total surplus generated by the technology. Non-adopting soybean producers in China and India suffered the largest losses, in aggregate more than US$1.5 billion.

Producer surplus for the producers of other oilseeds declined because they faced lower prices without any offsetting reductions in their variable costs. Combining sunflowers, rapeseed and palm oil, the total loss in producer surplus for these crops is US$2.4 billion. In contrast, producer surplus for other crops (for example, corn, cotton, wheat) increased by a total of US$1.6 billion because of price increases resulting from their small acreage reductions. As a result, the total net change in producer surplus from cross-commodity effects was a modest US$0.8 billion for the period of analysis.

All of the estimated changes in producer and consumer surplus reflect private gains and losses revealed in market transactions. In our calculations we have not captured any non-market impacts associated with environmental or other externalities that arise when adoption by particular individuals results in spillover benefits or costs to other individuals, for which they are not compensated through market transactions. Producers who grow RR soybean varieties may be seen as generating positive externalities if in doing so they reduce the total environmental burden of chemical pesticides. Negative externalities may arise if producers of RR soybeans impose costs on other producers and consumers, such as may be incurred when attempts are made to segregate RR from conventional soybeans, or when producer practices lead to build-up of weed resistance to glyphosate. These aspects that are not involved in farmers' perceived costs and benefits may still entail social costs and benefits to be added to the private measures calculated here.

All of the estimated benefits from the adoption of RR technology reflect the dynamic consequences of progressive adoption over time of technology that itself has evolved. Generally, the price impacts of RR soybeans increase, as expected, with the rate of adoption and the acreage planted to the new technology. Fast-growing acreage in the United States and Argentina in the early years, as well as in Brazil and Paraguay a few years later, led to soybean price declines of more than 3 per cent after 2001 (Figure 45.2). Yet, despite further increases in total adoption and acreage planted with RR soybeans, price declines were, in percentage terms, more muted in the later years, especially after 2006. Tighter global soybean stocks and overall demand and supply conditions, along with swelling commodity prices, led to proportionally smaller price effects from the innovation. Similar patterns of change are observed in the cross-commodity price impacts from the adoption of RR soybean technology (Figure 45.2).

With significant changes in the pattern of adoption across countries and over time, as well as in the price changes induced by the innovation across commodity markets, changes in consumer and producer surplus varied significantly from year to year. Figure 45.3 illustrates the year-to-year variations in the changes of the US producer and consumer surplus as well as in the fees paid to the technology provider. The illustration demonstrates that multi-year assessments are necessary in order to fully account for the economic impacts of new technologies with long and evolving product life cycles, like RR soybeans. Single-year assessments can be misleading, especially in the early stages of adoption and diffusion.

7 CONCLUSION

This chapter examined the global economic impacts of RR soybeans in their first 15 years of adoption. The analysis identifies substantial economic benefits, amounting to a total of about US$46 billion. These include benefits to consumers and producers all over the world (86 per cent of the total) as well as benefits to the innovator (14 per cent). Our estimates account for the surplus gained or lost by consumers and producers resulting from supply and demand shifts in all other major crop markets caused by the introduction of RR soybeans.

This analysis reinforces certain results and lessons learned from previous studies. First, early adopters benefit most from innovation. Out of the total world surplus,

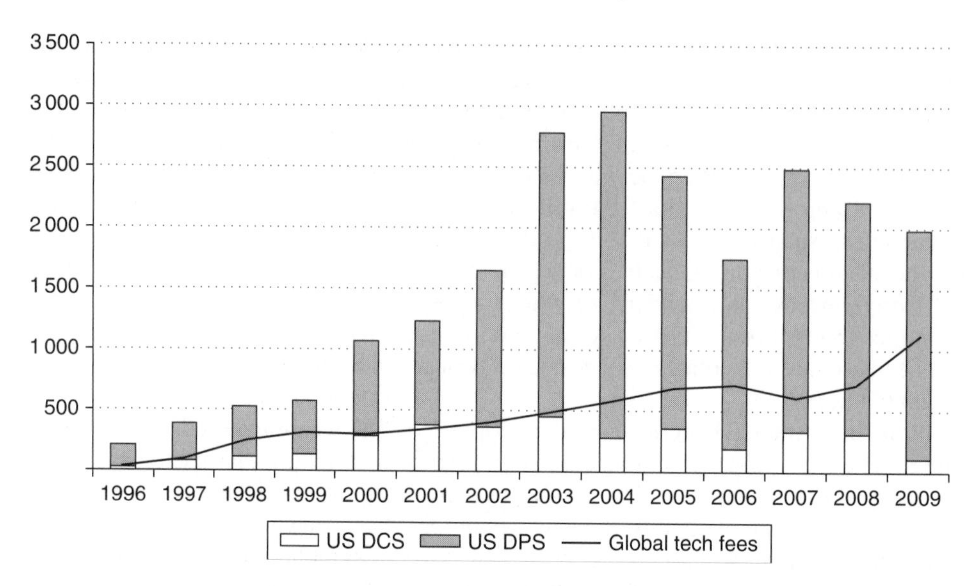

Figure 45.3 Changes in US producer surplus (DPS) and US consumer surplus (DCS) and global gross revenues of technology supplier (fees), in US$2009 million, 1996–2009

producers and consumers in the United States, Argentina and Brazil captured 75 per cent. Non-adopters and producers of competitive oilseeds that have not benefited from parallel innovations experienced economic losses. Second, through induced reductions in market prices, consumers (more precisely, buyers) have benefited from the technology almost as much as producers have. Because all major adopting countries are also major exporters, a large share of the benefits has been transferred to importing countries. Third, the aggregate economic impacts of RR soybeans are large and sustained but also quite dynamic as they are shaped by the patterns of adoption and conditions in global commodity markets. Tighter global stocks and demand conditions in recent years have meant lower proportional price impacts and hence smaller transfers from producers to consumers.

It should be noted, in closing, that as large as the measured benefits from the adoption of RR soybean technology are, they understate the total impacts of RR soybeans over the period of our analysis. The assumptions about the farm-level impacts of the technology on yields and cost savings are conservative and the measures do not account for the increases in the adoption of minimum tillage systems, the increase in double cropping, or the economic value of environmental and other non-pecuniary impacts that RR soybeans have facilitated. Hence, the estimates should be viewed as a lower bound of the total economic impacts of RR soybean technology for the period of analysis. Future studies could surely improve on these measures by firming up the accuracy of farm-level impacts across various countries and over time as well as by explicitly accounting for the effects of any positive and negative externalities.

NOTES

1. If a premium or discount applies for soybeans produced with the new technology, in some modeling approaches that too can be incorporated as an effective shift in supply.
2. When RR soybeans are grown, weeds can be controlled through over-the-top application of glyphosate, and multiple applications of selective post-emergence herbicides are typically replaced with one or two over-the-top applications of glyphosate.
3. In recent years, instances of weed resistance build-up to glyphosate have been reported. In such cases, over-the-top application of glyphosate alone may not be adequate to control all weeds, and the benefits from RR soybeans may not be clear. Furthermore, negative externalities may be present if such resistance build-up persists and covers large geographic areas. At this time, the extent of such resistance build-up is not clear. However, any such build-up of resistance will diminish the potential future benefits from RR soybeans compared with our estimates that apply under an assumption that the technology remains fully effective. Hence, care should be exercised in extrapolating our measures of past benefits forward far into the future.
4. This could entail fixed benefits from enhancements to farmer and farm-worker safety associated with the use of the technology as well as costs of risk or information costs associated with learning about the new technology (which could decline with experience, giving rise to progressive increases in adoption).
5. This measure does not account for any impacts on the suppliers of agricultural chemicals and others whose business may be reduced as a consequence of farmers shifting to the new technology. If the industries in question could be regarded as competitive and only earning 'normal' economic profits, then there would not be any net welfare impacts to consider. On the other hand, if the affected firms had been earning more than 'normal' economic profit, such as they would if they were exercising some market power in a patented technology, then they would experience net economic losses as a result of farmers adopting the new technology, and these economic losses ought to be considered as well in the estimate of net national benefits. Moschini et al. (2002) present a model of this kind of situation.
6. Welfare effects can be added up across all the affected markets and these measures will be valid and independent of the order of integration if integrability conditions are satisfied (see Just et al., 1982).
7. The model in Figure 45.1 is strictly illustrative. In particular, our simulation model does not entail an assumption of parallel shifts of linear supply functions associated with the introduction of RR soybeans.
8. No price discounts have been imposed on RR soybeans and, hence, they do not enter the calculations of the vertical shift. Yield changes associated with the use of RR technology are also not accounted for in the calculation of the vertical shift of the soybean supply in the model. As is discussed below, changes in soybean yields from the adoption of RR technology, both positive and negative, have been reported in a few studies but in almost all cases they have been rather small. The bulk of the literature has found no significant yield changes in the adoption of RR technology, and this assumption of 'no yield' impact is maintained in this study.
9. All reported sums are undiscounted sums of deflated nominal values.

REFERENCES

Alston, J.M. (2007), 'Benefits and beneficiaries from US farm subsidies' AEI Agricultural Policy Series: *The 2007 Farm Bill and Beyond*, Washington, DC: American Enterprise Institute, available at: (http://aic.ucdavis.edu/research/farmbill07/aeibriefs/20070515_alstonSubsidiesfinal.pdf).

Alston, J.M., G.W. Norton and P.G. Pardey (1995), *Science Under Scarcity: Principles and Practice for Agricultural Research Evaluation and Priority Setting*, Ithaca, NY: Cornell University Press.

Brookes, G. and P. Barfoot (2005), 'GM crops: the global economic and environmental impact: the first nine years 1996–2004', *AgBioForum*, **8**(2&3), 187–96.

Brookes, G. and P. Barfoot (2008), 'Global impact of biotech crops: socio-economic and environmental effects, 1996–2006', *AgBioForum*, **9**(3), 139–51.

Brookes, G. and P. Barfoot (2011), 'Global impact of biotech crops: environmental effects, 1996–2009', *GM Crops 2*, 1 January-March, pp. 1–16.

Brookes, G., T.H. Yu, S. Tokgoz and A. Elobeid (2010), 'The production and price impact of biotech corn, canola, and soybean crops', *AgBioForum*, **13**(1), 25–52.

Carpenter, J. (2010), 'Peer-reviewed surveys indicate positive impact of commercialized GM crops', *Nature Biotechnology*, **28**, 319–21.

Carpenter, J. and L. Gianessi (1999), 'Herbicide tolerant soybeans: why growers are adopting Roundup Ready varieties', *AgBioForum*, **2**(2), 65–72.

Carpenter, J. and L. Gianessi (2001), 'Agricultural biotechnology: updated benefit estimates', Washington, DC: National Center for Food and Agriculture Policy, available at: http://www.ncfap.org/reports/biotech/updatedbenefits.pdf.

Carpenter, J. and L. Gianessi (2003), 'Trends in pesticide use since the introduction of genetically engineered crops', in N. Kalaitzandonakes (ed.), *The Economic and Environmental Impacts of Agbiotech: A Global Perspective*, New York: Kluwer Academic Publishers.

Council for Biotechnology Information (2002), 'Agronomic, economic and environmental impacts of the commercial cultivation of Glyphosate tolerant soybeans in Ontario', unpublished report, Guelph, Ontario.

Couvillion, W. C., F. Kari, D. Hudson and A. Allen (2000), 'A preliminary economic assessment of Roundup Ready soybeans in Mississippi', Research Report no. 2000-005, Mississippi State University.

Davis, J.S. (1980) 'A note on the use of alternative lag structures for research expenditure in aggregate production function models', *Canadian Journal of Agricultural Economics*, **28**, 72–6.

Duffy, M. and M. Ernst (1999), 'Does planting GMO seed boost farmers' profits?', State University, Leopold Center for Sustainable Agriculture. Iowa *Leopold Letter*, **11**(3).

Ervin, D.E., S.S. Batie, R. Welsh, C.L. Carpentier, J.I. Fern, N.J. Richman and M.A. Schulz (2000), 'Transgenic crops: an environmental assessment', Henry Wallace Center for Agricultural & Environmental Policy at Winrock International, available at: *http://www.winrock.org/transgenic.pdf*.

Falck-Zepeda, J. B., G. Traxler and R.G. Nelson (2000a), 'Rent creation and distribution from biotechnology innovations: the case of Bt cotton and herbicide-tolerant soybeans in 1997', *Agribusiness*, **16**(1), 21–32.

Falck-Zepeda, J.B., G. Traxler and R.G. Nelson (2000b), 'Surplus distribution from the introduction of a biotechnology innovation', *American Journal of Agricultural Economics*, **82**(2), 360–69.

Fernandez-Cornejo, J. and W. McBride, with contributions from Cassandra Klotz-Ingram, Sharon Jans and Nora Brooks (2000), 'Genetically engineered crops for pest management in US agriculture', Agricultural Economics Report no. (AER786) May, available at: http://www.ers.usda.gov/publications/aer786/.

Fernandez-Cornejo, J. and W. McBride (2002), *Adoption of Bioengineered Crops*, Agricultural Economics Report no. 810, USDA, ERS.

Fernandez-Cornejo J., C. Klotz-Ingram and S. Jans (2002), 'Farm-level effects of adopting herbicide-tolerant soybeans in the USA', *Journal of Agricultural and Applied Economics*, **34**(1), 149–63.

Fernandez-Cornejo, J., C.K. Klotz-Ingram, R. Heimlich, M. Soule, W. McBride and S. Jans (2003), 'Economic and environmental impacts of herbicide tolerant and insect resistant crops in the United States', in N. Kalaitzandonakes (ed.), *The Economic and Environmental Impacts of Agbiotech: A Global Perspective*, New York: Kluwer Academic Publishers.

Freebairn, J.W. (1992) 'Evaluating the level and distribution of benefits from dairy industry research', *Australian Journal of Agricultural Economics*, **36**(2), 141–66.

Frisvold, G.B. (1997), 'Multimarket effects of agricultural research with technological spillovers', in T.W. Hertel (ed.), *Global Trade Analysis*, New York: Cambridge University Press.

Gianessi, L.P. and J.E. Carpenter (2000), 'Agricultural biotechnology: benefits of transgenic soybeans', Washington, DC: National Center for Food and Agricultural Policy.

Griliches, Z. (1958), 'Research costs and social returns: hybrid corn and related innovations', *Journal of Political Economy*, **66**(5), 419–31.

Hategekimana, B. (2002), 'Genetically modified grain corn and soybeans in Quebec and Ontario in 2000 and 2001', A. Division, Ottawa: Statistics Canada.

Just, R.E., D.L. Hueth and A. Schmitz (1982), *Applied Welfare Economics and Public Policy*, Englewood Cliffs, NJ: Prentice Hall.

Konduru, S., J. Kruse and N. Kalaitzandonakes (2008), 'Global economic impacts of Round-Up Ready soybeans', in G. Stacey (ed.), *Genetics and Genomics of Soybean*, New York: Springer.

Lin, W., G.K. Price and J. Fernandez-Cornejo (2001), 'Estimating farm-level effects of adopting herbicide-tolerant soybeans', in *Oil Crops Situation and Outlook Yearbook*, OCS-2001, USDA, ERS, Washington, DC.

Marra, M. (2001), 'Agricultural biotechnology: a critical review of the impact evidence to date', in Philip Pardey (ed.), *The Future of Food: Biotechnology Markets and Policies in an International Setting*, Washington, DC: International Food Policy Research Institute, pp. 155–84.

Marra, M.C. and N.E. Piggott (2006), 'The value of non-pecuniary characteristics of crop biotechnologies: a new look at the evidence', in R.E. Just, J.M. Alston and D. Zilberman (eds), *Regulating Agricultural Biotechnology: Economics and Policy*, New York: Springer Science and Business Media, pp. 145–78.

McBride, W., D. and N. Brooks (2000), 'Survey evidence on producer use and costs of genetically modified seed', *Agribusiness*, **16**, 6–20.

Moschini, G. and H. Lapan (1997), 'Intellectual property rights and the welfare effects of agricultural R&D', *American Journal of Agricultural Economics*, **79**, November, 1229–42.

Moschini, G., H. Lapan and A. Sobolevsky (2002), 'Roundup ready soybeans and welfare effects in the soybean complex', *Agribusiness*, **16**, 33–55.

Nelson, G. and D. Bullock (2003), 'Environmental effects of glyphosate resistant soybeans in the United States', in N. Kalaitzandonakes (ed.), *The Economic and Environmental Impacts of Agbiotech: A Global Perspective*, New York: Kluwer Academic Publishers.

Price, G.K., W. Lin, J.B. Falck-Zepeda and J. Fernandez-Cornejo (2003), 'Size and distribution of market benefits from adopting biotech crops', *USDA-ERS Technical Bulletin* no. 1906, Washington, DC.

Qaim, M. (2009), 'The economics of genetically modified crops', *Annual Review of Resource Economics*, **1**, 665–94.

Qaim, Matin and G. Traxler (2005), 'Roundup Ready soybeans in Argentina: farm level, environmental, and welfare effects', *Agricultural Economics*, **32**(1), 73–86.

Roberts, R.K. et al. (1999), 'Economic analysis of alternative herbicide regimes on Roundup Ready soybeans', *Journal of Production Agriculture*, **12**.

Sankula, S. (2006), 'Quantification of the impacts on US agriculture of biotechnology-derived crops planted in 2005', Washington DC: National Center for Food and Agricultural Policy.

Sankula, S. and E. Blumenthal (2004), 'Impacts on US agriculture of biotechnology-derived crops planted in 2003: an update of eleven case studies', Washington, DC: National Center for Food and Agriculture Policy, available at: http://www.ncfap.org/whatwedo/pdf/2004finalreport.pdf.

Sobolevsky, A., G. Moschini and H. Lapan (2005), 'Genetically modified crops and product differentiation: trade and welfare effects in the soybean complex', *American Journal of Agricultural Economics*, **87**(3), 621–44.

Sumner, D.A. (2005), 'Boxed in: conflicts between US farm policies and WTO obligations', *Cato Institute Trade Policy Analysis*, **32**, December, available at: http://www.freetrade.org/pubs/pas/pas.html.

Trigo, E.J. and E.J. Cap (2003), 'The impact of the introduction of transgenic crops in Argentinean agriculture', *AgBioForum*, **6**(3), 87–94.

Trigo, E.J. and E.J. Cap (2006), 'Ten years of genetically modified crops in Argentine agriculture', research report, Argentine Council for Information and Development of Biotechnology – ArgenBio.

Webster, Eric P. et al. (1999), 'Weed control and economics in non-transgenic and Glyphosate-resistant soybean (*Glycine max*)', *Weed Technology*, **13**.

Wohlgenant, M.K. (1997), 'The nature of the research-induced supply shift', *Australian Journal of Agricultural and Resource Economics*, **41**(3), 385–400.

Zhao, X., W.E. Griffiths, G.R. Griffith and J.D. Mullen (2000), 'Probability distributions for economic surplus changes: the case of technical change in the Australian wool industry', *Australian Journal of Agricultural and Resource Economics*, **44**, 83–106.

46 Wheat: status, outlook and implications

William W. Wilson

INTRODUCTION

Wheat is one of the next crops to be commercialized with genetically modified (GM) traits. Development of GM wheat is important for a number of reasons (see Wilson et al., 2003 for a comprehensive discussion of the issues related to GM wheat). First, wheat will be one of the first food grains in which GM traits are introduced and will likely be a precursor to similar developments in other food grains. Second, wheat is traded among many importing and exporting countries, many of which have very different mechanisms for regulating trade in GM crops and for marketing products with GM ingredients. Third, there is no doubt that there will be highly differentiated demand for products produced with or without GM ingredients and/or requirements to provide information to consumers among these countries.

There are numerous reasons for the recent increased investment in wheat technology. One is that wheat has been losing competitiveness to many other crops, due in part to the more rapid adoption of GM technology in corn, soybeans and canola. Second, wheat is a very large acreage crop worldwide and, as a result, some economies of trait development can be achieved. Third, expanded development has been encouraged by a political process in the United States. There is now overt support for GM development from major users, and internationally a tri-country process has emerged whereby grower organizations in each of the United States, Australia and Canada have supported the development of GM wheat (see National Association of Wheat Growers, 2010). As a result of this, the agbiotech companies have felt some of the risks associated with investments in GM wheat are mitigated.

GM wheat is currently being developed in a number of countries and by a number of companies. Traits under development in wheat using GM techniques include Fusarium resistance in the US (Huso and Wilson, 2005; Tollefson, 2011; Valliyodan and Nguyen, 2006), and drought resistance in Australia. Since about 2009 most of the major agbiotech companies have made announcements indicating their intentions to enter the GM wheat market, and in 2011 and 2012 there were field trials by several companies in the United States and Australia. Amongst these, the most common traits being pursued include yield, drought tolerance (DT) and nitrogen use efficiency.

The purpose of this chapter is to describe the outlook and issues related to developing GM technology for wheat. Specifically, the evolution of competitors in GM trait development is described. The chapter also discusses at some length some of the major issues related to development, uptake and use of the technology in wheat.

EVOLUTION OF NEW COMPETITORS IN GM WHEAT

Wheat is one of the world's largest acreage food crops, but has not been a recipient of the new GM technologies that have benefited corn, soybeans, canola and cotton. Compared to these crops, wheat has been losing its competitiveness for a number of reasons.[1] The area planted to wheat in the United States has declined by 30–40 per cent since the mid-1980s. During the same period, canola acreage in Canada has increased so that it now exceeds wheat acres. There have also been important geographical shifts in the composition of crops planted in these countries. Generally, the smaller wheat acreage has been matched with a gradual northerly and westerly shift to dryer areas.

Since 1996 a number of GM traits have been introduced in competing crops. For corn, Roundup Ready (RR), *Bacillus thuringiensis* (BT) and several other traits have been developed and widely adopted. Some of these are now stacked in multiples of three or four traits in a single variety. Looking forward, a large number of traits are under development and expected to be commercialized in the next 10 or more years (Wilson and Dahl, 2010a, 2010b). For corn, there are at least 21 new GM traits under development. There is a mix of producer, consumer and processor traits. Some of these traits are developed individually and some through joint initiatives. A comparable number and composition of traits is under development for soybeans.

Roundup Ready had earlier been developed by Monsanto as a GM trait for wheat, but was withdrawn in 2004 in part due to anticipated consumer resistance. While it was deregulated in the United States, it was not commercialized. Concurrently, Syngenta was developing a fusarium-resistant trait, but never pursued commercialization. Following a number of years in which wheat acres declined in North America, largely being shifted to corn, soybeans, canola and cotton, a number of events unfolded which helped spawn the recent interest in GM wheat. One was an international trilateral agreement amongst grower groups supporting development of GM wheat. The other was the sharp escalation in crop prices during 2008 which precipitated concerns by end-users about the longer-term supplies and competitiveness of wheat.

In 2009, Monsanto was the first to announce their intent to expand into GM wheat. This was followed within months by announcements from each of BASF, Bayer CropScience, Limagrain and Dow AgroSciences. Each of these companies is following work that had already been initiated in Australia by the Victoria Agrobiosciences Center (VABC) and CSIRO. Indeed much of the initial and early work was done in Australia, where the initial focus was on drought.[2] These are in addition to near-simultaneous development of initiatives on GM wheat in China (Xia et al., 2012). These firms and organizations have been pursuing varying strategies, including acquiring germplasm and creating public–private partnerships. In addition, to varying degrees they have each made claims about the traits they intend to develop using genetic modification.

Table 46.1 provides a summary of announcements to date. Each of the major firms has sought varying forms of alliances, acquisitions or partnerships to achieve their goals of technology improvement. The GM traits that are most common are yield, drought tolerance and nitrogen-use efficiency (NUE). The criteria for selecting these traits are not exactly clear. Most likely these choices are a result of experiences with other crops, evidence related to current plant stressors, and the anticipated changing geography of

Table 46.1 Partnerships and technology trait focus, 2012

Company	Acquisitions or partners (Year, Country)	GM trait targets
Bayer Crop Science	Athenix (2009, US)	Drought
	CSIRO (2009, AU)	NUE
	Evogene (2010, IL)	Yield
	NARDI (2011, RO)	
	RAGT (2011, FR)	
	SORT, EUROSPORT (2010, UA)	
	South Dakota State University (2011, US)	
	Texas A&M (2011, US)	
	University of Nebraska-Lincoln (2010, US)	
Dow Agrosciences	HRZ Wheat (2011, US)	
	Northwest Plant Breeding (2011, US)	
Limagrain	Arcadia (2010, US) (own a small share)	Disease
	Trident seed	
	Biogemma (FR)	Drought
	U of Idaho and crop quality at Ft. Collins	
	CISRO (2006, AU)	Quality
Monsanto	BASF (2010, US)	Drought – stress (incl of drought)
	Intergrain (2010, AU)	Herbicide tolerance
	Kansas State University (2010, US)	Yield
	Westbred (2009, US)	
	Virginia Tech (2010, US)	
	North Dakota State University (2012)	
Syngenta	CIMMYT (2010, MX)	Hybrid

production, in addition to concerns of future water availability and cost (James, 2011; Rice Today, 2012a, 2012b; Sindrich, 2012).

Drought tolerance (DT) is an example of a stress trait. It has been described in numerous articles and is the focus of extensive media promotion (*The Economist*, 2011a and 2011b; Tollefson, 2011; *Wall Street Journal*, 2012). DT is also the focus for trait development in other crops including corn and rice (Reyes, 2009). Genes are being identified to be activated by drought (that is, the efficiency gain by drought resistant gene would be realized when drought occurs) so as to avoid any yield penalty in normal conditions. Fatka (2008, p. 5) asserts that 'Drought tolerant crops look to be one of the most promising upcoming biotech traits in pipeline, providing ability to produce "more crop per drop" of water.' These crops would be designed to provide farmers' yield stability during periods when water supply is scarce by mitigating the effects of drought – or water stress – within the plant itself.

There have been fewer studies in the case of drought-tolerant wheat, in part because the trait discovery is just commencing (in 2010). In the case of GM drought tolerance, there have been four years of field trials in Australia. Results from those studies indicate that GM lines yield 20 per cent more than conventional wheat varieties under conditions of drought stress.

ISSUE 1: CONSUMER ACCEPTANCE

One of the issues confronting GM wheat is that of consumer acceptance. While this is an issue for most GM traits, it is perhaps more challenging in wheat, relative to non-food grains and oilseeds, for a couple of reasons. One is that wheat is broadly consumed across many countries and there are segments within each country as well as across countries with concerns about the composition of GM in food grains. Second, wheat is generally consumed as a food item (albeit, typically ground into flour and then baked in varying forms), as opposed to being used as animal feed or as an ingredient that goes through more intensive processing. Third, and probably most important, two of the larger consumers and importers of wheat are the EU and Japan, each of which has rather onerous regulatory mechanisms (discussed in the following section) that confront wheat.

While there have been a number of studies on consumer attitudes, generally the US industry has relied mostly upon survey work by the International Food Information Council (IFIC). Results of these surveys are broadly distributed. Some of the important points from recent surveys are noted. One is that in the United States about 70–80 per cent of the food products contain biotechnology ingredients. Second is that consumers are most concerned about issues related to sanitation, hygiene and foodborne illness. Less than half of 1 per cent of the respondents identified food biotechnology as a food safety concern. Third, regarding labeling, 76 per cent of respondents indicated there is no information they would like to see added to food labels. Further, less than 1 per cent of respondents wanted information contained on labels regarding biotechnology. Finally, support for the FDA policy to label food biotechnology has decreased.

While the surveys show US (and Canadian) consumers have a higher degree of trust in national regulatory authorities than do Europeans, notably both Japanese and European consumers prefer international regulatory agencies and have less trust in their national agencies.

Looking forward, it is important that in nearly all mature markets one should expect differentiated demands to arise. Indeed, one should expect that the relevant market segments for wheat-based products would be comprised of GM, non-GM and organic (as modeled in Wilson et al., 2008a). This is in part due to differentiated tastes, varying income levels, market maturity, and regulations. Market segmentation is a natural process of market maturity.

ISSUE 2: SEGREGATION

A second major issue confronting future commercialization of GM wheat relates to segregation. Ultimately, the reason this is important is due in part to the anticipated differentiation in demands as well as the multitude of regulatory regimes over GM content in the world wheat market. This section provides a summary of this problem and some of the recent studies on the topic.

Previous studies treated these regulatory regimes differently. Most regimes studied in the past involved speculation about what the import regime would be at the time of commercialization. While evolving regimes may or may not be in place by the time GM wheat is introduced (*Milling and Baking News*, 2004a), there are two regimes that stand out, are entrenched, and will likely dominate future trade. These are the ones in place in Japan and the EU, each of which are averse to GM content and have institutionalized import regimes that will apply to GM wheat.

Interpretation of the future administration of these regulatory regimes is prospective and tentative. Those in the European Union (EU) and in Japan are particularly important. Given the EU is a large importer from North America it is incumbent for wheat producers to develop and analyse a system that conforms to EU traceability requirements. EU legislation sets restrictive conditions for allowing import of grain from countries using GM seed. Regulation and control measures for GM products include testing, tolerance, shipping, and segregation strategies. Labeling is required for products containing more than 0.9 per cent GM material.[3] While EU traceability provisions[4] were initially optional, they have been mandatory since the implementation of Regulation 178/2002, which requires traceability 'one step forward one step backward'. This Regulation has been reinforced for GMOs by Regulation 1829/2003, which additionally requires the implementation of whole-chain traceability and the retention of traceability documents for a period of five years. Traceability is generally considered a means to restore consumer confidence in what was perceived as a 'broken food system'. Exporting wheat to meet EU traceability requirements would result in additional costs and risks for suppliers.

A number of studies have analysed various aspects of traceability, testing and coexistence (Wilson, 2008, Wilson and Dahl, 2005 and 2006; Wilson et al., 2007, 2008b). Hobbs (2004) described multiple forms of private and public traceability systems. Golan et al. (2004) examined existing traceability systems for fresh produce, grain and oilseeds, and cattle and beef in the US food supply. Though these and other studies point to the fact that contracts will be essential for executing traceability, we are not aware of other studies that have analysed contracting strategies designed to conform to traceability requirements. Meanwhile, numerous industry practices are emerging to facilitate traceability (*Milling and Baking News*, 2004b).

A series of studies have examined wheat marketing strategies designed to facilitate coexistence. These studies applied stylized models in order to estimate costs and evaluate alternative testing regimes in the United States (Wilson and Dahl, 2005), Canada (Wilson and Dahl, 2006), the EU (Wilson et al., 2007), and specifically in the EU import regime on traceability (Huso and Wilson 2006; Wilson et al., 2008b). These studies created models that used parameters based on data from the respective exporting country in order to accurately reflect the various forms of import regulations. In each case, the studies used stochastic optimization models that included random variables where appropriate.

The studies determined that optimal testing strategies would include a range of types of tests, frequency, and testing locations. One important output was the derived risk to both the buyer and the seller. Buyer risk is the risk of receiving GM content in a non-GM shipment. Seller risk is the risk of having a non-GM shipment rejected for containing GM content. Using the optimal testing strategy, the resulting probabilities were generally low and would be likely to be acceptable under the currently established regimes. The results also showed that traceability systems based on testing and segregation can efficiently assure buyers of the non-GMO status of their product at a relatively low cost. In addition, these costs are lower than the costs for implementing the different protocols involved in identity-preservation systems. A critical aspect revealed in these studies is that some form of information provided from growers to handlers (for example variety declaration) would reduce costs and risks substantially. Optimal testing strategies depend on GMO adoption rates. Specifically, greater GMO adoption requires more intense testing, yielding greater costs and greater risks of failing to conform to requirements. Finally, conforming to additional GMO-related traceability requirements for imported grains in the EU leads to a number of important added costs and may require the use of lower yielding varieties. The impact of implementing these traceability measures would raise the cost of conforming to the EU requirements to about US$18/tonne, but the buyer risk would be relatively low. This would be accrued largely by the importer.

Wilson et al. (2008b) used a game-theoretic approach to determine the optimal contracting strategies to procure wheat in a world of coexistence of GM and non-GM wheat. A non-GM wheat contracting strategy was defined to determine the premium that would be sufficient to motivate suppliers to conform to EU traceability requirements. Producers could enter into contracts with firms to grow a particular variety of grain or oilseed as a way to seek price premiums or agronomic benefits. Additional costs and risks due to segregation and tracking would then pose an additional problem for contracting. Results indicated in the base case equilibrium that the supplier would reject the contract offered by the buyer 53 per cent of the time. The rest of the time, the supplier accepts and offers a contract to farmers, who in turn reject it 95 per cent of the time. The acceptance probability for the supplier is high without a premium, because the supplier knows that the farmer will always reject the contract without a premium. His expected pay-off if the farmer accepts the contract is equal to his expected pay-off if the contract is rejected. This result shows that without a premium, the equilibrium violates the participation constraint. Specifically, it shows that the supply chain cannot use contracts without a premium. Premiums were then varied, and a mixed Nash equilibrium is reached when premiums are US$3.31/tonne and US$2.94/tonne for the farmer and handler, respectively. If players want to increase the probability of participation, premiums need to increase further. At a premium of US$5.14/tonne for both the supplier and the farmer, the probability that a contract will be accepted is 85 per cent for farmers and 72 per cent for suppliers. As a practical example, if Japanese buyers wanted to assure near complete production of non-GM wheat in a targeted procurement region or state, it could do so by increasing the premium.

ISSUE 3: PUBLIC GERMPLASM, TECHNOLOGY DEVELOPMENT AND PRIVATE TRAIT DEVELOPERS

There are numerous issues regarding the integration of germplasm and advanced breeding technologies. This is in part because most of the germplasm is controlled through the public sector, in varying ways, and most of the new investment in advanced breeding technologies is being developed by the private sector. The major issues relate to private sector firms wanting access to public germplasm and public breeders wanting access to traits developed by the private sector. Each is discussed.

Several technologies exist for improving wheat, including conventional breeding,[5] marker-assisted-selection (MAS) and genetic modification (GM), amongst others. As an example, drought tolerance has been a breeding goal for many years and has been described in the scientific literature for at least 40 years. Drought tolerance is felt to be difficult using conventional techniques due to the complexity of the plant's metabolic pathways. The appeal of using GM techniques for drought tolerance is that it may be more efficient in improving the crop.

Integration of these breeding technologies (conventional, MAS and GM) has brought about a paradigm shift in crop development – to 'seeds and traits' as integrated business functions. This involves combining novel genetic traits with elite germplasm to develop crops that thrive while expressing the desired trait. The steps include discovering novel genes, transferring them into the cells of plants, optimizing the expression of the genetic trait in plants in the correct plant tissues, at the appropriate time and in sufficient levels, and incorporating, through breeding, the genetic trait into commercially viable varieties or hybrids. As a business strategy, introduction of genetic traits via biotechnology does not reduce the importance of superior germplasm in the host plant, nor does it replace the need for plant science and plant breeding (Dow AgroSciences, n.d.; Kaehler, 2006).

Wheat germplasm has largely been controlled through the public sector, in varying ways and not exclusively. In the United States that takes the form of ownership through state agricultural experiment stations. In Canada, it is primarily through Agriculture and Agri-Food Canada. In Australia, the germplasm ownership is less predominantly public and more mixed between public and private ownership. Ownership varies across other countries. In addition to ownership, protocols exist in most countries that govern the relationship among breeding programs, particularly regarding germplasm exchange. Generally, in the US this has emerged from a system predominantly including multilateral exchange of germplasm among breeding programs (called reciprocal exchange) and facilitated through uniform regional nurseries. Essentially, this means that germplasm is available for multilateral exchange without financial obligation among parties of the exchange.

Most of the agbiotech companies have been pursuing strategies of either acquiring seed companies (for example Syngenta, Monsanto) or forming partnerships with public breeding companies (for example Monsanto and Bayer). The driving force to this is in part thought to be the breeding efficiency of the 'seeds and traits' concept (described above); moreover, there are a limited number of non-public breeding programs that would otherwise be available for acquisition. Hence, the industry is moving towards a mixed strategy encompassing some vertically integrated breeding-technology companies (for example Syngenta) and others forming varying public–private partnerships. This is having the effect of morphing the public breeding programs that had traditionally had multilateral reciprocal exchange without financial obligations into bilateral reciprocal exchange with financial obligations.

There is a long list of other emerging issues that is too voluminous to elaborate here. However, they include, without limits: patenting as a strategy to protect varieties or germplasm; pricing of public germplasm; mechanisms to assure quality control of released varieties; and distribution of seed varieties. Each of these will challenge the industry for years to come.

ISSUE 4: VALUE OF NEW GM TRAITS

The challenge in valuing GM traits is that development takes a long time, is highly risky as a result of uncertainties of numerous variables which are random, and is costly.

Typically, trait development including regulatory review takes about 10 years, costs about $100 million and consists of a number of distinct phases. Estimates of these costs are difficult since they ultimately are firm-level activities and information is not readily published. Earlier estimates were that development of a GM trait costs $60 million and can take 15 years (Goodman, 2004). Recent estimates for regulatory costs are in the $6 to $15 million range (Bradford et al., 2006; Just et al., 2006). Finally, one of the more recent studies that estimated these costs (Phillips McDougall, 2011) indicated the average cost of GM trait development is $136 million, and takes 13 years, though there is substantial variability across firms and traits on these estimates. Monsanto has indicated it will spend at least $150 million on its wheat initiative, though this includes costs of germplasm and breeding, in addition to MAS and GM. These costs reflect what are commonly referred to as discovery, proof of concept, early and advanced product development, and the regulatory phase, though the labels for these functions vary across firms.

Trait strategies are fraught with randomness and extended periods for development, which results in substantial risks. The trait pipeline typically is referred to as comprising phases ranging from proof of concept to regulatory approval. Finally, revenue streams from trait development do not ensue until a period following regulatory approval. For these reasons, trait development is highly risky and strategic, and as a result, real options are an appropriate methodology for valuing traits during the trait development period.

A number of recent studies have used the real options framework to analyze investment in R&D. The foundations are summarized in Schwartz and Trigeorgis (2004) who describe the evolution of this method and its role in valuing R&D. Features of the problem that are important include that uncertainties are resolved through time and managers have options that can be exercised throughout the duration of the project. Investments in R&D provide the option to continue, wait or abandon. Investing in R&D is equivalent to buying a call option and can be valued as a real option because it involves future opportunities, uncertainty and options. Earlier work on this methodology includes Luehrman (1997) and Morris et al., (1991) who describe why real options can be used to model R&D; and Brach and Paxson (2001), Jensen and Warren (2001) and Seppä and Laamanen (2001) who offers specific applications. Since the process of crop development is staged in discrete phases with measurable risks and uncertain outcomes for each stage, the real options approach lends itself well to use of this framework for valuation of GM traits.

Earlier studies used real options to analyze the value of GM traits in wheat (Furtan et al., 2003; Carter et al., 2005; Flagg, 2008). The two earlier studies analyzed decisions from a public sector perspective and were modeled as post-development timing options which were irreversible, and the values were derived using the Black–Scholes model. The approach of Flagg (2008) differs from these in several respects. While earlier studies modeled the public costs and benefits of releasing a GM trait, Flagg (2008) models the decision process of a private firm during the R&D process. Second, while early studies model a timing decision once the product is developed, Flagg (2008) models real options confronting management of biotechnology companies during the development process. The primary concern was the risks in the post-product development phase (that is, the commercialization phase).

In contrast, Shakya et al. (2012) developed a real-options model making R&D decisions through time and across phases of development. In this model, many of the parameters are random, and for this reason option values were derived at each phase using stochastic simulation.

Investing in R&D buys the option to abandon, wait, or continue to the next phase of development, which buys the option to continue to the commercialization phase. The option value, derived at each phase, can be either in-the-money (ITM), or out-of-the-money (OTM). If expected cash flows at an early development phase are positive, it is ITM and the developer would probably proceed to the next phase. If the value of the option is OTM, the developer can either wait, or abandon the project. Flagg (2008) modeled R&D as a compound call option using a stochastic binomial option specification. The 'continue' growth option represents the decision to continue to the next phase and make further investment to get to the next phase. So long as the option is ITM, management could choose to invest and continue.

This model was used to determine the *ex ante* value of random GM traits, specifically for corn and wheat. The empirical model comprised three steps. The value of drought-tolerant (DT) corn and wheat were determined at the farm level. Stochastic simulation and stochastic efficiency with respect to a function (SERF) were used to derive the risk premium for DT corn and wheat, which is interpreted as the expected value of the trait to growers. Farm budgets were simulated to measure risk and returns, with SERF used to determine certainty equivalents with and without the trait. These results were then used to derive the risk premium of the trait compared to no trait. The risk premium is defined as the value of the certainty equivalent that is required for a grower to be indifferent between the variety with and without the trait and is then used as the basis for pricing the new trait. Stochastic simulation and these values were used to derive the real option value of the R&D expenditure at each phase of development.

Trait efficiency is defined as the increase in yields as a result of the GM technology being inserted into conventional germplasm. While many companies are working on DT wheat, there are only a couple of public references indicating efficiency for this trait. In our base case we define trait efficiency using results from field trials in Australia (the only published reference to date on trait efficiency). Results from these studies indicate a trait-efficiency of .20 and apply this to the yield difference from maximum yields for the region. This value was from results of field trials in Australia and implies that DT wheat variety can recover 20 per cent of the yield loss attributable to drought. This is a critical value. There is also a subtle difference between corn and wheat in the interpretation of the probabilities. For corn, it is interpreted as 12 per cent of losses being able to be saved during drought years due to the GM trait, whereas the results for wheat are interpreted as yields being at least 20 per cent more than conventional varieties under stress conditions. It is also a critical value for competition amongst trait providers as they try to compete by increasing their trait efficiency in relation to their competitors. In the analysis, the base case assumes a trait efficiency of .20 and we conduct sensitivities at .25. Increases in trait efficiency have the impact of shifting the function rightwards, as illustrated in Figure 46.1.

The results indicate that the greatest value in the US on a per-acre basis in the base case, is for Heartland, Northern Crescent, Eastern Uplands, followed by Prairie Gateway and Northern Great Plains, with results of US$6.40, $6.37, $12.14, $9.20 and $8.58 per acre respectively. Strictly, a value of $8.58/acre is the value of the DT trait in the Northern Great Plains and reflects the value of the increased yield and reduced risk associated with

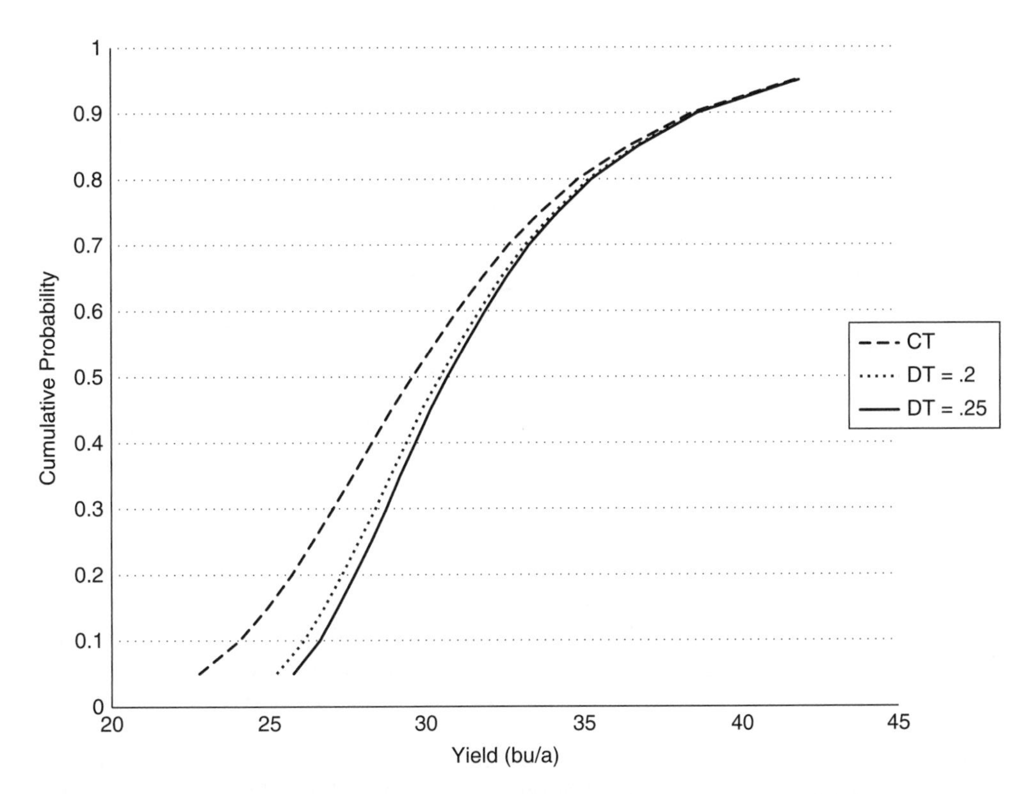

Figure 46.1 Comparison of distribution of wheat yields for conventional (CT) and drought tolerant (DT) with trait efficiency of .2 and .25: Prairie Gateway region

the DT trait, versus conventional technology. Thus, in the extreme case if the price of the trait exceeds this value, the grower will choose conventional technology. If the trait price is less than this value, the GM technology will be chosen.

The change in the risk premium with improvements in trait efficiency is important. Biotechnology companies compete on trait efficiency, ultimately trying to choose events that have greater efficiency than that of their competitors. We simulated impacts of a trait efficiency =.25. In this case, the value of the trait increases by $2 to $4/acre for the major wheat-producing regions. This is a very important figure and is an indicator of the prospective increased profitability associated with greater trait efficiency.

Finally, for comparison, values for DT corn using similar methodologies are lower than those shown here. In corn, the trait efficiency is in the area of .12, versus wheat at .20. Thus, for example in Northern Great Plains, the risk premium for DT corn is about US$6.39/acre, versus US$8.58/acre for wheat, indicating that value of improved resistance to drought is apparently much greater for wheat than for corn. This is largely driven by the measures of trait efficiency for the traits in the two crops.

Real option values were derived for each phase. Since this is a stochastic model, the distributional parameters are important. The results indicate that the real option value is in-the-money at each phase of development. It increases from US$12 million at the

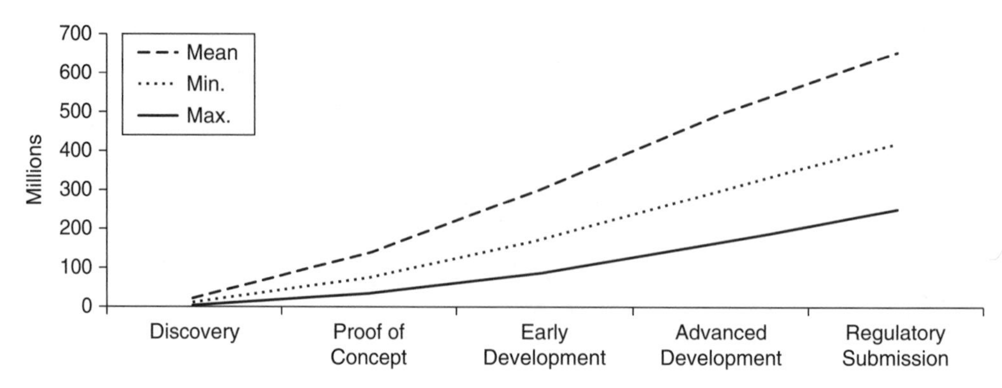

Figure 46.2 Option values of 'drought tolerance' in Hard Red Spring Wheat (HRSW) across stages of development (trait efficiency = .2; $ values in US$ millions)

discovery phase, to US$77 million during proof of concept, and ultimately to US$419 million at the point of regulatory submission. It is of interest that in all simulations the minimum option values exceed nil. Thus, the likelihood that the real option value in out-of-the-money is nil, which should provide substantial confidence about the future pay-off. Real option values and their distributions across phases are shown in Figure 46.2. The distribution of real option values increases as R&D moves through the development phases.

The results indicate that the value of drought-tolerant wheat using GM technology is in-the-money at each phase of development. Moreover, the value of GM drought-tolerant wheat exceeds that of drought-tolerant corn. These results explain why most of the agbiotech companies are developing this trait, amongst others. The greatest value would accrue to the Prairie Gateway and Northern Great Plains regions in the United States, though there would be similar value in numerous other countries which were not the focus of this study. The value of the trait has growing uncertainty throughout the trait development process. However, the variability in net present value (NPV) diminishes at each phase looking forward. For any probability of success at a developmental stage, the expected value of the trait increases with subsequent stages of development. Also, for a certain value of GM trait, there is less risk associated in later stages of development. A trait that is more likely to be discarded for development such as drought tolerance in initial stages due to high probability of being OTM becomes increasingly ITM as the developmental stages pass. The option tree provides the leverage to management to choose the option to wait by recognizing the need of market conditions and still to be able to get next best value for the GM trait by deciding to continue later on. Such flexibility is absent when the investment decision is made solely on NPV.

WELFARE STUDIES ON GM WHEAT

Four studies have analyzed impacts of GM traits in wheat (Furtan et al., 2003; Johnson et al., 2005; Berwald et al., 2006; and Wilson et al., 2008a). Furtan et al. (2003) analyzed the distribution of welfare from the release of RR wheat in Canada. Export markets that Canada risks losing due to the introduction of GM wheat were identified and segregation was not allowed. They concluded that while GM wheat would lower production costs, these agronomic benefits are more than outweighed by the costs of segregation of GM wheat, the impact of lost export markets, and lower wheat prices. Johnson et al. (2005) used a similar approach to buyer acceptance, conservative assumptions regarding cost savings and yield increases and assumed adoption would be 50 per cent for this biotech trait. That study indicated that introduction of GM wheat would lead to a small net loss of total economic welfare. However, they acknowledged the results were highly dependent on assumptions in the base case.

Berwald et al. (2006) developed an aggregate analysis of the world wheat market to explore impacts of introducing GM wheat. Their model allows GM introduction in major producing countries and differentiated demands by high and low-income countries. Results indicate that if Canada chooses not to adopt GM, producer welfare will decline.

These models differ, in terms of underlying assumptions, from each other and from the Wilson et al. (2008a) model. The aggregations were defined in each study to allow substitution with other lower protein classes to meet demands. Furtan et al. (2003) allowed for multiple segments of wheat based on low, medium and high quality and by GM segregations. Johnson et al. (2005) distinguished hard red winter (HRW) and hard red spring (HRS). Berwald et al. (2006) allowed for a global release of GM wheat that was potentially adopted in each of the US (for all classes including soft, hard and durum), Canada and Argentina. This is important as RR wheat was planned to be introduced into the HRS classes, and not the others, and Monsanto indicated that it would not be released in other countries due to their agronomics and also because of the inability to collect royalties in other countries. Thus, allowing other classes to benefit agronomically from this technology is highly speculative and allowing them to substitute with HRS higher protein wheat is not supported by empirical observations or trade practices.[6]

Second, each of these studies used models to analyze spatially separated markets in which suppliers compete for shipments and where shipping costs are important. Ignoring these conditions, along with the aggregation issues noted above, produces unrealistic impacts such as allowing lower-protein wheat from the southern US (or Argentina) to substitute in the EU market for higher-protein wheat shipped from Saskatchewan. Of course, this would never be observed in practice.

A third difference concerns treatment of adoption rates. Furtan et al. (2003) derived the optimal technology fee, from which they determined adoption rates. Berwald et al. (2006) and Johnson et al. (2005) assumed adoption rates, and calculated welfare changes conditional on these assumed rates. In practice, adoption rates are part of the equilibrium solution that is determined by a complex set of relationships including yield and cost-saving differentials, technology fees, and shipping costs relative to competing regions to the numerous segmented markets. Indeed, there should be less adoption in regions that are logistically and technologically more closely aligned to non-GM market segments.

A fourth difference involves treatment of the productivity gain associated with the GM trait and cost savings. Well documented information on yield gains from RR wheat (Kidnie et al., 2003; Blackshaw and Harker, 2002) suggests yield gains in the 5–15 per cent and 8–17 per cent range, respectively. However, Furtan et al. (2003) used a range of 0–5 per cent based on Holzman (2001, personal communication), whereas Johnson et al. (2005) used 3 per cent. These studies also under-represented on-farm cost savings: they used C$0–24.71/ha and Johnson et al. (2005) used US$11.86/ha, but in the case of HRS and CWRS, the more likely values are in the range US$15–26/ha and vary substantially across states and regions (Kalaitzandonakes et al., 2004). These savings are important and, taken together with the yield increases and convenience, are the major reasons for potential adoption of GM technology.

The final difference relates to segregation costs. Furtan et al. (2003) assumed that segregation cannot be done in a cost-effective manner and did not allow it. As a result, the price of wheat becomes the GM wheat price if GM is adopted. Johnson et al. (2005) used US$11–15/tonne based on the cost identity preservation (IP) procedures for food-grade non-GM soybeans. These procedures differ substantially from current practices for wheat segregation, as documented by Azziz (2007). Handlers in Canada and the US regularly and routinely segregate wheat and other grains. Further, IP practices are much more stringent and costly than required for segregating non-GM wheat, at least at tolerances in the 3–5 per cent range. These costs are relatively high in comparison to recent studies using simulation (Wilson and Dahl, 2005 and 2006; Wilson et al., 2007, 2008b) and industry practices, and they ignore that segregation costs escalate sharply with more stringent tolerances.

Wilson et al. (2008a) developed a spatial partial equilibrium model of the higher-protein hard wheat market and assessed the changes in the distribution of welfare associated with release of Roundup Ready® wheat. It incorporates segments for GM acceptance in each market and segregation costs for each segment. Suppliers are allowed to adopt or not adopt, depending on location and incentives, and handlers are allowed to segregate GM from non-GM at different tolerance levels at different costs. Other sources of cost saving, which vary geographically, are included. This study expands on methodologies used in welfare studies of GM traits by introducing spatial competition. In addition, it accounts for market segments and segregation costs which vary by tolerance, and determines equilibrium adoption rates by region. The latter are determined by a complicated set of relationships including regional differentials in yields, GM cost savings, segregation costs and relative shipping costs to segmented markets. It uses more appropriate measures of yield gains and cost savings, and analyzes welfare distribution inclusive of the impacts of transport, segments and segregation costs, and determines equilibrium prices, price differentials, adoption rates and trade flows.

The results indicate that in a likely scenario, producer and consumer welfare increase by US$301 and US$252 million, respectively. These are comparable to the estimated welfare gains on like traits introduced in soybeans and cotton. Producers of HRS wheat and CWRS wheat gain and HRW wheat growers lose. There is an overall price decline, which adversely impacts on HRW wheat because it does not capture any of the yield gains. Consumers in countries and segments allowing GM wheat gain in welfare and those with restrictions, notably the EU and Japan, have reduced welfare. Price differentials in each market and segment and are approximately equal to the cost differentials in production and marketing.

The total welfare gains are much greater than those found by Johnson et al. (2005) and Furtan et al. (2003), but much less than those in Berwald et al. (2006). The most important reasons for these differences include the different degrees of aggregation of wheat classes and countries in which GM is potentially adopted, whether segregation is allowed, the number of non-GM segments considered and associated segregation costs, cost savings and yield gains. The assumptions made by Furtan et al. (2003) and Johnson et al. (2005) have the effect of drastically reducing the welfare gains relative to those presented here.

There are many implications from these results, including for public and private policies and inter-country strategy. Due to the increased productivity and market segments, prices decline. The distribution of welfare gains is neither universal nor symmetric. Producers in regions with greater adoption have a greater gain in welfare than others. There are also differential impacts across consuming countries. Those with large segments which are GM tolerant benefit the most. However, those that restrict GM imports, notably the EU and Japan, suffer, due in part to the higher costs of production and segregation and to a minor extent the geographic shift in procurement.

There are two ways these results may influence public policy. One is to improve, however possible, acceptance of GM wheat. Welfare improvement of full versus segmented market acceptance is about US$234 million. This is sizable and results in improvements for all sectors. Hence, any efforts to improve consumer acceptance would result in further increases in welfare. The second relates to segregation costs. We allow for reasonable costs for segregation in determining our equilibrium results. However, a number of requisites are necessary to achieve these, including the availability of low-cost repeatable tests, certification, and mechanisms to facilitate variety declaration.

There are also several private sector implications that are important. With adoption, growers would be confronted with another production choice that has implications for farm management. These choices would affect marketing decisions and be impacted by price differentials, contracting mechanisms and obligations, and the prospective need to maintain segregations and assure improved variety purity. Handling firms and exporters would compete in this bifurcated market based on segregation costs and risks. Further, non-GM buyers with tight tolerances would be likely to require closer buyer–seller relationships and would use less transactional marketing. Finally, inter-segment competition amongst processors would be intensified with the introduction of GM wheat. Notably, processors of products requiring non-GM or limited amounts of GM for marketing purposes would face greater competition from those that do not, due in part to the lower-cost ingredients available to the latter.

SUMMARY

Wheat has been losing its competitiveness relative to competing crops, particularly those that have access to GM technology, notably corn, soybeans, cotton and canola. Partly in response to this, there has been an increase in grower support for developing GM technology in wheat. Many of the agbiotech companies have responded and are in the process of evaluating and developing traits for this crop. Important in this evaluation is that trait development takes a long period of time, there are many risks associated with development, and it is costly. Against this acceleration of research in wheat-breeding technology, there are a number of important issues.

One is consumer acceptance. Generally, consumers in North America are less averse to GM content than in other countries. In part this is due to greater confidence in the underlying regulatory mechanisms. However, in some other countries, notably the EU and Japan, there is greater aversion to GM content, as reflected partly in their regulatory regimes. Looking forward, one would expect in the future a more highly differentiated market for wheat products encompassing those non-averse to GM content, those averse to GM content and those seeking organic produce. Ultimately this means that commercialization of GM wheat will necessarily require fairly elaborate segregation systems which, while achievable, will need to be imitated by buyers through their contractual requirements. If so, the markets can be effective at segregation and at not unreasonable costs, though the incidence of the costs will vary across market segments and participants.

Another major issue confronting wheat is that most of the germplasm is or has been under control of the public sector. Hence as biotechnology companies seek to expand and pursue 'seeds and traits' strategies, they necessarily need to develop varying forms of public–private partnerships. Indeed, this is what is occurring as the industry evolves. Generally, the germplasm sector is morphing from one based on public ownership, with mechanisms that facilitate multiple reciprocal relationships with minimal financial obligations, into one with many bilateral reciprocal relations with financial obligations. This is a major change and challenge in the wheat sector.

Valuation of future traits is fraught with uncertainty, and is probably more problematic for wheat than other sectors. This was illustrated through the valuation of a drought-tolerant GM wheat and comparison through time and across regions. There are both public and private implications of these results. From a private perspective, the positive option values provide encouragement for further development of this trait. However, the option value is not as great as for other traits in other crops, which means that any variety of wheat would probably have to have a combination of stacked traits to be commercially acceptable. Finally, the fact that most of the big agbiotech companies are working on similar traits is important. Ultimately, competitive pressures will force companies to strive for the greatest trait efficiency as possible, which as illustrated here, has an important impact on trait valuation. These results differ from previous studies. Here the results indicate positive strategic option values of developing GM traits, in this case, drought tolerance, for wheat. This is encouraging such development, but, in the future as these become commercially available, many of the issues related to post-development commercialization may become more salient (for example, as discussed in Carter et al. 2005; Furtan et al., 2003). Strategies for segregating (Wilson and Dahl, 2005) and creation, capture and distribution of the overall welfare gains of GM wheat (Wilson et al., 2008a) will become key drivers.

NOTES

1. There have been numerous presentations to explain the extent of these changes. See in particular, Wilson (2008).
2. The status and outlook in Australia is described in several places, including: Agrifood Awareness, GM Wheat-Fact not Fiction: A 7–10 year program of consultation and collaboration, available at http://www.afaa.com.au/GM_wheat_2010/AFAA_GMWheatBrochure_WEB.pdf; and the most recent application for GM wheat trials in Australia are described at: International Service for the Acquisition of Biotech Applications, 2012, 'Australia's Gene Technology Regulator OKs Trial for GM Wheat and Barley', and, Pocket K No. 38 Biotech Wheat, which is available at http://www.isaaa.org/resources/publications/pocketk/38/default.asp. Finally, it was described previously at: http://www.financialexpress.com/news/

droughttolerant-gm-wheat-under-trial-down-under/329163/ and http://www.newscientist.com/article/mg19826623.500-droughtresistant-wheat-beats-australian-heat.html.
3. Or less, if GM presence is not fortuitous or technically unavoidable.
4. Already in place for all foodstuffs in the EU due to regulation (EC) 178/2002 ('General Food Law').
5. As an example, a salt-tolerant gene using non–genetically modified genes has recently been introduced into durum wheat by CSIRO and the University of Adelaide.
6. Specifically, the functional characteristics of HRS and Candian western red spring wheat (CWRS) differ substantially from Argentine wheat and HRW and hence their substitutability is nearly zero. As examples, water absorption is about 66 per cent for HRS and CWRS vs. 58 per cent for HRW; stability is about 19 minutes vs. 12 minutes for HRW and gluten content is 36 per cent vs. 29 per cent for HRW. These relationships are described in more detail elsewhere (e.g., *Wheat Marketing Center*, 2004). These unique characteristics are important to end-users and allow them to produce products that cannot be produced from lower protein wheat and/or that can blended with lower-protein, locally produced wheat. For these reasons, this class of wheat commands premiums. Allowing substitutability would imply that buyers forgo these functional requirements and ultimately produce other products.

REFERENCES

Azziz, E. (2007), 'The changing face of the U.S. grain system', Washington, DC: USDA, ERS Report No. 35.

Berwald, D., C. Carter and P. Gruere (2006), 'Rejecting new technology: the case of genetically modified wheat', *American Journal of Agricultural Economics*, **88**, 432–47.

Blackshaw, R.E. and K.N. Harker (2002), 'Selective weed control with glyphosate in glyphosate-resistant wheat (Triticum aestivum)', *Weed Technology*, **16**, 885–92.

Brach, M.A. and D.A. Paxson (2001), 'A gene to drug venture: Poisson options analysis', *R&D Management*, **31**(2), 203–14.

Bradford, K.J., J.M. Alston and N. Kalaitzandonakes (2006), 'Regulation of biotechnology for specialty crops', in R.E. Just, J.M. Alston, and D. Zilberman (eds), *Regulating Agricultural Biotechnology: Economics and Policy*, Vol. 30, Boston, MA: Springer US, pp. 683–97.

Carter, C.A., D. Berwald and A. Loyns (2005), 'Economics of genetically-modified wheat', Toronto: University of Toronto, Centre for Public Management.

Dow AgroSciences (n.d.), 'Product Information | Dow AgroSciences', available at: http://www.dowagro.com/stewardship/biotechnology/productprofiles.htm, accessed 27 February 2012.

Fatka, J. (2008), 'Biotech crop advancements key', *Feedstuffs*, **80**(30), 5.

Flagg, I.M. (2008), 'The valuation of agricultural biotechnology: the real options approach', Fargo, ND: North Dakota State University.

Furtan, W.H., R.S. Gray and J.J. Holzman (2003), 'The optimal time to license a biotech "lemon"', *Contemporary Economic Policy*, **21**(4), 433–44.

Golan, E., N. Krissoff, F. Kuchler, L. Calvin, K. Nelson and G. Price (2004), 'Traceability in the US food supply: economic theory and industry studies', Washington, DC: US Department of Agriculture, Agricultural Economics Report no. 830.

Goodman, M. (2004), 'Plant breeding requirements for applied molecular biology', *Crop Science*, **44**(6), 1913–14.

Hobbs, J. (2004), 'Information asymmetry and the role of traceability systems', *Agribusiness: An International Journal*, **20**(4), 397–415.

Holzman, J.J. (2001), 'The economics of herbicide tolerant wheat in Western Canadian crop rotations', master's thesis, Saskatoon, Saskatchewan Canada: University of Saskatchewan.

Huso, S.R. and W.W. Wilson (2005), 'Impacts of genetically modified (GM) traits on conventional technologies', Fargo, ND: Department of Agribusiness & Applied Economics, North Dakota State University, *Agribusiness & Applied Economics Report* no. 560, available at: http://purl.umn.edu/23491.

Huso, S.R. and W.W. Wilson (2006), 'Producer surplus distributions in GM crops: The ignored impacts in Roundup Ready® Wheat', *Journal of Agricultural and Resource Economics*, **31**(2), 339–54.

James, C. (2011), 'Global status of commercialized biotech/GM crops: 2011', *International Service for the Acquisition of Agri-Biotech Applications (ISAAA) Brief*, 43, Ithaca, NY: ISAAA, available at: http://www.isaaa.org/resources/publications/briefs/43/executivesummary/default.asp.

Jensen, K. and P. Warren (2001), 'The use of options theory to value research in the service sector', *R&D Management*, **31**(2), 173–80.

Johnson, D., W. Lin and G. Vocke (2005), 'Economic and welfare impacts of commercializing a herbicide-tolerant, biotech wheat', *Food Policy*, **30**, 162–84.

Just, R.E., J.M. Alston and D. Zilberman (2006), *Regulating Agricultural Biotechnology: Economics and Policy*, Amsterdam: Springer.
Kaehler, B (2006), 'Dow AgroSciences to expand its traits business', available at: http://findarticles.com/p/articles/mi_hb3147/is_4_44/ai_n29269452/, accessed 27 February 2012.
Kalaitzandonakes, N., P. Suntornpithug and P.W.B. Phillips (2004), 'Roundup Ready wheat in Canada and the US: A survey and analysis of potential farmer adoption', paper presented at the 8th ICABR International Conference on Agricultural Biotechnology: International Trade and Domestic Production, Ravello, Italy, 8–11 July.
Kidnie, M., R. Ripley, J. McNulty and R. Neyedley (2003), 'Performance of Roundup Transorb compared to commercial standard herbicides in Roundup Ready wheat', paper presented at the 2003 CWSS Annual Meeting, 3 December.
Luehrman, T.A. (1997), 'What's it worth? A general manager's guide to valuation', *Harvard Business Review*, **75**(3), 132–42.
Milling and Baking News (2004a), 'Monsanto withdraws applications', *Milling and Baking News*, 22 June, p. 8.
Milling and Baking News (2004b), 'Product traceability offers great potential benefits, faces hurdles', *Milling and Baking News*, **83**(7), 20 April, p. 13.
Morris, P.A., E.O. Teisberg and A.L. Kolbe (1991), 'When choosing R&D projects, go with long shots', *Research Technology Management*, **34**(1), 35.
National Association of Wheat Growers (2010), 'Wheat biotechnology commercialization: statement of Canadian, American and Australian Wheat Organizations', first published 14 May 2009, updated 6 December 2010, available at: http://www.wheatworld.org/issues/biotech/, accessed 14 May 2009.
Phillips McDougall (2011), 'The cost and time involved in the discovery, development and authorization of a new plant biotechnology derived trait', Consultancy Study for Crop Life International, Pathhead, Scotland, available at: http://www.croplife.org/PhillipsMcDougallStudy.
Reyes, L.C. (2009), 'Overcoming the toughest stress in rice: drought', International Rice Research Institute, *Rice Today*, **8**(3), July–September, 30–32.
Rice Today (2012a), 'Developing countries continue to embrace GM crops', *Rice Today*, **11**(2), April–June, 7–8.
Rice Today (2012b), 'Rice next to get salt tolerance gene?', *Rice Today*, **11**(2), April–June, 8.
Schwartz, E.S. and L. Trigeorgis (2004), *Real Options and Investment Under Uncertainty: Classical Readings and Recent Contributions*, Cambridge, MA: MIT Press.
Seppä, T.J. and T. Laamanen (2001), 'Valuation of venture capital investments: empirical evidence', *R&D Management*, **31**(2), 215–30.
Shakya, S., W. Wilson and B. Dahl (2012), 'Valuing new random GM traits: the case of drought tolerant wheat', *Agribusiness & Applied Economics Report* no. 691, Fargo, ND: Department of Agribusiness and Applied Economics Agricultural Experiment Station North Dakota State University.
Sindrich, J. (2012), 'Developing economies: new face of the biotech boom', *Farm Chemicals International*, February, available at: http://www.farmchemicalsinternational.com/magazine/?storyid=3428.
The Economist (2011a), 'The adoption of genetically modified crops: growth areas', *The Economist*, 23 February, available at: http://www.economist.com/blogs/dailychart/2011/02/adoption_genetically_modified_crops.
The Economist (2011b), 'Climate change and crop yields: one degree over', *The Economist*, 17 March, accessed at: http://www.economist.com/node/18386161.
Tollefson, J. (2011), 'Drought-tolerant maize gets US debut', *Nature News*, **469** (7329), 144–144.
Valliyodan, B. and H.T. Nguyen (2006), 'Understanding regulatory networks and engineering for enhanced drought tolerance in plants', *Current Opinion in Plant Biology*, **9**(2), 189–95.
Wall Street Journal (2012), 'Western Great Plains growers gearing up to plant Monsanto's new DroughtGard hybrids', *Wall Street Journal*, 21 February, available at: http://online.wsj.com/article/PR-CO-20120221-906840.html.
Wheat Marketing Center (2004), *Wheat and Flour Testing Methods: A Guide to Understanding Wheat and Flour Quality*, Portland, OR: Wheat Marketing Center.
Wilson, W.W. (2008), 'Challenges and strategies for commercializing GM wheat', paper presented at the National Association of Wheat Growers, 5 February, available at: http://www.wheatworld.org/wp-content/uploads/biotech-bill-wilson-presentation-20080205.pdf.
Wilson, W.W. and B.L. Dahl (2005), 'Costs and risks of testing and segregating GM wheat', *Review of Agricultural Economics*, **27**, 1–17.
Wilson, W.W. and B.L. Dahl (2006), 'Costs and risks of segregating GM wheat in Canada', *Canadian Journal of Agricultural Economics*, **54**, 341–59.
Wilson, W.W. and B. Dahl (2010a), 'Competition and dynamics in market structure in corn and soybean seed', *Competition Policy International–Antitrust Chronicle*, **4**(2), Spring, available at: https://www.competitionpolicyinternational.com/competition-and-dynamics-in-market-structure-in-corn-and-soybean-seed/.
Wilson, W.W. and B.L. Dahl (2010b), 'Dynamic changes in market structure and competition in the corn and soybean seed sector', Fargo, ND: North Dakota State University, Department of Agribusiness and Applied

Economics, Agribusiness & Applied Economics Report no. 657, available at: http://ageconsearch.umn.edu/handle/58487.
Wilson, W., B. Dahl and E. Jabs (2007), 'Optimal supplier testing and tolerance strategies for genetically modified (GM) wheat', *Agricultural Economics*, **36**, 39–48.
Wilson, W., X. Henry and B. Dahl (2008b), 'Costs and risks of conforming to EU traceability requirements; the case of hard red spring wheat', *Agribusiness: An International Journal*, **24**, 85–101.
Wilson, W.W., E.L. Janzen and B.L. Dahl (2003), 'Issues in development and adoption of genetically modified (GM) wheats', *AgBioforum*, **6**(3), 1–12.
Wilson, W.W., E.A. DeVuyst, R.D. Taylor, W.W. Koo and B.L. Dahl (2008a), 'Implications of biotech traits with segregation costs and market segments: the case of Roundup Ready® Wheat', *European Review of Agricultural Economics*, **35**(1), 51–73.
Xia, L., Y. Ma, Y. He and H.D. Jones (2012), 'GM wheat development in China: current status and challenges to commercialization', *Journal of Experimental Botany*, **63**(5), 1785–90.

47 Small grains: barley, oat and rye

Syed Masood H. Rizvi and Graham J. Scoles

1 INTRODUCTION

1.1 Contribution to Agriculture: Overview

The small-grain cereals are temperate annual grasses cultivated primarily for their grains. Barley (*Hordeum vulgare* L.), oat (*Avena sativa* L.) and rye (*Secale cereale* L.) are members of the sub-family Pooideae within the monocot grass family Poaceae. Phylogenetic studies have estimated that these grasses (along with wheat) diverged from a common ancestor (Gaut, 2002; Kellogg, 2001; Zohary and Hopf, 2000), although barley and rye (tribe Triticeae) are more closely related to each other than to oat (tribe Aveneae). All three crops are characterized by large genome sizes (compared to maize and rice) – 5500 Mb (Mega base pair) for barley, 8000 Mb for rye and 11 300 Mb for oat (Bennett and Smith, 1976). To date, no genetically modified lines of these three crops have been commercialized although biotechnology tools have been implemented into breeding programs of these crops to various degrees through molecular marker assisted selection (MMAS).

Barley (*Hordeum vulgare* L.) is an important cereal crop species, ranking after maize, wheat, rice and soybean in terms of area harvested worldwide (FAO, 2009, http://www.faostat.fao.org). The barley genome ($2n = 2x = 7$), comprising more than 5000 Mb, equals approximately 12 times the size of the rice genome and consists of about 80 per cent repetitive DNA (Flavell et al., 1974). Most oat cultivars belong to *Avena sativa* L., an allohexaploid species with $2n = 6x = 42$. Like barley, rye (*Secale cereale* L.) is also a diploid and also an important cereal crop in many countries around the world. More than 50 per cent of the annual rye harvest is used for making bread. Despite its relatively low production compared to other cereals, rye is of great importance due to its broad tolerance to biotic and abiotic stress, which is lacking in other temperate cereals. Therefore, rye remains an important grain crop species for cool temperate zones. This adaptation has also made it a valuable source of alien genes for wheat (*Triticum aestivum* L.) improvement. In addition, rye is important as the R-genome donor to triticale; the short arm of rye chromosome 1 (1RS) has been introgressed into more than 250 wheat cultivars (Schlegel, 2006). Many successful wheat cultivars carry the 1BL/1RS translocation as the presence of 1RS in the wheat genome increases both yield and protein content in grains. These characteristics make the study of the genetics of rye of considerable interest and importance.

1.2 History and Domestication

Barley was one of the first and earliest crops to diffuse from the Neolithic farming communities of the Middle East. The earliest remains of barley have been recovered alongside einkorn and emmer wheat at Neolithic sites in the so-called Fertile Crescent, dated about 8500 years BC. Barley was considered as a traditional staple food of Europe and Western Asia and is the principal crop that inspired Western agriculture in the Fertile Crescent (Salamini et al., 2002; Smith, 1998). The natural form of barley grows wild in the Middle East and central Asia; the main difference between domestic and wild barley (*H. vulgare ssp. spontaneum*) is that the domesticated form has non-brittle ears (Morrell and Clegg, 2007). When ripe, the undomesticated form of barley becomes brittle and shatters, allowing dispersion of its grains. An important step in the development of domesticated crops was the selection of mutant forms with larger grains and a tendency for the grain-containing spikelets to stay attached to the plant until harvested (Brown et al., 2009; Salamini et al., 2002). Because of the loss of their grain dispersal mechanism, all domesticated small grains have limited capacity to survive in the wild. However, wild ancestors of today's cultivated crops are still widespread in their areas of origin, can be crossed with their domesticated relatives without difficulty (Zohary and Hopf, 2000) and have been widely used as sources of valuable genes in plant breeding programs.

Whereas barley was domesticated directly as a crop, both oat and rye are thought to have been domesticated much later and indirectly. They probably occurred as weeds in early wheat and barley crops but in certain environments came to dominate local cropping systems, becoming crops in their own right (Zohary and Hopf, 2000). The non-shattering domesticated form of oat has only been cultivated for about 2000 years. It arose in Europe/Asia Minor and was probably used primarily to feed livestock. Eventually it gained favor as food for humans and became popular in northern Europe. Today, the majority of oat production occurs in North America and Europe. Like oat and barley, wild relatives of rye also grow in central and eastern Turkey and adjacent areas. Unlike oat and barley, rye is a cross-pollinating cereal. It was domesticated relatively late and has been found in small quantities at a number of Neolithic sites in Turkey, but is generally virtually absent from the archaeological record until the Bronze Age of central Europe, c. 1800–1500 BC. It is possible that rye traveled west from Turkey as a minor admixture of wheat and barley. Since the middle ages, rye has been widely cultivated in Central and Eastern Europe and is the main bread cereal in Poland, Germany, western Russia, Belarus and the Ukraine (Zohary and Hopf, 2000). Research into the evolution of small grains, particularly the transition from wild to domesticated forms of barley, oat and rye remains an area of great interest to anthropologists, evolutionary biologists and crop scientists (Brown et al., 2009).

Compared to the three major cereals used for human and/or animal food (wheat, maize and rice), barley, oat and rye are relatively minor crops on a world scale (see Table 47.1). In recent years the area devoted to all three crops has declined. In addition to being harvested for their grain, all three crops are also grown to a lesser extent as either green feed or hay for animals.

1.3 Utilization

Most barley is now grown either to produce malt for beer production or for animal feed, with small but slowly increasing amounts being used for direct human consumption (in areas such as Tibet, barley is a major component of the diet). To produce malt the barley

Table 47.1 Global cereals production

Cereals	Area harvested (1000 ha)	Yield (t/ha)	Production (1000 t)
Wheat *	225622	3.03	685614
Maize	158628	5.16	818823
Rice	158300	4.32	685240
Barley *	54059	2.81	152125
Sorghum	39969	1.40	56098
Millet	33692	0.79	26702
Oat *	10212	2.27	23258
Rye *	6559	2.76	18168
Triticale	4279	3.66	15669

Note: * Considered as small grain.

Source: FAOSTAT (2011) (table sorted on percentage of global area harvested).

grain is steeped in water under controlled conditions, allowing it to germinate or sprout, at which stage it is known as green malt. It is then dried or roasted in a kiln, cleaned and can then be stored for extended periods. By-products of the malting process such as screenings and sprouts are separated and used as animal feed. Malt is primarily an intermediate product and requires further processing. Most malt production is used to produce beer, the malt acting as a source of enzymes capable of breaking down starch to sugars which can then be fermented. Malt is also used by the distilling industry in the production of whisky and by the food industry in the production of some cake mixes and breads. Most barley used for food is either pearl barley or barley flour (with the exception of hulless barley, the majority of barley is surrounded by a fibrous hull or husk). Pearling consists of a polishing process, which removes the outer husk and part of the bran layer of the grain. Barley flour, a by-product of pearling, is used in North America for baby foods and other specialties. In addition to traditional 2-row and 6-row barley, hull-less barley cultivars have been developed for food application, where a minimal amount of cleaning is required prior to processing. Barley has nutritional advantages such as high soluble fiber (primarily beta-glucan) content, as well as being low in fat, thus having a low glycemic index. A number of current and ongoing research studies are concentrating on the potential health benefits of barley. Initial results demonstrating lowering of blood cholesterol are promising, but additional studies are necessary to confirm results.

Oat is used directly as livestock feed, primarily in horse and ruminant nutrition. Dehulled oat is also well suited for human nutrition and used in a variety of food products like oatmeal and breakfast cereals. Oat contains more soluble fiber than any other cereal, contributing to its high value in human diets. As with barley, high levels of beta-glucan and non-digestible polysaccharides (primarily located in the endosperm cell wall) have been found to have cholesterol-lowering properties.

Rye is the most hardy winter cereal (able to withstand temperatures down to −20°C). As mentioned earlier, rye bread is still a major human food in some parts of the world and also contains high levels of soluble fiber in the form of arabinoxylans. Rye can be used as animal feed and is also used for distillation into grain alcohol spirits, such as whisky.

2 GENOMIC STUDIES

Large-scale sequencing programs for the development of expressed sequence tags (ESTs) from various cDNA libraries have been initiated in the last 10 years. Major progress made in recent years has resulted in the generation of 556070 ESTs for barley, 25762 ESTs for oat and 10369 for rye, covering different cDNA libraries from various stages of plant development (http://www.ncbi.nlm.nih.gov/dbEST; 2011). As sequencing and other DNA technologies advance, the amount of genomic data for all species is growing almost exponentially.

2.1 Barley

Due to its importance as a staple crop, its diploidy and its inbreeding habit, barley initially served as a model species for genomic studies in the Triticeae, including wheat (*Triticum aestivum* L.) and rye (*Secale cereale* L.), and comprehensive genetic and genomic resources were established for barley prior to wheat and rye (Sreenivasulu et al., 2008b). These include a large number of genetic stocks and mutant collections (Bovina et al., 2011; Caldwell et al., 2004; Lundqvist et al., 1996), various genetic linkage maps (Sreenivasulu et al., 2008a; Varshney et al., 2004), large insert bacterial artificial chromosome (BAC) libraries (Isidore et al., 2005; Schulte et al., 2011), and a large collection of expressed sequence tag (EST) presently comprising more than 556000 entries in dbEST (expressed sequence tag database).

Several approaches have been pursued for detecting sequence polymorphisms in barley, relying initially on hybridization (Restriction Fragment Length Polymorphisms (RFLPs); for example, Graner et al., 1991), or PCR-based molecular marker systems like RAPD (Randomly Amplified Polymorphic DNA; for example, Weyen et al. 1996), simple sequence repeats (SSRs or microsatellites; for example, Pillen et al., 2000), amplified fragment length polymorphisms (AFLPs; for example, Waugh et al., 1997), and most recently, single nucleotide polymorphisms (SNPs; for example, Kota et al., 2008).

Traditionally barley linkage maps were constructed using RFLP or SSR markers (Kleinhofs et al., 1993; Ramsay et al., 2000; Varshney et al., 2007). These were seldom gene-based (with the exception of cDNA probes and EST-based SSRs). Unfortunately, both of these marker types suffer from the fact that they are implemented using serial assays (that is, they are low throughput) that generally depend upon the availability of a specialized laboratory. The high-throughput second-generation unified genotyping platform became a high priority recently for barley genomics because it was widely recognized that such a resource would enable the investigation of novel approaches for trait analysis such as association mapping. A dedicated platform for genotyping single nucleotide polymorphisms (SNPs) based on Beadarray technology was developed and commercialized by Illumina (Fan et al., 2003). A SNP-based integrated barley linkage map was developed by sequencing 1338 barley unigenes selected for association with various abiotic stresses in eight barley genotypes (Kota et al., 2001; Rostoks et al., 2005). The SNP frequency observed in this study was 1/200 bp, which is similar to that observed by others using different sets of germplasm and across different loci (Bundock et al., 2003; Bundock and Henry 2004; Kanazin et al., 2002; Kota et al., 2001). Computational algorithms have been developed for querying EST databases for the presence of SNPs (Kota et al., 2003), facilitating the systematic development of SNP markers, for which numerous assays have been developed (Rafalski, 2002). High density genetic maps of gene-based markers represent a powerful resource for enhanced genome analysis. They are essential for linking genetic and physical mapping information and allow for a detailed comparative genome analysis across both closely related and distantly related grass species. Moreover, gene-based markers, also termed 'functional markers', can be regarded as candidate genes in trait-mapping experiments. As a first step towards a comprehensive transcript map of barley, more than 330 EST-derived SNP markers were placed on a consensus map derived from three mapping populations (Rostoks et al., 2005). SNPs are the most abundant molecular markers identified in barley (Brookes, 1999). SNP discovery relies on the availability of sequencing data. In barley, SNP discovery has been achieved by searching assemblies of ESTs (Kota et al., 2003), sequencing selected sets of unigenes in several barley accessions (Rostoks et al., 2005), or exploiting the Affymetrix Barley1 GeneChip (Close et al., 2004).

In barley, 4596 SNPs have been arranged in three pilot GoldenGate assays. Barley oligonucleotide pool assays 1 and 2 (BOPA1 and BOPA2) were used to develop a consensus genetic linkage map composed of 2943 SNPs from Steptoe × Morex, OWB, Haruna Nijo × OHU602, and Morex × Barke doubled-haploid mapping populations (Close et al., 2009). The SNP-derived genetic map is available online (http://www.harvest-web.org). Other maps exploiting the Illumina GoldenGate SNP mapping tools and combining other molecular markers, particularly ESTs, are published (Table 47.2).

Table 47.2 Available barley maps

Reference	Marker	Loci
Rostoks et al. (2005)	SNP	1237
Wenzl et al. (2006)	DArT	2935
Varshney et al. (2007)	SSR	775
Marcel et al. (2007)	Multiple	3258
Stein et al. (2007)	EST	1255
Hearnden et al. (2007)	Multiple	1000
Potokina et al. (2008)	TDM	1596
Szucs et al. (2009)	Multiple	2383
Sato et al. (2009)	EST	2890
Sato and Takeda (2009)	SNP	2890
Close et al. (2009)	SNP	2943
Aghnoum et al. (2010)	Multiple	6990
Stein et al. (2011 unpublished data)	Multiple	1794

Notes: SNP (single nucleotide polymorphism); DArT (diversity array technology); SSR (simple sequence repeat); EST (expressed sequence tag); TDM (transcript derived markers).

2.2 Oat

The Quaker Oat Company Inc. supported the Oat Genomics Research Consortium consistently from 1985 to 2000, when oat was considered as an orphan crop for research and development. Modern genomics approaches in oat were first used in 1992 with the publication of the first RFLP map in diploid oat (O'Donoughue et al., 1992). This was followed by original and updated versions of hexaploid maps based on the 'Kanota' × 'Ogle' (K×O) recombinant inbred line (RIL) population (Wight et al., 2003) and by the addition of new sets of mapped markers (Locatelli et al., 2006; Orr and Molnar, 2007). Many additional maps, both partial and complete, have been published in hexaploid oat, as reviewed by Rines et al. (2006) and compiled in an online database (http://avena.agr.gc.ca/oatgenes/). However, most maps contain very few markers that are shared among other populations. The efforts in oat have not yet produced a true consensus map in which all linkage groups are assigned to the expected 21 oat chromosomes (Tinker et al., 2009). The current trend of molecular markers in oat is based on technologies that include Sequence Characterized Amplified Region (SCAR) SSR, AFLP and RFLP. Unfortunately, this diversity of technologies creates difficulties in performing comparative genomics within the oat community. The above factors highlight the urgent need for a set of molecular markers that provide complete genome coverage, that are based on a homogeneous technology and that can be scored readily in new germplasm by any member of the global oat research community. Furthermore, there is mounting evidence that whole genome association studies can yield informative results in an inbreeding species such as wheat, and this strategy has shown good potential in oat (Achleitner et al., 2008). There are only a few hundred microsatellite markers available for oat (Becher, 2007; Holland et al., 2001; Jannink and Gardner, 2005; Li et al., 2000; Pal et al., 2002; Zhu and Kaeppler, 2003), and only a few of them are integrated in the map of the oat reference population 'Kanota' × 'Ogle' (Pal et al., 2002; Wight et al., 2003).

In a recent attempt approximately 19000 genomic clones were isolated from complexity-reduced genomic representations of pooled DNA samples from 60 oat cultivars of global origin (Tinker et al., 2009) and screened on three discovery arrays, with more than 2000 polymorphic DArT markers being identified. DNA sequence was obtained for 2573 clones and assembled into a non-redundant set of 1770 contigs and singletons.

2.3 Rye

Despite the economic importance of rye, little is known about its genetic make-up at the DNA sequence level. The 1C DNA content of the rye genome is 9.5 pg (Bennett and Smith, 1976). Only 10–20 per cent of the genome can be assigned biochemically to the major part of the genome which belongs to the repeated sequence category (Flavell et al., 1977; Ranjekar et al., 1974). The kinetic analysis of genome organization has revealed that repeated sequences are, in general, interspersed among unrepeated sequences. The discovery of a very rapidly re-annealing class of DNA sequence provided useful information about specific regions of the genome. In rye this class of DNA constitutes 4–10 per cent of the genome and, although it is believed to be composed largely of sequences capable of renaturation, this class also contains long tandem arrays of simple, repeated sequences (Appels et al., 1978, 1982). Ranjekar et al. (1974) were the first to demonstrate several buoyant density components in a fraction of DNA restoration with a density of 10–12 per cent of the genome. This DNA was shown to contain several families of highly repeated sequence DNA. Two of them were purified, which resulted in a fraction restoration to a density of 1.701 g/cc and comprised 0.1 per cent of the total genome, and the other polypyrimidine tract DNA which comprised 0.1 per cent of the genome. Further hybridization studies between wheat, rye, barley and oat DNAs have shown that 22 per cent of rye DNA consists of species-specific repeated sequences (Rimpau et al., 1978) that have probably arisen by the amplification of single copy DNA since species divergence (Flavell, 1982).

At the time of writing there is no ongoing sequencing project in rye, and there are no plans to target gene-rich fractions of its genome. Rye is underrepresented in the sequence databases compared to wheat and barley, for which 1104K and 556K sequences respectively are deposited in GenBank. There are only 10000 rye sequences available, of which about 90 per cent are expressed sequence tags (ESTs). Molecular-marker technology has resulted in the creation of linkage maps in rye. Restriction fragment length polymorphisms (RFLPs) have been used to establish linkage maps because of their moderate level of polymorphism. In addition to RFLPs, PCR-based markers have also been used to improve mapping in cereals. Based on these two techniques, a series of rye genetic maps have been established (Korzun and Kunzel, 1996; Korzun et al., 1999). But many of the rye maps do not contain linkage maps of all seven chromosomes. The one rich map is that of Devos et al. (1993), which consists of 156 loci spanning about 1000 cM and covers all seven rye chromosomes. This map gives the most detailed description of rye chromosomes relative to their wheat homologues, but contains no rye genomic or cDNA markers.

The Philipp et al. (1994) map consists of 60 loci, in which most of the markers are rye genomic clones. The Loarce et al. (1996) map consists of 89 loci spanning 339.7 cM on all the rye chromosomes, except for 2R. The Senft and Wricke (1996) map consists of 127 loci, of which 15 showed dominant segregation, spanning about 760 cM. The Korzun et al. (1998) map is a joint map derived from two reciprocal crosses, spanning 660 cM consisting of 91 loci, of which 82 markers are integrated into previous rye, wheat, barley or oat maps, allowing for good comparisons between the different maps. The Wanous and Gustafson (1995) and Wanous et al. (1995) maps, which contain 68 markers covering about 702 cM on rye chromosomes 1R, 6R and 7R are the only maps to include cytological markers. As the rye genome comprises about 92 per cent repetitive sequences, rye genomic markers are capable of detecting multiple loci and, therefore, can appear non-syntenic in different populations (Gale, 1990; Harcourt and Gale, 1991; Devos et al., 1992). RFLP-based linkage mapping has revealed a non-uniform distribution of mapped loci in plants. The presence of intra- and inter-chromosomal duplications has also been observed through rye RFLP mapping (Liu et al., 1992).

Presently, 2050 biochemical, molecular and morphological markers are available in a database, out of which about 80 per cent are molecular markers gathered during the last decade, 12 per cent biochemical markers, and about 8 per cent mapped morphological features. Best investigated is chromosome 1R, followed by chromosomes 6R, 5R, 2R, 4R, 7R and 3R. An updated list of rye genes, markers and linkage data was created by Schlegel and Korzun (2013). The lack of sequence information is a major limitation for marker development and gene cloning in this species. End sequencing of BAC clones generates random sequence information distributed across the whole genome. Kelley et al. (1999) developed a protocol for high-throughput BAC end sequence (BES) generation using automated sequencers. This protocol is now routine in large sequencing centers, reducing cost and enabling the creation of large data sets (for example Paux et al., 2006) and could be applied to rye.

3 EXPERIENCE WITH TRANSFORMATION OF SMALL GRAINS

Like all grasses, barley, oat and rye are not as easy to manipulate in tissue culture as some other species. Tissue culture is a prerequisite for all genetic engineering as it is necessary to induce plant formation from individual cells into which DNA has been inserted. This factor, along with the fact that they are relatively minor crops, the significant costs of bringing transformed crops to the market and persistent concerns regarding the use of genetic engineering, has meant that to date genetic engineering has had no commercial impact on these three crops. Nevertheless, all three crops have been successfully transformed and in some cases experimental lines have been subjected to field trials.

3.1 Barley

Transformation of barley began in the early 1990s with the success of particle bombardment of the cultivar Golden Promise (see review by Lemaux et al., 1999). Since then, many reports of barley transformation using particle bombardment and *Agrobacterium* have been published. Cultivars such as Golden Promise, a spring barley, and Igri, a winter barley, have been popular choices for transformation because of the ease with which both cultivars form callus and regenerate plants. Initial research concentrated on developing transformation protocols using screenable marker genes such as GUS and selectable marker genes for antibiotic and herbicide tolerance. Various promoters such as those for the CaMV 35S, the maize ubiquitin and the rice actin genes were also tested. Once protocols were optimized, several agronomic and malting characteristics were successfully engineered into barley. In the last decade, significant advances were made in barley transformation (Dahleen and Manoharan, 2007).

Barley has been successfully transformed through various methods such as DNA uptake (Lazzeri et al., 1991; Funatsuki et al., 1995; Kihara et al., 1998; Nobre et al., 2000), electroporation (Salmenkallio-Marttila et al., 1995), microinjection (Holm et al., 2000), particle bombardment (Ritala et al., 1993, 1994; Jahne et al., 1994; Wan and Lemaux 1994; Hagio et al., 1995; Jensen et al., 1996; Koprek et al., 1996; Cho et al., 1998; Zhang et al., 2000; Manoharan and Dahleen, 2002) and *Agrobacterium tumefaciens* (Tingay et al., 1997; Horvath et al., 2000; Wang et al., 2001; Murray et al., 2004). Of all these methods, particle bombardment remains a favorite choice for barley transformation because of its ability to transform a wide variety of cultivars. Although *A. tumefaciens* transformation has been successfully demonstrated in barley, its success is limited mostly to cultivars such as Golden Promise that are easily re-generable. A substantial advantage of *Agrobacterium*-mediated transformation (Karami et al., 2009) compared to microprojectile bombardment is the possibility of transferring large segments of DNA with only minimal rearrangement (Hiei et al., 1997; Shibata and Liu, 2000). Other advantages include low copy number integration, higher transformation efficiency and a higher percentage of stable T-DNA inherited as a simple Mendelian trait. On the other hand, transgene silencing and rearrangements have been frequently observed in transformants produced by direct DNA delivery (Travella et al., 2005).

The main goal of barley transformation has been improvement of its quality traits by expression of new genes. The transformation procedure has been utilized in research focusing on improving characteristics such as malting quality (Wang et al., 2000, 2001; Manoharan et al., 2006), disease resistance (Nuutila et al., 1999; Kihara et al., 2000; Tull et al., 2003), amino acid composition (Hansen et al., 2007; Lange et al., 2007) and quality of grains used as feed (Xue et al., 2003). Interestingly, barley grains were chosen as a bioreactor for molecular farming (Schünmann et al., 2002; Joensuu et al., 2006). Companies like ORF Genetics (Reykjavik, Iceland) and Maltagen Forschung (Andernach, Germany) have started to produce pharmaceutical proteins (growth factors, cytokines, oral vaccines, food additives) in transgenic barley lines with robust endosperm-specific expression. In recent years, considerable progress in barley transformation has been achieved.

During the last decade, systematic efforts were made to improve grain quality traits, including those related to malting (reviewed in Von Wettstein, 2007). Malting improvement has been addressed by altering the expression of hydrolytic enzymes related to the degradation of storage products such as starch (α and β-amylases, Scheidig et al., 2002) and cell wall components. Since then, substantial efforts to improve methods and to insert genes of interest into barley have been made with varied success (for reviews, see Dahleen et al., 2001; Dahleen and Manoharan, 2007; Goedeke et al., 2007; Ganeshan et al., 2003; Harwood and Smedley, 2009; Harwood et al., 2009). Currently, no commercial transgenic barley cultivars are available because of government and industry policies and the sociological controversy of issues such as containment of transgenes and potential health effects. But transformation is a valuable research tool to determine gene effects and interactions.

3.2 Oat and Rye

Like barley, some (but much less) experimental work has been performed on oat and rye, with transformation of both of these crops. In both cases, transformants have been successfully produced although, unlike barley, these experiments have primarily been 'proof-of-concept' experiments rather than having the aim of producing transgenic plants with potentially useful traits for either greenhouse or field testing. As with barley, transformation initially was achieved through micro-bombardment of various tissues and more recently through *Agrobacterium*-based techniques. For reviews see: Somers, 1999; Gasparis et al., 2008; Popelka and Alpeter, 2003.

3.2.1 Oat

Genetic transformation of oat callus and regeneration of fertile, transgenic plants was first reported by Somers et al. (1992). Selection of transgenic oat was achieved later using an antibiotic marker (Torbert et al., 1995) and direct selection for expression of green fluorescent protein (GFP) with an epifluorescent microscope and no herbicide or antibiotic selection (Kaeppler et al., 2000). The ability to monitor the fluorescent phenotype of GFP during each stage of regeneration permitted the calculation of regeneration efficiencies and revealed the limiting steps in the protocol. Transgenics have been generated from both the leaf within the coleoptile (Torbert et al., 1998c) and from the mature leaf (Gless et al., 1998). Originally, this technique was limited to tissue culture amenable lines. However, it has since been expanded to include elite cultivar transformation (Torbert et al., 1998b; Zhang et al., 1999). Thus far oat has been transformed only via the biolistic method. Its genetic transformation was accomplished using embryogenic calli or cell suspensions derived from immature embryos (Kuai et al., 2001; Pawlowski and Somers, 1998; Perret et al., 2003; Somers et al., 1992; Torbert et al., 1995, 1998a), mature embryos (Kaeppler et al., 2000; Torbert et al., 1998b), leaf bases of young oat grainlings (Gless et al., 1998), highly regenerative cultures from grains (Cho et al., 1999) and shoot apical meristems (Cho et al., 1999; Maqbool et al., 2002; Zhang et al., 1999). However, transgenic plant production has been restricted to a small number of spring oat genotypes. Cultivar specificity of transformation is a severe obstacle to improvement of elite oat cultivars, and so far most of the plants produced with reasonably high efficiencies from embryo-derived callus cultures were from a non-commercial experimental genotype, GAF/Park (Torbert et al., 1995; McGrath et al., 1997; Torbert et al., 1998a, 1998c; Koev et al., 1998; Pawlowski et al., 1998; Cho et al., 1999), with the exception of a spring cutivar Belle (Torbert et al., 1998b). Other commercial spring cultivars have been successfully transformed using shoot meristem cultures (Zhang et al., 1999) or leaf base segments (Gless et al., 1998), target tissues which may be more amenable to transformation. However, transformation of winter and naked oat cultivars has yet to be reported.

3.2.2 Rye

Rye is known to be highly recalcitrant to tissue culture, making somatic cell regeneration difficult (Rybczyński and Zduńczyk, 1986; Rybczyński, 1990). De la Pena et al. (1987) reported a method for rye transformation without single cell culture and regeneration, by injecting DNA directly into immature floral tissues; however, this method was not substantiated. Castillo et al. (1994) used particle bombardment on embryonic callus tissue from immature embryos to create a few stable transformations. Rakoczy-Trojanowska and Malepszy (1995) determined that regeneration of rye *in vitro* is controlled by recessive genes. The recalcitrance of *Secale* sp. to somatic cell regeneration, which has restricted the development of transgenic rye using *Agrobacterium tumefaciens*, seems to have been surmounted. Herzfeld (2002) described a genetic transformation protocol for rye involving biolistic and *Agrobacterium*-mediated gene transfer. Popelka and Altpeter (2003), Popelka et al. (2003) and Altpeter (2006), using a time-critical strategy for embryo culture and avoiding the stumbling block of herbicide selection in the medium, succeeded in producing a high number of single-copy transgene inserts. The main hurdle for the establishment of a reproducible transformation protocol in rye is the non-availability of an efficient *in vitro* culture system, a basic requirement for genetic transformation.

4 MOLECULAR MARKER ASSISTED SELECTION (MMAS)

The first molecular markers to be employed in the development of molecular maps were RFLPs. Using sequences derived from either genomic DNA libraries or cDNA libraries as a source of DNA clones, mapping of the segregation of these clones within segregating populations allowed the development of the first molecular marker-based maps. If traits of interest were also segregating in those populations, and if the molecular map was sufficiently dense, then it was possible to identify an RFLP marker that segregated with that trait and could thus be used to select for that trait. Michelmore et al. (1991) developed the technique of bulked segregant analysis utilizing the polymerase chain reaction (PCR) which allowed molecular markers for phenotypic traits to be developed without the need of a map.

4.1 Barley

Almost two decades ago, RFLP markers were employed to develop the first comprehensive molecular marker maps in barley (Kleinhofs et al. 1993; Heun et al., 1991; Graner et al., 1991). Using those RFLP maps, a series of agronomic traits and characters including many quality traits and resistance against several diseases have been mapped (Friedt and Ordon, 2008; Hayes et al., 2003). Later, the availability of large numbers of ESTs facilitated the systematic development of functional markers, for example, by extracting ESTs containing simple sequence repeat (SSR) motifs using appropriate software tools (Thiel et al., 2003). Although EST-based SSR markers have been shown to be less polymorphic than their genomic counterparts, this drawback is more than compensated for by the ease of their development. Also, the availability of ESTs from multiple-genotypes/cultivars of barley provides the possibility to identify sequence polymorphisms (mainly single nucleotide polymorphisms and small InDels) in the corresponding EST alignments. These in turn can be exploited for the development of markers (Kota et. al., 2001, 2008). Kota et al. (2003) developed the computer algorithm SNiPping for discovery of functional markers through browsing EST assemblies in barley. Also an SNP2CAPS program has been published to facilitate the computational conversion of SNP markers into CAPS markers (Thiel et al., 2004). Information generated from the diverse mapping projects was further enhanced by the development of consensus maps (Karakousis et al., 2003; Varshney et al., 2007; Diab, 2006). These provide integrative genetic information by featuring high marker densities. Although the gel-based genotyping platforms offer the best quality marker systems, their low throughput encouraged researchers to explore high-throughput technologies that can simultaneously assay thousands of markers based on single nucleotide polymorphisms (SNP). Most recently, genome-wide scans using SNP-based genotyping platforms such as Illumina, GoldenGate and BeadArrays (Rostoks et al., 2006) and the diversity arrays technology (DArT), which do not require any sequence information (Wenzl et al., 2004), have been successfully established in barley. Although DArTs are not systematically interrogating expressed sequences, the choice of appropriate enzymes facilitates their enriched representation. Based on DArT technology, a high-density consensus map has recently been established (Wenzl et al., 2006). A number of recent studies also reported the use of the affymetrix Barley 1 GeneChip (Close et al., 2004) for identifying single-feature polymorphisms (SFPs), which cover not only SNPs but also indels and polymorphisms generated due to alternative splicing and polyadenylation (Rostocks et al., 2005).

The functional markers discussed above have found application in molecular marker-assisted selection (MMAS). MMAS is based on linking the DNA polymorphisms revealed by marker analysis with agronomical traits allowing for their rapid selection in routine breeding programs. MMAS can be performed at juvenile growth stages and before flowering and thus provides breeders with the opportunity to implement faster back-crossing strategies and allele enrichment in complex crosses, which eventually reduces the time and costs required for the development of improved cultivars. Despite its inherent advantages, the application of MAS in barley up to now has mainly been restricted to monogenic traits such as disease resistances. Here, one of the most widespread examples is the marker-assisted selection of the *rym4* gene giving resistance to the barley yellow mosaic virus complex. For this gene, several closely linked and easily scorable markers have been developed (Graner et al., 1999). More recently, cloning of the gene facilitated the exploitation of functional polymorphisms within the coding region of the resistance gene to differentiate between alleles (Stein et al., 2005). Using MAS, several genes providing full resistance could be readily combined in complex crosses without time-consuming progeny tests in the greenhouse or in the field. MMAS for quantitative traits suffers from two major limitations.

First, compared to monogenic traits, quantitative traits are characterized by lower heritabilities, impairing their accurate scoring and entailing a less accurately defined genetic position of the corresponding quantitative trait locus (QTL). As a result, a large chromosomal fragment needs to be selected for, resulting in the meiotic transfer of many potentially undesired genes. Meiotic purification of a QTL into a 'mendelian' locus, showing monogenic inheritance, provides a solution to this problem. The feasibility of downtracking a QTL to a single gene has been initially demonstrated in tomato, and requires the stepwise size reduction of a QTL fragment and its conversion into a near isogenic line by repeated backcrossing (for review see Salvi and Tuberosa, 2005). In barley, this approach has been successfully employed to isolate the *bot1* gene underlying a major QTL conferring boron tolerance (Sutton et al., 2007).

Secondly, many QTL alleles escape detection when transferred into a different genetic background. The reasons for the 'disappearance of QTLs' include epistatic interactions, QTL × environment effects, the allelic states of the parental lines or the small contribution of a single QTL to the overall variance. As a result, only few common QTLs were detected when the results of mapping studies that were performed in different crosses were compared (Rae et al., 2007). Although the number of successful examples for applying MAS in barley breeding is still rather limited, the recent implementation of high-throughput genotyping platforms (Illumina, DArT, and SNP identification by using Barley 1 GeneChip affymetrix array) in barley will significantly increase the identification of marker trait associations, and the subsequent identification of potential candidate genes. Finally, this will allow QTLs to be treated as monogenic traits and thus spurring their marker-assisted manipulation in breeding programs. In combination with a wide range of mapping populations developed for specific agronomic traits, this comprehensive resource of markers now allows the identification of polymorphisms in functionally defined sequences (Rostoks et al., 2005). Functional markers will also be useful for (i) association studies based on linkage disequilibrium; (ii) detection of *cis* and *trans*-acting regulators either based on genetical genomics studies using well-defined mapping populations or by investigating allelic imbalance (Doss et al. 2005); (iii) identification of alleles influencing agronomically important traits using TILLING/EcoTilling approaches (EcoTilling is a means to determine the extent of natural variation in selected genes); and (iv) genomics-assisted breeding.

4.2 Oat

The development of molecular maps (and thus MMAS) in oat occurred after such developments in barley. Unlike barley, oat is a hexaploid crop and this, combined with numerous translocations that appear to have occurred during oat domestication and cultivar development, made the development of a good molecular map of oat more difficult. Nevertheless MMAS is being applied for a few monogenic traits in some oat programs.

4.3 Rye

As previously mentioned, compared to barley and oat, the rye EST database is small. NCBI/GenBank listed 10369 ESTs of rye as of January 2012 compared to more than 1 million for wheat and more than half a million for barley. The Dana-Farber Cancer Institute (DFCI) Gene Index Project, which integrates research data from international EST sequencing and gene research projects, includes a Rye Gene Index (RyeGI) for *S. cereale* but presumably would include all rye sequences submitted. The DFCI RyeGI attempts to amass non-redundant sequences of all rye genes and data on their expression patterns, cellular roles, functions and evolutionary relationships. The total number of unique sequences for rye including singleton EST and tentative consensus sequences was 5587, which is the smallest number of sequences among all species included in the Gene Index Project (DFCI, 2008). Gene sequences of rye such as 5S ribosomal gene (Reddy and Appels, 1989) and the o-secalins gene (Hull et al., 1991) are also available. In addition, 661 gene sequences of rye can be obtained from NCBI. Ŝimková et al. (2008) constructed a deep coverage BAC library specific for the short arm of chromosome 1R, which will facilitate 1RS – wheat breeding. A BAC library of *S. cereale* cv. Blanco has been created that represents a valuable resource for *Secale* molecular studies. The library is composed of a 6 × genome coverage of S. cereale, which has a 8.1 Gb genome. The BAC clones have an average insert size of 131 kb, and the library is composed of ~2 per cent of empty or organelle derived clones. Shi et al. (2009) analyzed several BAC clones containing allelic forms of genes at or near the Alt4 locus, which revealed heterozygosity within cv. Blanco.

5 COMPARATIVE GENOMICS BETWEEN SMALL GRAIN AND OTHER CEREALS

In the early phase of genomics, comparative genomics was a science that exploited the results of comparative genetic mapping (Gale and Devos, 1998). However, due to the advent of next generation sequencing (NGS) technologies and bioinformatics tools, the scale of such analyses has now shifted to whole genomes, with comparison of genome structure, function and diversity now routine both within and between species.

In general, comparative genomics can be used to address four major research areas. First, all comparative genome analyses provide data that reveal both similarities and differences in genome structure on a macro- or micro-scale. Second, comparative genomics data are useful to identify both homologous genes and corresponding cis-regulatory elements based on their conservation between species. Third, comparative genomic data are useful to characterize genetic variation on a large scale, which has utility at the intraspecies level for development of new molecular markers and at the interspecies level for phylogenetic and evolutionary studies. Fourth, comparative genomics offers knowledge to understand the evolutionary and functional basis of novel traits, which has a great potential to impact the efficiency of breeding process. The utility of comparative studies is limited by the level of knowledge for the best-characterized species in any set of species under comparison. Thus, considerable energy has been directed towards understanding gene and genome function in model plant systems, such as Arabidopsis, rice, poplar and *Medicago*, based on the assumption that information obtained from one species can be applied to the tens of thousands of other plant species that are important for agriculture, the economy and natural environments. The validity of this assumption has been confirmed many times with respect to genome structure and gene function in plant growth and development. One manner in which to translate both structural and functional information from one gene or genome to another is through comparative mapping and sequence analysis (Table 47.3).

Table 47.3 Representative comparative genome analyses in cereals

Rice	Barley	Wheat	Sorghum	Maize	Millet	Oat	Rye	References	Method*
X	X	X	X					Mayer et al. (2011)	A
	X	X						Alm et al. (2011)	B
X			X	X	X	X	X	Devos (2005)	A & B
X				X				Keller and Feuillet (2000)	B
X			X		X	X		Gale and Devos (1998)	B
X		X		X				Bennetzen and Freeling (1997)	B

Note: * Method used for comparative study (A= sequenced based, B= map based).

The recent advances in next generation sequencing technology and bioinformatics analysis tools are accelerating the pace of comparative genomics and provide an opportunity to integrate structural and functional information across all taxonomic scales. From the perspective of agricultural species, this translational approach should facilitate molecular breeding through the identification of orthologous genes that are linked to, or control, important trait loci.

The Poaceae, also known as the grass family, is a large and diverse set of agronomically important crops and is the main source of dietary calories for both humans and livestock. The family includes more than 10 000 species and encompasses substantial morphological, physiological, ecological and genetic diversity (Kellogg, 1998). Grasses, in terms of genomic organization, represent a highly diverse family. Their chromosome number varies from 2n = 4 for *Colpodium vesicola* to 2n = 266 for the polyploidy grass *Poa litorosa*. Their genome sizes also vary greatly. For example, genomes of two typical grass crops, rice (430 Mbp) and bread wheat (17 000 Mbp), differ by a factor of 40. In the area of comparative genomic studies, the grass family has offered the most comprehensive dataset to date, which contains almost all the major cereal crops in general and small grains in particular. Such comprehensive analyses could be possible largely due to a long history of grass crop cultivation, availability of high-density genetic maps and large-scale genomic sequencing. Early studies mainly focused on identifying genomic regions that remained conserved over long evolutionary time periods. The use of common sets of low-copy number DNA markers, often coding sequences, in the mapping of grass genomes has confirmed that the gene content of different grass species does not vary greatly (Kurata et al., 1994; Bennetzen and Freeling, 1997). Genetic mapping of rice, wheat, maize and other grass species such as oat, with common DNA probes, revealed remarkable conservation of gene content (Gale and Devos, 1998; Keller and Feuillet. 2000; Devos et al., 2000).

Due to many genome sequence analyses in recent years, researchers could focus more on the discovery of sequence-level co-linearity, so-called micro-synteny and uncovered frequent small rearrangements disturbing the co-linearity in compared orthologous genomic regions. This analysis of sequence-based co-linearity is able to give deeper insights into whether gene order and content have remained conserved within the orthologous genomic segments, while a map-based approach may provide an overview of chromosomal rearrangements between related species. Since the first Crop Circles composed of eight species belonging to three subfamilies (Devos and Gale, 1997), it has been updated by adding more grass species such as rye (Devos, 2005).

6 TRENDS AND ISSUES AFFECTING THE SMALL GRAINS INDUSTRY

Per capita consumption of small grains has been declining for nearly a century, as demand for physical labor declined and human diets shifted and diversified as a result of increasing household incomes. Demand for small grain products reached a low in the early 1970s but rebounded when increasing health concerns began a shift from animal products to more grain-based diets. Consumption started to decrease again with the arrival of low carbohydrate diets in the early 2000s. More recently, this trend has been reversed in some markets, and over the last five years, domestic consumption of wheat food products – but also rye, oat and barley food products – is increasing. Globally, demand for small grains increased during the 1990s and 2000s due to a rising world population and increasing household incomes, particularly in emerging economies, where the demand for meat and dairy products is well above historic levels. The growing urbanization and middle class of developing countries is driving this trend to include more meat in diets. This, in turn, is increasing worldwide demand for grain products, particularly feed grains. As a result, world trade has reached record levels in recent years, with global stocks becoming low. Domestic demand for small grains is also driven by changes in the relative prices for corn and other grains, such as wheat, since livestock feed manufacturers can use a variety of feedstock for animal rations. Demand from livestock feed manufacturers, typically accounting for more than 40 per cent of the small grain market, has grown strongly over the last few years. The interest in alternative fuels, driven by recent trends in the energy sector and government policy, has contributed to the rise in demand. In particular, the interest in ethanol plays an important role in reducing the availability of corn for the feed industry. The diversion of part of the corn crop into ethanol production drives up prices for corn and forces livestock producers to substitute other feed grains. This, again, increases demand and, subsequently, prices for alternative feed grains, such as barley and oat, are increasing. In addition to these domestic conditions, the recent shortage of small grain production worldwide has resulted in low global inventories. In the most recent years, where world consumption of corn, sorghum, oat, barley and rye has exceeded production, global production of wheat has also remained below or equal to consumption, pushing up the price of alternative grains and limiting the extent to which grain users can substitute alternatives.

REFERENCES

Achleitner, A., N.A. Tinker, E. Zechner and H. Buerstmayr (2008), 'Genetic diversity among oat varieties of worldwide origin and associations of AFLP markers with quantitative traits', *TAG Theoretical and Applied Genetics*, **117**, 1041–53.

Aghnoum, R., T.C. Marcel, A. Johrde, N. Pecchioni, P. Schweizer and R.E. Niks (2010), 'Basal host resistance of barley to powdery mildew: connecting quantitative trait loci and candidate genes', *Molecular Plant-Microbe Interactions*, **23**, 91–102.

Alm, V., C.S. Busso, A. Ergon, H. Rudi, A. Larsen, M.W. Humphreys and O.A. Rognli (2011), 'QTL analyses and comparative genetic mapping of frost tolerance, winter survival and drought tolerance in meadow fescue (*Festuca pratensis Huds.*)', *Theoretical and Applied Genetics*, **123**, 369–82.

Altpeter, F. (2006), 'Rye (*Secale cereale* L.)', in K. Wang (ed.), *Methods in Molecular Biology*, 2nd edn, Totowa, NJ: Humana Press, pp. 223–32.

Appels, R., C. Driscoll and W.J. Peacock (1978), 'Heterochromatin and highly repeated DNA sequences in rye (Secale cereale)', Chromosoma **70**, 67–89.

Appels, R., J.P. Gustafson and C.E. May (1982), 'Structural variation in the heterochromatin of rye chromosomes in triticales', *Theoretical and Applied Genetics*, **63**, 235–44.

Becher, R. (2007), 'EST-derived microsatellites as a rich source of molecular markers for oats', *Plant Breeding*, **126**, 274–8.

Bennett, M.D. and J.B. Smith (1976), 'The nuclear DNA content of the egg, the zygote and young proembryo cells in Hordeum', *Caryologia*, **29**, 435–46.

Bennetzen, J.L. and M. Freeling (1997), 'The unified grass genome: synergy in synteny', *Genome Research*, **7**, 301–307.

Bovina, R., V. Talame, S. Silvio, M.C. Sanguineti, P. Trost, F. Sparla and R. Tuberosa (2011), 'Starch metabolism mutants in barley: a TILLING approach', *Plant Genetic Resources: Characterization and Utilization*, **9**, 170–73.

Brookes, A.J. (1999), 'The essence of SNPs', *Gene*, **234**, 177–86.

Brown, T.A., M.K. Jones, W. Powell and R.G. Allaby (2009), 'The complex origins of domesticated crops in the Fertile Crescent', *Trends in Ecology & Evolution*, **24**, 103–109.

Bundock, P.C. and Henry R.J. (2004), 'Single nucleotide polymorphism, haplotype diversity and recombination in the Isa gene of barley', *Theoretical and Applied Genetics*, **109**, 543–51.

Bundock, P.C., J.T. Christopher, P. Eggler, G. Ablett, R.J. Henry and T.A. Holton (2003), 'Single nucleotide polymorphisms in cytochrome P450 genes from barley', *Theoretical and Applied Genetics*, **106**, 676–82.

Caldwell, D.G., N. McCallum, P. Shaw, G.J. Muehlbauer, D.F. Marshall and R. Waugh (2004), 'A structured mutant population for forward and reverse genetics in barley (Hordeum vulgare L)', *Plant Journal*, **40**, 143–50.

Castillo, A.M., V. Vasil and I.K. Vasil (1994), 'Rapid production of fertile transgenic plants of rye (*Secale cereale* L.)', *Biotechnology*, **12**, 1366–71.

Cho, M-J., W. Jiang and P.G. Lemaux (1998), 'Transformation of recalcitrant cultivars through improvement in regenerability and decreased albinism', *Plant Science*, **138**, 229–44.

Cho, M.J., W. Jiang and P.G. Lemaux (1999), 'High-frequency transformation of oat via microprojectile bombardment of seed-derived highly regenerative cultures', *Plant Science*, **148**, 9–17.

Close, T.J., S.I. Wanamaker, R.A. Caldo, S.M. Turner, D.A. Ashlock, J.A. Dickerson, R.A. Wing, G.J. Muehlbauer, A. Kleinhofs and R.P. Wise (2004), 'A new resource for cereal genomics: 22K Barley GeneChip comes of age', *Plant Physiology*, **134**, 960–68.

Close, T.J., P.R. Bhat, S. Lonardi, Y.H. Wu, N. Rostoks, L. Ramsay, A. Druka, N. Stein, J.T. Svensson, S. Wanamaker, S. Bozdag, M.L. Roose, M.J. Moscou, S.M. Chao, R.K. Varshney, P. Szucs, K. Sato, P.M. Hayes, D.E. Matthews, A. Kleinhofs, G.J. Muehlbauer, J. DeYoung, D.F. Marshall, K. Madishetty,

R.D. Fenton and P. Condamine (2009), 'Development and implementation of high-throughput SNP genotyping in barley', *BMC Genomics*, **10**(4).
Dahleen, L.S. and M. Manoharan (2007), 'Recent advances in barley transformation', *In Vitro Cellular and Developmental Biology – Plant*, **43**, 493–506.
Dahleen, L.S., P.A. Okubara and A.E. Blechl (2001), 'Transgenic approaches to combat fusarium head blight in wheat and barley', *Crop Science*, **41**, 628–37.
De la Pena, A., H. Lörz and J. Schell (1987), 'Transgenic rye plants obtained by injecting DNA into young floral tillers', *Nature*, **325**, 274–6.
Devos, K.M. (2005), 'Updating the "Crop Circle"', *Current Opinion in Plant Biology*, **8**, 155–62.
Devos, K.M., T. Millan and M.D. Gale (1993), 'Comparative RFLP maps of the homoeologous group-2 chromosomes of wheat, rye and barley', *Theoretical and Applied Genetics*, **85**, 784–92.
Devos, K.M., T.S. Pittaway, A. Reynolds and M.D. Gale (2000), 'Comparative mapping reveals a complex relationship between the pearl millet genome and those of foxtail millet and rice', *Theoretical and Applied Genetics*, **100**, 190–98.
Devos, K.M., M.D. Atkinson, C.N. Chinoy, C.J. Liu and M.D. Gale (1992), 'RFLP-based genetic map of the homoeologous group-3 chromosomes of wheat and rye', *Theoretical and Applied Genetics*, **83**, 931–9.
DFCI (Dana-Farber Cancer Institute) (2008), Rye Gene Index (RyeGI) Release 4.0, available via DFCI, Computational Biology and Functional Genomics Laboratory, The Gene Index Project, http://compbio.dfci.harvard.edu/tgi/cgi-bin/tgi/gimain.pl?gudb=s_cereale.
Diab, A.A. (2006), 'Construction of barley consensus map showing chromosomal regions associated with economically important traits', *African Journal of Biotechnology*, **5**(3), 235–48.
Doss, S., E.E. Schadt, T.A. Drake and A.J. Lusis (2005), 'Cis-acting expression quantitative trait loci in mice', *Genome Research*, **15**, 681–91.
Fan, J.B., A. Oliphant, R. Shen, B.G. Kermani, F. Garcia, K.L. Gunderson, M. Hansen, F. Steemers, S.L. Butler, P. Deloukas, L. Galver, S. Hunt, C. McBride, M. Bibikova, T. Rubano, J. Chen E. Wickham, D. Doucet, W. Chang, D. Campbell, B. Zhang, S. Kruglyak, D. Bentley, J. Haas, P. Rigault, L. Zhou, J. Stuelpnagel and M.S. Chee (2003), 'Highly parallel SNP genotyping', *Cold Spring Harbor Symposium Quantitative Biology*, **68**, 69–78.
Flavell, B., J. Rimpau and D.B. Smith (1977), 'Repeated sequence DNA relationships in four cereal genomes', *Chromosoma*, **63**, 205–22.
Flavell, R.B. (1982), *Sequence Amplification, Deletion and Rearrangement: Major Sources of Variation During Species Divergence*, London: Academic Press.
Flavell, R.B., M.D. Bennett J.B. Smith and D.B. Smith (1974), 'Genome size and the proportion of repeated nucleotide sequence DNA in plants', *Biochemical Genetics*, **12**, 257–69.
Food and Agriculture Organization of United Nations (FAOSTAT) (2009), 'Rome', available at: http://www.faostat.fao.org.
Friedt, W. and F. Ordon (2008), 'Molecular markers for gene pyramiding and resistance breeding in barley', in R. Varshney and R. Tuberosa (eds), *Genomics-Assisted Crop Improvement, Vol 2: Genomics Applications in Crops*, Berlin: Springer.
Funatsuki, H., H. Kuroda, M. Kihara, P.A. Lazzeri, E. Muller, H. Lorz and I. Kishinami (1995), 'Fertile transgenic barley generated by direct DNA transfer to protoplasts', *Theoretical and Applied Genetics*, **91**, 707–12.
Gale, M.D. (1990), 'Comparative mapping in Triticeae genomes', in P.E. McGuire, H. Corke and C.O. Qualset (eds), *Genome Mapping of Wheat and Related Species*, Proceedings of the 1st Public Workshop of the International Triticeae Mapping Initiative, University of California, West Sacramento, California, pp. 17–19.
Gale, M.D. and K.M. Devos (1998), 'Comparative genetics in the grasses', *Proceedings of the National Academy of Sciences of the USA*, **95**, 1971–4.
Ganeshan, S., M. Baga, B.L. Harvey, B.G. Rossnagel, G.J. Scoles and R.N. Chibbar (2003), 'Production of multiple shoots from thidiazuron-treated mature embryos and leaf-base/apical meristems of barley (*Hordeum vulgare*)', *Plant Cell Tissue and Organ Culture*, **73**, 57–64.
Gasparis, S., C. Bregier, W. Orczyk and A. Nadolska-Orczyk (2008), 'Agrobacterium-mediated transformation of oat (*Avena sativa* L.) cultivars via immature embryo and leaf explants', *Plant Cell Reports*, **27**, 1721–9.
Gaut, B.S. (2002), 'Evolutionary dynamics of grass genomes', *New Phytologist*, **154**, 15–28.
Gless, C., H. Lörz and A. Jahne-Gartner (1998), 'Transgenic oat plants obtained at high efficiency by microprojectile bombardment of leaf base segments', *Journal of Plant Physiology*, **152**, 151–7.
Goedeke, S., G. Hensel, E. Kapusi, M. Gahrtz and J. Kumlehn (2007), 'Transgenic barley in fundamental research and biotechnology', *Transgenic Plant Journal*, **1**, 104–17.
Graner, A., A. Jahoor, J. Schondelmaier, H. Siedler, K. Pillen, G. Fischbeck, G. Wenzel and R.F. Herrmann (1991), 'Construction of an RFLP map of barley', *Theoretical and Applied Genetics*, **83**, 250–56.
Graner, A., S. Streng, A. Kellermann, A. Schiemann, E. Bauer, R. Waugh, B. Pellio and F. Ordon (1999), 'Molecular mapping and genetic fine-structure of the *rym5* locus encoding resistance to different strains of the barley yellow mosaic virus complex', *Theoretical and Applied Genetics*, **98**, 285–90.

Hagio, T., T. Hirabayashi, H. Machii and H. Tomotsune (1995), 'Production of fertile transgenic barley (*Hordeum vulgare* L) plants using the hygromycin-resistance marker', *Plant Cell Reports*, **14**, 329–34.
Hansen, M., M. Lange, C. Friis, G. Dionisio, P.B. Holm and E. Vincze (2007), 'Antisense-mediated suppression of C-hordein biosynthesis in the barley grain results in correlated changes in the transcriptome, protein profile, and amino acid composition', *Journal of Experimental Botany*, **58**, 3987–95.
Harcourt, R.L. and M.D. Gale (1991), 'A chromosome-specific DNA sequence which reveals a high level of RFLP in wheat', *Theoretical and Applied Genetics*, **81**, 397–400.
Harwood, W.A. and M. Smedley (2009), 'Barley transformation using biolistic techniques', in Huw D. Jones and Peter R. Shewry (eds), *Methods in Molecular Biology, Transgenic Wheat, Barley and Oats*, vol. **478**, pp. 125–36.
Harwood, W.A., J. Bartlett, S. Alves, M. Perry, M. Smedley, N. Leyland and J.W. Snape (2009), 'Barley transformation using Agrobacterium-mediated techniques', in H.D. Jones and P.R. Shewry (eds), *Methods in Molecular Biology, Transgenic Wheat, Barley and Oats*, Vol. **478**, pp. 137–47.
Hayes, P.M., A. Castro, L. Marquez-Cedillo et al. (2003), 'Genetic diversity for quantitatively inherited agronomic and malting quality traits', in R. von Bothmer, T. van Hintum, H. Knüpffer and K. Sato (eds), *Diversity in Barley (Hordeum vulgare)*, Amsterdam: Elsevier Science, pp. 201–26.
Hearnden, P.R.,P.J. Eckermann, G.L. McMichael, M.J. Hayden, J.K. Eglinton and K.J. Chalmers (2007), 'A genetic map of 1 000 SSR and DArT markers in a wide barley cross', *Theoretical and Applied Genetics*, **115**(3), 383–91.
Herzfeld, J.C.P. (2002), 'Development of a genetic transformation protocol for rye (*Secale cereale* L.) and characterisation of transgene expression after biolistic or *Agrobacterium*-mediated gene transfer', PhD thesis, Martin Luther University, Halle-Wittenberg, Germany.
Heun, M., A.E. Kennedy, J.A. Anderson, N.L.V. Lapitan, M.E. Sorrells and S.D. Tanksley (1991), 'Construction of a restriction fragment length polymorphism map for barley (*Hordeum vulgare*)', *Genome*, **34**(3), 437–47.
Hiei, Y., T. Komar and T. Kubo (1997), 'Transformation of rice mediated by Agrobacterium tumefaciens', *Plant Molecular Biology*, **35**, 205–18.
Holland, J.B., S.J. Helland, N. Sharopova and D.C. Rhyne (2001), 'Polymorphism of PCR-based markers targeting exons, introns, promoter regions, and SSRs in maize and introns and repeat sequences in oat', *Genome*, **44**, 1065–76.
Holm, P.B., O. Olsen, M. Schnorf, H. Brinch-Pederson and S. Knudsen (2000), 'Transformation of barley by microinjection into isolated zygote protoplasts', *Transgenic Research*, **9**, 21–32.
Horvath, H., J. Huang, O. Wong, E. Kohl, T. Okita, C.G. Kannangara and D. von Wettstein (2000), 'The production of recombinant proteins in transgenic barley grains', *Proceedings of the National Academy of Sciences of the USA*, **97**, 1914–19.
Hull, G.A., N.G. Halford, M. Kreis and P.R. Shewry (1991), 'Isolation and characterization of genes encoding rye prolamins containing a highly repetitive sequence motif', *Plant Molecular Biology*, **17**, 1111–15.
Isidore, E., B. Scherrer, A. Bellec, K. Budin, P. Faivre-Rampant, R. Waugh, B. Keller, M. Caboche, C. Feuillet and B. Chalhoub (2005), 'Direct targeting and rapid isolation of BAC clones spanning a defined chromosome region', *Functional & Integrative Genomics*, **5**, 97–103.
Jahne, A., D. Becker, R. Brettschneider and H. Lorz (1994), 'Generation of transgenic, microspore-derived, fertile barley', *Theoretical and Applied Genetics*, **89**, 525–33.
Jannink, J.L. and Gardner, S.W. (2005), 'Expanding the pool of PCR-based markers for oat', *Crop Science*, **45**, 2383–87.
Jensen, L.G., O. Olsen, O. Kops, N. Wolf and K.K. Thompsen (1996), 'Transgenic barley expressing a protein-engineered, thermostable (1,3–1,4)-beta glucanase during germination', *Proceedings of the National Academy of Sciences of the USA*, **93**, 3487–91.
Joensuu, J.J., M. Kotiaho, T.H. Teeri, L. Valmu, A.M. Nuutila, K-M. Oksman-Caldentey and V. Niklander-Teeri (2006), 'Glycosylated F4 (K88) fimbrial adhesin FaeG expressed in barley endosperm induces ETEC-neutralizing antibodies in mice', *Transgenic Research*, **15**, 359–73.
Kaeppler, H.F., G.K. Menon, R.W. Skadsen, A.M. Nuutila and A.R. Carlson (2000), 'Transgenic oat plants via visual selection of cells expressing green fluorescent protein', *Plant Cell Reports*, **19**, 661–6.
Kanazin, V., H. Talbert, D. See, P. DeCamp, E. Nevo and T. Blake (2002), 'Discovery and assay of single-nucleotide polymorphisms in barley (Hordeum vulgare)', *Plant Molecular Biology*, **48**, 529–37.
Karakousis, A., J.P. Gustafson, K.J. Chalmers, A.R. Barr and P. Langridge (2003), 'A consensus map of barley integrating SSR, RFLP, and AFLP markers', *Australian Journal of Agricultural Research*, **54**(11–12), 1173–85.
Karami, O., M. Esna-Ashari, G. Karimi Kurdistani and B. Aghavaisi (2009), 'Agrobacterium-mediated genetic transformation of plants: the role of host', *Biologia Plantarum*, **53**, 201–12.
Keller, B. and C. Feuillet (2000), 'Colinearity and gene density in grass genomes', *Trends in Plant Science*, **5**, 246–51.

Kelley, J.M., C.E. Field, M.B. Craven, D. Bocskai, U.J. Kim, S.D. Rounsley and M.D. Adams (1999), 'High throughput direct end sequencing of BAC clones', *Nucleic Acids Research*, **27**, 1539–46.
Kellogg, E.A. (1998), 'Relationships of cereal crops and other grasses', *Proceedings of the National Academy of Sciences of the USA*, **95**, 2005–10.
Kellogg, E.A. (2001), 'Evolutionary history of the grasses', *Plant Physiology*, **125**, 1198–205.
Kihara, M., K. Saek and K. Ito (1998), 'Rapid production of fertile transgenic barley (*Hordeum vulgare* L) by direct gene transfer to primary callus derived protoplasts', *Plant Cell Reports*, **17**, 937–40.
Kihara, M., Y. Okada, H. Kuroda, K. Saeki, N. Yoshigi and K. Ito (2000), 'Improvement of β-amylase thermostability in transgenic barley seeds and transgene stability in progeny', *Molecular Breeding*, **6**, 511–17.
Kleinhofs, A., A. Kilian, M.A. Saghai Maroof, R.M. Biyashev, P. Hayes, F.Q. Chen, N. Lapitan, A. Fenwick, T.K. Blake, V. Kanazin, E. Ananiev, L. Dahleen, D. Kudrna, J. Bollinger, S.J. Knapp, B. Liu, M. Sorrells, M. Heun, J.D. Franckowiak, D. Hoffman, R. Skadsen and B.J. Steffenson (1993), 'A molecular, isozyme and morphological map of the barley (Hordeum vulgare) genome', *Theoretical and Applied Genetics*, **86**, 705–12.
Koev, G., B.R. Mohan, S.D. Dinesh-Kumar, K.A. Torbert, D.A. Somers and W.A. Miller (1998), 'Extreme reduction of disease in oats transformed with the 5′ half of the barley yellow dwarf virus PAV genome', *Phytopathology*, **88**, 1013–1019.
Koprek, T., R. Haensch, A. Nerlich, R.R. Mendel and J. Schulze (1996), 'Fertile transgenic barley of different cultivars obtained by adjustment of bombardment conditions to tissue response', *Plant Science*, **119**, 79–91.
Korzun, L. and G. Kunzel (1996), 'The physical relationship of barley chromosome 5 (1H) to the linkage groups of rice chromosomes 5 and 10', *Molecular and General Genetics*, **253**, 225–31.
Korzun, V., S. Malyshev, N. Kartel, T. Westermann, W.E. Weber and A. Börner (1998), 'A genetic linkage map of rye (*Secale cereal* L.)', *Theoretical and Applied Genetics*, **96**, 203–208.
Korzun, V.V., A. Borner, R. Siebert, S. Malyshev, M. Hilpert, R. Kunze and H. Puchta (1999), 'Chromosomal location and genetic mapping of the mismatch repair gene homologs MSH2, MSH3, and MSH6 in rye and wheat', *Genome*, **42**, 1255–7.
Kota, R., R.K. Varshney, T. Thiel, K.J. Dehmer and A. Graner (2001), 'Generation and comparison of EST-derived SSRs and SNPs in barley (Hordeum vulgare L)', *Hereditas*, **135**(2–3), 145–51.
Kota, R., R.K. Varshney, M. Prasad, H. Zhang, N. Stein and A. Graner (2008), 'EST-derived single nucleotide polymorphism markers for assembling genetic and physical maps of the barley genome', *Functional & Integrative Genomics*, **8**, 223–33.
Kota, R., S. Rudd, A. Facius, G. Kolesov, T. Thiel, H. Zhang, N. Stein, K. Mayer and A. Graner (2003), 'Snipping polymorphisms from large EST collections in barley (Hordeum vulgare L)', *Molecular Genetics and Genomics*, **270**, 24–33.
Kuai, B., S. Perret, S.M. Wan, S.L. Dalton, A.J.E. Bettany and P. Morris (2001), 'Transformation of oat and inheritance of *bar* gene expression', *Plant Cell Tissue and Organ Culture*, **66**, 79–88.
Kurata, N., Y. Nagamura, K. Yamamoto, Y. Harushima, N. Sue, J. Wu, B.A. Antonio, A. Shomura, T. Shimizu, S.Y. Lin, T. Inoue, A. Fukuda, T. Shimano, Y. Kuboki, T. Toyama, Y. Miyamoto, T. Kirihara, K. Hayasaka, A. Miyao, L. Monna, H.S. Zhong, Y. Tamura, Z.X. Wang, T. Momma, Y. Umehara, M. Yano, T. Sasaki and Y. Minobe (1994), 'A 300-kb interval rice genetic map including 883 expressed sequences', *Nature Genetics*, **8**, 365–72.
Lange, M., E. Vincze, H. Wieser, J.K. Schjoerring and P.B. Holm (2007), 'Suppression of C-hordein synthesis in barley by antisense constructs results in a more balanced amino acid composition', *Journal of Agricultural and Food Chemistry*, 55, 6074–81.
Lazzeri, P.A., R. Brettschneider, R. Luhrs and H. Lorz (1991), 'Stable transformation of barley via PEG-induced DNA uptake into protoplasts', *Theoretical and Applied Genetics*, **18**, 437–44.
Lemaux, P.G., M-J. Cho, S. Zhang and P. Bregitzer (1999), 'Transgenic cereals: *Hordeum vulgare* L. (barley)', in I.K. Vasil (ed.), *Molecular Improvement of Cereal Crops*, Dordrecht: Kluwer Academic Publishers, pp. 255–316.
Li, C.D., B.G. Rossnagel and G.J. Scoles (2000), 'The development of oat microsatellite markers and their use in identifying relationships among Avena species and oat cultivars', *Theoretical and Applied Genetics*, **101**, 1259–68.
Liu, C.J., M.D. Atkinson, C.N. Chinoy, K.M. Devos and M.D. Gale (1992), 'Non-homoeologous translocations between group 4, 5 and 7 chromosomes in wheat and rye', *Theoretical and Applied Genetics*, **83**, 305–12.
Loarce, Y., G. Hueros and E. Ferrer (1996), 'A molecular linkage map of rye', *Theoretical and Applied Genetics*, **93**, 1112–18.
Locatelli, A.B., L.C. Federizzi, S.C.K. Milach, C.P. Wight, S.J. Molnar, J.T. Chapados and N.A. Tinker (2006), 'Loci affecting flowering time in oat under short-day conditions', *Genome*, **49**, 1528–38.
Lundqvist, U., J. Franckowiak and T. Konishi (1996), 'New and revised descriptions of barley genes', *Barley Genet Newsletter*, **26**, 22–43.

Manoharan, M. and L.S. Dahleen (2002), 'Genetic transformation of the commercial barley (*Hordeum vulgare* L.) cultivar Conlon by particle bombardment of callus', *Plant Cell Reports*, **21**, 76–80.

Manoharan, M., L.S. Dahleen, T.M. Hohn, S.M. Neate, X-H. Yu, N.J. Alexander, S.P. McCormick, P. Bregitzer, P.B. Schwarz and R.D. Horsley (2006), 'Expression of 3-OH trichothecene acetyltransferase in barley (*Hordeum vulgare* L.) and effects on deoxynivalenol', *Plant Science*, **171**, 699–706.

Marcel, T.C., R.K. Varshney, M. Barbieri, H. Jafary, M.J.D. de Kock, A. Graner and R.E. Niks (2007), 'A high-density consensus map of barley to compare the distribution of QTLs for partial resistance to *Puccinia hordei* and of defence gene homologues', *Theoretical and Applied Genetics*, **114**, 487–500.

Mayer, K.F.X., M. Martis, P.E. Hedley, H. Simkova, H. Liu, J.A. Morris, B. Steuernagel, S. Taudien, S. Roessner, H. Gundlach, M. Kubalakova, P. Suchankova, F. Murat, M. Felder, T. Nussbaumer, A. Graner, J. Salse, T. Endo, H. Sakai, T. Tanaka, T. Itoh, K. Sato, M. Platzer, T. Matsumoto, U. Scholz, J. Dolezel, R. Waugh and N. Stein (2011), 'Unlocking the barley genome by chromosomal and comparative genomics', *Plant Cell*, **23**, 1249–63.

Maqbool, S., H. Zhong, Y. El-Maghraby, A. Ahmad, B. Chai, W. Wang, R. Sabzikar and M. Sticklen (2002), 'Competence of oat (*Avena sativa* L.) shoot apical meristems for integrative transformation, inherited expression and osmotic tolerance of transgenic lines containing hva 1', *Theoretical and Applied Genetics*, **105**, 201–208.

McGrath, D.F., J.R. Vincent, C.H. Lei, W.P. Pawlowski, K.A. Torbert, W. Gu, H.F. Kaeppler, Y. Wan, P.G. Lemaux, H.R. Rines, D.A. Somers, B.A. Larkins and R.M. Lister (1997), 'Coat protein-mediated resistance to isolates of barley yellow dwarf in oats and barley', *European Journal of Plant Pathology*, **103**, 695–710.

Michelmore, R.W., I. Paran and R.V. Kesseli (1991), 'Identification of markers linked to disease-resistance genes by bulked segregant analysis: a rapid method to detect markers in specific genomic regions by using segregating populations', *Proceedings of the National Academy of Sciences of the USA*, **88**, 9828–32.

Morrell, P.L. and M.T. Clegg (2007), 'Genetic evidence for a second domestication of barley (Hordeum vulgare) east of the Fertile Crescent', *Proceedings of the National Academy of Sciences of the United States of America*, **104**, 3289–94.

Murray, F., R. Brettell, P. Matthews, D. Bishop and J. Jacobsen (2004), 'Comparison of Agrobacterium-mediated transformation of four barley cultivars using the GFP and GUS reporter genes', *Plant Cell Reports*, **22**, 397–402.

Nobre, J., M.R. Davey, P.A. Lazzeri and M.E. Cannell (2000), 'Transformation of barley protoplasts: regeneration of fertile transgenic plants', *Plant Cell Reports*, **19**, 1000–1005.

Nuutila, A.M., A. Ritala, R.W. Skadsen, L. Mannonen and V. Kauppinen (1999), 'Expression of fungal thermotolerant endo-1,4-B-glucanase in transgenic barley seeds during germination', *Plant Molecular Biology*, **41**, 777–83.

O'Donoughue, L.S., Z. Wang, M. Roder, B. Kneen, M. Leggett, M.E. Sorrells and S.D. Tanksley (1992), 'An RFLP-based linkage map of oats based on a cross between two diploid taxa (Avena atlantica X A hirtula)', *Genome*, **35**, 765–71.

Orr, W. and S.J. Molnar (2007), 'Development and mapping of PCR-based SCAR and CAPS markers linked to oil QTLs in oat', *Crop Science*, **47**, 848–52.

Pal, N., J.S. Sandhu, L.L. Domier and F L. Kolb (2002), 'Development and characterization of microsatellite and RFLP-derived PCR markers in oat', *Crop Science*, **42**, 912–18.

Paux, E., D. Roger, E. Badaeva, G. Gay, M. Bernard, P. Sourdille and C. Feuillet (2006), 'Characterizing the composition and evolution of homoeologous genomes in hexaploid wheat through BAC-end sequencing on chromosome 3B', *Plant Journal*, **48**, 463–74.

Pawlowski, W.P. and D.A. Somers (1998), 'Transgenic DNA integrated into the oat genome is frequently interspersed by host DNA', *Proceedings of the National Academy of Sciences of the USA*, **95**, 12106–10.

Pawlowski, W.P., K.A. Torbert, H.W. Rines and D.A. Somers (1998), 'Irregular patterns of transgene silencing in allohexaploid oats', *Plant Molecular Biology*, **38**, 597–607.

Perret, S., J. Valentine, J.M. Leggett and P. Morris (2003), 'Integration, expression and inheritance of transgenes in hexaploid oat (*Avena sativa* L.)', *Journal of Plant Physiology*, **160**, 931–43.

Philipp, U., Wehling, P. and Wricke, G. 1994. A linkage map of rye. Theoretical and Applied Genetics **88**, 243–248.

Pillen, K., A. Binder, B. Kreuzkam, L. Ramsay, R. Waugh, J. Forster and J. Leon (2000), 'Mapping new EMBL-derived barley microsatellites and their use in differentiating German barley cultivars', *Theoretical and Applied Genetics*, **101**, 652–60.

Popelka, J.C. and F. Altpeter (2003), 'Evaluation of rye (*Secale cereale* L) inbred lines and their crosses for tissue culture response and stable genetic transformation of homozygous rye inbred line L22 by biolistic gene transfer', *Theoretical and Applied Genetics*, **107**(4), 583–90.

Popelka, J.C., J.P. Xu and F. Altpeter (2003), 'Generation of rye (*Secale cereale* L.) plants with low transgene copy number after biolistic gene transfer and production of instantly marker-free transgenic rye', *Transgenic Research*, **12**, 587–96.

Potokina, E., A. Druka, Z. Luo, R. Wise, R. Waugh and M.J. Kearsey (2008), 'Gene expression quantitative trait locus analysis of 16000 barley genes reveals a complex pattern of genome-wide transcriptional regulation', *The Plant Journal*, **53**(1), 90–101.
Rae, S.J., M. Macaulay, L. Ramsay, F. Leigh, D. Matthews, D.M. O'Sullivan, P. Donini, P.C. Morris, W. Powell, D.F. Marshall, R. Waugh and W.T.B. Thomas (2007), 'Molecular barley breeding', *Euphytica*, **158**(3), 295–303.
Rafalski, J.A. (2002), 'Novel genetic mapping tools in plants: SNPs and LD-based approaches', *Plant Science*, **162**, 329–33.
Rakoczy-Trojanowska, M. and S. Malepszy (1995), 'Genetic factors influencing regeneration ability in rye (*Secale cereale* L.) II. Immature embryos', *Euphytica*, **83**, 233–9.
Ramsay, L., M. Macaulay, S. degli Ivanissevich, K. MacLean, L. Cardle, J. Fuller, K.J. Edwards, S. Tuvesson, M. Morgante, A. Massari, E. Maestri, N. Marmiroli, T. Sjakste, M. Ganal, W. Powell and R. Waugh (2000), 'A simple sequence repeat-based linkage map of barley', *Genetics*, **156**, 1997–2005.
Ranjekar, P.K., J.G. Lafontaine and D. Pallotta (1974), 'Characterization of repetitive DNA in rye (Secale cereale', *Chromosoma*, **48**, 427–40.
Reddy, P. and R. Appels (1989), 'A second locus for the 5S multigene family in *Secale* L.: sequence divergence in two lineages of the family', *Genome*, **32**, 456–67.
Rimpau, J., D. Smith and R. Flavell (1978), 'Sequence organisation analysis of the wheat and rye genomes by interspecies DNA/DNA hybridisation', *Journal of Molecular Biology*, **123**, 327–59.
Rines, H.W., S.J. Molnar, N.A. Tinker and R.L. Phillips (2006), 'Oat' in C. Kole (ed.), *Cereals and Millets*, Springer-Verlag GmbH: Heidelberg, pp. 211–42.
Ritala, A., L. Mannonen, K. Aspegren, M. Salmenkallio-Marttila, U. Kurten, R. Hannus, J.M. Lozano, T.H. Terri and V. Kauppinen (1993), 'Stable transformation of barley tissue culture by particle bombardment', *Plant Cell Reports*, **12**, 435–40.
Ritala, A., K. Aspegren, U. Kurten, M. Salmenkallio-Marttila, L. Mannonen, R. Hannus, V. Kauppinen, T.H. Terri and T. Enari (1994), 'Fertile transgenic barley by particle bombardment of immature embryos', *Plant Molecular Biology*, **24**, 317–25.
Rostoks, N., S. Mudie, L. Cardle, J. Russell, L. Ramsay, A. Booth, J.T. Svensson, S.I. Wanamaker, H. Walia, E.M. Rodriguez, P.E. Hedley, H. Liu, J. Morris, T.J. Close, D.F. Marshall, R. Waugh and H. Liu (2005), 'Genome-wide SNP discovery and linkage analysis in barley based on genes responsive to abiotic stress', *Molecular Genetics and Genomics*, **274**, 515–27.
Rostoks, N., L. Ramsay, K. MacKenzie, L. Cardle, P.R. Bhat, M.L. Roose, J.T. Svensson, N. Stein, R.K. Varshney, D.F. Marshall, A. Graner, T.J. Close and R. Waugh (2006), 'Recent history of artificial outcrossing facilitates whole-genome association mapping in elite inbred crop varieties', *Proceedings of the National Academy of Sciences of the United States of America*, **103**(49), 18656–61.
Rybczyński, J.J. and W. Zduńczyk (1986), 'Somatic embryogenesis and plantlet regeneration in the genus *Secale*. 1. Somatic embryogenesis and organogenesis from cultured immature embryos of five wild species of rye', *Theoretical and Applied Genetics*, **73**, 267–71.
Rybczyński, J.J. (1990), 'Plant tissue culture of *Secale* taxa', *Euphytica*, **46**, 57–70.
Salamini, F., H. Ozkan, A. Brandolini, R. Schafer-Pregl and W. Martin (2002), 'Genetics and geography of wild cereal domestication in the Near East', *Nature Reviews – Genetics*, **3**, 429–41.
Salmenkallio-Marttila, M., K. Aspergren, S. Akerman, U. Kurten, L. Mannonen, A. Ritala, T.H. Terri and V. Kauppinen (1995), 'Transgenic barley (*Hordeum vulgare* L) by electroporation of protoplasts', *Plant Cell Reports*, **15**, 301–304.
Salvi, S. and R. Tuberosa (2005), 'To clone or not to clone plant QTLs: present and future challenges', *Trends in Plant Science*, **10**(6), 297–304.
Sato, K. and K. Takeda (2009), 'An application of high-throughput SNP genotyping for barley genome mapping and characterization of recombinant chromosome substitution lines', *Theoretical and Applied Genetics*, **119**, 613–19.
Sato, K., N. Nankaku and K. Takeda (2009), 'A high-density transcript linkage map of barley derived from a single population', *Heredity*, **103**, 110–17.
Scheidig, A., A. Fröhlich, S. Schulze, J.R. Lloyd and J. Kossmann (2002), 'Downregulation of a chloroplast-targeted β-amylase leads to a starch-excess phenotype in leaves', *The Plant Journal*, **30**(5), 581–91.
Schlegel, R. (2006), 'Rye (Secale cereale L.) – a younger crop plant with bright future', in R.J. Singh and P. Jauhar (eds), *Genetic Resources, Chromosome Engineering, and Crop Improvement: Cereals*, Boca Raton: CRC Press, pp. 365–94.
Schlegel, R. and V. Korzun (2013), 'Genes, markers and linkage data of rye (*Secale cereale* L.)', 7th updated inventory, Version 09.13, pp. 1–111.
Schulte, D., R. Ariyadasa, B. Shi, D. Fleury, C. Saski, M. Atkins, P.d. Jong, C.C. Wu, A. Graner, P. Langridge, N. Stein, B.J. Shi and P. de Jong (2011), 'BAC library resources for map-based cloning and physical map construction in barley (Hordeum vulgare L), '*BMC Genomics*, **12**(19).

Schünmann, P.H.D., G. Coia and P.M. Waterhouse (2002), 'Biopharming the SimpliRedTM HIV diagnostic reagent in barley, potato and tobacco', *Molecular Breeding*, **9**, 113–21.

Senft, P. and G. Wricke (1996), 'An extended genetic map of rye (*Secale cereale* L.)', *Plant Breeding*, **115**(6), 508–10.

Shibata, D. and Y-G. Liu (2000), 'Agrobacterium-mediated plant transformation with large DNA fragments', *Trends in Plant Science*, **5**, 354–7.

Shi, B.-J., J.P. Gustafson, J. Button, J. Miyazaki, M. Pallotta, N. Gustafson, H. Zhou, P. Langridge and N.C. Collins (2009), 'Physical analysis of the complex rye (*Secale cereale* L.) Alt4 aluminium (aluminum) tolerance locus using a whole-genome BAC library of rye cv. Blanco', *Theoretical and Applied Genetics*, **119**, 695–704.

Šimková, H., J. Šafář, P. Suchánková, P. Kovářová, J. Bartoš, M. Kubaláková, J. Janda, J. Čihalíková, R. Mago, T. Lelley and J. Doležel (2008), 'A novel resource for genomics of Triticeae: BAC library specific for the short arm of rye (*Secale cereale* L.) chromosome 1R (1RS)', *BMC Genomics*, **9**, 237–45.

Smith, B.D. (1998), The Emergence of Agriculture, New York: Scientific American Library.

Somers, D.A. (1999), 'Genetic engineering of oat', in I. Vasil and R. Phillipes (eds), *Molecular Improvement of Cereal Crops*, Dordrecht: Kluwer Academic Publishers.

Somers, D.A., H.W. Rines, W. Gu, H.F. Kaeppler and W.R. Bushnell (1992), 'Fertile, transgenic oat plants', *Biotechnology*, **10**, 1589–94.

Sreenivasulu, N., A. Graner and U. Wobus (2008a), 'Barley genomics: an overview', *International Journal of Plant Genomics*, 2008, 486258.

Sreenivasulu, N., C. Pietsch, V. Radchuk, M. Strickert, M. Roder, W. Weschke and U. Wobus (2008b), 'Regulators determining grain maturation: a genetical genomics approach', *Options Méditerranéennes, Série A, Seminaires Méditerranéens*, 431–4.

Stein, N., D. Perovic, J. Kumlehn, B. Pellio, S. Stracke, S. Streng, F. Ordon and A. Graner (2005), 'The eukaryotic translation initiation factor 4E confers multiallelic recessive Bymovirus resistance in *Hordeum vulgare* (L.)', *The Plant Journal*, **42**(6), 912–22.

Stein, N., M. Prasad, U. Scholz et al. (2007), 'A 1 000-loci transcript map of the barley genome: new anchoring points for integrative grass genomics', *Theoretical and Applied Genetics*, **114**(5), 823–39.

Sutton, T., U. Baumann, J. Hayes, N.C. Collins, B.J. Shi, T. Schnurbusch, A. Hay, G. Mayo, M. Pallotta, M. Tester and P. Langridge (2007), 'Boron-toxicity tolerance in barley arising from efflux transporter amplification', *Science*, **318**(5855), 1446–9.

Szucs, P., V.C. Blake, P.R. Bhat, S. Chao, T.J. Close et al. (2009), 'An integrated resource for barley linkage map and malting quality QTL alignment', *The Plant Genome*, **2**, 134–40.

Thiel, T., W. Michalek, R.K. Varshney and A. Graner (2003), 'Exploiting EST databases for the development and characterization of gene-derived SSR-markers in barley (*Hordeum vulgare* L.)', *Theoretical and Applied Genetics*, **106**(3), 411–22.

Thiel, T., R. Kota, I. Grosse, N. Stein and A. Graner (2004), 'SNP2CAPS: a SNP and INDEL analysis tool for CAPS marker development', *Nucleic Acids Research*, **32**(1), e5.

Tingay, S., D. McElroy, R. Kalla, S. Fieg, M. Wang, S. Thornton and R. Brettell (1997), 'Agrobacterium tumefaciens-mediated barley transformation', *Plant Journal*, **11**, 1369–76.

Travella, S., S.M. Ross, J. Harden, C. Everett, J.W. Snape and W.A. Harwood (2005), 'A comparison of transgenic barley lines produced by particle bombardment and Agrobacterium-mediated techniques', *Plant Cell Reports*, **23**, 780–89.

Tinker, N.A. A. Kilian, C.P. Wight, K. Heller-Uszynska, P. Wenzl, H.W. Rines, A. Bjrnstad, C.J. Howarth, J.L. Jannink, J.M. Anderson, B.G. Rossnagel, D.D. Stuthman, M.E. Sorrells, E.W. Jackson, S. Tuvesson, F.L. Kolb, O. Olsson, L.C. Federizzi, M.L. Carson, H.W. Ohm, S.J. Molnar, G.J. Scoles, P.E. Eckstein, J.M. Bonman and A. Ceplitis (2009), 'New DArT markers for oat provide enhanced map coverage and global germplasm characterization', *BMC Genomics*, **10**(21).

Torbert, K.A., H.W. Rines and D.A. Somers (1995), 'Use of paromomycin as a selective agent for oat transformation', *Plant Cell Reports*, **14**, 635–40.

Torbert, K.A., M. Gopalraj, S.L. Medberry, N.E. Olszewski and D.A. Somers (1998a), 'Expression of the *Commelina* yellow mottle virus promoter in transgenic oat', *Plant Cell Reports*, **17**, 284–7.

Torbert, K.A., H.W. Rines, H.F. Kaeppler, G.K. Menon and D.A. Somers (1998b), 'Genetically engineering elite oat cultivars', *Crop Science*, **38**, 1–3.

Torbert, K.A., H.W. Rines and D.A. Somers (1998c), 'Transformation of oat using mature embryo-derived tissue cultures', *Crop Science*, **38**, 226–31.

Tull, D., B.A. Phillipson, B. Kramhoft, S. Knudsen, O. Olsen and B. Svensson (2003), 'Enhanced amylolytic activity in germinating barley through synthesis of a bacterial alpha-amylase', *Journal of Cereal Science*, **37**, 71–80.

Varshney, R.K., H. Zhang, E. Potokina, N. Stein, P. Langridge, A. Graner and H.N. Zhang (2004), 'A simple

hybridization-based strategy for the generation of non-redundant EST collections – a case study in barley (Hordeum vulgare L)', *Plant Science*, **167**, 629–34.
Varshney, R.K., T.C. Marcel, L. Ramsay, J. Russell, M.S. Roder, N. Stein, R. Waugh, P. Langridge, R.E. Niks and A. Graner (2007), 'A high density barley microsatellite consensus map with 775 SSR loci', *TAG Theoretical and Applied Genetics*, **114**, 1091–103.
Von Wettstein, D. (2007), 'From analysis of mutants to genetic engineering', *Annual Review of Plant Biology*, **58**, 1–19.
Wan, Y. and P.G. Lemaux (1994), 'Generation of large numbers of independently transformed fertile barley plants', *Plant Physiology*, **104**, 37–48.
Wang, M.-B., D.C. Abbott, N.M. Upadhyaya, J.V. Jacobsen and P.M. Waterhouse (2001), 'Agrobacterium tumefaciens-mediated transformation of an elite Australian barley cultivar with virus resistance and reporter genes', *Australian Journal of Plant Physiology*, **28**, 149–56.
Wang, M.-B., D.C. Abbott and P.M. Waterhouse (2000), 'A single copy of a virus-derived transgene encoding hairpin RNA gives immunity to barley yellow dwarf virus', *Molecular Plant Pathology*, **1**, 347–56.
Wanous, M.K. and J.P. Gustafson (1995), 'A genetic map of rye chromosome 1R integrating RFLP and cytogenetic loci', *Theoretical and Applied Genetics*, **91**, 720–26.
Wanous, M.K., P. Goicoechea and J.P. Gustafson (1995), 'RFLP maps of rye chromosomes 6R and 7R including terminal C-bands', *Genome*, **38**, 999–1004.
Waugh, R., N. Bonar, E. Baird, B. Thomas, A. Graner, P. Hayes and W. Powell (1997), 'Homology of AFLP products in three mapping populations of barley', *Molecular and General Genetics*, **255**, 311–21.
Wenzl, P., H. Li, J. Carling et al. (2006), 'A high-density consensus map of barley linking DArT markers to SSR, RFLP and STS loci and agricultural traits', *BMC Genomics*, **7**, article 206, 1–22.
Wenzl, P., J. Carling, D. Kudrna, D. Jaccoud, E. Huttner, A. Kleinhofs and A. Kilian (2004), 'Diversity Arrays Technology (DArT) for whole-genome profiling of barley', *Proceedings of the National Academy of Sciences of the USA*, **101**(26), 9915–20.
Wenzl, P., H. Li, J. Carling, M. Zhou, H. Raman, E. Paul, P. Hearnden, C. Maier, L. Xia, V. Caig, J. Ovesná, M. Cakir, D. Poulsen, J. Wang, R. Raman, K.P. Smith, G.J. Muehlbauer, K.J. Chalmers, A. Kleinhofs, E. Huttner and A.A. Kilian (2006), 'High-density consensus map of barley linking DArT markers to SSR, RFLP and STS loci and agricultural traits', *BMC Genomics*, **7**, article 206, 1–22.
Weyen, J., E. Bauer, A. Graner, W. Friedt and F. Ordon (1996), 'RAPD-mapping of the distal portion of chromosome 3 of barley, including the BaMMV/BaYMV resistance gene ym4', *Plant Breeding*, **115**, 285–7.
Wight, C.P., N.A. Tinker, S.F. Kianian, M.E. Sorrells, L.S. O'Donoughue, D.L. Hoffman, S. Groh, G.J. Scoles, C. Li, F.H. Webster, R.L. Phillips, H.W. Rines, S.M. Livingston, K.C. Armstrong, G. Fedak, S.J. Molnar and C.D. Li (2003), 'A molecular marker map in "Kanota" * "Ogle" hexaploid oat (Avena spp) enhanced by additional markers and a robust framework', *Genome*, **46**, 28–47.
Xue, G.P., M. Patel, J.S. Johnson, D.J. Smyth and C.E. Vickers (2003), 'Selectable marker-free transgenic barley producing a high level of cellulose (1,4-β-glucanase) in developing grains', *Plant Cell Reports*, **21**, 1088–94.
Zhang, S., M.-J. Cho, T. Koprek, R. Yun, P. Bregitzer and P.G. Lemaux (2000), 'Genetic transformation of commercial cultivars of oat (*Avena sativa* L.) and barley (*Hordeum vulgare* L.) using in vitro shoot meristematic cultivars derived from germinated seedlings', *Plant Cell Reports*, **18**, 959–66.
Zhang, S., M.-J. Cho, T. Koprek, R. Yun, P. Bregitzer and P.G. Lemaux (1999), 'Genetic transformation of commercial cultivars of oat (*Avena sativa* L.) and barley (*Hordeum vulgare* L.) using *in vitro* shoot meristematic cultures derived from germinated seedlings', *Plant Cell Reports*, **18**, 959–66.
Zhu, S. and H.F. Kaeppler (2003), 'A genetic linkage map for hexaploid, cultivated oat (Avena sativa L) based on an intraspecific cross "Ogle/MAM17-5"', *Theoretical and Applied Genetics*, **107**, 26–35.
Zohary, D. and M. Hopf (2000), '*Domestication of Plants in the Old World: the Origin and Spread of Cultivated Plants in West Asia, Europe and the Nile Valley*, Oxford: Oxford University Press.

48 Incremental benefits of genetically modified bananas in Uganda

Enoch M. Kikulwe, José Falck-Zepeda and Justus Wesseler

1 INTRODUCTION

Bananas (*Musa* spp) are the fourth most important food crop in the world, following rice, wheat and maize. They are grown in more than 150 countries, producing approximately 138.4 million tonnes of banana every year (FAOSTAT, 2010). Sub-Saharan Africa (SSA), where banana provides more than a quarter of the required dietary energy for over 100 million people, produces approximately 33 per cent of the global banana output. The East African region (including Rwanda, Uganda, Kenya, Tanzania and Burundi) is the major banana-producing and -consuming region in SSA. Uganda alone produces roughly 10.2 million tonnes (FAOSTAT, 2010) as the world's second producer after India, with the highest per capita consumption of cooking banana in the world (Clarke, 2003).

Banana production in Uganda is, however, limited by several productivity constraints such as pests, diseases, soil depletion and poor agronomic practices. To address those constraints, the country has invested significant resources in research and development and other publicly funded programmes, pursuing approaches over both the short and long term. Uganda formally initiated its short-term approach in the early 1990s, involving the collection of both local and foreign germplasm for the evaluation and selection of cultivars tolerant to productivity constraints. The long-term approach, launched in 1995, includes breeding for resistance to the productivity constraints using conventional breeding methods and genetic modification. Genetic modification projects in Uganda target the most popular and infertile cultivars that cannot be improved through conventional breeding. The main objective of genetic modification in Uganda is to develop genetically modified (GM) banana cultivars that are resistant to local pests and diseases, have improved agronomic attributes and are acceptable to consumers (Kikulwe et al., 2007).

However, GM bananas are currently a non-tradeable good in Uganda, as they are still undergoing confined field trial assessments. In fact, the introduction of GM banana in Uganda is likely to generate a wide range of concerns even if proven safe by scientists – as it has in other African countries. According to the Uganda National Council of Science and Technology (UNCST), for example, the main public concern is the safety of the technology for the environment and human health. Even if approved to be safe, the concerns about the GM crop compliance with biosafety regulations and the potential environmental and food safety risks can be important obstacles to public acceptance of biotechnology products in Africa (Paarlberg, 2008). Therefore, without the consent of the society at large, GM banana may fail in the Ugandan market.

In this chapter we present a real option model that shows how concerns about environmental risks can be considered within a cost–benefit analysis as a first step toward a socio-economic assessment of introducing a GM banana in Uganda. We estimate the economic welfare by considering the irreversible effects to see how the stream of incremental benefit will be affected under a longer planning horizon. This is the first study to show how much incremental benefit farmers (and consumers) would forgo if a GM banana is not introduced in Uganda even though the crop has passed the biosafety assessments. In a thought experiment we show the incremental benefits consumers would forgo if a safe GM banana is not accepted in Uganda. Some *ex ante* studies (Qaim, 1999; Kalyebara et al., 2007) have been conducted in the region to assess the economic benefits of biotechnology. Qaim (1999) assessed the welfare effects of adopting banana tissue culture planting materials in Kenya and Kalyebara et al., (2007) simulated gross economic benefits for banana that could be generated by a set of technology options – including current cultural practices, conventional improvement and genetic transformation – if they are successfully developed and adopted in Uganda. In these studies, authors estimated the welfare effects using the economic surplus framework considering a finite time period (that is, 20 years for Qaim, 1999 and 30 years for Kalyebara et al., 2007). In both studies, uncertainties about the benefits and costs as well as irreversible environmental concerns were not modelled explicitly, yet consumers are often more concerned about the unknown irreversible effects in the case of GM crops than the reversible benefits (for example, higher yields, resulting in increased farmers' profits) that may be generated in Africa and elsewhere (Paarlberg, 2008).

Irreversibilities and uncertainties have been considered within the literature on introducing GM crops (for example, Wesseler et al., 2007; Demont et al., 2004). Scatasta et al. (2006) introduced the term 'maximum incremental social tolerable irreversible costs' (MISTICs) to identify the threshold value for consumers' willingness-to-pay for not having a GM crop being introduced; the use of the concept within the biosafety debate is new.

Thus, we make two contributions to the knowledge concerning the relevance of socioeconomic analyses of GM crops. First, we present a general approach for assessing *ex ante* the economic benefits of introducing a GM banana in Uganda under uncertainty and irreversibility. Second, we discuss the main implications for biosafety regulations of GM crops in Uganda.

The chapter has five further sections. The next section presents the socio-economic aspects of a GM banana. Section 3 presents the theoretical approach, explaining and defining the different concepts used in the computation of the benefits and costs of GM banana. The data and its sources are presented in section 4. Section 5 reports and discusses the results. The final section draws conclusions and discusses implications for biosafety regulations of GM banana and GM crops in general in Uganda.

2 SOCIO-ECONOMIC ASPECTS OF A GM BANANA

The economic net benefits of introducing a GM banana depend on the reversible and irreversible benefits and costs the technology will generate. Reversible benefits and costs can be defined as those benefits and costs that can be reversed after the planting of the crop and do not result in additional *ex post* (after stopping production) benefits and costs. An illustrative example is the purchase of inorganic fertilizer. If the producer finds that producing a GM banana crop is no longer worthwhile – for instance if the price has drastically reduced, if consumers do not like the GM banana or if it is discovered that there are important negative effects of the crop and production is suddenly stopped by regulators – the purchased fertilizer can be used for other crops. Similarly, other variable costs can be considered as being reversible as well.

On the other hand, irreversible benefits and costs refer to those benefits and costs that will continue to occur even if GM bananas are no longer produced or those that cannot be fully reversed. Examples are sunk costs or chronic health damages from pesticide use. The reversible and irreversible benefits and costs can be further differentiated into private and public benefits and costs. This differentiation is useful for understanding the distribution of benefits and costs between, for example, farmers (private) and society at large (public) (see Wesseler, 2009). The non-private costs include the effects on non-target species, such as introduced genes in nematode-resistant GM banana that may affect beneficial non-target nematodes. Others include effects on human health such as antibiotic resistance and allergies, evolution of pests and disease resistant to the inserted genes (for example, Kendall et al., 1995) and loss of genetic diversity (FAO, 2001). Certainly, a net reduction in the use of insecticides and nematicides on GM banana will have positive impacts on human health, the environment and biodiversity: those can be considered irreversible benefits (Wesseler, 2003). Demont et al. (2004) provide a number of examples illustrating the difference between reversibility and irreversibility.

The different types of benefits and costs are summarized in Table 48.1, which shows a two-dimensional matrix differentiating between reversible and irreversible and private and non-private benefits and costs for an *ex ante* economic analysis of GM crops. The sum of quadrants 1 and 2 gives the value of the net social reversible benefits and that of quadrants 3 and 4 the net social irreversible costs. The irreversible costs are of critical importance for biosafety decision making. They are the major argument supporting biosafety regulations under the Cartagena Protocol on Biosafety (Secretariat of the Convention on Biological Diversity, 2000). However, it is not irreversibility itself that has been used exclusively to justify specific biosafety regulations for GM crops as well as to justify a delay in release to obtain additional knowledge and information on the new technology; rather, uncertainty about irreversible costs in combination with uncertainty about the economic benefits of GM crops has been put forward in the Cartagena Protocol and other regulatory processes to justify such interventions.

Table 48.1 The two dimensions of an ex ante *analysis of social benefits and costs of GM crops*

Reversibility	Scope	
	Private	Non-private
Reversible	**Quadrant 1** Private reversible benefits Private reversible costs	**Quadrant 2** External reversible benefits External reversible costs
Irreversible	**Quadrant 3** Private irreversible benefits Private irreversible costs	**Quadrant 4** External irreversible benefits External irreversible costs

Source: Demont et al. (2004).

In the context of the Cartagena Protocol, the introduction of a new GM crop becomes a decision-making process under uncertainty, irreversibility and flexibility. Analyzing decision-making under uncertainty, irreversibility and flexibility is not new to economists and has a tradition in environmental economics that originated in the early 1970s with papers published by Arrow and Fisher (1974) and Henry (1974), while in economics it can be traced back to Bachelier (1990) (Bernstein, 1992). Irreversible benefits and costs in combination with uncertainty and flexibility can be considered within a real option approach for the assessment of the adoption impacts of a GM crop. Examples are provided by Demont et al. (2004) and Scatasta et al. (2006) for the introduction of GM strains of sugar beet and corn in the European Union.

3 THEORETICAL MODEL

We followed a real option approach to analyze the social incremental benefits and costs, that is, the social incremental reversible benefits and maximum incremental social tolerable irreversible costs. In this approach, we begin with the assumption that incremental reversible net benefits follow a continuous-time, continuous-state process with trend, where GM crops may be released at a point in time. The social incremental reversible benefits W^* (the sign * indicates optimal threshold value) need to be greater than the difference between the social incremental irreversible costs (I) and the social incremental irreversible benefits (R), weighted by the size of the uncertainty and flexibility (or hurdle rate) associated with the introduction of the new technology. The hurdle rate is commonly expressed in the form $\frac{\beta}{\beta - 1}$, where $\beta > 1$ captures the uncertainty and flexibility effect and is a result of identifying the profit-maximizing decision rule under irreversibility, uncertainty and flexibility, if benefits follow a geometric Brownian motion.[1] The interpretation of the decision rule for the case of a GM banana is that as long as $W - \frac{\beta}{\beta - 1}(I - R) \leq 0$, Uganda should delay adoption of a GM banana until more information about the new technology is available.

In the context of GM crops, where people are more concerned about the not-so-well-known irreversible costs of the technology, it is feasible to estimate threshold values that indicate the maximum incremental social irreversible costs that an individual or society in general is willing to tolerate as compensation for the benefits of the technology. Scatasta et al. (2006) have called this threshold value the *maximum incremental social tolerable irreversible costs*, I^*, or MISTICs for short. In the specific case of Uganda, the estimated MISTICs can be interpreted as the maximum willingness to pay (WTP) for not having the GM banana approved for planting in the country. Actual incremental irreversible social costs, I, are to be no greater than the sum of incremental irreversible social benefits and incremental reversible social net benefits for introducing a GM banana, such that:

$$I < I^* = \frac{W}{\beta/(\beta - 1)} + R, \qquad (48.1)$$

Using equation (48.1) with parameter values generated for the case of GM banana can provide threshold values for the MISTICs. The values can be compared with information from secondary sources to identify whether the threshold value will be met in Uganda.

In practice, estimation of the MISTICs requires quantification of three factors: social incremental reversible benefits from GM crops (SIRBs or W); social incremental irreversible benefits (SIIBs or R) rate; and the hurdle rate, $\beta/(\beta - 1)$. All these factors can be estimated or calculated using econometric and mathematical modelling techniques following Demont et al. (2004), as presented in Appendix 48.1.

4 DATA AND DATA SOURCES

Secondary data have been used for the estimations of parameters in this chapter. Data are taken from the database of a NARO/IFPRI project conducted between 2003 and 2004 in Uganda. The data set is complemented by data from the Uganda Bureau of Statistics (UBOS) and the Food and Agriculture Organization. Table 48.2 lists the private and non-private reversible and irreversible benefits and costs directly and indirectly considered.

The social incremental reversible benefits (SIRBs) were estimated based on private net benefits. Private incremental reversible benefits can be defined as the difference between the gross margin from GM and non-GM bananas, excluding planting material. Table 48.3 shows the incremental benefits estimations for a GM banana in Uganda. The starting point for these estimations is the gross margin for a non-GM banana crop as reported by Bagamba (2007, p. 31). The annual variable costs for a non-GM banana crop include hired labour used mainly for weeding and crop sanitation. The use of other inputs such as fertilizer and pesticides is negligible (according to Bagamba, 2007). The average output in metric tons per year is about 10.6 per hectare with an average price of about 149 600 Uganda shillings (UGX) per metric ton. Under the current production practices, most farmers do not incur costs for planting materials. Most of the planting materials are exchanged for free between farmers (Kikulwe et al., 2007).

Table 48.2 Social benefits and costs for GM banana considered

Reversibility	Scope	
	Private	Non-private
Reversible	**Quadrant 1** *Benefits* • Higher yields *Costs* • Labour costs	**Quadrant 2** *Benefits* • Zero *Costs* • Zero
Irreversible	**Quadrant 3** *Benefits* • Negligible *Costs* • Planting material	**Quadrant 4** *Benefits* • Indirect: improved food safety and decreased vulnerability *Costs* • Indirect: possible health and environmental effects

Table 48.3 Incremental gross margin of cultivating one hectare of GM banana (20 per cent)

Variable	Non-GM banana (matooke)*	GM banana (matooke)**
Output (metric tons/year)	10.6	12.7
Price per ton (K)	149.6	149.6
Value of output (K)	1504.4	1902.9
Hired labour (hours)	232.8	256.1
Family labour (hours)	2295.8	2525.4
Total labour (hours)	2528.6	2781.5
Wage rate (UGX/hour)	400.1	400.1
Cost of hired labour (K)	93.2	102.5
Cost of planting materials (K)[a]	0.0	151.7
Gross margin (K)	1411.2	1648.7
Return to family labour (UGX/hour)	614.7	652.9
Expected average incremental gross margin (K)		237.5
Expected average private incremental reversible benefits (K)		389.2
Incremental average return per family labour (UGX/hour)		38.2
Total incremental labour income per hectare (K)		96.4

Notes:
Benefits and costs are valued in Uganda shillings (UGX 1750 ≈ US$1, by 2007); return to fixed resources (e.g., land) is not deducted from the gross margin in the computation of return to family labour.
[a] Tushemereirwe et al. (2003) recommend an average of about 1100 plantlets per hectare. Due to biosafety requirements, the cost of a GM banana plantlet may at least increase by 30% (UGX 1300), i.e., from the current UGX 1000 for a non-GM tissue cultured plantlet.

Sources: * Bagamba (2007); ** calculated by authors.

The average annual gross margin from producing one hectare of non-GM banana (traditional) is approximately UGX1 411 200 (US$800), excluding labour costs for planting. The main benefit of introducing a GM banana is an increase in banana yield through reduced biotic pressure. Assuming that planting a GM banana with a gene resistant to black Sigatoka[2] increases yield by 20 per cent and increases labour costs by about 10 per cent and that the average annual costs for planting material are UGX151 700, the gross margin per hectare would increase from about UGX1 411 200 to about UGX1 648 700, or by about UGX237 500. If the irreversible planting costs are not deducted, the expected average private incremental reversible benefits are about UGX389 200 per hectare (about US$222 per hectare).

The introduction of a GM banana will trigger an additional cost for planting materials. The total planting costs for about 1100 plantlets at a price of UGX1300 per plantlet are about UGX1 430 000 per hectare (about US$817). In our computations, we calculated the average annual cost of planting materials using a capital recovery factor for a 10 per cent interest rate and an expected GM banana plant life cycle of 30 years. Furthermore, we assumed no price discount for the GM banana at the farm gate and no other costs of adoption.

5 RESULTS AND DISCUSSION

5.1 Social Incremental Reversible Benefits (SIRBs)

The private incremental reversible benefits per hectare were used as the initial value for calculating the SIRBs. To obtain conservative estimates of SIRBs for Uganda, we assume that GM banana adoption follows a logistic function.[3] We used this function to predict what the incremental benefits would be if the GM banana were adopted according to the logistic adoption function. We estimated $\rho(t)$ based on the information provided by banana scientists of NARO. We asked experienced banana scientists to estimate the most likely adoption rate in the first four years after release of GM banana and computed averages for each year. We used an adoption ceiling rate (K) of about 50 per cent as a proxy for adopting any GM banana cultivar. This rate is based on the predicted demand for nakitembe (a commonly grown cultivar) after effective insertion of genes with 60 per cent resistance to both black Sigatoka and weevils, with supporting public investments in education, extension and market-related infrastructure as estimated by Edmeades and Smale (2006). The adoption curve for an adoption ceiling of 50 per cent and an estimated speed of adoption of 0.86 in linear form is $p(t) = 3.2 - 0.86t$. Figure 48.1 shows the assumed adoption curve.

In our analysis, we limit ourselves to the private incremental reversible benefits at the farm level, assuming all the rents from the new technology are captured by farmers. In the longer run, the rents will be distributed among farmers, the agents within the banana supply chain and banana consumers. Additional secondary benefits such as improved food security and reduced vulnerability to external shocks may be generated through higher farm income among banana growers. Assessing such benefits would require the use of a general equilibrium model for Uganda and be beyond the scope of this study. Thus, the computed SIRBs are equal to the private incremental reversible benefits (PIRBs) and reported in Table 48.4.

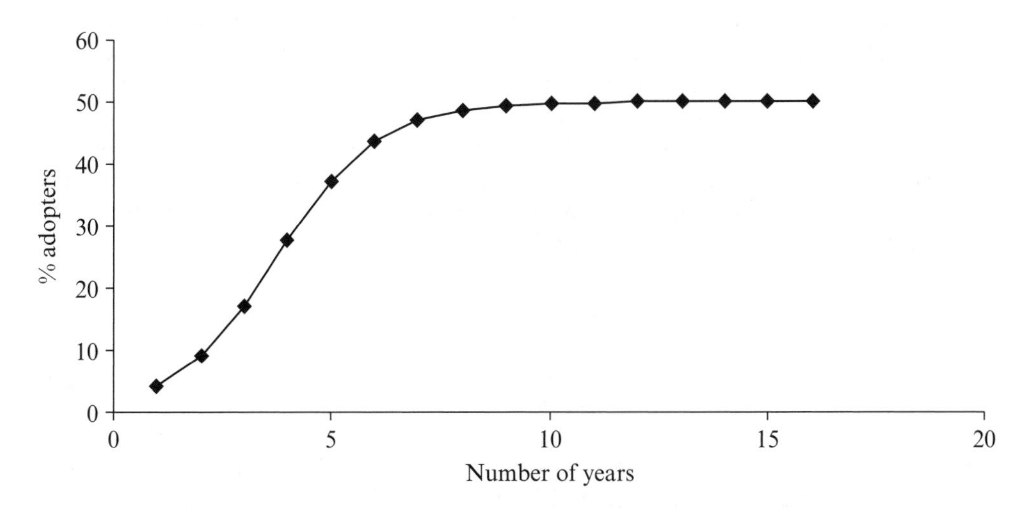

Figure 48.1 GM banana adoption rate over time (50 per cent adoption ceiling)

Table 48.4 Average annual SIRBs per banana-growing farm household per hectare at different risk-adjusted rates of return (μ)

	Risk-adjusted discount rates μ					
	0.04	0.06	0.08	0.10	0.12	0.14
SIRB (million $)	365	304	260	226	200	179
SIRB ($/ha)	459	399	356	326	303	287

Note: Exchange rate $1 = UGX1750.

Source: Calculated by authors.

The results in Table 48.4 show that the SIRBs, as expected, decrease with an increase in the risk-adjusted rate of return. The estimated SIRBs range between US$365 million and US$179 million per year, or US$459 and US$287 per hectare per year, for the range of risk-adjusted discount rates that varied from 4 per cent to 14 per cent.

We also tried to identify the social incremental irreversible benefits (SIIBs) on a per-hectare basis using information provided by Bagamba (2007). Most banana producers in Uganda do not use pesticides or fungicides to manage pests and diseases, as mentioned earlier. A small proportion (less than a quarter) of banana producers apply small amounts of pesticides.

5.2 Maximum Incremental Social Tolerable Irreversible Costs (MISTICs)

To estimate the MISTICs for introducing a GM banana, we first calculated the hurdle rate, a measure of irreversibility and uncertainty. Secondary time series data on banana yield per hectare (UBOS, 2006b) were used to estimate the drift and variance of the geometric Brownian motion as a proxy for the drift and variance rate from gross margin time series data. The geometric Brownian motion $U = (U_k(t), t \geq 0)$ is a continuous-time, continuous-state stochastic process in which the logarithm of the randomly varying quantity follows a Brownian motion: $U_k(t) = U_0 \exp[(\lambda - \frac{\sigma^2}{2})t + \sigma W(t)]$; where $W(t)$ is a Wiener process, U_0 is the initial real random number, t is the length of equally spaced intervals for all $t \in [0, T]$, and parameters λ and σ are constants.

The random variables $\log(U_k/U_0) \equiv g_k(t)$ are independently and identically distributed with mean $(\lambda - \sigma^2/2)\,t \equiv \alpha t$ (α is the expected growth rate or drift) and variance $\sigma^2 t$, where $k = 0,1,\ldots,n$. The maximum likelihood estimators for α and σ^2 were estimated as follows (see Campbell et al., 1997):

$$\alpha = \frac{1}{nt}\sum_{k=1}^{n} g_k(t) \tag{48.2}$$

$$\sigma^2 = \frac{1}{nt}\sum_{k=1}^{n}(g_k(t) - \alpha t)^2 \tag{48.3}$$

where t, the length of intervals, was one year ($t = 1$), and $n = 24$ years (1980 to 2004). The estimated parameter values were ultimately used to derive hurdle rates for Uganda.

From the Uganda Bureau of Statistics' data on total area and production of all types

Table 48.5 Hurdle rates, average annual MISTICs per hectare of GM banana, per household, and per banana-growing farm household at different risk-free rates of return (r) and risk-adjusted rates of return (μ)

Risk-free rate of return (r)		Risk-adjusted rates of return (μ)					
		0.04	0.06	0.08	0.10	0.12	0.14
	Hurdle rate	1.0169	1.0104	1.0075	1.0059	1.0048	1.0041
	MISTIC (million $)	359	301	258	225	199	178
0.00	MISTIC ($/ha)	451	394	353	324	302	285
	MISTIC ($/household)	69	58	50	43	38	34
	MISTIC ($/farmer)	239	201	172	150	133	119
	Hurdle rate	1.3298	1.0405	1.0166	1.0103	1.0075	1.0058
	MISTIC (million $)	274	293	256	224	198	178
0.04	MISTIC ($/ha)	345	383	350	322	301	285
	MISTIC ($/household)	53	56	49	43	38	34
	MISTIC ($/farmer)	183	195	170	149	132	119
	Hurdle rate				1.1386	1.0355	1.0161
	MISTIC (million $)				199	193	176
0.10	MISTIC ($/ha)				286	293	282
	MISTIC ($/household)				38	37	34
	MISTIC ($/farmer)				132	129	118

Note: Exchange rate $1 = UGX1750, in the year 2007.

Source: Calculated by authors.

of bananas, we estimated the average yield per hectare. Since cooking bananas contribute 80 per cent of total banana production in Uganda, this is a fairly good proxy for the yield of cooking bananas. We estimated a variance rate (σ^2) of 0.0328 and a drift rate (α) of 0.0083 for the yearly difference change for all years from 1980 to 2004. This implies that banana production in Uganda from 1980 to 2004 grew by 0.0083 per year, with a variance of approximately 0.03. Information about the risk-free rate of return and the risk-adjusted rate of return for farm household investments is rarely available and difficult to calculate.[4] Therefore, hurdle rates were calculated for different risk-free rates of return and risk-adjusted rates of return (0.04, 0.06, 0.08, 0.10, 0.12 and 0.14). Table 48.5 shows the computed annual MISTICs for a GM banana. The MISTICs are presented in total, on a per-hectare, per-household level, assuming 5 186 558 households as of November 2002 (UBOS, 2006a), and per banana-growing farmer, assuming 1 500 000 banana-planting farm households (Kalyebara et al. 2006).

The hurdle rates (Table 48.5) differ as the risk-free rate of return and risk-adjusted rate of return vary. For instance, at μ = 0.1 and r=0.04, the hurdle rate is about 1.01. In this case on average every US$1 of incremental social irreversible cost has to match with about 1.01 SIRBs to justify the immediate introduction of the GM banana. In general, the hurdle rates estimated in this chapter are very low compared with other estimates in the literature.

This indicates that the irreversibility effect is relatively small and much less important in comparison to other case studies, where the hurdle rates range between 1.04 and 3.69 (Demont et al., 2004) and 1.03 and 5.6 (Wesseler et al., 2007). Uganda's production data used to estimate the MISTICs are fairly smooth, in spite of observed biotic shocks in the 1990s and other years. Damage in a particular year may have been localized, yet heavy in those localized areas, so that national averages smooth out variations.

The annual MISTICs decrease as well with an increase in the risk-adjusted rate of return and with an increase in the risk-free rate of return. At $\mu = 0.10$ and $r = 0.04$, MISTICs are about US$224 million per year, or about US$322 per hectare per year. The MISTICs per banana-growing (farm) household and those per household (non-producing household) indicate a large difference between the two groups. The MISTICs per farm household are more than three times larger than the MISTICs per household. As indicated previously, the MISTICs can be interpreted as the maximum willingness-to-pay for not having a GM banana approved for planting in Uganda. Therefore, the difference in the MISTIC values between farm households and non-banana-producing households – both urban and rural – shows that in general the average banana-growing household may have a much larger interest than an average Ugandan household in having access to a GM banana even if the banana-growing household is concerned about the irreversible costs.

6 CONCLUSIONS AND POLICY IMPLICATIONS

In this study we have presented an approach for considering concerns about genetically modified crops within a socio-economic analysis of GM crops. We calculated the MISTICs associated with the adoption of a GM banana in Uganda. The MISTICs were presented for different risk-free and risk-adjusted rates of return. The results show the MISTICs to be between approximately US$176 million and US$359 million per year, or between US$282 and US$451 per hectare per year. In the scenario with a risk-adjusted rate of return of 12 per cent and a risk-free rate of interest of 4 per cent, which we consider to be a reasonable scenario based on the results of Mithöfer (2005), the annual MISTICs per household are about US$38. This result can be interpreted as follows: the immediate release of the GM banana should be postponed or abandoned only if the average household is willing to give up more than US$38 per year (or 3.2 per cent of the average annual household income) for not having such a banana introduced.

In the case where approval of the GM banana is delayed due to missing regulatory procedures and protocols, Uganda will forgo potential benefits (social incremental reversible benefits, or SIRBs) in the approximate range of US$179 million to US$365 million (or 7.5 to 15.3 per cent of the agricultural income) per year. This forgone benefit can be an indicator of how much Uganda can pay to compensate for potential damages. Additionally, the SIRBs provide a clue about the maximum costs farmers would endure in order to comply with biosafety regulations, including the cost of implementing coexistence policies and after deducting planting costs of US$101 per hectare. In a reasonable scenario, for instance, the average SIRBs total about US$303 per hectare. Adopters of the GM banana would not be willing to pay more than US$200 per hectare per year in transaction costs, that is, costs to comply with biosafety regulations, R&D costs and technology transfer costs. Assuming a maximum of 541 530 hectares that may be planted in GM banana in Uganda, this implies that the maximum total costs to bring the GM banana to Ugandan producers cannot exceed US$108 million. Otherwise, the GM banana is not a viable alternative.

The analyses in this chapter demonstrate the economic value and the effect of the forgone benefits as a result of waiting to release a GM banana. The results illustrate several implications to numerous stakeholders. First, the calculation of the MISTICs explicitly considers possible long-term effects of GM banana. The results indicate that with each year of delay in the introduction of a GM banana, Uganda loses about US$179 million to US$365 million to all households in Uganda. The MISTICs are in the order of about US$176 million or more. Only if the real average annual irreversible costs of planting a GM banana were to be as high as, or higher than, the irreversible benefits should the release be delayed. We have found no evidence yet that this will be the case. Given the potential and significant economic benefits from the introduction of a GM banana, NARO has to work harder to push the GM banana through the biosafety protocols as promptly and efficiently as possible.

Second, our findings indicate that a banana-growing household may have a much (three times) larger interest in having access to a GM banana than an average Ugandan household. This can be explained by the great losses experienced by farm households due to the prevailing banana constraints. The losses caused by banana constraints, therefore, make the opportunity cost to farmers of not using the GM banana technology extremely high. This implies that a farm household would naturally benefit disproportionately from a GM banana technology that is likely to ensure a return to sustainable production.

Third, biosafety regulatory assessment would be well advised to overcome the observed tendency of most regulatory processes globally of avoiding committing regulatory errors during decision-making and particularly stacking the odds in favour of not approving technologies that are safe against approving a technology that is not safe. In essence, decisions made by most regulatory bodies tend to be more precautionary than warranted by the evidence. To ensure a more balanced approach to decision-making, the literature suggests consideration of all benefits and costs – including opportunity and irreversible – supporting regulatory decision-making. This chapter proposes one alternative approach in this line of reasoning.

Lastly, the approach used here highlights how one can evaluate the socio-economic aspects of GM crops in general. To those stakeholders who are pessimistic about such technologies, it shows how much benefits are forgone as a result of a delayed release of a given technology. We have also indicated how one can consider long-term irreversible effects of GM crops. The approach can therefore be adapted to new GM crops requiring biosafety assessments prior to commercialization and can help to overcome one of the problems of establishing a biosafety system for Uganda or other African country.

NOTES

1. The geometric Brownian motion is a Wiener process with a geometric trend for which changes expressed as natural logarithms are normally distributed. The Wiener process is a continuous-time, continuous-state stochastic Markov process with three properties: (a) probability distributions of future values depend on the current value only; (b) the Wiener process grows at independent increments; and (c) changes are normally distributed. The assumption that the adoption of this technology follows a geometric Brownian motion accounts for the uncertainty of the technology (Cox and Miller, 1965).

2. In Uganda, black Sigatoka reduces yields by 30 to 50 per cent (Tushemereirwe et al., 2000) and greatly affects areas with medium and low productivity levels mostly lying in lowlands (below 1200 metres above sea level) of eastern and central Uganda. These areas contribute about two-thirds (64 per cent) of the total banana producing areas (Kalyebara et al., 2007).
3. Following Griliches (1957) and Feder et al. (1985), the adoption curve of a new technology is defined as $\rho(t) = K/1 + ae^{-bt}$, where $\rho(t)$ is the percentage planted with GM banana in a given year, K is the ceiling rate (the long-term upper bound of adoption), a is the constant, related to the time when adoption starts, b is the speed of adoption, and t is the time variable. We transformed the logistic adoption function into its log-linear form: $y(t) = \ln(\frac{\rho(t)}{K - \rho(t)}) = a + bt$, where $y(t)$ is the transformed function used to estimate parameters a and b using linear regression.
4. Mithöfer (2005) is a notable exception. The author estimated risk-adjusted rate of returns for farmers' investment in planting indigenous fruit trees in Zimbabwe ranging on average between 13.00 and 15.64 per cent.

REFERENCES

Arrow, K.J. and A.C. Fisher (1974), 'Environmental preservation, uncertainty and irreversibility', *Quarterly Journal of Economics*, **88**, 312–19.

Bachelier, L. (1900), 'Theory of speculation', Paris: Gauthier-Villars, trans. A.J. Boness, in P. Cootner (ed.), *The Random Character of Stock Market Prices*, Cambridge: MIT Press.

Bagamba, F. (2007), 'Market access and agricultural production: the case of banana production in Uganda', Doctoral thesis, Wageningen University, Wageningen, The Netherlands.

Bernstein, P.L. (1992), *Capital Ideas: The Improbable Origins of Modern Wall Street*, New York: The Free Press.

Campbell, J.Y., A.W. Lo and A.C. MacKinlay (1997), *The Econometrics of Financial Markets*, Princeton, NJ: Princeton University Press.

Clarke, T. (2003), 'Banana lab opens in Uganda: Genetic modification of clonal crop could soon follow', *Nature News*, 22 August, available at: http://www.bioedonline.org/news/news.cfm?art=430.

Cox, D.R. and H.D. Miller (1965), *The Theory of Stochastic Processes*, London: Chapman and Hall.

Demont, M., J. Wesseler and E. Tollens (2004), 'Biodiversity versus transgenic sugar beet: the one euro question', *European Review of Agricultural Economics*, **31**(1), 1–18.

Dixit, A.K. and R.S. Pindyck (1994), *Investment under Uncertainty*, Princeton, NJ: Princeton University Press.

Edmeades, S. and M. Smale (2006), 'A trait-based model of the potential demand for a genetically engineered food crop in a developing economy', *Agricultural Economics*, **35**, 351–61.

FAOSTAT (2010), 'FAOSTAT agriculture data', available at: http://faostat.fao.org/site/567/default. aspx#ancor.

Feder, G., R.E. Just and D. Zilberman (1985), 'Adoption of agricultural innovations in developing countries: a survey', *Economic Development and Cultural Change*, **33**, 255–98.

Food and Agriculture Organization (FAO) (2001), 'Biotechnology and banana production', Committee on commodity problems, Intergovernmental group on bananas and tropical fruits, San Jose, Costa Rica: FAO.

Griliches, Z. (1957), 'Hybrid corn: an exploration in the economics of technological change', *Econometrica*, **25**(4), 501–22.

Henry, C. (1974), 'Investment decision under uncertainty: the irreversibility effect', *American Economic Review*, **64**, 1006–12.

Kalyebara, M.R., P.E. Ragama, G.H. Kagezi, J. Kubiriba, F. Bagamba, C. Nankinga and W.K. Tushemereirwe (2006), 'Economic importance of the banana bacterial wilt in Uganda', *African Crop Science Journal*, **14**(2), 93–105.

Kalyebara, M.R., S. Wood and P.N. Abodi (2007), 'Assessing the potential impact of selected technologies on the banana industry in Uganda', in M. Smale and W.K. Tushemereirwe (eds), *An Economic Assessment of Banana Genetic Improvement and Innovation in the Lake Victoria Region of Uganda and Tanzania*, Washington, DC: International Food Policy Research Institute.

Kendall, H.W., R. Beachy, T. Eisner, F. Gould, R. Herdt, P.H. Raven, J.S. Schell and M.S. Swaminathan (1995), 'Bioengineering of crops, report of the World Bank panel on transgenic crops', Washington, DC: World Bank.

Kikulwe, M.E., K. Nowakunda, M.S.R. Byabachwezi, J.M. Nkuba, J. Namaganda, D. Talengera, E. Katungi and W.K. Tushemereirwe (2007), 'Development and dissemination of improved banana cultivars and management practices in Uganda and Tanzania', in M. Smale and W.K. Tushemereirwe (eds), *An Economic Assessment of Banana Genetic Improvement and Innovation in the Lake Victoria Region of Uganda and Tanzania*, Washington, DC: International Food Policy Research Institute, pp. 37–48.

Mithöfer, D. (2005), 'Economics of indigenous fruit tree crops in Zimbabawe', PhD thesis, Faculty of Economics, Hannover University, Germany.
Paarlberg, R. (2008), *Starved for Science: How Biotechnology is Being Kept out of Africa*, Cambridge, MA: Harvard University Press.
Qaim, M. (1999), 'Assessing the impact of banana biotechnology in Kenya', *ISAAA Briefs*, no. 10, Ithaca, NY: ISAAA.
Scatasta, S., J. Wesseler and M. Demont (2006), 'Irreversibility, uncertainty, and the adoption of transgenic crops: experiences from applications to Ht sugar beet, Ht corn, and Bt corn', in R.E. Just, J.M. Alston and D. Zilberman (eds), *Regulating Agricultural Biotechnology: Economics and Policy*, New York: Springer, pp. 327–52.
Secretariat of the Convention on Biological Diversity (2000), *Cartagena Protocol on Biosafety to the Convention on Biological Diversity: Text and Annexes*, Secretariat of the Convention on Biological Diversity, Montreal.
Tushemereirwe, W.K., I.N. Kashaija, W. Tinzaara, C. Nankinga and S. New (2003), *Banana Production Manual: A Guide to Successful Banana Production in Uganda*. 2nd edn, Kampala: Makerere University Printery.
Uganda Bureau of Statistics (UBOS) (2006a), '2002 Uganda population and housing census: main report, available at: http://www.ubos.org/2002%20Census%20Final%20Reportdoc.pdf.
Uganda Bureau of Statistics (UBOS) (2006b), National Statistics Databank, Agriculture and Fisheries, 'Area planted and production of selected food crops, 1980–2004', available at: http://www.ubos.org.
Wesseler, J. (2003), 'Resistance economics of transgenic crops: a real option approach', in R. Laxminarayan (ed.), *Battling Resistance to Antibiotics and Pesticides*, Washington, DC: Resources for the Future, pp. 214–37.
Wesseler, J. (2009), 'The Santaniello theorem of irreversible benefits', *AgBioForum*, **12**(1), 8–13.
Wesseler, J., S. Scatasta and E. Nillesen (2007), 'The maximum incremental social tolerable irreversible costs (MISTICs) and other benefits and costs of introducing transgenic maize in the EU-15', *Pedobiologia*, **51**(3), 261–9.

APPENDIX 48.1 DEFINING DIFFERENT CONCEPTS USED IN THE COMPUTATION OF THE BENEFITS AND COSTS OF GM BANANA

Social Incremental Reversible Benefits (SIRBs)

The social incremental reversible benefits (SIRBs) can be measured using a partial equilibrium model of the Ugandan banana industry. The estimation of the SIRBs is similar to what would be obtained if a traditional cost–benefit analysis based on a Ricardian rent model is used. But, since the $SIRB_{PV}$ (expected present value of SIRB) per hectare are uncertain, we estimated the value of the project under uncertainty by assuming annual SIRB follow a stochastic process based on a geometric Brownian motion. The incremental benefits, the expected future profit flow (*SIRB*), given by $\partial SIRB = (PQ^{GM} - C^{GM}) - (PQ^{non\text{-}GM} - C^{non\text{-}GM})$ follow a geometric Brownian motion. Where PQ^{GM} is the revenue from GM banana, C^{GM} is the cost of production of GM banana, $PQ^{non\text{-}GM}$ is the revenue from non-GM banana, and $C^{non\text{-}GM}$ is the cost of production for non-GM banana. If $SIRB = 0$, there is no extra gain from growing GM banana but the farmer gains income: the farmer gets the same income from GM banana as for non-GM banana. The use of the Ricardian rent model can be justified as the parameter estimates used for the calculations are based on time-series data reflecting the changes in prices and costs as a result of changes in demand and supply. Including additional demand and supply effects in this case would result in double counting. Further, effects on international trade of banana from Uganda can be ignored as the traditional varieties are not exported and gene transfer between the GM and non-GM banana varieties is not possible considering the biology of banana.

We computed the SIRBs at time t ($SIRB(t)$), as the *SIRBs* at complete adoption times the adoption rate at time t, $\rho(t)$, times the expected growth (or drift) at rate α: $SIRB(t) = SIRB \cdot \rho(t) \cdot e^{\alpha t}$. The discounted sum of *SIRBs*, $SIRB_{PV}$ for Uganda over time is calculated as:

$$SIRB_{PV} = \int_0^{\infty} SIRB(t)\, e^{-(\mu - \alpha)t} dt, \qquad (48A.1)$$

where μ is the risk-adjusted discount rate and α the drift rate of the geometric Brownian motion. The initial value for the calculation of the area for banana production is 1 670 000 hectares at full adoption.

Social Incremental Irreversible Benefits (SIIBs)

Similar to the SIRBs, the SIIBs function per hectare at time t can be computed as $SIIB(t) = SIIB \cdot \rho(t) \cdot e^{\alpha t}$. The discounted sum of *SIIBs*, $SIIB_{PV}$, over time are approximated by:

$$SIIB_{PV} = \int_0^{\infty} SIIB(t) e^{-(\mu - \alpha)t} dt \qquad (48A.2)$$

The Hurdle Rate

The hurdle rate, $\beta/(\beta - 1)$, depends on the uncertainty related to the expected SIRBs. The hurdle rates are calculated by defining β as follows (see Dixit and Pindyck, 1994, pp. 147–52):

$$\beta = \frac{1}{2} - \frac{r - \delta}{\sigma^2} + \sqrt{\left[\frac{r - \delta}{\sigma^2} - \frac{1}{2}\right]^2 + \frac{2r}{\sigma^2}} > 1 \tag{48A.3}$$

where r is the risk-free rate of return and δ is the convenience yield defined as the difference between the risk-adjusted discount rate μ and the drift rate α; that is, $\delta = \mu - \alpha > 0$, $\mu \geq r$, and α and σ^2 (variance rate) as before.

49 Biofuels and GM feedstocks

Alphanso Williams and William A. Kerr

1 INTRODUCTION

Biofuels are renewable energy used in transportation as a substitute and/or complement to fossil fuels. The application of these types of biofuels in transport may be in pure form, that is, 100 per cent of the fuel is bio-based and/or blended where a percentage of the fuel is renewable. For example, E15 or B15 means a 15 per cent blending of ethanol or biodiesel with fossil-based fuel. The two major types of biofuels currently produced are ethanol and biodiesel. These fuels are derived from biomass or waste. The major producers of biofuel are the United States (US), Brazil, the European Union (EU), China, Canada and India. Biofuels are expected to offer these countries improved energy security, a reduction in externalities that negatively impact the environment and rural development opportunities. Further, countries, particularly developing countries, may benefit from biofuels through an opportunity to supply a number of major nations that have mandated consumption, such as the US and EU.

One aspect of the production of biofuels is that it diverts productive agricultural land out of food production. As a result, food security may decline due to rising food prices, particularly for the very poor. Hence, there are potential negative externalities associated with biofuel production. A paradigm shift toward the encouragement of the development of biofuels industries took place in a number of countries before the negative externalities became apparent. As the effect on food prices became evident during the run-up to the food price spikes of 2008, the major producers of biofuels attempted to limit the effects of the externality through, for example, the US Energy Independence and Security Act of 2007 and the EU Directive 2009/28/EC. One way to reduce the fuel versus food conflict over the use of farmland is to increase the productivity of farmland. The use of modern biotechnology to improve crops is one way to improve the productivity of farmland, but the use of agricultural crops derived from biotechnology is an internationally contentious issue, with some countries firmly opposed to the planting of biotech crops while others have embraced it enthusiastically. Those who have not embraced biotechnology often object on the basis of the unknown risks these crops might pose for environmental sustainability and biodiversity. Hence, the fuel versus food controversy has become part of the much larger global debate over the acceptance of agricultural biotechnology.

2 THE GLOBAL BIOFUEL INDUSTRY

The US and Brazil are the two largest producers of ethanol while the EU is the largest producer of biodiesel. The standards used for quality control reflect the market conditions in the US, Brazil and the EU and product classification (Tripartite Task Force, 2007). For example, standards for biodiesel in the US and Brazil describe a product with

Table 49.1 Key differences and similarities in ethanol specification

Category A: Similar	Category B: Significant differences	Category C: Fundamental differences
Colour	Ethanol content	Water content
Appearance	Acidity	
Density	Phosphorus content	
Sulfate content	Chloride acid	
Copper content	Gum/evaportation residue	
Iron content	pHe	
Sodium content		
Electrolytic conductivity		

Source: Tripartite Task Force (2007).

a blending component in conventional hydrocarbon-based diesel fuel in contrast to the European description of a product that is either a blending component or a stand-alone substitute for petroleum-based diesel fuel. These three leading producers of biofuels have made an effort to reconcile standards through a White Paper on Internationally Compatible Biofuel Standards (Tripartite Task Force, 2007). The report shows significant differences in standards for ethanol among the US, the EU and Brazil, although these differences are not considered to significantly impede international trade in ethanol. Brazil and other producers wishing to supply the EU are required to provide additional drying and testing. The only fundamental difference in standards for ethanol among these three major producers is the water content requirement, with the EU having a limit of 0.24 per cent and the US and Brazil having 1 per cent and 0.4 per cent respectively. Table 49.1 summarizes the key similarities and differences in the ethanol specifications of the EU, US and Brazil.

The production of biofuels in China, Canada and India, the other major biofuel-producing countries, is skewed toward ethanol. In the case of China, Dong (2007) considers the domestic production of biofuels as a means to utilize any excess supply of grains as a way of supporting grain prices. The utilization of excess grain or stabilization of grain prices is correlated with rural development through improvements in farmers' income prospects.

The US is the major adopter of genetically modified (GM) corn, with almost total adoption. Thus, the crops used to produce the very large quantities of ethanol in the US are biotechnology based. Alternative ethanol inputs are sugar cane and wheat. For biodiesel, vegetable, palm and jatropha oils, rapeseed and soybeans are used as biomass inputs. The inputs used in production vary across nations based on climatic conditions, although the origin of the biomass is often crops initially developed as foods. Biofuels derived from food crops are categorized as first generation, whereas yet-to-be-commercialized, or even fully-developed, non-food alternative crops are denoted second-generation biomass sources for biofuel production. Only when, or if, these second-generation crops destined for use in biofuels production are profitably commercialized will the competition for farmland be reduced. The competition between food and fuel markets for crop and production resources is argued to be a contributory factor to high food prices (Hailu and

Weersink, 2010). In recognition, governments have moved towards alternative feedstock and/or limiting the use of food crops in the production of biofuels. Developing countries such as China and India, for example, are pursuing biofuels derived from sweet sorghum, cassava and jatropha. These inputs are expected eventually to ease the competition between food and renewable fuels markets. The key word is *eventually*. Until non-food-based crops are both technically feasible and economically profitable, biotechnology remains a major, maybe the major, mechanism to increase the productivity of agricultural land and assist in reducing the price of food.

2.1 The United States

The US Congress passed the Energy Independence and Security Act of 2007 (EISA, 2007) primarily to reduce dependence on foreign oil. In 2011, the US remained the number one producer of ethanol in the world with a reported production of 13.9 billion gallons, a 5 per cent increase over 2010 (Renewable Fuels Association, 2011). Biofuel development is supported through the EISA 2007. The mandate,[1] as set out by the Renewable Fuels Standards II (RFSII), requires that transportation fuel sold in the United States on an annual average basis contain at least the applicable volume of renewable fuel, advanced biofuel,[2] cellulosic biofuel or biomass-based diesel (EISA, 2007). A total of 36 billion gallons of renewable fuel is mandated to be blended with gasoline by 2022. Of the 36 billion gallons, 16 billion gallons should be cellulosic biofuel which reduces greenhouse gas emission (GHG) by at least 60 per cent. The applicable volume of advanced biofuel (not derived from corn starch) is 5 billion gallons and the conventional biofuel (derived from corn starch) requirement is residually capped at 15 billion gallons as of 2015. The GHG emission requirements for advanced and conventional biofuels are at least 50 per cent and 20 per cent respectively (EPA Regulatory February, 2010). Yano et al. (2010) argue that if no cheap alternative other than sugar cane-based ethanol is found, fuel based on sugar cane may be used to satisfy the requirements of the advanced biofuel mandate. The US is not a major sugar producer, thus raising the potential for imported sugar cane-based ethanol. The US mandate is depicted in Figure 49.1.

The US Environment Protection Agency (EPA) is responsible for developing and implementing regulations to ensure that transportation fuel sold in the US contains a minimum volume of renewable fuel on an annual basis. The regulatory requirements for RFSII apply to domestic and foreign producers of renewable fuels used in the US (EPA, 2010). The thresholds of 50 per cent and 60 per cent may be revised only if the levels are not commercially feasible for fuels made using a variety of feedstocks, technologies and processes to meet the applicable reduction (EISA, 2007, Sec 201 1E). Biodiesel from soy oil and renewable diesel from waste oils, fats and greases are expected to comply with the 50 per cent threshold reduction in greenhouse gas (GHG) emissions.

2.1.1 Requirements for feedstock producers

The feedstocks used to produce renewable fuels are required to be renewable biomass. Renewable biomass, as defined by Section 201 (I) of EISA 2007, may be planted crops, crop residue harvested from agricultural land, planted trees, tree residue, animal waste material and animal by-products. The EISA 2007 limits the types of biomass and lands dedicated to biofuel production. For both domestic and foreign non-agricultural sector feedstocks,

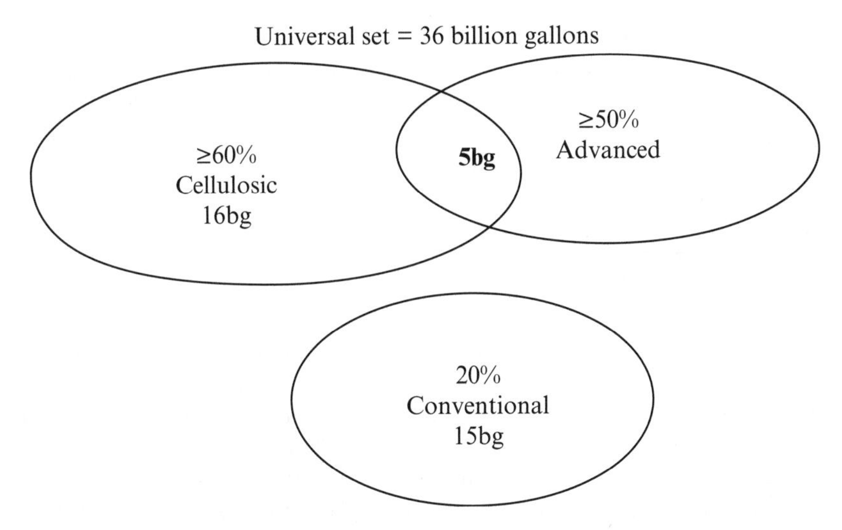

Note: The overlapping of cellulosic and advanced means that the ethanol mix under the RFSII can be made up of cellulosic that is not advanced, and advanced, which is cellulosic.

Figure 49.1 Ethanol mix of the US RFSII

renewable fuel producers are required to collect and maintain appropriate records from their feedstock suppliers to ensure compliance with the renewable biomass requirement. Furthermore, renewable fuel producers using agriculturally-based feedstocks grown in the US will be deemed compliant based on the EPA's aggregate compliance determination. Similarly, for foreign-based feedstocks derived from agricultural production used as an input to biofuels, the future aggregate option is available if the source region can provide sufficient data to support aggregate analysis and a monitoring program.

Even though the EISA requires 36 billion gallons (bg), of biofuels by 2022 for use in transportation, the US transport sector is faced with a technological constraint on demand known as the blend wall. In October 2010, the EPA waived a limitation on selling fuel that is more than 10 per cent ethanol. This limitation is referred to as the blend wall (BW), which restricts the volume of renewable fuel that can be blended with gasoline – primarily due to the technical capacity of automobile fuel systems to efficiently use non petroleum-based fuel sources. A BW of 15 per cent suggests that the maximum share for renewable fuel in total gasoline consumed is 15 per cent, or E15, and the remaining 85 per cent represents the share of gasoline produced from petroleum. The waiver increases the share of ethanol in the market for transportation fuel but remains a long way below the 25 per cent share required under RFSII by 2022.

2.2 Renewable Fuel Standard II

For 2012, the renewable fuel standard was 15.2 bg. The cellulosic standard was set at 8.65 million gallons (mg), biomass-based diesel at 1.5 and 2 bg for advanced biofuel (sugar cane) apart from biodiesel, as shown in Table 49.2. Generally, the cellulosic standard is subject to revision on a yearly basis in accordance with market conditions.

Table 49.2 Renewable fuel standards 2012

Fuel category	Percentage (%)	Volume (billion gallons)
Cellulosic biofuel	0.006	0.00865
Biomass-based diesel	0.91	1.5
Other advanced biofuel	1.21	2.0
Renewable fuel	9.23	15.2

Source: EPA (2011).

While the US is mandating non-corn-based inputs to ethanol production, biotechnology is not being used to improve the technical efficiency of these alternative crops to any significant degree. Where biotechnology is being used to increase the productivity of crop land – for example GM varieties of corn – the potential contribution of biotechnology is inhibited due to the cap in the mandate for corn-based ethanol. Of course, as the productivity of GM corn increases, the cap on corn-based ethanol will release some land to be re-applied to food production.

2.3 The European Union

In 2009, the EU Parliament issued Directive 2009/28/EC to member states promoting the use of renewable energy. The directive aims to reduce environmental degradation and enhance energy independence through the requirement that 20 per cent of overall EU energy consumption must come from renewable sources by the year 2020. Furthermore, there is a mandatory 10 per cent minimum target for all member states for the consumption share of renewable fuels in transportation. The EU mandate is a market share target provided in paragraph 18 of the Directive, defining the 10 per cent target as that share of final energy consumed in transport which is to be achieved from renewable sources as a whole but not solely from biofuels (EU Directive 2009/28/EC). In addition, the contribution to the target by second-generation biofuels, that is, biofuels made from wastes, residues, non-food cellulosic material and ligno-cellulosic material, is measured at twice the rate attributed to other biofuels (Article 21(2)).

The European Biodiesel Board (EBB) reports biodiesel production totaling approximately 2.4 billion gallons (US) in 2010 (EBB, 2012). In 2011, the EU produced 1.1 billion gallons of ethanol (RFA, 2011). In general, biofuel production is predominantly biodiesel, with Germany, France and Spain accounting for 54 per cent of total EU biodiesel production capacity. As in the US, however, the development of non-food biofuel crops remains largely at the conceptual or experimental level and, hence, they are unlikely to contribute to the renewable fuel target in any meaningful way by 2020. Unlike the US, however, there has been little acceptance or regulatory approval of GM crops (Viju et al., 2012). Basically, the improvements to agricultural productivity arising from biotechnology have been eschewed by the EU. Thus, in the EU the contribution that biotechnology could make in alleviating the competition between fuel and food has been effectively eliminated. Further, even if GM crops achieve regulatory approval, the likelihood of their adoption remains remote (Viju et al., 2012).

The Directive provides that sustainability criteria for biofuels as set out in Article 17[3] should be met for renewable fuels to be counted towards the blending target. However, biofuel produced from waste and residues, other than agriculture, fisheries and forestry residues, only need to satisfy sustainability criterion A (Article 17). The sustainability criteria are as follows:

A. The greenhouse gas emission (GHG) saving[4] is at least 35 per cent, increasing to 50 per cent effective 1 January 2017 and further increasing to 60 per cent,[5] effective 1 January 2018 (paragraph 2).
B. Biofuels are not produced from raw materials obtained from land with high biodiversity value[6] and high carbon stock (paragraphs 3 and 4).
C. Biofuels are not produced from raw materials on peatland in January 2008 unless evidence is provided that the cultivation and harvesting of that raw material does not involve drainage of previously undrained soil (paragraph 5).
D. The agricultural raw materials cultivated in the Community and used in the production of biofuels are obtained in accordance with the requirements and standards under the provisions referred to 'Environment' in part A and point 9 under Public, Animal and Plant Health of Council Regulation (EC) No. 73/2009 of 19 January 2009.

Criteria B and C can be viewed as capacity constraints or 'land specificity' requirements on biofuels producers willing to fulfill the target of 10 per cent (Williams and Kerr, 2011). In effect, the land-use criteria prohibit the growth of GM feedstocks produced on land which is being used to produce food. Given that the EU is not self-sufficient in biofuels and increases in farmland productivity are constrained through the virtual ban on the planting of GM crops, the EU is faced with meeting the mandate through imports. The major potential source of those imports is developing countries. Consequently, the restrictions on land use provide a potential advantage for countries with abundant land. The EU sustainability criteria apply equally to domestic and foreign-sourced biofuels. In other words, land used in foreign countries to produce biofuels destined for the EU must meet the sustainability criteria. Biofuels considered to have met the criteria are verified by the mass-balance method. The mass-balance system allows consignments of raw material or biofuel with differing sustainability characteristics to be mixed, but requires information about their sustainability characteristics (Article 18; paragraph 1 (a) and (b)). The Directive allows cooperation among member states in order to achieve the respective targets. The European Commission can independently rule a source of biofuel unfit and, hence, not allow a member state to take biofuel from that source (Article 25(3)[7] and (Article 18(8)). This clearly increases the risks associated with producing crops to be used as biomass.

Thus, just as in the EU itself, the contribution biotechnology can make to increasing productivity is limited because land currently used to produce food cannot be diverted to biofuel production. Further, while the EU does not explicitly restrict the import of biofuels derived from GM crops – for example, it imports corn-based ethanol from the US – the use of the same crops for food or fuel effectively limits the adoption of biotechnology-based food crops for the production of biofuels. This is because the EU bans the imports of GM-based foods.[8] As segregation of GM and non-GM crops is virtually impossible throughout the entire supply chain, if biotechnology crops can be grown, developing countries face the loss of the EU market for their conventional crops. As a result, a number of developing countries have not approved GM crops (see Chapter 12 in this volume by Paarlberg). Hence, GM crops cannot be grown as an input to biofuels or assist in increasing farmland productivity for those developing countries.

2.4 China

In the 1990s, biofuel policy was influenced by China moving from being a net exporter of oil to a net importer. The oil deficit was amplified by the dramatic increase in the number of automobiles along with rapid industrial growth. There was a significant increase in both gasoline and diesel consumption brought about by robust industrial growth. In conjunction with the rapid increase in consumption of fossil fuels came serious environmental problems, in particular air and water pollution in the densely populated parts of the country. A National Grain Reserve system is maintained in China which contains 17–18 per cent of grain produced every year. It is maintained to reduce the disruptions that can arise from poor harvests or natural disasters (Dong, 2007). China has a long history of famines and the government is very sensitive to the social disruptions and hardships they cause (Kerr, 2011). Li (2007) argues that the grain reserve became a burden to the government, and developing the fuel ethanol industry was a way to support alternative uses of grains such as corn and wheat to increase the efficiency of grains utilization. Liu (2005) posits that as the ethanol industry developed, the grain reserves were virtually exhausted. In 2011, ethanol production was 554 million gallons (RFA, 2011).

The Ministry of Agriculture of People's Republic of China (MAPRC) in the Agricultural biomass energy industry development plan (2007–2015), outlined its approach to the development of fuel ethanol in China, suggesting it should be guided by the rule of 'not competing with existing food crops' and 'not competing with land used for food production' (China Ministry of Agriculture, 2007; *Asia Times*, 2006; Dong, 2007). This guiding rule implies that even though the biofuel industry offers energy security and environmental advantages, these benefits should not undermine food security. As a result, China is moving towards using non-grain feedstocks as inputs for the production of biofuels. Land use is further limited by poor water supply in areas which are not used for food production. Given its huge population and relatively poor agricultural productivity, China had been forced to intensively use virtually all available agricultural land in the years when it pursued a policy of self-sufficiency – from approximately 1950 to the late 1970s. Non-food crops such as tuberous crops, sweet sorghum, cellulosic biomass and cassava are potential alternative feedstocks to grain. Tisdell (2009) highlights that currently all the feedstock used by China for ethanol production consists of food items, but possibly only those of a quality considered to be substandard as human food. Nonetheless, given China's large population, the threat to food security posed by food-based biofuels is quite credible. As such, the use of GM feedstocks in the development of biofuels which are a competitive alterative to fossil fuels is a potential option for lowering the risk of high food prices. Again, GM crops can contribute to alleviating the fuel versus food problem through increasing the productivity of agricultural land. China has implemented market policies, quality standards, laws and regulation to govern the biofuel industry (Loppacher and Kerr, 2004a). Table 49.3 shows that the General Administration of Quality Supervision, Inspection and Quarantine (AQSIQ) is the agency responsible

Table 49.3 Biofuel quality standards, market policies, laws and regulations

Title	Issuing date	Issuing ministries and commissions	Category
GB18350-2001 Denatured fuel ethanol	April 2001	AQSIQ*	Quality standards
GB18351-2001 Vehicle ethanol gasoline	April 2004	AQSIQ	Quality standards
National standards for biodiesel used in diesel/biodiesel blends (BD100)	May 2007	AQSIQ	Quality standards
Extended pilot scheme of fuel ethanol promotion rules	February 2004	Eight ministries and commissions	Policy of market admittance
Detailed extended pilot scheme of fuel ethanol promotion rules	February 2004	Eight ministries and commissions	Policy of market admittance
Renewable Energy Law	February 2005	The State Council of the People's Republic of China	Law

Note: * AQSIQ: General Administration of Quality Supervision, Inspection and Quarantine of the People's Republic of China.

Source: Extracted from Zhong et al. (2010, Table 1).

for quality standards. It issued standards for fuel ethanol in April 2001 and 2004 and May 2007 for biodiesel. In early 2004, policies were formulated concerning market admittance of biofuels, and the following year, 2005, the Renewable Energy Law was passed.

2.4.1 Production capacity

Table 49.4 shows the operational biofuel plants in China and their production capacity. Tianguan Group Co., Ltd and COFCO Bio-energy (Zhaodong) Co., Ltd, which are both supported by the government, have ventured into the production of cellulosic ethanol. The total annual biodiesel production capacity exceeds 1 million tons, while the actual annual production is close to 300000 tons as production is constrained by a limited supply of feedstocks (Zhong et al., 2010). Also, in contrast to fuel ethanol production, biodiesel production is provided with fewer incentives for development, at least in the short term (Dong, 2007).

2.4.2 Production

The three main feedstocks used for the production of biofuels are sugar, corn and cellulosic feedstocks, with production capacity regulated by the government. As of 2013, only GM corn has been developed. Dong (2007) asserts that current production costs of fuel ethanol from non-grain feedstocks are higher than those from corn or wheat. However, the technology for producing bioethanol from non-food-grain feedstock has reached the preliminary conditions needed for commercialization (NDRC PRC, 2007). Table 49.5 shows the feedstocks that are expected to be key inputs in biofuel production, implying a shift from food grains to non-food feedstocks.

Table 49.4 Biofuel (ethanol) plants operated in China

Plant	Year built	Location	Capacity (tons/year)	Feedstock
Huarun Ethanol Co. Ltd	1993	Heilongjiang	100 000	Corn
Jilin Fuel Ethanol Co. Ltd	2001	Jilin	400 000	Corn
Tianguan Fuel Ethanol Co. Ltd	2002	Henan	500 000	Wheat, tubers
Fengyuan Group	2002	Anhui	440 000	Corn
COFCO Bioenergy Co. Ltd	2005	Guangxi	200 000	Cassava

Source: Extracted from Tian et al. (2009, Table 2, p. S79).

Table 49.5 Crops for ethanol production to China and regions recommended for their cultivation

Region	Energy crops
North-east	Sweet sorghum
North China	Sweet sorghum, sweet potato
Loess plateau	Sweet sorghum
Inner Mongolia and Xingiang	Sugar beet, sweet sorghum
South China	Cassava
South-west	Sweet potato

Source: Extracted from Tian et al. (2009, Table 4, p. S80).

A sweet sorghum ethanol plant with a capacity of 3000 tons per year began operation in Jiangsu province in December 2007 (Sun et al., 2009) while a pilot plant has been built in Inner Mongolia to produce fuel ethanol from sweet sorghum (Zhang and Lv, 2008).

Sugar cane is another possible input. China is the third-largest producer of sugar cane. As of 2013, however, no GM sugar cane has been developed or commercialized. Although China ranks high in the production of this feedstock, sugar cane-based ethanol production has been crowded out by the high demand for sugar arising from its large population. Sugar cane is primarily cultivated in Guangxi and Yunnan provinces, with its cultivation area accounting for about 70 per cent of China's total sugar cane cultivation area (Zhong et al., 2010).

Corn could still be a major feedstock. Zhong et al. (2010) assert that government policies and new technologies resulted in the increased production of corn-based ethanol, but the surge peaked in 2005 with corn-based projects being spurned due to soaring food prices. Nonetheless, the technology for processing and the equipment used in the production of fuel ethanol from corn is relatively well developed compared to all other feedstocks.

Cassava is considered to be a reliable alternative to food crops for ethanol production, given the advantages of low-cost, reliable process technology and the fact that cassava does not compete with food production (Zhong et al., 2010). In 2008, the Food and Agriculture Organization (FAO) reported that an international network called the Global Cassava Partnership aims to increase research and investments in cassava to increase farm yields and to position cassava as an input in the production of biofuels (FAO, 2008).

Table 49.6 Cellulosic ethanol plants and production capacity

Enterprise	Production capacity (tons/year)	Date of operation	Note
COFCO*	500	2006	Pilot plant; corn stalk; steam explosion
Henan Tianguan Group Co. Ltd	3000	12/2007	Corn stalk; Acid pre-treatment
Xinjiang Jimsar Santai Brewery Group Co. Ltd	100000	12/2009	
Baicheng- Tingfeng Ethanol Company	30000		
Jilin Fuel Ethanol Co., Ltd	3000	12/2007	
Shandong Longlive Bio-technology Co., Ltd and Shandong University	3000		
PetroChina	3000	04/2008	
China Petrochemical Corporation and COFCO*	20 (plants) @ 100000		

Note: * COFCO, China National Cereals, Oils & Foodstuffs Import & Export Corporation.

Source: Extracted from Zhong et al. (2010), Table 2.

Furthermore, research by the FAO has shown that cassava is a competitive feedstock when crude oil price reaches US$45 per barrel.

Sweet potato also offers some opportunities. The breeding of sweet potato exclusively for energy production has been funded by the National '863 program', aimed at a new sweet potato species with high yield and high starch content as well as better pest resistance (Zhong et al., 2010). China accounts for more than 80 per cent of the world's sweet potato production with sweet potatoes outstripping other fuel ethanol feedstocks in both theoretical fuel ethanol yield and production cost per unit of cultivated area (Tian et al., 2009). It is unclear whether biotechnology is being used in the development of these varieties.

Finally, cellulosic feedstocks are significant. In 2005, the production of cellulosic waste from crop production (straw) was about 700 million tonnes with corn, wheat and rice straw accounting for 44 per cent, 22 per cent, and 17 per cent, respectively (Li et al., 2005). In addition, around 300 million tonnes are available that can be utilized for energy purposes (NDRC PRC, 2007). Table 49.6 shows the cellulosic ethanol plants operational in China with maximum and minimum capacity of 100000 tons and 3000 tons per year respectively.

Biodiesel is not likely to be a big part of the Chinese alternative fuels market. China is currently a net importer of all the major edible vegetable oils that could be used to produce biodiesel. Hence, production of biofuel is skewed towards ethanol due to the crops where potential surpluses can arise. Further, biodiesel production is provided with fewer incentives (Dong, 2007). The main feedstocks for biodiesel production include oil crops and algae along with waste oils and fats which serve as a supplementary feedstock for biodiesel production. Annual production capacity of biodiesel has increased from 85000 tons in 2005 to 1 million tons in 2007 (Sun and Liang, 2008). The GSI (2008) reported that China produced 300000 tons of biodiesel in 2007.

In China, it is the public sector that takes the lead in developing GM crops, making it different from the private sector-led research and development that characterizes most other major producers of new GM crops and varieties (Loppacher and Kerr, 2004b). Thus, whether or not attention is paid to the development of GM crops to support the biofuels industry depends on the priorities of the government. There is some evidence to suggest that the intellectual property system in China inhibits collaboration with other researchers working on biotechnology improvements across the globe (Loppacher and Kerr, 2004c). At present there seems to be no direct link between biotechnology and biofuels policy in China.

2.5 India

The Government of India formulated a National Policy on Biofuels in 2008 (Ministry of New and Renewable Energy, 2008) to ensure that a minimum level of biofuels is available in the market. The policy proposes a target of 20 per cent of fuels to be biofuels, both for biodiesel and bioethanol, by 2017.

2.5.1 Production

The production strategy for developing biofuels as an alternative to fossil fuels involves non-food feedstocks grown on degraded lands not suitable for agriculture. In 2009, India produced approximately 92 million gallons of ethanol (RFA, 2009). Production of biofuels is skewed to ethanol partly because of high production costs for biodiesel (US Department of Agriculture, 2011). Presently, ethanol is produced mainly from molasses, a by-product of sugar. India is the world's second largest producer of sugar cane (FAO, 2010). As a result, the sugar and distillery industries are being encouraged to augment production of ethanol to meet the requirements of the targets without constraining the supply of sugar or the availability of ethanol for industrial use.

India is a developing country with a large population similar to China and the utilization of agriculture land for biofuel production has the potential to threaten food security. As a result, the production of biodiesel involves utilizing waste and degraded forest and non-forest lands, and then only for the cultivation of shrubs and trees bearing non-edible oil seeds as inputs. Gonsalves (2006) asserts that the lack of assured supplies of vegetable oil feedstock has stymied efforts by the private sector to set up biodiesel plants in India. Given this constraint, non-edible oil from jatropha (*Jatropha curcus*) is used as feedstock for biodiesel (Gonsalves, 2006). Jatropha has a high (inedible) oil content and 'can grow in arid and semi-arid regions, tropical and subtropical areas and grow even on barren and wastelands, degraded soils having low fertility and moisture, but cannot stand heavy frost' (Punia, 2007). Punia (2007) estimates that 40 million hectacres (ha) of wasteland could be developed in India to grow jatropha. Biotechnology crops have not won easy acceptance in India. As a result, biotechnology is excluded from making a contribution to the development of the biofuel industry.

2.6 Brazil

Brazil is the world's second largest producer of ethanol, with a reported production of 6.9 billion gallons in 2011 (RFA, 2011). Walburger et al. (2006) posit that Brazil can be considered the most efficient producer with their production cost being similar to the cost of petroleum fuels, even though ethanol production has not been supported by a direct subsidy since the 1990s. Sugar cane is the primary feedstock for the production of ethanol. For biodiesel, the government supports production with a blending mandate of 5 per cent. Lendle and Schaus (2010) posit that soybean biodiesel from Brazil for the EU market meets the GHG emissions saving requirements, because low weight biodiesel is shipped rather than bulky soybeans. Furthermore, Brazil is an adopter of GM soybeans. Given that sugar cane is the overwhelming biofuel feedstock of choice in Brazil, any major contribution of biotechnology to biofuel production will have to await the commercialization of a GM sugar variety.

2.7 Canada

In 2011, Canada recorded a 30 per cent increase in ethanol production with an estimated 462 millions of gallons compared with 2010 (RFA, 2011). Biofuel production capacity, which is skewed towards ethanol, has grown significantly since 2005 due to substantial subsidies by both the federal and provincial governments. The industry is supported through federal and provincial mandates. The primary reason for 'going green' is to reduce GHG emissions as Canada is a net exporter of energy (Viju and Kerr, 2013). The main feedstocks used for producing ethanol are corn and wheat. The development of GM wheat has not gone forward due to concerns regarding the potential loss of the EU market for Canadian and American wheat. The inability to effectively segregate food wheat from industrial wheat destined to be an input to biofuel production has inhibited any research into industrial wheat varieties.

The Government of Canada has implemented various programs to encourage increased biofuel adoption. Some of these programs include the Agricultural Bioproducts Innovation Program and the Capital Formation Assistance Program for Renewable Fuels Production, which are designed for research and development and to provide incentives for participation in new renewable fuels production capacity respectively (Agriculture and Agri-Food Canada, 2007). Natural Resources Canada has funded two rounds of an Ethanol Expansion Program to provide long-term loans to biofuel producers as well as significant funding through the NextGen Biofuels Fund, designed to support the development and construction of cellulosic or 'second generation' biofuels (Fridfinsson and Rude, 2009). On this note, a concerted effort is being made to move away from first-generation biofuels due to the potential inflationary pressure on food prices. As yet, applications of biotechnology have not played a major role in the improvement of plants considered likely sources of feedstock for biofuels.

3 OVERVIEW OF BIOFUEL DEVELOPMENT

The EU and US have expanded their biofuel mandates as evident in the new directive of the European Parliament released in 2009 and the US Energy Independence and Security Act of 2007 (EISA, 2007). For Canada, federal legislation in 2008 resulted in the establishment of a mandated use of renewable fuel content in gasoline of 5 per cent by 2010 (GSI, 2009). Furthermore, the governments of the major producing

Table 49.7 Major producing nations mandates

Country	Blending/utilization mandates	GM feedstocks
US[a]	**36 billion** gallons of renewable fuels by **2022**.	Corn
Brazil[b]	All gasoline must contain between **20 per cent and 25 per cent** of anhydrous ethanol. **5 per cent** biodiesel.	Soybeans
EU[c]	**5 per cent** share in transport fuel by **2015** and **10 per cent** target by **2020**.	–
Canada[d]	**5 per cent** share in gasoline by **2010**. **2 per cent** share in diesel fuel and heating oil by **2012**.	Corn
China[e]	Ethanol: trial period of **10 per cent** blending mandates in some regions. Five Chinese provinces require **10 per cent** ethanol blends: Heilongjian, Jilin, Liaoning, Anhui, and Henan. **10 per cent** ethanol blending mandate by **2020**.	Corn
India	Ethanol blending; **5 per cent** in gasoline in designated states in **2008**, to increase to **20 per cent** by **2017**.	Corn

Notes:

[a] Energy Independence and Security Act 2007;

[b] http://www.ethanolproducer.com/article.jsp?article_id=2466;

[c] Directive 2009/28/EC promoting the use of energy from renewable sources;

[d] Report on Economic Assessment of Biofuel Support Policies, OECD Trade and Agriculture Directorate, July 2008;

[e] Tisdell (2009).

countries provide far-reaching assistance in the form of loans, subsidies, research and development (R&D) grants and tax incentives. The major producing nations have used blending or utilization mandates which can be considered another form of indirect subsidy offered by the governments as they are in effect guaranteeing a domestic market that is otherwise commercially unviable (Murphy, 2007). In addition, these measures bring stability and predictability for those considering investing in biofuel facilities (UNCTAD, 2009). Table 49.7 outlines the blending or utilization mandates for the major biofuels-producing nations and GM feedstocks used or available as biomass.

The long-term blending or utilization mandates can be viewed as ambitious given the substantial investments in infrastructure, including production capacity that will be required. With regard to the EU policy framework, it was proposed that the level of use of agro-fuels dependent on food crops be fixed on a yearly basis by the European Commission in consultation with the other agencies such as Food and Agriculture Organization (FAO) and possibly excluding this type of fuel from GHG reduction efforts (Inside US Trade, 2008). Furthermore, the technologies needed to achieve the utilization target (16 bg EISA) for second-generation biofuels (cellulosic) have yet to be developed. Hence, producing the mandated volume especially for the US RFSII appears to be more wishful thinking than grounded in the reality of the technical capacity (Williams and Kerr, 2011). Canada, China and India are the three smaller producers among the major six producing nations with production skewed to ethanol. The respective governments of these countries have committed to further developing the biofuel industry, notably, second-generation biofuels derived from cellulosic conversion. The potential new technologies are believed not to suffer from the adverse externalities associated with first-generation biofuels as they are expected to have a relatively small impact on agricultural markets (UNCTAD, 2009).

The respective policies guiding the development of the biofuel industry in the major producing countries, except for Brazil, continue to evolve with revised measures being implemented. For example, the mandate for ethanol derived from corn-based ethanol (conventional biofuel) under the US Energy Independence and Security Act of 2007 is a residual with a maximum of 15 billion gallons (bg) starting in 2015. Also, India and China made policy changes in 2008 in response to the apparent inflationary impact of biofuel production on food prices. The changes resulted in a shift away from using primarily food-based feedstock to using non-food-based feedstocks for the production of biofuels. China and India are both experiencing rapid economic growth, with the result that their oil deficit will be likely to continue to increase unless offset by biofuels through higher GM feedstock adoption. Although the benefits from engaging in biofuels production may be 'real', Asia faces several constraints in increasing their production of biofuels. Tisdell (2009) argues that opportunity costs are associated with the production of biofuels, which alludes to the food versus fuel debate and that the major expansion in biofuel production may result in biodiversity loss as a result of both agricultural intensification and expansion onto previously unused land. Furthermore, the area of land available for expanding biofuel feedstock production in Asia (including China and India) without reducing food supplies is limited.

Table 49.8 shows the potential demand of ethanol for the six major producing nations based on respective blending or utilization mandates and gasoline consumption estimates for 2006 as well as 2011 production totals. Opportunities for developing countries to supply the shortfall clearly exist, but the standards for biofuel required by potential importers may considerably inhibit the ability of these countries to realize those opportunities.

Table 49.8 Potential demand for ethanol until 2022

Country	Gasoline consumption (billion gallons) 2006	Potential demand for ethanol until 2022 (billion gallons)	Ethanol production* 2011 (billion gallons)
US	140	31	13.9
EU	39	1.2	1.1
China	18	2.7	0.55
Canada	10	0.52	0.46
Brazil	6.3	1.3	5.7
India	3.6	0.3	–

Note: * Renewable Fuels Association.

Source: Extracted from UNCTAD (2009).

3.1 Product Differentiation

The US requires corn-based ethanol produced by new technologies and at new facilities (or existing facilities that have increased capacity) to reduce GHG emissions by 20 per cent. For sugar cane-based ethanol, the requirement is at least 50 per cent (EISA, 2007). These may be difficult targets for developing countries to meet – or even to prove. In 2010, US soybean producers sought the support of the US Trade Representative to challenge the EU renewable fuel policy because of differing GHG emissions standards for like products and extensive monitoring expectations (Inside US Trade, 2010). Under the EU policy, soybean-based biodiesel does not meet the sustainability criteria when using default GHG emission savings values; therefore, exporters will need to prove compliance. The default GHG emission savings value for soybean-based biodiesel is 31 per cent (Directive 2009/28/EC) while the EU directive requires a minimum of 35 per cent. Ironically, the US standard is higher than the EU standard as biodiesel is expected to reduce GHG emissions by at least 50 per cent in contrast to the EU's minimum of 35 per cent. As a result, the US soybean producers claim that the EU erred in calculating that soy-based biodiesel only provides a 31 per cent reduction in GHG emissions. Hence, the issue of 'correct science' in the context of biofuels arises. For developing countries to mount similar challenges would require a major effort given their low level of institutional capacity to initiate trade disputes and the particular technical expertise required to make a credible case. There are significant domestic vested interests in all of the major potential biofuel markets for developing-country biofuels that advocate strongly for protection – particularly farm lobbies and those who have invested in domestic production capacity that may be underutilized.

As suggested throughout this chapter, the wider acceptance, regulatory and trade difficulties affecting the adoption of biotechnology have spilled over to inhibit the benefits that developing countries can expect to reap from biofuels. Trade in biofuels faces significant market access challenges across a range of issues including subsidization, product classification for tariff purposes and lack of international product characteristics standardization. The failure to agree internationally on a set of rules for trade for the products of modern biotechnology remains the major constraint (Loppacher and Kerr, 2005a). The direct risks to trade associated with the absence of a transparent internationally agreed set of trade rules for GM products are, however, dwarfed by the long-run effects such risks have on investments in the development of new, and improvements to existing, GM crops. There is little doubt that biotechnology could contribute to the development of crops, both food and non-food that could be tailored for use as biofuel feedstocks. The problems associated with segregation of GM crops make the development of biofuel-specific crops targeted to developing countries too risky for those considering investments in biotechnology. Of course other factors contribute to the unwillingness to invest in GM crops specifically for developing countries – in particular the poor protection provided to intellectual property in many developing countries (Cardwell and Kerr, 2008; Gaisford et al., 2007; Loppacher and Kerr, 2005b; Kerr and Isaac, 2005; Perdikis et al., 2004; Boyd et al., 2003; Gaisford et al., 2002). Hence, the impact of biotechnology's contribution to both the development of crops tailored to developing-country biofuel feedstock needs and to lessen the constraints that led to fuel versus food controversies is considerably reduced relative to its potential.

4 CONCLUSION

This overview of the global biofuel industry and the potential role for GM feedstocks in developing biofuels raises some awkward questions. In the first instance, the spectre of declining food security has prompted policymakers to revise their approach to biofuel development. The US (RFSII) and EU Directive 2009/28/EC require long-term use of renewable energy in transportation, subject to sustainability. These mandates create long-term demand for biofuels and an impetus for higher efficiency in production as non-food inputs such as cellulosic feedstocks are developed and become commercially viable. Technical development and commercial viability of non-food biofuel feedstock, however, remain elusive. A potential policy alternative to limiting food-based biofuels' contribution to the energy mix is the use of new GM feedstocks or higher adoption/ uptake of existing GM feedstocks for biofuels production. One of the main benefits of GM feedstocks is greater supply of crops, *ceteris paribus*, which would sustain supply to both food and fuel markets, particularly for developing countries with large populations. Coincidentally, the US is the largest producer of ethanol and a leading producer of GM corn. In light of the supply pressures brought by biofuel mandates, GM adoption may evolve from food security to energy security as long as second-generation technology is commercially infeasible. Therefore, GM feedstocks have the potential to reduce the competition between food and fuel markets over crop and production resources. The wider global controversy over the appropriateness of GM crops, however, inhibits this avenue, particularly for developing countries.

NOTES

1. In this case, a mandate is a government requirement to use a specific quantity of a biomass-based fuel in blending or stand-alone marketing of transportation fuels. In other words, firms producing or marketing biofuels must purchase the mandated quantities of inputs.
2. Advanced biofuel means renewable fuel other than ethanol derived from corn starch that has life-cycle greenhouse gas emissions that are at least 50 per cent less than baseline greenhouse gas (GHG) emissions. The type of fuels eligible for consideration as advanced biofuel include: ethanol derived from cellulose, hemi-cellulose or lignin, sugar or starch and waste material. Biodiesel, biogas, butanol and other fuel derived from cellulosic biomas may be considered (Sec. 201, Paragraph B, Energy Independence and Security Act 2007). According to Williams and Kerr (2011) these alternative fuels are not yet technically feasible on a commercially viable scale, and most are likely to be decades away from being at that stage.
3. The criteria are discussed below.
4. See Annex V of Directive 2009/ 28/EC for typical and default greenhouse gas emission saving values by production pathway if no net carbon emissions is from land use change (EU Directive, 2009).
5. A new proposal by the EU Commission if accepted by the EU Parliament will result in a 60 per cent minimum GHG saving threshold for new production by 1 January 2018.
6. A draft consultation document on the criteria and geographic ranges to determine which grassland can be considered to be highly biodiverse grassland is available at: http://ec.europa.eu/energy/renewables/consultations/doc/2010.
7. Further reference to Articles 3 and 7 of Decision 1999/468/EC applies having regard to the provisions of Article 8.
8. Recently the EU has approved for import a limited number of GM crops used for animal feed.

REFERENCES

Agriculture and Agri-Food Canada (2007), 'Agriculture and Agri-Food Canada's biofuels programs', Agricultural Co-operative Development Initiative Conference, 8 March, available at: http://www.coopzone.coop/files/AAFC%20Biofuels%20Programs,%20Chantale%20Courcy.pdf, accessed 8 August 2013.

Asia Times (2006), 'Biofuels eat into China's food stocks', *Asia Times*, 21 December, available at: http://www.atimes.com/atimes/China_Business/HL21Cb03.html, accessed 8 August 2013.

Boyd, S.L., W.A. Kerr and N. Perdikis (2003), 'Agricultural biotechnology innovations versus intellectual property rights – are developing countries at the mercy of multinationals?', *The Journal of World Intellectual Property*, **6**(2), 211–32.

Cardwell, R. and W.A. Kerr (2008), 'Protecting biotechnology IPRs in developing countries: simple analytics of a levy solution', *Journal of Agricultural Economics*, **59**(2), 217–36.

China Ministry of Agriculture (2007), 'Agricultural Bioenergy Industry Development Plan (2007–2015)'.

Dong, F. (2007), 'Food security and biofuels development: the case of China', *Briefing Paper no. 07-BP 52*, Center for Agricultural and Rural Development, Iowa State University.

Energy Independence and Security Act (2007), 'Energy Independence and Security Act of 2007', HR 6, United States, available at: http://www.gpo.gov/fdsys/pkg/BILLS-110 hr6enr/pdf/BILLS–110hr6enr.pdf, accessed 8 August 2013.

Environmental Protection Agency (2010), 'Regulatory announcement', February, available at: http://www.epa.gov/otaq/renewablefuels/420f10007.pdf, accessed 8 August 2013.

Environmental Protection Agency (2011), 'Regulatory announcement', December, available at: http://www.epa.gov/otaq/fuels/renewablefuels/documents/420f11044.pdf, accessed 8 August 2013.

EU Directive (2009), Directive 2009/28/EC of the European Parliament and of the Council, 23 April, available at: http://eur-lex.europa.eu/LexUriServ/LexUriServ.do? uri=Oj:L:2009:140:0016:0062:en.pdf, accessed 8 August 2013.

European Biodiesel Board (FBB) (2012), 'Statistics', available at: http://www.ebb-eu.org/stats.php and http://www.ebb-eu.org/prev_stats_production.php, accessed 8 August 2013.

Food and Agriculture Organization (2008), 'Cassava for food and energy security: investing in cassava research and development could boost yields and industrial uses', available at: http://www.fao.org/newsroom/en/news/2008/1000899/index.html.

Food and Agriculture Organization (2010), 'FAOSTAT–Brazil', available at: http://faostar.fao.org/site/339/default.aspx, accessed 8 August 2013.

Fridfinsson, B. and J. Rude (2009), 'The effects of biofuel policies on global commodity trade flows', working paper, Canada Agricultural Trade Policy Network (CATPRN).

Gaisford, J.D., J.E. Hobbs and W.A. Kerr (2007), 'Will the TRIPS Agreement foster appropriate biotechnologies for developing countries?', *Journal of Agricultural Economics*, **58**(2), 199–217.

Gaisford, J.D., R. Tarvydas, J.E. Hobbs and W.A. Kerr (2002), 'Biotechnology piracy: rethinking the international protection of intellectual property', *Canadian Journal of Agricultural Economics*, **50**(1), 1–14.

Global Subsidies Initiative (GSI) (2008), 'Biofuels – at what cost? Government support for ethanol and biodiesel in China', Geneva: International Institute for Sustainable Development, available at: http://www.iisd.org/gsi/sites/default/files/china_biofuels_subsidies.pdf, accessed 8 August 2013.

Global Subsidies Initiative (GSI) (2009), 'Biofuels – at what cost? Government support for ethanol and biodiesel in Canada', Geneva: International Institute for Sustainable Development, available at: http://www.globalsubsidies.org/files/assets/oecdbiofuels.pdf, accessed 8 August 2013.

Gonsalves, J. (2006), 'An assessment of the biofuels industry in India', *United Nations Conference on Trade and Development (UNCTAD) Report*, Geneva: UNCTAD.

Hailu, G. and A. Weersink (2010), 'Commodity Price Volatility: the impact of commodity index traders', CATPRN commissioned paper no. 2010–02, Canadian Agricultural Trade Policy and Competitiveness Network, available at: http://www.uoguelph.ca/catprn/PDF–CP/Commissioned_Paper_2010-2_Hailut Weersink. pdf, accessed 8 August 2013.

Inside US Trade (2008), 'EU developing strong biofuels law, criteria could be trade barrier', Inside US Trade, 23 May.

Inside US Trade (2010), 'Soybean producers lobby USTR, USDA to fight new fuel barrier in EU', *Daily News*, 15 November.

Kerr, W.A. (2011), 'Food sovereignty: old protectionism in somewhat recycled bottles', *African Technology Development Forum Journal*, **8**(1&2), 4–9.

Kerr, W.A. and G.E. Isaac (2005), 'The international treatment of biological material as intellectual property', *Journal of International Biotechnology Law*, **2**(3), 105–111.

Lendle, Andreas and Malorie Schaus, (2010), 'Sustainability criteria in the EU renewable energy directive: consistent with WTO rules?', ICTSD information note no. 2, September.

Li, G. (2007), 'Status, issues and plans for enhancing bio-ethanol production in China', unpublished paper, Institute of Crop Sciences, Chinese Academy of Agricultural Sciences, Beijing.

Li, J.F., R.Q. Hu, Y.Q. Song, J.L. Shi, S.C. Bhattacharya and P.A. Salam (2005), 'Assessment of sustainable energy potential of non-plantation biomass resources in China', *Biomass Bioenergy*, **29**, 167–77.

Liu, Y. (2005), 'Raw material shortage fetters China's ethanol ambition', *China Watch*, Beijing, 6 October.

Loppacher, L.J. and W.A. Kerr (2004a), 'China's regulation of biotechnology – does it conform to the WTO?', Saskatoon: Estey Centre for Law and Economics in International Trade, available at: www.esteycentre.com.

Loppacher, L.J. and W.A. Kerr (2004b), 'China's biotechnology industry and the international protection of intellectual property rights', *Journal of International Biotechnology Law*, **1**(5), 177–86.

Loppacher, L.J. and W.A. Kerr (2004c), 'Integrating China's biotechnology industry into global knowledge creation – intellectual property is the key', *Journal of World Intellectual Property*, **7**(4), 549–62.

Loppacher, L.J. and W.A. Kerr (2005a), 'Can biofuels become a global industry?: Government policies and trade constraints', *Energy Politics*, **5**, 7–27.

Loppacher, L.J. and W.A. Kerr (2005b), 'Developing countries' protection of intellectual property under TRIPS and innovation investment', *The Journal of World Intellectual Property Rights*, **1**(1–2), 1–17.

Ministry of New and Renewable Energy (2008), 'National policy on biofuels', New Delhi: Government of India, available at: http://mnre.gov.in/file-manager/UserFiles/biofuel_policy.pdf.

Murphy, S. (2007), 'The multilateral trade and investment context for biofuels: issues and challenges', London: IIED, and Minneapolis, USA: IATP, available at: http://pubs.iied.org/pdfs/15513IIED.pdf.

National Development and Reform Commission of the People's Republic of China (NDRCPRC) (2007), 'Medium and long term development plan for renewable energy', July, available at: http://www.chinaenvironmentallaw.com/wp-content/uploads/2008/04/medium-and-long-term-development-plan-for-renewable-energy.pdf.

Perdikis, N., S.L. Boyd and W.A. Kerr (2004), 'Multinationals, biotechnology, intellectual property and developing countries – should developing countries seek to be exploited?', in D. Meyer-Dinkgrafe (ed.), *European Culture in a Changing World: Between Nationalism and Globalism*, London: Cambridge Scholars Press, pp. 1–10.

Punia, M.S. (2007), 'Current status of research and development on jatropha (*Jatropha curcus*)', paper presented at USDA Global Conference on Agricultural Biofuels: Research and Economics, 20–22 August, Minneapolis, MN.

Renewable Fuels Association (2009), 'Industry statistics', available at: http://ethanolrfa.org/pages/World-Fuel-Ethanol-Production, accessed 8 August 2013.

Renewable Fuels Association (2011), 'Industry Statistics', available at: http://ethanolrfa.org/pages/World-Fuel-Ethanol-Production, accessed 8 August 2013.

Sun, C. and W. Liang (2008), 'Status of biodiesel development and production in China', *SinoGlobal Energy*, **13**, 23–8.

Sun, Z.M., L. Hou, J.B. Zhang, X. Zhou and L.P. Liu (2009), 'Present status and prospect of non-grains fuel ethanol industrialized production', *Liquor-Making Science & Technology*, **5**, 94–8.

Tian, Y.S., L.X. Zhao, H.B. Meng, L.Y. Sun and J.Y. Yan (2009), 'Estimation of unused potential for biofuels development in (the) People's Republic of China', *Applied Energy*, **86**(S1), 77–85.

Tisdell, C. (2009), 'The production of biofuels: welfare and environmental consequences for Asia', Economics, Ecology and the Environment, Working Paper no. 159, a paper presented at the International Workshop on Climate Change Growth in Asia, Brisbane, 8 September.

Tripartite Task Force (2007), '*White paper on internationally compatible biofuel standards*', Brazil, United States of America and European Union, available at: http://ec.europa.eu/energy/res/biofuels_standards/doc/white_paper_icbs_final.pdf.

United Nations Conference on Trade and Development (UNCTAD) (2009), *The Biofuels Market: Current Situation and Alternative Scenarios*, New York: UNCTAD.

United States Department of Agriculture (2011), 'Global Agriculture Information Network: India Biofuels Annual', Washington, DC: USDA.

Viju, C. and W.A. Kerr (2013), 'Taking an option on the future: subsidizing biofuels for energy security or reducing global warming', *Energy Policy*, **56**, 543–8.

Viju, C., M.T. Yeung and W.A. Kerr (2012), 'The trade implications of the post-moratorium European Union approval system for genetically modified organisms', *Journal of World Trade*, **46**(5), 1207–38.

Walburger, A.M., D. Le Roy, K.K. Kaushik and K.K. Klein (2006), 'Policies to stimulate biofuel production in Canada: lessons from Europe and the United States', A BIOCAP Research Integration Program Synthesis Paper, BIOCAP, available at: http://www.biocap.ca/rif/summary/Walburger_Allan.pdf.

Williams, A. and W.A. Kerr (2011), 'Wishful thinking in energy policy: biofuels in the US and EU', *Energy Politics*, **XXV**, 34–46.

Yano, Y., D. Blandford and Y. Surrey (2010), 'The implications of alternative US domestic and trade policies for

biofuels', paper presented at the 84th Annual Conference of the Agricultural Economics Society, Edinburgh, 29–31 March.
Zhang, M.H. and H.S. Lv (2008), 'Development of the production techniques of fuel alcohol by non-grain crops in China', *Liquor-Making Science & Technology*, **9**, 91–5.
Zhong, C., Cao Ying-Xiu, Li Bing-Zhi and Yuan Ying-Jin (2010), 'Biofuels in China: past, present and future', *Biofuels, Bioproducts and Biorefining*, **4**(3), 326–42.

50 Non-food GM crops: phytoremediation, industrial products and pharmaceuticals

George G. Khachatourians

1 INTRODUCTION

Contemporary biotechnology has altered how society perceives and utilizes natural resources like plants and microbes, creating large transformations in agriculture, the chemical industry and health care. Innovations in this sector have given rise to new knowledge, products and processes which have helped fuel the economy. The landscape of science and technology has changed and more changes lie ahead. It is in this context that plant manufactured proteins (PMPs) will be discussed and their function in environmental remediation considered, including addressing the rising demand for therapeutic drugs and industrial feedstocks. This will be undertaken by reviewing three applications of genetically modified (GM) plants, used for phytoremediation (PR), industrial products and plant manufactured pharmaceuticals (PMPh). All three areas have expanded the technological applications of GM plants to address societal needs. As a result these technologies build relationships between environmental sustainability, strategies to control infectious diseases and biotechnology development.

2 METHODS

The quantitative and qualitative approaches used in presentation of this chapter are primarily public domain literature relating progress in the molecular biology of plants and plant-based products. This domain has been a subject of increased focus by public and private funding institutions throughout the last three decades. Translation of this research into use has the potential to trigger intellectual property rights and a range of commercialization effects. As a result, publication of research findings, patent disclosures and industrially marketable products have all become tools for a systematic approach to developing aggregate forecasts of trends. These trends, in combination with their impact analyses, should lead to directed R&D and commercialization. Publicly available databases and their translational outcomes have become most valuable tools in this regard. In the sections that follow, a survey of the literature for all three facets of non-food GM crops is presented.

2.1 Phytoremediation (PR)

Industrial developments in some instances and locations have had significant and sometimes unavoidable effects in terms of environmental contamination of soil, water and air. Twentieth-century global human and industrial development has left us with a multitude of chemicals in the environment, whether agri-industrial, urban or rural. Physical removal of these materials in soil and water by conventional technologies is possible but often economically impractical. As a result, excavation and disposal is not a general course of action to resolve environmental contamination. Phytoremediation involves directing the natural ability of certain plants to bioaccumulate, degrade, or render harmless contaminants in soil or water. Contaminants such as metals, pesticides, solvents, explosives and crude oil and its derivatives have been managed by PR projects worldwide.

PR, a technology which affords uptake, sequestration, detoxification or volatilization of toxic chemicals, can be further enhanced through application of genetic manipulation. Research on transgenic plants aimed to accelerate PR generally uses three gene sources and approaches: (1) transformation with genes from other organisms (for example mammals or bacteria); (2) transformation with genes from other plant species; and (3) over-expression of genes from the same plant species (Maestri and Marmiroli, 2011). The outcome of these research and development (R&D) efforts remains commercially promising as the demand for environmental remediation becomes more urgent, due in part to: (1) increasing demand and unpredictability of the supplies of renewable and non-renewable resource extraction and transformation; (2) presence of residues of chemical warfare agents; and (3) explosives that contaminate millions of hectares of land globally. Regulating the transformation of many of these substrates by naturally occurring microorganisms or direct photo-chemical reactions is rather slow, especially with global climate change and unpredictable weather patterns. PR through GM plants is proposed to enhance adsorption, accumulation and transformation through inherent plant metabolic mechanisms.

Understanding the biology behind the metabolism of explosives by microorganisms and plants is well recognized. Rylott et al. (2011) have shown the biodegradation and biotransformation of explosives in plants occurs through genes involved in 2,4,6-trinitrotoluene (TNT) detoxification and the biochemical pathways of nitramine degradation in microorganisms. While identification of enzymes capable of degrading TNT is still elusive, promising PR research using transgenic poplar for field applicability is on the horizon. Two decades of research experience in PR has demonstrated that it can become a means for restoration of environments that were contaminated with chemicals deemed toxic. Therefore, GM plants offer additional options to excavation and disposal of the contaminant material, whether it is soil or water-borne, and in locales such as abandoned metal-mines or tailings and mine discharges.

During the last quarter-century, PR has undergone dramatic changes because of biotechnology advances and knowledge of genomic applications from microbial and plant species to engineered land or water-based plants. The translation ability from laboratory-scale studies to larger applications with advances from environmental engineering technologies has made PR useful for inaccessible sites, such as those from mining or contaminant spills. Thus, weeds, shrubs and trees have become increasingly popular at sub-optimal growth sites with soils contaminated with toxic heavy metals. A number of plants have a capacity to tolerate, accumulate and/or metabolize pollutants, creating the ideal plant for environmental clean-up. Plant properties important for metal PR are toxic metal tolerance and accumulation, which are determined by metal uptake, root–shoot translocation, intracellular sequestration, chemical modification and general stress resistance. Knowledge of which molecular mechanisms are involved in these tolerance and accumulation processes and which genes control these mechanisms, allows for the beneficial modification of these plants. The ideal plant for environmental clean-up can be envisioned as one with high biomass production, combined with superior capacity for pollutant tolerance, accumulation and/or degradation, depending on the type of pollutant and the PR technology of choice (Pilon-Smits and Pilon, 2002). In fact, both enhanced metal tolerance and accumulation have been achieved by overproducing metal chelating molecules or by the over-expression of metal transporter proteins. Mercury volatilization and tolerance was achieved by introduction of a bacterial pathway. The typical increase in metal accumulation as the result of these genetic engineering approaches is two to three fold.

Muratova et al. (2008) selected seven crop plants and three legume species that could be effective in the PR of a former oil-sludge pit. They showed that rye accelerated clean-up most effectively by a total of 52 per cent, yet alfalfa plants which had a lower PR but maintained larger numbers of polycyclic aromatic hydrocarbon degrading microorganisms in their rhizosphere. These two plants were used for a large-scale study to remediate an oil-sludge pit on the grounds of a petroleum refinery. Results have confirmed their effectiveness in decreasing oil-sludge content consistently for three years.

2.2 Phytoextraction (PE)

This involves extraction or accumulation of environmental contaminants into harvestable plant or algal biomass. These organisms take in larger-than-normal amounts of contaminants, especially heavy metals. GM plants with higher potential to absorb contaminants by their roots accumulate these metals, storing them in the roots, stems and leaves. Several plantation cycles achieve a significant extraction, permitting new cropping practices on previously contaminated sites. Examples of plant-based extraction are numerous, including the removal of arsenic, lead, cadmium, cesium and strontium and polychlorinated biphenyls, or PCBs. Plant species performing these tasks include sunflower, willow, mustard (*Brassica juncea*) and hemp (Adler, 1996; Meagher, 2000). Hyper-accumulation of some of the above metals in naturally adapted or GM plants is also noted. While a number of different plants can either break down a variety of organic contaminants or hyper-accumulate metals from the environment, even the most efficient of those plants is typically inhibited by the presence of the toxicant(s), which are responsible for stress that limits a plant's growth and ultimately its ability to phytoremediate. Glick and Stearns (2011) argue that the simple strategy of adding plant growth-promoting bacteria (preferably endophytes) that reduce plant ethylene levels can significantly increase both plant growth and PR activity in the presence of those toxicants.

Romih et al. (2012) conducted bioremediation with soil contaminated by heavy metals from the zinc industry at three locations in the Savinjska region of Slovenia. The soil was contaminated to a depth of 30 cm by cadmium (Cd), lead (Pb), and zinc (Zn). Two hybrids of *Brassica napus* were sown and after 277 days the mature plants' root, shoots and seeds were shown to have translocation, bioaccumulation and PE potential as the plants had higher shoot/root ratios and higher PE potential for Cd, Pb and Zn from the polluted site. By comparison, Neugschwandtner et al. (2012) show enhanced PE of agricultural soil contaminated with Cd and Pb by *Zea mays*. The two metals originated from smelting activities in Příbram (Czech Republic) were mobilized and taken up by the plants and were monitored for two additional years when *Triticum aestivum* was planted. However, less than 0.1 per cent of the metals were removed. Therefore, process efficiency and potential must be genetically enhanced for field application. Rivelli et al. (2012) show that in PR of Cd, Zn and Cu by sunflowers, concentrations in roots always exceeded those in the stem and leaves and there is a lower translocation from roots to shoots. These results suggest that in order to preserve younger tissue in sunflower, compartmentalization takes place in physiologically less active parts. Srivastava et al. (2012) reported that rice varieties that are hypertolerant and hyperaccumulators for arsenic (As), translocate it to its grains more efficiently than other cereal crops. Development of transgenic rice cultivars in contaminated areas for human consumption mitigating concerns for such residues is welcome news. Enhancement of PR efficiency can also utilize dual inoculation willow with ectomycorrhizal associated bacteria (*Micrococcus luteus* and *Sphingomonas* sp.) and ectomycorrhizal fungus (*Hebeloma crustuliniform*) on contaminated soil (Zimmer et al., 2010). The increased plant growth and total Cd and Zn accumulation in the shoot biomass was increased between 53 per cent and 62 per cent after inoculation with these microorganisms. Although this development did not use GM, these results suggest that new and promising approaches such as the use of microbial genes can improve the remediation of metal-contaminated soils by using willows.

Plants can also be used for naturally occurring radioisotopes found in mining and tailing situations. Pratas et al. (2012) have used several species of aquatic plants with tolerance to metals which can accumulate significant amounts of metals and be useful for phytofiltration of uranium (U) from contaminated water. These authors observed that three plants which grow in a uraniferous geochemical locale in Central Portugal accumulating significant amounts of U were; *Callitriche stagnalis* (1948.41 mg/kg of dry weight, DW), *Fontinalis antipyretica* (234.79 mg/kg DW), and *Lemna minor* (52.98 mg/kg DW). The particular genetic modification or transgenic potential of these plants should be of significant interest in U mining operations.

2.3 Phytotransformation (PT)

Phytotransformation uses plants for the management of environments containing a number of noxious industrial chemicals. PT occurs through plant metabolism and microbial activities in the rhizosphere. GM technology made it possible to insert bacterial nitroreductase genes from *Enterobacter cloacae* into tobacco plants for speedy removal and enhanced resistance to TNT (Hannink et al., 2001). The process of transformation for TNT has been shown to involve initial reactions, polarization and sequestration (Subramanian et al., 2006). Other types of PR have been reported in native wetland plants. Hoang et al. (2012) show in Vietnam's coastal wetlands, where quinolones are widely used, antibiotics in shrimp farming (frequently detected in sediments of former shrimp farms) are remediated through the roots. Also, the presence of resistant bacteria in these sediments suggests that bacteria could play a role in facilitating the phytodegradation. How these genetic traits could aid in transgenic (TG) plant-based technologies remains to be established.

A more dramatic application of the above technologies began in the 1990s. PR has been tested at more than 200 American sites. Two different examples are noteworthy. The Environmental Protection Agency (EPA) reports details of this research.[1] The Oregon

Poplar site comprises a three to four-acre abandoned grassy field within a vacant parcel located parallel to the small Mount Scott Creek stream in a primarily commercial and light industrial area with volatile organic compounds (VOCs), probably resulting from illegal dumping activities. The groundwater beneath the site is 2–10 feet below the surface, but in contact with the stream. These characteristics, along with low concentration of contaminants and little to no risk to human health, made the site a good candidate for PR. Hybrid poplar trees were planted on the site in 1998. By 30 July 2002, not only had the trees survived, but they had shown considerable growth. Tissue samples taken from four trees indicated active removal of VOCs from the groundwater and soil, with higher amounts in the trunk than the leaf tissue. The second example is that of J-Field at the Aberdeen Proving Ground located in Harford County, Maryland, a disposal site for chemical warfare agents, munitions and industrial chemicals from 1940 to the 1970s. The two most prevalent contaminants of concern in the groundwater at the site were trichloroethene (TCE) and 1,1,2,2-tetrachloroethan (TTCE). TTCE has been used as a refrigerant (R-130), a solvent and as an intermediate in the industrial production of TCE and derivatives. Chronic inhalation exposure in humans results in jaundice and an enlarged liver, headaches, neurological problems and cancer. Although no longer used in the USA, TCE and TTCE can be removed through PR using various tree species. Chosen for their rapid growth and high transpiration rates, 183 hybrid poplar trees were planted over a one-acre area in 1996. Breakdown products of VOCs (such as solvents, aliphatic and aromatic alcohols, organic acids) were detected in the leaf tissue and gas and water vapour expelled by the poplars. Sap flow rates and shallow ground water levels also indicate that the trees were intercepting and removing the contaminants from the site. It is estimated that within 30 years, contaminants at J-Field may be reduced by up to 85 per cent. The key limitation to applications of PR is the lack of a catabolic pathway for complete degradation/mineralization of, for example, TCE, TNT, atrazine, ethylene and hexahydro-1,3,5-trinitro-1,3,5-triazine (RDX), the latter being an explosive which presents an environmental hazard as a major land and groundwater contaminant (Eapen et al., 2007). These degradative pathways can, however, be introduced into plants by transgenic technology. Van Aken (2008) considers over-expression of a mammalian cytochrome P450, a family of enzymes commonly involved in the metabolism of toxic compounds including vinyl chloride, carbon tetrachloride, chloroform and benzene to create GM transgenic poplars. This type of strategy using stacked genes from bacteria, fungi and animals should yield transgenic plants with a wide range of catabolic genes leading to the wider and safer degradation of recalcitrant contaminants. Taking advantage of the advances in biotechnology and 'omics' technologies, development of novel transgenic plants for efficient PR of xenobiotic pollutants, field testing and commercialization will soon become a reality.

2.4 Plant-made Industrial and Pharmaceutical Products

Plants have had a long-standing history of providing pharmaceuticals and other feedstock for industrially useful products. Salicylic acid, the active extract from white willow (*Salix alba*) bark, was isolated and named by the German chemist Johann Andreas Buchner in 1826. Ever since, R&D in plant-made pharmaceuticals (PMPs) and plant-made industrial products (PMIPs) has entailed new inquiries, breakthroughs and commercial opportunities.

Studies on plant secondary metabolites have increased over the last 50 years. These molecules are known to play a major role in the adaptation of plants to their environment, but also represent an important source of active pharmaceuticals. Plant cell culture technologies were introduced at the end of the 1960s as a possible tool for both studying and producing plant secondary metabolites. Plant biotechnology efforts of the last 30 years have led to some commercial successes in the production of valuable secondary metabolic compounds. Different strategies, using *in vitro* systems, have been extensively studied with the objective of improving the production of secondary plant compounds. Biotechnology has opened a new field, metabolomics, with the possibility of directly modifying the expression of genes related to biosynthetic pathways that lead to secondary plant metabolites and metabolic engineering. The delivery of PMPs could occur through ingestion of raw plant material or from a full or possibly partial conventional processing method.

In 1983, the first proteins of human therapeutic value produced by recombinant DNA technology to be produced in bacteria and plants were insulin and interferon. In the early 1990s the concept of plant-made vaccines (PMV) as documented by Charles Arntzen at Arizona State University (Mason et al., 1992), Hilary Koprowski and Roy Curtiss III, at Washington University in St Louis (Curtiss and Cardineau, 1997). The idea of 'edible vaccine' was founded on the basis of expression of antigen-producing GM fruits or vegetables as safe and cost-effective PMVs. Thirty years after the first production of an antibody in a plant expression system (Hiatt et al., 1989), a special issue of 2010 *Plant Biotechnology Journal* presented six reviews which synthesize two decades of active research on PMPs. PMPs became the principal focus for the production of antibodies, enzymes and vaccines. Certainly there have been significant advances in transformation and expression systems for PMP, helping the maturation of this area of R&D. Research on this topic has produced at least 600 original publications in peer-reviewed journals and about 100 PMP-related patent families (Thangaraj et al., 2009; Faye and Gomord, 2010; Thomas et al., 2011).

There are ample demonstrations of plants and major food crops being efficiently transformed to produce, accumulate and store fully assembled and functional candidate transgene functional products (Khachatourians et al., 2002; Smyth et al., 2004; Daniell et al., 2009). Transgenic plants have also become the alternative system for increasing the production capacity and large-scale production and processing systems for antibodies and subunit vaccines (Faye et al., 2003; Daniell et al., 2009). The feasibility of commercial-scale production of PMPs and PMIPs has already been documented (Howard, 2005; Daniell et al., 2009). The starting point was avidin, a protein used as a diagnostic reagent, and trypsin, a protease, both produced in corn. Trypsin, produced from corn and developed by Prodigene using the bovine trypsin gene, is marketed by Sigma-Aldrich under the trademark name of *TrypZean* ®.

Although it seems feasible for edible PMVs in general to be introduced by TG technology, antigens of human pathogens have mostly been expressed in a few crops. It is clear that some plant-based products are not suitable for human consumption, for example uncooked potatoes. Cooking and boiling may reduce the immunogenicity of many antigens. In the last 10 years since the publication of one of the first compendia on transgenic crops (Khachatourians et al., 2002), the transformation of fruit and vegetable plants has been perfected towards edible PMVs or pharmaceuticals (Daniell et al., 2009; Hefferon,

2012; Wang and Ma, 2012). Antigens have been shown to be correctly processed in plants into forms that elicit immune responses when fed to animals or humans. Antigens expressed in many crops and fruits are particularly attractive since they can be conveniently delivered as food or feed to any animal that consumes grain, and for edible PMVs, they can be used for active and passive immunization.

The production of therapeutics through PMPs has two major advantages: first, the production can be in cell cultures, in greenhouses or open fields and is therefore inexpensive; and second, unlike animal tissue culture, they do not pose the risk of harbouring mammalian viruses or other pathogens. Additionally, there is reduced cost in comparison to the traditional fermentation technology-based manufacturing; furthermore, their development can also take place in the food and feed industries. Plant-manufactured pharmaceuticals are moving into the mainstream of biopharmaceutical manufacturing technologies. Daniell et al. (2001, 2009) review medical molecular farming, production of antibodies, biopharmaceuticals and edible PMVs in plants.

2.5 Regulatory Issues

A closer examination of the regulatory systems in Canada and the US reveals some surprising differences. Some would argue that the Canadian regulatory agencies have been more vigilant regarding transgenic crops than their American counterparts. Beginning in the early 1990s, Canadian regulators stated that all transgenic crops (as well as many mutagenic crops) would be treated as plants with novel traits (PNTs), and therefore subject to greater regulation than conventional crop varieties. Every new PNT requires mandatory oversight of their trials, efficacy and impact on safety of food, feed and the environment. Government agencies demand to see both the raw data and summaries of all tests performed and have the final say on every introduction. The Canadian system also has a formal system of contract registration for risky industrial crops and imposes criminal penalties for infractions. While the Canadian regulators have not completed their development of special rules for PMPs, they have been very influential in directing companies away from areas deemed to be of higher risk (for example canola) by simply reminding the developers that such products are unlikely to be approved.

The initial regulations in the early 1990s in the US were viewed by the industry as being too lax and therefore inadequate to establish trust with consumers. In response, the industry asked the regulators to strengthen the regulations for transgenic crops. Nevertheless, the American regulatory system has consistently been less rigorous in its approach to dealing with transgenic crops than regulators in Canada – for example most reviews are voluntary, non-transgenic novel traits are not reviewed and the regulatory agencies only see study summaries rather than raw data. As in Canada, the US regulators have not sorted out how to handle PMPs. The extra challenge they face is that they do not have the same powers and legal authority that Canadian regulators have to direct developers away from crops.

On 6 March 2003 the Animal and Plant Health Inspection Services division of the US Department of Agriculture announced that they would strengthen mandatory permit conditions for field-testing transgenic crops, including field trials for PMPs (APHIS, 2003). The number of site inspections will increase to five during the trial and two the following season. The permits for pharmaceutical trials will state that no corn can be grown within one mile of the trial site and that no food or feed crop can be grown on the site the following season. The size of the buffer zone was doubled from 25 to 50 feet. This strengthening of regulatory requirements, in part, can be seen as a method to address the concerns that arose following the regulatory violations between 2001 and 2003.

3 METRICS

How do the theory and methods that have been applied to other commodities covered in this book have a bearing on development and agriculture? How do ranges of impacts estimated and analyzed affect PMP-related scale and scope of current or future effects? How would an analysis of return on investments chart the prospects of a company? These are somewhat difficult questions, about which little data is available. Such an analysis helps to give an idea of future movements in price or the future impact of a newly implemented policy. One example of *ex ante* analysis in PMPs is when investments in a company determine *ex ante* stock values and then the predicted results are compared to the actual movement of the stock's price. However, there is uncertainty that needs to be resolved during the course of developmental events, necessitating that the *ex post* values (for example of expected gain) are calculated after the certainty has been attained.

The list of PMPs is steadily increasing. The number of products synthesized by plant nuclear genome include: 20 human vaccines; 35 antigens (25 viral, 9 bacterial and one protozoan); 36 pharmaceutically important proteins; and 24 transiently expressed bacterial and viral products (Daniell, 2009). Plant chloroplasts can also be engineered. Indeed they have been made to produce 8 vaccine antigens, 16 viral and 2 protozoal, and 14 biopharmaceuticals including two for diabetes. A number of these PMPs are used in the treatment of human diseases in clinical trials (Daniell, 2009). Recent developments in influenza vaccine production are presented.

Since 2000, the threat of an influenza pandemic has emphasized the need for a review of influenza vaccines and manufacturing practices. Between 1976 and 2007, the Center for Disease Control (CDC) estimated that the number of flu-associated deaths ranged from 3000 to 49 000, of which 90 per cent were among people 65 years or older. In 2012, a total of 135 million doses of influenza vaccine will be on hand. EvaluatePharma reports that Novartis ($NVS) will probably see the biggest jump in sales, with its OptaFlu vaccine; the company reported $36 million in 2011 sales, and EvaluatePharma projects $71 million for 2012. Sanofi ($SNY) and Sanofi Pasteur MSD's Fluzone (sold as Vaxigrip outside the US) will probably see a $10 million jump in sales, to $1.343 billion this year from $1.333 billion in 2011. Abbott Laboratories' ($ABT) Influvac will probably see a $10 million drop, to an estimated $188 million in 2012 from $198 million in 2011. Mitsubishi Tanabe Pharma's BIKEN HA vaccine is also expected to lose out, slumping by $6 million to a projected $108 million this year from $114 million in 2011.

In general terms the production of influenza vaccine uses inactivated influenza virus or its subunits. It is also possible to use virus-like particles (VLPs), non-infectious particles resembling the influenza virus. VLPs represent an alternative to traditional inactivated and split-influenza virions in inducing responses by both the humoral and cellular immune system (D'Aoust et al., 2010). Medicago Inc. CA is a biopharmaceutical company focused on developing PMVs based on VLPs through proprietary manufacturing technologies in plants. Shoji et al. (2011) also describe the large-scale production of recombinant hemagglutinin proteins from influenza viruses A/California/04/09 (H1N1) and A/Indonesia/05/05 (H5N1) in strains of *Nicotiana benthamiana*, a close relative of tobacco plants. They have shown immunogenicity and safety in animal models. These results support the deployment of our large-scale recombinant influenza PMV production.

A recent report illustrates the metrics associated with PMV. The report of a major development in VLPs was made on 25 September 2012, when Medicago and Philip Morris Products SA signed a commercial agreement for an exclusive license to develop, commercialize and manufacture Medicago's pandemic and seasonal influenza vaccines in China. Medicago was to receive up to US$12 million in upfront and milestone payments as well as royalties. Medicago also announced the signing of a licensing agreement with Philip Morris Products SA, a subsidiary of Philip Morris International Inc., the leading international tobacco product manufacturing company with products sold in approximately 180 countries. In addition, Medicago has signed an exclusive, worldwide license for a portfolio of plant-based protein development technologies from Philip Morris Products. In turn, Medicago is to receive an upfront payment of US$4.5 million from Philip Morris Products. Additionally, Medicago is eligible to receive payments totaling US$7.5 million upon meeting the development milestone, as well as royalty payments on any future sales of pandemic and seasonal influenza vaccines by Philip Morris Products in China. Medicago will pay US$0.7 million to Philip Morris Products, which is entitled to receive royalty payments on any future sales of Medicago products which utilize the technologies licensed from them.

The breakthrough technology in Medicago's PMP pipeline includes the initiation of a US Phase II clinical trial for a quadrivalent seasonal flu vaccine, with interim data expected in the first quarter of 2013. A Phase I clinical trial for a H5N1 VLP vaccine with a new adjuvant is planned in partnership with the Infectious Disease Research Institute (IDRI), with an expectation of interim data in the first quarter of 2013. Medicago's GMP process development and a GLP toxicology study for a rabies vaccine are ongoing, as is working with Mitsubishi Tanabe Pharma under a strategic alliance to develop a vaccine for rotavirus, and at least two additional vaccine candidates.

How did the VLP technology enter the highly competitive vaccine market? In the context of this chapter it becomes important to examine Medicago's development. Medicago is a clinical-stage biopharmaceutical company developing novel vaccines and therapeutic proteins to address a broad range of infectious diseases worldwide from its US and Canadian bases. It was established in Canada operating a 24 000-square-foot production facility located in the Technology Park in Quebec City. This facility includes a Biosafety level 2 greenhouse, an extraction and purification unit, and is cGMP compliant. Medicago has become a worldwide leader in the development of VLP vaccines using a transient expression system which produces recombinant PMVs with speed and cost advantages over competitive technologies. The company claims to enable the development of a PMV for testing in approximately one month after the identification and reception of genetic sequences from a pandemic strain. This production time frame has the potential to allow vaccination of the population before the first wave of a pandemic, and to supply large volumes of vaccine antigens to the world market.

Expanding on its Canadian base, Medicago entered into other areas such as biosimilars and biodefense products where the benefits of their technologies can make a significant competitive difference. The company established an operational base which expects full occupancy of its US operation by the end of October 2012. Medicago is under a $21 million obligation with the Defense Advanced Research Projects Agency (DARPA). A technology investment agreement provided for the development of the facility, a scaled-up and cGMP-capable version of the company's existing VLP vaccine factory. The DARPA initiative is part of the US government's Blue Angel project, which seeks to identify new ways to produce large amounts of vaccine grade protein in less than three months in response to emerging and novel biological threats. VLP is seen as a candidate. The manufacturing technology involves production of recombinant vaccine antigens in non-transgenic plants, offering potential speed and cost advantages over traditional egg-based vaccine production.

A second issue about the metrics of PMP development is that involving governance and regulatory bodies. Fischer et al. (2012) have examined the situation in some detail. From the regulatory perspective, those recombinant proteins which are PMPs and intended for medical use fall under the same regulatory guidelines for manufacturing that cover drugs. When such proteins enter clinical development this includes the GMP requirement. In principle, the well-characterized GMP regulations that apply to pharmaceutical proteins produced in bacteria and mammalian cells are directly transferrable to plants. In practice, the cell-specific terminology and the requirement for a contained, sterile environment mean that only plant cells in a bioreactor fully meet the original GMP criteria. In spite of these stringent criteria, recently the first GMP-compliant production processes for proteins produced in whole-plant systems have been presented and significant changes can only be expected (Fischer et al., 2012). Some of the previous research results of vaccines for Norwalk virus and Hepatitis B were derived from PMP production (Mason et al., 1996; Tacket et al., 1998, 2000). The CFIA and Health Canada are working to develop an appropriate regulatory framework that allows for the successful development of a commercial PMP industry in Canada.

4 CRITICAL ASSESSMENT

The trends in GM plant technologies, whether they apply to environmental clean-up, therapeutics or prevention of infectious diseases, is advancing at an increasing rate. The GM application in PR seemingly will continue towards its proper economic and social fruition. The case with PMPs and plant-derived vaccines (PDVs) is somewhat more complex as the science and technology requires more transdisciplinary perspective for the analyses of these issues (Gold et al., 2004). Notwithstanding, in general it is the case for PMPs and specifically PDVs that the proof of the concept has been established, promising products have been created, and all indications are that remaining scientific and technological challenges are surmountable (Das et al., 2008). Castle et al. (2009) examined the hypothetical case for commercialization of a PDV against hepatitis B in India. They developed a dynamic simulation model of technology diffusion including the role of regulatory burden for a variety of scenarios in which PDVs are produced and diffused. This model also relates the effects of the impact of the uptake of the PDV on mortality and morbidity and a variety of socio-economic indicators.

A forward-looking trend by assessment of discovery output and adoption remains optimistic and positive, yet the full manifestation of its disruptive impact is still lagging. A case in point is the Children's Vaccine Initiative (CVI), founded following the 1990 New York World Summit for Children. It aimed to save 2–3 million children annually from death or disability due to infectious diseases but was facing an uncertain future. However, past practices which relied mainly on public only financial support in current economic times will be more uncertain and necessitate various sorts of partnerships and philanthropic initiatives. In 2001, the Bill & Melinda Gates Foundation revisited two major global public health initiatives, the prevention of infectious disease and the rebuilding of the endangered CVI. PMPs in this regard could extend the lifeline of such support.

A major contributor to the changing of industry's assessment is the cost of production of a protein-based pharmaceutical requiring new manufacturing facilities costs close to US$100M and the fact that it can take up to four years for production. The size of the market for PMPs has justified the importance of exploring the potential of their cost in terms of total investment in R&D. Smyth et al. (2004) compared the costs of the various antibody production systems for the treatment of chronic human diseases. The use of mammalian cells to produce human antibodies costs in the range of US$105–175 per gram, but with PMP it would be one tenth the cost. The comparative cost of production of 1 kilogram of an PMP antibody through a conventional bioreactor would require an estimated US$3.75M, whereas the same product generated in corn would cost US$33 000 (Hileman, 2002) and would still require further focus process validation (Fahrner et al., 2001). PMPs that are already in clinical trials will change the economic aspects of production as the acreage and volume of GM crops are increasing, some of which can be dedicated to PMPs. The other challenge of PMPs is going to be to structure a fully integrated regulatory system that effectively evaluates, manages and communicates the risks of PMPs, and ultimately one that both enforces and is seen to enforce failures.

PMVs to ward off the threat of global pandemics, such as influenza virus, potential bioterrorism agents, such as anthrax, smallpox and the plague, require strategies of rapid production of vaccines in large quantities, stockpiled and on hand for potential future pandemics. Obviously, PMPs and PMVs will have a significant impact on this topic, which also should include assessments of risks and pose difficult questions for regulators. However, it is clear from the above presentation that the technology has matured over the last 20 years through obvious progress in technology (Twyman et al., 2003; Ribicky, 2010; Wang and Ma, 2012). Key drivers in the future of PMPs will be integrative, socio-economic-regulatory issues in the context of science-based processes in the production of high-quality products *in planta* as *in vitro* bioreactors. A continued focus on science-based regulations will be of major importance for this industry as several environmental organizations have already publicly expressed concern about the production and commercialization of PMP products. There must be a nexus for capacity and cost to meet the demand (Sparrow et al., 2007; Boehm, 2007; Penney et al., 2011).

Part of the future solutions to the conundrum of PMPs will be stricter regulation. Options such as (a) a closed loop system to confine PMPs to select authorized seed processors, certified growers, manufacturing sites and pharmaceutical firms; or (b) production of PMPs in regions where there are no major food crops are being examined by regulatory systems. It is encouraging to observe that industry–government dialogue and discussions relating to the development of a regulatory framework have been established and are continuing. While large-scale commercial PMP production may not occur in the immediate future, the development of clear, consistent and efficient regulations needs to continue moving forward. As the technology relating to molecular farming evolves and improves, additional issues will arise, and for this reason the establishment of a coordinated approach to the development of regulations will be a crucial determining factor in the successful growth of PMP-related industries.

NOTE

1. See: http://www.clu-in.org/products/newsltrs/tnandt/view.cfm?issue=0212.cfm, February 2012.

REFERENCES

Adler, T. (1996), 'Botanical cleanup crews: using plants to tackle polluted water and soil', *Science News*, **150**, 42–3.
Animal and Plant Health Inspection Services (APHIS) (2003), 'Field test releases in the US', available at: http://www.nbial.vt.edu/cfdocs/fieldtests1.cfm.
Boehm, R. (2007), 'Bioproduction of therapeutic proteins in the 21st century and the role of plants and plant cells as production platforms', *Annals of New York Academy of Science*, **1102**, 121–34.
Castle, D., K. Kumagai, C. Berard, M. Cloutier and R. Gold (2009), 'Regulatory burden in technology diffusion: the case of plant-derived vaccines', *AgbioForum*, **12**, 108–18.
Curtiss, R. and G.A. Cardineau (1997), 'Oral immunization by transgenic plants', US Patent 5679880, 21 October.
D'Aoust, M.A., M.M. Couture, N. Charland, S. Trépanier, N. Landry, F. Ors and L.P. Vézina (2010), 'The production of hemagglutinin-based virus-like particles in plants: a rapid, efficient and safe response to pandemic influenza', *Plant Biotechnology Journal*, **8**, 607–19.
Daniell, H., S.J. Streatfield and K. Wycoff (2001), 'Medical molecular farming: production of antibodies, biopharmaceuticals and edible vaccines in plants', *Trends in Plant Science*, **6**, 219–26.
Daniell, H., N.D. Singh, H. Mason and S.J. Streatfield (2009), 'Plant-made vaccine antigens and biopharmaceuticals', *Trends in Plant Science*, **14**(12), 669–79.
Das, V., W.B. Phillips and G.G. Khachatourians (2008), 'Commercialization of plant-derived vaccines in Canada: a distant dream?', *Health Law Review*, **16**, 25–9.
Eapen, S., S. Singh and S.F. D'Souza (2007), 'Advances in development of transgenic plants for remediation of xenobiotic pollutants', *Biotechnology Advances*, **25**, 442–51.
Fahrner, R.L., H.L. Knudsen, C.D. Basey, W. Galan, D. Feuerhelm, M. Vanderlaan and G.S. Blank (2001), 'Industrial purification of pharmaceutical antibodies: development, operation, and validation of chromatography process', *Biotechnology and Genetic Engineering Reviews*, **18**, 301–27.
Faye, L. and V. Gomord (2010), 'Editorial: success stories in molecular farming – a brief overview', *Plant Biotechnology Journal*, **8**, 525–8.
Faye, L., P. Lerouge and V. Gomord (2003), 'Production de molecules pharmaceutiques', *Annales de Pharmacologie*, **61**, 109–18.
Fischer, R., S. Schillberg, S. Hellwig, R.M. Twyman and J. Drossard (2012), 'GMP issues for recombinant plant-derived pharmaceutical proteins', *Biotechnology Advances*, **30**, 434–9.
Glick, B.R. and J.C. Stearns (2011), 'Making phytoremediation work better: maximizing a plant's growth potential in the midst of adversity', *International Journal of Phytoremediation*, **13**, 4–16.
Gold, E.R., W.A. Adams, D. Castle, G. Cleret de Langavant, L.M. Cloutier et al. (2004), 'Probing benefits: biotechnology innovation and the patent system', *Public Affairs Quarterly*, **18**, 299–344.
Hannink, N., S.J. Rosser, C.E. French, A. Basran, J.A. Murray, S. Nicklin and N.C. Bruce (2001), 'Phytodetoxification of TNT by transgenic plants expressing a bacterial nitroreductase', *Nature Biotechnology*, **19**, 168–72.
Hefferon, K.L. (2012), 'Recent advances in virus expression vector strategies for vaccine production in plants', *Virology and Mycology*, **1**, 174.

Hiatt, A., R. Cafferkey and K. Bowdish (1989), 'Production of antibodies in transgenic plants', *Nature*, **342**, 76–8.
Hileman, B. (2002), 'Drugs from plants stir debate', *Chemical Engineering News*, **80**, 22–5.
Hoang, T.T., T. Tu, L.T.C. Le and Q.P. Dao (2012), 'A preliminary study on the phytoremediation of antibiotic contaminated sediment', *International Journal of Phytoremediation*, **15**, 65–76.
Howard, J.A. (2005), 'Commercialization of biopharmaceutical and bioindustrial proteins from plants', *Crop Science*, **45**, 468–72.
Khachatourians, George G., Alan McHughen, Wai-Kit Nip, Ralph Scorza and Y.-H. Hui (eds) (2002), *Transgenic Plants and Crops*, New York: Marcel Dekker Inc.
Maestri, E. and N. Marmiroli (2011), 'Transgenic plants for phytoremediation', *International Journal of Phytoremediation*, **13**, 264–79.
Mason, H.S., D.M. Lamand and C.J. Arntzen (1992), 'Expression of hepatitis B surface antigen in transgenic plants', *Proceedings of the National Academy of Sciences of the United States of America*, **89**, 11745–9.
Mason, H.S., J.M. Ball, J.J. Shi, X. Jiang, M.K. Estes and C.J. Arntzen (1996), 'Expression of Norwalk virus capsid protein in transgenic tobacco and potato and its oral immunogenicity in mice', *Proceedings of the National Academy of Sciences of the United States of America*, **93**, 5335–40.
Meagher, R.B. (2000), 'Phytoremediation of toxic elemental and organic pollutants', *Current Opinions in Plant Biology*, **3**, 153–62.
Muratova, A.Y., T.V. Dmitrieva, L.V. Panchenko and O.V. Turkovskaya (2008), 'Phytoremediation of oil-sludge-contaminated soil', *International Journal of Phytoremediation*, **10**(6), 486–502.
Neugschwandtner, R.W. P. Tlustoš, M. Komárek, J. Száková and L. Lucie Jakoubková (2012), 'Chemically enhanced phytoextraction of risk elements from a contaminated agricultural soil using *Zea mays* and *Triticum aestivum*: performance and metal mobilization over a three year period', *International Journal of Phytoremediation*, **14**, 754–71.
Penney, C.A., D.R. Thomas, S.S. Deen and A.M. Walmsley (2011), 'Plant-made vaccines in support of the Millennium Development Goals', *Plant Cell Reports*, **30**, 789–98.
Pilon-Smits, E. and M. Pilon (2002), 'Phytoremediation of metals using transgenic plants', *Critical Reviews in Plant Science*, **21**, 439–56.
Pratas, J., P.J.C. Favas, C. Paulo, N. Rodrigues and M.N.V. Prasad (2012), 'Uranium accumulation by aquatic plants from uranium-contaminated water in central Portugal', *International Journal of Phytoremediation*, **14**, 221–34.
Ribicky, E.P. (2010), 'Plant-made vaccines for humans and animals', *Plant Biotechnology Journal*, **8**, 620–37.
Rivelli, A.R., S. De Maria, M. Puschenreiter and G. Piergiorgio (2012), 'Accumulation of cadmium, zinc, and copper by *Helianthus annuus* L.: impact on plant growth and uptake of nutritional elements', *International Journal of Phytoremediation*, **14**, 320–34.
Romih, N., B. Grabner, M. Lakota and C. Ribarič-Lasnika (2012), 'Distribution of Cd, Pb, Zn, Mo, and S in juvenile and mature *Brassica napus* L. var. *Napus*', *International Journal of Phytoremediation*, **14**, 282–301.
Rylott, E.R., A. Lorenz and N.C. Bruce (2011), 'Biodegradation and biotransformation of explosives', *Current Opinions in Biotechnology*, **22**, 434–40.
Shoji, Y., J.A. Chichester, M. Jones, S.D. Manceva, E. Damon, V. Mett et al. (2011), 'Plant-based rapid production of recombinant subunit hemagglutinin vaccines targeting H1N1 and H5N1 influenza', *Human Vaccine*, **7**, 41–50.
Smyth, S., G.G. Khachatourians and P.W.B. Phillips (2004), 'The liabilities from regulating gene flow in plant made pharmaceuticals', *Biotechnology and Genetic Engineering Reviews*, **10**, 277–97.
Sparrow, P.A.C., J.A. Irwin, P.J. Dale, R.M. Twyman and J.K.C. Ma (2007), 'Pharma-planta: road testing the developing regulatory guidelines for plant-made pharmaceuticals', *Transgenic Research*, **16**, 147–161.
Srivastava, S., S. Penna and S.F. D'Souza (2012), 'Mechanisms of arsenic tolerance and detoxification in plants and their application in transgenic technology: a critical appraisal', *International Journal of Phytoremediation*, **14**, 506–17.
Subramanian, M., D.J. Oliver and J.V. Shanks (2006), 'TNT phytotransformation pathway characteristics in *Arabidopsis*: role of aromatic hydroxylamines', *Biotechnology Progress*, **22**, 208–16.
Tacket, C.O., H.S. Mason, G. Losonsky, J.D. Clements, M.M. Levine and C.J. Arntzen (1998), 'Immunogenicity in humans of a recombinant bacterial antigen delivered in a transgenic potato', *Nature Medicine*, **4**, 607–609.
Tacket, C.O., H.S. Mason, G. Losonsky, M.M. Estes, M.M. Levine and C.J. Arntzen (2000), 'Human immune responses to a novel Norwalk Virus vaccine delivered in transgenic potatoes', *Journal of Infectious Diseases*, **182**, 302–305.
Thangaraj, H., C.J. van Dolleweerd, E.G. McGowan and J.K. Ma (2009), 'Dynamics of global disclosure through patent and journal publications for biopharmaceutical products', *Nature Biotechnology*, **27**, 614–18.
Thomas, D.R., C.A. Penney, A. Majumder and A.M. Walmsley (2011), 'Evolution of plant-made pharmaceuticals', *International Journal of Molecular Sciences*, **12**, 3220–36.

Twyman, R.M., E. Stoger, S. Schillberg, P. Christou and R. Fischer (2003), 'Molecular farming in plants: host systems and expression technology', *Trends in Biotechnology*, **21**, 570–78.
Van Aken, B. (2008), 'Transgenic plants for phytoremediation: helping nature to clean up environmental pollution', *Trends in Biotechnology*, **26**, 225–7.
Wang, Aiming and Shengwu Ma (2012), *Molecular Farming in Plants: Recent Advances and Future Prospects*, New York: Springer.
Zimmer, D., C. Baum, P. Leinweber, K. Hrynkiewicz and R. Meissner (2010), 'Associated bacteria increase the phytoextraction of cadmium and zinc from a metal-contaminated soil by mycorrhizal willows', *International Journal of Phytoremediation*, **11**, 200–213.

51 Tomatoes, potatoes and flax: exploring the cost of lost innovations

Camille D. Ryan and Alan McHughen

INTRODUCTION

The history of technological development and adoption is an interesting blend of 'eurekas' with unequivocal market 'failures'. That said, many good, viable technologies have entered the market at a full-scale run only to be usurped by competing technologies that are often only equivalent or, even in some cases, sub-par in nature. Take for instance MS-DOS and IBM, which came to dominate the corporate office suite market with Microsoft Office. The system was not superior to Apple's but IBM definitely had first-mover advantage (and buy-in from the corporate world) which led to their control of the market. The company also had the foresight to integrate spreadsheet technology into their product offering; an important and saleable feature for business. This is an example of a product adoption wherein 'path dependency', lock-in or network effects generated some form of 'market power' that plays a significant role in the lack of adoption of a good technology.

The history of agricultural biotechnology and food-related technologies has been even more unpredictable. Many products have been introduced to the market with the capacity to enhance agronomic traits and food production. They are backed by solid science and research and have navigated a costly and rigid regulatory approval process. They are, for all intents and purposes, superior products (when compared with contemporary products). But, in this highly politicized food environment, this matters little.

This chapter examines three cases where good, relevant food-related technologies were removed from the market due to anti-GM activism and political pressures: the Flavr Savr™ tomato; the NewLeaf™ potato; and CDC Triffid flax. These products were not removed due to safety issues. In fact, all products successfully received full government-mandated regulatory safety approval. These products were removed due to the pressures exerted by anti-biotechnology forces.

In section 1, we review the history of genetic engineering and innovation in food and crop technologies over the past couple of decades. In section 2, we introduce the three case studies of food/crop technologies that have been withdrawn from the market and outline the history of their development and ultimate market demise. In section 3, we review the role of activism in hindering the introduction of these technologies to the market and the economic and innovation-based implications of those constraints. In section 4, we finalize the chapter with some concluding thoughts.

1 THE EMERGENCE OF INNOVATIVE FOOD AND CROP TECHNOLOGIES

Agrobacterium tumefaciens, as a means of inserting genes into plants, has been extensively exploited in biotechnology. The first (transgenic) plants, genetically engineered using modified strains of *Agrobacterium* were reported in 1983 (Bevan et al., 1983; Fraley et al., 1983; Herrera-Estrella et al., 1983). This set off a race to extend the enabling technology to crop species from the widely-used experimental species (mainly tobacco and petunia). The engineering of bacteria *Agrobacterium tumefaciens* strains with pathogenic gene sequences deleted from the transfer DNA (T-DNA) which retained cellular transformation capability was the major breakthrough, at least for those plant species – mainly dicots – susceptible to transformation by *Agrobacterium*. In such plant species, the so-called 'disarmed', non-oncogenic *Agrobacterium* strain would attach and deliver the now impotent T-DNA into the susceptible host plant cells where the substitute DNA would integrate into the host cell's genome. The T-DNA would carry genes of interest, initially simple marker genes, to indicate when transfer and integration had been successful. Later, genes encoding traits of agronomic or economic value when expressed in the recipient host would be used. See Kerr (2011) for a brief history of employing *Agrobacterium* for plant genetic engineering.

Agrobacterium comprises a range of strains, collectively capable of transforming a correspondingly wide range of (mainly) dicotyledonous plants and all strains which could be rendered non-oncogenic or 'disarmed'. This naturally occurring cellular DNA-transferring genetic engineering agent opened a wide spectrum of plant species to the potential for genetic engineering.

However, having disarmed *Agrobacterium* deliver specific DNA into a plant cell is not by itself sufficient, as the transformed cell must then be coaxed into regenerating a whole fertile plant. That next step following cellular transformation, growing a complete plant from that single transformed cell, turned out to be no trivial matter. So, while the DNA delivery methods were largely mechanical or dependent upon the natural capability of *Agrobacterium* to transfer DNA to plant cells, the tissue culture technology of identifying and coaxing transformed cells to regenerate shoots and roots lagged behind. Fortunately, Solanaceous species like tobacco and petunia were not only highly susceptible to *Agrobacterium*, they were also amenable to the tissue culture and regeneration methods *in vitro*, so they were among the first transgenic plants produced.

Other crop species were less amenable to either cellular transformation or *in vitro* regeneration, or both, so considerable research effort was needed prior to delivery of some crop species, including the most important monocots such as rice, wheat and maize. Important dicot crops, such as grain legumes, although perhaps susceptible to transformation by *Agrobacterium*, proved difficult at whole-plant regeneration, so transgenic plants of soybeans and other grain legumes took longer.

2 CASE STUDIES

2.1 Calgene's Flavr Savr™ Tomato

The Flavr Savr™ (FS) tomato, commonly known as 'McGregor' was developed by the California company Calgene Inc. in partnership with Campbell Soup Company and was genetically engineered for slow ripening and extended shelf life. These 'antisense' tomatoes were developed through the Calgene/Campbell tomato fruit quality programme established in 1985. The tomatoes were engineered with an inverted polygalacturonase (PG) gene that significantly reduced the expression of the naturally occurring PG enzyme. PG is responsible for the breakdown of pectin, a primary component in the cell walls of fruit which affects shelf-life and the quality of produce. The slowed ripening process did not appear to negatively affect colour or flavour of the tomato fruit.

Until the introduction of the FS, the 'fresh tomato' market required that the fruit be picked 'green' in order that they survive transport to distributors and grocery stores. Product is then artificially ripened using ethylene gas (Martineau, 2001). The introduction of the Flavr Savr™ tomato changed all that. The new technology enabled vine ripening of tomatoes which resulted in a better tasting, better quality fruit with a longer shelf life. This represented huge benefits for the distributor, the retailer, as well as for the consumer.

The US patent for the tomato, which covered only the use of the gene in tomatoes, was granted in 1989. The EU patent was granted five years later. The Flavr Savr™ tomato was the first commercially grown, genetically engineered food to be granted a licence for human consumption in 1994 (18 May).[1] The product was launched nationally three days later.

From its introduction, Flavr Savr™ was extremely well received in the marketplace, 'reportedly selling like hot tomatoes' (Sugarman, 1994). By October of 1994, the tomatoes were being delivered to more than 700 stores in the West and Midwest of the US and making a huge impact on the then $4 billion per year US fresh tomato market. At the time, the Flavr Savr™ tomato was viewed as better than supermarket varieties but not as good as home-grown or farm-fresh (Martineau, 2001).

Competing varieties were hot on the heels of Calgene's FS tomato including Freshworld Farms VineSweet variety[2] and Monsanto's Premium Ripe vine variety. Zeneca, working with Dr Don Grierson and his lab at the University of Nottingham, also developed a tomato, but focused on the 'processed' rather than the 'fresh' market[3] (Harvey, 1999). Sales were eventually shut down for this product as well. This was due, in part, to a backlash from the anti-GM groups that opposed genetically modified foods. Additionally, reticence set in on the part of the company as approvals for cultivation in the EU were continually postponed.

The Flavr Savr™ tomato faced a similar fate. The 'halcyon days of market monopoly' for Calgene in the US ended soon after they started (Martineau, 2001). First, the company had trouble keeping up with national demand for the product. Additionally, there were bottom line problems. The tomatoes were priced at $1.99 per pound retail, a price that was deemed competitive with conventionally produced varieties. But the then estimated costs of production and distribution were estimated at $10 per pound and far outweighed those retail prices. A breakdown of those costs is not publicly available so it is unclear why costs associated with McGregor were so much higher than conventional varieties. It appeared, however, that the enthusiasm of getting this new, unique product to the market usurped efforts to properly strategize product 'roll out' and manage production. Short supply, the word that the technology may not work from the 'green to ripe' end as well as it did from the 'ripe to rotting' end of things and growing consumer opposition led to the eventual downfall of the Flavr Savr™ tomato (Martineau, 2001). The investment community caught wind of the Flavr Savr™ shortcomings and confidence in the product rapidly diminished. Monsanto, which owned more than 50 per cent of Calgene's share at the time, eventually bought out the company in 1997[4] as part of its strategy to expand further into biotechnology, a move that essentially saved the struggling company. Despite this, the Flavr Savr™ tomato was withdrawn from the market.

Consumer perceptions around food sovereignty, intense lobbying on the part of interest groups and media attention essentially pulled the Calgene tomato from its lofty position in the market. Poor management on the part of Calgene – a divide between science and business (Martineau, 2001) – was also a contributing factor.

2.2 NewLeaf™ Potato

The NewLeaf™ potato, developed by Monsanto and the first Bt crop variety approved for commercial use,[5] completed the US regulatory authorizations in 1995. This Russet Burbank variety was developed through biotechnological methods wherein a gene from a naturally occurring soil-based bacterium, *Bacillus thuringiensis* (*Bt*), was incorporated to protect the plant from the Colorado Potato Beetle (*Leptinotarsa decemlineata*). This pest was responsible for widespread damage to potatoes and related crops across the US and resistance to all available classes of conventional pesticides was spreading rapidly, making it very difficult to control. Thus, potato producers enthusiastically adopted the GM potato varieties to increase yields and to reduce environmental risks of pesticide use. By 1999, the GM share of the US and Canadian seed potato market had grown to 3 per cent.

Monsanto's subsidiary, NatureMark, trademarked and marketed NewLeaf™ potatoes, touting the new product as a key technology in reducing environmental risks. The strategic roll-out of the variety is cited as an example of effective proactive marketing of a biotechnology product (Phillips and Corkindale, 2002). In 1999, the company launched a campaign in the potato-producing province of Prince Edward Island (PEI).[6] GM fresh potatoes were made available in grocery stores in PEI and other grocery and wholesale outlets across Canada, complemented with good marketing tools such as strategic advertising, publicity, point-of-sale information and a toll-free phone line (Guenthner et al., 2011; Phillips and Corkindale, 2002). The market responded positively and, like the Flavr Savr™ tomatoes, the GM potatoes quickly sold out.

By the time that NewLeaf™ Plus – developed to be resistant to both the Colorado potato beetle and the potato leafroll virus – was introduced to the market in the late 1990s, plantings of GM potatoes had reached 55000 acres (Guenthner et al., 2011). This represented a huge increase in production from 1800 acres in 1995. Overall, things looked promising for Monsanto's new genetically modified 'brain child'. But the GM potato had entered a complex consumer arena in the mid to late 1990s; the market was now an environment with a strong and focused political anti-technology agenda where acceptance of food and ag-related technologies was tenuous at best. So, although producers were keen to adopt this new technology, fears of consumer resistance blocked further market development. In late 1999, the world's largest frozen potato processor and a buyer of PEI potatoes, McCain Foods, announced that it would not purchase GM potatoes for processing due to perceived consumer resistance. The company's President, Harrison McCain, was quoted as saying: 'We think genetically modified material is very good science [but] at the moment, very bad public relations' (Toronto Star, 1999). J.R. Simplot, a major supplier of French fries to McDonald's, instructed its producers to stop growing GM potato varieties: 'Virtually all the [fast food] chains have told us they prefer non-genetically modified potatoes' (Fred Zerza, J.R. Simplot spokesman, quoted in Killman, 2000).

Although GM potatoes appeared to be a commercial success in the fresh market, consumer interest groups, by targeting the quick service restaurant industry, convinced processors to cease production with GM potatoes. Despite their agronomic and environmental advantages, the destruction of markets for GM potatoes meant that producers were unwilling to plant them, thus leading to the disappearance of this valuable food technology (Guenthner et al., 2011).

2.3 Linseed Flax: 'CDC Triffid'

Common flax (*Linum usitatissimum* L.) was one of the first crops cultivated by man. Flax is used in the production of various industrial products including linen, fibre composites, paints, inks and linoleum. Over the past several years, flax has grown in popularity as a nutritional supplement for its value as an omega-3 fatty acid for both human and animal consumption. It is often consumed raw or is used in wholegrain products such as cereals and breads.

New flax cultivars in Canada must be grown in government-managed trials consisting of the experimental line, standard commercial check cultivars and candidate lines from other breeders. The trials are conducted at multiple locations and over at least two years. Two transgenic flax lines, designated FP967 and FP968, were grown in these 'co-op' trials in 1992 and 1993. One of the two lines, FP967, was presented for registration as a new commercial cultivar in February 1994, to the Prairie Registration Recommending Committee for Grain (PRRCG), the relevant government agency, who evaluated it based on scientific criteria and data generated primarily from the co-op field trials. A vote was taken and FP967 was approved unanimously, based on its meritorious performance. It became the first transgenic cultivar of any crop to gain such approval in Canada (transgenic canola was supported for registration the following year) (PRRCG, 2004; McHughen, 2000). The variety was designed to thrive in soil containing residues from sulfonylurea-type herbicides. Without Triffid, farmers would either have to continuous crop to cereals or summerfallow while waiting for the sulfonylurea in the soil to break down naturally. Neither of these practices was sustainable. At the time, there were no broad-leaf crops with sufficient sulfonylurea tolerance to introduce into a rotation. Development of the Triffid variety represented an innovative step in agricultural science and crop development given the variety's environmental benefits. It allowed farmers to practise sustainable agriculture.

FP967 was grown by commercial seed growers in the summer of 1994. However, unlike conventional cultivars, the transgenic flax had to pass additional regulatory scrutiny for food, feed and environmental safety prior to issuance of the formal Variety Registration

Certificate. That document was finally issued in May 1996. With the official certificate of registration, transgenic flax line FP967 was named 'CDC Triffid', a name registered and protected under the Seeds Act (R.S.C., 1985, c. S-8).

Seed multiplication continued in 1997 and by the end of the 1997 harvest there was an estimated 5000 tonnes of pedigreed CDC Triffid seed in Canada. That year turned out to be the final year for multiplication of Triffid flax. At that point, existing seed stocks of CDC Triffid were identified and contained in separate grain bins in compliance with pedigreed seed production regulations.

The imposition of the 1998 EU moratorium on GM crops and foods presented a formidable obstacle to the Canadian flax industry. Approximately half of the flax production in Canada at the time was being exported to Europe and the commercialization of a GM flax variety alarmed the European importing firms. While the canola industry was able to preserve the identity of their GM varieties effectively and continued to supply the European and Japanese markets from 1995–97 (Smyth and Phillips, 2001), the flax industry did not have this option because of the dominant role of the EU market. European importers were adamant, threatening that they would halt flax imports completely from Canada if any GM flax was grown commercially. Discussions about how to handle the situation were initiated by the Flax Council of Canada (FCC). It was determined that all of the existing seed stocks would remain sequestered from the export stream (marked with identifying confetti and sealed in bins on the farms of the commercial seed growers), until a suitable location could be found to crush the flax seed. Coordinated by the FCC, a CanAmera Foods crushing plant in Manitoba was ultimately contracted to crush the flax. The resulting flax meal was mixed into livestock feed and fed to Canadian livestock, while the oil from the crush was diverted into industrial applications in North America (where the product was fully approved for use). Any and all breeder seed stock held by the CDC was incinerated. This effectively removed all certified seed from seed growers that were contracted to multiply the seed.

The variety was deregistered in 2001 and it was thought that this represented an end to any trade issues associated with GM flax. Unfortunately, this was not to be the case. In July 2009, the EU reported that a Canadian shipment of flax had tested positive for the NPTII marker indicating a GM event and by September 2009 the EU's Rapid Alert System for Food and Feed (RASFF) was notified by a German company that its bakery/cereal products had tested positive for Triffid,[7] in spite of the fact that there was no Triffid-specific test in existence. The NPTII marker is common to many GM plants, so the test might have detected residual dust from GM canola, corn or soybeans mixed in with the flax, or low-level presence of one of the many other GM flax lines in open field trials.

The Canadian grain industry, however, was quick to respond to this initial notification. Industry stakeholders – the Flax Council of Canada, the Canadian Food Inspection Agency and the Canadian Grain Commission (CGC) – moved in quickly to try to mitigate the threat to market access for Canadian flax producers. On 19 October 2009, the FCC and the CFIA met together in Brussels to work with DG SANCO (Directorate General for Health and Consumer Affairs) and other EU stakeholders to develop a testing protocol to manage the situation in Canada (European Commission, 2009). The Protocol, launched immediately in the Canadian market, quickly established a system of sampling (one test for every 5000 bushels) and testing of flax stores. Samples that tested positive at levels greater than or equal to 0.01 per cent [8] would not be accepted for import into the EU market (FCC, 2009).[9] According to James (2010), Canadian flax averaged a failure rate of 20 per cent at EU ports during the last six months of 2009.

As part of the requirements for the EU and in keeping with the contractual obligations of the Protocol, testing of Canadian flax is ongoing. In fact, testing is conducted repeatedly all along the value chain, from farm-held stores of flax to the elevator and at ports where flax shipments await export. Early results suggested a widely distributed but extremely low-level presence of GM material in Canadian flax. As of September 2011, almost 26000 tests had been conducted on over 10000 seed lots.

The case of 'CDC Triffid' is enigmatic. The variety was deregistered and eliminated from the market in 2001 with pressure from producers and the grain industry in response to the EU moratorium. Despite this, GM flax detections became a problem for the Canadian industry only a few years later. Moreover, the positive assays did not follow the expected pattern for an inadvertent release, in that the detections came from a broad range of geographic origins, there were no 'hot spots' where detections were more frequent, and all positive test results were at the lowest limit of technical detection. Crucially, GM flax detections were reported in breeder seed, the highest purity of commercial seed, from non-GM varieties at breeding stations that had no GM flax breeding activity, and that were developed after Triffid was deregistered. Allegations that the tests were prone to false positives circulated early on. Lamb and Booker (2011) used a simple statistical approach on simulation modelling to explore the quantification of low levels of genetic modification contamination. Their results indicate that GM contamination is likely to be present at extremely low levels in breeder seed lots and that levels are virtually indistinguishable from zero, given the potential rates of false positive tests (Ryan and Smyth, 2012).

3 CONSUMER ACTIVISM AND THE EU MORATORIUM

Genetically-modified products entered a complex, dynamic market in the 1990s where consumer acceptance became a powerful force and interest groups aggressively pushed anti-technology agendas. There were a series of unrelated, non-GM food crises (including BSE) that occurred in that decade which led to consumer apprehension about food safety in general. According to Smyth et al. (2011), some critics viewed the use of biotechnology and related technologies as a '*slippery slope* to human cloning' or other unnatural extensions, or considered it to be a use of science 'outside of the boundaries of what is ethically acceptable' (Smyth et al., 2011, p. 11). Paarlberg and Pray (2008) suggest that the collective powerful voice behind these claims gain quick traction in society and often have direct impacts on government policy. This powerful anti-technology voice had many faces and several key events played a hand in the downfall of the GM products outlined in our case studies.

For example, Jeremy Rifkin, President of the Foundation on Economic Trends, marks the date of the FDA's approval of the Flavr Savr™ tomato as the start of what he calls the 'tomato war' (Martineau, 2001, p. 190; Barnum, 1994). Up until that point, Rifkin had, for the most part, ignored agricultural biotechnology. From thereon in, Rifkin vowed to lead efforts to picket markets, hand out anti-GM notices and organize boycotts. He quite effectively leveraged his already established Pure Food Campaign to give his anti-Flavr Savr™ tomato message 'legs' (Martineau, 2001). It is not clear just how much impact his lobby-lead had on the withdrawal of the Flavr Savr™ tomato from the market, but followers did picket and demonstrate at several retail outlets, receiving media attention, which undoubtedly affected consumer perceptions.

In August 1998, Dr Arpad Pusztai, a scientist with the Rowett Research Institute in Aberdeen, Scotland, announced on the popular BBC TV news programme *Panorama* in England that his unpublished research showed that rats that had been fed GM potatoes had suffered damage to their immune systems and showed underdevelopment of their vital organs. This was the first assertion by a scientist that GM crops could be hazardous to health (Firth, 1999). Within a few days of Pusztai's announcement, the Rowett Institute suspended him for publicly revealing research results before they were peer-reviewed. A debate ensued in the scientific community thereafter regarding the validity of Pusztai's study. Eventually, the editor of the scientific journal *The Lancet* invited Dr Pusztai to submit the study. It was sent to several reviewers, most of whom recommended that the paper be rejected due to a lack of appropriate controls used in the study (Miller, 2010). Despite this, portions of the study were published in the Research Letters section of the journal only a few months after the BBC programme. The article was eventually retracted due to concerns about its methodology but, by that time, the damage had already been done. Pusztai's controversial and inaccurate study made its rounds rapidly through the public sphere. The reputation of the GM potato – in fact, all genetically modified foods or crop varieties – was irrevocably damaged as a result of this very public issue.

Adding to all of this was the work of GM critics like Michael Pollan. In 2002, Pollan published a book entitled *Botany of Desire*, which focused on the relationship between humans and plants. In particular, he openly criticized Monsanto's NewLeaf™ potatoes. Pollan's work and others with similar anti-corporate personal views only serve to support anti-GM activists and their agendas.

The erosion of public trust in government agencies and oversight of the food industry pushed consumer concerns over the top, and the unofficial EU moratorium in 1998 resulted in the suspension of approvals of new genetically modified organisms in the European Union pending the adoption of rules to govern biotech products. This created huge entanglements in the EU regulatory system that only served to fuel the fire of anti-GM sentiments. The moratorium presented a huge obstacle to the North American agricultural industry given its role as a major trading partner for both Canada and the US.

The moratorium ended in 2003 when the European Food Safety Authority (EFSA) was established as the new regulatory body. But by this time the atmosphere in the EU and the related events had set the foundation for political control of the food and crop industries in the EU. The grassroots anti-GMO push from anti-GM activist groups is still apparent in the level of activism and eco-activism that continues today. The EU massively funds these worldwide activism efforts. In 2006 alone, the Friends of the Earth was earmarked (directly and through secondary or tertiary links) to receive roughly €790 million from European governments (Apel, 2010).

Adding fuel to the fire is the whole new generation of activism that has evolved to augment the anti-GM campaign. Interest groups rapidly adopt social media as a way to influence public opinion. Given its capacity to hyperlink across geographic boundaries and the relative low cost of access, the Internet has become a primary organizing tool for many non-government organizations. As more and more advocacy activities move online, the need for offline staffing and membership dollars to support these activities dwindles. Thus, even the smallest of interest groups can greatly impact public opinion on an issue with a well-executed online campaign strategy. They can quickly build coalitions and mobilize public interest around specific issues of interest at relatively low marginal costs (Ryan, 2010). These initiatives gain even more ground when augmented through non-expert celebrity endorsement.

Our complex world is one where good food and crop-related innovations – those that have the capacity to resolve production problems and/or feed the world – can be readily usurped or, at the very least, be over-regulated and kept from the market. Developers of new crop technologies have been unable to predict, with any confidence, how or in what way consumers may respond to genetically modified products in the marketplace.

3.1 Estimating the Cost of Lost Innovations

It is extremely difficult, if not impossible, to estimate the costs associated with products or technologies that are withdrawn from or never enter the market. In terms of our case studies, we can only extrapolate value from existing sector and market data. In the case of the tomato, according to Boriss and Brunke (2005), average annual per-capita consumption of the fruit steadily increased from 12.3 lbs in 1981 to 19.2 lbs in 2003. In 2009, the US fresh market tomato market had an estimated farm value of $1.3 billion (US). Sixty per cent of US world exports in fresh tomatoes went to Canada alone.[10] If the Flavr Savr™ tomato was on the market today, and conservatively speaking accounted for only 25 per cent of the total market, its share would easily be valued at US$325 million. Cost savings and other gains along the supply chain for distributors, retailers and consumers would only add to this value.

The genetically modified share of the US and Canadian seed potato market increased from 1 per cent in 1995 to 3 per cent in 1999, showing great promise for this new technology. By 2000, however, this share plummeted to 0.1 per cent (Guenthner, 2002). According to National Agricultural Statistic Services data, the fresh potato industry had an estimated value of US$1.1 billion in 2006. Taking into account the growth of the GM share between 1995 and 1999 and projecting a current market share at approximately 25 per cent, the value of the GM potato industry in the US could easily be worth US$275 million. In addition to lost revenues, there are other withdrawal costs to consider. Late blight is considered one of the most serious problems affecting the US potato industry, causing major losses in production and quality (Guenthner et al., 1999; Niederhauser, 1993). In research published in 2001, Guenthner et al. estimated the annual costs to US growers for late blight alone at US$287.8 million. This value includes both lost revenues to growers and fungicide costs. This does not include other costs sustained along the remainder of the value chain nor does it include lost revenues and costs associated with combating the Colorado Potato Beetle. One also has to consider the potato-related technologies that were in the pipeline at the time (and would also be considered 'lost innovations') when Monsanto withdrew GM potatoes from the market. These include a number of fungal, virus and blight resistance varieties. The loss of these follow-on innovations also represents huge opportunity costs to the agricultural industry and for society at large. According to Kaniewski and Thomas (2004), growers saved anywhere from $140 to $164 US per acre on their NewLeaf™ Plus plantings, which has been forgone.

The loss to the producer of Triffid's environmental benefits and the value-ad of the cultivar in the marketplace are difficult to quantify. Glean™, a sulfonylurea herbicide, has long since been pulled from the market. But equivalent products, such as Ally™, were still on the market at the time and Triffid could also have enjoyed a niche with producers that employed this product in their farming practices. Most certainly, the CDC Triffid flax variety would be obsolete by now. But this would be due primarily to improved cultivars that have since superseded the variety. Ryan and Smyth (2012) estimate the costs associated with Canada's management of the recent Triffid issue (2009 to 2011) to be almost C$30 million. This, of course, does not include historical costs associated with the management of the withdrawal and deregistration of Triffid (between 1998 and 2001) or the forgone benefits of a continued GM flax breeding programme.

Finally, the second-order effects of consumer activism have pushed regulatory costs through the roof which has and will represent a surcharge on any and all subsequent food or crop-related technologies. According to Jaffe (2005), it took both the FDA and APHIS twice as long to review biotech crops from 2000 to 2004 than it did in the 1990s with little to no justification for the longer review times. One might think that institutional knowledge and capacities would develop over time and such reviews and approval processes would be more expedient. Political, ethical and economic factors play a role in many risk assessments, thereby blurring the scientific focus (McHughen, 2007).

4 CONCLUSION

Despite numerous tests for safety of products passing through the rigorous regulatory approval processes, there appears to be no guarantee for success of GM food or crop technologies in today's market. The current environment into which these innovations are introduced is highly politically charged. Public perceptions of risk around technology are influenced substantially by public attitudes toward science and technology in general. The costs of science and research in the development of a product are sunk – and lost – whenever a good, useful, safe product wanted by customers is withdrawn from the market. As outlined in the previous section, there are other costs borne as well, including regulatory costs, lost value, opportunity costs and quarantine costs. This, of course, does not take into account the cascading or derived effects of the original 'lost innovation'. There is value in any second-order innovations that can potentially advance from product or technology development. These 'innovation progeny' are usually knocked out of the innovation race when the original product is eliminated from the market.

The introduction of innovative technologies (tomato, potato and flax) along with well-timed anti-technology activities leads to political pressures that essentially rendered some very good products and technologies null and void. It is extremely difficult to accurately calculate the value lost and costs of withdrawal of these products from the market. It is, however, evident that these costs are significant. Undoubtedly, consumer pressures and a politicized regulatory process are affecting society's ability to benefit from safe GM foods, feeds and crops. The reality is that food and crop technologies *will* develop over time but only if the unwarranted political and social constraints around acceptance of these technologies can be overcome.

NOTES

1. Calgene was granted a US patent for the Flavr Savr tomato in early 1989 and was subsequently granted a European patent for the technology in 1994.
2. This product was the result of the collaborative efforts between Meyer Tomatoes and DNA Plant Technology (DNAP).
3. The paste, with a prominent 'genetically modified' label, was sold in the UK in Sainsbury's and Safeway.
4. The remaining shares were bought for US$240 million.
5. In 1998 and 1999, Monsanto received full regulatory approval in both the US and Canada for other GM potatoes including: NewLeaf™ Plus for Russett Burbank, NewLeaf™ for the Russet Burbank and the Shepody varieties.
6. The GM potatoes were tested in confined sites in Prince Edward Island (PEI) (1992–95), New Brunswick (1992–95), Ontario (1993–95), Manitoba (1993–95), Saskatchewan (1995), Alberta (1994–95) and British Columbia (1994–95) (CFIA, 1996).
7. Notification on the RASFF system is equivalent to an air siren going off in the EU. It is an incredibly effective communication tool. This notification in September was the first of more than a hundred over the next several months that would report Triffid in bakeries, cereals and other products made by companies throughout the EU. Please refer to the initial (full) notification recorded as RASFF 2009.1171 at: https://webgate.ec.europa.eu/rasff-window/portal/index.cfm?event=notificationDetail&NOTIF_REFERENCE=2009.1171.
8. No detection at a 0.01 per cent level, 19 times out of 20.
9. There was, at that time, no Triffid-specific test in use.
10. Authors' calculations based on the United States Department of Agriculture's (USDA) National Agricultural Statistic Services data.

REFERENCES

Apel, A. (2010), 'The costly benefits of opposing agricultural biotechnology', forthcoming in *New Biotechnology*, **27**(5), available at: http://www.sciencedirect.com/science/article/B8JG4-504JYNT-1/2/8a13cbba83c6a95c4d584e09eedd26ee and http://www.ask-force.org/web/Vatican-PAS-NBT-publ/Apel-Costly-Benefits-Opposing-2010.pdf.

Barnum, A. (1994), 'Biotech tomato wins final ok for marketing', San Francisco Chronicle, 19 May.

Bevan, M.W., R.B. Flavell and M.D. Chilton (1983), 'A Chimeric antibiotic resistance gene as a selectable marker for plant cell transformation', *Nature*, **304**, 184–7.

Boriss, H. and H. Brunke (2005), 'Commodity profile: fresh market', Agriculture Issues Centre at the University of California, available at: http://aic.ucdavis.edu/profiles/FreshTomatoes-2005.pdf, accessed 18 November 2011.

Canadian Food Inspection Agency (CFIA) (1996), 'Determination of environmental safety of NatureMark potatoes' Colorado Potato Beetle (CPB) resistant potato (*Solanum tuberosum* L.)', Decision Document DD96-06, available at: http://www.inspection.gc.ca/plants/plants-with-novel-traits/approved-under-review/decision-documents/dd96-06/eng/1340663342030/1340663417992, accessed 31 July 2012.

European Commission (2009), 'Summary record of the Standing Committee on the Food Chain and Animal Health', available at: http://ec.europa.eu/food/Committees/regulatory/scfcan/modif_genet/sum_19102009_en.pdf, accessed 12 November 2009.

Firth, P. (1999), 'Leaving a bad taste: Scientific American on the Pusztai GE food safety controversy', *Scientific American*, June.

Flax Council of Canada (FCC) (2009), 'GMO flax update 19th October 2009', available at: http://www.flaxcouncil.ca/files/web/GM%20Flax%20update%2019%20October%202009.pdf,accessed 20 October 2009.

Fraley, R.T., S.G. Rogers, R.B. Horsch, P.R. Sanders, J.S. Flick, S.P. Adams, M.L. Bittner, L.A. Brand, C.L. Fink, J.S. Fry et al. (1983), 'Expression of bacterial genes in plant cells', *Proceedings of the National Academy of Sciences of the United States of America*, **80**, 4803–7.

Guenthner, J.F. (2002), 'Consumer acceptance of genetically modified potatoes', *American Journal of Potato Research*, **79**, 309–15.
Guenthner, J.F., K.C. Michael and P. Nolte (2001), 'The economic impact of potato late blight on US growers', *Potato Research*, **44**, 121–5.
Guenthner, J.F., Aaron J. Johnson, Christopher S. McIntosh and Michael K. Thornton (2011), 'An industry perspective of all-native and transgenic potatoes', *AgBioForum*, **14**(1), 14–19.
Guenthner, J.F., M.V. Wiese, A.D. Pavlista, J.B. Sieczka and J. Wyman (1999), 'Assessment of pesticide use in the US potato industry', *American Journal of Potato Research*, **76**, 25–9.
Harvey, M. (1999), 'Genetic modification as a bio-socio-economic process: one case of tomato purée', *CRIC Discussion Paper no. 31*, available at: http://www.cric.ac.uk/cric/Pdfs/dp31.pdf, accessed on 10 December 2012.
Herrera-Estrella, L., M. DeBlock, E. Messens, J.P. Hernalsteens, M. Van Montagu and J. Schell (1983), 'Chimeric genes as dominant selectable markers in plant cells', *European Molecular Biology Journal*, **2**, 987–96.
Jaffe, GA. (2005), *Withering on the Vine: Will Agricultural Biotech's Promises Bear Fruit?*, Washington, DC: Center for Science in the Public Interest.
James, T. (2010), 'Commodity perspectives: flax', paper presented at the Canada Grains Council's 41st annual meeting, Winnipeg, Canada, 19 April.
Kaniewski, W.K. and P.E. Thomas (2004), 'The potato story', *AgBioForum*, **7**(1&2), 41–6.
Kerr, A. (2011), 'GM crops – a mini-review', *Australasian Plant Pathology*, 40, 449–52.
Killman, S. (2000), 'Monsanto's biotech spud is being pulled from the fryer at fast-food chain', *Wall Street Journal*, 28 April.
Lamb, E. and H. Booker (2011), 'Quantification of low-level genetically modified (GM) seed presence in large seed lots: a case study of GM seed in Canadian flax breeder seed lots', *Seed Science Research*, July, pp. 1–7.
Martineau, B. (2001), *First Fruit: The Creation of the Flavr Savr™ Tomato and the Birth of Biotech Foods*, New York: McGraw-Hill Companies.
McHughen, A. (2000), *Pandora's Picnic Basket: the Potential and Hazards of Genetically Modified Foods*, Oxford: Oxford University Press.
McHughen, A. (2007), 'Fatal flaws in agbiotech regulatory policies', *Nature Biotechnology*, **25**, 725–7.
Miller, H. (2010), 'The *Lancet* pricks itself', Forbes Commentary, 5 February, available at: http://www.forbes.com/2010/02/05/lancet-vaccines-autism-opinions-contributors-henry-i-miller.html, accessed 11 December 2011.
Niederhauser, J.S. (1993), 'The role of the potato in the conquest of hunger', in J.F. Guenthner (ed.), *Past, Present and Future Uses of Potatoes*, proceedings of the Symposium, Potato Association of America, 76th Annual Meeting, College of Agriculture MS 164, University of Idaho, Moscow, ID.
Paarlberg, R. and C. Pray (2008), 'Political actors on the landscape', *AgBioForum*, **10**(3), article 3, available at: http://www.agbioforum.org/v10n3/v10n3a03-paarlberg.htm.
Phillips, P.W.B. and D. Corkindale (2002), 'Marketing GM foods: the way forward', *AgBioForum*, **5**(3), 113–21.
PRRCG, Prairie Registration Recommending Committee for Grain (2004), 'Oilseeds subcommittee report', Calgary, AB: PRRCG.
Ryan, C.D. and S.J. Smyth (2012), 'Economic implications of low level presence in a zero tolerance European import market: the case of Canadian Triffid flax', *AgBioForum*, **15**(1), 21–30.
Smyth, S. and P.W.B. Phillips (2001), 'Competitors co-operating: establishing a supply chain to manage genetically modified canola', *International Food and Agribusiness Management Review*, **4**, 51–66.
Smyth, S., W. Kerr and P.W.B. Phillips (2011), 'Recent trends in the scientific basis of sanitary and phytosanitary trade rules and their potential impact on investment', *The Journal of World Investment & Trade*, **12**(1), February, 5–26.
Sugarman, C. (1994), 'Tasting . . . 1, 2, 3, tasting', *Washington Post*, 8 June.
Toronto Star (1999), 'McCain won't use modified potatoes: move elates green groups, angers farmers', 30 November.

Index

Abbreviations used in the index include:

GM – genetically modified
HT – herbicide tolerant
IPRs – intellectual property rights
PMP – plant-made pharmaceuticals
PVP – plant varietal protection
RR – Roundup Ready®